# Statistical Foundations of Data Science

## CHAPMAN & HALL/CRC DATA SCIENCE SERIES

Reflecting the interdisciplinary nature of the field, this book series brings together researchers, practitioners, and instructors from statistics, computer science, machine learning, and analytics. The series will publish cutting-edge research, industry applications, and textbooks in data science.

The inclusion of concrete examples, applications, and methods is highly encouraged. The scope of the series includes titles in the areas of machine learning, pattern recognition, predictive analytics, business analytics, Big Data, visualization, programming, software, learning analytics, data wrangling, interactive graphics, and reproducible research.

Published Titles

**Feature Engineering and Selection**
A Practical Approach for Predictive Models
*Max Kuhn and Kjell Johnson*

**Probability and Statistics for Data Science**
Math + R + Data
*Norman Matloff*

**Introduction to Data Science**
Data Analysis and Prediction Algorithms with R
*Rafael A. Irizarry*

**Cybersecurity Analytics**
*Rakesh M. Verma and David J. Marchette*

**Basketball Data Science**
With Applications in R
*Paola Zuccolotto and Marcia Manisera*

**JavaScript for Data Science**
*Maya Gans, Toby Hodges, and Greg Wilson*

**Statistical Foundations of Data Science**
*Jianqing Fan, Runze Li, Cun-Hui Zhang and Hui Zou*

For more information about this series, please visit: https://www.crcpress.com/Chapman--HallCRC-Data-Science-Series/book-series/CHDSS

# Statistical Foundations of Data Science

By
Jianqing Fan
Runze Li
Cun-Hui Zhang
Hui Zou

CRC Press
Taylor & Francis Group
Boca Raton London New York

CRC Press is an imprint of the
Taylor & Francis Group, an **informa** business
A CHAPMAN & HALL BOOK

First edition published 2020
by CRC Press
6000 Broken Sound Parkway NW, Suite 300, Boca Raton, FL 33487-2742

and by CRC Press
2 Park Square, Milton Park, Abingdon, Oxon, OX14 4RN



CRC Press is an imprint of Taylor & Francis Group



ISBN: 978-1-466-51084-5 (hbk)

Visit the eResources: https://www.routledge.com/Statistical-Foundations-of-Data-Science/Fan-Li-Zhang-Zou/p/book/9781466510845

**TO THOSE**

**who educate us and love us;**

**whom we teach and we love;**

**with whom we collaborate and associate**

# Contents

# Preface

Big data are ubiquitous. They come in varying volume, velocity, and variety. They have a deep impact on systems such as storages, communications and computing architectures and analysis such as statistics, computation, optimization, and privacy. Engulfed by a multitude of applications, data science aims to address the large-scale challenges of data analysis, turning big data into smart data for decision making and knowledge discoveries. Data science integrates theories and methods from statistics, optimization, mathematical science, computer science, and information science to extract knowledge, make decisions, discover new insights, and reveal new phenomena from data. The concept of data science has appeared in the literature for several decades and has been interpreted differently by different researchers. It has nowadays become a multi-disciplinary field that distills knowledge in various disciplines to develop new methods, processes, algorithms and systems for knowledge discovery from various kinds of data, which can be either low or high dimensional, and either structured, unstructured or semi-structured. Statistical modeling plays critical roles in the analysis of complex and heterogeneous data and quantifies uncertainties of scientific hypotheses and statistical results.

This book introduces commonly-used statistical models, contemporary statistical machine learning techniques and algorithms, along with their mathematical insights and statistical theories. It aims to serve as a graduate-level textbook on the statistical foundations of data science as well as a research monograph on sparsity, covariance learning, machine learning and statistical inference. For a one-semester graduate level course, it may cover Chapters 2, 3, 9, 10, 12, 13 and some topics selected from the remaining chapters. This gives a comprehensive view on statistical machine learning models, theories and methods. Alternatively, a one-semester graduate course may cover Chapters 2, 3, 5, 7, 8 and selected topics from the remaining chapters. This track focuses more on high-dimensional statistics, model selection and inferences but both paths strongly emphasize sparsity and variable selections.

Frontiers of scientific research rely on the collection and processing of massive complex data. Information and technology allow us to collect big data of unprecedented size and complexity. Accompanying big data is the rise of dimensionality, and high dimensionality characterizes many contemporary statistical problems, from sciences and engineering to social science and humanities. Many traditional statistical procedures for finite or low-dimensional data are still useful in data science, but they become infeasible or ineffective for dealing with high-dimensional data. Hence, new statistical methods are indispensable. The authors have worked on high-dimensional statistics for two decades, and started to write the book on the topics of high-dimensional data analysis over a decade ago. Over the last decide, there have been surges in interest and exciting developments in high-dimensional and big data. This led us to concentrate mainly on statistical aspects of data science.

We aim to introduce commonly-used statistical models, methods and procedures in data science and provide readers with sufficient and sound theoretical justifications. It has been a challenge for us to balance statistical theories and methods and to choose the topics and works to cover since the number of publications in this emerging area is enormous. Thus, we focus on the foundational aspects that are related to sparsity, covariance learning, machine learning, and statistical inference.

Sparsity is a common assumption in the analysis of high-dimensional data. By sparsity, we mean that only a handful of features embedded in a huge pool suffice for certain scientific questions or predictions. This book introduces various regularization methods to deal with sparsity, including how to determine penalties and how to choose tuning parameters in regularization methods and numerical optimization algorithms for various statistical models. They can be found in Chapters 3–6 and 8.

High-dimensional measurements are frequently dependent, since these variables often measure similar things, such as aspects of economics or personal health. Many of these variables have heavy tails due to a large number of collected variables. To model the dependence, factor models are frequently employed, which exhibit low-rank plus sparse structures in data matrices and can be solved by robust principal component analysis from high-dimensional covariance. Robust covariance learning, principal component analysis, as well as their applications to community detection, topic modeling, recommender systems, etc. are also a feature of this book. They can be found in Chapters 9–11. Note that factor learning or more generally latent structure learning can also be regarded as unsupervised statistical machine learning.

Machine learning is critical in analyzing high-dimensional and complex data. This book also provides readers with a comprehensive account on statistical machine learning methods and algorithms in data science. We introduce statistical procedures for supervised learning in which the response variable (often categorical) is available and the goal is to predict the response based on input variables. This book also provides readers with statistical procedures for unsupervised learning, in which the responsible variable is missing and the goal concentrates on learning the association and patterns among a set of input variables. Feature creations and sparsity learning also arise in these problems. See Chapters 2, 12–14 for details.

Statistical inferences on high-dimensional data are another focus of this book. Statistical inferences require one to characterize the uncertainty, estimate the standard errors of the estimated parameters of primary interest and derive the asymptotic distributions of the resulting estimates. This is very challenging under the high-dimensional regime. See Chapter 7.

Fueled by the surging demands on processing high-dimensional and big data, there have been rapid and vast developments in high-dimensional statistics and machine learning over the last decade, contributed by data scientists from various fields such as statistics, computer science, information theory, applied and computational mathematics, and others. Even though we have narrowed the scope of the book to the statistical aspects of data science, the field is still too broad for us to cover. Many important contributions that do not fit our presentation have been omitted. Conscientious effort was made in the composition of the reference list and bibliographical notes, but they merely reflect our immediate interests. Omissions and discrepancies are inevitable. We apologize for their occurrence.

Although we all contribute to various chapters and share the responsibility for the whole book, Jianqing Fan was the lead author for Chapters 1, 3 and 9–11, 14 and some sections in other chapters, Runze Li for Chapters 5, and 8 and part of Chapters 6–7, Cun-Hui Zhang for Chapters 4 and 7, and Hui Zou for Chapters 2, 6, 11 and 12 and part of Chapter 5.

Many people have contributed importantly to the completion of this book. In particular, we would like to thank the editor, John Kimmel, who has been extremely helpful and patient with us for over 10 years! We greatly appreciate a set of around 10 anonymous reviewers for valuable comments that have led to the improvement of the book. We are particularly grateful to Cong Ma and Yiqiao Zhong for preparing a draft of Chapter 14, to Zhao Chen for helping us with putting our unsorted and non-uniform references into the present form, to Tracy Ke, Bryan Kelly, Dacheng Xiu and Jia Wang for helping us with constructing Figure 1.3, to Krishna Balasubramanian, Cong Ma, Lingzhou Xue, Boxiang Wang, Kaizheng Wang, Yi Yang, and Ziwei Zhu for producing some figures. Various people have carefully proof-read certain chapters of the book and made useful suggestions. They include Krishna Balasubramanian, Pierre Bayle, Alexander Chen, Elynn Chen, Wenyan Gong, Yongyi Guo, Bai Jiang, Cong Ma, Igor Silin, Qiang Sun, Francesca Tang, Bingyan Wang, Kaizheng Wang, Weichen Wang, Yuling Yan, Zhuoran Yang, Mengxin Yu, Wen-Xin Zhou, Yifeng Zhou, and Ziwei Zhu. We owe them many thanks.

We used a draft of this book as a textbook for a first-year graduate course at Princeton University in 2019 and 2020 and a senior graduate topic course at Pennsylvania State University in 2019. We would like to thank the graduate students in the classes for their careful readings. In particular, we are indebted to Cong Ma, Kaizheng Wang and Zongjun Tan for assisting in preparing the homework problems used for the Princeton course, most of which are now a part of our exercise at the end of each chapter. At Princeton, we covered chapters 2-3, 5, 8.1, 8.3, 9-12, 13.1-13.3, 14.

We are grateful to our teachers who educate us and to all of our collaborators for many enjoyable and stimulating collaborations. Finally, we would like to thank our families for their love and support.

Jianqing Fan
Runze Li
Cun-Hui Zhang
Hui Zou

January 2020.

*Book websites*:

http://personal.psu.edu/ril4/DataScience/

https://orfe.princeton.edu/~jqfan/DataScience/

Chapter 1

# Introduction

The first two decades of this century have witnessed the explosion of data collection in a blossoming age of information and technology. The recent technological revolution has made information acquisition easy and inexpensive through automated data collection processes. The frontiers of scientific research and technological developments have collected huge amounts of data that are widely available to statisticians and data scientists via internet dissemination. Modern computing power and massive storage allow us to process this data of unprecedented size and complexity. This provides mathematical sciences great opportunities with significant challenges. Innovative reasoning and processing of massive data are now required; novel statistical and computational methods are needed; insightful statistical modeling and theoretical understandings of the methods are essential.

## 1.1 Rise of Big Data and Dimensionality

Information and technology have revolutionized data collection. Millions of surveillance video cameras, billions of internet searches and social media chats and tweets produce massive data that contain vital information about security, public health, consumer preference, business sentiments, economic health, among others; billions of prescriptions, and an enormous amount of genetics and genomics information provide critical data on health and precision medicine; numerous experiments and observations in astrophysics and geosciences give rise to big data in science.

Nowadays, *Big Data* are ubiquitous: from the internet, engineering, science, biology and medicine to government, business, economy, finance, legal, and digital humanities. "There were 5 exabytes of information created between the dawn of civilization through 2003, but that much information is now created every 2 days", according to Eric Schmidt, the CEO of Google, in 2010; "Data are becoming the new raw material of business", according to Craig Mundie, Senior Advisor to the CEO at Microsoft; "Big data is not about the data", according to Gary King of Harvard University. The first quote is on the volume, velocity, variety, and variability of big data nowadays, the second is about the value of big data and its impact on society, and the third quote is on the importance of the smart analysis of big data.

Accompanying *Big Data* is rising dimensionality. Frontiers of scientific research depend heavily on the collection and processing of massive complex data. Big data collection and high dimensionality characterize many contemporary statistical problems, from sciences and engineering to social science and humanities. For example, in disease classification using microarray or proteomics data, tens of thousands of expressions of molecules or proteins are potential predictors; in genome-wide association studies, hundreds of thousands of single-nucleotide polymorphisms (SNPs) are potential covariates; in machine learning, millions or even billions of features are extracted from documents, images and other objects; in spatial-temporal problems in economics and earth sciences, time series of hundreds or thousands of regions are collected. When interactions are considered, the dimensionality grows much more quickly. Yet, interaction terms are needed for understanding the synergy of two genes, proteins or SNPs or the meanings of words. Other examples of massive data include high-resolution images, high-frequency financial data, e-commerce data, warehouse data, functional and longitudinal data, among others. See also Donoho (2000), Fan and Li (2006), Hastie, Tibshirani and Friedman (2009), Bühlmann and van de Geer (2011), Hastie, Tibshirani and Wainwright (2015), and Wainwright (2019) for other examples.

### *1.1.1 Biological sciences*

Bioimaging technology allows us to simultaneously monitor tens of thousands of genes or proteins as they are expressed differently in the tissues or cells under different experimental conditions. Microarray measures expression profiles of genes, typically in the order of tens of thousands, in a single hybridization experiment, depending on the microarray technology being used. For customized microarrays, the number of genes printed on the chip can be much smaller, giving more accurate measurements on the genes of focused interest. Figure 1.1 shows two microarrays using the Agilent microarray technology and cDNA micorarray technology. The intensity of each spot represents the level of expression of a particular gene. Depending on the nature of the studies, the sample sizes range from a couple to tens or hundreds. For cell lines, the individual variations are relatively small and the sample size can be very small, whereas for tissues from different human subjects, the individual variations are far larger and the sample sizes can be a few hundred.

RNA-seq (Nagalakshmi, et al., 2008), a methodology for RNA profiling based on next-generation sequencing (NGS, Shendure and Ji, 2008), has replaced microarrays for the study of gene expression. Next-generation sequencing is a term used to describe a number of different modern sequencing technologies that allow us to sequence DNA and RNA much more quickly and cheaply. RNA-seq technologies, based on assembling short reads 30$\sim$400 base pairs, offer advantages such as a wider range of expression levels, less noise, higher throughput, in addition to more information to detect allele-specific expression, novel promoters, and isoforms. There are a number of papers on

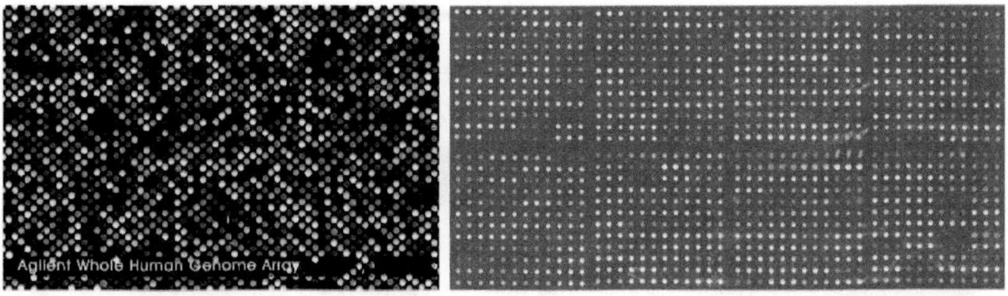


Figure 1.1: Gene expression profiles of microarrays. The intensity at each spot represents the gene expression profile (e.g. Agilent microarray, left panel) or relative profile (e.g. cDNA-microarray, right panel).

statistical methods for detecting differentially expressed genes across treatments/conditions; see Kvam, Liu and Si (2012) for an overview.

After the gene/RNA expression measurements have been properly normalized through RNA-seq or microarray technology, one can then select genes with different expressions under different experimental conditions (e.g. treated with cytokines) or tissues (e.g. normal versus tumor) and genes that express differently over time after treatments (time course experiments). See Speed (2003). This results in a lot of various literature on statistical analysis of controlling the *false discovery rate* in large scale hypothesis testing. See, for example, Benjamini and Hochberg (1995), Storey (2002), Storey and Tibshirani (2003), Efron (2007, 2010b), Fan, Han and Gu (2012), Barber and Candés (2015), Candés, Fan, Janson and Lv (2018), Fan, Ke, Sun and Zhou (2018), among others. The monograph by Efron (2010a) contains a comprehensive account on the subject.

Other aspects of analysis of gene/RNA expression data include association of gene/RNA expression profiles with clinical outcomes such as disease stages or survival time. In this case, the gene expressions are taken as the covariates and the number of variables is usually large even after preprocessing and screening. This results in high-dimensional regression and classification (corresponding to categorical responses, such as tumor types). It is widely believed that only a small group of genes are responsible for a particular clinical outcome. In other words, most of the regression coefficients are zero. This results in high-dimensional sparse regression and classification problems.

There are many other high throughput measurements in biomedical studies. In proteomics, thousands of proteins expression profiles, which are directly related to biological functionality, are simultaneously measured. Similar to genomics studies, the interest is to associate the protein expressions with clinical outcomes and biological functionality. In genomewide association studies, many common genetic variants (typically single-nucleotide polymorphisms or *SNPs*) in different individuals are examined to study if any variant is associated with a trait (heights, weights, eye colors, yields, etc.) or a disease. These

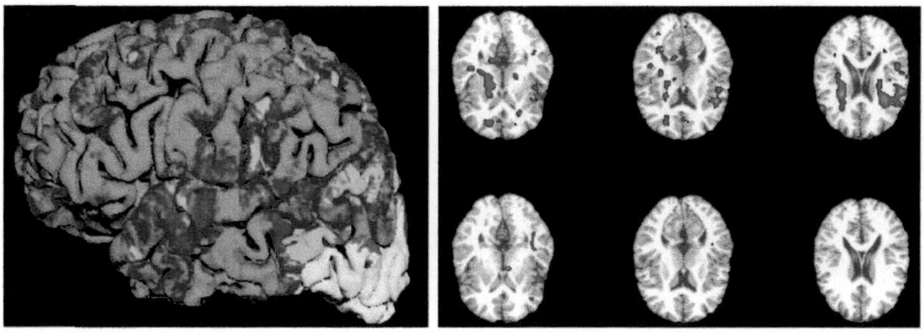

Figure 1.2: Schematic illustration of a brain response to a cognitive task and several slices of its associated fMRI measurements .

genetic variants are referred to as the *quantitative trait loci* (QTL) and hundreds of thousands or millions of SNPs are available for examination. The need for understanding pathophysiology has also led to investigating the so-called *eQTL* studies, the association between SNPs and the expressions of nearby genes. In this case, the gene expressions are regarded as the responses whereas the individual SNPs are taken as the covariates. This again results in high-dimensional regression problems.

High throughput measurements are also commonly used in neuroscience, astronomy, and agriculture and resource surveys using satellite and other imaging technology. In neuroscience, for example, *functional magnetic resonance imaging* (fMRI) technology is frequently applied to measure Blood Oxygenation Level-Dependent (*BOLD*) response to stimuli. This allows investigators to determine which areas of the brain are involved in a cognitive task, or more generally, the functionality of brains. Figure 1.2 gives a schematic illustration. fMRI data contain time-course measurements over tens or hundreds of thousand voxels, resulting in high-dimensional statistical problems.

### *1.1.2 Health sciences*

Health scientists employ many advanced bioinformatic tools to understand molecular mechanisms of disease initiation and progression, and the impact of genetic variations on clinical outcomes. Many health studies also collect a number of risk factors as well as clinical responses over a period of time: many covariates and responses of each subject are collected at different time points. These kinds of longitudinal studies can give rise to high-dimensional big data.

A famous example is the *Framingham Heart Study*, initiated in 1948 and sponsored by the National Heart, Lung and Blood Institute. Documentation of its first 55 years can be found at the website

http://www.framinghamheartstudy.org/.

More details on this study can be found from the website of the American Heart Association. Briefly, the study follows a representative sample of 5,209 adult residents and their offspring aged 28-62 years in Framingham, Massachusetts. These subjects have been tracked using standardized biennial cardiovascular examination, daily surveillance of hospital admissions, death information and information from physicians and other sources outside the clinic. In 1971, the study enrolled a second-generation group, consisting of 5,124 of the original participants' adult children and their spouses, to participate in similar examinations.

The aim of the Framingham Heart Study is to identify risk factors associated with heart disease, stroke and other diseases, and to understand the circumstances under which cardiovascular diseases arise, evolve and end fatally in the general population. In this study, there are more than 25,000 samples, each consisting of more than 100 variables. Because of the nature of this longitudinal study, some participants cannot be followed up due to their migrations. Thus, the collected data contain many missing values. During the study, cardiovascular diseases may develop for some participants, while other participants may never experience cardiovascular diseases. This implies that some data are censored because the event of particular interest never occurs. Furthermore, data between individuals may not be independent because data for individuals in a family are clustered and likely positively correlated. Missing, censoring and clustering are common features in health studies. These three issues make the data structure complicated and identification of important risk factors more challenging.

High-dimensionality is frequently seen in many other biomedical studies. It also arises in the studies of health costs, health care and health records.

### *1.1.3 Computer and information sciences*

The development of information and technology itself collects massive amounts of data. For example, there are billions of web pages on the internet, and an internet search engine needs to statistically learn the most likely outcomes of a query and fast algorithms need to evolve with empirical data. The input dimensionality of queries can be huge. In Google, Facebook and other social networks, algorithms are designed to predict the potential interests of individuals in certain services or products. A familiar example of this kind is amazon.com in which related books are recommended online based on user inputs. This kind of recommendation system applies to other types of services such as music and movies. These are just a few examples of statistical learning in which the data sets are huge and highly complex, and the number of variables is ultrahigh.

Machine learning algorithms have been widely applied to pattern recognition, search engines, computer vision, document and image classification, bioinformatics, medical diagnosis, natural language processing, knowledge graphs, automatic driving machines, internet doctors, among others. The development of these algorithms is based on high-dimensional statistical regres-

Figure 1.3: Some illustrations of machine learning. Top panel: the word clouds of sentiments of a company (Left: Negative Words; Right: Positive Words). The plots were constructed by using data used in Ke, Kelly and Xiu (2019). Bottom left: It is challenging for a computer to recognize the pavilion from the background in computer vision. Bottom right: Visualization of the friendship connections in Facebook.

sion and classification with a large number of predictors and a large amount of empirical data. For example, in text and document classification, the data of documents are summarized by word-document information matrices: the frequencies of the words and phrases $x$ in document $y$ are computed. This step of *feature extraction* is very important for the accuracy of classification. A specific example of document classification is E-mail spam in which there are only two classes of E-mails, junk or non-junk. Clearly, the number of features should be very large in order to find important features for accurate document classifications. This results in high-dimensional classification problems.

Similar problems arise for image or object classifications. Feature extractions play critical roles. One approach for such a feature extrapolation is the classical *vector quantization* technique, in which images are represented by many small subimages or *wavelet* coefficients, which are further reduced by summary statistics. Again, this results in high-dimensional predictive variables. Figure 1.3 illustrates a few problems that arise in machine learning.

### *1.1.4 Economics and finance*

Thanks to the revolution in information and technology, high-frequency financial data have been collected for a host of financial assets, from stocks, bonds, and commodity prices to foreign exchange rates and financial derivatives. The asset correlations among 500 stocks in the S&P500 Index already involve over a hundred thousand parameters. This poses challenges in accurately measuring the financial risks of the portfolios, systemic risks in the financial systems, bubble migrations, and risk contagions, in additional to portfolio allocation and management (Fan, Zhang and Yu, 2012; Brownlees and Engle, 2017). For an overview of high-dimensional economics and finance, see, for example, Fan, Lv and Qi (2012).

To understand the dynamics of financial assets, large panels of financial time series are widely available within asset classes (e.g. components of Russell 3000 stocks) and across asset classes (e.g. stocks, bonds, options, commodities, and other financial derivatives). This is important for understanding the dynamics of price co-movements, time-dependent large volatility matrices of asset returns, systemic risks, and bubble migrations.

Large panel data also arise frequently in economic studies. To analyze the joint evolution of macroeconomic time series, hundreds of macroeconomic variables are compiled to better understand the impact of government policies and to gain better statistical accuracy via, for example, the vector autoregressive model (Sims, 1980). The number of parameters is very large since it grows quadratically with the number of predictors. To enrich the model information, Bernanke et al. (2005) propose to augment standard VAR models with estimated factors (FAVAR) to measure the effects of monetary policy. Factor analysis also plays an important role in prediction using large dimensional data sets (for reviews, see Stock and Watson (2006), Bai and Ng (2008)). A comprehensive collection of 131 macroeconomics time series (McCracken and Ng, 2015) with monthly updates can be found in the website

https://research.stlouisfed.org/econ/mccracken/fred-databases/ .

Spatial-temporal data also give rise to big data in economics. Unemployment rates, housing price indices and sale data are frequently collected in many regions, detailed up to zip code level, over a period of time. The use of spatial correlation enables us to better model the joint dynamics of the data and forecast future outcomes. In addition, exploring homogeneity enables us to aggregate a number of homogeneous regions to reduce the dimensionality, and hence statistical uncertainties, and to better understand heterogeneity across spatial locations. An example of this in prediction of housing appreciation was illustrated in the paper by Fan, Lv, and Qi (2012). See Figure 1.4 and Section 3.9.

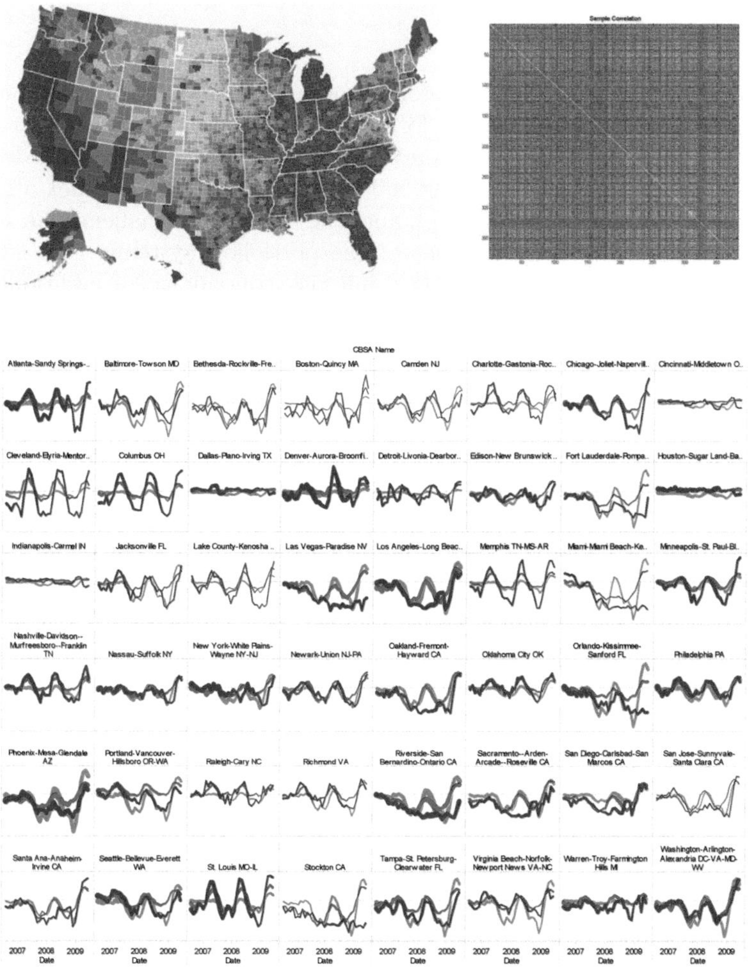


Figure 1.4: Prediction of monthly housing appreciation. Top panel-left: Choropleth map for the 2009 U.S. unemployment rate by county. Top panel-right: Spatial correlation of monthly housing price appreciation among 352 largest counties in the United States from January 2000 to December 2009 (from Fan, Lv, and Qi, 2012). Bottom panel: Prediction of monthly housing pricing appreciation in 48 regions from January 2006 to December 2009 using a large sparse econometrics model with 352 monthly time series from January 2000 to December 2005. Blue: OLS. Red: PLS. Black: Actual. Thickness: Proportion to repeated sales. Adapted from Fan, Lv, and Qi (2012).

### 1.1.5 *Business and program evaluation*

Big data arises frequently in marketing and program evaluation. Multi-channel strategies are frequently used to market products, such as drugs and medical devices. Data from hundreds of thousands of doctors are collected with different marketing strategies over a period of time, resulting in big data. The design of marketing strategies and the evaluation of a program's effectiveness are important to corporate revenues and cost savings. This also applies to online advertisements and AB-tests.

Similarly, to evaluate government programs and policies, large numbers of confounders are collected, along with many individual responses to the treatment. This results in big and high-dimensional data.

### 1.1.6 *Earth sciences and astronomy*

Spatial-temporal data have been widely available in the earth sciences. In meteorology and climatology studies, measurements such as temperatures and precipitations are widely available across many regions over a long period of time. They are critical for understanding climate changes, local and global warming, and weather forecasts, and provide an important basis for energy storage and pricing weather based financial derivatives.

In astronomy, sky surveys collect a huge amount of high-resolution imaging data. They are fundamental to new astronomical discoveries and to understanding the origin and dynamics of the universe.

## 1.2 Impact of Big Data

The arrival of *Big Data* has had deep impact on data systems and analysis. It poses great challenges in terms of storage, communication and analysis. It has forever changed many aspects of computer science, statistics, and computational and applied mathematics: from hardware to software; from storage to super-computing; from data base to data security; from data communication to parallel computing; from data analysis to statistical inference and modeling; from scientific computing to optimization. The efforts to provide solutions to these challenges gave birth to a new disciplinary science, data science. Engulfed by the applications in various disciplines, *data science* consists of studies on data acquisition, storage and communication, data analysis and modeling, and scalable algorithms for data analysis and artificial intelligence. For an overview, see Fan, Han, and Liu (2014).

Big Data powers the success of statistical prediction and artificial intelligence. Deep *artificial neural network* models have been very successfully applied to many *machine learning* and prediction problems, resulting in a discipline called *deep learning* (LeCun, Bengio and Hinton, 2015; Goodfellow, Bengio and Courville, 2016). Deep learning uses a family of over parameterized models, defined through deep neural networks, that have small modeling biases. Such an over-parameterized family of models typically has large variances, too big to be useful. It is the big amount of data that reduces the variance to an acceptable level, achieving bias and variance trade-offs in prediction. Similarly, such an over-parameterized family of models typically is too hard to find reasonable local minima, and it is modern computing power and cheap GPUs that make the implementation possible. It is fair to say that today's success of deep learning is powered by the arrivals of big data and modern computing power. These successes will be further carried into the future, as we collect even bigger data and become even better computing architecture.

As Big Data are typically collected by automated process and by different generations of technologies, the quality of data is low and measurement errors are inevitable. Since data are collected from various sources and populations, the problem of *heterogeneity* of big data arises. In addition, since the number of variables is typically large, many variables have high kurtosis (much higher than the normal distribution). Moreover, *endogeneity* occurs incidentally due to high-dimensionality that has huge impacts on model selection and statistical inference (Fan and Liao, 2014). These intrinsic features of Big Data have significant impacts on the future developments of big data analysis techniques, from heterogeneity and heavy tailedness to endogeneity and measurement errors. See Fan, Han, and Liu (2014).

Big data are often collected at multiple locations and owned by different parties. They are often too big and unsafe to be stored in one single machine. In addition, the processing power required to manipulate big data is not satisfied by standard computers. For these reasons, big data are often distributed in multiple locations. This creates the issues of communications, privacy and owner issues.

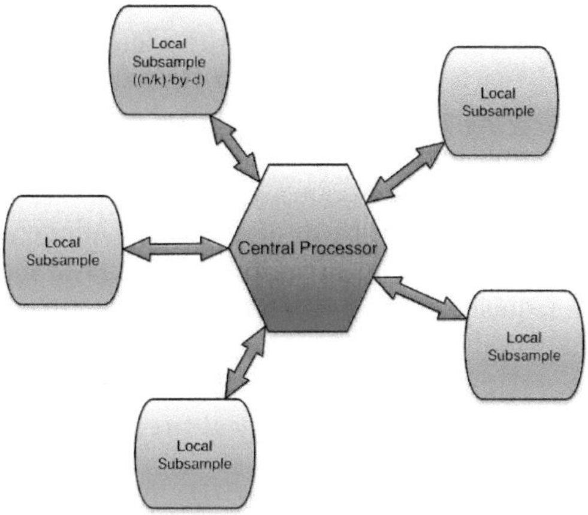


Figure 1.5: Schematic illustration of the distributed data analysis and computing architecture.

A simple architecture that tackles simultaneously the storage, communication, privacy and ownership issues is the *distributed data analysis* in Figure 1.5. Here, each node analyzes the local data and communicates only the results to the central machine. The central machine then aggregates the results and reports the final results (one-shot analysis) or communicates the results back to each node machine for further analysis (multi-shot analysis). For recent developments on this subject, see Shamir, Srebro and Zhang (2014), Zhang, Duchi and Wainwright (2015), Jordan, Lee and Yang (2018) for low-dimensional regression; Chen and Xie (2014), Lee, Liu, Sun and Taylor (2017), Battey, Fan, Liu, Lu and Zhu (2018) for high-dimensional sparse regression and inference, and El Karoui and d'Aspremont (2010), Liang, et al. (2014), Bertrand and Moonen (2014), Schizas and Aduroja (2015), Garber, Shamir and Srebro (2017), and Fan, Wang, Wang and Zhu (2019) for *principal component analysis*.

As mentioned before, big data are frequently accompanied by high-dimensionality. We now highlight the impacts of dimensionality on data analysis.

## 1.3 Impact of Dimensionality

What makes high-dimensional statistical inference different from traditional statistics? High-dimensionality has a significant impact on computation, spurious correlation, noise accumulation, and theoretical studies. We now briefly touch these topics.

### *1.3.1 Computation*

Statistical inferences frequently involve numerical optimization. Optimizations in millions and billions of dimensional spaces are not unheard of and arise easily when interactions are considered. High-dimensional optimization is not only expensive in computation, but also slow in convergence. It also creates numerical instability. Algorithms can easily get trapped at local minima. In addition, algorithms frequently use iteratively the inversions of large matrices, which causes many instability issues in addition to large computational costs and memory storages. Scalable and stable implementations of high-dimensional statistical procedures are very important to statistical learning.

Intensive computation comes also from the large number of observations, which can be in the order of millions or even billions as in marketing and machine learning studies. In these cases, computation of summary statistics such as correlations among all variables is expensive, yet statistical methods often involve repeated evaluations of summation of loss functions. In addition, when new cases are added, it is ideal to only update some of the summary statistics, rather than to use the entire updated data set to redo the computation. This also saves considerable data storage and computation. Therefore, scalability of statistical techniques to both dimensionality and the number of cases is paramountly important.

The high dimensionality and the availability of big data have reshaped statistical thinking and data analysis. Dimensionality reduction and feature extraction play pivotal roles in all high-dimensional statistical problems. This helps reduce computation costs as well as improve statistical accuracy and scientific interpretability. The intensive computation inherent in these problems has altered the course of methodological developments. Simplified methods have been developed to address the large-scale computational problems. Data scientists are willing to trade statistical efficiencies with computational expediency and robust implementations. Fast and stable implementations of optimization techniques are frequently used.

### *1.3.2 Noise accumulation*

High-dimensionality has significant impact on statistical inference in at least two important aspects: *noise accumulation* and *spurious correlation*. Noise accumulation refers to the fact that when a statistical rule depends on many parameters, each estimated with stochastic errors, the estimation errors in the rule can accumulate. For high-dimensional statistics, noise accumulation is more severe, and can even dominate the underlying signals. Consider, for example, a linear classification rule which classifies a new data point $\mathbf{x}$ to class 1 if $\mathbf{x}^T\boldsymbol{\beta} > 0$. This rule can have high discrimination power when $\boldsymbol{\beta}$ is known. However, when an estimator $\widehat{\boldsymbol{\beta}}$ is used instead, due to accumulation of errors in estimating the high-dimensional vector $\widehat{\boldsymbol{\beta}}$, the classification rule can be as bad as random guessing.

To illustrate the above point, let us assume that we have random samples $\{\mathbf{X}_i\}_{i=1}^n$ and $\{\mathbf{Y}_i\}_{i=1}^n$ from class 0 and class 1 with the population distributions $N(\boldsymbol{\mu}_0, \mathbf{I}_p)$ and $N(\boldsymbol{\mu}_1, \mathbf{I}_p)$, respectively. To mimic the gene expression data, we take $p = 4500$, $\boldsymbol{\mu}_0 = 0$ without loss of generality, and $\boldsymbol{\mu}_1$ from a realization of $0.98\delta_0 + 0.02 * \mathrm{DE}$, a mixture of point mass 0 with probability 0.98 and the standard double exponential distribution with probability 0.02. The realized $\boldsymbol{\mu}_1$ is shown in Figure 1.6, which should have about 90 non-vanishing components and is taken as true $\boldsymbol{\mu}_1$. The components that are considerably different from zero are numbered far less than 90, around 20 to 30 or so.

Unlike high-dimensional regression problems, high-dimensional classification does not have implementation issues if the Euclidian distance based classifier is used; see Figure 1.6. It classifies $\mathbf{x}$ to class 1 if

$$\|\mathbf{x} - \boldsymbol{\mu}_1\|^2 \leq \|\mathbf{x} - \boldsymbol{\mu}_0\|^2 \qquad \text{or} \qquad \boldsymbol{\beta}^T(\mathbf{x} - \boldsymbol{\mu}) \geq 0, \tag{1.1}$$

where $\boldsymbol{\beta} = \boldsymbol{\mu}_1 - \boldsymbol{\mu}_0$ and $\boldsymbol{\mu} = (\boldsymbol{\mu}_0 + \boldsymbol{\mu}_1)/2$. For the particular setting in the last paragraph, the distance-based classifier is the Fisher classifier and is the optimal Bayes classifier if prior probability of class 0 is 0.5. The misclassification probability for $\mathbf{x}$ from class 1 into class 0 is $\Phi(-\|\boldsymbol{\mu}_1 - \boldsymbol{\mu}_0\|/2)$. This reveals the fact that components with large differences contribute more to differentiating

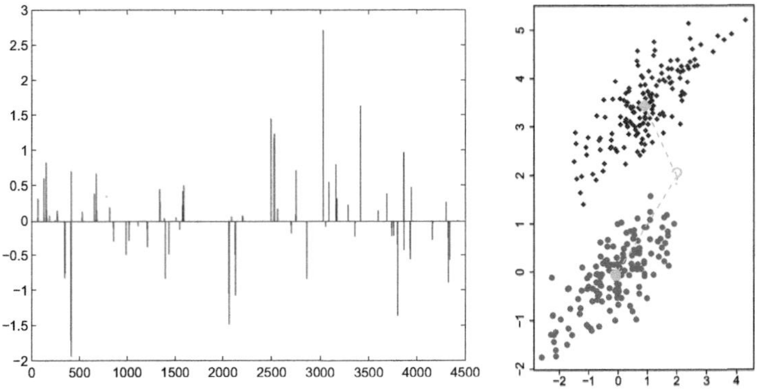


Figure 1.6: Illustration of Classification. Left panel: a realization of $\{\mu_j\}_{j=1}^{4500}$ from the mixture distribution $0.98\delta_0 + 0.02 * \text{DE}$, where DE stands for the standard Double Exponential distribution. Right panel: Illustration of the Euclidian distance based classifier, which classifies the query to a class according to its distances to the centroids.

the two classes, and the more components the smaller the discrimination error. In other words, $\Delta_p = \|\boldsymbol{\mu}_1 - \boldsymbol{\mu}_0\|$ is a nondecreasing function of $p$. Let $\Delta_{(m)}$ be the distance computed based on the $m$ largest components of the difference vector $\boldsymbol{\mu}_1 - \boldsymbol{\mu}_0$. For our particular specification in the last paragraph, the misclassification rate is around $\Phi(-\sqrt{2^2 + 2.5^2}/2) = 0.054$ when the two most powerful components are used ($m = 2$). In addition, $\Delta_{(m)}$ stops increasing noticeably when $m$ reaches 30 and will be constant when $m \geq 100$.

The practical implementation requires estimates of the parameters such as $\widehat{\boldsymbol{\beta}}$. The actual performance of the classifiers can differ from our expectation due to the noise accumulation. To illustrate the noise accumulation phenomenon, let us assume that the rank of the importance of the $p$ features is known to us. In this case, if we use only two features, the classification power is very high. This is shown in Figure 1.7(a). Since the dimensionality is low, the noise in the estimated parameters is negligible. Now, if we take $m = 100$, the signal strength $\Delta_m$ increases. On the other hand, we need to estimate 100 coefficients $\boldsymbol{\beta}$, which accumulate stochastic noises in the classifier. To visualize this, we project the observed data onto the first two principal components of these 100-dimensional selected features. From Figure 1.7(b), it is clear that signal and noise effect cancel. We still have classification power to differentiate the two classes. When $m = 500$ and 4500, there is no further increase of signals and noise accumulation effect dominates. The performance is as poorly as random guessing. Indeed, Fan and Fan (2008) show that almost all high-dimensional classifiers can perform as poorly as random guessing unless the signal is excessively strong. See Figure 1.7(c) and (d).

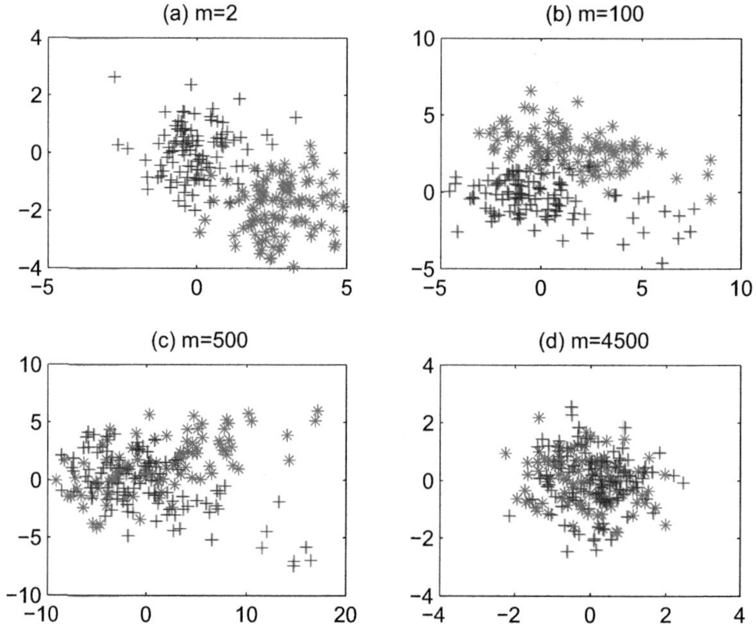


Figure 1.7: Illustration of noise accumulation. Left panel: Projection of observed data ($n = 100$ from each class) onto the first two principal components of $m$-dimensional selected feature space. The $m$ most important features are extracted before applying the principal component analysis.

Fan and Fan (2008) quantify explicitly the price paid with use of more features. They demonstrate that the classification error rate depends on $\Delta_m/\sqrt{m}$. The numerator shows the benefit of the dimensionality through the increase of signals $\Delta_m$, whereas the denominator represents the noise accumulation effect due to estimation of the unknown parameters. In particular, when $\Delta_p/\sqrt{p} \to \infty$ as $p \to \infty$, Hall, Pittelkow and Ghosh (2008) show that the problem is perfectly classifiable (error rate converges to zero).

The above illustration of the noise accumulation phenomenon reveals the pivotal role of feature selection in high dimensional statistical endeavors. Not only does it reduce the prediction error, but also improves the interpretability of the classification rule. In other words, the use of sparse $\boldsymbol{\beta}$ is preferable.

### *1.3.3 Spurious correlation*

Spurious correlation refers to the observation that two variables which have no population correlation have a high sample correlation. The analogy is that two persons look alike but have no genetic relation. In a small village, spurious correlation rarely occurs. This explains why spurious correlation is not an issue in traditional low-dimensional statistics. In a moderate sized city, however, spurious correlations start to occur. One can find two similar looking persons with no genetic relation. In a large city, one can easily find two persons with similar appearances who have no genetic relation. In the same vein, high dimensionality easily creates issues of spurious correlation.

To illustrate the above concept, let us generate a random sample of size $n = 50$ of $p+1$ independent standard normal random variables $Z_1, \cdots, Z_{p+1} \sim_{i.i.d.} N(0,1)$. Theoretically, the sample correlation between any of two random variables is small. When $p$ is small, say $p = 10$, this is indeed the case and the issue of spurious correlation is not severe. However, when $p$ is large, the spurious correlation starts to be noticeable. To illustrate this, let us compute

$$\widehat{r} = \max_{j \geq 2} \widehat{\text{cor}}(Z_1, Z_j) \tag{1.2}$$

where $\widehat{\text{cor}}(Z_1, Z_j)$ is the sample correlation between the variables $Z_1$ and $Z_j$. Similarly, let us compute

$$\widehat{R} = \max_{|\mathcal{S}|=5} \widehat{\text{cor}}(Z_1, \mathbf{Z}_{\mathcal{S}}) \tag{1.3}$$

where $\widehat{\text{cor}}(Z_1, \mathbf{Z}_{\mathcal{S}})$ is the multiple correlation between $Z_1$ and $\mathbf{Z}_{\mathcal{S}}$, namely, the correlation between $Z_1$ and its best linear predictor using $\mathbf{Z}_{\mathcal{S}}$. To avoid computing all $\binom{p}{5}$ multiple $R^2$ in (1.3), we use the forward selection algorithm to compute $\widehat{R}$. The actual value of $\widehat{R}$ is larger than what we present here. We repeat this experiment 200 times and present the distributions of $\widehat{r}$ and $\widehat{R}$ in Figure 1.8.

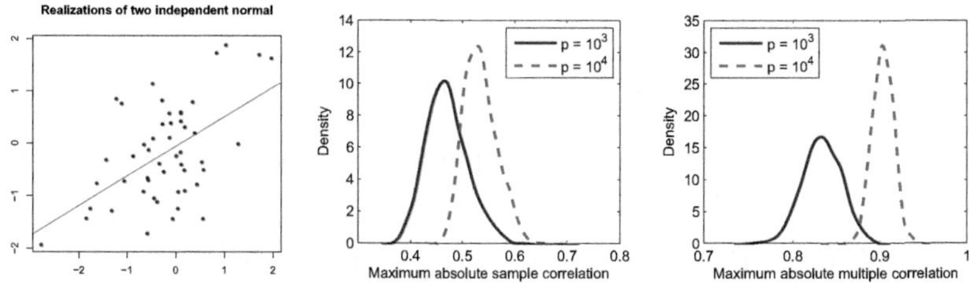


Figure 1.8: Illustration of spurious correlation. Left panel: a typical realization of $Z_1$ with its mostly spuriously correlated variable ($p = 1000$); middle and left panels: distributions of $\widehat{r}$ and $\widehat{R}$ for $p = 1,000$ and $p = 10,000$, respectively. The sample size is $n = 50$.

The maximum spurious correlation $\widehat{r}$ is around 0.45 for $p = 1000$ and 0.55 for $p = 10,000$. They become 0.85 and 0.91 respectively when multiple correlation $\widehat{R}$ in (1.3) is considered. Theoretical results on the order of these spurious correlations can be found in Cai and Jiang (2012) and Fan, Guo and

Hao (2012), and more comprehensively in Fan, Shao, and Zhou (2018) and Fan and Zhou (2016).

The impact of *spurious correlation* includes false scientific discoveries and false statistical inferences. Since the correlation between $Z_1$ and $\mathbf{Z}_{\widehat{S}}$ is around 0.9 for a set $\widehat{S}$ with $|\widehat{S}| = 5$ (Figure 1.8), $Z_1$ and $\mathbf{Z}_{\widehat{S}}$ are practically indistinguishable given $n = 50$. If $Z_1$ represents the gene expression of a gene that is responsible for a disease, we will also discover 5 genes $\widehat{S}$ that have a similar predictive power although they have no relation to the disease.

To further appreciate the concept of spurious correlation, let us consider the neuroblastoma data used in Oberthuer et al. (2006). The study consists of 251 patients, aged from 0 to 296 months at diagnosis with a median age of 15 months, of the German Neuroblastoma Trials NB90-NB2004, diagnosed between 1989 and 2004. Neuroblastoma is a common pediatric solid cancer, accounting for around 15% of pediatric cancers. 251 neuroblastoma specimens were analyzed using a customized oligonucleotide microarray with $p = 10,707$ gene expressions available after preprocessing. The clinical outcome is taken as the indicator of whether a neuroblastoma child has a 3 year event-free survival. 125 cases are taken at random as the training sample (with 25 positives) and the remaining data are taken as the testing sample. To illustrate the spurious correlation, we now replace the gene expressions by artificially simulated Gaussian data. Using only $p = 1000$ artificial variables along with the traditional forward selection, we can easily find 10 of those artificial variables that perfectly classify the clinical outcomes. Of course, these 10 artificial variables have no relation with the clinical outcomes. When the classification rule is applied to the test samples, the classification result is the same as random guessing.

To see the impact of spurious correlation on statistical inference, let us consider a linear model

$$Y = \mathbf{X}^T \boldsymbol{\beta} + \varepsilon, \qquad \sigma^2 = \mathrm{Var}(\varepsilon). \tag{1.4}$$

Let $\widehat{S}$ be a selected subset and we compute the residual variances based on the selected variables $\widehat{S}$:

$$\widehat{\sigma}^2 = \mathbf{Y}^T(I_n - \mathbf{P}_{\widehat{S}})\mathbf{Y}/(n - |\widehat{S}|), \qquad \mathbf{P}_{\widehat{S}} = \mathbf{X}_{\widehat{S}}(\mathbf{X}_{\widehat{S}}^T \mathbf{X}_{\widehat{S}})^{-1}\mathbf{X}_{\widehat{S}}^T. \tag{1.5}$$

In particular, when $\boldsymbol{\beta} = 0$, all selected variables are spurious. In this case, $\mathbf{Y} = \boldsymbol{\varepsilon}$ and

$$\widehat{\sigma}^2 \approx (1 - \gamma_n^2)\|\boldsymbol{\varepsilon}\|^2/n \approx (1 - \gamma_n^2)\sigma^2, \tag{1.6}$$

when $|\widehat{S}|/n \to 0$, where $\gamma_n^2 = \boldsymbol{\varepsilon}^T \mathbf{P}_{\widehat{S}}\boldsymbol{\varepsilon}/\|\boldsymbol{\varepsilon}\|^2$. Therefore, $\sigma^2$ is underestimated by a factor of $\gamma_n^2$

Suppose that we select only one spurious variable, then that variable must be mostly correlated with $\mathbf{Y}$. Since the spurious correlation is high, the bias is large. The two left panels of Figure 1.9 depicts the distribution of $\gamma_n$ along with the associated estimates of $\widehat{\sigma}^2$ for different choices of $p$. Clearly, the bias increases with the dimensionality $p$.

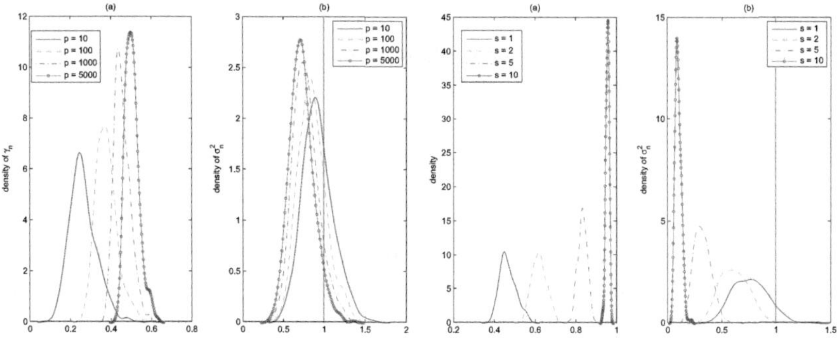


Figure 1.9: Distributions of spurious correlations. Left panel: Distributions of $\gamma_n$ for the null model when $|\widehat{S}| = 1$ and their associated estimates of $\sigma^2 = 1$ for various choices of $p$. Right panel: Distributions of $\gamma_n$ for the model $Y = 2X_1 + 0.3X_2 + \varepsilon$ and their associated estimates of $\sigma^2 = 1$ for various choices of $|\widehat{S}|$ but fixed $p = 1000$. The sample size $n = 50$. Adapted from Fan, Guo, and Hao (2012).

Spurious correlation gets larger when more than one spurious variables are selected, as seen in Figure 1.8. To see this, let us consider the linear model $Y = 2X_1 + 0.3X_2 + \varepsilon$ and use forward selection methods to recruit variables. Again, the spurious variables are selected mainly due to their spurious correlation with $\varepsilon$, the unobservable but realized random noises. As shown in the right panel of Figure 1.9, the spurious correlation is very large and $\widehat{\sigma}^2$ gets notably more biased when $|\widehat{S}|$ gets larger.

Underestimate of residual variance leads to further wrong statistical inferences. More variables will be called statistically significant and that further leads to wrong scientific conclusions. There is active literature on selective inference for dealing with such kinds of issues, starting from Lockhart, Taylor, Tibshirani and Tibshirani (2014); see also Taylor and Tibshirani (2015) and Tibshirani, Taylor, Lockhart and Tibshirani (2016).

### *1.3.4 Statistical theory*

High dimensionality has a strong impact on statistical theory. The traditional asymptotic theory assumes that sample size $n$ tends to infinity while keeping $p$ fixed. This does not reflect the reality of the high dimensionality and cannot explain the observed phenomena such as noise accumulation and spurious correlation. A more reasonable framework is to assume $p$ grows with $n$ and investigate how high the dimensionality $p_n$ a given procedure can handle given the sample size $n$. This new paradigm is now popularly used in the literature.

High dimensionality gives rise to new statistical theory. Many new insights have been unveiled and many new phenomena have been discovered. Subsequent chapters will unveil some of these.

## 1.4 Aim of High-dimensional Statistical Learning

As shown in Section 1.1, high-dimensional statistical learning arises from various different scientific contexts and has very different disciplinary goals. Nevertheless, its statistical endeavor can be abstracted as follows. The main goals of high dimensional inferences, according to Bickel (2008), are

**(a)** to construct a method as effective as possible to predict future observations and

**(b)** to gain insight into the relationship between features and responses for scientific purposes, as well as, hopefully, to construct an improved prediction method.

This view is also shared by Fan and Li (2006). The former appears in problems such as text and document classifications or portfolio optimizations, in which the performance of the procedure is more important than understanding the features that select spam e-mail or stocks that are chosen for portfolio construction. The latter appears naturally in many genomic studies and other scientific endeavors. In these cases, scientists would like to know which genes are responsible for diseases or other biological functions, to understand the molecular mechanisms and biological processes, and predict future outcomes. Clearly, the second goal of high dimensional inferences is more challenging.

The above two objectives are closely related. However, they are not necessarily the same and can be decisively different. A procedure that has a good mean squared error or, more generally risk properties, might not have model selection consistency. For example, if an important variable is missing in a model selection process, the method might find 10 other variables, whose linear combination acts like the missing important variable, to proxy it. As a result, the procedure can still have good prediction power. Yet, the absence of that important variable can lead to false scientific discoveries for objective (b).

As will be seen in Sec. 3.3.2, Lasso (Tibshirani, 1996) has very good risk properties under mild conditions. Yet, its model selection consistency requires the restricted *irrepresentable condition* (Zhao and Yu, 2006; Zou, 2006; Meinshausen and Bühlmann, 2006). In other words, one can get optimal rates in mean squared errors, and yet the selected variables can still differ substantially from the underlying true model. In addition, the estimated coefficients are biased. In this view, Lasso aims more at objective (a). In an effort to resolve the problems caused by the $L_1$-penalty, a class of *folded-concave* penalized least-squares or likelihood procedures, including SCAD, was introduced by Fan and Li (2001), which aims more at objective (b).

## 1.5 What Big Data Can Do

Big Data hold great promise for the discovery of heterogeneity and search for personalized treatments and precision marketing. An important aim for big data analysis is to understand heterogeneity for personalized medicine or services from large pools of variables, factors, genes, environments and their interactions as well as latent factors. Such a kind of understanding is only possible when sample size is very large, particularly for rare diseases.

Another important aim of big data is to discover the commonality and weak patterns, such as the impact of drinking teas and wines on health, in the presence of large variations. Big data allow us to reduce large variances of complexity models such as deep neural network models, as discussed in Section 1.2. The successes of *deep learning* technologies rest to quite an extent on the variance reduction due to big data so that a stable model can be constructed.

## 1.6 Scope of the Book

This book will provide a comprehensive and systematic account of theories and methods in high-dimensional data analysis. The statistical problems range from high-dimensional sparse regression, compressed sensing, sparse likelihood-based models, supervised and unsupervised learning, large covariance matrix estimation and graphical models, high-dimensional survival analysis, robust and quantile regression, among others. The modeling techniques can either be parametric, semi-parametric or nonparametric. In addition, variable selection via regularization methods and sure independent feature screening methods will be introduced.

Chapter 2

# Multiple and Nonparametric Regression

## 2.1 Introduction

In this chapter we discuss some popular linear methods for regression analysis with continuous response variable. We call them linear regression models in general, but our discussion is not limited to the classical multiple linear regression. They are extended to multivariate nonparametric regression via the kernel trick. We first give a brief introduction to multiple linear regression and least-squares, presenting the basic and important ideas such as inferential results, Box-Cox transformation and basis expansion. We then discuss linear methods based on regularized least-squares with ridge regression as the first example. We then touch on the topic of nonparametric regression in a reproducing kernel Hilbert space (RKHS) via the kernel trick and kernel ridge regression. Some basic elements of the RKHS theory are presented, including the famous representer theorem. Lastly, we discuss the leave-one-out analysis and generalized cross-validation for tuning parameter selection in regularized linear models.

## 2.2 Multiple Linear Regression

Consider a *multiple linear regression* model:

$$Y = \beta_1 X_1 + \cdots + \beta_p X_p + \varepsilon, \tag{2.1}$$

where $Y$ represents the *response* or *dependent variable* and the $X$ variables are often called *explanatory variables* or *covariates* or *independent variables*. The intercept term can be included in the model by including 1 as one of the covariates, say $X_1 = 1$. Note that the term "random error" $\varepsilon$ in (2.1) is a generic name used in statistics. In general, the "random error" here corresponds to the part of the response variable that cannot be explained or predicted by the covariates. It is often assumed that "random error" $\varepsilon$ has zero mean, uncorrelated with covariates $X$, which is referred to as *exogenous* variables. Our goal is to estimate these $\beta$'s, called *regression coefficients*, based on a random sample generated from model (2.1).

Suppose that $\{(X_{i1}, \cdots, X_{ip}, Y_i)\}, i = 1, \cdots, n$ is a random sample from model (2.1). Then, we can write

$$Y_i = \sum_{j=1}^{p} X_{ij}\beta_j + \varepsilon_i. \tag{2.2}$$

The method of least-squares is a standard and popular technique for data fitting. It was advanced early in the nineteenth century by Gauss and Legendre. In (2.2) we have the residuals ($r_i$'s)

$$r_i = Y_i - \sum_{j=1}^{p} X_{ij}\beta_j.$$

Assume that random errors $\varepsilon_i$'s are *homoscedastic*, i.e., they are uncorrelated random variables with mean 0 and common variance $\sigma^2$. The *least-squares method* is to minimize the residual sum-of-squares (RSS):

$$\mathrm{RSS}(\boldsymbol{\beta}) = \sum_{i=1}^{n} r_i^2 = \sum_{i=1}^{n}(Y_i - \sum_{j=1}^{p} X_{ij}\beta_j)^2. \tag{2.3}$$

with respect to $\boldsymbol{\beta}$. Since (2.3) is a nice quadratic function of $\boldsymbol{\beta}$, there is a closed-form solution. Denote by

$$\mathbf{Y} = \begin{pmatrix} Y_1 \\ \vdots \\ Y_n \end{pmatrix}, \mathbf{X}_j = \begin{pmatrix} X_{1j} \\ \vdots \\ X_{nj} \end{pmatrix}, \mathbf{X} = \begin{pmatrix} X_{11} & \cdots & X_{1p} \\ \vdots & \cdots & \vdots \\ X_{n1} & \cdots & X_{np} \end{pmatrix}, \boldsymbol{\beta} = \begin{pmatrix} \beta_1 \\ \vdots \\ \beta_p \end{pmatrix}, \boldsymbol{\varepsilon} = \begin{pmatrix} \varepsilon_1 \\ \vdots \\ \varepsilon_n \end{pmatrix}.$$

Then (2.2) can be written in the matrix form

$$\mathbf{Y} = \mathbf{X}\boldsymbol{\beta} + \boldsymbol{\varepsilon}.$$

The matrix $\mathbf{X}$ is known as the *design matrix* and is of crucial importance to the whole theory of linear regression analysis. The $\mathrm{RSS}(\boldsymbol{\beta})$ can be written as

$$\mathrm{RSS}(\boldsymbol{\beta}) = \|\mathbf{Y} - \mathbf{X}\boldsymbol{\beta}\|^2 = (\mathbf{Y} - \mathbf{X}\boldsymbol{\beta})^T(\mathbf{Y} - \mathbf{X}\boldsymbol{\beta}).$$

Differentiating $\mathrm{RSS}(\boldsymbol{\beta})$ with respect to $\boldsymbol{\beta}$ and setting the gradient vector to zero, we obtain the *normal equations*

$$\mathbf{X}^T\mathbf{Y} = \mathbf{X}^T\mathbf{X}\boldsymbol{\beta}.$$

Here we assume that $p < n$ and $\mathbf{X}$ has rank $p$. Hence $\mathbf{X}^T\mathbf{X}$ is invertible and the normal equations yield the least-squares estimator of $\boldsymbol{\beta}$

$$\widehat{\boldsymbol{\beta}} = (\mathbf{X}^T\mathbf{X})^{-1}\mathbf{X}^T\mathbf{Y}. \tag{2.4}$$

In this chapter $\mathbf{X}^T\mathbf{X}$ is assumed to be invertible unless specifically mentioned otherwise.

The fitted $Y$ value is

$$\widehat{\mathbf{Y}} = \mathbf{X}\widehat{\boldsymbol{\beta}} = \mathbf{X}(\mathbf{X}^T\mathbf{X})^{-1}\mathbf{X}^T\mathbf{Y},$$

and the regression residual is

$$\widehat{\mathbf{r}} = \mathbf{Y} - \widehat{\mathbf{Y}} = (\mathbf{I} - \mathbf{X}(\mathbf{X}^T\mathbf{X})^{-1}\mathbf{X}^T)\mathbf{Y}.$$

**Theorem 2.1** *Define* $\mathbf{P} = \mathbf{X}(\mathbf{X}^T\mathbf{X})^{-1}\mathbf{X}^T$. *Then we have*

$$\mathbf{P}\mathbf{X}_j = \mathbf{X}_j, \quad j = 1, 2, \cdots, p;$$

$$\mathbf{P}^2 = \mathbf{P} \quad \text{or} \quad \mathbf{P}(\mathbf{I}_n - \mathbf{P}) = \mathbf{0},$$

*namely* $\mathbf{P}$ *is a projection matrix onto the space spanned by the columns of* $\mathbf{X}$.

**Proof.** It follows from the direct calculation that

$$\mathbf{P}\mathbf{X} = \mathbf{X}(\mathbf{X}^T\mathbf{X})^{-1}\mathbf{X}^T\mathbf{X} = \mathbf{X}.$$

Taking the $j$ column of the above equality, we obtain the first results. Similarly,

$$\mathbf{P}\mathbf{P} = \mathbf{X}(\mathbf{X}^T\mathbf{X})^{-1}\mathbf{X}^T\mathbf{X}(\mathbf{X}^T\mathbf{X})^{-1}\mathbf{X}^T = \mathbf{X}(\mathbf{X}^T\mathbf{X})^{-1}\mathbf{X}^T = \mathbf{P}.$$

This completes the proof. ∎

By Theorem 2.1 we can write

$$\widehat{\mathbf{Y}} = \mathbf{P}\mathbf{Y}, \quad \widehat{\mathbf{r}} = (\mathbf{I}_n - \mathbf{P})\mathbf{Y} \tag{2.5}$$

and we see two simple identities:

$$\mathbf{P}\widehat{\mathbf{Y}} = \widehat{\mathbf{Y}}, \quad \widehat{\mathbf{Y}}^T\widehat{\mathbf{r}} = 0.$$

This reveals an interesting geometric interpretation of the method of least-squares: the least-squares fit amounts to projecting the response vector onto the linear space spanned by the covariates. See Figure 2.1 for an illustration with two covariates.

### *2.2.1 The Gauss-Markov theorem*

We assume the linear regression model (2.1) with

- *exogeneity*: $\mathrm{E}(\varepsilon|X) = 0$;
- *homoscedasticity*: $\mathrm{Var}(\varepsilon|X) = \sigma^2$.

**Theorem 2.2** *Under model (2.1) with exogenous and homoscedastic error, it follows that*

**(i)** (unbiasedness) $\mathrm{E}(\widehat{\boldsymbol{\beta}}|\mathbf{X}) = \boldsymbol{\beta}$.

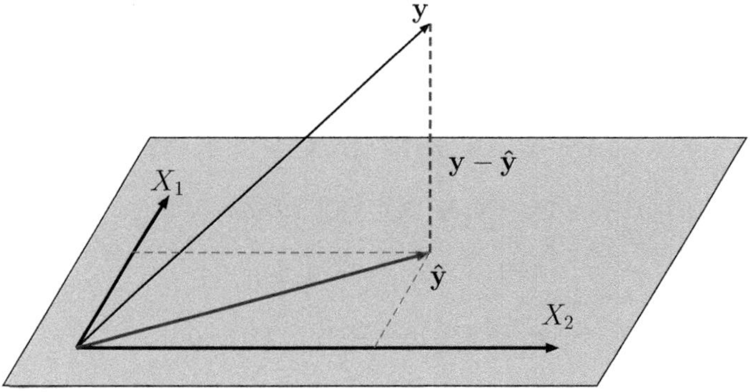


Figure 2.1: Geometric view of least-squares. The fitted value is the blue arrow, which is the projection of $\mathbf{Y}$ on the plane spanned by $X_1$ and $X_2$.

**(ii)** (conditional standard errors) $\mathrm{Var}(\widehat{\boldsymbol\beta}|\mathbf{X}) = \sigma^2(\mathbf{X}^T\mathbf{X})^{-1}$.

**(iii)** (BLUE) *The least-squares estimator* $\widehat{\boldsymbol\beta}$ *is the best linear unbiased estimator (BLUE). That is, for any given vector* $\mathbf{a}$, $\mathbf{a}^T\widehat{\boldsymbol\beta}$ *is a linear unbiased estimator of the parameter* $\theta = \mathbf{a}^T\boldsymbol\beta$. *Further, for any linear unbiased estimator* $\mathbf{b}^T\mathbf{Y}$ *of* $\theta$, *its variance is at least as large as that of* $\mathbf{a}^T\widehat{\boldsymbol\beta}$.

**Proof.** The first property follows directly from $\mathrm{E}(\mathbf{Y}|\mathbf{X}) = \mathbf{X}\boldsymbol\beta$ and

$$\mathrm{E}(\widehat{\boldsymbol\beta}|\mathbf{X}) = (\mathbf{X}^T\mathbf{X})^{-1}\mathbf{X}^T(\mathbf{X}\boldsymbol\beta) = \boldsymbol\beta.$$

To prove the second property, note that for any linear combination $\mathbf{AY}$, its variance-covariance matrix is given by

$$\mathrm{Var}(\mathbf{AY}|\mathbf{X}) = \mathbf{A}\,\mathrm{Var}(\mathbf{Y}|\mathbf{X})\mathbf{A}^T = \sigma^2\mathbf{A}\mathbf{A}^T. \tag{2.6}$$

Applying this formula to the least-squares estimator with $\mathbf{A} = (\mathbf{X}^T\mathbf{X})^{-1}\mathbf{X}^T$, we obtain the property (ii).

To prove property (iii), we first notice that $\mathbf{a}^T\widehat{\boldsymbol\beta}$ is an unbiased estimator of the parameter $\theta = \mathbf{a}^T\boldsymbol\beta$, with the variance

$$\mathrm{Var}(\mathbf{a}^T\widehat{\boldsymbol\beta}|\mathbf{X}) = \mathbf{a}^T\,\mathrm{Var}(\widehat{\boldsymbol\beta}|\mathbf{X})\mathbf{a} = \sigma^2\mathbf{a}^T(\mathbf{X}^T\mathbf{X})^{-1}\mathbf{a}.$$

Now, consider any linear unbiased estimator, $\mathbf{b}^T\mathbf{Y}$, of the parameter $\theta$. The unbiasedness requires that

$$\mathbf{b}^T\mathbf{X}\boldsymbol\beta = \mathbf{a}^T\boldsymbol\beta,$$

namely $\mathbf{X}^T\mathbf{b} = \mathbf{a}$. The variance of this linear estimator is

$$\sigma^2\mathbf{b}^T\mathbf{b}.$$

To prove (iii), we need only to show that

$$\mathbf{a}^T(\mathbf{X}^T\mathbf{X})^{-1}\mathbf{a} \le \mathbf{b}^T\mathbf{b}.$$

Note that

$$\mathbf{a}^T(\mathbf{X}^T\mathbf{X})^{-1}\mathbf{a} = \mathbf{b}^T\mathbf{X}(\mathbf{X}^T\mathbf{X})^{-1}\mathbf{X}^T\mathbf{b} = \mathbf{b}^T\mathbf{P}\mathbf{b}.$$

$\mathbf{P} = \mathbf{P}^2$ means that the eigenvalues of $\mathbf{P}$ are either 1 or 0 and hence $\mathbf{I}_n - \mathbf{P}$ is a semi-positive matrix. Thus,

$$\mathbf{b}^T(\mathbf{I}_n - \mathbf{P})\mathbf{b} \ge 0,$$

or equivalently $\mathbf{b}^T\mathbf{b} \ge \mathbf{b}^T\mathbf{P}\mathbf{b}$. ∎

Property (ii) of Theorem 2.2 gives the variance-covariance matrix of the least-squares estimate. In particular, the conditional standard error of $\widehat{\beta}_i$ is simply $\sigma a_{ii}^{1/2}$ and the covariance between $\widehat{\beta}_i$ and $\widehat{\beta}_j$ is $\sigma^2 a_{ij}$, where $a_{ij}$ is the $(i,j)$-th element of matrix $(\mathbf{X}^T\mathbf{X})^{-1}$.

In many applications $\sigma^2$ is often an unknown parameter of the model in addition to the regression coefficient vector $\boldsymbol{\beta}$. In order to use the variance-covariance formula, we first need to find a good estimate of $\sigma^2$. Given the least-squares estimate of $\boldsymbol{\beta}$, RSS can be written as

$$\text{RSS} = \sum_{i=1}^{n}(Y_i - \widehat{Y}_i)^2 = (\mathbf{Y} - \widehat{\mathbf{Y}})^T(\mathbf{Y} - \widehat{\mathbf{Y}}). \tag{2.7}$$

Define

$$\widehat{\sigma}^2 = \text{RSS}\,/(n-p).$$

We will show in Theorem 2.3 that $\widehat{\sigma}^2$ is an unbiased estimator of $\sigma^2$.

**Theorem 2.3** *Under the linear model (2.1) with homoscedastic error, it follows that*

$$\text{E}(\widehat{\sigma}^2|\mathbf{X}) = \sigma^2.$$

**Proof.** First by Theorem 2.1 we have

$$\text{RSS} = \|(\mathbf{I}_n - \mathbf{P})\mathbf{Y}\|^2 = \|(\mathbf{I}_n - \mathbf{P})(\mathbf{Y} - \mathbf{X}\boldsymbol{\beta})\|^2 = \boldsymbol{\varepsilon}^T(\mathbf{I}_n - \mathbf{P})\boldsymbol{\varepsilon}.$$

Let $\text{tr}(\mathbf{A})$ be the trace of the matrix $\mathbf{A}$. Using the property that $\text{tr}(\mathbf{AB}) = \text{tr}(\mathbf{BA})$, we have

$$\text{RSS} = \text{tr}\{(\mathbf{I}_n - \mathbf{P})\boldsymbol{\varepsilon}\boldsymbol{\varepsilon}^T\}.$$

Hence,

$$\text{E}(\text{RSS}\,|\mathbf{X}) = \sigma^2\,\text{tr}(\mathbf{I}_n - \mathbf{P}).$$

Because the eigenvalues of $\mathbf{P}$ are either 1 or 0, its trace is equal to its rank which is $p$ under the assumption that $\mathbf{X}^T\mathbf{X}$ is invertible. Thus,

$$E(\widehat{\sigma}^2|\mathbf{X}) = \sigma^2(n-p)/(n-p) = \sigma^2.$$

This completes the proof. ∎

*2.2.2 Statistical tests*

After fitting the regression model, we often need to perform some tests on the model parameters. For example, we may be interested in testing whether a particular regression coefficient should be zero, or whether several regression coefficients should be zero at the same time, which is equivalent to asking whether these variables are important in the presence of other covariates. To facilitate the discussion, we focus on the fixed design case where $\mathbf{X}$ is fixed. This is essentially the same as the random design case but conditioned upon the given realization $\mathbf{X}$.

We assume a homoscedastic model (2.1) with normal error. That is, $\varepsilon$ is a Gaussian random variable with zero mean and variance $\sigma^2$, written as $\varepsilon \sim N(0, \sigma^2)$. Note that

$$\widehat{\boldsymbol{\beta}} = \boldsymbol{\beta} + (\mathbf{X}^T\mathbf{X})^{-1}\mathbf{X}^T\boldsymbol{\varepsilon}. \tag{2.8}$$

Then it is easy to see that

$$\widehat{\boldsymbol{\beta}} \sim N(\boldsymbol{\beta}, (\mathbf{X}^T\mathbf{X})^{-1}\sigma^2). \tag{2.9}$$

If we look at each $\widehat{\beta}_j$ marginally, then $\widehat{\beta}_j \sim N(\beta_j, v_j\sigma^2)$ where $v_j$ is the $j$th diagonal element of $(\mathbf{X}^T\mathbf{X})^{-1}$. In addition,

$$(n-p)\widehat{\sigma}^2 \sim \sigma^2\chi^2_{n-p} \tag{2.10}$$

and $\widehat{\sigma}^2$ is independent of $\widehat{\boldsymbol{\beta}}$. The latter can easily be shown as follow. By (2.7), $\widehat{\sigma}^2$ depends on $\mathbf{Y}$ through $\mathbf{Y} - \widehat{\mathbf{Y}} = (\mathbf{I}_n - \mathbf{P})\boldsymbol{\varepsilon}$ whereas $\widehat{\boldsymbol{\beta}}$ depends on $\mathbf{Y}$ through (2.8) or $\mathbf{X}^T\boldsymbol{\varepsilon}$. Note that both $(\mathbf{I}_n - \mathbf{P})\boldsymbol{\varepsilon}$ and $\mathbf{X}^T\boldsymbol{\varepsilon}$ are jointly normal because they are linear transforms of normally distributed random variables, and therefore their independence is equivalent to their uncorrelatedness. This can easily be checked by computing their covariance

$$E(\mathbf{I}_n - \mathbf{P})\boldsymbol{\varepsilon}(\mathbf{X}^T\boldsymbol{\varepsilon})^T = E(\mathbf{I}_n - \mathbf{P})\boldsymbol{\varepsilon}\boldsymbol{\varepsilon}^T\mathbf{X} = \sigma^2(\mathbf{I}_n - \mathbf{P})\mathbf{X} = 0.$$

If we want to test the hypothesis that $\beta_j = 0$, we can use the following $t$ test statistic

$$t_j = \frac{\widehat{\beta}_j}{\sqrt{v_j}\widehat{\sigma}} \tag{2.11}$$

which follows a $t$-distribution with $n-p$ degrees of freedom under the null hypothesis $H_0 : \beta_j = 0$. A level $\alpha$ test rejects the null hypothesis if $|t_j| > t_{n-p,1-\alpha/2}$, where $t_{n-p,1-\alpha/2}$ denotes the $100(1-\alpha/2)$ percentile of the $t$-distribution with $n-p$ degrees of freedom.

In many applications the null hypothesis is that a subset of the covariates have zero regression coefficients. That is, this subset of covariates can be deleted from the regression model: they are unrelated to the response variable given the remaining variables. Under such a null hypothesis, we can reduce the model to a smaller model. Suppose that the reduced model has $p_0$ many regression coefficients. Let RSS and $\text{RSS}_0$ be the residual sum-of-squares based on the least-squares fit of the full model and the reduced smaller model, respectively. If the null hypothesis is true, then these two quantities should be similar: The RSS reduction by using the full model is small, in relative terms. This leads to the *F-statistic*:

$$F = \frac{(\text{RSS}_0 - \text{RSS})/(p-p_0)}{\text{RSS}/(n-p)}. \tag{2.12}$$

Under the null hypothesis that the reduced model is correct, $F \sim F_{p-p_0,n-p}$.

The normal error assumption can be relaxed if the sample size $n$ is large. First, we know that $(\mathbf{X}^T\mathbf{X})^{\frac{1}{2}}(\widehat{\boldsymbol{\beta}}-\boldsymbol{\beta})/\sigma$ always has zero mean and an identity variance-covariance matrix. On the other hand, (2.8) gives us

$$(\mathbf{X}^T\mathbf{X})^{\frac{1}{2}}(\widehat{\boldsymbol{\beta}}-\boldsymbol{\beta})/\sigma = (\mathbf{X}^T\mathbf{X})^{-\frac{1}{2}}\mathbf{X}^T\boldsymbol{\varepsilon}/\sigma.$$

Observe that $(\mathbf{X}^T\mathbf{X})^{-\frac{1}{2}}\mathbf{X}^T\boldsymbol{\varepsilon}/\sigma$ is a linear combination of $n$ i.i.d. random variables $\{\varepsilon_i\}_{i=1}^n$ with zero mean and variance 1. Then the central limit theorem implies that under some regularity conditions,

$$\widehat{\boldsymbol{\beta}} \xrightarrow{D} N(\boldsymbol{\beta}, (\mathbf{X}^T\mathbf{X})^{-1}\sigma^2). \tag{2.13}$$

Consequently, when $n$ is large, the distribution of the $t$ test statistic in (2.11) is approximately $N(0,1)$, and the distribution of the $F$ test statistic in (2.12) is approximately $\chi^2_{p-p_0}/(p-p_0)$.

## 2.3 Weighted Least-Squares

The method of least-squares can be further generalized to handle the situations where errors are *heteroscedastic* or correlated. In the linear regression model (2.2), we would like to keep the assumption $\mathrm{E}(\boldsymbol{\varepsilon}|\mathbf{X}) = 0$ which means there is no structure information left in the error term. However, the constant variance assumption $\mathrm{Var}(\varepsilon_i|\mathbf{X}_i) = \sigma^2$ may not likely hold in many applications. For example, if $y_i$ is the average response value of the $i$th subject in a study in which $k_i$ many repeated measurements have been taken, then it would be more reasonable to assume $\mathrm{Var}(\varepsilon_i|\mathbf{X}_i) = \sigma^2/k_i$.

Let us consider a modification of model (2.1) as follows

$$Y_i = \sum_{j=1}^{p} X_{ij}\beta_j + \varepsilon_i; \quad \mathrm{Var}(\varepsilon_i|\mathbf{X}_i) = \sigma^2 v_i \tag{2.14}$$

where $v_i$s are known positive constants but $\sigma^2$ remains unknown. One can still use the ordinary least-squares (OLS) estimator $\widehat{\boldsymbol{\beta}} = (\mathbf{X}^T\mathbf{X})^{-1}\mathbf{X}^T\mathbf{Y}$. It is easy to show that the OLS estimator is unbiased but no longer BLUE. In fact, the OLS estimator can be improved by using the *weighted least-squares* method.

Let $Y_i^* = v_i^{-1/2}Y_i$, $X_{ij}^* = v_i^{-1/2}X_{ij}$, $\varepsilon_i^* = v_i^{-1/2}\varepsilon_i$. Then the new model (2.14) can be written as

$$Y_i^* = \sum_{j=1}^{p} X_{ij}^*\beta_j + \varepsilon_i^* \tag{2.15}$$

with $\text{Var}(\varepsilon_i^*|\mathbf{X}_i^*) = \sigma^2$. Therefore, the working data $\{(X_{i1}^*, \cdots, X_{ip}^*, Y_i^*)\}_{i=1}^n$ obey the standard *homoscedastic* linear regression model. Applying the standard least-squares method to the working data, we have

$$\widehat{\boldsymbol{\beta}}^{wls} = \text{argmin}_{\boldsymbol{\beta}} \sum_{i=1}^{n}\left(Y_i^* - \sum_{j=1}^{p} X_{ij}^*\beta_j\right)^2 = \text{argmin}_{\boldsymbol{\beta}} \sum_{i=1}^{n} v_i^{-1}\left(Y_i - \sum_{j=1}^{p} X_{ij}\beta_j\right)^2.$$

It follows easily from Theorem 2.2 that the weighted least-squares estimator is the BLUE for $\boldsymbol{\beta}$.

In model (2.14) the errors are assumed to be uncorrelated. In general, the method of least-squares can be extended to handle heteroscedastic and correlated errors.

Assume that

$$\mathbf{Y} = \mathbf{X}\boldsymbol{\beta} + \boldsymbol{\varepsilon}.$$

and the variance-covariance matrix of $\boldsymbol{\varepsilon}$ is given

$$\text{Var}(\boldsymbol{\varepsilon}|\mathbf{X}) = \sigma^2\mathbf{W}, \tag{2.16}$$

in which $\mathbf{W}$ is a known positive definite matrix. Let $\mathbf{W}^{-1/2}$ be the square root of $\mathbf{W}^{-1}$, i.e.,

$$(\mathbf{W}^{-1/2})^T\mathbf{W}^{-1/2} = \mathbf{W}^{-1}.$$

Then

$$\text{Var}(\mathbf{W}^{-1/2}\boldsymbol{\varepsilon}) = \sigma^2\mathbf{I},$$

which are homoscedastic and uncorrelated.

Define the working data as follows:

$$\mathbf{Y}^* = \mathbf{W}^{-1/2}\mathbf{Y}, \quad \mathbf{X}^* = \mathbf{W}^{-1/2}\mathbf{X}, \quad \boldsymbol{\varepsilon}^* = \mathbf{W}^{-1/2}\boldsymbol{\varepsilon}.$$

Then we have

$$\mathbf{Y}^* = \mathbf{X}^*\boldsymbol{\beta} + \boldsymbol{\varepsilon}^*. \tag{2.17}$$

Thus, we can apply the standard least-squares to the working data. First, the residual sum-of-squares (RSS) is

$$\text{RSS}(\boldsymbol{\beta}) = ||\mathbf{Y}^* - \mathbf{X}^*\boldsymbol{\beta}||^2 = (\mathbf{Y} - \mathbf{X}\boldsymbol{\beta})^T\mathbf{W}^{-1}(\mathbf{Y} - \mathbf{X}\boldsymbol{\beta}). \tag{2.18}$$

Then the *general least-squares* estimator is defined by

$$\begin{aligned}\widehat{\boldsymbol{\beta}} &= \operatorname{argmin}_{\boldsymbol{\beta}} \mathrm{RSS}(\boldsymbol{\beta}) \\ &= (\mathbf{X}^{*T}\mathbf{X}^{*})^{-1}\mathbf{X}^{*T}\mathbf{Y}^{*} \\ &= (\mathbf{X}^{T}\mathbf{W}^{-1}\mathbf{X})^{-1}\mathbf{X}^{T}\mathbf{W}^{-1}\mathbf{Y}.\end{aligned} \tag{2.19}$$

Again, $\widehat{\boldsymbol{\beta}}$ is the BLUE according to Theorem 2.2.

In practice, it is difficult to know precisely the $n \times n$ covariance matrix $\mathbf{W}$; the misspecification of $\mathbf{W}$ in the general least-squares seems hard to avoid. Let us examine the robustness of the general least-squares estimate. Assume that $\mathrm{Var}(\boldsymbol{\varepsilon}) = \sigma^2 \mathbf{W}_0$, where $\mathbf{W}_0$ is unknown to us, but we employ the general least-squares method (2.19) with the wrong covariance matrix $\mathbf{W}$. We can see that the general least-squares estimator is still unbiased:

$$\mathrm{E}(\widehat{\boldsymbol{\beta}}|\mathbf{X}) = (\mathbf{X}^{T}\mathbf{W}^{-1}\mathbf{X})^{-1}\mathbf{X}^{T}\mathbf{W}^{-1}\mathbf{X}\boldsymbol{\beta} = \boldsymbol{\beta}.$$

Furthermore, the variance-covariance matrix is given by

$$\mathrm{Var}(\widehat{\boldsymbol{\beta}}) = (\mathbf{X}^{T}\mathbf{W}^{-1}\mathbf{X})^{-1}(\mathbf{X}^{T}\mathbf{W}^{-1}\mathbf{W}_0\mathbf{W}^{-1}\mathbf{X})(\mathbf{X}^{T}\mathbf{W}^{-1}\mathbf{X})^{-1},$$

which is of order $O(n^{-1})$ under some mild conditions. In other words, using the wrong covariance matrix would still give us a root-$n$ consistent estimate. So even when errors are heteroscedastic and correlated, the ordinary least-squares estimate with $\mathbf{W} = \mathbf{I}$ and the weighted least-squares estimate with $\mathbf{W} = \mathrm{diag}(\mathbf{W}_0)$ still give us an unbiased and $n^{-1/2}$ consistent estimator. Of course, we still prefer using a working $\mathbf{W}$ matrix that is identical or close to the true $\mathbf{W}_0$.

## 2.4 Box-Cox Transformation

In practice we often take a transformation of the response variable before fitting a linear regression model. The idea is that the transformed response variable can be modeled by the set of covariates via the classical multiple linear regression model. For example, in many engineering problems we expect $Y \propto X_1^{\beta_1} X_2^{\beta_2} \cdot X_p^{\beta_p}$ where all variables are positive. Then a linear model seems proper by taking logarithms: $\log(Y) = \sum_{j=1}^{p} \beta_j X_j + \varepsilon$. If we assume $\varepsilon \sim N(0, \sigma^2)$, then in the original scale the model is $Y = (\prod_j^p X_j^{\beta_j})\varepsilon^*$ where $\varepsilon^*$ is a log-normal random variable: $\log \varepsilon^* \sim N(0, \sigma^2)$.

Box and Cox (1964) advocated the variable transformation idea in linear regression and also proposed a systematic way to estimate the transformation function from data. Their method is now known as the *Box-Cox transform* in the literature. Box and Cox (1964) suggested a parametric family for the transformation function. Let $Y^{(\lambda)}$ denote the transformed response where $\lambda$ parameterizes the transformation function:

$$Y^{(\lambda)} = \begin{cases} \frac{Y^{\lambda}-1}{\lambda} & \text{if } \lambda \neq 0 \\ \log(Y) & \text{if } \lambda = 0 \end{cases}.$$

The Box-Cox model assumes that

$$Y^{(\lambda)} = \sum_{j=1}^{p} X_j\beta_j + \varepsilon$$

where $\varepsilon \sim N(0, \sigma^2)$.

The likelihood function of the Box-Cox model is given by

$$L(\lambda, \boldsymbol{\beta}, \sigma^2) = (\frac{1}{\sqrt{2\pi}\sigma})^n e^{-\frac{1}{2\sigma^2}\|\mathbf{Y}^{(\lambda)} - \mathbf{X}\boldsymbol{\beta}\|^2} \cdot J(\lambda, \mathbf{Y})$$

where $J(\lambda, \mathbf{Y}) = \prod_{i=1}^n |\frac{dy_i^{(\lambda)}}{dy_i}| = (\prod_{i=1}^n |y_i|)^{\lambda-1}$. Given $\lambda$, the maximum likelihood estimators (MLE) of $\boldsymbol{\beta}$ and $\sigma^2$ are obtained by the ordinary least-squares:

$$\widehat{\boldsymbol{\beta}}(\lambda) = (\mathbf{X}^T\mathbf{X})^{-1}\mathbf{X}^T\mathbf{Y}^{(\lambda)}, \quad \widehat{\sigma}^2(\lambda) = \frac{1}{n}\|\mathbf{Y}^{(\lambda)} - \mathbf{X}(\mathbf{X}^T\mathbf{X})^{-1}\mathbf{X}^T\mathbf{Y}^{(\lambda)}\|^2.$$

Plugging $\widehat{\boldsymbol{\beta}}(\lambda), \widehat{\sigma}^2(\lambda)$ into $L(\lambda, \boldsymbol{\beta}, \sigma^2)$ yields a likelihood function of $\lambda$

$$\log L(\lambda) = (\lambda - 1)\sum_{i=1}^{n} \log(|y_i|) - \frac{n}{2}\log\widehat{\sigma}^2(\lambda) - \frac{n}{2}.$$

Then the MLE of $\lambda$ is

$$\widehat{\lambda}_{mle} = \text{argmax}_\lambda \log L(\lambda),$$

and the MLE of $\boldsymbol{\beta}$ and $\sigma^2$ are $\widehat{\boldsymbol{\beta}}(\widehat{\lambda}_{mle})$ and $\widehat{\sigma}^2(\widehat{\lambda}_{mle})$, respectively.

## 2.5 Model Building and Basis Expansions

Multiple linear regression can be used to produce nonlinear regression and other very complicated models. The key idea is to create new covariates from the original ones by adopting some transformations. We then fit a multiple linear regression model using augmented covariates.

For simplicity, we first illustrate some useful transformations in the case of $p = 1$, which is closely related to the curve fitting problem in *nonparametric regression*. In a nonparametric regression model

$$Y = f(X) + \varepsilon,$$

we do not assume a specific form of the regression function $f(x)$, but assume only some qualitative aspects of the regression function. Examples include that $f(\cdot)$ is continuous with a certain number of derivatives or that $f(\cdot)$ is convex. The aim is to estimate the function $f(x)$ and its derivatives, without a specific parametric form of $f(\cdot)$. See, for example Fan and Gijbels (1996), Li and Racine (2007), Hastie, Tibshirani and Friedman (2009), among others.

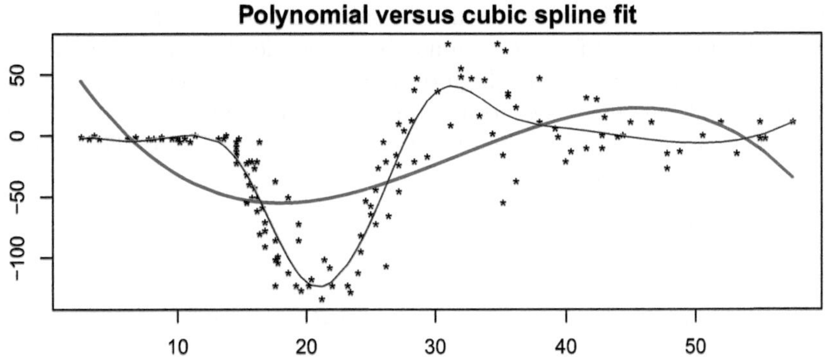


Figure 2.2: Scatter plot of time (in milliseconds) after a simulated impact on motorcycles against the head acceleration of a test object. Red = cubic polynomial fit, blue = cubic spline fit.

### 2.5.1 *Polynomial regression*

Without loss of generality, assume $X$ is bounded on $[0,1]$ for simplicity. The Weierstrass approximation theorem states that any continuous $f(x)$ can be uniformly approximated by a polynomial function up to any precision factor. Let us approximate the model by

$$Y = \underbrace{\beta_0 + \beta_1 X + \cdots + \beta_d X^d}_{\approx f(X)} + \varepsilon$$

This *polynomial regression* is a multiple regression problem by setting $X_0 = 1, X_1 = X, \cdots, X_d = X^d$. The design matrix now becomes

$$\mathbf{B}_1 = \begin{pmatrix} 1 & x_1 & \cdots & x_1^d \\ \vdots & \vdots & \cdots & \vdots \\ 1 & x_n & \cdots & x_n^d \end{pmatrix}.$$

We estimate $f(x)$ by

$$\widehat{f}(x) = \widehat{\beta}_0 + \sum_{m=1}^{d} \widehat{\beta}_m x^m,$$

where $\widehat{\boldsymbol{\beta}} = (\mathbf{B}_1^T \mathbf{B}_1)^{-1} \mathbf{B}_1^T \mathbf{Y}$ is the least-squares estimate.

Polynomial functions have derivatives everywhere and are global functions. They are not very flexible in approximating functions with local features such as functions with various degrees of smoothness at different locations. Figure 2.2 shows the cubic polynomial fit to a motorcycle data. Clearly, it does not fit the data very well. Increasing the order of the polynomial fits will help reduce the bias issue, but will not solve the lack of fit issue. This is because the underlying function cannot be economically approximated by a polynomial function. It requires high-order polynomials to reduce approximation biases, but this increases both variances and instability of the fits. This leads to the introduction of spline functions that allow for more flexibility in function approximation.

### 2.5.2 *Spline regression*

Let $\tau_0 < \tau_1 < \cdots < \tau_{K+1}$. A *spline function* of degree $d$ on $[\tau_0, \tau_{K+1}]$ is a piecewise polynomial function of degree $d$ on intervals $[\tau_j, \tau_{j+1})$ ($j = 0, \cdots, K$), with continuous first $d-1$ derivatives. The points where the spline function might not have continuous $d^{th}$ derivatives are $\{\tau_j\}_{j=1}^K$, which are called *knots*. Thus, a cubic spline function is a piecewise polynomial function with continuous first two derivatives and the points where the third derivative might not exist are called knots of the cubic spline. An example of a cubic fit is given by Figure 2.2.

All spline functions of degree $d$ form a linear space. Let us determine its basis functions.

**Linear Splines:** A continuous function on $[0, 1]$ can also be approximated by a piecewise constant or linear function. We wish to use a continuous function to approximate $f(x)$. Since a piecewise constant function is not continuous unless the function is a constant in the entire interval, we use a continuous piecewise linear function to fit $f(x)$. Suppose that we split the interval $[0, 1]$ into three regions: $[0, \tau_1], [\tau_1, \tau_2], [\tau_2, 1]$ with given knots $\tau_1, \tau_2$. Denote by $l(x)$ the continuous piecewise linear function. In the first interval $[0, \tau_1]$ we write

$$l(x) = \beta_0 + \beta_1 x, \; x \in [0, \tau_1],$$

as it is linear. Since $l(x)$ must be continuous at $\tau_1$, the newly added linear function must have an intercept 0 at point $\tau_1$. Thus, in $[\tau_1, \tau_2]$ we must have

$$l(x) = \beta_0 + \beta_1 x + \beta_2 (x - \tau_1)_+, \; x \in [\tau_1, \tau_2],$$

where $z_+$ equals $z$ if $z > 0$ and zero otherwise. The function is linear in $[\tau_1, \tau_2]$ with slope $\beta_1 + \beta_2$. Likewise, in $[\tau_2, 1]$ we write

$$l(x) = \beta_0 + \beta_1 x + \beta_2 (x - \tau_1)_+ + \beta_3 (x - \tau_2)_+, \; x \in [\tau_2, 1].$$

The function is now clearly a piecewise linear function with possible different slopes on different intervals. Therefore, the basis functions are

$$B_0(x) = 1, B_1(x) = x, B_2(x) = (x - \tau_1)_+, B_3(x) = (x - \tau_2)_+; \tag{2.20}$$

which are called a *linear spline* basis. We then approximate the nonparametric regression model as

$$Y = \underbrace{\beta_0 B_0(X) + \beta_1 B_1(X) + \beta_2 B_2(X) + \beta_3 B_3(X)}_{\approx f(X)} + \varepsilon.$$

This is again a multiple regression problem where we set $X_0 = B_0(X), X_1 = B_1(X), X_2 = B_2(X), X_3 = B_3(X)$. The corresponding design matrix becomes

$$\mathbf{B}_2 = \begin{pmatrix} 1 & x_1 & (x_1-\tau_1)_+ & (x_1-\tau_2)_+ \\ \vdots & \vdots & \vdots & \vdots \\ 1 & x_n & (x_n-\tau_1)_+ & (x_n-\tau_2)_+ \end{pmatrix},$$

and we estimate $f(x)$ by

$$\widehat{f}(x) = \widehat{\beta}_0 + \widehat{\beta}_1 x + \widehat{\beta}_2 (x-\tau_1)_+ + \widehat{\beta}_3 (x-\tau_2)_+,$$

where $\widehat{\boldsymbol{\beta}} = (\mathbf{B}_2^T\mathbf{B}_2)^{-1}\mathbf{B}_2^T\mathbf{Y}$. The above method applies more generally to a multiple knot setting for the data on any intervals.

**Cubic Splines:** We can further consider fitting piecewise polynomials whose derivatives are also continuous. A popular choice is the so-called cubic spline that is a piecewise cubic polynomial function with continuous first and second derivatives. Again, we consider two knots and three regions: $[0,\tau_1], [\tau_1,\tau_2], [\tau_2,1]$. Let $c(x)$ be a cubic spline. In $[0,\tau_1]$ we write

$$c(x) = \beta_0 + \beta_1 x + \beta_2 x^2 + \beta_3 x^3, \ x \le \tau_1.$$

And $c(x) = \beta_0 + \beta_1 x + \beta_2 x^2 + \beta_3 x^3 + \delta(x)$ in $[\tau_1,\tau_2]$. By definition, $\delta(x)$ is a cubic function in $[\tau_1,\tau_2]$ and its first and second derivatives equal zero at $x = \tau_1$. Then we must have

$$\delta(x) = \beta_4 (x-\tau_1)_+^3, \ x \in [\tau_1,\tau_2]$$

which means

$$c(x) = \beta_0 + \beta_1 x + \beta_2 x^2 + \beta_3 x^3 + \beta_4 (x-\tau_1)_+^3, \ x \in [\tau_1,\tau_2].$$

Likewise, in $[\tau_2, 1]$ we must have

$$c(x) = \beta_0 + \beta_1 x + \beta_2 x^2 + \beta_3 x^3 + \beta_4 (x-\tau_1)_+^3 + \beta_5 (x-\tau_2)_+^3, \ x > \tau_2.$$

Therefore, the basis functions are

$$\begin{aligned} B_0(x) = 1, B_1(x) = x, B_2(x) = x^2, B_3(x) = x^3 \\ B_4(x) = (x-\tau_1)_+^3, B_5(x) = (x-\tau_2)_+^3. \end{aligned}$$

The corresponding transformed design matrix becomes

$$\mathbf{B}_3 = \begin{pmatrix} 1 & x_1 & x_1^2 & x_1^3 & (x_1-\tau_1)_+^3 & (x_1-\tau_2)_+^3 \\ \vdots & \vdots & \vdots & \vdots & \vdots & \vdots \\ 1 & x_n & x_n^2 & x_n^3 & (x_n-\tau_1)_+^3 & (x_n-\tau_2)_+^3 \end{pmatrix},$$

and we estimate $f(x)$ by

$$\widehat{f}(x) = \widehat{\beta}_0 + \widehat{\beta}_1 x + \widehat{\beta}_2 x^2 + \widehat{\beta}_3 x^3 + \widehat{\beta}_4 (x - \tau_1)_+^3 + \widehat{\beta}_5 (x - \tau_2)_+^3,$$

where $\widehat{\boldsymbol{\beta}} = (\mathbf{B}_3^T \mathbf{B}_3)^{-1} \mathbf{B}_3^T \mathbf{Y}$ is the least-squares estimate of the coefficients.

In general, if there are $K$ knots $\{\tau_1, \cdots, \tau_K\}$, then the *basis functions* of *cubic splines* are

$$B_0(x) = 1, B_1(x) = x, B_2(x) = x^2, B_3(x) = x^3$$
$$B_4(x) = (x - \tau_1)_+^3, \cdots, B_{K+3}(x) = (x - \tau_K)_+^3.$$

By approximating the nonparametric function $f(X)$ by the spline function with knots $\{\tau_j\}_{j=1}^K$, we have

$$Y = \underbrace{\beta_0 B_0(X) + \beta_1 B_1(X) + \cdots + \beta_{K+3} B_{K+3}(X)}_{\approx f(X)} + \varepsilon \tag{2.21}$$

This *spline regression* is again a multiple regression problem.

**Natural Cubic Splines:** Extrapolation is always a serious issue in regression. It is not wise to fit a cubic function to a region where the observations are scarce. If we must, extrapolation with a linear function is preferred. A *natural cubic spline* is a special cubic spline with additional constraints: the cubic spline must be linear beyond two end knots. Consider a natural cubic spline, NC$(x)$, with knots at $\{\tau_1, \cdots, \tau_K\}$. By its cubic spline representation, we can write

$$\mathrm{NC}(x) = \beta_0 + \beta_1 x + \beta_2 x^2 + \beta_3 x^3 + \sum_{j=1}^{K} \beta_{3+j} (x - \tau_j)_+^3.$$

First, NC$(x)$ is linear for $x < \tau_1$, which implies that

$$\beta_2 = \beta_3 = 0.$$

Second, NC$(x)$ is linear for $x > \tau_K$, which means that

$$\sum_{j=1}^{K} \beta_{3+j} = 0, \qquad \sum_{j=1}^{K} \tau_j \beta_{3+j} = 0,$$

corresponding to the coefficients for the cubic and quadratic term of the polynomial $\sum_{j=1}^K \beta_{3+j}(x - \tau_j)^3$ for $x > \tau_K$. We solve for $\beta_{K+2}, \beta_{K+3}$ from the above equations and then write NC$(x)$ as

$$\mathrm{NC}(x) \quad = \quad \sum_{j=0}^{K-1} \beta_j B_j(x),$$

where the *natural cubic spline* basis functions are given by

$$B_0(x) = 1, B_1(x) = x,$$
$$B_{j+1}(x) = \frac{(x-\tau_j)_+^3 - (x-\tau_K)_+^3}{\tau_j - \tau_K} - \frac{(x-\tau_{K-1})_+^3 - (x-\tau_K)_+^3}{\tau_{K-1} - \tau_K}$$
for $j = 1, \cdots, K-2$.

Again, by approximating the nonparametric function with the natural cubic spline, we have

$$Y = \sum_{j=0}^{K-1} \beta_j B_j(X) + \varepsilon. \tag{2.22}$$

which can be solved by using multiple regression techniques.

### 2.5.3 Multiple covariates

The concept of polynomial regression extends to multivariate covariates. The simplest example is the bivariate regression model

$$Y = \beta_0 + \beta_1 X_1 + \beta_2 X_2 + \beta_3 X_1^2 + \beta_4 X_1 X_2 + \beta_5 X_2^2 + \varepsilon.$$

The term $X_1 X_2$ is called the *interaction*, which quantifies how $X_1$ and $X_2$ work together to contribute to the response. Often, one introduces interactions without using the quadratic term, leading to a slightly simplified model

$$Y = \beta_0 + \beta_1 X_1 + \beta_2 X_2 + \beta_3 X_1 X_2 + \varepsilon.$$

More generally, the multivariate quadratic regression is of the form

$$Y = \sum_{j=1}^{p} \beta_j X_j + \sum_{j \le k} \beta_{jk} X_j X_k + \varepsilon \tag{2.23}$$

and the multivariate regression with main effects (the linear terms) and interactions is of the form

$$Y = \sum_{j=1}^{p} \beta_j X_j + \sum_{j < k} \beta_{jk} X_j X_k + \varepsilon. \tag{2.24}$$

This concept can also be extended to the multivariate spline case. The basis function can be the tensor of the univariate spline basis function for not only unstructured $f(\mathbf{x})$, but also other basis functions for structured $f(\mathbf{x})$. Unstructured nonparametric functions are not very useful: If each variable uses 100 basis functions, then there are $100^p$ basis functions in the tensor products, which is prohibitively large for say, $p = 10$. Such an issue is termed the "curse-of-dimensionality" in literature. See Hastie and Tibshirani (1990) and Fan and Gijbels (1996). On the other hand, for the structured multivariate model, such as the following additive model (Stone, 1985, 1994; Hastie and Tibshirani, 1990),

$$Y = f_1(X_1) + \cdots + f_p(X_p) + \varepsilon \tag{2.25}$$

the basis functions are simply the collection of all univariate basis functions for approximating $f_1, \cdots, f_p$. The total number grows only linearly with $p$.

In general, let $B_m(\mathbf{x})$ be the basis functions $m = 1, \cdots, M$. Then, we approximate the multivariate nonparametric regression model $Y = f(\mathbf{X}) + \varepsilon$ by

$$Y = \sum_{m=1}^{M} \beta_j B_j(\mathbf{X}) + \varepsilon. \tag{2.26}$$

This can be fit using a multiple regression technique. The new design matrix is

$$\mathbf{B} = \begin{pmatrix} B_1(\mathbf{X}_1) & \cdots & B_M(\mathbf{X}_1) \\ \vdots & \cdots & \vdots \\ B_1(\mathbf{X}_n) & \cdots & B_M(\mathbf{X}_n) \end{pmatrix}$$

and the least-squares estimate is given by

$$\widehat{f}(\mathbf{x}) = \sum_{m=1}^{M} \widehat{\beta}_m B_m(\mathbf{x}),$$

where

$$\widehat{\beta} = (\mathbf{B}^T\mathbf{B})^{-1}\mathbf{B}^T\mathbf{Y}.$$

The above fitting implicitly assumes that $M \ll n$. This condition in fact can easily be violated in unstructured multivariate nonparametric regression. For the additive model (2.25), in which we assume $f(\mathbf{x}) = \sum_{j=1}^{p} f_j(x_j)$ where each $f_j(x_j)$ is a smooth univariate function of $x_j$, the univariate basis expansion ideas can be readily applied to approximation of each $f_j(x_j)$:

$$f_j(x_j) \approx \sum_{m=1}^{M_j} B_{jm}(x_j)\beta_{jm}$$

which implies that the fitted regression function is

$$f(\mathbf{x}) \approx \sum_{j=1}^{p}\sum_{m=1}^{M_j} B_{jm}(x_j)\beta_{jm}.$$

In Section 2.6.5 and Section 2.7 we introduce a fully nonparametric multiple regression technique which can be regarded as a basis expansion method where the basis functions are given by kernel functions.

## 2.6 Ridge Regression

### *2.6.1 Bias-variance tradeoff*

Recall that the ordinary least squares estimate is defined by $\widehat{\boldsymbol{\beta}} = (\mathbf{X}^T\mathbf{X})^{-1}\mathbf{X}^T\mathbf{Y}$ when $\mathbf{X}$ is of full rank. In practice, we often encounter highly correlated covariates, which is known as the *collinearity* issue. As a result, although $\mathbf{X}^T\mathbf{X}$ is still invertible, its smallest eigenvalue can be very small. Under the homoscedastic error model, the variance-covariance matrix of the OLS estimate is $\mathrm{Var}(\widehat{\boldsymbol{\beta}}) = (\mathbf{X}^T\mathbf{X})^{-1}\sigma^2$. Thus, the collinearity issue makes $\mathrm{Var}(\widehat{\boldsymbol{\beta}})$ large.

Hoerl and Kennard (1970) introduced the *ridge regression* estimator as follows:

$$\widehat{\boldsymbol{\beta}}_\lambda = (\mathbf{X}^T\mathbf{X} + \lambda\mathbf{I})^{-1}\mathbf{X}^T\mathbf{Y}, \tag{2.27}$$

where $\lambda > 0$ is a regularization parameter. In the usual case ($\mathbf{X}^T\mathbf{X}$ is invertible), ridge regression reduces to OLS by setting $\lambda = 0$. However, ridge regression is always well defined even when $\mathbf{X}$ is not full rank.

Under the assumption $\mathrm{Var}(\boldsymbol{\varepsilon}) = \sigma^2\mathbf{I}$, it is easy to show that

$$\mathrm{Var}(\widehat{\boldsymbol{\beta}}_\lambda) = (\mathbf{X}^T\mathbf{X} + \lambda\mathbf{I})^{-1}\mathbf{X}^T\mathbf{X}(\mathbf{X}^T\mathbf{X} + \lambda\mathbf{I})^{-1}\sigma^2. \tag{2.28}$$

We always have $\mathrm{Var}(\widehat{\boldsymbol{\beta}}_\lambda) < (\mathbf{X}^T\mathbf{X})^{-1}\sigma^2$. The ridge regression estimator reduces the estimation variance by paying a price in estimation bias:

$$\mathrm{E}(\widehat{\boldsymbol{\beta}}_\lambda) - \boldsymbol{\beta} = (\mathbf{X}^T\mathbf{X} + \lambda\mathbf{I})^{-1}\mathbf{X}^T\mathbf{X}\boldsymbol{\beta} - \boldsymbol{\beta} = -\lambda(\mathbf{X}^T\mathbf{X} + \lambda\mathbf{I})^{-1}\boldsymbol{\beta}. \tag{2.29}$$

The overall estimation accuracy is gauged by the mean squared error (MSE). For $\widehat{\boldsymbol{\beta}}_\lambda$ its MSE is given by

$$\mathrm{MSE}(\widehat{\boldsymbol{\beta}}_\lambda) = \mathrm{E}(\|\widehat{\boldsymbol{\beta}}_\lambda - \boldsymbol{\beta}\|^2). \tag{2.30}$$

By (2.28) and (2.29) we have

$$\begin{aligned}
\mathrm{MSE}(\widehat{\boldsymbol{\beta}}_\lambda) &= \mathrm{tr}\left((\mathbf{X}^T\mathbf{X} + \lambda\mathbf{I})^{-1}\mathbf{X}^T\mathbf{X}(\mathbf{X}^T\mathbf{X} + \lambda\mathbf{I})^{-1}\sigma^2\right) \\
&\quad + \lambda^2\boldsymbol{\beta}^T(\mathbf{X}^T\mathbf{X} + \lambda\mathbf{I})^{-2}\boldsymbol{\beta} \\
&= \mathrm{tr}\left((\mathbf{X}^T\mathbf{X} + \lambda\mathbf{I})^{-2}[\lambda^2\boldsymbol{\beta}\boldsymbol{\beta}^T + \sigma^2\mathbf{X}^T\mathbf{X}]\right).
\end{aligned} \tag{2.31}$$

It can be shown that $\frac{d\mathrm{MSE}(\widehat{\boldsymbol{\beta}}_\lambda)}{d\lambda}|_{\lambda=0} < 0$, which implies that there are some proper $\lambda$ values by which ridge regression improves OLS.

### 2.6.2 $\ell_2$ penalized least squares

Define a penalized residual sum-of-squares (PRSS) as follows:

$$\mathrm{PRSS}(\boldsymbol{\beta}|\lambda) = \sum_{i=1}^{n}(Y_i - \sum_{j=1}^{p} X_{ij}\beta_j)^2 + \lambda \sum_{j=1}^{p} \beta_j^2. \tag{2.32}$$

Then let

$$\widehat{\boldsymbol{\beta}}_\lambda = \mathrm{argmin}_{\boldsymbol{\beta}}\, \mathrm{PRSS}(\boldsymbol{\beta}|\lambda). \tag{2.33}$$

Note that we can write it in a matrix form

$$\mathrm{PRSS}(\boldsymbol{\beta}|\lambda) = \|\mathbf{Y} - \mathbf{X}\boldsymbol{\beta}\|^2 + \lambda\|\boldsymbol{\beta}\|^2.$$

The term $\lambda\|\boldsymbol{\beta}\|^2$ is called the $\ell_2$-penalty of $\boldsymbol{\beta}$. Taking derivatives with respect to $\boldsymbol{\beta}$ and setting it to zero, we solve the root of the following equation

$$-\mathbf{X}^T(\mathbf{Y} - \mathbf{X}\boldsymbol{\beta}) + \lambda\boldsymbol{\beta} = 0,$$

which yields

$$\widehat{\boldsymbol{\beta}}_\lambda = (\mathbf{X}^T\mathbf{X} + \lambda\mathbf{I})^{-1}\mathbf{X}^T\mathbf{Y}.$$

The above discussion shows that ridge regression is equivalent to the $\ell_2$ penalized least-squares.

We have seen that ridge regression can achieve a smaller MSE than OLS. In other words, the $\ell_2$ penalty term helps regularize (reduce) estimation variance and produces a better estimator when the reduction in variance exceeds the induced extra bias. From this perspective, one can also consider a more general $\ell_q$ penalized least-squares estimate

$$\min_{\boldsymbol{\beta}} \|\mathbf{Y} - \mathbf{X}\boldsymbol{\beta}\|^2 + \lambda \sum_{j=1}^{p} |\beta_j|^q \tag{2.34}$$

where $q$ is a positive constant. This is referred to as the Bridge estimator (Frank and Friedman, 1993). The $\ell_q$ penalty is strictly concave when $0 < q < 1$, and strictly convex when $q > 1$. For $q = 1$, the resulting $\ell_1$ penalized least-squares is also known as the Lasso (Tibshirani, 1996). Chapter 3 covers the Lasso in great detail. Among all Bridge estimators only the ridge regression has a nice closed-form solution with a general design matrix.

### 2.6.3 Bayesian interpretation

Ridge regression has a neat Bayesian interpretation in the sense that it can be a formal Bayes estimator. We begin with the homoscedastic Gaussian error model:

$$Y_i = \sum_{j=1}^{p} X_{ij}\beta_j + \varepsilon_i$$

and $\varepsilon_i|\mathbf{X}_i \sim N(0, \sigma^2)$. Now suppose that $\beta_j$'s are also independent $N(0, \tau^2)$ variables, which represent our knowledge about the regression coefficients before seeing the data. In Bayesian statistics, $N(0, \tau^2)$ is called the prior distribution of $\beta_j$. The model and the prior together give us the posterior distribution of $\boldsymbol{\beta}$ given the data (the conditional distribution of $\boldsymbol{\beta}$ given $\mathbf{Y}, \mathbf{X}$). Straightforward calculations yield

$$P(\boldsymbol{\beta}|\mathbf{Y}, \mathbf{X}) \propto \exp(-\frac{1}{2\sigma^2}\|\mathbf{Y} - \mathbf{X}\boldsymbol{\beta}\|^2)\exp(-\frac{1}{2\tau^2}\|\boldsymbol{\beta}\|^2). \tag{2.35}$$

A maximum posteriori probability (MAP) estimate is defined as

$$\begin{aligned} \widehat{\boldsymbol{\beta}}^{\text{MAP}} &= \text{argmax}_{\boldsymbol{\beta}} P(\boldsymbol{\beta}|\mathbf{Y}, \mathbf{X}) \\ &= \text{argmax}_{\boldsymbol{\beta}} \left\{ -\frac{1}{2\sigma^2}\|\mathbf{Y} - \mathbf{X}\boldsymbol{\beta}\|^2 - \frac{1}{2\tau^2}\|\boldsymbol{\beta}\|^2 \right\}. \end{aligned} \tag{2.36}$$

It is easy to see that $\widehat{\boldsymbol{\beta}}^{\text{MAP}}$ is ridge regression with $\lambda = \frac{\sigma^2}{\tau^2}$. Another popular Bayesian estimate is the posterior mean. In this model, the posterior mean and posterior mode are the same.

From the Bayesian perspective, it is easy to construct a generalized ridge regression estimator. Suppose that the prior distribution for the entire $\boldsymbol{\beta}$ vector is $N(0, \boldsymbol{\Sigma})$, where $\boldsymbol{\Sigma}$ is a general positive definite matrix. Then the posterior distribution is computed as

$$P(\boldsymbol{\beta}|\mathbf{Y}, \mathbf{X}) \propto \exp(-\frac{1}{2\sigma^2}\|\mathbf{Y} - \mathbf{X}\boldsymbol{\beta}\|^2)\exp(-\frac{1}{2}\boldsymbol{\beta}^T\boldsymbol{\Sigma}^{-1}\boldsymbol{\beta}). \tag{2.37}$$

The corresponding MAP estimate is

$$\begin{aligned} \widehat{\boldsymbol{\beta}}^{\text{MAP}} &= \text{argmax}_{\boldsymbol{\beta}} P(\boldsymbol{\beta}|\mathbf{Y}, \mathbf{X}) \\ &= \text{argmax}_{\boldsymbol{\beta}} \left\{ -\frac{1}{2\sigma^2}\|\mathbf{Y} - \mathbf{X}\boldsymbol{\beta}\|^2 - \frac{1}{2}\boldsymbol{\beta}^T\boldsymbol{\Sigma}^{-1}\boldsymbol{\beta} \right\}. \end{aligned} \tag{2.38}$$

It is easy to see that

$$\widehat{\boldsymbol{\beta}}^{\text{MAP}} = (\mathbf{X}^T\mathbf{X} + \sigma^2\boldsymbol{\Sigma}^{-1})^{-1}\mathbf{X}^T\mathbf{Y}. \tag{2.39}$$

This generalized ridge regression can take into account different scales of covariates, by an appropriate choice of $\boldsymbol{\Sigma}$.

### *2.6.4 Ridge regression solution path*

The performance of ridge regression heavily depends on the choice of $\lambda$. In practice we only need to compute ridge regression estimates at a fine grid of $\lambda$ values and then select the best from these candidate solutions. Although ridge regression is easy to compute for a $\lambda$ owing to its nice closed-form solution expression, the total cost could be high if the process is repeated many times.

Through a more careful analysis, one can see that the solutions of ridge regression at a fine grid of $\lambda$ values can be computed very efficiently via singular value decomposition.

Assume $n > p$ and $\mathbf{X}$ is full rank. The singular value decomposition (SVD) of $\mathbf{X}$ is given by

$$\mathbf{X} = \mathbf{U}\mathbf{D}\mathbf{V}^T$$

where $\mathbf{U}$ is a $n \times p$ orthogonal matrix, $\mathbf{V}$ is a $p \times p$ orthogonal matrix and $\mathbf{D}$ is a $p \times p$ diagonal matrix whose diagonal elements are the ordered (from large to small) singular values of $\mathbf{X}$. Then

$$\mathbf{X}^T\mathbf{X} = \mathbf{V}\mathbf{D}\mathbf{U}^T\mathbf{U}\mathbf{D}\mathbf{V}^T = \mathbf{V}\mathbf{D}^2\mathbf{V}^T,$$

$$\mathbf{X}^T\mathbf{X} + \lambda\mathbf{I} = \mathbf{V}\mathbf{D}^2\mathbf{V}^T + \lambda\mathbf{I} = \mathbf{V}(\mathbf{D}^2 + \lambda\mathbf{I})\mathbf{V}^T,$$

$$(\mathbf{X}^T\mathbf{X} + \lambda\mathbf{I})^{-1} = \mathbf{V}(\mathbf{D}^2 + \lambda\mathbf{I})^{-1}\mathbf{V}^T.$$

The ridge regression estimator $\widehat{\boldsymbol\beta}_\lambda$ can now be written as

$$\begin{aligned}\widehat{\boldsymbol\beta}_\lambda &= (\mathbf{X}^T\mathbf{X} + \lambda\mathbf{I})^{-1}\mathbf{X}^T\mathbf{Y} \\ &= \mathbf{V}(\mathbf{D}^2 + \lambda\mathbf{I})^{-1}\mathbf{D}\mathbf{U}^T\mathbf{Y} \\ &= \sum_{j=1}^{p} \frac{d_j}{d_j^2 + \lambda}\langle \mathbf{U}_j, \mathbf{Y}\rangle \mathbf{V}_j, \qquad (2.40)\end{aligned}$$

where $d_j$ is the $j^{th}$ diagonal element of $\mathbf{D}$ and $\langle \mathbf{U}_j, \mathbf{Y}\rangle$ is the inner product between $\mathbf{U}_j$ and $\mathbf{Y}$ and $\mathbf{U}_j$ ($\mathbf{V}_j$ are respectively the $j^{th}$ column of $\mathbf{U}$ and $\mathbf{V}$). In particular, when $\lambda = 0$, ridge regression reduces to OLS and we have

$$\widehat{\boldsymbol\beta}^{\mathrm{OLS}} = \mathbf{V}D^{-1}\mathbf{U}^T\mathbf{Y} = \sum_{j=1}^{p} \frac{1}{d_j}\langle \mathbf{U}_j, \mathbf{Y}\rangle \mathbf{V}_j. \qquad (2.41)$$

Based on (2.40) we suggest the following procedure to compute ridge regression at a fine grid $\lambda_1, \cdots, \lambda_M$:

1. Compute the SVD of $\mathbf{X}$ and save $\mathbf{U}, \mathbf{D}, \mathbf{V}$.
2. Compute $\mathbf{w}_j = \frac{1}{d_j}\langle \mathbf{U}_j \cdot \mathbf{Y}\rangle \mathbf{V}_j$ for $j = 1, \cdots, p$ and save $\mathbf{w}_j$s.
3. For $m = 1, 2 \cdots, M$,

   (i). compute $\gamma_j = \frac{d_j^2}{d_j^2 + \lambda_m}$

   (ii). compute $\widehat{\boldsymbol\beta}_{\lambda_m} = \sum_{j=1}^{p} \gamma_j \mathbf{w}_j$.

The essence of the above algorithm is to compute the common vectors $\{\mathbf{w}_j\}_{j=1}^p$ first and then utilize (2.40).

### 2.6.5 *Kernel ridge regression*

In this section we introduce a nonparametric generalization of ridge regression. Our discussion begins with the following theorem.

**Theorem 2.4** *Ridge regression estimator is equal to*

$$\widehat{\boldsymbol{\beta}}_\lambda = \mathbf{X}^T(\mathbf{X}\mathbf{X}^T + \lambda\mathbf{I})^{-1}\mathbf{Y} \tag{2.42}$$

*and the fitted value of* $Y$ *at* $\mathbf{x}$ *is*

$$\widehat{y} = \mathbf{x}^T\widehat{\boldsymbol{\beta}}_\lambda = \mathbf{x}^T\mathbf{X}^T(\mathbf{X}\mathbf{X}^T + \lambda\mathbf{I})^{-1}\mathbf{Y} \tag{2.43}$$

**Proof**. Observe the following identity

$$(\mathbf{X}^T\mathbf{X} + \lambda\mathbf{I})\mathbf{X}^T = \mathbf{X}^T\mathbf{X}\mathbf{X}^T + \lambda\mathbf{X}^T = \mathbf{X}^T(\mathbf{X}\mathbf{X}^T + \lambda\mathbf{I}).$$

Thus, we have

$$\mathbf{X}^T = (\mathbf{X}^T\mathbf{X} + \lambda\mathbf{I})^{-1}\mathbf{X}^T(\mathbf{X}\mathbf{X}^T + \lambda\mathbf{I})$$

and

$$\mathbf{X}^T(\mathbf{X}\mathbf{X}^T + \lambda\mathbf{I})^{-1} = (\mathbf{X}^T\mathbf{X} + \lambda\mathbf{I})^{-1}\mathbf{X}^T.$$

Then by using (2.27) we obtain (2.42) and hence (2.43). ∎

It is important to see that $\mathbf{X}\mathbf{X}^T$ and not $\mathbf{X}^T\mathbf{X}$ appears in the expression for $\widehat{\boldsymbol{\beta}}_\lambda$. Note that $\mathbf{X}\mathbf{X}^T$ is a $n \times n$ matrix and its $ij$ elements are $\langle \mathbf{x}_i, \mathbf{x}_j \rangle$. Similarly, $\mathbf{x}^T\mathbf{X}^T$ is an $n$-dimensional vector with the $i$th element being $\langle \mathbf{x}, \mathbf{x}_i \rangle$ $i = 1, \cdots, n$. Therefore, the prediction by ridge regression boils down to computing the inner product between $p$-dimensional covariate vectors. This is the foundation of the so-called "*kernel trick*".

Suppose that we use another "inner product" to replace the usual inner product in Theorem 2.4; then we may end up with a new ridge regression estimator. To be more specific, let us replace $\langle \mathbf{x}_i, \mathbf{x}_j \rangle$ with $K(\mathbf{x}_i, \mathbf{x}_j)$ where $K(\cdot, \cdot)$ is a known function:

$$\mathbf{x}^T\mathbf{X}^T \rightarrow (K(\mathbf{x}, \mathbf{X}_1), \cdots, K(\mathbf{x}, \mathbf{X}_n)),$$

$$\mathbf{X}\mathbf{X}^T \rightarrow \mathbf{K} = (K(\mathbf{X}_i, \mathbf{X}_j))_{1 \le i,j \le n}.$$

By doing so, we turn (2.43) into

$$\widehat{y} = (K(\mathbf{x}, \mathbf{X}_1), \cdots, K(\mathbf{x}, \mathbf{X}_n))(\mathbf{K} + \lambda\mathbf{I})^{-1}\mathbf{Y} = \sum_{i=1}^n \widehat{\alpha}_i K(\mathbf{x}, \mathbf{X}_i), \tag{2.44}$$

where $\widehat{\boldsymbol{\alpha}} = (\mathbf{K} + \lambda\mathbf{I})^{-1}\mathbf{Y}$. In particular, the fitted $\mathbf{Y}$ vector is

$$\widehat{\mathbf{Y}} = \mathbf{K}(\mathbf{K} + \lambda\mathbf{I})^{-1}\mathbf{Y}. \tag{2.45}$$

The above formula gives the so-called kernel ridge regression. Because $\mathbf{X}\mathbf{X}^T$ is at least positive semi-definite, it is required that $\mathbf{K}$ is also positive semi-definite. Some widely used kernel functions (Hastie, Tibshirani and Friedman, 2009) include

- *linear kernel*: $K(\mathbf{x}_i, \mathbf{x}_j) = \langle \mathbf{x}_i, \mathbf{x}_j \rangle$,
- *polynomial kernel*: $K(\mathbf{x}_i, \mathbf{x}_j) = (1 + \langle \mathbf{x}_i, \mathbf{x}_j \rangle)^d$, $d = 2, 3, \cdots$,
- *radial basis kernel*: $K(\mathbf{x}_i, \mathbf{x}_j) = e^{-\gamma\|\mathbf{x}_i - \mathbf{x}_j\|^2}$, $\gamma > 0$, which is the *Gaussian kernel*, and $K(\mathbf{x}_i, \mathbf{x}_j) = e^{-\gamma\|\mathbf{x}_i - \mathbf{x}_j\|}$, $\gamma > 0$, which is the *Laplacian kernel*.

To show how we get (2.45) more formally, let us consider to approximate the multivariate regression by using the kernel basis functions $\{K(\cdot, \mathbf{x}_j)\}_{j=1}^n$ so that our observed data are now modeled as

$$Y_i = \sum_{j=1}^{n} \alpha_j K(\mathbf{X}_i, \mathbf{X}_j) + \varepsilon_i$$

or in matrix form $\mathbf{Y} = \mathbf{K}\boldsymbol{\alpha} + \boldsymbol{\varepsilon}$. If we apply the ridge regression

$$\frac{1}{2}\|\mathbf{Y} - \mathbf{K}\boldsymbol{\alpha}\|^2 + \lambda\boldsymbol{\alpha}\mathbf{K}\boldsymbol{\alpha},$$

the minimizer of the above problem is

$$\widehat{\boldsymbol{\alpha}} = (\mathbf{K}^T\mathbf{K} + \lambda\mathbf{K})^{-1}\mathbf{K}^T\mathbf{Y} = \{\mathbf{K}(\mathbf{K} + \lambda\mathbf{I})\}^{-1}\mathbf{K}\mathbf{Y},$$

where we use the fact that $\mathbf{K}$ is symmetric. Assuming $\mathbf{K}$ is invertible, we easily get (2.45).

So far we have only derived the kernel ridge regression based on heuristics and the kernel trick. In Sec. 2.7 we show that the kernel ridge regression can be formally derived based on the theory of function estimation in a reproducing kernel Hilbert space.

## 2.7 Regression in Reproducing Kernel Hilbert Space

A *Hilbert space* is an abstract vector space endowed by the structure of an inner product. Let $\mathcal{X}$ be an arbitrary set and $\mathcal{H}$ be a Hilbert space of real-valued functions on $\mathcal{X}$, endowed by the inner product $\langle \cdot, \cdot \rangle_{\mathcal{H}}$. The evaluation functional over the Hilbert space of functions $\mathcal{H}$ is a linear functional that evaluates each function at a point $x$:

$$L_x : f \to f(x), \forall f \in \mathcal{H}.$$

A Hilbert space $\mathcal{H}$ is called a *reproducing kernel Hilbert space* (RKHS) if, for all $x \in \mathcal{X}$, the map $L_x$ is continuous at any $f \in \mathcal{H}$, namely, there exists some $C > 0$ such that

$$|L_x(f)| = |f(x)| \le C\|f\|_{\mathcal{H}}, \qquad \forall f \in \mathcal{H}.$$

By the Riesz representation theorem, for all $x \in \mathcal{X}$, there exists a unique element $K_x \in \mathcal{H}$ with the reproducing property

$$f(x) = L_x(f) = \langle f, K_x \rangle_{\mathcal{H}}, \qquad \forall f \in \mathcal{H}.$$

Since $K_x$ is itself a function in $\mathcal{H}$, it holds that for every $x' \in \mathcal{X}$, there exists a $K_{x'} \in \mathcal{H}$ such that

$$K_x(x') = \langle K_x,\ K_{x'}\rangle_{\mathcal{H}}.$$

This allows us to define the *reproducing kernel* $K(x, x') = \langle K_x,\ K_{x'}\rangle_{\mathcal{H}}$. From the definition, it is easy to see that the reproducing kernel $K$ is a symmetric and semi-positive function:

$$\sum_{i,j=1}^{n} c_i c_j K(x_i, x_j) = \sum_{i,j=1}^{n} c_i c_j \langle K_{x_i},\ K_{x_j}\rangle_{\mathcal{H}} = \Big\| \sum_{i=1}^{n} c_i K_{x_i} \Big\|_{\mathcal{H}}^2 \geq 0,$$

for all $c's$ and $x's$. The reproducing Hilbert space is a class of *nonparametric functions*, satisfying the above properties.

Let $\mathcal{H}_K$ denote the reproducing kernel Hilbert space (RKHS) with kernel $K(\mathbf{x}, \mathbf{x}')$ (Wahba, 1990; Halmos, 2017). Then, the kernel $K(\mathbf{x}, \mathbf{x}')$ admits the eigen-decomposition

$$K(\mathbf{x}, \mathbf{x}') = \sum_{j=1}^{\infty} \gamma_j \psi_j(\mathbf{x}) \psi_j(\mathbf{x}'). \tag{2.46}$$

where $\gamma_j \geq 0$ are eigen-values and $\sum_{j=1}^{\infty} \gamma_j^2 < \infty$. Let $g$ and $g'$ be any two functions in $\mathcal{H}_K$ with expansions in terms of these eigen-functions

$$g(\mathbf{x}) = \sum_{j=1}^{\infty} \beta_j \psi_j(\mathbf{x}), \qquad g'(\mathbf{x}) = \sum_{j=1}^{\infty} \beta_j' \psi_j(\mathbf{x})$$

and their inner product is defined as

$$\langle g, g' \rangle_{\mathcal{H}_K} = \sum_{j=1}^{\infty} \frac{\beta_j \beta_j'}{\gamma_j}. \tag{2.47}$$

The functional $\ell_2$ norm of $g(\mathbf{x})$ is equal to

$$\|g\|_{\mathcal{H}_K}^2 = \langle g, g \rangle_{\mathcal{H}_K} = \sum_{j=1}^{\infty} \frac{\beta_j^2}{\gamma_j}. \tag{2.48}$$

The first property shows the *reproducibility* of the kernel $K$.

**Theorem 2.5** *Let $g$ be a function in $\mathcal{H}_K$. The following identities hold:*

*(i). $\langle K(\cdot, \mathbf{x}'), g \rangle_{\mathcal{H}_K} = g(\mathbf{x}')$,*

*(ii). $\langle K(\cdot, \mathbf{x}_1), K(\cdot, \mathbf{x}_2) \rangle_{\mathcal{H}_K} = K(\mathbf{x}_1, \mathbf{x}_2)$.*

*(iii). If $g(\mathbf{x}) = \sum_{i=1}^{n} \alpha_i K(\mathbf{x}, \mathbf{x}_i)$, then $\|g\|_{\mathcal{H}_K}^2 = \sum_{i=1}^{n} \sum_{j=1}^{n} \alpha_i \alpha_j K(\mathbf{x}_i, \mathbf{x}_j)$.*

**Proof**. Write $g(\mathbf{x}) = \sum_{j=1}^{\infty} \beta_j \psi_j(\mathbf{x})$, by (2.46) we have $K(\mathbf{x}, \mathbf{x}') = \sum_{j=1}^{\infty} (\gamma_j \psi_j(\mathbf{x}')) \psi_j(\mathbf{x})$. Thus

$$\langle K(\cdot, \mathbf{x}'), g \rangle_{\mathcal{H}_K} = \sum_{j=1}^{\infty} \frac{\beta_j \gamma_j \psi_j(\mathbf{x}')}{\gamma_j} = \sum_{j=1}^{\infty} \beta_j \psi_j(\mathbf{x}') = g(\mathbf{x}').$$

This proves part (i). Now we apply part (i) to get part (ii) by letting $g(\mathbf{x}) = K(\mathbf{x}, \mathbf{x}_2)$.

For part (iii) we observe that

$$\begin{aligned} \|g\|^2_{\mathcal{H}_K} &= \langle \sum_{i=1}^n \alpha_i K(\mathbf{x}, \mathbf{x}_i), \sum_{j=1}^n \alpha_j K(\mathbf{x}, \mathbf{x}_j) \rangle_{\mathcal{H}_K} \\ &= \sum_{i=1}^n \sum_{j=1}^n \alpha_i \alpha_j \langle K(\mathbf{x}, \mathbf{x}_i), K(\mathbf{x}, \mathbf{x}_j) \rangle_{\mathcal{H}_K} \\ &= \sum_{i=1}^n \sum_{j=1}^n \alpha_i \alpha_j K(\mathbf{x}_i, \mathbf{x}_j), \end{aligned}$$

where we have used part (ii) in the final step.

■

Consider a general regression model

$$Y = f(\mathbf{X}) + \varepsilon \tag{2.49}$$

where $\varepsilon$ is independent of $\mathbf{X}$ and has zero mean and variance $\sigma^2$. Given a realization $\{(\mathbf{X}_i, Y_i)\}_{i=1}^n$ from the above model, we wish to fit the regression function in $\mathcal{H}_K$ via the following penalized least-squares:

$$\widehat{f} = \text{argmin}_{f \in \mathcal{H}_K} \sum_{i=1}^n [Y_i - f(\mathbf{X}_i)]^2 + \lambda \|f\|^2_{\mathcal{H}_K}, \quad \lambda > 0. \tag{2.50}$$

Note that without the $\|f\|^2_{\mathcal{H}_K}$ term there are infinitely many functions in $\mathcal{H}_K$ that can fit the observations perfectly, i.e., $Y_i = f(\mathbf{X}_i)$ for $i = 1, \cdots, n$. By using the eigen-function expansion of $f$

$$f(\mathbf{x}) = \sum_{j=1}^{\infty} \beta_j \psi_j(\mathbf{x}), \tag{2.51}$$

an equivalent formulation of (2.50) is

$$\min_{\{\beta_j\}_{j=1}^{\infty}} \sum_{i=1}^n [Y_i - \sum_{j=1}^{\infty} \beta_j \psi_j(\mathbf{X}_i)]^2 + \lambda \sum_{j=1}^{\infty} \frac{1}{\gamma_j} \beta_j^2. \tag{2.52}$$

Define $\beta_j^* = \frac{\beta_j}{\sqrt{\gamma_j}}$ and $\psi_j^* = \sqrt{\gamma_j}\psi_j$ for $j = 1, 2, \cdots$. Then (2.52) can be rewritten as

$$\min_{\{\beta_j^*\}_{j=1}^{\infty}} \sum_{i=1}^n [Y_i - \sum_{j=1}^{\infty} \beta_j^* \psi_j^*(\mathbf{X}_i)]^2 + \lambda \sum_{j=1}^{\infty} (\beta_j^*)^2. \tag{2.53}$$

The above can be seen as a ridge regression estimate in an infinite dimensional space. Symbolically, our covariate vector is now $(\psi_1^*(\mathbf{x}), \psi_2^*(\mathbf{x}), \cdots)$ and the enlarged design matrix is

$$\mathbf{\Psi} = \begin{pmatrix} \psi_1^*(\mathbf{X}_1) & \cdots & \psi_j^*(\mathbf{X}_1) & \cdots \\ \vdots & \cdots & \vdots & \cdots \\ \psi_1^*(\mathbf{X}_n) & \cdots & \psi_j^*(\mathbf{X}_n) & \cdots \end{pmatrix}.$$

Because Theorem 2.4 is valid for any finite dimensional covariate space, it is not unreasonable to extrapolate it to the above infinite dimensional setting. The key assumption is that we can compute the inner product in the enlarged space. This is indeed true because

$$\text{inner product} = \sum_{j=1}^{\infty} \psi_j^*(\mathbf{x}_i)\psi_j^*(\mathbf{x}_{i'}) = \sum_{j=1}^{\infty} \gamma_j \psi_j(\mathbf{x}_i)\psi_j(\mathbf{x}_{i'}) = K(\mathbf{x}_i, \mathbf{x}_j).$$

Now we can directly apply the kernel ridge regression formula from Section 2.6.5 to get

$$\widehat{f}(\mathbf{x}) = \sum_{i=1}^{n} \widehat{\alpha}_i K(\mathbf{x}, \mathbf{X}_i), \tag{2.54}$$

where $\mathbf{K} = (K(\mathbf{X}_i, \mathbf{X}_j))_{1 \le i,j \le n}$ and

$$\widehat{\boldsymbol{\alpha}} = (\mathbf{K} + \lambda \mathbf{I})^{-1}\mathbf{Y}. \tag{2.55}$$

We have derived (2.54) by extrapolating Theorem 2.4 to an infinite dimensional space. Although the idea seems correct, we still need a rigorous proof. Moreover, Theorem 2.4 only concerns ridge regression, but it turns out that (2.54) can be made much more general.

**Theorem 2.6** *Consider a general loss function $L(y, f(\mathbf{x}))$ and let*

$$\widehat{f} = \operatorname{argmin}_{f \in \mathcal{H}_K} \sum_{i=1}^{n} L(y_i, f(\mathbf{x}_i)) + P_\lambda(\|f\|_{\mathcal{H}_K}), \quad \lambda > 0.$$

*where $P_\lambda(t)$ is a strictly increasing function on $[0, \infty)$. Then we must have*

$$\widehat{f}(\mathbf{x}) = \sum_{i=1}^{n} \widehat{\alpha}_i K(\mathbf{x}, \mathbf{X}_i) \tag{2.56}$$

*where $\widehat{\boldsymbol{\alpha}} = (\widehat{\alpha}_1, \cdots, \widehat{\alpha}_n)$ is the solution to the following problem*

$$\min_{\boldsymbol{\alpha}} \sum_{i=1}^{n} L\left(y_i, \sum_{j=1}^{n} \alpha_j K(\mathbf{x}, \mathbf{x}_j)\right) + P_\lambda(\sqrt{\boldsymbol{\alpha}^T \mathbf{K} \boldsymbol{\alpha}}). \tag{2.57}$$

**Proof.** Any function $f$ in $\mathcal{H}_K$ can be decomposed as the sum of two functions: one is in the span $\{K(\cdot, \mathbf{X}_1), \cdots, K(\cdot, \mathbf{X}_n)\}$ and the other is in the orthogonal complement. In other words, we write

$$f(\mathbf{x}) = \sum_{i=1}^{n} \alpha_i K(\mathbf{x}, \mathbf{X}_i) + r(\mathbf{x})$$

where $\langle r(\mathbf{x}), K(\mathbf{x}, \mathbf{X}_i)\rangle_{\mathcal{H}_K} = 0$ for all $i = 1, 2, \cdots, n$. By part (i) of Theorem 2.5 we have

$$r(\mathbf{x}_i) = \langle r, K(\cdot, \mathbf{X}_i)\rangle_{\mathcal{H}_K} = 0, \quad 1 \le i \le n.$$

Denote by $g(\mathbf{x}) = \sum_{i=1}^{n} \alpha_i K(\mathbf{x}, \mathbf{X}_i)$. Then we have $g(\mathbf{X}_i) = f(\mathbf{X}_i)$ for all $i$, which implies

$$\sum_{i=1}^{n} L(Y_i, f(\mathbf{X}_i)) = \sum_{i=1}^{n} L(Y_i, g(\mathbf{X}_i)). \tag{2.58}$$

Moreover, we notice

$$\begin{aligned} \|f\|^2_{\mathcal{H}_K} &= \langle g + r, g + r\rangle_{\mathcal{H}_K} \\ &= \langle g, g\rangle_{\mathcal{H}_K} + \langle r, r\rangle_{\mathcal{H}_K} + 2\langle g, r\rangle_{\mathcal{H}_K} \end{aligned}$$

and

$$\langle g, r\rangle_{\mathcal{H}_K} = \sum_{i=1}^{n} \alpha_i \langle K(\cdot, \mathbf{X}_i), r\rangle_{\mathcal{H}_K} = 0.$$

Thus $\|f\|^2_{\mathcal{H}_K} = \|g\|^2_{\mathcal{H}_K} + \|r\|^2_{\mathcal{H}_K}$. Because $P_\lambda(\cdot)$ is a strictly increasing function, we then have

$$P_\lambda(\|f\|_{\mathcal{H}_K}) \ge P_\lambda(\|g\|_{\mathcal{H}_K}) \tag{2.59}$$

and the equality holds if and only if $f = g$. Combining (2.58) and (2.59) we prove (2.56).

To prove (2.57), we use (2.56) and part (iii) of Theorem 2.5 to write

$$\|f\|^2_{\mathcal{H}_K} = \boldsymbol{\alpha}^T \mathbf{K} \boldsymbol{\alpha}. \tag{2.60}$$

Hence $P_\lambda(\|f\|_{\mathcal{H}_K}) = P_\lambda(\sqrt{\boldsymbol{\alpha}^T \mathbf{K} \boldsymbol{\alpha}})$ under (2.56) . ■

Theorem 2.6 is known as the *representer theorem* (Wahba, 1990). It shows that for a wide class of statistical estimation problems in a RKHS, although the criterion is defined in an infinite dimensional space, the solution always has a finite dimensional representation based on the kernel functions. This provides a solid mathematical foundation for the kernel trick without resorting to any optimization/computational arguments.

Let the loss function in Theorem 2.6 be the squared error loss and $P_\lambda(t) = \lambda t^2$. Then Theorem 2.6 handles the problem defined in (2.50) and (2.57) reduces to

$$\min_{\boldsymbol{\alpha}} \|\mathbf{Y} - \mathbf{K}\boldsymbol{\alpha}\|^2 + \lambda \boldsymbol{\alpha}^T \mathbf{K} \boldsymbol{\alpha}. \tag{2.61}$$

Table 2.1: A list of commonly used kernels.

| | |
|---|---|
| Linear kernel | $K(\mathbf{x}_i, \mathbf{x}_j) = \langle \mathbf{x}_i, \mathbf{x}_j \rangle$ |
| Polynomial kernel | $K(\mathbf{x}_i, \mathbf{x}_j) = (1 + \langle \mathbf{x}_i, \mathbf{x}_j \rangle)^d$ |
| Gaussian kernel | $K(\mathbf{x}_i, \mathbf{x}_j) = e^{-\gamma \|\mathbf{x}_i - \mathbf{x}_j\|^2}$ |
| Laplacian kernel | $K(\mathbf{x}_i, \mathbf{x}_j) = e^{-\gamma \|\mathbf{x}_i - \mathbf{x}_j\|}$ |

It is easy to see the solution is

$$\widehat{\alpha} = (\mathbf{K} + \lambda \mathbf{I})^{-1} \mathbf{Y}.$$

which is identical to (2.55). The fitted multivariate nonparametric regression function is given by (2.56). In practice, one takes a kernel function from the list of linear, polynomial, Gaussian or Laplacian kernels given in Table 2.1. It remains to show how to choose the regularization parameter $\lambda$ (and $\gamma$ for Gaussian and Laplacian kernels) to optimize the prediction performance. This can be done by cross-validation methods outlined in the next section.

## 2.8 Leave-one-out and Generalized Cross-validation

We have seen that both ridge regression and the kernel ridge regression use a *tuning parameter* $\lambda$. In practice, we would like to use the data to pick a data-driven $\lambda$ in order to achieve the "best" estimation/prediction performance. This problem is often called tuning parameter selection and is ubiquitous in modern statistics and machine learning. A general solution is $k$-fold *cross-validation* (CV), such as 10-fold or 5-fold CV. $k$-fold CV estimates prediction errors as follows.

- Divide data randomly and evenly into $k$ subsets.
- Use one of the subsets as the *testing set* and the remaining $k-1$ subsets of data as a *training set* to compute testing errors.
- Compute testing errors for each of $k$ subsets of data and average these testing errors.

An interesting special case is the $n$-fold CV, which is also known as the *leave-one-out CV*.

In this section we focus on regression problems under the squared error loss. Following the above scheme, the leave-one-out CV error, using the quadratic loss, is defined as

$$\mathrm{CV} = \frac{1}{n} \sum_{i=1}^{n} (Y_i - \widehat{f}^{(-i)}(\mathbf{X}_i))^2, \tag{2.62}$$

where $\widehat{f}^{(-i)}(\mathbf{X}_i)$ is the predicted value at $\mathbf{x}_i$ computed by using all the data except the $i$th observation. So in principle we need to repeat the same data

Table 2.2: A list of commonly used regression methods and their $\mathbf{S}$ matrices. $d_j$s are the singular values of $\mathbf{X}$ and $\gamma_i$s are the eigenvalues of $\mathbf{K}$.

| Method | $\mathbf{S}$ | $\operatorname{tr}\mathbf{S}$ |
|---|---|---|
| Multiple Linear Regression | $\mathbf{X}(\mathbf{X}^T\mathbf{X})^{-1}\mathbf{X}^T$ | $p$ |
| Ridge Regression | $\mathbf{X}(\mathbf{X}^T\mathbf{X}+\lambda\mathbf{I})^{-1}\mathbf{X}^T$ | $\sum_{j=1}^p \frac{d_j^2}{d_j^2+\lambda}$ |
| Kernel Regression in RKHS | $\mathbf{K}(\mathbf{K}+\lambda\mathbf{I})^{-1}$ | $\sum_{i=1}^n \frac{\gamma_i}{\gamma_i+\lambda}$ |

fitting process $n$ times to compute the leave-one-out CV. Fortunately, we can avoid much computation for many popular regression methods.

A fitting method is called a *linear smoother* if we can write

$$\widehat{\mathbf{Y}} = \mathbf{S}\mathbf{Y} \tag{2.63}$$

for any dataset $\{(\mathbf{X}_i, Y_i\}_1^n$ where $\mathbf{S}$ is a $n \times n$ matrix that only depends on $\mathbf{X}$. Many regression methods are linear smoothers with different $\mathbf{S}$ matrices. See Table 2.2.

Assume that a linear smoother is fitted on $\{\mathbf{X}_i, Y_i\}_{i=1}^n$. Let $\mathbf{x}$ be a new covariate vector and $\widehat{f}(\mathbf{x})$ be its the predicted value by using the linear smoother. We then augment the dataset by including $(\mathbf{x}, \widehat{f}(\mathbf{x}))$ and refit the linear smoother on this augmented dataset. The linear smoother is said to be *self-stable* if the fit based on the augmented dataset is identical to the fit based on the original data regardless of $\mathbf{x}$.

It is easy to check that the three linear smoothers in Table 2.2 all have the self-stable property.

**Theorem 2.7** *For a linear smoother $\widehat{\mathbf{Y}} = \mathbf{S}\mathbf{Y}$ with the self-stable property, we have*

$$Y_i - \widehat{f}^{(-i)}(\mathbf{X}_i) = \frac{Y_i - \widehat{Y}_i}{1 - S_{ii}}, \tag{2.64}$$

*and its leave-one-out CV error is equal to* $\frac{1}{n}\sum_{i=1}^n \left(\frac{Y_i - \widehat{Y}_i}{1-S_{ii}}\right)^2$.

**Proof.** We first apply the linear smoother to all the data except the $i$th to compute $\widehat{f}^{(-i)}(\mathbf{X}_i)$. Write $\widetilde{y}_j = y_j$ for $j \neq i$ and $\widetilde{y}_i = \widehat{f}^{(-i)}(\mathbf{X}_i)$. Then we apply the linear smoother to the following working dataset:

$$\{(\mathbf{X}_j, Y_j), j \neq i, (\mathbf{X}_i, \widetilde{Y}_i)\}$$

The self-stable property implies that the fit stays the same. In particular,

$$\widetilde{Y}_i = \widehat{f}^{(-i)}(\mathbf{X}_i) = (\mathbf{S}\widetilde{\mathbf{Y}})_i = S_{ii}\widetilde{Y}_i + \sum_{j \neq i} S_{ij} Y_j \tag{2.65}$$

and

$$\widehat{Y}_i = (\mathbf{SY})_i = S_{ii}Y_i + \sum_{j \neq i} S_{ij}Y_j. \tag{2.66}$$

Combining (2.65) and (2.66) yields

$$\widetilde{Y}_i = \frac{\widehat{Y}_i - S_{ii}Y_i}{1 - S_{ii}}.$$

Thus,

$$Y_i - \widetilde{Y}_i = Y_i - \frac{\widehat{Y}_i - S_{ii}Y_i}{1 - S_{ii}} = \frac{Y_i - \widehat{Y}_i}{1 - S_{ii}}.$$

The proof is now complete. ∎

Theorem 2.7 shows a nice shortcut for computing the leave-one-out CV error of a self-stable linear smoother. For some smoothers $\operatorname{tr}\mathbf{S}$ can be computed more easily than its diagonal elements. To take advantage of this, *generalized cross-validation* (GCV) (Golub, Heath and Wahba, 1979) is a convenient computational approximation of the leave-one-out CV error. Suppose that we approximate each diagonal elements of $\mathbf{S}$ by their average which equals $\frac{\operatorname{tr}\mathbf{S}}{n}$, then we have

$$\frac{1}{n}\sum_{i=1}^{n}\left(\frac{Y_i - \widehat{Y}_i}{1 - S_{ii}}\right)^2 \approx \frac{1}{n}\frac{\sum_{i=1}^{n}(Y_i - \widehat{Y}_i)^2}{(1 - \frac{\operatorname{tr}\mathbf{S}}{n})^2} := \text{GCV}.$$

In the literature $\operatorname{tr}\mathbf{S}$ is called the *effective degrees of freedom* of the linear smoother. Its rigorous justification is based on Stein's unbiased risk estimation theory (Stein, 1981; Efron, 1986). In Table 2.2 we list the degrees of freedom of three popular linear smoothers.

Now we are ready to handle the tuning parameter selection issue in the linear smoother. We write $\mathbf{S} = \mathbf{S}_\lambda$ and

$$\text{GCV}(\lambda) = \frac{1}{n}\frac{\mathbf{Y}^T(\mathbf{I} - \mathbf{S}_\lambda)^2\mathbf{Y}}{(1 - \frac{\operatorname{tr}\mathbf{S}_\lambda}{n})^2}.$$

According to GCV, the best $\lambda$ is given by

$$\lambda^{\text{GCV}} = \operatorname{argmin}_\lambda \frac{1}{n}\frac{\mathbf{Y}^T(\mathbf{I} - \mathbf{S}_\lambda)^2\mathbf{Y}}{(1 - \frac{\operatorname{tr}\mathbf{S}_\lambda}{n})^2}.$$

## 2.9 Exercises

2.1 Suppose that $\{\mathbf{X}_i, Y_i\}$, $i = 1, \cdots, n$ is a random sample from linear regression model (2.1). Assume that the random error $\varepsilon \sim N(0, \sigma^2)$ and is independent of $\mathbf{X} = (X_1, \cdots, X_p)^T$.

(a) Show that the maximum likelihood estimate of $\boldsymbol{\beta}$ is the same as its least squares estimator, while the maximum likelihood estimate of $\sigma^2$ is $\mathrm{RSS}/n$, where $\mathrm{RSS} = \sum_{i=1}^n (Y_i - \mathbf{X}_i^T\widehat{\boldsymbol{\beta}})^2$ is the residual sum-of-squares and $\widehat{\boldsymbol{\beta}}$ is the least squares estimator.

(b) Assume that $\mathbf{X}$ is of full rank. Show that $\mathrm{RSS} \sim \sigma^2\chi^2_{n-p}$.

(c) Prove that $1-\alpha$ CI for $\beta_j$ is $\widehat{\beta}_j \pm t_{n-p}(1-\alpha/2)\sqrt{v_j\,\mathrm{RSS}/(n-p)}$, where $v_j$ is the $j^{th}$ diagonal element of $(\mathbf{X}^T\mathbf{X})^{-1}$.

(d) Dropping the normality assumption, if $\{\mathbf{X}_i\}$ are independent and identically distributed from a population with $\mathrm{E}\,\mathbf{X}_1\mathbf{X}_1^T = \boldsymbol{\Sigma}$ and independent of $\{\varepsilon_i\}_{i=1}^n$, which are independent and identically distributed from a population with $\mathrm{E}\,\varepsilon = 0$ and $\mathrm{Var}(\varepsilon) = \sigma^2$, show that

$$\sqrt{n}(\widehat{\boldsymbol{\beta}} - \boldsymbol{\beta}) \xrightarrow{d} N(0, \sigma^2\boldsymbol{\Sigma}^{-1}).$$

2.2 Suppose that a random sample of size $n$ from linear regression model (2.1), where the random error $\varepsilon \sim N(0,\sigma^2)$ and is independent of $(X_1, \cdots, X_p)$. Consider a general linear hypothesis $H_0 : \mathbf{C}\boldsymbol{\beta} = \mathbf{h}$ versus $H_0 : \mathbf{C}\boldsymbol{\beta} \neq \mathbf{h}$, where $\mathbf{C}$ is a $q \times p$ constant matrix with rank $q$ $(\leq p)$, and $\mathbf{h}$ is a $q \times 1$ constant vector.

(a) Derive the least squares estimator of $\boldsymbol{\beta}$ under $H_0$, denoted by $\widehat{\boldsymbol{\beta}}_0$.

(b) Define $\mathrm{RSS}_1 = \|\mathbf{Y} - \mathbf{X}\widehat{\boldsymbol{\beta}}\|^2$ and $\mathrm{RSS}_0 = \|\mathbf{Y} - \mathbf{X}\widehat{\boldsymbol{\beta}}_0\|^2$, the residual sum-of-squares under $H_1$ and $H_0$. Show that $\mathrm{RSS}_1/\sigma^2 \sim \chi^2_{n-p}$. Further, under the null hypothesis $H_0$, $(\mathrm{RSS}_0 - \mathrm{RSS}_1)/\sigma^2 \sim \chi^2_q$ and is independent of $\mathrm{RSS}_1$.

(c) Show that under $H_0$, $F = \{(\mathrm{RSS}_0 - \mathrm{RSS}_1)/q\}/\{\mathrm{RSS}_1/(n-p)\}$ follows an $F_{q,n-p}$ distribution.

(d) Show that the $F$-test for $H_0$ is equivalent to the likelihood ratio test for $H_0$.

2.3 Suppose that we have $n$ independent data $Y_i \sim N(\mu, \sigma_i^2)$, where $\sigma_i = \sigma^2 v_i$ with known $v_i$. Use the weighted least-squares method to find an estimator of $\mu$. Show that it is the best linear unbiased estimator. Compare the variance of the sample mean $\bar{y}$ with that of the weighted least-squares estimator $v_i^2 = \log(i+1)$ when $n = 20$.

2.4 Consider the linear model $\mathbf{Y} = \mathbf{X}\boldsymbol{\beta} + \boldsymbol{\varepsilon}$, where $\boldsymbol{\varepsilon} \sim N(0, \boldsymbol{\Sigma})$, and $\mathbf{X}$ is of full rank.

(a) Show that the general least-squares estimator, which minimizes $(\mathbf{Y} - \mathbf{X}\boldsymbol{\beta})^T\boldsymbol{\Sigma}^{-1}(\mathbf{Y} - \mathbf{X}\boldsymbol{\beta})$, is the best linear unbiased estimator. More precisely, for any vector $\mathbf{c} \neq 0$, $\mathbf{c}^T\widehat{\boldsymbol{\beta}}$ is the best linear estimator of $\boldsymbol{\beta}$. Do we need the normality assumption?

(b) Deduce from part (a) that the weighted least-squares estimator is the best linear unbiased estimator, when the error distribution is uncorrelated.

(c) If $\mathbf{\Sigma}$ is the equi-correlation matrix with unknown correlation $\rho$, what is the solution to part (a)?

2.5 Suppose that $Y_1, \cdots, Y_n$ are random variables with common mean $\mu$ and covariance matrix $\sigma^2\mathbf{V}$, where $\mathbf{V}$ is of the form $v_{ii} = 1$ and $v_{ij} = \rho$ $(0 < \rho < 1)$ for $i \neq j$.

(a) Find the generalized least squares estimate of $\mu$.

(b) Show that it is the same as the ordinary least squares estimate.

2.6 Suppose that data $\{X_{i1}, \cdots, X_{ip}, Y_i\}, i = 1, \cdots, n$, are an independent and identically distributed sample from the model

$$Y = f(X_1\beta_1 + \cdots + X_p\beta_p + \varepsilon),$$

where $\varepsilon \sim N(0, \sigma^2)$ with unknown $\sigma^2$, and $f(\cdot)$ is a known, differentiable, strictly increasing, non-linear function.

(a) Consider transform $Y_i^* = h(Y_i)$, where $h(\cdot)$ is a differentiable function yet to be determined. Show that $\text{Var}(Y_i^*)$ =constant for all $i$ leads to the equation: $[h'\{f(u)\}]^2\{f'(u)\}^2$ =constant for all $u$.

(b) Let $f(x) = x^p$ $(p > 1)$. Find the corresponding $h(\cdot)$ using the equation in (a).

(c) Let $f(x) = \exp(x)$. Find the corresponding $h$ transform.

2.7 The data set 'hkepd.txt' consists of daily measurements of levels of air pollutants and the number of total hospital admissions for circulatory and respiratory problems from January 1, 1994 to December 31, 1995 in Hong Kong. This data set can be downloaded from this book website. Of interest is to investigate the association between the number of total hospital admissions and the levels of air pollutants.
We set the $Y$ variable to be the number of total hospital admissions and the $X$ variables the levels of air pollutants. Define

$X_1$ = the level of sulfur dioxide ($\mu g/m^3$);
$X_2$ = the level of nitrogen dioxide ($\mu g/m^3$);
$X_3$ = the level of dust ($\mu g/m^3$).

(a) Fit the data to the following linear regression model

$$Y = \beta_0 + \beta_1 X_1 + \beta_2 X_2 + \beta_3 X_3 + \varepsilon, \tag{2.67}$$

and test whether the level of each air pollutant has significant impact on the number of total hospital admissions.

(b) Construct residual plots and examine whether the random error approximately follows a normal distribution.

(c) Take $Z = \log(Y)$ and fit the data to the following linear regression model

$$Z = \beta_0 + \beta_1 X_1 + \beta_2 X_2 + \beta_3 X_3 + \varepsilon, \tag{2.68}$$

and test whether the level of each air pollutant has significant impact on the logarithm of the number of total hospital admissions.

(d) Construct residual plots based on model (2.68) and compare this residual plot with the one obtained in part (b).

(e) Since the observations in this data set were collected over time, it is of interest to explore the potential seasonal trends in the total number of hospital admissions. For simplicity, define $t$ to be the day on which data were collected. This corresponds to the first column of the data set. Consider the time-varying effect model

$$Z = \beta_0(t) + \beta_1(t)X_1 + \beta_2(t)X_2 + \beta_3(t)X_3 + \varepsilon, \tag{2.69}$$

which allows the effects of predictors varying over time. Model (2.69) indeed is a nonparametric regression model. Fit model (2.69) to the data by using nonparametric regression techniques introduced in Section 2.5.

(f) For model (2.69), construct an $F$-type test for $H_0 : \beta_3(\cdot) \equiv 0$ (i.e., the level of dust not significant) by comparing their residual sum of squares under $H_0$ and $H_1$.

2.8 The data set 'macroecno.txt' consists of 129 macroecnomic time series and can be downloaded from this book website. Let the response $Y_t = \Delta\log(\text{PCE}_t)$ be the changes in personal consumption expenditure. Define covariates as follows.

$$X_{t,1} = \text{Unrate}_{t-1},\ X_{t,2} = \Delta\log(\text{IndPro}_{t-1}),\ X_{t,3} = \Delta\log(\text{M2Real}_{t-1}),$$
$$X_{t,4} = \Delta\log(\text{CPI}_{t-1}),\quad X_{t,5} = \Delta\log(\text{SPY}_{t-1}),\quad X_{t,6} = \text{HouSta}_{t-1},$$
$$X_{t,7} = \text{FedFund}_{t-1}$$

Set the last 10 years data as testing data and the remaining as training data. Conduct linear regression analysis and address the following questions.

(a) What are $\widehat{\sigma}^2$, adjusted $R^2$ and insignificant variables?

(b) Conduct the stepwise deletion, eliminating one least significant variable at a time (by looking at the smallest $|t|$-statistic) until all variables are statistically significant. Name this model as model $\widehat{\mathcal{M}}$. (The function `step` can do the job automatically)

(c) Using model $\widehat{\mathcal{M}}$, what are root mean-square prediction error and mean absolute deviation prediction error for the test sample?

(d) Compute the standardized residuals. Present the time series plot of the residuals, fitted values versus the standardized residuals, and QQ plot for the standardized residuals.

2.9 Zillow is an online real estate database company that was founded in 2006. The most important task for Zillow is to predict the house price. However, their accuracy has been criticized a lot. According to Fortune,

"Zillow has Zestimated the value of 57 percent of U.S. housing stock, but only 65 percent of that could be considered 'accurate' by its definition, within 10 percent of the actual selling price. And even that accuracy isn't equally distributed". Therefore, Zillow needs your help to build a housing pricing model to improve their accuracy. Download the data from the book website, and read the data (training data: 15129 cases, testing data: 6484 cases)

```
train.data <- read.csv('train.data.csv', header=TRUE)
test.data <- read.csv('test.data.csv', header=TRUE)
train.data$zipcode <- as.factor(train.data$zipcode)
test.data$zipcode <- as.factor(test.data$zipcode)
```

where the last two lines make sure that zip code is treated as a factor. Let $\mathcal{T}$ be a test set, define out-of-sample $R^2$ as of a prediction method $\{\widehat{y}_i^{pred}\}$ as

$$R^2 = 1 - \frac{\sum_{i\in\mathcal{T}}(y_i - \widehat{y}_i^{pred})^2}{\sum_{i\in\mathcal{T}}(y_i - \bar{y}^{pred})^2},$$

where $\bar{y}^{pred} = \text{ave}(\{y_i\}_{i\in\mathcal{T}_0})$ and $\mathcal{T}_0$ is the training set.

(a) Calculate out-of-sample $R^2$ using variables "bedrooms", "bathrooms", "sqft_living", and "sqft_lot".

(b) Calculate out-of-sample $R^2$ using the 4 variables above along with interaction terms.

(c) Compare the result with the nonparametric model using the Gaussian kernel with $\gamma = 0.3^2/2$ and $\gamma = 0.1^2/2$ (standardize predictors first) and $\lambda$ chosen by 5-fold CV or GCV. **Hint**: To speed up computation, please divide data randomly into 10 pieces and get 10 predicted values based on 10 fitted kernel models. Use the median of these 10 predicted values as your final prediction.

(d) Add the factor zipcode to (b) and compute out-of-sample $R^2$.

(e) Add the following additional variables to (d): $X_{12} = I(view == 0)$, $X_{13} = L^2$, $X_{13+i} = (L - \tau_i)_+^2$, $i = 1, \cdots, 9$, where $\tau_i$ is $10 * i^{th}$ percentile and $L$ is the size of the living area ("sqft_living"). Compute out-of-sample $R^2$.

(f) Why do you see the increased out-of-sample $R^2$ with modeling complexity?

Chapter 3

# Introduction to Penalized Least-Squares

Variable selection is vital to high-dimensional statistical learning and inference, and is essential for scientific discoveries and engineering innovation. Multiple regression is one of the most classical and useful techniques in statistics. This chapter introduces *penalized least-squares* approaches to variable selection problems in multiple regression models. They provide fundamental insights and the basis for *model selection* problems in other more sophisticated models.

## 3.1 Classical Variable Selection Criteria

In this chapter, we will follow the notation and model introduced in Chapter 2. To reduce noise accumulation and to enhance interpretability, variable selection techniques have been popularly used even in traditional statistics. When the number of predictors $p$ is larger than the sample size $n$, the model parameters in the linear model (2.2) are not identifiable. What makes them estimable is the *sparsity* assumption on the regression coefficients $\{\beta_j\}_{j=1}^p$: many of them are too small to matter, so they are ideally regarded as zero. Throughout this chapter, we assume the linear model (2.1):

$$Y = \beta_1 X_1 + \cdots + \beta_p X_p + \varepsilon,$$

unless otherwise stated.

### *3.1.1 Subset selection*

One of the most popular and intuitive variable selection techniques is the *best subset* selection. Among all models with $m$ variables, pick the one with the smallest residual sum of squares (2.3), which is denoted by $\text{RSS}_m$. This is indeed very intuitive: among the models with the same complexity, a better fit is preferable. This creates a sequence of submodels $\{\mathcal{M}_m\}_{m=0}^p$ indexed by the model size $m$. The choice of the model size $m$ will be further illuminated in Section 3.1.3.

Computation of the best subset method is expensive even when $p$ is moderately large. At each step, we compare the goodness-of-fit among $\binom{p}{m}$ models of size $m$ and there are $2^p$ submodels in total. Intuitive and greedy algorithms have been introduced to produce a sequence of submodels with different numbers of variables. These include *forward selection* also called *stepwise addition*, *backward elimination* also named *stepwise deletion*, and *stepwise regression*. See for example, Weisberg (2005).

Forward selection recruits one additional regressor at a time to optimize the fit. Starting with $\mathcal{M}_0^a$ as the empty set, at step $m$ one chooses a variable not in $\mathcal{M}_{m-1}^a$ along with the $m-1$ variables in $\mathcal{M}_{m-1}^a$ to minimize the RSS. At step $m$, only $p-m+1$ submodels instead of $\binom{p}{m}$ of size $m$ are fitted and compared. The total number of regressions is only $p(p+1)/2$, which is considerably less than $2^p$. Of course, the sequence of submodels $\{\mathcal{M}_m^a\}_{m=0}^p$ may not have been as good of a fit as the models produced by the best subset selection.

Backward elimination deletes one least statistically significant variable in each fitted model. Starting from the full model $\mathcal{M}_p$, denoted also by $\mathcal{M}_p^d$, one eliminates the least statistically significant variable in the full model, resulting in a model $\mathcal{M}_{p-1}^d$. We then use the variables in $\mathcal{M}_{p-1}^d$ to fit the data again and delete the least statistically significant variable to obtain model $\mathcal{M}_{p-2}^d$, continuing this process to yield a sequence of models $\{\mathcal{M}_m^d\}_{m=0}^p$.

In classical statistics, one does not produce the full sequence of the models in the forward selection and backward elimination methods. One often sets a very simple stopping criterion such as when all variables are statistically significant (e.g. P-values for each fitted coefficient is smaller than 0.05).

When $p$ is larger than $n$, backward elimination cannot be applied since we cannot fit the full model. Yet, the forward selection can still be used to select a sequence of submodels. When $p < n$ or when $p$ is relatively large compared to $n$, backward elimination cannot produce a stable selection process, but forward selection can as long as it is stopped early enough. These are the advantages of the forward selection algorithm.

Stepwise regression is a combination of backward elimination and forward selection procedures. We omit its details. Other greedy algorithms include *matching pursuit* (Mallot and Zhang, 1993), which picks the most correlated variable with the residuals from the previous step of fitting, also referred to as *partial residuals*, and runs the univariate regression to fit the partial residuals. See Section 3.5.11 for additional details.

### *3.1.2 Relation with penalized regression*

Best subset selection can be regarded as *penalized least-squares* (PLS). Let $\|\boldsymbol{\beta}\|_0$ be the $L_0$-norm of the vector $\boldsymbol{\beta}$, which counts the number of non-vanishing components of $\boldsymbol{\beta}$. Consider the penalized least-squares with $L_0$ penalty:

$$\|\mathbf{Y} - \mathbf{X}\boldsymbol{\beta}\|^2 + \lambda\|\boldsymbol{\beta}\|_0. \tag{3.1}$$

The procedure is also referred to as *complexity* or *entropy* based PLS. Clearly, given the model size $\|\boldsymbol{\beta}\|_0 = m$, the solution to the penalized least-squares (3.1) is the best subset selection. The computational complexity is NP-hard.

The stepwise algorithms in the last subsection can be regarded as greedy (approximation) algorithms for penalized least-squares (3.1).

Recall $\text{RSS}_m$ is the smallest residual sum of squares among models with size $m$. With this definition, minimization of (3.1) can be written as

$$\text{RSS}_m + \lambda m. \tag{3.2}$$

The optimal model size is obtained by minimizing (3.2) with respect to $m$. Clearly, the regularization parameter $\lambda$ dictates the size of the model. The larger the $\lambda$, the larger the penalty on the model complexity $m$, and the smaller the selected model.

### *3.1.3 Selection of regularization parameters*

The best subset technique does not tell us the choice of model size $m$. The criterion used to compare two models is usually the prediction error. For a completely new observation $(\mathbf{X}^{*T}, Y^*)$, the *prediction error* of using model $\mathcal{M}_m$ is

$$\text{PE}(\mathcal{M}_m) = \text{E}(Y^* - \widehat{\boldsymbol{\beta}}_m^T \mathbf{X}^*_{\mathcal{M}_m})^2,$$

where $\widehat{\boldsymbol{\beta}}_m$ is the fitted regression coefficient vector with $m$ variables, $\mathbf{X}^*_{\mathcal{M}_m}$ is the subvector of $\mathbf{X}^*$ with the selected variables, and the expectation is taken only with respect to the new random variable $(\mathbf{X}^{*T}, Y^*)$.

An unbiased estimation of the prediction error $n\text{PE}(\mathcal{M}_m)$ was derived by Mallows (1973) (after ignoring a constant; see Section 3.6.1 for a derivation):

$$C_p(m) = \text{RSS}_m + 2\sigma^2 m. \tag{3.3}$$

This corresponds to taking $\lambda = 2\sigma^2$ in the penalized least-squares problem (3.1) or (3.2). The parameter $m$ is chosen to minimize (3.3), which is often referred to as *Mallow's $C_p$ criterion.*

Akaike (1973, 1974) derived an approximately unbiased estimate of the prediction error (in terms of the Kullback-Leibler divergence) in a general likelihood based model. His work is regarded as one of the important breakthroughs in statistics in the twentieth century. Translating his criterion into the least-squares setting, it becomes

$$\text{AIC}(m) = \log(\text{RSS}_m/n) + 2m/n,$$

which is called the *Akaike information criterion* (AIC). Note that when $\text{RSS}_m/n \approx \sigma^2$, which is correct when $\mathcal{M}_m$ contains the true model, by Taylor's expansion,

$$\begin{aligned} \log(\text{RSS}_m/n) &= \log\sigma^2 + \log(1 + \text{RSS}_m/(n\sigma^2) - 1) \\ &\approx \log\sigma^2 + (\text{RSS}_m/(n\sigma^2) - 1). \end{aligned}$$

Therefore,

$$\text{AIC}(m) \approx [\text{RSS}_m + 2\sigma^2 m]/(n\sigma^2) + \log\sigma^2 - 1,$$

which is approximately the same as the $C_p$ criterion (3.3) after ignoring the affine transformation.

Many information criteria have been derived since the pioneering work of Akaike and Mallow. They correspond to different choices of $\lambda$ in

$$\mathrm{IC}(m) = \log(\mathrm{RSS}_m/n) + \lambda m/n. \tag{3.4}$$

Examples include

- *Bayesian information criterion* (BIC, Schwarz, 1978): $\lambda = \log(n)\sigma^2$;
- *$\phi$-criterion* (Hannan and Quinn, 1979; Shibata, 1984): $\lambda = c(\log\log n)$;
- Risk inflation criterion (RIC, Foster and George, 1994): $\lambda = 2\log(p)$.

Using the Taylor expansion above, the information criteria (3.4) is asymptotically equivalent to (3.2). An advantage of using these information criteria over criterion (3.2) is that they do not need to estimate $\sigma^2$. But this also creates the bias issue. In particular, when a submodel contains modeling biases, AIC is no longer an approximately unbiased estimator. The issue of model selection consistency has been thoroughly studied in Shao (1997).

In summary, the best subset method along with an information criterion corresponds to the $L_0$-penalized least-squares with penalty parameters $\lambda$ being a multiple 2, $\log(n)$, $c\log\log n$, and $2\log p$ of $\sigma^2$, respectively for the AIC, BIC, $\phi$-criterion, and RIC.

Cross-validation (Allen 1974; Stone, 1974) is a novel and widely applicable idea for estimating the prediction error of a model. It is one of the most widely used and innovative techniques in statistics. It involves partitioning a sample of data into a *training set* used to estimate model parameters and a *testing set* reserved for validating the analysis of the fitted model. See Section 2.8.

In $k$-fold cross-validation, the original sample is randomly partitioned into $k$ approximately equal-sized subsamples with index sets $\{\mathcal{S}_j\}_{j=1}^k$. Of the $k$ subsamples, a single subsample $\mathcal{S}_k$ is retained as the validation set, and the remaining data $\{\mathcal{S}_j\}_{j\neq k}$ are used as a training set. The cross-validation process is then repeated $k$ times, with each of the $k$ subsamples used exactly once as the validation data. The prediction error of the $k$-fold cross-validation is computed as

$$\mathrm{CV}_k(m) = n^{-1}\sum_{j=1}^{k}\Big\{\sum_{i\in\mathcal{S}_j}(Y_i - \widehat{\boldsymbol{\beta}}_{m,-\mathcal{S}_j}^T\mathbf{X}_{i,\mathcal{M}_m})^2\Big\}, \tag{3.5}$$

where $\widehat{\boldsymbol{\beta}}_{m,-\mathcal{S}_j}$ is the fitted coefficients of the submodel $\mathcal{M}_m$ without using the data indexed in $\mathcal{S}_j$. The number of fittings is $k$, which is much smaller than $n$. In practice, the popular choice of $k$ is 5 or 10. An interesting choice of $k$ is $n$, which is called the leave-one-out cross-validation. The leave-one-out CV error of the submodel $\mathcal{M}_m$ is

$$\mathrm{CV}(m) = n^{-1}\sum_{i=1}^{n}(Y_i - \widehat{\boldsymbol{\beta}}_{m,-i}^T\mathbf{X}_{i,\mathcal{M}_m})^2. \tag{3.6}$$

In general, the leave-one-out CV error is expensive to compute. For multiple linear regression and other linear smoothers with a *self-stable* property, there is a neat formula for computing $\mathrm{CV}(m)$ without fitting the model $n$ times. See Theorem 2.7 in Section 2.8 of Chapter 2. Another simplification of $\mathrm{CV}(m)$ is to use Generalized Cross-Validation (GCV, Craven and Wahba, 1979), defined by

$$\mathrm{GCV}(m) = \frac{\mathrm{RSS}_m}{n(1-m/n)^2}. \tag{3.7}$$

By using a simple Taylor expansion,

$$(1-m/n)^{-2} = 1 + 2m/n + o(m/n)$$

and $\mathrm{RSS}_m/n \approx \sigma^2$, one can easily see that $\mathrm{GCV}(m)$ is approximately the same as $C_p(m)/n$.

A classical choice of $m$ is to maximize the adjusted multiple $R^2$, defined by

$$R^2_{adj,m} = 1 - \frac{n-1}{n-m}\frac{\mathrm{RSS}_m}{\mathrm{RSS}_0}, \tag{3.8}$$

where $\mathrm{RSS}_0$ is the sample standard deviation of the response variable $\{Y_i\}$. This is equivalent to minimizing $\mathrm{RSS}_m/(n-m)$. Derived the same way as the GCV, it corresponds to approximately $\lambda = \sigma^2$ in (3.2).

## 3.2 Folded-concave Penalized Least Squares

The complexity based PLS (3.1) possesses many nice statistical properties, as documented in the paper by Barron, Birgé and Massart (1999). However, its minimization problem is impossible to carry out when the dimensionality is high. A natural relaxation is to replace the discontinuous $L_0$-penalty by more regular functions. This results in penalized least-squares

$$\begin{aligned} Q(\boldsymbol{\beta}) &= \frac{1}{2n}\|\mathbf{Y}-\mathbf{X}\boldsymbol{\beta}\|^2 + \sum_{j=1}^{p} p_\lambda(|\beta_j|) \\ &\equiv \frac{1}{2n}\|\mathbf{Y}-\mathbf{X}\boldsymbol{\beta}\|^2 + \|p_\lambda(|\boldsymbol{\beta}|)\|_1, \end{aligned} \tag{3.9}$$

where $p_\lambda(\cdot)$ is a penalty function in which the *regularization* or *penalization parameters* $\lambda$ are the same for convenience of presentation.

A natural choice is $p_\lambda(\theta) = \lambda\theta^2/2$, whose solution is ridge regression

$$\widehat{\boldsymbol{\beta}}_{\text{ridge}} = (\mathbf{X}^T\mathbf{X} + n\lambda\mathbf{I}_p)^{-1}\mathbf{X}^T\mathbf{Y}, \tag{3.10}$$

which is also called the *Tikhonov regularization* (Tikhonov, 1943). The estimator shrinks all components toward zero, but none of them are actually zero. It does not have a model selection property and creates biases for large parameters. In order to reduce the bias, Frank and Friedman (1993) propose to use

$p_\lambda(\theta) = \lambda|\theta|^q$ for $0 < q < 2$, called the *bridge regression*, which bridges the best subset selection (penalized $L_0$) and ridge regression (penalized $L_2$). Donoho and Johnstone (1994), Tibshirani (1996) and Chen, Donoho and Sanders (1998) observe that penalized $L_1$ regression leads to a sparse minimizer and hence possesses a variable selection property. The procedure is called *Lasso* by Tibshirani (1996), for 'least absolute shrinkage and selection operator'. Unlike the complexity penalty $p_\lambda(|\theta|) = \lambda I(|\theta| \neq 0)$, Lasso solves a convex optimization problem. This gives the Lasso huge computational advantages.

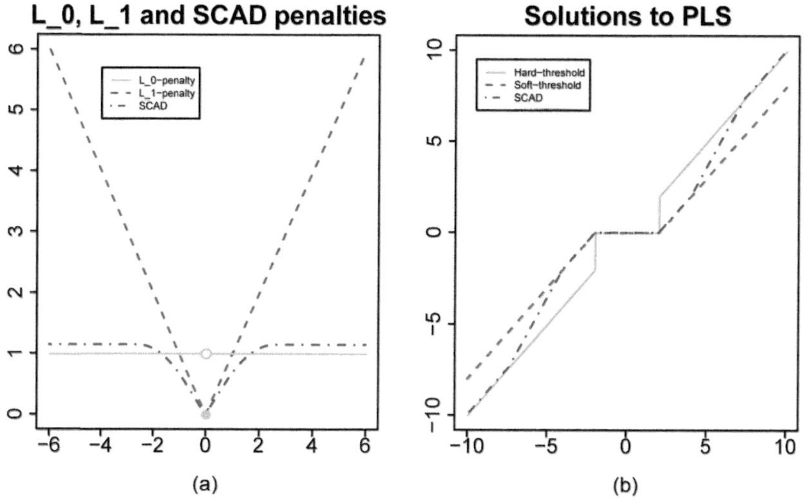


Figure 3.1: (a) The entropy or complexity penalty $L_0$ (green solid), $L_1$-penalty (blue dash), and the smoothly clipped $L_1$-penalty (SCAD, red dotdash). Clearly, SCAD inherits desired properties from $L_0$ and $L_1$ penalty at tails and the origin. (b) Solution to the penalized least-squares (3.12) (a=3.7, $\lambda = 2$ for SCAD)

As shown in Figure 3.1, the $L_1$-penalty differs substantially from the $L_0$-penalty. It penalizes the large parameters too much. To further reduce the bias in the estimation, Antoniadis and Fan (2001) and Fan and Li (2001) introduce folded concave penalized least-squares, in which $p_\lambda(\theta)$ is symmetric and concave on each side. In particular, the *smoothly clipped absolute deviation* (SCAD) penalty [see (3.14) below] is introduced to improve the bias property. As shown in Figure 3.1, SCAD behaves like the $L_1$-penalty at the origin in order to keep the variable selection property and acts like the $L_0$-penalty at the tails in order to improve the bias property of the $L_1$-penalty. The smoothness of the penalty function is introduced to ensure the continuity of the solution for *model stability*.

Let us examine what kind of penalty functions are desirable for variable selection. Significant insights can be gained by studying a specific case in which the design matrix is orthonormal.

### 3.2.1 Orthonormal designs

For an orthonormal design in which the design matrix multiplied by $n^{-1/2}$ is orthonormal (i.e., $\mathbf{X}^T\mathbf{X} = nI_p$, which implies $p \le n$), (3.9) reduces to

$$\frac{1}{2n}\|\mathbf{Y} - \mathbf{X}\widehat{\boldsymbol{\beta}}\|^2 + \frac{1}{2}\|\widehat{\boldsymbol{\beta}} - \boldsymbol{\beta}\|^2 + \|p_\lambda(|\boldsymbol{\beta}|)\|_1, \tag{3.11}$$

where $\widehat{\boldsymbol{\beta}} = n^{-1}\mathbf{X}^T\mathbf{Y}$ is the ordinary least-squares estimate. Noticing that the first term is constant, minimizing (3.11) becomes minimizing

$$\sum_{j=1}^{p}\left\{\frac{1}{2}(\widehat{\beta}_j - \beta_j)^2 + p_\lambda(|\beta_j|)\right\},$$

which is a componentwise regression problem: each component consists of the univariate PLS problem of the form

$$\widehat{\theta}(z) = \arg\min_{\theta}\left\{\frac{1}{2}(z-\theta)^2 + p_\lambda(|\theta|)\right\}. \tag{3.12}$$

Fan and Li (2001) advocate penalty functions that give estimators with the following three properties:

1) *Sparsity*: The resulting estimator automatically sets small estimated coefficients to zero to accomplish variable selection and reduce model complexity.
2) *Unbiasedness*: The resulting estimator is nearly unbiased, especially when the true coefficient $\beta_j$ is large, to reduce model bias.
3) *Continuity*: The resulting estimator is continuous in the data to reduce instability in model prediction (Breiman, 1996).

The third property is nice to have, but not necessarily required.

Let $p_\lambda(t)$ be nondecreasing and continuously differentiable on $[0,\infty)$. Assume that the function $-t - p'_\lambda(t)$ is strictly unimodal on $(0,\infty)$ with the convention $p'_\lambda(0) = p'_\lambda(0+)$. Antoniadis and Fan (2001) characterize the properties of $\widehat{\theta}(z)$ as follows:

(1) Sparsity if $\min_{t\ge 0}\{t + p'_\lambda(t)\} > 0$, which holds if $p'(0+) > 0$;

(2) Approximate unbiasedness if $p'_\lambda(t) = 0$ for large $t$;

(3) Continuity if and only if $\arg\min_{t\ge 0}\{t + p'_\lambda(t)\} = 0$.

Note that properties 1) and 2) require the penalty functions to be folded-concave. Fan and Li (2001) advocate to use a family of folded-concave penalized likelihoods as a viable variable selection technique. They do not expect the form of the penalty functions to play a particularly important role provided that it satisfies properties 1) – 3). Antoniadis and Fan (2001) show further that $\widehat{\theta}(z)$ is an anti-symmetric shrinkage function:

$$|\widehat{\theta}(z)| \leq |z|, \quad \text{and} \quad \widehat{\theta}(-z) = -\widehat{\theta}(z). \tag{3.13}$$

The approximate unbiasedness requires $\theta(z)/z \to 1$, as $|z| \to \infty$.

### *3.2.2 Penalty functions*

From the above discussion, singularity at the origin (i.e., $p'_\lambda(0+) > 0$) is sufficient for generating sparsity in variable selection and concavity is needed to reduce the estimation bias. This leads to a family of folded concave penalty functions with singularity at the origin. The $L_1$ penalty can be regarded as both a concave and convex function. It falls on the boundary of the family of the folded-concave *penalty functions*.

The $L_q$ penalty with $q > 1$ is convex. It does not satisfy the sparsity condition, whereas the $L_1$ penalty does not satisfy the unbiasedness condition. The $L_q$ penalty with $0 \leq q < 1$ is concave but does not satisfy the continuity condition. In other words, none of the $L_q$ penalties possesses all three aforementioned properties simultaneously. For this reason, Fan (1997) introduces the smoothly clipped absolute deviation (SCAD), whose derivative is given by

$$p'_\lambda(t) = \lambda \left\{ I\left(t \leq \lambda\right) + \frac{(a\lambda - t)_+}{(a-1)\,\lambda} I\left(t > \lambda\right) \right\} \quad \text{for some } a > 2, \tag{3.14}$$

where $p_\lambda(0) = 0$ and often $a = 3.7$ is used (suggested by a Bayesian argument in Fan and Li, 2001). Now this satisfies the aforementioned three properties. Note that when $a = \infty$, SCAD reduces to the $L_1$-penalty.

In response to Fan (1997), Antoniadis (1997) proposes the penalty function

$$p_\lambda(t) = \frac{1}{2}\lambda^2 - \frac{1}{2}(\lambda - t)_+^2, \tag{3.15}$$

which results in the hard-thresholding estimator

$$\widehat{\theta}_H(z) = zI(|z| > \lambda). \tag{3.16}$$

Fan and Li (2001) refer to this penalty function as the hard thresholding penalty, whose derivative function is $p'_\lambda(t)/2 = (\lambda - t)_+$. An extension of this penalty function, derived by Zhang (2010) from a minimax point of view, is the *minimax concave penalty (MCP)*, whose derivative is given by

$$p'_\lambda(t) = (\lambda - t/a)_+ \,. \tag{3.17}$$

Note that the hard thresholding penalty corresponds to $a = 1$ and the MCP

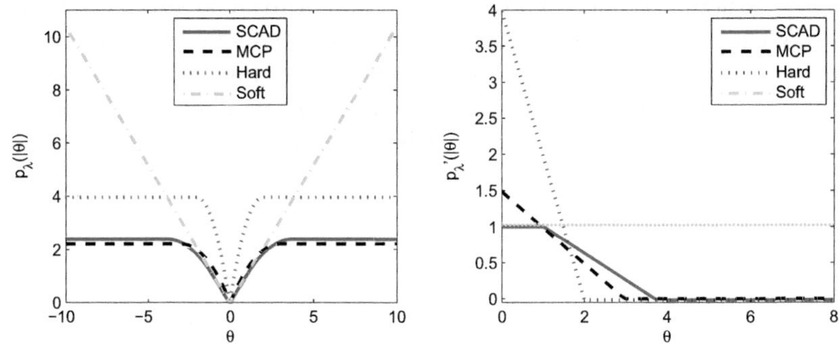


Figure 3.2: Some commonly used penalty functions (left panel) and their derivatives (right panel). They correspond to the risk functions shown in the right panel of Figure 3.3. More precisely, $\lambda = 2$ for hard thresholding penalty, $\lambda = 1.04$ for $L_1$-penalty, $\lambda = 1.02$ for SCAD with $a = 3.7$, and $\lambda = 1.49$ for MCP with $a = 2$. Taken from Fan and Lv (2010).

does not satisfy the continuity property. But this is not that important as noted before. Figure 3.2 depicts some of those commonly used penalty functions.

### *3.2.3 Thresholding by SCAD and MCP*

We now look at the PLS estimator $\widehat{\theta}(z)$ in (3.12) for some penalties. The entropy penalty ($L_0$ penalty) and the hard thresholding penalty (3.15) yield the hard thresholding rule (3.16) (Donoho and Johnstone, 1994) and the $L_1$ penalty gives the soft thresholding rule (Bickel, 1983; Donoho and Johnstone, 1994):

$$\widehat{\theta}_{\text{soft}}(z) = \text{sgn}(z)(|z| - \lambda)_+. \tag{3.18}$$

The SCAD and MCP give rise to analytical solutions to (3.12), each of which is a linear spline in $z$. For the SCAD penalty, the solution is

$$\widehat{\theta}_{\text{SCAD}}(z) = \begin{cases} \text{sgn}(z)(|z| - \lambda)_+, & \text{when } |z| \le 2\lambda; \\ \text{sgn}(z)[(a-1)|z| - a\lambda]/(a-2), & \text{when } 2\lambda < |z| \le a\lambda; \\ z, & \text{when } |z| \ge a\lambda. \end{cases} \tag{3.19}$$

See Fan (1997) and Figure 3.1(b). Note that when $a = \infty$, the SCAD estimator becomes the *soft-thresholding* estimator (3.18).

For the MCP with $a \ge 1$, the solution is

$$\widehat{\theta}_{\text{MCP}}(z) = \begin{cases} \text{sgn}(z)(|z| - \lambda)_+/(1 - 1/a), & \text{when } |z| < a\lambda; \\ z, & \text{when } |z| \ge a\lambda. \end{cases} \tag{3.20}$$

It has discontinuity points at $|z| = \lambda$, which can create model instability. In particular, when $a = 1$, the solution is the hard thresholding function $\widehat{\theta}_H(z)$ (3.16). When $a = \infty$, it also becomes a soft-thresholding estimator.

In summary, SCAD and MCP are folded concave functions. They are generalizations of the soft-thresholding and hard-thresholding estimators. The former is continuous whereas the latter is discontinuous.

### 3.2.4 Risk properties

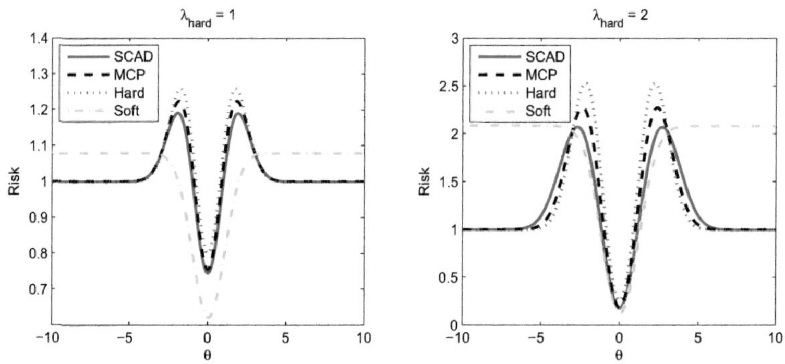


Figure 3.3: The risk functions for penalized least squares under the Gaussian model with the hard-thresholding penalty, $L_1$-penalty, SCAD ($a = 3.7$), and MCP ($a = 2$). The left panel corresponds to $\lambda = 1$ and the right panel corresponds to $\lambda = 2$ for the hard-thresholding estimator, and the rest of the parameters are chosen so that their risks are the same at the point $\theta = 3$. Adapted from Fan and Lv (2010).

We now numerically compare the risk property of several commonly thresholded-shrinkage estimators under the fundamental model $Z \sim N(\theta, 1)$. Let

$$R(\theta) = \mathrm{E}(\widehat{\theta}(Z) - \theta)^2$$

be the risk function for the estimator $\widehat{\theta}(Z)$. Figure 3.3 depicts $R(\theta)$ for some commonly used penalty functions. To make them comparable, we chose $\lambda = 1$ and 2 for the hard thresholding penalty, and for other penalty functions the values of $\lambda$ are selected to make their risks at $\theta = 3$ the same as that of the hard thresholding estimator $\widehat{\theta}_H(z)$.

Figure 3.3 shows that the PLS estimators improve the ordinary least squares estimator $Z$ in the region where $\theta$ is near zero, and have the same risk as the ordinary least squares estimator when $\theta$ is far away from zero (e.g., 4 standard deviations away). An exception to this is the Lasso estimator. The Lasso estimator has a bias approximately of size $\lambda$ for large $\theta$, and this causes higher risk as shown in Figure 3.3. The better risk property at the origin is the payoff that we earn for exploring sparsity.

When $\lambda_{\text{hard}} = 2$, Lasso has a higher risk than the SCAD estimator except in a small region. Lasso prefers smaller $\lambda$ due to its bias. For $\lambda_{\text{hard}} = 1$, Lasso outperforms other methods near the origin. As a result, when $\lambda$ is chosen automatically by data, Lasso has to choose a smaller $\lambda$ in order to have a desired mean squared error (to reduce the modeling bias). Yet, a smaller value of $\lambda$ yields a more complex model. This explains why Lasso tends to have many false positive variables in selected models.

### *3.2.5 Characterization of folded-concave PLS*

Folded-concave penalized least-squares (3.9) is in general a non-convex function. It is challenging to characterize the global solution so let us first characterize its local minimizers.

From Lv and Fan (2009) and Zhang (2010), the local concavity of the penalty $p_\lambda(\cdot)$ at $\mathbf{v} = (v_1, \cdots, v_q)^T$ is defined as

$$\kappa(p_\lambda; \mathbf{v}) = \lim_{\epsilon \to 0+} \max_{1 \le j \le q} \sup_{t_1 < t_2 \in (|v_j| - \epsilon, |v_j| + \epsilon)} - \frac{p'_\lambda(t_2) - p'_\lambda(t_1)}{t_2 - t_1}. \tag{3.21}$$

By the concavity of $p_\lambda$ on $[0, \infty)$, $\kappa(p_\lambda; \mathbf{v}) \ge 0$. It is easy to see by the mean-value theorem that $\kappa(p_\lambda; \mathbf{v}) = \max_{1 \le j \le q} -p''_\lambda(|v_j|)$ when the second derivative of $p_\lambda(\cdot)$ is continuous. For the $L_1$ penalty, $\kappa(p_\lambda; \mathbf{v}) = 0$ for any $\mathbf{v}$. For the SCAD penalty, $\kappa(p_\lambda; \mathbf{v}) = 0$ unless some component of $|\mathbf{v}|$ takes values in $[\lambda, a\lambda]$. In the latter case, $\kappa(p_\lambda; \mathbf{v}) = (a-1)^{-1}\lambda^{-1}$.

Let $\lambda_{\min}(\mathbf{A})$ be the minimum eigenvalue of a symmetric matrix $\mathbf{A}$ and $\|\mathbf{a}\|_\infty = \max_j |a_j|$. Lv and Fan (2009) prove the following result. The gap between the necessary condition for local minimizer and sufficient condition for strict local minimizer is tiny (non-strict versus strict inequalities).

**Theorem 3.1 (Characterization of PLSE)** *Assume that $p_\lambda(|\theta|)$ is folded concave. Then a necessary condition for $\widehat{\boldsymbol\beta} \in R^p$ being a local minimizer of $Q(\boldsymbol\beta)$ defined by (3.9) is*

$$n^{-1}\mathbf{X}_1^T(\mathbf{Y} - \mathbf{X}\widehat{\boldsymbol\beta}) - p'_\lambda(|\widehat{\boldsymbol\beta}_1|)\,\mathrm{sgn}(\widehat{\boldsymbol\beta}_1) = \mathbf{0}, \tag{3.22}$$

$$\|n^{-1}\mathbf{X}_2^T(\mathbf{Y} - \mathbf{X}\widehat{\boldsymbol\beta})\|_\infty \le p'_\lambda(0+), \tag{3.23}$$

$$\lambda_{\min}(n^{-1}\mathbf{X}_1^T\mathbf{X}_1) \ge \kappa(p_\lambda; \widehat{\boldsymbol\beta}_1), \tag{3.24}$$

*where $\mathbf{X}_1$ and $\mathbf{X}_2$ are respectively the submatrices of $\mathbf{X}$ formed by columns indexed by* $\mathrm{supp}(\widehat{\boldsymbol\beta})$ *and its complement, and $\widehat{\boldsymbol\beta}_1$ is a vector of all non-vanishing components $\widehat{\boldsymbol\beta}$. On the other hand, if (3.22) – (3.24) hold with inequalities replaced by strict inequalities, then $\widehat{\boldsymbol\beta}$ is a strict local minimizer of $Q(\boldsymbol\beta)$.*

Conditions (3.22) – (3.24) can be regarded as the *Karush-Kuhn-Tucker conditions*. They can also be derived by using subgradient calculus. Conditions (3.22) and (3.24) are respectively the first and second order conditions for $(\widehat{\boldsymbol{\beta}}_1, \mathbf{0})$ to be the local minimizer of $Q(\boldsymbol{\beta}_1, \mathbf{0})$, the local minimizer on the restricted coordinate subspace. Condition (3.23) guarantees the local minimizer on the restricted coordinate subspace is also the local minimizer of the whole space $R^p$.

When $Q(\boldsymbol{\beta})$ is strictly convex, there exists at most one local minimizer. In this case, the local minimizer is also the unique global minimizer. For a folded concave penalty function, let $\kappa(p_\lambda)$ be the maximum concavity of the penalty function $p_\lambda$ defined by

$$\kappa(p_\lambda) = \sup_{t_1 < t_2 \in (0,\infty)} -\frac{p'_\lambda(t_2) - p'_\lambda(t_1)}{t_2 - t_1}. \tag{3.25}$$

For the $L_1$ penalty, SCAD and MCP, we have $\kappa(p_\lambda) = 0$, $(a-1)^{-1}$, and $a^{-1}$, respectively. Thus, the maximum concavity of SCAD and MCP is small when $a$ is large. When

$$\lambda_{\min}(n^{-1}\mathbf{X}^T\mathbf{X}) > \kappa(p_\lambda), \tag{3.26}$$

the function $Q(\boldsymbol{\beta})$ is strictly convex, as the convexity of the quadratic loss dominates the maximum concavity of the penalty in (3.9). Hence, the global minimum is unique. Note that condition (3.26) requires $p \le n$.

In general, the global minimizer of the folded-concave penalized least-squares is hard to characterize. Fan and Lv (2011) are able to give conditions under which a solution is global optimal on the union of all $m$-dimensional coordinate subspaces:

$$\mathbb{S}_m = \{\boldsymbol{\beta} \in R^p : \|\boldsymbol{\beta}\|_0 \le m\}. \tag{3.27}$$

## 3.3 Lasso and $L_1$ Regularization

Lasso gains its popularity due to its convexity and computational expedience. The predecessor of Lasso is the negative garrote. The study of Lasso also leads to the Dantzig selector, the *adaptive Lasso* and the *elastic net.* This section touches on the basis of these estimators in which the $L_1$-norm regularization plays a central role.

### *3.3.1 Nonnegative garrote*

The nonnegative garrote estimator, introduced by Breiman (1995), is the first modern statistical method that uses the $L_1$-norm regularization to do variable selection in multiple linear regression. Consider the usual setting with $p < n$ and let $\widehat{\boldsymbol{\beta}} = (\widehat{\beta}_1, \cdots, \widehat{\beta}_p)^T$ be the OLS estimator. When $p > n$, $\widehat{\boldsymbol{\beta}}$ can be the ridge regression estimator (Yuan and Lin, 2005), the main idea stays the same. Then, the fitted model becomes

$$\widehat{Y} = \widehat{\beta}_1 X_1 + \cdots + \widehat{\beta}_p X_p$$

The above model uses all variables. To do variable selection, we introduce the nonnegative shrinkage parameter $\boldsymbol{\theta} = (\theta_1, \cdots, \theta_p)^T$ and regard $\widehat{\boldsymbol{\beta}}$ as fixed; a new fitted model becomes

$$\widehat{Y} = \theta_1 Z_1 + \cdots + \theta_p Z_p, \qquad Z_j = \widehat{\beta}_j X_j.$$

If $\theta_j$ is zero, then variable $X_j$ is excluded from the fitted model. The *nonnegative garrote* estimates $\boldsymbol{\theta}$ via the following $L_1$ regularized least squares:

$$\min_{\boldsymbol{\theta} \geq 0} \frac{1}{2n} \|\mathbf{Y} - \mathbf{Z}\boldsymbol{\theta}\|^2 + \lambda \sum_j \theta_j. \tag{3.28}$$

Note that under the nonnegative constraints $\sum_{j=1}^p \theta_j = \|\boldsymbol{\theta}\|_1$.

By varying $\lambda$, the nonnegative garrote automatically achieves model selection. Many components of the minimizer of (3.28), $\widehat{\boldsymbol{\theta}}$, will be zero. This can be easily seen when $\mathbf{X}$ is scaled orthonormal $\mathbf{X}^T\mathbf{X} = n\mathbf{I}_p$, as in Section 3.2.1. In this case, $\{\mathbf{Z}_j\}$ are still orthogonal and $\mathbf{Z}^T\mathbf{Z}$ is diagonal. The ordinary least-squares estimator is given by

$$\widehat{\boldsymbol{\theta}}_0 = (\mathbf{Z}^T\mathbf{Z})^{-1}\mathbf{Z}^T\mathbf{Y} = (\mathbf{Z}_1/\|\mathbf{Z}_1\|^2, \cdots, \mathbf{Z}_p/\|\mathbf{Z}_p\|^2)^T\mathbf{Y}.$$

Note that by the orthogonality of the least-squares fit to its residuals,

$$\begin{aligned} \|\mathbf{Y} - \mathbf{Z}\boldsymbol{\theta}\|^2 &= \|\mathbf{Y} - \mathbf{Z}\widehat{\boldsymbol{\theta}}_0\|^2 + \|\mathbf{Z}(\boldsymbol{\theta} - \widehat{\boldsymbol{\theta}}_0)\|^2 \\ &= \|\mathbf{Y} - \mathbf{Z}\widehat{\boldsymbol{\theta}}_0\|^2 + \sum_{j=1}^p \|\mathbf{Z}_j\|^2(\theta_j - \widehat{\theta}_{j0})^2, \end{aligned}$$

where $\widehat{\theta}_{j0} = \mathbf{Z}_j^T\mathbf{Y}/\|\mathbf{Z}_j\|^2$ is the $j^{th}$ component of $\boldsymbol{\theta}_0$ and the last equality uses the orthogonality of $\mathbf{Z}_j$. Therefore, problem (3.28) becomes

$$\min_{\boldsymbol{\theta} \geq 0} \frac{1}{2} \sum_{j=1}^p \|\mathbf{Z}_j\|^2(\theta_j - \widehat{\theta}_{j0})^2 + \lambda \sum_{j=1}^p \theta_j.$$

This reduces to the componentwise minimization problem

$$\min_{\theta \geq 0} \frac{1}{2}(\theta - \theta_0)^2 + \lambda\theta,$$

whose minimizer is clearly $\widehat{\theta} = (\theta_0 - \lambda)_+$ by taking the first derivative and setting it to zero. Applying this to our scenario and noticing $\|\mathbf{Z}_j\|^2 = n\widehat{\beta}_j^2$ and $\widehat{\theta}_{j0} = n^{-1}\mathbf{X}_j^T\mathbf{Y}/\widehat{\beta}_j$ we have

$$\widehat{\theta}_j = \left(\widehat{\theta}_{j0} - \frac{\lambda}{\|\mathbf{Z}_j\|^2}\right)_+ = \left(\frac{\mathbf{X}_j^T\mathbf{Y}}{n\widehat{\beta}_j} - \frac{\lambda}{n\widehat{\beta}_j^2}\right)_+.$$

In particular, if $\widehat{\boldsymbol\beta} = n^{-1}\mathbf{X}^T\mathbf{Y}$ is the ordinary least-squares estimate, then

$$\widehat{\theta}_j = \Big(1 - \frac{\lambda}{n\widehat{\beta}_j^2}\Big)_+.$$

Model selection of the negative garrote now becomes clear. When $|\widehat{\beta}_j| \le \sqrt{\lambda/n}$, it is shrunk to zero. The larger the original estimate, the smaller the shrinkage. Furthermore, the shrinkage rule is continuous. This is in the same spirit as the folded concave PLS such as SCAD introduced in the last section.

### *3.3.2 Lasso*

*Lasso*, the term coined by Tibshirani (1996), estimates the sparse regression coefficient vector $\boldsymbol\beta$ by minimizing

$$\frac{1}{2n}\|\mathbf{Y} - \mathbf{X}\boldsymbol\beta\|^2 + \lambda\|\boldsymbol\beta\|_1. \tag{3.29}$$

This corresponds to (3.9) by taking $p_\lambda(\theta) = \lambda\theta$ for $\theta \ge 0$. Comparing (3.29) and (3.28), we see that the Lasso does not need to use a preliminary estimator of $\boldsymbol\beta$, although both use the $L_1$-norm to achieve variable selection.

The KKT conditions (3.22)–(3.24) now become

$$n^{-1}\mathbf{X}_1^T(\mathbf{Y} - \mathbf{X}_1\widehat{\boldsymbol\beta}_1) - \lambda\,\mathrm{sgn}(\widehat{\boldsymbol\beta}_1) = \mathbf{0}, \tag{3.30}$$

and

$$\|(n\lambda)^{-1}\mathbf{X}_2^T(\mathbf{Y} - \mathbf{X}_1\widehat{\boldsymbol\beta}_1)\|_\infty \le 1, \tag{3.31}$$

since (3.24) is satisfied automatically. This first condition says that the signs of nonzero components of Lasso are the same as the correlations of the covariates with the current residual. The equations (3.30) and (3.31) imply that

$$\|n^{-1}\mathbf{X}^T(\mathbf{Y} - \mathbf{X}\widehat{\boldsymbol\beta})\|_\infty \le \lambda. \tag{3.32}$$

Note that condition (3.31) holds for $\widehat{\boldsymbol\beta} = 0$ when

$$\lambda > \|n^{-1}\mathbf{X}^T\mathbf{Y}\|_\infty. \tag{3.33}$$

Since the condition is imposed with a strict inequality, it is a sufficient condition (Theorem 3.1). In other words, when $\lambda > \|n^{-1}\mathbf{X}^T\mathbf{Y}\|_\infty$, $\widehat{\boldsymbol\beta} = \mathbf{0}$ is the unique solution and hence Lasso selects no variables. Therefore, we need only to consider $\lambda$ in the interval $[0, \|n^{-1}\mathbf{X}^T\mathbf{Y}\|_\infty]$.

We now look at the model selection consistency of Lasso. Assuming the invertibility of $\mathbf{X}_1^T\mathbf{X}_1$, solving equation (3.30) gives

$$\widehat{\boldsymbol\beta}_1 = (\mathbf{X}_1^T\mathbf{X}_1)^{-1}(\mathbf{X}_1^T\mathbf{Y} - n\lambda\,\mathrm{sgn}(\widehat{\boldsymbol\beta}_1)), \tag{3.34}$$

and substituting this into equation (3.31) yields

$$\|(n\lambda)^{-1}\mathbf{X}_2^T(\mathbf{I}_n - \mathbf{P}_{\mathbf{X}_1})\mathbf{Y} + \mathbf{X}_2^T\mathbf{X}_1(\mathbf{X}_1^T\mathbf{X}_1)^{-1}\,\mathrm{sgn}(\widehat{\boldsymbol\beta}_1)\|_\infty \le 1, \tag{3.35}$$

where $\mathbf{P}_{\mathbf{X}_1} = \mathbf{X}_1(\mathbf{X}_1^T\mathbf{X}_1)^{-1}\mathbf{X}_1^T$ is the projection matrix onto the linear space spanned by the columns of $\mathbf{X}_1$. For the true parameter, let $\text{supp}(\boldsymbol{\beta}_0) = \mathcal{S}_0$ so that

$$\mathbf{Y} = \mathbf{X}_{\mathcal{S}_0}\boldsymbol{\beta}_0 + \boldsymbol{\varepsilon}. \tag{3.36}$$

If $\text{supp}(\widehat{\boldsymbol{\beta}}) = \mathcal{S}_0$, i.e., *model selection consistency* holds, then $\mathbf{X}_{\mathcal{S}_0} = \mathbf{X}_1$ and $(\mathbf{I}_n - \mathbf{P}_{\mathbf{X}_1})\mathbf{X}_{\mathcal{S}_0} = 0$. By substituting (3.36) into (3.35), we have

$$\|(n\lambda)^{-1}\mathbf{X}_2^T(\mathbf{I}_n - \mathbf{P}_{\mathbf{X}_1})\boldsymbol{\varepsilon} + \mathbf{X}_2^T\mathbf{X}_1(\mathbf{X}_1^T\mathbf{X}_1)^{-1}\text{sgn}(\widehat{\boldsymbol{\beta}}_1)\|_\infty \le 1. \tag{3.37}$$

Note that this condition is also sufficient if the inequality is replaced with the strict one (see Theorem 3.1).

Typically, the first term in (3.37) is negligible. This will be formally shown in Chapter 4. A specific example is the case $\boldsymbol{\varepsilon} = 0$ as in the compressed sensing problem. In this case, condition (3.37) becomes

$$\|\mathbf{X}_2^T\mathbf{X}_1(\mathbf{X}_1^T\mathbf{X}_1)^{-1}\text{sgn}(\widehat{\boldsymbol{\beta}}_1)\|_\infty \le 1.$$

This condition involves $\text{sgn}(\widehat{\boldsymbol{\beta}}_1)$. If we require a stronger consistency $\text{sgn}(\widehat{\boldsymbol{\beta}}) = \text{sgn}(\boldsymbol{\beta}_0)$, called the *sign consistency*, then the condition becomes

$$\|\mathbf{X}_2^T\mathbf{X}_1(\mathbf{X}_1^T\mathbf{X}_1)^{-1}\text{sgn}(\boldsymbol{\beta}_{\mathcal{S}_0})\|_\infty \le 1. \tag{3.38}$$

The above condition does not depend on $\lambda$. It appeared in Zou (2006) and Zhao and Yu (2006) who coined the name *the irrepresentable condition*.

Note that $(\mathbf{X}_1^T\mathbf{X}_1)^{-1}\mathbf{X}_1^T\mathbf{X}_2$ in (3.38) is the matrix of the regression coefficients of each 'unimportant' variable $X_j$ ($j \notin \mathcal{S}_0$) regressed on the important variables $\mathbf{X}_1 = \mathbf{X}_{\mathcal{S}_0}$. The irrepresentable condition is a condition on how strongly the important and unimportant variables can be correlated. Condition (3.38) states that the sum of the signed regression coefficients of each unimportant variable $X_j$ for $j \notin \mathcal{S}_0$ on the important variables $\mathbf{X}_{\mathcal{S}_0}$ cannot exceed 1. The more the unimportant variables, the harder the condition is to meet. The irrepresentable condition is in general very restrictive. Using the regression intuition, one can easily construct an example when it fails. For example, if an unimportant variable is generated by

$$X_j = \rho s^{-1/2}\sum_{k\in\mathcal{S}_0}\text{sgn}(\beta_k)X_k + \sqrt{1-\rho^2}\varepsilon_k, \qquad s = |\mathcal{S}_0|,$$

for some given $|\rho| \le 1$ (all normalization is to make $\text{Var}(X_j) = 1$), where all other random variables are independent and standardized, then the $L_1$-norm of the signed regression coefficients of this variable is $|\rho|s^{1/2}$, which can easily exceed 1. The larger the 'important variable' set $\mathcal{S}_0$, the more easily the irrepresentable condition fails. In addition, we need only one such unimportant predictor that has such a non-negligible correlation with important variables to make the condition fail. See also Corollary 1 in Zou (2006) for a counterexample.

The moral of the above story is that Lasso can have sign consistency, but this happens only in very specific cases. The irrepresentable condition (3.38) is independent of $\lambda$. When it fails, Lasso does not have sign consistency and this cannot be rescued by using a different value of $\lambda$.

We now look at the risk property of Lasso. It is easier to explain it under the constrained form:

$$\min_{\|\boldsymbol{\beta}\|_1 \le c} \|\mathbf{Y} - \mathbf{X}\boldsymbol{\beta}\|^2 \tag{3.39}$$

for some constant $c$, as in Tibshirani (1996). Define the *theoretical risk* and *empirical risk* respectively as

$$R(\boldsymbol{\beta}) = \mathrm{E}(Y - \mathbf{X}^T\boldsymbol{\beta})^2 \quad \text{and} \quad R_n(\boldsymbol{\beta}) = n^{-1}\sum_{i=1}^n (Y_i - \mathbf{X}_i^T\boldsymbol{\beta})^2,$$

which are prediction errors using the parameter $\boldsymbol{\beta}$. The best prediction error is $R(\boldsymbol{\beta}_0)$. Note that

$$R(\boldsymbol{\beta}) = \boldsymbol{\gamma}^T\boldsymbol{\Sigma}^*\boldsymbol{\gamma} \quad \text{and} \quad R_n(\boldsymbol{\beta}) = \boldsymbol{\gamma}^T\mathbf{S}_n^*\boldsymbol{\gamma},$$

where $\boldsymbol{\gamma} = (-1, \boldsymbol{\beta}^T)^T$, $\boldsymbol{\Sigma}^* = \mathrm{Var}((Y, \mathbf{X}^T)^T)$, and $\mathbf{S}_n^*$ is the sample covariance matrix based on the data $\{(Y_i, \mathbf{X}_i^T)^T\}_{i=1}^n$. Thus, for any $\boldsymbol{\beta}$, we have the following risk approximation:

$$\begin{aligned} |R(\boldsymbol{\beta}) - R_n(\boldsymbol{\beta})| &= |\boldsymbol{\gamma}^T(\boldsymbol{\Sigma}^* - \mathbf{S}_n^*)\boldsymbol{\gamma}| \\ &\le \|\boldsymbol{\Sigma}^* - \mathbf{S}_n^*\|_{\max}\|\boldsymbol{\gamma}\|_1^2 \\ &= (1 + \|\boldsymbol{\beta}\|_1)^2\|\boldsymbol{\Sigma}^* - \mathbf{S}_n^*\|_{\max}, \end{aligned} \tag{3.40}$$

where $\|\mathbf{A}\|_{\max} = \max_{i,j}|a_{ij}|$ for a matrix $\mathbf{A}$ with elements $a_{ij}$. On the other hand, if the true parameter $\boldsymbol{\beta}_0$ is in the feasible set, namely, $\|\boldsymbol{\beta}_0\|_1 \le c$, then $R_n(\widehat{\boldsymbol{\beta}}) - R_n(\boldsymbol{\beta}_0) \le 0$. Using this,

$$0 \le R(\widehat{\boldsymbol{\beta}}) - R(\boldsymbol{\beta}_0) \le \{R(\widehat{\boldsymbol{\beta}}) - R_n(\widehat{\boldsymbol{\beta}})\} + \{R_n(\boldsymbol{\beta}_0) - R(\boldsymbol{\beta}_0)\}.$$

By (3.40) along with $\|\widehat{\boldsymbol{\beta}}\|_1 \le c$ and $\|\boldsymbol{\beta}_0\|_1 \le c$, we conclude that

$$|R(\widehat{\boldsymbol{\beta}}) - R(\boldsymbol{\beta}_0)| \le 2(1 + c)^2\|\boldsymbol{\Sigma}^* - \mathbf{S}_n^*\|_{\max}. \tag{3.41}$$

When $\|\boldsymbol{\Sigma}^* - \mathbf{S}_n^*\|_{\max} \to 0$, the risk converges. Such a property is called *persistency* by Greenshtein and Ritov (2004). Further details on the rates of convergence for estimating large covariance matrices can be found in Chapter 11. The rate is of order $O(\sqrt{(\log p)/n})$ for the data with Gaussian tails. The above discussion also reveals the relationship between covariance matrix estimation and sparse regression. A robust covariance matrix estimation can also reveal a robust sparse regression.

Persistency requires that the risk based on $\widehat{\boldsymbol{\beta}}$ is approximately the same as that of the optimal parameter $\boldsymbol{\beta}_0$, i.e.,

$$R(\widehat{\boldsymbol{\beta}}) - R(\boldsymbol{\beta}_0) = o_P(1).$$

By (3.41), this requires only $\boldsymbol{\beta}_0$ sparse in the sense that $\|\boldsymbol{\beta}_0\|_1$ does not grow too quickly (recalling $\|\boldsymbol{\beta}_0\|_1 \le c$) and the large covariance matrix $\boldsymbol{\Sigma}^*$ can be uniformly consistently estimated. For data with Gaussian tails, since $\|\boldsymbol{\Sigma}^* - \mathbf{S}_n^*\|_{\max} = O_P(\sqrt{(\log p)/n})$ (see Chapter 11), we require

$$\|\boldsymbol{\beta}_0\|_1 \le c = o((n/\log p)^{1/4})$$

for Lasso to possess persistency. Furthermore, the result (3.41) does not require having a true underlying linear model. As long as we define

$$\boldsymbol{\beta}_0 = \operatorname{argmin}_{\|\boldsymbol{\beta}\|_1 \le c} R(\boldsymbol{\beta}),$$

the risk approximation inequality (3.41) holds by using the same argument above. In conclusion, Lasso has a good risk property when $\boldsymbol{\beta}_0$ is sufficiently sparse.

### *3.3.3 Adaptive Lasso*

The irrepresentable condition indicates restrictions on the use of the Lasso as a model/variable selection method. Another drawback of the Lasso is its lack of unbiasedness for large coefficients, as explained in Fan and Li (2001). This can be seen from (3.34). Even when the signal is strong so that $\operatorname{supp}(\widehat{\boldsymbol{\beta}}) = \mathcal{S}_0$, by substituting (3.36) into (3.34), we have

$$\widehat{\boldsymbol{\beta}}_1 = \boldsymbol{\beta}_0 + (\mathbf{X}_1^T\mathbf{X}_1)^{-1}\mathbf{X}_1^T\boldsymbol{\varepsilon} - n\lambda(\mathbf{X}_1^T\mathbf{X}_1)^{-1}\operatorname{sgn}(\widehat{\boldsymbol{\beta}}_1).$$

The last term is the bias due to the $L_1$ penalty. Unless $\lambda$ goes to 0 sufficiently fast, the bias term is not negligible. However, $\lambda \approx \frac{1}{\sqrt{n}}$ is needed in order to make the Lasso estimate root-$n$ consistent under the fixed $p$ large $n$ setting. For $p \gg n$, the Lasso estimator uses $\lambda \approx \sqrt{\log(p)/n}$ to achieve the optimal rate $\sqrt{|\mathcal{S}_0|\log(p)/n}$. See Chapter 4 for more details. So now, it is clear that the optimal Lasso estimator has non-negligible biases.

Is there a nice fix for these two problems? Zou (2006) proposes to use the adaptively weighted $L_1$ penalty (a.k.a. *adaptive Lasso*) to replace the $L_1$ penalty in penalized linear regression and penalized generalized linear models. With the weighted $L_1$ penalty, (3.9) becomes

$$\frac{1}{2n}\|\mathbf{Y} - \mathbf{X}\boldsymbol{\beta}\|^2 + \lambda\sum_{j=1}^{p} w_j|\beta_j|. \tag{3.42}$$

To keep the convexity property of the Lasso, $w_j$ should be nonnegative. It is important to note that if the weights are deterministic, then they cannot fix the aforementioned two problems of the Lasso. Suppose that some deterministic weights can make the Lasso gain sign consistency. Then no $w_j$ should be zero, otherwise the variable $X_j$ is always included, which will violate the sign consistency of the Lasso if the underlying model does not include $X_j$, i.e. $X_j$

is not an important variable. Hence, all $w_j$s are positive. Then we redefine the regressors as $X_j^{\mathrm{w}} = X_j/w_j$ and $\theta_j = w_j\beta$, $1 \le j \le p$. The underlying regression model can be rewritten as

$$Y = \sum_{j=1}^{p} X_j^{\mathrm{w}}\theta_j + \epsilon$$

and (3.42) becomes

$$\frac{1}{2n}\|\mathbf{Y} - \mathbf{X}^{\mathrm{w}}\boldsymbol{\theta}\|^2 + \lambda\sum_{j=1}^{p}|\theta_j|.$$

Its corresponding irrepresentable condition is

$$\|(\mathbf{X}_2^{\mathrm{w}})^T\mathbf{X}_1^{\mathrm{w}}[(\mathbf{X}_1^{\mathrm{w}})^T\mathbf{X}_1^{\mathrm{w}}]^{-1}\,\mathrm{sgn}(\boldsymbol{\beta}_{\mathcal{S}_0})\|_\infty \le 1.$$

We write $\mathbf{W} = (w_1, \ldots, w_p)^T = (\mathbf{W}_1, \mathbf{W}_2)^T$. and express the irrepresentable conditions using the original variables, we have

$$\|[\mathbf{X}_2^T\mathbf{X}_1(\mathbf{X}_1^T\mathbf{X}_1)^{-1}\mathbf{W}_1 \circ \mathrm{sgn}(\boldsymbol{\beta}_{\mathcal{S}_0})] \circ \mathbf{W}_2^{-1}\| \le 1$$

Observe that if $\max \mathbf{W}_1/\inf \mathbf{W}_2 \to 0$, then this representable condition can be satisfied for general $\mathbf{X}_1, \mathbf{X}_2$ and $\mathrm{sgn}(\boldsymbol{\beta}_{\mathcal{S}_0})$. This condition can only be achieved using a data-driven scheme, as we do not know the set $\mathcal{S}_0$.

Zou (2006) proposes to use a preliminary estimate $\widehat{\beta}_j$ to construct $w_j$. For example, $w_j = |\widehat{\beta}_j|^{-\gamma}$ for some $\gamma > 0$, and $\gamma = 0.5$, 1 or 2. In the case of fixed $p$ large $n$, the preliminary estimate can be the least-squares estimate. When $p \gg n$, the preliminary estimate can be the Lasso estimate and $w_j = p_\lambda'(|\widehat{\beta}_j^{\mathrm{lasso}}|)/\lambda$ with a folded concave penalty $p_\lambda(\cdot)$.

As will be seen in Section 3.5.5, the adaptive Lasso is connected to the penalized least-squares estimator (3.9) via the *local linear approximation* with $p_\lambda'(\theta) = \lambda\theta^{-\gamma}$ or $L_{1-\gamma}$ penalty. Since the derivative function is decreasing, the spirit of the adaptive Lasso is the same as the folded-concave PLS. Hence, the adaptive Lasso is able to fix the bias caused by the $L_1$ penalty. In particular, the adaptive Lasso estimator for $\boldsymbol{\beta}_{\mathcal{S}_0}$ shares the asymptotical normality property of the oracle OLS estimator for $\boldsymbol{\beta}_{\mathcal{S}_0}$, i.e., $\widehat{\boldsymbol{\beta}}_{\mathcal{S}_0}^{\mathrm{oracle}} = (\mathbf{X}_1^T\mathbf{X}_1)^{-1}\mathbf{X}_1^T\mathbf{Y}$.

### *3.3.4 Elastic Net*

In the early 2000s, the Lasso was applied to regression with micrarrays to do gene selection. The results were concerning because of high variability. This is mainly caused by the spurious correlation in high-dimensional data, as illustrated in Section 1.3.3 of Chapter 1. How to handle the strong (empirical) correlations among high-dimensional variables while keeping the continuous shrinkage and selection property of the Lasso? Zou and Hastie (2005) propose the *Elastic Net* regularization that uses a convex combination of $L_1$ and $L_2$

penalties. For the penalized least squares, the Elastic Net estimator is defined as

$$\arg\min_{\boldsymbol{\beta}} \left\{ \frac{1}{n} \|\mathbf{Y} - \mathbf{X}\boldsymbol{\beta}\|^2 + \lambda_2 \|\boldsymbol{\beta}\|^2 + \lambda_1 \|\boldsymbol{\beta}\|_1 \right\}, \tag{3.43}$$

where $p_{\lambda_1,\lambda_2}(t) = \lambda_1|t| + \lambda_2 t^2$ is called the Elastic Net penalty. Another form of the Elastic Net penalty is

$$p_{\lambda,\alpha}(t) = \lambda J(t) = \lambda[(1-\alpha)t^2 + \alpha|t|],$$

with $\lambda = \lambda_1 + \lambda_2$ and $\alpha = \frac{\lambda_1}{\lambda_1+\lambda_2}$. The Elastic Net is a pure ridge regression when $\alpha = 0$ and a pure Lasso when $\alpha = 1$. The advantage of using $(\lambda, \alpha)$ parametrization is that $\alpha$ has a natural range $[0, 1]$. In practice, we can use CV to choose $\alpha$ over a grid such as $0.1k$, $k = 1, \ldots, 10$. For the penalized least squares problem, using $(\lambda_2, \lambda_1)$ parametrization is interesting because it can be shown that for a fixed $\lambda_2$ the solution path is piecewise linear with respect to $\lambda_1$. Zou and Hastie (2005) exploit this property to derive an efficient path-following algorithm named LARS-EN for computing the entire solution path of the Elastic Net penalized least squares (for each fixed $\lambda_2$).

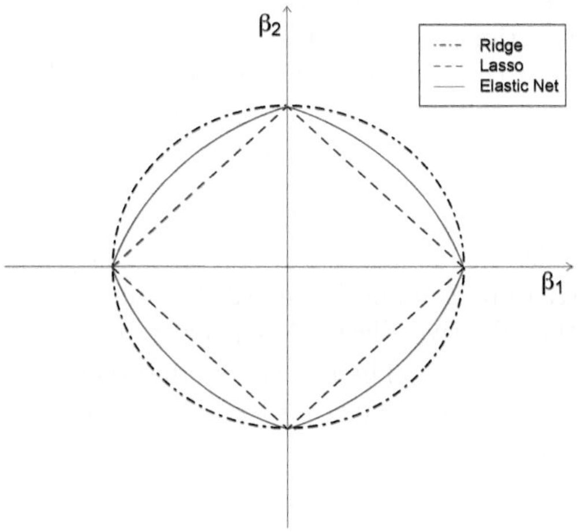


Figure 3.4: Geometric comparison of Lasso, Ridge and Elastic Net in two-dimensions. $\alpha = 0.5$ in the Elastic Net penalty. The Lasso and Elastic Net share four sharp vertices–sparsity. Similar to Ridge, Elastic Net has 'round' edges–strict convexity.

It is well known that $L_2$ regularization gracefully handles collinearity and achieves a good bias-variance tradeoff for prediction. The Elastic Net inherits the ability to handle collinearity from its $L_2$ component and keeps the sparsity property of the Lasso through its $L_1$ component. Figure 3.4 shows a geometric comparison of the two penalty functions in two-dimensions. With high-dimensional data the Elastic Net often generates a more accurate predictive model than the Lasso. Zou and Hastie (2005) also reveal the group effect of the Elastic Net in the sense that highly correlated variables tend to enter or exit the model together, while the Lasso tends to randomly pick one variable and ignore the rest.

To help visualize the fundamental differences between Lasso and Elastic Net, let us consider a synthetic model as follows. Let $Z_1$ and $Z_2$ be two independent unif(0, 20) variables. Response $\mathbf{Y}$ is generated by $\mathbf{Y} = Z_1 + 0.1 \cdot Z_2 + \epsilon$, with $\epsilon \sim N(0,1)$ and the observed regressors are generated by

$$\begin{aligned} X_1 &= Z_1 + \epsilon_1, \quad X_2 = -Z_1 + \epsilon_2, \quad X_3 = Z_1 + \epsilon_3, \\ X_4 &= Z_2 + \epsilon_4, \quad X_5 = -Z_2 + \epsilon_5, \quad X_6 = Z_2 + \epsilon_6, \end{aligned}$$

where $\epsilon_i$ are iid $N(0, \frac{1}{16})$. $X_1, X_2, X_3$ form a group whose underlying factor is $Z_1$, and $X_4, X_5, X_6$ form the other group whose underlying factor is $Z_2$. The within group correlations are almost 1 and the between group correlations are almost 0. Ideally, we would want to only identify the $Z_1$ group $(X_1, X_2, X_3)$ as the important variables. We generated two independent datasets with sample size 100 from this model. Figure 3.5 displays the solution paths of Lasso and Elastic Net; see Section 3.5 for details. The two Lasso solution paths are very different, suggesting the high instability of the Lasso under strong correlations. On the other hand, the two Elastic Net solution paths are almost identical. Moreover, the Elastic Net identifies the corrected variables.

The Elastic Net relies on its $L_1$ component for sparsity and variable selection. Similar to the Lasso case, the Elastic Net also requires a restrictive condition on the design matrix for selection consistency (Jia and Yu, 2010). To bypass this restriction, Zou and Zhang (2009) follow the adaptive Lasso idea and introduce the adaptive Elastic Net penalty $p(|\beta_j|) = \lambda_1 w_j |\beta_j| + \lambda_2 |\beta_j|^2$ where $w_j = |\widehat{\beta}^{\text{enet}} + 1/n|^{-\gamma}$. The numeric studies therein shows the very competitive performance of the adaptive Elastic Net in terms of variable selection and model estimation.

### *3.3.5 Dantzig selector*

The *Dantzig selector*, introduced by Candés and Tao (2007), is a novel idea of casting the regularization problem into a linear program. Recall that Lasso satisfies (3.32), but it might not have the smallest $L_1$ norm. One can find the estimator to minimize its $L_1$-norm:

$$\min_{\boldsymbol{\beta} \in R^p} \|\boldsymbol{\beta}\|_1, \qquad \text{subject to} \quad \|n^{-1}\mathbf{X}^T(\mathbf{Y} - \mathbf{X}\boldsymbol{\beta})\|_\infty \le \lambda. \tag{3.44}$$

The target function and constraints in (3.44) are linear. The problem can be

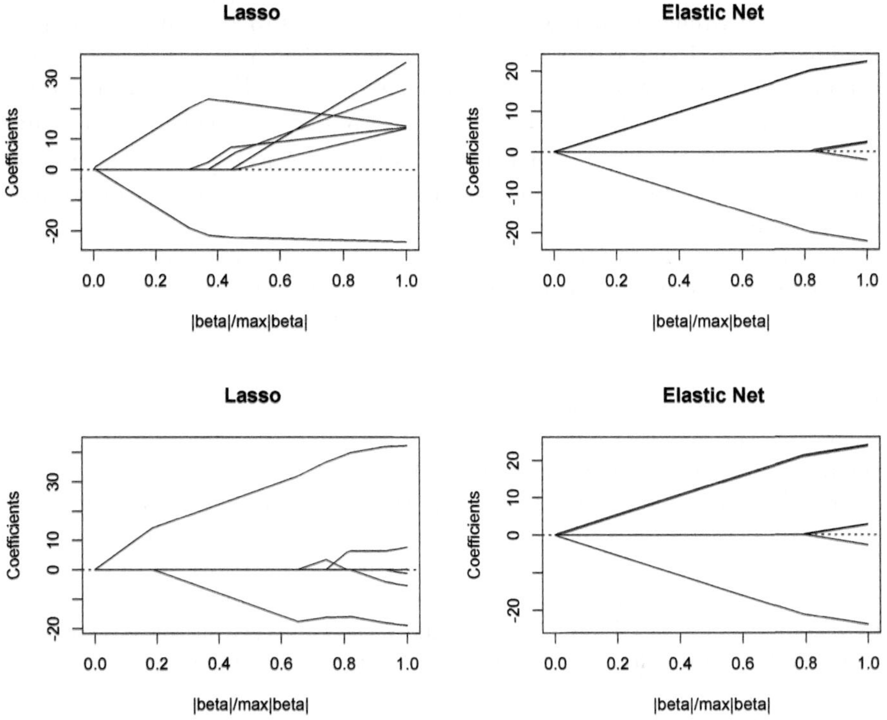


Figure 3.5: A toy example to illustrate the instability of the Lasso and how this is improved by the Elastic Net. Left two panels show the Lasso solution path on two independent datasets generated from the same model; the right two panels are the Elastic Net solution path on the same two datasets. The Lasso paths are very different, but the Elastic Net paths are almost the same. The x-axis is the fraction of $L_1$ norm defined as $\|\widehat{\boldsymbol{\beta}}(\lambda)\|_1 / \max_{\lambda \le \lambda_{\max}} \|\widehat{\boldsymbol{\beta}}(\lambda)\|_1$.

formulated as a linear program by expressing it as

$$\min_{\mathbf{u}} \sum_{i=1}^{p} u_i, \qquad \mathbf{u} \ge 0, \quad -\mathbf{u} \le \boldsymbol{\beta} \le \mathbf{u}, \quad -\lambda \mathbf{1} \le n^{-1} \mathbf{X}^T (\mathbf{Y} - \mathbf{X}\boldsymbol{\beta}) \le \lambda \mathbf{1}.$$

The name "Dantzig selector" was coined by Emmanuel Candès and Terence Tao to pay tribute to George Dantzig, the father of linear programming who passed away while their manuscript was finalized.

Let $\widehat{\boldsymbol{\beta}}_{\mathrm{DZ}}$ be the solution. A necessary condition for $\widehat{\boldsymbol{\beta}}_{\mathrm{DZ}}$ to have model selection consistency is that $\boldsymbol{\beta}_0$ is in the feasible set of (3.44), with probability tending to one. Using model (3.36), this implies that $\lambda \geq n^{-1}\|\mathbf{X}^T\boldsymbol{\varepsilon}\|_\infty$. For example, in the case when $\boldsymbol{\varepsilon} \sim N(0, \sigma^2 I_n)$ and columns $\{\mathbf{X}_j\}_{j=1}^p$ of $\mathbf{X}$ are standardized so that $n^{-1}\|\mathbf{X}_j\|^2 = 1$, then $\mathbf{X}_j^T\boldsymbol{\varepsilon} \sim N(0, \sigma^2/n)$. Then, it can easily be shown (see Section 3.3.7) that it suffices to take $\lambda$ as $\sigma\sqrt{2(1+\delta)n^{-1}\log p}$ for any $\delta > 0$, by using the union bound and the tail probability of normal distribution.

The Dantzig selector opens a new chapter for sparse regularization. Since the value $\lambda$ is chosen so that the true parameter $\boldsymbol{\beta}_0$ falls in the constraint:

$$P\{\|n^{-1}\mathbf{X}^T(\mathbf{Y} - \mathbf{X}\boldsymbol{\beta}_0)\|_\infty \leq \lambda\} \to 1, \tag{3.45}$$

Fan (2014) interprets the set $\{\boldsymbol{\beta} : \|n^{-1}\mathbf{X}^T(\mathbf{Y} - \mathbf{X}\boldsymbol{\beta})\|_\infty \leq \lambda\}$ as the high confidence set and this high confidence set summarizes the information on the parameter $\boldsymbol{\beta}_0$ provided by the data. He argues that this set is too big to be useful in high-dimensional spaces and that we need some additional prior about $\boldsymbol{\beta}_0$. If the prior is that $\boldsymbol{\beta}_0$ is sparse, one naturally combines these two pieces of information. This leads to finding the *sparsest solution in the high-confidence set* as a natural solution to the sparse regulation. This idea applies to *quasi-likelihood* based models and includes the Dantzig selector as a specific case. See Fan (2014) for details. The idea is reproduced by Fan, Han, and Liu (2014).

To see how norm-minimization plays a role, let us assume that $\boldsymbol{\beta}_0$ is in the feasible set by taking a large enough value $\lambda$, i.e., $\lambda \geq n^{-1}\|\mathbf{X}^T\boldsymbol{\varepsilon}\|_\infty$ as noted above. This is usually achieved by a probabilistic statement. Let $\widehat{\boldsymbol{\Delta}} = \widehat{\boldsymbol{\beta}}_{\mathrm{DZ}} - \boldsymbol{\beta}_0$. From the norm minimization, we have

$$\|\boldsymbol{\beta}_0\|_1 \geq \|\widehat{\boldsymbol{\beta}}_{\mathrm{DZ}}\|_1 = \|\boldsymbol{\beta}_0 + \widehat{\boldsymbol{\Delta}}\|_1. \tag{3.46}$$

Noticing $\mathcal{S}_0 = \mathrm{supp}(\boldsymbol{\beta}_0)$, we have

$$\begin{aligned} \|\boldsymbol{\beta}_0 + \widehat{\boldsymbol{\Delta}}\|_1 &= \|(\boldsymbol{\beta}_0 + \widehat{\boldsymbol{\Delta}})_{\mathcal{S}_0}\|_1 + \|(\mathbf{0} + \widehat{\boldsymbol{\Delta}})_{\mathcal{S}_0^c}\|_1 \\ &\geq \|\boldsymbol{\beta}_0\|_1 - \|\widehat{\boldsymbol{\Delta}}_{\mathcal{S}_0}\|_1 + \|\widehat{\boldsymbol{\Delta}}_{\mathcal{S}_0^c}\|_1. \end{aligned} \tag{3.47}$$

This together with (3.46) entails that

$$\|\widehat{\boldsymbol{\Delta}}_{\mathcal{S}_0}\|_1 \geq \|\widehat{\boldsymbol{\Delta}}_{\mathcal{S}_0^c}\|_1, \tag{3.48}$$

or that $\widehat{\boldsymbol{\Delta}}$ is sparse (the $L_1$-norm of $\widehat{\boldsymbol{\Delta}}$ on a much bigger set is controlled by that on a much smaller set) or 'restricted'. For example, with $s = |\mathcal{S}_0|$,

$$\|\widehat{\boldsymbol{\Delta}}\|_2 \geq \|\widehat{\boldsymbol{\Delta}}_{\mathcal{S}_0}\|_2 \geq \|\widehat{\boldsymbol{\Delta}}_{\mathcal{S}_0}\|_1/\sqrt{s} \geq \|\widehat{\boldsymbol{\Delta}}\|_1/(2\sqrt{s}), \tag{3.49}$$

where the last inequality utilizes (3.48). At the same time, since $\widehat{\boldsymbol{\beta}}$ and $\boldsymbol{\beta}_0$ are in the feasible set (3.44), we have $\|n^{-1}\mathbf{X}^T\mathbf{X}\widehat{\mathbf{\Delta}}\|_\infty \le 2\lambda$, which implies further that

$$\|\mathbf{X}\widehat{\mathbf{\Delta}}\|_2^2 = \widehat{\mathbf{\Delta}}^T(\mathbf{X}^T\mathbf{X}\widehat{\mathbf{\Delta}}) \le \|\mathbf{X}^T\mathbf{X}\widehat{\mathbf{\Delta}}\|_\infty \|\widehat{\mathbf{\Delta}}\|_1 \le 2n\lambda\|\widehat{\mathbf{\Delta}}\|_1.$$

Using (3.49), we have

$$\|\mathbf{X}\widehat{\mathbf{\Delta}}\|_2^2 \le 4n\lambda\sqrt{s}\|\widehat{\mathbf{\Delta}}\|_2. \tag{3.50}$$

The regularity condition on $\mathbf{X}$ such as the *restricted eigenvalue condition* (Bickel, Ritov and Tsybakov, 2009)

$$\min_{\|\widehat{\mathbf{\Delta}}_{S_0}\|_1 \ge \|\widehat{\mathbf{\Delta}}_{S_0^c}\|_1} n^{-1}\|\mathbf{X}\widehat{\mathbf{\Delta}}\|_2^2/\|\widehat{\mathbf{\Delta}}\|_2^2 \ge a$$

implies a convergence in $L_2$. Indeed, from (3.50), we have

$$a\|\widehat{\mathbf{\Delta}}\|_2^2 \le 4\lambda\sqrt{s}\|\widehat{\mathbf{\Delta}}\|_2, \quad \text{or} \quad \|\widehat{\mathbf{\Delta}}\|_2^2 \le 16a^{-2}\lambda^2 s,$$

which is of order $O(sn^{-1}\log p)$ by choosing the smallest feasible $\lambda = O(\sqrt{2n^{-1}\log p})$ as noted above. Note that the squared error of each nonsparse term is $O(n^{-1})$ and we have to estimate at least $s$ terms of nonsparse parameters. Therefore, $\|\widehat{\mathbf{\Delta}}\|_2^2$ should be at least of order $O(s/n)$. The price that we pay for searching the unknown locations of nonsparse elements is merely a factor of $\log p$. In addition, Bickel, Ritov and Tsybakov (2009) show that the Dantzig selector and Lasso are asymptotically equivalent. James, Radchenko and Lv (2009) develop the explicit condition under which the Dantzig selector and Lasso will give identical fits.

The restricted eigenvalue condition basically imposes that the *condition number* (the ratio of the largest to the smallest eigenvalue) is bounded for any matrix $n^{-1}\mathbf{X}_S^T\mathbf{X}_S$ with $|\mathcal{S}| = s$. This requires that the variables in $\mathbf{X}$ are weakly correlated. It does not allow covariates to share common factors and can be very restrictive. A method to weaken this requirement and to adjust for latent *factors* is given by Kneip and Sarda (2011) and Fan, Ke and Wang (2016).

### *3.3.6 SLOPE and sorted penalties*

The Sorted L-One ($\ell_1$) Penalized Estimation (*SLOPE*) is introduced in Bogdan *et al.* (2015) to control the *false discovery rate* in variable selection. Given a sequence of penalty levels $\lambda_1 \ge \lambda_2 \ge \cdots \ge \lambda_p \ge 0$, it finds the solution to the sorted $\ell_1$ penalized least squares problem

$$\frac{1}{2n}\|\mathbf{Y} - \mathbf{X}\beta\|^2 + \sum_{j=1}^{p} \lambda_j |\beta|_{(j)} \tag{3.51}$$

where $|\beta|_{(1)} \ge \cdots \ge |\beta|_{(p)}$ are the *order statistics* of $\{|\boldsymbol{\beta}|_j\}_{j=1}^p$, namely the decreasing sequence of $\{|\boldsymbol{\beta}|_j\}_{j=1}^p$.

For orthogonal design as in Section 3.2.1 with $\boldsymbol{\varepsilon} \sim N(0, \sigma^2\mathbf{I}_p)$, Bogdan *et al.* (2015) show that the *false discovery rate* for variable selection is controlled at level $q$ if $\lambda_j = \Phi^{-1}(1 - jq/2p)\sigma/\sqrt{n}$, the rescaled critical values used by Benjamini and Hochberg (1995) for *multiple testing*. They also provide a fast computational algorithm. Su and Candés (2016) demonstrate that it achieves adaptive minimaxity in prediction and coefficient estimation for high-dimensional linear regression. Note that using the tail property of the standard normal distribution (see (3.53)), it is not hard to see that $\lambda_j \approx \sigma\sqrt{(2/n)\log(p/j)}$.

From the bias reduction point of view, the SLOPE is not satisfactory as it is still an $\ell_1$-based penalty. This motivates Feng and Zhang (2017) to introduce *sorted folded concave penalties* that combine the strengths of concave and sorted penalties. Given a family of univariate penalty functions $p_\lambda(t)$ indexed by $\lambda$, the associated estimator is defined as

$$\frac{1}{2n}\|\mathbf{Y} - \mathbf{X}\boldsymbol{\beta}\|^2 + \sum_{j=1}^{p} p_{\lambda_j}\big(|\beta|_{(j)}\big).$$

This is a direct extension of (3.51) as it automatically reproduces it with $p_\lambda(t) = \lambda|t|$. The properties of SLOPE and its generalization will be thoroughly investigated in Section 4.5 under a unified framework.

### 3.3.7 *Concentration inequalities and uniform convergence*

The uniform convergence appears in a number of occasions for establishing consistency of regularized estimators. See, for example, (3.32), (3.37) and (3.45). It is fundamental to high-dimensional analysis. Let us illustrate the technique to prove (3.45), which is equivalent to showing

$$P\Big\{\|n^{-1}\mathbf{X}\boldsymbol{\varepsilon}\|_\infty \le \lambda\Big\} = 1 - P\Big\{\max_{1\le j\le p} |n^{-1}\sum_{i=1}^{n} X_{ij}\epsilon_i| > \lambda\Big\} \to 1. \qquad (3.52)$$

If we assume $\varepsilon_i \sim N(0,\sigma^2)$, the conditional distribution of $n^{-1}\sum_{i=1}^n X_{ij}\epsilon_i \sim N(0,\sigma^2/n)$ under the standardization $n^{-1}\|\mathbf{X}_j\|^2 = 1$. Therefore, for any $t > 0$, we have

$$\begin{aligned} P\Big\{|n^{-1}\sum_{i=1}^{n} X_{ij}\epsilon_i| > t\sigma/\sqrt{n}\Big\} &= 2\int_t^\infty \frac{1}{\sqrt{2\pi}}\exp(-x^2/2)dx \\ &\le \frac{2}{\sqrt{2\pi}}\int_t^\infty \frac{x}{t}\exp(-x^2/2)dx \\ &= \frac{2}{\sqrt{2\pi}}\exp(-t^2/2)/t. \qquad (3.53) \end{aligned}$$

In other words, the probability of the average of random variables at least $t$ standard deviations from its mean converges to zero as $t$ goes to $\infty$ exponentially fast. It is highly concentrated, and such a kind of inequality is called a *concentration inequality*.

Now, by the union bound, (3.52) and (3.53), we have

$$
\begin{aligned}
P\Big\{\|n^{-1}\mathbf{X}\boldsymbol{\varepsilon}\|_\infty > \frac{t\sigma}{\sqrt{n}}\Big\} &\le \sum_{j=1}^{p} P\Big\{|n^{-1}\sum_{i=1}^{n} X_{ij}\epsilon_i| > \frac{t\sigma}{\sqrt{n}}\Big\} \\
&\le p\frac{2}{\sqrt{2\pi}}\exp(-t^2/2)/t.
\end{aligned}
$$

Taking $t = \sqrt{2(1+\delta)\log p}$, the above probability is $o(p^{-\delta})$. In other words, with probability at least $1 - o(p^{-\delta})$,

$$
\|n^{-1}\mathbf{X}\boldsymbol{\varepsilon}\|_\infty \le \sqrt{2(1+\delta)}\sigma\sqrt{\frac{\log p}{n}}. \tag{3.54}
$$

The essence of the above proof relies on the concentration inequality (3.53) and the union bound. Note that the concentration inequalities in general hold for the sum of independent random variables with *sub-Gaussian* tails or weaker conditions (see Lemma 4.2). They will appear in later chapters. See Boucheron, Lugosi and Massart (2013), Tropp (2015), Chapters 2 and 3 Wainwright (2019) for general treatments. Below, we give a few of them so that readers can get an idea of these inequalities. These types of inequalities began with Hoeffding's work in 1963.

**Theorem 3.2 (Concentration inequalities)** *Assume that $Y_1, \cdots, Y_n$ are independent random variables with mean zero (without loss of generality). Let $S_n = \sum_{i=1}^n Y_i$ be the sum of the random variables.*

*a) Hoeffding's inequality: If $Y_i \in [a_i, b_i]$, then*

$$
P(|S_n| \ge t) \le 2\exp\Big(-\frac{2t^2}{\sum_{i=1}^n (b_i - a_i)^2}\Big).
$$

*b) Bounded difference inequality: For independent random vectors $\{\mathbf{X}_i\}_{i=1}^n$, if $Z_n = g(\mathbf{X}_1, \cdots, \mathbf{X}_n)$ with $|g(\mathbf{x}_1, \cdots, \mathbf{x}_n) - g(\mathbf{x}_1, \cdots, \mathbf{x}_{i-1}, \mathbf{x}_i', \cdots, \mathbf{x}_n)| \le c_i$ for all data $\{\mathbf{x}_i\}_{i=1}^n$ and $\mathbf{x}_i'$ (changing only one data point from $\mathbf{x}_i$ to $\mathbf{x}_i'$), then*

$$
P(|Z_n - EZ_n| > t) \le 2\exp(-\frac{2t^2}{c_1^2 + \cdots c_n^2}).
$$

*c) Berstein's inequality. If $E|Y_i|^m \le m!M^{m-2}v_i/2$ for every $m \ge 2$ and all $i$ and some positive constants $M$ and $v_i$, then*

$$
P(|S_n| \ge t) \le 2\exp\Big(-\frac{t^2}{2(v_1 + \cdots + v_n + Mt)}\Big).
$$

*See Lemma 2.2.11 of van der Vaart and Wellner (1996).*

*d) Sub-Gaussian case: If $E\exp(aY_i) \le exp(v_i a^2/2)$ for all $a > 0$ and some $v_i > 0$, then, for any $t > 0$,*

$$
P(|S_n| \ge t) \le 2\exp\Big(-\frac{t^2}{2(v_1 + \cdots + v_n)}\Big).
$$

e) *Bounded second moment– Adaptive Huber loss: Assume that $Y_i$ are i.i.d. with mean $\mu$ and variance $\sigma^2$. Let*

$$\widehat{\mu}_\tau = \operatorname{argmin} \sum_{i=1}^{n} \rho_\tau(Y_i - \mu), \qquad \rho_\tau(x) = \begin{cases} x^2, & \text{if } |x| \le \tau \\ \tau(2|x| - \tau), & \text{if } |x| > \tau \end{cases}$$

*be the adaptive Huber estimator. Then, for $\tau = \sqrt{n}c/t$ with $c \ge SD(Y)$ (standard deviation of $Y$), we have (Fan, Li, and Wang, 2017)*

$$P(|\widehat{\mu}_\tau - \mu| \ge t\frac{c}{\sqrt{n}}) \;\; \le \;\; 2\exp(-t^2/16), \quad \forall t \le \sqrt{n/8},$$

f) *Bounded second moment – Winsorized data: Set $\widetilde{Y}_i = sgn(Y_i)\min(|Y_i|, \tau)$. When $\tau \asymp \sqrt{n}\sigma$, then*

$$P\Big(\Big|\frac{1}{n}\sum_{i=1}^{n}\widetilde{Y}_i - \mu\Big| \ge t\frac{\sigma}{\sqrt{n}}\Big) \le 2\exp\Big(-ct^2\Big)$$

*for some universal constant $c$. See Fan, Wang, and Zhu (2016).*

**Proof**. The Hoeffding inequality is a specific case of the bounded difference inequality, whose proof can be found in Corollary 2.21 of Wainwright (2019). We give a proof of the sub-Gaussian case to illustrate the simple idea. By Makov's inequality, independence, sub-Gaussianity, we have for any $a > 0$

$$P(S_n \ge t) \le \exp(-at)E\exp(aS_n) \le \exp(-at)\prod_{i=1}^{n}\exp(v_i a^2/2).$$

By taking the optimal $a = t/(v_1 + \cdots + v_n)$, we obtain

$$P(S_n \ge t) \le \exp\Big(-\frac{t^2}{2(v_1 + \cdots + v_n)}\Big).$$

This is called the *Chernoff bound*. Now, applying the above inequality to $\{-Y_i\}$, we obtain that

$$P(S_n \le -t) \le \exp\Big(-\frac{t^2}{2(v_1 + \cdots + v_n)}\Big).$$

Combining the last two inequalities, we obtain the result. ■

The common theme of the above results is that the probability of $S_n$ deviating from its mean more than $t$ times of its standard deviation converges to zero in the rate $\exp(-ct^2)$ for some positive constant $c$. Theorem 3.2(a) is for bounded random variables, whereas Cases b) and c) extend it to the case with sub-Gaussian moments or tails out. They all yield the same rate of convergence. Additional results and derivations can be found in Chapters 2–5 of Wainwright (2019). Cases d) and e) extend the results further to the case only with bounded second moment. This line of work began with Catoni (2012). See also Devroye, Lerasle, Lugosi and Oliveira (2016).

### 3.3.8 A brief history of model selection

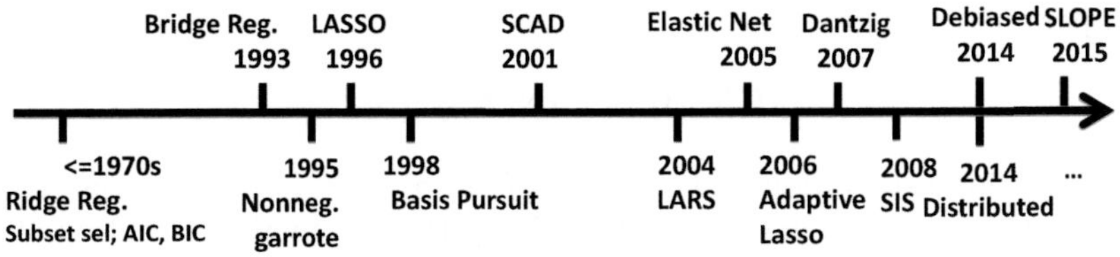


Figure 3.6: A snapshot of history on the major developments of the model selection techniques.

Figure 3.6 summarizes the important developments in model selection techniques. Particular emphasis is given to the development of the penalized least-squares methods. The list is far from complete. For example, Bayesian model selection is not even included. It intends only to give readers a snapshot on some historical developments. For example, the SCAD penalty function was actually introduced by Fan (1997) but its systematic developments were given by Fan and Li (2001), who studied the properties and computation of the whole class of the folded-concave penalized least-squares, not just SCAD. As discussed in Section 3.1.2, AIC and BIC criteria can be regarded as penalized $L_0$ regression. The idea of ridge regression and subset selection appear long before the 1970s (see, e.g., Tikhonov, 1943; Hoerl, 1962).

Sure independence screening, which selects variables based on marginal utilities such as their marginal correlations with the response variable, is not a penalized method. It was introduced by Fan and Lv (2008) to reduce the dimensionality for high-dimensional problems with massive data. It will be systematically introduced in Chapter 8. Because of its importance in analysis of big data and that it can be combined with PLS, we include it here for completeness.

Debiased Lasso was proposed by Zhang and Zhang (2014), which is further extended by van de Geer, Bühlmann, Ritov, and Dezeure (2014) and improved by Javanmard and Montanari (2014). For *distributed estimation* of high-dimensional problem, see Chen and Xie (2014), Shamir, Srebro and Zhang (2014), Lee, Liu, Sun and Taylor (2017), Battey, Fan, Liu, Lu and Zhu (2018), Jordan, Lee and Yang (2018), among others.

## 3.4 Bayesian Variable Selection

### 3.4.1 Bayesian view of the PLS

Sparse penalized regression can be put in the Bayesian framework. One can regard the parameters $\{\beta_j\}_{j=1}^p$ as a realization from a prior distribution having a density $\pi(\cdot)$ with the mode at the origin. If the observed data $\mathbf{Y}$ has a density $p_Y(\mathbf{Y}|\mathbf{X}\boldsymbol{\beta})$ (conditioned on $\mathbf{X}$), then the joint density of the data and parameters is given by

$$f(\mathbf{Y};\boldsymbol{\beta}) = p_Y(\mathbf{Y}|\mathbf{X}\boldsymbol{\beta})\pi(\boldsymbol{\beta}).$$

The posterior distribution of $\boldsymbol{\beta}$ given $\mathbf{Y}$ is $f(\mathbf{Y};\boldsymbol{\beta})/g(\mathbf{Y})$, which is proportional to $f(\mathbf{Y};\boldsymbol{\beta})$ as a function of $\boldsymbol{\beta}$, where $g(\mathbf{Y})$ is the marginal distribution of $\mathbf{Y}$. Bayesian inference is based on the posterior distribution of $\boldsymbol{\beta}$ given $\mathbf{X}$ and $\mathbf{Y}$.

One possible estimator is to use the *posterior mean* $E(\boldsymbol{\beta}|\mathbf{X},\mathbf{Y})$ to estimate $\boldsymbol{\beta}$. Another is the *posterior mode*, which finds

$$\widehat{\boldsymbol{\beta}} = \text{argmax}_{\boldsymbol{\beta}\in R^p} \log p_Y(\mathbf{Y}|\mathbf{X}\boldsymbol{\beta}) + \log \pi(\boldsymbol{\beta}).$$

It is frequently taken as a *Bayesian estimator*. In particular, when $\mathbf{Y} \sim N(\mathbf{X}\boldsymbol{\beta}, \mathbf{I}_n)$ (the standard deviation is taken to be one for convenience), then

$$\log p_Y(\mathbf{Y}|\mathbf{X}\boldsymbol{\beta}) = -\frac{n}{2}\log(2\pi) - \frac{1}{2}\|\mathbf{Y} - \mathbf{X}\boldsymbol{\beta}\|^2.$$

Thus, finding the posterior mode reduces to minimizing

$$\frac{1}{2}\|\mathbf{Y} - \mathbf{X}\boldsymbol{\beta}\|^2 - \log \pi(\boldsymbol{\beta}).$$

Typically, the prior distributions are taken to be independent: $\pi(\boldsymbol{\beta}) = \prod_{j=1}^p \pi_j(\boldsymbol{\beta}_j)$ where $\pi_j(\cdot)$ is the marginal prior for $\beta_j$, though this is not mandatory. In this case, the problem becomes the penalized least-squares

$$\frac{1}{2}\|\mathbf{Y} - \mathbf{X}\boldsymbol{\beta}\|^2 - \sum_{j=1}^p \log \pi_j(\beta_j). \tag{3.55}$$

When $\beta_j \sim_{i.i.d.} \exp(-p_\lambda(|\beta_j|))$, where we hide the normalization constant, (3.55) becomes the penalized least-squares (3.9). In particular, when $\beta_j \sim_{i.i.d.} \lambda\exp(-\lambda|\beta_j|)/2$, the double exponential distribution with scale parameter $\lambda$, the above minimization problem becomes the Lasso problem (3.29). Note that when $p_\lambda(|\beta_j|)$ is flat (constant) at the tails, the function $\exp(-p_\lambda(|\beta_j|))$ cannot be scaled to be a density function as it is not integrable. Such a prior is called an *improper prior*, one with very heavy tails. The SCAD penalty corresponds to an improper prior.

The prior $\pi_j(\theta)$ typically involves some parameters $\gamma$, called *hyper parameters*. An example is the scale parameter $\lambda$ in the double exponential distribution. One can regard them as the parameters generated from some other prior distributions. Such methods are called *hierarchical Bayes*. They can also be regarded as fixed parameters and are estimated through maximum likelihood, maximizing the marginal density $g(\mathbf{Y})$ of $\mathbf{Y}$ with respect to $\gamma$. In other words, one estimates $\boldsymbol{\gamma}$ by the maximum likelihood and employs a Bayes rule to estimate parameters $\boldsymbol{\beta}$ for the given estimated $\boldsymbol{\gamma}$. This procedure is referred to as *empirical Bayes*. Park and Casella (2008) discuss the use of empirical Bayes in the *Bayesian Lasso* by exploiting a hierarchical representation of the double exponential distribution as a scale mixture of normals (Andrews and Mallows 1974):

$$\frac{a}{2}e^{-a|z|} = \int_0^\infty \frac{e^{-z^2/(2s)}}{\sqrt{2\pi s}} \frac{a^2}{2} e^{-a^2 s/2} ds, \ a > 0.$$

They develop a nice Gibbs sampler for sampling the posterior distribution. As demonstrated in Efron (2010), the empirical Bayes plays a very prominent role in large-scale statistical inference.

### *3.4.2 A Bayesian framework for selection*

The Bayesian inference of $\boldsymbol{\beta}$ and model selection are related but not identical problems. Bayesian model selection can be more complex. To understand this, let us denote $\{\mathcal{S}\}$ as all possible models; each model has a prior probability $p(\mathcal{S})$. For example, $\mathcal{S}$ is equally likely among models with the same size, and one assigns the probability proportion to $|\mathcal{S}|^{-\gamma}$. We can also assign the prior probability $p_j$ to models with size $j$ such that models of size $j$ are all equally likely. Within each model $\mathcal{S}$, there is a parameter vector $\boldsymbol{\beta}_\mathcal{S}$ with prior $\pi_\mathcal{S}(\cdot)$. In this case, the "joint density" of the models, the model parameters, and the data is

$$p(\mathcal{S})\pi_\mathcal{S}(\boldsymbol{\beta}_\mathcal{S})p_Y(\mathbf{Y}|\mathbf{X}_\mathcal{S}\boldsymbol{\beta}_\mathcal{S}).$$

There is a large amount of literature on *Bayesian model selection*. The posterior modes are in general computed by using the *Markov Chain Monte Carlo*. See for example Andrieu, De Freitas, Doucet and Jordan (2003) and Liu (2008). Bayesian model selection techniques are very powerful in many applications, where disciplinary knowledge can be explicitly incorporated into the prior. See, for example, Raftery (1995).

A popular Bayesian idea for variable selection is to introduce $p$ latent binary variables $Z = (z_1, \ldots, z_p)$ such that $z_j = 1$ means variable $x_j$ should be included in the model and $z_j = 0$ means excluding $x_j$. Given $z_j = 1$, the distribution of $\beta_j$ has a flat tail (slab), but the distribution of $\beta_j$ given $z_j = 0$ is concentrated at zero (spike). The marginal distribution of $\beta_j$ is a *spike and slab prior*. For example, assume that $\beta_j$ is generated from a mixture of the point mass at 0 and a distribution $\pi_j(\beta)$ with probability $\alpha_j$:

$$\beta_j \sim \alpha_j\delta_0 + (1 - \alpha_j)\pi_j(\beta).$$

See Johnstone and Silverman (2005) for an interesting study of this in wavelet regularization. For computation considerations, the spike distribution is often chosen to be a normal distribution with mean zero and a small variance. The slab distribution is another normal distribution with mean zero and a much bigger variance. See, for example, George and McCulloch (1993), Ishwaran and Rao (2005) and Narisetty and He (2014), among others. A working Bayesian selection model with the *Gaussian spike and slab prior* is given as follows:

$$
\begin{aligned}
\mathbf{Y}|(\mathbf{X}, \boldsymbol{\beta}, \sigma^2) &\sim N(\mathbf{X}\boldsymbol{\beta}, \sigma^2 \mathbf{I}_n), \\
\beta_j|(Z_j = 0, \sigma^2) &\sim N(0, \sigma^2 v_0), \\
\beta_j|(Z_j = 1, \sigma^2) &\sim N(0, \sigma^2 v_1), \\
P(Z_j = 1) &= q, \\
\sigma^2 &\sim IG(\alpha_1, \alpha_2).
\end{aligned}
$$

where $IG$ denotes the inverse Gamma distribution. The data generating process is bottom up in the above representation. The joint posterior distribution $P(\boldsymbol{\beta}, Z, \sigma^2|\mathbf{Y}, \mathbf{X})$ can be sampled by a neat Gibbs sampler. Model selection is based on the marginal posterior probabilities $P(Z_j = 1|\mathbf{Y}, \mathbf{X})$. According to Barbieri and Berger (2004), $x_j$ is selected if $P(Z_j = 1|\mathbf{Y}, \mathbf{X}) \geq 0.5$. This selection method leads to the *median probability model* which is shown to be predictive optimal. For the high-dimension setting $p \gg n$, Narisetty and He (2014) establish the frequentist selection consistency of the Bayesian approach by using dimension-varying prior parameters: $v_0 = v_0(n, p) = o(n^{-1})$, $v_1 = v_1(n, p) = O(\frac{p^{2+\delta}}{n})$ and $q = q(n, p) \approx p^{-1}$.

## 3.5 Numerical Algorithms

This section introduces some early developed algorithms to compute the folded-concave penalized least-squares. We first present the algorithms for computing the Lasso as it is more specific. We then develop algorithms for more general folded concave PLS such as SCAD and MCP. In particular, the connections between the folded-concave PLS and iteratively reweighted adaptive Lasso are made. These algorithms provide us not only a way to implement PLS but also statistical insights on the procedures.

In many applications, we are interested in finding the solution to PLS (3.9) for a range of values of $\lambda$. The solutions $\widehat{\boldsymbol{\beta}}(\lambda)$ to the PLS as a function of $\lambda$ are called *solution paths* or *coefficient paths* (Efron, Hastie, Johnstone and Tibshirani, 2004). This allows one to examine how the variables enter into the solution as $\lambda$ decreases. Figure 3.7 gives an example of coefficient paths.

Each section below is independent, where some sections are harder than others, and can be skipped without significant impact on understanding the other sections.

### 3.5.1 *Quadratic programs*

There are several algorithms for computing Lasso: Quadratic programming, least-angle regression, and coordinate descent algorithm. The first two algorithms are introduced in this and the next sections, and the last one will be introduced in Section 3.5.6.

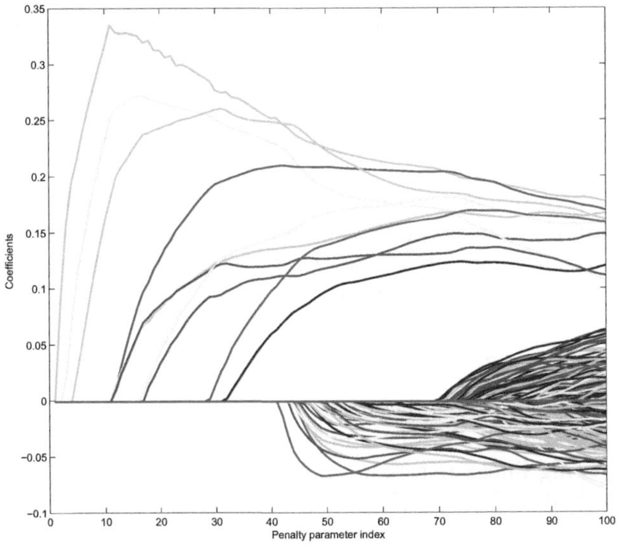


Figure 3.7: Solution paths $\widehat{\boldsymbol{\beta}}(\lambda)$ as a function of $1/\lambda$. For each given value of $\lambda$, only non-vanishing coefficients are presented. As $\lambda$ decreases, we can examine how variables enter into the regression. Each curve shows $\widehat{\beta}_j(\lambda)$ as a function of $\lambda$ for an important regressor $X_j$.

First of all, as in Tibshirani (1996), a convenient alternative is to express the penalized $L_1$-regression (3.29) into its dual problem (3.39). Each $\lambda$ determines a constant $c$ and vice versa. The relationship depends on the data $(\mathbf{X}, \mathbf{Y})$.

The quadratic program, employed by Tibshirani (1996), is to regard the constraints $\|\boldsymbol{\beta}\|_1 \leq c$ as $2^p$ linear constraints $\mathbf{b}_j^T \boldsymbol{\beta} \leq c$ for all $p$-tuples $\mathbf{b}_j$ of form $(\pm 1, \pm 1, \cdots, \pm 1)$. A simple solution is to write (3.39) as

$$\begin{array}{rl} & \min_{\boldsymbol{\beta}^+, \boldsymbol{\beta}^-} \|\mathbf{Y} - \mathbf{X}(\boldsymbol{\beta}^+ - \boldsymbol{\beta}^-)\|^2 \\ \text{s.t.} & \sum_{i=1}^p \beta_i^+ + \sum_{i=1}^p \beta^- \leq c, \quad \beta_i^+ \geq 0, \quad \beta_i^- \geq 0. \end{array} \tag{3.56}$$

This is a $2p$-variable convex optimization problem and the constraints are linear in those variables. Therefore, the standard *convex optimization* algorithms and solvers (Boyd and Vandenberghe, 2004) can be employed. An alternative expression to the optimization problem (3.39) is

$$\begin{array}{rl} & \min_{\boldsymbol{\beta}, \boldsymbol{\gamma}} \|\mathbf{Y} - \mathbf{X}\boldsymbol{\beta}\|^2 \\ \text{s.t.} & -\gamma_i \leq \beta_i \leq \gamma_i, \quad \sum_{i=1}^p \gamma_i \leq c, \quad \gamma_i \geq 0. \end{array} \tag{3.57}$$

This is again a $2p$-variable convex optimization problem with linear constraints.

To find the solution paths, one needs to repeatedly solve a quadratic programming problem for a grid values of $c$. This is very inefficient and does not offer statistical insights.

Osborne, Presnell and Turlach (2000) expressed the $L_1$ constraint as

$$\text{sgn}(\boldsymbol{\beta})^T \boldsymbol{\beta} \le c.$$

They treated the problem as a quadratic program with $\text{sgn}(\boldsymbol{\beta})$ taken from the previous step of the iteration and developed a "homotopy method" based on this linearized constraint. Their homotopy method is related to the solution-path algorithm of Efron *et al.* (2004).

### *3.5.2 Least angle regression**

Efron *et al.* (2004) introduce the Least-Angle Regression (LARS) to explain the striking but mysterious similarity between the Lasso regression path and the $\epsilon$-boosting linear regression path observed by Hastie, Friedman and Tibshirani in 2001. Efron *et al.* (2004) show that the Lasso regression and $\epsilon$-boosting linear regression are two variants of *LARS* with different small modifications, thus explaining their similarity and differences. LARS itself is also an interesting procedure for variable selection and model estimation.

LARS is a forward stepwise selection procedure but operates in a less greedy way than the standard *forward selection* does. Assuming that all variables have been standardized so that they have mean-zero and unit variance, we now describe the LARS algorithm for the constrained least-squares problem (3.39). Let $\mathbf{z} = \mathbf{X}^T\mathbf{Y}/n$ and $\chi_j = \mathbf{X}^T\mathbf{X}_j/n$, where $\mathbf{X}_j$ is the $j^{th}$ column of $\mathbf{X}$. Then, necessary conditions for minimizing the Lasso problem (3.29) are [see (3.30) and (3.31)]

$$\begin{cases} \tau z_j - \chi_j^T \mathbf{b} = \text{sgn}(b_j) & \text{if } b_j \neq 0, \\ |\tau z_j - \chi_j^T \mathbf{b}| \le 1 & \text{if } b_j = 0, \end{cases} \tag{3.58}$$

where $\tau = 1/\lambda$ and $\mathbf{b} = \tau\widehat{\boldsymbol{\beta}}$. The solution path is given by $\widehat{\mathbf{b}}(\tau)$ that solves (3.58). When $\tau \le 1/\|n^{-1}\mathbf{X}^T\mathbf{Y}\|_\infty$, as noted in Section 3.3.2, the solution is $\widehat{\mathbf{b}}(\tau) = 0$. We now describe the LARS algorithm for the constrained least-squares problem (3.39). First of all, when $c = 0$ in (3.57), no variables are selected. This corresponds to $\widehat{\boldsymbol{\beta}}(\lambda) = 0$ for $\lambda > \|n^{-1}\mathbf{X}^T\mathbf{Y}\|_\infty$, as noted in Section 3.3.2.

As soon as $c$ moves slightly away from zero, one picks only one variable ($\mathbf{X}_1$, say) that has the maximum absolute correlation (least angle) with the response variable $\mathbf{Y}$. Then, $\widehat{\boldsymbol{\beta}}_c = (\text{sgn}(r_1)c, 0, \cdots, 0)^T$ is the solution to problem (3.39) for sufficiently small $c$, where $r_1$ is the correlation between $\mathbf{X}_1$ and $\mathbf{Y}$. Now, as $c$ increases, the absolute correlation between the current residual

$$\mathbf{R}_c = \mathbf{Y} - \mathbf{X}\widehat{\boldsymbol{\beta}}_c$$

and $\mathbf{X}_1$ decreases until a (smallest) value $c_1$ at which there exists a second variable $\mathbf{X}_2$, say, that has the same absolute correlation (equal angle) with $\mathbf{R}_{c_1}$:

$$|\text{cor}(\mathbf{X}_1, \mathbf{R}_{c_1})| = |\text{cor}(\mathbf{X}_2, \mathbf{R}_{c_1})|.$$

Then, $\widehat{\boldsymbol{\beta}}_c$ is the solution to problem (3.39) for $0 \leq c \leq c_1$ and the value $c_1$ can easily be determined, as in (3.62) below.

LARS then proceeds equiangularly between $\mathbf{X}_1$ and $\mathbf{X}_2$ until a third variable, $\mathbf{X}_3$ (say), joins the ranks of "most correlated variables" with the current residuals. LARS then proceeds equiangularly between $\mathbf{X}_1$, $\mathbf{X}_2$ and $\mathbf{X}_3$ and so on. As we proceed down this path, the maximum of the absolute correlation of covariates with the current residual keeps decreasing until it becomes zero.

The equiangular direction of a set of variables $\mathbf{X}_\mathcal{S}$ is given by

$$\mathbf{u}_\mathcal{S} = \mathbf{X}_\mathcal{S}(\mathbf{X}_\mathcal{S}^T\mathbf{X}_\mathcal{S})^{-1}\mathbf{1}/w_\mathcal{S} \equiv \mathbf{X}_\mathcal{S}\widehat{\boldsymbol{\beta}}_\mathcal{S} \tag{3.59}$$

where $\mathbf{1}$ is a vector of $1'$s and $w_\mathcal{S}^2 = \mathbf{1}^T(\mathbf{X}_\mathcal{S}^T\mathbf{X}_\mathcal{S})^{-1}\mathbf{1}$ is a normalization constant. The equiangular property can easily be seen:

$$\mathbf{X}_\mathcal{S}^T\mathbf{u}_\mathcal{S} = \mathbf{1}/w_\mathcal{S}, \qquad \|\mathbf{u}_\mathcal{S}\| = 1.$$

We now furnish some details of LARS. Assume that $\mathbf{X}$ is of full rank. Start from $\boldsymbol{\mu}_0 = 0$, $\mathcal{S} = \varnothing$, the empty set, and $\boldsymbol{\beta}_\mathcal{S} = 0$. Let $\mathcal{S}$ be the current active set of variables and $\widehat{\boldsymbol{\mu}}_\mathcal{S} = \mathbf{X}_\mathcal{S}\widehat{\boldsymbol{\beta}}_\mathcal{S}$ be its current fitted value of $\mathbf{Y}$. Compute the marginal correlations of covariates $\mathbf{X}$ with the current residual (except a normalization constant)

$$\widehat{\mathbf{c}} = \mathbf{X}^T(\mathbf{Y} - \widehat{\boldsymbol{\mu}}_\mathcal{S}). \tag{3.60}$$

Define $s_j = \text{sgn}(\widehat{c}_j)$ for $j \in \mathcal{S}$. Note that the absolute correlation does not change if the columns of $\mathbf{X}$ are multiplied by $\pm 1$. Update the active set of variables by taking the most correlated set

$$\mathcal{S}_{\text{new}} = \{j : |\widehat{c}_j| = \|\widehat{\mathbf{c}}\|_\infty\} \tag{3.61}$$

and compute the equiangular direction $\mathbf{u}_{\mathcal{S}_{\text{new}}}$ by (3.59) using variables $\{\text{sgn}(\widehat{c}_j)\mathbf{X}_j, j \in \mathcal{S}\}$. For the example in the beginning of this section, if $\mathcal{S} = \phi$, an empty set, then $\widehat{\boldsymbol{\mu}}_\mathcal{S} = 0$ and $\mathcal{S}_{\text{new}} = \{1\}$. If $\mathcal{S} = \{1\}$, then $\widehat{\boldsymbol{\mu}}_\mathcal{S} = \mathbf{X}\widehat{\boldsymbol{\beta}}_{c_1} = \text{sgn}(r_1)c_1\mathbf{X}_1$ and $\mathcal{S}_{\text{new}} = \{1, 2\}$.

Now compute

$$\gamma_\mathcal{S} = \min_{j \in \mathcal{S}^c}{}^+ \left\{ \frac{\|\mathbf{c}\|_\infty - \widehat{c}_j}{w_\mathcal{S}^{-1} - a_j}, \frac{\|\mathbf{c}\|_\infty + \widehat{c}_j}{w_\mathcal{S}^{-1} + a_j} \right\}, \qquad \mathbf{a} = \mathbf{X}^T\mathbf{u}_\mathcal{S}. \tag{3.62}$$

where "$\min^+$" is the minimum taken only over positive components. It is not hard to show that this step size $\gamma_\mathcal{S}$ is the smallest positive constant $\gamma$ such that some new indices will join the active set (see Efron *et al.*, 2004). For example, if $\mathcal{S} = \{1\}$, this $\gamma_\mathcal{S}$ is $c_1$ in the first step. Update the fitted value along the equiangular direction by

$$\widehat{\boldsymbol{\mu}}_{\mathcal{S}_{\text{new}}} = \widehat{\boldsymbol{\mu}}_\mathcal{S} + \gamma_{\mathcal{S}_{\text{new}}}\mathbf{u}_{\mathcal{S}_{\text{new}}}. \tag{3.63}$$

The solution path for $\gamma \in (0, \gamma_{\mathcal{S}_{\text{new}}})$ is

$$\widehat{\boldsymbol{\mu}}_{\mathcal{S}_{\text{new}},\gamma} = \widehat{\boldsymbol{\mu}}_{\mathcal{S}} + \gamma \mathbf{u}_{\mathcal{S}_{\text{new}}}.$$

Note that $\mathcal{S}_{\text{new}}$ is always a bigger set than $\mathcal{S}$. Write $\widehat{\boldsymbol{\mu}}_{\mathcal{S}} = \mathbf{X}\widehat{\boldsymbol{\beta}}_{\mathcal{S}}$, in which $\widehat{\boldsymbol{\beta}}_{\mathcal{S}}$ has support $\mathcal{S}$ so that $\widehat{\boldsymbol{\mu}}_{\mathcal{S}}$ is in the linear space spanned by columns of $\mathbf{X}_{\mathcal{S}}$. By (3.59), we have

$$\widehat{\boldsymbol{\mu}}_{\mathcal{S}_{\text{new}},\gamma} = \mathbf{X}(\widehat{\boldsymbol{\beta}}_{\mathcal{S}} + \gamma \boldsymbol{\beta}_{\mathcal{S}_{\text{new}}}).$$

Note that by (3.59), $\widehat{\boldsymbol{\beta}}_{\mathcal{S}} + \gamma \boldsymbol{\beta}_{\mathcal{S}_{\text{new}}}$ has a support $\mathcal{S}_{\text{new}}$. In terms of coefficients, we have updated the coefficients from $\widehat{\boldsymbol{\beta}}_{\mathcal{S}}$ for variables $\mathbf{X}_{\mathcal{S}}$ to

$$\widehat{\boldsymbol{\beta}}_{\mathcal{S}_{\text{new}},\gamma} = \widehat{\boldsymbol{\beta}}_{\mathcal{S}} + \gamma \boldsymbol{\beta}_{\mathcal{S}_{\text{new}}} \tag{3.64}$$

for variables $\mathbf{X}_{\mathcal{S}_{\text{new}}}$, expressed in $R^p$. Some modifications of the signs in the second term in (3.64) are needed since we use the variables $\{\text{sgn}(\widehat{c}_j)\mathbf{X}_j, j \in \mathcal{S}\}$ rather than $\mathbf{X}_{\mathcal{S}}$ to compute the equiangular direction $\mathbf{u}_{\mathcal{S}_{\text{new}}}$.

The LARS algorithm is summarized as follows.

- **Initialization**: Set $\mathcal{S} = \phi$, $\widehat{\boldsymbol{\mu}}_{\mathcal{S}} = 0$, $\widehat{\boldsymbol{\beta}}_{\mathcal{S}} = \mathbf{0}$.
- **Step 1**: Compute the current correlation vector $\widehat{\mathbf{c}}$ by (3.60), the new subset $\mathcal{S}_{\text{new}}$, the least angular covariates with the current residual $\mathbf{Y} - \widehat{\boldsymbol{\mu}}_{\mathcal{S}}$, by (3.61), and the stepsize $\gamma_{\mathcal{S}_{\text{new}}}$, the largest stepsize along the equiangular direction, by (3.62).
- **Step 2**: Update $\mathcal{S}$ with $\mathcal{S}_{\text{new}}$, $\widehat{\boldsymbol{\mu}}_{\mathcal{S}}$ with $\widehat{\boldsymbol{\mu}}_{\mathcal{S}_{\text{new}}}$ in (3.63), and $\widehat{\boldsymbol{\beta}}_{\mathcal{S}}$ with $\widehat{\boldsymbol{\beta}}_{\mathcal{S}_{\text{new}}}$ in (3.64) with $\gamma = \gamma_{\mathcal{S}_{\text{new}}}$.
- **Iterations**: Iterate between Steps 1 and 2 until all variables are included in the model and the solution reaches the OLS estimate.

The entire LARS solution path simply connects $p$-dimensional coefficients linearly at each discrete step above. However, it is not necessarily the solution to the Lasso problem (3.39). The LARS model size is enlarged by one after each step, but the Lasso may also drop a variable from the current model as $c$ increases. Technically speaking, (3.30) shows that Lasso and the current correlation must have the same sign, but the LARS solution path does not enforce this. Efron *et al.* (2004) show that this sign constraint can easily be enforced in the LARS algorithm: during the ongoing LARS update step, if the $\tilde{j}$th variable in $\mathcal{S}$ has a sign change before the new variable enters $\mathcal{S}$, stop the ongoing LARS update, drop the $\tilde{j}$th variable from the model and recalculate the new equiangular direction for doing the LARS update. Efron *et al.* (2004) prove that the modified LARS path is indeed the Lasso solution path under a "one at a time" condition, which assumes that at most one variable can enter or leave the model at any time.

Other modifications of the LARS algorithm are also possible. For example, by modifying LARS shrinkage, James and Radchenko (2008) introduce *variable inclusion and shrinkage algorithms* (VISA) that intend to attenuate the over-shrinkage problem of Lasso. James, Radchenko and Lv (2009) develop an algorithm called *Dasso* that allows one to fit the entire path of regression coefficients for different values of the Dantzig selector tuning parameter.

The key argument in the LARS algorithm is the piecewise linearity property of the Lasso solution path. This property is not unique to the Lasso PLS. Many statistical models can be formulated as $\min\{\text{Loss} + \lambda\text{Penalty}\}$. In Rosset and Zhu (2007) it is shown that if the loss function is almost quadratic and the penalty is $L_1$ (or piecewise linear), then the solution path is piecewise linear as a function of $\lambda$. Examples of such models include the $L_1$ penalized Huber regression (Rosset and Zhu, 2007) and the $L_1$ penalized support vector machine (Zhu, Rosset, Hastie and Tibshirani, 2004). Interestingly, if the loss function is $L_1$ (or piecewise linear) and the penalty function is quadratic, we can switch their roles when computing and the solution path is piecewise linear as a function of $1/\lambda$. See, for example, the solution path algorithms for the support vector machine (Hastie, Rosset, Tibshirani and Zhu, 2003) and support vector regression (Gunter and Zhu, 2007).

### *3.5.3 Local quadratic approximations*

Local quadratic approximation (LQA) was introduced by Fan and Li (2001) before LARS-Lasso or other effective methods that are available in statistics for computing Lasso. It allows statisticians to implement folded concave penalized likelihood and to compute the standard error of estimated nonzero components. Given an initial value $\beta_0$, approximate the function $p_\lambda(|\beta|)$ locally at this point by a quadratic function $q(\beta|\beta_0)$. This quadratic function is required to be symmetric around zero, satisfying

$$q(\beta_0|\beta_0) = p_\lambda(|\beta_0|) \quad \text{and} \quad q'(\beta_0|\beta_0) = p'_\lambda(|\beta_0|).$$

These three conditions determine uniquely the quadratic function

$$q(\beta|\beta_0) = p_\lambda(|\beta_0|) + \frac{1}{2}\frac{p'_\lambda(|\beta_0|)}{|\beta_0|}(\beta^2 - \beta_0^2). \tag{3.65}$$

See Figure 3.8.

Given the current estimate $\boldsymbol{\beta}_0$, by approximating each folded concave function in PLS (3.9) by its LQA, our target becomes minimizing

$$Q(\boldsymbol{\beta}|\boldsymbol{\beta}_0) = \frac{1}{2n}\|\mathbf{Y} - \mathbf{X}\boldsymbol{\beta}\|^2 + \sum_{j=1}^{p} q(\beta_j|\beta_{j0}). \tag{3.66}$$

Minimizing (3.66) is the same as minimizing

$$\frac{1}{2n}\|\mathbf{Y} - \mathbf{X}\boldsymbol{\beta}\|^2 + \sum_{j=1}^{p} \frac{p'_\lambda(|\beta_{j0}|)}{2|\beta_{j0}|}\beta_j^2.$$

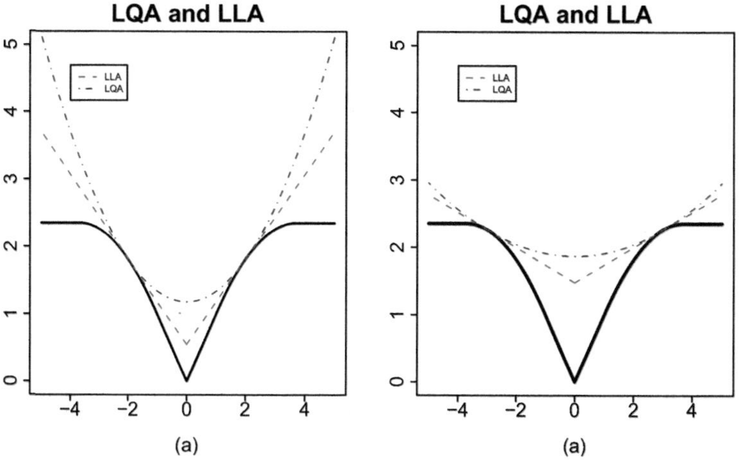


Figure 3.8: Local quadratic and local linear approximations to SCAD with $\lambda = 1$ and $a = 3.7$ at at point $x$. (a) $x = 2$ and (b) $x = 3$.

This is a ridge regression problem with solution computed analytically as

$$\widehat{\boldsymbol{\beta}}_{\text{new}} = (\mathbf{X}^T\mathbf{X} + n\,\text{diag}\{p'_\lambda(|\beta_{j0}|)/|\beta_{j0}|\})^{-1}\mathbf{X}^T\mathbf{Y}. \tag{3.67}$$

The LQA is to iteratively use (3.67), starting from an initial value (e.g. univariate marginal regression coefficients). Fan and Li (2001) note that the approximation (3.65) is not good when $|\beta_{j0}| \le \varepsilon_0$, a tolerance level. When this happens, delete variables from the model before applying (3.67). This speeds up the computation. Furthermore, they proposed to compute the standard error for surviving variables using (3.67), as if $p'_\lambda(\beta_{j0})/|\beta_{j0}|$ were non-stochastic. They validated the accuracy of the estimated standard error. See Fan and Peng (2004) for a theoretical proof.

Does the algorithm converge? and if so, in what sense? Hunter and Li (2005) realized that the local quadratic approximation is a specific case of the *majorization-minimization* (MM) algorithm (Hunter and Lange, 2000). First of all, as shown in Figure 3.8, thanks to the folded-concaveness,

$$q(\beta|\beta_0) \ge p_\lambda(\beta) \quad \text{and} \quad q(\beta_0|\beta_0) = p_\lambda(\beta_0),$$

namely $q(\beta|\beta_0)$ is a convex majorant of $p_\lambda(\cdot)$ with $q(\beta_0|\beta_0) = p_\lambda(|\beta_0|)$. This entails that

$$Q(\boldsymbol{\beta}|\boldsymbol{\beta}_0) \ge Q(\boldsymbol{\beta}) \quad \text{and} \quad Q(\boldsymbol{\beta}_0|\boldsymbol{\beta}_0) = Q(\boldsymbol{\beta}_0), \tag{3.68}$$

namely, $Q(\boldsymbol{\beta}|\boldsymbol{\beta}_0)$ is a convex majorization of the folded-concave PLS $Q(\boldsymbol{\beta})$ defined by (3.9). Let $\boldsymbol{\beta}_{\text{new}}$ minimize $Q(\boldsymbol{\beta}|\boldsymbol{\beta}_0)$. Then, it follows from (3.68) that

$$Q(\boldsymbol{\beta}_{\text{new}}) \le Q(\boldsymbol{\beta}_{\text{new}}|\boldsymbol{\beta}_0) \le Q(\boldsymbol{\beta}_0|\boldsymbol{\beta}_0) = Q(\boldsymbol{\beta}_0), \tag{3.69}$$

where the second inequality follows from the definition of the minimization. In other words, the target function decreases after each iteration and will converge.

### 3.5.4 Local linear algorithm

With LARS and other efficient algorithms for computing Lasso, the *local linear approximation* (LLA) approximates $p_\lambda(|\beta|)$ at $\beta_0$ by

$$l(\beta|\beta_0) = p_\lambda(|\beta_0|) + p'_\lambda(|\beta_0|)(|\beta| - |\beta_0|),$$

which is the first-order Taylor expansion of $p_\lambda(|\beta|)$ at the point $\beta_0$. Clearly, as shown in Figure 3.8, $l(\beta|\beta_0)$ is a better approximation than LQA $q(\beta|\beta_0)$. Indeed, it is the minimum convex majorant of $p_\lambda(|\beta|)$ with $l(\beta_0|\beta_0) = p_\lambda(|\beta_0|)$.

With the local linear approximation, given the current estimate $\boldsymbol{\beta}_0$, the folded concave penalized least-squares (3.9) now becomes

$$\frac{1}{2n}\|\mathbf{Y} - \mathbf{X}\boldsymbol{\beta}\|^2 + \sum_{j=1}^{p} p'_\lambda(|\beta_{j0}|)|\beta_j|, \tag{3.70}$$

after ignoring the constant term. This is now an adaptively weighted Lasso problem and can be solved by using algorithms in Sections 3.5.1 and 3.5.2. The algorithm was introduced by Zou and Li (2008). As it is also a specific MM algorithm, as shown in (3.69), the LLA algorithm also enjoys a decreasing target value property (3.69).

Unlike LQA, if one component hits zero at a certain step, it will not always be stuck at zero. For example, if $\boldsymbol{\beta}_0 = 0$, (3.70) reduces to Lasso. In this view, even when the initial estimator is very crude, LLA gives a good one-step estimator.

Through LLA approximation (3.70), the folded-concave PLS can be viewed as an iteratively reweighted penalized $L_1$ regression. The weights depend on where the current estimates are. The larger the magnitude, the smaller the weighted penalty. This reduces the biases for estimating large true coefficients. Lasso is the one-step estimator of the folded concave PLS with initial estimate $\widehat{\boldsymbol{\beta}} = \mathbf{0}$. Lasso puts a full stop yet the folded concave PLS iterates further to reduce the bias due to the Lasso shrinkage.

It is now clear that the adaptive Lasso is a specific example of the LLA implementation of the folded-concave PLS with $p'_\lambda(|\beta|) = |\beta|^{-\gamma}$. This function is explosive near 0 and is therefore inappropriate to use in the iterative application of (3.70): once a component $\beta_j$ hits zero at a certain iteration, its weights cannot be computed or the variable $X_j$ is eliminated forever.

The LLA implementation of folded concave PLS has a very nice theoretical property. Fan, Xue and Zou (2014) show that with Lasso as the initial estimator, with high probability, the LLA implementation (3.70) produces the oracle estimator in one step. The result holds in a very general likelihood-based context under some mild conditions. This gives additional endorsement of the folded-concave PLS implemented by LLA (3.70).

The implementation of LLA is available in the R package called "SIS" (function: *scadglm*), contributed by Fan, Feng, Samworth, and Wu.

### 3.5.5 *Penalized linear unbiased selection**

The penalized linear unbiased selection (PLUS) algorithm, introduced by Zhang (2010), finds multiple local minimizers of folded-concave PLS in a branch of the graph (indexed by $\tau = \lambda^{-1}$) of critical points determined by (3.22) and (3.23). The PLUS algorithm deals with the folded concave-penalized functions $p_\lambda(t) = \lambda^2 \rho(t/\lambda)$, in which $\rho(\cdot)$ is a quadratic spline. This includes $L_1$, SCAD (3.14), hard-threhsolding penalty (3.15) and MCP (3.17) as specific examples. Let $t_1 = 0, \cdots, t_m$ be the knots of the quadratic spline $\rho(\cdot)$. Then, the derivative of $\rho(\cdot)$ can be expressed as

$$\rho'(t) = \sum_{i=1}^{m} (u_i - v_i t) I(t_i < t \le t_{i+1}), \tag{3.71}$$

for some constants $\{v_i\}_{i=1}^m$, in which $u_1 = 1$ (normalization), $u_m = v_m = 0$ (flat tail), and $t_{m+1} = \infty$. For example, the $L_1$ penalty corresponds to $m = 1$; MCP corresponds to $m = 2$ with $t_2 = a$, $v_1 = 1/a$; SCAD corresponds to $m = 3$ with $t_2 = 1$, $t_3 = a$, $v_1 = 0$, $u_2 = a/(a-1)$ and $v_2 = 1/(a-1)$.

Let $\mathbf{z} = \mathbf{X}^T\mathbf{Y}/n$ and $\chi_j = \mathbf{X}^T\mathbf{X}_j/n$, where $\mathbf{X}_j$ is the $j^{th}$ column of $\mathbf{X}$. Then, estimating equations (3.22) and (3.23) can be written as

$$\begin{cases} \tau z_j - \chi_j^T \mathbf{b} = \text{sgn}(b_j)\rho'(|b_j|) & \text{if } b_j \neq 0, \\ |\tau z_j - \chi_j^T \mathbf{b}| \le 1 & \text{if } b_j = 0, \end{cases} \tag{3.72}$$

where $\tau = 1/\lambda$ and $\mathbf{b} = \tau\boldsymbol{\beta}$. (3.72) can admit multiple solutions for each given $\lambda$. For example, $\mathbf{b} = \mathbf{0}$ is a local solution to (3.72), when $\lambda \ge \|\mathbf{X}^T\mathbf{Y}/n\|_\infty$. See also (3.33). Unlike Lasso, there can be other local solutions to (3.72). PLUS computes the main branch $\widehat{\boldsymbol{\beta}}(\tau)$ starting from $\widehat{\boldsymbol{\beta}}(\tau) = \mathbf{0}$, where $\tau = 1/\|\mathbf{X}^T\mathbf{Y}/n\|_\infty$.

Let us characterize the solution set of (3.72). The component $b_j$ of a solution $\mathbf{b}$ falls in one of the intervals $\{(t_i, t_{i+1}]\}_{i=1}^m$, or 0, or in one of the intervals $\{[-t_{i+1}, -t_i)\}_{i=1}^m$. Let us use $i_j \in \{-m, \cdots, m\}$ to indicate such an interval and $\mathbf{i} \in \{-m, \cdots, m\}^p$ be the the vector of indicators. Then, by (3.71), (3.72) can be written as

$$\begin{cases} \tau z_j - \chi_j^T \mathbf{b} = \text{sgn}(i_j)(u_{i_j} - b_j v_{i_j}), & \bar{t}_{i_j} \le b_j \le \bar{t}_{i_j+1}, & i_j \neq 0, \\ -1 \le \tau z_j - \chi_j^T \mathbf{b} \le 1, & b_j = 0, & i_j = 0, \end{cases} \tag{3.73}$$

where $u_{-k} = u_k$, $v_{-k} = v_k$, and $\bar{t}_i = t_i$ for $0 < i \le m+1$ and $-t_{|i|+1}$ for $-m \le i \le 0$. Let $\mathcal{S}_\tau(\mathbf{i})$ be the set of $(\tau\mathbf{z}^T, \mathbf{b}^T)^T$ in $R^{2p}$, whose coordinates satisfy (3.73). Note that the solution $\mathbf{b}$ is piecewise linear $\tau$.

Let $H = R^p$ represent the data $\mathbf{z}$ and its dual $H^* = R^p$ represent the solution $\mathbf{b}$, and $\mathbf{z} \oplus \mathbf{b}$ be members of $H \oplus H^* = R^{2p}$. The set $\mathcal{S}_\tau(\mathbf{i})$ in $R^{2p}$ is more compactly expressed as

$$\mathcal{S}_\tau(\mathbf{i}) = \{\tau\mathbf{z} \oplus \mathbf{b} : \ \tau\mathbf{z} \text{ and } \mathbf{b} \text{ satisfy (3.73)}\}.$$

For each given $\tau$ and $\mathbf{i}$, the set $\mathcal{S}_\tau(\mathbf{i})$ is a parallelepiped in $R^{2p}$ and $\mathcal{S}_\tau = \tau\mathcal{S}_1$. The solution $\mathbf{b}$ is the projection of $\mathcal{S}_\tau(\mathbf{i})$ onto $H^*$, denoted by $\mathcal{S}_\tau(\mathbf{i}|\mathbf{z})$. Clearly, all solutions to (3.72) are a $p$-dimensional set given by

$$\mathcal{S}_\tau(\mathbf{z}) = \cup\{\mathcal{S}_\tau(\mathbf{i}|\mathbf{z}) : \mathbf{i} \in \{-m, -m+1, \cdots, m\}^p\}. \tag{3.74}$$

Like LARS, the PLUS algorithm computes a solution $\boldsymbol{\beta}(\tau)$ from $\mathcal{S}_\tau(\mathbf{z})$. Starting from $\tau_0 = 1/\|n^{-1}\mathbf{X}^T\mathbf{Y}\|_\infty$, $\widehat{\boldsymbol{\beta}}(\tau_0) = 0$ and $\mathbf{i} = \mathbf{0}$, PLUS updates the active set of variables as well as the branch $\mathbf{i}$, determines the step size $\tau$ and solution $\mathbf{b}$. The solutions between two turning points are connected by lines. We refer to Zhang (2010) for additional details. In addition, Zhang (2010) gives the conditions under which the solution becomes the oracle estimator, and derives the risk and model selection properties of the PLUS estimators.

### 3.5.6 *Cyclic coordinate descent algorithms*

Consider the sparse penalized least squares; the computation difficulty comes from the nonsmoothness of the penalty function. Observe that the penalty function part is the sum of $p$ univariate nonsmooth functions. Then, we can employ *cyclic coordinate descent algorithms* (Tseng, 2001; Tseng and Yun, 2009) that successively optimize one coefficient (coordinate) at a time. Let

$$L(\beta_1, \ldots, \beta_p) = \frac{1}{2n}\|\mathbf{Y} - \mathbf{X}\boldsymbol{\beta}\|^2 + \sum_{j=1}^p p_\lambda(|\beta_j|)$$

and the cyclic coordinate descent (CCD) algorithm proceeds as follows:

1. choose an initial value of $\widehat{\boldsymbol{\beta}}$
2. for $j = 1, 2, \ldots, p, 1, 2, \ldots$, update $\widehat{\beta}_j$ by solving a univariate optimization problem of $\beta_j$:

$$\widehat{\beta}_j^{\text{update}} \leftarrow \text{argmin}_{\beta_j} L(\widehat{\beta}_1, \ldots, \widehat{\beta}_{j-1}, \beta_j, \widehat{\beta}_{j+1}, \widehat{\beta}_p). \tag{3.75}$$

3. Repeat (2) till convergence.

Let $\mathbf{R}_j = \mathbf{Y} - \mathbf{X}_{-j}\widehat{\boldsymbol{\beta}}_{-j}$ be the current residual, where $\mathbf{X}_{-j}$ and $\widehat{\boldsymbol{\beta}}_{-j}$ are respectively $\mathbf{X}$ and $\widehat{\boldsymbol{\beta}}$ with the $j^{th}$ column and $j^{th}$ component removed. Then, the target function in (3.75) becomes, after ignoring a constant,

$$Q_j(\beta_j) \quad \equiv \quad \frac{1}{2n}\|\mathbf{R}_j - \mathbf{X}_j\beta_j\|^2 + p_\lambda(|\beta_j|),$$

Recall that $\|\mathbf{X}_j\|^2 = n$ by standardization and $\widehat{c}_j = n^{-1}\mathbf{X}_j^T\mathbf{R}_j$ is the current covariance [c.f. (3.60)]. Then, after ignoring a constant,

$$Q_j(\beta_j) = \frac{1}{2}(\beta_j - \widehat{c}_j)^2 + p_\lambda(|\beta_j|). \tag{3.76}$$

This is the same problem as (3.12). For $L_1$, SCAD and MCP penalty, (3.76) admits an explicit solution as in (3.18)–(3.20). In this case, the *CCD* algorithm is simply an iterative thresholding method.

The CCD algorithm for the Lasso regression is the same as the shooting algorithm introduced by Fu (1998). Friedman, Hastie, Höfling and Tibshirani (2007) implement the CCD algorithm by using several tricks such as warm start, active set update, etc. As a result, they were able to show that the coordinate descent algorithm is actually very effective in computing the Lasso solution path, proving to be even faster than the LARS algorithm. Fan and Lv (2011) extend the CCD algorithm to the penalized likelihood.

The user needs to be careful when applying the coordinate descent algorithm to solve the concave penalized problems because the algorithm converges to a local minima but this solution may not be the statistical optimal one. The choice of initial value becomes very important. In Fan, Xue and Zou (2014) there are simulation examples showing that the solution by CCD is suboptimal compared with the LLA solution in SCAD and MCP penalized regression and logistic regression. It is beneficial to try multiple initial values when using CCD to solve nonconvex problems.

### 3.5.7 *Iterative shrinkage-thresholding algorithms*

The iterative shrinkage-thresholding algorithm (ISTA, Daubechies *et al.*, 2004) is developed to optimize the functions of form $Q(\boldsymbol{\beta}) = f(\boldsymbol{\beta}) + g(\boldsymbol{\beta})$, in which $f$ is smooth whereas $g(\boldsymbol{\beta})$ is non-smooth. Note that the *gradient descent algorithm*

$$\boldsymbol{\beta}_k = \boldsymbol{\beta}_{k-1} - s_k f'(\boldsymbol{\beta}_{k-1}),$$

for a suitable stepsize $s_k$ is the minimizer to the *local isotropic quadratic* approximation of $f$ at $\boldsymbol{\beta}_{k-1}$:

$$f_A(\boldsymbol{\beta}|\boldsymbol{\beta}_{k-1}, s_k) = f(\boldsymbol{\beta}_{k-1}) + f'(\boldsymbol{\beta}_{k-1})^T(\boldsymbol{\beta} - \boldsymbol{\beta}_{k-1}) + \frac{1}{2s_k}\|\boldsymbol{\beta} - \boldsymbol{\beta}_{k-1}\|^2. \quad (3.77)$$

The local isotropic approximation avoids computing the Hessian matrix, which is expensive and requires a lot of storage for high-dimensional optimization. Adapting this idea to minimizing $Q(\cdot)$ yields the algorithm

$$\boldsymbol{\beta}_k = \operatorname{argmin}\{f_A(\boldsymbol{\beta}|\boldsymbol{\beta}_{k-1}, s_k) + g(\boldsymbol{\beta})\}.$$

In particular, when $g(\boldsymbol{\beta}) = \sum_{j=1}^p p_\lambda(|\beta_j|)$, the problem becomes a componentwise optimization after ignoring a constant

$$\boldsymbol{\beta}_k = \operatorname{argmin}\left\{\frac{1}{2s_k}\|\boldsymbol{\beta} - (\boldsymbol{\beta}_{k-1} - s_k f'(\boldsymbol{\beta}_{k-1}))\|^2 + \sum_{j=1}^p p_\lambda(|\beta_j|)\right\},$$

for each component of the form (3.12). Let us denote

$$\theta_s(z) = \operatorname{argmin}_\theta \left\{\frac{1}{2}(z-\theta)^2 + sp_\lambda(|\theta|)\right\}.$$

Then, ISTA is to iteratively apply

$$\boldsymbol{\beta}_k = \theta_{s_k}\big(\boldsymbol{\beta}_{k-1} - s_k f'(\boldsymbol{\beta}_{k-1})\big). \tag{3.78}$$

In particular, for the Lasso problem (3.29), the ISTA becomes

$$\boldsymbol{\beta}_k = \big(\boldsymbol{\beta}_{k-1} - s_k n^{-1}\mathbf{X}^T(\mathbf{Y} - \mathbf{X}\boldsymbol{\beta}_{k-1}) - s_k\lambda\big)_+. \tag{3.79}$$

Similar iterative formulas can be obtained for SCAD and MCP. This kind of algorithm is called a *proximal gradient method* in the optimization literature.

Note that when $\|f'(\boldsymbol{\beta}) - f'(\boldsymbol{\theta})\| \leq \|\boldsymbol{\beta} - \boldsymbol{\theta}\|/s_k$ for all $\boldsymbol{\beta}$ and $\boldsymbol{\theta}$, we have $f_A(\boldsymbol{\beta}|\boldsymbol{\beta}_{k-1}, s_k) \geq f(\boldsymbol{\beta})$. This holds when the largest eigenvalue of the Hessian matrix $f''(\boldsymbol{\beta})$ is bounded by $1/s_k$. Therefore, the ISTA algorithm is also a specific implementation of the MM algorithm, when the condition is met.

The above isotropic quadratic majorization requires strong conditions regarding to the function $f$. Inspecting the proof in (3.69) for the MM algorithm, we indeed do not require majorization but only the *local majorization* $Q(\boldsymbol{\beta}_{\text{new}}) \leq Q(\boldsymbol{\beta}_{\text{new}}|\boldsymbol{\beta}_0)$. This can be achieved by using the *backtracking rule* to choose the step size $s_k$ as follows. Take an initial step size $s_0 > 0$, $\delta < 1$, and the initial value $\boldsymbol{\beta}_0$. Find the smallest nonnegative integer $i_k$ such that with $s = \delta^{i_k} s_{k-1}$,

$$Q(\boldsymbol{\beta}_{k,s}) \leq Q_A(\boldsymbol{\beta}_{k,s}) \equiv f_A(\boldsymbol{\beta}_{k,s}|\boldsymbol{\beta}_{k-1}, s) + g(\boldsymbol{\beta}_{k,s}), \tag{3.80}$$

where $\boldsymbol{\beta}_{k,s} = \theta_s\big(\boldsymbol{\beta}_{k-1} - sf'(\boldsymbol{\beta}_{k-1})\big)$ is the same as the above with emphasis on its dependence on $s$. Set $s_k = \delta^{i_k} s_{k-1}$ and compute

$$\boldsymbol{\beta}_k = \theta_{s_k}\big(\boldsymbol{\beta}_{k-1} - s_k f'(\boldsymbol{\beta}_{k-1})\big).$$

Note that the requirement (3.80) is really the local majorization requirement. It can easily hold since $s_k \to 0$ exponentially fast as $i_k \to \infty$. According to (3.69), the sequence of objective values $\{Q(\boldsymbol{\beta}_k)\}$ is non-increasing. The above choice of the step size of $s_k$ can be very small as $k$ gets large. Another possible scheme is to use $s = \delta^{i_k} s_0$ rather than $s = \delta^{i_k} s_{k-1}$ in (3.80) in choosing $s_k$.

The fast iterative shrinkage-thresholding algorithm (FISTA, Beck and Teboulle, 2009) is proposed to improve the convergence rate of ISTA. It employs the Nesterov acceleration idea (Nesterov, 1983). The algorithm runs as follows. Input the step size $s$ such that $s^{-1}$ is the upper bound of the Lipchitz constant of $f'(\cdot)$. Take $\mathbf{x}_1 = \boldsymbol{\beta}_0$ and $t_1 = 1$. Compute iteratively for $k \geq 1$

$$\begin{aligned}
\boldsymbol{\beta}_k &= \theta_s(\mathbf{x}_k - sf'(\mathbf{x}_k)), \quad t_{k+1} = (1 + \sqrt{1 + 4t_k^2}\,)/2, \\
\mathbf{x}_{k+1} &= \boldsymbol{\beta}_k + \frac{t_k - 1}{t_{k+1}}(\boldsymbol{\beta}_k - \boldsymbol{\beta}_{k-1}).
\end{aligned}$$

The algorithm utilizes a constant "stepsize" $s$. The backtracking rule can also be employed to make the algorithm more practical. Beck and Teboulle (2009) show that the FISTA has a quadratic convergence rate whereas the ISTA has only a linear convergence rate.

### 3.5.8 *Projected proximal gradient method*

Agarwal, Negahban and Wainwright (2012) propose a projected proximal gradient descent algorithm to solve the problem

$$\min_{R(\boldsymbol{\beta})\le c} \{f(\boldsymbol{\beta}) + g(\boldsymbol{\beta})\}. \tag{3.81}$$

Given the current value $\boldsymbol{\beta}_{k-1}$, approximate the smooth function $f$ by isotropic quadratic (3.77). The resulting unconstrained solution is given by (3.78). Now, project $\boldsymbol{\beta}_k$ onto the set $\{\boldsymbol{\beta} : R(\boldsymbol{\beta}) \le c\}$ and continue with the next iteration by taking the projected value as the initial value. When $\|R(\boldsymbol{\beta})\| = \|\boldsymbol{\beta}\|_1$, the projection admits an analytical solution. If $\|\boldsymbol{\beta}_k\|_1 \le c$, then the projection is just itself; otherwise, it is the soft-thresholding at level $\lambda_n$ so that the constraint $\|\boldsymbol{\beta}_k\|_1 = c$. The threshold level $\lambda_n$ can be computed as follows: (1) sort $\{|\beta_{k,j}|\}_{j=1}^p$ into $b_1 \ge b_2 \ge \ldots \ge b_p$; (2) find $J = \max\{1 \le j \le p : b_j - (\sum_{r=1}^j b_r - c)/j > 0\}$ and let $\lambda_n = (\sum_{r=1}^J b_j - c)/J$.

### 3.5.9 *ADMM*

The *alternating direction method of multipliers* (ADMM) (Douglas and Rachford (1956), Eckstein and Bertsekas (1992)) has a number of successful applications in modern statistical machine learning. Boyd *et al.* (2011) give a comprehensive review on *ADMM*. Solving the Lasso regression problem is a classical application of ADMM. Consider the Lasso penalized least square

$$\min_{\boldsymbol{\beta}} \frac{1}{2n}\|\mathbf{Y} - \mathbf{X}\boldsymbol{\beta}\|^2 + \lambda\|\boldsymbol{\beta}\|_1$$

which is equivalent to

$$\min_{\boldsymbol{\beta},\mathbf{z}} \frac{1}{2n}\|\mathbf{Y} - \mathbf{X}\boldsymbol{\beta}\|^2 + \lambda\|\mathbf{z}\|_1 \quad \text{subject to} \quad \mathbf{z} = \boldsymbol{\beta}.$$

The augmented Lagrangian is

$$\mathcal{L}_\eta(\boldsymbol{\beta}, \mathbf{z}, \boldsymbol{\theta}) = \frac{1}{2n}\|\mathbf{Y} - \mathbf{X}\boldsymbol{\beta}\|^2 + \lambda\|\mathbf{z}\|_1 - \boldsymbol{\theta}^T(\mathbf{z} - \boldsymbol{\beta}) + \frac{\eta}{2}\|\mathbf{z} - \boldsymbol{\beta}\|_2^2,$$

where $\eta$ can be a fixed positive constant set by the user, e.g. $\eta = 1$. The term $\boldsymbol{\theta}^T(\mathbf{z} - \boldsymbol{\beta})$ is the Lagrange multiplier and the term $\frac{\eta}{2}\|\mathbf{z} - \boldsymbol{\beta}\|_2^2$ is its augmentation. The choice of $\eta$ can affect the convergence speed. ADMM is an iterative procedure. Let $(\boldsymbol{\beta}^k, \mathbf{z}^k, \boldsymbol{\theta}^k)$ denote the $k$th iteration of the ADMM algorithm for $k = 0, 1, 2, \ldots$. Then the algorithm proceeds as follows:

$$\begin{aligned}
\boldsymbol{\beta}^{k+1} &= \operatorname{argmin}_{\boldsymbol{\beta}} \mathcal{L}_\eta(\boldsymbol{\beta}, \mathbf{z}^k, \boldsymbol{\theta}^k), \\
\mathbf{z}^{k+1} &= \operatorname{argmin}_{\mathbf{z}} \mathcal{L}_\eta(\boldsymbol{\beta}^{k+1}, \mathbf{z}, \boldsymbol{\theta}^k), \\
\boldsymbol{\theta}^{k+1} &= \boldsymbol{\theta}^k - (\mathbf{z}^{k+1} - \boldsymbol{\beta}^{k+1}).
\end{aligned}$$

It is easy to see that $\boldsymbol{\beta}^{k+1}$ has a close form expression and $\mathbf{z}^{k+1}$ is obtained by solving $p$ univariate $L_1$ penalized problems. More specifically, we have

$$\begin{aligned}
\boldsymbol{\beta}^{k+1} &= (\mathbf{X}^T\mathbf{X}/n + \eta\mathbf{I})^{-1}(\mathbf{X}^T\mathbf{Y}/n + \eta\mathbf{z}^k - \eta\boldsymbol{\theta}^k), \\
z_j^{k+1} &= \operatorname{sgn}(\beta_j^{k+1} + \theta_j^k)(|\beta_j^{k+1} + \theta_j^k| - \lambda/\eta), j = 1, \ldots, p.
\end{aligned}$$

### *3.5.10 Iterative local adaptive majorization and minimization*

Iterative local adaptive majorization and minimization is an algorithmic approach to solve the folded concave penalized least-squares problem (3.9) or more generally the *penalized quasi-likelihood* of the form:

$$f(\boldsymbol{\beta}) + \sum_{j=1}^p p_\lambda(|\beta_j|) \tag{3.82}$$

with both algorithmic and statistical guaranteed, proposed and studied by Fan, Liu, Sun, and Zhang (2018). It combines the local linear approximation (3.70) and the proximal gradient method (3.78) to solve the problem (3.82). More specifically, starting from the initial value $\boldsymbol{\beta}^{(0)} = 0$, we use LLA to case problem (3.82) into the sequence of problems:

$$\widehat{\boldsymbol{\beta}}^{(1)} = \arg\min\Big\{f(\boldsymbol{\beta}) + \sum_{j=1}^d \lambda_j^{(0)}|\beta_j|\Big\}, \quad \text{with } \lambda_j^{(0)} = p'_\lambda(|\widehat{\beta}_j^{(0)}|) \tag{3.83}$$

..............................................

$$\widehat{\boldsymbol{\beta}}^{(t)} = \arg\min\Big\{f(\boldsymbol{\beta}) + \sum_{j=1}^d \lambda_j^{(t-1)}|\beta_j|\Big\}, \text{ with } \lambda_j^{(t-1)} = p'_\lambda(|\widehat{\beta}_j^{(t-1)}|) \tag{3.84}$$

Within each problem (3.83) or (3.84) above, we apply the proximal gradient method. More specifically, by (3.79), starting from the initial value $\widehat{\boldsymbol{\beta}}_{t,0} = \widehat{\boldsymbol{\beta}}_{t-1}$, the algorithm used to solve (3.84) utilizes the iterations

$$\widehat{\boldsymbol{\beta}}_{t,k} = \big(\widehat{\boldsymbol{\beta}}_{t,k-1} - s_{t,k}n^{-1}\mathbf{X}^T(\mathbf{Y} - \mathbf{X}\widehat{\boldsymbol{\beta}}_{t,k-1}) - s_{t,k}\lambda\big)_+, \tag{3.85}$$

for $k = 1, \cdots, k_t$, where the step size is computed by using $s = \delta^{i_k} s_0$ to check (3.80) in choosing $s_k$. The flowchat of the algorithm can be summarized in Figure 3.9. This algorithmic approach of the statistical estimator is called *I-LAMM* by Fan *et al.* (2018).

Note that the problem (3.83) is convex but not strongly convex. It converges only at a sublinear rate. Hence, it takes longer to get to a consistent

$$\begin{array}{llllll}
\boldsymbol{\lambda}^{(0)}: & \boldsymbol{\beta}^{(1,0)}=\mathbf{0} & \xrightarrow{\text{LAMM}} \boldsymbol{\beta}^{(1,1)} \xrightarrow{\text{LAMM}} & \cdots \xrightarrow{\text{LAMM}} & \boldsymbol{\beta}^{(1,k_1)} = \widetilde{\boldsymbol{\beta}}^{(1)}, & k_1 \lesssim \varepsilon_c^{-2}; \\
\boldsymbol{\lambda}^{(1)}: & \boldsymbol{\beta}^{(2,0)}=\widetilde{\boldsymbol{\beta}}^{(1)} & \xrightarrow{\text{LAMM}} \boldsymbol{\beta}^{(2,1)} \xrightarrow{\text{LAMM}} & \cdots \xrightarrow{\text{LAMM}} & \boldsymbol{\beta}^{(2,k_2)} = \widetilde{\boldsymbol{\beta}}^{(2)}, & k_2 \lesssim \log(\varepsilon_t^{-1}); \\
\vdots & & & \vdots & & \vdots \\
\boldsymbol{\lambda}^{(T-1)}: & \boldsymbol{\beta}^{(T,0)}=\widetilde{\boldsymbol{\beta}}^{(T-1)} & \xrightarrow{\text{LAMM}} \boldsymbol{\beta}^{(T,1)} \xrightarrow{\text{LAMM}} & \cdots \xrightarrow{\text{LAMM}} & \boldsymbol{\beta}^{(T,k_T)} = \widetilde{\boldsymbol{\beta}}^{(T)}, & k_T \lesssim \log(\varepsilon_t^{-1}).
\end{array}$$

Figure 3.9: Flowchart of iterative local majorization and minorization algorithm. For Gaussian noise, $\varepsilon_c \asymp \sqrt{n^{-1}\log p}$ and $\varepsilon_t \asymp \sqrt{n^{-1}}$. Taken from Fan, Liu, Sun, and Zhang (2018)

neighborhood. Once the estimate is in a consistent neighborhood, from step 2 and on, the solutions are sparse and therefore the function (3.84) is strongly convex in this restricted neighborhood and the algorithmic convergence is exponentially fast (at a linear rate). This leads Fan *et al.* (2018) to take $k_1 \asymp n/\log p$ and $k_2 \asymp \log n$. In addition, they show that when the number of outer loop $T \asymp \log(\log(p))$, the estimator achieves statistical optimal rates and further iteration will not improve nor deteriorate the statistical errors.

### 3.5.11 Other methods and timeline

There are many other algorithms for computing the penalized least-squares problem. For example, *matching pursuit*, introduced by Mallot and Zhang (1993), is similar to the forward selection algorithm for subset selection. As in the forward selection and LARS, the most correlated variable $\mathbf{X}_j$ (say) with the current residual $\mathbf{R}$ is selected and the univariate regression

$$\mathbf{R} = \beta_j \mathbf{X}_j + \varepsilon$$

is fitted. This is an important deviation from the forward selection in high-dimensional regression as the matching pursuit does not compute multiple regression. It is similar but more greedy than the coordinate descent algorithm, as only the most correlated coordinate is chosen. With fitted univariate coefficient $\widehat{\beta}_j$, we update the current residual by $\mathbf{R} - \widehat{\beta}_j \mathbf{X}_j$. The variables selected as well as coefficients used to compute $\mathbf{R}$ can be recorded along the fit.

Iterated SIS (sure independence screening) introduced in Fan and Lv (2008) and extended by Fan, Samworth and Wu (2009) can be regarded as another greedy algorithm for computing folded concave PLS. The basic idea is to iteratively use large scale screening (e.g. marginal screening) and moderate scale selection by using the penalized least-squares. Details will be introduced in Chapter 8.

The *DC algorithm* (An and Tao, 1997) is a general algorithm for minimizing the difference of two convex functions. Suppose that $Q(\boldsymbol{\beta}) = Q_1(\boldsymbol{\beta}) - Q_2(\boldsymbol{\beta})$, where $Q_1$ and $Q_2$ are convex. Given the current value $\boldsymbol{\beta}_0$, linearize $Q_2(\boldsymbol{\beta})$ by

$$Q_{2,L}(\boldsymbol{\beta}) = Q_2(\boldsymbol{\beta}_0) + Q_2'(\boldsymbol{\beta}_0)^T(\boldsymbol{\beta} - \boldsymbol{\beta}_0).$$

Now update the minimizer by the convex optimization problem

$$\widehat{\boldsymbol{\beta}} = \text{argmin}_{\boldsymbol{\beta}}\{Q_1(\boldsymbol{\beta}) - Q_{2,L}(\boldsymbol{\beta})\}.$$

Note that for any convex function

$$Q_2(\boldsymbol{\beta}) \geq Q_{2,L}(\boldsymbol{\beta}) \quad \text{with} \quad Q_2(\boldsymbol{\beta}_0) = Q_{2,L}(\boldsymbol{\beta}_0).$$

Thus, the DC algorithm is a special case of the MM-algorithm. Hence, its target value should be non-increasing $Q(\boldsymbol{\beta}_{\text{new}}) \leq Q(\boldsymbol{\beta}_0)$ [c.f. (3.69)]. The algorithm has been implemented to support vector machine classifications by Liu, Shen and Doss (2005) and Wu and Liu (2007). It was used by Kim, Choi and Oh (2008) to compute SCAD in which the SCAD penalty function is decomposed as

$$p_\lambda(|\beta|) = \lambda|\beta| - [\lambda|\beta| - p_\lambda(|\beta|)].$$

Agarwal, Negahban and Wainwright (2012) propose the composite gradient descent algorithm. Liu, Yao and Li (2016) propose a mixed integer programming-based global optimization (MIPGO) to solve the class of folded concave penalized least-squares that find a provably global optimal solution. Fan, Liu, Sun and Zhang (2018) propose I-LAMM to simultaneously control of algorithmic complexity and statistical error.

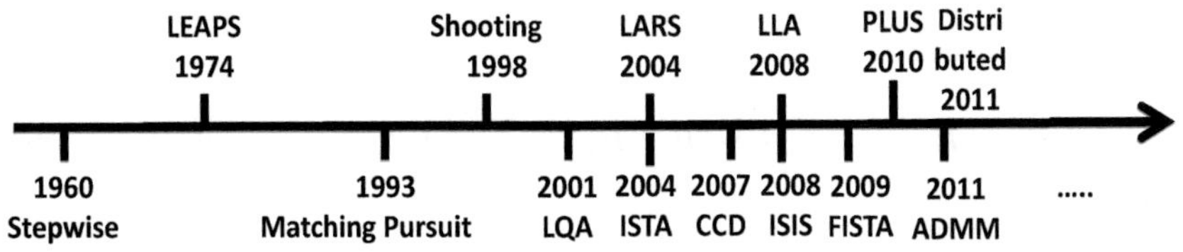


Figure 3.10: A snapshot of the history of the algorithms for computing penalized least-squares.

## 3.6 Regularization Parameters for PLS

In applications of the folded concave PLS (3.9), one needs to determine the regularization parameter $\lambda$. The solution paths such as in Figure 3.7 can help us to choose a model. For example, it is not unreasonable to select a model with $1/\lambda$ somewhat larger than 40 in Figure 3.7. After that point, the model complexity increases substantially and there will be no more variables with large coefficients.

In many situations, one would also like to have a data-driven choice of

$\lambda$. The choice of $\lambda$ for the $L_0$-penalty was addressed in Section 3.1.3. The basic idea of choosing regularization parameters to minimize the estimated prediction error continues to apply. For example, one can choose $\lambda$ in the folded concave PLS by using cross-validation (3.5). However, other criteria such as AIC and BIC utilize the model size $m$ that is specific to $L_0$-penalty. We need to generalize this concept of model size, which will be called the degrees of freedom.

### 3.6.1 Degrees of freedom

To help motivate the definition of *degrees of freedom*, following Efron (1986) and Efron *et al.* (2004), we assume that given the covariates $\mathbf{X}$, $\mathbf{Y}$ has the conditional mean vector $\boldsymbol{\mu}(\mathbf{X})$ (also called regression function) that depends on $\mathbf{X}$ and homoscedastic variance $\sigma^2$. The conditional mean vector $\boldsymbol{\mu}$ (whose dependence on $\mathbf{X}$ is suppressed) is unknown and estimated by $\widehat{\boldsymbol{\mu}}$, a function of the data $(\mathbf{X}, \mathbf{Y})$. Note that

$$\|\boldsymbol{\mu} - \widehat{\boldsymbol{\mu}}\|^2 = \|\mathbf{Y} - \widehat{\boldsymbol{\mu}}\|^2 - \|\mathbf{Y} - \boldsymbol{\mu}\|^2 + 2(\widehat{\boldsymbol{\mu}} - \boldsymbol{\mu})^T(\mathbf{Y} - \boldsymbol{\mu}). \tag{3.86}$$

Thus, we have *Stein's identity*: the *mean squared error*

$$\mathrm{E}\,\|\boldsymbol{\mu} - \widehat{\boldsymbol{\mu}}\|^2 = \mathrm{E}\left\{\|\mathbf{Y} - \widehat{\boldsymbol{\mu}}\|^2 - n\sigma^2\right\} + 2\sum_{i=1}^n \mathrm{cov}(\widehat{\mu}_i, Y_i). \tag{3.87}$$

and the *prediction error*

$$\mathrm{E}\,\|\mathbf{Y}^{\mathrm{new}} - \widehat{\boldsymbol{\mu}}\|^2 = n\sigma^2 + \mathrm{E}\,\|\boldsymbol{\mu} - \widehat{\boldsymbol{\mu}}\|^2 = \mathrm{E}\left\{\|\mathbf{Y} - \widehat{\boldsymbol{\mu}}\|^2\right\} + 2df_{\widehat{\boldsymbol{\mu}}}\sigma^2 \tag{3.88}$$

with

$$df_{\widehat{\boldsymbol{\mu}}} = \sigma^{-2}\sum_{i=1}^n \mathrm{cov}(\widehat{\mu}_i, Y_i) \tag{3.89}$$

as the *degrees of freedom.*

If $df_{\widehat{\boldsymbol{\mu}}}$ is known and $\sigma^2$ is given, a $C_p$-type of unbiased risk estimation is given by

$$C_p(\widehat{\boldsymbol{\mu}}) = \|\mathbf{Y} - \widehat{\boldsymbol{\mu}}\|^2 + 2\sigma^2\, df_{\widehat{\boldsymbol{\mu}}}. \tag{3.90}$$

The above formula shows that $df_{\widehat{\boldsymbol{\mu}}}$ plays the same role as the number of parameters in (3.3).

For many linear smoothers, their degrees of freedom are indeed known quantities. A linear estimator has the form $\widehat{\boldsymbol{\mu}} = \mathbf{S}\mathbf{Y}$ with $\mathbf{S}$ being a smoother matrix that only depends on $\mathbf{X}$. See examples given in Section 2.8 of Chapter 2. By independence among $\mathbf{Y}_i$s, $\mathrm{cov}(\widehat{\mu}_i, \mathbf{Y}_i) = \mathbf{S}_{ii}\sigma^2$. From (3.86), it follows that

$$df_{\widehat{\boldsymbol{\mu}}} = \frac{1}{\sigma^2}\sum_{i=1}^n \mathbf{S}_{ii}\sigma^2 = \mathrm{tr}(\mathbf{S}).$$

We mentioned $\mathrm{tr}(\mathbf{S})$ as the degrees of freedom of the linear smoother $\mathbf{S}$ in

Chapter 2. Here, a formal justification is provided. In particular, when $\mathbf{S} = \mathbf{X}_1(\mathbf{X}_1^T\mathbf{X}_1)^{-1}\mathbf{X}_1^T$, the projection matrix using $m$ variables of the full model, we have

$$df_{\widehat{\boldsymbol{\mu}}} = \mathrm{tr}(\mathbf{X}_1(\mathbf{X}_1^T\mathbf{X}_1)^{-1}\mathbf{X}_1^T) = m.$$

Therefore, the degrees of freedom formula is an extension of the number of variables used in the classical linear model.

Degrees of freedom can be much more complex for nonlinear model fitting procedures. For example, let us consider the best subset selection. For a given subset size $m$, the final model always has $m$ variables and one may naively think the degrees of freedom is $m$. This is in general wrong unless $m = 0$ or $m = p$. This is because the final subset is obtained by exclusively searching over $\binom{p}{m}$ many candidate models. We can not ignore the stochastic nature of the search unless $m = 0$ or $m = p$. A simulation study in Lucas, Fithian and Hastie (2015) shows that the degree of freedom is larger than $m$ and can be even larger than $p$. Another interesting and counter-intuitive finding is that the degrees of freedom is not a monotonic increasing function of $m$, which again reflects the complexity due to the stochastic search over $\binom{p}{m}$ many submodels. The same phenomenon is also observed for the degrees of freedom of forward selection.

For least angle regression, Efron *et al.* (2004) show that under the orthogonal design assumption, the degree of freedom in the $m^{th}$ step of the LARS algorithm is $m$. This matches our intuition, as at the $m^{th}$ step of the LARS algorithm, $m$ variables are effectively recruited. For a general design matrix, let $\widehat{\boldsymbol{\beta}}_\lambda^{\text{lasso}}$ be the Lasso penalized least square estimator with penalization parameter $\lambda$. Let $df_\lambda^{\text{lasso}}$ denote its degrees of freedom. Zou, Hastie and Tibshirani (2007) prove a surprising result:

$$df_\lambda^{\text{lasso}} = \mathrm{E}[\|\widehat{\boldsymbol{\beta}}_\lambda^{\text{lasso}}\|_0]. \tag{3.91}$$

Therefore, the number of nonzero estimated coefficients is an exact unbiased estimator of the degrees freedom of the Lasso. The estimation consistency is also established. In theory we view the $L_1$ PLS as a convex relaxation of $L_0$ PLS, but their degrees of freedom (model complexity) has very different properties. For the $L_0$ PLS, the number of nonzero estimated coefficients can severely underestimate the true degrees of freedom. The final model of $L_1$ PLS is also obtained via a stochastic search, but (3.91) implies that on average the complexity due to stochastic search is zero.

The unbiasedness result is good enough for constructing a $C_p$ type statistic for the Lasso:

$$C_p^{\text{lasso}} = \|\mathbf{Y} - \mathbf{X}\widehat{\boldsymbol{\beta}}_\lambda^{\text{lasso}}\|^2 + 2\sigma^2\|\widehat{\boldsymbol{\beta}}_\lambda^{\text{lasso}}\|_0, \tag{3.92}$$

which is an exact unbiased estimator of the prediction risk of the Lasso.

### 3.6.2 *Extension of information criteria*

Suppose that $\widehat{\boldsymbol{\mu}}(\lambda)$ is constructed by using a regularization parameter $\lambda$. An extension of the $C_p$ criterion (3.3) and the information criterion (3.4) is

$$C_p(\lambda) = \|\mathbf{Y} - \widehat{\boldsymbol{\mu}}(\lambda)\|^2 + \gamma\sigma^2 \, df_{\widehat{\boldsymbol{\mu}}(\lambda)}, \tag{3.93}$$

and

$$\mathrm{IC}(\lambda) = \log(\|\mathbf{Y} - \widehat{\boldsymbol{\mu}}(\lambda)\|^2/n) + \gamma \, df_{\widehat{\boldsymbol{\mu}}(\lambda)}/n. \tag{3.94}$$

As shown in (3.87), $C_p(\lambda)$ with $\gamma = 2$ is an unbiased estimation of the risk $E\|\widehat{\boldsymbol{\mu}}(\lambda) - \boldsymbol{\mu}\|^2$ except a constant term $-n\sigma^2$. When $\gamma = 2$, $\log(n)$ and $2\log(p)$, the criteria (3.93) will be called respectively the AIC, BIC, and RIC criterion. With an estimate of $\sigma^2$ (see Section 3.7), one can choose $\lambda$ to minimize (3.93). Similarly, we can choose $\lambda$ to minimize (3.94).

Similarly, one can extend the *generalized cross-validation* criterion (3.7) to this framework by

$$\mathrm{GCV}(\lambda) = \frac{\|\mathbf{Y} - \widehat{\boldsymbol{\mu}}(\lambda)\|^2}{(1 - df_{\widehat{\boldsymbol{\mu}}(\lambda)}/n)^2}. \tag{3.95}$$

In particular, when the linear estimator $\widehat{\boldsymbol{\mu}}(\lambda) = \mathbf{H}(\lambda)\mathbf{Y}$ is used, by (3.90), we have

$$C_p(\lambda) = \|\mathbf{Y} - \widehat{\boldsymbol{\mu}}(\lambda)\|^2 + \gamma\sigma^2 \, \mathrm{tr}(\mathbf{H}(\lambda)), \tag{3.96}$$

$$\mathrm{IC}(\lambda) = \log(\|\mathbf{Y} - \widehat{\boldsymbol{\mu}}(\lambda)\|^2/n) + \gamma \, \mathrm{tr}(\mathbf{H}(\lambda))/n. \tag{3.97}$$

and

$$\mathrm{GCV}(\lambda) = \frac{\|\mathbf{Y} - \widehat{\boldsymbol{\mu}}(\lambda)\|^2}{[1 - \mathrm{tr}(\mathbf{H}(\lambda))/n]^2}. \tag{3.98}$$

As mentioned in Section 3.1.3, an advantage of the information criterion and GCV is that no estimation of $\sigma^2$ is needed, but this can lead to inaccurate estimation of prediction error.

### 3.6.3 *Application to PLS estimators*

For PLS estimator (3.9), $\widehat{\boldsymbol{\mu}}(\lambda)$ is not linear in $\mathbf{Y}$. Some approximations are needed. For example, using the LQA approximation, Fan and Li (2001) regard (3.67) as a linear smoother with (recalling $\boldsymbol{\mu}(\lambda) = \mathbf{X}\widehat{\boldsymbol{\beta}}(\lambda)$)

$$\mathbf{H}(\lambda) = \mathbf{X}(\mathbf{X}^T\mathbf{X} + n \, \mathrm{diag}\{p'_\lambda(\widehat{\beta}_j(\lambda))/|\widehat{\beta}_j(\lambda)|\})^{-1}\mathbf{X}^T,$$

and choose $\lambda$ by GCV (3.98).

For the LARS-Lasso algorithm, as mentioned at the end of Section 3.6.1, Zou, Hastie and Tibshirani (2007) demonstrate that the degree of freedom is the same as the number of variables used in the LARS algorithm. This motivates Wang, Li and Tsai (2007) and Wang and Leng (2007) to use directly $\|\widehat{\boldsymbol{\beta}}\|_0$ as the degree of freedom. This leads to the definition of modified information criterion as

$$\mathrm{IC}^*(\lambda) = \log(\|\mathbf{Y} - \mathbf{X}\widehat{\boldsymbol{\beta}}_\lambda\|^2/n) + \gamma\frac{\|\widehat{\boldsymbol{\beta}}_\lambda\|_0}{n}C_n \tag{3.99}$$

for a sequence of constants $C_n$. It has been shown by Wang, Li and Tsai (2007) that SCAD with AIC ($\gamma = 2$, $C_n = 1$) yields an inconsistent model (too many false positives) while BIC ($\gamma = \log n$, $C_n = 1$) yields a consistent estimation of the model when $p$ is fixed. See also Wang and Leng (2007) for similar model selection results. Wang, Li and Leng (2009) show that the modified BIC (3.99) with $\gamma = \log n$ and $C_n \to \infty$ produces consistent model selection when SCAD is used. For high-dimensional model selection, Chen and Chen (2008) propose an extended BIC, which adds a multiple of the logarithm of the prior probability of a submodel to BIC. Here, they successfully establish its model selection consistency.

## 3.7 Residual Variance and Refitted Cross-validation

Estimation of noise variance $\sigma^2$ is fundamental in statistical inference. It is prominently featured in the statistical inference of regression coefficients. It is also important for variable selection using the $C_p$ criterion (3.96). It provides a benchmark for forecasting error when an oracle actually knows the underlying regression function. It also arises from genomewise association studies (see Fan, Han and Gu, 2012). In the classical linear model as in Chapter 2, the noise variance is estimated by the residual sum of squares divided by $n - p$. This is not applicable to the high-dimensional situations in which $p > n$. In fact, as demonstrated in Section 1.3.3 (see Figure 1.9 there), the impact of spurious correlation on residual variance estimation can be very large. This leads us to introducing the refitted cross-validation.

In this section, we introduce methods for estimating $\sigma^2$ in the high-dimensional framework. Throughout this section, we assume the linear model (2.2) with homoscedastic variance $\sigma^2$.

### *3.7.1 Residual variance of Lasso*

A natural estimator of $\sigma^2$ is the residual variance of penalized least-squares estimators. As demonstrated in Section 3.3.2, Lasso has a good risk property. We therefore examine when its residual variance gives a consistent estimator of $\sigma^2$.

Recall that the *theoretical risk* and *empirical risk* are defined by

$$R(\boldsymbol{\beta}) = \mathrm{E}(Y - \mathbf{X}^T\boldsymbol{\beta})^2 \quad \text{and} \quad R_n(\boldsymbol{\beta}) = n^{-1}\sum_{i=1}^{n}(Y_i - \mathbf{X}_i^T\boldsymbol{\beta})^2,$$

Let $\widehat{\boldsymbol{\beta}}$ be the solution to the Lasso problem (3.39) and $c$ be sufficiently large so that $\|\boldsymbol{\beta}_0\|_1 \le c$. Then, $R_n(\boldsymbol{\beta}_0) \ge R_n(\widehat{\boldsymbol{\beta}})$. Using this, we have

$$\begin{aligned} R(\boldsymbol{\beta}_0) - R_n(\widehat{\boldsymbol{\beta}}) &= [R(\boldsymbol{\beta}_0) - R_n(\boldsymbol{\beta}_0)] + [R_n(\boldsymbol{\beta}_0) - R_n(\widehat{\boldsymbol{\beta}})] \\ &\ge R(\boldsymbol{\beta}_0) - R_n(\boldsymbol{\beta}_0) \\ &\ge -\sup_{\|\boldsymbol{\beta}\|_1 \le c} |R(\boldsymbol{\beta}) - R_n(\boldsymbol{\beta})|. \end{aligned}$$

On the other hand, by using $R(\boldsymbol{\beta}_0) \le R(\widehat{\boldsymbol{\beta}})$, we have

$$R(\boldsymbol{\beta}_0) - R_n(\widehat{\boldsymbol{\beta}}) \le R(\widehat{\boldsymbol{\beta}}) - R_n(\widehat{\boldsymbol{\beta}}) \le \sup_{\|\boldsymbol{\beta}\|_1 \le c} |R(\boldsymbol{\beta}) - R_n(\boldsymbol{\beta})|.$$

Therefore,

$$|R(\boldsymbol{\beta}_0) - R_n(\widehat{\boldsymbol{\beta}})| \le \sup_{\|\boldsymbol{\beta}\|_1 \le c} |R(\boldsymbol{\beta}) - R_n(\boldsymbol{\beta})|.$$

By (3.40), we conclude that

$$|R(\boldsymbol{\beta}_0) - R_n(\widehat{\boldsymbol{\beta}})| \le (1+c)^2 \|\boldsymbol{\Sigma}^* - \mathbf{S}_n^*\|_\infty, \tag{3.100}$$

provided that $\|\boldsymbol{\beta}_0\|_1 \le c$. In other words, the average residual sum of squares of Lasso

$$\widehat{\sigma}^2_{\text{Lasso}} = n^{-1}\|\mathbf{Y} - \mathbf{X}\widehat{\boldsymbol{\beta}}\|^2$$

provides a consistent estimation of $\sigma^2$, if the right hand side of (3.100) goes to zero and $\|\boldsymbol{\beta}_0\|_1 \le c$

$$R(\boldsymbol{\beta}_0) = \sigma^2 \qquad \text{and} \qquad R_n(\widehat{\boldsymbol{\beta}}) = \widehat{\sigma}^2_{\text{Lasso}}.$$

As shown in Chapter 11, $\|\boldsymbol{\Sigma} - \widehat{\mathbf{S}}_n\|_\infty = O_P(\sqrt{(\log p)/n})$ for the data with Gaussian tails. That means that $\widehat{\sigma}^2_{\text{Lasso}}$ is consistent when

$$\|\boldsymbol{\beta}_0\|_1 \le c = o\big((n/\log p)^{1/4}\big). \tag{3.101}$$

Condition (3.101) is actually very restrictive. It requires the number of significantly nonzero components to be an order of magnitude smaller than $(n/\log p)^{1/4}$. Even when that condition holds, $\widehat{\sigma}^2_{\text{Lasso}}$ is only a consistent estimator and can be biased or not optimal. This leads us to consider refitted cross-validation.

### *3.7.2 Refitted cross-validation*

Refitted cross-validation (RCV) was introduced by Fan, Guo and Hao (2012) to deal with the spurious correlation induced by data-driven model selection. In high-dimensional regression models, model selection consistency is very hard to achieve. When some important variables are missed in the selected model, they create a non-negligible bias in estimating $\sigma^2$. When spurious variables are selected into the model, they are likely to predict the realized but unobserved noise $\varepsilon$. Hence, the residual variance will seriously underestimate $\sigma^2$ as shown in Section 1.3.3.

Note that our observed data follow

$$\mathbf{Y} = \mathbf{X}\boldsymbol{\beta} + \boldsymbol{\varepsilon}, \text{ and } \operatorname{Var}(\boldsymbol{\varepsilon}) = \sigma^2 \mathbf{I}_n.$$

Even though we only observe $(\mathbf{X}, \mathbf{Y})$, $\boldsymbol{\varepsilon}$ is a realized vector in $R^n$. It can have a spurious correlation with a subgroup of variables $\mathbf{X}_{\mathcal{S}}$, namely, there exists a vector $\boldsymbol{\beta}_{\mathcal{S}}$ such that $\mathbf{X}_{\mathcal{S}}\boldsymbol{\beta}_{\mathcal{S}}$ and $\boldsymbol{\varepsilon}$ are highly correlated. This can occur easily when the number of predictors $p$ is large as shown in Section 1.3.3. In this case, $\mathbf{X}_{\mathcal{S}}$ can be seen by a model selection technique as important variables. A way to validate the model is to collect new data to see whether the variables in the set $\mathcal{S}$ still correlated highly with the newly observed $Y$. But this is infeasible in many studies and is often replaced by the data splitting technique.

RCV splits data evenly at random into two halves, where we use the first half of the data along with a model selection technique to get a submodel. Then, we fit this submodel to the second half of the data using the ordinary least-squares and get the residual variance. Next we switch the role of the first and the second half of the data and take the average of the two residual variance estimates. The idea differs importantly from cross-validation in that the refitting in the second stage reduces the influence of the spurious variables selected in the first stage.

We now describe the procedure in detail. Let datasets $(\mathbf{Y}^{(1)}, \mathbf{X}^{(1)})$ and $(\mathbf{Y}^{(2)}, \mathbf{X}^{(2)})$ be two randomly split data. Let $\widehat{\mathcal{S}}_1$ be the set of selected variables using data $(\mathbf{Y}^{(1)}, \mathbf{X}^{(1)})$. The variance $\sigma^2$ is then estimated by the residual variance of the least-squares estimate using the second dataset along with variables in $\widehat{\mathcal{S}}_1$ (only the selected model, not the data, from the first stage is carried to the fit in the second stage), namely,

$$\widehat{\sigma}_1^2 = \frac{(\mathbf{Y}^{(2)})^T (\mathbf{I}_{n/2} - \mathbf{P}^{(2)}_{\widehat{\mathcal{S}}_1}) \mathbf{Y}^{(2)}}{n/2 - |\widehat{\mathcal{S}}_1|}, \qquad \mathbf{P}^{(2)}_{\widehat{\mathcal{S}}_1} = \mathbf{X}^{(2)}_{\widehat{\mathcal{S}}_1} (\mathbf{X}^{(2)T}_{\widehat{\mathcal{S}}_1} \mathbf{X}^{(2)}_{\widehat{\mathcal{S}}_1})^{-1} \mathbf{X}^{(2)T}_{\widehat{\mathcal{S}}_1}. \tag{3.102}$$

Compare residual variance estimation in (2.7). Switching the role of the first and second half, we get a second estimate

$$\widehat{\sigma}_2^2 = \frac{(\mathbf{Y}^{(1)})^T (\mathbf{I}_{n/2} - \mathbf{P}^{(1)}_{\widehat{\mathcal{S}}_2}) \mathbf{Y}^{(1)}}{n/2 - |\widehat{\mathcal{S}}_2|}.$$

We define the final estimator as the simple average

$$\widehat{\sigma}^2_{\mathrm{RCV}} = (\widehat{\sigma}_1^2 + \widehat{\sigma}_2^2)/2,$$

or the weighted average defined by

$$\widehat{\sigma}^2_{\mathrm{wRCV}} = \frac{(\mathbf{Y}^{(2)})^T (\mathbf{I}_{n/2} - \mathbf{P}^{(2)}_{\widehat{\mathcal{S}}_1}) \mathbf{Y}^{(2)} + (\mathbf{Y}^{(1)})^T (\mathbf{I}_{n/2} - \mathbf{P}^{(1)}_{\widehat{\mathcal{S}}_2}) \mathbf{Y}^{(1)}}{n - |\widehat{\mathcal{S}}_1| - |\widehat{\mathcal{S}}_2|}. \tag{3.103}$$

The latter takes into account the degrees of freedom used in fitting the linear model in the second stage. We can now randomly divide the data multiple times and take the average of the resulting RCV estimates.

The point of refitting is that even though $\widehat{\mathcal{S}}_1$ may contain some unimportant variables that are highly correlated with $\boldsymbol{\varepsilon}^{(1)}$, they play minor roles in estimating $\sigma^2$ in the second stage since they are unrelated with the realized noise vector $\boldsymbol{\varepsilon}^{(2)}$ in the second half of data set. Furthermore, even when some important variables are missed in $\widehat{\mathcal{S}}_1$, they still have a good chance of being well approximated by the other variables in $\widehat{\mathcal{S}}_1$. Thanks to the refitting in the second stage, the best linear approximation of those selected variables is used to reduce the biases in (3.102).

Unlike cross-validation, the second half of data also plays an important role in fitting. Therefore, its size can not be too small. For example, it should be bigger than $|\widehat{\mathcal{S}}_1|$. Yet, a larger set $\mathcal{S}_1$ gives a better chance of sure screening (no false negatives) and hence reduces the bias of estimator (3.102). RCV is applicable to any variable selection rule, including the marginal screening procedure in Chapter 8. Fan, Guo and Hao (2012) show that under some mild conditions, the method yields an asymptotic efficient estimator of $\widehat{\sigma}^2$. In particular, it can handle intrinsic model size $s = o(n)$, much higher than (3.101), when folded concave PLS is used. They verified the theoretical results by numerous simulations. See Section 8.7 for further developments.

## 3.8 Extensions to Nonparametric Modeling

The fundamental ideas of the penalized least squares can be easily extended to the more flexible *nonparametric models*. This section illustrates the versatility of high-dimensional linear techniques.

### 3.8.1 Structured nonparametric models

A popular modeling strategy is the *generalized additive model* (GAM):

$$Y = \mu + f_1(X_1) + \cdots + f_p(X_p) + \varepsilon, \tag{3.104}$$

This model was introduced by Stone (1985) to deal with the "*curse of dimensionality*" in multivariate nonparametric modeling and was thoroughly treated in the book by Hastie and Tibshirani (1990). A simple way to fit the additive model (3.104) is to expand the regression function $f_j(x)$ into a basis:

$$f_j(x) = \sum_{k=1}^{K_j} \beta_{jk} B_{jk}(x), \tag{3.105}$$

where $\{B_{jk}(x)\}_{k=1}^{K_j}$ are the basis functions (e.g. a B-spline basis with certain number of knots) for variable $X_j$. See Section 2.5.2. Substituting the expansion into (3.104) yields

$$\begin{aligned} Y &= \mu + \{\beta_{1,1}B_{1,1}(X_1) + \cdots + \beta_{1,K_1}B_{1,K_1}(X_1)\} + \cdots \\ &\quad + \{\beta_{p,1}B_{p,1}(X_p) + \cdots + \beta_{p,K_p}B_{p,K_p}(X_p)\} + \varepsilon. \end{aligned} \tag{3.106}$$

Treating the basis functions $\{B_{j,k}(X_j) : k = 1, \cdots, K_j, j = 1, \cdots, p\}$ as predictors, (3.106) is a high-dimensional linear model. By imposing a sparsity assumption, we assume that only a few $f_j$ functions actually enter the model. So, many $\beta$ coefficients are zero. Therefore, we can employ penalized folded concave PLS (3.9) to solve this problem. Another selection method is via the *group penalization*. See, for example, the PLASM algorithm in Baskin (1999) and the SpAM algorithm in Ravikumar, Liu, Lafferty and Wasserman (2007).

The *varying-coefficient model* is another widely-used nonparametric extension of the multiple linear regression model. Conditioning on an exposure variable $U$, the response and covariates follow a linear model. In other words,

$$Y = \beta_0(U) + \beta_1(U)X_1 + \cdots + \beta_p(U)X_p + \varepsilon. \tag{3.107}$$

The model allows regression coefficients to vary with the level of exposure $U$, which is an observed covariate variable such as age, time, or gene expression. For a survey and various applications, see Fan and Zhang (2008). Expanding the coefficient functions similar to (3.105), we can write

$$Y = \sum_{j=0}^{p}\{\sum_{k=1}^{K_j} \beta_{j,k}B_{j,k}(U)X_j\} + \varepsilon, \tag{3.108}$$

where $X_0 = 1$. By regarding variables $\{B_{j,k}(U)X_j, k = 1, \cdots, K_j, j = 0, \cdots, p\}$ as new predictors, model (3.108) is a high-dimensional linear model. The sparsity assumption says that only a few variables should be in the model (3.107) which implies many zero $\beta$ coefficients in (3.108). Again, we can employ penalized folded concave PLS (3.9) to do variable selection or use the group selection method.

### 3.8.2 *Group penalty*

The penalized least-squares estimate to the nonparametric models in Section 3.8 results in term-by-term selection of the basis functions. In theory, when the folded concave penalty is employed, the selection should be fine. On the other hand, the term-by-term selection does not fully utilize the sparsity assumption of the functions. In both the additive model and varying coefficient model, a zero function implies that the whole group of its associated coefficients in the basis expansion is zero. Therefore, model selection techniques should ideally keep or kill a group of coefficients at the same time.

Group penalty was proposed in Antoniadis and Fan (2001, page 966) to keep or kill a block of wavelets coefficients. It was employed by Lin and Zhang (2006) for component selection in smoothing spline regression models, including the additive model as a special case. Their COSSO algorithm iterates between a smoothing spline fit and a non-negative garrote shrinkage and selection. A special case of COSSO becomes a more familiar group lasso regression formulation considered in Yuan and Lin (2006) who named the group penalty *group-Lasso*.

Let $\{\mathbf{x}_j\}_{j=1}^p$ be $p$ groups of variables, each consisting of $K_j$ variables. Consider a generic linear model

$$Y = \sum_{j=1}^p \mathbf{x}_j^T \boldsymbol{\beta}_j + \varepsilon. \tag{3.109}$$

Two examples of (3.109) are (3.106) and (3.108) in which $\mathbf{x}_j$ represents $K_j$ spline bases and $\boldsymbol{\beta}_j$ represents their associated coefficients. In matrix form, the observed data based on a sample of size $n$ follow the model

$$\mathbf{Y} = \sum_{j=1}^p \mathbf{X}_j \boldsymbol{\beta}_j + \boldsymbol{\varepsilon}, \tag{3.110}$$

where $\mathbf{X}_j$ is $n \times K_j$ design matrix of variables $\mathbf{x}_j$.

The *group penalized least-squares* is to minimize

$$\frac{1}{2n}\|\mathbf{Y} - \sum_{j=1}^p \mathbf{X}_j\boldsymbol{\beta}_j\|^2 + \sum_{j=1}^p p_\lambda(\|\boldsymbol{\beta}_j\|_{W_j}) \tag{3.111}$$

where $p_\lambda(\cdot)$ is a penalty function and

$$\|\boldsymbol{\beta}_j\|_{W_j} = \sqrt{\boldsymbol{\beta}_j^T \mathbf{W}_j \boldsymbol{\beta}_j}$$

is a generalized norm with a semi-definite matrix $\mathbf{W}_j$. In many applications, one takes $\mathbf{W}_j = \mathbf{I}_{K_j}$, resulting in

$$\frac{1}{2n}\|\mathbf{Y} - \sum_{j=1}^p \mathbf{X}_j\boldsymbol{\beta}_j\|^2 + \sum_{j=1}^p p_\lambda(\|\boldsymbol{\beta}_j\|). \tag{3.112}$$

For example, the group-Lasso is defined as

$$\frac{1}{2n}\|\mathbf{Y} - \sum_{j=1}^p \mathbf{X}_j\boldsymbol{\beta}_j\|^2 + \lambda\sum_{j=1}^p K_j^{1/2}\|\boldsymbol{\beta}_j\|. \tag{3.113}$$

The extra factor $K_j^{1/2}$ is included to balance the impact of group size.

The group-Lasso (3.113) was proposed by Baskin (1999) for variable selection in the additive model. Turlach, Venables and Wright (2005) also used the group-Lasso for simultaneous variable selection in multiple responses linear regression, an example of multi-task learning.

Assuming a group-wise orthogonality condition, that is, $\mathbf{X}_j^T\mathbf{X}_j = n\mathbf{I}_{K_j}$ for all $j$, Yuan and Lin (2006) used a group descent algorithm to solve (3.112). Similar to coordinate descent, we update the estimate one group at a time. Consider the coefficients of group $j$ while holding all other coefficients fixed. Then, by $\mathbf{X}_j^T\mathbf{X}_j = n\mathbf{I}_{K_j}$ (3.112) can be written as

$$\frac{1}{2n}\|\mathbf{Y}_{-j} - \mathbf{X}_j\widehat{\boldsymbol{\beta}}_{-j}\|^2 + \frac{1}{2}\|\widehat{\boldsymbol{\beta}}_{-j} - \boldsymbol{\beta}_j\|^2 + \sum_{k=1}^p p_\lambda(\|\boldsymbol{\beta}_k\|), \tag{3.114}$$

where $\mathbf{Y}_{-j} = \mathbf{Y} - \sum_{k \neq j} \mathbf{X}_k \boldsymbol{\beta}_k$ and $\widehat{\boldsymbol{\beta}}_{-j} = n^{-1}\mathbf{X}_j^T \mathbf{Y}_{-j}$. This problem was solved by Antoniadis and Fan (2001, page 966). They observed that

$$\min_{\boldsymbol{\beta}_j} \frac{1}{2}\|\widehat{\boldsymbol{\beta}}_{-j} - \boldsymbol{\beta}_j\|^2 + p_\lambda(\|\boldsymbol{\beta}_j\|) = \min_r \left\{ \frac{1}{2} \min_{\|\boldsymbol{\beta}_j\|=r} \|\widehat{\boldsymbol{\beta}}_{-j} - \boldsymbol{\beta}_j\|^2 + p_\lambda(r) \right\}. \tag{3.115}$$

The inner bracket is minimized at $\widehat{\boldsymbol{\beta}}_{j,r} = r\widehat{\boldsymbol{\beta}}_{-j}/\|\widehat{\boldsymbol{\beta}}_{-j}\|$. Substituting this into (3.115), the problem becomes

$$\min_r \left\{ \frac{1}{2}(\|\widehat{\boldsymbol{\beta}}_{-j}\| - r)^2 + p_\lambda(r) \right\}. \tag{3.116}$$

Problem (3.116) is identical to problem (3.12), whose solution is denoted by $\widehat{\theta}(\|\widehat{\boldsymbol{\beta}}_{-j}\|)$. For the $L_1$, SCAD and MCP, the explicit solutions are given respectively by (3.18)–(3.20). With this notation, we have

$$\widehat{\boldsymbol{\beta}}_j = \frac{\widehat{\theta}(\|\widehat{\boldsymbol{\beta}}_{-j}\|)}{\|\widehat{\boldsymbol{\beta}}_{-j}\|} \widehat{\boldsymbol{\beta}}_{-j}. \tag{3.117}$$

In particular, for the $L_1$-penalty,

$$\widehat{\boldsymbol{\beta}}_j = \left(1 - \frac{\lambda}{\|\widehat{\boldsymbol{\beta}}_{-j}\|}\right)_+ \widehat{\boldsymbol{\beta}}_{-j},$$

and for the hard-thresholding penalty

$$\widehat{\boldsymbol{\beta}}_j = I(\|\widehat{\boldsymbol{\beta}}_{-j}\| \geq \lambda)\widehat{\boldsymbol{\beta}}_{-j}.$$

These formulas were given by Antoniadis and Fan (2001, page 966). They clearly show that the strength of the group estimates is pulled together to decide whether or not to keep a group of coefficients.

The groupwise orthogonality condition is in fact not natural and necessary to consider. Suppose that the condition holds for the data $(\mathbf{Y}_i, \mathbf{X}_i), 1 \leq i \leq n$. If we bootstrap the data or do cross-validation to selection $\lambda$, the groupwise orthogonality condition easily fails on the perturbed dataset. For computational considerations, the groupwise orthogonality condition is not needed for using the group descent algorithm. Several algorithms for solving the Lasso regression, such as ISTA, FISTA and ADMM, can be readily used to solve the group-Lasso regression with a general design matrix. We omit the details here.

## 3.9 Applications

We now illustrate high-dimensional statistical modeling using the monthly house price appreciations (HPA) for 352 counties in the United States. The housing price appreciation is computed based on monthly repeated sales.

These 352 counties have the largest repeated sales and hence their measurements are more reliable. The spatial correlations of these 352 HPAs, based on the data in the period from January 2000 to December 2009, are presented in Figure 1.4.

To take advantage of the spatial correlation in their prediction, Fan, Lv and Qi (2011) utilize the following high-dimensional time-series regression. Let $Y_t^i$ be the HPA in county $i$ at time $t$ and $\mathbf{X}_{i,t}$ be the observable factors that drive the market. In the application below, $\mathbf{X}_{i,t}$ will be taken as the national HPA, the returns of the national house price index that drives the overall housing markets. They used the following $s$-period ahead county-level forecast model:

$$Y_{t+s}^i = \sum_{j=1}^{p} b_{ij} Y_t^j + \mathbf{X}_{i,t}^T \boldsymbol{\beta}_i + \varepsilon_{t+s}^i, \quad i = 1, \ldots, p, \tag{3.118}$$

where $p = 352$ and $b_{ij}$ and $\boldsymbol{\beta}_i$ are regression coefficients. In this model, we allow neighboring HPAs to influence the future housing price, but we do not know which counties have such prediction power. This leads to the following PLS problem: For each given county $i$,

$$\min_{\{b_{ij}, j=1,\ldots,p, \boldsymbol{\beta}_i\}} \sum_{t=1}^{T-s} \Big( Y_{t+s}^i - \mathbf{X}_{i,t}^T \boldsymbol{\beta}_i - \sum_{j=1}^{p} b_{ij} Y_t^j \Big)^2 + \sum_{j=1}^{p} w_{ij} p_\lambda(|b_{ij}|),$$

where the weights $w_{ij}$ are chosen according to the geographical distances between counties $i$ and $j$. The weights are used to discourage HPAs from remote regions from being used in the prediction. The non-vanishing coefficients represent the selected neighbors that are useful for predicting HPA at county $i$.

Monthly HPA data from January 2000 to December 2009 were used to fit model (3.118) for each county with $s = 1$. The top panel of Figure 3.11 highlights the selected neighborhood HPAs used in the prediction. For each county $i$, only 3-4 neighboring counties are chosen on average, which is reasonable. Figure 3.11 (bottom left) presents the spatial correlations of the residuals using model (3.118). No pattern can be found, which indicates the spatial correlations have already been explained by the neighborhood HPAs. In contrast, if we ignore the neighborhood selection (namely, setting $b_{ij} = 0, \forall i \neq j$), which is a lower-dimensional problem and will be referred to as the OLS estimate, the spatial correlations of the residuals are visible (bottom right). This provides additional evidence on the effectiveness of the neighborhood selection by PLS.

We now compare the forecasting power of the PLS with OLS. The training sample covers the data for 2000.1-2005.12, and the test period is 2006.1-2009.12. Fan, Lv and Qi (2011) carried out prediction throughout the next 3 years in the following manner. For the short-term prediction horizons $s$ from 1 to 6 months, each month is predicted separately using model (3.118); for

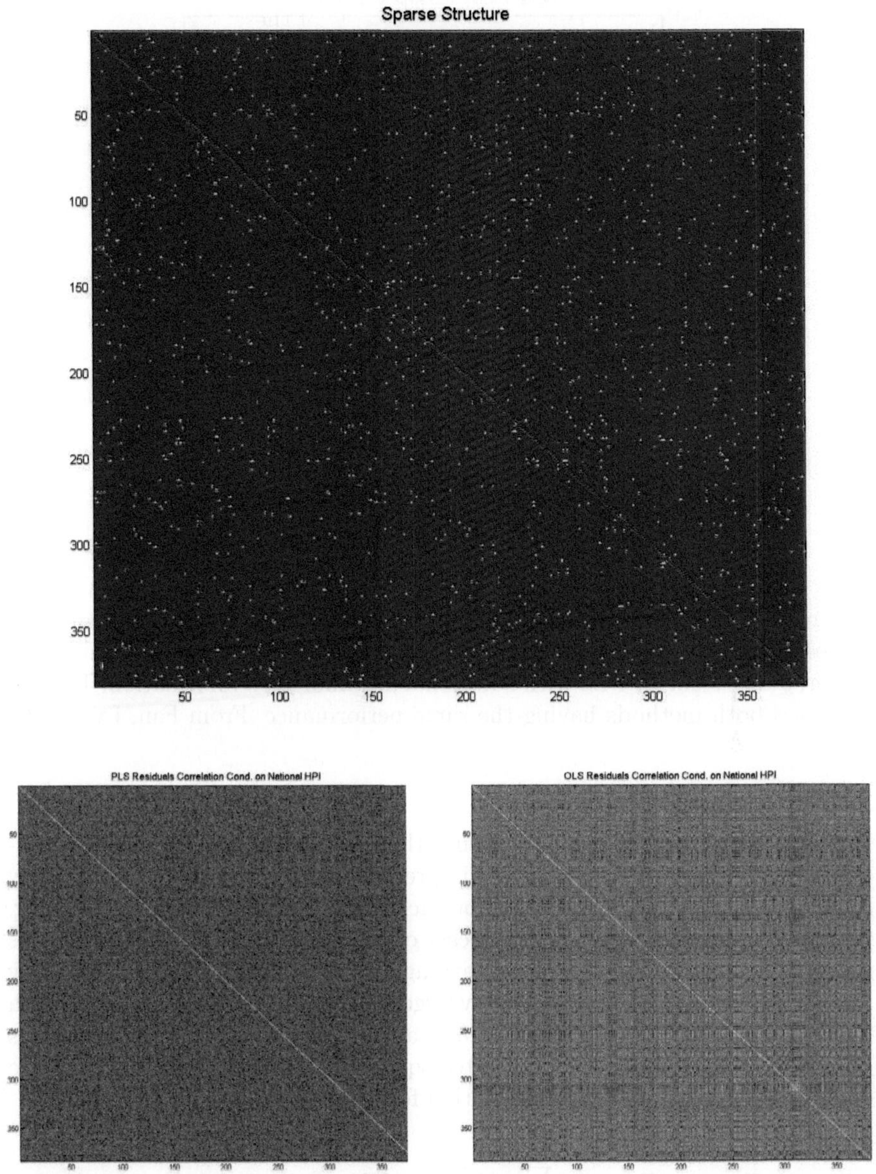


Figure 3.11: Top panel: Neighborhoods with non-zero regression coefficients: for each county $i$ in the $y$-axis, each row, i.e. $x$-axis indicates the neighborhood that has impact on the HPA in county $i$. Bottom left: Spatial-correlation of residuals with national HPA and neighborhood selection. Bottom right: Spatial-correlation of residuals using only national HPA as the predictor. Adapted from Fan, Lv and Qi (2011).

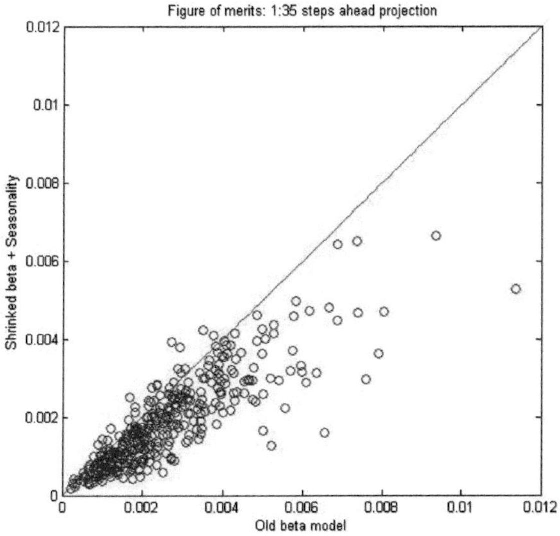


Figure 3.12: Aggregated forecast errors (3.119) in 36 months over 352 counties. For each dot, the $x$-axis represents prediction errors by OLS with only national factor, $y$-axis error by PLS with additional neighborhood information. The line indicates both methods having the same performance. From Fan, Lv and Qi (2011).

the time horizon of 7-36 months, only the average HPA over 6-month periods (e.g. months 7-12, 13-18, etc) is predicted. This increases the stability of the prediction. More precisely, for each of the 6 consecutive months (e.g. months 13-18), they obtained a forecast of average HPA during the 6 months using PLS with historical 6-month average HPAs as a training sample. They treated the (annualized) 6-month average as a forecast of the middle month of the 6-month period (e.g. month 15.5) and linearly interpolated the months in between. The discounted aggregated squared errors were used as a measure of overall performance of the prediction for county $i$:

$$\text{Forecast Error}_i = \sum_{s=1}^{\tau} \rho^s (\widehat{Y}^i_{T+s} - Y^i_{T+s})^2, \quad \rho = 0.95, \tag{3.119}$$

where $\tau$ is the time horizon to be predicted.

The results in Figure 3.12 show that over 352 counties, the sparse regression model (3.118) with neighborhood information performs on average 30% better in terms of prediction error than the model without using the neighborhood information. Figure 1.4 compares forecasts using OLS with only the national

HPA (blue) and PLS with additional neighborhood information (red) for the largest counties with the historical HPAs (black).

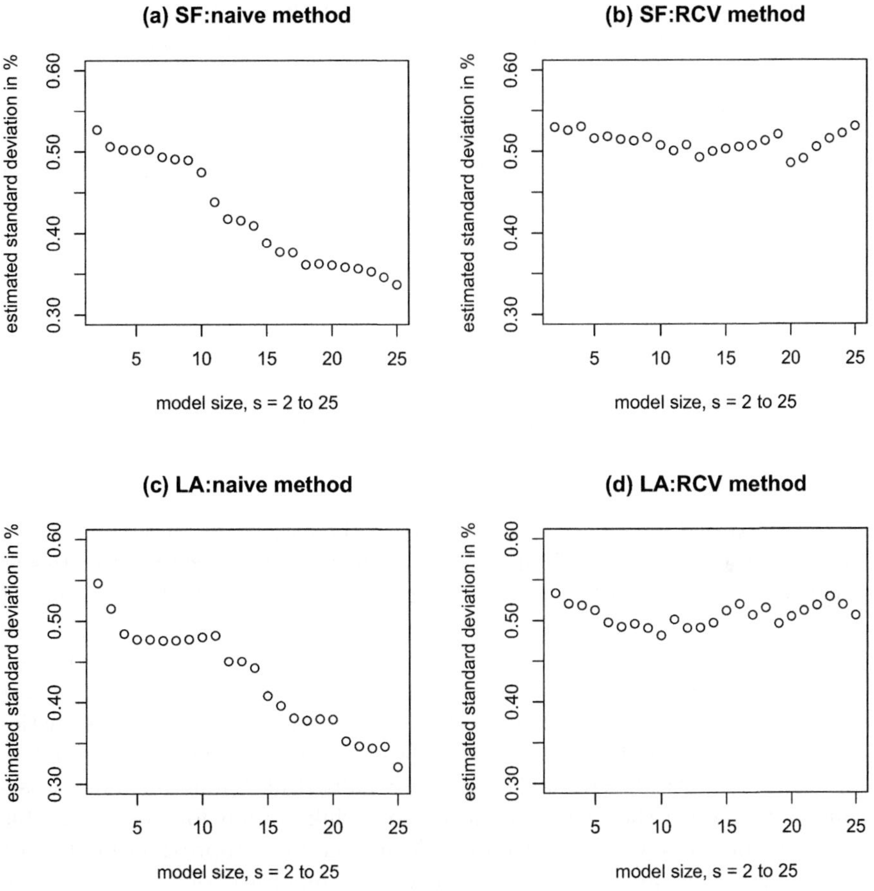


Figure 3.13: Estimated standard deviation $\sigma$ for one-step ahead forecast as a function of selected model size $s$ in both San Francisco (top panel) and Los Angeles (bottom panel) using both naive (left panel) and RCV (right panel) methods. Taken from Fan, Lv and Qi (2011).

How good is a prediction method? The residual standard deviation $\sigma$ provides a benchmark measure when the ideal prediction rule is used. To illustrate this, we estimate $\sigma$ for a one-step forecast in San Francisco and Los Angeles, using the HPA data from January 1998 to December 2005 (96 months). The RCV estimates, as a function of the selected model size $s$, are shown in Figure

3.13. The naive estimates, which compute directly the residual variances, decrease with $s$ due to spurious correlation. On the other hand, the RCV gives reasonably stable estimates for a range of selected models. The benchmarks of prediction errors for both San Francisco and Los Angeles regions are about .53%, comparing the standard deviations of month over month variations of HPAs 1.08% and 1.69%, respectively. In contrast, the rolling one-step prediction errors over 12 months in 2006 are .67% and .86% for the San Francisco and Los Angeles areas, respectively. They are clearly larger than the benchmark as expected, but considerably smaller than the standard deviations, which used no variables to forecast. They also show that some small room for improvements is the PLS are possible.

## 3.10 Bibliographical Notes

There are many other exciting developments on variable selection. We have no intention to give a comprehensive survey here. Instead, we focus only on some important inventions on penalized least-squares that led to the vast literature today.

The idea of $L_2$-regularization appears in the early work of Tikhonov regularization in the 1940's (Tikhonov, 1943). It was introduced to statistics as ridge regression by Hoerl (1962) and Hoerl and Kennard (1970). The concept of sparsity and $L_1$ penalty appeared in a series of work by David Donoho and Iain Johnstone (see e.g. Donoho and Johnstone, 1994). Penalized $L_1$-regression was employed by David Donoho and Shaobing Chen, in a technical report on "basis pursuit" in 1994, to select a basis from an over complete dictionary. It was then used by Tibshirani (1996) to conduct variable selection. Fan and Li (2001) introduced folded concave penalized likelihood including least-squares to reduce the biases in the Lasso shrinkage and for better variable selection consistency. They introduced local quadratic approximation to cast the optimization problem into a sequence of a quadratic optimization problems and established the oracle property. LARS was introduced by Efron *et al.* (2004) to efficiently compute the Lasso path. An early work on the asymptotic study of penalized least-squares (indeed, penalized likelihood) with diverging dimensionality was given by Fan and Peng (2004). Zou and Zhang (2009) introduced the adaptive Elastic Net and studied its properties under diverging dimensions. Zhao and Yu (2006), Meinshausen and Bühlmann (2006) and Zou (2006) gave irrepresentable conditions for model selection consistency of Lasso. Candés and Tao (2007) proposed the Dantzig selector, which can be cast as a linear program. Zhang (2010) introduced the PLUS algorithm for computing a solution path of a specific class of folded concave PLS including MCP and SCAD and established a strong oracle property. A family of folded concave penalties that bridge the $L_0$ and $L_1$ penalties was studied by Lv and Fan (2009). A thorough investigation of the properties of folded concave PLS when dimensionality diverges was given by Lv and Fan (2010). Meinshausen and Bühlmann (2010) proposed stability selection based on subsampling.

Belloni, Chernozhukov, and Wang (2011) proposed square-root Lasso for sparse regression. Negahban, Ravikumar, Wainwright and Yu (2012) proposed a unified analysis of high-dimensional $M$-estimators with decomposable regularizers. Agarwal, Negahban and Wainwright (2012) proposed the composite gradient descent algorithm and developed the sampling properties by taking computational error into consideration. Belloni, Chen, Chernozhukov, and Hansen (2012) investigated optimal instruments selection. The focussed GMM was proposed by Fan and Liao (2014) to solve endogeneity problems pandemic in high-dimensional sparse regression. Belloni and Chernozhukov (2013) investigated post-model selection estimators. Fan, Xue and Zou (2014) showed that the one-step LLA algorithm produces a strong oracle solution as long as the problem is localizable and regular. Loh and Wainwright (2014) developed statistical and algorithmic theory for local optima of regularized M-estimators with nonconvexity penalty. They showed surprisingly that all local optima will lie within statistical precision of the sparse true parameter vector. Support recovery without incoherence was investigated by Loh and Wainwright (2017) using nonconvex regularization.

There were many developments on robust regularization methods and quantile regression. For fixed dimensionality variable selection, see, for example, Wang, Li and Jiang (2007), Li and Zhu (2008), Zou and Yuan (2008), and Wu and Liu (2009). The penalized composite likelihood method was proposed in Bradic, Fan, and Wang (2011) for improvement of the efficiency of Lasso in high dimensions. Belloni and Chernozhukov (2011) showed that the $L_1$-penalized quantile regression admits the near-oracle rate and derived bounds on the size of the selected model, uniformly in a compact set of quantile indices. Bounds on the prediction error were derived in van de Geer and Müller (2012) for a large class of $L_1$ penalized estimators, including quantile regression. Wang, Wu and Li (2012) showed that the oracle estimate belongs to the set of local minima of the nonconvex penalized quantile regression. Fan, Fan and Barut (2014) proposed and studied the adaptive robust variable selection. Fan, Li, and Wang (2017) considered estimating high-dimensional mean regression using adaptive Huber loss, which assumes only the second moment condition on the error distribution. Sun, Zhou, and Fan (2017) weakened the second moment condition to $(1+\delta)$-moment and unveiled optimality and phase transition for the adaptive Lasso. Loh (2017) investigated theoretical properties of regularized robust M-estimators, applicable for data contaminated by heavy-tailed distributions and/or outliers in the noises and covariates.

## 3.11 Exercises

3.1 Let $g(\theta|z,\lambda) = \frac{1}{2}(z-\theta)^2 + p_\lambda(|\theta|)$ for a given $\lambda$ and $z$, and denote $\widehat{\theta}(z|\lambda) = \operatorname{argmin}_\theta g(\theta|z,\lambda)$.

(a) Let $p_\lambda(|\theta|) = \frac{\lambda^2}{2} I(|\theta| \neq 0)$, the $L_0$-penalty. Show that $\widehat{\theta}_H(z|\lambda) = zI(|z| \geq \lambda)$, the hard thresholding rule.

(b) Let $p_\lambda(|\theta|) = \frac{1}{2}\lambda^2 - \frac{1}{2}(\lambda - \theta)_+^2$, the hard-thresholding penalty defined in (3.15). Show that $\widehat{\theta}_H(z|\lambda) = zI(|z| \geq \lambda)$. This implies that different penalty functions may result in the same penalized least squares solution.

(c) Comment upon the advantages of the hard-thresholding penalty over the $L_0$-penalty.

(d) Let $p_\lambda(|\theta|) = \lambda|\theta|$, the $L_1$-penalty. Show that $\widehat{\theta}_S(z|\lambda) = \mathrm{sgn}(z)(|z| - \lambda)_+$, the soft-thresholding rule. Compare with the $L_0$-penalty; the regularization parameter $\lambda$s in different penalty functions may be in different scale.

(e) Let $p_\lambda(\theta) = \lambda\{(1-\alpha)\theta^2 + \alpha|\theta|\}$ with a fixed $\alpha$. Derive the close-form solution of $\widehat{\theta}(z|\lambda)$, which is the Elastic Net thresholding rule.

3.2 Let $g(\theta|z, \lambda) = \frac{1}{2}(z - \theta)^2 + p_\lambda(|\theta|)$ for a given $\lambda$ and $z$, and denote $\widehat{\theta}(z|\lambda) = \mathrm{argmin}_\theta\, g(\theta|z, \lambda)$. Following the convention, let $p'_\lambda(0) = p'_\lambda(0+)$. Assume that $p_\lambda(\theta)$ is nondescreasing and continuously differentiable on $[0, \infty)$, and the function $-\theta - p'_\lambda(\theta)$ is strictly unimodal on $[0, \infty)$.

(a) Show that if $t_0 = \min_{\theta \geq 0}\{\theta + p'_\lambda(\theta)\} > 0$, then $\widehat{\theta}(z|\lambda) = 0$ when $z \leq t_0$. This leads to the sparsity of $\widehat{\theta}(z|\lambda)$.

(b) Show that if $p'_\lambda(|\theta|) = 0$ for $|\theta| \geq t_1$, then $\widehat{\theta}(z|\lambda) = z$ for $|\theta| \geq t_1$ with large $t_1$. This leads to the unbiasedness of $\widehat{\theta}(z|\lambda)$.

(c) Show that $\widehat{\theta}(z|\lambda)$ is continuous in $z$ if and only if $\mathrm{argmin}_{\theta \geq 0}\{\theta + p'_\lambda(\theta)\} = 0$. This leads to the continuity of $\widehat{\theta}(z|\lambda)$.

3.3 Let $g(\theta|z, \lambda, \sigma) = \frac{1}{2\sigma^2}(z - \theta)^2 + p_\lambda(|\theta|)$ for a given $\lambda$, $z$ and $\sigma$, and denote $\widehat{\theta}(z|\lambda, \sigma) = \mathrm{argmin}_\theta\, g(\theta|z, \lambda, \sigma)$.

(a) Take the penalty function to be the SCAD penalty whose derivative is given in (3.14). Derive the expressive form solution of $\widehat{\theta}(z|\lambda, \sigma)$.

(b) Take the penalty function to be the MCP whose derivative is given in (3.17). Derive the closed form solution $\widehat{\theta}(z|\lambda, \sigma)$.

(c) Comment upon how $\lambda$ relates to $\sigma$ so that the solutions in (a) and (b) still have sparsity, unbiasedness and continuity.

3.4 Suppose that $Z \sim N(\theta, \sigma^2)$. Derive the closed form of risk function $R(\theta) = E(\widehat{\theta}(Z) - \theta)^2$ for the hard-thresholding rule (3.16), the soft-thresholding rule (3.18), the SCAD thresholding rule given (3.19) and the MCP thresholding rule (3.20), respectively. Plot $R(\theta)$ against $\theta$ with $\sigma = 2$ and 3, and compare your plot with Figure 3.3.

3.5 Consider the Lasso problem $\min_{\boldsymbol{\beta}} \frac{1}{2n}\|\mathbf{Y} - \mathbf{X}\boldsymbol{\beta}\|_2^2 + \lambda\|\boldsymbol{\beta}\|_1$, where $\lambda > 0$ is a tuning parameter.

(a) If $\widehat{\boldsymbol{\beta}}_1$ and $\widehat{\boldsymbol{\beta}}_2$ are both minimizers of the Lasso problem, show that they have the same prediction, i.e., $\mathbf{X}\widehat{\boldsymbol{\beta}}_1 = \mathbf{X}\widehat{\boldsymbol{\beta}}_2$. **Hint:** Consider the vector $\alpha\widehat{\boldsymbol{\beta}}_1 + (1-\alpha)\widehat{\boldsymbol{\beta}}_2$ for $\alpha \in (0,1)$.

(b) Let $\widehat{\boldsymbol{\beta}}$ be a minimizer of the Lasso problem with $j^{th}$ component $\widehat{\beta}_j$. Denote $\mathbf{X}_j$ to be the $j$-th column of $\mathbf{X}$. Show that

$$\begin{cases} \lambda = n^{-1}\mathbf{X}_j^T(\mathbf{Y} - \mathbf{X}\widehat{\boldsymbol{\beta}}) & \text{if } \widehat{\beta}_j > 0; \\ \lambda = -n^{-1}\mathbf{X}_j^T(\mathbf{Y} - \mathbf{X}\widehat{\boldsymbol{\beta}}) & \text{if } \widehat{\beta}_j < 0; \\ \lambda \geq |n^{-1}\mathbf{X}_j^T(\mathbf{Y} - \mathbf{X}\widehat{\boldsymbol{\beta}})| & \text{if } \widehat{\beta}_j = 0. \end{cases}$$

(c) If $\lambda > \|n^{-1}\mathbf{X}^T\mathbf{Y}\|_\infty$, prove that $\widehat{\boldsymbol{\beta}}_\lambda = \mathbf{0}$, where $\widehat{\boldsymbol{\beta}}_\lambda$ is the minimizer of the Lasso problem with regularization parameter $\lambda$.

3.6 Verify the KKT conditions in (3.22), (3.23) and (3.24) for penalized least squares.

3.7 Consider the Elastic-Net penalty $p(\theta) = \lambda_1|\theta| + \lambda_2\theta^2$ with $\lambda_2 > 0$. Let $\widehat{\boldsymbol{\beta}}$ be the minimizer of $\frac{1}{2n}\|\mathbf{Y} - \mathbf{X}\boldsymbol{\beta}\|_2^2 + \sum_{j=1}^p p(|\beta_j|)$.

(a) Show that $\widehat{\boldsymbol{\beta}}$ is unique.

(b) Give the necessary and sufficient conditions for $\widehat{\boldsymbol{\beta}}$ being the penalized least-squares solution.

(c) If $\lambda_1 > \|n^{-1}\mathbf{X}^T\mathbf{Y}\|_\infty$, show that $\widehat{\boldsymbol{\beta}} = \mathbf{0}$.

3.8 Concentration inequalities.

(a) The random vector $\boldsymbol{\varepsilon} \in \mathbb{R}^n$ is called $\sigma$-sub-Gaussian if $E\exp\left(\mathbf{a}^T\boldsymbol{\varepsilon}\right) \leq \exp\left(\|\mathbf{a}\|_2^2\sigma^2/2\right)$, $\forall \mathbf{a} \in \mathbb{R}^n$. Show that $E\boldsymbol{\varepsilon} = \mathbf{0}$ and $\operatorname{Var}(\boldsymbol{\varepsilon}) \leq \sigma^2\mathbf{I}_n$. **Hint:** Expand exponential functions as infinite series.

(b) For $\mathbf{X} \in \mathbb{R}^{n\times p}$ with the $j$-th column denoted by $\mathbf{X}_j \in \mathbb{R}^n$, suppose that $\|\mathbf{X}_j\|_2^2 = n$ for all $j$, and $\boldsymbol{\varepsilon} \in \mathbb{R}^n$ is a $\sigma$-sub-Gaussian random vector. Show that there exists a constant $C > 0$ such that

$$P\Big(\|n^{-1}\mathbf{X}^T\boldsymbol{\varepsilon}\|_\infty > \sqrt{2(1+\delta)}\sigma\sqrt{\frac{\log p}{n}}\Big) \leq Cp^{-\delta}, \qquad \forall \delta > 0.$$

3.9 The goal is to show the concentration inequality for the median-of-means estimator when the random variable only has finite second moment. We divide the problem into three simple steps.

(a) Let $X$ be a random variable with $\operatorname{E}X = \mu < \infty$ and $\operatorname{Var}(X) = \sigma^2 < \infty$. Suppose that we have $m$ i.i.d. random samples $\{X_i\}_{i=1}^m$ with the same distribution as $X$. Let $\widehat{\mu}_m = \frac{1}{m}\sum_{i=1}^m X_i$. Show that

$$P\big(|\widehat{\mu}_m - \mu| \geq \frac{2\sigma}{\sqrt{m}}\big) \leq \frac{1}{4}.$$

(b) Given $k$ i.i.d. Bernoulli random variables $\{B_j\}_{j=1}^k$ with $\mathrm{E}\,B_j = p < \frac{1}{2}$, use the moment generating function of $B_j$, i.e., $\mathrm{E}(\exp(tB_j))$, to show that

$$P\Big(\frac{1}{k}\sum_{j=1}^{k} B_j \geq \frac{1}{2}\Big) \leq (4p(1-p))^{\frac{k}{2}}.$$

(c) Suppose that we have $n$ i.i.d. random samples $\{X_i\}_{i=1}^n$ from a population with mean $\mu < \infty$ and variance $\sigma^2 < \infty$. For any positive integer $k$, we randomly and uniformly divide all the samples into $k$ subsamples, each having size $m = n/k$ (for simplicity, we assume $n$ is always divisible by $k$). Let $\widehat{\mu}_j$ be the sample average of the $j^{th}$ subsample and $\widetilde{m}$ be the median of $\{\widehat{\mu}_j\}_{j=1}^k$. Apply the previous two results to show that

$$P\Big(|\widetilde{m} - \mu| \geq 2\sigma\sqrt{\frac{k}{n}}\Big) \leq \Big(\frac{\sqrt{3}}{2}\Big)^k.$$

Hint: Consider the Bernoulli random variable $B_j = \mathbb{1}\{|\widehat{\mu}_j - \mu| \geq 2\sigma\sqrt{\frac{k}{n}}\}$ for $j = 1, ..., k$.

3.10 This problem intends to show that the gradient descent method for a convex function $f(\cdot)$ is a member of majorization-minimization algorithms and has a sublinear rate of convergence in terms of function values. From now on, the function $f(\cdot)$ is convex and let $\mathbf{x}^* \in \mathrm{argmin} f(\mathbf{x})$. Here we implicitly assume the minimum can be attained at some point $\mathbf{x}^\star \in \mathbb{R}^p$.

(a) Suppose that $f''(\mathbf{x}) \leq L\mathbf{I}_p$ and $\delta \leq 1/L$. Show that the quadratic function $g(\mathbf{x}) = f(\mathbf{x}_{i-1}) + f'(\mathbf{x}_{i-1})^T(\mathbf{x} - \mathbf{x}_{i-1}) + \frac{1}{2\delta}\|\mathbf{x} - \mathbf{x}_{i-1}\|^2$ is a majorization of $f(\mathbf{x})$ at point $\mathbf{x}_{i-1}$, i.e., $g(\mathbf{x}) \geq f(\mathbf{x})$ for all $\mathbf{x}$ and also $g(\mathbf{x}_{i-1}) = f(\mathbf{x}_{i-1})$.

(b) Show that gradient step $\mathbf{x}_i = \mathbf{x}_{i-1} - \delta f'(\mathbf{x}_{i-1})$ is the minimizer of the majorized quadratic function $g(\mathbf{x})$ and hence the gradient descent method can be regarded as a member of MM-algorithms.

(c) Use (a) and the convexity of $f(\cdot)$ to show that

$$f(\mathbf{x}_i) \leq f(\mathbf{x}^*) + \frac{1}{2\delta}(\|\mathbf{x}_{i-1} - \mathbf{x}^*\|^2 - \|\mathbf{x}^* - \mathbf{x}_i\|^2).$$

(d) Conclude using (c) that $f(\mathbf{x}_k) - f(\mathbf{x}^*) \leq \|\mathbf{x}_0 - \mathbf{x}^*\|^2/(2k\delta)$, namely gradient descent converges at a sublinear rate. (**Note**: The gradient descent method converges linearly if $f(\cdot)$ is strongly convex.)

3.11 Conduct a numerical comparison among the quadratic programming algorithm (3.57), the LARS algorithm, the cyclic coordinate descent algorithms and the ADMM algorithm for Lasso.

3.12 Conduct a numerical comparison between the Lasso and the one-step SCAD estimator using LLA with Lasso initial values.

3.13 Show that when the dimension of $\mathbf{x}$ is finite and fixed, the SCAD with GCV-tuning parameter selector defined in (3.98) leads to an overfitted model. That is, the selected model contains all important variables, but with a positive probability, the selected model contains some unimportant variables. See, Wang, Li and Tsai (2007).

3.14 Show that when the dimension of $\mathbf{x}$ is finite and fixed, the SCAD with BIC-tuning parameter selector defined in (3.99) yields a consistent estimation of the model.

3.15 Show that $\widehat{\sigma}_1^2$ defined in (3.102) is a root-$n$ consistent estimator of $\sigma^2$.

3.16 Extend the ADMM algorithm in Section 3.5.9 for group-Lasso regression.

3.17 Extend the ISTA algorithm in Section 3.5.7 for penalized least squares with group penalty, and further apply the new algorithm for variable selection in varying coefficient models in (3.107).

3.18 Let us consider the 128 macroeconomic time series from Jan. 1959 to Dec. 2018, which can be downloaded from the book website. In this problem, we will explore what macroeconomic variables are associated with the unemployment rate contemporarily and which macroeconomic variables lead the unemployment rates.

(a) Extract the data from Jan. 1960 to Oct. 2018 (in total 706 months) and remove the feature named "sasdate". Then, remove the features with missing entries and report their names.

(b) The column with name "UNRATE" measures the difference in unemployment rate between the current month and the previous month. Take this column as the response and take the remaining variables as predictors. To conduct contemporary association studies, do the following steps for Lasso (using R package `glmnet`) and SCAD (using R package `ncvreg`): Set a random seed by `set.seed(525)`; Plot the regularization paths as well as the mean squared errors estimated by 10-fold cross-validation; Choose a model based on cross-validation, report its in-sample $R^2$, and point out two most important macroeconomic variables that are correlated with the current change of unemployment rate and explain why you choose them.

(c) In this sub-problem, we are going to study which macroeconomic variables are leading indicators for the changes of future unemployment rate. To do so, we will pair each row of predictors with the next row of response. The last row of predictors and the first element in the response are hence discarded. After this change, do the same exercise as (b).

(d) Consider the setting of (c). Leave the last 120 months as testing data and use the rest as training data. Set a random seed by `set.seed(525)`. Run Lasso and SCAD on the training data using `glmnet` and `ncvreg`, respectively, and choose a model based on 10-fold cross-validation. Compute the out-of-sample $R^2$'s for predicting the changes of the future unemployment rates.

3.19 Consider the Zillow data analyzed in Exercise 2.9. We drop the first 3 columns ("(empty)", "id", "date") and treat "zipcode" as a factor variable. Now, consider the variables

(a) "bedrooms", "bathrooms", "sqft_living", and "sqft_lot" and their interactions and the remaining 15 variables in the data, including "zipcode".

(b) "bedrooms", "bathrooms", "sqft_living", "sqft_lot" and "zipcode", and their interactions and the remaining 14 variables in the data. (We can use *model.matrix* to expand factors into a set of dummy variables.)

(c) Add the following additional variables to (b): $X_{12} = I(view == 0)$, $X_{13} = L^2$, $X_{13+i} = (L - \tau_i)_+^2$, $i = 1, \cdots, 9$, where $\tau_i$ is $10 * i^{th}$ percentile and $L$ is the size of living area ("sqft_living").

Compute and compare out-of-sample $R^2$ using ridge regression, Lasso, SCAD with regularization parameter chosen by 10 fold cross-validation.

Chapter 4

# Penalized Least Squares: Properties

This chapter describes properties of PLS methods in linear regression models with a large number of covariate variables. We will study the performance of such methods in prediction, coefficient estimation, and variable selection under proper regularity conditions on the noise level, the sparsity of regression coefficients, and the covariance of covariate variables. To make reading easier, we defer some lengthier proofs to the end of each section and make this chapter a self-contained chapter, despite some repetition and slightly modified notation.

## 4.1 Performance Benchmarks

This section provides a general description of theoretical objectives of penalized least squares estimation along with some basic concepts and terminologies for studying such methods in high-dimension.

Suppose we observe covariates $X_{ij}, 1 \le j \le p$, and a response variable $Y_i$ from the $i$-th data point in the sample, $i = 1, \ldots, n$. As in the *linear regression model* (2.1), the covariates and response variable satisfy the relationship

$$Y_i = \sum_{j=1}^{p} X_{ij}\beta_j + \varepsilon_i,$$

In vector notation, it is written as

$$\mathbf{Y} = \mathbf{X}\boldsymbol{\beta} + \boldsymbol{\varepsilon}. \tag{4.1}$$

See Section 2.1. Unless otherwise stated, the design matrix $\mathbf{X}$ is considered as deterministic. Of course, in the case of random design, the noise vector $\boldsymbol{\varepsilon}$ is assumed to be independent of $\mathbf{X}$.

We are interested in the performance of PLS in the case of large $p$, including $p \gg n$. Thus, unless otherwise stated, $\{p, \mathbf{X}, \boldsymbol{\beta}, \boldsymbol{\varepsilon}\}$ are all allowed to depend on $n$ and $p = p_n \to \infty$ as $n \to \infty$.

We denote by $\mathbf{X}_j$ the $j$-th column $(X_{1j}, \ldots, X_{nj})^T$ of the design, $\mathbf{X}_A = (\mathbf{X}_j, j \in A)$ the sub-matrix of the design with variables in $A$ for subsets $A \subseteq \{1, \ldots, p\}$, $\overline{\boldsymbol{\Sigma}} = \mathbf{X}^T\mathbf{X}/n$ the normalized Gram matrix, $\overline{\boldsymbol{\Sigma}}_{A,B} = \mathbf{X}_A^T\mathbf{X}_B/n$

its subblocks, and $\mathbf{P}_A$ the orthogonal projection to the range of $\mathbf{X}_A$. Likewise, we denote by $\mathbf{v}_A = (v_j, j \in A)^T$ the subvector of $\mathbf{v} = (v_1, \ldots, v_p)^T$ with indices in $A$. We denote by $\|\mathbf{v}\|_q = \{\sum_{i=1}^n |v_i|^q\}^{1/q}$ the $\ell_q$ norm for $1 \le q < \infty$, with the usual extension $\|\mathbf{v}\|_\infty = \max_i |v_i|$, $\mathrm{supp}(\mathbf{v}) = \{j : v_j \neq 0\}$ the support of a vector $\mathbf{v}$, and $\|\mathbf{v}\|_0 = \#\{j : v_j \neq 0\}$ the size of the support. Denote by $\phi_{\min}(\mathbf{M})$ and $\phi_{\max}(\mathbf{M})$ the smallest and largest singular values of a matrix $\mathbf{M}$, respectively.

Let $\mathbf{A} = (a_{ij})$ be an $m \times n$ matrix. We use $\|\mathbf{A}\|_{q,r}$ to denote the $\ell_q$ to $\ell_r$ operator norm:

$$\|\mathbf{A}\|_{q,r} = \max_{\|\mathbf{u}\|_q = 1} \|\mathbf{A}\mathbf{u}\|_r. \tag{4.2}$$

When $q = r$, it is denoted as $\|\mathbf{A}\|_q$. In particular, $\|\mathbf{A}\|_{2,2} = \phi_{\max}(\mathbf{A}^T\mathbf{A})^{1/2}$ and will also be denoted as $\|\mathbf{A}\|$ and

$$\|\mathbf{A}\|_1 = \max_{1\le j\le n} \sum_{i=1}^m |a_{ij}| \quad \text{and} \quad \|\mathbf{A}\|_\infty = \max_{1\le i\le m} \sum_{j=1}^n |a_{ij}| \tag{4.3}$$

are the maximum $L_1$-norm of columns and rows, respectively. In particular, for a symmetric matrix $\mathbf{A}$, $\|\mathbf{A}\|_1 = \|\mathbf{A}\|_\infty$. It also holds that

$$\|\mathbf{A}\|_2 \le \sqrt{\|\mathbf{A}\|_1 \|\mathbf{A}\|_\infty}. \tag{4.4}$$

### *4.1.1 Performance measures*

Here we describe performance measures for prediction, coefficient estimation and variable selection. The goal of prediction is to estimate the response $Y$ at a future design point $(X_1, \ldots, X_p)$, where $(X_1, \ldots, X_p, Y)$ is assumed to be independent of the data $(\mathbf{X}, \mathbf{Y})$ in (4.1) but follows the same model,

$$Y = \sum_{j=1}^p X_j \beta_j + \varepsilon, \qquad \text{with} \quad \mathrm{E}\,\varepsilon = 0.$$

We measure the performance of a predictor $\widehat{\boldsymbol\beta}$ by the *mean squared prediction error,*

$$\mathrm{E}\left[\left(Y - \sum_{j=1}^p X_j \widehat\beta_j\right)^2 \middle| \widehat{\boldsymbol\beta}\right].$$

It follows from the independence between the future observations and the current data that this error measure can be decomposed as

$$\mathrm{E}\left[\left(Y - \sum_{j=1}^p X_j \widehat\beta_j\right)^2 \middle| \widehat{\boldsymbol\beta}\right] = \mathrm{E}\,\varepsilon^2 + \mathrm{E}\left[\left(\sum_{j=1}^p X_j \widehat\beta_j - \sum_{j=1}^p X_j \beta_j\right)^2 \middle| \widehat{\boldsymbol\beta}\right]$$

Thus, the true $\boldsymbol{\beta}$ is an optimal predictor, and minimizing the mean squared prediction error is equivalent to minimizing

$$R_{\text{pred}}(\widehat{\boldsymbol{\beta}}, \boldsymbol{\beta}; X_1, \ldots, X_p) = \left( \sum_{j=1}^{p} X_j \widehat{\beta}_j - \sum_{j=1}^{p} X_j \beta_j \right)^2.$$

The above quantity can be viewed as *prediction regret* of not knowing $\boldsymbol{\beta}$ at design point $(X_1, \ldots, X_p)$, because it is the difference between the mean squared prediction errors of a predictor $\widehat{\boldsymbol{\beta}}$ and the unknown optimal predictor $\boldsymbol{\beta}$.

For simplicity, we typically further assume that the future design point resembles the design points in the current data in the following sense. For deterministic designs, we assume unless otherwise specified that $(X_1, \ldots, X_p)$ is equally likely to be any of the $n$ current design points $(X_{i1}, \ldots, X_{ip})$ in (4.1), so that the expected prediction regret given the data $(\mathbf{X}, \mathbf{Y})$ is

$$R_{\text{pred}}(\widehat{\boldsymbol{\beta}}, \boldsymbol{\beta}) = \frac{1}{n} \sum_{i=1}^{n} R_{\text{pred}}(\widehat{\boldsymbol{\beta}}; X_{i1}, \ldots, X_{ip}) = \|\mathbf{X}\widehat{\boldsymbol{\beta}} - \mathbf{X}\boldsymbol{\beta}\|_2^2 / n. \tag{4.5}$$

We will simply call this quantity (4.5) *prediction error*. For *random designs*, we assume that $\mathbf{X}$ has independent and identically distributed (*iid*) rows from a population with a second moment structure $\boldsymbol{\Sigma} = (\mathrm{E}\, X_{ij} X_{ik})_{p \times p}$. In this case, we assume that the future design point comes from the same population, so that the expected prediction regret given $(\mathbf{X}, \mathbf{Y})$ is

$$\begin{aligned} R_{\text{pred}}(\widehat{\boldsymbol{\beta}}, \boldsymbol{\beta}) &= \mathrm{E}\left[ R_{\text{pred}}(\widehat{\boldsymbol{\beta}}, \boldsymbol{\beta}; X_1, \ldots, X_p) \Big| \widehat{\boldsymbol{\beta}} \right] \\ &= (\widehat{\boldsymbol{\beta}} - \boldsymbol{\beta})^T \boldsymbol{\Sigma} (\widehat{\boldsymbol{\beta}} - \boldsymbol{\beta}). \end{aligned} \tag{4.6}$$

As in Section 3.3.2, an estimator $\widehat{\boldsymbol{\beta}}$ is *persistent* if (4.5) or (4.6) converges to zero respectively for deterministic and random designs (Greenshtein and Ritov, 2004).

We can also measure estimation performance of $\widehat{\boldsymbol{\beta}}$ with the $\ell_q$ *estimation error*

$$\|\widehat{\boldsymbol{\beta}} - \boldsymbol{\beta}\|_q = \left( \sum_{j=1}^{p} |\widehat{\beta}_j - \beta_j|^q \right)^{1/q}.$$

This quantity is closely related to the prediction regret $R_{\text{pred}}(\widehat{\boldsymbol{\beta}}, \boldsymbol{\beta}; X_1, \ldots, X_p)$ through the duality between the $\ell_q$ and $\ell_{q/(q-1)}$ norms as

$$\|\widehat{\boldsymbol{\beta}} - \boldsymbol{\beta}\|_q^2 = \max\left\{ \left( \sum_{j=1}^{p} X_j \widehat{\beta}_j - \sum_{j=1}^{p} X_j \beta_j \right)^2 : \sum_{j=1}^{p} |X_j|^{q/(q-1)} \le 1 \right\} \tag{4.7}$$

is the maximum prediction regret for a deterministic future design vector in the unit $\ell_{q/(q-1)}$ ball. In this sense, the $\ell_q$ estimation error is a conservative prediction error without assuming the resemblance of the future and current design points. For random designs, the persistency property with (4.6) is equivalent to the convergence of the $\ell_2$ estimation error to zero, $\|\widehat{\boldsymbol\beta} - \boldsymbol\beta\|_2 \to 0$, when the eigenvalues of the population covariance matrix $\boldsymbol\Sigma$ are uniformly bounded away from zero and infinity.

The problem of variable selection is essentially the estimation of the *support set* of the true $\boldsymbol\beta$, or equivalently the *true model*,

$$\text{supp}(\boldsymbol\beta) = \big\{j : \beta_j \neq 0\big\}.$$

Here are some commonly used performance measures for variable selection.

**Definition 4.1** *Loss functions for variable selection.*

- *Incorrect sign selection:* $I\{\text{sgn}(\widehat{\boldsymbol\beta}) \neq \text{sgn}(\boldsymbol\beta)\}$ *with* $\text{sgn}(0) = 0$
- *Incorrect selection:* $I\{\text{supp}(\widehat{\boldsymbol\beta}) \neq \text{supp}(\boldsymbol\beta)\}$
- *False positive:* $\text{FP}(\widehat{\boldsymbol\beta}) = |\text{supp}(\widehat{\boldsymbol\beta}) \setminus \text{supp}(\boldsymbol\beta)|$
- *False negative:* $\text{FN}(\widehat{\boldsymbol\beta}) = |\text{supp}(\boldsymbol\beta) \setminus \text{supp}(\widehat{\boldsymbol\beta})|$
- *Total miss:* $\text{TM}(\widehat{\boldsymbol\beta}) = \text{FP}(\widehat{\boldsymbol\beta}) + \text{FN}(\widehat{\boldsymbol\beta})$
- *Model size:* $\|\widehat{\boldsymbol\beta}\|_0 = |\text{supp}(\widehat{\boldsymbol\beta})|$
- *Family wise error rate* (FWER)*:* $P\{\text{FP}(\widehat{\boldsymbol\beta}) > 0\} = P\{\exists j : \widehat\beta_j \neq 0 = \beta_j\}$
- *Per comparison error rate* (PCER)*:* $E\{\text{FP}(\widehat{\boldsymbol\beta})\}/p = \sum_{\{j:\beta_j\neq 0\}} P\{\widehat\beta_j \neq 0\}/p$
- *False discovery rate* (FDR)*:* $\text{E}\{\text{FP}(\widehat{\boldsymbol\beta})/\max(1, \|\widehat{\boldsymbol\beta}\|_0)\}$

The expectation of the false positive, false negative, and total miss can be all expressed as sums of the Type-I or Type-II errors of tests $I\{\widehat\beta_j \neq 0\}$ for the hypotheses $H_j : \beta_j = 0$. As the error probabilities of these individual tests may not be easy to track, the false positive and false negative are considered instead. Among other performance measures above, the most stringent one is the correct sign selection. It follows easily from their definitions that

$$\begin{aligned}\text{PCER} &\leq \text{FDR} \\ &\leq \text{FWER} \\ &\leq P\big\{\text{supp}(\widehat{\boldsymbol\beta}) \neq \text{supp}(\boldsymbol\beta)\big\} \\ &\leq P\big\{\text{sgn}(\widehat{\boldsymbol\beta}) \neq \text{sgn}(\boldsymbol\beta)\big\}.\end{aligned}$$

As the three smaller error measures only control Type-I errors in testing $H_j$, we will focus on the two larger error measures. An estimator $\widehat{\boldsymbol\beta}$ is *sign-consistent* if

$$\lim_{n\to\infty} P\Big\{\text{sgn}(\widehat{\boldsymbol\beta}) = \text{sgn}(\boldsymbol\beta)\Big\} = 1. \tag{4.8}$$

A slightly weaker criterion is the *variable selection consistency*:

$$\lim_{n\to\infty} P\Big\{\text{supp}(\widehat{\boldsymbol\beta}) = \text{supp}(\boldsymbol\beta)\Big\} = 1. \tag{4.9}$$

A concept closely related to variable selection consistency (4.9) is *oracle property*. When the true $\boldsymbol{\beta}$ is sparse with a small support set $\mathcal{S} = \text{supp}(\boldsymbol{\beta})$, the *oracle LSE* (oracle least squares estimator), denoted by $\widehat{\boldsymbol{\beta}}^o$, is defined by

$$\widehat{\boldsymbol{\beta}}^o_{\mathcal{S}} = (\mathbf{X}_{\mathcal{S}}^T\mathbf{X}_{\mathcal{S}})^{-1}\mathbf{X}_{\mathcal{S}}^T\mathbf{Y}, \quad \widehat{\boldsymbol{\beta}}^o_{\mathcal{S}^c} = 0. \tag{4.10}$$

This "estimator" is constructed with the aim of an oracle expert with the knowledge of $\mathcal{S}$, but is not available to the statistician (Fan and Li, 2001). We may say that an estimator $\widehat{\boldsymbol{\beta}}$ has the oracle property if it is selection consistent, as statistical procedures based on the selected model would have a high probability of being identical to the same procedure in the true model $\mathcal{S}$. We may also say that an estimator $\widehat{\boldsymbol{\beta}}$ has the oracle property if

$$\sup_{\mathbf{a}} P\Big\{|\mathbf{a}^T(\widehat{\boldsymbol{\beta}} - \widehat{\boldsymbol{\beta}}^o)|^2 > \epsilon_n^2 \operatorname{Var}(\mathbf{a}^T\widehat{\boldsymbol{\beta}}^o)\Big\} \le \epsilon_n \tag{4.11}$$

for a certain $\epsilon_n \to 0$. This oracle property, amongst the weakest version of such, implies that for the estimation of linear functions of $\boldsymbol{\beta}$, confidence intervals and other commonly used statistical inference procedures based on $\widehat{\boldsymbol{\beta}}$ would have nearly the same performance as those based on the oracle estimator $\widehat{\boldsymbol{\beta}}^o$.

### *4.1.2 Impact of model uncertainty*

What is the cost of not knowing the true model $\mathcal{S} = \text{supp}(\boldsymbol{\beta})$? This can be measured by comparing the best possible performance without knowing $\mathcal{S}$ with the performance of the oracle LSE $\widehat{\boldsymbol{\beta}}^o$ in (4.10) at a different noise level. If the best possible performance without knowing $\mathcal{S}$ at the true noise level $\sigma$ is comparable with the performance of the oracle LSE at an inflated noise level $\sigma'$, the ratio $\sigma'/\sigma$ is often used to measure the cost of not knowing the oracle model $\mathcal{S}$. This ratio can be called the *noise inflation factor*. It is convenient to measure the cost of not knowing the true model $\mathcal{S}$ with the noise inflation factor as the performance of the oracle LSE is well understood.

Let $s = |\mathcal{S}|$ be the size of the true model. We will show in this section that to a large extent, the noise inflation factor is at least of the order $\sqrt{\log(p/s)}$ for prediction and coefficient estimation and $\sqrt{\log(p-s)}$ for variable selection in high-dimensional regression models with Gaussian noise. The impact of this noise inflation may decline or diminish when the signal is strong. When $\log s$ is of smaller order than $\log p$, the noise inflation factor is of the order $\sqrt{\log p}$.

We will justify the above summary statements with lower and upper performance bounds of matching order in prediction, coefficient estimation, and variable selection. The lower performance bounds, presented in the rest of this section, are applicable to all estimators. The upper performance bounds are derived for PLS estimators in other sections of this chapter.

For simplicity, we assume throughout this discussion of lower bounds for the noise inflation factor that the errors are iid $N(0, \sigma^2)$, i.e. $\boldsymbol{\varepsilon} \sim N(0, \sigma^2\mathbf{I}_{n\times n})$

in (4.1). Moreover, we assume that the true noise level $\sigma$ is known and positive as lower bounds in the more difficult case of unknown $\sigma$ are not smaller.

#### 4.1.2.1 Bayes lower bounds for orthogonal design

We will start with a simpler case of *orthogonal design*. A linear regression model has an orthogonal design if $\mathbf{X}_j^T\mathbf{X}_k = 0$ for all $j \neq k$. This does allow $p \to \infty$. However, since rank($\mathbf{X}$) $= p$ for orthogonal designs, $p \leq n$ is required. For simplicity, we consider here linear regression models with *orthonormal designs*: $\mathbf{X}^T\mathbf{X}/n = \mathbf{I}_{p\times p}$, or equivalently $\mathbf{X}_j^T\mathbf{X}_k/n = I\{j = k\}$. In this case, $Z_j = \mathbf{X}_j^T\mathbf{Y}/n$ are *sufficient statistics* as $\sigma$ is assumed known here. Moreover, these sufficient statistics are independent of each other and

$$Z_j \sim N(\beta_j, \sigma_n^2),$$

where $\sigma_n = \sigma/n^{1/2}$. This is called the *Gaussian sequence model.*

For orthonormal designs, we have

$$\|\mathbf{X}(\widehat{\boldsymbol\beta} - \boldsymbol\beta)\|_2^2/n = \|\widehat{\boldsymbol\beta} - \boldsymbol\beta\|_2^2 = \sum_{j=1}^{p}(\widehat{\beta}_j - \beta_j)^2,$$

so that the expected prediction regret is identical to the $\ell_2$ estimation error.

The following theorem provides a lower bound for the maximum *$\ell_q$ estimation risk* of the Bayes rule, including the maximum prediction risk via its equivalence to the $\ell_2$ estimation risk using (4.7) with $q = 2$, when the regression coefficients, $\beta_j, j = 1, \ldots, p$, follow a coin tossing model.

**Theorem 4.1** *Suppose $\mathbf{X}^T\mathbf{X}/n = \mathbf{I}_{p\times p}$ and $\boldsymbol\varepsilon \sim N(0, \sigma^2\mathbf{I}_{n\times n})$. Let $1 \leq q < \infty$, $0 < \pi_0 < \epsilon_0 \leq 1 - \pi_0$, $\sigma_n = \sigma/\sqrt{n}$, $\mu_0 = \sqrt{2\log(1/\pi_0)} - \sqrt{2\log(1/\epsilon_0)}$, and $G_0$ be the prior under which $\beta_j$ are iid random variables with $P_{G_0}\{\beta_j \neq 0\} = P_{G_0}\{\beta_j = \mu_0\sigma_n\} = \pi_0$. Then,*

$$\inf_{\widehat{\boldsymbol\beta}} \mathrm{E}_{G_0} \|\widehat{\boldsymbol\beta} - \boldsymbol\beta\|_q^q \geq (1 - \epsilon_0/2)(1 - \epsilon_1)(\pi_0 p)(\sigma_n\mu_0)^q,$$

*where $\epsilon_1 = 1 - 1/[\{\epsilon_0/(1 - \pi_0)\}^{1/(q-1)} + 1]^{q-1}$ for $q > 1$ and $\epsilon_1 = 0$ for $q = 1$.*

**Proof**. The proof is based on the following simple lower bound of the posterior Bayes risk in the estimation of a single $\beta_j$.

Let $\delta_1$ be an unknown $\{0,1\}$-valued random variable, whose (posterior) mean is $\pi_1$, and $\omega_1 = \pi_1/(1 - \pi_1)$. Let $R_q(\pi_1)$ be the minimum (posterior) risk for the estimation of $\delta_1$ under the $\ell_q$ loss. Then,

$$R_q(\pi_1) = \min_{0\leq x\leq 1}\{\pi_1(1 - x)^q + (1 - \pi_1)x^q\},$$

whose minimum is attained when $x = I\{\pi_1 > 1/2\}$ for $q = 1$ and $x/(1-x) = \omega_1^{1/(q-1)}$ for $q > 1$. Thus

$$R_q(\pi_1) = \Big\{\pi_1 \wedge (1-\pi_1)\Big\} I_{\{q=1\}} + \Big\{\pi_1/\{\omega_1^{1/(q-1)} + 1\}^{q-1}\Big\} I_{\{q>1\}}.$$

Let $\zeta = Z_1/\sigma_n$ and $\delta_1 = I\{\beta_1 \neq 0\}$. Then, $\zeta|\delta_1 \sim N(\mu_0\delta_1, 1)$ and $\delta_1 \sim$Bernoulli$(\pi_0)$ under $P_{G_0}$. As $(Z_j, \beta_j)$ are iid under $P_{G_0}$, we have

$$\begin{aligned}\inf_{\widehat{\boldsymbol{\beta}}} E_{G_0}\|\widehat{\boldsymbol{\beta}} - \boldsymbol{\beta}\|_q^q &= p \inf_{\widehat{\beta}_1} E_{G_0}|\widehat{\beta}_1 - \beta_1|^q \\ &= p\sigma_n^q \mu_0^q \min_{f(\cdot)} \mathrm{E}_{G_0}|f(\zeta) - \delta_1|^q.\end{aligned}$$

The posterior probability for $\delta_1 = 1$ is

$$\begin{aligned}\pi_1(\zeta) &= P_{G_0}\{\delta_1 = 1|\zeta\} \\ &= \pi_0\varphi(\zeta-\mu_0)/\{\pi_0\varphi(\zeta-\mu_0) + (1-\pi_0)\varphi(\zeta)\},\end{aligned}$$

where $\varphi(x) = e^{-t^2/2}/\sqrt{2\pi}$ is the standard normal density. Let $\omega_1(\zeta) = \pi_1(\zeta)/\{1-\pi_1(\zeta)\}$ be the posterior odds. Because $\pi_0\varphi(\zeta-\mu_0) + (1-\pi_0)\varphi(\zeta)$ is the density of $\zeta$, the formula for $R_q(\pi_1)$ gives

$$\begin{aligned}\min_{f(\cdot)} E_{G_0}|f(\zeta) - \delta_1|^q &= E_{G_0} R_q(\pi_1(\zeta)) \\ &= \int \frac{\pi_0\varphi(\zeta-\mu_0)}{\{\omega_1^{1/(q-1)}(\zeta) + 1\}^{q-1}} d\zeta,\end{aligned}$$

where $\{\omega_1^{1/(q-1)}(\zeta) + 1\}^{q-1}$ is treated as $\max\{1, \omega_1(\zeta)\}$ for $q = 1$.

Let $a_0 = \sqrt{2\log(1/\epsilon_0)} = \sqrt{2\log(1/\pi_0)} - \mu_0$ and $\pi_1^* = \epsilon_0/(1-\pi_0)$. As the likelihood ratio $\varphi(\zeta-\mu_0)/\varphi(\zeta)$ is increasing in $\zeta$, for $\zeta \le \mu_0 + a_0$ the posterior odds is bounded by

$$\begin{aligned}\omega_1(\zeta) &= \pi_0\varphi(\zeta-\mu_0)/\{(1-\pi_0)\varphi(\zeta)\} \\ &\le \pi_0\varphi(a_0)/\{(1-\pi_0)\varphi(\mu_0+a_0)\} \\ &= \pi_0 e^{-a_0^2/2 + (a_0+\mu_0)^2/2}/(1-\pi_0) \\ &= \epsilon_0/(1-\pi_0).\end{aligned}$$

By assumption, we have $\epsilon_0/(1-\pi_0) \le 1$, so that

$$1/\{\omega_1^{1/(q-1)}(\zeta) + 1\}^{q-1} \ge 1/\{(\epsilon_0/(1-\pi_0))^{1/(q-1)} + 1\}^{q-1} = 1 - \epsilon_1.$$

Thus,

$$\begin{aligned}\min_{f(\cdot)} \mathrm{E}_{G_0}|f(\zeta) - \delta_1|^q &\ge (1-\epsilon_1)\int_{-\infty}^{\mu_0+a_0} \pi_0\varphi(\zeta-\mu_0) d\zeta \\ &\ge (1-\epsilon_1)\pi_0(1-\epsilon_0/2),\end{aligned}$$

due to $P\{N(\mu_0, 1) > \mu_0 + a_0\} = P\{N(0,1) > a_0\} \le e^{-a_0^2/2}/2 = \epsilon_0/2$ [see

(3.52)]. The conclusion follows by applying this inequality to the identity at the beginning of the proof. ■

This theorem is at the heart of the lower bound results in Donoho and Johnstone (1994) where the minimax $\ell_q$ risk in $\ell_r$ balls is studied. The interpretation of this theorem is the clearest when both $\pi_0$ and $\epsilon_0$ are small and $\log(1/\epsilon_0)$ is of smaller order than $\log(1/\pi_0)$. In this case, Theorem 4.1 asserts that in the case where the support of $\boldsymbol{\beta}$ is unknown, the $\ell_q$ risk is bounded from below by

$$\mathrm{E}_{G_0} \|\widehat{\boldsymbol{\beta}} - \boldsymbol{\beta}\|_q^q \geq (1+o(1))\{\, \mathrm{E}_{G_0} \|\boldsymbol{\beta}\|_0\} \left\{\sigma_n \sqrt{2\log\left(p/(\mathrm{E}_{G_0}\|\boldsymbol{\beta}\|_0)\right)}\right\}^q, \quad (4.12)$$

noticing $\mathrm{E}_{G_0} \|\boldsymbol{\beta}\|_0 = \pi_0 p$. Under the assumptions of Theorem 4.1, the oracle estimator (4.10) becomes

$$\widehat{\boldsymbol{\beta}}^o = (Z_j I\{\beta_j \neq 0\}, j \leq p)^T,$$

with $Z_j \sim N(\beta_j, \sigma_n^2)$. It then follows that the oracle estimator has the $\ell_q$ risk

$$\mathrm{E}_{G_0} \|\widehat{\boldsymbol{\beta}}^o - \boldsymbol{\beta}\|_q^q = \mathrm{E}_{G_0} \sum_{j=1}^p |Z_j - \boldsymbol{\beta}_j|^q I\{\beta_j \neq 0\} = \mathrm{E}_{G_0} \|\boldsymbol{\beta}\|_0 \sigma_n^q \,\mathrm{E}\, |N(0,1)|^q.$$

Thus, Theorem 4.1 has the interpretation

$$\left(\frac{\inf_{\widehat{\boldsymbol{\beta}}} \mathrm{E}_{G_0} \|\widehat{\boldsymbol{\beta}} - \boldsymbol{\beta}\|_q^q}{\mathrm{E}_{G_0} \|\widehat{\boldsymbol{\beta}}^o - \boldsymbol{\beta}\|_q^q}\right)^{1/q} \geq (1+o(1)) \frac{\sqrt{2\log\left(p/(\mathrm{E}_{G_0}\|\boldsymbol{\beta}\|_0)\right)}}{\{\mathrm{E}\,|N(0,1)|^q\}^{1/q}}. \quad (4.13)$$

In particular, for $q = 2$, the above inequality asserts that in a Bayes setting, the Bayes rule, or the best one can do without knowing the support of $\boldsymbol{\beta}$, has a mean squared error no smaller than that of the oracle LSE inflated by approximately a factor of $\sqrt{2\log(p/\|\boldsymbol{\beta}\|_0)}$. This also gives a noise inflation factor for prediction as the prediction risk is identical to the $\ell_2$ estimation error.

The lower bound in Theorem 4.1 is obtained by setting the *signal strength*, or equivalently the magnitude of the nonzero $\beta_j$, at approximately the least informative level. The following example demonstrates that the *Bayes estimator* may outperform the oracle LSE when the signal is strong and a true prior can be specified.

**Example 4.1** *Let $\{\mathbf{X}, \boldsymbol{\beta}, \boldsymbol{\varepsilon}\}$ be as in Theorem 4.1 with possibly different positive parameters $\{\pi_0, \mu_0, a_0, \epsilon_0\}$ satisfying $\mu_0^2/2 \geq a_0\mu_0 + \log\{(1/\pi_0 - 1)(\mu_0^2/\epsilon_0 - 1)\}$, $0 < \pi_0 \leq 1/2$ and $P\{N(0,1) > a_0\} \leq \epsilon_0/\mu_0^2 < 1$. Then,*

$$\mathrm{E}_{G_0} \|\widehat{\boldsymbol{\beta}} - \boldsymbol{\beta}\|_2^2 \leq 2\epsilon_0 \,\mathrm{E}_{G_0} \|\widehat{\boldsymbol{\beta}}^o - \boldsymbol{\beta}\|_2^2,$$

*where $\widehat{\beta}$ is the Bayes rule under $G_0$ and $\widehat{\boldsymbol{\beta}}^o$ is the oracle LSE in (4.10). This can be seen as follows. Because $\varphi(a_0)/\varphi(\mu_0 - a_0) = e^{\mu_0^2/2 - a_0\mu_0} \ge (1/\pi_0 - 1)(\mu_0^2/\epsilon_0 - 1)$ and $\pi_1(\zeta)$ is increasing in $\zeta$, $\zeta \ge \mu_0 - a_0$ implies*

$$\pi_1(\zeta) = \frac{\varphi(\zeta - \mu_0)/\varphi(\zeta)}{\varphi(\zeta - \mu_0)/\varphi(\zeta) + (1/\pi_0 - 1)} \ge \pi_1(\mu_0 - a_0) \ge 1 - \epsilon_0/\mu_0^2.$$

*Thus, $\int\{1 - \pi_1(\zeta)\}\varphi(\zeta - \mu_0)d\zeta \le \epsilon_0/\mu_0^2 + P\{N(\mu_0, 1) < \mu_0 - a_0\}$, which is bounded by $2\epsilon_0/\mu_0^2$. Let $\sigma_n = \sigma/n^{1/2}$. As in the proof of Theorem 4.1, the $\ell_2$ Bayes risk of the Bayes estimator, $p\sigma_n^2\mu_0^2\int\{1 - \pi_1(\zeta)\}\varphi(\zeta - \mu_0)d\zeta$ is no greater than $2\epsilon_0 p\sigma_n^2 = 2\epsilon_0 \, \mathrm{E}_{G_0} \|\widehat{\boldsymbol{\beta}}^o - \boldsymbol{\beta}\|_2^2$.*

For variable selection, the noise inflation is expressed in terms of required signal strength. Consider testing $H_j : \beta_j = 0$ with a common one-sided rejection rule $Z_j > \lambda$ for a certain constant $\lambda$, where $Z_j \sim N(\beta_j, \sigma_n^2)$ are the independent sufficient statistics for $\beta_j$. This class of separable tests includes all Bayes rules for maximizing $P_G\{\mathrm{supp}(\widehat{\boldsymbol{\beta}}) = \mathrm{supp}(\boldsymbol{\beta})\}$ when $\beta_j$ are iid nonnegative variables *a priori*. The following theorem provides necessary conditions on $\lambda$ and the *minimum signal strength* $\beta_{\min} = \min_{\beta_j \ne 0} |\beta_j|$ for the selection consistency of such methods. Let $\Phi^{-1}(t)$ be the standard normal quantile function, or equivalently the inverse function of $\Phi(t) = \int_{-\infty}^t \varphi(z)dz$. It is well known that for all $t > 0$

$$\frac{t\varphi(t)}{1 + t^2} \le \Phi(-t) = e^{-t^2/2}\int_0^\infty e^{-xt}\varphi(x)dx \le \min\{\varphi(t)/t, e^{-t^2/2}/2\}.$$

**Theorem 4.2** *Suppose $Z_j$ are independent $N(\beta_j, \sigma_n^2)$ variables conditionally on $\boldsymbol{\beta} = (\beta_1, \ldots, \beta_p)^T$ under a probability measure $P_G$. Let $1 > \alpha' > \alpha > 0$, $0 < s < p$ and $\beta_* > 0$ be constants satisfying*

$$1 - P_G\{\beta_{\min} \le \beta_*, \|\boldsymbol{\beta}\|_0 \le s\} = \alpha' - \alpha.$$

*Let $\lambda$ be a positive constant and $\widehat{\boldsymbol{\beta}} = (\widehat{\beta}_1, \ldots, \widehat{\beta}_p)^T$ be an estimator of $\boldsymbol{\beta}$ such that $\widehat{\beta}_j = 0$ iff $Z_j \le \lambda$. If $P_G\{\mathrm{supp}(\widehat{\boldsymbol{\beta}}) = \mathrm{supp}(\boldsymbol{\beta})\} \ge 1 - \alpha$, then*

$$\lambda \ge \sigma_n\Phi^{-1}\{(1 - \alpha')^{1/(p-s)}\}, \quad \beta_* \ge \lambda + \sigma_n\Phi^{-1}(1 - \alpha').$$

**Proof**. From the assumption, we have

$$P_G\{\mathrm{supp}(\widehat{\boldsymbol{\beta}}) = \mathrm{supp}(\boldsymbol{\beta}), \beta_{\min} \le \beta_*, \|\boldsymbol{\beta}\|_0 \le s\} \ge 1 - \alpha' > 0.$$

Using the independence of $Z_j$ given $\boldsymbol{\beta}$, the right-hand side is bounded from above by

$$\begin{aligned} 1 - \alpha' &\le P(|Z_j| \le \lambda, j \in \mathrm{supp}(\boldsymbol{\beta}), |Z_j| \ge \lambda, j \notin \mathrm{supp}(\boldsymbol{\beta}), \|\boldsymbol{\beta}\|_0 \le s) \\ &\le \Phi^{p-s}(\lambda/\sigma_n)\Phi\big((\beta_* - \lambda)/\sigma_n\big) \end{aligned}$$

$$\leq \min\Big\{\Phi^{p-s}(\lambda/\sigma_n), \Phi\big((\beta_* - \lambda)/\sigma_n\big)\Big\}.$$

Thus, $\lambda/\sigma_n \geq \Phi^{-1}\{(1-\alpha')^{1/(p-s)}\}$ and $(\beta_* - \lambda)/\sigma_n \geq \Phi^{-1}(1-\alpha')$. The conclusion follows as $(1-\alpha')^{1/(p-s_2)} \geq 1 + (p-s)^{-1}\log(1-\alpha')$. ■

Suppose $\beta_j$ are iid variables under $P_G$ and $p - s \to \infty$, where $s$ is the median of $\|\boldsymbol{\beta}\|_0$. Theorem 4.2 asserts that the Bayes rule under the loss function $I\{\text{supp}(\widehat{\boldsymbol{\beta}}) \neq \text{supp}(\boldsymbol{\beta})\}$, which is a thresholding rule with a special $\lambda$, requires $\lambda \geq \sigma_n\Phi^{-1}((1-\alpha')^{1/(p-s)}) \approx \sigma_n\sqrt{2\log(p-s)}$ and $\beta_{\min} \geq \lambda + O(\sigma_n)$ to achieve selection consistency. In comparison, for testing an individual hypothesis $\beta_j = 0$ against the alternative $\beta_j \geq \beta_{\min}$, both Type I and Type II error probability are bounded by $\alpha$ when $\beta_{\min} \geq 2\sigma_n\Phi^{-1}(1-\alpha)$. Thus, due to the multiplicity of the testing problem, an increase of the signal strength $\beta_{\min}$ by a factor of the order $\sqrt{\log(p-s)}$ is required to control the noise. Again, this can be viewed as a noise inflation factor due to model uncertainty.

#### *4.1.2.2 Minimax lower bounds for general design*

We consider here lower performance bounds for general design matrices. This is important since orthogonal designs require $p \leq n$ but we are mainly interested in the case of $p > n$.

In addition to the normality assumption $\boldsymbol{\varepsilon} \sim N(0, \sigma^2\mathbf{I}_{n\times n})$, we assume for simplicity that the design vectors are normalized to $\|\mathbf{X}_j\|_2^2/n = \sum_{i=1}^n X_{ij}^2/n = 1$ for deterministic designs and $\mathrm{E}\,\|\mathbf{X}_j\|_2^2 = n$ for random designs.

We first consider a general *compound loss function*, defined as a sum of losses for decisions about individual $\beta_j$. The following theorem asserts that when $\beta_j$ are iid random variables *a priori*, the regression problem is always more difficult than the Gaussian sequence model for any compound loss on $\boldsymbol{\beta}$.

**Theorem 4.3** *Let $\mathscr{D}_{\text{reg}}$ be the set of all estimators based on data $(\mathbf{X}, \mathbf{Y})$ in a linear regression model with Gaussian error $\boldsymbol{\varepsilon} \sim N(0, \sigma^2\mathbf{I}_{n\times n})$ and design vectors normalized to $\|\mathbf{X}_j\|_2^2 = n\ \forall j$. Let $\mathscr{D}_{\text{seq}}$ be the set of all estimators based on independent observations $Z_j \sim N(\beta_j, \sigma^2/n), j \leq p$, in a Gaussian sequence model. Assume that $\sigma$ is known. Let $G$ be a prior under which $\beta_j$ are independent random variables. Then, for any given nonnegative loss function $L$,*

$$\inf_{\widehat{\boldsymbol{\beta}}\in\mathscr{D}_{\text{reg}}} \mathrm{E}_G \sum_{j=1}^p L(\widehat{\beta}_j, \beta_j) \geq \inf_{\widehat{\boldsymbol{\beta}}\in\mathscr{D}_{\text{seq}}} \mathrm{E}_G \sum_{j=1}^p L(\widehat{\beta}_j, \beta_j).$$

**Proof.** Let $\widetilde{Z}_j = \mathbf{X}_j^T(\mathbf{Y} - \sum_{k\neq j}\mathbf{X}_k\beta_k)/n$. Since $\boldsymbol{\varepsilon} \sim N(0, \sigma^2\mathbf{I}_{n\times n})$, $\widetilde{Z}_j \sim N(\beta_j, \sigma^2/n)$ is sufficient for $\beta_j$ in the linear regression model when all other parameters, $\beta_k, k \neq j$, are known. It follows that

$$\begin{aligned}
\inf_{\widehat{\beta}\in\mathscr{D}_{\rm reg}} \mathrm{E}_G \sum_{j=1}^p L(\widehat{\beta}_j, \beta_j) &= \sum_{j=1}^p \mathrm{E}_G\Big\{\inf_{\widehat{\beta}\in\mathscr{D}_{\rm reg}} \mathrm{E}_G\Big[L(\widehat{\beta}_j,\beta_j)\Big|\mathbf{X},\mathbf{Y}\Big]\Big\} \\
&\geq \sum_{j=1}^p \mathrm{E}_G\Big\{\inf_{\widehat{\beta}\in\mathscr{D}_{\rm reg}} \mathrm{E}_G\Big[L(\widehat{\beta}_j,\beta_j)\Big|\mathbf{X},\mathbf{Y},\beta_k, k\neq j\Big]\Big\} \\
&= \sum_{j=1}^p \mathrm{E}_G\Big\{\inf_{\widehat{\beta}\in\mathscr{D}_{\rm reg}} \mathrm{E}_G\Big[L(\widehat{\beta}_j,\beta_j)\Big|\widetilde{Z}_j\Big]\Big\} \\
&= \sum_{j=1}^p \mathrm{E}_G\Big\{\inf_{\widehat{\beta}\in\mathscr{D}_{\rm seq}} \mathrm{E}_G\Big[L(\widehat{\beta}_j,\beta_j)\Big|Z_j\Big]\Big\}.
\end{aligned}$$

The proof is complete due to the sufficiency of $Z_j$ for $\beta_j$ and the independence of $(Z_j, \beta_j)$ in the Gaussian sequence model. ∎

An immediate consequence of Theorems 4.1 and 4.3 is the following lower bound for the minimum $\ell_q$ Bayes risk under the prior $G_0$.

**Corollary 4.1** *Suppose* $\|\mathbf{X}_j\|_2^2 = n$ *for all* $j \leq p$ *and* $\boldsymbol{\varepsilon} \sim N(0, \sigma^2\mathbf{I}_{n\times n})$. *Let* $\{q, \pi_0, \epsilon_0, \epsilon_1\}$ *be as in Theorem 4.1,* $\sigma_n = \sigma/\sqrt{n}$, $\mu_0 = \sqrt{2\log(1/\pi_0)} - \sqrt{2\log(1/\epsilon_0)}$, *and* $G_0$ *be the prior under which* $\beta_j$ *are iid random variables with* $P_{G_0}\{\beta_j \neq 0\} = P_{G_0}\{\beta_j = \mu_0\sigma/\sqrt{n}\} = \pi_0$. *Then,*

$$\inf_{\widehat{\boldsymbol{\beta}}} \mathrm{E}_{G_0} \|\widehat{\boldsymbol{\beta}} - \boldsymbol{\beta}\|_q^q \geq (1-\epsilon_0/2)(1-\epsilon_1)(\pi_0 p)(\sigma_n\mu_0)^q.$$

Corollary 4.1 indicates a lower bound of the order $\sqrt{2\log(p/\|\boldsymbol{\beta}\|_0)}$ for the noise inflation factor for the estimation of $\boldsymbol{\beta}$. This can be seen as follows. For submodels of no greater dimension than real number $t$, define the *lower sparse eigenvalue* of a nonnegative matrix $\boldsymbol{\Sigma}$ as

$$\phi_-(t;\boldsymbol{\Sigma}) = \min_{|A|<t+1} \phi_{\min}(\boldsymbol{\Sigma}_{A,A}), \tag{4.14}$$

Namely, it is the smallest eigenvalue among all sub-matrices of size no larger than $t$. Let $E_{\boldsymbol{\beta}}$ be the expectation under which the linear model (4.1) has Gaussian noise $\boldsymbol{\varepsilon} \sim N(0, \sigma^2\mathbf{I}_{n\times n})$. Let $\mathcal{S} = \mathrm{supp}(\boldsymbol{\beta})$. Under $E_{\boldsymbol{\beta}}$, the oracle LSE (4.10) satisfies $\widehat{\boldsymbol{\beta}}^o \sim N\big(\boldsymbol{\beta}, \sigma^2(\mathbf{X}_{\mathcal{S}}^T\mathbf{X}_{\mathcal{S}})^{-1}\big)$ when $\mathrm{rank}(\mathbf{X}_{\mathcal{S}}) = |\mathcal{S}|$. Thus, for $\|\boldsymbol{\beta}\|_0 = |\mathcal{S}| \leq s^*$ and $1 \leq q < \infty$,

$$\begin{aligned}
&\mathrm{E}_{\boldsymbol{\beta}} \|\mathbf{X}\widehat{\boldsymbol{\beta}}^o - \mathbf{X}\boldsymbol{\beta}\|_2^2 = \sigma_n^2\|\boldsymbol{\beta}\|_0, \\
&\mathrm{E}_{\boldsymbol{\beta}} \|\widehat{\boldsymbol{\beta}}^o - \boldsymbol{\beta}\|_q^q \leq \sigma_n^q\|\boldsymbol{\beta}\|_0\, \mathrm{E}\,|N(0,1)|^q/\phi_-^{q/2}(s^*;\overline{\boldsymbol{\Sigma}}),
\end{aligned} \tag{4.15}$$

where $\overline{\boldsymbol{\Sigma}} = \mathbf{X}^T\mathbf{X}/n$ is the (normalized) Gram matrix. Since $\pi_0 p\mu_0^q \approx \mathrm{E}_{G_0}\|\boldsymbol{\beta}\|_0\{2\log(p/\mathrm{E}_{G_0}\|\boldsymbol{\beta}\|_0)\}^{q/2}$ in Corollary 4.1, the noise inflation factor, or the $q$-root of the ratio of the $\ell_q$ risks of the Bayes and oracle estimators, is of the order $\sqrt{\log(p/\mathrm{E}_{G_0}\|\boldsymbol{\beta}\|_0)}$ when $\phi_-(s^*;\overline{\boldsymbol{\Sigma}})$ can be treated as a constant.

Although the above comparison provides some rough idea about the noise inflation factor based on relatively simple arguments, several issues arise. Firstly, the noise inflation factor for prediction is unclear since the prediction risk is no longer equivalent to the $\ell_2$ estimation risk for non-orthonormal designs. Secondly, the risks are evaluated with different expectations $\mathrm{E}_{G_0}$ and $\mathrm{E}_{\boldsymbol{\beta}}$. Thirdly, since $\|\boldsymbol{\beta}\|_0$ is unbounded under $G_0$, Corollary 4.1 does not provide a lower bound for the estimation error under commonly imposed sparsity conditions expressed in terms of $\|\boldsymbol{\beta}\|_0$ or other complexity measures of $\boldsymbol{\beta}$. The next theorem addresses these issues by finding a least favorable configuration of sparse $\boldsymbol{\beta}$ in the following intersection of $\ell_0$ and $\ell_\infty$ balls:

$$\Theta_{n,p,s^*} = \big\{\mathbf{v} \in I\!\!R^p : \|\mathbf{v}\|_0 \le s^*, \|\mathbf{v}\|_\infty \le \lambda_n\big\},$$

where $\lambda_n = \sigma\sqrt{(2/n)\log(p/s^*)}$.

**Theorem 4.4** *Let* $\mathbf{X}$ *be a deterministic design matrix with* $\|\mathbf{X}_j\|_2^2 = n$ *for all* $j$ *and* $P_{\boldsymbol{\beta}}$ *be probability measures under which* $\mathbf{Y} = \mathbf{X}\boldsymbol{\beta} + \boldsymbol{\varepsilon}$ *with* $\boldsymbol{\varepsilon} \sim N(0, \sigma^2\mathbf{I}_{n\times n})$. *Let* $p \gg s^* \to \infty$ *and* $0 < \epsilon < 1 \le q < \infty$. *Then,*

$$\inf_{\widehat{\boldsymbol{\beta}}} \sup_{\boldsymbol{\beta}\in\Theta_{n,p,s^*}} \mathrm{E}_{\boldsymbol{\beta}}\|\widehat{\boldsymbol{\beta}} - \boldsymbol{\beta}\|_q^q \ge (1+o(1))s^*\lambda_n^q,$$

*and*

$$\inf_{\widehat{\boldsymbol{\beta}}} \sup_{\boldsymbol{\beta}\in\Theta_{n,p,s^*}} P_{\boldsymbol{\beta}}\Big\{\|\widehat{\boldsymbol{\beta}} - \boldsymbol{\beta}\|_q^q \ge (1-\epsilon)^q s^*\lambda_n^q\Big\} \ge \frac{1-(1-\epsilon)^q + o(1)}{(3-\epsilon)^q - (1-\epsilon)^q}.$$

*Moreover, for the expected prediction regret in (4.5),*

$$\inf_{\widehat{\boldsymbol{\beta}}} \sup_{\boldsymbol{\beta}\in\Theta_{n,p,s^*}} \mathrm{E}_{\boldsymbol{\beta}}\|\mathbf{X}\widehat{\boldsymbol{\beta}} - \mathbf{X}\boldsymbol{\beta}\|_2^2/n \ge (1+o(1))s^*\lambda_n^q\phi_-(2s^*;\overline{\boldsymbol{\Sigma}})/4,$$

*and*

$$\inf_{\widehat{\boldsymbol{\beta}}} \sup_{\boldsymbol{\beta}\in\Theta_{n,p,s^*}} P_{\boldsymbol{\beta}}\Big\{\|\mathbf{X}\widehat{\boldsymbol{\beta}} - \mathbf{X}\boldsymbol{\beta}\|_2^2/n \ge \frac{(1-\epsilon)^2 s^*\lambda_n^2}{4/\phi_-(2s^*;\overline{\boldsymbol{\Sigma}})}\Big\} \ge \epsilon/4 + o(1),$$

*where* $\phi_-(\cdot;\cdot)$ *is the lower sparse eigenvalue in (4.14) and* $\overline{\boldsymbol{\Sigma}} = \mathbf{X}^T\mathbf{X}/n$.

Theorem 4.4 provides $s^*\sigma^2(2/n)\log(p/s^*)$ as a lower bound for the *minimax* prediction and $\ell_2$ estimation errors. In view of the prediction and $\ell_2$ risks of the oracle LSE, it implies that the noise inflation factor due to model uncertainty is at least of the order $\sqrt{\log(p/s^*)}$ under the *$\ell_0$ sparsity assumption* $\|\boldsymbol{\beta}\|_0 \le s^*$.

Theorem 4.4 is a slight improvement of a special case of the results in Ye and Zhang (2010), where the minimax rate of $\ell_q$ risk and loss in $\ell_r$ balls were obtained for $0 \le r \le q \le \infty$. Their more general lower bounds for the tail probability of the $\ell_q$ error can be stated as follows: Let $0 < r \vee 1 \le q$, $\sigma_n = \sigma/\sqrt{n}$ and $\lambda_n = \sigma_n\{2\log(p(\sigma_n/C_r)^r)\}^{1/2}$. If $\min\{\lambda_n/\sigma_n, (C_r/\lambda_n)^r\} \to \infty$, then

$$\inf_{\widehat{\boldsymbol{\beta}}} \sup_{\|\boldsymbol{\beta}\|_r \le C_r} P_{\boldsymbol{\beta}}\Big\{\|\widehat{\boldsymbol{\beta}} - \boldsymbol{\beta}\|_q^q \ge (1-\epsilon)^q C_r^r \lambda_n^{q-r}\Big\} \ge \epsilon/2 + o(1). \tag{4.16}$$

**Proof of Theorem 4.4** Let $\{\pi_0, \epsilon_0, \epsilon_1, \mu_0\}$ and $P_{G_0}$ be defined as in Theorem 4.1 such that $\pi_0 \vee \epsilon_0 \vee \epsilon_1 \to 0$. Let $s^* = (1+\epsilon_0)\pi_0 p$, $\mu = \mu_0\sigma/\sqrt{n}$ and $k = \pi_0 p$. For $1/\epsilon_0 \ge 1+\epsilon_0$,

$$\begin{aligned} & \sqrt{n/2}(\lambda_n - \mu)/\sigma \\ = \; & \sqrt{\log(p/s^*)} + \sqrt{\log(1/\epsilon_0)} - \sqrt{\log(p/s^*) + \log(1+\epsilon_0)} \\ > \; & 0. \end{aligned}$$

Let $N = \|\boldsymbol{\beta}\|_0$. Under $P_{G_0}$ and the $\ell_q$ loss, the Bayes rule given $N \le s^*$ is

$$\begin{aligned} \boldsymbol{\delta}^* &= \delta^*(\mathbf{X}, \mathbf{Y}, s^*) \\ &= \arg\min_{\mathbf{b}} \mathrm{E}_{G_0}\big[\|\boldsymbol{\beta} - \mathbf{b}\|_q^q \big| \mathbf{X}, \mathbf{Y}, N \le s^*\big]. \end{aligned}$$

We note that $s^* = (1+\epsilon_0)k$ and $\|\boldsymbol{\beta}\|_q^q = N\mu^q$. As the risk of $\delta^*$ is no smaller than the Bayes rule without conditioning, Corollary 4.1 implies

$$\begin{aligned} & (1-\epsilon_0/2)(1-\epsilon_1)k\mu^q \\ \le \; & \mathrm{E}_{G_0}\big[\|\boldsymbol{\delta}^* - \boldsymbol{\beta}\|_q^q \big| N \le s^*\big] + \mathrm{E}_{G_0}\|\boldsymbol{\delta}^* - \boldsymbol{\beta}\|_q^q I\{N > s^*\}. \end{aligned}$$

Because $\|\boldsymbol{\beta}\|_q^q \le s^*\mu^q$ when $N \le s^*$, $\|\boldsymbol{\delta}^*\|_q^q \le s^*\mu^q$, so that $\|\boldsymbol{\delta}^* - \boldsymbol{\beta}\|_q^q \le 2^{q-1}\mu^q(s^*+N)$. It follows that

$$\begin{aligned} & \mathrm{E}_{G_0}\|\boldsymbol{\delta}^* - \boldsymbol{\beta}\|_q^q I\{N > s^*\} \\ \le \; & 2^{q-1}\mu^q \, \mathrm{E}_{G_0}(s^* + N) I\{N > (1+\epsilon_0)k\}. \end{aligned}$$

Since $N$ is a binomial variable with $\mathrm{E}_{G_0} N = k \to \infty$, the right-hand side above is of smaller order than $k = s^*/(1+\epsilon_0)$. Thus, letting $\epsilon_0 \to 0+$ and $\epsilon_1 \to 0+$ slowly, we find

$$\begin{aligned} \mathrm{E}_{G_0}\big[\|\boldsymbol{\delta}^* - \boldsymbol{\beta}\|_q^q \big| N \le s^*\big] &\ge (1+o(1))k\mu^q \\ &= (1+o(1))s^*\lambda_n^q. \end{aligned}$$

This yields the lower bound for the $\ell_q$ minimax risk as the minimax risk is no smaller than the Bayes risk.

Consider the loss function $L(\widehat{\boldsymbol{\beta}}, \boldsymbol{\beta}) = I\{\|\widehat{\boldsymbol{\beta}} - \boldsymbol{\beta}\|_q \ge c(s^*)^{1/q}\lambda_n\}$ for a given $\widehat{\boldsymbol{\beta}}$. Define

$$\widetilde{\boldsymbol{\beta}} = \widehat{\boldsymbol{\beta}} I\Big\{\|\widehat{\boldsymbol{\beta}}\|_q \le (1+c)(s^*)^{1/q}\lambda_n\Big\}.$$

For $N \le s^*$, we have $\|\boldsymbol{\beta}\|_q^q \le s^*\mu^q \le s^*\lambda_n^q$, so that

$$\|\widetilde{\boldsymbol{\beta}} - \boldsymbol{\beta}\|_q^q \le c^q s^* \lambda_n^q \{1 - L(\widehat{\boldsymbol{\beta}}, \boldsymbol{\beta})\} + (2+c)^q s^* \lambda_n^q L(\widehat{\boldsymbol{\beta}}, \boldsymbol{\beta}).$$

Since $\|\boldsymbol{\beta}\|_q^q = N\mu^q \le N\lambda_n^q$, it follows that

$$\begin{aligned} & \mathrm{E}_{G_0} \|\widetilde{\boldsymbol{\beta}} - \boldsymbol{\beta}\|_q^q \\ \le\ & c^q s^* \lambda_n^q + \{(2+c)^q - c^q\} s^* \lambda_n^q \mathrm{E}_{G_0} L(\widehat{\boldsymbol{\beta}}, \boldsymbol{\beta}) I\{N \le s^*\} \\ & + 2^{q-1} s^* \lambda_n^q \mathrm{E}_{G_0} \Big(N/s^* + (1+c)^q\Big) I\{N > (1+\epsilon_0)k\}. \end{aligned}$$

Since $\mathrm{E}_{G_0} \big(N/s^* + (1+c)^q\big) I\{N > (1+\epsilon_0)k\} \to 0$, Corollary 4.1 gives

$$\begin{aligned} & \{(2+c)^q - c^q\} \sup_{\boldsymbol{\beta} \in \Theta_{n,p,s^*}} \mathrm{E}_{\boldsymbol{\beta}} L(\widehat{\boldsymbol{\beta}}, \boldsymbol{\beta}) \\ \ge\ & \mathrm{E}_{G_0} \|\widetilde{\boldsymbol{\beta}} - \boldsymbol{\beta}\|_q^q / (s^* \lambda_n^q) - c^q + o(1) \\ \ge\ & 1 - c^q + o(1). \end{aligned}$$

This gives the lower bound for $\sup_{\boldsymbol{\beta} \in \Theta_{n,p,s^*}} \mathrm{E}_{\boldsymbol{\beta}} L(\widehat{\boldsymbol{\beta}}, \boldsymbol{\beta})$ with $c = 1 - \epsilon$.

Now consider the prediction performance of a given $\widehat{\boldsymbol{\beta}}$. Define

$$\widetilde{\boldsymbol{\beta}} = \arg\min_{\mathbf{b}} \Big\{ \big\|\mathbf{X}\widehat{\boldsymbol{\beta}} - \mathbf{X}\mathbf{b}\big\|_2 : b_j \ne 0 \Rightarrow b_j = \mu, \|\mathbf{b}\|_0 \le s^* \Big\}.$$

For $\|\boldsymbol{\beta}\|_0 \le s^*$, we have $\|\widetilde{\boldsymbol{\beta}} - \boldsymbol{\beta}\|_0 \le 2s^*$, so that

$$\phi_-(2s^*; \boldsymbol{\Sigma}) \|\widetilde{\boldsymbol{\beta}} - \boldsymbol{\beta}\|_2^2 \le \big\|\mathbf{X}\widetilde{\boldsymbol{\beta}} - \mathbf{X}\boldsymbol{\beta}\big\|_2^2 / n$$

by the definition of the sparse eigenvalue. Since $\widetilde{\boldsymbol{\beta}}$ is the minimum distance estimator in the prediction loss based on $\widehat{\boldsymbol{\beta}}$, $\big\|\mathbf{X}\widetilde{\boldsymbol{\beta}} - \mathbf{X}\boldsymbol{\beta}\big\|_2 \le 2\big\|\mathbf{X}\widehat{\boldsymbol{\beta}} - \mathbf{X}\boldsymbol{\beta}\big\|_2$. Consequently,

$$\frac{4\big\|\mathbf{X}\widehat{\boldsymbol{\beta}} - \mathbf{X}\boldsymbol{\beta}\big\|_2^2}{\phi_-(2s^*; \boldsymbol{\Sigma}) n} \ge \|\widetilde{\boldsymbol{\beta}} - \boldsymbol{\beta}\|_2^2,$$

and the lower bound for the expected prediction regret follows from that of the $\ell_2$ estimation error. ■

Next we use the Fano (1961) lemma to derive a necessary uniform signal strength condition for variable selection. Fano's lemma can be regarded as the average correct classification rate when data $\mathbf{Z}$ are from one of the $m$ classes. We omit its proof.

**Lemma 4.1** *Let $P_1, \ldots, P_m$ be probability measures for data $\mathbf{Z}$, $\delta(\mathbf{z})$ a $\{1, \ldots, m\}$-valued statistic, and $K(P, Q) = \mathrm{E}_P \log(dP/dQ)$ the Kullback-Leibler information. Then,*

$$\frac{1}{m} \sum_{j=1}^m P_j \Big\{\delta(\mathbf{Z}) = j\Big\} \le \frac{1}{m^2} \sum_{1 \le j,k \le m} \frac{K(P_j, P_k) + \log 2}{\log(m-1)}.$$

**Theorem 4.5** *Consider linear regression models with a design matrix* $\mathbf{X}$ *and a Gaussian error vector* $\boldsymbol{\varepsilon} \sim N(0, \sigma^2 \mathbf{I}_{n\times n})$. *Suppose either* $\|\mathbf{X}_j\|_2^2 = n\ \forall j$ *for deterministic design or* $\mathrm{E}\,\|\mathbf{X}_j\|_2^2 = n\ \forall j$ *for random design. Let* $\mathscr{S}(s, \beta_{\min})$ *be the set of all coefficient vectors* $\boldsymbol{\beta}$ *satisfying* $\|\boldsymbol{\beta}\|_0 = s$ *and* $\min_{\beta_j \neq 0} |\beta_j| \geq \beta_{\min}$, *with* $s < p - 1$. *Then,*

$$\inf_{\widehat{\boldsymbol{\beta}}} \sup_{\boldsymbol{\beta} \in \mathscr{S}(s,\beta_{\min})} P\big\{\mathrm{supp}(\widehat{\boldsymbol{\beta}}) \neq \mathrm{supp}(\boldsymbol{\beta})\big\} \geq 1 - \frac{2\beta_{\min}^2 + (\sigma^2/n)\log 2}{(\sigma^2/n)\log(p-s)}.$$

Theorem 4.5 is a simplified version of a result in Zhang (2007), while a similar result was proved earlier in Wainwright (2009) for the random design with iid $N(0,1)$ entries. It asserts that selection consistency among models with $\boldsymbol{\beta} \in \mathscr{S}(s, \beta_{\min})$ requires the "beta-min condition"

$$\beta_{\min} \geq (1 + o(1))\sigma\sqrt{(2n)^{-1}\log(p-s)},$$

which can be viewed as a uniform signal strength condition. Compared with the direct proof for orthonormal designs in Theorem 4.2, Theorem 4.5 applies to all column-standardized design matrices but loses approximately a factor of 2 in the uniform signal strength condition.

**Proof of Theorem 4.5.** Let $m = p - s + 1$. For $j = 1, \ldots, m$ let $P_j$ be the probability under which $\beta_j = \beta_{p-s+2} = \cdots = \beta_p = \beta_{\min}$ and $\beta_k = 0$ for all $k \neq j$ and $k \leq p - s + 1$. Clearly $\boldsymbol{\beta} \in \mathscr{S}(s, \beta_{\min})$ under $P_j$. Let us calculate the Kullback-Leibler information between $P_u = N(\mathbf{u}, \sigma^2 \mathbf{I}_{n\times n})$ and $P_v = N(\mathbf{v}, \sigma^2 \mathbf{I}_{n\times n})$. Notice that $dP_u/dP_v$ is just the density ratio of these two distributions. Let $\mathbf{X} \sim P_u$. Then

$$K(P_u, P_v) = \mathrm{E}_{P_u} \frac{1}{2\sigma^2}(\|\mathbf{X}-\mathbf{v}\|^2 - \|\mathbf{X}-\mathbf{u}\|^2) = \frac{1}{2\sigma^2}\mathrm{E}_{P_u}(\mathbf{u}-\mathbf{v})^T(2\mathbf{X}-\boldsymbol{\mu}-\mathbf{v}),$$

which is $\|\mathbf{u}-\mathbf{v}\|_2^2/(2\sigma^2)$ by using $\mathrm{E}_{P_u} X = \mathbf{u}$. Hence,

$$K(P_j, P_k) = \mathrm{E}\,\|\mathbf{X}_j\beta_{\min} - \mathbf{X}_k\beta_{\min}\|_2^2/(2\sigma^2) \leq 2n\beta_{\min}^2/\sigma^2.$$

Let $\mathbf{Z} = (\mathbf{X}, \mathbf{Y})$ and $\delta(\mathbf{Z}) = \min\{j : \widehat{\beta}_j \neq 0\}$. Under model $P_j$, $\mathrm{supp}(\boldsymbol{\beta}) = \{j, m+1, \cdots, p\}$ and on the event $\big\{\mathrm{supp}(\widehat{\boldsymbol{\beta}}) = \mathrm{supp}(\boldsymbol{\beta})\big\}$, $\delta(\mathbf{Z}) = j$, which implies that

$$P_j\big\{\mathrm{supp}(\widehat{\boldsymbol{\beta}}) = \mathrm{supp}(\boldsymbol{\beta})\big\} \leq P_j\{\delta(\mathbf{Z}) = j\}.$$

It follows from Fano's lemma that

$$\begin{aligned}\min_j P_j\big\{\mathrm{supp}(\widehat{\boldsymbol{\beta}}) = \mathrm{supp}(\boldsymbol{\beta})\big\} &\leq \frac{1}{m}\sum_{j=1}^m P_j\{\delta(\mathbf{Z}) = j\} \\ &\leq \frac{2\beta_{\min}^2/(\sigma^2/n) + \log 2}{\log(m-1)}\end{aligned}$$

This completes the proof. ■

### 4.1.3 Performance goals, sparsity and sub-Gaussian noise

We have provided lower performance bounds in the previous subsection for prediction, estimation, and variable selection in the linear regression model (4.1). A natural question is whether these lower bounds are attained by penalized least squares estimators. More precisely, for $\boldsymbol{\beta}$ satisfying $\|\boldsymbol{\beta}\|_0 \le s^*$ or a comparable sparsity condition such as $\|\boldsymbol{\beta}\|_q \le C_q$ with $s^* = (C_q/\lambda)^q$, we are interested in finding methodologies for which

$$\begin{aligned} \|\mathbf{X}\widehat{\boldsymbol{\beta}} - \mathbf{X}\boldsymbol{\beta}\|_2^2/n &\le C_{\text{pred}}\, s^*\lambda^2, \\ \|\widehat{\boldsymbol{\beta}} - \boldsymbol{\beta}\|_q^q &\le C_{\text{est},q}^q s^*\lambda^q, \end{aligned} \tag{4.17}$$

with large probability under proper regularity conditions on $\mathbf{X}$, where $\lambda = \sigma\sqrt{(2/n)\log(p/s^*)}$. For variable selection we are interested in methodologies providing selection consistency

$$P\{\text{supp}(\widehat{\boldsymbol{\beta}}) = \text{supp}(\boldsymbol{\beta})\} \to 1$$

under an additional *minimum signal strength condition* of the form

$$\beta_{\min} = \min_{\beta_j \ne 0} |\beta_j| \;\ge\; C_{\text{select}}\,\lambda \tag{4.18}$$

with $\lambda = \sigma\sqrt{(2/n)\log(p - s^*)}$. Here $\{C_{\text{pred}}, C_{\text{est},q}, C_{\text{select}}\}$ are constants which may depend on the design $\mathbf{X}$, but they should be bounded as $n \wedge p \to \infty$. It follows from Theorems 4.4 and 4.5 that (4.17) and (4.18) imply the rate optimality of $\widehat{\boldsymbol{\beta}}$ in a minimax sense.

As $s^*$ is typically unknown, it is often convenient to take the *universal threshold level* $\lambda = \sigma\sqrt{(2/n)\log p}$ in (4.17) and (4.18). The use of this larger threshold level does not change the rate minimaxity implication of (4.17) and (4.18) when $p \ge n^\gamma$ or $p \ge \gamma n$, $\gamma > 1$, respectively.

We will prove in the rest of the chapter that the performance bounds in (4.17) and the information bound in (4.18) indeed hold for PLS under proper regularity conditions, although we do not have a universal "winner" which is easy to compute and achieves all the above performance bounds with the sharpest constant factors and under minimum regularity conditions.

We use the rest of the subsection to discuss regularity conditions on the coefficient vector and the noise. Regularity conditions on the design can be imposed in many forms and will be discussed as needed after their appearance.

We will consider two types of regularity conditions on the regression coefficients, respectively on the sparsity of $\boldsymbol{\beta}$ and the magnitude of nonzero $|\beta_j|$. For the sparsity of $\boldsymbol{\beta}$, we have so far focused on the so called $\ell_0$ sparsity condition $\|\boldsymbol{\beta}\|_0 \le s^*$. The uniform signal strength condition has already been given in (4.18). It is restrictive but necessary for variable selection consistency (4.9) and oracle property (4.11) in view of the lower bound in Theorem 4.5.

As the rank of the design matrix $\mathbf{X}$ is no greater than the sample size $n$ in linear regression, the coefficient vector $\boldsymbol{\beta}$ is not identifiable when $p > n$.

However, in many applications, it is often practical to impose a sparsity assumption on $\boldsymbol{\beta}$. A vector is sparse if it has many zero or near zero components. Among the following sparsity conditions, the first two are commonly imposed.

**Definition 4.2** *If one of the following conditions holds for a vector $\boldsymbol{\beta}$, we say that $\boldsymbol{\beta}$ is $s^*$-complex under the corresponding sparsity criterion.*

- *$\ell_0$ sparsity or hard sparsity:* $\|\boldsymbol{\beta}\|_0 = |\mathrm{supp}(\boldsymbol{\beta})| \le s^*$.
- *$\ell_r$ sparsity:* $\|\boldsymbol{\beta}\|_r \le C_r$, $0 < r < 2$. *Typically,* $0 < r < 1$ *is needed.*
- *Capped-$\ell_1$ sparsity: For a certain threshold level $\lambda$,*

$$\sum_{j=1}^{p} \min\{\|\mathbf{X}_j \beta_j\|_2/(\lambda\sqrt{n}), 1\} \le s^*.$$

*If the noise variables are iid $N(0, \sigma^2)$ in (4.1), we may take $\lambda = \sigma\lambda_{\mathrm{univ}}$ with the universal threshold level*

$$\lambda_{\mathrm{univ}} = \sqrt{(2/n)\log p}.$$

*If $\mathbf{X}$ is column normalized to $\|\mathbf{X}_j\|_2^2 = n$, the condition becomes* $\sum_j \min\{|\beta_j|/\lambda, 1\} \le s^*$.

- *Approximate sparsity: For certain $\lambda > 0$ and $\overline{\boldsymbol{\beta}} = (\overline{\beta}_1, \ldots, \overline{\beta}_p)^T$,*

$$\|\mathbf{X}\boldsymbol{\beta} - \mathbf{X}\overline{\boldsymbol{\beta}}\|_2^2/(\lambda^2 n) + \sum_{j=1}^{p} \min\{\|\mathbf{X}_j \overline{\beta}_j\|_2/(\lambda\sqrt{n}), 1\} \le s^*.$$

*Again, we may take $\lambda = \sigma\lambda_{\mathrm{univ}}$ for Gaussian errors.*

Compared with the $\ell_0$ sparsity condition, the $\ell_r$ and capped-$\ell_1$ sparsity conditions are more practical since they allow small $|\beta_j|$s. The capped-$\ell_1$ sparsity condition is weaker than the simpler $\ell_0$ sparsity condition, as the summands in the definition of the former are bounded by 1. In the column normalized case $\|\mathbf{X}_j\|_2^2 = n$, the $\ell_r$ sparsity condition for $0 < r < 1$ implies the capped-$\ell_1$ sparsity condition with $s^* = (C_r/\lambda)^r$, due to

$$\sum_j \min(|\beta_j|/\lambda, 1) \le \sum_j |\beta_j|^r/\lambda^r \le C_r^r/\lambda^r. \tag{4.19}$$

The approximate sparsity condition is even weaker than the capped-$\ell_1$ condition as it allows bias; It requires only that $\mathbf{X}\boldsymbol{\beta}$ be close to some $\mathbf{X}\overline{\boldsymbol{\beta}}$ with a capped-$\ell_1$ sparse $\overline{\boldsymbol{\beta}}$. Under the approximate sparsity condition, it is still sensible to consider the estimation of $\mathbf{X}\boldsymbol{\beta}$ (prediction) and $\overline{\boldsymbol{\beta}}$, but the estimation of $\boldsymbol{\beta}$ is in general not tractable, especially in the $p > n$ setting.

In our derivation of lower performance bounds in this section, the noise vector $\boldsymbol{\varepsilon}$ in (4.1) is assumed to be Gaussian $N(0, \sigma^2 \mathbf{I}_{n\times n})$. We may still impose this condition in our derivation of upper performance bounds in the rest of the chapter. However, in many instances such upper error bounds hold nearly verbatim for *sub-Gaussian* noise vectors satisfying

$$\mathrm{E}\exp\left(\mathbf{a}^T\varepsilon\right) \le \exp\left(\|\mathbf{a}\|_2^2\sigma^2/2\right),\ \forall \mathbf{a} \in I\!R^n. \tag{4.20}$$

By Taylor expansion of the exponential function around 0, it follows that condition (4.20) implies $\mathrm{E}\,\varepsilon = \mathbf{0}$ and $\mathrm{E}(\mathbf{a}^T\varepsilon)^2 \le \|\mathbf{a}\|_2^2\sigma^2$. When $E\,\varepsilon = \mathbf{0}$ and the components $\varepsilon_i$ of $\varepsilon$ are independent, (4.20) holds if $\varepsilon_i$ are symmetric and uniformly bounded by $\sigma$ or asymmetric and uniformly bounded by $\sigma/2$. When $\varepsilon \sim N(0, \sigma^2\mathbf{I}_{n\times n})$, $\mathbf{a}^T\varepsilon \sim N(0, \|\mathbf{a}\|_2^2\sigma^2)$, so that (4.20) holds with equality. The following lemma provides some tail probability bounds to match conditions on the noise and design imposed in the rest of the chapter.

**Lemma 4.2** *Let* $\mathbf{P}$ *be a deterministic orthogonal projection in* $I\!R^n$.
*(i) Suppose (4.20) holds for a random vector* $\varepsilon$. *Then, for all vectors* $\mathbf{a} \neq 0$ *and* $t > 0$, $\mathrm{E}\,\mathbf{a}^T\varepsilon = 0$, $\mathrm{E}(\mathbf{a}^T\varepsilon)^2 \le \|\mathbf{a}\|_2^2\sigma^2$ *and*

$$\sup_{\|\mathbf{a}\|_2=1} P\left\{\mathbf{a}^T\varepsilon > \sigma t\right\} \le e^{-t^2/2}. \tag{4.21}$$

*(ii) Suppose (4.21) holds. Let* $\mathbf{X}$ *be a design matrix with* $\max_{j\le p}\|\mathbf{X}_j\|_2^2 \le n$ *and* $\lambda_{\text{univ}}(\epsilon) = \sqrt{(2/n)\log(p/\epsilon)}$ *with* $0 < \epsilon < p/e$. *Then,*

$$P\left\{\|\mathbf{X}^T\mathbf{P}\varepsilon/n\|_\infty \ge \sigma\lambda_{\text{univ}}(\epsilon/2)\right\} \le \epsilon.$$

*(iii) Suppose (4.21) holds. Let* $r$ *be the rank of* $\mathbf{P}$. *Then,*

$$P\left\{\|\mathbf{P}\varepsilon\|_2^2/\sigma^2 \ge ra_t\right\} \le e^{-rt},\ \forall t > 0,$$

*where* $a_t = \min_{0<\epsilon<1/2}\{2t + 2\log(1+1/\epsilon)\}/(1-2\epsilon)^2$. *Moreover,* $a_t \le 2.5t + 7.5$ *for all* $t > 0$ *and* $a_t = 2t + 2\log t + O(1)$ *for large* $t$.

The weaker condition (4.21) can be treated as an alternative definition of the sub-Gaussian condition. As we only need (4.21) in Lemma 4.2 (ii) and (iii), condition (4.20) can be often replaced by condition (4.21) in the rest of this chapter. If $\varepsilon \sim N(0, \sigma^2\mathbf{I}_{n\times n})$, then the inequality in Lemma 4.2 (ii) holds with $\lambda_{\text{univ}}(\epsilon/2)$ replaced by the slightly smaller quantity $-\Phi^{-1}(\epsilon/(2p))/\sqrt{n}$. The tail probability bound in Lemma 4.2 (iii) is comparable to that of the chi-square distribution with $r$ degrees of freedom.

**Proof of Lemma 4.2.** (i) Consider $\|\mathbf{a}\|_2 = 1$ without loss of generality. Since the moment generating function of $\mathbf{a}^T\varepsilon$ is bounded from above by $e^{t^2\sigma^2/2}$, its Taylor expansion $1 + \mu_1 t + \mu_2 t^2/2 + \cdots$ must have $\mu_1 = \mathrm{E}\,\mathbf{a}^T\varepsilon = 0$ and $\mu_2 = E(\mathbf{a}^T\varepsilon)^2 \le \sigma^2$. The tail probability bound follows from

$$P\left\{\mathbf{a}^T\varepsilon/\sigma > t\right\} \le e^{-t^2}\,\mathrm{E}\exp(t\mathbf{a}^T\varepsilon/\sigma) \le e^{-t^2/2}.$$

(ii) Since $\|\mathbf{P}\mathbf{X}_j/\sqrt{n}\|_2 \le 1$, the conclusion follows from (4.21) and the union bound.

(iii) The random vector $\mathbf{P}\boldsymbol{\varepsilon}$ is a sub-Gaussian vector living in an $r$-dimensional linear space. Since the sub-Gaussian condition does not depend on the coordinate system, it suffices to consider $\mathbf{P} = \mathbf{I}$ and $n = r$.

The proof of the concentration bound for the tail of $\|\boldsymbol{\varepsilon}\|_2$ involves the packing number for balls in $I\!R^r$. Let $\epsilon \in (0, 1/2)$ and $N$ be the maximum number of balls of radius $\epsilon$ that can be packed into the ball $\{\mathbf{u} : \|\mathbf{u}\|_2 \le 1+\epsilon\}$. Let $\mathbf{u}_1, \ldots, \mathbf{u}_N$ be the centers of these $N$ balls, with $\|\mathbf{u}_j\|_2 \le 1$ necessarily. Because $N$ balls of radius $\epsilon$ are packed into the larger ball of radius $1 + \epsilon$, volume comparison yields $N\epsilon^r \le (1+\epsilon)^r$,, which means $N \le (1 + 1/\epsilon)^r$.

For $\|\mathbf{a}\|_2 \le 1$, the ball with center $\mathbf{a}$ and radius $\epsilon$ must intersect with one of these $N$ balls since one can no longer pack another ball into the larger ball of radius $1 + \epsilon$. This implies $\min_{j \le N} \|\mathbf{a} - \mathbf{u}_j\|_2 \le 2\epsilon$. It follows that

$$\begin{aligned}\max_{\|\mathbf{a}\|_2 \le 1} \mathbf{a}^T\boldsymbol{\varepsilon} - \max_{1 \le j \le N} \mathbf{u}_j^T\boldsymbol{\varepsilon} &\le \max_{\|\mathbf{a}-\mathbf{u}_j\|_2 \le 2\epsilon} (\mathbf{a} - \mathbf{u}_j)^T\boldsymbol{\varepsilon} \\ &\le 2\epsilon \max_{\|\mathbf{a}\|_2 \le 1} \mathbf{a}^T\boldsymbol{\varepsilon}.\end{aligned}$$

Thus, $\max_{1 \le j \le N} \mathbf{u}_j^T\boldsymbol{\varepsilon} \ge (1 - 2\epsilon) \max_{\|\mathbf{a}\|_2 \le 1} \mathbf{a}^T\boldsymbol{\varepsilon} = (1 - 2\epsilon)\|\boldsymbol{\varepsilon}\|_2$. By (4.21),

$$\begin{aligned}& P\Big\{(1-2\epsilon)^2\|\boldsymbol{\varepsilon}\|_2^2/\sigma^2 \ge 2rt + 2r\log(1+1/\epsilon)\Big\} \\ \le\; & N \max_{j \le N} P\Big\{\mathbf{u}_j^T\boldsymbol{\varepsilon}/\sigma \ge \sqrt{2rt + 2r\log(1+1/\epsilon)}\Big\} \\ \le\; & (1+1/\epsilon)^r \exp\big(-(2rt + 2r\log(1+1/\epsilon))/2\big) \\ =\; & e^{-rt}.\end{aligned}$$

This yields the tail probability bound by taking the optimal $\epsilon$. Taking $\epsilon = 0.052$ we find that $a_t \le 2.5t + 7.5$. We omit the calculation of the approximate value of $a_t$ for large $t$. ■

## 4.2 Penalized $L_0$ Selection

We provide a concrete picture of what PLS can achieve by studying penalized subset selection. As we have discussed in Chapter 3, the penalty here is imposed on the number of nonzero estimated coefficients, or the $\ell_0$ norm of the estimator. Among PLS methods, the best subset selection was introduced the earliest (Akaike, 1973; Mallows, 1973; Schwarz, 1978) and has motivated many other proposals. Moreover, the $\ell_0$ PLS attains the rate optimality as in (4.17) and (4.18) under relatively simple conditions on the design. However, the $\ell_0$ PLS is hard to compute because of the discontinuity of the penalty at zero.

Let $p_\lambda(k)$ be the penalty for estimators with $k$ nonzero entries. The *penalized subset selector*, or equivalently *$\ell_0$ PLS estimator*, is defined by

$$\widehat{\boldsymbol\beta}^{(\ell_0)} = \mathop{\arg\min}_{\mathbf{b}\in I\!R^p} \Big\{ \|\mathbf{X}\mathbf{b} - \mathbf{Y}\|_2^2/(2n) + p_\lambda(\|\mathbf{b}\|_0) \Big\}. \tag{4.22}$$

Throughout this section, we assume that $p_\lambda(0) = 0$ and $p_\lambda(k)$ is increasing and sub-additive in $k$; For any $k_1 \le k_2$,

$$p_\lambda(k_1) \le p_\lambda(k_2) \le p_\lambda(k_1) + p_\lambda(k_2 - k_1).$$

We prove in this section that under mild regularity conditions, the prediction, coefficient estimation and selection performance of the $\ell_0$ PLS in (4.22) matches the lower bounds given in Theorems 4.4 and 4.5 up to certain constant factors. Thus, the estimator is rate optimal and the lower bounds in these theorems are also rate optimal.

We first describe the performance of the $\ell_0$ PLS in prediction and coefficient estimation, and its sparsity property. The following theorem is an extension of Theorem 3 of Zhang and Zhang (2012) from the specific $p_\lambda(k) = (\lambda^2/2)k$ to the general *sub-additive penalty*. For nonnegative-definite $\boldsymbol\Sigma$ and $\mathcal{S} \subset \{1, \ldots, p\}$, define a relaxation of the *lower sparse eigenvalue* in (4.14) as

$$\phi_*(m; \mathcal{S}, \boldsymbol\Sigma) = \min_{|B\setminus\mathcal{S}|\le m} \phi_{\min}(\boldsymbol\Sigma_{B,B}). \tag{4.23}$$

This quantity, which is no smaller than (4.14) for $t = m+|\mathcal{S}|$, is often sufficient for studying high-dimensional regression as the inclusion of a true model or a target model is often required anyway. Consider the event

$$\max_{|B|=r} \|\mathbf{P}_B\boldsymbol\varepsilon\|_2 \le \eta\sqrt{2np_\lambda(r)}, \ \forall\, 1 \le r \le n. \tag{4.24}$$

**Theorem 4.6** *Suppose (4.24) holds with $\eta \in (0,1)$. Let $s = \|\boldsymbol\beta\|_0$ and define*

$$m = \max\{j \le n : p_\lambda(j) \le p_\lambda(s)(1+\eta^2)/(1-\eta^2)\}.$$

*Then, we have $\|\widehat{\boldsymbol\beta}^{(\ell_0)}\|_0 \le m$,*

$$\|\mathbf{X}\widehat{\boldsymbol\beta}^{(\ell_0)} - \mathbf{X}\boldsymbol\beta\|_2^2/n \le 2p_\lambda(s)(1+\eta)/(1-\eta),$$

*and with $\overline{\boldsymbol\Sigma} = \mathbf{X}^T\mathbf{X}/n$*

$$\|\widehat{\boldsymbol\beta}^{(\ell_0)} - \boldsymbol\beta\|_2^2 \le \frac{2(1+\eta)p_\lambda(s)}{(1-\eta)\phi_*(m; \mathcal{S}, \overline{\boldsymbol\Sigma})}.$$

**Proof.** For notation simplicity let $\widehat{\boldsymbol\beta} = \widehat{\boldsymbol\beta}^{(\ell_0)}$, $\widehat{s} = \|\widehat{\boldsymbol\beta}\|_0$, $\mathcal{S} = \mathrm{supp}(\boldsymbol\beta)$, $\widehat{\mathcal{S}} = \mathrm{supp}(\widehat{\boldsymbol\beta}^{(\ell_0)})$ and $\mathbf{h} = \widehat{\boldsymbol\beta} - \boldsymbol\beta$. Then, $\mathbf{Y} - \mathbf{X}\widehat{\boldsymbol\beta} = \boldsymbol\varepsilon - \mathbf{X}\mathbf{h}$.

The optimality of $\widehat{\boldsymbol\beta}$ for (4.22) implies its target value no larger than that at $\boldsymbol\beta_0$:

$$\begin{aligned} 0 &\le \|\boldsymbol\varepsilon\|_2^2/(2n) + p_\lambda(s) - \|\boldsymbol\varepsilon - \mathbf{Xh}\|_2^2/(2n) - p_\lambda(\widehat{s}) \\ &= p_\lambda(s) + \boldsymbol\varepsilon^T\mathbf{Xh}/n - \|\mathbf{Xh}\|_2^2/(2n) - p_\lambda(\widehat{s}). \end{aligned}$$

Let $\mathbf{B}$ be the columns corresponding to those in supp($\mathbf{h}$). Then, $\mathbf{Xh} = \mathbf{Bh} = \mathbf{P}_B\mathbf{h}$. By condition (4.24), we have

$$|\boldsymbol\varepsilon^T\mathbf{Xh}| = |\boldsymbol\varepsilon^T\mathbf{P_B}\mathbf{Bh}| \le \|\boldsymbol\varepsilon^T\mathbf{P_B}\|\|\mathbf{Bh}\| \le \eta\sqrt{2np_\lambda(\|\mathbf{h}\|_0)}\|\mathbf{Xh}\|_2.$$

Combining the last two results, we have

$$\begin{aligned} 0 &\le p_\lambda(s) + \eta\sqrt{2p_\lambda(\|\mathbf{h}\|_0)/n}\|\mathbf{Xh}\|_2 - \|\mathbf{Xh}\|_2^2/(2n) - p_\lambda(\widehat{s}) \qquad (4.25)\\ &= p_\lambda(s) + \eta^2 p_\lambda(\|\mathbf{h}\|_0) - p_\lambda(\widehat{s}) - \frac{1}{2n}\Big(\|\mathbf{Xh}\| - \eta\sqrt{2np_\lambda(\|\mathbf{h}\|_0)}\Big)^2. \end{aligned}$$

Since $p_\lambda(\|\mathbf{h}\|_0) \le p_\lambda(s) + p_\lambda(\widehat{s})$ by the sub-additivity of the penalty, we have

$$0 \le p_\lambda(s) + \eta^2[p_\lambda(s) + p_\lambda(\widehat{s})] - p_\lambda(\widehat{s})$$

and the dimension bound follows. Now applying $\sqrt{2ab} \le a + b$ into (4.25), we have

$$0 \le p_\lambda(s) + \eta p_\lambda(\|\mathbf{h}\|_0) - (1-\eta)\|\mathbf{Xh}\|_2^2/(2n) - p_\lambda(\widehat{s}).$$

This yields the prediction performance bound by the sub-additivity of the penalty. Finally, as $\|\mathbf{h}\|_0 \le \|\boldsymbol\beta\|_0 + m$, the $\ell_2$ error bound follows from the prediction error bound and the definition of the lower sparse eigenvalue. ■

Note that the result in Theorem 4.6 is deterministic. As there are at most $\binom{p}{r}$ models of size $r$, the probability of the event in (4.24) is no smaller than $1 - \binom{p}{r}e^{-rt}$ when $2\eta^2 np_\lambda(r) = \sigma^2 r a_t$ with $a_t \approx 2t$ as given in Lemma 4.2 (iii). This leads to the following specific choice:

$$p_\lambda(k) = (A^2\sigma^2/n)\log\binom{p}{k \wedge \lfloor p/2\rfloor}, \qquad \text{for some } A > 1/\eta > 1. \qquad (4.26)$$

For this choice, $p_\lambda(k+1) - p_\lambda(k) = (A^2\sigma^2/n)\log((p-k)/(k+1))$ is decreasing in $k$ for $k+1 \le p/2$, so that $p_\lambda(k)$ is concave in $k \ge 0$ and thus sub-additive. It follows from the Stirling formula that for positive integers $k \le p/2$

$$\begin{aligned} \frac{p_\lambda(k)}{A^2(\sigma^2/n)} &= (1 + o(1))k\log(p/k) \\ &\le k\log(ep/k) - 2^{-1}\log(2\pi k), \end{aligned}$$

and that for positive integers $s$ and $k$ with $s + k \le p/2$,

$$k\log\Big(\frac{p-1}{s+k} - 1\Big) \le \frac{p_\lambda(k+s) - p_\lambda(s)}{A^2(\sigma^2/n)} \le k\log\Big(\frac{p-s}{s+1}\Big).$$

**Corollary 4.2** *(i) Suppose the sub-Gaussian condition (4.21) holds and the* $p_\lambda(k)$ *in (4.26) is taken in (4.22). Then, for large* $p$*, the conclusions of Theorem 4.6 hold with at least probability* $1 - p^{(1-A^2\eta^2)/2}$*. Moreover, when* $p/s$ *is large,* $m = (1+o(1))s(1+\eta^2)/(1-\eta^2)$*.*
*(ii) The conclusions of part (i) also hold for* $p_\lambda(k) = A^2k(\sigma^2/n)\log p$ *with* $A > 1/\eta > 1$*.*

For regression models (4.1) with sub-Gaussian error, the simple dimension penalty $p_\lambda(k) = A^2k(\sigma^2/n)\log p$ is often considered, although the penalty in (4.26) produces sharper results.

It follows from Corollary 4.2 that the prediction and $\ell_2$ estimation losses of the *best subset selector* $\widehat{\boldsymbol{\beta}}^{(\ell_0)}$ in (4.22) and the corresponding lower bounds in Theorem 4.4 are all rate optimal, as they match up to constant factors. Moreover, the size of the selected model is of the right order. An interesting feature of this theory of $\ell_0$ regularization is that the bounds for the model size and prediction error do not require conditions on the design matrix. However, the rate optimality result for the estimation of $\boldsymbol{\beta}$ requires that the involved lower sparse eigenvalues be greater than a fixed constant.

**Proof of Corollary 4.2.** We only prove part (i) as the proof of part (ii) is simpler and is left as an exercise (Exercise 4.5). For each $r$, we set $e^{rt} = \binom{p}{r}^{1+(A^2\eta^2-1)(2/3)}$ in Lemma 4.2 (iii). For large $p$, $\sigma^2 r a_t = (2 + o(1))\sigma^2 rt = (2+o(1))\sigma^2\{1+(A^2\eta^2-1)(2/3)\}\log\binom{p}{r} \le 2\eta^2 n p_\lambda(r)$ as $t$ is large. Thus, as $p$ is large, condition (4.24) holds with at least probability $1 - \sum_{k=1}^{p\wedge n}\binom{p}{k}^{(1-A^2\eta^2)(2/3)} \ge 1 - p^{(1-A^2\eta^2)/2}$. As $p_\lambda(k)$ is nondecreasing and sub-additive, the conclusions of Theorem 4.6 hold. The approximation formula for $m$ follows from the Stirling approximation of $p_\lambda(k)$. ∎

Next, we describe the performance of the best subset selector (4.22) in variable selection. The following theorem improves upon Theorem 4 of Zhang and Zhang (2012) by allowing sub-additive penalty functions and penalty levels not dependent on the lower sparse eigenvalue defined in (4.14). Let $\mathcal{S} = \text{supp}(\boldsymbol{\beta})$, $\widehat{\boldsymbol{\beta}}^o$ be the oracle estimator in (4.10) and $\varepsilon^o = \mathbf{Y} - \mathbf{X}\widehat{\boldsymbol{\beta}}^o$. Consider the condition that

$$\max_{|B|=r, B\subset\mathcal{S}^c} \|\mathbf{P}_B\varepsilon^o\|_2^2/(2n) \le \eta^2\Big\{p_\lambda(r+|\mathcal{S}|) - p_\lambda(|\mathcal{S}|)\Big\} \tag{4.27}$$

for all integers $1 \le r \le p - |\mathcal{S}|$ with certain constants $\eta \in (0,1)$. Define

$$\theta(m) = \max_{|B|\le m, B\subset\mathcal{S}^c} \phi_{\max}\big(\mathbf{P}_B\mathbf{P}_{\mathcal{S}}\big).$$

The largest singular value $\phi_{\max}\big(\mathbf{P}_B\mathbf{P}_{\mathcal{S}}\big)$ has the geometric interpretation as the cosine of the smallest principal angle between the columns spaces of $\mathbf{X}_B$ and $\mathbf{X}_{\mathcal{S}}$. It is highly relevant to the least squares estimation as the method is about orthogonal projections.

**Theorem 4.7** *Suppose $p_\lambda(k)$ is nondecreasing and concave in $k \ge 0$ with $p_\lambda(0) = 0$ and that (4.24) and (4.27) hold. Let $\{\widehat{\boldsymbol\beta}^{(\ell_0)}, \eta, m, \overline{\boldsymbol\Sigma}, \phi_*(m; \mathcal{S}, \overline{\boldsymbol\Sigma})\}$ be as in Theorem 4.6, $\mathcal{S} = \mathrm{supp}(\boldsymbol\beta)$ and $\widehat{\mathcal{S}} = \mathrm{supp}(\widehat{\boldsymbol\beta}^{(\ell_0)})$. Suppose $\theta^2(m) < 1-\eta$ for the $\eta \in (0,1)$ in (4.27). Let $\mathbf{h} = \widehat{\boldsymbol\beta}^{(\ell_0)} - \widehat{\boldsymbol\beta}^{o}$ with the oracle estimator in (4.10) and*

$$s_* = \#\left\{ j \in \mathcal{S} : |\widehat{\beta}_j^o|^2 \le \frac{2(2-\eta)p_\lambda(1)}{\phi_{\min}(\overline{\boldsymbol\Sigma}_{\mathcal{S},\mathcal{S}})\{1-\theta^2(m)/(1-\eta)\}_+} \right\}.$$

*Then, $|\widehat{\mathcal{S}}| \le m$,*

$$|\mathcal{S} \setminus \widehat{\mathcal{S}}| + \{p_\lambda(|\mathcal{S} \cup \widehat{\mathcal{S}}|) - p_\lambda(|\mathcal{S}|)\}/p_\lambda(1) \le s_*(2-\eta)/(1-\eta)$$

*and the prediction and $\ell_2$ estimation errors of $\widehat{\boldsymbol\beta}^{(\ell_0)}$ are bounded by*

$$\frac{\|\mathbf{h}\|_2^2}{\phi_*(m; \mathcal{S}, \overline{\boldsymbol\Sigma})} \le \|\mathbf{X}\mathbf{h}\|_2^2/n \le \frac{2s_* p_\lambda(1)(2-\eta)}{(1-\eta)\eta/(1+\eta)}.$$

For sub-Gaussian errors, Lemma 4.2 (iii) can be used to verify condition (4.27) in the same way as the verification of (4.24) for both the penalty functions in (4.26) and $p_\lambda(k) = k(A^2\sigma^2/n)\log p$. We formally state the result and its implication on selection consistency in the following corollary.

**Corollary 4.3** *Suppose the sub-Gaussian condition (4.21) holds and the $p_\lambda(k) = k(A^2\sigma^2/n)\log p$ is taken with the subset selector in (4.22), with $A > 1/\eta > 1$. Then, for large p, the conclusions of Corollary 4.2 and Theorem 4.7 hold with at least probability $1 - p^{(1-A^2\eta^2)/2}$. In particular,*

$$\min_{j\in\mathcal{S}} |\beta_j| \ge C\sigma\sqrt{(2/n)\log p} \;\; \text{impies} \;\; P\{\widehat{\mathcal{S}} = \mathcal{S}\} \ge 1 - 2p^{(1-A^2\eta^2)/2} \qquad (4.28)$$

*with $\mathcal{S} = \mathrm{supp}(\boldsymbol\beta)$, $\widehat{\mathcal{S}} = \mathrm{supp}(\widehat{\boldsymbol\beta}^{(\ell_0)})$ and*

$$C = \sqrt{\frac{1+A^2\eta^2}{\phi_*(0; \mathcal{S}, \overline{\boldsymbol\Sigma})}} + A\sqrt{\frac{(2-\eta)/(\phi_*(0; \mathcal{S}, \overline{\boldsymbol\Sigma})}{\{1-\theta^2(m)/(1-\eta)\}_+}}.$$

For selection consistency (4.8), the minimum signal strength condition in Corollary 4.3 matches the order of the lower bound in Theorem 4.5 so that both the upper and lower bounds are rate optimal, provided that the involved lower sparse eigenvalues and $1-\theta^2(m)/(1-\eta)$ can be treated as positive constants. Thus, in view of Corollary 4.2, the *best subset selector* (4.22) is rate optimal in prediction, coefficient estimation and variable selection.

As the penalty is imposed on the number of selected variables, $\widehat{\boldsymbol\beta}^{(\ell_0)}$ is the least squares estimator satisfying $\mathbf{X}\widehat{\boldsymbol\beta}^{(\ell_0)} = \mathbf{P}_{\widehat{\mathcal{S}}}\mathbf{Y}$. Thus, when the model is correctly selected, $\widehat{\mathcal{S}} = \text{supp}(\boldsymbol{\beta})$, the estimator (4.22) is identical to the oracle LSE $\widehat{\boldsymbol{\beta}}^o$ in (4.10).

Theorem 4.7 and Corollary 4.3 also assert that the $\ell_2$ estimation and prediction errors beyond those of the oracle $\widehat{\boldsymbol{\beta}}^o$ should be in proportion to $s_*$, the number of small nonzero signals, instead of in proportion to $|\mathcal{S}|$. Thus, the estimator (4.22) may have estimation and prediction errors of smaller order than the lower bounds given in Theorem 4.4 when the fraction $s_*/|\mathcal{S}|$ of small signals is small.

It is worthwhile to mention here that while subset selection enjoys great optimality properties under quite reasonable conditions on the sparsity of $\boldsymbol{\beta}$ and regularity of the design matrix $\mathbf{X}$, it is notoriously difficult to compute or approximate. This has motivated many computationally more efficient regularization methods as discussed and studied in the rest of this chapter.

**Proof of Theorem 4.7.** Let $\mathbf{h} = \widehat{\boldsymbol{\beta}}^{(\ell_0)} - \widehat{\boldsymbol{\beta}}^o$, $B = \widehat{\mathcal{S}} \setminus \mathcal{S}$, $\widehat{s} = |\widehat{\mathcal{S}}|$, $s = |\mathcal{S}|$, $k^* = |\widehat{\mathcal{S}} \cup \mathcal{S}|$ and $k_* = |\widehat{\mathcal{S}} \cap \mathcal{S}|$. We have $\widehat{s} + s = k^* + k_*$ and $|B| \le \widehat{k} \le m$ by Theorem 4.6. As in the proof of Theorem 4.6, the optimality of $\widehat{\boldsymbol{\beta}}^{(\ell_0)}$ implies

$$0 \le p_\lambda(s) + (\boldsymbol{\varepsilon}^o)^T \mathbf{X}\mathbf{h}/n - \|\mathbf{X}\mathbf{h}\|_2^2/(2n) - p_\lambda(\widehat{s}).$$

Because $(\boldsymbol{\varepsilon}^o)^T \mathbf{X}_{\mathcal{S}} = \mathbf{0}$ and $\mathbf{0} = \mathbf{X}_B^T(\mathbf{Y} - \mathbf{X}\widehat{\boldsymbol{\beta}}^{(\ell_0)}) = \mathbf{X}_B^T(\boldsymbol{\varepsilon}^o - \mathbf{X}_{\mathcal{S}}\mathbf{h}_{\mathcal{S}} - \mathbf{X}_B\widehat{\boldsymbol{\beta}}_B^{(\ell_0)})$,

$$\begin{aligned}
& 2(\boldsymbol{\varepsilon}^o)^T \mathbf{X}\mathbf{h} - \|\mathbf{X}\mathbf{h}\|_2^2 \\
= \;& 2(\boldsymbol{\varepsilon}^o)\mathbf{X}_B\widehat{\boldsymbol{\beta}}_B^{(\ell_0)} - 2(\mathbf{X}_{\mathcal{S}}\mathbf{h}_{\mathcal{S}})^T\mathbf{X}_B\widehat{\boldsymbol{\beta}}_B^{(\ell_0)} - \|\mathbf{X}_B\widehat{\boldsymbol{\beta}}_B^{(\ell_0)}\|_2^2 - \|\mathbf{X}_{\mathcal{S}}\mathbf{h}_{\mathcal{S}}\|_2^2 \\
= \;& (\boldsymbol{\varepsilon}^o - \mathbf{X}_{\mathcal{S}}\mathbf{h}_{\mathcal{S}})^T\mathbf{X}_B\widehat{\boldsymbol{\beta}}_B^{(\ell_0)} - \|\mathbf{X}_{\mathcal{S}}\mathbf{h}_{\mathcal{S}}\|_2^2 \\
= \;& \|\mathbf{P}_B(\boldsymbol{\varepsilon}^o - \mathbf{X}_{\mathcal{S}}\mathbf{h}_{\mathcal{S}})\|_2^2 - \|\mathbf{X}_{\mathcal{S}}\mathbf{h}_{\mathcal{S}}\|_2^2 \\
\le \;& \|\mathbf{P}_B\boldsymbol{\varepsilon}^o\|_2^2/\eta + \|\mathbf{P}_B\mathbf{X}_{\mathcal{S}}\mathbf{h}_{\mathcal{S}}\|_2^2/(1-\eta) - \|\mathbf{X}_{\mathcal{S}}\mathbf{h}_{\mathcal{S}}\|_2^2,
\end{aligned}$$

due to $(a+b)^2 \le a^2/\eta + b^2/(1-\eta)$ via Cauchy-Schwarz. By assumption, $\|\mathbf{P}_B\boldsymbol{\varepsilon}^o\|_2^2/(2n\eta) \le \eta\{p_\lambda(k^*) - p_\lambda(s)\}$, so that

$$\begin{aligned}
0 \;\le\; & p_\lambda(s) - p_\lambda(\widehat{s}) + \eta\{p_\lambda(k^*) - p_\lambda(s)\} \\
& + \{\|\mathbf{P}_B\mathbf{X}_{\mathcal{S}}\mathbf{h}_{\mathcal{S}}\|_2^2/(1-\eta) - \|\mathbf{X}_{\mathcal{S}}\mathbf{h}_{\mathcal{S}}\|_2^2\}/(2n).
\end{aligned}$$

As $|B| \le \widehat{k} \le m$, $\|\mathbf{P}_B\mathbf{P}_{\mathcal{S}}\|^2/(1-\eta) \le \theta^2(m)/(1-\eta) < 1$. Thus,

$$0 \le p_\lambda(s) - p_\lambda(\widehat{s}) + \eta\{p_\lambda(k^*) - p_\lambda(s)\} - \{1 - \theta^2(m)/(1-\eta)\}\|\mathbf{X}_{\mathcal{S}}\mathbf{h}_{\mathcal{S}}\|_2^2/(2n).$$

By the definition of $\phi_*(0; \mathcal{S}, \overline{\boldsymbol{\Sigma}})$ and concavity of $p_\lambda(\cdot)$, we have

$$\begin{aligned}
& \{1 - \theta^2(m)/(1-\eta)\}_+ \phi_*(0; \mathcal{S}, \overline{\boldsymbol{\Sigma}})\|\mathbf{h}_{\mathcal{S}}\|_2^2/2 \\
\le \;& \{1 - \theta^2(m)/(1-\eta)\}_+ \|\mathbf{X}_{\mathcal{S}}\mathbf{h}_{\mathcal{S}}\|_2^2/(2n) \\
\le \;& p_\lambda(s) - p_\lambda(\widehat{s}) + \eta\{p_\lambda(k^*) - p_\lambda(s)\} \\
= \;& p_\lambda(k^*) - p_\lambda(\widehat{s}) - (1-\eta)\{p_\lambda(k^*) - p_\lambda(s)\} \\
\le \;& p_\lambda(1)(k^* - \widehat{s}) - (1-\eta)\{p_\lambda(k^*) - p_\lambda(s)\}.
\end{aligned}$$

As $h_j = -\widehat{\beta}_j^o$ for $j \in \mathcal{S} \setminus \widehat{\mathcal{S}}$, it follows from the definition of $s_*$ that

$$\phi_*(0; \mathcal{S}, \overline{\mathbf{\Sigma}})\{1 - \theta^2(m)/(1-\eta)\}_+ \|\mathbf{h}_{\mathcal{S}\setminus\widehat{\mathcal{S}}}\|_2^2 \geq 2(2-\eta)(|\mathcal{S} \setminus \widehat{\mathcal{S}}| - s_*)p_\lambda(1).$$

As $|\mathcal{S} \setminus \widehat{\mathcal{S}}| = k^* - \widehat{s}$, we have

$$(2-\eta)(k^* - \widehat{s} - s_*)p_\lambda(1) \leq p_\lambda(1)(k^* - \widehat{s}) - (1-\eta)\{p_\lambda(k^*) - p_\lambda(s)\}.$$

This yields the selection error bound by simple algebra.

For the prediction bound, we slightly change the calculation via

$$\begin{aligned} 0 &\leq p_\lambda(s) + (\varepsilon^o)^T \mathbf{X}\mathbf{h}/n - \|\mathbf{X}\mathbf{h}\|_2^2/(2n) - p_\lambda(\widehat{s}). \\ &\leq p_\lambda(s) - p_\lambda(\widehat{s}) + \{\|\mathbf{P}_B \varepsilon^o\|_2^2/\eta^2 + \|\mathbf{P}_B \mathbf{X}_{\mathcal{S}} \mathbf{h}_{\mathcal{S}}\|_2^2/(1-\eta^2) - \|\mathbf{X}_{\mathcal{S}}\mathbf{h}_{\mathcal{S}}\|_2^2\}/(2n) \\ &\leq p_\lambda(k^*) - p_\lambda(\widehat{s}) + \|\mathbf{X}_{\mathcal{S}}\mathbf{h}_{\mathcal{S}}\|_2^2/(1+\eta) - \|\mathbf{X}_{\mathcal{S}}\mathbf{h}_{\mathcal{S}}\|_2^2. \end{aligned}$$

As $\mathbf{X}\mathbf{h} = \mathbf{X}_B \mathbf{h}_B + \mathbf{X}_{\mathcal{S}}\mathbf{h}_{\mathcal{S}} = \mathbf{P}_B(\varepsilon^o - \mathbf{X}_{\mathcal{S}}\mathbf{h}_{\mathcal{S}}) + \mathbf{X}_{\mathcal{S}}\mathbf{h}_{\mathcal{S}}$, we have

$$\begin{aligned} & \{\eta/(1+\eta)\}\|\mathbf{X}\mathbf{h}\|_2^2/(2n) \\ = \; & \{\eta/(1+\eta)\}\|\mathbf{P}_B\varepsilon^o\|_2^2/(2n) + \{\eta/(1+\eta)\}\|\mathbf{P}_B^\perp \mathbf{X}_{\mathcal{S}}\mathbf{h}_{\mathcal{S}}\|_2^2/(2n) \\ \leq \; & \{\eta/(1+\eta)\}\eta^2\{p_\lambda(k^*) - p_\lambda(s)\} + p_\lambda(k^*) - p_\lambda(\widehat{s}) \\ \leq \; & p_\lambda(k^*) - p_\lambda(s) + p_\lambda(1)(k^* - \widehat{s}) \\ \leq \; & s_* p_\lambda(1)(2-\eta)/(1-\eta). \end{aligned}$$

The $\ell_2$ error bound follows immediately from the prediction error bound and the definition of $\phi_*(m; \mathcal{S}, \overline{\mathbf{\Sigma}})$ in (4.23). ■

## 4.3 Lasso and Dantzig Selector

The *Lasso* is defined as

$$\widehat{\boldsymbol{\beta}}^{(\ell_1)} \equiv \widehat{\boldsymbol{\beta}}(\lambda) = \arg\min_{\mathbf{b}} \Big\{ \|\mathbf{Y} - \mathbf{X}\mathbf{b}\|_2^2/(2n) + \lambda\|\mathbf{b}\|_1 \Big\}, \tag{4.29}$$

where $\mathbf{X}$ and $\mathbf{Y}$ are the design matrix and response vector in (4.1) and $\lambda$ is a penalty level. The name Lasso is an acronym that stands for the "least absolute shrinkage and selection operator" Tibshirani (1996).

We have shown in the previous section that the penalized subset selector (4.22) attains rate optimality in prediction, coefficient estimation and variable selection by matching its error bounds with the lower bounds derived in Section 4.1. However, best subset selection is not feasible in many applications due to its computational complexity, as the evaluation of $\binom{p}{k}$ least squares estimators is required to find the $k$-dimensional sub-model with the best fit. To alleviate this computational burden, the $\ell_1$ penalty can be used as a surrogate for the $\ell_0$ penalty, resulting in a *convex minimization* problem. What is the cost of this computational convenience in terms of statistical performance? We answer this question here by studying theoretical properties of the Lasso.

The *Dantzig selector*, proposed by Candés and Tao (2007), is an $\ell_1$ regularization method closely related to the Lasso. It is defined as

$$\widehat{\boldsymbol{\beta}}^{(DS)} \equiv \widehat{\boldsymbol{\beta}}^{(DS)}(\lambda) = \arg\min_{\mathbf{b}} \Big\{ \|\mathbf{b}\|_1 : \|\mathbf{X}^T(\mathbf{Y} - \mathbf{X}\mathbf{b})/n\|_\infty \le \lambda \Big\}. \tag{4.30}$$

We will also study properties of the Dantzig selector in this section as the analyses of the Lasso and Dantzig selector are parallel to each other.

The properties of these two estimators have been prelimilarily studied in Section 3.3.

### *4.3.1 Selection consistency*

We begin with a careful statement of the Karush-Kuhn-Tucker (KKT) conditions, which characterize the Lasso. As outlined in Section 3.3.2, conditions for the selection consistency of the Lasso can be directly derived from the KKT conditions for the procedure.

As both the loss function $\|\mathbf{Y}-\mathbf{X}\mathbf{b}\|_2^2/(2n)$ and penalty function $\lambda\|\mathbf{b}\|_1$ are convex in $\mathbf{b}$, the Lasso is a convex minimization problem. As such, a vector $\widehat{\boldsymbol{\beta}}$ is a solution of (4.29) if and only if the penalized loss is supported by a constant hyperplane at $\mathbf{b} = \widehat{\boldsymbol{\beta}}$. Moreover, the solution is unique if and only if the intersection of the support hyperplane and the graph of the penalized loss, which is a nonempty convex set, contains exactly one point. The KKT conditions, stated in the following proposition, are in essence an algebraic expression of the geometric interpretation of the Lasso solution.

**Proposition 4.1** *A vector $\widehat{\boldsymbol{\beta}} = \widehat{\boldsymbol{\beta}}(\lambda)$ is a solution of the Lasso, or equivalently a minimizer of the penalized loss in (4.29), if and only if the negative gradient of the loss, $\mathbf{g} = \mathbf{g}(\lambda) = (g_1, \ldots, g_p)^T = \mathbf{X}_j^T(\mathbf{Y} - \mathbf{X}\widehat{\boldsymbol{\beta}})/n$, satisfies*

$$g_j = \begin{cases} \lambda \mathrm{sgn}(\widehat{\beta}_j), & \widehat{\beta}_j \neq 0 \\ \in [-\lambda, \lambda], & \widehat{\beta}_j = 0. \end{cases} \tag{4.31}$$

*Moreover, $\widehat{\boldsymbol{\beta}}$ is the unique solution of the Lasso if and only if $\|\mathbf{X}\mathbf{u}\|_2^2 > 0$ for all $\mathbf{u} \neq 0$ satisfying the conditions $\mathrm{supp}(\mathbf{u}) \subseteq \{j : |g_j| = \lambda\}$ and $g_j u_j \geq 0$ for all $j$ with $\widehat{\beta}_j = 0$. In particular, the Lasso solution is unique if $\mathrm{rank}(\mathbf{X}_{\mathcal{S}_1}) = |\mathcal{S}_1|$, where $\mathcal{S}_1 = \{j : |g_j| = \lambda\}$.*

**Proof of Proposition 4.1.** Let $L_\lambda(\mathbf{b}) = \|\mathbf{Y} - \mathbf{X}\mathbf{b}\|_2^2/(2n) + \lambda\|\mathbf{b}\|_1$. For each $\mathbf{u} \in I\!R^p$, there exists a real number $t^* > 0$ (depending on $\mathbf{u}$) such that for all $0 < t < t^*$

$$\begin{aligned} & (\partial/\partial t)L_\lambda(\widehat{\boldsymbol{\beta}} + t\mathbf{u}) \\ = \; & t\|\mathbf{X}\mathbf{u}\|_2^2/n + \sum_{\widehat{\beta}_j \neq 0} (-g_j + \lambda \mathrm{sgn}(\widehat{\beta}_j))u_j + \sum_{\widehat{\beta}_j = 0} (-g_j u_j + \lambda |u_j|). \end{aligned}$$

Because $L_\lambda(\mathbf{b})$ is convex in $\mathbf{b}$, it is minimized at $\mathbf{b} = \widehat{\boldsymbol{\beta}}$ if and only if

$(\partial/\partial t)L_\lambda(\widehat{\boldsymbol\beta}+t\mathbf{u})\big|_{t=0+} \ge 0$ for all $\mathbf{u} \in I\!R^p$. Since this holds for all $\mathbf{u} \in I\!R^p$, it is equivalent to (4.31).

For the uniqueness of the solution, it suffices to consider directions $\mathbf{u} \neq 0$ in which $(\partial/\partial t)L_\lambda(\widehat{\boldsymbol\beta}+t\mathbf{u})\big|_{t=0+} = 0$. That is the collection of all $\mathbf{u}$ satisfying $\mathrm{supp}(\mathbf{u}) \subseteq \{j : |g_j| = \lambda\}$ and $g_j u_j \ge 0$ for $\widehat{\beta}_j = 0$. As the second derivative of $L_\lambda(\widehat{\boldsymbol\beta}+t\mathbf{u})$ is $\|\mathbf{X}\mathbf{u}\|_2^2/n$ for $0 < t < t^*$, the necessity and sufficiency of the uniqueness condition follow. ■

The Lasso is selection consistent if and only if a solution of the KKT conditions has support $\mathcal{S} = \mathrm{supp}(\boldsymbol\beta)$. However, this statement is not far from the definition of selection consistency itself. A better understanding of the selection consistency of the Lasso requires a more explicit statement of the conditions for the property in terms of the design matrix, the coefficient vector and the noise. To this end, we define

$$\kappa_{\text{select}} = \|\overline{\boldsymbol\Sigma}_{\mathcal{S}^c,\mathcal{S}}\overline{\boldsymbol\Sigma}_{\mathcal{S},\mathcal{S}}^{-1}\,\mathrm{sgn}(\boldsymbol\beta_{\mathcal{S}})\|_\infty,$$

where $\mathcal{S} = \mathrm{supp}(\boldsymbol\beta)$ and $\overline{\boldsymbol\Sigma} = \mathbf{X}^T\mathbf{X}/n$. Selection consistency of the Lasso hinges on the following two conditions. The first one, called neighborhood stability condition in Meinshausen and Bühlmann (2006), the exact recovery coefficient condition in Tropp (2006), and strong irrepresentable condition in Zhao and Yu (2006), can be written as

$$\kappa_{\text{select}} < 1. \tag{4.32}$$

The second one, a uniform signal strength condition, can be written as

$$\min_{j\in\mathcal{S}}\Big(|\beta_j| - \mathrm{sgn}(\beta_j)\lambda(\overline{\boldsymbol\Sigma}_{\mathcal{S},\mathcal{S}}^{-1}\mathrm{sgn}(\boldsymbol\beta_{\mathcal{S}}))_j - \lambda'(\overline{\boldsymbol\Sigma}_{\mathcal{S},\mathcal{S}}^{-1})_{j,j}^{1/2}\Big) > 0, \tag{4.33}$$

with the penalty level $\lambda$ in (4.29) and a suitable $\lambda' > 0$.

**Theorem 4.8** *Let* $\mathcal{S} = \mathrm{supp}(\boldsymbol\beta)$ *and* $\mathbf{P}_{\mathcal{S}}$ *be the orthogonal projection to the column space of* $\mathbf{X}_{\mathcal{S}}$. *Suppose* $\mathrm{rank}(\mathbf{X}_{\mathcal{S}}) = |\mathcal{S}|$. *Let* $\widehat{\boldsymbol\beta} = \widehat{\boldsymbol\beta}(\lambda)$ *be the Lasso estimator in (4.29) and* $\widehat{\boldsymbol\beta}^{o}$ *the oracle LSE in (4.10).*
*(i) Suppose* $\max_{j\notin\mathcal{S}}|\mathbf{X}_j^T\mathbf{P}_{\mathcal{S}}^{\perp}\boldsymbol\varepsilon/n| \le (1-\kappa_{\text{select}})\lambda$ *and* $\mathrm{sgn}(\boldsymbol\beta_{\mathcal{S}}) = \mathrm{sgn}(\widehat{\boldsymbol\beta}^{o}_{\mathcal{S}} - \lambda\overline{\boldsymbol\Sigma}_{\mathcal{S},\mathcal{S}}^{-1}\mathrm{sgn}(\boldsymbol\beta_{\mathcal{S}}))$. *Then, a Lasso solution is sign consistent:*

$$\mathrm{sgn}(\widehat{\boldsymbol\beta}) = \mathrm{sgn}(\boldsymbol\beta).$$

*Moreover, the solution is unique if the first condition holds strictly.*
*(ii) Suppose* $\boldsymbol\varepsilon$ *is a symmetric continuous random vector. If either (4.33) fails to hold for* $\lambda' = 0$ *or if* $\kappa_{\text{select}} \ge 1$, *then*

$$P\Big\{\mathrm{sgn}(\widehat{\boldsymbol\beta}) = \mathrm{sgn}(\boldsymbol\beta)\Big\} \le 1/2.$$

*Moreover, if (4.33) holds and* $\kappa_{\text{select}} \ge 1$*, then*

$$\begin{aligned} & P\Big\{\text{supp}(\widehat{\boldsymbol\beta}) = \text{supp}(\boldsymbol\beta)\Big\} \\ \le\ & 1/2 + P\Big\{\min_{j\in\mathcal{S}} \text{sgn}(\beta_j)(\widehat\beta_j^o - \beta_j)/(\overline{\boldsymbol\Sigma}_{\mathcal{S},\mathcal{S}}^{-1})_{j,j}^{1/2} < -\lambda'\Big\}. \end{aligned}$$

*(iii) Suppose* $\boldsymbol\varepsilon$ *satisfies the sub-Gaussian condition (4.21),* $\max_j \|\mathbf{P}_{\mathcal{S}}^{\perp}\mathbf{X}_j\|_2^2 \le n$*,* $\lambda \ge (1-\kappa_{\text{select}})_+^{-1}\sigma\sqrt{(2/n)\log((p-|\mathcal{S}|)/\epsilon)}$*, and (4.33) holds with* $\lambda' = \sigma\sqrt{(2/n)\log(|\mathcal{S}|/\epsilon)}$*. Then,*

$$P\Big\{\text{sgn}(\widehat{\boldsymbol\beta}) = \text{sgn}(\boldsymbol\beta)\Big\} \ge 1 - 3\epsilon.$$

The probability of selection consistency is bounded by 1/2 in Theorem 4.8 (ii) as we only tested the KKT conditions in one worst coordinate with the symmetry assumption. This probability could be much smaller in the setting of Theorem 4.8 (ii) as the KKT conditions involve $p$ coordinates.

Among the conditions of Theorem 4.8, (4.32) and (4.33) are most significant and most likely to limit the scope of application of the Lasso for the purpose of variable selection. And yet these conditions are necessary by part (ii) of the theorem. Since $\|\text{sgn}(\boldsymbol\beta_{\mathcal{S}})\|_2 = |\mathcal{S}|^{1/2}$, one cannot bound $\kappa_{\text{select}}$ or $\text{sgn}(\beta_j)\lambda(\overline{\boldsymbol\Sigma}_{\mathcal{S},\mathcal{S}}^{-1}\text{sgn}(\boldsymbol\beta_{\mathcal{S}}))_j$ by $\ell_2$-type quantities such as the eigenvalues of sub-matrices of $\overline{\boldsymbol\Sigma}$ of proper size. This is natural as both conditions (4.32) and (4.33) are of the $\ell_\infty$ type. In fact, if the maximum is taken over $\text{sgn}(\boldsymbol\beta_{\mathcal{S}})$ given $\mathcal{S}$, condition (4.32) can be written as $\|\overline{\boldsymbol\Sigma}_{\mathcal{S}^c,\mathcal{S}}\overline{\boldsymbol\Sigma}_{\mathcal{S},\mathcal{S}}^{-1}\|_\infty < 1$ and a slightly strengthened version of (4.33) can be written as

$$\min_{\beta_j\ne 0} |\beta_j| \ge \lambda\|\overline{\boldsymbol\Sigma}_{\mathcal{S},\mathcal{S}}^{-1}\|_\infty + \lambda'\|\operatorname{diag}(\overline{\boldsymbol\Sigma}_{\mathcal{S},\mathcal{S}}^{-1})\|_\infty.$$

For Gaussian designs $\mathbf{X}$ with iid $N(0,\boldsymbol\Sigma)$ rows and given $\mathcal{S}$, Wainwright (2009) proved that $\kappa_{\text{select}}$ and $\|\overline{\boldsymbol\Sigma}_{\mathcal{S},\mathcal{S}}^{-1}\text{sgn}(\boldsymbol\beta_{\mathcal{S}})\|_\infty$ are within a small fraction of their population versions with $\boldsymbol\Sigma$ in place of $\overline{\boldsymbol\Sigma}$ when $\|\boldsymbol\beta\|_0(\log p)/n$ is sufficiently small. For random $\boldsymbol\beta$ with uniformly distributed $\text{sgn}(\boldsymbol\beta)$ given $\{|\beta_j|, j\le p\}$, Candés and Plan (2009) proved $\|\overline{\boldsymbol\Sigma}_{\mathcal{S},\mathcal{S}}^{-1}\text{sgn}(\boldsymbol\beta_{\mathcal{S}})\|_\infty \le 2$ and $\kappa_{\text{select}} \le 1 - 1/\sqrt{2}$ with large probability when $\|\boldsymbol\beta\|_0\{\|\mathbf{X}\mathbf{X}^T/p\|\log p\}/n$ and $\max_{j\ne k}|\mathbf{X}_j\mathbf{X}_k/n|(\log p)$ are bounded.

It will be shown in our study of folded concave penalties that condition (4.32) is not necessary for variable selection consistency by PLS algorithms with computation complexity similar to or slightly higher than the Lasso. The main reason for the requirement of this irrepresentable condition for the selection consistency of the Lasso is the bias of the Lasso, as the gradient does not diminish for large signals by the KKT conditions (4.31). However, we will prove in Subsection 4.3.3 that the magnitude of the selection error of the Lasso can still be controlled at a reasonable level when (4.32) is replaced with certain $\ell_2$-type conditions on the design matrix.

Condition (4.33) is a consequence of the uniform signal strength condition

$$\beta_{\min} = \min_{j\in\mathcal{S}} |\beta_j| \ge \lambda\|\overline{\boldsymbol{\Sigma}}_{\mathcal{S},\mathcal{S}}^{-1}\mathrm{sgn}(\boldsymbol{\beta}_{\mathcal{S}})\|_\infty + \lambda'\|\,\mathrm{diag}(\overline{\boldsymbol{\Sigma}}_{\mathcal{S},\mathcal{S}}^{-1})\|_\infty^{1/2}.$$

When $\log(|\mathcal{S}|) = o(1)\log((p-|\mathcal{S}|))$, this condition can be written as

$$\beta_{\min} \ge (1+o(1))\lambda\|\overline{\boldsymbol{\Sigma}}_{\mathcal{S},\mathcal{S}}^{-1}\mathrm{sgn}(\boldsymbol{\beta}_{\mathcal{S}})\|_\infty/(1-\kappa_{\mathrm{select}}) \tag{4.34}$$

with the $\lambda$ in (4.18). It follows from Theorem 4.5 that this uniform signal strength condition is rate optimal for variable selection consistency. The problem with (4.34) is that the boundedness of the factor $\|\overline{\boldsymbol{\Sigma}}_{\mathcal{S},\mathcal{S}}^{-1}\mathrm{sgn}(\boldsymbol{\beta}_{\mathcal{S}})\|_\infty/(1-\kappa_{\mathrm{select}})$ is restrictive and hard to interpret. Again it will be shown in our study of folded concave PLS that the factor can be replaced with a constant when a certain restricted eigenvalue of the Gram matrix is bound away from zero.

**Proof of Theorem 4.8.** (i) It follows from Proposition 4.1 that $\widehat{\boldsymbol{\beta}}$ is a solution of (4.29) satisfying $\mathrm{sgn}(\widehat{\boldsymbol{\beta}}) = \mathrm{sgn}(\boldsymbol{\beta})$ if and only if

$$\begin{aligned} &\mathbf{X}_j^T(\mathbf{Y}-\mathbf{X}_{\mathcal{S}}\widehat{\boldsymbol{\beta}}_{\mathcal{S}})/n = \lambda\,\mathrm{sgn}(\beta_j) = \lambda\,\mathrm{sgn}(\widehat{\beta}_j),\ j\in\mathcal{S},\\ &|\mathbf{X}_j^T(\mathbf{Y}-\mathbf{X}_{\mathcal{S}}\widehat{\boldsymbol{\beta}}_{\mathcal{S}})/n| \le \lambda \text{ and } \widehat{\beta}_j = 0,\ j\notin\mathcal{S}. \end{aligned} \tag{4.35}$$

Moreover, since $\mathrm{rank}(\mathbf{X}_{\mathcal{S}}) = |\mathcal{S}|$, the uniqueness of this solution follows from Proposition 4.1 if the inequality in (4.35) holds strictly for all $j\notin\mathcal{S}$.

Since the number of equations matches the number of $\widehat{\beta}_j$ for $j\in\mathcal{S}$ in (4.35), the equations can be solved to obtain

$$\begin{aligned} \widehat{\boldsymbol{\beta}}_{\mathcal{S}} &= (\mathbf{X}_{\mathcal{S}}^T\mathbf{X}_{\mathcal{S}})^{-1}\{\mathbf{X}_{\mathcal{S}}^T\mathbf{Y} - n\lambda\mathrm{sgn}(\boldsymbol{\beta}_{\mathcal{S}})\}\\ &= \widehat{\boldsymbol{\beta}}_{\mathcal{S}}^o - \lambda\overline{\boldsymbol{\Sigma}}_{\mathcal{S},\mathcal{S}}^{-1}\mathrm{sgn}(\boldsymbol{\beta}_{\mathcal{S}}). \end{aligned}$$

Moreover, since $\mathbf{X}_{\mathcal{S}}\widehat{\boldsymbol{\beta}}_{\mathcal{S}}^o = \mathbf{P}_{\mathcal{S}}\mathbf{Y}$ and $\mathbf{P}_{\mathcal{S}}^\perp\mathbf{Y} = \mathbf{P}_{\mathcal{S}}^\perp\boldsymbol{\varepsilon}$, (4.35) can be written as

$$\begin{aligned} &\mathrm{sgn}(\widehat{\boldsymbol{\beta}}_{\mathcal{S}}^o - \lambda\overline{\boldsymbol{\Sigma}}_{\mathcal{S},\mathcal{S}}^{-1}\mathrm{sgn}(\boldsymbol{\beta}_{\mathcal{S}})) = \mathrm{sgn}(\boldsymbol{\beta}_{\mathcal{S}}),\\ &\|\mathbf{X}_{\mathcal{S}^c}^T\mathbf{P}_{\mathcal{S}}^\perp\boldsymbol{\varepsilon}/n + \lambda\overline{\boldsymbol{\Sigma}}_{\mathcal{S}^c,\mathcal{S}}\overline{\boldsymbol{\Sigma}}_{\mathcal{S},\mathcal{S}}^{-1}\mathrm{sgn}(\boldsymbol{\beta}_{\mathcal{S}})\|_\infty \le \lambda. \end{aligned} \tag{4.36}$$

The conclusions follow from (4.36) with an application of the triangle inequality, $\|\mathbf{X}_{\mathcal{S}^c}^T\mathbf{P}_{\mathcal{S}}^\perp\boldsymbol{\varepsilon}/n + \lambda\overline{\boldsymbol{\Sigma}}_{\mathcal{S}^c,\mathcal{S}}\overline{\boldsymbol{\Sigma}}_{\mathcal{S},\mathcal{S}}^{-1}\mathrm{sgn}(\boldsymbol{\beta}_{\mathcal{S}})\|_\infty \le \|\mathbf{X}_{\mathcal{S}^c}^T\mathbf{P}_{\mathcal{S}}^\perp\boldsymbol{\varepsilon}/n\|_\infty + \lambda\kappa_{\mathrm{select}}$.

(ii) The first condition in (4.36) can be written as

$$\mathrm{sgn}(\beta_j)(\widehat{\beta}_j^o - \beta_j) + |\beta_j| - \mathrm{sgn}(\beta_j)\lambda(\overline{\boldsymbol{\Sigma}}_{\mathcal{S},\mathcal{S}}^{-1}\mathrm{sgn}(\boldsymbol{\beta}_{\mathcal{S}}))_j > 0,\ \forall\, j\in\mathcal{S}. \tag{4.37}$$

Note that $P\{\mathrm{sgn}(\beta_j)(\widehat{\beta}_j^o - \beta_j) \le 0\} = 1/2$ by the symmetry of $\widehat{\beta}_j^o - \beta_j$. If (4.33) fails to hold for $\lambda' = 0$, then $|\beta_j| - \mathrm{sgn}(\beta_j)\lambda(\overline{\boldsymbol{\Sigma}}_{\mathcal{S},\mathcal{S}}^{-1}\mathrm{sgn}(\boldsymbol{\beta}_{\mathcal{S}}))_j \le 0$ for some $j\in\mathcal{S}$. Namely, the second term in (4.37) is also non-positive. Thus, (4.37) fails with at least 1/2 probability by the symmetry of $\widehat{\beta}_j^o - \beta_j$. Similarly, when $\kappa_{\text{select}} \ge 1$, $|\overline{\mathbf{\Sigma}}_{j,\mathcal{S}}\overline{\mathbf{\Sigma}}_{\mathcal{S},\mathcal{S}}^{-1}\text{sgn}(\boldsymbol{\beta}_{\mathcal{S}})| \ge 1$ for some $j \in \mathcal{S}^c$, so that the second condition in (4.36) fails with at least 1/2 probability, due to the symmetry of $\mathbf{X}_j^T\mathbf{P}_{\mathcal{S}}^{\perp}\boldsymbol{\varepsilon}$. When (4.33) holds and $\text{sgn}(\beta_j)(\widehat{\beta}_j^o - \beta_j) \ge -\lambda'(\overline{\mathbf{\Sigma}}_{\mathcal{S},\mathcal{S}}^{-1})_{j,j}^{1/2}$ for all $j \in \mathcal{S}$, the first condition in (4.36) holds. In this case, $\text{supp}(\widehat{\boldsymbol{\beta}}) = \text{supp}(\boldsymbol{\beta})$ if and only if $\text{sgn}(\widehat{\boldsymbol{\beta}}) = \text{sgn}(\boldsymbol{\beta})$, if and only if the second condition in (4.36) holds. Again, the probability for the second condition in (4.36) to hold is no greater than 1/2 when $\kappa_{\text{select}} \ge 1$.

(iii) We claim that for certain vectors $\mathbf{a}_j$ satisfying $\|\mathbf{a}_j\|_2^2 \le 1$,

$$\text{sgn}(\beta_j)(\widehat{\beta}_j^o - \beta_j)/(\overline{\mathbf{\Sigma}}_{\mathcal{S},\mathcal{S}}^{-1})_{j,j}^{1/2} = \mathbf{a}_j^T\boldsymbol{\varepsilon}/n^{1/2},\ j \in \mathcal{S},$$
$$\mathbf{X}_j^T\mathbf{P}_{\mathcal{S}}^{\perp}\boldsymbol{\varepsilon}/n = \mathbf{a}_j^T\boldsymbol{\varepsilon}/n^{1/2},\ j \notin \mathcal{S}.$$

It suffices to verify this claim in the Gaussian case $\boldsymbol{\varepsilon} \sim N(0, \sigma^2\mathbf{I}_{n\times n})$, and this can be easily seen via the variance calculation $\text{Var}(\widehat{\beta}_j^o) = (\overline{\mathbf{\Sigma}}_{\mathcal{S},\mathcal{S}}^{-1})_{j,j}\sigma^2/n$ and $\text{Var}(\mathbf{X}_j^T\mathbf{P}_{\mathcal{S}}^{\perp}\boldsymbol{\varepsilon}/n) = \sigma^2\|\mathbf{X}_j\mathbf{P}_{\mathcal{S}}^{\perp}\|_2^2/n^2 \le \sigma^2/n$. By the sub-Gaussian condition, $P\{|\mathbf{a}_j^T\boldsymbol{\varepsilon}| \ge \sigma t\} \le e^{-t^2/2}$ for all $j$, so that by the choice of $\lambda'$,

$$P\Big\{\text{sgn}(\beta_j)(\widehat{\beta}_j^o - \beta_j) \le \lambda'(\overline{\mathbf{\Sigma}}_{\mathcal{S},\mathcal{S}}^{-1})_{j,j}^{1/2}\forall j \in \mathcal{S}\Big\} \le |\mathcal{S}|e^{-n\lambda'^2/2} \le \epsilon.$$

Thus, condition (4.33) implies that the first part of (4.36) holds with at least probability $1-\epsilon$. Similarly, by the choice of $\lambda$,

$$P\Big\{\|\mathbf{X}_{\mathcal{S}^c}^T\mathbf{P}_{\mathcal{S}}^{\perp}\boldsymbol{\varepsilon}/n\|_\infty + \lambda\kappa_{\text{select}} > \lambda\Big\} \le 2\epsilon,$$

so that the second part of (4.36) holds with at least probability $1-2\epsilon$. ■

### 4.3.2 *Prediction and coefficient estimation errors*

To study prediction and coefficient estimation, we consider a general target predictor $\boldsymbol{\beta}^*$. The choice of $\boldsymbol{\beta}^*$ is quite flexible; possible candidates include the true $\boldsymbol{\beta}$ in (4.1), the oracle estimator (4.10), and its extension given by

$$\widehat{\boldsymbol{\beta}}_{\mathcal{S}}^o = \boldsymbol{\beta}_{\mathcal{S}} + (\mathbf{X}_{\mathcal{S}}^T\mathbf{X}_{\mathcal{S}})^{-1}\mathbf{X}_{\mathcal{S}}^T\boldsymbol{\varepsilon},\ \widehat{\boldsymbol{\beta}}_{\mathcal{S}^c}^o = \boldsymbol{\beta}_{\mathcal{S}^c}, \tag{4.38}$$

allowing large $\|\boldsymbol{\beta}\|_0$ with $\mathcal{S} \ne \text{supp}(\boldsymbol{\beta})$.

The target $\boldsymbol{\beta}^*$ is not required to be sparse. However, we assume the existence of a sparse $\overline{\boldsymbol{\beta}}$ such that $\mathbf{X}\overline{\boldsymbol{\beta}}$ approximates $\mathbf{X}\boldsymbol{\beta}^*$ well. This condition is essential for the Lasso to perform well since the method aims to produce a sparse predictor $\widehat{\boldsymbol{\beta}}$. For $\boldsymbol{\beta}^* = \boldsymbol{\beta}$, a quantitative description of this condition is given as the approximate sparsity condition in Definition 4.2. We will show that the Lasso achieves rate optimal performance in prediction by approximating such $\overline{\boldsymbol{\beta}}$ under mild conditions. As the error bounds provided here allow for the comparison of the performance of the Lasso with that of the oracle estimator, they are sometimes referred to as oracle inequalities.

We need to define some quantities involving a set $\mathcal{S} \subset \{1, \ldots, p\}$, the sign of $\overline{\boldsymbol{\beta}}$ on $\mathcal{S}$, and the Gram matrix $\overline{\boldsymbol{\Sigma}} = \mathbf{X}^T\mathbf{X}/n$. We may view $\mathcal{S}$ as an index set of variables with high prediction power although the choice of $\mathcal{S}$ is quite arbitrary. For $\overline{\Delta} \geq 0$, $0 \leq \eta_1$ and $\eta_2 \leq 1$, define a constant factor for bounding the prediction error of the Lasso as

$$\begin{aligned} \overline{C}^{(\ell_1)}_{\text{pred}}(\mathcal{S}, \overline{\Delta}) &\equiv \overline{C}^{(\ell_1)}_{\text{pred}}(\mathcal{S}, \overline{\Delta}; \eta_1, \eta_2) \\ &= \sup \left\{ \frac{\left[\psi(\mathbf{u}) + \{\psi^2(\mathbf{u}) + 2\overline{\Delta}\}^{1/2}\right]^2}{4(1 \vee |\mathcal{S}|)} : \mathbf{u}^T \overline{\boldsymbol{\Sigma}} \mathbf{u} = 1 \right\}, \end{aligned} \tag{4.39}$$

where $\psi(\mathbf{u}) = \eta_1 \|\mathbf{u}_{\mathcal{S}}\|_1 - \text{sgn}(\overline{\boldsymbol{\beta}}_{\mathcal{S}})^T \mathbf{u}_{\mathcal{S}} - (1 - \eta_2)\|\mathbf{u}_{\mathcal{S}^c}\|_1$ with the convention $\text{sgn}(0) = 0$. Similarly, for any seminorm $\|\cdot\|$ as a loss function, define a constant factor for bounding the estimation error of the Lasso as

$$\begin{aligned} \overline{C}^{(\ell_1)}_{\text{est}}(\mathcal{S}, \overline{\Delta}; \|\cdot\|) &\equiv \overline{C}^{(\ell_1)}_{\text{est}}(\mathcal{S}, \overline{\Delta}; \|\cdot\|, \eta_1, \eta_2) \\ &= \sup \left\{ \|\mathbf{u}\| \left[\psi(\mathbf{u}) + \{\psi^2(\mathbf{u}) + 2\overline{\Delta}\}^{1/2}\right]/2 : \mathbf{u}^T \overline{\boldsymbol{\Sigma}} \mathbf{u} = 1 \right\}. \end{aligned} \tag{4.40}$$

The normalization factor $1 \vee |\mathcal{S}|$ in (4.39) is tailored to match the $\ell_2$-type scaling $\mathbf{u}^T \overline{\boldsymbol{\Sigma}} \mathbf{u} = 1$, so that $\overline{C}^{(\ell_1)}_{\text{pred}}(\mathcal{S}, \overline{\Delta})$ can be reasonably treated as a constant independent of $\{n, p, |\mathcal{S}|\}$, at least in the $p < n$ setting. However, as the loss $\|\cdot\|$ is arbitrary, (4.40) is not properly dimension normalized. For the $\ell_q$ estimation loss, we define dimension-normalized version of (4.40) as

$$\begin{aligned} \overline{C}^{(\ell_1)}_{\text{est},q}(\mathcal{S}, \overline{\Delta}) &\equiv \overline{C}^{(\ell_1)}_{\text{est},q}(\mathcal{S}, \overline{\Delta}; \eta_1, \eta_2) \\ &= \overline{C}^{(\ell_1)}_{\text{est}}(\mathcal{S}, \overline{\Delta}; \|\cdot\|_q, \eta_1, \eta_2)/(1 \vee |\mathcal{S}|)^{1/(q \wedge 2)}. \end{aligned} \tag{4.41}$$

We note that due to $(x + \sqrt{x^2 + 2y})^2 \leq 4x^2 + 4y$ and $\psi(\mathbf{u}) \leq (1 + \eta_1)\|\mathbf{u}\|_1$,

$$\begin{aligned} \overline{C}^{(\ell_1)}_{\text{pred}}(\mathcal{S}, \overline{\Delta}) &\leq \overline{C}^{(\ell_1)}_{\text{pred}}(\mathcal{S}, 0) + \overline{\Delta}/(1 \vee |\mathcal{S}|) \\ &\leq (1 + \eta_1)\overline{C}^{(\ell_1)}_{\text{est},1}(\mathcal{S}, 0) + \overline{\Delta}/(1 \vee |\mathcal{S}|), \end{aligned} \tag{4.42}$$

and that $\overline{C}^{(\ell_1)}_{\text{pred}}(\mathcal{S}, 0) = \overline{C}^{(\ell_1)}_{\text{est}}(\mathcal{S}, 0; \|\cdot\|) = 0$ when $|\mathcal{S}| = 0$.

In the special case where $|\mathcal{S}| = p$ is fixed and $\overline{\boldsymbol{\Sigma}} = \mathbf{I}_{p \times p}$, we have

$$\overline{C}^{(\ell_1)}_{\text{pred}}(\mathcal{S}, 0) \leq (1 + \eta_1)^2, \quad \overline{C}^{(\ell_1)}_{\text{est,q}}(\mathcal{S}, 0) \leq 1 + \eta_1,$$

so that the constant factors are properly dimension normalized in the most favorable scenario. For $\eta_2 < 1$, it is still reasonable to use the same dimension normalization in general as $\|\mathbf{u}_{\mathcal{S}^c}\|_1$ is penalized through $\psi(\mathbf{u})$.

The above constant factors are closely related to the restricted eigenvalue

(Bickel, Ritov and Tsybakov, 2009; Koltchinskii, 2009), the compatibility factor (van de Geer and Bühlmann, 2009), and the weak cone invertibility factor (Ye and Zhang, 2010). Further discussion of these and other quantities related to (4.39), (4.40) and (4.41) can be found in Subsection 4.3.5. In particular, we will provide proper sufficient conditions under which

$$\begin{aligned}(1\vee|\mathcal{S}|)\overline{C}^{(\ell_1)}_{\text{pred}}(\mathcal{S},\overline{\Delta}) &\lesssim |\mathcal{S}|+\overline{\Delta},\\ (1\vee|\mathcal{S}|)^{1/(q\wedge 2)}\overline{C}^{(\ell_1)}_{\text{est},q}(\mathcal{S},\overline{\Delta}) &\lesssim (|\mathcal{S}|+\overline{\Delta})^{1/(q\wedge 2)},\end{aligned} \tag{4.43}$$

so that the error bounds for prediction and coefficient estimation with $\overline{\Delta}\lesssim|\mathcal{S}|$ are of the same order as the simplest case of $\overline{\Delta}=0$.

**Theorem 4.9** *Let $\widehat{\boldsymbol{\beta}}=\widehat{\boldsymbol{\beta}}(\lambda)$ be the Lasso estimator in (4.29), $\{\boldsymbol{\beta}^*,\overline{\boldsymbol{\beta}}\}\subset I\!R^p$, $\mathcal{S}\subseteq\{1,\ldots,p\}$, $\eta_1\ge 0$, and $\eta_2\le 1$. Suppose*

$$\|\mathbf{X}_{\mathcal{S}}^T(\mathbf{Y}-\mathbf{X}\boldsymbol{\beta}^*)/n\|_\infty\le\eta_1\lambda,\ \ \|\mathbf{X}_{\mathcal{S}^c}^T(\mathbf{Y}-\mathbf{X}\boldsymbol{\beta}^*)/n\|_\infty\le\eta_2\lambda. \tag{4.44}$$

*Let $\overline{\Delta}=\|\mathbf{X}\overline{\boldsymbol{\beta}}-\mathbf{X}\boldsymbol{\beta}^*\|_2^2/(\lambda^2 n)+4\|\overline{\boldsymbol{\beta}}_{\mathcal{S}^c}\|_1/\lambda$. Then,*

$$\|\mathbf{X}\widehat{\boldsymbol{\beta}}-\mathbf{X}\boldsymbol{\beta}^*\|_2^2/n\le\begin{cases}\lambda^2\overline{\Delta}-\|\mathbf{X}\widehat{\boldsymbol{\beta}}-\mathbf{X}\overline{\boldsymbol{\beta}}\|_2^2/n, & \mathcal{S}=\emptyset,\\ \lambda^2\overline{\Delta}+\lambda^2|\mathcal{S}|\overline{C}^{(\ell_1)}_{\text{pred}}(\mathcal{S},0), & \\ \lambda^2(1\vee|\mathcal{S}|)\overline{C}^{(\ell_1)}_{\text{pred}}(\mathcal{S},\overline{\Delta}), & \overline{\boldsymbol{\beta}}=\boldsymbol{\beta}^*,\end{cases} \tag{4.45}$$

*and for all seminorms $\|\cdot\|$,*

$$\|\widehat{\boldsymbol{\beta}}-\overline{\boldsymbol{\beta}}\|\le\begin{cases}2\lambda\overline{C}^{(\ell_1)}_{\text{est}}(\mathcal{S},\overline{\Delta}/2;\|\cdot\|), & \\ \lambda\overline{C}^{(\ell_1)}_{\text{est}}(\mathcal{S},\overline{\Delta};\|\cdot\|), & \overline{\boldsymbol{\beta}}=\boldsymbol{\beta}^*.\end{cases} \tag{4.46}$$

*Moreover, when $\eta_1=\eta_2$, the minimum over $\mathcal{S}$ and/or $\overline{\boldsymbol{\beta}}$ can be taken in the right-hand side of (4.45) and (4.46).*

Theorem 4.9 is a refinement of Theorem 4 of Sun and Zhang (2012). It imposes an analytical condition (4.44) on the noise $\mathbf{Y}-\mathbf{X}\boldsymbol{\beta}^*$. When the noise vector $\boldsymbol{\varepsilon}$ in (4.1) is sub-Gaussian and the design vectors are normalized to $\|\mathbf{X}_j\|_2^2=n$, this analytical condition holds with large probability for the true or oracle $\boldsymbol{\beta}^*$ and

$$\lambda=(1/\eta)\sigma\sqrt{(2/n)\log(p/\epsilon)},\quad\text{where } 0<\eta\le 1 \text{ is a constant.} \tag{4.47}$$

This statement and its consequences are formally stated as follows.

**Corollary 4.4** *Suppose that the sub-Gaussian condition (4.21) holds and the penalty level (4.47) is used with the Lasso in (4.29).*
*(i) Suppose $\max_{j\le p}\|\mathbf{X}_j\|_2^2/n\le 1$. For $\boldsymbol{\beta}^*=\boldsymbol{\beta}$, condition (4.44) holds for $\eta_1=\eta_2=\eta$ with at least probability $1-2\epsilon$. Alternatively, the condition holds for the same $\eta_2=\eta$ and smaller $\eta_1=\eta_2\sqrt{\log(|\mathcal{S}|/\epsilon)/\log(p/\epsilon)}$ with at least*

*probability* $1-4\epsilon$. *Consequently, the conclusions of Theorem 4.9 hold for the respective* $(\eta_1,\eta_2)$ *with at least probability* $1-2\epsilon$ *or* $1-4\epsilon$.
*(ii) Suppose* $\max_{j\in\mathcal{S}^c}\|\mathbf{P}_{\mathcal{S}}^{\perp}\mathbf{X}_j\|_2^2/n\le 1$. *For* $\boldsymbol{\beta}^*=\widehat{\boldsymbol{\beta}}^{o}$ *with the oracle estimator in (4.38), (4.44) holds for* $\eta_1=0$ *and* $\eta_2=\eta$ *with at least probability* $1-2\epsilon$. *Consequently, the conclusions of Theorem 4.9 hold with at least probability* $1-2\epsilon$.

Theorem 4.9 and Corollary 4.4 cover an array of oracle inequalities as the choice of vectors $\{\boldsymbol{\beta}^*,\overline{\boldsymbol{\beta}}\}$, the set $\mathcal{S}$ and the norm $\|\cdot\|$ are all arbitrary. In general $\eta_2<1$ is required when $\mathcal{S}\ne\emptyset$ and $\mathrm{rank}(\overline{\boldsymbol{\Sigma}})\ll p$, due to the need to penalize $\|\mathbf{u}_{\mathcal{S}^c}\|_1$ through (4.39) and (4.40). We state a selected subset of these oracle inequalities in the following corollary under different sparsity criteria given in Definition 4.2. The proof of the corollary follows that of Theorem 4.9.

**Corollary 4.5** *Suppose the conditions of Theorem 4.9 or Corollary 4.4 hold. Let* $\boldsymbol{\varepsilon}^*=\mathbf{Y}-\mathbf{X}\boldsymbol{\beta}^*$ *and* $0\le\eta_0\le\eta_1\le\eta_2=\eta<1$.
*(i) For all choices of* $\boldsymbol{\beta}^*$ *fulfilling condition (4.44),*

$$\|\mathbf{X}\widehat{\boldsymbol{\beta}}-\mathbf{X}\boldsymbol{\beta}^*\|_2^2/n\le 2\lambda\|\boldsymbol{\beta}^*\|_1.$$

*(ii) Let* $\overline{\boldsymbol{\beta}}=\boldsymbol{\beta}^*=\boldsymbol{\beta}$. *Suppose* $\boldsymbol{\beta}$ *is* $s^*$*-complex under the capped-*$\ell_1$ *sparsity criterion. If* $\overline{C}_{\text{pred}}^{(\ell_1)}(\mathcal{S},0)=O(1)$, *then*

$$\|\mathbf{X}\widehat{\boldsymbol{\beta}}-\mathbf{X}\boldsymbol{\beta}\|_2^2/n\le\lambda^2|\mathcal{S}|\overline{C}_{\text{pred}}^{(\ell_1)}(\mathcal{S},0)+4\lambda\|\boldsymbol{\beta}_{\mathcal{S}^c}\|_1\lesssim\lambda^2 s^*.$$

*If* $\|\mathbf{X}_{\mathcal{S}}^T\boldsymbol{\varepsilon}^*/n\|_\infty\le(\eta_1-\eta_0)\lambda$ *and* $\|\mathbf{X}_{\mathcal{S}^c}^T\boldsymbol{\varepsilon}^*/n\|_\infty\le(\eta_2-\eta_0)\lambda$ *with* $\eta_0>0$, *then*

$$\|\widehat{\boldsymbol{\beta}}-\boldsymbol{\beta}\|_1\le\eta_0^{-1}\Big\{\lambda|\mathcal{S}|\overline{C}_{\text{pred}}^{(\ell_1)}(\mathcal{S},0)/4+2\|\boldsymbol{\beta}_{\mathcal{S}^c}\|_1\Big\}\lesssim\lambda s^*.$$

*If (4.43) holds, then*

$$\|\widehat{\boldsymbol{\beta}}-\boldsymbol{\beta}\|_q\le\lambda(1\vee|\mathcal{S}|)^{1/(q\wedge 2)}\overline{C}_{\text{est},q}^{(\ell_1)}(\mathcal{S},4\|\boldsymbol{\beta}_{\mathcal{S}^c}\|_1/\lambda)\lesssim\lambda(s^*)^{1/(q\wedge 2)}.$$

*(iii) Let* $\boldsymbol{\beta}^*=\boldsymbol{\beta}$. *Suppose* $\boldsymbol{\beta}$ *is* $s^*$*-complex under the approximate sparsity criterion for certain* $\overline{\boldsymbol{\beta}}$ *and* $\mathcal{S}$. *If* $\overline{C}_{\text{pred}}^{(\ell_1)}(\mathcal{S},0)=O(1)$, *then*

$$\|\mathbf{X}\widehat{\boldsymbol{\beta}}-\mathbf{X}\boldsymbol{\beta}\|_2^2/n\le\|\mathbf{X}\boldsymbol{\beta}-\mathbf{X}\overline{\boldsymbol{\beta}}\|_2^2/n+4\lambda\|\overline{\boldsymbol{\beta}}_{\mathcal{S}^c}\|_1+\lambda^2|\mathcal{S}|\overline{C}_{\text{pred}}^{(\ell_1)}(\mathcal{S},0)\lesssim\lambda^2 s^*.$$

*If* $\|\mathbf{X}_{\mathcal{S}}^T\boldsymbol{\varepsilon}^*/n\|_\infty\le(\eta_1-\eta_0)\lambda$ *and* $\|\mathbf{X}_{\mathcal{S}^c}^T\boldsymbol{\varepsilon}^*/n\|_\infty\le(\eta_2-\eta_0)\lambda$ *with* $\eta_0>0$, *then*

$$\|\widehat{\boldsymbol{\beta}}-\overline{\boldsymbol{\beta}}\|_1\le(2\eta_0)^{-1}\Big\{\lambda|\mathcal{S}|\overline{C}_{\text{pred}}^{(\ell_1)}(\mathcal{S},0)+\overline{\Delta}\Big\}\lesssim\lambda s^*.$$

*If (4.43) holds, then*

$$\|\widehat{\boldsymbol{\beta}}-\overline{\boldsymbol{\beta}}\|_q\le 2\lambda(1\vee|\mathcal{S}|)^{1/(q\wedge 2)}\overline{C}_{\text{est},q}^{(\ell_1)}(\mathcal{S},\overline{\Delta}/2)\lesssim\lambda(s^*)^{1/(q\wedge 2)}.$$

When $\log(p/s^*) \asymp \log p$, Corollary 4.5 (ii) and (iii) assert the rate optimality of the Lasso in the sense of (4.17) for prediction and coefficient estimation with $\ell_q$ losses, $1 \le q \le 2$. This is under the capped-$\ell_1$ and approximate sparsity conditions respectively on $\boldsymbol{\beta}$ and proper conditions on the noise and design. As we will mention below in Definition 4.2, the estimation of $\boldsymbol{\beta}$ does not make sense under the approximate sparsity condition when $p > n$ so the estimation of $\overline{\boldsymbol{\beta}}$ is considered in part (iii).

**Remark 4.1** The prediction error bound in Corollary 4.5 (i), which is proportional to $\lambda \asymp \sqrt{(\log p)/n}$, is of the so called "slow rate" as the optimal rate is proportional to $\lambda^2$. It becomes comparable with the optimal rate only when $\|\boldsymbol{\beta}^*\|_1 \asymp \lambda s^*$. We note that $\lambda s^* \le \|\boldsymbol{\beta}\|_1$ is always feasible under the capped-$\ell_1$ or approximate sparsity conditions on $\boldsymbol{\beta}$, so that Corollary 4.5 (ii) and (iii) are not subsumed in (i). A significant advantage of part (i) is that the slow rate is achieved without imposing a condition on the design $\mathbf{X}$. In fact, under condition (4.44) with $\eta_1 \vee \eta_2 \le 1$, a somewhat sharper bound

$$\|\mathbf{X}\widehat{\boldsymbol{\beta}} - \mathbf{X}\boldsymbol{\beta}^*\|_2^2/n \le (1+\eta_1)\lambda\|\boldsymbol{\beta}^*_{\mathcal{S}}\|_1 + (1+\eta_2)\lambda\|\boldsymbol{\beta}^*_{\mathcal{S}^c}\|_1 \tag{4.48}$$

can be obtained by a simplification of the proof of Lemma 4.3 below.

**Remark 4.2** An interesting phenomenon with the prediction and $\ell_1$ error bounds in Corollary 4.5 (ii) and (iii), in view of (4.42), is that they guarantee rate optimality in the respective losses under the same regularity condition $\overline{C}^{(\ell_1)}_{\text{pred}}(\mathcal{S}, 0) = O(1)$ on the design. Another interesting aspect of Corollary 4.5 (iii) is the following interpretation of the prediction error bound: The Lasso performs at least as well as the optimal choice of a predictor $\overline{\boldsymbol{\beta}}$ if the cost of choosing $\overline{\boldsymbol{\beta}}$ is $4\lambda\|\overline{\boldsymbol{\beta}}_{\mathcal{S}^c}\|_1 + \lambda^2|\mathcal{S}|\overline{C}^{(\ell_1)}_{\text{pred}}(\mathcal{S}, 0)$. This cost is of the same order as the product of $\lambda^2$ and the capped-$\ell_1$ complexity of $\overline{\boldsymbol{\beta}}$.

The following lemma is needed in the proof of Theorem 4.9.

**Lemma 4.3** *Let $\widehat{\boldsymbol{\beta}} = \widehat{\boldsymbol{\beta}}(\lambda)$ be the Lasso estimator in (4.29), $\{\boldsymbol{\beta}^*, \overline{\boldsymbol{\beta}}\} \subset I\!R^p$, $\mathcal{S} \subset \{1, \ldots, p\}$, and $\psi(\mathbf{u}) = \eta_1\|\mathbf{u}_{\mathcal{S}}\|_1 - \text{sgn}(\overline{\boldsymbol{\beta}}_{\mathcal{S}})^T\mathbf{u}_{\mathcal{S}} - (1-\eta_2)\|\mathbf{u}_{\mathcal{S}^c}\|_1$ as in (4.39). Suppose (4.44) holds. Let $\overline{\Delta}$ be as in Theorem 4.9. Then,*

$$\|\mathbf{X}\widehat{\boldsymbol{\beta}} - \mathbf{X}\boldsymbol{\beta}^*\|_2^2/n + \|\mathbf{X}\widehat{\boldsymbol{\beta}} - \mathbf{X}\overline{\boldsymbol{\beta}}\|_2^2/n \le \lambda^2\overline{\Delta} + 2\lambda\psi(\widehat{\boldsymbol{\beta}} - \overline{\boldsymbol{\beta}}). \tag{4.49}$$

**Proof.** Let $\overline{\mathbf{h}} = (\overline{h}_1, \ldots, \overline{h}_p)^T = \widehat{\boldsymbol{\beta}} - \overline{\boldsymbol{\beta}}$, $\mathbf{h}^* = \widehat{\boldsymbol{\beta}} - \boldsymbol{\beta}^*$, and $\mathbf{g} = (g_1, \ldots, g_p)^T = \mathbf{X}^T(\mathbf{Y} - \mathbf{X}\widehat{\boldsymbol{\beta}})/n$. By the KKT conditions (4.31),

$$-\overline{h}_j g_j \le \begin{cases} -\overline{h}_j\lambda\text{sgn}(\widehat{\beta}_j) \le -\lambda\overline{h}_j\text{sgn}(\overline{\beta}_j), & j \in \mathcal{S}, \\ \lambda(|\overline{\beta}_j| - |\widehat{\beta}_j|) \le 2\lambda|\overline{\beta}_j| - \lambda|\overline{h}_j|, & j \in \mathcal{S}^c. \end{cases}$$

Note that the convention $\text{sgn}(0) = 0$ is used here. As

$$\overline{h}_j g_j + \overline{h}_j\mathbf{X}_j^T\mathbf{X}\mathbf{h}^*/n = \overline{h}_j\mathbf{X}_j^T(\mathbf{Y} - \mathbf{X}\boldsymbol{\beta}^*)/n,$$

condition (4.44) implies that

$$\overline{h}_j \mathbf{X}_j^T \mathbf{X}\mathbf{h}^*/n \le \begin{cases} \eta_1\lambda|\overline{h}_j| - \lambda\overline{h}_j\mathrm{sgn}(\overline{\beta}_j), & j \in \mathcal{S}, \\ -(1-\eta_2)\lambda|\overline{h}_j| + 2\lambda|\overline{\beta}_j|, & j \in \mathcal{S}^c. \end{cases} \quad (4.50)$$

Summing the above inequalities over $j$, we find by the definition of $\psi(\mathbf{u})$ that

$$\begin{aligned} & (\mathbf{X}\overline{\mathbf{h}})^T \mathbf{X}\mathbf{h}^*/n \\ \le\ & \eta_1\lambda\|\overline{\mathbf{h}}_{\mathcal{S}}\|_1 - \lambda\overline{\mathbf{h}}_{\mathcal{S}}^T\mathrm{sgn}(\overline{\boldsymbol{\beta}}_{\mathcal{S}}) - (1-\eta_2)\lambda\|\overline{\mathbf{h}}_{\mathcal{S}^c}\|_1 + 2\lambda\|\overline{\boldsymbol{\beta}}_{\mathcal{S}^c}\|_1. \\ =\ & \lambda\psi(\overline{\mathbf{h}}) + 2\lambda\|\overline{\boldsymbol{\beta}}_{\mathcal{S}^c}\|_1. \end{aligned}$$

As $ab = b^2/2 - (a-b)^2/2 + a^2/2$, the above inequality can be written as

$$\|\mathbf{X}\mathbf{h}^*\|_2^2/(2n) + \|\mathbf{X}\overline{\mathbf{h}}\|_2^2/(2n) \le \|\mathbf{X}\overline{\boldsymbol{\beta}} - \mathbf{X}\boldsymbol{\beta}^*\|_2^2/(2n) + \lambda\psi(\overline{\mathbf{h}}) + 2\lambda\|\overline{\boldsymbol{\beta}}_{\mathcal{S}^c}\|_1,$$

which gives (4.49). ∎

We take an optimization approach by asking the following question: What are the sharpest upper bounds for the prediction and estimation errors one can possibly prove based on Lemma 4.3? This problem is solved in the proof of Theorem 4.9 below, so that the oracle inequalities in the theorem are the sharpest possible respectively on the inequality in (4.49).

**Proof of Theorem 4.9.** Let $\overline{\mathbf{h}} = \widehat{\boldsymbol{\beta}} - \overline{\boldsymbol{\beta}}$ and $\overline{\boldsymbol{\Sigma}} = \mathbf{X}^T\mathbf{X}/n$. For $\mathcal{S} = \emptyset$, we have $\psi(\overline{\mathbf{h}}) \le 0$, so that (4.45) follows directly from (4.49). For $\mathcal{S} \ne \emptyset$ and $\mathbf{X}\overline{\mathbf{h}} = \mathbf{0}$, $\lambda^2|\mathcal{S}|\overline{C}_{\mathrm{pred}}^{(\ell_1)}(\mathcal{S}, \overline{\Delta}) \ge \lambda^2\overline{\Delta} \ge \|\mathbf{X}\overline{\boldsymbol{\beta}} - \mathbf{X}\boldsymbol{\beta}^*\|_2^2/n = \|\mathbf{X}\widehat{\boldsymbol{\beta}} - \mathbf{X}\boldsymbol{\beta}^*\|_2^2/n$, so that (4.45) also holds. For $\mathcal{S} \ne \emptyset$ and $\mathbf{X}\overline{\mathbf{h}} \ne \mathbf{0}$, $\overline{\mathbf{h}} = t\mathbf{u}$ for some $t \ge 0$ and $\mathbf{u}$ satisfying $\mathbf{u}^T\overline{\boldsymbol{\Sigma}}\mathbf{u} = 1$, so that (4.49) can be written as

$$\|\mathbf{X}\widehat{\boldsymbol{\beta}} - \mathbf{X}\boldsymbol{\beta}^*\|_2^2/n \le c + 2bt - t^2$$

with $b = \lambda\psi(\mathbf{u})$ and $c = \lambda^2\overline{\Delta}$. Consequently, (4.45) follows from

$$c + 2bt - t^2 \le c + b^2 = c + \lambda^2\psi^2(\mathbf{u}) \le \lambda^2\overline{\Delta} + \lambda^2|\mathcal{S}|\overline{C}_{\mathrm{pred}}^{(\ell_1)}(\mathcal{S}, 0)$$

as $\max\{0, \psi(\mathbf{u})\} = \left[\psi(\mathbf{u}) + \{\psi^2(\mathbf{u}) + 2\overline{\Delta}\}^{1/2}\right]/2$ for $\overline{\Delta} = 0$ in (4.39). For $\overline{\boldsymbol{\beta}} = \boldsymbol{\beta}^*$, (4.45) is trivial for $\mathbf{X}\overline{\mathbf{h}} = \mathbf{0}$, and for $\mathbf{X}\overline{\mathbf{h}} \ne \mathbf{0}$ (4.49) gives

$$t^2 = \|\mathbf{X}\overline{\mathbf{h}}\|_2^2/n = \|\mathbf{X}\mathbf{h}^*\|_2^2/n \le \lambda^2\overline{\Delta}/2 + \lambda\psi(\overline{\mathbf{h}}) = c/2 + bt,$$

which again implies (4.45) because

$$t^2 \le \left\{\left(b + \sqrt{b^2 + 2c}\right)/2\right\}^2 \le \lambda^2(1 \vee |\mathcal{S}|)\overline{C}_{\mathrm{pred}}^{(\ell_1)}(\mathcal{S}, \overline{\Delta})$$

by the definition of $\overline{C}_{\mathrm{pred}}^{(\ell_1)}(\mathcal{S}, \overline{\Delta})$ in (4.39). The proof of (4.45) is complete.

To prove (4.46), we first consider $\mathbf{X}\overline{\mathbf{h}} \neq \mathbf{0}$. As in the proof of (4.45), $\|\overline{\mathbf{h}}\| = t\|\mathbf{u}\|$ with $0 \le c + 2bt - t^2$, so that (4.46) follows from

$$\begin{aligned} \|\overline{\mathbf{h}}\| &\le \|\mathbf{u}\|(b + \sqrt{b^2 + c}) \\ &= \lambda\|\mathbf{u}\|\big[\psi(\mathbf{u}) + \{\psi^2(\mathbf{u}) + \overline{\Delta}\}^{1/2}\big] \\ &\le 2\lambda\overline{C}^{(\ell_1)}_{\rm est}(\mathcal{S}, \overline{\Delta}/2; \|\cdot\|) \end{aligned}$$

by the definition of $\overline{C}^{(\ell_1)}_{\rm est}(\mathcal{S}, \overline{\Delta}; \|\cdot\|)$ in (4.40). For $\overline{\boldsymbol{\beta}} = \boldsymbol{\beta}^*$, (4.46) follows from $\|\mathbf{h}^*\| = \|\overline{\mathbf{h}}\| = t\|\mathbf{u}\| \le \|\mathbf{u}\|(b + \sqrt{b^2 + 2c})/2 \le \lambda\overline{C}^{(\ell_1)}_{\rm est}(\mathcal{S}, \overline{\Delta}; \|\cdot\|)$.

For $\mathbf{X}\overline{\mathbf{h}} = \mathbf{0}$, we perturb a single component of $\overline{\boldsymbol{\beta}} : \overline{\beta}_j \to \overline{\beta}_j + \epsilon$ with a $j \notin \mathcal{S}$, so that $\mathbf{X}\overline{\mathbf{h}} \to \epsilon\mathbf{X}_j \neq \mathbf{0}$ after the perturbation. Because the perturbation does not change $\psi(\mathbf{u})$, the function $\overline{C}^{(\ell_1)}_{\rm est}(\mathcal{S}, x; \|\cdot\|)$ is also unchanged. The conclusion follows from the lower semicontinuity of $\overline{C}^{(\ell_1)}_{\rm est}(\mathcal{S}, x; \|\cdot\|)$ in $x$ as (4.46) holds whenever $\mathbf{X}\overline{\mathbf{h}} \neq \mathbf{0}$. ■

**Proof of Corollary 4.5.** When $\eta_1$ and $\eta_2$ is reduced to $\eta_1 - \eta_0$ and $\eta_2 - \eta_0$, $\widetilde{\psi}(\mathbf{u}) = \psi(\mathbf{u}) - \eta_0\|\mathbf{u}\|_1 = (\eta_1 - \eta_0)\|\mathbf{u}_{\mathcal{S}}\|_1 - \mathrm{sgn}(\overline{\boldsymbol{\beta}}_{\mathcal{S}})^T\mathbf{u}_{\mathcal{S}} - (1 - \eta_2 + \eta_0)\|\mathbf{u}_{\mathcal{S}^c}\|_1$ is used in (4.40) instead of $\psi(\mathbf{u})$. As $(a - x)(x + \sqrt{x^2 + b}) \le (a^2 + b)/2$ with the maximum attained at $x = (a - b/a)/2$,

$$\begin{aligned} &\eta_0\|\mathbf{u}\|_1\big[\widetilde{\psi}(\mathbf{u}) + \{\widetilde{\psi}^2(\mathbf{u}) + 2\overline{\Delta}\}^{1/2}\big]/2 \\ \le\ &\{\psi(\mathbf{u}) - \widetilde{\psi}(\mathbf{u})\}\big[\widetilde{\psi}(\mathbf{u}) + \{\widetilde{\psi}^2(\mathbf{u}) + 2\overline{\Delta}\}^{1/2}\big]/2 \\ \le\ &\{(\psi(\mathbf{u}))_+^2 + 2\overline{\Delta}\}/4. \end{aligned}$$

Thus, as $\psi(\mathbf{u}) \le 0$ for $\mathcal{S} = \emptyset$, it follows from the definitions of the constant factors in (4.39) and (4.40) that

$$\eta_0(1 \vee |\mathcal{S}|)\overline{C}^{(\ell_1)}_{\rm est,1}(\mathcal{S}, \overline{\Delta}; \eta_1 - \eta_0, \eta_2 - \eta_0) \le |\mathcal{S}|\overline{C}^{(\ell_1)}_{\rm pred}(\mathcal{S}, 0; \eta_1, \eta_2)/4 + \overline{\Delta}/2.$$

The rest of the corollary follows directly from Theorem 4.9 and (4.42). ■

As we have mentioned above in the proof of Theorem 4.9, all the oracle inequalities in the theorem are the sharpest possible based on Lemma 4.3. Still, Theorem 4.9 does not provide rate optimality for coefficient estimation under the $\ell_q$ loss with $q > 2$. The main reason for this analytical deficiency is a gap introduced in the proof of Lemma 4.3 when the summation of the more basic inequality (4.50) is taken over the index $j \in \{1, \ldots, p\}$. In the rest of this section, we derive even sharper oracle inequalities in the case of $\overline{\boldsymbol{\beta}} = \boldsymbol{\beta}^*$ based on (4.50) and an additional set of inequalities from the KKT conditions. This sharper analysis will provide sufficient conditions for the rate optimality of the Lasso estimation under the $\ell_q$ loss for all $1 \le q \le \infty$.

Recall that the Gram matrix is $\overline{\boldsymbol{\Sigma}} = \mathbf{X}^T\mathbf{X}/n$. For $\overline{\boldsymbol{\beta}} = \boldsymbol{\beta}^*$, (4.50) becomes

$$h_j^*\big(\overline{\boldsymbol{\Sigma}}\mathbf{h}^*\big)_j \le \begin{cases} \eta_1\lambda|h_j^*| - \lambda h_j^*\mathrm{sgn}(\beta_j^*), & j \in \mathcal{S}, \\ -(1 - \eta_2)\lambda|h_j^*| + 2\lambda|\beta_j^*|, & j \in \mathcal{S}^c, \end{cases}$$

with $\mathbf{h}^* = (h_1^*, \ldots, h_p^*)^T = \widehat{\boldsymbol{\beta}} - \boldsymbol{\beta}^*$, provided that $\mathcal{S} \subseteq \text{supp}(\boldsymbol{\beta}^*)$. Let $\mathbf{u} = \mathbf{h}^*/\lambda$ and $\Delta_j^* = 2|\beta_j^*|/\lambda$. The above inequalities can be written as

$$u_j \big(\overline{\boldsymbol{\Sigma}}\mathbf{u}\big)_j \le \begin{cases} \eta_1 |u_j| - u_j \text{sgn}(\beta_j^*), & j \in \mathcal{S}, \\ -(1-\eta_2)|u_j| + \Delta_j^*, & j \in \mathcal{S}^c. \end{cases} \tag{4.51}$$

Moreover, the noise bound (4.44) and the KKT conditions (4.31) imply that

$$\|\big(\overline{\boldsymbol{\Sigma}}\mathbf{u}\big)_{\mathcal{S}}\|_\infty \le (1+\eta_1), \quad \|\big(\overline{\boldsymbol{\Sigma}}\mathbf{u}\big)_{\mathcal{S}^c}\|_\infty \le (1+\eta_2). \tag{4.52}$$

For sets $\mathcal{S} \subset \{1, \ldots, p\}$ and vectors $\boldsymbol{\Delta}^* = (\Delta_j^*, j \in \mathcal{S}^c)^T$, define a class of vectors $\mathbf{u} = (u_1, \ldots, u_p)^T$ by

$$\begin{aligned} \mathscr{U}(\mathcal{S}, \boldsymbol{\Delta}^*) &\equiv \mathscr{U}(\mathcal{S}, \boldsymbol{\Delta}^*; \eta_1, \eta_2) \\ &= \Big\{\text{all } \mathbf{u} \text{ satisfying (4.51) and (4.52)}\Big\}. \end{aligned} \tag{4.53}$$

For any semi-norm $\|\cdot\|$ as a loss function, including $\|\mathbf{b}\|_{\text{pred}} = \big(\mathbf{b}^T\overline{\boldsymbol{\Sigma}}\mathbf{b}\big)^{1/2}$ for prediction, define

$$C^{(\ell_1)}(\mathscr{U}; \|\cdot\|) = \sup\big\{\|\mathbf{u}\| : \mathbf{u} \in \mathscr{U}\big\}. \tag{4.54}$$

**Theorem 4.10** *Let $\widehat{\boldsymbol{\beta}} = \widehat{\boldsymbol{\beta}}(\lambda)$ be the Lasso estimator in (4.29), $\boldsymbol{\beta}^* \in I\!\!R^p$, and $\boldsymbol{\Delta}^* = (2|\beta_j^*|/\lambda, j \in \mathcal{S}^c)^T$. Suppose (4.44) holds. Then,*

$$(\widehat{\boldsymbol{\beta}} - \boldsymbol{\beta}^*)/\lambda \in \mathscr{U}(\mathcal{S}, \boldsymbol{\Delta}^*), \tag{4.55}$$

*and for any semi-norm $\|\cdot\|$ as a loss function*

$$\|\widehat{\boldsymbol{\beta}} - \boldsymbol{\beta}^*\| \le \lambda C\big(\mathscr{U}(\mathcal{S}, \boldsymbol{\Delta}^*); \|\cdot\|\big). \tag{4.56}$$

*If in addition $\lambda$ is as in (4.47) and the sub-Gaussian condition (4.21) holds as in Corollary 4.4, then (4.55) and (4.56) hold with probability $1-2\epsilon$ or $1-4\epsilon$ in the respective options for $\{\boldsymbol{\beta}^*, \eta_1, \eta_2\}$ as in Corollary 4.4.*

Theorem 4.10 sharpens Theorem 4.9 and Corollary 4.4 in the case of $\overline{\boldsymbol{\beta}} = \boldsymbol{\beta}^*$, for example, when $\boldsymbol{\beta}$ is capped-$\ell_1$ sparse. By definition, (4.56) is the sharpest possible oracle inequality based on (4.51) and (4.52). As (4.51) is equivalent to (4.50) on which Theorem 4.9 is entirely based, Theorem 4.9 and Corollary 4.4 are subsumed in Theorem 4.10. Moreover, Theorem 4.10 provides proper conditions for the rate optimality of the Lasso in the $\ell_q$ estimation loss for $q > 2$ as we will state in Theorem 4.11 below. This closes the gap between the upper bound in Corollary 4.5 and the lower bound in Theorem 4.4 for $q > 2$, at least in the case of $\log(p/|\mathcal{S}|) \asymp \log p$ for sub-Gaussian noise.

The constant $C\big(\mathscr{U}(\mathcal{S},\mathbf{\Delta}^*);\|\cdot\|\big)$ in (4.56) is still quite abstract. The following lemma provides a more explicit expression of the quantity in the case of $\mathbf{\Delta}^*=\mathbf{0}$, which is equivalent to $\text{supp}(\boldsymbol{\beta}^*)\subseteq\mathcal{S}$. For any $\mathbf{u}\in I\!R^p$, define

$$T_+(\mathbf{u})=\min\left[\frac{1+\eta_1}{\|(\overline{\mathbf{\Sigma}}\mathbf{u})_{\mathcal{S}}\|_\infty},\frac{1+\eta_2}{\|(\overline{\mathbf{\Sigma}}\mathbf{u})_{\mathcal{S}^c}\|_\infty},\min_{j\in\mathcal{S}}\frac{\{\eta_1|u_j|-u_j\text{sgn}(\beta_j^*)\}_+}{(u_j\big(\overline{\mathbf{\Sigma}}\mathbf{u}\big)_j)_+}\right]$$

with the convention $(a)_+/(b)_+=0$ for $a<0=b$ and $0/0=\infty$ otherwise, and

$$T_-(\mathbf{u})=\max\left[\max_{j\in\mathcal{S}^c}\frac{(1-\eta_2)|u_j|}{(u_j\big(\overline{\mathbf{\Sigma}}\mathbf{u}\big)_j)_-},\max_{j\in\mathcal{S}}\frac{\{\eta_1|u_j|-u_j\text{sgn}(\beta_j^*)\}_-}{(u_j\big(\overline{\mathbf{\Sigma}}\mathbf{u}\big)_j)_-}\right]$$

with the convention $0/0=0$.

**Lemma 4.4** *For $\mathbf{\Delta}^*=\mathbf{0}$, the quantity defined in (4.56) can be written as*

$$C^{(\ell_1)}(\mathscr{U}(\mathcal{S},\mathbf{0});\|\cdot\|)=\sup_{T_+(\mathbf{u})\ge T_-(\mathbf{u})}\Big(\|\mathbf{u}\|T_+(\mathbf{u})\Big)=\sup_{\mathbf{u}\in\mathscr{U}(\mathcal{S},\mathbf{0})}\Big(\|\mathbf{u}\|T_+(\mathbf{u})\Big),$$

*with the convention $\|\mathbf{u}\|T_+(\mathbf{u})=0$ for $\mathbf{u}=\mathbf{0}$ and $T_+(\mathbf{u})=\infty$.*

The formula in Lemma 4.4 allows for various more explicit relaxations by weakening, dropping and/or combining the components of $T_\pm(\mathbf{u})$. It will be clear from the proof of Lemma 4.4 below that a similar expression of $C^{(\ell_1)}(\mathscr{U}(\mathcal{S},\mathbf{\Delta}^*);\|\cdot\|)$ can be also obtained for $\mathbf{\Delta}^*\neq\mathbf{0}$, in which $T_+(\mathbf{u})$ and $T_-(\mathbf{u})$ will respectively be the minimum of the larger roots of quadratic equations and maximum of smaller roots. We omit this calculation. We note that $C^{(\ell_1)}(\mathscr{U}(\mathcal{S},\mathbf{0});\|\cdot\|)=0$ when $\mathcal{S}=\emptyset$ and $\eta_2<1$, as $\mathscr{U}(\mathcal{S},\mathbf{0})=\{\mathbf{0}\}$ in this special case.

**Proof of Lemma 4.4.** In the direction $\{\mathbf{u}=t\mathbf{v}:t\ge 0\}$, (4.51) can be written as

$$tv_j\big(\overline{\mathbf{\Sigma}}\mathbf{v}\big)_j\le\begin{cases}\eta_1|v_j|-v_j\text{sgn}(\beta_j^*), & j\in\mathcal{S},\\ -(1-\eta_2)|v_j|, & j\in\mathcal{S}^c.\end{cases}$$

For each $j$, this inequality can be written as $b_jt\le a_j$, which can be also written as $c_j\le t\le d_j$ unless $a_j<0\le b_j$. If $a_j<0\le b_j$ for some $j$, then (4.51) does not hold for $\mathbf{u}$. In this case, $T_+(\mathbf{v})=T_+(\mathbf{u})=0$ as desired, so that $\mathbf{u}$ does not contribute to the computation of the maximum of $\|\mathbf{u}\|T_+(\mathbf{u})$.

Suppose $a_j<0\le b_j$ does not happen for any $j$. The inequalities $b_jt\le a_j$ hold simultaneously if and only if $\max_j c_j\le t\le\min_j d_j$. As $t\ge 0$, $d_j=\infty=(a_j)_+/(b_j)_+$ when $b_j\le 0$ and $d_j=a_j/b_j=(a_j)_+/(b_j)_+$ otherwise. Thus, $\min_j d_j=\min_{j\in\mathcal{S}}(a_j)+/(b_j)_+$. The combination of this upper bound with (4.52) yield $t\le T_+(\mathbf{v})$. Similarly, $\max_j c_i=\max_j(a_j)_-/(b_j)_-=T_-(\mathbf{v})\le t$. Thus $\mathbf{u}$ is feasible if and only if $T_-(\mathbf{v})\le T_+(\mathbf{v})$. As $\|\mathbf{u}\|=t\|\mathbf{v}\|\le\|\mathbf{v}\|T_+(\mathbf{v})$, the conclusion follows. ■

We now consider oracle inequalities for prediction and $\ell_q$ estimation with $\mathbf{\Delta}^* = \|\boldsymbol{\beta}^*_{\mathcal{S}^c}\|_1 = 0$. Such error bounds are most useful when the true $\boldsymbol{\beta}$ is $\ell_0$ sparse, for example with $\beta^*_j = \beta_j$ for large $|\beta_j|$. Define

$$C^{(\ell_1)}_{\mathrm{pred}}(\mathcal{S}) \equiv C^{(\ell_1)}_{\mathrm{pred}}(\mathcal{S};\eta_1,\eta_2) = \sup_{T_+(\mathbf{u})\ge T_-(\mathbf{u})} \frac{(\mathbf{u}^T\overline{\mathbf{\Sigma}}\mathbf{u})T^2_+(\mathbf{u})}{1\vee|\mathcal{S}|}$$

and

$$C^{(\ell_1)}_{\mathrm{est},q}(\mathcal{S}) \equiv C^{(\ell_1)}_{\mathrm{est},q}(\mathcal{S};\eta_1,\eta_2) = \sup_{T_+(\mathbf{u})\ge T_-(\mathbf{u})} \frac{\|\mathbf{u}\|_q T_+(\mathbf{u})}{1\vee|\mathcal{S}|^{1/q}}.$$

By Lemma 4.4, the above quantities are the normalized version of (4.54) respectively with the prediction loss $\|\mathbf{u}\| = \|\mathbf{u}\|_{\mathrm{pred}} = (\mathbf{u}^T\overline{\mathbf{\Sigma}}\mathbf{u})^{1/2}$ and the $\ell_q$ loss. The normalization factors are tailored to match the scaling of $\|\mathbf{u}\|$ by the $\ell_\infty$ norm of the components of $\overline{\mathbf{\Sigma}}\mathbf{u}$ in the denominator of $T_+(\mathbf{u})$. For $\overline{\mathbf{\Sigma}}_{\mathcal{S},\mathcal{S}} = \mathbf{I}_{|\mathcal{S}|\times|\mathcal{S}|}$ and $\|\mathbf{u}_{\mathcal{S}^c}\|_1 = 0 < \|\mathbf{u}_{\mathcal{S}}\|_1$, $\|\mathbf{u}\|_q T_+(\mathbf{u})/(1+\eta_1) \le \|\mathbf{u}_{\mathcal{S}}\|_q/\|\mathbf{u}_{\mathcal{S}}\|_\infty \le |\mathcal{S}|^{1/q}$ and $(\mathbf{u}^T\overline{\mathbf{\Sigma}}\mathbf{u})T^2_+(\mathbf{u}) \le (1+\eta_1)^2|\mathcal{S}|$. Thus, as $\mathbf{u}_{\mathcal{S}^c}$ is penalized through the requirement $T_-(\mathbf{u}) \le T_+(\mathbf{u})$, $C^{(\ell_1)}_{\mathrm{pred}}(\mathcal{S})$ and $C^{(\ell_1)}_{\mathrm{est},q}(\mathcal{S})$, $1 \le q \le \infty$, are all properly dimension normalized.

**Theorem 4.11** *Under the conditions of Theorem 4.10, the following specializations of (4.56) hold when* $\mathrm{supp}(\boldsymbol{\beta}^*) \subseteq \mathcal{S}$*:*

$$\begin{aligned}\|\mathbf{X}\widehat{\boldsymbol{\beta}} - \mathbf{X}\boldsymbol{\beta}^*\|_2^2/n &\le \lambda^2|\mathcal{S}|C^{(\ell_1)}_{\mathrm{pred}}(\mathcal{S}) \\ &= \lambda^2 \sup_{T_-(\mathbf{u})\le T_+(\mathbf{u})} (\mathbf{u}^T\overline{\mathbf{\Sigma}}\mathbf{u})T^2_+(\mathbf{u})\end{aligned}$$

*and*

$$\|\widehat{\boldsymbol{\beta}} - \boldsymbol{\beta}^*\|_q \le \lambda|\mathcal{S}|^{1/q}C^{(\ell_1)}_{\mathrm{est},q}(\mathcal{S}) = \lambda \sup_{T_-(\mathbf{u})\le T_+(\mathbf{u})} \|\mathbf{u}\|_q T_+(\mathbf{u}).$$

The constant factors in Theorem 4.11, which are smaller than the corresponding constant factors in Theorem 4.9 in the case of $\overline{\Delta} = 0$, are closely related to the (sign-restricted) cone invertibility factor (Ye and Zhang, 2010). The relationship among these and other related quantities and their upper bounds are discussed in Subsections 4.3.3, 4.3.4 and 4.3.5.

As Theorem 4.11 is a more explicit statement of Theorem 4.10 under the $\ell_0$ sparsity condition $\|\boldsymbol{\beta}^*_{\mathcal{S}^c}\|_1 = 0$, it provides the sharpest possible error bounds for prediction and $\ell_q$ estimation based on basic inequalities (4.51) and (4.52) under the condition. Possible choices of $\boldsymbol{\beta}^*$ include the true $\boldsymbol{\beta}^* = \boldsymbol{\beta}$ and the oracle $\boldsymbol{\beta}^* = \widehat{\boldsymbol{\beta}}^o$ in (4.38). Under proper conditions on $\overline{\mathbf{\Sigma}}$ and $\mathcal{S}$, the quantities $C^{(\ell_1)}_{\mathrm{pred}}(\mathcal{S})$ and $C^{(\ell_1)}_{\mathrm{est},q}(\mathcal{S})$ are uniformly bounded. When $\log(p/|\mathcal{S}|) \asymp \log p$, this gives the rate optimality of the Lasso in prediction and coefficient estimation in the sense of (4.17), especially in the $\ell_q$ loss for $q > 2$.

The $\ell_\infty$ error bound provides theoretical justification for thresholding the Lasso. In fact, as $|\mathcal{S}|^{1/q} \le e$ for $q = \log|\mathcal{S}|$, the $\ell_q$ bound for the moderately large $q = \log|\mathcal{S}|$ suffices. Let $\lambda^* > 0$ and define

$$\widehat{\boldsymbol{\beta}}^{(\text{thresh})} = \Big(\widehat{\beta}_j I\{|\widehat{\beta}_j| > \lambda^*\} : j = 1, \ldots, p\Big)^T$$

as a hard-thresholded Lasso.

**Corollary 4.6** *Suppose* $\lambda^* \ge e\lambda C^{(\ell_1)}_{\text{est},\log|\mathcal{S}|}(\mathcal{S})$. *Under the conditions and notation of Theorem 4.11,*

$$\{j : |\beta_j^*| > 2\lambda^*\} \subseteq \text{supp}\big(\widehat{\boldsymbol{\beta}}^{(\text{thresh})}\big) \subseteq \mathcal{S}$$

*and*

$$\|\widehat{\boldsymbol{\beta}}^{(\text{thresh})} - \boldsymbol{\beta}^*\|_2 \le 2|\mathcal{S}|^{1/2}\lambda^*.$$

This corollary asserts that when $C^{(\ell_1)}_{\text{est},\log|\mathcal{S}|}(\mathcal{S})$ is uniformly bounded, thresholding the Lasso at level $\lambda^*$ removes all false positives and retains all signals greater than $2\lambda^*$. It has a similar spirit to the sure screening concept in Fan and Lv (2008). See also Chapter 8. The selection consistency of this thresholded Lasso requires a *minimum signal strength* $\min_{j\in\mathcal{S}}|\beta_j^*| \ge 2\lambda^* \asymp \lambda$, which matches the order of the lower bound in Theorem 4.5. However, this result has two drawbacks: the requirement of the $\ell_\infty$ type condition $C^{(\ell_1)}_{\text{est},\log|\mathcal{S}|}(\mathcal{S}) = O(1)$ and the lack of the knowledge of $\lambda^*$ in typical applications.

As a majority of the existing analyses of the prediction and estimation properties of the Lasso have been based on basic inequalities in (4.49) in Lemma 4.3, (4.51) or (4.52), Theorems 4.9, 4.10 and 4.11 summarize this literature respectively under the approximate, capped-$\ell_1$ and hard sparsity conditions on the regression coefficients. Throughout this analysis of the Lasso, the required analytical condition on the penalty level $\lambda$ is (4.44) and the required probabilistic condition on $\lambda$ is (4.47) for sub-Gaussian noise. Regularity conditions on the design matrix $\mathbf{X}$, or equivalently on the Gram matrix $\overline{\boldsymbol{\Sigma}}$, are expressed in terms of constant factors in (4.39), (4.40), (4.41), (4.54) and Lemma 4.4. Sufficient conditions for the stability of these constant factors and related quantities for the analysis of the Lasso and Dantzig selector will be provided in Subsection 4.3.5 and Corollary 4.7 of Theorem 4.12 in Subsection 4.3.3.

It is worthwhile to mention here that the rate optimality results in this subsection focus on the worst case scenario under the respective sparsity assumptions. When the signal to noise ratio is high for a significant fraction of $\text{supp}(\boldsymbol{\beta})$ as in Theorem 4.7, the prediction and squared $\ell_2$ estimation losses of the Lasso are still of the order $\lambda^2|\mathcal{S}|$, while the losses for the subset selector are of the smaller order $\lambda^2 s_* + |\mathcal{S}|\sigma^2/n$, where $s_*$ is the number of small signals in $\mathcal{S}$. The main reason for the sub-optimality of the Lasso in this scenario is again its bias. This gap is clear for orthogonal designs where the Lasso and subset selector become the soft and hard threshold estimators respectively.

### *4.3.3 Model size and least squares after selection*

In this subsection, we study the size of the model selected by the Lasso and the least squares estimation after Lasso selection.

As the Lasso is designed to have sparse solutions, the size of the selected model is a quantity of considerable interest. Let $\mathcal{S} = \text{supp}(\boldsymbol{\beta})$ be the true model and $\widehat{\mathcal{S}} = \text{supp}(\widehat{\boldsymbol{\beta}})$ be the model selected by the Lasso. Let $\text{FP}(\widehat{\boldsymbol{\beta}}) = |\widehat{\mathcal{S}} \setminus \mathcal{S}|$ be the size of false positive and $\text{FN}(\widehat{\boldsymbol{\beta}}) = |\mathcal{S} \setminus \widehat{\mathcal{S}}|$ be the size of the false negative as discussed in Subsection 4.1.1. As $|\widehat{\mathcal{S}}| = |\mathcal{S}| + \text{FP}(\widehat{\boldsymbol{\beta}}) - \text{FN}(\widehat{\boldsymbol{\beta}})$, the model size of the Lasso is closely related to $\text{FP}(\widehat{\boldsymbol{\beta}})$ and $\text{FN}(\widehat{\boldsymbol{\beta}})$, especially the false positive.

It follows from Theorem 4.8 that the selection consistency of the Lasso requires the irrepresentable condition (4.32). While this condition is among the strongest on the design matrix in the analysis of PLS, it is not necessary for controlling the magnitude of the number of false positives.

Another topic of interest is the least squares estimation after selection, which is commonly used for possible improvements in prediction and coefficient estimation. As the expected prediction error of the least squares estimator in deterministic linear models is proportional to the dimension of the model when the noise is homoscedastic, least squares after Lasso is not expected to provide a good predictor when the selected model is too large. While least squares estimation *post Lasso selection* may remove the bias of the Lasso within the selected model, it may also amplify both the bias and variability of the Lasso due to false positive selection.

To provide an upper bound for the false positive, we define for $\mathscr{U} \subset \mathbb{R}^p$

$$\text{FP}(\mathscr{U}, \mathcal{S}) = \max\left\{|\text{supp}(\mathbf{u}) \setminus \mathcal{S}| : \mathbf{u} \in \mathscr{U}\right\}. \tag{4.57}$$

According to Theorem 4.10, $(\widehat{\boldsymbol{\beta}} - \boldsymbol{\beta}^*)/\lambda \in \mathscr{U}(\mathcal{S}, \boldsymbol{\Delta}^*)$ under condition (4.44) on the noise with the set $\mathscr{U}(\mathcal{S}, \boldsymbol{\Delta}^*) \subset \mathbb{R}^p$ in (4.53). Thus, by definition $\text{FP}(\mathscr{U}(\mathcal{S}, \boldsymbol{\Delta}^*), \mathcal{S})$ is the smallest upper bound for the false positive based on inequalities (4.51) and (4.52). Theorem 4.12 below provides a more explicit upper bound of $\text{FP}(\mathscr{U}(\mathcal{S}, \boldsymbol{\Delta}^*), \mathcal{S})$ under the *sparse Riesz condition* (SRC; Zhang and Huang, 2008). For nonnegative-definite $\boldsymbol{\Sigma}$ and $\mathcal{S} \subset \{1, \ldots, p\}$, define

$$\phi_{\text{cond}}(m; \mathcal{S}, \boldsymbol{\Sigma}) = \max_{|B \setminus \mathcal{S}| \le (1 \vee m)} \left\{\phi_{\max}(\boldsymbol{\Sigma}_{B,B}) / \phi_{\min}(\boldsymbol{\Sigma}_{B,B})\right\} \tag{4.58}$$

as a relaxed *sparse condition number* corresponding to the lower sparse eigenvalue in (4.23). We use the version of the SRC given in Zhang (2010a) which can be stated as the existence of a positive integer $m^* \le p - |\mathcal{S}|$ satisfying

$$|\mathcal{S}| < \frac{2(1-\eta_2)^2 m^*}{(1+\eta_1)^2\{\phi_{\rm cond}(m^*;\mathcal{S},\overline{\mathbf{\Sigma}})-1\}}, \tag{4.59}$$

where $\overline{\mathbf{\Sigma}} = \mathbf{X}^T\mathbf{X}/n$ is the Gram matrix. The SRC can be viewed as an $\ell_2$-type condition as it depends only on the eigenvalue of sub-Gram matrices. Define

$$C_{\rm FP}^{(\ell_1)}(\mathcal{S},\overline{\mathbf{\Sigma}}) = \min\left\{t : t = \frac{\phi_{\rm cond}(m;\mathcal{S},\overline{\mathbf{\Sigma}})-1}{2(1-\eta_2)^2/(1+\eta_1)^2} < \frac{m}{|\mathcal{S}|}\right\}.$$

**Theorem 4.12** *Let $\widehat{\boldsymbol\beta} = \widehat{\boldsymbol\beta}(\lambda)$ be the Lasso estimator in (4.29), $\{\boldsymbol\beta^*, \boldsymbol\beta\} \subset I\!R^p$, $\mathcal{S} \supseteq \text{supp}(\boldsymbol\beta) \cup \text{supp}(\boldsymbol\beta^*)$, $\eta_1 \ge 0$ and $\eta_2 \le 1$, and $\widehat{\mathcal{S}} = \text{supp}(\widehat{\boldsymbol\beta})$. Suppose (4.59) holds with an integer $m^* \ge 1$. If (4.44) holds, then*

$$\begin{aligned} |\widehat{\mathcal{S}}\setminus\mathcal{S}| &\le \text{FP}(\mathscr{U}(\mathcal{S},\mathbf{0}),\mathcal{S}) \le |\mathcal{S}|C_{\rm FP}^{(\ell_1)}(\mathcal{S},\overline{\mathbf{\Sigma}}) \le m^*, \\ |\mathcal{S}\setminus\widehat{\mathcal{S}}| &\le \#\{j\in\mathcal{S} : |\beta_j^*| \le e\lambda C_{{\rm est},\log|\mathcal{S}|}^{(\ell_1)}(\mathcal{S})\}, \end{aligned} \tag{4.60}$$

*where $C_{{\rm est},\log|\mathcal{S}|}^{(\ell_1)}(\mathcal{S})$ is as in Corollary 4.6. Moreover, if the penalty level $\lambda$ is as in (4.47) and the sub-Gaussian condition (4.21) holds as in Corollary 4.4, then (4.60) holds with probability $1-2\epsilon$ or $1-4\epsilon$ in the respective options of $\{\boldsymbol\beta^*, \eta_1, \eta_2\}$ in Corollary 4.4.*

The main result of Theorem 4.12 is the upper bound on false positive. This automatically implies that the size of the model selected by the Lasso is of the same order as the size of the true model. This bound is obtained under the $\ell_2$-type SRC on the design. A stronger version of the bound will be given in Lemma 4.5 below. Such bounds on the false positive are quite useful in further analysis of the Lasso and estimation after the Lasso selection. In Corollary 4.7 below, (4.60) is used to bound the constant factors in Theorem 4.11. Further discussion of these constant factors can be found in Subsections 4.3.4 and 4.3.5.

The upper bound of the false negative is a direct consequence of Theorem 4.11. It asserts that the false negative is limited to the set where the absolute value of the target coefficient $\beta_j^*$ is smaller than a threshold of order $\lambda$. As the condition on the noise holds when $\lambda \approx \sigma\sqrt{(2/n)\log p}$ under the sub-Gaussian condition, such requirement on the signal strength is of the optimal order in view of Theorem 4.4. However, as we will discuss in Subsection 4.3.5, the condition $C_{{\rm est},\log|\mathcal{S}|}^{(\ell_1)}(\mathcal{S}) = O(1)$, which is essentially of the $\ell_\infty$ type, may limit the scope of application of the upper bound for false negative.

For nonnegative-definite matrices $\mathbf{\Sigma}$, define an $\ell_q \to \ell_r$ extension of the lower sparse eigenvalue in (4.23) as

$$\phi_{*,q,r}(m;\mathcal{S},\mathbf{\Sigma}) = \min_{|B\setminus\mathcal{S}|\le m}\ \min_{\|\mathbf{u}\|_q=1} \|\mathbf{\Sigma}_{B,B}\mathbf{u}\|_r. \tag{4.61}$$

We note that $1/\phi_{*,q,r}(m;\mathcal{S},\mathbf{\Sigma}) = \|\mathbf{\Sigma}_{B,B}^{-1}\|_{r,q}$ is the $\ell_r \to \ell_q$ operator norm of the inverse $\mathbf{\Sigma}_{B,B}^{-1}$ of the subblock. The lower sparse eigenvalue can be written as $\phi_*(m;\mathcal{S},\mathbf{\Sigma}) = \phi_{*,2,2}(m;\mathcal{S},\mathbf{\Sigma})$.

**Corollary 4.7** *Let* $\{C_{\text{pred}}^{(\ell_1)}(\mathcal{S}), C_{\text{est},q}^{(\ell_1)}(\mathcal{S})\}$ *and* $C_{\text{FP}}^{(\ell_1)}(\mathcal{S},\overline{\mathbf{\Sigma}})$ *be the constant factors in Theorems 4.11 and 4.12 respectively. Let* $m_1 = \lfloor |\mathcal{S}| C_{\text{FP}}^{(\ell_1)}(\mathcal{S},\overline{\mathbf{\Sigma}})\rfloor$. *Then, the conclusions of Theorems 4.11 and 4.12 hold with*

$$C_{\text{pred}}^{(\ell_1)}(\mathcal{S}) \le \frac{(1+\eta_1)^2}{\phi_*(m_1;\mathcal{S},\overline{\mathbf{\Sigma}})}, \tag{4.62}$$

*and with* $\xi_{1,2} = \{1+(\eta_1-\eta_2)/2\}^2/(1-\eta_2)$

$$C_{\text{est},q}^{(\ell_1)}(\mathcal{S}) \le \frac{\xi_{1,2}^{2/q-1}(1+\eta_1)^{2-2/q}}{\phi_*(m_1;\mathcal{S},\overline{\mathbf{\Sigma}})}, \quad 1 \le q \le 2. \tag{4.63}$$

*Furthermore, for all* $1 \le q \le \infty$,

$$\begin{aligned} C_{\text{est},q}^{(\ell_1)}(\mathcal{S}) &\le \frac{1+\eta_1 \vee \eta_2}{(1 \vee |\mathcal{S}|^{1/q})\phi_{*,q,\infty}(m_1;\mathcal{S},\overline{\mathbf{\Sigma}})} \\ &\le \frac{\{1+C_{\text{FP}}^{(\ell_1)}(\mathcal{S},\overline{\mathbf{\Sigma}})\}^{1/q}(1+\eta_1\vee\eta_2)}{\phi_{*,q,q}(m_1;\mathcal{S},\overline{\mathbf{\Sigma}})}. \end{aligned} \tag{4.64}$$

Theorem 4.12 extends Theorem 6 of Zhang (2010a) to the special case of the $\ell_1$ penalty by allowing $\eta_1 > 0$ for $\boldsymbol{\beta}^* \ne \widehat{\boldsymbol{\beta}}^o$. The bound for the false positive is actually proved in more general vector classes of the form

$$\begin{aligned} &\mathscr{U}_0(\mathcal{S},\mathbf{\Sigma};\eta_1,\eta_2) \\ = \; &\big\{\mathbf{u} : \big|u_j(\mathbf{\Sigma}\mathbf{u})_j + |u_j|\big| \le \eta_2|u_j| \ \forall\, j \notin \mathcal{S}, \|(\mathbf{\Sigma}\mathbf{u})_{\mathcal{S}}\|_\infty \le 1+\eta_1\big\}, \end{aligned} \tag{4.65}$$

where $\mathcal{S} \subset \{1,\ldots,p\}$, $\mathbf{\Sigma}$ is a $p\times p$ nonnegative-definite matrix, $0 \le \eta_1 < \infty$ and $\eta_2 \in [0,1)$. It is clear from (4.53) that $\mathscr{U}(\mathcal{S},\mathbf{0}) \subseteq \mathscr{U}_0(\mathcal{S},\overline{\mathbf{\Sigma}};\eta_1,\eta_2)$. We state this stronger version of the upper bound in the following lemma, which immediately implies Theorem 4.12 and will be used again when we study concave PLS. As the inequalities in Corollary 4.7 are nontrivial, we provide a proof of the corollary below the proof of Lemma 4.5.

**Lemma 4.5** *Suppose the SRC holds with* $\overline{\mathbf{\Sigma}}$ *in (4.59) replaced by* $\mathbf{\Sigma}$. *Then,*

$$\frac{\|\mathbf{u}_{\mathcal{S}^c}\|_1\phi_{\max}(\mathbf{\Sigma}_{\mathcal{S},\mathcal{S}})}{1-\eta_2} + \big|\text{supp}(\mathbf{u})\setminus\mathcal{S}\big| \le \frac{\{\phi_{\text{cond}}(m^*;\mathcal{S},\mathbf{\Sigma})-1\}|\mathcal{S}|}{2(1-\eta_2)^2/(1+\eta_1)^2}.$$

*for all* $\mathbf{u} \in \mathscr{U}_0(\mathcal{S},\mathbf{\Sigma};\eta_1,\eta_2)$.

Lemma 4.5 improves upon Lemma 1 of Zhang (2010) in the special case of the Lasso by including the term $\|\mathbf{u}_{\mathcal{S}^c}\|_1\phi_{\max}(\mathbf{\Sigma}_{\mathcal{S},\mathcal{S}})/(1-\eta_2)$ on the left-hand side and allowing $\eta_1 > 0$. The proof there, which covers concave penalties as well as the Lasso, is modified to keep the two following additional items.

**Proof of Lemma 4.5.** Let $\mathscr{U}_1 = \mathscr{U}_0(\mathcal{S}, \mathbf{\Sigma}; \eta_1, \eta_2)$. For each $\mathbf{u} \in \mathscr{U}_1$, there exists a small $\epsilon > 0$ for which $\mathbf{u} \in \mathscr{U}_0(\mathcal{S}, \mathbf{\Sigma} + \epsilon^2 \mathbf{I}_{p\times p}; \eta_1 + \epsilon, \eta_2 + \epsilon)$. Thus, as $\phi_{\text{cond}}(m^*; \mathcal{S}, \mathbf{\Sigma} + \epsilon^2 \mathbf{I}_{p\times p}) \le \phi_{\text{cond}}(m^*; \mathcal{S}, \mathbf{\Sigma})$ and the conclusion is continuous in $(\eta_1, \eta_2)$, we assume without loss of generality that $\mathbf{\Sigma}$ is positive definite.

Let $B_{\mathbf{u}} = \{j \in \mathcal{S}^c : |(\mathbf{\Sigma u})_j| \ge 1 - \eta_2\}$. We have $\text{supp}(\mathbf{u}) \setminus \mathcal{S} \subseteq B_{\mathbf{u}}$. Define

$$k^* = \max\Big\{|B_{\mathbf{u}}| : \mathbf{u} \in \mathscr{U}_1\Big\}, \quad t^* = \frac{\{\phi_{\text{cond}}(m^*; \mathcal{S}, \mathbf{\Sigma}) - 1\}|\mathcal{S}|}{2(1-\eta_2)^2/(1+\eta_1)^2}.$$

We split the proof into two-steps. In the first step, we prove that for any integer $k \in [0, k^*]$, there exists a vector $\mathbf{u} \in \mathscr{U}_1$ and $A$ satisfying

$$k = |A \setminus \mathcal{S}|, \quad \mathcal{S} \cup \text{supp}(\mathbf{u}) \subseteq A \subseteq \mathcal{S} \cup \{j : |(\mathbf{\Sigma u})_j| \ge 1 - \eta_2\}. \tag{4.66}$$

In the second step, we prove that when (4.66) holds with $k \le m^*$,

$$\|\mathbf{u}_{\mathcal{S}^c}\|_1 \phi_{\max}(\mathbf{\Sigma}_{\mathcal{S},\mathcal{S}})/(1-\eta_2) + k \le t^*. \tag{4.67}$$

As $t^* < m^*$ by the SRC, $k \le m^*$ implies $k < m^*$ by the second step, so that $m^* \notin [0, k^*]$ by the first step. The conclusion follows as $|\text{supp}(\mathbf{u}) \setminus \mathcal{S}| \le k^*$.

*Step 1.* Let $\mathbf{u}^* \in \mathscr{U}_1$ with $|B_{\mathbf{u}^*}| = k^*$. Let $B = B_{\mathbf{u}^*}$. Define a vector $\mathbf{z}$ by

$$\mathbf{z}_B = (\mathbf{\Sigma u}^*)_B + \text{sgn}(\mathbf{u}^*_B), \quad \mathbf{z}_{\mathcal{S}} = (\mathbf{\Sigma u}^*)_{\mathcal{S}}.$$

As $\mathbf{u}^* \in \mathscr{U}_1$, we have $\|\mathbf{z}_B\|_\infty \le \eta_2$ and $\|\mathbf{z}_{\mathcal{S}}\|_\infty \le 1 + \eta_1$. Consider an auxiliary optimization problem

$$\mathbf{b}(\lambda) = \arg\min_{\mathbf{b}} \Big\{\mathbf{b}^T \mathbf{\Sigma b}/2 - \mathbf{z}^T \mathbf{b} + \lambda \|\mathbf{b}_B\|_1 : \text{supp}(\mathbf{b}) \subseteq \mathcal{S} \cup B\Big\}.$$

Similar to Proposition 4.1, the KKT conditions for $\mathbf{b}(\lambda)$ can be written as

$$\begin{cases} \big|\big(\mathbf{\Sigma b}(\lambda) - \mathbf{z}\big)_j\big| \le \lambda, & j \in B, \\ b_j(\lambda)\big(\mathbf{\Sigma b}(\lambda) - \mathbf{z}\big)_j + \lambda\,|b_j(\lambda)| = 0, & j \in \mathcal{S}^c, \\ \big(\mathbf{\Sigma b}(\lambda) - \mathbf{z}\big)_j = 0, & j \in \mathcal{S}, \\ b_j(\lambda) = 0, & j \notin \mathcal{S} \cup B. \end{cases}$$

Note that $b_j$ is penalized only for $j \in B$, but $j \in B$ does not guarantee $b_j(\lambda) \ne 0$. Due to the positive-definiteness of $\mathbf{\Sigma}$, the objective function of the auxiliary minimization problem is strictly convex, so that $\mathbf{b}(\lambda)$ is uniquely defined by the KKT conditions and continuous in $\lambda$.

Let $\mathbf{u}(\lambda) = \mathbf{b}(\lambda)/\lambda$. For $\lambda \ge 1$, the KKT conditions imply

$$\big|u_j(\lambda)\big(\mathbf{\Sigma u}(\lambda)\big)_j + |u_j(\lambda)|\big| = |b_j(\lambda) z_j|/\lambda^2 \le \eta_2 |u_j(\lambda)|, \ \forall\, j \in B,$$

and $\|(\mathbf{\Sigma u}(\lambda))_{\mathcal{S}}\|_\infty = \|\mathbf{z}_{\mathcal{S}}\|_\infty/\lambda \le 1+\eta_1$, so that $\mathbf{u}(\lambda) \in \mathscr{U}_1$. Let

$$B(\lambda) = \{j \in B : |(\mathbf{\Sigma u}(\lambda))_j| \ge 1-\eta_2\}.$$

For $\lambda = 1$, the KKT conditions yield $\mathbf{b}(1) = \mathbf{u}^*$. Let

$$\lambda^* = \|\mathbf{\Sigma}_{B,\mathcal{S}}\mathbf{\Sigma}_{\mathcal{S},\mathcal{S}}^{-1}\mathbf{z}_{\mathcal{S}} - \mathbf{z}_B\|_\infty.$$

For $\lambda \ge \lambda^*$, the solution is given by

$$\mathbf{b}_{\mathcal{S}}(\lambda) = \mathbf{\Sigma}_{\mathcal{S},\mathcal{S}}^{-1}\mathbf{z}_{\mathcal{S}}, \quad \mathbf{b}_B(\lambda) = 0.$$

Thus, $\mathbf{u}(\lambda)$ is a continuous path in $\mathscr{U}_1$ with $\mathrm{supp}(\mathbf{u}(\lambda^*)) = \mathcal{S}$, $\mathbf{u}(1) = \mathbf{u}^*$ and $\mathrm{supp}(\mathbf{u}(\lambda)) \subseteq \mathcal{S} \cup B(\lambda)$. Let $k \in [0, k^*]$ and

$$\lambda_k = \sup\{\lambda \in [1,\lambda^*] : |B(\lambda)| \ge k \ \text{ or } \ \lambda = \lambda^*\}.$$

If $B(\lambda^*) \ge k$, then (4.66) is feasible with $\mathbf{u} = \mathbf{u}(\lambda^*)$ due to $\mathrm{supp}(\mathbf{u}(\lambda^*)) = \mathcal{S}$. Otherwise, $\lambda_k \in [1,\lambda^*)$, $\mathrm{supp}(\mathbf{u}(\lambda_k)) \subseteq \mathcal{S} \cup B(\lambda_k+)$, $|B(\lambda_k+)| < k$, and $|B(\lambda_k)| \ge k$ due to the continuity of $\mathbf{u}(\lambda)$ and the fact that $k \le k^* = |B(1)|$. Thus, (4.66) is feasible with $\mathbf{u} = \mathbf{u}(\lambda^*)$.

*Step 2.* Suppose (4.66) holds for certain $A$, $k \le m^*$ and $\mathbf{u} \in \mathscr{U}_1$. We need to prove (4.67). Let $B = A \setminus \mathcal{S}$, $\mathbf{v} = (\mathbf{\Sigma u})_A \in \mathbb{R}^A$, $\mathbf{v}_{(\mathcal{S})} = (v_j I\{j \in \mathcal{S}\}, j \in A) \in \mathbb{R}^A$ and $\mathbf{v}_{(B)} = \mathbf{v} - \mathbf{v}_{(\mathcal{S})}$. By algebra,

$$\mathbf{v}^T\mathbf{\Sigma}_{A,A}^{-1}\mathbf{v} + \mathbf{v}_{(B)}^T\mathbf{\Sigma}_{A,A}^{-1}\mathbf{v}_{(B)} - \mathbf{v}_{(\mathcal{S})}^T\mathbf{\Sigma}_{A,A}^{-1}\mathbf{v}_{(\mathcal{S})} = 2\mathbf{v}^T\mathbf{\Sigma}_{A,A}^{-1}\mathbf{v}_{(B)}.$$

Because $\mathbf{v}^T\mathbf{\Sigma}_{A,A}^{-1}\mathbf{v}_{(B)} = (\mathbf{v}^T\mathbf{\Sigma}_{A,A}^{-1})_B\mathbf{v}_B = \mathbf{u}_B^T(\mathbf{\Sigma u})_B \le -(1-\eta_2)\|\mathbf{u}_B\|_1$,

$$\begin{aligned}\frac{\|\mathbf{v}_{(B)}\|_2^2 + \|\mathbf{v}\|_2^2}{\phi_{\max}(\mathbf{\Sigma}_{A,A})} &\le \mathbf{v}_{(\mathcal{S})}^T\mathbf{\Sigma}_{A,A}^{-1}\mathbf{v}_{(\mathcal{S})} - 2(1-\eta_2)\|\mathbf{u}_B\|_1 \\ &\le \frac{\|\mathbf{v}_{(\mathcal{S})}\|_2^2}{\phi_{\min}(\mathbf{\Sigma}_{A,A})} - 2(1-\eta_2)\|\mathbf{u}_B\|_1.\end{aligned}$$

Since $\|\mathbf{v}_{(B)}\|_2^2 \ge (1-\eta_2)^2|B|$ and $\|\mathbf{v}\|_2^2 - \|\mathbf{v}_{(B)}\|_2^2 = \|\mathbf{v}_{(\mathcal{S})}\|_2^2 \le (1+\eta_1)^2|\mathcal{S}|$,

$$\begin{aligned}&2(1-\eta_2)\|\mathbf{u}_B\|_1\phi_{\max}(\mathbf{\Sigma}_{A,A}) + 2(1-\eta_2)^2|B| \\ \le\ &(1+\eta_1)^2|\mathcal{S}|\Big(\frac{\phi_{\max}(\mathbf{\Sigma}_{A,A})}{\phi_{\min}(\mathbf{\Sigma}_{A,A})} - 1\Big).\end{aligned}$$

As $|A \setminus \mathcal{S}| = k \le m^*$ and $\mathcal{S} \subseteq A$, $\phi_{\max}(\mathbf{\Sigma}_{A,A})/\phi_{\min}(\mathbf{\Sigma}_{A,A}) \le \phi_{\mathrm{cond}}(m^*; \mathcal{S}, \mathbf{\Sigma})$ and $\phi_{\max}(\mathbf{\Sigma}_{A,A}) \ge \phi_{\max}(\mathbf{\Sigma}_{\mathcal{S},\mathcal{S}})$. It follows that

$$\|\mathbf{u}_B\|_1\phi_{\max}(\mathbf{\Sigma}_{\mathcal{S},\mathcal{S}})/(1-\eta_2) + |B| \le \frac{|\mathcal{S}|\big(\phi_{\mathrm{cond}}(m^*; \mathcal{S}, \mathbf{\Sigma}) - 1\big)}{2(1-\eta_2)^2/(1+\eta_1)^2} = t^*.$$

This completes Step 2 and thus the proof of the lemma. ∎

**Proof of Corollary 4.7.** Since $C_{\rm pred}^{(\ell_1)}(\mathcal{S})$ and $C_{{\rm est},q}^{(\ell_1)}(\mathcal{S})$ are the constant factors in the sharpest oracle inequalities based on the basic inequalities in (4.51) and (4.52) and $\overline{C}_{\rm pred}^{(\ell_1)}(\mathcal{S},\overline{\Delta})$ and $\overline{C}_{{\rm est},q}^{(\ell_1)}(\mathcal{S},\overline{\Delta})$ are based on the weaker basic inequality in (4.49), it suffices to bound the expressions for $\overline{C}_{\rm pred}^{(\ell_1)}(\mathcal{S},0)$ and $\overline{C}_{{\rm est},q}^{(\ell_1)}(\mathcal{S},0)$ under the additional constraint $|{\rm supp}(\mathbf{u})\setminus\mathcal{S}|\le m_1$.

Consider $\mathbf{u}$ satisfying $\mathbf{u}^T\overline{\mathbf{\Sigma}}\mathbf{u}=1$. Let $\psi_+(\mathbf{u})=\max\{0,\psi(\mathbf{u})\}$ with the $\psi(\mathbf{u})$ in (4.39). By definition, $\psi_+^2(\mathbf{u})\le(1+\eta_1)^2\|\mathbf{u}_{\mathcal{S}}\|_1^2\le(1+\eta_1)^2|\mathcal{S}|\|\mathbf{u}\|_2^2$. As $|{\rm supp}(\mathbf{u})\setminus\mathcal{S}|\le m_1$, $\|\mathbf{u}\|_2^2\phi_*(m_1;\mathcal{S},\overline{\mathbf{\Sigma}})\le\mathbf{u}^T\overline{\mathbf{\Sigma}}\mathbf{u}=1$. Thus,

$$\frac{\psi_+^2(\mathbf{u})}{1\vee|\mathcal{S}|}\le\frac{|\mathcal{S}|(1+\eta_1)^2}{(1\vee|\mathcal{S}|)\phi_*(m_1;\mathcal{S},\overline{\mathbf{\Sigma}})}.$$

This yields (4.62) by the expression of $\overline{C}_{\rm pred}^{(\ell_1)}(\mathcal{S},0)$ in (4.39).

For (4.63), we notice that $\|\mathbf{u}\|_1\psi(\mathbf{u})\le\|\mathbf{u}_{\mathcal{S}}\|_1^2(1+t)\{(1+\eta_1)-(1-\eta_2)t\}$ with $t=\|\mathbf{u}_{\mathcal{S}^c}\|_1/\|\mathbf{u}_{\mathcal{S}}\|_1$. As $(1+t)\{(1+\eta_1)-(1-\eta_2)t\}\le\xi_{1,2}$ with the maximum attained at $t=(\eta_1+\eta_2)/\{2(1-\eta_2)\}$, $\|\mathbf{u}\|_1\psi_+(\mathbf{u})\le\xi_{1,2}\|\mathbf{u}_{\mathcal{S}}\|_1^2$. By the Hölder inequality, $\|\mathbf{u}\|_q\le\|\mathbf{u}\|_1^{2/q-1}\|\mathbf{u}\|_2^{2-2/q}$ for $1\le q\le 2$. It follows that

$$\begin{aligned}\|\mathbf{u}\|_q\psi_+(\mathbf{u}) &\le \Big(\|\mathbf{u}\|_1\psi_+(\mathbf{u})\Big)^{2/q-1}\Big(\|\mathbf{u}\|_2\psi_+(\mathbf{u})\Big)^{2-2/q}\\ &\le \Big(\xi_{1,2}\|\mathbf{u}_{\mathcal{S}}\|_1^2\Big)^{2/q-1}\Big(\|\mathbf{u}\|_2(1+\eta_1)\|\mathbf{u}_{\mathcal{S}}\|_1\Big)^{2-2/q}\\ &\le \xi_{1,2}^{2/q-1}(1+\eta_1)^{2-2/q}|\mathcal{S}|^{1/q}\|\mathbf{u}\|_2^2\\ &\le \xi_{1,2}^{2/q-1}(1+\eta_1)^{2-2/q}|\mathcal{S}|^{1/q}/\phi_*(m_1;\mathcal{S},\overline{\mathbf{\Sigma}}).\end{aligned}$$

This yields (4.63) by the expression of $\overline{C}_{{\rm est},q}^{(\ell_1)}(\mathcal{S},0)$ in (4.40).

Now we prove (4.64). By the definition of $T_+(\mathbf{u})$, we have

$$\begin{aligned}\|\mathbf{u}\|_qT_+(\mathbf{u}) &\le \|\mathbf{u}\|_q(1+\eta_1\vee\eta_2)/\|\overline{\mathbf{\Sigma}}\mathbf{u}\|_\infty\\ &\le (1+\eta_1\vee\eta_2)/\phi_{*,q,\infty}(m_1;\mathcal{S},\overline{\mathbf{\Sigma}}).\end{aligned}$$

Moreover, as $\|\overline{\mathbf{\Sigma}}_{B,B}\mathbf{u}\|_q\le|B|^{1/q}\|\overline{\mathbf{\Sigma}}_{B,B}\mathbf{u}\|_\infty$, we have $\phi_{*,q,\infty}(m_1;\mathcal{S},\overline{\mathbf{\Sigma}})\ge\phi_{*,q,q}(m_1;\mathcal{S},\overline{\mathbf{\Sigma}})/(m_1+|\mathcal{S}|)^{1/q}$. This yields (4.64) by the definition of $C_{{\rm est},q}^{(\ell_1)}(\mathcal{S})$ above the statement of Theorem 4.11. ∎

Now we consider least squares estimation after Lasso selection. Given $\mathcal{M}\subset\{1,\ldots,p\}$, the least squares estimator in model $\mathcal{M}$ is defined by

$$\widehat{\boldsymbol{\beta}}_{\mathcal{M}}^{(\mathcal{M})}=\Big(\mathbf{X}_{\mathcal{M}}^T\mathbf{X}_{\mathcal{M}}\Big)^{-1}\mathbf{X}_{\mathcal{M}}\mathbf{Y},\quad \widehat{\boldsymbol{\beta}}_{\mathcal{M}^c}^{(\mathcal{M})}=\mathbf{0}. \tag{4.68}$$

**Theorem 4.13** *Suppose $\mathcal{S}\supseteq{\rm supp}(\boldsymbol{\beta})$, the sub-Gaussian condition (4.21) holds, and the SRC (4.59) holds for the design. Let $\widehat{\boldsymbol{\beta}}(\lambda)$ be the Lasso with the penalty level $\lambda=(1/\eta)\sigma\sqrt{(2/n)\log(p/\epsilon)}$, where $\eta\le\eta_2<1$. Let $\boldsymbol{\beta}^*=\boldsymbol{\beta}$ with*

*$\eta_1 = \eta_2$ or $\boldsymbol{\beta}^* = \widehat{\boldsymbol{\beta}}^{(\mathcal{S})} = \widehat{\boldsymbol{\beta}}^{o}$ as in (4.10) with $\eta_1 = 0$. Let $\widehat{\mathcal{S}} = \mathrm{supp}(\widehat{\boldsymbol{\beta}}(\lambda))$, $m_1 = \lfloor |\mathcal{S}| C_{\mathrm{FP}}^{(\ell_1)}(\mathcal{S}, \overline{\boldsymbol{\Sigma}})\rfloor \le m^*$ with the $C_{\mathrm{FP}}^{(\ell_1)}(\mathcal{S}, \overline{\boldsymbol{\Sigma}})$ in Theorem 4.12, and $\lambda_1 = \sigma\sqrt{(2/n)\log(ep/(\epsilon m_1))}$. Then,*

$$\begin{aligned} & |\mathcal{S}|^{-1}\|\widehat{\boldsymbol{\beta}}^{(\widehat{\mathcal{S}})} - \boldsymbol{\beta}^*\|_1^2/\{C_{\mathrm{FP}}^{(\ell_1)}(\mathcal{S}, \overline{\boldsymbol{\Sigma}})\phi_*(m_1; \mathcal{S}, \overline{\boldsymbol{\Sigma}})\} \\ \le\ & \|\widehat{\boldsymbol{\beta}}^{(\widehat{\mathcal{S}})} - \boldsymbol{\beta}^*\|_2^2/\phi_*(m_1; \mathcal{S}, \overline{\boldsymbol{\Sigma}}) \\ \le\ & \|\mathbf{X}\widehat{\boldsymbol{\beta}}^{(\widehat{\mathcal{S}})} - \mathbf{X}\boldsymbol{\beta}^*\|_2^2/n \\ \le\ & |\mathcal{S}|\Big\{C_{\mathrm{pred}}^{(\ell_1)}(\mathcal{S})(\lambda^2 + 7.5/n) + 2.5 C_{\mathrm{FP}}^{(\ell_1)}(\mathcal{S}, \overline{\boldsymbol{\Sigma}})(\lambda_1^2 + 3/n)\Big\} \end{aligned}$$

*with at least probability $1 - 2\epsilon - \epsilon^{m_1}$, where $\phi_*(m; \mathcal{S}, \overline{\boldsymbol{\Sigma}})$ is the lower sparse eigenvalue in (4.23) and $C_{\mathrm{pred}}^{(\ell_1)}(\mathcal{S})$ is as in Theorem 4.11.*

Theorem 4.13 asserts that under the $\ell_0$ sparsity condition, the least squares estimator in the model selected by the Lasso is rate optimal in prediction and coefficient estimation, provided that both $C_{\mathrm{pred}}^{(\ell_1)}(\mathcal{S})$ and $C_{\mathrm{FP}}^{(\ell_1)}(\mathcal{S}, \overline{\boldsymbol{\Sigma}})$ are bounded.

**Proof of Theorem 4.13.** As $\mathbf{X}\widehat{\boldsymbol{\beta}}^{(\widehat{\mathcal{S}})}$ is the projection of $\mathbf{Y}$ to the column space of $\mathbf{X}_{\widehat{\mathcal{S}}}$, we have

$$\begin{aligned} \|\mathbf{X}\widehat{\boldsymbol{\beta}}^{(\widehat{\mathcal{S}})} - \mathbf{X}\boldsymbol{\beta}\|_2^2 &= \|\mathbf{P}_{\widehat{\mathcal{S}}}^{\perp}\mathbf{X}\boldsymbol{\beta}\|_2^2 + \|\mathbf{P}_{\widehat{\mathcal{S}}}\boldsymbol{\varepsilon}\|_2^2 \\ &\le \|\mathbf{X}\widehat{\boldsymbol{\beta}} - \mathbf{X}\boldsymbol{\beta}\|_2^2 + \|\mathbf{P}_{\widehat{\mathcal{S}}\cup\mathcal{S}}\boldsymbol{\varepsilon}\|_2^2. \end{aligned}$$

By Theorem 4.12, the second term above is no greater than $|\mathcal{S}|\lambda^2 C_{\mathrm{pred}}^{(\ell_1)}(\mathcal{S})$ with at least probability $1 - 2\epsilon$. As there are $\binom{p-|\mathcal{S}|}{m_1}$ models $\mathcal{M}$ with $|\mathcal{M}\setminus\mathcal{S}| \le m_1$, Lemma 4.2 (iii) implies

$$P\Big\{\|\mathbf{P}_{\widehat{\mathcal{S}}\cup\mathcal{S}}\boldsymbol{\varepsilon}\|_2^2 \ge (|\mathcal{S}| + m_1)(2.5t + 7.5)\Big\} \le \binom{p - |\mathcal{S}|}{m_1} e^{-(|\mathcal{S}|+m_1)t}.$$

As $\log\binom{p-|\mathcal{S}|}{m_1} \le m_1\log(ep/m_1)$, The conclusion follows by taking $(|\mathcal{S}|+m_1)t = m_1\lambda_1^2$. ∎

### *4.3.4 Properties of the Dantzig selector*

In this subsection, we study the Dantzig selector (4.30) under the condition

$$\|\mathbf{X}^T(\mathbf{Y} - \mathbf{X}\boldsymbol{\beta}^*)/n\|_\infty \le \eta\lambda,\ \eta \le 1, \tag{4.69}$$

where $\boldsymbol{\beta}^*$ is a target coefficient vector. This condition is identical to (4.44) when $\eta_1 = \eta_2 = \eta$. We will show that for prediction and coefficient estimation, the performance of the Dantzig selector is guaranteed by oracle inequalities similar to those for the Lasso in Theorems 4.9 and 4.11. Moreover, we will provide some upper bounds for constant factors in the oracle inequalities for both the Lasso and Dantzig selector.

It follows from (4.69) and the $\ell_\infty$ constraint in (4.30) that

$$\left\|\mathbf{X}^T\mathbf{X}(\widehat{\boldsymbol{\beta}}^{(DS)} - \boldsymbol{\beta}^*)/n\right\|_\infty \le (1+\eta)\lambda.$$

Moreover, as $\boldsymbol{\beta}^*$ is a feasible solution satisfying the constraint, the optimality of $\widehat{\boldsymbol{\beta}}^{(DS)}$ implies $\|\widehat{\boldsymbol{\beta}}^{(DS)}\|_1 \le \|\boldsymbol{\beta}^*\|_1$. Letting $\widehat{\boldsymbol{\Delta}} = \widehat{\boldsymbol{\beta}}^{(DS)} - \boldsymbol{\beta}^*$, following the same arguments as those for obtaining (3.48), we have

$$\begin{aligned}\|\boldsymbol{\beta}^*\|_1 &\ge \|\boldsymbol{\beta}^* + \widehat{\boldsymbol{\Delta}}\|_1 \\ &= \|(\boldsymbol{\beta}^* + \widehat{\boldsymbol{\Delta}})_{\mathcal{S}}\|_1 + \|(\boldsymbol{\beta}^* + \widehat{\boldsymbol{\Delta}})_{\mathcal{S}^c}\|_1 \\ &\ge \|\boldsymbol{\beta}^*_{\mathcal{S}}\|_1 - \|\widehat{\boldsymbol{\Delta}}_{\mathcal{S}}\|_1 + \|\widehat{\boldsymbol{\Delta}}_{\mathcal{S}^c}\|_1 - \|\boldsymbol{\beta}^*_{\mathcal{S}^c}\|_1\end{aligned}$$

Therefore,

$$\|\widehat{\boldsymbol{\Delta}}_{\mathcal{S}^c}\|_1 \le \|\widehat{\boldsymbol{\Delta}}_{\mathcal{S}}\|_1 + 2\|\boldsymbol{\beta}^*_{\mathcal{S}^c}\|_1.$$

Thus, letting $\mathbf{u} = (\widehat{\boldsymbol{\beta}}^{(DS)} - \boldsymbol{\beta}^*)/\lambda$ and $\overline{\boldsymbol{\Sigma}} = \mathbf{X}^T\mathbf{X}/n$, we have

$$\|\overline{\boldsymbol{\Sigma}}\mathbf{u}\|_\infty \le 1+\eta, \quad \|\mathbf{u}_{\mathcal{S}^c}\|_1 \le \xi\|\mathbf{u}_{\mathcal{S}}\|_1 + s_1^*, \tag{4.70}$$

with $\xi = 1$ and $s_1^* = 2\|\boldsymbol{\beta}^*_{\mathcal{S}^c}\|_1/\lambda$. In the analysis below, the general $\xi$ allows simultaneous study of the Lasso and Dantzig selector and a connection between the cases of $s_1^* > 0$ and $s_1^* = 0$. In the same spirit of the analysis of the Lasso in Subsection 4.3.2, Theorem 4.14 below provides the sharpest error bounds for prediction and coefficient estimation based on (4.70).

Let

$$\mathscr{U}^{(DS)}(\mathcal{S}, s_1^*; \eta, \xi) = \Big\{\text{all vectors } \mathbf{u} \text{ satisfying (4.70)}\Big\}. \tag{4.71}$$

With the convention $0/0 = \infty$, define

$$T^{(DS)}(\mathbf{u}) = \min\left\{\frac{1+\eta}{\|\overline{\boldsymbol{\Sigma}}\mathbf{u}\|_\infty}, \frac{s_1^*}{(\|\mathbf{u}_{\mathcal{S}^c}\|_1 - \xi\|\mathbf{u}_{\mathcal{S}}\|_1)_+}\right\}.$$

Note that $T^{(DS)}(\mathbf{u}) = \{(1+\eta)/\|\overline{\boldsymbol{\Sigma}}\mathbf{u}\|_\infty\} I\{\|\mathbf{u}_{\mathcal{S}^c}\|_1 \le \xi\|\mathbf{u}_{\mathcal{S}}\|_1\}$ for $s_1^* = 0$. Define

$$C_{\text{pred}}^{(DS)}(\mathcal{S}, s_1^*; \eta, \xi) = \max\left\{\frac{\left(T^{(DS)}(\mathbf{u})\right)^2}{1 \vee |\mathcal{S}|} : \mathbf{u}^T\overline{\boldsymbol{\Sigma}}\mathbf{u} = 1\right\} \tag{4.72}$$

as a constant factor for bounding the prediction error, and

$$C_{\text{est},q}^{(DS)}(\mathcal{S}, s_1^*; \eta, \xi) = \max\left\{\frac{T^{(DS)}(\mathbf{u})}{1 \vee |\mathcal{S}|^{1/q}} : \|\mathbf{u}\|_q = 1\right\} \tag{4.73}$$

as a constant factor for bounding the $\ell_q$ estimation error.

**Theorem 4.14** *Let $\widehat{\boldsymbol\beta}^{(DS)} = \widehat{\boldsymbol\beta}^{(DS)}(\lambda)$ be the Dantzig selector in (4.30), $\boldsymbol\beta^*$ a target coefficient vector in $\mathbb{R}^p$, $\mathcal{S} \subset \{1, \ldots, p\}$, and $s_1^* = 2\|\boldsymbol\beta^*_{\mathcal{S}^c}\|_1/\lambda$. Let $0 \le \eta \le 1$. If (4.69) holds, then,*

$$(\widehat{\boldsymbol\beta}^{(DS)} - \boldsymbol\beta^*)/\lambda \in \mathscr{U}^{(DS)}(\mathcal{S}, s_1^*; \eta, 1), \tag{4.74}$$

$$\begin{aligned}\|\mathbf{X}\widehat{\boldsymbol\beta}^{(DS)} - \mathbf{X}\boldsymbol\beta^*\|_2^2/n &\le \lambda^2(1 \vee |\mathcal{S}|)C_{\text{pred}}^{(DS)}(\mathcal{S}, s_1^*; \eta, 1) \qquad (4.75)\\ &\le \lambda^2(1 \vee |\mathcal{S}|)(1+\eta)C_{\text{est},1}^{(DS)}(\mathcal{S}, s_1^*; \eta, 1),\end{aligned}$$

*and*

$$\|\widehat{\boldsymbol\beta}^{(DS)} - \boldsymbol\beta^*\|_q \le \lambda(1 \vee |\mathcal{S}|^{1/q})C_{\text{est},q}^{(DS)}(\mathcal{S}, s_1^*; \eta, 1). \tag{4.76}$$

*Alternatively, if $\lambda = (1/\eta)\sigma\sqrt{(2/n)\log(p/\epsilon)}$ and the sub-Gaussian condition (4.21) holds, then (4.74), (4.75) and (4.76) hold with probability $1 - 2\epsilon$ for either $\boldsymbol\beta^* = \boldsymbol\beta$ or $\boldsymbol\beta^* = \widehat{\boldsymbol\beta}^{o}$, where $\widehat{\boldsymbol\beta}^{o}$ is the oracle estimator in (4.38).*

**Remark 4.3** For $\mathcal{S} = \emptyset$ and $\xi = 1$, $\big(T^{(DS)}(\mathbf{u})\big)^2 \le (1+\eta)s_1^*/(\|\mathbf{u}\|_1\|\overline{\boldsymbol\Sigma}\mathbf{u}\|_\infty) \le (1+\eta)s_1^*$ subject to $\mathbf{u}^T\overline{\boldsymbol\Sigma}\mathbf{u} = 1$, so that (4.75) implies

$$\|\mathbf{X}\widehat{\boldsymbol\beta}^{(DS)} - \mathbf{X}\boldsymbol\beta^*\|_2^2/n \le \lambda^2(1+\eta)s_1^* = 2(1+\eta)\lambda\|\boldsymbol\beta^*\|_1. \tag{4.77}$$

This slow rate for prediction is nearly identical to (4.48).

**Proof of Theorem 4.14.** We have already proved (4.74) before the statement of the theorem. For any seminorm $\|\cdot\|$ as a loss function, we want to prove

$$\max\Big\{\|\mathbf{v}\| : \mathbf{v} \in \mathscr{U}^{(DS)}(\mathcal{S}, s_1^*; \eta, \xi)\Big\} = \max_{\|\mathbf{u}\|=1} T^{(DS)}(\mathbf{u}). \tag{4.78}$$

We write $\mathbf{v} = t\mathbf{u}$ with $\|\mathbf{u}\| = 1$, so that (4.74) is equivalent to $t\|\overline{\boldsymbol\Sigma}\mathbf{u}\|_\infty \le 1+\eta$ and $t(\|\mathbf{u}_{\mathcal{S}^c}\|_1 - \xi\|\mathbf{u}_{\mathcal{S}}\|_1)_+ \le s_1^*$, or equivalently $\|\mathbf{v}\| = t \le T^{(DS)}(\mathbf{u})$. The relationship between the prediction and $\ell_1$ losses in (4.75) follows from $\mathbf{u}^T\overline{\boldsymbol\Sigma}\mathbf{u} \le \|\mathbf{u}\|_1\|\overline{\boldsymbol\Sigma}\mathbf{u}\|_\infty$. ■

The oracle inequalities (4.75) and (4.76) are the sharpest based on (4.74). For $s_1^* = 0$, Candés and Tao (2007), formulated $C_{\text{est},2}^{(DS)}(\mathcal{S}, 0; 1, 1)$ as an $\ell_2$ optimization problem and derived upper bounds of the quantity under a restricted isometry property (*RIP*) of the design matrix. Their results were sharpened and extended in Bickel, Ritov and Tsybakov (2009), van de Geer and Bühlmann (2009), Zhang (2009), Cai, Wang and Xu (2010), Ye and Zhang

(2010) and Cai and Zhang (2013) among others, using the restricted eigenvalue, compatibility factor, cone invertibility factor and similar quantities. As we have mentioned above Theorem 4.9, results related to these quantities are discussed in detail in Subsection 4.3.5.

Theorem 4.15 below presents an analysis along the line started in Candés and Tao (2007) of the constant factors in Theorems 4.11 and 4.14, including those for the $\ell_q$ loss with $2 < q \le \infty$. These upper bounds are expressed in terms of the matrix operator norms in sparse diagonal and off-diagonal sub-blocks.

For $1 \le q \le \infty$, $\mathcal{S} \subset \{1, \ldots, p\}$, $s \ge 0$, integer $\ell \ge 0$, $\xi \ge 0$, and nonnegative-definite matrices $\mathbf{\Sigma}$, define

$$\begin{aligned} & C^*_{\text{est},q}(\mathcal{S}, s, \ell, \xi; \mathbf{\Sigma}) \qquad (4.79) \\ = & \min_r \max_{A,B} \frac{\{1 + \xi(a_q s/\ell)^{1-1/q}\} s^{-1/q} \|\mathbf{\Sigma}^{-1}_{A,A}\|_{\infty,q}}{\{1 - \xi s^{1-1/q}(a_r/\ell)^{1-1/r} \|\mathbf{\Sigma}^{-1}_{A,A}\mathbf{\Sigma}_{A,B}\|_{r,q}\}_+}, \end{aligned}$$

where $a_\infty = 1$, $a_q = (1 - 1/q)/q^{1/(q-1)}$ for $1 \le q < \infty$, $\|\cdot\|_{r,q}$ is the $\ell_r$ to $\ell_q$ operator norm, the minimum is taken over $1 \le r \le \infty$, and the maximum is taken with $|A \setminus \mathcal{S}| = \ell$ and $|B| \le \lceil \ell/a_r \rceil$ with $A \cap B = \emptyset$. As $|A| \le s + \ell$,

$$s^{-1/q}\|\mathbf{\Sigma}^{-1}_{A,A}\|_{\infty,q} \le (1 + \ell/s)^{1/q}\|\mathbf{\Sigma}^{-1}_{A,A}\|_{q,q}.$$

This inequality and the choice $r = q$ yields

$$\begin{aligned} & C^*_{\text{est},q}(\mathcal{S}, s, \ell, \xi; \mathbf{\Sigma}) \qquad (4.80) \\ \le & \max_{A,B} \frac{\{1 + \xi(a_q s/\ell)^{1-1/q}\}(1 + \ell/s)^{1/q}\|\mathbf{\Sigma}^{-1}_{A,A}\|_{q,q}}{\{1 - \xi(a_q s/\ell)^{1-1/q}\|\mathbf{\Sigma}^{-1}_{A,A}\mathbf{\Sigma}_{A,B}\|_{q,q}\}_+}. \end{aligned}$$

**Theorem 4.15** *Let* $1 \le q \le \infty$ *and* $C^*_{\text{est},q}(\mathcal{S}, s, \ell, \xi; \mathbf{\Sigma})$ *be as in (4.79).*
*(i) For any* $\{\eta_0, s^*, \ell\}$ *satisfying* $\eta_0(a_1 s_1^*/\ell)^{1-1/q} < 1$ *and* $s^* \ge \max(s_1^*, |\mathcal{S}|)$,

$$\begin{aligned} & (1 \vee |\mathcal{S}|^{1/q}) C^{(DS)}_{\text{est},q}(\mathcal{S}, s_1^*; \eta, \xi) \\ \le & (s^*)^{1/q} \max\left\{1/\eta_0, (1+\eta) C^*_{\text{est},q}(\mathcal{S}, s^*, \ell, \xi_0; \overline{\mathbf{\Sigma}})\right\} \qquad (4.81) \end{aligned}$$

*with* $\xi_0 = (\xi + \eta_0)/\{1 - \eta_0(a_q s_1^*/\ell)^{1-1/q}\}$.
*(ii) Let* $\eta = \eta_1 \vee \eta_2$, $\xi = (1+\eta_1)/(1-\eta_2)$ *and* $s_1^* = 2\|\boldsymbol{\beta}^*_{\mathcal{S}^c}\|_1/\{\lambda(1-\eta_2)\}$. *For the* $\ell_q$ *loss, the constant factor in Theorem 4.10 is bounded by (4.52)*

$$C^{(\ell_1)}(\mathscr{U}(\mathcal{S}, \mathbf{\Delta}^*); \|\cdot\|_q) \le (1 \vee |\mathcal{S}|^{1/q}) C^{(DS)}_{\text{est},q}(\mathcal{S}, s_1^*; \eta, \xi). \qquad (4.82)$$

*Moreover, the prediction loss of the Lasso is bounded by*

$$\left\{C^{(\ell_1)}(\mathscr{U}(\mathcal{S}, \mathbf{\Delta}^*); \|\cdot\|_{\text{pred}})\right\}^2 \le (1 \vee |\mathcal{S}|) C^{(DS)}_{\text{est},1}(\mathcal{S}, s_1^*; \eta, \xi) \qquad (4.83)$$

*with* $\|\mathbf{u}\|_{\text{pred}} = (\mathbf{u}^T\overline{\mathbf{\Sigma}}\mathbf{u})^{1/2}$. *Consequently, the upper bound in (4.81) is also applicable to the prediction and* $\ell_q$ *loss of the Lasso.*

The main features of Theorem 4.15 are the proper dimension normalization for all $1 \le q \le \infty$, the applicability under the capped-$\ell_1$ sparsity assumption and the possibility of taking a large $\ell$ to maintain a positive denominator in (4.79) and (4.80). It follows from Theorems 4.10, 4.14 and 4.15 that both the Lasso and Dantzig selector are rate optimal in prediction and coefficient estimation in the sense of (4.17), provided the capped-$\ell_1$ sparsity of $\boldsymbol{\beta}^*$ with $\log(p/s^*) \asymp \log p$, uniformly bounded $1/\eta_0 + C^*_{\text{est},q}(\mathcal{S}, s^*, \ell, \xi_0; \overline{\boldsymbol{\Sigma}})$ for the design, and sub-Gaussian condition (4.21) on the noise.

**Remark 4.4** Let $1 \le q' \le q \le \infty$. We will show in the proof of Theorem 4.15 that the left-hand side of (4.81) can be replaced by

$$(1 \vee |\mathcal{S}|^{1/q'}) C^{(DS)}_{\text{est},q'}(\mathcal{S}, s^*_1; \eta, \xi) / C^{1/q'-1/q}_{q,\xi,s,\ell}$$

where $C_{q,\xi,s,\ell} = (1+\xi)^{1/(1-1/q)} / \{1 + \xi(a_q s/\ell)^{1-1/q}\}^{1/(1-1/q)}$. Thus, the prediction and $\ell_q$ estimation loss with $1 \le q \le 2$ are of the optimal rate in the sense of (4.17) when Theorem 4.15 implies the rate optimality of the $\ell_2$ loss.

The proof of Theorem 4.15 is based on the following two lemmas. For vectors $\mathbf{w} \in I\!\!R^p$ and $\mathcal{S} \subset \{1, \ldots, p\}$, let $w_1^\# \ge \cdots \ge w^\#_{p-|\mathcal{S}|}$ be the ordered values of $|w_j|, j \notin \mathcal{S}$, and define

$$f_q(\mathbf{w}, t, \mathcal{S}) = \Big\{ (t - \lfloor t \rfloor)(w^\#_{\lfloor t \rfloor + 1})^q + \sum_{1 \le j \le t} (w_j^\#)^q \Big\}^{1/q}$$

for real $t \le p - |\mathcal{S}|$, and $f_q(\mathbf{w}, t, \mathcal{S}) = f_q(\mathbf{w}, p - |\mathcal{S}|, \mathcal{S})$ for $t > p - |\mathcal{S}|$. When $t$ is an integer no greater than $|\mathcal{S}^c|$, $f_q(\mathbf{w}, t, \mathcal{S}) = \max_{B \cap \mathcal{S} = \emptyset, |B| \le t} \|\mathbf{w}_B\|_q$.

**Lemma 4.6** *Let $1 \le q \le \infty$ and $a_q = (1 - 1/q)/q^{1/(q-1)}$ with $a_\infty = 1$. Let $\mathbf{u} \in \mathbb{R}^p$, $\mathcal{S} \subset \{1, \ldots, p\}$ and $A$ be the union of $\mathcal{S}$ and the indices of the $\ell$ largest $|u_j|$ with $j \notin \mathcal{S}$, $1 \le \ell \le p - |\mathcal{S}|$. Then, $\|\mathbf{u}_{A^c}\|_q \le (a_q/\ell)^{1-1/q} \|\mathbf{u}_{\mathcal{S}^c}\|_1$. Moreover, for any vector $\mathbf{w} \in \mathbb{R}^p$,*

$$\sum_{j \notin A} w_j u_j \le \|\mathbf{u}_{\mathcal{S}^c}\|_1 \Big( \frac{a}{\ell} \vee \frac{a_q}{\ell} \Big)^{1-1/q} f_{q/(q-1)}(\mathbf{w}, \ell/a, \mathcal{S}).$$

**Proof.** Let $h(t)$ be a nonnegative non-increasing function. We first prove that

$$\int_a^{1+a} h^q(x)dx \le \max\big\{1, (a_q/a)^{q-1}\big\} \Big\{ \int_0^1 h(x)dx \Big\}^q.$$

It suffices to consider $0 < a \le a_q$ and $\int_0^1 h(x)dx = 1$. As $\int_a^{1+a} h^q(x)dx$ is convex in $h$, it suffices to consider $h(x) = uI\{x \le w\} + vI\{x > w\}$ for certain $u \ge 1 \ge v$ and $a \le w \le 1$. As $\int_0^1 h(x)dx = 1$, $v = (1 - uw)/(1 - w)$. Thus, for fixed $w$, $h(x)$ is linear in $u$. As $\int_a^{1+a} h^q(x)dx$ is convex in $u$, its maximum is attained at the extreme points $u \in \{1, 1/w\}$. For $u = 1/w$, we have $v = 0$ and $\int_a^{1+a} h^q(x)dx = u^{q-1} - au^q$, so that the optimal $u$ satisfies $au = (q-1)/q$, resulting in the maximum $\{(q-1)/(qa)\}^{q-1}/q = (a_q/a)^{q-1}$. For $u = 1$, we have $v = 1$ and $\int_a^{1+a} h^q(x)dx = 1$. This yields the inequality as the right-hand side is the maximum at the two extreme points.

Let $c_0 = \ell/a$, $x_m = \ell + m\ell/a$ and $u(t) = h(t/c_0)$. We have

$$\begin{aligned}\Big(\int_\ell^\infty u^q(t)dt\Big)^{1/q} &\le \sum_{m=0}^\infty \Big(\int_{\ell+m\ell/a}^{\ell+(m+1)\ell/a} u^q(t)dt\Big)^{1/q} \\ &= \sum_{m=0}^\infty c_0^{1/q}\Big(\int_{x_m/c_0}^{x_m/c_0+1} h^q(x)dx\Big)^{1/q} \\ &\le c_0^{1/q}\max\big\{1, (a_q/a)^{1-1/q}\big\}\int_{x_0/c_0-a}^\infty h(x)dx \\ &= \max\big\{(a/\ell)^{1-1/q}, (a_q/\ell)^{1-1/q}\big\}\int_0^\infty u(t)dt.\end{aligned}$$

Assume without loss of generality that $\mathcal{S} = \emptyset$ and $u_j$ is nonnegative and decreasing in $j$. We have $A = \{1, \ldots, \ell\}$. Define $u(t)$ by $u(t) = u_j$ in $[j-1, j)$ for $1 \le j \le p$ and $u(t) = 0$ for $t > p$. It follows that with $a = a_q$,

$$\|\mathbf{u}_{A^c}\|_q = \Big(\int_\ell^\infty u^q(t)dt\Big)^{1/q} \le (a_q/\ell)^{1-1/q}\|\mathbf{u}\|_1.$$

Finally, define $w(t)$ by $w(t) = w_j^\#$ in $[j-1, j)$ for $1 \le j \le p$ and $w(t) = 0$ for $t > p$. We have $\big(\int_x^{x+c_0} |w(t)|^q dt\big)^{1/q} \le f_q(\mathbf{w}, \ell/a, \emptyset)$, so that by the Hölder inequality

$$\begin{aligned}\sum_j w_j u_j &= \int_0^\infty w(t)u(t)dt \\ &\le f_{q/(q-1)}(\mathbf{w}, \ell/a, \emptyset)\sum_{m=0}^\infty \Big(\int_{\ell+m\ell/a}^{\ell+(m+1)\ell/a} u^q(t)dt\Big)^{1/q}.\end{aligned}$$

The conclusion follows as $\|\mathbf{u}\|_1 = \int_0^\infty u(t)dt$. ∎

For $1 \le q \le \infty$, $\mathcal{S} \subset \{1, \ldots, p\}$, $s \ge |\mathcal{S}|$, integer $\ell \ge 0$ and $\xi \ge 0$, define

$$\mathscr{C}_q(\mathcal{S}, s, \ell, \xi) = \Big\{\mathbf{u} : \|\mathbf{u}_{\mathcal{S}^c}\|_1 \le \xi s^{1-1/q}\|\mathbf{u}_A\|_q,\ \exists\ |A \setminus \mathcal{S}| = \ell\Big\}, \tag{4.84}$$

**Lemma 4.7** *Let $\mathscr{C}_q(\mathcal{S}, s, \ell, \xi)$ and $C^*_{\mathrm{est},q}(\mathcal{S}, s, \ell, \xi; \mathbf{\Sigma})$ be as in (4.84) and (4.79) respectively with $s \ge |\mathcal{S}|$. For all $1 \le q \le \infty$, $\ell \ge 0$ and $\mathbf{0} \ne \mathbf{u} \in \mathscr{C}_q(\mathcal{S}, s, \ell, \xi)$,*

$$\frac{\|\mathbf{u}\|_q}{s^{1/q}\|\overline{\mathbf{\Sigma}}\mathbf{u}\|_\infty} \le \max_{|A \setminus \mathcal{S}| \le \ell}\frac{\{1 + \xi(a_q s/\ell)^{1-1/q}\}\|\mathbf{u}_A\|_q}{s^{1/q}\|\overline{\mathbf{\Sigma}}\mathbf{u}\|_\infty} \le C^*_{\mathrm{est},q}(\mathcal{S}, s, \ell, \xi; \overline{\mathbf{\Sigma}}).$$

**Proof of Lemma 4.7.** Let $A$ be the union of $\mathcal{S}$ and the index set of the $\ell$ largest $|u_j|$ with $j \notin \mathcal{S}$, $A = \text{argmax}_{\mathcal{S}}\{\|\mathbf{u}_J\|_1 : |J \setminus \mathcal{S}| = \ell\}$. It follows from Lemma 4.6 and (4.84) that for all $\mathbf{u} \in \mathscr{C}_q(\mathcal{S}, s, \ell, \xi)$

$$\|\mathbf{u}\|_q \le \|\mathbf{u}_A\|_q + (a_q/\ell)^{1-1/q}\|\mathbf{u}_{\mathcal{S}^c}\|_1 \le \|\mathbf{u}_A\|_q\{1 + \xi(a_q s/\ell)^{1-1/q}\}.$$

Let $\mathbf{u}^*$ be the $\ell_q$ dual of $\mathbf{u}$ on $A$ satisfying $\|\mathbf{u}^*\|_{q/(q-1)} = 1$, $(\mathbf{u}_A^*)^T\mathbf{u}_A = 1$ and $\mathbf{u}_{A^c}^* = \mathbf{0}$. Define $\mathbf{v}$ by $\mathbf{v}_A = \overline{\boldsymbol{\Sigma}}_{A,A}^{-1}\mathbf{u}_A^*$ and $\mathbf{v}_{A^c} = \mathbf{0}$, and $\mathbf{w}$ by $\mathbf{w}_A = \mathbf{0}$ and $\mathbf{w}_{A^c} = \overline{\boldsymbol{\Sigma}}_{A^c,A}\mathbf{v}_A$. Again, it follows from Lemma 4.6 and (4.84) that

$$\begin{aligned}
\mathbf{v}^T\overline{\boldsymbol{\Sigma}}\mathbf{u} &= \mathbf{v}_A^T\overline{\boldsymbol{\Sigma}}_{A,A}\mathbf{u}_A + \mathbf{v}_A^T\overline{\boldsymbol{\Sigma}}_{A,A^c}\mathbf{u}_{A^c} \\
&= (\mathbf{u}_A^*)^T\mathbf{u}_A + \mathbf{w}_{A^c}^T\mathbf{u}_{A^c} \\
&\ge \|\mathbf{u}_A\|_q - \|\mathbf{u}_{\mathcal{S}^c}\|_1(a_r/\ell)^{1-1/r} f_{r/(r-1)}(\mathbf{w}, \ell/a_r, \mathcal{S}) \\
&\ge \|\mathbf{u}_A\|_q\Big\{1 - \xi s^{1-1/q}(a_r/\ell)^{1-1/r} f_{r/(r-1)}(\mathbf{w}, \ell/a_r, \mathcal{S})\Big\}
\end{aligned}$$

for all $\mathbf{u} \in \mathscr{C}_q(\mathcal{S}, s, \ell, \xi)$. It follows that

$$\begin{aligned}
& \frac{\|\mathbf{u}_A\|_q\{1 + \xi(a_q s/\ell)^{1-1/q}\}}{s^{1/q}\|\overline{\boldsymbol{\Sigma}}\mathbf{u}\|_\infty} \\
\le\ & \frac{\{1 + \xi(a_q s/\ell)^{1-1/q}\}\|\mathbf{u}_A\|_q\|\mathbf{v}\|_1}{s^{1/q}\mathbf{v}^T\overline{\boldsymbol{\Sigma}}\mathbf{u}} \qquad (4.85)\\
\le\ & \frac{\{1 + \xi(a_q s/\ell)^{1-1/q}\}\|\mathbf{v}\|_1 s^{-1/q}}{\{1 - \xi s^{1-1/q}(a_r/\ell)^{1-1/r} f_{r/(r-1)}(\mathbf{w}, \ell/a_r, \mathcal{S})\}_+}.
\end{aligned}$$

Since $\mathbf{v}_A = \overline{\boldsymbol{\Sigma}}_{A,A}^{-1}\mathbf{u}_A^*$ with $\|\mathbf{u}_A^*\|_{q/(q-1)} = 1$, we have

$$\|\mathbf{v}\|_1 = \|\overline{\boldsymbol{\Sigma}}_{A,A}^{-1}\mathbf{u}_A^*\|_1 \le \|\overline{\boldsymbol{\Sigma}}_{A,A}^{-1}\|_{q/(1-q),1} = \|\overline{\boldsymbol{\Sigma}}_{A,A}^{-1}\|_{\infty,q}.$$

Similarly, as $\mathbf{w}_{A^c} = \overline{\boldsymbol{\Sigma}}_{A^c,A}\mathbf{v}_A = \overline{\boldsymbol{\Sigma}}_{A^c,A}\overline{\boldsymbol{\Sigma}}_{A,A}^{-1}\mathbf{u}_A^*$, we have

$$\begin{aligned}
f_{r/(r-1)}(\mathbf{w}, \ell/a_r, A) &\le \max_{|B| \le \lceil \ell/a_r \rceil} \|\overline{\boldsymbol{\Sigma}}_{B,A}\overline{\boldsymbol{\Sigma}}_{A,A}^{-1}\mathbf{u}_A^*\|_{r/(r-1)} \\
&\le \max_{|B| \le \lceil \ell/a_r \rceil} \Big\|\overline{\boldsymbol{\Sigma}}_{B,A}\overline{\boldsymbol{\Sigma}}_{A,A}^{-1}\Big\|_{q/(q-1),r/(r-1)} \\
&= \max_{|B| \le \lceil \ell/a_r \rceil} \Big\|\overline{\boldsymbol{\Sigma}}_{A,A}^{-1}\overline{\boldsymbol{\Sigma}}_{A,B}\Big\|_{r,q}.
\end{aligned}$$

Inserting the above bounds for $\|\mathbf{v}\|_1$ and $f_{r/(r-1)}(\mathbf{w}, \ell/a_r, A)$ to (4.85) yields the conclusion. ■

**Proof of Theorem 4.15.** (i) Consider the maximization of $T^{(DS)}(\mathbf{u})$ in the definition of $C_{\text{est},q}^{(DS)}(\mathcal{S}, s_1^*; \eta, \xi)$ in (4.73). As there is nothing to prove when $T^{(DS)}(\mathbf{u}) \le (s^*)^{1/q}/\eta_0$, it suffices to consider $T^{(DS)}(\mathbf{u}) > (s^*)^{1/q}/\eta_0$. As the scale is set with $\|\mathbf{u}\|_q = 1$ in (4.73), we have $(s^*)^{1/q} < \eta_0\|\mathbf{u}\|_q T^{(DS)}(\mathbf{u}) \le \eta_0\|\mathbf{u}\|_q s_1^*/(\|\mathbf{u}_{\mathcal{S}^c}\|_1 - \xi\|\mathbf{u}_{\mathcal{S}}\|_1)$, so that

$$\begin{aligned}\|\mathbf{u}_{\mathcal{S}^c}\|_1 - \xi\|\mathbf{u}_{\mathcal{S}}\|_1 &\le \eta_0(s_1^*)^{1-1/q}\|\mathbf{u}\|_q \\ &\le \eta_0(s_1^*)^{1-1/q}\{\|\mathbf{u}_A\|_q + (a_q/\ell)^{1-1/q}\|\mathbf{u}_{\mathcal{S}^c}\|_1\}\end{aligned}$$

by Lemma 4.6. It follows that

$$\begin{aligned}&\{1-\eta_0(a_q s_1^*/\ell)^{1-1/q}\}\|\mathbf{u}_{\mathcal{S}^c}\|_1 \\ \le\ & \xi\|\mathbf{u}_{\mathcal{S}}\|_1 + \eta_0(s_1^*)^{1-1/q}\|\mathbf{u}_A\|_q \\ \le\ & \{\xi|\mathcal{S}|^{1-1/q} + \eta_0(s_1^*)^{1-1/q}\}\|\mathbf{u}_A\|_q,\end{aligned}$$

which gives $\|\mathbf{u}_{\mathcal{S}^c}\|_1 \le \xi_0(s^*)^{1-1/q}\|\mathbf{u}_A\|_q$. Thus, $\mathbf{u} \in \mathscr{C}_q(\mathcal{S}, s^*, \ell, \xi_0)$ and

$$T^{(DS)}(\mathbf{u})\|\mathbf{u}\|_q \le \frac{(1+\eta)\|\mathbf{u}\|_q}{\|\overline{\mathbf{\Sigma}}\mathbf{u}\|_\infty} \le (1+\eta)(s^*)^{1/q}C^*_{\text{est},q}(\mathcal{S}, s^*, \ell, \xi_0; \overline{\mathbf{\Sigma}})$$

by Lemma 4.7.

(ii) Let $\mathbf{u} \in \mathscr{U}(\mathcal{S}, \mathbf{\Delta}^*)$ as in (4.53), so that (4.51) and (4.52) hold. By taking the sum over all $j$ in (4.51), we have

$$\mathbf{u}^T\overline{\mathbf{\Sigma}}\mathbf{u} \le \psi(\mathbf{u}) + 2\|\boldsymbol{\beta}^*_{\mathcal{S}^c}\|_1/\lambda,$$

where $\psi(\mathbf{u}) = \eta_1\|\mathbf{u}_{\mathcal{S}}\|_1 - \text{sgn}(\boldsymbol{\beta}^*_{\mathcal{S}})^T\mathbf{u}_{\mathcal{S}} - (1-\eta_2)\|\mathbf{u}_{\mathcal{S}^c}\|_1$ with the convention $\text{sgn}(0) = 0$ as in (4.39). This and (4.52) imply (4.70) with the specified $\{s^*, \eta, \xi\}$. Because $\mathbf{u}$ is an arbitrary member of $\mathscr{U}(\mathcal{S}, \mathbf{\Delta}^*)$,

$$\mathscr{U}(\mathcal{S}, \mathbf{\Delta}^*) \subseteq \mathscr{U}^{(DS)}(\mathcal{S}, s_1^*; \eta, \xi).$$

The conclusion follows from (4.54), (4.72) and (4.73).

Finally, we prove the assertion in Remark 4.4. By the Hölder inequality,

$$\|\mathbf{u}\|_{q'} \le \|\mathbf{u}\|_1^{(1/q'-1/q)/(1-1/q)}\|\mathbf{u}\|_q^{(1-1/q')/(1-1/q)}.$$

Because $\|\mathbf{u}\|_1 \le \|\mathbf{u}_{\mathcal{S}}\|_1 + \xi s^{1-1/q}\|\mathbf{u}_A\|_q \le (1+\xi)s^{1-1/q}\|\mathbf{u}_A\|_q$ and $\|\mathbf{u}\|_q \le \{1+\xi(a_q s/\ell)^{1-1/q}\}\|\mathbf{u}_A\|_q$ for $\mathbf{u} \in \mathscr{C}_q(\mathcal{S}, s, \ell, \xi)$,

$$\frac{\|\mathbf{u}\|_{q'}s^{1/q-1/q'}}{\{1+\xi(a_q s/\ell)^{1-1/q}\}\|\mathbf{u}_A\|_q} \le C_{q,\xi,s,\ell}^{1/q'-1/q}\{1+\xi(a_q s/\ell)^{1-1/q}\}.$$

This completes the proof. ■

### *4.3.5 Regularity conditions on the design matrix*

In this subsection, we describe the relationship between different forms of regularity conditions on the design matrix for controlling the prediction and estimation errors of the Lasso and Dantzig selector, as we have already discussed the conditions for selection consistency in Subsection 4.3.1. We will focus on the relationship of conditions in the previous subsections and the literature. At the end of this subsection, we will prove that for sub-Gaussian designs, the regularity conditions for the prediction and $\ell_q$ estimation, $1 \leq q \leq 2$, largely hold under matching conditions on the population covariance matrix.

Regularity conditions on the design matrix $\mathbf{X}$, or equivalently on the Gram matrix $\overline{\mathbf{\Sigma}} = \mathbf{X}^T\mathbf{X}/n$, are required for the estimation of $\boldsymbol{\beta}$ and variable selection since $\boldsymbol{\beta}$ is related to the response vector $\mathbf{Y}$ through $\mathbf{X}\boldsymbol{\beta}$ in (4.1). For prediction or equivalently the estimation of $\mathbf{X}\boldsymbol{\beta}$, the $\ell_0$ PLS achieves rate optimality in Theorem 4.6 and $\ell_1$ regularized methods achieve the slow rate in (4.48) and (4.77) without condition on the design. However, to achieve rate optimality in prediction in the form (4.17), regularity conditions on $\mathbf{X}$ are still required in the analyses here and in the literature as the $\ell_1$ penalty is related to $\mathbf{Y}$ through $\mathbf{X}$. It is worthwhile to mention here that in addition to the $\ell_0$ penalty, *folded concave PLS*, discussed in the next section, may also require weaker conditions on $\mathbf{X}$ than $\ell_1$ regularized methods.

The prediction and estimation performance of the Lasso and Dantzig selector has been studied in Knight and Fu (2000), Greenshtein and Ritov (2004), Candés and Tao (2007), Bunea, Tsybakov and Wegkamp (2007), Zhang and Huang (2008), Meinshausen and Yu (2009), Bickel, Ritov and Tsybakov (2009), Koltchinskii (2009), Zhang (2009), van de Geer and Bühlmann (2009), Ye and Zhang (2010), Zhang and Zhang (2012) among many other papers in the literature. As Theorems 4.9, 4.10, 4.11 and 4.14 give the sharpest oracle inequalities based on basic inequalities (4.49), (4.51), (4.52) and (4.70), results based on the same basic inequalities, including a majority of these cited above, can be viewed as weaker versions of these theorems with larger and hopefully more explicit constant factors. In this sense, Theorems 4.9, 4.10, 4.11 and 4.14 provide a baseline from which the merits of different rate-optimal results can be compared by inspecting the corresponding constant factors.

While Theorems 4.9, 4.10, 4.11 and 4.14 provide the sharpest error bounds based on the basic inequalities for the Lasso and Dantzig selector, the constant factors in these theorems are also the most abstract. Among regularity conditions on the design, the most intuitive ones involve sparse eigenvalues and sparse operator norms. Analysis based on such quantities, which provides rate optimality in prediction and $\ell_q$ estimation for both the Lasso and Dantzig selector, and the false positive for the Lasso, are carried out in the previous subsections as stated in Theorem 4.12, Corollary 4.7 and Theorem 4.15. Here we provide a brief review of the literature.

Sparse eigenvalue-based regularity conditions were first introduced in the compressed sensing literature as the *RIP* in Candés and Tao (2005) for $\ell_1$ regularized sparse recovery under random projection. Suppose the design vectors are normalized to $\|\mathbf{X}_j\|_2^2 = n$. Define the restricted isometry constant as

$$\delta_t = \max\left\{\phi_+(t;\overline{\mathbf{\Sigma}}) - 1, 1 - \phi_-(t;\overline{\mathbf{\Sigma}})\right\}$$

with the Gram matrix $\overline{\mathbf{\Sigma}} = \mathbf{X}^T\mathbf{X}/n$, the lower sparse eigenvalue $\phi_-(k;\mathbf{\Sigma})$ in (4.14) and the corresponding upper sparse eigenvalue

$$\phi_+(t;\mathbf{\Sigma}) = \max_{|B|<t+1} \phi_{\max}(\mathbf{\Sigma}_{B,B}). \tag{4.86}$$

Define the off-diagonal version of the restricted isometry constant as

$$\theta_{s,t} \quad = \quad \max\left\{\|\overline{\mathbf{\Sigma}}_{A,B}\|_S : |A| < s+1, |B| < t+1, A\cap B = \emptyset\right\}.$$

Let $s = |\text{supp}(\boldsymbol{\beta})|$. The original RIP, which can be stated as the sufficiency of $\delta_s + \delta_{2s} + \delta_{3s} < 1$ for the recovery of $\boldsymbol{\beta}$ by minimizing $\|\boldsymbol{\beta}\|_1$ subject to $\mathbf{Y} = \mathbf{X}\boldsymbol{\beta}$, which is equivalent to $\widehat{\boldsymbol{\beta}}^{(DS)}(0) = \boldsymbol{\beta}$ with noiseless data. Candés and Tao (2005) actually proved that $\delta_s + \theta_{s,s} + \theta_{s,2s} < 1$ is sufficient. Candés and Tao (2007) sharpened the RIP to $\delta_{2s} + \theta_{s,2s} < 1$ and called the weaker condition the *uniform uncertainty principle* (UUP). Moreover, they proved under the UUP an $\ell_2$ error bound of the form (4.17) for the Dantzig selector which can be also stated as

$$C_{\text{est},2}^{(DS)}(\mathcal{S}, 0) = O(1)/(1 - \delta_{2s} - \theta_{s,2s})_+^2$$

in Theorem 4.14. The RIP and UUP were further sharpened and generalized to a condition of the form (4.79) with $r = \infty$ in Zhang (2009), to $\delta_{1.25s}+\theta_{1.25s,s} < 1$ in Cai, Wang and Xu (2010), to Theorem 4.15 (i) in Ye and Zhang (2010) and to $\min(3\delta_s, 2\delta_{2s}) < 1$ in Cai and Zhang (2013). For example, by Theorem 4.15 (i), the rate optimality of the $\ell_2$ loss of both the Lasso and Dantzig selector is guaranteed by $1/\phi_-(1.25s) = O(1)$ and

$$\max\left\{\|\overline{\mathbf{\Sigma}}_{A,A}^{-1}\overline{\mathbf{\Sigma}}_{A,B}\|_S : |A| < 1.25s+1, |B| \le s, A\cap B = \emptyset\right\} < 1,$$

with $\|\overline{\mathbf{\Sigma}}_{A,A}^{-1}\overline{\mathbf{\Sigma}}_{A,B}\|_S \le \theta_{1.25s,s}/\delta_{1.25s}$.

For the Lasso (4.29), prediction and estimation error bounds of the form (4.17) were established in Zhang and Huang (2008) under an SRC. Their SRC was sharpened in Zhang (2010a) to

$$\frac{s}{m^*} < \left(\frac{\phi_+(m^*;\overline{\mathbf{\Sigma}})}{\phi_-(m^*;\overline{\mathbf{\Sigma}})} + \widetilde{\kappa}^2 - 1\right)^{-1}\left(2 - \widetilde{\kappa}^2\right)_+ \tag{4.87}$$

for possibly non-convex penalties with a certain small $\widetilde{\kappa} > 0$. For the Lasso, this can be more precisely stated as (4.59), leading to error bounds in (4.60),

(4.62) and (4.63). Compared with the RIP, the SRC is more general. For example, the SRC allows the upper sparse eigenvalue to be greater than 2. The RIP condition implies the SRC in the sense of

$$\delta_{ks} < (k-1)/k \ \Rightarrow \ (4.87) \text{ with } m^* = (k-1)s.$$

For example, $\delta_{2s} < 1/2$ implies the SRC with $m^* = s$ and $\delta_{3s} < 2/3$ implies the SRC with $m^* = 2s$. Thus, such RIP conditions are also sufficient for the rate optimality of the Lasso in (4.62) and (4.63). However, the sharpest UUP and SRC seem not to imply each other.

While the regularity conditions discussed above provide rate optimality for the prediction and $\ell_q$ loss for all $1 \le q \le 2$ for both the Lasso and Dantzig selector, they require an upper bound for the Gram matrix, at least for the off-diagonal block $\overline{\mathbf{\Sigma}}_{A,B}$ in (4.79) and (4.80). As the invertibility of $\overline{\mathbf{\Sigma}}$ suffices for the estimation of $\boldsymbol{\beta}$ in the low-dimensional case, it is natural to ask whether the upper bound condition on $\overline{\mathbf{\Sigma}}_{A,B}$ is essential. We answer this question by studying the restricted eigenvalue and the constant factors in Theorem 4.9.

For simplicity, we confine our discussion in the rest of this subsection to the hard sparsity scenario: $\text{supp}(\boldsymbol{\beta}) = \mathcal{S}$, $\boldsymbol{\beta}^* = \overline{\boldsymbol{\beta}} = \boldsymbol{\beta}$ and $\eta_1 = \eta_2 = \eta$, where $\overline{\Delta} = 0$. We write the constant factors (4.39) and (4.41) in Theorem 4.9 as

$$\overline{C}^{(\ell_1)}_{\text{pred}}(\mathcal{S};\eta) = \sup_{\mathbf{u}\neq\mathbf{0}} \frac{\{[\max\{\psi(\mathbf{u}),0\}]^2}{(1\vee|\mathcal{S}|)\mathbf{u}^T\overline{\mathbf{\Sigma}}\mathbf{u}}, \tag{4.88}$$

with $\psi(\mathbf{u}) = \eta\|\mathbf{u}_\mathcal{S}\|_1 - \text{sgn}(\boldsymbol{\beta}_\mathcal{S})^T\mathbf{u}_\mathcal{S} - (1-\eta)\|\mathbf{u}_{\mathcal{S}^c}\|_1$, and

$$\overline{C}^{(\ell_1)}_{\text{est},q}(\mathcal{S};\eta) = \sup_{\mathbf{u}\neq\mathbf{0}} \frac{\|\mathbf{u}\|_q\max\{\psi(\mathbf{u}),0\}}{(1\vee|\mathcal{S}|^{1/q})\mathbf{u}^T\overline{\mathbf{\Sigma}}\mathbf{u}}. \tag{4.89}$$

These quantities are closely related to the *restricted eigenvalue* (*RE*). For possibly asymmetric $\mathbf{M} \in \mathbb{R}^{p\times p}$, an RE of general form, with options explicitly matching the original definition of the RE, can be defined as

$$\text{RE}_k(\mathcal{S};\eta,\mathbf{M},\psi_1,\psi_2) = \inf_{\mathbf{u}\in\mathscr{C}(\mathcal{S};\eta,\psi_1)} \left\{\frac{(\mathbf{u}^T\mathbf{M}\mathbf{u})_+|\mathcal{S}|^{1-k/2}}{\|\mathbf{u}\|_2^k\psi_2^{2-k}(\mathbf{u}_\mathcal{S})}\right\}^{1/2}, \tag{4.90}$$

where $k \in \{0,1,2\}$, $\psi_j(\cdot)$ satisfies $\psi_j(t\mathbf{u}_\mathcal{S}) = t\psi_j(\mathbf{u}_\mathcal{S})$ for $t > 0$ and $\psi_j(\mathbf{u}_\mathcal{S}) \le |\mathcal{S}|^{1/2}\|\mathbf{u}_\mathcal{S}\|_2$, and $\mathscr{C}(\mathcal{S};\eta,\psi_1) = \{\mathbf{u} : (1-\eta)\|\mathbf{u}_{\mathcal{S}^c}\|_1 < (1+\eta)\psi_1(\mathbf{u}_\mathcal{S})\}$. We omit the reference to $\mathbf{M}$ and $\psi_j$ in (4.90) and $\mathscr{C}(\mathcal{S};\eta,\psi_1)$ when $\mathbf{M} = \overline{\mathbf{\Sigma}}$ and $\psi_j(\mathbf{u}_\mathcal{S}) = \|\mathbf{u}_\mathcal{S}\|_1$, and we denote $\text{RE}_0(\mathcal{S};\eta,\overline{\mathbf{\Sigma}},\|\cdot\|_1,|\mathcal{S}|^{1/2}\|\cdot\|_2)$ by $\text{RE}_{2,0}(\mathcal{S};\eta)$. We note that $\text{RE}_{2,0}(\mathcal{S};\eta)$ and $\text{RE}_2(\mathcal{S};\eta)$ are the RE introduced in Bickel, Ritov and Tsybakov (2009) respectively for prediction and $\ell_2$ estimation, and $\text{RE}_2(\mathcal{S};\eta)$ in Koltchinskii (2009). Moreover, $\text{RE}_0(\mathcal{S};\eta)$ is the *compatibility factor* in van de Geer (2007) and van de Geer and Bühlmann (2009), and $\max_{|A|=s}\text{RE}_2(A;\eta,\overline{\mathbf{\Sigma}},\|\cdot\|_1,s^{1/2}\|\cdot\|_2)$ is equivalent to the *restricted strong convexity* (*RSC*) coefficient in Negahban *et al.*(2012). The quantity $\text{RE}_0(\mathcal{S};\eta)$

is also called the $\ell_1$ restricted eigenvalue (van de Geer and Bühlmann, 2009), and sometimes denoted as $\mathrm{RE}_1$ in the literature. Quantities of the form (4.90) will play an important role in our analysis of concave PLS in Section 4.4.

The relationship of some existing RE error bounds to Theorem 4.11, Theorem 4.9 and Corollary 4.5 can be written as

$$C^{(\ell_1)}_{\mathrm{pred}} \le \overline{C}^{(\ell_1)}_{\mathrm{pred}} \le \frac{(1+\eta)^2}{\mathrm{RE}_0^2} \le \frac{(1+\eta)^2}{\mathrm{RE}_{2,0}^2} \tag{4.91}$$

with the dependence on $\mathcal{S}$ and $\eta$ omitted. For the $\ell_2$ loss, we have

$$C^{(\ell_1)}_{\mathrm{est},2} \le \overline{C}^{(\ell_1)}_{\mathrm{est},2} \le \frac{1+\eta}{\mathrm{RE}_1^2} \le \frac{1+\eta}{\mathrm{RE}_0\mathrm{RE}_2} \le \frac{1+\eta}{\mathrm{RE}_2^2}. \tag{4.92}$$

The inequalities in (4.91) and (4.92) can be easily seen from the definition of the quantities, especially that of $\psi(\mathbf{u})$ in (4.88), as Theorem 4.11 is always sharper than Theorem 4.9. Similar inequalities hold for the Dantzig selector. The inequalities in (4.91) and (4.92) are nontrivial as the ratio of the two-sides of each inequality is unbounded in the worst configuration of $\{\mathcal{S}, \eta, \overline{\mathbf{\Sigma}}\}$ (van de Geer and Bühlmann, 2009).

The sparse and restricted eigenvalues and the quantities $\overline{C}^{(\ell_1)}_{\mathrm{pred}}$ and $\overline{C}^{(\ell_1)}_{\mathrm{est},q}$ in Theorem 4.9 and Corollary 4.5 can all be viewed as of the $\ell_2$ type (Zhang and Zhang, 2012) due to their association with $\mathbf{u}^T\overline{\mathbf{\Sigma}}\mathbf{u}$, which is of the same order as $\|\mathbf{u}\|_2^2$ when $\mathrm{RE}_2$ can be treated as constant.

A problem with $\ell_2$ type quantities is that for $2 < q \le \infty$, they do not control the $\ell_q$ estimation loss at the optimal rate, as the quantity $\max_{\|\mathbf{u}\|_2=1} \|\mathbf{u}_{\mathcal{S}}\|_q \|\mathbf{u}_{\mathcal{S}}\|_1/|\mathcal{S}|^{1/q}$ is unbounded for large $|\mathcal{S}|$. Zhang (2009) and Ye and Zhang (2010) uses the sparse operator norm and *cone invertibility factor* (CIF) to address this issue. We have already studied in Theorem 4.15 error bounds based on the sparse operator norm. With the same cone $\mathscr{C}(\mathcal{S};\eta) = \mathscr{C}(\mathcal{S};\eta, \|\cdot\|_1)$ as in (4.90), the CIF is defined as

$$\mathrm{CIF}_{q,\ell}(\mathcal{S};\eta) = \inf_{\mathbf{u}\in\mathscr{C}(\mathcal{S};\eta)} \min_{|A\setminus S|\le\ell} \frac{|\mathcal{S}|^{1/q}\|\overline{\mathbf{\Sigma}}\mathbf{u}\|_\infty}{\|\mathbf{u}_A\|_q}. \tag{4.93}$$

For studying the Lasso, a sign-restricted CIF is defined as

$$\mathrm{SCIF}_{q,\ell}(\mathcal{S};\eta) = \inf_{\mathbf{u}\in\mathscr{C}_-(\mathcal{S};\eta)} \min_{|A\setminus S|\le\ell} \frac{|\mathcal{S}|^{1/q}\|\overline{\mathbf{\Sigma}}\mathbf{u}\|_\infty}{\|\mathbf{u}_A\|_q}, \tag{4.94}$$

where $\mathscr{C}_-(\mathcal{S};\eta) = \{\mathbf{u} : (1-\eta)\|\mathbf{u}_{\mathcal{S}^c}\|_1 \le (1+\eta)\|\mathbf{u}_{\mathcal{S}}\|_1, u_j(\overline{\mathbf{\Sigma}}\mathbf{u})_j \le 0\}$. The CIF and SCIF can be viewed as of the $\ell_q$ type as they can both be bounded from below by functions of the sparse $\ell_q$ operator norms of the Gram matrix with an application of Theorem 4.15.

The CIF and SCIF provide rate optimal $\ell_q$ bounds for both the Lasso and

Dantzig selector for all $1 \le q \le \infty$. In fact, when $\boldsymbol{\beta}^*$ is hard sparse with $s_1^* = 0$, the CIF is the sharpest possible based on (4.70) as

$$C_{\mathrm{est},q}^{(DS)}(\mathcal{S}, 0; \eta, \xi) = \frac{1+\eta}{\mathrm{CIF}_{q,p}(\mathcal{S};\eta)} \le \sup_{\mathbf{u}\in\mathscr{C}(\mathcal{S};\eta)} \frac{(1+\eta)\|\mathbf{u}\|_1\|\mathbf{u}\|_q}{|\mathcal{S}|^{1/q}\mathbf{u}^T\overline{\boldsymbol{\Sigma}}\mathbf{u}} \tag{4.95}$$

with $\xi = (1+\eta)/(1-\eta)$, so that the both the CIF and and SCIF are bounded from below via Theorem 4.15. For the Lasso, the SCIF-based $\ell_q$ error bond can be written as

$$C_{\mathrm{est},q}^{(\ell_1)}(\mathcal{S};\eta,\eta) \le \frac{1+\eta}{\mathrm{SCIF}_{q,p}(\mathcal{S};\eta)} \le \sup_{\mathbf{u}\in\mathscr{C}_-(\mathcal{S};\eta)} \frac{(1+\eta)\|\mathbf{u}_{\mathcal{S}}\|_1\|\mathbf{u}\|_q}{|\mathcal{S}|^{1/q}\mathbf{u}^T\overline{\boldsymbol{\Sigma}}\mathbf{u}}. \tag{4.96}$$

In addition to the above, the CIF are related to the RE through the following statements. As the cone $\mathscr{C}_-(\mathcal{S};\eta)$ in (4.94) is contained in the set $\mathscr{U}(\mathcal{S},0;\eta,\eta)$ in (4.53), $\overline{C}_{\mathrm{pred}}^{(\ell_1)}$ can be replaced by $(1+\eta)^2/\mathrm{SCIF}_{1,0}$ in (4.91), and $\overline{C}_{\mathrm{est},2}^{(\ell_1)}$ by $(1+\eta)/\mathrm{SCIF}_{2,p}$ in (4.92).

Although $\ell_2$-type regularity conditions do not provide rate optimal error bounds for the $\ell_q$ loss for $q > 2$, they are simpler to study. In fact, a lower sparse eigenvalue condition guarantees all the $\ell_2$-type regularity conditions discussed in this section, and for sub-Gaussian designs the sample version of the RE in (4.90) is within a small fraction of its population version when $(|S|/n)\log(p/|S|)$ is small. We prove these statements for the general RE defined in (4.90) in the following two theorems, for deterministic and random designs, respectively. These two theorems slightly extend the results of Rudelson and Zhou (2013). Moreover, we prove in Theorem 4.17 that a population version of the SRC implies its sample version (4.59) with high probability for Gaussian designs.

Define a sparse version of (4.90) by

$$\begin{aligned} & R_{k,\ell,m}^2(\mathcal{S};\eta,\mathbf{M},\psi_1,\psi_2) \qquad\qquad (4.97)\\ = & \sup_{\mathbf{u}\in\mathscr{C}(\mathcal{S};\eta,\psi_1)} \left\{ \frac{(\mathbf{u}^T\mathbf{M}\mathbf{u})_+|\mathcal{S}|^{1-k/2}}{\|\mathbf{u}_A\|_2^k\psi_1^{2-k}(\mathbf{u}_{\mathcal{S}})} : |A\setminus\mathcal{S}| \le \ell, |\mathrm{supp}(\mathbf{u})\setminus A| \le m \right\}. \end{aligned}$$

**Theorem 4.16** *Let $\mathcal{S} \subset \{1,\ldots,p\}$ and $\mathbf{M}$ be a $p\times p$ matrix and $\delta_0 \in \{0,1\}$ be the indicator for the nonnegative-definiteness of $\mathbf{M}$. Let $k \in \{0,1,2\}$, $\ell = 0$ for $k = 0$ and $\ell = \lceil|\mathcal{S}|/4\rceil$ for $k > 0$. Let $m$ be a positive integer satisfying*

$$\frac{4^{1-\delta_0}\|\mathbf{M}_{\mathcal{S},\mathcal{S}}\|_{\max}\xi^2|\mathcal{S}|}{mR_{k,\ell,m}^2(\mathcal{S};\eta,\mathbf{M},\psi_1,\psi_2)} \le \eta_0 < 1.$$

*with $\psi_1(\mathbf{u}_{\mathcal{S}}) \le \psi_2(\mathbf{u}_{\mathcal{S}})$ $\forall\mathbf{u}$. Let $R_k(\mathcal{S};\eta,\mathbf{M},\psi_1,\psi_2)$ be as in (4.90). Then,*

$$\frac{R_{k,\ell,m}^2(\mathcal{S};\eta,\mathbf{M},\psi_1,\psi_2)}{R_k^2(\mathcal{S};\eta,\mathbf{M},\psi_1,\psi_2)} \le \frac{(1+\xi)^k}{1-\eta_0}. \tag{4.98}$$

*In particular, if* $\|\mathbf{X}_j\|_2^2/n \le 1$ *and* $\xi^2|\mathcal{S}| \le \eta_0 m\phi_*(\ell+m;\mathcal{S},\overline{\mathbf{\Sigma}})$ *for the Gram matrix* $\overline{\mathbf{\Sigma}} = \mathbf{X}^T\mathbf{X}/n$ *and the lower sparse eigenvalue* $\phi_*$ *in (4.23), then*

$$\frac{\phi_*(\ell+m;\mathcal{S},\overline{\mathbf{\Sigma}})}{R_k^2(\mathcal{S};\eta,\overline{\mathbf{\Sigma}},\psi_1,\psi_2)} \le \frac{(1+\xi)^k}{(1-\eta_0)}. \tag{4.99}$$

For $k=2$, Theorem 4.16 and the comparisons in (4.91), (4.92), (4.95) and (4.96) yield

$$\begin{aligned}&\min\left\{\frac{(1+\eta)^2}{C_{\rm pred}^{(\ell_1)}}, \frac{(1+\eta)^2}{(C_{{\rm est},2}^{(\ell_1)})^2}, {\rm SCIF}_{2,p}^2, {\rm CIF}_{2,p}^2, {\rm RE}_0^2, {\rm RE}_2^2\right\}\\ \ge\ & \frac{(1-\eta_0)\phi_*(\ell+m;\mathcal{S},\overline{\mathbf{\Sigma}})}{(1+\xi)^2}.\end{aligned}$$

under the sparsity condition $\xi^2|\mathcal{S}| \le \eta_0 m\phi_*(\ell+m;\mathcal{S},\overline{\mathbf{\Sigma}})$, as the squared restricted eigenvalue ${\rm RE}_2^2$ is the smallest on the left.

The proof of Theorem 4.16 uses Maurey's empirical method as in the following lemma.

**Lemma 4.8** *Let* $A \subset \{1,\ldots,p\}$, $\mathbf{M} \in I\!R^{p\times p}$, *and* $m$ *be a positive integer. For any vector* $\mathbf{u} \in I\!R^p$ *there exists a vector* $\mathbf{v} \in I\!R^p$ *such that*

$$\mathbf{v}_A = \mathbf{u}_A,\ {\rm supp}(\mathbf{v}) \subseteq {\rm supp}(\mathbf{u}),\ \|\mathbf{v}_{A^c}\|_0 \le m,\ \|\mathbf{v}_{A^c}\|_1 = \|\mathbf{u}_{A^c}\|_1, \tag{4.100}$$

*and with* $\delta_0 = I\{\mathbf{u}_{A^c}^T\mathbf{M}_{A^c,A^c}\mathbf{u}_{A^c} \ge 0\}$,

$$\mathbf{v}^T\mathbf{M}\mathbf{v} - \mathbf{u}^T\mathbf{M}\mathbf{u} \le 4^{1-\delta_0}\|\mathbf{M}_{A^c,A^c}\|_{\max}\|\mathbf{u}_{A^c}\|_1^2/m. \tag{4.101}$$

**Proof.** Assume without loss of generality that $\|\mathbf{u}_{A^c}\|_1 > 0$. Let $w_j = |u_j|/\|\mathbf{u}_{A^c}\|_1$, $\mathbf{e}_j$ be the canonical basis vectors in $I\!R^p$, and $\mathbf{Z} \in I\!R^p$ and $\mathbf{Z}^i, i=1,\ldots,m$, be a iid discrete random vectors with distribution

$$P\{\mathbf{Z} = \mathbf{u}_A + \|\mathbf{u}_{A^c}\|_1{\rm sgn}(u_j)\mathbf{e}_j\} = w_j,\ j \notin A.$$

Let $m_j = \#\{i : \mathbf{Z}^i = \mathbf{u}_A + {\rm sgn}(u_j)\mathbf{e}_j\|\mathbf{u}_{A^c}\|_1\}$. We have

$$\frac{1}{m}\sum_{i=1}^m \mathbf{Z}^i = \mathbf{u}_A + \|\mathbf{u}_{A^c}\|_1\sum_{j\notin A}(m_j/m){\rm sgn}(u_j)\mathbf{e}_j.$$

As $m_j = 0$ for $u_j = 0$ and $\sum_{j\notin A} m_j/m = 1$, (4.100) holds for all realizations $\mathbf{v}$ of $\sum_{i=1}^m \mathbf{Z}^i/m$. Moreover, as ${\rm E}\,\mathbf{Z} = \mathbf{u}$,

$${\rm E}\Big(\frac{1}{m}\sum_{i=1}^m \mathbf{Z}^i\Big)^T\mathbf{M}\Big(\frac{1}{m}\sum_{i=1}^m \mathbf{Z}^i\Big) - \mathbf{u}^T\mathbf{M}\mathbf{u}$$

$$
\begin{aligned}
&= \frac{1}{m}\mathrm{E}(\mathbf{Z}-\mathbf{u})_{A^c}^T \mathbf{M}_{A^c,A^c}(\mathbf{Z}-\mathbf{u})_{A^c} \\
&\le \frac{\|\mathbf{u}_{A^c}\|_1^2}{m}\|\mathbf{M}_{A^c,A^c}\|_{\max}\sum_{j\notin A^c} 4^{1-\delta_0} w_j(1-w_j)^2.
\end{aligned}
$$

Thus, one of the realizations of $\sum_{i=1}^m \mathbf{Z}^i/m$, say $\mathbf{v}$, must satisfy (4.101). ■

**Proof of Theorem 4.16.** Consider a given $\mathbf{u}$ satisfying $\|\mathbf{u}_{\mathcal{S}^c}\|_1 < \psi_1(\mathbf{u}_{\mathcal{S}})$. It follows from Lemma 4.6 that for $4\ell \ge |\mathcal{S}|$

$$
\|\mathbf{u}\|_2 \le \|\mathbf{u}_A\|_2 + \|\mathbf{u}_{\mathcal{S}^c}\|_1/\sqrt{4\ell} \le (1+\xi)\|\mathbf{u}_A\|_2,
$$

where $A$ is the union of $\mathcal{S}$ and the index set of the $\ell$ largest $\{|u_j|, j \notin \mathcal{S}\}$.

Let $\mathbf{v}$ be as in Lemma 4.8 with $\mathbf{v}_A = \mathbf{u}_A$. By the condition on $\psi_1(\cdot)$,

$$
\begin{aligned}
\|\mathbf{u}_{\mathcal{S}^c}\|_1^2 &< \xi^2\psi_1^2(\mathbf{v}_{\mathcal{S}}) \\
&\le \xi^2(|\mathcal{S}|^{1/2}\|\mathbf{v}_A\|_2)^k\psi_2^{2-k}(\mathbf{v}_{\mathcal{S}}) \\
&\le \xi^2|\mathcal{S}|(\mathbf{v}^T\mathbf{M}\mathbf{v})_+/R_{k,\ell,m}^2(\mathcal{S};\eta,\mathbf{M},\psi_1,\psi_2),
\end{aligned}
$$

which implies $\mathbf{v}^T\mathbf{M}\mathbf{v} \ge 0$. Moreover, Lemma 4.8 and the above inequality imply

$$
\begin{aligned}
\mathbf{v}^T\mathbf{M}\mathbf{v} - \mathbf{u}^T\mathbf{M}\mathbf{u} &\le 4^{1-\delta_0}\|\mathbf{M}_{\mathcal{S},\mathcal{S}}\|_{\max}\|\mathbf{u}_{\mathcal{S}^c}\|_1^2/m \\
&\le \frac{4^{1-\delta_0}\|\mathbf{M}_{\mathcal{S},\mathcal{S}}\|_{\max}\xi^2|\mathcal{S}|(\mathbf{v}^T\mathbf{M}\mathbf{v})}{mR_{k,\ell,m}^2(\mathcal{S};\eta,\mathbf{M},\psi_1,\psi_2)} \\
&\le \eta_0\mathbf{v}^T\mathbf{M}\mathbf{v}.
\end{aligned}
$$

Consequently, as $\eta_0 < 1$, $\mathbf{u}^T\mathbf{M}\mathbf{u} \ge (1-\eta_0)\mathbf{v}^T\mathbf{M}\mathbf{v} \ge 0$, so that

$$
\frac{\|\mathbf{u}\|_2^k\psi_2^{2-k}(\mathbf{u}_{\mathcal{S}})}{(\mathbf{u}^T\mathbf{M}\mathbf{u})_+|\mathcal{S}|^{1-k/2}} \le \frac{(1+\xi)^k\|\mathbf{v}_A\|_2^k\psi_2^{2-k}(\mathbf{v}_{\mathcal{S}})}{(1-\eta_0)(\mathbf{v}^T\mathbf{M}\mathbf{v})_+|\mathcal{S}|^{1-k/2}} \le \frac{(1+\xi)^k/(1-\eta_0)}{R_{k,\ell,m}^2(\mathcal{S};\eta,\mathbf{M},\psi_1,\psi_2)}
$$

as $\mathbf{v}_A = \mathbf{u}_A$ and $A \supseteq \mathcal{S}$. This gives (4.98). For (4.99), we have $\mathbf{M} = \overline{\mathbf{\Sigma}}$ and

$$
R_{k,\ell,m}^2(\mathcal{S};\eta,\overline{\mathbf{\Sigma}},\psi_1,\psi_2) \ge \inf_{|B\setminus\mathcal{S}|<\ell+m}\frac{\mathbf{u}_B^T\overline{\mathbf{\Sigma}}_{B,B}\mathbf{u}_B}{\|\mathbf{u}_B\|_2^2} \ge \phi_*(\ell+m;\mathcal{S},\overline{\mathbf{\Sigma}}),
$$

as the condition on $\{m,\eta_0\}$ holds with $\delta_0 = 1$ and $\|\mathbf{M}\|_{\max} \le 1$. ■

Finally, consider random designs $\mathbf{X} = (X_{ij})_{n\times p}$ with $\mathbf{\Sigma} = \mathrm{E}\overline{\mathbf{\Sigma}} = \mathrm{E}[\mathbf{X}^T\mathbf{X}/n]$ and independent rows satisfying the following sub-Gaussian condition,

$$
\mathrm{E}\Big(\sum_{j=1}^p X_{ij}u_j\Big)^{2k} \le \frac{k!}{2}\Big(V_0\mathbf{u}^T\mathbf{\Sigma}\mathbf{u}\Big)^{k-1}\mathrm{E}\Big(\sum_{j=1}^p X_{ij}u_j\Big)^2 \tag{4.102}
$$

for all integers $k \ge 2$, $1 \le i \le n$, and $\mathbf{u}$ with $|\mathrm{supp}(\mathbf{u}) \setminus \mathcal{S}| \le m_1$. It is a generalization of centralized Gaussian distribution, in which $\sum_{j=1}^p X_{ij}u_j \sim N(0,\mathbf{u}^T\mathbf{\Sigma}\mathbf{u})$ and $V_0 = 3$ if $\mathbf{X}$ has iid Gaussian rows.

**Theorem 4.17** *Suppose* $\mathbf{X}$ *is a sub-Gaussian matrix with independent rows satisfying (4.102). Let* $s = |\mathcal{S}|$, $\{\ell, m\}$ *be as in Theorem 4.16,* $m_1 = m + \ell$ *and* $m_2 = s + m_1$. *Suppose* $m_1 \log(e(p-s)/m_1) \le a_0 n$. *Let* $\epsilon_1$ *and* $\epsilon_2$ *be positive constants and* $\epsilon_0 = \{\epsilon_1 + \epsilon_2(1+\epsilon_2)\}/\{1 - \epsilon_2(1+\epsilon_2)\}$. *Then,*

$$P\left\{\left|\frac{\mathbf{u}^T\overline{\mathbf{\Sigma}}\mathbf{u}}{\mathbf{u}^T\mathbf{\Sigma}\mathbf{u}} - 1\right| \ge \epsilon_0 \ \forall \mathbf{u} \text{ with } |\text{supp}(\mathbf{u}) \setminus \mathcal{S}| \le m_1\right\} \le 2e^{-na_1}, \tag{4.103}$$

*when* $a_1 \le \epsilon_1^2/\{2V_0(1+\epsilon_1)\} - (m_2/n)\log(1+2/\epsilon_2) - a_0$. *Consequently,* $\mathbf{M} = \overline{\mathbf{\Sigma}}$,

$$P\left\{\frac{R^2_{k,\ell,m}(\mathcal{S};\eta,\mathbf{\Sigma},\psi_1,\psi_2)}{R^2_k(\mathcal{S};\eta,\overline{\mathbf{\Sigma}},\psi_1,\psi_2)} \le \frac{(1+\xi)^k}{(1-\epsilon_0)(1-\eta_0)}\right\} \ge 1 - 2e^{-na_1} \tag{4.104}$$

*where* $R_k(\mathcal{S};\eta,\mathbf{M},\psi_1,\psi_2)$ *is as in (4.90) and* $R_{k,\ell,m}(\mathcal{S};\eta,\mathbf{M},\psi_1,\psi_2)$ *as in (4.97). Moreover, under the following population SRC condition,*

$$|\mathcal{S}| \le \frac{2m^*(1-\eta_2)^2/(1+\eta_1)^2}{\phi_{\text{cond}}(m^*;\mathcal{S},\overline{\mathbf{\Sigma}})(1+\epsilon_0)/(1-\epsilon_0) - 1}, \tag{4.105}$$

*the sample version (4.59) holds with at least probability* $1 - e^{-na_1}$.

As we commented below Theorem 4.16,

$$\begin{aligned} &\min\left\{\frac{(1+\eta)^2}{C^{(\ell_1)}_{\text{pred}}}, \frac{(1+\eta)^2}{(C^{(\ell_1)}_{\text{est},2})^2}, \text{SCIF}^2_{2,p}, \text{CIF}^2_{2,p}, \text{RE}^2_0, \text{RE}^2_2\right\} \\ \ge\ & \phi_*(\ell+m;\mathcal{S},\mathbf{\Sigma})(1-\epsilon_0)(1-\eta_0)/(1+\xi)^2 \end{aligned}$$

with at least probability $1 - e^{-na_1}$ under conditions of Theorem 4.17.

**Proof of Theorem 4.17.** Let $\mu_{2,i}(\mathbf{u}) = \mathrm{E}\left(\sum_{j=1}^p X_{ij}u_j\right)^2$. For vectors $\mathbf{u}$ satisfying $\mathbf{u}\mathbf{\Sigma}\mathbf{u} = 1$, $\sum_{i=1}^n \mu_{2,i}(\mathbf{u}) = n$ and the Bernstein inequality is given by

$$\begin{aligned} & P\left\{\mathbf{u}^T\overline{\mathbf{\Sigma}}\mathbf{u} \ge 1 + \epsilon_1\right\} \\ \le\ & e^{-nt(1+\epsilon_1)}\prod_{i=1}^n\left\{1 + \mu_{2,i}(\mathbf{u})\sum_{k=1}^\infty \frac{t^k V_0^{k-1}}{2\wedge k}\right\} \\ \le\ & \exp\left\{-\frac{n\epsilon_1^2}{2V_0(1+\epsilon_1)}\right\} \end{aligned} \tag{4.106}$$

under condition (4.102), where $t = V_0^{-1}\epsilon_1/(1+\epsilon_1)$. Similarly

$$P\left\{\mathbf{u}^T\overline{\mathbf{\Sigma}}\mathbf{u} \le 1 - \epsilon_1\right\} \le \exp\left\{-\frac{n\epsilon_1^2}{2V_0(1+\epsilon_1)}\right\}. \tag{4.107}$$

Let $\mathscr{U}_A = \{\mathbf{u} : \text{supp}(\mathbf{u}) \subseteq A, \mathbf{u}^T\mathbf{\Sigma u} = 1\}$ with $\mathcal{S} \subseteq A \subseteq \{1, \ldots, p\}$ satisfying $|A \setminus \mathcal{S}| = m_1$. Let $\mathbf{u}_1, \ldots, \mathbf{u}_N$ be vectors in $\mathscr{U}_A$ such that

$$\max_{\mathbf{u}\in\mathscr{U}_A} \min_{1\le j\le N} (\mathbf{u} - \mathbf{u}_j)^T \mathbf{\Sigma}(\mathbf{u} - \mathbf{u}_j) \le \epsilon_2^2$$

and $\min_{1\le j<k\le N}(\mathbf{u}_k - \mathbf{u}_j)^T\mathbf{\Sigma}(\mathbf{u}_k - \mathbf{u}_j) \ge \epsilon_2^2$. As in the proof of Lemma 4.2 (iii), $N \le (1 + 2/\epsilon_2)^{m_1}$ by volume comparison. We have

$$\begin{aligned}
&\max_{\mathbf{u}\in\mathscr{U}_A} \left|\mathbf{u}^T\overline{\mathbf{\Sigma}}\mathbf{u} - 1\right| - \max_{1\le j\le N}\left|\mathbf{u}_j^T\overline{\mathbf{\Sigma}}\mathbf{u}_j - 1\right| \\
\le\ & \max_{\mathbf{u}\in\mathscr{U}_A} \min_{1\le j\le N} \left|\mathbf{u}^T\overline{\mathbf{\Sigma}}\mathbf{u} - \mathbf{u}_j^T\overline{\mathbf{\Sigma}}\mathbf{u}_j\right| \\
\le\ & \max_{\mathbf{u}\in\mathscr{U}_A} \min_{1\le j\le N} \left|(\mathbf{u} - \mathbf{u}_j)^T\overline{\mathbf{\Sigma}}(\mathbf{u} + \mathbf{u}_j)\right| \\
\le\ & \epsilon_2(1 + \epsilon_2) \max_{\mathbf{u}\in\mathscr{U}_A} \mathbf{u}^T\overline{\mathbf{\Sigma}}\mathbf{u}
\end{aligned}$$

with an application of the Cauchy-Schwarz inequality. It follows that

$$\max_{\mathbf{u}\in\mathscr{U}_A} \left|\mathbf{u}^T\overline{\mathbf{\Sigma}}\mathbf{u} - 1\right| \le \frac{\max_{1\le j\le N}\left|\mathbf{u}_j^T\overline{\mathbf{\Sigma}}\mathbf{u}_j - 1\right| + \epsilon_2(1+\epsilon_2)}{1 - \epsilon_2(1 + \epsilon_2)}.$$

Thus, as $\epsilon_0 = \{\epsilon_1 + \epsilon_2(1 + \epsilon_2)\}/\{1 - \epsilon_2(1 + \epsilon_2)\}$, (4.106) and (4.107) yield

$$P\Big\{\max_{\mathbf{u}\in\mathscr{U}_A} \left|\mathbf{u}^T\overline{\mathbf{\Sigma}}\mathbf{u} - 1\right| \ge \epsilon_0\Big\} \le 2N \exp\Big\{-\frac{n\epsilon_1^2}{2V_0(1 + \epsilon_1)}\Big\}.$$

As there are totally $\binom{p-s}{m_1} \le e^{a_0 n}$ models $A$ with $A \supset \mathcal{S}$ and $|A \setminus \mathcal{S}| = m_1$, the left-hand side of (4.103) is bounded by

$$2\binom{p-s}{m^*}(1 + 2/\epsilon_2)^{m^*} \exp\Big\{-\frac{n\epsilon_1^2}{2V_0(1 + \epsilon_1)}\Big\} \le e^{-na_1}.$$

Inequality (4.104) follows from (4.103) and Theorem 4.16.

## 4.4 Properties of Concave PLS

In this section, we consider general penalized loss functions of the form

$$Q(\mathbf{b}) = \|\mathbf{Y} - \mathbf{Xb}\|_2^2/(2n) + \text{Pen}(\mathbf{b}). \tag{4.108}$$

We have already studied the best subset selection and Lasso, respectively corresponding to the $\ell_0$ penalty $\text{Pen}(\mathbf{b}) = p_\lambda(\|\mathbf{b}\|_0)$ and the $\ell_1$ penalty $\text{Pen}(\mathbf{b}) = \lambda\|\mathbf{b}\|_1$. As we have mentioned in Chapter 3, the purpose of considering the more general penalty function is to simultaneously facilitate the sparsity, unbiasedness, continuity, computational feasibility and other desirable properties for the PLS. Our objective is to prove that under an $\ell_2$ restricted eigenvalue condition on the design matrix, these goals can be achieved with separable folded concave penalties discussed in Subsection 4.4.1 below.

The subset selection, which becomes the hard threshold estimator for orthogonal designs, is not continuous. For general designs, this discontinuity means the existence of too many local minima for (4.108) and high computational complexity, as the stepwise regression and other iterative algorithms may get stuck at a bad local minimum.

The Lasso, which can be viewed as a convex surrogate of the best subset selection, greatly alleviates the computational burden of the procedure. However, the Lasso is biased. This can be easily seen in the orthogonal designs where the Lasso is the soft threshold estimator. For correlated designs and large $p$, the bias of the Lasso may cancel out true effects and create spurious effects, leading to variable selection inconsistency. This is expressed as the requirement of the much stronger $\ell_\infty$-type conditions for the Lasso selection consistency in Theorem 4.8, compared to more scalable $\ell_2$-type conditions such as sparse and restricted eigenvalues. For prediction and coefficient estimation, the Lasso does achieve rate optimality under $\ell_2$-type conditions. Still, the Lasso garners limited benefit from signal strength, as the bias of the Lasso does not diminish as signal strengthens.

The $\ell_\alpha$ (bridge) penalty (Frank and Friedman, 1993) was introduced to interpolate the $\ell_0$ and $\ell_2$ penalty. For $0 < \alpha < 1$, it provides a continuous transition from $\ell_0$ to $\ell_1$ penalties. However, for $0 < \alpha < 1$, the bridge penalty still has too many local minima as its derivative is unbounded near zero.

Fan and Li (2001) introduced the *SCAD* penalty, indeed a class of *folded concave penalty* function to remedy these problems. The SCAD penalty is a folded concave penalty with both sparseness and unbiasedness properties as well as a bounded derivative. This allows gradient descent and other continuous algorithms to search for proper solutions in large subspaces. However, as the penalty is necessarily nonconvex, the PLS (4.108) may still have many local minimum. The *MCP* (Zhang, 2010) was introduced to minimize the concavity subject to given levels of sparseness and unbiasedness, so that the sparse convexity of the PLS is maximized. This sparse convexity provides the uniqueness of sparse local minimizers so that the local minimizer produced by iterative algorithms can be identified with oracle estimators to achieve variable selection consistency. The results in this section will provide proper conditions for the existence of an oracle local solution of the PLS problem, the oracle and other optimality properties of the local minima produced by path-following and gradient descent algorithms, and the benefits of signal strength in prediction and estimation even when variable selection is infeasible. All these are achieved under $\ell_2$-type regularity conditions on the design matrix.

As the penalized loss function in (4.108) is not guaranteed to be convex, the minimization problem may have many local solutions, and local solutions can be further classified by their properties or computational algorithms. We provide below more precise definitions of some local and global solutions.

**Definition 4.3** *For the minimization of* $Q(\boldsymbol{\beta})$, *a vector* $\widehat{\boldsymbol{\beta}}$ *is*

- *a critical point if*

$$\liminf_{t\to 0_+} t^{-1}\Big\{Q(\widehat{\boldsymbol\beta}+t\mathbf{u})-Q(\widehat{\boldsymbol\beta})\Big\}\ge 0 \ \forall\ \mathbf{u}\in \mathbb{R}^p \tag{4.109}$$

- *a local minimizer if there exists $t_0>0$ such that*

$$Q(\mathbf{b})\ge Q(\widehat{\boldsymbol\beta}),\quad \forall 0<\|\mathbf{b}-\widehat{\boldsymbol\beta}\|_2\le t_0 \tag{4.110}$$

- *a strict local minimizer if there exists $t_0>0$ such that*

$$Q(\mathbf{b})> Q(\widehat{\boldsymbol\beta}),\quad \forall 0<\|\mathbf{b}-\widehat{\boldsymbol\beta}\|_2\le t_0 \tag{4.111}$$

- *a strong local minimizer if (4.110) holds and*

$$Q(\mathbf{b}+t\mathbf{e}_j)\ge Q(\widehat{\boldsymbol\beta})\quad \forall\ j\le p,\ t\in \mathbb{R} \tag{4.112}$$

- *a restricted minimizer under the condition $\boldsymbol\beta\in\mathscr{B}$ if*

$$\widehat{\boldsymbol\beta}=\arg\min_{\mathbf{b}\in\mathscr{B}} Q(\mathbf{b}) \tag{4.113}$$

- *a global minimizer if (4.113) holds with $\mathscr{B}=\mathbb{R}^p$,*

$$\widehat{\boldsymbol\beta}=\arg\min_{\mathbf{b}\in\mathbb{R}^p} Q(\mathbf{b}) \tag{4.114}$$

*In general a minimizer of $Q(\boldsymbol\beta)$, whether local or global, is considered as a solution of (4.108). However, we may call a critical point a local solution.*

We will describe in Subsection 4.4.1 some basic and optional conditions imposed on the penalty function in our analysis and their implication on the resulting PLS in the simplest setting. We will then study the existence of oracle local solutions in Subsection 4.4.2, properties of general local solutions in Subsection 4.4.3, and global and approximate global solutions in Subsection 4.4.4.

### *4.4.1 Properties of penalty functions*

We describe in this subsection properties of separable penalties of the form

$$\mathrm{Pen}(\mathbf{b})=\|p_\lambda(\mathbf{b})\|_1=\sum_{j=1}^p p_\lambda(b_j). \tag{4.115}$$

As the penalty is separable, we focus on properties of the function $p_\lambda(t)$ and their relation to the properties of the PLS in the simplest case of $p=1$.

Define the left- and right-derivative of the penalty by

$$p'_\lambda(t\pm)=\lim_{\epsilon\to 0+}\frac{p_\lambda(t\pm\epsilon)-p_\lambda(t)}{\pm\epsilon}\in[-\infty,\infty],$$

and denote by $p'_\lambda(t)$ any real number satisfying

$$\min\{p'_\lambda(t+), p'_\lambda(t-)\} \le p'_\lambda(t) \le \max\{p'_\lambda(t+), p'_\lambda(t-)\}. \tag{4.116}$$

Thus, by writing $p'_\lambda(t) = x$ we imply that both the left- and right-derivatives of $p_\lambda(\cdot)$ are defined at $t$ and the real number $x$ is between the two one-sided derivatives. We note that $p'_\lambda(t\pm)$ is in general different from the limit of $p'_\lambda(t)$ from the left and right, and that $p'_\lambda(0\pm) = \pm\infty$ for the $\ell_\alpha$ penalties, $\alpha \in [0, 1)$, and any penalty with discontinuity at $t = 0$.

Unless otherwise stated, we assume throughout this section that the following basic properties hold for the penalty function in (4.115):

- *Zero baseline penalty*: $p_\lambda(0) = 0$;
- *Symmetry*: $p_\lambda(-t) = p_\lambda(t)$;
- *Sparsity*: $p'(0+) = \lim_{t\to 0+} p_\lambda(t)/t > 0$;
- *Monotonicity*: $p_\lambda(x) \le p_\lambda(y)$ for all $0 \le x < y < \infty$;
- *One-sided differentiability*: $p'_\lambda(t\pm)$ are defined for all $t > 0$;
- *Subadditivity*: $p_\lambda(x + y) \le p_\lambda(x) + p_\lambda(y)$ for all $0 \le x \le y < \infty$.

This class of penalty functions was introduced in Antoniadis and Fan (2001) and Fan and Li (2001). It includes all nondecreasing concave functions of $|t|$ with $p_\lambda(0) = 0$ and $\|p_\lambda\|_\infty > 0$, and is closed under the operations of summation and maximization. It is convenient to include the one-sided differentiability condition, which is not much stronger than the monotonicity condition, although the one-sided differentiability condition is not required in our analysis of the global solution of the PLS. In addition to the $\ell_0$, $\ell_1$, SCAD and MCP penalties mentioned at the beginning of the section, examples of penalty functions satisfying the basic properties include the bridge penalty with $p_\lambda(t) \propto |t|^\alpha$ for $0 < \alpha < 1$ and the capped-$\ell_1$ penalty $p_\lambda(t) = \min(\lambda|t|, a\lambda^2)$. However, the sub-additivity, which promotes sparsity, excludes strictly convex penalties such as the Elastic Net.

To ensure that the PLS has desired properties under proper conditions on the design matrix, we may also impose additional conditions on the penalty based on quantities defined below.

We define the *concavity* of the penalty at $t$ as

$$\overline{\kappa}(p_\lambda; t) = \sup_{\epsilon \ne 0} \frac{p'_\lambda(t-\epsilon) - p'_\lambda(t)}{\epsilon}, \tag{4.117}$$

the one-sided *local concavity* at $t$ as

$$\kappa(p_\lambda; t\pm) = \lim_{\epsilon \to 0+} \frac{p'_\lambda(t \pm \epsilon) - p'_\lambda(t\pm)}{\mp\epsilon}, \tag{4.118}$$

and the *maximum concavity* as

$$\kappa(p_\lambda) = \sup_t \kappa(p_\lambda; t+). \tag{4.119}$$

When $p_\lambda(t)$ is concave in $|t|$, the quantities defined above are all nonnegative. However, this may be true for a general penalty. In this case, negative concavity actually means convexity. Due to the monotonicity and symmetry of $p_\lambda(t)$, $tp'_\lambda(t) \ge 0$, so that $\overline{\kappa}(p_\lambda; t)$ is always attained when $t-\epsilon$ and $t$ have the same sign. Thus, the maximum concavity in (4.119) can be written as

$$\kappa(p_\lambda) = \sup_{0<t_1<t_2<\infty} \frac{p'_\lambda(t_2) - p'_\lambda(t_1)}{t_1 - t_2},$$

as defined in (3.21), and in the case of $p = 1$, the function $\kappa(p_\lambda; t) = \max\{\kappa(p_\lambda; t+), \kappa(p_\lambda; t-)\}$ is the local concavity $\kappa(p_\lambda; t)$ in Chapter 3.

Define a penalized loss for a univariate data point $z$ as

$$Q_\lambda(b|z) = (z-b)^2/2 + p_\lambda(b)$$

and let $b_\lambda(z)$ be the global minimizer of $Q_\lambda(b|z)$,

$$b_\lambda(z) = \arg\min_b \Big\{(z-b)^2/2 + p_\lambda(b)\Big\} \in \Big\{b : b = z - p'_\lambda(b)\Big\}. \tag{4.120}$$

When $p_\lambda(t)$ is convex in $t$, $b_\lambda(z)$ is the associated *proximal mapping*. For the $\ell_1$ penalty $p_\lambda(t) = \lambda|t|$, $b_\lambda(z) = \text{sgn}(z)(|z| - \lambda)_+$ is the soft threshold estimator. The condition $\kappa(p_\lambda) \le 1$ guarantees the convexity of $Q_\lambda(b|z)$ in $b$. As $z = b_\lambda(z) + p'_\lambda(b_\lambda(z))$, $b_\lambda(z)$ is unique and continuous in $z$ iff $t + p'_\lambda(t)$ is a strictly increasing mapping from $(0,\infty)$ onto $(\lambda,\infty)$ under the convention (4.116). Moreover, the strict convexity condition $\kappa(p_\lambda) < 1$ guarantees that $b_\lambda(z)$ has the Lipschitz coefficient $1/\{1 - \kappa(p_\lambda)\}$.

When $z \sim N(\beta, \sigma_n^2)$ with small $\sigma_n$, the first order approximation of the bias of the univariate estimator $\widehat{\beta} = b_\lambda(z)$ is

$$\mathrm{E}\,\widehat{\beta} - \beta \approx p'_\lambda(\beta) \approx p'_\lambda(z).$$

Thus, for the estimation of the coefficient vector $\boldsymbol{\beta}$ in (4.1) and separable penalty (4.115), the bias of the PLS (4.108) can be measured by

$$\text{bias} \asymp \big\|p'_\lambda(\boldsymbol{\beta}_S)\big\|_2 \approx \big\|p'_\lambda(\widehat{\boldsymbol{\beta}}^o_S)\big\|_2 \tag{4.121}$$

with the oracle LSE $\widehat{\boldsymbol{\beta}}^o$ in (4.10). This is expected from the univariate case when the PLS is nearly selection consistent, at least for nearly orthogonal designs. A formal justification of (4.121) can be found in Theorem 4.21 in Subsection 4.4.3. For selection consistency, we define the *bias threshold* of the penalty by

$$a_\lambda = \inf\Big\{t > 0 : p_\lambda(x) = p_\lambda(t)\ \forall\ |x| \ge t\Big\}, \tag{4.122}$$

from which point to $\infty$, the function $p_\lambda(\cdot)$ is flat. Thus, $b_\lambda(z) = z$ for all

$|z| > a_\lambda$ and $\|p'_\lambda(\boldsymbol{\beta}_{\mathcal{S}})\|_2 = 0$ under the *minimum signal strength* condition (4.18) with $C_{\text{select}} > a_\lambda$.

We measure the *global threshold level* of the penalty by

$$\lambda_* = \lambda_*(p_\lambda) = \inf_{t>0} \{t/2 + p_\lambda(t)/t\}, \tag{4.123}$$

as $b = 0$ is the global minimizer of $Q_\lambda(b|z)$ if and only if $|z| \le \lambda_*$, namely

$$\min_b Q_\lambda(b|z) = Q_\lambda(0|z) \;\Leftrightarrow |z| \le \lambda_*.$$

In other words, when $|z| \le \lambda_*$, the solution to the penalized least-squares is zero. This can be seen as follows: Assume WOLG that $z \ge 0$. Then, the minimizer is obtained only on $[0, \infty)$. The result follows from

$$\begin{aligned} Q_\lambda(b|z) &= z^2/2 + b(-z + b/2 + p_\lambda(b)/b) \ge z^2/2 + b(-z + \lambda_*) \\ &\ge z^2/2 = Q_\lambda(0|z), \end{aligned}$$

for $z \le \lambda_*$. Similarly, we measure the *local threshold level* of the penalty by $p'_\lambda(0+)$, since when $|z| \le p'_\lambda(0+)$,

$$\liminf_{|b|\to 0+} \frac{Q_\lambda(b|z) - Q_\lambda(0|z)}{|b|} \ge 0$$

so that the minimizer is again zero. While the local threshold level is allowed to be infinity, the global threshold level, which is no greater, is always finite. Thus, the global threshold level is a more meaningful measure of the threshold level of the penalty when $p'_\lambda(0+) = \infty$.

To simplify the notation, we assume that the penalty function $p_\lambda(t)$ is indexed by a real number $\lambda$ and standardized as follows:

- *Standardized local threshold level*: $p'_\lambda(0+) = \lambda$ if $p'_\lambda(0+) < \infty$;
- *Standardized global threshold level*: $\lambda_*(p_\lambda) = \lambda$ if $p'_\lambda(0+) = \infty$;
- *Standardized bias threshold*: $a = a_\lambda/\lambda$;
- *Standardized local threshold level*: $a^* = p'_\lambda(0+)/\lambda$;
- *Standardized global threshold level*: $a_* = \lambda_*/\lambda$;
- *Standardized maximum penalty*: $\gamma^* = \|p_\lambda(t)\|_\infty/\lambda^2$.

Note that we have either $a^* = 1 \ge a_*$ or $(a^*, a_*) = (\infty, 1)$. The quantities $a$, $\kappa(p_\lambda)$, $a^*$, $a_*$ and $\gamma^*$ do not depend on $\lambda$ if the penalty function is *scale invariant* in the sense of

$$p_\lambda(t) = \lambda^2 p_1(t/\lambda). \tag{4.124}$$

We list the characterizations of a number of scale invariant penalty functions in Table 4.1 below along with their formulas for $\lambda = 1$.

We note that the global threshold level of the penalty functions in Table 4.1 are all equal to $\lambda$ as those with finite local threshold level happen to have

Table 4.1: Characteristics of Penalties: $C_\alpha = \{2(1-\alpha)\}^{1-\alpha}/(2-\alpha)^{2-\alpha}, 0 < \alpha < 1$

| Penalty | $p_1(t)$ | Bias $a$ | Maximum Concavity $\kappa(p_\lambda)$ | Threshold Levels $a^*/a_*$ | Maximum Penalty $\gamma^*$ |
|---|---|---|---|---|---|
| $\ell_0$ | $2^{-1}I\{t \neq 0\}$ | 0 | $\infty$ | $\infty/1$ | $1/2$ |
| Bridge | $C_\alpha \lvert t\rvert^\alpha$ | $\infty$ | $\infty$ | $\infty/1$ | $\infty$ |
| Lasso | $\lvert t\rvert$ | $\infty$ | 0 | $1/1$ | $\infty$ |
| SCAD | $\int_0^{\lvert t\rvert}\left(1-\frac{(x-1)_+}{a-1}\right)_+ dx$ | $a \geq 2$ | $1/(a-1)$ | $1/1$ | $a/2+1/2$ |
| MCP | $\int_0^{\lvert t\rvert}\left(1-x/a\right)_+ dx$ | $a \geq 1$ | $1/a$ | $1/1$ | $a/2$ |
| Capped-$\ell_1$ | $\min(\lvert t\rvert, a)$ | $a \geq 1/2$ | $\infty$ | $1/1$ | $a$ |

the same global threshold level due to the specified limitation in $a$. Given the threshold level, we want to have small bias threshold and small maximum concavity. In addition to the basic properties and the characteristics in Table 4.1, we may further impose optional conditions listed below:

- *Concavity*: $p'_\lambda(t)$ is decreasing in $t$ for $t > 0$;
- *Approximate unbiasedness*: $\lim_{t\to\infty} p'_\lambda(t) = 0$;
- Monotonicity of $p_\lambda(t)/t$ for $t > 0$;
- Continuity of $p_\lambda(t)$ at $t = 0$.

The above optional assumptions, if invoked, will always be explicitly stated alongside the statement of the result. Note that the concavity condition is slightly stronger than the monotonicity of $p_\lambda(t)/t$ for $t > 0$, which is slightly stronger than the basic sub-additivity property.

We conclude this subsection with the following proposition summarizing some useful bounds for all the penalty functions under consideration.

**Proposition 4.2** *For any penalty function with the basic properties,*

$$\min\left\{|t|\lambda_*/2, \lambda_*^2/2\right\} \leq p_\lambda(t) \leq |t|\lambda_* + (\lambda_* - |t|/2)_+^2/2.$$

*Consequently, for any $A \subset \{1, \ldots, p\}$ and vector $\mathbf{v}_A$,*

$$\|p_\lambda(\mathbf{v}_A)\|_1 < |A|\lambda^2 \min\left\{a_*\|\mathbf{v}_A\|_1/(\lambda|A|) + a_*^2/2, \gamma^*\right\}.$$

**Proof of Proposition 4.2.** By (4.123), $p_\lambda(t) \geq |t|(\lambda_* - |t|/2)$. This and the monotonicity of $p_\lambda(t)$ give the lower bound in both cases $|t| \leq \lambda_*$ and $|t| > \lambda_*$.

To prove the upper bound, we consider $x \to t_0$ with $x/2 + p_\lambda(x)/x \to \lambda_*$. Let $t > 0$ and $q = \lfloor t/x \rfloor$. Because $p_\lambda(t)$ is nondecreasing and sub-additive,

$$p_\lambda(t) \leq p_\lambda(qx) + p_\lambda(t - qx) \leq (q+1)p_\lambda(x) \leq (t+x)p_\lambda(x)/x.$$

Let $x \to t_0$. We find that

$$\begin{aligned} p_\lambda(t) &\le (t+t_0)(\lambda_* - t_0/2) \\ &\le \max_{x>0}(t+x)(\lambda_* - x/2) \\ &= \lambda_* t + (\lambda_* - t/2)_+^2/2, \end{aligned}$$

as the maximum is reached at $x = (\lambda_* - t/2)_+$. The bound for $\|p_\lambda(\mathbf{v}_A)\|_1$ follows directly from the upper bound and the definitions of $a_*$ and $\gamma^*$. ■

### *4.4.2 Local and oracle solutions*

We first study necessary and sufficient conditions for local solutions of the general PLS (4.108) and then provide a set of sufficient conditions for the existence of an oracle local solution with the selection-consistency property. Throughout this subsection, we shall confine our discussion to separable folded penalties of the form (4.115) and satisfying the basic properties and conventions stated in the previous subsection.

With the basic assumption on the one-sided differentiability of the penalty and the convention for the derivative in (4.116), the partial derivative of the penalized loss in (4.108) is defined as

$$Q_j'(\boldsymbol{\beta}) = -\mathbf{X}_j^T(\mathbf{Y} - \mathbf{X}\boldsymbol{\beta})/n + p_\lambda'(\beta_j).$$

The following theorem is a sharper version of Theorem 3.1 with more details.

**Theorem 4.18** *Consider separable penalties in (4.115).*
*(i) Condition (4.109) holds at $\widehat{\boldsymbol{\beta}}$ iff it holds uniformly in $\mathbf{u}$ in the sense of*

$$\liminf_{t\to 0+} \inf_{0<\|\mathbf{u}\|_2\le 1} t^{-1}\Big\{Q(\widehat{\boldsymbol{\beta}} + t\mathbf{u}) - Q(\widehat{\boldsymbol{\beta}})\Big\} \ge 0.$$

*In this case $Q_j'(\widehat{\boldsymbol{\beta}}) = 0$ for all $j = 1, \ldots, p$, or equivalently*

$$\begin{cases} \mathbf{X}_j^T(\mathbf{Y} - \mathbf{X}\widehat{\boldsymbol{\beta}})/n = p_\lambda'(\widehat{\beta}_j) & \widehat{\beta}_j \ne 0, \\ |\mathbf{X}_j^T(\mathbf{Y} - \mathbf{X}\widehat{\boldsymbol{\beta}})/n| \le p_\lambda'(0+) & \widehat{\beta}_j = 0. \end{cases} \tag{4.125}$$

*Conversely, if (4.125) holds and $p_\lambda'(t)$ is continuous at $t \in \{\widehat{\beta}_j : \widehat{\beta}_j \ne 0\}$, then $\widehat{\boldsymbol{\beta}}$ is a critical point in the sense of (4.109).*
*(ii) Suppose $\kappa(p_\lambda) < \infty$. A critical point $\widehat{\boldsymbol{\beta}}$ is a strict local minimizer as defined in (4.111) if*

$$\|\mathbf{X}\mathbf{u}\|_2^2/n > \sum_{u_j>0} \kappa(p_\lambda; \widehat{\beta}_j+)u_j^2 + \sum_{u_j<0} \kappa(p_\lambda; \widehat{\beta}_j-)u_j^2 \tag{4.126}$$

*for all* $\mathbf{u}$ *with* $\{j : u_j > 0\} \subseteq \mathcal{S}_+$ *and* $\{j : u_j < 0\} \subseteq \mathcal{S}_-$*, where* $\mathcal{S}_\pm = \{j : \mathbf{X}_j^T(\mathbf{Y} - \mathbf{X}\widehat{\boldsymbol{\beta}})/n = p'_\lambda(\widehat{\beta}_j \pm)\}$.

*(iii) Suppose* $p_\lambda(t)$ *is a quadratic spline in* $[0, \infty)$*. Then, a critical point* $\widehat{\boldsymbol{\beta}}$ *is a strict local minimizer if and only if (4.126) holds, and it is a local minimizer as defined in (4.110) if and only if*

$$\|\mathbf{X}\mathbf{u}\|_2^2/n \geq \sum_{u_j>0} \kappa(p_\lambda; \widehat{\beta}_j+)u_j^2 + \sum_{u_j<0} \kappa(p_\lambda; \widehat{\beta}_j-)u_j^2 \tag{4.127}$$

*for the same set of* $\mathbf{u}$ *as in (4.126).*

In what follows we may also call $\widehat{\boldsymbol{\beta}}$ a *local solution* for the PLS problem with penalty (4.115) when condition (4.125) is satisfied. Theorem 4.18 is an extension of Proposition 4.1 as it includes the Lasso as a special case with $p'_\lambda(t) = \lambda\,\mathrm{sgn}(t)$ for $t \neq 0$ and $\kappa(p_\lambda; t\pm) = 0$ for all $t$. Theorem 4.18 (iii) is applicable to the SCAD and MCP as they are folded quadratic splines. However, the capped-$\ell_1$ penalty is only a linear spline due to the discontinuity of its derivative when $|t| = a$. Anyway, parts (ii) and (iii) of the theorem are not applicable to penalty functions with $\kappa(p_\lambda) = \infty$.

**Proof of Theorem 4.18.** (i) Condition (4.109) holds uniformly in $\mathbf{u}$ iff

$$\liminf_{t\to 0+} \frac{Q(\widehat{\boldsymbol{\beta}} + t\mathbf{u}) - Q(\widehat{\boldsymbol{\beta}})}{t} \geq 0 \geq \limsup_{t\to 0+} \frac{Q(\widehat{\boldsymbol{\beta}}) - Q(\widehat{\boldsymbol{\beta}} - t\mathbf{u})}{t},$$

uniformly in $\mathbf{u}$ given each choice of $\mathrm{sgn}(\mathbf{u}) \in \{-1, 0, 1\}^p$. As $p'_\lambda(t)$ is one-sided differentiable due to its basic properties, the above one-sided derivatives are both uniform in $\mathbf{u}$ given $\mathrm{sgn}(\mathbf{u})$. As $p'_\lambda(t)$ can be chosen as any value between the one-sided derivatives and $\mathbf{u}$ is arbitrary, e.g. $\mathbf{u} = \mathbf{e}_j$, the above inequalities can be written as $Q'_j(\widehat{\boldsymbol{\beta}}) = 0$, $j = 1, \ldots, p$ under the convention (4.116). Conversely, if (4.125) holds and $p'_\lambda(t)$ is continuous at $t = \widehat{\beta}_j$ for all $\widehat{\beta}_j \neq 0$, then $Q'_j(\widehat{\boldsymbol{\beta}}\pm) = 0$ for $\widehat{\beta}_j \neq 0$ and $\mathbf{X}_j^T(\mathbf{Y} - \mathbf{X}\widehat{\boldsymbol{\beta}})/n = p'_\lambda(0) \in p'_\lambda(0)[-1, 1]$ for $\widehat{\beta}_j = 0$, so that

$$\liminf_{t\to 0+} \frac{Q(\widehat{\boldsymbol{\beta}} + t\mathbf{u}) - Q(\widehat{\boldsymbol{\beta}})}{t} = \sum_{\widehat{\beta}_n = 0} \Big\{ -u_j \mathbf{X}_j^T(\mathbf{Y} - \mathbf{X}\widehat{\boldsymbol{\beta}})/n + p'_\lambda(0+)|u_j| \Big\} \geq 0.$$

(ii) Suppose $\kappa(p_\lambda) < \infty$ and $\widehat{\boldsymbol{\beta}}$ is a critical point. It suffices to consider directions $\mathbf{u}$ with $(d/dt)Q(\widehat{\boldsymbol{\beta}} + t\mathbf{u})\big|_{t=0+} = 0$, due to the uniformity of the directional differentiation by the proof of part (i). Given $\mathrm{sgn}(\mathbf{u})$, the directional derivative of $Q(\cdot)$ can be written as

$$\begin{aligned} &(d/dt)Q(\widehat{\boldsymbol{\beta}} + t\mathbf{u})\Big|_{t=0+} \\ &= \sum_{u_j>0} \Big\{p'_\lambda(\widehat{\beta}_j+) - \mathbf{X}_j^T(\mathbf{Y} - \mathbf{X}\widehat{\boldsymbol{\beta}})/n\Big\}u_j \end{aligned}$$

$$+\sum_{u_j<0}\Big\{p'_\lambda(\widehat{\beta}_j-)-\mathbf{X}_j^T(\mathbf{Y}-\mathbf{X}\widehat{\boldsymbol{\beta}})/n\Big\}u_j$$

Because $(d/dt)Q(\widehat{\boldsymbol{\beta}}+t\,\mathrm{sgn}(u_j)\mathbf{e}_j)\geq 0$ at $t=0$, the sign of individual terms in the above sums must all be nonnegative. It follows that

$$\begin{aligned}
&(d/dt)Q(\widehat{\boldsymbol{\beta}}+t\mathbf{u})\Big|_{t=0+}\\
=&\sum_{u_j>0}\Big\{p'_\lambda(\widehat{\beta}_j+)-\mathbf{X}_j^T(\mathbf{Y}-\mathbf{X}\widehat{\boldsymbol{\beta}})/n\Big\}_+u_j\\
&+\sum_{u_j<0}\Big\{p'_\lambda(\widehat{\beta}_j-)-\mathbf{X}_j^T(\mathbf{Y}-\mathbf{X}\widehat{\boldsymbol{\beta}})/n\Big\}_-|u_j|.
\end{aligned}$$

Thus, $(d/dt)Q(\widehat{\boldsymbol{\beta}}+t\mathbf{u})\Big|_{t=0+}=0$ happens only when $\{j:u_j>0\}\subseteq\mathcal{S}_+$ and $\{j:u_j<0\}\subseteq\mathcal{S}_-$. In this case,

$$\begin{aligned}
&t^{-1}(d/dt)Q(\widehat{\boldsymbol{\beta}}+t\mathbf{u})\\
=&\ t^{-1}\sum_{u_j>0}\Big\{p'_\lambda(\widehat{\beta}_j+tu_j)-\mathbf{X}_j^T(\mathbf{Y}-\mathbf{X}\widehat{\boldsymbol{\beta}}-t\mathbf{X}\mathbf{u}))/n\Big\}u_j\\
&+t^{-1}\sum_{u_j<0}\Big\{p'_\lambda(\widehat{\beta}_j+tu_j)-\mathbf{X}_j^T(\mathbf{Y}-\mathbf{X}\widehat{\boldsymbol{\beta}}-t\mathbf{X}\mathbf{u})/n\Big\}u_j\\
=&\ \|\mathbf{X}\mathbf{u}\|_2^2+t^{-1}\sum_{u_j>0}\Big\{p'_\lambda(\widehat{\beta}_j+tu_j)-p'_\lambda(\widehat{\beta}_j+)\Big\}u_j\\
&+t^{-1}\sum_{u_j<0}\Big\{p'_\lambda(\widehat{\beta}_j+tu_j)-p'_\lambda(\widehat{\beta}_j-)\Big\}u_j\\
\geq&\ \|\mathbf{X}\mathbf{u}\|_2^2-(1+o(1))\Big\{\sum_{u_j>0}\kappa(p_\lambda,\widehat{\beta}_j+)u_j^2+\sum_{u_j<0}\kappa(p_\lambda,\widehat{\beta}_j-)u_j^2\Big\},
\end{aligned}$$

which is positive for small $t$.

(iii) When $p_\lambda(t)$ is a quadratic spline for $t>0$, the prove of part (ii) yields

$$\begin{aligned}
&t^{-1}(d/dt)Q(\widehat{\boldsymbol{\beta}}+t\mathbf{u})\\
=&\ \|\mathbf{X}\mathbf{u}\|_2^2-\Big\{\sum_{u_j>0}\kappa(p_\lambda,\widehat{\beta}_j+)u_j^2+\sum_{u_j<0}\kappa(p_\lambda,\widehat{\beta}_j-)u_j^2\Big\}
\end{aligned}$$

for sufficiently small $t>0$ whenever $(d/dt)Q(\widehat{\boldsymbol{\beta}}+t\mathbf{u})=0$ at $t=0$. The conclusions follow. ■

We now turn our attention to the existence of a local solution $\widehat{\boldsymbol{\beta}}$ satisfying $\mathrm{supp}(\widehat{\boldsymbol{\beta}})=\mathrm{supp}(\boldsymbol{\beta})$. Following Fan and Li (2001), we say that the PLS has the *oracle property* if such a selection consistent local solution exists, which we call the *oracle solution*. In addition to Fan and Li (2001), the oracle property of PLS and more general penalized losses have been studied by Fan and Peng (2004), Huang, Horowitz and Ma (2008), Kim, Choi and Oh (2008), Zou and

Zhang (2009), Xie and Huang (2009), Bradic, Fan and Wang (2011), Fan and Lv (2011), Zhang and Zhang (2012), and many more. We present a general result, which can be viewed as a variation of Theorem 6 (ii) of Zhang and Zhang (2012), based on the Brouwer fixed point theorem.

We first need to define two quantities depending on $\{\mathbf{X}, \boldsymbol{\beta}, p_\lambda\}$:

$$\theta_{\text{select}}(p_\lambda, \boldsymbol{\beta}) = \inf\left\{\theta : \frac{\|\mathbf{v}_\mathcal{S} - \boldsymbol{\beta}_\mathcal{S}\|_\infty}{\theta\lambda + \lambda'} \le 1 \Rightarrow \|\overline{\boldsymbol{\Sigma}}_{\mathcal{S},\mathcal{S}}^{-1} p'_\lambda(\mathbf{v}_\mathcal{S})\|_\infty \le \theta\lambda\right\}$$

with $\mathcal{S} = \text{supp}(\boldsymbol{\beta})$ and an additional threshold level $\lambda'$, and

$$\kappa_{\text{select}}(p_\lambda, \boldsymbol{\beta}) = \sup\left\{\left\|\overline{\boldsymbol{\Sigma}}_{\mathcal{S}^c,\mathcal{S}}\overline{\boldsymbol{\Sigma}}_{\mathcal{S},\mathcal{S}}^{-1}\Big(\frac{p'_\lambda(\mathbf{v}_\mathcal{S})}{\lambda}\Big)\right\|_\infty : \frac{\|\mathbf{v}_\mathcal{S} - \boldsymbol{\beta}_\mathcal{S}\|_\infty}{\theta_{\text{select}}(p_\lambda, \boldsymbol{\beta})\lambda + \lambda'} \le 1\right\}.$$

Consider conditions

$$\kappa_{\text{select}}(p_\lambda, \boldsymbol{\beta}) < 1 \tag{4.128}$$

as an extension of the strong irrepresentable condition (4.32), and

$$\min_{j\in\mathcal{S}} |\beta_j| \ge \theta_{\text{select}}(p_\lambda, \boldsymbol{\beta})\lambda + \lambda' \tag{4.129}$$

as a variation of the signal strength condition (4.33).

**Theorem 4.19** *Let $\mathcal{S} = \text{supp}(\boldsymbol{\beta})$. Suppose $p'_\lambda(\cdot)$ is continuous in the region*

$$\cup_{j\in\mathcal{S}}[\beta_j - \theta_{\text{select}}(p_\lambda, \boldsymbol{\beta})\lambda - \lambda', \beta_j + \theta_{\text{select}}(p_\lambda, \boldsymbol{\beta})\lambda + \lambda'].$$

*Suppose* $\text{rank}(\overline{\boldsymbol{\Sigma}}_{\mathcal{S},\mathcal{S}}) = |\mathcal{S}|$. *Let $\widehat{\boldsymbol{\beta}}^o$ be the oracle LSE in (4.10). Suppose (4.128) and (4.129) hold, and*

$$\|\widehat{\boldsymbol{\beta}}^o - \boldsymbol{\beta}\|_\infty \le \lambda', \quad \frac{\|\mathbf{X}^T(\mathbf{Y} - \mathbf{X}\widehat{\boldsymbol{\beta}}^o)/n\|_\infty}{(1 - \kappa_{\text{select}}(p_\lambda, \boldsymbol{\beta}))_+} \le \lambda. \tag{4.130}$$

*Then, there exists a local solution $\widehat{\boldsymbol{\beta}}$ in the sense of (4.109) such that*

$$\text{sgn}(\widehat{\boldsymbol{\beta}}) = \text{sgn}(\boldsymbol{\beta}), \quad \|\widehat{\boldsymbol{\beta}} - \boldsymbol{\beta}\|_\infty \le \theta_{\text{select}}(p_\lambda, \boldsymbol{\beta})\lambda + \lambda'. \tag{4.131}$$

*Moreover, if in addition $\theta_{\text{select}}(p_\lambda, \boldsymbol{\beta}) = 0$, then $\kappa_{\text{select}}(p_\lambda, \boldsymbol{\beta}) = 0$ and*

$$\widehat{\boldsymbol{\beta}}^o \text{ is a local minimizer of the penalized loss in (4.108).} \tag{4.132}$$

When $p_\lambda(t) = \lambda|t|$ is the $\ell_1$ penalty and $\min_{j\in\mathcal{S}} |\beta_j| \ge \theta_{\text{select}}(p_\lambda, \boldsymbol{\beta})\lambda + \lambda'$, (4.128) is identical to the irrepresentable condition (4.32) for the selection consistency of the Lasso. Thus, Theorem 4.19 can be viewed as an extension of Theorem 4.8.

Compared with the sufficient and nearly necessary conditions for the selection consistency of the Lasso in Theorem 4.8, Theorem 4.19 demonstrates the benefits of using a penalty function with potentially small $p'_\lambda(\boldsymbol{\beta}_{\mathcal{S}})$, especially those with finite bias threshold such as SCAD and MCP. Let $a_\lambda$ be the bias threshold in (4.122). Under condition (4.130) with $\kappa_{\text{select}}(p_\lambda, \boldsymbol{\beta}) = 0$,

$$\min_{j\in\mathcal{S}} |\beta_j| \ge a_\lambda + \lambda' \;\Rightarrow\; \theta_{\text{select}}(p_\lambda, \boldsymbol{\beta}) = 0 \;\Rightarrow\; (4.132). \tag{4.133}$$

This provides the simplest signal strength condition for the oracle property.

Of course, a significant drawback of Theorem 4.19 is the need to identify the oracle solution with a computationally feasible one. This issue will be addressed in Subsection 4.4.3 and Section 4.5.

**Proof of Theorem 4.19.** Let $B = \{\mathbf{h}_{\mathcal{S}} : \|\mathbf{h}_{\mathcal{S}}\|_\infty \le \theta_{\text{select}}(p_\lambda, \boldsymbol{\beta})\lambda\}$. Consider a given vector $\mathbf{u}_{\mathcal{S}}$ with $\|\mathbf{u}_{\mathcal{S}}\|_\infty \le \lambda'$. For any $\mathbf{h}_{\mathcal{S}} \in B$, the vector $\mathbf{v}_{\mathcal{S}} = \boldsymbol{\beta}_{\mathcal{S}} + \mathbf{u}_{\mathcal{S}} + \mathbf{h}_{\mathcal{S}}$ satisfies $\|\mathbf{v}_{\mathcal{S}} - \boldsymbol{\beta}_{\mathcal{S}}\|_\infty \le \theta_{\text{select}}(p_\lambda, \boldsymbol{\beta})\lambda + \lambda'$, so that $\|\overline{\boldsymbol{\Sigma}}_{\mathcal{S},\mathcal{S}}^{-1} p'_\lambda(\boldsymbol{\beta}_{\mathcal{S}} + \mathbf{u}_{\mathcal{S}} + \mathbf{h}_{\mathcal{S}})\|_\infty \le \theta_{\text{select}}(p_\lambda, \boldsymbol{\beta})\lambda$. Thus,

$$\mathbf{h}_{\mathcal{S}} \to -\overline{\boldsymbol{\Sigma}}_{\mathcal{S},\mathcal{S}}^{-1} p'_\lambda(\boldsymbol{\beta}_{\mathcal{S}} + \mathbf{u}_{\mathcal{S}} + \mathbf{h}_{\mathcal{S}})$$

is a continuous map from $B$ into $B$. It follows from the Brouwer fixed point theorem that the mapping must have an equilibrium $\mathbf{h}_{\mathcal{S}} \in B$ for the given $\mathbf{u}_{\mathcal{S}}$.

When $\|\widehat{\boldsymbol{\beta}}^o - \boldsymbol{\beta}\|_\infty \le \lambda'$, we may choose $\mathbf{u}_{\mathcal{S}} = \widehat{\boldsymbol{\beta}}^o_{\mathcal{S}} - \boldsymbol{\beta}_{\mathcal{S}}$ and define a vector $\widehat{\boldsymbol{\beta}}_{\mathcal{S}} = \widehat{\boldsymbol{\beta}}^o_{\mathcal{S}} + \mathbf{h}_{\mathcal{S}} = \boldsymbol{\beta}_{\mathcal{S}} + \mathbf{u}_{\mathcal{S}} + \mathbf{h}_{\mathcal{S}}$ satisfying $\|\mathbf{u}_{\mathcal{S}} + \mathbf{h}_{\mathcal{S}}\|_\infty \le \theta_{\text{select}}(p_\lambda, \boldsymbol{\beta})\lambda + \lambda'$ and

$$\widehat{\boldsymbol{\beta}}_{\mathcal{S}} - \widehat{\boldsymbol{\beta}}^o_{\mathcal{S}} = \mathbf{h}_{\mathcal{S}} = -\overline{\boldsymbol{\Sigma}}_{\mathcal{S},\mathcal{S}}^{-1} p'_\lambda(\boldsymbol{\beta}_{\mathcal{S}} + \mathbf{u}_{\mathcal{S}} + \mathbf{h}_{\mathcal{S}}) = -\overline{\boldsymbol{\Sigma}}_{\mathcal{S},\mathcal{S}}^{-1} p'_\lambda(\widehat{\boldsymbol{\beta}}_{\mathcal{S}}).$$

As $\overline{\boldsymbol{\Sigma}} = \mathbf{X}^T\mathbf{X}/n$ and $\mathbf{X}_{\mathcal{S}}^T\mathbf{Y}/n = \overline{\boldsymbol{\Sigma}}_{\mathcal{S},\mathcal{S}}\widehat{\boldsymbol{\beta}}^o_{\mathcal{S}}$, we have

$$\mathbf{X}_{\mathcal{S}}^T(\mathbf{Y} - \mathbf{X}_{\mathcal{S}}\widehat{\boldsymbol{\beta}}_{\mathcal{S}})/n = p'_\lambda(\widehat{\boldsymbol{\beta}}_{\mathcal{S}}).$$

Let $\widehat{\boldsymbol{\beta}}_{\mathcal{S}^c} = \mathbf{0}$ so that the entire $\widehat{\boldsymbol{\beta}}$ is defined. Because

$$\begin{aligned} \mathbf{X}_{\mathcal{S}^c}^T(\mathbf{Y} - \mathbf{X}_{\mathcal{S}}\widehat{\boldsymbol{\beta}}_{\mathcal{S}})/n &= \mathbf{X}_{\mathcal{S}^c}^T(\mathbf{Y} - \mathbf{X}\widehat{\boldsymbol{\beta}}^o)/n - \overline{\boldsymbol{\Sigma}}_{\mathcal{S}^c,\mathcal{S}}(\widehat{\boldsymbol{\beta}}_{\mathcal{S}} - \widehat{\boldsymbol{\beta}}^o_{\mathcal{S}}) \\ &= \mathbf{X}_{\mathcal{S}^c}^T(\mathbf{Y} - \mathbf{X}\widehat{\boldsymbol{\beta}}^o)/n + \overline{\boldsymbol{\Sigma}}_{\mathcal{S}^c,\mathcal{S}}\overline{\boldsymbol{\Sigma}}_{\mathcal{S},\mathcal{S}}^{-1} p'_\lambda(\widehat{\boldsymbol{\beta}}_{\mathcal{S}}) \end{aligned}$$

with $\|\widehat{\boldsymbol{\beta}}_{\mathcal{S}} - \boldsymbol{\beta}_{\mathcal{S}}\|_\infty \le \theta_{\text{select}}(p_\lambda, \boldsymbol{\beta})\lambda + \lambda'$, we have

$$\big\|\mathbf{X}_{\mathcal{S}^c}^T(\mathbf{Y} - \mathbf{X}_{\mathcal{S}}\widehat{\boldsymbol{\beta}}_{\mathcal{S}})/n\big\|_\infty \le \big\|\mathbf{X}_{\mathcal{S}^c}^T(\mathbf{Y} - \mathbf{X}\widehat{\boldsymbol{\beta}}^o)/n\big\|_\infty + \kappa_{\text{select}}(p_\lambda, \boldsymbol{\beta})\lambda \le \lambda.$$

Thus, by Theorem 4.18 (i), $\widehat{\boldsymbol{\beta}}$ is a local solution satisfying (4.131).

If in addition $\theta_{\text{select}}(p_\lambda, \boldsymbol{\beta}) = 0$, then $\kappa_{\text{select}}(p_\lambda, \boldsymbol{\beta}) = 0$ by definition and we may set $\widehat{\boldsymbol{\beta}} = \widehat{\boldsymbol{\beta}}^o$ with $\mathbf{v}_{\mathcal{S}} = \mathbf{0}$. ■

### 4.4.3 Properties of local solutions

In this subsection, we study properties of local solutions (4.125) for minimization of penalized loss (4.108) with separable penalties (4.115) and approximate local solutions $\widehat{\boldsymbol{\beta}}$ of the form

$$\begin{cases} \big(\lambda - \kappa_0|\widehat{\beta}_j|\big)_+ \le \mathrm{sgn}(\widehat{\beta}_j)\mathbf{X}_j^T(\mathbf{Y} - \mathbf{X}\widehat{\boldsymbol{\beta}})/n \le \lambda(1+\nu), & \widehat{\beta}_j \ne 0, \\ \big|\mathbf{X}_j^T(\mathbf{Y} - \mathbf{X}\widehat{\boldsymbol{\beta}})/n\big| \le (1+\nu)\lambda, & \widehat{\beta}_j = 0, \end{cases} \tag{4.134}$$

with $\lambda \ge \lambda_0$ for a certain minimum penalty level $\lambda_0 > 0$, where $\nu \ge 0$ measures approximation or computational errors.

Condition (4.134) certainly holds for exact local solutions (4.125) when $\|p'_\lambda(t)\|_\infty = p'_\lambda(0+) = \lambda$ and the maximum concavity $\kappa(p_\lambda)$ of the penalty is no greater than $\kappa_0$. For $\nu = 0$, the upper penalty level for $\widehat{\beta}_j \ne 0$ is attained by the Lasso, and the lower penalty level by the MCP. We note that the collection of solutions satisfying (4.134) is identical to the collection of solutions satisfying

$$\mathbf{X}_j^T(\mathbf{Y} - \mathbf{X}\widehat{\boldsymbol{\beta}})/n = p'_{\lambda_j}(\widehat{\beta}_j; j) \quad \forall j \tag{4.135}$$

with different folded concave penalties for different $j$, where $\lambda \le \lambda_j \le (1+\nu)\lambda$ and $p_\lambda(.;j)$ are penalties with penalty level $\lambda = p'_\lambda(0+;j)$ and maximum concavity bounded by $\kappa_0$. For example, when

$$\mathrm{sgn}(\widehat{\beta}_j)\mathbf{X}_j^T(\mathbf{Y} - \mathbf{X}\widehat{\boldsymbol{\beta}})/n = \big(\lambda_j - \kappa_j|\widehat{\beta}_j|\big)_+ \ \forall\, \widehat{\beta}_j \ne 0, \quad 0 \le \kappa_j \le \kappa_0,$$

we may take $p_{\lambda_j}(b_j; j)$ as the MCP penalty with concavity $\kappa_j$. Analogous to (4.125), (4.135) can be viewed as a local solution for minimizing the penalized loss (4.108) with $\mathrm{Pen}(\mathbf{b}) = \sum_{j=1}^p p_{\lambda_j}(b_j; j)$.

More precisely, we study in this subsection prediction, coefficient estimation and variable selection properties of solutions of (4.134) computable through iterative algorithms. We assume that the iteration begins with the origin and in each iteration of the algorithm, an approximate local solution of the form (4.134) is produced such that

$$\big\|\widehat{\boldsymbol{\beta}}^{(k)} - \widehat{\boldsymbol{\beta}}^{(k-1)}\big\|_1 \le a_0\lambda^{(k)}, \quad k = 0, 1, \ldots, k^*, \quad \widehat{\boldsymbol{\beta}}^{(k^*)} = \widehat{\boldsymbol{\beta}}, \tag{4.136}$$

for a sufficiently small $a_0$, where $\lambda^{(k)}$ is the penalty level for $\widehat{\boldsymbol{\beta}}^{(k)}$ and $a_0 > 0$ is as in Proposition 4.3 below. Condition (4.136) certainly holds if the solution of (4.134) under consideration is connected to the origin through a continuous path of such approximate local solutions with potentially different penalty level $\lambda$, including the LARS path for the Lasso and the PLUS path for quadratic spline concave penalties such as the SCAD and MCP. Moreover, the iterative algorithm may begin from any solution, such as the Lasso, known to be connected to the origin in such a way.

For $\mathcal{S} \subseteq \{1, \ldots, p\}$ and $\eta \in [0, 1)$, define restricted eigenvalues (*RE*)

$$\mathrm{RE}_k = \mathrm{RE}_k(\mathcal{S}; \eta) = \inf_{\mathbf{0} \ne \mathbf{u} \in \mathscr{C}(\mathcal{S}; \eta, \|\cdot\|_1)} \left\{ \frac{\|\mathbf{X}\mathbf{u}\|_2^2 |\mathcal{S}|^{1-k/2}}{n\|\mathbf{u}\|_2^k \|\mathbf{u}_{\mathcal{S}}\|_1^{2-k}} \right\}^{1/2}, \tag{4.137}$$

as in (4.90), where $\mathscr{C}(\mathcal{S};\eta,\|\cdot\|_1)=\{\mathbf{u}:(1-\eta)\|\mathbf{u}_{\mathcal{S}^c}\|_1\le(1+\eta)\|\mathbf{u}_{\mathcal{S}}\|_1\}$. By (4.92), $\mathrm{RE}_2$ guarantees the $\ell_2$ estimation error bound for the Lasso as in Bickel, Ritov and Tsybakov (2009). More importantly, Theorems 4.16 and 4.17 can be used to bound $\mathrm{RE}_2$ from below by the lower sparse eigenvalue.

We study the performance of $\widehat{\boldsymbol{\beta}}$ by comparing it with an oracle solution $\widehat{\boldsymbol{\beta}}^o$ under a restricted eigenvalue condition. Let $\mathcal{S}=\mathrm{supp}(\boldsymbol{\beta})$. We shall consider oracle solutions $\widehat{\boldsymbol{\beta}}^o$ satisfying conditions $\mathrm{supp}(\widehat{\boldsymbol{\beta}}^o)\subseteq\mathcal{S}$ and

$$\widehat{\beta}_j^o\mathbf{X}_j^T(\mathbf{Y}-\mathbf{X}\widehat{\boldsymbol{\beta}}^o)/n\ge 0,\ \forall\, j\in\mathcal{S}, \tag{4.138}$$

$$\big\|\mathbf{X}^T(\mathbf{Y}-\mathbf{X}\widehat{\boldsymbol{\beta}}^o)/n\big\|_\infty\le(\eta_0-\epsilon_0)\lambda_0, \tag{4.139}$$

with $0<\epsilon_0<\eta_0<1$ satisfying $(1+\nu)(1-\eta)=(1-\eta_0)(1+\eta)$ for some $\eta\in(0,1)$. For the oracle LSE in (4.10), (4.138) holds automatically and (4.139) can be verified by Lemma 5.2 (ii) under the sub-Gaussian condition (4.20). Conditions (4.138) and (4.139) also hold for the oracle solution in (4.131) with a penalty level $\lambda\le(\eta_0-\epsilon_0)\lambda_0$, if the oracle solution exists. The analytical approach in this subsection aims to study approximate local solutions (4.134) for which the estimation error $\widehat{\boldsymbol{\beta}}-\widehat{\boldsymbol{\beta}}^o$ belongs to the cone $\mathscr{C}(\mathcal{S};\eta,\|\cdot\|_1)$, so that the restricted eigenvalues in (4.137) can be used. The role of the chain (4.136) of approximate solutions is to guarantee this cone membership for the solutions under consideration, as the following proposition asserts.

**Proposition 4.3** *Suppose (4.138) and (4.139) hold with* $\mathrm{supp}(\widehat{\beta}^o)\subseteq\mathcal{S}$. *Let* $\widehat{\boldsymbol{\beta}}$ *and* $\widetilde{\boldsymbol{\beta}}$ *be approximate solutions satisfying (4.134), respectively with penalty levels* $\lambda$ *and* $\widetilde{\lambda}$ *no smaller than* $\lambda_0$. *Let* $a_0=\epsilon_0^2/\{\kappa_0(2-\eta_0+\nu+2\epsilon_0)\}$. *Suppose* $\kappa_0\le\mathrm{RE}_2^2$ *for the RE in (4.137). If* $\widehat{\boldsymbol{\beta}}-\widehat{\boldsymbol{\beta}}^o\in\mathscr{C}(\mathcal{S};\eta,\|\cdot\|_1)$ *and* $\|\widetilde{\boldsymbol{\beta}}-\widehat{\boldsymbol{\beta}}\|_1\le a_0\widetilde{\lambda}$, *then* $\widetilde{\boldsymbol{\beta}}-\widehat{\boldsymbol{\beta}}^o\in\mathscr{C}(\mathcal{S};\eta,\|\cdot\|_1)$.

As the chain of approximate solutions (4.136) begins with the origin and the origin belongs to the cone $\mathscr{C}(\mathcal{S};\eta,\|\cdot\|_1)$, Proposition 4.3 asserts by induction that all members of the chain (4.136) belong to the cone.

**Proof**. Let $\mathbf{h}=\widehat{\boldsymbol{\beta}}-\widehat{\boldsymbol{\beta}}^o$, $\widetilde{\mathbf{h}}=\widetilde{\boldsymbol{\beta}}-\widehat{\boldsymbol{\beta}}^o$ and

$$\mathbf{g}^o=\mathbf{X}^T(\mathbf{Y}-\mathbf{X}\widehat{\boldsymbol{\beta}}^o)/n,\quad \widetilde{\mathbf{g}}=\mathbf{X}^T(\mathbf{Y}-\mathbf{X}\widetilde{\boldsymbol{\beta}})/n.$$

For the $\eta_0$ in (4.139), $\widetilde{\mathbf{h}}\in\mathscr{C}(\mathcal{S};\eta,\|\cdot\|_1)$ iff $(1-\eta_0)\|\widetilde{\mathbf{h}}_{\mathcal{S}^c}\|_1\le(1+\nu)\|\widetilde{\mathbf{h}}_{\mathcal{S}}\|_1$. By condition (4.134) on $\{\widetilde{\boldsymbol{\beta}},\widetilde{\lambda}\}$ and (4.138) on $\widehat{\boldsymbol{\beta}}^o$, $(\widetilde{\beta}_j\widetilde{g}_j)\wedge(\widehat{\beta}_j^o g_j^o)\ge 0$, so that

$$\widetilde{h}_j\big(\overline{\boldsymbol{\Sigma}}\widetilde{\mathbf{h}}\big)_j=(\widetilde{\beta}_j-\widehat{\beta}_j^o)(g_j^o-\widetilde{g}_j)\le|\widetilde{h}_j|\big(|g_j^o|\vee|\widetilde{g}_j|\big)\le(1+\nu)\widetilde{\lambda}|\widetilde{h}_j|$$

for $j\in\mathcal{S}$. Moreover, for $j\in\mathcal{S}^c$, we have

$$\begin{aligned}|\widetilde{h}_j|(1-\kappa_0|\widetilde{h}_j|) &\le \widetilde{h}_j\mathbf{X}_j^T(\mathbf{Y}-\mathbf{X}\widehat{\boldsymbol{\beta}}^o)/n-\widetilde{h}_j\big(\overline{\boldsymbol{\Sigma}}\widetilde{\mathbf{h}}\big)_j\\ &\le |\widetilde{h}_j|(\eta_0-\epsilon_0)\widetilde{\lambda}-\widetilde{h}_j\big(\overline{\boldsymbol{\Sigma}}\widetilde{\mathbf{h}}\big)_j.\end{aligned}$$

Summing up the above inequalities over $j$, we have

$$\widetilde{\mathbf{h}}^T\overline{\mathbf{\Sigma}}\widetilde{\mathbf{h}} + (1-\eta_0+\epsilon_0)\widetilde{\lambda}\big\|\widetilde{\mathbf{h}}_{\mathcal{S}^c}\big\|_1 \le (1+\nu)\widetilde{\lambda}\|\widetilde{\mathbf{h}}_{\mathcal{S}}\|_1 + \kappa_0\|\widetilde{\mathbf{h}}_{\mathcal{S}^c}\big\|_2^2. \quad (4.140)$$

When $\|\mathbf{h}-\widetilde{\mathbf{h}}\|_1 \le a_0\widetilde{\lambda}$ and $\|\mathbf{h}\|_1 \le a_1\widetilde{\lambda}$ with $a_1 = \epsilon_0/\kappa_0 - a_0$, we have $\kappa_0\|\widetilde{\mathbf{h}}_{\mathcal{S}^c}\|_2 \le \kappa_0\|\widetilde{\mathbf{h}}\|_1 \le \kappa_0(a_1+a_1)\widetilde{\lambda} \le \epsilon_0\widetilde{\lambda}$, so that $(1-\eta_0)\big\|\widetilde{\mathbf{h}}_{\mathcal{S}}\big\|_1 \le (1+\nu)\|\widetilde{\mathbf{h}}_{\mathcal{S}}\|_1$. On the other hand, as $(1-\eta_0)\|\mathbf{h}_{\mathcal{S}^c}\|_1 \le (1+\nu)\|\mathbf{h}_{\mathcal{S}}\|_1$ and $\mathrm{RE}_2^2 \ge \kappa_0$, $\mathbf{h}^T\overline{\mathbf{\Sigma}}\mathbf{h} \ge \kappa_0\|\mathbf{h}\|_2^2$, so that the $\mathbf{h}$ version of (4.140) gives

$$(1-\eta_0+\epsilon_0)\big\|\mathbf{h}_{\mathcal{S}^c}\big\|_1 \le (1+\nu)\|\mathbf{h}_{\mathcal{S}}\|_1.$$

When $\|\mathbf{h}-\widetilde{\mathbf{h}}\|_1 \le a_0\widetilde{\lambda}$ and $\|\mathbf{h}\|_1 > a_1\widetilde{\lambda}$, we have by algebra

$$\frac{\|\widetilde{\mathbf{h}}-\mathbf{h}\|_1}{\|\mathbf{h}\|_1} \le \frac{a_0}{a_1} = \frac{a_3}{1+\nu}, \quad a_3 = \frac{a_0(1+\nu)}{\epsilon_0/\kappa_0 - a_0} = \frac{\epsilon_0(1+\nu)}{2-\eta_0+\nu+\epsilon_0}.$$

Thus, by triangular inequality and some more algebra,

$$\begin{aligned}
& (1-\eta_0)\|\widetilde{\mathbf{h}}_{\mathcal{S}^c}\|_1 - (1+\nu)\|\widetilde{\mathbf{h}}_{\mathcal{S}}\|_1 \\
\le\ & (1-\eta_0)\|\mathbf{h}_{\mathcal{S}^c}\|_1 - (1+\nu)\|\mathbf{h}_{\mathcal{S}}\|_1 + (1+\nu)\|\widetilde{\mathbf{h}}-\mathbf{h}\|_1 \\
\le\ & (1-\eta_0+a_3)\|\mathbf{h}_{\mathcal{S}^c}\|_1 - (1+\nu-a_3)\|\mathbf{h}_{\mathcal{S}}\|_1 \\
\propto\ & (1-\eta_0+\epsilon_0)\|\mathbf{h}_{\mathcal{S}^c}\|_1 - (1+\nu)\|\mathbf{h}_{\mathcal{S}}\|_1 \\
\le\ & 0.
\end{aligned}$$

Thus, $(1-\eta_0)\|\widetilde{\mathbf{h}}_{\mathcal{S}^c}\|_1 \le (1+\nu)\|\widetilde{\mathbf{h}}_{\mathcal{S}}\|_1$ in both cases. □

We are now ready to study properties of the following class of solutions,

$$\mathscr{B}_0 = \Big\{\widehat{\boldsymbol{\beta}}\text{: (4.134) and (4.136) hold with } \lambda \ge \lambda_0\Big\}. \quad (4.141)$$

As we have discussed earlier, a solution $\widehat{\boldsymbol{\beta}}$ in $\mathscr{B}_0$ must also satisfy (4.135) for some penalty functions $p_{\lambda_j}(\cdot;j)$ as specified. Given an oracle $\widehat{\boldsymbol{\beta}}^o$ satisfying (4.138) and (4.139), define for every $\widehat{\boldsymbol{\beta}} \in \mathscr{B}_0$ a vector

$$\mathbf{w}^o = \frac{\mathbf{X}^T(\mathbf{Y}-\mathbf{X}\widehat{\boldsymbol{\beta}}^o)/n - \big(p'_{\lambda_j}(\widehat{\beta}_j^o;j), j=1,\ldots,p\big)^T}{(1+\nu)\lambda} \quad (4.142)$$

with the penalty functions $p_{\lambda_j}(\cdot;j)$ fulfilling (4.135). When $p'_{\lambda_j}(\cdot;j) = p_\lambda(\cdot)$ for all $j$ and $\widehat{\boldsymbol{\beta}}^o$ is the oracle LSE in (4.10),

$$(1+\nu)\lambda\|\mathbf{w}_{\mathcal{S}}^o\|_2 = \|p'_\lambda(\widehat{\boldsymbol{\beta}}^o)\|_2,$$

which is used in (4.121) to measures the magnitude of the bias.

**Theorem 4.20** *Suppose (4.138) and (4.139) hold for an oracle solution* $\widehat{\beta}^o$ *and* $\{\nu, \epsilon_0, \eta_0, \eta\}$ *with* $\epsilon_0 \geq \nu$. *Let* $\mathbf{w}^o$ *be as in (4.142) and* $\mathrm{RE}_k$ *as in (4.137). Suppose* $\mathrm{RE}_2^2 \geq \kappa_0$ *for the* $\ell_2$ *restricted eigenvalue. Then, for* $\widehat{\beta} \in \mathscr{B}_0$

$$\frac{\|\widehat{\beta} - \widehat{\beta}^o\|}{(1+\eta_0)\lambda} \leq \sup\left\{\frac{\|\mathbf{u}\|}{\|\overline{\Sigma}\mathbf{u}\|_\infty} : \frac{\|\mathbf{u}_{\mathcal{S}^c}\|_1}{1+\eta} \leq \frac{(\mathbf{u}_{\mathcal{S}}^T \mathbf{w}_{\mathcal{S}}^o) \wedge \|\mathbf{u}_{\mathcal{S}}\|_1}{1-\eta}\right\} \tag{4.143}$$

*for any seminorm* $\|\cdot\|$ *as loss function. In particular, (4.143) implies*

$$\begin{aligned} &\max\left\{\frac{\|\widehat{\beta} - \widehat{\beta}^o\|_2}{\mathrm{RE}_1^{-2}}, \frac{\mathrm{RE}_0^2 \|\widehat{\beta} - \widehat{\beta}^o\|_1}{2(1 \vee |\mathcal{S}|)^{1/2}/(1-\eta)}, \frac{\|\mathbf{X}\widehat{\beta} - \mathbf{X}\widehat{\beta}^o\|_2}{n^{1/2}/\mathrm{RE}_0}\right\} \\ \leq\ & 2\lambda |\mathcal{S}|^{1/2}(1+\eta_0)/(1-\eta). \end{aligned} \tag{4.144}$$

*Moreover, when* $\phi_{\min}(\overline{\Sigma}_{\mathcal{S},\mathcal{S}}) > \kappa_0$,

$$\mathbf{w}_{\mathcal{S}}^o = \mathbf{0} \Rightarrow \widehat{\beta} = \widehat{\beta}^o. \tag{4.145}$$

**Corollary 4.8** *Let* $\mathcal{S} = \mathrm{supp}(\beta)$. *Suppose an oracle solution* $\widehat{\beta}^o$ *of (4.125) exists with* $\mathrm{supp}(\widehat{\beta}^o) = \mathcal{S}$. *Suppose* $\mathrm{RE}_2^2 \geq \kappa_0$ *and* $\phi_{\min}(\overline{\Sigma}_{\mathcal{S},\mathcal{S}}) > \kappa_0$. *If* $\widehat{\beta} \in \mathscr{B}_0$ *is a solution of (4.125) for the same penalty, then* $\widehat{\beta} = \widehat{\beta}^o$. *In particular, the oracle LSE* $\widehat{\beta}^o$ *in (4.10) is a solution of (4.125) when* $\min_{j \in \mathcal{S}} |\widehat{\beta}_j^o| \geq a\lambda$, *where* $a = \inf\{c > 0 : p'_\lambda(t) = 0\ \forall t > c\}$.

Theorem 4.20 asserts that the prediction and estimation errors of the folded concave PLS are rate optimal in the sense of (4.17) under the $\ell_2$ restricted eigenvalue condition on the design, which is comparable with the condition imposed in Bickel, Ritov and Tsybakov (2009) for similar bounds for the Lasso. For the Lasso, the condition $\mathrm{RE}_2^2 \geq \kappa_0$ holds automatically due to $\kappa_0 = 0$, so that Theorem 4.20 produces a slightly weaker version of Theorem 4.11. When the bias of the PLS is small, $\|p'_\lambda(\widehat{\beta}_{\mathcal{S}}^o)\|_2/\lambda = (1+\nu)\|\mathbf{w}_{\mathcal{S}}^o\|_2 \ll |\mathcal{S}|^{1/2}$, the CIF based error bound in (4.143) is of sharper order than (4.144) as (4.145) attests. This signal strength condition $\|\mathbf{w}_{\mathcal{S}}^o\|_2 \ll |\mathcal{S}|^{1/2}$ holds for the concave PLS when the number of small signals is of smaller order than $|\mathcal{S}|$. This advantage of the concave PLS is stated explicitly in Theorem 4.21 below. Another significant advantage of the folded concave PLS, compared to the strong irrepresentable condition (4.32) for the Lasso, is the fact that no additional regularity condition on the design is required for selection consistency in (4.145), provided a uniform signal strength condition of the form (4.18) as in Corollary 4.8.

**Theorem 4.21** *Let* $\widehat{\beta} \in \mathscr{B}_0$ *be a solution of (4.125) for a certain penalty* $p_\lambda(\cdot)$ *with penalty level* $\lambda \geq \lambda_0$, *maximum concavity* $\kappa(p_\lambda) \leq \kappa_0$ *as in (4.119), and bias threshold* $a_\lambda$ *as in (4.122). Let* $\widehat{\beta}^o$ *be the oracle LSE in (4.10). Suppose*

*(4.139) holds with the specified* $\{\epsilon_0, \eta_0, \eta\}$. *Let* $\mathrm{RE}_k$ *as in (4.137). Suppose* $\mathrm{RE}_2^2(1 - 1/C_0) \ge \kappa_0$ *for some constant* $C_0$. *Then,*

$$\frac{\|\widehat{\boldsymbol\beta} - \widehat{\boldsymbol\beta}^o\|}{(1+\nu)\lambda} \le C_0 \sup_{\|\mathbf{u}\|=1} \frac{\mathbf{u}_{\mathcal{S}}^T\mathbf{w}_{\mathcal{S}}^o - \|\mathbf{u}_{\mathcal{S}^c}\|_1(1-\eta)/(1+\eta)}{\mathbf{u}^T\overline{\boldsymbol\Sigma}\mathbf{u}} \tag{4.146}$$

*for any seminorm* $\|\cdot\|$. *In particular, for* $s_1 = \{j \in \mathcal{S} : |\widehat\beta_j^o| \le a_\lambda\}$,

$$\max\left\{\|\widehat{\boldsymbol\beta} - \widehat{\boldsymbol\beta}^o\|_2, \frac{\|\mathbf{X}\widehat{\boldsymbol\beta} - \mathbf{X}\widehat{\boldsymbol\beta}^o\|_2}{n^{1/2}\mathrm{RE}_2}\right\} \le \frac{C_0\|p'_\lambda(\widehat{\boldsymbol\beta}_{\mathcal{S}}^o)\|_2}{\mathrm{RE}_2^2} \le \frac{C_0\lambda s_1^{1/2}}{\mathrm{RE}_2^2}. \tag{4.147}$$

**Proof of Theorems 4.20 and 4.21.** Let $\mathbf{h} = \widehat{\boldsymbol\beta} - \widehat{\boldsymbol\beta}^o$ and $\mathbf{w}^o$ be as in (4.142). By (4.135) and the concavity bound $\kappa_0$ for the penalties,

$$h_j\big(\overline{\boldsymbol\Sigma}\mathbf{h}\big)_j = h_j w_j^o(1+\nu)\lambda + h_j\{p'_{\lambda_j}(\widehat\beta_j) - p'_{\lambda_j}(\widehat\beta_j^o)\} \le h_j w_j^o(1+\nu)\lambda + \kappa_0 h_j^2$$

for all $j \in \mathcal{S}$. This and the $\{\boldsymbol\beta, \lambda\}$ version of the inequality above (4.140) yield

$$\mathbf{h}^T\overline{\boldsymbol\Sigma}\mathbf{h} + (1 - \eta_0 + \epsilon_0)\lambda\big\|\mathbf{h}_{\mathcal{S}^c}\big\|_1 \le (1+\nu)\lambda\mathbf{h}_{\mathcal{S}}^T\mathbf{w}_{\mathcal{S}}^o + \kappa_0\|\mathbf{h}\|_2^2. \tag{4.148}$$

By Proposition 4.3, $\mathbf{h} \in \mathscr{C}(\mathcal{S}; \eta, \|\cdot\|_1)$, so that $\mathbf{h}^T\overline{\boldsymbol\Sigma}\mathbf{h} \ge \mathrm{RE}_2^2\|\mathbf{h}\|_2^2$. As $\mathrm{RE}_2^2 \ge \kappa_0$, we have $(1 - \eta_0 + \epsilon_0)\lambda\big\|\mathbf{h}_{\mathcal{S}^c}\big\|_1 \le (1+\nu)\lambda\mathbf{h}_{\mathcal{S}}^T\mathbf{w}_{\mathcal{S}}^o$, which implies

$$(1-\eta)\big\|\mathbf{h}_{\mathcal{S}^c}\big\|_1 \le (1+\eta)\min\{\mathbf{h}_{\mathcal{S}}^T\mathbf{w}_{\mathcal{S}}^o, \|\mathbf{h}_{\mathcal{S}}\|_1\}. \tag{4.149}$$

On the other hand, by (4.134) and (4.139),

$$\|\overline{\boldsymbol\Sigma}\mathbf{h}\|_\infty \le (1 + \nu + \eta_0 - \epsilon_0)\lambda \le (1+\eta_0)\lambda,$$

so that $\|\mathbf{h}\| \le (1+\eta_0)\lambda\|\mathbf{h}\|/\|\overline{\boldsymbol\Sigma}\mathbf{h}\|_\infty$. This and (4.149) imply (4.143). Moreover, (4.144) follows from

$$\frac{2\|\overline{\boldsymbol\Sigma}\mathbf{u}\|_\infty}{(1-\eta)\|\mathbf{u}\|} \ge \frac{\mathbf{u}^T\overline{\boldsymbol\Sigma}\mathbf{u}}{\|\mathbf{u}_{\mathcal{S}}\|_1\|\mathbf{u}\|} \ge \begin{cases} \mathrm{RE}_1^2, & \|\mathbf{u}\| = \|\mathbf{u}\|_2, \\ \mathrm{RE}_0^2(1-\eta)/2, & \|\mathbf{u}\| = \|\mathbf{u}\|_1, \\ \mathrm{RE}_0, & \|\mathbf{u}\| = (\mathbf{u}^T\overline{\boldsymbol\Sigma}\mathbf{u})^{1/2} \end{cases}$$

for $\mathbf{u} \ne \mathbf{0}$ satisfying $(1-\eta)\big\|\mathbf{u}_{\mathcal{S}^c}\big\|_1 \le (1+\eta)\|\mathbf{u}_{\mathcal{S}}\|_1$, When $\mathbf{w}_{\mathcal{S}}^o = \mathbf{0}$, we have $\big\|\widehat{\boldsymbol\beta}_{\mathcal{S}^c}\big\|_1 = \big\|\mathbf{h}_{\mathcal{S}^c}\big\|_1 = 0$ and $\mathbf{h}^T\overline{\boldsymbol\Sigma}\mathbf{h} \le \kappa_0\|\mathbf{h}\|_2^2$ by (4.148) and (4.149), so that (4.145) follows.

When $(1 - 1/C_0)\mathrm{RE}_2^2 \ge \kappa_0$, (4.148) gives

$$C_0^{-1}\mathbf{h}^T\overline{\boldsymbol\Sigma}\mathbf{h} + (1 - \eta_0 + \epsilon_0)\lambda\big\|\mathbf{h}_{\mathcal{S}^c}\big\|_1 \le (1+\nu)\lambda\mathbf{h}_{\mathcal{S}}^T\mathbf{w}_{\mathcal{S}}^o.$$

Let $\mathbf{u} = \mathbf{h}/\|\mathbf{h}\|$. As $(1+\nu)(1-\eta) = (1-\eta_0)(1+\eta)$, we find that

$$\|\mathbf{h}\|C_0^{-1}\mathbf{u}^T\overline{\boldsymbol\Sigma}\mathbf{u} \le (1+\nu)\lambda\mathbf{u}_{\mathcal{S}}^T\mathbf{w}_{\mathcal{S}}^o - (1 - \eta_0 + \epsilon_0)\lambda\big\|\mathbf{u}_{\mathcal{S}^c}\big\|_1$$

$$\le \; (1+\nu)\lambda\|\mathbf{u}\|\big\{\mathbf{u}_{\mathcal{S}}^T\mathbf{w}_{\mathcal{S}}^o - \big\|\mathbf{u}_{\mathcal{S}^c}\big\|_1(1-\eta)/(1+\eta)\big\}.$$

This yields (4.146). Finally (4.147) follows from $\|p'_\lambda(\widehat{\boldsymbol{\beta}}^o_{\mathcal{S}})\|_2 = \lambda(1+\nu)\|\mathbf{w}^o_{\mathcal{S}}\|_2 \le \lambda s_1^{1/2}$ as $\mathbf{X}_{\mathcal{S}}^T(\mathbf{Y}-\mathbf{X}\widehat{\boldsymbol{\beta}}^o) = \mathbf{0}$ in (4.142) for the $\widehat{\boldsymbol{\beta}}^o$ in (4.10). ∎

### *4.4.4 Global and approximate global solutions*

As the global minimizer in (4.114) is numerically optimal in the PLS problem $Q(\boldsymbol{\beta})$ defined in (4.108), a natural question is whether it also has statistical optimality properties such as those in (4.17). Suppose that the answer to the above question is affirmative as the theorems in this subsection will demonstrate the next natural question is whether such optimality properties also hold for some approximately global solutions which are easier to find. We address the second question by considering solutions $\widehat{\boldsymbol{\beta}}$ satisfying

$$Q(\widehat{\boldsymbol{\beta}}) \le Q(\boldsymbol{\beta}^*) + \nu \tag{4.150}$$

for certain small $\nu \ge 0$ and target coefficient vector $\boldsymbol{\beta}^*$. Such a $\widehat{\boldsymbol{\beta}}$ has been called an $\{\nu, \boldsymbol{\beta}^*\}$ *approximate global solution* by Zhang and Zhang (2012).

A significant practical implication of this analysis of approximate global solution is that once (4.150) is attained in a gradient descent algorithm as discussed in Chapter 3, it remains valid throughout further iterations of the algorithm. Thus, if such an algorithm eventually reaches a local solution, the final output is a local solution as well as an approximate global solution.

For simplicity, we assume throughout this subsection that the design is column standardized with $\|\mathbf{X}_j\|_2^2 = n$ for all $j \le p$.

The main technical assumption in our analysis of the global and approximate global solutions is the following null consistency condition. Let $\eta \in (0,1]$. We say that the penalized loss (4.108) satisfies the $\eta$ *null-consistency* ($\eta$-NC) condition if the following equality holds:

$$\min_{\mathbf{b}\in \mathbb{R}^p}\Big(\|\boldsymbol{\varepsilon}/\eta - \mathbf{X}\mathbf{b}\|_2^2/(2n) + \|p_\lambda(\mathbf{b})\|_1\Big) = \|\boldsymbol{\varepsilon}/\eta\|_2^2/(2n). \tag{4.151}$$

The $\eta$ null-consistency holds if and only if $\widehat{\boldsymbol{\beta}} = \mathbf{0}$ is a global solution when $\mathbf{Y} = \boldsymbol{\varepsilon}/\eta$, or equivalently when $\boldsymbol{\beta} = \mathbf{0}$ is the truth but the noise is inflated by a factor $1/\eta$. For $\eta = 1$, the NC condition is necessary for selection consistency or for rate optimality in prediction and estimation with $s^* = \|\boldsymbol{\beta}\|_0$ in (4.17).

Zhang and Zhang (2012) provided sufficient conditions on the design and noise for the $\eta$-NC of the PLS. It was shown that the null-consistency condition holds for the $\ell_0$ and $\ell_1$ penalties with at least probability $1-\epsilon$ under the sub-Gaussian condition (4.21) on the noise when $\lambda = (\sigma/\eta)\sqrt{(2/n)\log(p/\epsilon)}$ as in (4.47). For more general penalty functions, their result requires an upper sparse eigenvalue condition when $\lambda \asymp \sigma\sqrt{(\log p)/n}$ for sub-Gaussian noise.

In addition to the column standardization, we assume that the following *restricted invertibility factor* (RIF) of the design matrix is positive:

$$\mathrm{RIF}_{\|\cdot\|}(\mathcal{S};\eta) = \inf\Big\{\frac{\|\overline{\mathbf{\Sigma}}\mathbf{u}\|_\infty}{\|\mathbf{u}\|} : \frac{\|p_\lambda(\mathbf{u}_{\mathcal{S}^c})\|_1}{1+\eta} < \frac{\|p_\lambda(\mathbf{u}_{\mathcal{S}})\|_1}{1-\eta}\Big\}, \tag{4.152}$$

where $\|\cdot\|$ is a norm representing the loss function of concern and $\overline{\mathbf{\Sigma}} = \mathbf{X}^T\mathbf{X}/n$ is the Gram matrix. Similar to (4.46) and (4.54), $\mathrm{RIF}_{\|\cdot\|}$ is not dimension scaled. For the rate optimality in prediction and $\ell_q$ estimation, we need

$$\frac{|\mathcal{S}|^{-1/2}}{\mathrm{RIF}_{\|\cdot\|_{\mathrm{pred}}}(\mathcal{S};\eta)} = O(1) \;\text{ and }\; \frac{|\mathcal{S}|^{-1/q}}{\mathrm{RIF}_{\|\cdot\|_q}(\mathcal{S};\eta)} = O(1) \tag{4.153}$$

respectively, where $\|\mathbf{u}\|_{\mathrm{pred}} = (\mathbf{u}^T\overline{\mathbf{\Sigma}}\mathbf{u})^{1/2}$. This is comparable with the regularity conditions for the $\ell_1$ regularization as the RIF is equivalent to the CIF in the following sense: If $t/p_\lambda(t)$ is increasing in $t \in (0,\infty)$, then

$$\mathrm{RIF}_{\|\cdot\|}(\mathcal{S};\eta) \ge \inf_{|A|=|\mathcal{S}|} \mathrm{CIF}_{\|\cdot\|}(A;\eta), \tag{4.154}$$

where $\mathrm{CIF}_{\|\cdot\|}(\mathcal{S};\eta)$ is the $\mathrm{RIF}_{\|\cdot\|}(\mathcal{S};\eta)$ when $p_\lambda(t) = \lambda|t|$. We note that this definition of the CIF is equivalent to the CIF in (4.93) as $|\mathcal{S}|^{1/q}\mathrm{CIF}_{\|\cdot\|_q}(\mathcal{S};\eta) = \mathrm{CIF}_{q,p}(\mathcal{S};\eta)$, and that in view of (4.95), the $\mathrm{CIF}_{q,p}(\mathcal{S};\eta)$ corresponds to the sharpest possible constant factor for the analysis of the Dantzig selector based on (4.70). In view of the discussion in Subsection 4.3.5, it is reasonable to expect that (4.153) holds for $1 \le q \le 2$. A proof of (4.154) will be included as Step 5 of the proof of Theorem 4.22 below.

We are now ready to state the first theorem of this subsection on the prediction and estimation errors of the global solution in (4.114) and the approximate global solution in (4.150). Consider $\widehat{\boldsymbol{\beta}}$ satisfying the following condition:

$$\widehat{\boldsymbol{\beta}} = \begin{cases} \text{a local solution satisfying (4.109) when } \lambda = p'_\lambda(0+), \\ \text{a strong local solution satisfying (4.112) when } \lambda = \lambda_*, \end{cases} \tag{4.155}$$

where $\lambda_*$ is the global threshold level defined in (4.123). Recall that we standardize the penalty with $\lambda = p'_\lambda(0+)$ when $p'_\lambda(0+) < \infty$ and $\lambda = \lambda_*$ when $p'_\lambda(0+) = \infty$. Recall that $a_* = \lambda_*/\lambda \le 1$ and $\gamma^* = \|p_\lambda(t)\|_\infty/\lambda^2$. Define

$$C_{\mathrm{pred}}^{(p_\lambda)}(\mathcal{S};\eta) = \frac{2(1+\eta)}{1-\eta}\min\Big\{\frac{a_*(1+\eta)}{(1\vee|\mathcal{S}|)\mathrm{RIF}_{\|\cdot\|_1}(\mathcal{S};\eta)} + \frac{a_*^2}{2}, \gamma^*\Big\}.$$

As in (4.23) define a relaxed upper sparse eigenvalue as

$$\phi^*(m;\mathcal{S},\mathbf{\Sigma}) = \max_{|B\setminus\mathcal{S}|\le m} \phi_{\max}\big(\mathbf{\Sigma}_{B,B}\big). \tag{4.156}$$

**Theorem 4.22** *Let $\mathcal{S} = \mathrm{supp}(\boldsymbol{\beta})$, $\boldsymbol{\varepsilon} = \mathbf{Y} - \mathbf{X}\boldsymbol{\beta}$, and $\eta \in (0,1)$.*
*(i) Suppose $\|p'_\lambda(\cdot)\|_\infty = p'_\lambda(0+)$ when $p'_\lambda(0+) < \infty$. Let $\widehat{\boldsymbol{\beta}}$ be a vector satisfying*

*both (4.155) and (4.150) with* $\boldsymbol{\beta}^* = \boldsymbol{\beta}$, *and* $\mathrm{RIF}_{\|\cdot\|}(\mathcal{S};\eta)$ *as in (4.152). Suppose the* $\eta$*-NC condition (4.151) holds. Then,*

$$2\|p_\lambda(\widehat{\boldsymbol{\beta}}_{\mathcal{S}^c})\|_1 + \|\mathbf{X}\widehat{\boldsymbol{\beta}} - \mathbf{X}\boldsymbol{\beta}\|_2^2/n < \lambda^2|\mathcal{S}|C_{\text{pred}}^{(p_\lambda)}(\mathcal{S};\eta) + \frac{2\nu}{1-\eta}, \tag{4.157}$$

*and for any seminorm* $\|\cdot\|$, $\|\mathbf{X}\widehat{\boldsymbol{\beta}} - \mathbf{X}\boldsymbol{\beta}\|_2^2/n \geq 2\nu/(1-\eta)$ *implies*

$$\|\widehat{\boldsymbol{\beta}} - \boldsymbol{\beta}\| \leq (1+\eta)\lambda/\mathrm{RIF}_{\|\cdot\|}(\mathcal{S};\eta). \tag{4.158}$$

*(ii) Suppose* $\|\mathbf{X}^T\boldsymbol{\varepsilon}/n\|_\infty \leq \eta\lambda$ *and* $p_\lambda(t) \leq \lambda|t|$. *For any given constants* $C_{\text{pred}}$ *and* $C_{\text{est},1}$ *set* $\nu = |\mathcal{S}|\lambda^2\{C_{\text{pred}}/2 + (1+\eta)C_{\text{est},1}\}$. *Then, the Lasso estimator* $\widehat{\boldsymbol{\beta}}^{(\ell_1)}$ *in (4.29) is a* $\{\nu,\boldsymbol{\beta}\}$ *approximate global solution with* $C_{\text{pred}} = C_{\text{pred}}^{(\ell_1)}(\mathcal{S};\eta,\eta)$ *and* $C_{\text{est},1} = C_{\text{est},1}^{(\ell_1)}(\mathcal{S};\eta,\eta)$ *in Theorem 4.11, and the Dantzig selector* $\widehat{\boldsymbol{\beta}}^{(DS)}$ *in (4.30) is a* $\{\nu,\boldsymbol{\beta}\}$ *approximate global solution with* $C_{\text{pred}} = C_{\text{pred}}^{(DS)}(\mathcal{S},|\mathcal{S}|;\eta,1)$ *and* $C_{\text{est},1} = C_{\text{est},1}^{(DS)}(\mathcal{S},|\mathcal{S}|;\eta,1)$ *in Theorem 4.14.*

*(iii) Let* $\widehat{\boldsymbol{\beta}}$ *be a local solution in the sense of (4.109). Suppose* $\|\mathbf{X}^T\boldsymbol{\varepsilon}/n\|_\infty \leq \eta\lambda$,

$$\frac{\|f_0(|\widehat{\boldsymbol{\beta}}_{\mathcal{S}^c}|)\|_1}{f_0(t_0\lambda)} \leq |\mathcal{S}|C_{\text{est},f_0}, \ \|\mathbf{X}\widehat{\boldsymbol{\beta}} - \mathbf{X}\boldsymbol{\beta}\|_2^2/n < |\mathcal{S}|\lambda^2 C_{\text{pred}}, \tag{4.159}$$

*for certain nonnegative constants* $\{t_0, C_{\text{est},f_0}\}$ *and nondecreasing function* $f_0$, $\inf_{0<s<t_0\lambda} p'_\lambda(s) = (1-\eta_0)\lambda$ *for a certain* $\eta_0 \geq 0$, *and*

$$|\mathcal{S}|C_{\text{pred}} \leq (1-\eta_0-\eta)_+^2 m_0/\phi^*(m_0;\mathcal{S},\overline{\boldsymbol{\Sigma}}) \tag{4.160}$$

*for some integer* $m_0 > 0$. *Then,*

$$|\widehat{\mathcal{S}} \setminus \mathcal{S}| < |\mathcal{S}|C_{\text{est},f_0} + \frac{\phi^*(m_0;\mathcal{S},\overline{\boldsymbol{\Sigma}})|\mathcal{S}|C_{\text{pred}}}{(1-\eta_0-\eta)_+^2}. \tag{4.161}$$

**Remark 4.5** When $p_\lambda(t)$ is a folded concave penalty, $p'_\lambda(0+) = \|p'_\lambda(\cdot)\|_\infty$ and $(1-\eta_0) = p'_\lambda(t_0\lambda)/\lambda$. It follows from Proposition 4.2 that $p_\lambda(t_0\lambda)/\lambda^2 \geq \min(t_0 a_*/2, a_*^2/2)$, so that (4.157) implies (4.159) with $f_0(t) = p_\lambda(t)$ when $\nu \lesssim \lambda^2|\mathcal{S}|$. Another convenient choice of $f_0(t)$ is $t_+^q$ when a proper $\ell_q$ error bound is available. For penalty functions with $p_\lambda(0+) > 0$ such as the $\ell_0$ penalty, we may take $m_0 = t_0 = 0$. For the Lasso, a more explicit version of (4.161) is provided in Theorem 4.12. We note that the NC condition is replaced by the simpler $\|\mathbf{X}^T\boldsymbol{\varepsilon}/n\|_\infty \leq \eta\lambda$ in Theorem 4.22 (ii) and (iii).

**Remark 4.6** In Theorem 4.22 (ii), the condition $p_\lambda(t) \leq \lambda|t|$, which holds for all folded concave penalties with $p'_\lambda(0+) < \infty$, including the SCAD and MCP, can be removed for the Lasso if Proposition 4.2 and Theorem 4.12 are invoked, with

$$\nu = |\mathcal{S}|\lambda^2\left[C_{\text{pred}}^{(\ell_1)} + 2\min\left\{C_{\text{est},1}^{(\ell_1)}a_* + (C_{\text{FP}}^{(\ell_1)}+1)a_*^2/2, (C_{\text{FP}}^{(\ell_1)}+1)\gamma^*\right\}\right].$$

Theorem 4.22, adapted from Theorems 1, 2 and 7 of Zhang and Zhang (2012) asserts that under suitable assumptions, the prediction error and the penalty on the error are bounded in a proper scale for any $\{\nu, \boldsymbol{\beta}\}$ approximate global solution when it is also a local solution satisfying (4.155), and that the Lasso is a $\{\nu, \boldsymbol{\beta}\}$ approximate global solution with $\nu \lesssim \lambda^2|\mathcal{S}|$. Moreover, all local solutions satisfying such prediction and penalty error bounds are sparse. These results have significant implications as we remark below.

**Remark 4.7** Since the Lasso and Dantzig selector are $\{\nu, \boldsymbol{\beta}\}$ approximate global solutions with $\nu \lesssim \lambda^2|\mathcal{S}|$, all the gradient descent algorithms initializing from such an $\ell_1$ regularized estimator, such as LQA, LLA, coordinate descent and iterative shrinkage-thresholding algorithms discussed in Chapter 3, produce $\{\nu, \boldsymbol{\beta}\}$ approximate global solutions with no greater $\nu$ in all iterations. If such an algorithm converges to a local solution satisfying (4.155), then Theorem 4.22 (i) can be applied to obtain prediction and estimation oracle inequalities and Theorem 4.22 (iii) can be applied to prove the sparsity property $\|\widehat{\boldsymbol{\beta}}\|_0 \lesssim |\mathcal{S}|$ for the local solution in the limit.

**Remark 4.8** Let $\kappa(p_\lambda)$ be the maximum concavity of the penalty as defined in (4.119). Suppose the relaxed lower sparse eigenvalue in (4.23) satisfies

$$\phi_*(m; \mathcal{S}, \overline{\boldsymbol{\Sigma}}) > \kappa(p_\lambda).$$

The penalized loss $Q(\boldsymbol{\beta})$ in (4.108) is convex in all models $\text{supp}(\mathbf{b}) = A$ with $|A \setminus \mathcal{S}| \leq m$. This is called the sparse convexity condition. If $m_1$ is the integer part of the right-hand side of (4.161) and $\widetilde{\boldsymbol{\beta}}$ is a local solution in the sense of (4.125) with $\#\{j \notin \mathcal{S} : \widetilde{\boldsymbol{\beta}}_j \neq 0\} + m_1 \leq m$, then the local solution must be identical to the global solution. For example, under the conditions of Theorem 4.22, an approximate global solution such as a local minimizer produced by gradient descent as discussed in Remark 4.7 is identical to the global solution, and it is also identical to the oracle solution if the conditions of Theorem 4.19 are satisfied as well.

**Remark 4.9** Consider local solutions in the class $\mathscr{B}_0$ in (4.141). It follows from Theorem 4.22 (iii) and the oracle inequalities in Theorem 4.20 that the estimated coefficient vector is sparse, $\|\widehat{\boldsymbol{\beta}}\|_0 \lesssim |\mathcal{S}|$, if $\widehat{\boldsymbol{\beta}}$ is also a solution of (4.125). Because the global solution is sparse by Theorem 4.22, $\widehat{\boldsymbol{\beta}}$ must be identical to the global solution for the same penalty under the sparse convexity condition.

**Proof of Theorem 4.22.** We divide the proof into six steps, deriving some basic inequalities in steps 1 and 2, proving the three parts of the theorem in steps 3, 4 and 5, and linking RIF and CIF in step 6.

*Step 1.* Prove that under condition (4.155),

$$\|\mathbf{X}^T(\mathbf{Y} - \mathbf{X}\widehat{\boldsymbol{\beta}})/n\|_\infty \leq \lambda. \tag{4.162}$$

This also implies $\|\mathbf{X}^T\boldsymbol{\varepsilon}/n\|_\infty \le \eta\lambda$ under the $\eta$-NC condition (4.151) as $\widehat{\boldsymbol{\beta}} = \mathbf{0}$ is a global solution when $\mathbf{Y} = \boldsymbol{\varepsilon}$.

When $\widehat{\boldsymbol{\beta}}$ is a local solution and $\lambda = \|p'_\lambda(\cdot)\|_\infty$, (4.162) follows from (4.125) in Theorem 4.20 (i). When $\widehat{\boldsymbol{\beta}}$ is a strong local solution and $\lambda = \lambda_*$,

$$\|\mathbf{Y} - \mathbf{X}\widehat{\boldsymbol{\beta}}\|_2^2/(2n) + p_\lambda(\widehat{\beta}_j) \le \|\mathbf{Y} - \mathbf{X}\widehat{\boldsymbol{\beta}} - \mathbf{X}_j t\|_2^2/(2n) + p_\lambda(\widehat{\beta}_j + t)$$

for all real $t$. Since $p_\lambda(t)$ is sub-additive in $t$,

$$t\mathbf{X}_j^T(\mathbf{Y} - \mathbf{X}\widehat{\boldsymbol{\beta}})/n \le t^2\|\mathbf{X}_j\|_2^2/(2n) + p_\lambda(\widehat{\beta}_j + t) - p_\lambda(\widehat{\beta}_j) \le t^2/2 + p_\lambda(t).$$

Since $t$ is arbitrary, (4.162) follows from the definition of $\lambda_*$ in (4.123).

*Step 2.* Prove that for all $\{\nu, \boldsymbol{\beta}\}$ approximate solutions $\widehat{\boldsymbol{\beta}}$ as defined in (4.150), the NC condition (4.151) implies

$$\begin{aligned} & (1-\eta)\|\mathbf{X}\boldsymbol{\Delta}\|_2^2/(2n) \\ \le \;& (\eta+1)\|p_\lambda(\boldsymbol{\Delta}_{\mathcal{S}})\|_1 + (\eta-1)\|p_\lambda(\boldsymbol{\Delta}_{\mathcal{S}^c})\|_1 + \nu, \end{aligned} \tag{4.163}$$

where $\mathcal{S} = \mathrm{supp}(\boldsymbol{\beta})$ and $\boldsymbol{\Delta} = \widehat{\boldsymbol{\beta}} - \boldsymbol{\beta}$.

As $\widehat{\boldsymbol{\beta}}$ is an $\{\nu, \boldsymbol{\beta}\}$ approximate solution, we have

$$\begin{aligned} -\nu &\le \|\mathbf{Y} - \mathbf{X}\boldsymbol{\beta}\|_2^2/(2n) + \|p_\lambda(\boldsymbol{\beta})\|_1 - \|\mathbf{Y} - \mathbf{X}\widehat{\boldsymbol{\beta}}\|_2^2/(2n) - \|p_\lambda(\widehat{\boldsymbol{\beta}})\|_1 \\ &= -\|\mathbf{X}\boldsymbol{\Delta}\|_2^2/(2n) + \boldsymbol{\varepsilon}^T\mathbf{X}\boldsymbol{\Delta}/n + \|p_\lambda(\boldsymbol{\beta})\|_1 - \|p_\lambda(\boldsymbol{\beta} + \boldsymbol{\Delta})\|_1. \end{aligned}$$

By (4.151), $\|\boldsymbol{\varepsilon}/\eta\|_2^2/(2n) \le \|\boldsymbol{\varepsilon}/\eta - \mathbf{X}\boldsymbol{\Delta}\|_2^2/(2n) + \|p_\lambda(\boldsymbol{\Delta})\|_1$, so that

$$\boldsymbol{\varepsilon}^T\mathbf{X}\boldsymbol{\Delta}/n \le \eta\|\mathbf{X}\boldsymbol{\Delta}\|_2^2/(2n) + \eta\|p_\lambda(\boldsymbol{\Delta})\|_1.$$

The above two displayed inequalities yield

$$(1-\eta)\|\mathbf{X}\boldsymbol{\Delta}\|_2^2/(2n) \le \eta\|p_\lambda(\boldsymbol{\Delta})\|_1 + \|p_\lambda(\boldsymbol{\beta})\|_1 - \|p_\lambda(\boldsymbol{\beta} + \boldsymbol{\Delta})\|_1 + \nu.$$

As $\boldsymbol{\beta}_{\mathcal{S}^c} = 0$ and $p_\lambda(t)$ is sub-additive, we have

$$\|p_\lambda(\boldsymbol{\beta})\|_1 - \|p_\lambda(\boldsymbol{\beta} + \boldsymbol{\Delta})\|_1 \le \|p_\lambda(\boldsymbol{\Delta}_{\mathcal{S}})\|_1 - \|p_\lambda(\boldsymbol{\Delta}_{\mathcal{S}^c})\|_1.$$

The above two displayed inequalities imply (4.163).

*Step 3.* Part (i). For (4.158), (4.163) and $\|\mathbf{X}\boldsymbol{\Delta}\|_2^2/n \ge 2\nu/(1-\eta)$ yield

$$(1-\eta)\|p_\lambda(\boldsymbol{\Delta}_{\mathcal{S}^c})\|_1 \le (1+\eta)\|p_\lambda(\boldsymbol{\Delta}_{\mathcal{S}})\|_1. \tag{4.164}$$

It follows from (4.162) that

$$\|\overline{\boldsymbol{\Sigma}}\boldsymbol{\Delta}\|_\infty \le \|\mathbf{X}^T(\mathbf{Y} - \mathbf{X}\widehat{\boldsymbol{\beta}})/n\|_\infty + \|\mathbf{X}^T\boldsymbol{\varepsilon}/n\|_\infty \le \lambda + \eta\lambda.$$

Thus, the definition of the RIF in (4.152) yields (4.158) as

$$\|\boldsymbol{\Delta}\| \le \frac{\|\overline{\boldsymbol{\Sigma}}\boldsymbol{\Delta}\|_\infty}{\mathrm{RIF}_{\|\cdot\|}(\mathcal{S};\eta)} \le \frac{\lambda + \eta\lambda}{\mathrm{RIF}_{\|\cdot\|}(\mathcal{S};\eta)}.$$

The prediction bound (4.157) follows from (4.163) and Proposition 4.2:

$$\begin{aligned}
&\|\mathbf{X}\mathbf{\Delta}\|_2^2/n + 2\|p_\lambda(\mathbf{\Delta}_{\mathcal{S}^c})\|_1 \\
\le\ & \frac{2}{1-\eta}\Big\{(\eta+1)\|p_\lambda(\mathbf{\Delta}_{\mathcal{S}})\|_1 + \nu\Big\} \\
\le\ & \frac{2(\eta+1)}{1-\eta}|\mathcal{S}|\lambda^2 \min\Big\{\frac{a_*\|\mathbf{\Delta}_{\mathcal{S}}\|_1}{\lambda(|\mathcal{S}|\vee 1)} + \frac{a_*^2}{2}, \gamma^*\Big\} + \frac{2\nu}{1-\eta}
\end{aligned}$$

with the $\ell_1$ bound $\|\mathbf{\Delta}_{\mathcal{S}}\|_1 \le \lambda(1+\eta)/\mathrm{RIF}_{\|\cdot\|_1}(\mathcal{S};\eta)$.

*Step 4.* Part (ii). As $\|\mathbf{X}^T\varepsilon/n\|_\infty \le \eta\lambda$, the condition of Theorem 4.11 holds with $\eta_1 = \eta_2 = \eta$ and $\boldsymbol{\beta}^* = \boldsymbol{\beta}$. Let $\mathbf{\Delta}^{(\ell_1)} = \widehat{\boldsymbol{\beta}}^{(\ell_1)} - \boldsymbol{\beta}$. As $|\varepsilon^T\mathbf{X}\mathbf{\Delta}^{(\ell_1)}/n| \le \eta\lambda\|\mathbf{\Delta}^{(\ell_1)}\|_1$ and $p_\lambda(t) \le \lambda|t|$, we have by Theorem 4.11

$$\begin{aligned}
\nu &= \|\mathbf{X}\mathbf{\Delta}^{(\ell_1)}\|_2^2/(2n) - \varepsilon^T\mathbf{X}\mathbf{\Delta}^{(\ell_1)}/n + \|p_\lambda(\widehat{\boldsymbol{\beta}}^{(\ell_1)})\|_1 - \|p_\lambda(\boldsymbol{\beta})\|_1 \\
&\le \|\mathbf{X}\mathbf{\Delta}^{(\ell_1)}\|_2^2/(2n) + \eta\lambda\|\mathbf{\Delta}^{(\ell_1)}\|_1 + \|p_\lambda(\mathbf{\Delta}^{(\ell_1)})\|_1 \\
&\le \lambda^2|\mathcal{S}|C_{\mathrm{pred}}^{(\ell_1)}(\mathcal{S};\eta,\eta)/2 + (1+\eta)\lambda^2|\mathcal{S}|C_{\mathrm{est},1}^{(\ell_1)}(\mathcal{S};\eta,\eta).
\end{aligned}$$

For the Dantzig selector, Theorem 4.14 is used instead of Theorem 4.11.

*Step 5.* Part (iii). Let

$$\widehat{\mathcal{S}}_1 = \{j \in \widehat{\mathcal{S}} \setminus \mathcal{S} : |\widehat{\boldsymbol{\beta}}_j| \ge \lambda t_0\}, \quad \widehat{\mathcal{S}}_2 = \{j \in \widehat{\mathcal{S}} \setminus \mathcal{S} : |\widehat{\boldsymbol{\beta}}_j| < \lambda t_0\}.$$

Since $f_0$ is a nondecreasing function, (4.159) yields

$$|\widehat{\mathcal{S}}_1| \le \frac{\|f(|\widehat{\boldsymbol{\beta}}_{\mathcal{S}^c}|)\|_1}{f(\lambda t_0)} \le |\mathcal{S}|C_{\mathrm{est},f_0}. \tag{4.165}$$

For $j \in \widehat{\mathcal{S}}_2$, (4.125) and the assumption $\|\mathbf{X}^T\varepsilon/n\|_\infty \le \eta\lambda$ imply that

$$\begin{aligned}
|(\overline{\mathbf{\Sigma}}\mathbf{\Delta})_j| &= \Big|\mathbf{X}_j^T(\mathbf{Y} - \mathbf{X}\widehat{\beta} - \varepsilon)/n\Big| \\
&\ge \inf_{0<s<t_0\lambda} p'_\lambda(s) - \eta\lambda \\
&\ge (1-\eta-\eta_0)\lambda.
\end{aligned}$$

For any set $A \subset \widehat{\mathcal{S}}_2$ with $|A| \le m_0$, we have

$$(1-\eta_0-\eta)_+^2\lambda^2|A| \le \|(\overline{\mathbf{\Sigma}}\mathbf{\Delta})_A\|_2^2 \le \phi^*(m_0;\mathcal{S},\overline{\mathbf{\Sigma}})\|\mathbf{X}\mathbf{\Delta}\|_2^2/n.$$

It follows from conditions (4.159) and (4.160) that

$$\begin{aligned}
(1-\eta_0-\eta)_+^2|A| &< \phi^*(m_0;\mathcal{S},\overline{\mathbf{\Sigma}})|\mathcal{S}|C_{\mathrm{pred}} \\
&\le (1-\eta_0-\eta)_+^2 m_0
\end{aligned}$$

Thus, $\max_{A\subset\widehat{\mathcal{S}}_2,|A|\le m_0}|A| < m_0$, which implies that $|\widehat{\mathcal{S}}_2| < m_0$. Combining this estimate with (4.165), we obtain the desired bound on $|\widehat{\mathcal{S}} \setminus \mathcal{S}|$.

*Step 6.* Prove (4.154). Suppose $(1-\eta)\|p_\lambda(\mathbf{u}_{\mathcal{S}^c})\|_1 < (1+\eta)\|p_\lambda(\mathbf{u}_{\mathcal{S}})\|_1$. It suffices to prove $(1-\eta)\|\mathbf{u}_{A^c}\|_1 < (1+\eta)\|\mathbf{u}_A\|_1$. Let $f(t) = t/p_\lambda(t)$ and $A$ be the index set of the $|S|$ largest $|u_j|$. Since $p_\lambda(t)$ is nondecreasing in $|t|$, we have $(1-\eta)\|p_\lambda(\mathbf{u}_{A^c})\|_1 < (1+\eta)\|p_\lambda(\mathbf{u}_A)\|_1$. Since $f(t)$ is nondecreasing in $t$,

$$\begin{aligned}(1-\eta)\|\mathbf{u}_{A^c}\|_1 &\le (1-\eta)\|p_\lambda(\mathbf{u}_{A^c})\|_1 f(\|\mathbf{u}_{A^c}\|_\infty)\\ &< (1+\eta)\|p_\lambda(\mathbf{u}_A)\|_1 f(\|\mathbf{u}_{A^c}\|_\infty)\\ &\le (1+\eta)\|\mathbf{u}_A\|_1.\end{aligned}$$

■

## 4.5 Smaller and Sorted Penalties

As we have discussed in Section 4.4, concave PLS was introduced to reduce the bias of the Lasso. The effect of this bias reduction can be seen from the prediction and $\ell_2$ estimation error bounds in (4.147) of Theorem 4.21, which justify the use of $\|p'_\lambda(\widehat{\boldsymbol{\beta}}^o_S)\|_2$ to measure the bias of the PLS as in (4.121). The bias-reduction effect can be also seen from the removal of the strong irrepresentable condition for selection consistency in Theorem 4.20.

For folded concave penalties with $p'_\lambda(0+) = \lambda$, $p'_\lambda(t)$ is increasing in the penalty level $\lambda$ and decreasing in the concavity $-p''_\lambda(t)$. Thus, bias reduction can be achieved by taking a concave penalty and/or a small penalty level. As we have already studied the benefits of folded concave penalization in Section 4.4, we shall focus on properties of PLS with smaller penalty levels than those considered earlier, e.g. in (4.44), (4.69) and (4.139), in this section.

For sub-Gaussian noise and standardized design, the $\ell_\infty$ bounds in (4.44), (4.69) and (4.139) all require a penalty level

$$\lambda \ge (\sigma/\eta)\sqrt{(2/n)\log p} \tag{4.166}$$

in view of Lemma 4.2 (ii) for some $\eta < 1$, at least for $\log p = (1+o(1))\log(p-s)$ and nearly orthogonal designs, where $s = \|\boldsymbol{\beta}\|_0$. Such penalty or threshold levels are commonly used in literature to study regularized methods in high-dimensional regression. The lower bound (4.166) for the penalty level is also within a small fraction of being necessary for selection consistency. However, (4.166) is conservative and may yield poor numerical results. Moreover, in view of the benchmark error bounds in (4.17), rate minimaxity in prediction and coefficient estimation dictates smaller penalty levels satisfying

$$\lambda \ge \lambda_0 = (\sigma/\eta)\sqrt{(2/n)\log(p/s)}, \tag{4.167}$$

which can be of smaller order than (4.166). We shall prove this rate minimaxity of the Lasso and concave PLS by extending the prediction and estimation error bounds in Theorems 4.20 and 4.21 to the smaller penalty level (4.167).

A problem with such rate optimalility results is the dependence of the penalty level (4.167) on $s = \|\boldsymbol{\beta}\|_0$, which is typically unknown. This has led to *sorted PLS* and aggregation methods for adaptively achieving the benchmark rate (4.17) with smaller $\lambda$, and *local concave approximation* (*LCA*) schemes for the computation of the sorted concave PLS.

We shall study in separate subsections *sorted concave penalties* and their local convex approximation *LCA*, properties of approximate local solutions for smaller and sorted penalties, and statistical properties of the *LLA* and *LCA*.

### *4.5.1 Sorted concave penalties and their local approximation*

The Sorted L-One ($\ell_1$) Penalized Estimation (*SLOPE*) was introduced in Bogdan *et al.* (2015) and Su and Candés (2016) to control the *FDR* in variable selection and to achieve adaptive minimaxity in prediction and coefficient estimation in high-dimensional linear regression. Given a sequence of penalty levels $\lambda_1 \ge \lambda_2 \ge \cdots \ge \lambda_p \ge 0$, the sorted $\ell_1$ penalty is defined as

$$\mathrm{Pen}(\mathbf{b}) = \textstyle\sum_{j=1}^{p} \lambda_j b_j^{\#}, \tag{4.168}$$

where $b_j^{\#}$ is the j-th largest value among $|b_1|, \ldots, |b_p|$.

The basic idea can be described as follows. Let $\lambda_j = (\sigma/\eta)\sqrt{(2/n)\log(p/j)}$ for some $\eta < 1$ (see Section 3.3.6). Assume the SLOPE is selection consistent. The penalty level for variables in $S^c$ is approximately $\lambda_s$ as $\lambda_j$ decays slowly. On the other hand, according to (4.121), the squared bias of the SLOPE can be measured by $\sum_{j=1}^{s} \lambda_j^2 = (1 + o(1))s\lambda_s^2$. Thus, without the knowledge of $s = \|\boldsymbol{\beta}\|_0$, the SLOPE matches the Lasso with fixed penalty level $\lambda_s$ in both the penalty level and bias. This hopefully leads to the minimax rate $s\lambda_s^2$ in prediction and the squared estimation errors under proper conditions on the design without requiring selection consistency.

From the bias reduction point of view, the SLOPE is not completely satisfying as the ideal squared bias for the concave PLS, $\|p'_\lambda(\widehat{\boldsymbol{\beta}}^o)\|_2^2$, may still be of smaller order than $s\lambda_s^2$ due to signal strength. The sorted concave penalty was introduced in Feng and Zhang (2017) to combine the strengths of concave and sorted penalties. Given a family of univariate penalty functions $p_\lambda(t)$ and a vector $\boldsymbol{\lambda} = (\lambda_1, \ldots, \lambda_p)^T$ with non-increasing nonnegative elements, the associated sorted penalty is defined as

$$\mathrm{Pen}_{\#}(\mathbf{b}) = \mathrm{Pen}_{\#}(\mathbf{b}; \boldsymbol{\lambda}) = \textstyle\sum_{j=1}^{p} p_{\lambda_j}(b_j^{\#}). \tag{4.169}$$

This is a direct extension of (4.168) as it automatically reproduces it with $p_\lambda(t) = \lambda|t|$. In the next subsection, we prove that under an RE condition, the sorted concave penalty inherits the benefits of both concave and sorted penalties, namely bias reduction from signal strength and adaptation to the smaller penalty level $\lambda_s$. However, what are the penalty level and concavity of such sorted penalties? How do we compute sorted concave penalized estimators?

We shall begin by studying properties of general penalty functions through their sub-differential and maximum concavity. A vector $\mathbf{g}$ is a (local) *sub-gradient* of a function $f : I\!R^p \to I\!R$ at a point $\mathbf{b} \in I\!R^p$ if

$$\liminf_{t\to 0+} t^{-1}\{f(\mathbf{b}+t\mathbf{u}) - f(\mathbf{b})\} \ge \mathbf{g}^T\mathbf{u}, \quad \forall\ \mathbf{u} \in I\!R^p. \tag{4.170}$$

The (local) sub-differential of $f(\mathbf{b})$ at $\mathbf{b}$, denoted by $\partial f(\mathbf{b})$, can be defined as the set of all sub-gradients $\mathbf{g}$ satisfying (4.170). As $\mathbf{g}^T\mathbf{u}$ is continuous in $\mathbf{g}$, $\partial f(\mathbf{b})$ is always a closed convex set of $I\!R^p$. We say that $f(\cdot)$ is sub-differentiable at $\mathbf{b}$ if $\partial f(\mathbf{b})$ is nonempty, and denote by $\nabla f(\mathbf{b})$ members of $\partial f(\mathbf{b})$. For functions $f(\cdot)$ sub-differentiable at $\mathbf{b}$ and $\boldsymbol{\beta}$, define

$$D_f(\mathbf{b},\boldsymbol{\beta}) = (\mathbf{b}-\boldsymbol{\beta})^T\{\nabla f(\mathbf{b}) - \nabla f(\boldsymbol{\beta})\}.$$

When $f(\mathbf{b})$ is convex, $D_f(\mathbf{b},\boldsymbol{\beta})$, called the *Bregman divergence*, is always nonnegative. Similarly to how we did in (4.119), we define the maximum concavity of $\mathrm{Pen}(\mathbf{b})$ as

$$\kappa(\mathrm{Pen}) = \inf\Big\{\kappa : \mathrm{Pen}(\mathbf{b}) + \kappa\|\mathbf{b}\|_2^2/2 \ \text{ is convex in } \mathbf{b}\Big\}. \tag{4.171}$$

We decompose the penalty as the difference of two convex functions,

$$\mathrm{Pen}(\mathbf{b}) = \mathrm{Pen}_+(\mathbf{b}) - \kappa\|\mathbf{b}\|_2^2/2, \tag{4.172}$$

when $\kappa(\mathrm{Pen}) \le \kappa < \infty$, where $\mathrm{Pen}_+(\mathbf{b}) = \mathrm{Pen}(\mathbf{b}) + \kappa\|\mathbf{b}\|_2^2/2$. The following lemma provides some basic properties of multivariate penalties.

**Lemma 4.9** *Let $L_0(\mathbf{b})$ be a Fréchet differentiable function with derivative $\nabla L_0(\mathbf{b})$. Let $Q(\mathbf{b}) = L_0(\mathbf{b}) + \mathrm{Pen}(\mathbf{b})$. Then, $\widehat{\boldsymbol{\beta}}$ is a critical point of $Q(\mathbf{b})$ in the sense of (4.109) iff*

$$-\nabla L_0(\widehat{\boldsymbol{\beta}}) = \nabla \mathrm{Pen}(\widehat{\boldsymbol{\beta}}) \in \partial\,\mathrm{Pen}(\widehat{\boldsymbol{\beta}}). \tag{4.173}$$

*Consequently, for any target vector $\boldsymbol{\beta}^*$ and any $\nabla\mathrm{Pen}(\boldsymbol{\beta}^*) \in \partial\,\mathrm{Pen}(\boldsymbol{\beta}^*)$,*

$$\begin{aligned} & D_{L_0}(\widehat{\boldsymbol{\beta}},\boldsymbol{\beta}^*) + (\boldsymbol{\beta}^* - \widehat{\boldsymbol{\beta}})^T\nabla\mathrm{Pen}(\boldsymbol{\beta}^*) \\ \le\ & D_{L_0}(\widehat{\boldsymbol{\beta}},\boldsymbol{\beta}^*) + D_{\mathrm{Pen}_+}(\widehat{\boldsymbol{\beta}},\boldsymbol{\beta}^*) + (\boldsymbol{\beta}^* - \widehat{\boldsymbol{\beta}})^T\nabla\mathrm{Pen}(\boldsymbol{\beta}^*) \\ =\ & -(\widehat{\boldsymbol{\beta}} - \boldsymbol{\beta}^*)^T\nabla L_0(\boldsymbol{\beta}^*) + \kappa\|\widehat{\boldsymbol{\beta}} - \boldsymbol{\beta}^*\|_2^2/2. \end{aligned} \tag{4.174}$$

As (4.173) characterizes local solutions for the minimization of $Q(\mathbf{b})$, it can be viewed as an extension of the KKT condition (4.31) to general penalized loss. Similarly, as $D_{L_0}(\widehat{\boldsymbol{\beta}},\boldsymbol{\beta}^*) = \|\mathbf{X}\widehat{\boldsymbol{\beta}} - \mathbf{X}\boldsymbol{\beta}^*\|_2^2/n$ for $L_0(\mathbf{b}) = \|\mathbf{Y} - \mathbf{X}\mathbf{b}\|_2^2/(2n)$, (4.173) can be viewed as an extension of the basic inequality (4.49) with $\overline{\boldsymbol{\beta}} = \boldsymbol{\beta}^*$.

**Proof of Lemma 4.9.** As $L_0(\cdot)$ is Fréchet differentiable at $\mathbf{b}$,

$$\liminf_{t\to 0+} \frac{Q(\mathbf{b}+t\mathbf{u}) - Q(\mathbf{b})}{t} = \nabla L_0(\widehat{\boldsymbol{\beta}}) + \liminf_{t\to 0+} \frac{\mathrm{Pen}(\mathbf{b}+t\mathbf{u}) - \mathrm{Pen}(\mathbf{b})}{t}$$

as long as one of the lim inf is well defined. This gives us (4.173). The identity in (4.174) then follows after taking the inner product of $\widehat{\boldsymbol{\beta}} - \boldsymbol{\beta}^*$ and (4.173) on both sides with some algebra. ■

Next, we study $\nabla \text{Pen}_{\#}(\boldsymbol{\beta})$ and $\kappa(\text{Pen}_{\#})$ for the sorted penalty (4.169) as they play key roles in the application of Lemma 4.9 to sorted penalties. We shall consider throughout this section penalty families $\{p_\lambda(\cdot)\}$ with finite local threshold level, increasing $|p'_\lambda(t)|$ in $\lambda$ and maximum concavity:

$$p'_\lambda(0+) = \lambda, \quad p'_\lambda(t) \uparrow \text{ in } \lambda \ \forall t > 0, \quad \kappa(p_\lambda) \le \kappa. \tag{4.175}$$

Given $\lambda_1 \ge \cdots \ge \lambda_p \ge 0$ and integer $s \in [0, p)$, define sorted $\ell_1$ norm

$$\|\mathbf{b}\|_{\#,s} = \sum_{j=1}^{p-s} (\lambda_{s+j}/\lambda_s) b_j^{\#}, \quad \forall\, \mathbf{b} \in I\!R^{p-s}, \tag{4.176}$$

where $b_j^{\#}$ is the $j$-th largest value among $|b_1|, \ldots, |b_{p-s}|$.

**Lemma 4.10** *Suppose (4.175) holds. Let* $\text{Pen}_{\#}(\boldsymbol{\beta})$ *be as in (4.169).*
*(i) The maximum concavity of* $\text{Pen}_{\#}(\cdot)$ *is bounded by* $\kappa(\text{Pen}_{\#}) \le \kappa$ *and*

$$\text{Pen}_{\#}(\mathbf{b}; \boldsymbol{\lambda}) = \max\Big\{ \textstyle\sum_{j=1}^{p} p_{\lambda_{k_j}}(b_j) : \mathbf{k} \in \text{perm}(p) \Big\}, \tag{4.177}$$

*where* $\text{perm}(p)$ *is the set of* $\mathbf{k} \in I\!R^p$ *generated by permuting* $(1, \ldots, p)^T$.
*(ii) If* $\text{supp}(\mathbf{b}) = \mathcal{S}$ *with* $|\mathcal{S}| = s$*, then*

$$\partial\, \text{Pen}_{\#}(\mathbf{b}) = \overline{\big\{ \mathbf{v} : v_{k_j} = p'_{\lambda_j}(b_{k_j}), \mathbf{k} \in K_{\mathcal{S}}, \|\mathbf{v}_{\mathcal{S}^c}\|_{\#,s,*} \le \lambda_s \big\}}, \tag{4.178}$$

*where* $K_{\mathcal{S}}$ *is the set of all vectors* $\mathbf{k} = (k_1, \ldots, k_s)^T$ *satisfying* $\{k_1, \ldots, k_s\} = \mathcal{S}$ *and* $|b_{k_j}| \ge |b_{k_{j+1}}|$ *for* $1 \le j < s$, $\overline{C}$ *denotes the closure of the convex hull of* $C$, *and* $\|\cdot\|_{\#,s,*}$ *is the dual norm of* $\|\cdot\|_{\#,s}$ *in (4.176). Consequently,*

$$\sup_{\mathbf{g} \in \partial \text{Pen}_{\#}(\mathbf{b})} \mathbf{h}^T \mathbf{g} = \lambda_s \big\|\mathbf{h}_{\mathcal{S}^c}\big\|_{\#,s} + \max_{\mathbf{k} \in K_{\mathcal{S}}} \sum_{j=1}^{s} p'_{\lambda_j}(b_{k_j}) h_{k_j}. \tag{4.179}$$

Condition (4.175) holds for many commonly used penalty families, including the SCAD penalty and MCP. For families of the form $p_\lambda(t) = \lambda^2 p_1(t/\lambda)$, $\kappa(p_\lambda)$ does not depend on $\lambda$, so that by Lemma 4.10 (i), the sorted penalty $\text{Pen}_{\#}$ has the same maximum concavity as the original $p_\lambda$. Lemma 4.10 (ii) provides an explicit formula for the sub-differential of sorted penalties. We note that $\big\|\cdot\big\|_{\#,s}$ is the $\ell_1$ norm when $\lambda_p = \lambda_s$.

**Proof of Lemma 4.10.** (i) For $\lambda \ge \lambda'$ and $t \ge t' \ge 0$,

$$p_\lambda(t) + p_{\lambda'}(t') - p_{\lambda'}(t) - p_\lambda(t') = \int_{t'}^{t} \{p'_\lambda(x) - p'_{\lambda'}(x)\} dx \ge 0,$$

so that (4.177) holds. As $p_{\lambda_j}(t) + \kappa t^2/2$ is convex in $t$,

$$\mathrm{Pen}_{\#}(\mathbf{b}) + \kappa\|\mathbf{b}\|_2^2/2 = \max\left[\sum_{j=1}^{p}\left\{p_{\lambda_{k_j}}(b) + \kappa|b_j|^2/2\right\} : \mathbf{k} \in \mathrm{perm}(p)\right]$$

is convex, as the maximum of convex functions is convex.

(ii) For sufficiently small $t > 0$,

$$\begin{aligned}\frac{\mathrm{Pen}_{\#}(\mathbf{b}+t\mathbf{u}) - \mathrm{Pen}_{\#}(\mathbf{b})}{t} &= \max_{\mathbf{k}\in K_S}\sum_{j=1}^{s} p'_{\lambda_j}(b_{k_j})u_{k_j} + \lambda_s\|\mathbf{u}_{\mathcal{S}^c}\|_{\#,s} \\ &\geq \min_{\mathbf{k}\in K_S}\sum_{j=1}^{s} p'_{\lambda_j}(b_{k_j})u_{k_j} + \lambda_s\|\mathbf{u}_{\mathcal{S}^c}\|_{\#,s}.\end{aligned}$$

The conclusion follows. ■

In the rest of the subsection, we describe the local convex approximation (LCA) (Feng and Zhang, 2017). Consider a penalized loss

$$Q(\mathbf{b}) = L_0(\mathbf{b}) + \mathrm{Pen}(\mathbf{b}) \tag{4.180}$$

where $L_0(\cdot)$ is convex and continuously differentiable. The simplest version of the LCA can be written as

$$\widehat{\boldsymbol{\beta}}^{(new)} = \arg\min_{\mathbf{b}}\left\{Q(\mathbf{b}) + (\kappa/2)\|\mathbf{b} - \widehat{\boldsymbol{\beta}}^{(old)}\|_2^2\right\}. \tag{4.181}$$

when the maximum concavity of the penalty in (4.172) is bounded by $\kappa(\mathrm{Pen}) \leq \kappa$. Similar to the LQA (Fan and Li, 2001) and LLA (Zou and Li, 2008), the LCA is a *majorization-minimization* (*MM*) algorithm. Given a tentative solution $\widehat{\boldsymbol{\beta}}^{(old)}$, the new penalty

$$\mathrm{Pen}^{(new)}(\mathbf{b}) = \mathrm{Pen}(\mathbf{b}) + (\kappa/2)\|\mathbf{b} - \widehat{\boldsymbol{\beta}}^{(old)}\|_2^2$$

is a convex majorization of $\mathrm{Pen}(\mathbf{b})$, so that the loss function $Q(\mathbf{b})$ is non-increasing in the LCA iteration (4.181): $Q(\widehat{\boldsymbol{\beta}}^{(new)}) \leq Q(\widehat{\boldsymbol{\beta}}^{(old)})$; see (3.69).

Figure 4.1 demonstrates that for $p = 1$ the LCA also majorizes the LLA

$$\mathrm{Pen}^{(new)}(b) = p_\lambda(|\widehat{\beta}^{(old)}|) + p'_\lambda(|\widehat{\beta}^{(old)}|)(|b| - |\widehat{\beta}^{(old)}|).$$

However, the LLA is not feasible when $\lambda_1 > \lambda_p$. As the LCA majorizes the LLA, it imposes larger penalty on solutions with larger step size, but this does not change our analysis of their statistical properties.

The LCA can be computed by proximal gradient algorithms such as the FISTA (Beck and Teboulle, 2009) as described in Chapter 3. The proximal mapping

$$\mathrm{prox}(\mathbf{x}; f) = \arg\min_{\mathbf{b}}\left\{\|\mathbf{b} - \mathbf{x}\|_2^2/2 + f(\mathbf{b})\right\} \tag{4.182}$$

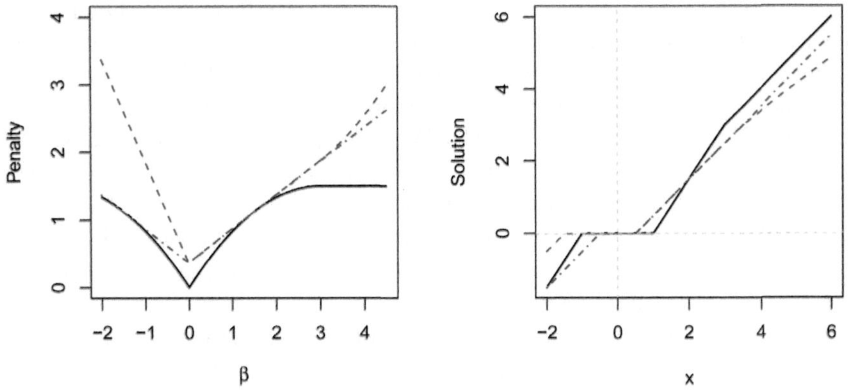


Figure 4.1: LCA (red dashed), LLA (blue mixed) and original penalty (black solid) for MCP with $\lambda = 1$ and $\kappa = 1/3$ at $\widehat{\beta}^{(old)} = 1.5$. Left: penalty function and its approximations; Right: $\arg\min_b \{(x-b)^2/2 + \mathrm{Pen}(b)\}$.

is not separable for $f(\mathbf{b}) \propto \mathrm{Pen}_+(\mathbf{b}) = \mathrm{Pen}_\#(\mathbf{b}) + (\kappa/2)\|\mathbf{b}\|_2^2$ for truly sorted penalties with $\lambda_1 > \lambda_p$. However, it still preserves the sign and ordering in absolute value of the input. Thus, after removing the sign and sorting $\mathbf{x}$ according to absolute value, (4.182) can be solved by the following algorithm:

**Input:** $\boldsymbol{\lambda}\downarrow$, $\boldsymbol{\lambda} \in [0,\infty)^p$, $\mathbf{x}\downarrow$, $\mathbf{x} \in [0,\infty)^p$, $t > 0$
**Compute** $b_j = \arg\min_b\{(x_j - b)^2/2 + t(p_{\lambda_j}(b) + \kappa b^2/2)\}$, $1 \le j \le p$
**While b** is not nonincreasing **do**
Identify blocks of violators of the monotonicity constraint,
$b_{j'-1} > b_{j'} \le b_{j'+1} \le \cdots \le b_{j''} > b_{j''+1}$, $b_{j'} < b_{j''}$
Replace $b_j$, $j' \le j \le j''$, with the solution of
$\arg\min_b \sum_{j=j'}^{j''}\{(x_j - b)^2/2 + t(p_{\lambda_j}(b) + \kappa b^2/2)\}$
**Output:** $\mathbf{b}$

### *4.5.2 Approximate PLS with smaller and sorted penalties*

In this subsection, we provide prediction and estimation error bounds for approximate local solutions of the PLS problem (4.108) for smaller and sorted penalties. Error bounds of the same type were established by Sun and Zhang (2013) and Bellec, Lecué and Tsybakov (2016) for the Lasso with smaller penalty levels, by Su and Candés (2016) and Bellec, Lecué and Tsybakov (2016) for the SLOPE, and by Feng and Zhang (2017) in the general setting considered here.

We write an approximate solution for the general PLS problem (4.108) as

$$\mathbf{X}^T(\mathbf{Y} - \mathbf{X}\widehat{\boldsymbol{\beta}})/n = \nabla\mathrm{Pen}(\widehat{\boldsymbol{\beta}}) + \boldsymbol{\nu}_{approx}. \tag{4.183}$$

Solutions of the form (4.183) are called approximate local solutions in Zhang and Zhang (2012) where their uniqueness, variable selection properties and relationship to the global solution were studied. It follows from Lemma 4.9 that a local solution (4.109) must satisfy (4.183) with $\boldsymbol{\nu}_{approx} = \mathbf{0}$. If we are confined to separable penalties with penalty level $(1+\nu/2)\lambda$ and no greater concavity than $\kappa_0$ and approximation errors satisfying $\|\boldsymbol{\nu}_{approx}\|_\infty \le \nu\lambda/2$, then (4.183) is equivalent to (4.134). We shall present a unified analysis of separable and sorted penalties that allows the penalty level to be smaller than (4.166) and the computational error $\boldsymbol{\nu}_{approx}$ to be a sum of $\ell_\infty$ and $\ell_2$ components. For simplicity, we focus on the difference between $\widehat{\boldsymbol{\beta}}$ and the oracle solution $\widehat{\boldsymbol{\beta}}^o$ in (4.10), as the deviation of $\widehat{\boldsymbol{\beta}}^o$ from the true coefficient vector is already well understood.

Let $\kappa(\mathrm{Pen})$ be the maximum concavity (4.171) of the penalty in (4.183). An application of the calculation in (4.174) to (4.183) yields

$$\mathbf{h}^T\overline{\boldsymbol{\Sigma}}\mathbf{h} + \sup_{\mathbf{g}\in\partial\,\mathrm{Pen}(\widehat{\boldsymbol{\beta}}^o)} \mathbf{h}^T\mathbf{g} \le \mathbf{h}^T(\mathbf{Z}-\boldsymbol{\nu}_{approx}) + \kappa(\mathrm{Pen})\|\mathbf{h}\|_2^2,$$

where $\mathbf{h} = \widehat{\boldsymbol{\beta}} - \widehat{\boldsymbol{\beta}}^o$, $\overline{\boldsymbol{\Sigma}} = \mathbf{X}^T\mathbf{X}/n$ and $\mathbf{Z} = \mathbf{X}^T(\mathbf{Y}-\mathbf{X}\widehat{\boldsymbol{\beta}}^o)/n$. Thus, by (4.179),

$$\mathbf{h}^T\overline{\boldsymbol{\Sigma}}\mathbf{h} + \lambda_s\|\mathbf{h}_{\mathcal{S}^c}\|_{\#,s} + \lambda_s\mathbf{h}_{\mathcal{S}}^T\mathbf{w}_{\mathcal{S}}^o \le \mathbf{h}^T(\mathbf{Z}-\boldsymbol{\nu}_{approx}) + \kappa(\mathrm{Pen})\|\mathbf{h}\|_2^2 \quad (4.184)$$

with $s = |\mathcal{S}|$, the $\|\cdot\|_{\#,s}$ in (4.176) and $(\mathbf{w}^o)_j = p'_{\lambda_j}(\widehat{\beta}^o_{k_j})/\lambda_s$, where $\{k_1,\ldots,k_s\} = \mathcal{S}$ satisfies $|\widehat{\beta}^o_{k_j}| \ge |\widehat{\beta}^o_{k_{j+1}}|$ for $j = 1,\ldots,s-1$. This is a basic inequality in our analysis of (4.183). For separable penalties $\mathrm{Pen}(\mathbf{b}) = \|p_\lambda(\mathbf{b})\|_1$, or equivalently sorted penalties with $\lambda_1 = \cdots = \lambda_p = \lambda$, (4.184) holds with $\|\mathbf{h}_{\mathcal{S}^c}\|_{\#,s} = \|\mathbf{h}_{\mathcal{S}^c}\|_1$. However, in general, $\|\mathbf{h}_{\mathcal{S}^c}\|_{\#,s} \le \|\mathbf{h}_{\mathcal{S}^c}\|_1$, so that sorted penalties provide weaker control of the noise. This can be viewed as a the cost of adaptation to $\lambda_s$.

Our analysis of (4.184) can be outlined as follows. For positive constants $\{r,\eta,\gamma\}$ with $\eta\in(0,1)$ and vectors $\mathbf{w}$ and $\boldsymbol{\nu}$, define

$$\Delta(r,\eta,\mathbf{w},\boldsymbol{\nu}) = \sup_{\mathbf{u}\neq 0} \frac{\mathbf{u}^T(\mathbf{Z}-\boldsymbol{\nu})/\lambda_s - \eta\|\mathbf{u}_{\mathcal{S}^c}\|_{\#,s} - \mathbf{u}_{\mathcal{S}}^T\mathbf{w}_{\mathcal{S}}}{r\max\left\{\|\mathbf{u}\|_2, \|\mathbf{X}\mathbf{u}\|_2\sqrt{\gamma/n}\right\}} \quad (4.185)$$

with $\mathbf{Z} = \mathbf{X}^T(\mathbf{Y}-\mathbf{X}\widehat{\boldsymbol{\beta}}^o)/n$. We summarize the required conditions on the noise and the approximation error by

$$\Delta(r,\eta,\mathbf{w}^o,\boldsymbol{\nu}_{approx}) < 1 \quad \text{for certain} \quad r < (1-\eta)\xi s^{1/2} \quad (4.186)$$

with a constant $\xi > 0$. This condition and (4.184) yield

$$\begin{aligned} &\mathbf{h}^T\overline{\boldsymbol{\Sigma}}\mathbf{h} + (1-\eta)\lambda\big\|\mathbf{h}_{\mathcal{S}^c}\big\|_{\#,s} - \kappa(\mathrm{Pen})\|\mathbf{h}\|_2^2 \\ \le\ & r\lambda\max\left\{\|\mathbf{h}\|_2, \|\mathbf{X}\mathbf{h}\|_2\sqrt{\gamma/n}\right\} \quad (4.187) \\ \le\ & (1-\eta)\xi s^{1/2}\lambda\max\left\{\|\mathbf{h}\|_2, \|\mathbf{X}\mathbf{h}\|_2\sqrt{\gamma/n}\right\}. \end{aligned}$$

Similar to Subsection 4.4.3, we shall impose an RE condition on the design. However, because (4.183) specifies a larger class of estimators, the RE condition is imposed on a larger cone

$$\mathscr{C}_{\#}(\mathcal{S};\xi,\gamma) = \left\{\mathbf{u} : \|\mathbf{u}_{\mathcal{S}^c}\|_{\#,s} \le \xi s^{1/2} \max\left(\|\mathbf{u}\|_2, (\gamma \mathbf{u}^T \overline{\mathbf{\Sigma}} \mathbf{u})^{1/2}\right)\right\} \quad (4.188)$$

with $s = |\mathcal{S}|$. The corresponding $\ell_2$ restricted eigenvalue is

$$\mathrm{RE}_{\#}(S;\xi,\gamma) = \inf\left\{\frac{(\mathbf{u}^T\overline{\mathbf{\Sigma}}\mathbf{u})^{1/2}}{\|\mathbf{u}\|_2} : \mathbf{0} \ne \mathbf{u} \in \mathscr{C}_{\#}(S;\xi,\gamma)\right\}. \quad (4.189)$$

As the cone $\mathscr{C}_{\#}(S;\xi,\gamma)$ in (4.188) depends on the sorted $\boldsymbol{\lambda} = (\lambda_1,\ldots,\lambda_p)^T$, the infimum in (4.189) is also taken over all $\boldsymbol{\lambda}$ under consideration. We note that

$$\mathrm{RE}^2_{\#}(S;\xi,\gamma) \ge \left[\inf\left\{\frac{\mathbf{u}^T\overline{\mathbf{\Sigma}}\mathbf{u}}{\|\mathbf{u}\|_2^2} : \|\mathbf{u}_{\mathcal{S}^c}\|_{\#,s} < \xi|\mathcal{S}|^{1/2}\|\mathbf{u}\|_2\right\}\right] \wedge \frac{1}{\gamma}.$$

When the cone is confined to $\lambda_1 = \lambda_p$, i.e. $\|\mathbf{u}_{\mathcal{S}^c}\|_{\#,s} = \|\mathbf{u}_{\mathcal{S}^c}\|_1$, the RE condition on $\mathrm{RE}^2_{\#}(S;\xi,\gamma)$ is equivalent to the restricted strong convexity condition Negahban *et al.*(2012) as $\|\mathbf{u}_{\mathcal{S}^c}\|_1 < \xi|\mathcal{S}|^{1/2}\|\mathbf{u}\|_2$ implies $\|\mathbf{u}\|_1 < (\xi+1)|\mathcal{S}|^{1/2}\|\mathbf{u}\|_2$. Compared with the $\mathrm{RE}_2$ in (4.137), the quantity in (4.189) is smaller due to the use of a larger cone. However, this is hard to avoid because the smaller penalty does not control the $\ell_\infty$ measure of the noise as in (4.139) and we do not wish to impose a uniform bound on $\boldsymbol{\nu}_{approx}$ here.

Still, membership of $\mathbf{h} = \widehat{\boldsymbol{\beta}} - \widehat{\boldsymbol{\beta}}^o$ in the cone (4.188) is not guaranteed for all approximate local solutions satisfying condition (4.187). When $\kappa(\mathrm{Pen})\|\mathbf{h}\|_2^2 \le \mathbf{h}^T\overline{\mathbf{\Sigma}}\mathbf{h}$, (4.187) implies $\mathbf{h} \in \mathscr{C}_{\#}(\mathcal{S};\xi,\gamma)$. When $\mathbf{h} \in \mathscr{C}_{\#}(\mathcal{S};\xi,\gamma)$, we have $\kappa(\mathrm{Pen})\|\mathbf{h}\|_2^2 \le \mathbf{h}^T\overline{\mathbf{\Sigma}}\mathbf{h}$ under the RE condition $\mathrm{RE}^2_{\#}(S;\xi,\gamma) \ge \kappa(\mathrm{Pen})$. The following proposition, an extension of Proposition 4.3 for more general approximate solutions (4.183), asserts that $\widehat{\boldsymbol{\beta}} - \widehat{\boldsymbol{\beta}}^o$ is in the cone (4.188) if $\widehat{\boldsymbol{\beta}}$ can be computed iteratively through a chain of solutions as follows:

$$\begin{aligned} &\widehat{\boldsymbol{\beta}} = \widehat{\boldsymbol{\beta}}^{(k^*)},\ \|(\widehat{\boldsymbol{\beta}}^{(0)} - \widehat{\boldsymbol{\beta}}^o)_{\mathcal{S}^c}\|_{\#,s}^{(k-1)} \le \xi s^{1/2} F(\widehat{\boldsymbol{\beta}}^{(0)} - \widehat{\boldsymbol{\beta}}^o), \qquad (4.190)\\ &\|\widehat{\boldsymbol{\beta}}^{(k)} - \widehat{\boldsymbol{\beta}}^{(k-1)}\|_1 \le a_1^{(k)} s\lambda_s^{(k)},\ F(\widehat{\boldsymbol{\beta}}^{(k)} - \widehat{\boldsymbol{\beta}}^{(k-1)}) \le a_2^{(k)} s^{1/2}\lambda_s^{(k)},\\ &\|\cdot\|_{\#,s}^{(k)} \le \|\cdot\|_{\#,s}^{(k-1)},\ \kappa(\mathrm{Pen}^{(k)}) \le \left\{\mathrm{RE}_{\#}^{(k)}(\mathcal{S};\xi,\gamma)\right\}^2, \end{aligned}$$

where $F(\mathbf{u}) = \max\left\{\|\mathbf{u}\|_2, \|\mathbf{X}\mathbf{u}\|_2\sqrt{\gamma/n}\right\}$, $\mathrm{Pen}^{(k)}$ is the penalty for $\widehat{\boldsymbol{\beta}}^{(k)}$, $\|\cdot\|_{\#,s}^{(k)}$ is the sorted $\ell_1$ norm (4.176) corresponding to the penalty levels $\lambda_1^{(k)} \ge \cdots \ge \lambda_p^{(k)}$ for $\mathrm{Pen}^{(k)}$, and $\mathrm{RE}_{\#}^{(k)}(\mathcal{S};\xi,\gamma)$ is the restricted eigenvalue (4.189) for $\|\cdot\|_{\#,s}^{(k)}$.

**Proposition 4.4** *Let* $\mathcal{S} \supseteq \mathrm{supp}(\boldsymbol{\beta})$, $\widehat{\boldsymbol{\beta}}^o$ *be the oracle LSE in (4.10) and* $s = |\mathcal{S}|$. *Let* $\eta \in (0,1)$, $0 < \epsilon_0 < 1-\eta$, $\epsilon_1 = (\epsilon_0/2)/(1-\eta-\epsilon_0)$ *and* $\widehat{\boldsymbol{\beta}}^{(k)}$ *be as in (4.190) with* $a_1^{(k)} = \epsilon_1\epsilon_0\xi^2/\kappa(\mathrm{Pen}^{(k)})$ *and* $a_2^{(k)} = \epsilon_1\epsilon_0\xi/\kappa(\mathrm{Pen}^{(k)})$. *Suppose*

*(4.187) holds with* $\mathbf{h} = \widehat{\boldsymbol{\beta}}^{(k)} - \widehat{\boldsymbol{\beta}}^{o}$, $\|\cdot\|_{\#,s} = \|\cdot\|_{\#,s}^{(k)}$, *and some* $r \le (1-\eta-\epsilon_0)\xi s^{1/2}$. *Then,*

$$\big\|(\widehat{\boldsymbol{\beta}}^{(k)} - \widehat{\boldsymbol{\beta}}^{o})_{\mathcal{S}^c}\big\|_{\#,s}^{(k)} \le \xi s^{1/2} F(\widehat{\boldsymbol{\beta}}^{(k)} - \widehat{\boldsymbol{\beta}}^{o}),\ k = 0, 1, \ldots, k^*.$$

The initialization condition of (4.190) holds with $\widehat{\boldsymbol{\beta}}^{(0)} = \mathbf{0}$. Compared with Proposition 4.3, Proposition 4.4 asserts the membership of $\widehat{\boldsymbol{\beta}} - \boldsymbol{\beta}^o$ in a larger cone. However, (4.190) also allows the larger step size. The restricted eigenvalue condition is comparable with $\mathrm{RE}_2^2 \ge \kappa_0$ in Theorem 4.20. In practice, Proposition 4.4 suggests iterations satisfying

$$\big\|\widehat{\boldsymbol{\beta}}^{(k)} - \widehat{\boldsymbol{\beta}}^{(k-1)}\big\|_1 \le a_1 s\lambda_{*,s}, \qquad F(\widehat{\boldsymbol{\beta}}^{(k)} - \widehat{\boldsymbol{\beta}}^{(k-1)}) \le a_2 s^{1/2}\lambda_{*,s}, \quad (4.191)$$

with $\widehat{\boldsymbol{\beta}}^{(0)} = \mathbf{0}$ or $\kappa(\mathrm{Pen}^{(0)}) = 0$, small $a_1$ and $a_2$ and lower bounds $\lambda_{*,j} \le \lambda_j^{(k)}$, where $s\lambda_{*,s}$ and $s^{1/2}\lambda_{*,s}$ represent acceptable levels of the $\ell_1$ and prediction errors.

**Proof of Proposition 4.4.** Let $\mathbf{h} = \widehat{\boldsymbol{\beta}}^{(k-1)} - \widehat{\boldsymbol{\beta}}^{o}$, $\widetilde{\mathbf{h}} = \widehat{\boldsymbol{\beta}}^{(k)} - \widehat{\boldsymbol{\beta}}^{o}$, $\widetilde{\lambda} = \lambda^{(k)}$ and $\kappa_0 = \kappa(\mathrm{Pen}^{(k)})$. As the conclusion holds for $k = 0$, it suffices to prove $\|\widetilde{\mathbf{h}}_{\mathcal{S}^c}\|_{\#,s}^{(k)} \le \xi s^{1/2} F(\widetilde{\mathbf{h}})$ under the induction assumption $\|\mathbf{h}_{\mathcal{S}^c}\|_{\#,s}^{(k-1)} \le \xi s^{1/2} F(\mathbf{h})$. As (4.187) holds with $r = (1-\eta-\epsilon_0)s^{1/2}\xi$, we have

$$\begin{aligned} & \widetilde{\mathbf{h}}^T \overline{\boldsymbol{\Sigma}} \widetilde{\mathbf{h}} + (1-\eta)\widetilde{\lambda}\|\widetilde{\mathbf{h}}_{\mathcal{S}^c}\|_{\#,s}^{(k)} \\ \le\ & r\widetilde{\lambda} F(\widetilde{\mathbf{h}}) + \kappa_0\|\widetilde{\mathbf{h}}\|_2^2 \\ \le\ & (1-\eta-\epsilon_0)\widetilde{\lambda}\xi s^{1/2} F(\widetilde{\mathbf{h}}) + \kappa_0 \|\widetilde{\mathbf{h}}\|_2^2. \end{aligned} \quad (4.192)$$

When $F(\widetilde{\mathbf{h}}) \le \epsilon_0 \widetilde{\lambda}\xi s^{1/2}/\kappa_0$, we have $\kappa_0\|\widetilde{\mathbf{h}}\|_2^2 \le \kappa_0 F^2(\widetilde{\mathbf{h}}) \le \epsilon_0\widetilde{\lambda}\xi s^{1/2} F(\widetilde{\mathbf{h}})$, so that $\|\widetilde{\mathbf{h}}_{\mathcal{S}^c}\|_{\#,s}^{(k)} \le \xi s^{1/2} F(\widetilde{\mathbf{h}})$ by (4.192). It remains to consider the case where

$$\begin{aligned} & F(\widetilde{\mathbf{h}}) \ge \epsilon_0\widetilde{\lambda}\xi s^{1/2}/\kappa_0 = (a_2^{(k)}/\epsilon_1)\widetilde{\lambda} s^{1/2}, \\ & \|\widetilde{\mathbf{h}}_{\mathcal{S}^c}\|_{\#,s}^{(k)} \ge (\xi s^{1/2})(\epsilon_0\widetilde{\lambda}\xi s^{1/2}/\kappa_0) = (a_1^{(k)}/\epsilon_1)\widetilde{\lambda} s. \end{aligned}$$

It follows from our condition on $\widetilde{\mathbf{h}} - \mathbf{h} = \widehat{\boldsymbol{\beta}}^{(k)} - \widehat{\boldsymbol{\beta}}^{(k-1)}$ that

$$\frac{F(\widetilde{\mathbf{h}})}{F(\mathbf{h})} \ge \frac{1}{1 + F(\widetilde{\mathbf{h}} - \mathbf{h})/F(\widetilde{\mathbf{h}})} \ge \frac{1}{1 + a_2^{(k)}/(a_2^{(k)}/\epsilon_1)} = \frac{1}{1+\epsilon_1}$$

and

$$\frac{\|\mathbf{h}_{\mathcal{S}^c}\|_{\#,s}^{(k-1)}}{\|\widetilde{\mathbf{h}}_{\mathcal{S}^c}\|_{\#,s}^{(k)}} \ge \frac{\|\mathbf{h}_{\mathcal{S}^c}\|_{\#,s}^{(k)}}{\|\widetilde{\mathbf{h}}_{\mathcal{S}^c}\|_{\#,s}^{(k)}} \ge 1 - \frac{\|(\widetilde{\mathbf{h}} - \mathbf{h})_{\mathcal{S}^c}\|_{\#,s}^{(k)}}{\|\widetilde{\mathbf{h}}_{\mathcal{S}^c}\|_{\#,s}^{(k)}} \ge 1 - \frac{\|\widetilde{\mathbf{h}} - \mathbf{h}\|_1}{\|\widetilde{\mathbf{h}}_{\mathcal{S}^c}\|_{\#,s}^{(k)}} \ge 1 - \epsilon_1.$$

As $\epsilon_1 = (\epsilon_0/2)/(1-\eta-\epsilon_0)$, we have

$$\frac{F(\widetilde{\mathbf{h}})}{\|\widetilde{\mathbf{h}}_{\mathcal{S}^c}\|_{\#,s}^{(k)}} \geq \frac{(1-\epsilon_1)F(\mathbf{h})}{(1+\epsilon_1)\|\mathbf{h}_{\mathcal{S}^c}\|_{\#,s}^{(k-1)}} \geq \frac{(1-\epsilon_1)(1-\eta-\epsilon_0)}{(1+\epsilon_1)(1-\eta)\xi} = \frac{1}{\xi}.$$

The last inequality above follows from the $\mathbf{h}$ version of (4.192). □

We are now ready to present the main results of this subsection. Let

$$\mathscr{B}_1 = \left\{ \begin{matrix} \text{all } \widehat{\boldsymbol{\beta}} \text{ computable by (4.190) through a chain of solutions} \\ \text{of (4.183) with penalties satisfying condition (4.186)} \end{matrix} \right\}. \quad (4.193)$$

Here, the conditions on the maximum concavity of the penalty and the restricted eigenvalue of the design are imposed in (4.190), which can be understood as $\kappa(\text{Pen}) \leq \text{RE}^2$. The penalty level condition is imposed through (4.186), which also involves the approximation error in (4.183).

**Theorem 4.23** *Let $\mathcal{S} \supseteq \text{supp}(\boldsymbol{\beta})$, $\widehat{\boldsymbol{\beta}}^o$ be the oracle LSE in (4.10), $s = |\mathcal{S}|$ and $F(\mathbf{u}) = \max\{\|\mathbf{u}\|_2, (\gamma\mathbf{u}^T\overline{\boldsymbol{\Sigma}}\mathbf{u})^{1/2}\}$. Let $\widehat{\boldsymbol{\beta}} \in \mathscr{B}_1$ be a specific solution of (4.183) in the class (4.193) with penalty levels $\lambda_1 \geq \cdots \geq \lambda_p$ such that (4.186) holds and $\kappa(\text{Pen}) \leq (1-1/C_0)\text{RE}_{\#}^2(\mathcal{S};\xi,\gamma)$ for the specific $\kappa(\text{Pen})$ and $\|\mathbf{u}_{\mathcal{S}^c}\|_{\#,s}$ associated with the penalty for $\widehat{\boldsymbol{\beta}}$ as in (4.171), (4.176) and (4.188). Then, for any semi-norm $\|\cdot\|$,*

$$\|\mathbf{h}\| \leq C_0\lambda_s \sup_{\mathbf{u}\neq 0} \frac{\|\mathbf{u}\|\{rF(\mathbf{u}) - (1-\eta)\|\mathbf{u}_{\mathcal{S}^c}\|_{\#,s}\}}{\mathbf{u}^T\overline{\boldsymbol{\Sigma}}\mathbf{u}} \quad (4.194)$$

*with $\mathbf{h} = \widehat{\boldsymbol{\beta}} - \widehat{\boldsymbol{\beta}}^o$ and the value $r$ in (4.186). In particular,*

$$\frac{\|\mathbf{h}_{\mathcal{S}}\|_1 + \|\mathbf{h}_{\mathcal{S}^c}\|_{\#,s}}{(1+\xi)|\mathcal{S}|^{1/2}} \leq F(\mathbf{h}) \leq \frac{\|\mathbf{X}\mathbf{h}\|_2/\sqrt{n}}{\text{RE}_{\#}(\mathcal{S};\xi,\gamma)} \leq \frac{C_0(r\lambda_s)}{\text{RE}_{\#}^2(\mathcal{S};\xi,\gamma)}. \quad (4.195)$$

**Proof.** It follows from the $\mathbf{h} = \widehat{\boldsymbol{\beta}} - \widehat{\boldsymbol{\beta}}^o$ version of (4.192) and the condition on the specific $\kappa(\text{Pen})$ that

$$C_0^{-1}\mathbf{h}^T\overline{\boldsymbol{\Sigma}}\mathbf{h} + (1-\eta)\lambda_s\|\mathbf{h}_{\mathcal{S}^c}\|_{\#,s}^{(k)} \leq r\lambda_s F(\mathbf{h}).$$

The conclusions follow in view of the definition of $\text{RE}_{\#}(\mathcal{S};\xi,\gamma)$ in (4.189). ■

Theorem 4.23 asserts that under proper conditions on the penalty level and concavity and the restricted eigenvalue, the prediction and $\ell_2$-estimation errors are of order $(r\lambda_s)^2$, with the small $r$ in (4.186). This rate can be understood as

$$(r\lambda_s)^2 \asymp \left\{p'_{\lambda_j}(\widehat{\beta}^o_{k_j})\right\}^2 + (\sigma^2/n)s \lesssim (\sigma^2/n)\left\{s_1\log(p/s_1) + s\right\}, \quad (4.196)$$

as we shall discuss in Theorem 4.24 below, where $s_1$ can be understood as the number of small nonzero signals. We note that $\sigma^2 s/n$ is the expected prediction error of the oracle LSE $\widehat{\beta}^o$ in (4.10). Moreover, (4.178) shows that the sorted concave PLS automatically picks the penalty level $\lambda_s$ from the sequence $\{\lambda_j\}$ without requiring the knowledge of $s$.

To understand condition (4.186) and thus the rate statement (4.196), we consider penalty levels satisfying

$$\lambda_j \ge \lambda_{*,j} = A_0\sigma\sqrt{(2/n)\log(p/(\alpha j))},\ \ \forall\, j=1,\ldots,p, \tag{4.197}$$

$$\textstyle\sum_{j=1}^s \lambda_j^2 \le B_0 s\lambda_s^2,\quad \sum_{j=1}^{p-s} u_j^{\#}\lambda_{s+j}/\lambda_s \ge \|\mathbf{u}\|_{\#,s,*}\ \forall\, s,\mathbf{u}, \tag{4.198}$$

for certain $A_0 > 1 > \alpha$, and approximation errors satisfying

$$\mathbf{u}^T\boldsymbol{\nu}_{approx} \le \eta_1\lambda_s\|\mathbf{u}_{\mathcal{S}^c}\|_{\#,s} + r_2\lambda_s\|\mathbf{u}\|_2\ \forall\ \mathbf{u}\in I\!R^p \tag{4.199}$$

with $\eta_1 \in (0,\eta)$ and $r_2 > 0$. The first part of (4.198) holds uniformly when $\lambda_j \propto \lambda_{*,j}$ for all penalties under consideration. The second part of (4.198) holds with $\|\mathbf{u}\|_{\#,s,*} = B_1^{-1}\sum_{j=1}^{p-s} u_j^{\#}\lambda_{*,s+j}/\lambda_{*,s}$ when $B_1\lambda_{s+j}/\lambda_s \ge \lambda_{*,s+j}/\lambda_{*,s}$ for all penalties under consideration. Condition (4.199) is fulfilled when

$$\boldsymbol{\nu}_{approx} = \boldsymbol{\nu}_1 + \boldsymbol{\nu}_2 \text{ with } \nu_{1,j}^{\#} \le \eta_1\lambda_j \text{ and } \|\boldsymbol{\nu}_2\|_2 + \eta_1\|\lambda_{1:s}\|_2 \le \lambda_s r_2,$$

as this decomposition provides

$$\begin{aligned}\mathbf{u}^T\boldsymbol{\nu}_{approx} &\le \eta_1\sum_{j=1}^p u_j^{\#}\lambda_j + \|\mathbf{u}\|_2\|\boldsymbol{\nu}_2\|_2\\ &\le \eta_1\lambda_s\|\mathbf{u}_{\mathcal{S}^c}\|_{\#,s} + (\eta_1\|\lambda_{1:s}\|_2 + \|\boldsymbol{\nu}_2\|_2)\|\mathbf{u}\|_2.\end{aligned}$$

Under these conditions, we consider a more specific class of solutions

$$\mathscr{B}_2 = \left\{\begin{array}{c}\text{all } \widehat{\boldsymbol{\beta}} \text{ computable by (4.191) through a chain}\\ \text{of solutions of (4.183) with sorted penalties}\\ \text{satisfying (4.197), (4.198) and } \kappa(\text{Pen}) \le \text{RE}^2\\ \text{and approximation errors satisfying (4.199)}\end{array}\right\}, \tag{4.200}$$

where $\text{RE} = \text{RE}_{\#,*}(\mathcal{S};\xi,\gamma)$ be the restricted eigenvalue (4.189) for the norm $\|\cdot\|_{\#,s} = \|\cdot\|_{\#,s,*}$ in (4.198).

For $\alpha\in(0,1)$, $A = A_0(\eta-\eta_1) > 1$ and integers $s < j_1 \le p$, define

$$L_x = \sqrt{2\log(p/(\alpha x))},\ p_{\alpha,A} = \sum_{k=1}^{k^*}\alpha^{1+(A-1)A^{k-1}},\ q_{\alpha,A} = 1 - \frac{2p_{\alpha,A}}{1-s/j_1},$$

where $k^* = \lceil(\log A)^{-1}\log(L_{j_1}^2/L_p^2)\rceil$. Define

$$\mu_{\#,s} = \left\{\frac{8p(\alpha j_1/p)^{A^2}/q_{\alpha,A}}{A^2L_s^2(A^2L_{j_1}^2+2)}\right\}^{1/2} I_{\{s>0\}}. \tag{4.201}$$

For fixed $\{\alpha, A, c_0\}$ with $s/j_1 \le c_0 < 1$, $\mu_{\#,s}^2 \lesssim (s/p)^{A^2-1}s/L_s^2$.

**Theorem 4.24** *Let $\mathcal{S} \supseteq \text{supp}(\boldsymbol{\beta})$, $\widehat{\boldsymbol{\beta}}^o$ be the oracle LSE in (4.10), $s = |\mathcal{S}|$ and $F(\mathbf{u}) = \max\{\|\mathbf{u}\|_2, (\gamma \mathbf{u}^T\boldsymbol{\Sigma}\mathbf{u})^{1/2}\}$. Let $a_0 \in (0,1)$, $B_0$ be as in (4.198), $\{\eta_1, r_2\}$ as in (4.199), $\mu_{\#,s}$ as in (4.201), $r_1 = (\eta - \eta_1)\mu_{\#,s}/(1-a_0)$, and $\mathscr{B}_2$ be as in (4.200) with $\xi = \{(r_1+r_2)/s^{1/2} + B_0^{1/2}\}/(1-\eta)$ in the definition of RE. Let $r = (1-a_0)^{-1}(r_1+w+r_2)$ and $\widehat{\boldsymbol{\beta}} \in \mathscr{B}_2$ be a specific solution of (4.183) with penalty levels $\lambda_1 \ge \cdots \ge \lambda_p$ and $\kappa(\text{Pen}) \le (1-1/C_0)\text{RE}^2_{\#}(\mathcal{S};\xi_1,\gamma)$ where $\xi_1 = r/\{(1-\eta)s^{1/2}\}$. Suppose $\boldsymbol{\varepsilon} \sim N(\mathbf{0}, \sigma^2\mathbf{I})$ and $\max_{j\le p}\|\mathbf{X}_j\|_2^2 \le n$. Then,*

$$\begin{aligned} & P\Big\{ (4.194) \text{ and } (4.195) \text{ hold for } \widehat{\boldsymbol{\beta}} \Big\} \\ \ge\ & \Phi\big(a_0 r_1 A_0 L_s \sqrt{\gamma}\big) - P\Big\{ \|\mathbf{w}^o_{\mathcal{S}}\|_2 > w \Big\}. \end{aligned} \tag{4.202}$$

*In particular, with $s_1 = \#\{j : 0 < (\widehat{\boldsymbol{\beta}}^o)^{\#}_j < a\lambda_j\}$ for the penalties and constant $a$ in Table 4.1, $\lambda_s^2\|\mathbf{w}^o_{\mathcal{S}}\|_2^2 \le B_0 s_1 \lambda_{s_1}^2$ and with $\mathbf{h} = \widehat{\boldsymbol{\beta}} - \widehat{\boldsymbol{\beta}}^o$*

$$\|\mathbf{h}\|_2^2 \le \frac{\|\mathbf{X}\mathbf{h}\|_2^2/n}{\text{RE}^2_{\#}(\mathcal{S};\xi_1,\gamma)} \lesssim s_1\lambda_{s_1}^2 + r_2^2\lambda_s^2 + \frac{(s/p)^{A^2-1}s\lambda_s^2}{\log(p/s)}. \tag{4.203}$$

**Corollary 4.9** *Under the conditions of Theorem 4.24,*

$$\text{E}\,\big\|\mathbf{X}\widehat{\boldsymbol{\beta}} - \mathbf{X}\widehat{\boldsymbol{\beta}}^o\big\|_2^2/n \le \frac{(1-\eta)^2\xi^2 s\lambda_s^2}{\text{RE}^2_{\#}(\mathcal{S};\xi,0)} + \frac{(1-\eta)^2\xi^2 s\lambda_s^2}{2(a_0 r_1 A_0 L_s)^2}.$$

Prediction and estimation risk bounds of PLS were established in Ye and Zhang (2010) for the Lasso and Dantzig selector under an $\ell_q$ condition on $\boldsymbol{\beta}$, and in Bellec, Lecué and Tsybakov (2016) for the Lasso and SLOPE. Bellec, Lecué and Tsybakov (2016) introduced the idea of including $\|\mathbf{X}\mathbf{u}\|_2$ in the denominator in (4.185), which is crucial in the analysis of expected loss.

We note that $\|\widehat{\boldsymbol{\beta}}^o - \boldsymbol{\beta}\|_2^2 \lesssim \sigma^2 s/n$, so that for $r_2^2\log(p/s) \lesssim s_1\log(p/s_1) + s$ and $\lambda_s \lesssim \lambda_{*,s}$, (4.203) yields

$$\big\|\widehat{\boldsymbol{\beta}} - \boldsymbol{\beta}\big\|_2^2 + \|\mathbf{X}\widehat{\boldsymbol{\beta}} - \mathbf{X}\boldsymbol{\beta}\|_2^2/n \lesssim (\sigma^2/n)\big\{s_1\log(p/s_1) + s\big\}. \tag{4.204}$$

Thus, the sorted concave PLS inherits the advantages of both the concave penalty (when $s_1 \ll s$) and sorted penalty (with adaptive choice of $\lambda_s$ without requiring the knowledge of $s$). In the worst case scenario where $s_1 \asymp s$, (4.203) matches the lower bound (4.17) and thus attains the minimax rate.

We first need the following lemma in the proof of Theorem 4.24.

**Lemma 4.11** *Let $\{s, A, L_x, q_{\alpha,A}, x_1, \mu_{\#,s}\}$ be as in (4.201).*
*(i) Let $z_j \sim N(0,\sigma_n^2)$. Then, for $s > 0$ and $q_{\alpha,A} > 0$*

$$P\bigg\{ \max_{x_1 \le j \le p} \frac{z^{\#}_{j-s}}{A\sigma_n L_j} < 1, \sum_{s<j<x_1} \frac{\big(z^{\#}_{j-s} - A\sigma_n L_j\big)_+^2}{(A\sigma_n L_s)^2} < \mu^2_{\#,s} \bigg\} > 1/2,$$

*and the right-hand side is greater than* $(1+q_{\alpha,A})/2$ *for* $s=0$.
*(ii) Let* $\lambda_{*,j}$ *be as in (4.197),* $\eta'=\eta-\eta_1$, $A=\eta' A_0$,

$$\widetilde{\Delta}(r,\eta') = \sup_{\|\mathbf{u}\|_2=1} \frac{\{\mathbf{u}_{\mathcal{S}^c}^T \mathbf{Z}_{\mathcal{S}^c} - \eta' \sum_{j=1}^{p-s} (\mathbf{u}_{\mathcal{S}^c})_j^{\#} \lambda_{*,s+j}\}_+}{r\lambda_{*,s} \max\{\|\mathbf{u}\|_2, \|\mathbf{X}\mathbf{u}\|_2 \sqrt{\gamma/n}\}},$$

*and* $r_1 = \eta' \mu_{\#,s}/(1-a_0)$. *If* $\boldsymbol{\varepsilon} \sim N(\mathbf{0}, \sigma^2 \mathbf{I})$ *and* $\max_{j\le p} \|\mathbf{X}_j\|_2^2 \le n$, *then*

$$P\Big\{\widetilde{\Delta}(r,\eta') \ge 1\Big\} \le \Phi\big(-a_0 r_1 A_0 L_s \sqrt{\gamma}\big).$$

**Proof of Lemma 4.11.** Set $\sigma_n = 1$ without loss of generality. We have

$$P\{z_j^{\#} > t\} \le \frac{2}{j} \sum_{k=1}^{p} \mathrm{E}\, e^{t(z_k - t)} I_{\{z_k > t\}} = (p/j) e^{-t^2/2}.$$

Define $x_k$ and $j_k$ by $L_{x_k}^2 = A^{-1} L_{j_{k-1}}^2$ and $j_k = \lceil x_k \rceil$, $k = 2, \ldots, k^*+1$. We have $L_{x_{k^*+1}}^2 \le A^{-k^*} L_{j_1}^2 \le L_p^2$. Because $z_j^{\#} \ge z_{j+1}^{\#}$ and $j_k/(j_k - s) \le 1/(1 - s/j_1)$,

$$\begin{aligned}
& P\Big\{\max_{j_1 \le j \le p} z_{j-s}^{\#}/L_j \ge A\Big\} \\
\le\ & \sum_{1\le k\le k^*} P\{z_{j_k - s}^{\#} \ge A L_{x_{k+1}} = A^{1/2} L_{j_k}\} \\
\le\ & \sum_{1\le k\le k^*} \frac{p}{j_k - s} \exp\Big(-A L_{j_k}^2/2\Big) \qquad (4.205) \\
\le\ & \sum_{1\le k\le k^*} \frac{p}{j_k - s} \Big(\frac{\alpha j_k}{p}\Big) \exp\Big(-(A-1)A^{k^*-k} L_p^2/2\Big) \\
\le\ & \frac{j_1}{j_1 - s} \sum_{1\le k\le k^*} \alpha^{1+(A-1)A^{k^*-k}} \\
=\ & (1 - q_{\alpha,A})/2.
\end{aligned}$$

This completes the proof for $s = 0$, which gives $x_1 = 1$. For $s > 0$,

$$\mathrm{E} \sum_{s<j<j_1} \frac{(z_{j-s}^{\#} - A L_j)_+^2}{A^2 L_s^2} \le \frac{p\,\mathrm{E}(|z_1| - A L_{j_1 - 1})_+}{A^2 L_s^2}.$$

Let $J_k(t) = \int_0^\infty x^k e^{-x - x^2/(2t^2)} dx$. Because

$$J_{k+1}(t) + J_{k+2}(t)/t^2 = -\int_0^\infty x^{k+1} d e^{-x - x^2/(2t^2)} = (k+1) J_k(t),$$

we have

$$\frac{t^2 E(z_1 - t)_+^2}{\Phi(-t)} = \frac{J_2(t)}{J_0(t)} = \frac{J_2(t)}{\{J_2(t) + J_3(t)/t^2\}/2 + J_2(t)/t^2} \le \frac{2t^2}{t^2+2}.$$

As $\Phi(-AL_{x_1}) \le e^{-(AL_x)^2/2} \le (\alpha x_1/p)^{A^2}$, it follows that

$$\mathrm{E}\sum_{s<j<j_1} \frac{\left(z^{\#}_{j-s} - AL_j\right)_+^2}{A^2 L_s^2} \le \frac{4p(\alpha x_1/p)^{A^2}}{A^2 L_s^2 (A^2 L_{x_1}^2 + 2)} = \mu^2_{\#,s} q_{\alpha,A}/2.$$

Thus, by the Markov inequality

$$P\Big\{ \sum_{s<j<x_1} \left(z^{\#}_{j-s} - AL_j\right)_+^2 > A^2 L_s^2 \mu^2_{\#,s} \Big\} \le q_{\alpha,A}/2.$$

This and (4.205) complete the proof of part (i) by the union bound.

By the definition of the sorted norm $\|\cdot\|_{\#,s,*}$,

$$\mathbf{u}^T_{\mathcal{S}^c}\mathbf{Z}_{\mathcal{S}^c} - \eta' \sum_{j=1}^{p-s} (\mathbf{u}_{\mathcal{S}^c})^{\#}_j \lambda_{*,s+j} \le \sum_{j=1}^{p-s} (\mathbf{u}_{\mathcal{S}^c})^{\#}_j \left(z^{\#}_j - \eta' \lambda_{*,s+j}\right)_+.$$

Thus, as $\eta' \lambda_{*,s} = A(\sigma/n^{1/2})L_j$ and $Z_j \sim N(0, \sigma^2/n)$, Lemma 4.11 yields

$$\mathrm{median}\Big(\widetilde{\Delta}(r_1, \eta')\Big) \le \eta' \mu_{\#,s}/r_1 = 1 - a_0.$$

As $\mathrm{supp}(\mathbf{Z}) \subseteq \mathcal{S}^c$ and $\mathbf{u}^T_{\mathcal{S}^c}\mathbf{Z}_{\mathcal{S}^c} = (\mathbf{X}\mathbf{u})^T(\mathbf{Y} - \mathbf{X}\widehat{\boldsymbol\beta}^o)/n$, $\widetilde{\Delta}(r_1, \eta')$ is convex in $\mathbf{Y} - \mathbf{X}\boldsymbol\beta^o$ with Lipschitz coefficient $1/(r_1 \lambda_{*,s}\sqrt{n\gamma}) = 1/(r_1 A_0 \sigma L_s \sqrt{\gamma})$. Therefore, part (ii) follows from the Gaussian concentration inequality. □

**Proof of Theorem 4.24.** Let $\eta' = \eta - \eta_1$ and $\lambda_{*,j}$ be as in (4.197). As $\sum_{j=1}^{p-s}(u_j z_{s+j} - \eta u^{\#}_j \lambda_{s+j}) \le \sum_{j=1}^{p-s}(u_j z_{s+j} - \eta u^{\#}_j \lambda_{*,s+j})$ and $\mathbf{Z}_{\mathcal{S}} = \mathbf{0}$,

$$\Delta(r, \eta, \mathbf{w}, \boldsymbol\nu_{approx}) \le \frac{r_1}{r}\widetilde{\Delta}(r_1, \eta') + \frac{\|\mathbf{w}_{\mathcal{S}}\|_2 + r_2}{r}$$

under conditions (4.197) and (4.199), in view of the definition of $\Delta(r, \eta, \mathbf{w}, \boldsymbol\nu)$ in (4.185). It follows that when $\Delta(r_1, \eta - \eta_1, \mathbf{0}, \mathbf{0}) < 1$

$$\Delta((1-\eta)\xi s^{1/2}, \eta, \mathbf{w}^o, \boldsymbol\nu_{approx}) < 1 \ \forall \widehat{\boldsymbol\beta} \in \mathscr{B}_2$$

due to the choice of $\xi$ and the bound $\|\mathbf{w}^o_{\mathcal{S}}\|_2^2 \le \sum_{i=1}^s \lambda_j^2/\lambda_s^2 \le B_0 s$ by (4.198). Moreover, for any specific $\widehat{\boldsymbol\beta} \in \mathscr{B}_2$, $\Delta(r, \eta, \mathbf{w}^o, \boldsymbol\nu_{approx}) < (r_1 + w + r_2)/r$ when $\|\mathbf{w}^o_{\mathcal{S}}\|_2 \le w$. This verifies condition (4.186) for the whole class $\mathscr{B}_2$ with the larger $r = (1-\eta)\xi s^{1/2}$ and for the specific $\widehat{\boldsymbol\beta} \in \mathscr{B}_2$ with the specific $r$. Hence, in view of the condition $\kappa(\mathrm{Pen}) \le \mathrm{RE}^2$ for the class $\mathscr{B}_2$, the conclusion follows from Theorem 4.23. ■

**Proof of Corollary 4.9.** When the penalty function is convex,

$$\mathbf{h}^T\overline{\boldsymbol\Sigma}\mathbf{h} + (1-\eta)\lambda_s \|\mathbf{h}_{\mathcal{S}^c}\|^{(k)}_{\#,s} \le (1-\eta)\xi s^{1/2}\lambda_s F(\mathbf{h})$$

with probability at least $\Phi\big(a_0 r_1 A_0 L_s\sqrt{\gamma}\big)$. In this event,

$$(\mathbf{h}^T\overline{\boldsymbol{\Sigma}}\mathbf{h})^{1/2} \le (1-\eta)\xi s^{1/2}\lambda_s \max\left\{\frac{1}{\mathrm{RE}_{\#}(\mathcal{S};\xi,0)}, \gamma^{1/2}\right\}$$

as $\|\mathbf{h}\|_2^2 \ge \gamma\mathbf{h}^T\overline{\boldsymbol{\Sigma}}\mathbf{h}$ implies

$$\|\mathbf{h}\|_2\mathrm{RE}_{\#}(\mathcal{S};\xi,0)(\mathbf{h}^T\overline{\boldsymbol{\Sigma}}\mathbf{h})^{1/2} \le \mathbf{h}^T\overline{\boldsymbol{\Sigma}}\mathbf{h} \le (1-\eta)\xi s^{1/2}\lambda_s\|\mathbf{h}\|_2$$

and $\|\mathbf{h}\|_2^2 \le \gamma\mathbf{h}^T\overline{\boldsymbol{\Sigma}}\mathbf{h}$ implies $\mathbf{h}^T\overline{\boldsymbol{\Sigma}}\mathbf{h} \le (1-\eta)\xi s^{1/2}\lambda_s(\gamma\mathbf{h}^T\overline{\boldsymbol{\Sigma}}\mathbf{h})^{1/2}$. It follows that

$$(\mathbf{h}^T\overline{\boldsymbol{\Sigma}}\mathbf{h})^{1/2} \le (1-\eta)\xi s^{1/2}\lambda_s \max\left\{\frac{1}{\mathrm{RE}_{\#}(\mathcal{S};\xi,0)}, \frac{(N(0,1))_+}{a_0 r_1 A_0 L_s}\right\}$$

for a certain standard Gaussian variable. The conclusion follows. ∎

### *4.5.3 Properties of LLA and LCA*

In this subsection, we provide prediction and estimation error bounds for local solutions computable by a logarithmic number of recursive applications of the local linear approximation (*LLA*) or local convex approximation (*LCA*), respectively for separable and sorted penalties. The error bounds are comparable with those in Theorems 4.23 and Theorems 4.24 and hold under a similar restricted eigenvalue condition on the design. Error bounds of the same type were established under various sets of regularity conditions in Zhang (2010), Agarwal, Negahban and Wainwright (2012), Huang and Zhang (2012), Negahban *et al.* (2012), Wang, Liu and Zhang (2014), Loh and Wainwright (2015), Fan *et al.* (2015), and Feng and Zhang (2017) among others. However, most of these studies have focused on separable penalties with relatively high penalty levels such as (4.166). The analysis presented here is adapted from Feng and Zhang (2017).

As we have discussed respectively in Chapter 3 and Subsection 4.5.1, both the LLA and LCA of the PLS problem (4.108) can be written as *MM* algorithm

$$\widehat{\boldsymbol{\beta}}^{(new)} = \arg\min_{\mathbf{b}}\left\{\|\mathbf{Y}-\mathbf{X}\mathbf{b}\|_2^2/(2n) + \mathrm{Pen}^{(new)}(\mathbf{b})\right\} \tag{4.206}$$

with $\mathrm{Pen}^{(new)}(\mathbf{b}) \ge \mathrm{Pen}(\mathbf{b})$ for all $\mathbf{b}$ and $\mathrm{Pen}^{(new)}(\widehat{\boldsymbol{\beta}}^{(old)}) = \mathrm{Pen}(\widehat{\boldsymbol{\beta}}^{(old)})$. In the case of the LLA, $\mathrm{Pen}(\mathbf{b}) = \sum_{j=1}^p p_\lambda(|b_j|)$ is separable and possibly concave, and

$$\mathrm{Pen}^{(new)}(\mathbf{b}) = \sum_{j=1}^p \left\{p_\lambda\big(|\widehat{\beta}_j^{(old)}|\big) + p_\lambda'\big(|\widehat{\beta}_j^{(old)}|\big)\Big(|b_j| - |\widehat{\beta}_j^{(old)}|\Big)\right\}.$$

In the case of the LCA, the maximum concavity $\kappa(\text{Pen})$ of the penalty (4.171) is required to be no greater than $\kappa_0$ and as in (4.181)

$$\text{Pen}^{(new)}(\mathbf{b}) = \text{Pen}(\mathbf{b}) + (\kappa_0/2)\big\|\mathbf{b} - \widehat{\boldsymbol{\beta}}^{(old)}\big\|_2^2.$$

In both cases, (4.206) is a convex minimization problem. As (4.206) is typically applied recursively in practice, it is wasteful to compute its exact solution. Thus, we shall study its approximate solutions of the form

$$\mathbf{X}^T(\mathbf{Y} - \mathbf{X}\widehat{\boldsymbol{\beta}}^{(new)})/n = \nabla\text{Pen}^{(new)}(\widehat{\boldsymbol{\beta}}^{(new)}) + \boldsymbol{\nu}_{approx}. \tag{4.207}$$

As $\text{Pen}^{(new)}(\mathbf{b})$ is convex in $\mathbf{b}$ for both the LLA and LCA, an application of the calculation in (4.174) to (4.207) yields

$$\begin{aligned}\mathbf{h}^T\overline{\boldsymbol{\Sigma}}\mathbf{h} &\le \mathbf{h}^T(\mathbf{Z} - \boldsymbol{\nu}_{approx}) - \mathbf{h}^T\nabla\text{Pen}^{(new)}(\widehat{\boldsymbol{\beta}}^o) \\ &= \mathbf{h}^T(\mathbf{Z} - \boldsymbol{\nu}_{approx} + \boldsymbol{\nu}_{carry}) - \mathbf{h}^T\nabla\text{Pen}(\widehat{\boldsymbol{\beta}}^o),\end{aligned}$$

where $\mathbf{h} = \widehat{\boldsymbol{\beta}}^{(new)} - \widehat{\boldsymbol{\beta}}^o$, $\overline{\boldsymbol{\Sigma}} = \mathbf{X}^T\mathbf{X}/n$, $\mathbf{Z} = \mathbf{X}^T(\mathbf{Y} - \mathbf{X}\widehat{\boldsymbol{\beta}}^o)/n$, and $\boldsymbol{\nu}_{carry} = \nabla\text{Pen}(\widehat{\boldsymbol{\beta}}^o) - \nabla\text{Pen}^{(new)}(\widehat{\boldsymbol{\beta}}^o)$ can be viewed as the carryover error from $\widehat{\boldsymbol{\beta}}^{(old)}$. Similar to the derivation of (4.184), this inequality and Lemma 4.10 yield

$$\mathbf{h}^T\overline{\boldsymbol{\Sigma}}\mathbf{h} + \lambda_s\|\mathbf{h}_{\mathcal{S}^c}\|_{\#,s} + \lambda_s\mathbf{h}_{\mathcal{S}}^T\mathbf{w}_{\mathcal{S}}^o \le \mathbf{h}^T(\mathbf{Z} - \boldsymbol{\nu}_{approx} + \boldsymbol{\nu}_{carry}). \tag{4.208}$$

We note that $\|\cdot\|_{\#,s} = \|\cdot\|_1$ when $\lambda_1 = \cdots = \lambda_p$, and that the term $\kappa_0\|\mathbf{h}\|_2^2$ in (4.184) is replaced by the carryover term $\mathbf{h}^T\boldsymbol{\nu}_{carry}$ here. Interestingly,

$$\big\|\boldsymbol{\nu}_{carry}\big\|_2 \le \kappa_0\big\|\widehat{\boldsymbol{\beta}}^{(old)} - \widehat{\boldsymbol{\beta}}^o\big\|_2 \tag{4.209}$$

when $\kappa(\text{Pen}) \le \kappa_0$, as $\big|p'_\lambda(|\widehat{\beta}_j^{(old)}|) - p'_\lambda(|\widehat{\beta}_j^o|)\big| \le \kappa_0\big|\widehat{\beta}_j^{(old)} - \widehat{\beta}_j^o\big|$ for the LLA and $\boldsymbol{\nu}_{carry} = \kappa_0\big(\widehat{\boldsymbol{\beta}}^{(old)} - \widehat{\boldsymbol{\beta}}^o\big)$ for the LCA.

It can be clearly seen by inspecting (4.184) and (4.208) that the analysis in Subsection 4.5.2 can also be applied to obtain error bounds for the approximation solutions (4.207) for the LLA and LCA. Here we apply this analysis to recursive applications of the algorithm:

$$\widehat{\boldsymbol{\beta}}^{(k)} \leftarrow \text{LA}\big(\widehat{\boldsymbol{\beta}}^{(k-1)}, \text{Pen}^{(k)}, \boldsymbol{\nu}_{approx}^{(k)}\big) \tag{4.210}$$

where $\text{Pen}^{(k)}$ are sorted penalties (4.169) with penalty levels $\lambda_1^{(k)} \ge \cdots \ge \lambda_p^{(k)}$ and maximum concavity $\kappa(\text{Pen}^{(k)}) \le \kappa_0$, $\widehat{\boldsymbol{\beta}}^{(new)} \leftarrow \text{LA}\big(\widehat{\boldsymbol{\beta}}^{(old)}, \text{Pen}, \boldsymbol{\nu}\big)$ is the one-step mapping in (4.207) with $\text{Pen}^{(new)}$ depending on $\widehat{\boldsymbol{\beta}}^{(old)}$ and Pen. This includes separable penalties with $\lambda_1^{(k)} = \cdots = \lambda_p^{(k)}$.

**Theorem 4.25** *Let $\mathcal{S} \supseteq \text{supp}(\boldsymbol{\beta})$, $\widehat{\boldsymbol{\beta}}^o$ be the oracle LSE in (4.10) and $s = |\mathcal{S}|$. Let $\widehat{\boldsymbol{\beta}}^{(k)}$ be generated by (4.210) with penalty levels and concavity as specified.*

*Let* $\theta_0 = \kappa_0/\mathrm{RE}^2_{\#}(\mathcal{S};\xi,\gamma)$, $\lambda_s^{(0)} = \lambda_s^{(1)}$, $r^{(0)} > 0$, *and* $r^{(k)}$, $1 \le k \le k^*$, *be real numbers satisfying* $\sum_{\ell=0}^{k}\theta_0^{k-\ell}(\lambda_s^{(\ell)}/\lambda_s^{(k)})r^{(\ell)} \le (1-\eta)\xi s^{1/2}$. *Let* $\mathbf{w}^{(k)} = \nabla \mathrm{Pen}^{(k)}(\widehat{\boldsymbol\beta}^{o})/\lambda_s^{(k)}$. *Suppose* $\lambda_s^{(0)} r^{(0)} \ge \mathrm{RE}^2_{\#}(\mathcal{S};\xi,\gamma)\big\|\widehat{\boldsymbol\beta}^{(0)} - \widehat{\boldsymbol\beta}^{o}\big\|_2$ *and*

$$\Delta\big(r^{(k)}, \mathbf{w}^{(k)}, \boldsymbol\nu^{(k)}_{approx}\big) \le 1,\ k = 1,\ldots,k^*. \tag{4.211}$$

*Let* $\psi^{(k)}(\mathbf{u}) = F(\mathbf{u})\sum_{\ell=0}^{k}\theta_0^{k-\ell}(\lambda_s^{(\ell)}/\lambda_s^{(k)})r^{(\ell)} - (1-\eta)\lambda_s^{(k)}\|\mathbf{u}_{\mathcal{S}^c}\|_{\#,s}$ *and* $\mathbf{h}^{(k)} = \widehat{\boldsymbol\beta}^{(k)} - \boldsymbol\beta^o$. *Then, for any seminorm* $\|\cdot\|$,

$$\big\|\mathbf{h}^{(k)}\big\| \le \lambda_s^{(k)} \sup_{\mathbf{u}} \frac{\|\mathbf{u}\|\psi^{(k)}(\mathbf{u})}{\mathbf{u}^T\overline{\boldsymbol\Sigma}\mathbf{u}},\quad k = 1,\ldots,k^*. \tag{4.212}$$

*In particular, for* $F(\mathbf{u}) = \max\big\{\|\mathbf{u}\|_2, (\gamma\mathbf{u}^T\overline{\boldsymbol\Sigma}\mathbf{u})^{1/2}\big\}$,

$$F\big(\mathbf{h}^{(k)}\big) \le \frac{r^{(k)}\lambda_s^{(k)}}{\mathrm{RE}^2_{\#}(\mathcal{S};\xi,\gamma)} + \theta_0\big\|\mathbf{h}^{(k-1)}\big\|_2 \le \sum_{\ell=0}^{k}\frac{\theta_0^{k-\ell}r^{(\ell)}\lambda_s^{(\ell)}}{\mathrm{RE}^2_{\#}(\mathcal{S};\xi,\gamma)}. \tag{4.213}$$

Theorem 4.25 is most useful when $\theta_0 < 1$. For the computation of an approximate solution with a specific sorted penalty, (4.213) yields

$$F\big(\mathbf{h}^{(k)}\big) \le \frac{(1+o(1))C_0 r\lambda_s}{\mathrm{RE}^2_{\#}(\mathcal{S};\xi,\gamma)} \tag{4.214}$$

with the same $r\lambda_s$ in Theorem 4.23 and $C_0 = 1/(1-\theta_0)$, when $k$ is of the order $\log(1+r^{(0)}\lambda_s^{(0)}/(r\lambda_s))$. This error bound is nearly identical to the corresponding one in (4.195). A main advantage here is that the algorithm and the number of required iterations are more explicitly given. However, a stronger RE condition is required here as $\xi \approx C_0 r/\{(1-\eta)s^{1/2}\}$ has an extra factor $C_0$, compared to (4.186). As we have discussed below Theorem 4.24, the right-hand side of (4.214) is of the order $s_1\lambda_{s_1}^2 + r_2^2\lambda_s^2 + o(1)s\lambda_s^2/\log(p/s)$, where $s_1$ is the number of small nonzero $\beta_j^o$ and $r_2$ is the size of the $\ell_2$ component of the approximation error.

Next, we derive the counterpart of Theorem 4.24 for the LLA and LCA.

**Theorem 4.26** *Let* $\{a_0, A_0, B_0, \eta, \eta_1, r_1, r_2, \mu_{\#,s}, L_s\}$ *be as in Theorem 4.24 and* $\{\mathcal{S}, \widehat{\boldsymbol\beta}^{o}, s, F(\mathbf{u})\}$ *as in Theorem 4.25. Suppose (4.197) and (4.198) hold for* $\boldsymbol\lambda = \boldsymbol\lambda^{(k)}$, $k \ge 1$, *with sorted penalty levels* $\boldsymbol\lambda^{(k)} = \big(\lambda_1^{(k)},\ldots,\lambda_p^{(k)}\big)^T$. *Let* $\lambda_s^{(0)} = \lambda_s^{(1)}$, $r^{(0)} > 0$, $r^{(k)} = r_1 + r_2 + w^{(k)}$ *for* $k \ge 1$. *Suppose for all* $1 \le k \le k^*$ *and certain* $\kappa_0 > 0$, $\gamma \ge 0$ *and* $\xi > 0$,

$$\sum_{\ell=0}^{k}\left\{\frac{\kappa_0}{\mathrm{RE}^2_{\#,*}(\mathcal{S};\xi,\gamma)}\right\}^{k-\ell}\frac{(\lambda_s^{(\ell)}/\lambda_s^{(k)})r^{(\ell)}}{(1-a_0)(1-\eta)s^{1/2}} \le \xi \tag{4.215}$$

*for the restricted eigenvalue in (4.200). Let* $\mathrm{Pen}^{(k)}$ *be sorted (concave) penalties with penalty level* $\boldsymbol\lambda^{(k)}$ *and maximum concavity* $\kappa\big(\mathrm{Pen}^{(k)}\big) \le \kappa_0$. *Suppose*

$\widehat{\boldsymbol{\beta}}^{(k)}$ *are generated by (4.210) such that (4.199) holds for* $\boldsymbol{\nu}_{approx} = \boldsymbol{\nu}^{(k)}_{approx}$, $1 \le k \le k^*$. *Suppose* $\boldsymbol{\varepsilon} \sim N(\mathbf{0}, \sigma^2\mathbf{I})$ *and* $\max_{j\le p}\|\mathbf{X}_j\|_2^2 \le n$. *Then,*

$$\begin{aligned} & P\Big\{ (4.212) \text{ and } (4.213) \text{ hold for all } 1 \le k \le k^* \Big\} \qquad (4.216) \\ \ge\ & P\Big\{ \max_{1\le k\le k^*} \|\mathbf{w}_S^{(k)}\|_2/w^{(k)} \le 1,\ \mathrm{RE}^2_{\#,*}(\mathcal{S};\xi,\gamma)\big\|\widehat{\boldsymbol{\beta}}^{(0)} - \widehat{\boldsymbol{\beta}}^{o}\big\|_2 \le \lambda_s^{(0)} r^{(0)} \Big\} \\ & -\Phi\big(a_0 r_1 A_0 L_s\sqrt{\gamma}\big). \end{aligned}$$

Theorems 4.25 and 4.26 are applicable to a rich spectrum of possible implementations of the LLA and LCA, and provide guidelines on the choice of penalties and constraints on the approximation error in intermediate steps. For the PLS with separable concave penalties, the number of steps needed is of the order $\log(1 + r^{(0)}\lambda/(r\lambda)) \asymp \log(1 + s/s_1)$ if we initialize from the Lasso, where $s_1$ is the number of small nonzero coefficients. For sorted concave penalties, the number of steps needed is of the order $\log(1 + r^{(0)}\lambda_s^{(0}/(r\lambda)) \asymp \log(1 + s\log(p/s)/(s_1\log(p/s_1))$ if we initialize from the SLOPE.

**Proof of Theorem 4.25.** We prove the theorem by induction on

$$\big\|\mathbf{h}^{(k)}\big\|_2 \le \sum_{\ell=0}^{k} \theta_0^{k-\ell}\lambda_s^{(\ell)} r^{(\ell)}/\mathrm{RE}^2_{\#}(\mathcal{S};\xi,\gamma)$$

as in (4.213) since $r^{(0)}\lambda_s^{(0)}/\mathrm{RE}^2_{\#}(\mathcal{S};\xi,\gamma)$ is assumed to bound $\|\mathbf{h}^{(0)}\|_2$. By the induction assumption,

$$\begin{aligned} r^{(k)} + \kappa_0\big\|\mathbf{h}^{(k-1)}\big\|_2/\lambda_s^{(k)} &\le r^{(k)} + \frac{\kappa_0}{\lambda_s^{(k)}}\sum_{\ell=0}^{k-1}\frac{\theta_0^{k-1-\ell}\lambda_s^{(\ell)} r^{(\ell)}}{\mathrm{RE}^2_{\#}(\mathcal{S};\xi,\gamma)} \\ &= \sum_{\ell=0}^{k}\theta_0^{k-\ell}\big(\lambda_s^{(\ell)}/\lambda_s^{(k)}\big) r^{(\ell)}. \end{aligned}$$

It follows from (4.208), (4.209) and (4.211) that

$$\begin{aligned} & \big(\mathbf{h}^{(k)}\big)^T\overline{\boldsymbol{\Sigma}}\mathbf{h}^{(k)} + (1-\eta)\lambda_s^{(k)}\|\mathbf{h}^{(k)}_{\mathcal{S}^c}\|_{\#,s} \\ \le\ & \lambda_s^{(k)} r^{(k)} F\big(\mathbf{h}^{(k)}\big) + \kappa_0\big\|\mathbf{h}^{(k-1)}\big\|_2\big\|\mathbf{h}^{(k)}\big\|_2 \\ \le\ & \big\{\lambda_s^{(k)} r^{(k)} + \kappa_0\big\|\mathbf{h}^{(k-1)}\big\|_2\big\}F\big(\mathbf{h}^{(k)}\big) \\ \le\ & \lambda_s^{(k)} F\big(\mathbf{h}^{(k)}\big)\sum_{\ell=0}^{k}\theta_0^{k-\ell}\big(\lambda_s^{(\ell)}/\lambda_s^{(k)}\big) r^{(\ell)} \\ \le\ & \lambda_s^{(k)}(1-\eta)\xi s^{1/2} F\big(\mathbf{h}^{(k)}\big), \end{aligned}$$

so that $\mathbf{h}^{(k)}$ is inside the cone (4.188). Thus,

$$\mathrm{RE}^2_{\#}(\mathcal{S};\xi,\gamma)F\big(\mathbf{h}^{(k)}\big) \le \lambda_s^{(k)} r^{(k)} + \kappa_0\big\|\mathbf{h}^{(k-1)}\big\|_2 \le \sum_{\ell=0}^{k}\theta_0^{k-\ell} r^{(\ell)}\lambda_s^{(\ell)}$$

by (4.189) and the induction assumption. This completes the induction, and the conclusions follow. ■

**Proof of Theorem 4.26.** The proof of Theorem 4.26 is nearly identical to that of Theorem 4.24 as it suffices to verify that (4.211) holds with large probability. As $w^{(k)}$ changes with $k$, we use the bound

$$\Delta\big(r^{(k)}, \mathbf{w}^{(k)}, \boldsymbol{\nu}^{(k)}_{approx}\big) I\big\{\|\mathbf{w}^{(k)}_{S}\|_2 \le w^{(k)}\big\} \le \frac{r_1}{r^{(k)}}\widetilde{\Delta}(r_1, \eta') + \frac{w^{(k)} + r_2}{r^{(k)}}.$$

Thus, as (4.211) holds for the approximation error $\boldsymbol{\nu}_{approx}$,

$$P\Big\{\Delta\big(r^{(k)}, \mathbf{w}^{(k)}, \boldsymbol{\nu}^{(k)}_{approx}\big) \le 1,\ 1 \le k \le k^*,\ \|\mathbf{h}^{(0)}\|_2 \le \lambda^{(0)}_{s} r^{(0)}\Big\}$$

is no smaller than the right-hand side of (4.216). The conclusion follows from Theorem 4.25. ■

## 4.6 Bibliographical Notes

Properties of regularized least squares methods have been studied by many. See Bibliographical Notes in Chapter 3 for additional notes. As a comprehensive discussion of the literature is infeasible, we shall focus on some important contributions, for example the establishment of rate optimal bounds such as (4.17) and (4.18).

For $\ell_1$ regularization, prediction error bounds of type (4.17) were established for the Lasso in Greenshtein and Ritov (2004) under an $\ell_1$ condition on the coefficient coefficient, selection consistency of the Lasso was established by Meinshausen and Bühlmann (2006), Tropp (2006) and Zhao and Yu (2006) under (4.18) and the strong irrepresentable condition on the design, and the prediction and coefficient estimation error bounds of type (4.17) were established in Candés and Tao (2007) for the Dantzig selector under the *RIP* condition on the design and in Zhang and Huang (2008) for the Lasso under the *SRC*. As the RIP and SRC are imposed on both the upper and lower sparse eigenvalues of the design, these conditions were significantly weakened in Bickel, Ritov and Tsybakov (2009) which was later proved to depend on the lower sparse eigenvalue only in Rudelson and Zhou (2013). van de Geer and Bühlmann (2009) proved that for prediction and $\ell_1$ error bounds, the RE condition can be further weakened to the compatibility condition. For smaller penalty levels, error bounds of type (4.17) were established in Sun and Zhang (2013) with nonadaptive penalty level and in Su and Candés (2016) and Bellec, Lecué and Tsybakov (2016) adaptively with the SLOPE.

Lower performance bounds for selection, estimation and prediction, comparable with (4.17) and (4.18), were first obtained in Wainwright (2009), Ye and Zhang (2010) and Raskutti, Wainwright and Yu (2011).

For concave PLS, Fan and Li (2001) and Fan and Peng (2004) established the existence of the oracle local solution, and thus the oracle property of the concave PLS. Comparable results were obtained in Zou (2006) for adaptive Lasso. Under the SRC, Zhang (2010a) proved the selection consistency of concave PLS under (4.18) and estimation error bounds of type (4.17) for a computed local solution. This significantly weakens the strong irrepresentable condition for selection consistency to the SRC. Similar theoretical results for the LLA were established in Zhang (2010). Fan, Xue and Zou (2014) showed that the two-step LLA algorithm produces a strong oracle solution as long as the problem is localizable and regular. Lv and Fan (2010) investigated the properties of folded concave penalized least-squares when dimensionality diverges and Fan and Lv (2011) established a uniform rate of convergence and the oracle property of folded concave penalized likelihood. Recently, the SRC was further reduced to the RE condition in Feng and Zhang (2017) for selection, prediction and coefficient estimation.

## 4.7 Exercises

4.1 Verify (4.4)

4.2 Show that (4.7) with $q = 2$.

4.3 Verify (4.12), and further show (4.13).

4.4 For all $t > 0$, prove

$$\frac{t\varphi(t)}{1+t^2} \le \Phi(-t) = e^{-t^2/2}\int_0^\infty e^{-xt}\varphi(x)dx \le \min\left\{\varphi(t)/t, e^{-t^2/2}/2\right\}.$$

where $\varphi(t)$ and $\Phi(t)$ are the probability density and cumulative distribution function of the standard normal distribution $N(0,1)$.

4.5 Prove Corollary 4.1, Corollary 4.2(ii). and Corollary 4.3.

Chapter 5

# Generalized Linear Models and Penalized Likelihood

This chapter extends techniques related to the least squares regression to regression models with binary, categorial and count responses. We begin by introducing generalized linear models which provide a unified framework for modeling binary, categorial and count outcomes, and then move to introduce the penalized likelihood method and discuss further the numerical optimization algorithms and tuning parameter selection for the penalized likelihood. Asymptotic properties for both low-dimensional and high-dimensional problems are also presented. We conclude this chapter by presenting the penalized $M$-estimation with decomposable penalty functions.

## 5.1 Generalized Linear Models

*Generalized linear models* (*GLIM*) have numerous applications in diverse research fields including medicine, psychology, engineering, among others. McCullagh and Nelder (1989) provides a comprehensive account of the generalized linear models. This section gives a brief introduction of GLIM based on the theory of exponential families.

### 5.1.1 *Exponential family*

The distribution of a random variable $Y$ belongs to a canonical exponential family, if the density or probability mass function of $Y$ can be written as

$$f(y) = \exp[\{\theta y - b(\theta)\}/a(\phi) + c(y, \phi)] \tag{5.1}$$

for some known functions $a(\cdot)$, $b(\cdot)$ and $c(\cdot, \cdot)$. The parameter $\theta$ is called a *canonical parameter* and $\phi$ is called a *dispersion parameter*. As illustrated below, the exponential family includes many commonly-used distributions as special cases, including normal distributions, binomial distributions, Poisson distributions and gamma distributions.

**Example 5.1** (*Normal distribution*) Rewrite the density of normal distribution $N(\mu, \sigma^2)$ as

$$f(y; \mu, \sigma^2) = \exp\left\{\frac{y\mu - \mu^2/2}{\sigma^2} - \frac{y^2}{2\sigma^2} - \log(\sqrt{2\pi}\sigma)\right\}.$$

Let $\theta = \mu$ and $\phi = \sigma^2$. By setting $a(\phi) = \sigma^2$, $b(\theta) = \theta^2/2$ and $c(y, \phi) = -y^2/(2\sigma^2) - \log(\sqrt{2\pi}\sigma)$, the normal distribution belongs to the exponential family. ■

**Example 5.2** (*Binomial distribution*) The probability mass function of binomial distribution $b(m, p)$ with $p \in (0, 1)$, is

$$P(Y = y) = \binom{m}{y} p^y (1-p)^{m-y}.$$

We can rewrite the probability function as

$$P(Y = y) = \exp\left\{y \log(\frac{p}{1-p}) + m \log(1-p) + \log\binom{m}{y}\right\}.$$

The canonical parameter $\theta = \log(\frac{p}{1-p})$, and $\phi = 1$. Take $a(\phi) = 1$, $b(\theta) = m \log(1-p) = -m \log\{1 + \exp(\theta)\}$ and $c(y, \phi) = \log\binom{m}{y}$. Thus, binomial distribution also belongs to the exponential family. ■

**Example 5.3** (*Poisson distribution*) The probability mass function of a Poisson distribution with mean $\lambda$ is

$$P(Y = y) = \frac{\lambda^y \exp(-\lambda)}{y!} = \exp\big(y \log(\lambda) - \lambda - \log(y!)\big).$$

Here, the canonical parameter $\theta = \log(\lambda)$ and dispersion parameter $\phi = 1$, taking $a(\phi) = 1$, $b(\theta) = \lambda = \exp(\theta)$ and $c(y, \phi) = \log(y!)$. Thus, the Poisson distribution belongs to the exponential family. The Poisson distribution is useful for modelling count data. It is particularly useful for situations in which the outcome is a counting variable with approximately the same mean and variance. ■

**Example 5.4** (*Gamma distribution*) The gamma distribution with mean $\mu$ and variance $\mu^2/\alpha$ has the following density function

$$\frac{1}{\Gamma(\alpha)}\left(\frac{\alpha y}{\mu}\right)^{\alpha} \exp\left(-\frac{\alpha y}{\mu}\right)\frac{1}{y},$$

where $\Gamma(\cdot)$ is the gamma function. The density function can be expressed in the following form

$$f(y;\mu,\alpha) = \exp\{\alpha(-y/\mu - \log\mu) + (\alpha-1)\log(y) + \alpha\log\alpha - \log\Gamma(\alpha)\}.$$

For the gamma distribution, the canonical parameter $\theta = -\frac{1}{\mu}$ and dispersion parameter $\phi = 1/\alpha$. Set $a(\phi) = 1/\alpha$, $b(\theta) = \log(\mu) = -\log(-\theta)$ and $c(y,\phi) = (\alpha-1)\log(y) + \alpha\log\alpha - \log\Gamma(\alpha)$. The value of $\alpha$ determines the shape of the distribution. In particular, the distribution with $\alpha = 1$ corresponds to the exponential distribution. ■

Denote $\ell(\theta)$ as the logarithm of likelihood function. Under some regularity conditions, it is easy to show that

$$\mathrm{E}\,\ell'(\theta) = 0, \tag{5.2}$$
$$\mathrm{E}\,\ell''(\theta) + \mathrm{E}\{\ell'(\theta)\}^2 = 0. \tag{5.3}$$

To see this, we let $f(\mathbf{x};\theta)$ be the density or probability mass function, then $\ell(\theta) = \log f(\mathbf{x};\theta)$. If the integral and derivative are exchangeable, then

$$\mathrm{E}\,\ell'(\theta) = \mathrm{E}\,\frac{\partial f(\mathbf{X};\theta)}{\partial\theta}/f(\mathbf{X};\theta) = \int \frac{\partial f(\mathbf{x};\theta)}{\partial\theta}d\mathbf{x} = \frac{\partial}{\partial\theta}\int f(\mathbf{x};\theta)d\mathbf{x} = 0,$$

since the integral of the density is one. Taking one more derivative with respect to $\theta$, we get

$$0 = \frac{\partial}{\partial\theta}\mathrm{E}\,\ell'(\theta) = \frac{\partial}{\partial\theta}\int \ell'(\theta)f(\mathbf{x};\theta)dx = \mathrm{E}[\ell''(\theta) + \{\ell'(\theta)\}^2],$$

which is (5.3). Equations (5.2) and (5.3) have been referred to as the *Bartlett first and second identities*. The Bartlett first identity gives us the consistency of the maximum likelihood estimate of $\theta$, and the second identity provides us with the efficiency of the maximum likelihood estimate of $\theta$. The above identity holds for multivariate parameters. In this case $\ell'(\boldsymbol{\theta})$ is the gradient of vector and $\ell''(\boldsymbol{\theta})$ is its associated Hessian matrix and the last term in (5.3) should be replaced by $\mathrm{E}\,\ell'(\boldsymbol{\theta})\ell'(\boldsymbol{\theta})^T$. The proof is identical to what is given above.

For the canonical exponential family,

$$\ell(\theta) = \{Y\theta - b(\theta)\}/a(\phi) + c(y,\phi).$$

It follows from (5.2) that

$$\mathrm{E}\,\ell'(\theta) = \mathrm{E}\{Y - b'(\theta)\}/a(\phi) = 0$$

which yields that

$$\mathrm{E}(Y) = b'(\theta). \tag{5.4}$$

Table 5.1: Canonical link functions for some exponential family models

| Distribution | $b(\theta)$ | Canonical Link | Range |
|---|---|---|---|
| Normal | $\theta^2/2$ | $g(\mu) = \mu$ | $(-\infty, +\infty)$ |
| Poisson | $\exp(\theta)$ | $g(\mu) = \log \mu$ | $(-\infty, +\infty)$ |
| Binomial | $-\log(1 - e^\theta)$ | $g(\mu) = \log \frac{\mu}{1-\mu}$ | $(-\infty, +\infty)$ |
| Gamma | $-\log(-\theta)$ | $g(\mu) = -\frac{1}{\mu}$ | $(-\infty, 0)$ |

Using (5.3), it follows that

$$\mathrm{E}\{-b''(\theta)/a(\phi)\} + \mathrm{E}\{Y - b'(\theta)\}^2/a^2(\phi) = 0.$$

That is,

$$\mathrm{Var}(Y) = a(\phi)b''(\theta). \tag{5.5}$$

Thus, $b''(\theta)$ is called the variance function, denoted by $V(\theta)$.

Denote by the mean $\mu = \mathrm{E}(Y)$. From (5.4) $\mu = b'(\theta)$, so that

$$\theta = (b')^{-1}(\mu).$$

The function $g = (b')^{-1}$ that links the mean $\mu$ to the canonical parameter is called the *canonical link*

$$g(\mu) = \theta.$$

Because $b''$ is positive by (5.5), $b'$ is a strictly increasing function, so is $g$. Moreover, the likelihood function $\ell(\theta)$ is strictly concave since $\ell''(\theta) = -b''(\theta)/a(\phi) < 0$ by (5.3). This is a very nice property for theoretical and computational considerations. For instance, the MLE is uniquely defined with an exponential family model.

From Examples 5.1–5.4, we can directly derive the canonical link functions for normal, binomial, Poisson and gamma distributions. Table 5.1 presents the canonical link functions of these distributions.

### 5.1.2 *Elements of generalized linear models*

Generalized linear models are developed for regression analysis when the response variable is binary, categorical or counts. Recall that the normal linear regression model is

$$Y = \mathbf{X}^T\boldsymbol{\beta} + \epsilon, \quad \epsilon \sim N(0, \sigma^2),$$

in which the regression function $\mu(\mathbf{x}) = \mathrm{E}(Y|\mathbf{X} = \mathbf{x})$ equals $\mathbf{x}^T\boldsymbol{\beta}$, and the response ranges from $-\infty$ to $+\infty$. Therefore, this model does not fit well with discrete outcomes such as binary, categorial and count outcomes. It is necessary to extend the normal linear model to handle the discrete response.

In doing so, we assume the regression function to be a (monotonic) function of $\mathbf{x}^T\boldsymbol{\beta}$ in order to retain the nice interpretability of the normal linear model. The resulting model is called a *generalized linear model* (GLIM).

In generalized linear models, it is assumed that the conditional density or probability mass function of $Y$ given $\mathbf{X} = \mathbf{x}$ belongs to an exponential family:

$$f(y|\mathbf{x}) = \exp\left([\theta(\mathbf{x})y - b\{\theta(\mathbf{x})\}]/a(\phi) + c(y, \phi)\right). \qquad (5.6)$$

Here we use $\theta(\mathbf{x})$ to emphasize that $\theta$ depends on $\mathbf{x}$. We have shown that this distribution family includes the normal linear model as a special case. Each *GLIM* consists of three components: response distribution, linear predictor and link function. The latter two are used to model the regression function. The assumption regarding conditional distribution of response given predictors is referred to as the *random component* of the generalized linear model in literature. The second and third component of the generalized linear model are regarded to as the *linear predictor* and *link function*, respectively:

$$g\{\mu(\mathbf{x})\} = \mathbf{x}^T\boldsymbol{\beta},$$

where $\mu(\mathbf{x}) = \mathrm{E}(Y|\mathbf{X} = \mathbf{x})$ for a given function $g(\cdot)$, called a link function. The link function bridges the random component and linear predictor: It models the conditional mean function by

$$\mu(\mathbf{x}) = g^{-1}(\mathbf{x}^T\boldsymbol{\beta}). \qquad (5.7)$$

The normal linear model uses an identical link function $g$. With the link function $g$, we can write $\theta(\mathbf{x})$ as

$$\theta(\mathbf{x}) = (b')^{-1}(\mu(\mathbf{x})) = (g \circ b')^{-1}(\mathbf{x}^T\boldsymbol{\beta}) \widehat{=} h(\mathbf{x}^T\boldsymbol{\beta}).$$

If $g$ is the canonical link, $h$ becomes the identical function and $\theta(\mathbf{x}) = \mathbf{x}^T\boldsymbol{\beta}$.

We shall introduce several most commonly-used generalized linear models in the next section.

### 5.1.3 *Maximum likelihood*

Suppose that $(\mathbf{X}_i, Y_i)$, $i = 1, \cdots, n$, is an independent random sample from a generalized linear model with link $g(\cdot)$, and the conditional distribution of response given the covariates is

$$f(Y_i|\mathbf{X}_i, \theta_i, \phi) = \exp[\{Y_i\theta_i - b(\theta_i)\}/a_i(\phi) + c(Y_i, \phi)].$$

Denote by $\mu_i = \mu(\mathbf{X}_i) = \mathrm{E}(Y|\mathbf{X}_i)$. Then

$$\theta_i = (b')^{-1}(\mu_i) = h(\mathbf{X}_i^T\boldsymbol{\beta}).$$

The likelihood function of $\boldsymbol{\beta}$ and $\phi$ is

$$\ell_n(\boldsymbol{\beta}, \phi) = \sum_{i=1}^n [Y_i h(\mathbf{X}_i^T\boldsymbol{\beta}) - b\{h(\mathbf{X}_i^T\boldsymbol{\beta})\}]/a_i(\phi) + \sum_{i=1}^n c(Y_i, \phi), \qquad (5.8)$$

and the MLE for $\boldsymbol{\beta}$ and $\phi$ are

$$(\widehat{\boldsymbol{\beta}}^{\text{mle}}, \widehat{\phi}^{\text{mle}}) = \arg\max_{\boldsymbol{\beta},\phi} \ell_n(\boldsymbol{\beta}, \phi).$$

Under some mild regularity conditions, the asymptotic normality for $\widehat{\boldsymbol{\beta}}$ and $\widehat{\phi}$ can be established. From the fact that $\mathrm{E}(Y_i|\mathbf{X}_i) = b'(\theta_i)$, we have that

$$\mathrm{E}\left\{\frac{\partial^2 \ell_n(\boldsymbol{\beta}, \phi)}{\partial \boldsymbol{\beta} \partial \phi}\right\} = 0,$$

which implies that $\widehat{\boldsymbol{\beta}}^{\text{mle}}$ and $\widehat{\phi}^{\text{mle}}$ are asymptotically independent. In practice, $\phi$ is treated as a nuisance parameter, and is estimated by using its moment estimate rather than its MLE.

To accommodate for heteroscedasticity, it is typically assumed that

$$a_i(\phi) = \phi / w_i. \tag{5.9}$$

For example, if $Y_i$ is the average of $n_i$ data points with covariate $\mathbf{X}_i$, we take $w_i = n_i$. The assumption (5.9) is very mild and valid for many commonly used distributions such as the binomial, Poisson, normal and gamma distributions, as shown in Examples 5.1 – 5.4. Thus, we impose the assumption (5.9) for the rest of this section. One can also simply take $w_i = 1$ for measurements that are not averaged.

By (5.5), the moment estimate of $\phi$ is

$$\widetilde{\phi} = \frac{1}{n-d} \sum_{i=1}^{n} \frac{w_i (Y_i - \widehat{\mu}_i)^2}{b''(\widehat{\theta}_i)},$$

where $d$ is the dimension of $\boldsymbol{\beta}$, $\widehat{\theta}_i = h(\mathbf{X}_i^T \widehat{\boldsymbol{\beta}})$ and $\widehat{\mu}_i = b'(\widehat{\theta}_i)$, the fitted values of $\theta_i$ and $\mu_i$, respectively.

### 5.1.4 *Computing MLE: Iteratively reweighed least squares*

For ease of presentation, let $\widehat{\boldsymbol{\beta}}$ be the MLE of $\boldsymbol{\beta}$. Under (5.9), we have

$$\ell_n(\boldsymbol{\beta}, \phi) = \sum_{i=1}^{n} w_i [Y_i h(\mathbf{X}_i^T \boldsymbol{\beta}) - b\{h(\mathbf{X}_i^T \boldsymbol{\beta})\}]/\phi + \sum_{i=1}^{n} c(Y_i, \phi).$$

Note that the dispersion parameter $\phi$ is a positive constant. The MLE $\widehat{\boldsymbol{\beta}}$ is the maximizer of

$$\ell(\boldsymbol{\beta}) = \sum_{i=1}^{n} w_i [Y_i h(\mathbf{X}_i^T \boldsymbol{\beta}) - b\{h(\mathbf{X}_i^T \boldsymbol{\beta})\}].$$

It is easy to see that $\ell(\boldsymbol{\beta})$ is a concave function under the canonical link:

$$\ell''(\boldsymbol{\beta}) = -\sum_{i=1}^{n} w_i b''\{\mathbf{X}_i^T \boldsymbol{\beta}\} \mathbf{X}_i \mathbf{X}_i^T \leq 0,$$

using $b''(\theta) > 0$ as implied by (5.5).

The *Newton-Raphson algorithm* may be used to compute the MLE. Given an initial value $\boldsymbol{\beta}^{(0)}$, we iteratively update the value of $\boldsymbol{\beta}$ by

$$\boldsymbol{\beta}^{(k+1)} = \boldsymbol{\beta}^{(k)} - [\ell''(\boldsymbol{\beta}^{(k)})]^{-1}\ell'(\boldsymbol{\beta}^{(k)}) \tag{5.10}$$

for $k = 0, 1, \cdots$, until the algorithm converges. For canonical links, as noted above, $-\ell''(\boldsymbol{\beta})$ is nonnegative definite. However, this is not guaranteed for non-canonical links. On the other hand, $\mathrm{E}\{-\ell''(\boldsymbol{\beta})\}$ is proportional to the Fisher information matrix and therefore is positive definite under mild conditions. The Fisher scoring algorithm replaces $\ell''(\boldsymbol{\beta}^{(k)})$ by $\mathrm{E}\{\ell''(\boldsymbol{\beta}^{(k)})\}$ conditioning on $\mathbf{x}_1, \cdots, \mathbf{x}_n$. From $b'(\theta) = \mu$, we have $d\mu/d\theta = b''(\theta)$. By the chain rule, it follows that

$$\ell'(\boldsymbol{\beta}) = \mathbf{X}^T\mathbf{W}\mathbf{G}(\mathbf{Y} - \boldsymbol{\mu}),$$

where $\boldsymbol{\mu} = (\mu_1, \cdots, \mu_n)^T$ and $\mathbf{W}$ and $\mathbf{G}$ is the $n \times n$ diagonal matrix with $i$-th diagonal element being $w_i[b''(\theta_i)\{g'(\mu_i)\}^2]^{-1}$ and $g'(\mu_i)$, respectively. Furthermore

$$-\mathrm{E}\,\ell''(\boldsymbol{\beta}) = \mathbf{X}^T\mathbf{W}\mathbf{X}.$$

Then

$$\begin{aligned}\boldsymbol{\beta}^{(k+1)} &= \boldsymbol{\beta}^{(k)} + [\mathbf{X}^T\mathbf{W}\mathbf{X}]^{-1}\mathbf{X}^T\mathbf{W}\mathbf{G}(\mathbf{Y} - \boldsymbol{\mu}) \\ &= [\mathbf{X}^T\mathbf{W}\mathbf{X}]^{-1}\mathbf{X}^T\mathbf{W}\{\mathbf{G}(\mathbf{Y} - \boldsymbol{\mu}) + \mathbf{X}\boldsymbol{\beta}^{(k)}\}.\end{aligned}$$

We observe that the above update formula is the same as the weighted least squares solution with weight matrix $\mathbf{W}$ and the *working response* variable $\mathbf{z} \widehat{=} \mathbf{G}(\mathbf{Y} - \boldsymbol{\mu}) + \mathbf{X}\boldsymbol{\beta}^{(k)}$. The elements of $\mathbf{G}(\mathbf{Y} - \boldsymbol{\mu})$ are called *working residuals* because they represent the difference between the working response and their predictor $\mathbf{X}\boldsymbol{\beta}^{(k)}$. Note that the conditional covariance matrix of the working response is $\phi\mathbf{W}^{-1}$. So the weight $\mathbf{W}$ provides the right weighting matrix. The maximization of $\ell(\boldsymbol{\beta})$ can be carried out by using an *iteratively reweighted least squares* (IRLS) algorithm, which is described as follows.

**IRLS algorithm**:

Step 1: Set initial value $\boldsymbol{\beta}^{(0)}$, and let $k = 0$;

Step 2: For given $k$, calculate $\mathbf{G}$ and $\boldsymbol{\mu}$ by replacing $\boldsymbol{\beta}$ with $\boldsymbol{\beta}^{(k)}$, and further construct the working response variable

$$\mathbf{z} = \mathbf{G}(\mathbf{Y} - \boldsymbol{\mu}) + \mathbf{X}\boldsymbol{\beta}^{(k)}.$$

Step 3: For given $k$, calculate $\mathbf{W}$ by replacing $\boldsymbol{\beta}$ with $\boldsymbol{\beta}^{(k)}$, and update $\boldsymbol{\beta}^{(k)}$ by using the weighted least squares

$$\boldsymbol{\beta}^{(k+1)} = (\mathbf{X}^T\mathbf{W}\mathbf{X})^{-1}\mathbf{X}^T\mathbf{W}\mathbf{z}.$$

Step 4: Let $k = 1, \cdots$, iterate between Step 2 and Step 3 until the sequence of $\boldsymbol{\beta}^{(k)}$ converges.

When the algorithm converges, it follows that

$$\widehat{\boldsymbol{\beta}} = (\mathbf{X}^T\widehat{\mathbf{W}}\mathbf{X})^{-1}\mathbf{X}^T\widehat{\mathbf{W}}\mathbf{z}.$$

Note that $\mathrm{cov}(\mathbf{z}|\mathbf{X}) \approx \mathrm{cov}\{\mathbf{G}(\mathbf{Y}-\boldsymbol{\mu})|\mathbf{X}\} = \phi\mathbf{W}^{-1}$. The covariance matrix of $\widehat{\boldsymbol{\beta}}$ can be estimated by a *sandwich formula*

$$\widehat{\mathrm{cov}}(\widehat{\boldsymbol{\beta}}|\mathbf{X}) = \widehat{\phi}(\mathbf{X}^T\widehat{\mathbf{W}}\mathbf{X})^{-1}\mathbf{X}^T\widehat{\mathbf{W}}\mathbf{X}(\mathbf{X}^T\widehat{\mathbf{W}}\mathbf{X})^{-1} = \widehat{\phi}(\mathbf{X}^T\widehat{\mathbf{W}}\mathbf{X})^{-1}.$$

Here $\widehat{\phi}$ is the MLE or the moment estimator of $\phi$, if $\phi$ is unknown. The MLE of $\phi$ can be obtained by maximizing $\ell_n(\widehat{\boldsymbol{\beta}}, \phi)$ with respect to $\phi$.

We can obtain estimated standard errors of $\widehat{\boldsymbol{\beta}}$ using the sandwich formula, and make statistical inferences on $\boldsymbol{\beta}$. The asymptotic normality of $\widehat{\boldsymbol{\beta}}$ enables us to construct an asymptotic confidence interval of each $\beta_j$ and confidence region of the entire regression coefficient vector $\boldsymbol{\beta}$.

The second order method, namely the Newton-Raphson algorithm or its Fisher scoring version, implicitly assumes that the Hessian matrix in (5.10) is invertible. This implicitly assumes that $p \le n$ or even $p \ll n$, namely, a low-dimensional region. For a high-dimensional regime, computing an $O(p^2)$ Hessian matrix and its inverse is very computationally expensive, not to mention the ill-conditioning of the Hessian matrix. For this reason, we replace $[\ell''(\boldsymbol{\beta}^{(k)})]^{-1}$ in (5.10) with $\lambda_k$, resulting in the *gradient ascent algorithm*

$$\boldsymbol{\beta}^{(k+1)} = \boldsymbol{\beta}^{(k)} - \lambda_k \ell'(\boldsymbol{\beta}^{(k)}), \tag{5.11}$$

where $\lambda_k$ is called the *step size* or *learning rate*. See Section 3.5.7 for a related idea.

### 5.1.5 Deviance and analysis of deviance

With the MLE $\widehat{\boldsymbol{\beta}}$ we can calculate the fitted value of $\widehat{\theta}$:

$$\widehat{\theta}_i = h(\mathbf{X}_i^T\widehat{\boldsymbol{\beta}}).$$

Without putting any model restriction on $\theta_i$, we would have to maximize

$$\ell_n(\boldsymbol{\theta}, \phi; \mathbf{Y}, \mathbf{X}) = \sum_{i=1}^n w_i\{Y_i\theta_i - b(\theta_i)\}/\phi$$

with respect to $\boldsymbol{\theta} = (\theta_1, \cdots, \theta_n)^T$. The maximization can be done by maximizing $Y_i\theta_i - b(\theta_i)$ for $i = 1, \cdots, n$. So the unrestricted maximizer $\widetilde{\theta}_i = (b')^{-1}(Y_i)$, by taking the derivative and setting it to zero: $Y_i - b'(\widetilde{\theta}_i) = 0$. Then the difference is the lack-of-fit due to the model restriction. Define *deviance*, denoted by $D(\mathbf{Y}; \widehat{\boldsymbol{\mu}})$, to be

$$D(\mathbf{Y}; \widehat{\boldsymbol{\mu}}) = 2\sum_{i=1}^n w_i\{Y_i(\widetilde{\theta}_i - \widehat{\theta}_i) - b(\widetilde{\theta}_i) + b(\widehat{\theta}_i)\}.$$

In contrast,

$$D^*(\mathbf{Y};\widehat{\boldsymbol{\mu}}) = D(\mathbf{Y};\widehat{\boldsymbol{\mu}})/\phi$$

is called the *scaled deviance*.

For normal linear models, $\theta(\mu) = \mu$ and $b(\theta) = \theta^2/2$. Then

$$D(\mathbf{Y};\widehat{\boldsymbol{\mu}}) = 2\sum_{i=1}^{n} w_i\{Y_i(Y_i - \widehat{Y}_i) - \frac{Y_i^2}{2} + \frac{\widehat{Y}_i^2}{2}\} = \sum_{i=1}^{n} w_i(Y_i - \widehat{Y}_i)^2.$$

This is the weighted residual sum of squares under a weighted linear model (see (2.3)). Table 5.2 depicts the deviances of several commonly used distributions.

Table 5.2: Deviances for commonly used distributions

| Distribution | deviance ($w_i = 1$) |
|---|---|
| Normal | $\sum(Y_i - \widehat{\mu}_i)^2$ |
| Binomial | $2\sum(Y_i \log(Y_i/\widehat{\mu}_i) - (m - Y_i)\log\{(m - Y_i)/(m - \widehat{Y}_i)\})$ |
| Poisson | $2\sum\{Y_i \log(Y_i/\widehat{\mu}_i) - (Y_i - \widehat{\mu}_i)\}$ |
| Gamma | $2\sum\{-\log(Y_i/\widehat{\mu}_i) + (Y_i - \widehat{\mu}_i)/\widehat{\mu}_i\}$ |

The value of deviance is usually not scaled, however, the reduction of deviance provides useful information. Suppose that we are comparing a larger model with a smaller one. Let $\Theta_1$ and $\Theta_0$ be the corresponding parameter space, and $\Theta$ be the set consisting of all possible (no constraint) models. By the definition,

$$\begin{aligned} & \text{Deviance(smaller model)}-\text{Deviance(larger model)} \\ = \; & 2\phi\{(\max_{\theta\in\Theta}\ell_n(\theta) - \max_{\theta\in\Theta_0}\ell_n(\theta)) - (\max_{\theta\in\Theta}\ell_n(\theta) - \max_{\theta\in\Theta_1}\ell_n(\theta))\} \\ = \; & 2\phi\{\max_{\theta\in\Theta_1}\ell_n(\theta) - \max_{\theta\in\Theta_0}\ell_n(\theta)\} \\ \to \; & \phi\chi^2_{\text{df}} \end{aligned}$$

in distribution by the theory of likelihood ratio test, where df $= \dim(\Theta_1) - \dim(\Theta_0)$.

The deviance plays a very similar role to that of the sum of squared residuals for the normal linear model. Like the table of ANalysis of Variance (*ANOVA*), we may construct a table of *analysis of deviance*. Suppose that we have factors $A$ and $B$. The deviance can be decomposed as

$$\begin{aligned} \text{Dev}(I) &= \underbrace{\text{Dev}(I) - \text{Dev}(A)}_{R_A} + \underbrace{\text{Dev}(A) - \text{Dev}(A,B)}_{R_{B|A}} \\ &\quad + \underbrace{\text{Dev}(A,B) - \text{Dev}(A*B)}_{R_{A*B|(A,B)}}. \end{aligned}$$

where $\text{Dev}(I)$ stands for the deviance of the null model with intercept only,

Dev($A$) for the deviance of a model with intercept and Factor A only, Dev($A, B$) for the deviance of a two-way model without interaction, and Dev($A * B$) for the deviance of a two-way model with interaction. Therefore, the analysis of deviance table can be constructed as follows:

Table 5.3: Table for analysis of deviance

| Resource | DF | Deviance Reduction | Deviance | P-value |
|---|---|---|---|---|
| Intercept | | | Dev($I$) | |
| $A$ | $\text{df}_A - 1$ | $R_A$ | Dev(A) | |
| $B\|A$ | $\text{df}_B - 1$ | $R_{B\|A}$ | Dev(A,B) | |
| $A * B\|(A, B)$ | $(\text{df}_A - 1) * (\text{df}_B - 1)$ | $R_{A*B\|(A,B)}$ | Dev(A*B) | |

For binomial and Poisson models without over-dispersion, we can take $\phi = 1$. Thus, $R_A$, $R_{B|A}$ and $R_{A*B|(A,B)}$ follow asymptotic $\chi^2$-distributions. Then, the P-value can be calculated by using the corresponding $\chi^2$-distribution. For other models, it may be required to estimate $\phi$ in order to obtain the $P$-value.

The idea of constructing the analysis of deviance for a two-way model can be directly extended to multi-way models and analysis of covariance models in the presence of continuous covariates. In general, the terms in the model will not be orthogonal any more, so the conclusion should be drawn with caution. The decomposition of deviance depends on the order.

### 5.1.6 *Residuals*

For generalized linear models, there are several ways to define residuals. In this section, we introduce deviance residuals, *Pearson residuals* and *Anscombe residuals*.

Denoted by $\widehat{\mu}_i$ the fitted value:

$$\widehat{\mu}_i = g^{-1}(\mathbf{X}_i^T \widehat{\boldsymbol{\beta}})$$

in generalized linear models. Define $d_i^2 = Y_i(\widetilde{\theta}_i - \widehat{\theta}_i) - \{b(\widetilde{\theta}_i) - b(\widehat{\theta}_i)\}$ as the discrepancy of fit in the $i$th unit, and then we can write $D(\mathbf{Y}, \widehat{\mu}) = \sum_{i=1}^n w_i d_i^2$. Define *deviance residual* as

$$r_{D,i} = d_i \text{sgn}(Y_i - \widehat{\mu}_i).$$

To study the property of the deviance residual, we drop the subscript $i$ for ease of presentation. Let $z = y - \widehat{\mu}$. Regard $\theta$ as a function of $\mu$, denoted by $\theta(\mu)$. Letting $f(z) = r_D^2$, it follows that

$$\begin{aligned} f(z) &= y(\widetilde{\theta} - \widehat{\theta}) - \{b(\widetilde{\theta}) - b(\widehat{\theta})\} \\ &= y(\widetilde{\theta} - \theta(\widehat{\mu})) - [b(\widetilde{\theta}) - b\{\theta(\widehat{\mu})\}] \\ &= y\{\widetilde{\theta} - \theta(y - z)\} - b(\widetilde{\theta}) + b\{\theta(y - z)\}. \end{aligned}$$

Recall that $b'(\theta) = \mu$, which implies that $b'\{\theta(y-z)\} = y-z$ and $d\mu = b''(\theta)d\theta$. Thus, $d\theta/d\mu = 1/b''(\theta) > 0$ and

$$f'(z) = y\theta'(y-z) - b'\{\theta(y-z)\}\theta'(y-z) = z\theta'(y-z).$$

Hence the sign of $f'(z)$ is the same as that of $z$. Thus $f(z)$ is increasing when $z > 0$ and decreasing when $z < 0$.

In summary, the deviance residual has the following property. For each given $Y_i$, the deviance residual $r_{D,i}$ is an increasing function of $Y_i - \widehat{\mu}_i$. Deviance residuals provide useful raw materials for model diagnostics, in a similar way to the residuals for linear regression models. The deviance, the sum of squares of deviance residuals, plays a very similar role as the sum of squared residuals.

The summands in Table 5.2 give the formulas for $d_i^2$. In particular, for the Bernoulli model ($m = 1$ in Table 5.2 with convention $0 \log 0 = 0$), we have

$$d_i^2 = -2 \log \widehat{p}_i I(Y_i = 1) - 2 \log \widehat{q}_i I(Y_i = 0),$$

where $\widehat{q}_i = 1 - \widehat{p}_i$. Thus

$$r_{D,i} = \sqrt{-2 \log \widehat{p}_i} I(Y_i = 1) - \sqrt{-2 \log \widehat{q}_i} I(Y_i = 0).$$

We next introduce the *Pearson residual.* Let $\mathrm{E}(Y) = \mu$ and $\mathrm{Var}(Y) = V(\mu)$. For generalized linear models, $V(\cdot)$ is known and given by (5.5). The *Pearson residual* is defined as

$$\widehat{r}_{P,i} = \frac{Y_i - \widehat{\mu}_i}{\sqrt{V(\widehat{\mu}_i)}}.$$

The Pearson residual is the same as the ordinary residual for the normal distributed error, and is skewed for non-normal distributed responses.

Pearson's method was invented before the time of modern computers. It has a simple computation advantage and has been used for the goodness of fit test. The Pearson goodness-of-fit statistic is

$$X^2 = \sum_{i=1}^{n} \widehat{r}_{p,i}^2.$$

This is also referred to as the Pearson chi-square statistic since it can be shown that under some mild conditions, $X^2$ follows an asymptotic $\chi^2$-distribution with degrees of freedom $(n - d)$, where $d$ is the dimension of $\boldsymbol{\beta}$. The residual defined above plays a similar role as the usual one. It can be used to construct a residual plot for examining an overall pattern, detecting heteroscedasticity and suggesting alternative models.

We next describe the *Anscombe residual.* For non-normal data $y$, a traditional approach is to transform $y$ so that the distribution of $A(y)$ is as close to the normal distribution as possible in some sense. It was shown by Wedderburn (unpublished, but see Barndorff-Nielsen 1978) that the function is given by

$$A(z) = \int_a^z \frac{d\mu}{V^{1/3}(\mu)},$$

where $a$ is the lower limit of the range of $\mu$. The residual

$$A(y) - A(\widehat{\mu}) \approx A'(\mu)(y - \widehat{\mu})$$

when $\widehat{\mu}$ is close to $y$. Hence the variance of the "residual" $A(y) - A(\widehat{\mu})$ is approximate

$$A'(\mu)^2 V(\mu) = V^{-2/3}(\mu)V(\mu) = V(\mu)^{1/3}.$$

This leads to defining the *Anscombe residual* as

$$r_A = \frac{A(y) - A(\widehat{\mu})}{V(\widehat{\mu})^{1/6}}.$$

We now use the Poisson distribution to illustrate the three residuals.

$$A(z) = \int_0^z \frac{d\mu}{\mu^{1/3}} = \frac{3}{2} z^{2/3}.$$

Hence

$$r_A = \frac{\frac{3}{2}(y^{2/3} - \widehat{\mu}^{2/3})}{\widehat{\mu}^{1/6}}.$$

In contrast, the Pearson residual is

$$r_p = \frac{Y - \widehat{\mu}}{\widehat{\mu}^{1/2}}$$

and the deviance residual

$$r_D = 2\{y \log(y/\widehat{\mu}) - (y - \widehat{\mu})\}\text{sgn}(y - \widehat{\mu}).$$

## 5.2 Examples

To illustrate the generality of the generalized linear models, we introduce several most useful examples of generalized linear models in practice.

### *5.2.1 Bernoulli and binomial models*

Binary response is a common feature of many studies. For example, it is of primary interest in medical studies whether a patient has recovered after receiving a treatment. The response takes only two possible values, $Y = 1$ for a recovered patient and $Y = 0$ otherwise. Similarly, for classification, it is also natural to label one class as $Y = 0$ and the other class as $Y = 1$ and the same can be applied to many problems with binary response such as "success" and "failure".

When the response $Y$ is binary, it is natural to assume that given $\mathbf{X} = \mathbf{x}$, $Y$ follows a Bernoulli distribution with success probability $p(\mathbf{x})$. That is, for $y = 0$ and 1,

$$P(Y = y|\mathbf{x}) = p(\mathbf{x})^y\{1 - p(\mathbf{x})\}^{1-y}.$$

Its regression function is $\mu(\mathbf{x}) = \mathrm{E}(Y|\mathbf{x}) = p(\mathbf{x}) \in [0, 1]$. From Example 5.2, the canonical link for the Bernoulli distribution is the *logit* link

$$g(\mu) = \log\left(\frac{\mu}{1-\mu}\right).$$

This yields the logistic regression:

$$\log\left(\frac{p(\mathbf{x})}{1-p(\mathbf{x})}\right) = \mathbf{x}^T\boldsymbol{\beta}.$$

Equivalently,

$$p(\mathbf{x}) = \frac{\exp(\mathbf{x}^T\boldsymbol{\beta})}{1+\exp(\mathbf{x}^T\boldsymbol{\beta})}.$$

The logit link is the canonical link for the Bernoulli distribution. Therefore it is important in modeling binary response data. In many fields, researchers are interested in examining odds ratios of different treatments. Thus the logit link becomes the most popular link function in modeling binary responses as it directly gives us *odds*, defined as $p/(1-p)$, which is equal to $\exp(\mathbf{x}^T\boldsymbol{\beta})$. This offers us an easy interpretation of the coefficient. For instance, suppose that $x_1$ is the treatment variable $T$: $T = 1$ stands for a treatment being received, and $T = 0$ otherwise. Let $\mathbf{x}_2$ consist of other covariates. For any given covariate value $\mathbf{x}_2$, denote $p_1 = p(T = 1, \mathbf{x}_2)$ and $p_2 = p(T = 0, \mathbf{x}_2)$. Then it follows that

$$\beta_1 = \log\left\{\frac{p_1/(1-p_1)}{p_2/(1-p_2)}\right\},$$

which is the log odds ratio of the event that a patient receives the treatment given the other covariates.

The logistic regression model is the most popular model for binary response. Other link functions have been used in the literature. For the Bernoulli model, the range of $\mu(\mathbf{x})$ is [0,1]. So any appropriate link function for the Bernoulli model should be defined as a function from [0,1] onto $(-\infty, +\infty)$. Intuitively, any continuous, strictly increasing distribution can be a link function for the Bernoulli model. In fact, suppose that latent variable $L$ follows the linear model

$$L = -\mathbf{x}^T\boldsymbol{\beta} + \epsilon,$$

where $\epsilon \sim F$. We observe $Y = I(L \le \tau)$ according to certain thresholding parameter $\tau$. Then given $\mathbf{x}$, $Y$ follows a Bernoulli distribution with $p(\mathbf{x}) =$

$F(\tau + \mathbf{x}^T\boldsymbol{\beta})$. It is convenient to assume that $\tau = 0$ by including an intercept term in the linear model (one column of $\mathbf{x}$ is 1). This results in

$$p(\mathbf{x}) = F(\mathbf{x}^T\boldsymbol{\beta}).$$

If $\epsilon$ follows a logistic distribution with

$$F(x) = \frac{\exp(x)}{1 + \exp(x)}$$

then

$$p(\mathbf{x}) = \frac{\exp(\mathbf{x}^T\boldsymbol{\beta})}{1 + \exp(\mathbf{x}^T\boldsymbol{\beta})}.$$

This coincides with the logistic regression.

If $\epsilon$ follows the normal distribution $N(0, \sigma^2)$, then

$$p(\mathbf{x}) = \Phi(\mathbf{x}^T\boldsymbol{\beta}/\sigma),$$

where $\Phi(\cdot)$ is the cumulative distribution of $N(0, 1)$. Note that both $\boldsymbol{\beta}$ and $\sigma$ are not identifiable. However, $\boldsymbol{\alpha} = \boldsymbol{\beta}/\sigma$ is identifiable. Thus we reparameterize the above model by considering

$$p(\mathbf{x}) = \Phi(\mathbf{x}^T\boldsymbol{\alpha}).$$

This leads to *probit* link: $\Phi^{-1}\{\mu(\mathbf{x})\} = \mathbf{x}^T\boldsymbol{\alpha}$.

Let $R$ follow an exponential distribution with mean 1, and $\epsilon = \exp(R)$. Then the distribution of $\epsilon$ is $F(x) = 1 - \exp(-\exp(x))$ and

$$p(\mathbf{x}) = 1 - \exp\{-\exp(\mathbf{x}^T\boldsymbol{\beta})\}$$

which is equivalent to

$$\eta = \log\{-\log(1 - \mu)\} = \mathbf{x}^T\boldsymbol{\beta}$$

This coincides with the *complementary log-log link*, another commonly used link function.

Figure 5.1 depicts the plot of the logit, probit and complementary log-log links. From Figure 5.1, it can be seen that both logit and probit links are symmetric about (0.5, 0), while the complementary log-log link is not.

In many social and behavioral studies, the sample unit typically is a group of subjects such as a class or a school instead of an individual person. The response is often recorded as the number of subjects having a specific characterization of interest out of the group size $m_i$. It is more convenient to model the response by using a binomial distribution rather than a Bernoulli distribution. Suppose that given $\mathbf{X} = \mathbf{x}$, $Y$ follows a binomial distribution with index $m$ and success probability $p(\mathbf{x})$. Thus, for $y = 0, \cdots, m$,

$$P(Y = y|\mathbf{x}) = \exp\left(y \log[p(x)/\{1 - p(x)\}] + m \log\{1 - p(\mathbf{x})\} + \log\binom{m}{y}\right).$$

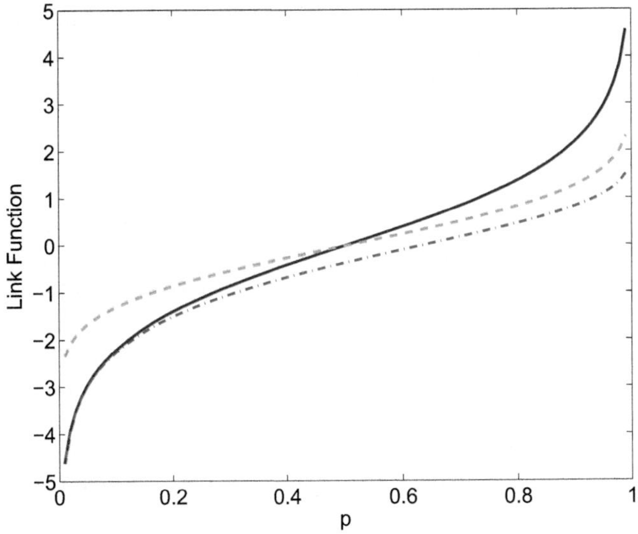


Figure 5.1: Plot of link functions. Solid line is for logit link, dashed line is for probit link and dash-dotted line is for complementary log-log link.

From Example 5.2, the canonical link for a binomial distribution is the logit link, which yields the logistic regression for binomial response:

$$\log\left(\frac{p(\mathbf{x})}{1-p(\mathbf{x})}\right) = \mathbf{x}^T\boldsymbol{\beta}.$$

Parallel to the Bernoulli model, the probit link or complementary log-log link can be considered for binomial models. Chapter 4 of McCullagh and Nelder (1989) provides a detailed introduction to analysis of data with binary responses. One may directly apply the IRLS algorithm in Section 5.1.4 for Bernoulli and binomial models (Exercise 5.2).

### *5.2.2 Models for count responses*

In various scientific studies, the outcome can be the count of a specific event of interest. For instance, the outcome may be the number of new cases of HIV infection in a community in an AIDS study, or the number of a certain type of cancer in a population in cancer studies. It is natural to use the Poisson distribution for count responses. The canonical link for the Poisson distribution is the log-link. The Poisson *log-linear model* is referred to as that given $\mathbf{X} = \mathbf{x}$, $Y$ has a Poisson distribution with mean $\mu(\mathbf{x})$ and

$$\eta = \log(\mu(\mathbf{x})) = \mathbf{x}^T\boldsymbol{\beta}.$$

Equivalently,

$$\lambda(\mathbf{x}) = \exp(\mathbf{x}^T\boldsymbol{\beta}).$$

since $\mu(\mathbf{x}) = \lambda(\mathbf{x})$ for Poisson distribution. The Poisson log-linear model implies that the response has variance equal to the mean for any given values of the covariates, the changes in the mean function from combined effects of covariates are multiplicative, and the logarithm of mean function changes linearly with equal increment increases in the predictors.

In many health survey studies, surveys are conducted on the general population or a community sample and the symptom count measure tends to have a high frequency of zero values. In quality control studies, the response typically is the number of defective products in a sample unit, and therefore, tends to have a high frequency of zero values. In such situations, a Poisson log-linear regression model may not be appropriate. Mullahy (1986) proposed a hurdle model for zero-inflated count responses, while Lambert (1992) proposed a zero-inflated Poisson (ZIP) model for zero-inflated count responses.

The hurdle model assumes that the response $Y_i$ given $\mathbf{X}_i$ is from a finite mixture: with probability $p_i = p(\mathbf{X}_i)$, $Y_i = 0$, and with probability $1 - p_i$, $Y_i$ follows a truncated Poisson with mean $\lambda_i = \lambda(\mathbf{X}_i)$. Thus, the probability mass function can be written as

$$\begin{aligned} P(Y_i = 0|\mathbf{X}_i) &= p_i, \\ P(Y_i = k|\mathbf{X}_i) &= (1 - p_i)\frac{\lambda(\mathbf{X}_i)^k}{k!}\frac{\exp\{-\lambda(\mathbf{X}_i)\}}{[1 - \exp\{-\lambda(\mathbf{X}_i)\}]}. \end{aligned}$$

Similar to the logistic regression model and Poisson log-linear model, the probability $p(\mathbf{X}_i)$ and $\lambda(\mathbf{X}_i)$ can be modeled by

$$\text{logit}\{p(\mathbf{X}_i)\} = \mathbf{X}_i^T\gamma, \qquad \text{and} \qquad \log\{\lambda(\mathbf{X}_i)\} = \mathbf{X}_i^T\boldsymbol{\beta}.$$

The ZIP model is similar to the hurdle model, but it assumes that the response $Y_i$ given $\mathbf{X}_i$ follows a zero-inflated Poisson distribution:

$$Y_i \sim p(\mathbf{X}_i)\text{Poisson}(\lambda_0) + \{1 - p(\mathbf{X}_i)\}\text{Poisson}(\lambda(\mathbf{x})),$$

with $\lambda_0 = 0$. We further assume that the probability $p(\mathbf{X}_i)$ and $\mu(\mathbf{X}_i)$ can be modeled by

$$\text{logit}\{p(\mathbf{X}_i)\} = \mathbf{X}_i^T\gamma, \qquad \text{and} \qquad \log\{\lambda(\mathbf{X}_i)\} = \mathbf{X}_i^T\boldsymbol{\beta}.$$

Denote $p_i = p(\mathbf{X}_i)$ and $\lambda_i = \lambda(\mathbf{X}_i)$. Then

$$\begin{aligned} P(Y_i = 0|\mathbf{X}_i) &= p_i + (1 - p_i)\exp(-\lambda_i), \\ P(Y_i = k|\mathbf{X}_i) &= (1 - p_i)\exp(-\lambda_i)\lambda_i^k/k!. \end{aligned}$$

The major difference between the hurdle model and the ZIP model is how each one models the probability of the event $Y_i = 0$. The hurdle model may be used to model zero-deflated count response by considering $p_i < \exp(-\lambda_i)$, while the ZIP model is not appropriate for zero-deflated count response since

$P(Y_i = 0|\mathbf{X}_i) = p_i + (1 - p_i)\exp(-\lambda_i) = \exp(-\lambda_i) + p_i\{1 - \exp(-\lambda_i)\}$, which is always greater than $\exp(-\lambda_i)$.

### *5.2.3 Models for nonnegative continuous responses*

For modeling a nonnegative response with a constant *coefficient of variation* (standard deviation divided by mean), it is natural to consider using a gamma distribution. Given $\mathbf{X} = \mathbf{x}$, $Y$ has a density given Example 5.4 with $\mu(\mathbf{x})$ and constant shape parameter $\alpha$. The canonical link for the gamma distribution is

$$\eta = -\frac{1}{\mu}.$$

Then the gamma reciprocal linear model is

$$\mathrm{E}(Y|\mathbf{x}) = \frac{-1}{\mathbf{x}^T\boldsymbol{\beta}} = \frac{1}{\mathbf{x}^T\boldsymbol{\alpha}},$$

where $\boldsymbol{\alpha} = -\boldsymbol{\beta}$. The reciprocal linear model is useful for modeling the rate of a process (Nelder, 1966), but it does not map $(0, \infty)$, the range of $\mu$, to $(-\infty, \infty)$. This implies restrictions on the parameters $\boldsymbol{\beta}$ in any linear model. For this reason, the log-link, $\eta = \log(\mu)$ has been proposed in the literature. The identity link, $\theta = \mu$, has also been used.

### *5.2.4 Normal error models*

Suppose that the conditional density of $Y$ given $\mathbf{x}$ is a normal distribution with mean $\mu(\mathbf{x})$ and variance $\sigma^2$. The normal linear model corresponds to the case with the canonical link. That is, $\mu(\mathbf{x}) = \eta = \mathbf{x}^T\boldsymbol{\beta}$.

For other link function $g(\cdot)$, let $h = g^{-1}$. This yields a nonlinear normal error model:

$$Y = h(\mathbf{X}^T\boldsymbol{\beta}) + \epsilon, \tag{5.12}$$

where $\epsilon \sim N(0, \sigma^2)$. If $h$ is a pre-specified link function, the nonlinear least squares method can be used to estimate the regression coefficient $\boldsymbol{\beta}$.

When $h(\cdot)$ is an unknown nonparametric function, model (5.12) is referred to as a *single index model* (Carroll *et al.*, 1997), a commonly-used semiparametric regression model.

## 5.3 Sparest Solution in High Confidence Set

This section is mainly taken from Fan (2014). Suppose that we have data $\{(\mathbf{X}_i, Y_i)\}$ from a statistical model with the true parameter vector $\boldsymbol{\beta}_0$. The information on $\boldsymbol{\beta}_0$ provided by the data can be summarized by the confidence sets of $\boldsymbol{\beta}_0$ in $\mathbb{R}^p$. However, this confidence set is too big to be useful when the number of parameters is large. Thus, we frequently impose the sparsity assumption on the underlying parameter $\boldsymbol{\beta}_0$. Combining these two pieces of information, a general solution to high-dimensional statistics is naturally the sparsest solution in a high-confidence set.

### 5.3.1 A general setup

Suppose that our task is to find an estimate of the sparse vector $\boldsymbol{\beta}_0 \in \mathbb{R}^p$ such that it minimizes $L(\boldsymbol{\beta}) = \mathrm{E}\{L(\mathbf{X}^T\boldsymbol{\beta}, Y)\}$. The loss function is assumed convex in the first argument so that $L(\boldsymbol{\beta})$ is convex. The setup encompasses the generalized linear models (5.1) with $L(\theta, y) = b(\theta) - \theta y$ under the canonical link (negative log-likelihood with $\phi = 1$), robust regression with $L(\theta, y) = |y - \theta|$, the hinge loss $L(\theta, y) = (1 - \theta y)_+$ in the support vector machine and exponential loss $L(\theta, y) = \exp(-\theta y)$ in AdaBoost. Let

$$L_n(\boldsymbol{\beta}) = \frac{1}{n}\sum_{i=1}^{n} L(\mathbf{X}_i^T\boldsymbol{\beta}, Y_i)$$

be the empirical loss and $\nabla L_n(\boldsymbol{\beta})$ be its gradient. Given that $\nabla L(\boldsymbol{\beta}_0) = 0$, a natural confidence set is of form

$$\mathcal{C}_n = \{\boldsymbol{\beta} \in \mathbb{R}^p : \|\nabla L_n(\boldsymbol{\beta})\|_\infty \le \gamma_n\}$$

for some given $\gamma_n$ that is related to the confidence level.

In principle, any norm can be used in constructing a confidence set. However, we take the $L_\infty$-norm as it is the conjugate norm to the $L_1$-norm in Hölder's inequality (see below). The tuning parameter $\gamma_n$ is chosen so that the set $\mathcal{C}_n$ has confidence level $1 - \delta_n$, namely,

$$\Pr(\boldsymbol{\beta}_0 \in \mathcal{C}_n) = \Pr\{\|\nabla L_n(\boldsymbol{\beta}_0)\|_\infty \le \gamma_n\} \ge 1 - \delta_n. \tag{5.13}$$

The confidence region $\mathcal{C}_n$ is called a high confidence set because $\delta_n \to 0$.

The set $\mathcal{C}_n$ is the summary of the data information about $\boldsymbol{\beta}_0$. If in addition we assume that $\boldsymbol{\beta}_0$ is sparse, then a natural solution is the intersection of these two pieces of information, namely, finding the *sparsest solution in the high-confidence set*:

$$\min_{\boldsymbol{\beta} \in \mathcal{C}_n} \|\boldsymbol{\beta}\|_1 = \min_{\|\nabla L_n(\boldsymbol{\beta})\|_\infty \le \gamma_n} \|\boldsymbol{\beta}\|_1. \tag{5.14}$$

This is a convex optimization problem. Here, the sparsity is measured by the $L_1$-norm, but it can also be measured by other norms such as the weighted $L_1$-norm (Zou, 2006).

### 5.3.2 Examples

The Danzig selector (Candés and Tao, 2007) is a specific case of problem (5.14). It corresponds to using the quadratic loss: $L(x, y) = (x - y)^2$. This provides an alternative view to the Danzig selector. In this case, the confidence set implied by the data is

$$\mathcal{C}_n = \{\boldsymbol{\beta} \in \mathbb{R}^p : \|\mathbf{X}^T(\mathbf{Y} - \mathbf{X}\boldsymbol{\beta})\|_\infty \le \gamma_n\}$$

and the sparsest solution in the high confidence set is now given by

$$\min \|\boldsymbol{\beta}\|_1, \quad \text{subject to } \|\mathbf{X}^T(\mathbf{Y}-\mathbf{X}\boldsymbol{\beta})\|_\infty \le \gamma_n,$$

which is the Danzig selector.

Similarly, in estimating sparse precision matrix $\boldsymbol{\Theta} = \boldsymbol{\Sigma}^{-1}$ for the Gaussian graphic model, if $L(\boldsymbol{\Theta}, \mathbf{S}_n) = \|\boldsymbol{\Theta}\mathbf{S}_n - \mathbf{I}_p\|_F^2$ where $\mathbf{S}_n$ is the sample covariance matrix and $\|\cdot\|_F$ is the Frobenius norm, then the high confidence set provided by the data is

$$\mathcal{C}_n = \{\boldsymbol{\Theta} : \|\mathbf{S}_n \cdot (\boldsymbol{\Theta}\mathbf{S}_n - \mathbf{I}_p)\|_\infty \le \gamma_n\},$$

where $\cdot$ denotes the componentwise product (a factor 2 of off-diagonal elements is ignored). If we construct the high-confidence set based directly on the estimation equations $\boldsymbol{\Theta}\mathbf{S}_n - \mathbf{I}_p = 0$, then the sparse high-confidence set becomes

$$\min_{\|\boldsymbol{\Theta}\mathbf{S}_n - \mathbf{I}_p\|_\infty \le \gamma_n} \|\text{vec}(\boldsymbol{\Theta})\|_1.$$

If the matrix $L_1$-norm is used in (5.14) to measure the sparsity, then the resulting estimator is the CLIME estimator of Cai, Liu, and Luo (2011), which solves

$$\min_{\|\boldsymbol{\Theta}\mathbf{S}_n - \mathbf{I}_p\|_\infty \le \gamma_n} \|\boldsymbol{\Theta}\|_1.$$

If we use the Gaussian log-likelihood

$$L_n(\boldsymbol{\Theta}) = -\ln(|\boldsymbol{\Theta}|) + \text{tr}(\boldsymbol{\Theta}\mathbf{S}_n),$$

then $\nabla L_n(\boldsymbol{\Theta}) = -\boldsymbol{\Theta}^{-1} + \mathbf{S}_n$ and

$$\mathcal{C}_n = \{\|\boldsymbol{\Theta}^{-1} - \mathbf{S}_n\|_\infty \le \gamma_n\}.$$

The sparsest solution is then given by

$$\min_{\|\boldsymbol{\Theta}^{-1} - \mathbf{S}_n\|_\infty \le \gamma_n} \|\boldsymbol{\Theta}\|_1.$$

### 5.3.3 *Properties*

Let $\widehat{\boldsymbol{\beta}}$ be a solution to (5.14) and $\widehat{\boldsymbol{\Delta}} = \widehat{\boldsymbol{\beta}} - \boldsymbol{\beta}_0$. As in the Danzig selection, the feasibility of $\boldsymbol{\beta}_0$ implied by (5.13) entails that

$$\|\boldsymbol{\beta}_0\|_1 \ge \|\widehat{\boldsymbol{\beta}}\|_1 = \|\boldsymbol{\beta}_0 + \widehat{\boldsymbol{\Delta}}\|_1.$$

Letting $\mathcal{S}_0 = \text{supp}(\boldsymbol{\beta}_0)$ and $s = |\mathcal{S}_0|$, using the same argument as in (3.48) and (3.49), we have

$$\|\widehat{\boldsymbol{\Delta}}_{\mathcal{S}_0}\|_1 \ge \|\widehat{\boldsymbol{\Delta}}_{\mathcal{S}_0^c}\|_1 \quad \text{and} \quad \|\widehat{\boldsymbol{\Delta}}\|_2 \ge \|\widehat{\boldsymbol{\Delta}}\|_1/(2\sqrt{s}) \tag{5.15}$$

At the same time, since $\widehat{\boldsymbol{\beta}}$ and $\boldsymbol{\beta}_0$ are in the feasible set (5.13), we have

$$\|\nabla L_n(\widehat{\boldsymbol{\beta}}) - \nabla L_n(\boldsymbol{\beta}_0)\|_\infty \le 2\gamma_n$$

with probability at least $1-\delta_n$. By Hölder's inequality, we get

$$|[L_n'(\widehat{\beta}) - \nabla L_n(\boldsymbol{\beta}_0)]^T \widehat{\boldsymbol{\Delta}}| \leq 2\gamma_n \|\widehat{\boldsymbol{\Delta}}\|_1 \leq 4\sqrt{s}\gamma_n \|\widehat{\boldsymbol{\Delta}}\|_2 \tag{5.16}$$

with probability at least $1-\delta_n$, where the last inequality utilizes (5.15). Under the *restricted strong convexity* assumption that

$$\inf_{\|\boldsymbol{\Delta}_{S_0}\|_1 \geq \|\boldsymbol{\Delta}_{S_0^c}\|_1} \left| [\nabla L_n(\boldsymbol{\beta}_0 + \Delta) - \nabla L_n(\boldsymbol{\beta}_0)]^T \Delta \right| \geq a \|\Delta\|^2, \tag{5.17}$$

for a constant $a > 0$, we conclude from (5.16) that

$$\|\widehat{\boldsymbol{\Delta}}\|_2 \leq 4a^{-1}\sqrt{s}\gamma_n \quad \text{and} \quad \|\widehat{\boldsymbol{\Delta}}\|_1 \leq 8a^{-1} s\gamma_n, \tag{5.18}$$

where the last inequality follows from (5.15).

## 5.4 Variable Selection via Penalized Likelihood

As discussed in Chapter 1, sparsity is necessary from a scientific and statistical perspective when modeling high-dimensional data. The sparse penalization techniques discussed in Chapter 3 can be applied to GLIM via the penalized maximum likelihood method.

Suppose that the observed data $(\mathbf{X}_i, Y_i)_{i=1}^n$ is a random sample from a model where the conditional distribution of $Y_i$ depends on $\mathbf{X}_i$ through a linear combination $\mathbf{X}_i^T\boldsymbol{\beta}$ and a possible nuisance parameter $\phi$. Let $p$ be the dimension of the covariates $\mathbf{x}$. Denote the conditional log-likelihood as $\ell_i(\boldsymbol{\beta}, \phi)$. For some distributions such as the Poisson and Bernoulli distributions, no such nuisance parameter $\phi$ exists. Some GLIMs may have a nuisance parameter $\phi$ such as the dispersion parameter in the canonical exponential family. The primary goal is to infer $\boldsymbol{\beta}$. For ease of presentation, we suppress $\phi$ and write $\ell_i(\boldsymbol{\beta})$ for $\ell_i(\boldsymbol{\beta}, \phi)$ throughout this section. Furthermore, denote

$$\ell(\boldsymbol{\beta}) = \sum_{i=1}^n \ell_i(\boldsymbol{\beta}),$$

the likelihood of the observed data. However, we should also mention that the estimation of $\phi$ can be important for high-dimensional learning. For example, error variance is the nuisance parameter in the normal-linear model and Section 3.7 discussed the estimation of variance in penalized linear regression.

Similar to penalized least squares introduced in Chapter 3, we define *penalized likelihood* as follows

$$Q(\boldsymbol{\beta}) = \ell(\boldsymbol{\beta}) - n\sum_{j=1}^p p_{\lambda_j}(|\beta_j|), \tag{5.19}$$

and so the *penalized MLE* is

$$\widehat{\boldsymbol{\beta}} = \arg\max_{\boldsymbol{\beta}} Q(\boldsymbol{\beta}),$$

where $p_{\lambda_j}(\cdot)$ is a given nonnegative penalty function with a tuning parameter $\lambda_j$. Without the penalty function, $\widehat{\boldsymbol{\beta}}$ is the usual MLE. Popularly used *folded concave penalty* functions include $L_1$-penalty, and SCAD and MCP penalties (see Section 3.2). The use of an appropriate penalty function yields a sparse $\widehat{\boldsymbol{\beta}}$, and therefore achieves the purpose of variable selection.

In the case of the normal linear model, the penalized MLE becomes the penalized least squares. To see this, notice that the corresponding penalized likelihood is

$$-\frac{1}{2\sigma^2}\sum_{i=1}^{n}(Y_i - \mathbf{X}_i^T\boldsymbol{\beta})^2 - n\sum_{j=1}^{p} p_{\lambda_j}(|\beta_j|)$$

after dropping a constant related to $\sigma^2$. Maximizing the penalized likelihood is equivalent to minimizing the following *penalized least-squares*

$$\frac{1}{2n}\sum_{i=1}^{n}(Y_i - \mathbf{X}_i^T\boldsymbol{\beta})^2 + \sum_{j=1}^{p}\sigma^2 p_{\lambda_j}(|\beta_j|).$$

The unknown $\sigma^2$ can be absorbed into the penalty function. This implies that the penalized likelihood is a natural extension of the penalized least squares introduced in Chapter 3. Tools and techniques for analyzing the penalized least squares may be applicable for the penalized likelihood.

The previous definition applies to all likelihood-based models. Now we focus on GLIM. Under model (5.7), using the likelihood function (5.8), after suppressing the dispersion parameter $\phi$ as aforementioned and we define a *penalized likelihood* function as

$$Q(\boldsymbol{\beta}) = \sum_{i=1}^{n}[Y_i h(\mathbf{X}_i^T\boldsymbol{\beta}) - b(h(\mathbf{X}_i^T\boldsymbol{\beta}))] - n\sum_{j=1}^{p} p_{\lambda}(|\beta_j|). \tag{5.20}$$

With proper choice of penalty function and tuning parameter $\lambda$, maximizing $Q(\boldsymbol{\beta})$ results in variable selection for generalized linear models.

For canonical link, $\theta_i = \mathbf{X}_i^T\boldsymbol{\beta}$, the penalized likelihood function becomes

$$Q(\boldsymbol{\beta}) = \sum_{i=1}^{n}[Y_i\mathbf{X}_i^T\boldsymbol{\beta} - b(\mathbf{X}_i^T\boldsymbol{\beta})] - n\sum_{j=1}^{p} p_{\lambda}(|\beta_j|). \tag{5.21}$$

Denote $\boldsymbol{\theta} = (\theta_1, \cdots, \theta_n)^T = \mathbf{X}\boldsymbol{\beta}$, and define

$$\begin{aligned}\boldsymbol{\mu}(\boldsymbol{\beta}) &= (b'(\mathbf{X}_1^T\boldsymbol{\beta}), \cdots, b'(\mathbf{X}_n^T\boldsymbol{\beta}))^T,\\ \mathbf{W}(\boldsymbol{\beta}) &= \operatorname{diag}\{b''(\mathbf{X}_1^T\boldsymbol{\beta}), \cdots, b''(\mathbf{X}_n^T\boldsymbol{\beta})\}.\end{aligned}$$

As shown in Section 5.1, $\mathrm{E}(\mathbf{Y}|\mathbf{X}) = \boldsymbol{\mu}(\boldsymbol{\beta})$ and $\mathrm{cov}(\mathbf{Y}|\mathbf{X}) = a(\phi)\mathbf{W}(\boldsymbol{\beta})$, the Hessian of the log-likelihood is $-\mathbf{X}^T\mathbf{W}(\boldsymbol{\beta})\mathbf{X}$, which implies the log-likelihood function is concave if $\mathbf{X}$ has rank $p$.

When the Hessian matrix of $Q(\boldsymbol{\beta})$ is negative definite, the target function $Q(\cdot)$ is strictly concave and there is a unique maximizer. Recall the log-likelihood function $\ell(\boldsymbol{\beta}) = \sum_{i=1}^n [Y_i \mathbf{X}_i^T \boldsymbol{\beta} - b(\mathbf{X}_i^T \boldsymbol{\beta})]$. Denote that $\mathcal{L}_c = \{\boldsymbol{\beta} : \ell(\boldsymbol{\beta}) \geq c\}$ is a sublevel set of $-\ell(\boldsymbol{\beta})$ for some $c < \ell(\mathbf{0})$. The following proposition due to Fan and Lv (2011) provides us with a sufficient condition for $\widehat{\boldsymbol{\beta}}$ to be a global maximizer of $Q(\boldsymbol{\beta})$.

**Proposition 5.1** *Suppose that* $\mathbf{X}$ *has rank* $p$ *and satisfies that*

$$\min_{\boldsymbol{\beta} \in \mathcal{L}_c} \lambda_{\min}\{\frac{1}{n}\mathbf{X}^T \mathbf{W}(\boldsymbol{\beta})\mathbf{X}\} \geq \kappa\{p_{\lambda_n}(\cdot)\}, \tag{5.22}$$

*where* $\lambda_{\min}(\mathbf{A})$ *stands for the minimum eigenvalue of a symmetric matrix* $\mathbf{A}$ *and* $\kappa(\cdot)$ *is the maximum concavity, defined by* (3.25). *Then the* $\widehat{\boldsymbol{\beta}}$ *is a global maximizer of* $Q(\boldsymbol{\beta})$ *if* $\widehat{\boldsymbol{\beta}} \in \mathcal{L}_c$ *for some* $c < \ell(\mathbf{0})$.

For the $L_1$ penalty, the condition (5.22) holds automatically since $\kappa\{p_{\lambda_n}(\cdot)\} = 0$. For the SCAD and the MCP penalty, the condition holds for sufficiently large $a$ when correlation between covariates is not too strong. We leave the details as an exercise (Exercise 5.4). Of course, this argument is not valid when $p > n$ because $\frac{1}{n}\mathbf{X}^T\mathbf{W}(\boldsymbol{\beta})\mathbf{X}$ is singular.

The penalized likelihood can be used for variable selection beyond generalized linear models. Suppose that we have the heteroscedastic model $\mathrm{Var}(Y_i|\mathbf{X}_i) \propto V(\mu_i)$. This includes GLIM (5.1) by using (5.5). A direct extension is to consider the *penalized quasi-likelihood* as follows:

$$\sum_{i=1}^n \int_{Y_i}^{\mu_i} \frac{Y_i - t}{V(t)}\, dt - n \sum_{j=1}^p p_\lambda(|\beta_j|),$$

where $\mu_i = \mathrm{E}(Y_i|\mathbf{X}_i) = g^{-1}(\mathbf{X}_i^T\boldsymbol{\beta})$ with a link function $g(\cdot)$ and readers are welcome to verify that it reduces to penalized likelihood (5.20) for GLIM.

Penalized likelihood has been used for variable selection in finite mixture regression models in Khalili and Chen (2007) and Städler, Bühlmann, van de Geer (2010). The finite mixture regression model is defined as follows. Let $C$ be a latent class variable with possible values $k = 1, \cdots, K$. Conditioning on $C = k$ and given $\mathbf{X}_i$, the conditional density function (with respect to a measure $\nu$) of $Y_i$ is $f(Y_i; \theta_k(\mathbf{X}_i), \phi_k)$, where, just like the generalized linear models, $\theta_k(\mathbf{X}_i) = h(\mathbf{X}_i^T\boldsymbol{\beta}_k)$, and $\phi_k$'s are nuisance parameters. Thus, the conditional density function of $Y_i$ given $\mathbf{X}_i$ is

$$f(Y_i; \mathbf{X}_i, \boldsymbol{\beta}_1, \cdots, \boldsymbol{\beta}_K, \phi_1, \cdots, \phi_k, \pi_1, \cdots, \pi_k) = \sum_{k=1}^K \pi_k f(Y_i; \theta_k(\mathbf{X}_i), \phi_k).$$

The penalized likelihood is

$$\sum_{i=1}^n \log\left\{\sum_{k=1}^K \pi_k f(Y_i; \theta_k(\mathbf{X}_i), \phi_k)\right\} - \sum_{k=1}^K \pi_k \sum_{j=1}^p p_\lambda(|\beta_{kj}|),$$

where $\boldsymbol{\beta}_k = (\beta_{k1}, \cdots, \beta_{kd})^T$. Khalili and Chen (2007) studied the theory of this penalized likelihood and demonstrated that with certain penalty functions, such as the SCAD penalty, this penalized likelihood can be used for variable selection in finite mixture regression. If interested, refer to Khalili and Chen (2007) and Städler, Bühlmann and van de Geer (2010) for details.

Buu, Johnson, Li and Tan (2011) extended the penalized likelihood with the SCAD penalty for variable selection in Zero-Inflated Poisson (ZIP) regression models. They further applied the technique for an empirical analysis of a substance abuse data set which identifies a subset of important neighborhood risk factors for alcohol use disorder. Interested readers can refer to Buu *et al.* (2011) for more details.

## 5.5 Algorithms

The high-dimensionality and non-differentiability of the penalty function make the computation of penalized likelihood a challenging task. In Chapter 3, we have discussed several algorithms for fitting the penalized least squares model, including the *local quadratic approximation* (LQA) algorithm, the *local linear approximation* (LLA) algorithm and the *cyclic coordinate descent* algorithm, etc. One may extend other algorithms in Section 3.5 for penalized likelihood. We leave these as exercises (Exercises 5.5). We can apply these algorithms to the penalized likelihood problem in conjunction with the Newton-Raphson idea or gradient method.

### 5.5.1 *Local quadratic approximation*

The local quadratic approximation (LQA) algorithm introduced in Section 3.5.3 for penalized least squares can be naturally extended to the penalized likelihood problem. Denote by $p_\lambda'(|\beta|+)$ the limit of $p_\lambda'(x)$ as $x \to |\beta|$ from above. Assume $p_\lambda'(|\beta|+)$ exists for all $\beta$ including zero. This assumption holds for all popular penalty functions such as the Lasso, the SCAD and the MCP.

Let $\ell(\boldsymbol{\beta})$ be the logarithm of likelihood function. Given the current estimate $\boldsymbol{\beta}^{(k)}$, it follows from the Taylor expansion that

$$\ell(\boldsymbol{\beta}) \approx \ell(\boldsymbol{\beta}^{(k)}) + \ell'(\boldsymbol{\beta}^{(k)})^T(\boldsymbol{\beta} - \boldsymbol{\beta}^{(k)}) + \frac{1}{2}(\boldsymbol{\beta} - \boldsymbol{\beta}^{(k)})^T \ell''(\boldsymbol{\beta}^{(k)})(\boldsymbol{\beta} - \boldsymbol{\beta}^{(k)}).$$

In the LQA algorithm we locally approximate the penalty function $p_\lambda(|\beta_j|)$ by the following quadratic function

$$q(\beta_j|\beta_j^{(k)}) = p_\lambda(|\beta_j^{(k)}|) + \frac{1}{2}\frac{p_\lambda'(|\beta_j^{(k)}|)}{|\beta_j^{(k)}|}(\beta_j^2 - (\beta_j^{(k)})^2).$$

Then, we consider the following quadratic problem

$$\max_{\boldsymbol{\beta}}\{\ell'(\boldsymbol{\beta}^{(k)})^T(\boldsymbol{\beta}-\boldsymbol{\beta}^{(k)})+\frac{1}{2}(\boldsymbol{\beta}-\boldsymbol{\beta}^{(k)})^T\ell''(\boldsymbol{\beta}^{(k)})(\boldsymbol{\beta}-\boldsymbol{\beta}^{(k)})-n\sum_{j=1}^{p}q(\beta_j|\beta_j^{(k)})\}$$

The updating formula is given by

$$\boldsymbol{\beta}^{(k+1)}=\boldsymbol{\beta}^{(k)}-\{\ell''(\boldsymbol{\beta}^{(k)})-n\mathbf{Q}\}^{-1}\{\ell'(\boldsymbol{\beta}^{(k)})-n\mathbf{Q}\boldsymbol{\beta}^{(k)}\},$$

where $\mathbf{Q}$ is a diagonal matrix with diagonal element $p'_\lambda(|\beta_j^{(k)}|+)/|\beta_j^{(k)}|$. If $|\beta_j^{(k)}|<\epsilon_0$ (a small threshold), truncate $\beta_j^{(k)}$ to be zero and remove variable $X_j$ from the model.

In some situations, other algorithms other than the Newton-Raphson algorithm are used for maximizing the likelihood function. We may modify the updating of $\boldsymbol{\beta}^{(k+1)}$ so that the corresponding LQA algorithm can be carried out in a similar manner. As discussed in Section 5.1, the IRLS algorithm can be used for maximizing the corresponding likelihood function for generalized linear models. Applying the LQA to the IRLS algorithm, we obtain the following iteratively reweighted ridge regression (IRRR) algorithm for variable selection in generalized linear models as follows. Update $\mathbf{W}$, $\mathbf{G}$ and $\boldsymbol{\mu}$ defined in Section 5.1. Construct the working response vector $\mathbf{G}(\mathbf{Y}-\boldsymbol{\mu})+\mathbf{X}\boldsymbol{\beta}^{(k)}$. With the unpenalized likelihood estimate as the initial value $\boldsymbol{\beta}^{(0)}$, we iteratively update

$$\boldsymbol{\beta}^{(k+1)}=\{\mathbf{X}^T\mathbf{W}\mathbf{X}+n\mathbf{Q}\}^{-1}\mathbf{X}^T\mathbf{W}\{\mathbf{G}(\mathbf{Y}-\boldsymbol{\mu})+\mathbf{X}\boldsymbol{\beta}^{(k)}\},$$

only for non-vanishing $\beta_j^{(k)}$. If $|\beta_j^{(k)}|<\epsilon_0$ (a small threshold), truncate $\beta_j^{(k)}$ to be zero and remove variable $j$ from the model.

For logistic regression, $\mu_i=\pi_i=\exp(\mathbf{X}_i^T\boldsymbol{\beta})/\{1+\exp(\mathbf{X}_i^T\boldsymbol{\beta})\}$, $\mathbf{W}=\mathrm{diag}\{\pi_1(1-\pi_1),\cdots,\pi_n(1-\pi_n)\}$ and $\mathbf{G}=\mathbf{W}^{-1}$. For Poisson log-linear regression, $\mu_i=\exp(\mathbf{X}_i^T\boldsymbol{\beta})$, $\mathbf{W}=\mathrm{diag}\{\mu_1,\cdots,\mu_n\}$ and $\mathbf{G}=\mathbf{W}^{-1}$. Thus, the IRRR algorithm can be easily carried out and used for variable selection in both logistic regression models and Poisson log-linear models.

### *5.5.2 Local linear approximation*

There are two ways to carry out the LLA algorithm for fitting the penalized likelihood. Given the current estimate $\boldsymbol{\beta}^{(k)}$, the LLA algorithm introduced in Section 3.5.4 uses a linear majorization function of the penalty function:

$$l(\beta_j|\beta_j^{(k)})=p_\lambda(|\beta_j^{(k)}|)+p'_\lambda(|\beta_j^{(k)}|)(|\beta_j|-|\beta_j^{(k)}|).$$

The first approach updates the estimate by solving the following weighted $L_1$ penalized likelihood problem

$$\boldsymbol{\beta}^{(k+1)}=\arg\max_{\boldsymbol{\beta}}\{\ell(\boldsymbol{\beta})-n\sum_{j=1}^{p}l(\beta_j|\beta_j^{(k)})\}.$$

Note that if we initially take $\boldsymbol{\beta}^{(0)} = \mathbf{0}$, then $p'_\lambda(0) = \lambda$ and $l_j(\beta_j|\beta_j^{(0)}) = c_j + \lambda|\beta_j|$ for a constant $c_j$. Hence, the first-step is simply the Lasso estimator.

After LLA, the problem becomes a convex optimization and can be efficiently solved by using a solution path algorithm in Park and Hastie (2007) or by using the coordinate descent algorithm. The second approach is more often used where the log-likelihood function is approximated by its second order Taylor expansion as is done in the usual MLE calculation. We solve

$$\max_{\boldsymbol{\beta}}\{\ell'(\boldsymbol{\beta}^{(k)})^T(\boldsymbol{\beta}-\boldsymbol{\beta}^{(k)}) + \frac{1}{2}(\boldsymbol{\beta}-\boldsymbol{\beta}^{(k)})^T\ell''(\boldsymbol{\beta}^{(k)})(\boldsymbol{\beta}-\boldsymbol{\beta}^{(k)}) - n\sum_{j=1}^p l(\beta_j|\beta_j^{(k)})\}$$

which is equivalent to solving

$$\min_{\boldsymbol{\beta}}\{-\ell'(\boldsymbol{\beta}^{(k)})^T(\boldsymbol{\beta}-\boldsymbol{\beta}^{(k)}) + \frac{1}{2}(\boldsymbol{\beta}-\boldsymbol{\beta}^{(k)})^T[-\ell''(\boldsymbol{\beta}^{(k)})](\boldsymbol{\beta}-\boldsymbol{\beta}^{(k)}) + n\sum_{j=1}^p d_j|\beta_j|\}, \tag{5.23}$$

where $d_j = p'_\lambda(|\beta_j^{(k)}|)$.

Now let us examine the above algorithm for the penalized GLIM with a canonical link. We have

$$\ell(\boldsymbol{\beta}) = \sum_{i=1}^n w_i[Y_i\mathbf{X}_i^T\boldsymbol{\beta} - b(\mathbf{X}_i^T\boldsymbol{\beta})].$$

Then direct calculation shows that (5.23) becomes

$$\min_{\boldsymbol{\beta}}\{\sum_{i=1}^n \frac{1}{2}w_i b''(\mathbf{X}_i^T\boldsymbol{\beta}^{(k)})(z_i - \mathbf{X}_i^T\boldsymbol{\beta})^2 + n\sum_{j=1}^p d_j|\beta_j|\}, \tag{5.24}$$

where

$$z_i = \mathbf{X}_i^T\boldsymbol{\beta}^{(k)} + \frac{Y_i - b'(\mathbf{X}_i^T\boldsymbol{\beta}^{(k)})}{b''(\mathbf{X}_i^T\boldsymbol{\beta}^{(k)})}.$$

Note that (5.24) is a standard $L_1$ penalized weighted least squares problem. There are many efficient algorithms for solving (5.24) such as *coordinate descent*, *FISTA* and *ADMM*, as discussed in Chapter 3. This scheme can be directly applied for both the logistic regression and the Poisson log-linear model.

### *5.5.3 Coordinate descent*

We can also just apply the Newton-Raphson idea to approximate the log-likelihood function while keeping the original penalty function. Given the current estimate $\boldsymbol{\beta}^{(k)}$, the update is given by minimizing

$$-\ell'(\boldsymbol{\beta}^{(k)})^T(\boldsymbol{\beta}-\boldsymbol{\beta}^{(k)}) + \frac{1}{2}(\boldsymbol{\beta}-\boldsymbol{\beta}^{(k)})^T[-\ell''(\boldsymbol{\beta}^{(k)})](\boldsymbol{\beta}-\boldsymbol{\beta}^{(k)}) + n\sum_{j=1}^p p_\lambda(|\beta_j|). \tag{5.25}$$

one component at a time.

In the case of a GLIM with the canonical link, we have

$$\sum_{i=1}^{n} \frac{1}{2} w_i b''(\mathbf{X}_i^T \boldsymbol{\beta}^{(k)})(z_i - \mathbf{X}_i^T \boldsymbol{\beta})^2 + n \sum_{j=1}^{p} p_\lambda(|\beta_j|). \tag{5.26}$$

Coordinate descent has been used to solve (5.26) even when the penalty function is concave. Some details about the implementation of this algorithm can be found in Zhang and Li (2011) and Fan and Lv (2011).

### 5.5.4 *Iterative local adaptive majorization and minimization*

The LQA and LLA utilize the second order approximation and involve inversions of $p \times p$ Hessian matrices. This is very expensive to compute in ultrahigh-dimensional settings, if not infeasible. *I-LAMM*, introduced in Section 3.5.10 is the first-order based algorithmic estimator used to solve general *penalized quasi-likelihood* (3.82). Thus, it is also applicable to our specific GLIM. In this case,

$$f(\boldsymbol{\beta}) = -\sum_{i=1}^{n} w_i [Y_i \mathbf{X}_i^T \boldsymbol{\beta} - b(\mathbf{X}_i^T \boldsymbol{\beta})]. \tag{5.27}$$

In addition, Fan, Liu, Sun, and Zhang (2018) proved such a method achieve s statistically optimal rate of convergence.

## 5.6 Tuning Parameter Selection

In applications we often need to use a data-driven method to select a proper tuning (or penalization) parameter $\lambda$, which controls the model complexity and the goodness of fit. In Section 3.6, we have discussed this issue for the penalized least squares where degree of freedom plays an important role in describing the model complexity. In this section, we extend the discussion to the penalized likelihood setting.

**Measure of model fitting**. For linear regression models with homogeneous errors, the sum of squared residuals can be used to measure how well a candidate model fits the data. For generalized linear models, there are several choices for the measure. One may consider the *deviance*, the *sum of squared Pearson residuals* or the *sum of squared Anscombe residuals* as a measure of model fitting. For a given $\lambda$, denote by $\widehat{\boldsymbol{\beta}}_\lambda$ the corresponding penalized likelihood estimate. Let $G(y, \mathbf{x}^T \widehat{\boldsymbol{\beta}}_\lambda)$ be a measure of how well the model associated with $\widehat{\boldsymbol{\beta}}_\lambda$ fits a generic observation $(y, \mathbf{x})$.

**Cross-validation**. The leave-one-out *cross-validation* method can be described as follows. For a given $\lambda$, denote by $\widehat{\boldsymbol{\beta}}_\lambda^{(-i)}$ the corresponding penalized likelihood estimate without including the $i$-th observation, $i = 1, \cdots, n$. Define a cross-validation score as follows

$$\mathrm{CV}(\lambda) = \sum_{i=1}^{n} G(Y_i, \mathbf{X}_i^T \boldsymbol{\beta}_\lambda^{(-i)}). \tag{5.28}$$

The cross-validation tuning parameter selector is to choose the tuning parameter by minimizing $\mathrm{CV}(\lambda)$ over a set of finite grid points of $\lambda$.

When the sample size is large, the leave-one-out cross-validation method usually demands prohibitive computation. One may consider using $K$-fold cross-validation. Randomly partition the whole data set into $K$ almost equal-sized subsets, denoted by $T^{(k)}$, $k = 1, \cdots, K$. For each $\lambda$ and $k$, let $\widehat{\boldsymbol{\beta}}_\lambda^{(k)}$ be the penalized likelihood estimate based on the training data excluding the $k$th folder. Then define the $K$-fold cross-validation score as

$$\mathrm{CV}_K(\lambda) = \sum_{k=1}^{K} \sum_{(\mathbf{X}_i, Y_i) \in T^{(k)}} G(Y_i, \mathbf{X}_i^T \boldsymbol{\beta}_\lambda^{(k)}). \tag{5.29}$$

One chooses the tuning parameter by minimizing $\mathrm{CV}_K(\lambda)$ over a set of finite grid points of $\lambda$. The value of $K$ is usually set to be 5 or 10 in practical implementation.

In doing cross-validation, we need to specify the range for candidate values of $\lambda$. A practical guideline is to plot the cross-validation score versus $\lambda$, and adjust the range of $\lambda$ so that the minimum of $\mathrm{CV}(\lambda)$ is reached around the middle of the range of $\lambda$.

**Generalized cross-validation**. Another popular method for tuning parameter selection is the generalized cross-validation (GCV). Define the *generalized cross-validation score*:

$$\mathrm{GCV}(\lambda) = \frac{1}{n} \frac{G(\mathbf{Y}, \widehat{\boldsymbol{\beta}}_\lambda)}{\{1 - \mathrm{df}(\lambda)/n\}^2}, \tag{5.30}$$

where $\mathrm{df}(\lambda)$ is the effective number of parameters corresponding to $\widehat{\boldsymbol{\beta}}_\lambda$. The GCV tuning parameter selector chooses $\lambda$ to be the minimizer of $\mathrm{GCV}(\lambda)$.

As discussed in Section 3.6, the number of nonzero estimates in the Lasso penalized least squares is an exact unbiased estimate of the degrees of freedom (Zou, Hastie and Tibshirani, 2007). The penalized likelihood estimate can be obtained by iteratively solving the corresponding $L_1$ penalized least squares. Based on this observation, Park and Hastie (2007) argued that the number of nonzero coefficients is still a good estimate for the degree of freedom in the context of generalized linear models. Fan and Li (2001) consider another way to define the effective number of parameters. Under some mild conditions, Zhang, Li and Tsai (2010) showed that with probability tending to one, the effective number of parameters defined in Fan and Li (2001) is the same as the number of nonzero coefficients.

**Generalized information criterion method**. Wang, Li and Tsai (2007) studied the issue of selecting a tuning parameter for the penalized least squares with the SCAD penalty. Zhang, Li and Tsai (2010) proposed a *generalized information criterion* (GIC) tuning parameter selector for penalized likelihood with a generic penalty. Define the GIC score as follows.

$$\mathrm{GIC}_{\kappa_n}(\lambda) = \frac{1}{n}\{G(\mathbf{Y}, \widehat{\boldsymbol{\beta}}_\lambda) + \kappa_n \mathrm{df}(\lambda)\}, \tag{5.31}$$

where $\kappa_n$ is a positive number that controls the properties of variable selection and $\mathrm{df}(\lambda)$ is the effective degrees of freedom of the corresponding candidate model. For a given $\kappa_n$, the GIC selector chooses $\lambda$ that minimizes $\mathrm{GIC}_{\kappa_n}(\lambda)$.

Under some conditions given in Section 5.8, Zhang, Li and Tsai (2010) showed that if there exists a positive constant $M$ such that $\kappa_n < M$, with zero probability, the penalized likelihood estimate with $\lambda$ selected by the GIC selector produces an underfitted model that excludes at least one significant predictors. Furthermore, with a positive probability, the penalized likelihood estimate with $\lambda$ selected by GIC selector results in an overfitted model that includes all significant predictors and at least one inactive predictors. They further showed that if $\kappa_n \to \infty$ and $\kappa_n/\sqrt{n} \to 0$ as $n \to \infty$, then with probability tending to 1, the penalized likelihood estimate with $\lambda$ selected by GIC selector yields a model exactly containing all truly significant predictors. Theoretical properties of the GIC selector are discussed in Section 5.8.

Analogous to the classical AIC and BIC variable selection criteria, we referred to the GIC selector with $\kappa_n = 2$ as AIC selector, and name the GIC with $\kappa_n \to 2$ the AIC-type selector. In contrast, we call GIC with $\kappa_n \to \infty$ and $\kappa_n/\sqrt{n} \to 0$ the BIC-type selector as the factor $\log(n)$ in the classical BIC variable selection criterion satisfies that $\log(n) \to \infty$ and $\log(n)/\sqrt{n} \to 0$ as $n \to \infty$. When the true model lies in the pools of candidate models, Zhang, Li and Tsai (2010) suggested using the BIC-type selector.

## 5.7 An Application

This section is devoted to an empirical analysis of a mammographic mass data set, which is available at the UCI Machine Learning Repository. This data set consists of 516 benign and 445 malignant instances, among which 131 are missing values and 15 are coding errors. As a result, this analysis will be based on a total of 815 cases. It is known that mammography is the most effective method to screen the presence of breast cancer. However, there is about a 70% rate of unnecessary biopsies with benign outcomes in mammogram interpretation. Thus, researchers have developed several computer aided diagnosis (CAD) systems to assist physicians in predicting breast biopsy outcomes from the findings of the breast imaging reporting and data system (BI-RADS). Elter, Schulz-Wendtland and Wittenberg (2007) analyzed this data set to examine the capability of some novel CAD approaches.

This data set has been used as an illustration in Zhang, Li and Tsai (2010) for tuning parameter selection. The following definitions of response and covariates are extracted from Zhang, Li and Tsai (2010). Consider the binary response $Y$, $Y = 0$ for benign cases and $Y = 1$ for malignant cases and 10 covariates as follows:

$X_1$: birads, BI-RADS assessment assigned in a double review process by physicians. This variable takes five possible values from 1 to 5, in which $X_1 = 1$ stands for definitely benign, and $X_1 = 5$ for highly suggestive of malignancy.

$X_2$: patient's age in years.

$X_3$: mass density. high = 1, iso = 2, low = 3 and fat containing = 4.

$X_4$: dummy variable, $X_4 = 1$ stands for round mass shape, and 0 otherwise.

$X_5$: dummy variable, $X_5 = 1$ stands for oval mass shape, and 0 otherwise

$X_6$: dummy variable, $X_6 = 1$ stands for lobular mass shape, and 0 otherwise. For mass shape, irregular shape is used as the baseline. That is, if all $X_4$, $X_5$ and $X_6$ equal 0, the shape is irregular shape.

$X_7$: dummy variable for mass margins, $X_7 = 1$ stands for circumscribed mass margins, and 0 otherwise.

$X_8$: dummy variable for mass margins, $X_8 = 1$ stands for microlobulated mass margins, and 0 otherwise.

$X_9$: dummy variable for mass margins, $X_9 = 1$ stands for obscured mass margins, and 0 otherwise.

$X_{10}$: dummy variable for mass margins, $X_{10} = 1$ stands for ill-defined mass margins, and 0 otherwise. In the dummy variables $X_7$-$X_{10}$, a spiculated mass margin is used as the baseline.

Let $\pi(\mathbf{X}^T\boldsymbol{\beta})$ be the probability of the case being classified as malignant, $\mathbf{X} = (X_0, X_1, \cdots, X_{10})^T$ with $X_0 = 1$, and $\boldsymbol{\beta} = (\beta_0, \cdots, \beta_{10})$. We consider the following logistic regression model:

$$\text{logit}\{\pi(\mathbf{X}^T\boldsymbol{\beta})\} = \beta_0 + \beta_1 X_1 \cdots + \beta_{10} X_{10}. \tag{5.32}$$

For the purpose of comparison, the best subset selection with the classical AIC and BIC is first considered. The resulting estimates along with their standard errors are reported in columns labeled 'BS-AIC' and 'BS-BIC' in Table 5.4. Penalized logistic regression with the SCAD penalty was further applied for this model. The AIC and BIC-tuning parameter selectors correspond to $\kappa_n = 2$ and $\log(n)$, respectively. The selected $\lambda$s are 0.0332 and 0.1512 based on the AIC and BIC selectors for the SCAD, respectively. The resulting estimates along with their standard errors are depicted in the columns labeled 'SCAD-AIC' and 'SCAD-BIC' of Table 5.4. Comparing the BS-AIC and the SCAD-AIC, we find that the SCAD-AIC includes $x_9$, while the BS-AIC does not. Comparing the BS-BIC and the SCAD-BIC, it can be seen that BS-BIC excludes $x_4$, while SCAD-BIC includes $x_4$, which has been also selected by both BS-AIC and SCAD-AIC. Although these four procedures result in four different models, they are nested. Thus, we may conduct analysis of deviance. Table 5.5 depicts the corresponding analysis of deviance table. From Table 5.5,

Table 5.4: Estimated coefficients for Model (5.32) with standard errors in parentheses (Adapted from Zhang, Li and Tsai (2010))

| | MLE | BS-AIC | BS-BIC | SCAD-AIC | SCAD-BIC |
|---|---|---|---|---|---|
| $\beta_0$ | -11.04(1.48) | -11.40(1.14) | -11.98(1.13) | -11.15(1.13) | -11.16(1.07) |
| $\beta_1$ | 2.18(0.23) | 2.21(0.23) | 2.29(0.23) | 2.19(0.23) | 2.25(0.23) |
| $\beta_2$ | 0.05(0.01) | 0.05(0.01) | 0.05(0.01) | 0.04(0.01) | 0.04(0.01) |
| $\beta_3$ | -0.04(0.29) | 0 (−) | 0 (−) | 0 (−) | 0 (−) |
| $\beta_4$ | -0.98(0.37) | -1.00(0.36) | 0 (−) | -0.99(0.37) | -0.80(0.34) |
| $\beta_5$ | -1.21(0.32) | -1.23(0.32) | -0.74(0.27) | -1.22(0.32) | -1.07(0.30) |
| $\beta_6$ | -0.53(0.35) | -0.59(0.34) | 0 (−) | -0.54(0.34) | 0 (−) |
| $\beta_7$ | -1.05(0.42) | -0.87(0.31) | -1.42(0.24) | -0.98(0.32) | -1.01(0.30) |
| $\beta_8$ | -0.03(0.65) | 0 (−) | 0 (−) | 0 (−) | 0 (−) |
| $\beta_9$ | -0.48(0.39) | 0 (−) | 0 (−) | -0.42(0.30) | 0 (−) |
| $\beta_{10}$ | -0.09(0.33) | 0 (−) | 0 (−) | 0 (−) | 0 (−) |

Table 5.5: Table of analysis of deviance

| Resource | DF | Deviance reduction | Residual Deviance | P-value |
|---|---|---|---|---|
| Intercept | | | 1128.6505 | |
| BS-BIC | 4 | 510.2339 | 618.4166 | 0.0000 |
| SCAD-BIC\|BS-BIC | 1 | 5.2700 | 613.1466 | 0.0217 |
| AIC\|SCAD-BIC | 1 | 3.0731 | 610.0735 | 0.0796 |
| SCAD-AIC\|AIC | 1 | 1.8554 | 608.2181 | 0.1732 |
| Full Model\|SCAD-AIC | 3 | 0.0985 | 608.1196 | 0.9920 |

Deviance(BS-BIC)−Deviance(SCAD-BIC)=5.2700 with P-value 0.0217. This implies that $x_4$ significantly reduces deviance at level 0.05, and should not be excluded. This table also indicates that it is unnecessary to retain $x_6$ and $x_9$. Deviance(SCAD-BIC)−Deviance(Full model)= 5.0269 with P-value 0.4126.

## 5.8 Sampling Properties in Low-dimension

The asymptotic properties of the penalized likelihood have been studied under different settings. The theory in Fan and Li (2001) was established under the settings in which the dimension $p$ is fixed and the sample size $n$ goes to $\infty$. This section introduces some of these classical results to give readers some ideas on the developments. More thorough treatments for modern high-dimensional problems will be given in the next two sections. For this reason, we omit the proofs.

### 5.8.1 *Notation and regularity conditions*

To present the asymptotic properties of the penalized likelihood estimate, it needs the following regularity conditions. For ease of presentation, we write $\mathbf{v}_i = (\mathbf{X}_i^T, Y_i)$, $i = 1, \cdots, n$, and denote by $\Omega$ the parameter space for $\boldsymbol{\beta}$.

**Regularity Conditions:**

**(C1)** The observations $\mathbf{v}_i$, $i = 1, \cdots, n$, are independent and identically distributed with probability density $f(\mathbf{v}, \boldsymbol{\beta})$ with respect to some measure $\nu$. $f(\mathbf{v}, \boldsymbol{\beta})$ has a common support and the model is identifiable. Furthermore, the first and second logarithmic derivatives of $f$ satisfy the *Bartlett first and second identities* (see (5.2) and (5.3)):

$$\mathrm{E}_{\boldsymbol{\beta}}\left[\frac{\partial \log f(\mathbf{v}, \boldsymbol{\beta})}{\partial \beta_j}\right] = 0 \quad \text{for } j = 1, \cdots, p$$

and

$$I_{jk}(\boldsymbol{\beta}) = \mathrm{E}_{\boldsymbol{\beta}}\left[\frac{\partial}{\partial \beta_j} \log f(\mathbf{v}, \boldsymbol{\beta}) \frac{\partial}{\partial \beta_k} \log f(\mathbf{v}, \boldsymbol{\beta})\right] = E_{\boldsymbol{\beta}}\left[-\frac{\partial^2}{\partial \beta_j \partial \beta_k} \log f(\mathbf{v}, \boldsymbol{\beta})\right].$$

**(C2)** The Fisher information matrix

$$\mathbf{I}(\boldsymbol{\beta}) = \mathrm{E}\left\{\left[\frac{\partial}{\partial \boldsymbol{\beta}} \log f(\mathbf{v}, \boldsymbol{\beta})\right]\left[\frac{\partial}{\partial \boldsymbol{\beta}} \log f(\mathbf{v}, \boldsymbol{\beta})\right]^T\right\}$$

is finite and positive definite at $\boldsymbol{\beta} = \boldsymbol{\beta}_0$, the true value of $\boldsymbol{\beta}$.

**(C3)** There exists an open subset $\omega$ of $\Omega$ containing the true parameter point $\boldsymbol{\beta}_0$ such that for almost all $\mathbf{v}$ the density $f(\mathbf{v}, \boldsymbol{\beta})$ admits all third derivatives $\frac{\partial f(\mathbf{v}, \boldsymbol{\beta})}{\partial \beta_j \partial \beta_k \partial \beta_l}$ for all $\boldsymbol{\beta} \in \omega$. Further there exist functions $M_{jkl}$ such that

$$\left|\frac{\partial^3}{\partial \beta_j \partial \beta_k \partial \beta_l} \log f(\mathbf{v}, \boldsymbol{\beta})\right| \leq M_{jkl}(\mathbf{v}) \quad \text{for all} \quad \boldsymbol{\beta} \in \omega,$$

where $m_{jkl} = \mathrm{E}_{\boldsymbol{\beta}_0}[M_{jkl}(\mathbf{v})] < \infty$ for $j, k, l$.

Conditions (C1)-(C3) are used to establish the asymptotic normality of the ordinary maximum likelihood estimates (see, for example, Lehmann (1983)). These regularity conditions are mild conditions. For the generalized linear models with a canonical link, $\mathbf{v}$ has the following density:

$$f(\mathbf{v}, \boldsymbol{\beta}) = \exp\left\{\frac{y\mathbf{x}^T\boldsymbol{\beta} - b(\mathbf{x}^T\boldsymbol{\beta})}{a(\phi)} + c(y, \phi)\right\} f(\mathbf{x}).$$

where $f(\mathbf{x})$ is the density of $\mathbf{X}$ with respect to some measure. The Bartlett's first and second identities guarantee Condition (A). The Fisher information matrix is

$$\mathbf{I}(\boldsymbol{\beta}) = \mathrm{E}\{b''(\mathbf{X})\mathbf{X}\mathbf{X}^T\}/a(\phi).$$

Therefore if $\mathrm{E}\{b''(\mathbf{X}^T\boldsymbol{\beta}_0)\mathbf{X}\mathbf{X}^T\}$ is finite and positive definite, then Condition (C2) holds. If for all $\boldsymbol{\beta}$ in some neighborhood of $\boldsymbol{\beta}_0$, $|b^{(3)}(\mathbf{x}^T\boldsymbol{\beta})| \le M_0(\mathbf{x})$ for some function $M_0(\mathbf{x})$ satisfying $\mathrm{E}_{\boldsymbol{\beta}_0}\{M_0(\mathbf{x})X_jX_kX_l\} < \infty$ for all $j$, $k$, $l$, then Condition (C3) holds.

Write

$$\boldsymbol{\beta}_0 = (\beta_{10}, \cdots, \beta_{p0})^T = (\boldsymbol{\beta}_{10}^T, \boldsymbol{\beta}_{20}^T)^T.$$

Without loss of generality, assume that $\boldsymbol{\beta}_{20} = \mathbf{0}$. Let $I(\boldsymbol{\beta}_0)$ be the Fisher information matrix and $I_1(\boldsymbol{\beta}_{10}) = I_1(\boldsymbol{\beta}_{10}, \mathbf{0})$ be the Fisher information knowing $\boldsymbol{\beta}_{20} = \mathbf{0}$. Throughout this section, we write $p_{\lambda_n}(\cdot)$ for $p_\lambda(\cdot)$ to emphasize $\lambda$ depending on the sample size $n$. We further denote by $s$ the dimension of $\boldsymbol{\beta}_{10}$ and set

$$a_n = \max\{p'_{\lambda_n}(|\beta_{j0}|) : \beta_{j0} \neq 0\}, \tag{5.33}$$

$$\boldsymbol{\Sigma} = \mathrm{diag}\left\{p''_{\lambda_n}(|\beta_{10}|), \cdots, p''_{\lambda_n}(|\beta_{s0}|)\right\}, \tag{5.34}$$

$$\mathbf{b} = \left(p'_{\lambda_n}(|\beta_{10}|)\mathrm{sgn}(\beta_{10}), \cdots, p'_{\lambda_n}(|\beta_{s0}|)\mathrm{sgn}(\beta_{s0})\right)^T. \tag{5.35}$$

Define an *oracle MLE* as follows: an MLE that knows in advanced the true subset of the important variables. So we have $\widehat{\boldsymbol{\beta}}^{\mathrm{ora}} = (\widehat{\boldsymbol{\beta}}_1^{\mathrm{ora}}, 0)$ and

$$\widehat{\boldsymbol{\beta}}_1^{\mathrm{ora}} = \arg\max_{\boldsymbol{\beta}_2 = 0} \ell(\boldsymbol{\beta}). \tag{5.36}$$

That is, the oracle 'estimator' only provides the MLE of the nonzero elements of $\boldsymbol{\beta}$, knowing the true subset model. The classical MLE theory implies that

$$\sqrt{n}(\widehat{\boldsymbol{\beta}}_1^{\mathrm{ora}} - \boldsymbol{\beta}_{10}) \xrightarrow{D} N(\mathbf{0}, \mathbf{I}_1(\boldsymbol{\beta}_{10})).$$

The oracle 'estimator' is not a genuine estimator because it assumes knowing the true support of $\boldsymbol{\beta}_0$. However, it can be used as a benchmark for theoretical analysis of an estimator. Fan and Li (2001) introduced the concept of an *oracle property*. An estimator $\widehat{\boldsymbol{\beta}}$ is said to have the oracle property if (i) with probability going to one, $\widehat{\boldsymbol{\beta}}_2 = 0$, and (ii) $\widehat{\boldsymbol{\beta}}_1$ has the same asymptotic distribution as $\widehat{\boldsymbol{\beta}}_1^{\mathrm{ora}}$.

### *5.8.2 The oracle property*

Recall $\ell(\boldsymbol{\beta})$ to be the log-likelihood function of the observations $\mathbf{v}_1, \cdots, \mathbf{v}_n$, and

$$Q(\boldsymbol{\beta}) = \ell(\boldsymbol{\beta}) - n\sum_{j=1}^{p} p_{\lambda_n}(|\beta_j|).$$

The rate of convergence of penalized likelihood estimate was established in Fan and Li (2001). Specifically,

**Theorem 5.1** *Under Conditions (C1)-(C3), suppose that* $a_n \to 0$ *and* $\max\{|p''_{\lambda_n}(|\beta_{j0}|)| : \beta_{j0} \neq 0\} \to 0$ *as* $n \to \infty$. *Then there exists a local maximizer* $\widehat{\boldsymbol\beta}$ *of* $Q(\boldsymbol\beta)$ *such that*

$$\|\widehat{\boldsymbol\beta} - \boldsymbol\beta_0\| = O_P(n^{-1/2} + a_n). \tag{5.37}$$

Since $a_n$ depends on both penalty function and $\lambda_n$, (5.37) demonstrates how the convergent rate of penalized likelihood estimate depends on $\lambda_n$. In general, we should choose an appropriate convergent rate of $\lambda_n$ so that $a_n = O_P(1/\sqrt{n})$ in order for $\widehat{\boldsymbol\beta}$ to be root-$n$ consistent. For the $L_1$ penalty, $a_n = \lambda_n$. Thus, the $L_1$-penalized likelihood estimate is root-$n$ consistent if $\lambda_n = O_P(1/\sqrt{n})$. For the SCAD and MCP penalties, $a_n = 0$ if $\lambda_n \to 0$. Therefore, the SCAD- and MCP-penalized likelihood estimates are root-$n$ consistent if $\lambda_n = o_P(1)$ as $n \to \infty$. Proofs of Theorem 5.1 and Theorem 5.2 below are good exercise (Exercise 5.7).

**Theorem 5.2** *Suppose that* $p_{\lambda_n}(|\beta|)$ *satisfies*

$$\liminf_{n\to\infty} \liminf_{\beta\to 0+} \lambda_n^{-1} p'_{\lambda_n}(\beta) > 0. \tag{5.38}$$

*If* $\lambda_n \to 0$ *and* $\sqrt{n}\lambda_n \to \infty$ *as* $n \to \infty$, *then under Conditions (C1)-(C3), with probability tending to 1, the root n consistent local maximizers* $\widehat{\boldsymbol\beta} = (\widehat{\boldsymbol\beta}_1^T, \widehat{\boldsymbol\beta}_2^T)^T$ *in (5.37) must satisfy:*

*(i) (***Sparsity***)* $\widehat{\boldsymbol\beta}_2 = \mathbf{0}$*;*

*(ii) (***Asymptotic normality***)*

$$\sqrt{n}(\mathbf{I}_1(\boldsymbol\beta_{10}) + \boldsymbol\Sigma)\left\{\widehat{\boldsymbol\beta}_1 - \boldsymbol\beta_{10} + (\mathbf{I}_1(\boldsymbol\beta_{10}) + \boldsymbol\Sigma)^{-1}\mathbf{b}\right\} \xrightarrow{D} N\{\mathbf{0}, \mathbf{I}_1(\boldsymbol\beta_{10})\}$$

Theorem 5.2 indicates that with a proper tuning parameter $\lambda_n$, and for certain penalty functions, the corresponding penalized likelihood can correctly identify zero components in $\boldsymbol\beta_0$ and estimate its nonzero components $\boldsymbol\beta_1$ as efficiently as the maximum likelihood estimates for estimating $\boldsymbol\beta_1$ knowing $\boldsymbol\beta_{20} = \mathbf{0}$, if $\boldsymbol\Sigma = 0$ and $\mathbf{b} = 0$. In this case, $\widehat{\boldsymbol\beta}_1$ is more efficient than the maximum likelihood estimates of $\boldsymbol\beta_1$ based on the full model.

For the SCAD and MCP penalties, when $\lambda_n \to 0$, $a_n = 0$, $\boldsymbol\Sigma = 0$ and $\mathbf{b} = 0$. Hence, by Theorem 5.2, if $\sqrt{n}\lambda_n \to \infty$, their corresponding penalized likelihood estimators enjoy *the oracle property*:

$$\widehat{\boldsymbol\beta}_2 = 0 \quad \text{and} \quad \sqrt{n}(\widehat{\boldsymbol\beta}_1 - \boldsymbol\beta_{10}) \xrightarrow{D} N(\mathbf{0}, \mathbf{I}_1(\boldsymbol\beta_{10})).$$

However, for the $L_1$ penalty, $a_n = \lambda_n$. Hence, the root-$n$ consistency requires that $\lambda_n = O_P(1/\sqrt{n})$. On the other hand, the conditions in Theorem 5.2 require that $\sqrt{n}\lambda_n \to \infty$. These two conditions for the Lasso cannot be satisfied simultaneously. Zou (2006) showed that in general, the Lasso does not have the oracle property. But for the $L_q$ penalty with $0 < q < 1$, the oracle property continues to hold with a suitable choice of $\lambda_n$.

From Theorem 5.2, the asymptotic covariance matrix of $\sqrt{n}(\widehat{\boldsymbol\beta}_1 - \boldsymbol\beta_{10})$ is

$$\{\mathbf{I}_1(\boldsymbol\beta_{10}) + \boldsymbol\Sigma\}^{-1}\mathbf{I}_1(\boldsymbol\beta_{10})\{\mathbf{I}_1(\boldsymbol\beta_{10}) + \boldsymbol\Sigma\}^{-1}.$$

Replacing $\mathbf{I}_1(\widehat{\boldsymbol\beta}_{10})$ and $\boldsymbol\Sigma$ with their estimates, we can estimate the asymptotic covariance matrix consistently. The consistency of the resulting estimate was established under a more general setting in Fan and Peng (2004); see (5.43).

Theorem 5.2 shows that in principle, the folded-concave penalized estimator is theoretically equivalent to the oracle estimator. However, it does not explicitly describe which estimator has that nice theoretical property, as there may be multiple local solutions. To resolve this issue, we study the local solution to the *folded concave penalized likelihood* estimator when using the LLA algorithm. In Zou and Li (2008), they took the ordinary MLE as $\boldsymbol\beta^{(0)}$, the initial value in the LLA algorithm. Interestingly, they showed that even after one LLA iteration, $\boldsymbol\beta^{(1)}$ has a strong theoretical property. They call $\boldsymbol\beta^{(1)}$ the *one-step estimator*, denoted by $\widehat{\boldsymbol\beta}^{\text{os}}$.

There are two classes of folded concave penalty functions in the literature:

**(P1)** There exists a constant $m$ such that penalty function $p_{\lambda_n}(|\beta|)$ satisfies that $p'_{\lambda_n}(0+) > 0$ and $p'_{\lambda_n}(|\beta|) = 0$ for $|\beta| > m\lambda$.

**(P2)** The penalty function has the following form: $p_{\lambda_n}(|\beta|) = \lambda_n p(|\beta|)$. Suppose $p(\cdot)$ has continuous first derivatives on $(0, \infty)$ and there is a constant $q > 0$ such that $p'(|\beta|) = O(|\beta|^{-q})$ as $|\beta| \to 0+$.

Penalty functions in (P1) include the SCAD, MCP and hard-thresholding penalty functions, while penalty functions in (P2) include the $L_r$-penalty with $r \in (0, 1)$ and the logarithm penalty described in Zou and Li (2008).

**Theorem 5.3** *Suppose that Conditions (C1)-(C3) hold. For penalty in (P1), assume that $\sqrt{n}p'_{\lambda_n}(0+) \to \infty$ and $\lambda_n \to 0$ as $n \to \infty$, while for penalty in (P2), assume that $n^{(1+q)/2}\lambda_n \to \infty$ and $\sqrt{n}\lambda_n \to 0$ as $n \to \infty$. Then the one-step estimator $\widehat{\boldsymbol\beta}^{\text{os}}$ must satisfy*

*(a) Sparsity: with probability tending to one, $\widehat{\boldsymbol\beta}_2^{\text{os}} = 0$.*

*(b) Asymptotic normality: $\sqrt{n}(\widehat{\boldsymbol\beta}_1^{\text{os}} - \boldsymbol\beta_{10}) \xrightarrow{D} N(0, \mathbf{I}_1^{-1}(\boldsymbol\beta_{10}))$.*

### *5.8.3 Sampling properties with diverging dimensions*

Fan and Peng (2004) studied the asymptotic behavior of penalized likelihood with a diverging number of parameters. We use $p_n$ to emphasize $p$ growing with $n$. In this section, it is assumed that $p_n = O(n^\alpha)$ for $\alpha \in (0, 1)$. Under the setting of diverging number of predictors, it is typical to assume that the true value of the parameter depends on the sample size. To emphasize this, one would write $\boldsymbol\beta_n$ for $\boldsymbol\beta$, and denote by $\boldsymbol\beta_{n0}$ the true value of $\boldsymbol\beta$.

Assume that there exist positive constants $C_j$, $j = 1, \cdots, 5$ such that

$$0 < C_1 < \lambda_{\min}\{\mathbf{I}(\boldsymbol{\beta}_n)\} \le \lambda_{\infty}\{\mathbf{I}(\boldsymbol{\beta}_n)\} < C_2 < \infty \qquad \text{for all } n, \tag{5.39}$$

for $j, k = 1, 2, \cdots, p_n$,

$$\mathrm{E}_{\boldsymbol{\beta}_n}\left\{\frac{\partial \log f(\mathbf{v}_n, \boldsymbol{\beta}_n)}{\partial \beta_{nj}}\frac{\partial \log f(\mathbf{v}_n, \boldsymbol{\beta}_n)}{\partial \beta_{nk}}\right\}^2 < C_3 < \infty \tag{5.40}$$

and

$$\mathrm{E}_{\boldsymbol{\beta}_n}\left\{\frac{\partial^2 \log f_n(\mathbf{v}_n, \boldsymbol{\beta}_n)}{\partial \beta_{nj}\partial \beta_{nk}}\right\}^2 < C_4 < \infty, \tag{5.41}$$

and $\mathrm{E}_{\boldsymbol{\beta}_n}[M_{jkl}^2(\mathbf{v}_n)] < C_5 < \infty$ for $j, k, l$.

Under Conditions (C1)—(C3) and the moment conditions (5.39)—(5.41), Fan and Peng (2004) showed that then there is a local maximizer $\widehat{\boldsymbol{\beta}}_n$ of $Q(\boldsymbol{\beta}_n)$ such that

$$\|\widehat{\boldsymbol{\beta}}_n - \boldsymbol{\beta}_{n0}\| = O_P\{\sqrt{p_n}(n^{-1/2} + a_n)\},$$

if $p_n^4/n \to 0$ as $n \to \infty$.

Parallel to the notation in Section 5.8.1, write

$$\boldsymbol{\beta}_{n0} = (\beta_{n10}, \cdots, \beta_{nd_n0})^T = (\boldsymbol{\beta}_{n10}^T, \boldsymbol{\beta}_{n20}^T)^T.$$

Without loss of generality, assume that $\boldsymbol{\beta}_{n20} = \mathbf{0}$, and all elements of $\boldsymbol{\beta}_{n10}$ do not equal 0. Denote by $s_n$ the dimension of $\boldsymbol{\beta}_{n10}$. Let $\mathbf{I}(\boldsymbol{\beta}_{n0})$ be the Fisher information matrix and $\mathbf{I}_1(\boldsymbol{\beta}_{n10}) = \mathbf{I}_1(\boldsymbol{\beta}_{n10}, \mathbf{0})$ be the Fisher information knowing $\boldsymbol{\beta}_{n20} = \mathbf{0}$. We also assume the following *minimum signal assumption*:

$$\min_{1 \le j \le s_n} |\beta_{n0j}|/\lambda_n \to \infty, \qquad n \to \infty,$$

which is used for the variable selection consistency of penalized likelihood.

Under Conditions (C1)—(C3), the moment conditions (5.39)—(5.41) and the minimal signal condition, Fan and Peng (2004) showed that if $\lambda_n \to 0$, and $\sqrt{n/p_n}\lambda_n \to \infty$, and $p_n^5/n \to 0$ as $n \to \infty$, then with probability tending to one, $\widehat{\boldsymbol{\beta}}_{n2} = \mathbf{0}$. Furthermore, for any $s_n$-dimensional vector $\mathbf{c}_n$ such that $\mathbf{c}_n^T\mathbf{c}_n = 1$, it follows that

$$\sqrt{n}\mathbf{c}_n^T\mathbf{I}_1^{-1/2}(\boldsymbol{\beta}_{n10})(\mathbf{I}_{n1}(\boldsymbol{\beta}_{10}) + \boldsymbol{\Sigma})\left\{\widehat{\boldsymbol{\beta}}_{n1} - \boldsymbol{\beta}_{n10} + (\mathbf{I}_1(\boldsymbol{\beta}_{n10}) + \boldsymbol{\Sigma})^{-1}\mathbf{b}\right\} \xrightarrow{D} N(0,1) \tag{5.42}$$

where $\boldsymbol{\Sigma}$ and $\mathbf{b}$ are defined in (5.34) and (5.35), respectively. Thus, the asymptotic covariance matrix of $\widehat{\boldsymbol{\beta}}_{n1}$ is

$$\mathbf{A}_n = \{\mathbf{I}_n(\boldsymbol{\beta}_{n10}) + \boldsymbol{\Sigma}\}^{-1}\mathbf{I}_n(\boldsymbol{\beta}_{n10})\{\mathbf{I}_n(\boldsymbol{\beta}_{n10}) + \boldsymbol{\Sigma}\}^{-1}.$$

Thus, its sandwich estimator for the covariance matrix of $\widehat{\boldsymbol{\beta}}_{n1}$ is

$$\widehat{\mathbf{A}}_n = n\{\ell''(\widehat{\boldsymbol{\beta}}_{n1}) - n\widehat{\boldsymbol{\Sigma}}\}^{-1}\widehat{\text{cov}}\{\ell'(\widehat{\boldsymbol{\beta}}_{n1})\}\{\ell''(\widehat{\boldsymbol{\beta}}_{n1}) - n\widehat{\boldsymbol{\Sigma}}\}^{-1}$$

with an estimate for $\text{cov}\{\ell'(\widehat{\boldsymbol{\beta}}_{n1})\}$ being

$$\widehat{\text{cov}}\{\ell'(\widehat{\boldsymbol{\beta}}_{n1})\} = \frac{1}{n}\sum_{i=1}^{n}\ell'_{ni}(\widehat{\boldsymbol{\beta}}_{n1})\ell'_{ni}(\widehat{\boldsymbol{\beta}}_{n1})^T - \{\frac{1}{n}\ell'(\widehat{\boldsymbol{\beta}}_{n1})\}\{\frac{1}{n}\ell'(\widehat{\boldsymbol{\beta}}_{n1})\}^T,$$

where $\ell_{ni}(\boldsymbol{\beta}_n) = \log f(\mathbf{v}_{ni}, \boldsymbol{\beta}_n)$.

Fan and Peng (2004) further established the consistency of the sandwich estimator. If $p_n^5/n \to 0$ as $n \to \infty$, then for any $s_n$-dimensional vector $\mathbf{c}_n$ such that $\mathbf{c}_n^T\mathbf{c}_n = 1$,

$$\mathbf{c}_n^T(\widehat{\mathbf{A}}_n - \mathbf{A}_n)\mathbf{c}_n \to 0 \tag{5.43}$$

in probability as $n \to \infty$.

Fan and Peng (2004) also studied the following hypothesis testing problem:

$$H_0 : \mathbf{C}_n\boldsymbol{\beta}_{n10} = 0 \quad \text{versus} \quad H_1 : \mathbf{C}_n\boldsymbol{\beta}_{n10} \neq 0, \tag{5.44}$$

where $\mathbf{C}_n$ is a $q \times s_n$ matrix and $\mathbf{C}_n\mathbf{C}_n^T = \mathbf{I}_q$ for a fixed $q$. Many testing hypothesis problems can be formulated as the above *linear hypotheses*. For diverging dimensionality, the likelihood ratio test can still be used for the hypothesis testing. Define the likelihood ratio test

$$T_{n1} = 2[\ell\{\widehat{\boldsymbol{\beta}}(H_1)\} - \ell\{\widehat{\boldsymbol{\beta}}(H_0)\}],$$

where $\widehat{\boldsymbol{\beta}}(H_0)$ and $\widehat{\boldsymbol{\beta}}(H_1)$ are the maximum likelihood estimate under $H_0$ and $H_1$, respectively. Alternatively, we may consider the penalized likelihood ratio test

$$T_{n2} = 2[Q\{\widehat{\boldsymbol{\beta}}(H_1)\} - Q\{\widehat{\boldsymbol{\beta}}(H_0)\}],$$

where $\widehat{\boldsymbol{\beta}}(H_0)$ and $\widehat{\boldsymbol{\beta}}(H_1)$ are the penalized likelihood estimate under $H_0$ and $H_1$, respectively. Fan and Peng (2004) showed that $T_{n2}$ follows an asymptotic $\chi^2$-distribution with $q$ degrees of freedom under certain regularity conditions.

### *5.8.4 Asymptotic properties of GIC selectors*

In the applications of these selectors, the tuning parameter $\lambda$ needs to be chosen by a data-driven method. Zhang, Li and Tsai (2010) established some interesting theoretical results for the GIC selector. We introduce their results in this section.

**Definition 5.1** *Let $\bar{\alpha} = \{1, ..., p\}$, and $\alpha$ be a subset of $\bar{\alpha}$. A candidate model $\alpha$ means that the corresponding predictors labelled by $\alpha$ are included in the model. Accordingly, $\bar{\alpha}$ is the full model. In addition, denote by $\mathcal{A}$ the collection of all candidate models.*

Denote by $d_\alpha$ and $\boldsymbol{\beta}_\alpha$ the size of model $\alpha$ (i.e., the number of nonzero parameters in $\alpha$) and the coefficients associated with the predictors in model $\alpha$, respectively. For a penalized estimator $\widehat{\boldsymbol{\beta}}_\lambda$ that minimizes the objective function (5.19), let $\alpha_\lambda$ stand for the model associated with $\widehat{\boldsymbol{\beta}}_\lambda$ .

To establish the model selection consistency of variable selection, it is typically assumed that the set of candidate models contains the unique true model. We further assume that the number of parameters in the full model is finite. This assumption enables one to define underfitted and overfitted models as follows.

**Definition 5.2** *Suppose that there is a unique true model $\alpha_0$ in $\mathcal{A}$, whose corresponding coefficients are nonzero. A candidate model $\alpha$ is said to be an underfitted model if $\alpha \not\supset \alpha_0$. A candidate model $\alpha$ is said to be an overfitted model if $\alpha \supset \alpha_0$ but $\alpha \neq \alpha_0$.*

Based on the above definitions, one may partition the tuning parameter interval $[0, \lambda_\infty]$ into the underfitted, true, and overfitted subsets, respectively,

$$\begin{aligned}\Lambda_- &= \{\lambda : \alpha_\lambda \not\supset \alpha_0\},\\ \Lambda_0 &= \{\lambda : \alpha_\lambda = \alpha_0\}, \quad \text{and}\\ \Lambda_+ &= \{\lambda : \alpha_\lambda \supset \alpha_0 \quad \text{and} \quad \alpha_\lambda \neq \alpha_0\}.\end{aligned}$$

**Some Additional Regularity Conditions**:

**(C4)** For any candidate model $\alpha \in \mathcal{A}$, there exists $c_\alpha > 0$ such that $\frac{1}{n}G(\mathbf{Y}, \widehat{\boldsymbol{\beta}}^*_\alpha) \xrightarrow{P} c_\alpha$, where $\widehat{\boldsymbol{\beta}}^*_\alpha$ is the non-penalized maximum likelihood estimator of $\boldsymbol{\beta}$. In addition, for any underfitted model $\alpha \not\supset \alpha_0$, $c_\alpha > c_{\alpha_0}$, where $c_{\alpha_0}$ is the limit of $\frac{1}{n}G(\mathbf{Y}, \boldsymbol{\beta}_{\alpha_0})$ and $\boldsymbol{\beta}_{\alpha_0}$ is the parameter vector of the true model $\alpha_0$.

**(C5)** Assume that $\lambda_\infty$ depends on $n$ and satisfied $\lambda_\infty \to 0$ as $n \to \infty$.

**(C6)** There exists a constant $m$ such that the penalty $p_{\lambda_n}(|\beta|)$ satisfies $p'_{\lambda_n}(|\beta|) = 0$ for $|\beta| > m\lambda$.

**(C7)** If $\lambda_n \to 0$ as $n \to \infty$, then the penalty function satisfies

$$\lim_{n\to\infty} \inf \liminf_{\theta\to 0+} \sqrt{n}p'_{\lambda_n}(\theta) \to \infty.$$

Condition (C4) assures that the underfitted model yields a larger measure of model fitting than that of the true model.

For a generalized linear model $\alpha_\lambda$, let us use the scaled deviance $D(\mathbf{Y}; \widehat{\boldsymbol{\mu}}_\lambda)$ as the goodness-of-fit measure $G(\mathbf{Y}, \widehat{\boldsymbol{\beta}}_\lambda)$. The resulting GIC for the generalized linear model is

$$\mathrm{GIC}_{\kappa_n}(\lambda) = \frac{1}{n}\{D(\mathbf{Y}, \widehat{\boldsymbol{\mu}}_\lambda) + \kappa_n \mathrm{df}(\lambda)\}. \tag{5.45}$$

The asymptotic performances of GIC for generalized linear models is given in the following theorem due to Zhang, Li and Tsai (2010).

**Theorem 5.4** *Suppose that* $(\mathbf{X}_i^T, Y_i)$, $i = 1, \cdots, n$, *is a random sample from a generalized linear model, and Conditions (C1)-(C4) hold.*

**(A)** *If there exists a positive constant* $M$ *such that* $\kappa_n < M$, *then the tuning parameter* $\widehat{\lambda}$ *selected by minimizing* $GIC_{\kappa_n}(\lambda)$ *in (5.45) satisfies*

$$P\left\{\widehat{\lambda} \in \Lambda_-\right\} \to 0, \text{ and } P\left\{\widehat{\lambda} \in \Lambda_+\right\} \geq \pi,$$

*where* $\pi$ *is a nonzero probability.*

**(B)** *Assume that Conditions (C5)-(C7) are valid. If* $\kappa_n \to \infty$ *and* $\kappa_n/\sqrt{n} \to 0$, *then the tuning parameter* $\widehat{\lambda}$ *selected by minimizing* $GIC_{\kappa_n}(\lambda)$ *in (5.45) satisfies* $P\left\{\alpha_{\widehat{\lambda}} = \alpha_0\right\} \to 1$.

Theorem 5.4 provides guidance on the choice of the regularization parameter. Theorem 5.4(A) implies that the *AIC-type selector* tends to overfit without regard to which penalty function is being used. In contrast, Theorem 5.4(B) indicates that the *BIC-type selector* identifies the true model consistently. To achieve the oracle property, the BIC-type selector should be used.

In applications, the assumption that the true model is included in a family of candidate models may not hold. Some work has been done to assess the efficiency of the classical AIC procedure (see Shibata, 1981, 1984 and Li, 1987) in linear regression models. Zhang, Li and Tsai (2010) studied the asymptotic efficiency of the AIC-type selector. Under some regularity conditions, they established the asymptotic average quadratic loss efficiency for linear regression models, and asymptotic Kullback-Leibler loss efficiency for generalized linear models. Interested readers can refer to their work for more details.

## 5.9 Properties under Ultrahigh Dimensions

Suppose that $(\mathbf{X}_i^T, Y_i)$, $i = 1, \cdots, n$, is a sample from a generalized linear model with the canonical link. Let $\mathbf{Y} = (Y_1, \cdots, Y_n)^T$, the response vector, and let $\mathbf{X}$ be the corresponding $n \times p$ design matrix, which is assumed to be fixed. Ultrahigh dimensionality means that $p$ can grow with $n$ of exponential order: $\log p \asymp n^\alpha$ for some $\alpha > 0$. The model parameter is $\boldsymbol{\beta}_0$ and its support set is $\mathcal{S}_0 = \{j : \beta_{0j} \neq 0\}$ with cardinality $s = |\mathcal{S}_0|$. Without loss of generality, let $\mathcal{S}_0 = \{1, 2, \ldots, s\}$, and write $\boldsymbol{\beta}_0 = (\boldsymbol{\beta}_{10}^T, 0^T)^T$. Note that $\boldsymbol{\beta}_{10}, \mathcal{S}_0, s$ may depend on $n$, but we do not explicitly use $n$ to index these quantities for ease of presentation. Define $b_n = \min_{j \in \mathcal{S}_0} |\beta_{0j}|$ as the minimum signal strength. Let $\|\mathbf{v}\|_{\max} = \max_j |v_j|$ for a vector $\mathbf{v}$.

### 5.9.1 *The Lasso penalized estimator and its risk property*

As the Lasso estimator can be regarded as the one-step estimator of the folded concave penalized estimator, let us begin by considering its properties.

The Lasso penalized likelihood estimator is

$$\widehat{\boldsymbol{\beta}}_L = \arg\min_{\boldsymbol{\beta}} \{L(\boldsymbol{\beta}) + \lambda_{\mathrm{L}} \sum_{j=1}^{p} |\beta_j|\} \tag{5.46}$$

where, for simplicity of notation, $L(\boldsymbol{\beta}) = -\frac{1}{n}\ell(\boldsymbol{\beta})$ is the *log-likelihood function* with a $1/n$ factor. The following proposition shows that the estimation error is sparse.

**Proposition 5.2** *Let* $\widehat{\boldsymbol{\Delta}} = \widehat{\boldsymbol{\beta}}_{\mathrm{L}} - \boldsymbol{\beta}_0$. *Suppose that* $L(\boldsymbol{\beta})$ *is convex. Then, in the event*

$$\{\|\nabla L(\boldsymbol{\beta}_0)\|_\infty \le \frac{1}{2}\lambda_{\mathrm{L}}\}, \tag{5.47}$$

*we have*

$$\|\widehat{\boldsymbol{\Delta}}_{\mathcal{S}_0{}^c}\|_1 \le 3\|\widehat{\boldsymbol{\Delta}}_{\mathcal{S}_0}\|_1. \tag{5.48}$$

The proposition states that $\widehat{\boldsymbol{\Delta}}$ falls in the cone $\mathcal{H} = \{\mathbf{u} : \|\mathbf{u}_{\mathcal{S}_0{}^c}\|_1 \le 3\|\mathbf{u}_{\mathcal{S}_0}\|_1\}$. This shows that $\widehat{\boldsymbol{\Delta}}$ is sparse: its $L_1$ norm on a high-dimensional space $\mathcal{S}_0^c$ is controlled by that of a much smaller space $\mathcal{S}_0$.

**Proof of Proposition 5.2**. By definition, we have

$$L(\widehat{\boldsymbol{\beta}}_{\mathrm{L}}) + \lambda_{\mathrm{L}}\|\widehat{\boldsymbol{\beta}}_{\mathrm{L}}\|_1 \le L(\boldsymbol{\beta}_0) + \lambda_{\mathrm{L}}\|\boldsymbol{\beta}_0\|_1.$$

By the convexity of negative log-likelihood $L(\boldsymbol{\beta})$, we obtain

$$(\nabla L(\boldsymbol{\beta}_0))^T\widehat{\boldsymbol{\Delta}} + \lambda_{\mathrm{L}}\|\widehat{\boldsymbol{\beta}}_{\mathrm{L}}\|_1 \le \lambda_{\mathrm{L}}\|\boldsymbol{\beta}_0\|_1. \tag{5.49}$$

In the event (5.47), we have

$$|\nabla L(\boldsymbol{\beta}_0)^T\widehat{\boldsymbol{\Delta}}| \le \frac{1}{2}\lambda_{\mathrm{L}}\|\widehat{\boldsymbol{\Delta}}\|_1$$

and hence by (5.49)

$$-\frac{1}{2}\|\widehat{\boldsymbol{\Delta}}\|_1 + \|\widehat{\boldsymbol{\beta}}_{\mathrm{L}}\|_1 \le \|\boldsymbol{\beta}_0\|_1.$$

Now using (3.47), we complete the proof. ■

Similar to the penalized least squares case, the *restricted eigenvalue condition* is defined as follows:

$$\kappa = \min_{\mathbf{u}\neq\mathbf{0}:\|\mathbf{u}_{\mathcal{S}_0{}^c}\|_1\le 3\|\mathbf{u}_{\mathcal{S}_0}\|_1} \frac{\mathbf{u}^T[\nabla^2 L(\boldsymbol{\beta}_0)]\mathbf{u}}{\mathbf{u}^T\mathbf{u}} \in (0,\infty). \tag{5.50}$$

In the rest of our discussion, we focus on the logistic regression model. We use this to illustrate how the risk bound is derived. The idea of the proof is fundamental and can be generalized to a more general case in Section 5.10. We begin with a simple lemma.

**Lemma 5.1** *Suppose $F(\mathbf{x})$ is a convex function in $R^p$ with $F(\mathbf{0}) = 0$ and the set $\mathcal{C}$ is a cone with the vertex at the origin, namely, if $\mathbf{x} \in \mathcal{C}$, then $a\mathbf{x} \in \mathcal{C}$ for any $a \geq 0$. If $F(\mathbf{x}) > 0$, $\forall \mathbf{x} \in \mathcal{C} \bigcap \{\|\mathbf{x}\| = \delta\}$, then the minimizer $\widehat{\mathbf{x}}$ of $F(\cdot)$ over $\mathcal{C}$ must have $\|\widehat{\mathbf{x}}\| < \delta$.*

**Proof.** If $\|\widehat{\mathbf{x}}\| > \delta$, then $a\widehat{\mathbf{x}} \in \mathcal{C} \bigcap \{\|\mathbf{x}\| = \delta\}$ for $a = \delta/\|\widehat{\mathbf{x}}\| < 1$. Note that $F(\widehat{\mathbf{x}}) \leq F(\mathbf{0}) = 0$. It follows by the convexity

$$F(a\widehat{\mathbf{x}}) = F\big(a\widehat{\mathbf{x}} + (1-a)\mathbf{0}\big) \leq aF(\widehat{\mathbf{x}}) + (1-a)F(\mathbf{0}) \leq 0,$$

which is a contradiction. Hence, $\|\widehat{\mathbf{x}}\| < \delta$. ■

Let $m = \max_{ij} |x_{ij}|$ and $M = \max_j \frac{1}{n}\|\mathbf{X}_j\|_2^2$.

**Theorem 5.5** *For logistic regression, if $\lambda_L \leq \frac{\kappa}{20ms}$, with probability at least $1 - 2p \cdot \exp(-\frac{n}{2M}\lambda_L^2)$,*

$$\|\widehat{\boldsymbol{\beta}}_L - \boldsymbol{\beta}_0\|_2 \leq 5\kappa^{-1}s^{1/2}\lambda_L,$$

*and*

$$\|\widehat{\boldsymbol{\beta}}_L - \boldsymbol{\beta}_0\|_1 \leq 20\kappa^{-1}s\lambda_L,$$

*where $\kappa$ is the restricted eigenvalue defined by* (5.50) *for logistic regression.*

The result is nonasymptotic. In order for the probability to be at least $1 - 2p^{-a+1}$, which goes to 1 when $a > 1$, we take $\lambda_L = \sqrt{\frac{2aM\log p}{n}}$. Then the condition $\lambda_L \leq \frac{\kappa}{20ms}$ of Theorem 5.5 is satisfied as long as $s < b\sqrt{n/\log p}$ for sufficient small constant $b$, and we have the rate of convergence

$$\|\widehat{\boldsymbol{\beta}}_L - \boldsymbol{\beta}_0\|_2 = O_P\Big(\sqrt{\frac{s\log p}{n}}\Big).$$

**Proof of Theorem 5.5.** Let $\widehat{\boldsymbol{\Delta}} = \widehat{\boldsymbol{\beta}}_L - \boldsymbol{\beta}_0$ and

$$F(\boldsymbol{\Delta}) = L(\boldsymbol{\beta}_0 + \boldsymbol{\Delta}) - L(\boldsymbol{\beta}_0) + \lambda_L(\|\boldsymbol{\beta}_0 + \boldsymbol{\Delta}\|_1 - \|\boldsymbol{\beta}_0\|_1).$$

By definition $F(\mathbf{0}) = 0$ and $F(\widehat{\boldsymbol{\Delta}}) \leq 0$. By Proposition 5.2, in the event (5.47), we need only to consider the optimization of $F(\boldsymbol{\Delta})$ on the cone:

$$\|\boldsymbol{\Delta}_{\mathcal{S}_0^c}\|_1 \leq 3\|\boldsymbol{\Delta}_{\mathcal{S}_0}\|_1.$$

Define $\mathcal{D} = \{\boldsymbol{\Delta} \in \mathbb{R}^p : \|\boldsymbol{\Delta}_{\mathcal{S}_0^c}\|_1 \leq 3\|\boldsymbol{\Delta}_{\mathcal{S}_0}\|_1 \text{ and } \|\boldsymbol{\Delta}\|_2 = 5\kappa^{-1}s^{1/2}\lambda_L\}$. By Lemma 5.1, if $F(\boldsymbol{\Delta}) > 0$ for any $\boldsymbol{\Delta} \in \mathcal{D}$, then $\|\widehat{\boldsymbol{\Delta}}\|_2 \leq 5\kappa^{-1}s^{1/2}\lambda_L$. For the $L_1$-norm, the result follows from the same calculation as in (5.51) below.

For $\boldsymbol{\Delta} \in \mathcal{D}$

$$\|\boldsymbol{\Delta}\|_1 = \|\boldsymbol{\Delta}_{\mathcal{S}_0}\|_1 + \|\boldsymbol{\Delta}_{\mathcal{S}_0^c}\|_1 \leq 4\|\boldsymbol{\Delta}_{\mathcal{S}_0}\|_1 \leq 4s^{1/2}\|\boldsymbol{\Delta}\|_2 = z_0 \qquad (5.51)$$

where $z_0 = 20s\kappa^{-1}\lambda_L$.

We derive a lower bound for terms in $F(\mathbf{\Delta})$. Using again the arguments for obtaining (3.47), we have

$$\begin{aligned}\|\boldsymbol{\beta}_0 + \mathbf{\Delta}\|_1 - \|\boldsymbol{\beta}_0\|_1 &= \|\boldsymbol{\beta}_{\mathcal{S}_0} + \mathbf{\Delta}_{\mathcal{S}_0}\|_1 + \|\mathbf{\Delta}_{\mathcal{S}_0^c}\|_1 - \|\boldsymbol{\beta}_{\mathcal{S}_0}\|_1 \\ &\geq \|\mathbf{\Delta}_{\mathcal{S}_0^c}\|_1 - \|\mathbf{\Delta}_{\mathcal{S}_0}\|_1 \end{aligned} \tag{5.52}$$

We now deal with the first difference term in $F(\mathbf{\Delta})$, which is $G(1)-G(0)$ where $G(u) = L(\boldsymbol{\beta}_0 + u\mathbf{\Delta})$. Recall the negative log-likelihood for logistic regression is

$$L(\boldsymbol{\beta}) = -\frac{1}{n}\sum_i \{Y_i \mathbf{X}_i^T\boldsymbol{\beta} - \psi(\mathbf{X}_i^T\boldsymbol{\beta})\}$$

with $\psi(t) = \log(1 + \exp(t))$. It can easily be calculated that

$$\begin{aligned} G''(u) &= \frac{1}{n}\sum_i \psi''(\mathbf{X}_i^T(\boldsymbol{\beta}_0 + u\mathbf{\Delta})) \cdot (\mathbf{X}_i^T\mathbf{\Delta})^2 \\ G'''(u) &= \frac{1}{n}\sum_i \psi'''(\mathbf{X}_i^T(\boldsymbol{\beta}_0 + u\mathbf{\Delta})) \cdot (\mathbf{X}_i^T\mathbf{\Delta})^3 \end{aligned}$$

where with $\theta(t) = (1 + \exp(t))^{-1}$,

$$\psi''(t) = \theta(t)(1 - \theta(t)) \quad \text{and} \quad \psi'''(t) = \theta(t)(1 - \theta(t))(2\theta(t) - 1).$$

By $|\psi'''(t)| \leq \psi''(t)$, we have

$$|G'''(u)| \leq \max_i |\mathbf{X}_i^T\mathbf{\Delta}| \cdot G''(u) \leq m\|\mathbf{\Delta}\|_1 \cdot G''(u) \leq mz_0 \cdot G''(u).$$

Then, by Lemma 1 of Bach (2010), we have

$$G(1) - G(0) - G'(0) \geq G''(0) \cdot h(mz_0),$$

with $h(t) = t^{-2}(\exp(-t) + t - 1)$. Observe that $h(t)$ is a decreasing function in $t > 0$. We pick $\lambda_L$ such that $mz_0 \leq 1$, that is,

$$\lambda_L \leq \frac{\kappa}{20ms},$$

and $h(mz_0) \geq h(1) = \exp(-1) > 1/3$. We rewrite $G(1) - G(0) - G'(0) \geq G''(0) \cdot h(mz_0) \geq 1/3$ as

$$G(1) - G(0) > (\nabla L(\boldsymbol{\beta}_0))^T\mathbf{\Delta} + \frac{1}{3}\mathbf{\Delta}^T[\nabla^2 L(\boldsymbol{\beta}_0)]\mathbf{\Delta}.$$

Note that $|(\nabla L(\boldsymbol{\beta}_0))^T\mathbf{\Delta}| \leq \|(\nabla L(\boldsymbol{\beta}_0))\|_\infty\|\mathbf{\Delta}\|_1 \leq \frac{1}{2}\lambda_L\|\mathbf{\Delta}\|_1$ in the event (5.47). Using the restricted eigenvalue condition and $\mathbf{\Delta} \in \mathcal{D}$, it follows that

$$G(1) - G(0) \geq -\frac{1}{2}\lambda_L\|\mathbf{\Delta}\|_1 + \frac{1}{3}\kappa\|\mathbf{\Delta}\|_2. \tag{5.53}$$

Combining (5.52) and (5.53) we have

$$\begin{aligned} F(\mathbf{\Delta}) &\geq -\frac{1}{2}\lambda_{\rm L}\|\mathbf{\Delta}\|_1 + \frac{1}{3}\kappa\|\mathbf{\Delta}\|_2 + \lambda_{\rm L}(\|\mathbf{\Delta}_{\mathcal{S}_0^c}\|_1 - \|\mathbf{\Delta}_{\mathcal{S}_0}\|_1) \\ &\geq \frac{1}{3}\kappa\|\mathbf{\Delta}\|_2 - \frac{3}{2}\lambda_{\rm L}\|\mathbf{\Delta}_{\mathcal{S}_0}\|_1 \\ &\geq \frac{1}{3}\kappa\|\mathbf{\Delta}\|_2^2 - \frac{3}{2}\lambda_{\rm L}\cdot s^{1/2}\|\mathbf{\Delta}\|_2 \\ &= \frac{5s\lambda_{\rm L}^2}{6\kappa}, \end{aligned} \tag{5.54}$$

where in (5.54) we plugged in $\|\mathbf{\Delta}\|_2 = 5s^{1/2}\lambda_{\rm L}/\kappa$. This shows $F(\mathbf{\Delta}) > 0$ for $\mathbf{\Delta} \in \mathcal{D}$.

It remains to calculate the probability of the event (5.47). By the union bound, we have

$$P(\|\frac{1}{n}\mathbf{X}^T(\mathbf{Y} - \boldsymbol{\mu}(\boldsymbol{\beta}_0))\|_\infty > \frac{1}{2}\lambda_{\rm L}) \leq \sum_{j=1}^p P(\|\frac{1}{n}\mathbf{X}_j^T(\mathbf{Y} - \boldsymbol{\mu}(\boldsymbol{\beta}_0))\| > \frac{1}{2}\lambda_{\rm L}),$$

where $\mathbf{X}_j$ is the $j$-th column of $\mathbf{X}$. Note that $\mathrm{E}(Y_i) = \mu_i(\boldsymbol{\beta}_0)$ and $Y_i - \mu_i(\boldsymbol{\beta}_0) \in [-\mu_i(\boldsymbol{\beta}_0), 1 - \mu_i(\boldsymbol{\beta}_0)]$ for the binary response. By Hoeffding's inequality (Hoeffding 1963; Theorem 3.2) we have

$$\begin{aligned} P(\|\frac{1}{n}\mathbf{X}_j^T(\mathbf{Y} - \boldsymbol{\mu}(\boldsymbol{\beta}_0))\| > \frac{1}{2}\lambda_{\rm L}) &\leq 2\exp(-\frac{2n^2(\lambda_{\rm L}/2)^2}{\sum_{i=1}^n |x_{ij}|^2}) \\ &\leq 2\exp(-\frac{n}{2M}\lambda_{\rm L}^2), \end{aligned}$$

with $M = \max_j n^{-1}\|\mathbf{X}_j\|_2^2$, which is less than $m^2$ by the assumption on $|x_{ij}|$. Thus,

$$P(\|\frac{1}{n}\mathbf{X}^T(\mathbf{Y} - \boldsymbol{\mu}(\boldsymbol{\beta}_0))\|_\infty \leq \frac{1}{2}\lambda_{\rm L}) \geq 1 - 2p\cdot\exp(-\frac{n}{2M}\lambda_{\rm L}^2).$$

This completes the proof. ■

### 5.9.2 *Strong oracle property*

The folded concave penalized likelihood estimator is defined as

$$\min_{\boldsymbol{\beta}}\{L(\boldsymbol{\beta}) + \sum_j p_\lambda(|\beta_j|)\}$$

where $p_\lambda(\cdot)$ is a folded concave penalty and $L(\boldsymbol{\beta}) = -\frac{1}{n}\ell(\boldsymbol{\beta})$ is the negative log-likelihood function (with a $1/n$ factor). In the case of a generalized linear model with the canonical link (see (5.8) after ignoring the dispersion)

$$L(\boldsymbol{\beta}) = \frac{1}{n}\sum_{i=1}^{n}\{-Y_i\mathbf{X}_i^T\boldsymbol{\beta} + b(\mathbf{X}_i^T\boldsymbol{\beta})\}.$$

Fan and Lv (2011) established the *oracle property* of the penalized likelihood estimator when $p = O(\exp(n^\alpha))$ for some $\alpha \in (0,1)$. When an estimator equals the oracle "estimator" with probability going to one, we say that this estimator has *strong oracle property*. Obviously, the strong oracle property implies the oracle property. Fan, Xue and Zou (2014) established an explicit procedure under which the strong oracle property of the folded concave penalized estimator holds. Their theory covers the penalized likelihood as an example.

Throughout this section, we assume that the penalty $p_\lambda(|t|)$ is a general *folded concave penalty* function defined on $t \in (-\infty, \infty)$ satisfying

(i) $p_\lambda(t)$ is increasing and concave in $t \in [0, \infty)$ with $p_\lambda(0) = 0$;

(ii) $p_\lambda(t)$ is differentiable in $t \in (0, \infty)$ with $p'_\lambda(0) := p'_\lambda(0+) \geq a_1\lambda$;

(iii) $p'_\lambda(t) \geq a_1\lambda$ for $t \in (0, a_2\lambda]$;

(iv) $p'_\lambda(t) = 0$ for $t \in [a\lambda, \infty)$ with the pre-specified constant $a > a_2$, where $a_1$ and $a_2$ are two fixed positive constants.

The derivative of the SCAD penalty is

$$p'_\lambda(t) = \lambda I_{\{t\leq\lambda\}} + \frac{(a\lambda - t)_+}{a-1} I_{\{t>\lambda\}}, \quad \text{for some } a > 2,$$

and the derivative of the MCP is $p'_\lambda(t) = (\lambda - \frac{t}{a})_+$, for some $a > 1$. It is easy to see that $a_1 = a_2 = 1$ for the SCAD, and $a_1 = 1 - a^{-1}$, $a_2 = 1$ for the MCP.

We consider the local solution of the folded penalized likelihood estimator by using the LLA algorithm. Recall that in the LLA iteration, the $k$ step update is given by

$$\widehat{\boldsymbol{\beta}}^{(k)} = \min_{\boldsymbol{\beta}}\{L(\boldsymbol{\beta}) + \sum_j \widehat{d}_j^{(k-1)} \cdot |\beta_j|\},$$

where $\widehat{d}_j^{(k-1)} = p'_\lambda(|\widehat{\beta}_j^{(m-1)}|)$. The initial estimator $\widehat{\boldsymbol{\beta}}^{(0)} = \widehat{\boldsymbol{\beta}}^{\text{ini}}$ fully determines the LLA sequence. Let $\widehat{\boldsymbol{\beta}}^{\text{ora}}$ be the oracle estimator, which is constructed using the knowledge of $\mathcal{S}_0$, satisfying

$$\widehat{\boldsymbol{\beta}}_{\mathcal{S}_0} = \arg\min_{\boldsymbol{\beta}:\boldsymbol{\beta}_{\mathcal{S}_0^c}=0} L(\boldsymbol{\beta}), \qquad \widehat{\boldsymbol{\beta}}_{\mathcal{S}_0^c} = \mathbf{0}. \tag{5.55}$$

See also (5.36).

The following Theorem is due to Fan, Xue and Zou (2014).

**Theorem 5.6** *Suppose the minimal signal strength $\beta_{\min} = \min_{j\in\mathcal{S}_0}|\beta_{0j}|$ satisfies $\beta_{\min} > (a+1)\lambda$. Given a folded concave penalty $p_\lambda(\cdot)$ satisfying (i)–(iv), let $a_0 = \min\{1, a_2\}$. Under the event*

$$\mathcal{E}_1 = \{\|\widehat{\boldsymbol\beta}^{\text{ini}} - \boldsymbol\beta_0\|_\infty \le a_0\lambda\} \cap \{\|\nabla_{\mathcal{S}_0{}^c} L(\widehat{\boldsymbol\beta}^{\text{ora}})\|_\infty < a_1\lambda\},$$

*the LLA algorithm initialized by $\widehat{\boldsymbol\beta}^{\text{ini}}$ finds $\widehat{\boldsymbol\beta}^{\text{ora}}$ after one iteration. Therefore, with probability at least $1-\delta_0-\delta_1$, the one-step LLA estimator $\widehat{\boldsymbol\beta}^{(1)}$ equals the oracle estimator $\widehat{\boldsymbol\beta}^{\text{ora}}$, where*

$$\delta_0 = \Pr\big(\|\widehat{\boldsymbol\beta}^{\text{ini}} - \boldsymbol\beta_0\|_\infty > a_0\lambda\big)$$

$$\delta_1 = \Pr\big(\|\nabla_{\mathcal{S}_0{}^c} L(\widehat{\boldsymbol\beta}^{\text{ora}})\|_\infty \ge a_1\lambda\big).$$

**Proof.** Write $\widehat{\boldsymbol\beta}^{(0)} = \widehat{\boldsymbol\beta}^{\text{ini}}$. Under the event $\{\|\widehat{\boldsymbol\beta}^{(0)} - \boldsymbol\beta_0\|_\infty \le a_0\lambda\}$, for $j \in \mathcal{S}_0{}^c$ we have

$$|\widehat{\beta}_j^{(0)}| \le \|\widehat{\boldsymbol\beta}^{(0)} - \boldsymbol\beta_0\|_\infty \le a_0\lambda \le a_2\lambda$$

and for $j \in \mathcal{S}_0$.

$$|\widehat{\beta}_j^{(0)}| \ge \beta_{\min} - \|\widehat{\boldsymbol\beta}^{(0)} - \boldsymbol\beta_0\|_\infty > a\lambda,$$

where we use the assumption $\beta_{\min} \ge (1+a)\lambda$. By property (iv), $p'_\lambda(|\widehat{\beta}_j^{(0)}|) = 0$ for all $j \in \mathcal{S}_0$. Thus $\widehat{\boldsymbol\beta}^{(1)}$ is the solution to the problem

$$\widehat{\boldsymbol\beta}^{(1)} = \arg\min_{\boldsymbol\beta} L(\boldsymbol\beta) + \sum_{j\in\mathcal{S}_0{}^c} p'_\lambda(|\widehat{\beta}_j^{(0)}|)\cdot|\beta_j|. \tag{5.56}$$

By property (ii) & (iii), $p'_\lambda(|\widehat{\beta}_j^{(0)}|) \ge a_1\lambda$ holds for $j \in \mathcal{S}_0{}^c$. We now show that $\widehat{\boldsymbol\beta}^{\text{ora}}$ is the unique global solution to (5.56) under the additional condition $\{\|\nabla_{\mathcal{S}_0{}^c} L(\widehat{\boldsymbol\beta}^{\text{ora}})\|_\infty < a_1\lambda\}$. By convexity, we have

$$\begin{aligned} L(\boldsymbol\beta) &\ge L(\widehat{\boldsymbol\beta}^{\text{ora}}) + \sum_j \nabla_j L(\widehat{\boldsymbol\beta}^{\text{ora}})(\beta_j - \widehat{\beta}_j^{\text{ora}}) \\ &= L(\widehat{\boldsymbol\beta}^{\text{ora}}) + \sum_{j\in\mathcal{S}_0{}^c} \nabla_j L(\widehat{\boldsymbol\beta}^{\text{ora}})\beta_j. \end{aligned}$$

where the last equality used $\nabla_j L(\widehat{\boldsymbol\beta}^{\text{ora}}) = 0$ for $j \in \mathcal{S}_0$ and $\widehat{\boldsymbol\beta}_j^{\text{ora}} = 0$ for $j \in \mathcal{S}_0{}^c$. Thus, for any $\boldsymbol\beta$, we have

$$\begin{aligned} &\{L(\boldsymbol\beta) + \sum_{j\in\mathcal{S}_0{}^c} p'_\lambda(|\widehat{\beta}_j^{(0)}|)|\beta_j|\} - \{L(\widehat{\boldsymbol\beta}^{\text{ora}}) + \sum_{j\in\mathcal{S}_0{}^c} p'_\lambda(|\widehat{\beta}_j^{(0)}|)|\underbrace{\widehat{\beta}_j^{\text{ora}}}_{=0}|\} \\ \ge\; &\sum_{j\in\mathcal{S}_0^c} \{p'_\lambda(|\widehat{\beta}_j^{(0)}|) + \nabla_j L(\widehat{\boldsymbol\beta}^{\text{ora}})\cdot \text{sign}(\beta_j)\}\cdot|\beta_j| \\ \ge\; &\sum_{j\in\mathcal{S}_0^c} \{a_1\lambda + \nabla_j L(\widehat{\boldsymbol\beta}^{\text{ora}})\cdot \text{sign}(\beta_j)\}\cdot|\beta_j| \ge 0. \end{aligned}$$

The strict inequality holds unless $\beta_j = 0, \forall j \in \mathcal{S}_0{}^c$. In either case, we must have $\widehat{\boldsymbol{\beta}}^{(1)} = \widehat{\boldsymbol{\beta}}^{\text{ora}}$. ■

When $\delta_0$ and $\delta_1$ in Theorem 5.6 go to zero as $n, p$ diverge, the one-step LLA iteration equals the oracle estimator with overwhelming probability. The two quantities $\delta_0$ and $\delta_1$ have explicit meanings. First, $\delta_0$ represents the *localizability* of the underlying model. To apply Theorem 5.6 we need to have an appropriate initial estimator to make $\delta_0$ go to zero as $n$ and $p$ diverge to infinity, namely the underlying problem is localizable. Second, $\delta_1$ represents the *regularity* behavior of the oracle estimator, i.e., its closeness to the true parameter measured by the score function. Note that $\nabla_{\mathcal{S}_0{}^c} L(\widehat{\boldsymbol{\beta}}^{\text{ora}})$ is concentrated around zero. Thus, $\delta_1$ is usually small. Theorem 5.6 states that as long as the penalized likelihood problem is *localizable and regular*, we can find an oracle estimator by using the one-step LLA update. This is a generalization of the one-step estimation idea in Zou and Li (2008) to the ultrahigh dimensional setting.

Fan, Xue and Zou (2014) further reveals a sufficient condition under which the LLA algorithm actually converges after two iterations.

**Theorem 5.7** *Assume $p_\lambda(\cdot)$ is a folded concave penalty satisfying (i)–(iv). Under the event*

$$\mathcal{E}_2 = \big\{\|\nabla_{\mathcal{S}_0{}^c} L(\widehat{\boldsymbol{\beta}}^{\text{ora}})\|_\infty < a_1\lambda\big\} \cap \big\{\|\widehat{\boldsymbol{\beta}}^{\text{ora}}_{\mathcal{S}_0}\|_{\min} > a\lambda\big\},$$

*if $\widehat{\boldsymbol{\beta}}^{\text{ora}}$ is obtained, the LLA algorithm will find $\widehat{\boldsymbol{\beta}}^{\text{ora}}$ again in the next iteration, i.e., it converges to $\widehat{\boldsymbol{\beta}}^{\text{ora}}$ in the next iteration and is a fixed point. Therefore, with probability at least $1 - \delta_0 - \delta_1 - \delta_2$, the LLA algorithm initialized by $\widehat{\boldsymbol{\beta}}^{\text{ini}}$ converges to $\widehat{\boldsymbol{\beta}}^{\text{ora}}$ after two iterations, where*

$$\delta_2 = \Pr\big(\|\widehat{\boldsymbol{\beta}}^{\text{ora}}_{\mathcal{S}_0}\|_{\min} \le a\lambda\big).$$

**Proof.** Given that the LLA algorithm finds $\widehat{\boldsymbol{\beta}}^{\text{ora}}$ at the current iteration, we denote $\widehat{\boldsymbol{\beta}}$ as the solution to the convex optimization problem in the next iteration of the LLA algorithm. Note that $\widehat{\boldsymbol{\beta}}^{\text{ora}}_{\mathcal{S}_0{}^c} = 0$ and $p'_\lambda(|\widehat{\beta}^{\text{ora}}_j|) = 0$ for $j \in \mathcal{S}_0$. Under the event $\{\|\widehat{\boldsymbol{\beta}}^{\text{ora}}_{\mathcal{S}_0}\|_{\min} > a\lambda\}$, we have

$$\widehat{\boldsymbol{\beta}} = \arg\min_{\boldsymbol{\beta}} L(\boldsymbol{\beta}) + \sum_{j \in \mathcal{S}_0{}^c} \gamma \cdot |\beta_j|, \tag{5.57}$$

where $\gamma = p'_\lambda(0) \ge a_1\lambda$. This problem is very similar to (5.56). We can follow the proof of Theorem 5.6 to show that under the additional condition $\{\|\nabla_{\mathcal{S}_0{}^c} L(\widehat{\boldsymbol{\beta}}^{\text{ora}})\|_\infty < a_1\lambda\}$, $\widehat{\boldsymbol{\beta}}^{\text{ora}}$ is the unique solution to (5.57). Hence the LLA algorithm converges to $\widehat{\boldsymbol{\beta}}^{\text{ora}}$. ■

Note that $\delta_1$ and $\delta_2$ only depend on the model and the oracle estimator and $\delta_0$ depends on the initial estimator. Fan, Xue and Zou (2014) suggested two initial estimators. The first one is the $L_1$ penalized likelihood estimator. The second one is $\widehat{\boldsymbol\beta}^{\text{ini}} = 0$. It is interesting to observe that for the SCAD and MCP penalized likelihood, if we start with zero, the first LLA update gives the $L_1$ penalized likelihood estimator.

We now apply the general theorem to study the penalized logistic regression.

$$\min_{\boldsymbol\beta} \frac{1}{n}\sum_i \{-Y_i \mathbf{X}_i^T \boldsymbol\beta + \psi(\mathbf{X}_i^T\boldsymbol\beta)\} + \sum_j p_\lambda(|\beta_j|),$$

with $\psi(t) = \log(1+\exp(t))$.

For ease of presentation, we define $\boldsymbol\mu(\boldsymbol\beta) = (\psi'(\mathbf{X}_1^T\boldsymbol\beta), \ldots, \psi'(\mathbf{X}_n^T\boldsymbol\beta))^T$ and $\boldsymbol\Sigma(\boldsymbol\beta) = \text{diag}\{\psi''(\mathbf{X}_1^T\boldsymbol\beta), \ldots, \psi''(\mathbf{X}_n^T\boldsymbol\beta)\}$. We also define three useful quantities:

$$Q_1 = \max_j \gamma_{\max}(\frac{1}{n}\mathbf{X}_{\mathcal{S}_0}^T \,\text{diag}\{|\mathbf{x}_{(j)}|\}\mathbf{X}_{\mathcal{S}_0}),$$

$$Q_2 = \|(\frac{1}{n}\mathbf{X}_{\mathcal{S}_0}^T \boldsymbol\Sigma(\boldsymbol\beta_0)\mathbf{X}_{\mathcal{S}_0})^{-1}\|_{\ell_\infty},$$

and

$$Q_3 = \|\mathbf{X}_{\mathcal{S}_0^c}^T \boldsymbol\Sigma(\boldsymbol\beta_0)\mathbf{X}_{\mathcal{S}_0}(\mathbf{X}_{\mathcal{S}_0}^T \boldsymbol\Sigma(\boldsymbol\beta_0)\mathbf{X}_{\mathcal{S}_0})^{-1}\|_{\ell_\infty},$$

where $\text{diag}\{|\mathbf{x}_{(j)}|\}$ is a diagonal matrix with elements $\{|x_{ij}|\}_{i=1}^n$. $\gamma_{\max}(\mathbf{G})$ denotes the largest eigenvalue of the matrix $\mathbf{G}$, and $\|\mathbf{G}\|_{\ell_\infty}$ denotes the matrix $L_\infty$ norm of a matrix $\mathbf{G}$.

When using Theorem 5.6 we do not need to compute $\delta_0$ and $\delta_1$. A good upper bound is sufficient. In the case of the logistic regression, we can bound $\delta_1$ and $\delta_2$ as follows (See Theorem 4 of Fan, Xue and Zou (2014) )

$$\delta_1^{\text{logi}} = 2s \cdot \exp(-\frac{n}{M}\min\{\frac{2}{Q_1^2 Q_2^4 s^2}, \frac{a_1^2\lambda^2}{2(1+2Q_3)^2}\}) + 2(p-s)\cdot\exp(-\frac{a_1^2 n\lambda^2}{2M})$$

and

$$\delta_2^{\text{logi}} = 2s \cdot \exp(-\frac{n}{MQ_2^2}\min\{\frac{2}{Q_1^2 Q_2^2 s^2}, \frac{1}{2}(\beta_{\min} - a\lambda)^2\}).$$

For $\delta_0$ Theorem 5.5 gave an automatic upper bound by noticing that the max norm is bounded by the $L_2$ norm.

$$\delta_0^{\text{logi}} = P(\|\widehat{\boldsymbol\beta}_{\text{L}} - \boldsymbol\beta_0\|_\infty > a_0\lambda) \le P(\|\widehat{\boldsymbol\beta}_{\text{L}} - \boldsymbol\beta_0\|_2 > a_0\lambda)$$

By Theorem 5.5, if we choose $\lambda$ such that

$$\lambda \ge \frac{5s^{1/2}\lambda_L}{a_0\kappa_{\text{logi}}}, \tag{5.58}$$

we have $\delta_0^{\text{logi}} \le 2p\exp(-\frac{n}{2M}\lambda_{\text{L}}^2)$.

To sum up the above discussion, the LLA algorithm initialized by $\widehat{\boldsymbol{\beta}}_{\mathrm{L}}$ finds $\widehat{\boldsymbol{\beta}}^{\mathrm{ora}}$ after one iteration with a high probability. If we use zero as the initial estimator, a similar conclusion holds with an extra condition. Note that after the first LLA update, the solution is a Lasso penalized logistic regression with $\lambda_{\mathrm{L}} = \lambda$. The condition in (5.58) becomes

$$\lambda \geq \frac{5s^{1/2}\lambda}{a_0 \kappa_{\mathrm{logi}}},$$

or

$$\kappa_{\mathrm{logi}} \geq 5a_0^{-1}s^{1/2}. \tag{5.59}$$

Assuming (5.59), the two-step LLA solution from zero equals the oracle estimator with high probability.

### 5.9.3 *Numeric studies*

We conducted a simulation study to check the theory of folded concave penalized likelihood. We designed a logistic regression model. $Y|\mathbf{X} \sim \mathrm{Bernoulli}(\frac{\exp(\mathbf{X}^T\boldsymbol{\beta}_0)}{1+\exp(\mathbf{X}^T\boldsymbol{\beta}_0)})$, where $\mathbf{X} \sim N_p(0, \boldsymbol{\Sigma})$ with $\boldsymbol{\Sigma} = (0.5^{|i-j|})$ and $\boldsymbol{\beta}_0 = (3, 1.5, 0, 0, 2, 0_{p-5})$. For tuning, we generated an independent validation set. The validation error of a generic estimator $\widehat{\boldsymbol{\beta}}$ is defined as

$$\sum_{i \in \text{validation set}} \{-Y_i \mathbf{X}_i^T \widehat{\boldsymbol{\beta}} + \log(1 + \exp(\mathbf{X}_i^T \widehat{\boldsymbol{\beta}}))\}.$$

We picked the penalization parameter $\lambda$ by minimizing the validation error.

The following estimators were considered:

- Lasso penalized logistic regression
- SCAD-CD: coordinate descent solution of the SCAD penalized logistic regression
- SCAD-LLA*: the fully converged LLA solution of the SCAD penalized logistic regression with the Lasso solution as the initial estimator
- SCAD-2SLLA*: the two-step LLA solution of the SCAD penalized logistic regression with the Lasso solution as the initial estimator
- SCAD-LLA0: the fully converged LLA solution of the SCAD penalized logistic regression with zero as the initial estimator
- SCAD-3SLLA0: the three-step LLA solution of the SCAD penalized logistic regression with zero as the initial estimator

We can make several remarks based on the simulation study. First, coordinate descent and LLA yield different local solutions. The LLA with different initial estimators end up with slightly different solutions. This shows that the

Table 5.6: Comparison for penalized logistic regression estimators for simulated data. $\ell_1$ loss is $\mathrm{E}[\|\widehat{\boldsymbol{\beta}} - \boldsymbol{\beta}_0\|_1]$, $\ell_w$ loss is $\mathrm{E}[\|\widehat{\boldsymbol{\beta}} - \boldsymbol{\beta}_0\|_2]$. FP and FN denote the false positive rate and the false negative rate, respectively. Adapted from Fan, Xue and Zou (2014).

| Method | $(n,p) = (200, 500)$ | | | | $(n,p) = (200, 1000)$ | | | |
|---|---|---|---|---|---|---|---|---|
| | $\ell_1$ loss | $\ell_2$ loss | # FP | # FN | $\ell_1$ loss | $\ell_2$ loss | # FP | # FN |
| Lasso | 5.27 | 2.25 | 20.30 | 0.01 | 5.67 | 2.37 | 24.02 | 0.04 |
| | (0.05) | (0.02) | (0.39) | (0.01) | (0.05) | (0.02) | (0.44) | (0.01) |
| SCAD-CD | 4.09 | 2.01 | 10.79 | 0.04 | 4.50 | 2.13 | 13.99 | 0.08 |
| | (0.05) | (0.02) | (0.25) | (0.01) | (0.06) | (0.02) | (0.31) | (0.01) |
| SCAD-3SLLA0 | 1.85 | 1.20 | 0.34 | 0.10 | 2.26 | 1.36 | 0.32 | 0.24 |
| | (0.09) | (0.05) | (0.04) | (0.02) | (0.11) | (0.06) | (0.05) | (0.02) |
| SCAD-LLA0 | 1.85 | 1.17 | 0.31 | 0.09 | 2.16 | 1.32 | 0.31 | 0.22 |
| | (0.09) | (0.05) | (0.04) | (0.02) | (0.11) | (0.06) | (0.05) | (0.02) |
| SCAD-2SLLA$^\star$ | 1.85 | 1.15 | 0.26 | 0.07 | 2.17 | 1.33 | 0.31 | 0.23 |
| | (0.09) | (0.06) | (0.04) | (0.02) | (0.10) | (0.06) | (0.04) | (0.02) |
| SCAD-LLA$^\star$ | 1.82 | 1.17 | 0.24 | 0.10 | 2.08 | 1.28 | 0.26 | 0.19 |
| | (0.09) | (0.06) | (0.04) | (0.02) | (0.10) | (0.06) | (0.04) | (0.02) |

multiple local solutions issue is real in folded concave penalized logistic regression. Secondly, although the local solution by coordinate descent is better than the Lasso estimator, the local solution by LLA is much better. Finally, there is little difference between the fully converged LLA solution and the two-step or three-step LLA solutions.

## 5.10 Risk Properties

This section is devoted to studying the risk of penalized likelihood estimators under the general framework of M-estimation with decomposable regularization. This section deals with the convex penalty. Section 6.6 treats further the penalized $M$-estimator with folded concave penalty.

Many statistical problems can be abstracted as M-estimation problems by minimizing their empirical loss function $L(\boldsymbol{\theta})$. This can be robust regression, in which the loss can be the $L_1$-loss or the *Huber loss*. It can also be *quantile regression*, in which the loss function is a check function, or Cox's *proportional hazards model*, in which the loss function is the negative *partial likelihood function*. Chapter 6 considers these problems in detail. For the high-dimensional setting, one naturally imposes the sparsity on the parameter $\boldsymbol{\theta}$ and uses the *penalized $M$-estimator* of the form

$$\widehat{\boldsymbol{\theta}} \in \operatorname*{arg\,min}_{\boldsymbol{\theta} \in I\!R^p} \{L(\boldsymbol{\theta}) + \lambda \sum_{j=1}^{p} p_\lambda(|\theta_j|)\}. \tag{5.60}$$

This is a natural generalization of the *penalized likelihood* method (5.49).

In many applied mathematics and engineering problems, the parameters are more naturally regarded as a matrix. Examples include matrix completion, reduced-rank or multi-task regression, compressed sensing, and image reconstructions. An example that includes matrix completion, reduced-rank or multi-task regression, and compress sensing is the following trace regression problem:

$$Y_i = \mathrm{tr}(\mathbf{X}_i^T \boldsymbol{\Theta}) + \varepsilon_i, \qquad i = 1, \cdots, n. \tag{5.61}$$

See Negahban, Ravikumar, Wainwright and Yu (2012) and Fan, Wang, and Zhu (2016) and references therein. Other problems that involve matrix parameters include estimating a structured high-dimensional covariance matrix and a precision matrix, discussed in Chapter 11. In both classes of problems on estimating a high-dimensional matrix, the sparsity concept of a vector extends naturally to low-rankness of a matrix. In this case, a natural extension of the $L_1$-penalty for vectors becomes the *nuclear norm* for a matrix $\boldsymbol{\Theta}$:

$$\|\boldsymbol{\Theta}\|_N = \sum_{i=1}^{\min(d_1,d_2)} \sigma_i(\boldsymbol{\Theta}),$$

where $\{\sigma_i(\boldsymbol{\Theta})\}$ are the singular values of the $d_1 \times d_2$ matrix $\boldsymbol{\Theta}$, namely, they are the square-root of the eigenvalues of $\boldsymbol{\Theta}^T\boldsymbol{\Theta}$.

The regularized M-estimation problem can be abstracted as follows

$$\widehat{\boldsymbol{\theta}} \in \arg\min_{\boldsymbol{\theta} \in I\!\!R^p} \{L(\boldsymbol{\theta}) + \lambda_n R(\boldsymbol{\theta})\}. \tag{5.62}$$

It targets at estimating

$$\boldsymbol{\theta}^* = \arg\min_{\boldsymbol{\theta} \in I\!\!R^p} E\{L(\boldsymbol{\theta})\} \tag{5.63}$$

with a certain structure *a priori* on $\boldsymbol{\theta}^*$. Negahban *et al.*(2012) gives various other examples that fit the above framework. In this section, we plan to extend the techniques used in the proof of Theorem 5.5 to a more general setting. It follows closely the development of Negahban *et al.*(2012).

The first requirement of the penalty function is its decomposability. Given a pair of subspaces $\mathcal{M} \subseteq \overline{\mathcal{M}}$. Define

$$\overline{\mathcal{M}}^{\perp} = \{\mathbf{v} \in I\!\!R^p : \langle \mathbf{u}, \mathbf{v} \rangle = 0 \text{ for all } \mathbf{u} \in \overline{\mathcal{M}}\}.$$

**Definition 5.3** *A norm based regularizer $R(\boldsymbol{\theta})$ is said decomposable with respect to $(\mathcal{M}, \overline{\mathcal{M}}^{\perp})$ if*

$$R(\boldsymbol{\theta} + \boldsymbol{\gamma}) = R(\boldsymbol{\theta}) + R(\boldsymbol{\gamma}), \quad \textit{for all } \boldsymbol{\theta} \in \mathcal{M} \textit{ and } \boldsymbol{\gamma} \in \overline{\mathcal{M}}^{\perp} \tag{5.64}$$

For the $L_1$-regularization $R(\boldsymbol{\theta}) = \|\boldsymbol{\theta}\|_1$, we take $\mathcal{M} = \{\boldsymbol{\theta} \in I\!\!R^p : \boldsymbol{\theta}_{S^c} = 0\}$ and $\overline{\mathcal{M}} = \mathcal{M}$ so that $\overline{\mathcal{M}}^{\perp} = \{\boldsymbol{\gamma} \in I\!\!R^p : \boldsymbol{\gamma}_S = 0\}$ and it satisfies (5.64).

For the nuclear norm $R(\mathbf{\Theta}) = \|\mathbf{\Theta}\|_N$, define

$$\mathcal{M} = \{\mathbf{\Theta} \in I\!R^{d_1 \times d_2} : \text{row}(\mathbf{\Theta}) \subseteq \mathbf{V}, \text{col}(\mathbf{\Theta}) \subseteq \mathbf{U}\}.$$

for a given pair of $r$-dimensional subspaces $\mathbf{U} \in I\!R^{d_1}$ and $\mathbf{V} \in I\!R^{d_2}$. These pairs represent the left and right singular vectors of the true matrix $\mathbf{\Theta}^*$. Here, row($\mathbf{\Theta}$) and col($\mathbf{\Theta}$) are respectively the row and column space of $\mathbf{\Theta}$. Define

$$\overline{\mathcal{M}}^{\perp} = \{\mathbf{\Theta} \in I\!R^{d_1 \times d_2} : \text{row}(\mathbf{\Theta}) \subseteq \mathbf{V}^{\perp}, \text{col}(\mathbf{\Theta}) \subseteq \mathbf{U}^{\perp}\}.$$

Then, we have that (5.64) holds.

**Definition 5.4** *The dual norm of $R$ is defined as*

$$R^*(\mathbf{v}) = \sup_{\mathbf{u} \in I\!R^p} \langle \mathbf{u}, \mathbf{v} \rangle / R(\mathbf{u}),$$

*where we used the convention* $0/0 = 0$.

As a result of this definition, we have

$$\langle \mathbf{u}, \mathbf{v} \rangle \le R(\mathbf{u}) R^*(\mathbf{v}). \tag{5.65}$$

For the $L_1$-norm $R(\mathbf{u}) = \|\mathbf{u}\|_1$, its dual norm is $R^*(\mathbf{v}) = \|\mathbf{v}\|_\infty$.

We first extend the result of Proposition 5.2, which shows that we need only to restrict our attention to a star set.

**Proposition 5.3** *Let $\widehat{\mathbf{\Delta}} = \widehat{\boldsymbol{\theta}} - \boldsymbol{\theta}^*$. Suppose that $L(\boldsymbol{\beta})$ is convex. Then, in the event*

$$\{R^*(\nabla L(\boldsymbol{\theta}^*)) \le \frac{1}{2}\lambda_n\}, \tag{5.66}$$

*for any pair $(\mathcal{M}, \overline{\mathcal{M}}^{\perp})$, we have*

$$R(\widehat{\mathbf{\Delta}}_{\bar{\mathcal{M}}^\perp}) \le 3R(\widehat{\mathbf{\Delta}}_{\bar{\mathcal{M}}}) + 4R(\boldsymbol{\theta}^*_{\mathcal{M}^\perp}). \tag{5.67}$$

**Proof.** Let $F(\mathbf{\Delta}) = \underbrace{L(\boldsymbol{\theta}^* + \mathbf{\Delta}) - L(\boldsymbol{\theta}^*)}_{(1)} + \lambda_n \underbrace{\{R(\boldsymbol{\theta}^* + \mathbf{\Delta}) - R(\boldsymbol{\theta}^*)\}}_{(2)}$. Then $F(\widehat{\mathbf{\Delta}}) \le 0$. The first term (1) can be bounded from below as

$$(1) \ge_{(a)} -|\langle \nabla L(\boldsymbol{\theta}^*),\ \mathbf{\Delta} \rangle| \ge_{(b)} -R^*(\nabla L(\boldsymbol{\theta}^*)) R(\mathbf{\Delta}) \ge_{(c)} -\frac{\lambda_n}{2} R(\mathbf{\Delta})$$

The inequality (a) is due to the convexity of $L(\cdot)$, the inequalities (b) and (c) are due to (5.65) and (5.66).

We now bound the second term (2) as

$$\begin{aligned}
(2) \quad &\ge_{(d)} \quad R(\boldsymbol{\theta}^*_{\mathcal{M}} + \mathbf{\Delta}_{\bar{\mathcal{M}}^\perp}) - R(\boldsymbol{\theta}^*_{\mathcal{M}^\perp}) - R(\mathbf{\Delta}_{\bar{\mathcal{M}}}) - R(\boldsymbol{\theta}^*) \\
&\ge_{(e)} \quad R(\boldsymbol{\theta}^*_{\mathcal{M}}) + R(\mathbf{\Delta}_{\bar{\mathcal{M}}^\perp}) - R(\boldsymbol{\theta}^*_{\mathcal{M}^\perp}) - R(\mathbf{\Delta}_{\bar{\mathcal{M}}}) - \{R(\boldsymbol{\theta}^*_{\mathcal{M}}) + R(\boldsymbol{\theta}^*_{\mathcal{M}^\perp})\} \\
&= \quad R(\mathbf{\Delta}_{\bar{\mathcal{M}}^\perp}) - R(\mathbf{\Delta}_{\bar{\mathcal{M}}}) - 2R(\boldsymbol{\theta}^*_{\mathcal{M}^\perp}).
\end{aligned}$$

The inequality (d) is implied by the triangle inequality of the norm $R$:

$$\begin{aligned} R(\boldsymbol{\theta}^* + \boldsymbol{\Delta}) &= R(\boldsymbol{\theta}^*_{\mathcal{M}} + \boldsymbol{\theta}^*_{\mathcal{M}^\perp} + \boldsymbol{\Delta}_{\bar{\mathcal{M}}} + \boldsymbol{\Delta}_{\bar{\mathcal{M}}^\perp}) \\ &\geq R(\boldsymbol{\theta}^*_{\mathcal{M}} + \boldsymbol{\Delta}_{\bar{\mathcal{M}}^\perp}) - R(\boldsymbol{\theta}^*_{\mathcal{M}^\perp}) - R(\boldsymbol{\Delta}_{\bar{\mathcal{M}}}). \end{aligned}$$

The inequality (e) is also due to the triangle inequality that $R(\boldsymbol{\theta}^*) \leq R(\boldsymbol{\theta}^*_{\mathcal{M}}) + R(\boldsymbol{\theta}^*_{\mathcal{M}^\perp})$.

Combining (1) and (2), along with $R(\boldsymbol{\Delta}) \leq R(\boldsymbol{\Delta}_{\bar{\mathcal{M}}}) + R(\boldsymbol{\Delta}_{\bar{\mathcal{M}}^\perp})$, we obtain

$$F(\Delta) \geq \frac{\lambda_n}{2}\Big[R(\boldsymbol{\Delta}_{\bar{\mathcal{M}}^\perp}) - 3R(\boldsymbol{\Delta}_{\bar{\mathcal{M}}}) - 4R(\boldsymbol{\theta}^*_{\mathcal{M}^\perp})\Big].$$

Hence, from $F(\widehat{\boldsymbol{\Delta}}) \leq 0$, we obtain the desired conclusion. ∎

Proposition 5.3 holds for any decomposition pairs $(\mathcal{M}, \overline{\mathcal{M}}^\perp)$. But it is useful when the pair is taken appropriately. For the sparse vector $\boldsymbol{\theta}^*$, if we take $\mathcal{M} = \mathcal{S}_0 = \text{supp}(\boldsymbol{\theta}^*)$, then $R(\boldsymbol{\theta}^*_{\mathcal{M}^\perp}) = 0$ and the result reduces to Proposition 5.2.

Proposition 5.3 states that we can restrict the optimization to the star-shape set

$$\mathcal{C} = \{\boldsymbol{\Delta} \in R^p : R(\boldsymbol{\Delta}_{\bar{\mathcal{M}}^\perp}) \leq 3R(\boldsymbol{\Delta}_{\bar{\mathcal{M}}}) + 4R(\boldsymbol{\theta}^*_{\mathcal{M}^\perp})\}$$

We would like the function to be strongly convex over this set with some tolerance. This motivates the following definition.

**Definition 5.5** *The loss function satisfies a restricted strong convexity (RSC) condition with curvature $\kappa_L$ and tolerance function $\tau_L$ if*

$$L(\boldsymbol{\theta}^* + \boldsymbol{\Delta}) - L(\boldsymbol{\theta}^*) - \langle \nabla L(\boldsymbol{\theta}^*), \boldsymbol{\Delta}\rangle \geq \kappa_L \|\boldsymbol{\Delta}\|^2 - \tau_L^2(\boldsymbol{\theta}^*), \tag{5.68}$$

*for all $\boldsymbol{\Delta} \in \mathcal{C}$.*

We are now ready to state the main result of this section, due to Negahban, Ravikumar, Wainwright and Yu (2012).

**Theorem 5.8** *Let $\widehat{\boldsymbol{\Delta}} = \widehat{\boldsymbol{\theta}} - \boldsymbol{\theta}^*$. Suppose that $L(\boldsymbol{\theta})$ is convex, satisfying* (5.68). *Assume $R(\boldsymbol{\theta})$ is decomposable with respect to $(\mathcal{M}, \overline{\mathcal{M}}^\perp)$. Then, in the event* (5.66), *we have*

$$\|\widehat{\boldsymbol{\Delta}}\|^2 \leq \frac{9\lambda_n^2}{4\kappa_L^2}\psi^2(\overline{\mathcal{M}}) + \frac{4\lambda_n R(\boldsymbol{\theta}^*_{\mathcal{M}^\perp})}{\kappa_L} + \frac{2\tau_L^2(\boldsymbol{\theta}^*)}{\kappa_T} \tag{5.69}$$

*where $\psi(\overline{\mathcal{M}}) = \sup_{\mathbf{u} \in \overline{\mathcal{M}}} R(\mathbf{u})/\|\mathbf{u}\|$. In addition,*

$$R(\widehat{\boldsymbol{\Delta}}) \leq 4\psi(\overline{\mathcal{M}})\|\widehat{\boldsymbol{\Delta}}\| + 4R(\boldsymbol{\theta}^*_{\mathcal{M}^\perp}). \tag{5.70}$$

**Proof of Theorem 5.8**. We now use Lemma 5.1 to derive a rate of convergence. To do this, we need to lower bound the objective function $F$. By using (5.68) and (2), we have

$$F(\boldsymbol{\Delta}) \geq \langle \nabla L(\boldsymbol{\theta}^*), \boldsymbol{\Delta}\rangle + \kappa_L \|\boldsymbol{\Delta}\|^2 - \tau_L^2(\boldsymbol{\theta}^*) + \lambda_n \{R(\boldsymbol{\Delta}_{\bar{\mathcal{M}}^\perp}) - R(\boldsymbol{\Delta}_{\bar{\mathcal{M}}}) - 2R(\boldsymbol{\theta}^*_{\mathcal{M}^\perp})\}.$$

By (1) and the triangular inequality, we have

$$|\langle \nabla L(\boldsymbol{\theta}^*), \boldsymbol{\Delta}\rangle| \leq \frac{\lambda_n}{2} R(\boldsymbol{\Delta}) \leq \frac{\lambda_n}{2}\{R(\boldsymbol{\Delta}_{\bar{\mathcal{M}}^\perp}) + R(\boldsymbol{\Delta}_{\bar{\mathcal{M}}})\}.$$

Hence, we bound further

$$F(\boldsymbol{\Delta}) \geq \kappa_L \|\boldsymbol{\Delta}\|^2 - \tau_L^2(\boldsymbol{\theta}^*) - \frac{\lambda_n}{2}\{3R(\boldsymbol{\Delta}_{\bar{\mathcal{M}}}) + 4R(\boldsymbol{\theta}^*_{\mathcal{M}^\perp})\}$$

By using the definition of $\psi(\overline{\mathcal{M}})$, we have $R(\boldsymbol{\Delta}_{\bar{\mathcal{M}}}) \leq \psi(\overline{\mathcal{M}})\|\boldsymbol{\Delta}_{\bar{\mathcal{M}}}\|$. Hence, we conclude that $F(\boldsymbol{\Delta})$ is bounded by the quadratic function (in $\|\boldsymbol{\Delta}\|$):

$$F(\boldsymbol{\Delta}) \geq \kappa_L \|\boldsymbol{\Delta}\|^2 - \tau_L^2(\boldsymbol{\theta}^*) - \frac{\lambda_n}{2}\{3\psi(\overline{\mathcal{M}})\|\boldsymbol{\Delta}\| + 4R(\boldsymbol{\theta}^*_{\mathcal{M}^\perp})\}.$$

Solving the quadratic function, it is easy to see the right-hand side is non-negative, when

$$\|\boldsymbol{\Delta}\|^2 \geq \frac{9\lambda_n^2}{4\kappa_L^2}\psi^2(\overline{\mathcal{M}}) + \frac{4\lambda_n R(\boldsymbol{\theta}^*_{\mathcal{M}^\perp})}{\kappa_L} + \frac{2\tau_L^2(\boldsymbol{\theta}^*)}{\kappa_\tau}:$$

Finally, since $\widehat{\boldsymbol{\Delta}} \in \mathcal{C}$, we have by Proposition 5.3 that

$$R(\widehat{\boldsymbol{\Delta}}) \leq R(\widehat{\boldsymbol{\Delta}}_{\bar{\mathcal{M}}^\perp}) + R(\widehat{\boldsymbol{\Delta}}_{\bar{\mathcal{M}}}) \leq 4R(\widehat{\boldsymbol{\Delta}}_{\bar{\mathcal{M}}}) + 4R(\boldsymbol{\theta}^*_{\mathcal{M}^\perp}) \leq 4\psi(\overline{\mathcal{M}})\|\widehat{\boldsymbol{\Delta}}\| + 4R(\boldsymbol{\theta}^*_{\mathcal{M}^\perp}).$$

This completes the proof. ■

## 5.11 Bibliographical Notes

Penalized likelihood for variable selection traces back to the AIC (Akaike, 1974) and BIC (Schwarz, 1978). During the 1980s, most work on variable selection focused on linear regression models. Bayesian variable selection is closely related to penalized likelihood. Modern Bayesian variable selection was pioneered by Mitchell and Beauchamp (1988), where they proposed a Bayesian variable selection procedure with "spike and slab" prior distribution for regression coefficients in normal linear models. A related paper is George and McCulloch (1993). Berger and Pericchi (1996) proposed the intrinsic Bayes factor (IBF) for model selection (more precisely, model comparison).

Tibshirani (1996) proposed the Lasso method for variable selection in linear regression models and briefly discussed its potential use for generalized linear models. Fan and Li (2001) is the first paper that systematically studied penalized likelihood by using a folded concave penalty, where the concept of oracle estimator, the oracle property and the LQA algorithm were formulated. Fan and Peng(2004) studied the asymptotical property of penalized likelihood under a diverging dimension setting. van de Geer (2008) studied properties of high-dimensional Lasso. Fan and Lv (2011) studied the theoretical properties of penalized likelihood for ultrahigh dimensional generalized linear models. Fan, Xue and Zou (2014) developed a general theoretical framework to prove the strong oracle property of penalized likelihood under ultrahigh dimensions. Negahban, Ravikumar, Wainwright and Yu (2012) gave a unified treatment of convex penalized $M$ estimators.

There are rich computational developments for penalized likelihood. Fan and Li (2001) proposed the LQA algorithm for nonconcave penalized likelihood. Hunter and Li (2005) studied the convergence property of the LQA algorithm. Zou and Li (2008) proposed the LLA algorithm for nonconcave penalized likelihood, and suggested a one-step sparse estimator. Park and Hastie (2007) proposed to use a predictor-corrector algorithm for computing the Lasso penalized GLIM. Friedman, Hastie, Hoefling and Tibshirani (2007) revealed the efficiency of coordinate descent for solving penalized linear regression. Later, Friedman, Hastie and Tibshirani (2010) applied the coordinate descent algorithm to compute the regularization paths of penalized generalized linear models such as logistic regression. Fan and Lv (2011) proposed a modified version of the coordinate ascent algorithm for penalized likelihood. Wang, Kim and Li (2013) studied the difference convex algorithm for penalized logistic regression. Gradient-based algorithms (Nesterov, 2007) were considered in Wang, Liu and Zhang (2014) and Loh and Wainwright (2015) for computing one of the local solutions. Loh and Wainwright (2014) showed that any stationary point of folded-concave penalized $M$-estimation lies within statistical precision of the underlying parameter vector and Loh and Wainwright (2017) showed that any stationary point of the SCAD or MCP penalized likelihood can be used to recover the support of the true model. The asymptotic normality for high-dimensional robust $M$-estimators is established in Loh(2017).

Tuning parameter selection of the penalized likelihood for generalized linear models was first studied by Zhang, Li and Tsai (2010) under finite dimensional settings, and further studied by Fan and Tang under high dimensional settings. Wang, Kim (2013) and Li (2014) proposed HBIC for ultrahigh dimensional linear regression and extended it to logistic regression. Ninomiya and Kawano (2016) derived the AIC for the Lasso estimator under the framework of generalized linear models.

## 5.12 Exercises

5.1 Suppose a latent variable $Z$ (e.g. severity of autism) follows the linear model

$$Z = \boldsymbol{\beta}^T \mathbf{X} + \varepsilon,$$

where $\mathbf{X}$ is the covariate and $\varepsilon$ has a cumulative distribution function $F(t)$. That is, $\mathbb{P}(\varepsilon \le t) = F(t)$ for all $t \in \mathbb{R}$. Instead of observing $Z$, one observes $Y = 0$ if $Z \le c_1$, $Y = 1$ if $Z \in (c_1, c_2]$ and $Y = 2$ if $Z > c_2$ for some unknown parameters $c_1 < c_2$.

(a) What is the conditional distribution of $Y$ given $\mathbf{X}$?

(b) Suppose that $\{(\mathbf{X}_i, Y_i)\}_{i=1}^n$ is a random sample from the above model. Write down the log-likelihood function.

(c) How do we generalize the logistic regression model to categorical data with more than 2 categories? (**Hint**: Use multiple $\beta$'s.)

5.2 Derive the IRLS algorithm for generalized linear models

(a) Derive the IRLS algorithm for Bernoulli and binomial models in Section 5.2.1.

(b) Derive the IRLS algorithm for Poisson loglinear models in Section 5.2.2.

(c) Derive the IRLS algorithm for generalized linear models with nonnegative continuous response in Section 5.2.3 and normal error models in Section 5.2.4.

5.3 Consider the generalized linear model $f(y|\mathbf{x}) = \exp\{\frac{y\theta - b(\theta)}{\phi} + c(y, \phi)\}$ with the canonical link. Let $\ell_n(\boldsymbol{\beta})$ denote the negative log-likelihood of the data $\{(\mathbf{X}_i, Y_i)\}_{i=1}^n$.

(a) The formula for estimating the variance of the MLE $\widehat{\boldsymbol{\beta}}$ is $\widehat{\mathrm{Var}}(\widehat{\boldsymbol{\beta}}) = [\nabla^2 \ell_n(\widehat{\boldsymbol{\beta}})]^{-1}$. Show that

$$\widehat{\mathrm{Var}}(\widehat{\boldsymbol{\beta}}) = \phi[\sum_{i=1}^n b''(\widehat{\theta}_i)\mathbf{X}_i\mathbf{X}_i^T]^{-1}, \qquad \text{where} \qquad \widehat{\theta}_i = \mathbf{X}_i^T\widehat{\boldsymbol{\beta}}.$$

(b) Deduce $\widehat{\mathrm{Var}}(\widehat{\boldsymbol{\beta}})$ for logistic regression and Poisson regression.

(c) Write the optimization problem for finding the sparsest solution in a high-confidence set.

(d) For logistic regression, what does the formulation above look like?

5.4 Regard to condition (5.22) with the MCP and the SCAD penalty.

(a) For linear regression models, derive a sufficient condition for condition (5.22), and construct an example such that (5.22) is invalid.

(b) For logistic regression models, derive a sufficient condition for condition (5.22), and construct an example such that (5.22) is invalid.

(c) For Poisson loglinear regression models, derive a sufficient condition for condition (5.22), and construct an example such that (5.22) is invalid.

5.5 Derive iterative shrinkage-thresholding (IST) algorithms for the penalized likelihood with folded concave penalty.

(a) Extend the IST algorithm in Section 3.5.7 for the penalized logistic regression with the MCP and the SCAD penalty, and provide a sufficient condition under which the IST algorithm possesses the ascent property.

(b) Extend the IST algorithm in Section 3.5.7 for the penalized Poisson loglinear regression model with the MCP and the SCAD penalty.

5.6 Suppose that $f(\mathbf{x})$ is smooth and strongly convex in the sense that there exist some $L > \sigma > 0$ such that

$$f''(\mathbf{x}) \le L\mathbf{I}_p \qquad \text{and} \qquad f(\mathbf{x}) \ge f(\mathbf{x}_0) + f'(\mathbf{x}_0)^T(\mathbf{x}-\mathbf{x}_0) + \frac{\sigma}{2}\|\mathbf{x}-\mathbf{x}_0\|^2$$

for any $\mathbf{x}_0$ and $\mathbf{x}$. Let

$$f_Q(\mathbf{x}|\mathbf{x}_0) = f(\mathbf{x}_0) + f'(\mathbf{x}_0)^T(\mathbf{x}-\mathbf{x}_0) + \frac{1}{2\delta}\|\mathbf{x}-\mathbf{x}_0\|^2$$

be a quadratic majorization at point $\mathbf{x}_0$ with $\delta \le 1/L$. Let $F(\mathbf{x}) = f(\mathbf{x}) + \lambda\|\mathbf{x}\|_1$ be the objective function to be minimized and $G(\mathbf{x}|\mathbf{x}_0) = f_Q(\mathbf{x}|\mathbf{x}_0) + \lambda\|\mathbf{x}\|_1$ be its penalized quadratic majorization at the point $\mathbf{x}_0$.

(a) Show that $G(\mathbf{x}|\mathbf{x}_0) \le F(\mathbf{x}) + \frac{1}{2\delta}\|\mathbf{x}-\mathbf{x}_0\|^2$. **Hint**: $f(\mathbf{x}) \ge f(\mathbf{x}_0) + f'(\mathbf{x}_0)^T(\mathbf{x}-\mathbf{x}_0)$.

(b) To find $\mathbf{x}^* = \arg\min F(\mathbf{x})$,[1] consider the iteration

$$\mathbf{x}_i = \arg\min G(\mathbf{x}|\mathbf{x}_{i-1})$$

whose solution is given by component-wise thresholding. Show that

$$F(\mathbf{x}_i) \le F(\mathbf{x}_{i-1}) + \min_w\{-w(F(\mathbf{x}_{i-1}) - F(\mathbf{x}^*)) + \frac{w^2}{2\delta}\|\mathbf{x}^* - \mathbf{x}_{i-1}\|^2\}.$$

**Hint**: Use $F(\mathbf{x}_i) \le \min_{\mathbf{x}} G(\mathbf{x}|\mathbf{x}_{i-1})$, part (a) and consider minimization only on the line $w\mathbf{x}^* + (1-w)\mathbf{x}_{i-1}$.

(c) Use the optimality condition of $\mathbf{x}^*$ to show first $F(\mathbf{x}_{i-1}) - F(\mathbf{x}^*) \ge \frac{\sigma}{2}\|\mathbf{x}_{i-1} - \mathbf{x}^*\|^2$.

(d) Use (b) and (c) to show the linear rate convergence: $F(\mathbf{x}_i) - F(\mathbf{x}^*) \le (1 - \frac{\delta\sigma}{4})[F(\mathbf{x}_{i-1}) - F(\mathbf{x}^*)]$.

[1]Please verify by yourself that the minimizer exists and is unique. You don't need to prove this.

5.7 Establish the oracle property of the folded concave penalized likelihood.

(a) Prove Theorem 5.1.

(b) Prove Theorem 5.2.

5.8 Establish the oracle property of the sparse one-step estimator by proving Theorem 5.3.

5.9 Prove Theorem 5.7.

5.10 Let $\ell_n(\boldsymbol{\beta}) = n^{-1}\sum_{i=1}^n [b(\mathbf{X}_i^T\boldsymbol{\beta}) - Y_i\mathbf{X}_i^T\boldsymbol{\beta}]$ be the (normalized) negative log-likelihood of the generalized linear model with $\phi = 1$. Consider the penalized likelihood estimator

$$\widehat{\boldsymbol{\beta}} \in \arg\min_{\boldsymbol{\beta}}\{\ell_n(\boldsymbol{\beta}) + \lambda_n\|\boldsymbol{\beta}\|_1\}.$$

(a) Show that $\nabla\ell_n(\boldsymbol{\beta}^*) = n^{-1}\sum_{i=1}^n \varepsilon_i\mathbf{X}_i$, where $\varepsilon_i = b'(\mathbf{X}_i^T\boldsymbol{\beta}^*) - Y_i$.

(b) Let $\boldsymbol{\Delta} = \widehat{\boldsymbol{\beta}} - \boldsymbol{\beta}^*$ where $\boldsymbol{\beta}^*$ is the true parameter. If $\lambda_n \geq 2\|\nabla\ell_n(\boldsymbol{\beta}^*)\|_\infty$, then for any set $\mathcal{S} \subseteq \{1, 2, \cdots, p\}$ we have

$$\|\boldsymbol{\Delta}_{\mathcal{S}^c}\|_1 \leq 3\|\boldsymbol{\Delta}_{\mathcal{S}}\|_1 + 4\|\boldsymbol{\beta}^*_{\mathcal{S}^c}\|_1.$$

**Hint**: Apply Proposition 5.2.

(c) Let $\mathcal{S}_0 = \text{supp}(\boldsymbol{\beta}^*)$ and $s_n = |\mathcal{S}_0|$. Suppose that $\|\boldsymbol{\beta}^*_{\mathcal{S}_0^c}\|_1 \lesssim \lambda_n$ (here $p_n \lesssim q_n$ means there exists some constant $C > 0$ such that $p_n \leq Cq_n$ holds for sufficiently large $n$), and the restricted strong convexity holds with $\tau_L = 0$ and $\kappa_L$ being a positive constant. Given $\lambda_n \geq 2\|\nabla\ell_n(\boldsymbol{\beta}^*)\|_\infty$, show that

$$\|\boldsymbol{\Delta}\|_2^2 \lesssim s_n\lambda_n^2 \qquad \text{and} \qquad \|\boldsymbol{\Delta}\|_1 \lesssim s_n\lambda_n.$$

**Hint**: Apply Theorem 5.5.

5.11 Let us consider the 128 macroeconomic time series from Jan. 1959 to Dec. 2018, which can be downloaded from the book website (see the "transformed macroeconomic data" at the book website. As before, we extract the data from Jan. 1960 to Oct. 2018 (in total 706 months) and remove the feature named "sasdate" and the features with missing entries. Suppose that we only observe the binary response $Y = I(\text{UNRATE} \geq 0)$ rather than "UNRATE" (up or down rather than the actual unemployment rate changes).

(a) Take the remaining variables as predictors. To conduct contemporary association studies, do the following steps for Lasso (using R package `glmnet`) and SCAD (using R package `ncvreg`): Set a random seed by `set.seed(525)`; Plot the regularization paths as well as the prediction error estimated by 10-fold cross-validation; Choose a model based on cross-validation, report the model, and point out two most important macroeconomic variables (largest coefficients in the standardized variables) that are correlated with the current change of unemployment rate.

(b) In this sub-problem, we are going to study which macroeconomic variables are leading indicators for driving future unemployment rates up and down. To do so, we will pair each row of predictors with the indicator of the next row of response. The last row of predictors and the first element in the response are hence discarded. After this change, do the same exercise as (a).

(c) Consider the setting in (b) and set a random seed by `set.seed(525)`. Let us take the variables selected by the Lasso with absolute coefficient $> 0.01$ (under standardized variables), calling it model 1. Run the logistic regression and summarize the fit. Now pick the significant variables whose absolute z-statistics are larger than 1.96. Run the model again with the subset of the variable, calling this model 2. Are models 1 and 2 statistically significant by running a likelihood ratio test? Are two models practically different by plotting the residuals of model 1 against model 2? Also, plot the residuals of model 2 to see if there are any patterns. (You can use the function `glm` and `anova`. It is a good practice to do the ANOVA part of the calculation by hand for understanding)

(d) Consider the setting of (b). Leave the last 120 months as testing data and use the rest as training data. Set a random seed by `set.seed(525)`. Run Lasso and SCAD on the training data using `glmnet` and `ncvreg`, respectively, and choose a model based on 10-fold cross-validation. Compute the out-of-sample proportion of prediction error.

5.12 Upright Human Detection in Photos

Go to the book's class website, download the image data `pictures.zip` and its associated preliminary codes `human.r`. In this problem, we are going to create a human detector that tells us whether there is an upright human in a given photo. We treat this as a classification problem with two classes: having humans or not in a photo. You are provided with two datasets `POS` and `NEG` that have photos with and without upright humans respectively.

(a) Load Pictures and Extract Features (The code has already been written for you).

The tutorial below explains the data loading and feature extraction in `human.r`. If you do not want to read it, you can just execute the code to get the extracted features. Remember to install the package `png` by using `install.packages("png")` and to change the working directory to yours.

i. Install the package `png` by using `install.packages("png")`, and use the function `readPNG` to load photos. The function `readPNG()` will return the grayscale matrix of the picture.

ii. Use the function `grad` to obtain the gradient field of the central $128 \times 64$ part of the grayscale matrix.

iii. Use the function hog (Histograms of Oriented Gradient) to extract a feature vector from the gradient field obtained in the previous step. Your feature vector should have 96 components. Please see the appendix for parameter configuration of this function.

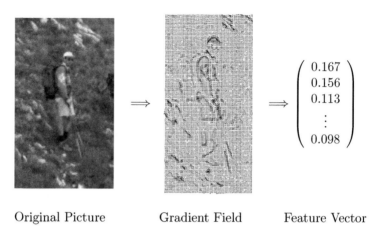

$$\Longrightarrow \begin{pmatrix} 0.167 \\ 0.156 \\ 0.113 \\ \vdots \\ 0.098 \end{pmatrix}$$

Original Picture    Gradient Field    Feature Vector

Figure 5.2: Illustration of feature extraction for a positive example from `POS`.

(b) Logistic Regression (You are responsible for writing the code).

In this question, we will apply logistic regression to the data and use the cross validation to test the classification accuracy of the fitted model. Please finish the following steps:

i. Set the random seed to be 525, i.e., type in `set.seed(525)` in your code. Randomly divide your whole data into five parts. Let the first four parts be the training data and the rest be your testing data.

ii. Feed the training data to the function `glm` and store the fitted model as `fitted1` in R.

iii. Use the function `predict` to apply your fitted model `fitted1` to do classification on the testing data and report the misclassification rate (prediction error).

iv. Let us now select the features using stepwise selection, by issuing the R-function `step(fitted1)`. Let us call the selected model (the last model in the output) `fitted2`. Report the model `fitted2` and the misclassification rate of `fitted2`.

v. Report the misclassification rate by adding a Lasso penalty with $\lambda$ chosen by 10-fold cross-validation. (You can use the function `glmnet`.)

## Appendix

*A. Histogram of Oriented Gradient*

Here we give a brief introduction of what `hog(xgrad, ygrad, hn, wn, an)` does. First of all, it uniformly partitions the whole picture into `hn*wn` small parts with `hn` partitions of the height and `wn` partitions of the width. For each small part, it counts the gradient direction whose angle falls in the intervals $[0, 2\pi/\mathtt{an}), [2\pi/\mathtt{an}, 4\pi/\mathtt{an}), ..., [2(\mathtt{an}-1)\pi/\mathtt{an}, 2\pi)$ respectively. So `hog` can get `an` frequencies for each small picture. Applying the same procedure to all the small parts, `hog` will have `hn*wn*an` frequencies that constitute the final feature vector for the given gradient field.

*B. Useful Functions*

(a) `crop.r(X, h, w)` randomly crops a sub-picture that has height $h$ and width $w$ from `X`. The output is therefore a sub-matrix of `X` with `h` rows and `w` columns.

(b) `crop.c(X, h, w)` crops a sub-picture that has height $h$ and width $w$ at the center of `X`. The output is therefore a sub-matrix of `X` with `h` rows and `w` columns. This function helps the `hog(...)` function. For, the cropping in your assignment, use `crop.r()`.

(c) `grad(X, h, w, pic)` yields the gradient field at the center part of the given grayscale matrix `X`. The center region it examines has height `h` and width `w`. It returns a list of two matrices `xgrad` and `ygrad`. The parameter `pic` is a boolean variable. If it is `TRUE`, the generated gradient filed will be plotted. Otherwise the plot will be omitted.

(d) `hog(xgrad, ygrad, hn, wn, an)` returns a feature vector in the length of `hn*wn*an` from the given gradient field. `(xgrad[i,j], ygrad[i,j])` gives the grayscale gradient at the position `(i,j)`. `hn` and `wn` are the partition number of height and width respectively. `an` is the partition number of the angles (or the interval $[0, 2\pi)$ equivalently).

Chapter 6

# Penalized M-estimators

In Chapters 3 and 5, we introduce the methodology of penalized least squares and penalized likelihood, respectively. This chapter is devoted to introducing regularization methods in quantile regression, composite quantile regression, robust regression and rank regression under the general framework of penalized M-estimators. This chapter also offers a brief introduction to penalized partial likelihood for survival data analysis.

## 6.1 Penalized Quantile Regression

Quantile regression was first proposed by Koenker and Basset (1978), and has many applications in various research areas such as growth charts (Wei and He, 2006), survival analysis (Koenker and Geling, 2001) and economics (Koenker and Hallock, 2001), and so on. Quantile regression has become a popular data analytic tool in the literature of both statistics and econometrics. A comprehensive account on quantile regression can be found in Koenker (2005).

### 6.1.1 *Quantile regression*

Let us begin with the univariate unconditional quantile. Let $F_Y(y) = \Pr(Y \le y)$ be the cumulative distribution function (cdf) of $Y$ and $Q_Y(\tau) = \inf\{y\colon F_Y(y) \ge \tau\}$ be the $\tau$th quantile of $Y$ for a given $\tau \in (0,1)$. Koenker and Basset (1978) observed the following result

$$Q_Y(\tau) = \arg\min_t \mathrm{E}[\rho_\tau(Y - t)], \tag{6.1}$$

where $\rho_\tau(u) = u\{\tau - I(u < 0)\}$ is the so-called check loss function. Figure 6.1 shows the *check loss* function for $\tau = 0.75$. When $\tau = 0.5$, the check loss becomes the absolute value $L_1$-loss.

We now consider the *quantile regression*. Consider a response variable $Y$ and a vector of covariates $\mathbf{X} = (X_1, \ldots, X_p)^T$. Given $\mathbf{X}$, denote by $F_Y(y|\mathbf{X})$ the conditional cumulative distribution function, and let $Q_Y(\tau|\mathbf{x}) = \inf\{y\colon F_Y(y|\mathbf{X} = \mathbf{x}) \ge \tau\}$ to be the $\tau$th *conditional quantile*. For the conditional quantiles, (6.1) can be naturally generalized as

$$Q_Y(\tau|\mathbf{x}) = \arg\min_t \mathrm{E}[\rho_\tau(Y - t)|\mathbf{X} = \mathbf{x}]. \tag{6.2}$$

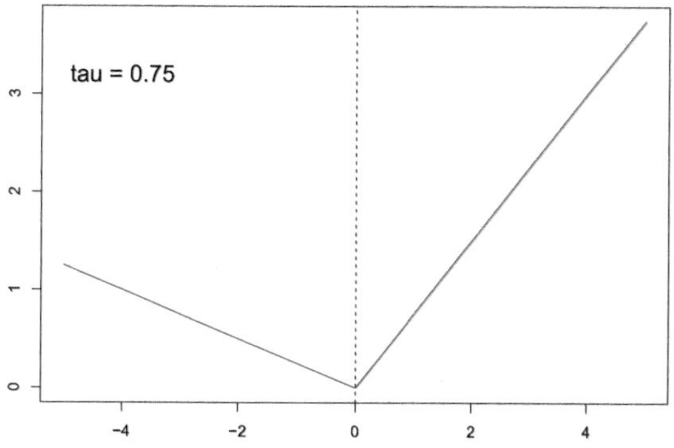


Figure 6.1: The check loss function with $\tau = 0.75$.

Quantile regression is based on (6.2). Consider the linear quantile regression in which $Q_Y(\tau|\mathbf{x})$ is a linear function of $\mathbf{x}$. Without loss of generality, let $Q_Y(\tau|\mathbf{x}) = \mathbf{x}^T\boldsymbol{\beta}^*$. Then, the quantile regression estimate of $\boldsymbol{\beta}^*$ is

$$\widehat{\boldsymbol{\beta}}_\tau = \arg\min_{\boldsymbol{\beta}} \frac{1}{n}\sum_{i=1}^{n} \rho_\tau(Y_i - \mathbf{X}_i^T\boldsymbol{\beta}). \tag{6.3}$$

This is due to Koenker and Basset (1978) and allows us to estimate the conditional quantile for each given $\mathbf{X}$.

Theoretical properties of linear quantile regression have been well studied. See, for example, Koenker (2005). Suppose that $\{\mathbf{X}_i, Y_i\}_{i=1}^n$, is a random sample from a linear model

$$Y = \mathbf{X}^T\boldsymbol{\beta}^* + \epsilon,$$

where $\epsilon$ is a random error with a positive probability density function $f(\cdot)$, and independent of the covariate $\mathbf{X}$. Let $\mathbf{X}$ be the design matrix and assume $\lim_{n\to\infty}\frac{1}{n}\mathbf{X}^T\mathbf{X} = \boldsymbol{\Sigma}$, where $\boldsymbol{\Sigma}$ is a $p \times p$ positive definite matrix. Under this assumption along with other weak regularity conditions, it has been shown that

$$\sqrt{n}(\widehat{\boldsymbol{\beta}}_\tau - \boldsymbol{\beta}^*) \xrightarrow{D} N\left(0, \frac{\tau(1-\tau)}{f^2(b_\tau^*)}\boldsymbol{\Sigma}^{-1}\right), \tag{6.4}$$

where $b_\tau^*$ is the $\tau$-quantile of the error distribution. Under the same setting, if $E\varepsilon = 0$ and $\mathrm{Var}(\epsilon) = \sigma^2$, it is known that the corresponding least squares estimator has the following asymptotic normality:

$$\sqrt{n}(\widehat{\boldsymbol{\beta}}^{\mathrm{ols}} - \boldsymbol{\beta}^*) \xrightarrow{D} N\left(0, \sigma^2\boldsymbol{\Sigma}^{-1}\right).$$

It is interesting to note that the asymptotic normality of the quantile regression estimator is still valid even when the error distribution does not have a finite variance, provided that $f(b_\tau^*)$ is not zero. This implies that the quantile regression may perform well for heavy tail distributions such as Cauchy distributions. Thus, quantile regression is viewed as a robust statistical tool for data analysis. Comparison of asymptotic efficiency between quantile regression and the least squares method would be an interesting exercise (Exercise 6.2).

Since the check loss is not differential at the origin, the Newton-Raphson algorithm and its variants cannot be directly used to solve (6.3). Hunter and Lange (2000) proposed a majorization-minimization (MM) algorithm for linear quantile regression. Alternatively, the minimization problem in (6.3) can be recast into a linear programming problem. Thus, one employs algorithms such as the interior point algorithm for linear programming to solve the quantile regression (Koenker and Ng, 2005). Details are left as an exercise (Exercise 6.1).

### *6.1.2 Variable selection in quantile regression*

The penalized least squares method can be naturally extended to *penalized quantile regression*

$$\frac{1}{n}\sum_{i=1}^{n}\rho_\tau(Y_i - \mathbf{X}_i^T\boldsymbol{\beta}) + \sum_{j=1}^{p}p_\lambda(|\beta_j|), \tag{6.5}$$

where $p_\lambda(\cdot)$ is a penalty function with a regularization parameter $\lambda$. Minimizing (6.5) results in a penalized quantile regression estimate. With an appropriate choice of penalty function and $\lambda$, the penalized quantile regression may achieve the purpose of variable selection.

By taking the penalty to the $\ell_1$-penalty, the Lasso method can be naturally applied to the quantile regression setting for variable selection with high-dimensional data. Belloni and Chernozhukov (2011) studied the $\ell_1$ penalized quantile regression when $p \gg n$, and showed that the Lasso penalized quantile regression estimator is consistent at a near-oracle rate. Following their notation, the $\ell_1$ quantile regression is defined as

$$\widehat{\boldsymbol{\beta}}^{\text{lasso}} = \arg\min_{\boldsymbol{\beta}} \frac{1}{n}\sum_{i=1}^{n}\rho_\tau(Y_i - \mathbf{X}_i^T\boldsymbol{\beta}) + \frac{\lambda}{n}\sum_{j=1}^{n}\widehat{\sigma}_j|\beta_j|, \tag{6.6}$$

where $\widehat{\sigma}_j^2 = \frac{1}{n}\sum_{i=1}^{n}X_{ij}^2$. In practice, it is typical to standardize each covariate so that its sample mean and variance equal 0 and 1, respectively. Thus, (6.6) can be viewed as a special case of (6.5).

Suppose that $Q_Y(\tau|\mathbf{x}) = \mathbf{x}^T\boldsymbol{\beta}^*$ with $\|\boldsymbol{\beta}^*\|_0 = s$. Let $\mathcal{S}$ denote the support of $\boldsymbol{\beta}^*$. Assume that the conditional density $f(y|\mathbf{x})$ is continuously differentiable in $y$ and upper bounded by a constant $\bar{f}$. In addition, assume that $f(y|\mathbf{x})$ evaluated at its $\tau$-quantile is bounded away from zero, i.e., $f(\mathbf{x}^T\boldsymbol{\beta}^*|\mathbf{x}) >$

$\underline{f} > 0$ uniformly, and $\frac{\partial f(y|\mathbf{x})}{\partial y}$ is upper bounded by a constant $\bar{f}'$. These are the standard regularity conditions for quantile regression. Without loss of generality, assume that all covariates are centered and have variance 1. Assume further that

$$P(\max_j |\widehat{\sigma}_j - 1| \leq 1/2) \geq 1 - \gamma \to 0. \tag{6.7}$$

For a vector $\mathbf{v}$ in $\mathbb{R}^p$ and $\mathcal{S}'$ any subset of $\{1, \ldots, p\}$, $\mathbf{v}_{\mathcal{S}'}$ is a vector in $\mathbb{R}^p$ such that $(\mathbf{v}_{\mathcal{S}'})_j = v_j$ if $j \in \mathcal{S}'$ and $(\mathbf{v}_{\mathcal{S}'})_j = 0$ if $j \notin \mathcal{S}'$. A restricted set is defined as

$$\mathcal{A} = \{\mathbf{v} \in \mathbb{R}^p : \|\mathbf{v}_{\mathcal{S}^c}\|_1 \leq c_0 \|\mathbf{v}_{\mathcal{S}}\|_1, \|\mathbf{v}_{\mathcal{S}^c}\|_0 \leq n\}. \tag{6.8}$$

It can be shown that with a high probability $\widehat{\boldsymbol{\beta}}^{\text{lasso}} - \boldsymbol{\beta}^* \in \mathcal{A}$. To bound $\widehat{\boldsymbol{\beta}}^{\text{lasso}} - \boldsymbol{\beta}^*$ additional regularity conditions are required. Define $\bar{\mathcal{S}}(\mathbf{v}, m) \subset \{1, \ldots, p\} \setminus \mathcal{S}$ as the support of the $m$ largest in absolute value elements of $\mathbf{v}$ outside $\mathcal{S}$. Assume that for some $m > 0$ we have

$$\kappa_m^2 = \inf_{\mathbf{v} \in \mathcal{A}, \mathbf{v} \neq 0} \frac{\mathbf{v}^T \mathbf{E}[\mathbf{X}\mathbf{X}^T]\mathbf{v}}{\|\mathbf{v}_{\mathcal{S} \cup \bar{\mathcal{S}}(\mathbf{v}, m)}\|^2} > 0 \tag{6.9}$$

$$q_m = \frac{3}{8} \frac{\underline{f}^{3/2}}{\bar{f}'} \inf_{\mathbf{v} \in \mathcal{A}, \mathbf{v} \neq 0} \frac{\mathrm{E}[|\mathbf{X}^T \mathbf{v}|^2]^{3/2}}{\mathrm{E}[|\mathbf{X}^T \mathbf{v}|^3]} > 0 \tag{6.10}$$

The condition (6.9) is similar to the restricted eigenvalue condition for the $\ell_1$ least squares estimator.

**Theorem 6.1** *If we choose $\lambda_0(\alpha) \leq \lambda \leq C\sqrt{s \log(p)/n}$ for some constant $C$, where $\lambda_0(\alpha)$ only depends on $\alpha$, with probability at least* $1 - 4\gamma - \alpha - 3p^{-b^2}$,

$$\|\widehat{\boldsymbol{\beta}}^{\text{lasso}} - \boldsymbol{\beta}^*\| \leq \frac{1 + c_0\sqrt{s/m}}{\kappa_m} \frac{8C(1 + c_0)}{\underline{f}\kappa_0\sqrt{\tau(1-\tau)}} b\sqrt{\frac{s \log p}{n}},$$

*where $b$ is any constant $b > 1$.*

This is a direct result of Theorem 2 of Belloni and Chernozhukov (2011) with $p \geq n$. By taking the penalty function in (6.5) to be the SCAD penalty, we obtain the SCAD penalized quantile regression. Comparing with the $\ell_1$ quantile regression, the SCAD penalized quantile regression is expected to eliminate the estimation bias introduced by the $\ell_1$ penalty. See, for example, Wu and Liu (2009) and Wang, Wu and Li (2012). Fan, Fan and Barut (2014) introduced weighted robust Lasso and adaptive robust Lasso to reduce the biases caused by the $L_1$-penalty and established the oracle property and asymptotic normality in the ultrahigh dimensional setting. Fan, Xue and Zou (2014) proved the strong oracle property of the SCAD penalized quantile regression estimator:

$$\widehat{\boldsymbol{\beta}}^{\text{scad}} = \arg\min_{\boldsymbol{\beta}} \frac{1}{n} \sum_{i=1}^{n} \rho_\tau(Y_i - \mathbf{X}_i^T \boldsymbol{\beta}) + \sum_{j=1}^{p} p_\lambda(|\beta_j|), \tag{6.11}$$

where $p_\lambda(\cdot)$ is the SCAD penalty function with the regularization parameter $\lambda$, via the *local linear approximation* (LLA). Details are left as an exercise (Exercise 6.3).

With the aid of the LLA approximation (Zou and Li, 2008), the minimization problem of the SCAD penalized quantile regression in (6.11) can be solved by the iteratively weighted $\ell_1$ penalized quantile regression problem in which the weights come from the derivative of the penalty function. In practice, we can initialize the LLA algorithm with $\widehat{\boldsymbol\beta}^{\text{lasso}}$. Let $\widehat{\boldsymbol\beta}^{k}$ denote the solution at the $k$th iteration. For $k = 1, 2, \ldots,$, repeat

(i) Compute the weights $w_j = \lambda^{-1} p'_\lambda(|\widehat{\beta}_j^{k-1}|)$, $j = 1, \ldots, p$.

(ii) Solve $\widehat{\boldsymbol\beta}^{k} = \arg\min_{\boldsymbol\beta} \frac{1}{n}\sum_{i=1}^n \rho_\tau(Y_i - \mathbf{X}_i^T\boldsymbol\beta) + \lambda\sum_{j=1}^p w_j|\beta_j|$.

Define the oracle quantile regression estimator $\widehat{\boldsymbol\beta}^{\text{oracle}} = (\widehat{\boldsymbol\beta}_{\mathcal{S}}^{\text{oracle}}, 0)$ where

$$\widehat{\boldsymbol\beta}_{\mathcal{S}}^{\text{oracle}} = \arg\min_{\boldsymbol\beta} \frac{1}{n}\sum_{i=1}^n \rho_\tau(Y_i - \sum_{j\in\mathcal{S}} X_{ij}\beta_j).$$

As shown in Fan, Xue and Zou (2014), we have the following result.

**Theorem 6.2** *Under the regularity conditions for the $\ell_1$ penalized quantile regression for Theorem 6.1, with a high probability, the LLA algorithm initialized by $\widehat{\boldsymbol\beta}^{\text{lasso}}$ converges to $\widehat{\boldsymbol\beta}^{\text{oracle}}$ after two iterations.*

To compare the performance of the $\ell_1$ and the SCAD penalized quantile regression, we generate a random sample of size 100 from the simulation model

$$Y = \mathbf{X}^T\boldsymbol\beta^* + \varepsilon$$

with $p = 400$, where $\boldsymbol\beta^*$ is constructed by randomly choosing 10 elements in $\boldsymbol\beta^*$ as $t_1 s_1, \ldots, t_{10}s_{10}$ and setting the other $p - 10$ elements as zero, where $t_j$'s are independently drawn from Unif(1, 2), and $s_j$'s are independent Bernoulli samples with $\Pr(s_j = 1) = \Pr(s_j = -1) = 0.5$. The error $\varepsilon$ follows the standard Cauchy distribution. The simulation results are summarized in Table 6.1. It is clear that the SCAD penalized quantile regression performs much better than the Lasso penalized quantile regression. Because the error distribution is Cauchy, the SCAD or Lasso penalized least square estimator breaks down. Their corresponding results are not presented here.

### *6.1.3 A fast algorithm for penalized quantile regression*

The minimization problem associated with penalized quantile regression can be more challenging than solving the penalized least squares problem.

Table 6.1: Estimation accuracy is measured by the $\ell_1$ loss and the $\ell_2$ loss, and selection accuracy is measured by counts of false negative (#FN) or false positive (#FP). Each metric is averaged over 100 replications with its standard error shown in the parenthesis. Values in this table are extracted from Table 1 in Fan, Xue and Zou (2014).

| Method | $\ell_1$ loss | $\ell_2$ loss | # FP | # FN | $\ell_1$ loss | $\ell_2$ loss | # FP | # FN |
|---|---|---|---|---|---|---|---|---|
| | $\tau = 0.3$ | | | | $\tau = 0.5$ | | | |
| LASSO | 14.33 | 2.92 | 39.31 | 1.03 | 13.09 | 2.62 | 41.42 | 0.61 |
| | (0.35) | (0.07) | (1.29) | (0.14) | (0.33) | (0.06) | (1.18) | (0.09) |
| SCAD | 5.92 | 1.93 | 8.89 | 1.63 | 3.96 | 1.27 | 10.18 | 0.69 |
| | (0.37) | (0.11) | (0.68) | (0.19) | (0.39) | (0.08) | (0.74) | (0.11) |

With the aid of LLA, the computing of penalized quantile regression boils down to solving

$$\min_{\boldsymbol{\beta}} \frac{1}{n}\sum_{i=1}^{n} \rho_\tau(Y_i - \mathbf{X}_i^T\boldsymbol{\beta}) + \lambda \sum_{j=1}^{p} w_j|\beta_j|. \tag{6.12}$$

For the $\ell_1$ penalized quantile regression, we solve (6.12) once with $w_j = 1$ for all $j$. For the SCAD or MCP penalized quantile regression, we solve (6.12) twice in two LLA iterations with different $w_j$s, according to Theorem 6.2. More iterations will help reduce the optimization errors.

A standard method for solving (6.12) is to transform the corresponding optimization problem into a linear programming, which can then be solved by many existing optimization software packages. Koenker and Ng (2005) proposed an interior-point method for quantile regression and the $\ell_1$ penalized quantile regression. However, this interior-point method is only efficient for solving small scale problems and is very inefficient when $p$ is large.

By following the LARS algorithm for the Lasso linear regression (Efron, Hastie, Johnstone and Tibshirani 2004), Li and Zhu (2008) proposed a piecewise linear path following the algorithm for computing the solution path of the Lasso penalized quantile regression. Although the LARS-type algorithm is very efficient for solving the Lasso linear regression, it loses its efficiency when the squared error loss is replaced with the nonsmooth check loss. More recently, Peng and Wang (2015) suggested using a coordinate descent algorithm for solving sparse penalized quantile regression. Coordinate descent is very successful in solving the Lasso penalized least squares (Friedman *et al.* 2010). However, when the loss function is nonsmooth and non-separable (which is the case for quantile regression), the coordinate descent algorithm does not have a theoretical guarantee to give the right solution. Here is a simple example to demonstrate this point. Consider quantile regression ($\tau = 0.5$) with a 'fake' dataset:

| $y$ | $x_1$ | $x_2$ |
|---|---|---|
| 0 | 1 | 1 |
| 0 | 1 | -1 |
| 0 | 1 | -1 |

and the corresponding optimization problem is

$$\arg\min_{\beta_1,\beta_2} \frac{1}{2}|\beta_1+\beta_2| + |\beta_1-\beta_2|.$$

Obviously, the minimizer is $\beta_1 = \beta_2 = 0$. It can be directly shown that the cyclic coordinate descent algorithm is trapped at the initial value. Thus, unless the initial value is 0, the cyclic coordinate descent algorithm does not provide the right solution.

Gu, Fan, Kong, Ma and Zou (2018) offered an efficient algorithm for solving (6.12). Their algorithm is based on the *alternating direction method of multipliers (ADMM)*. For ease of notation, introduce new variables $\mathbf{z} = \mathbf{Y} - \mathbf{X}\boldsymbol{\beta}$, and denote $\mathbb{Q}_\tau(\mathbf{z}) = (1/n)\sum_{i=1}^n \rho_\tau(z_i)$ for $\mathbf{z} = (z_1, \ldots, z_n)^T$. Problem (6.12) is equivalent to

$$\begin{array}{ll} \min_{\boldsymbol{\beta},\mathbf{z}} & \mathbb{Q}_\tau(\mathbf{z}) + \lambda\sum_{j=1}^p w_j|\beta_j| \\ \text{subject to} & \mathbf{X}\boldsymbol{\beta} + \mathbf{z} = \mathbf{Y}. \end{array} \tag{6.13}$$

The augmented Lagrangian function of (6.13) is

$$\mathcal{L}_\eta(\boldsymbol{\beta}, \mathbf{z}, \theta) = \mathbb{Q}_\tau(\mathbf{z}) + \lambda\sum_{j=1}^p w_j|\beta_j| - \langle\theta, (\mathbf{X}\boldsymbol{\beta} + \mathbf{z} - \mathbf{Y})\rangle + \frac{\eta}{2}\|\mathbf{X}\boldsymbol{\beta} + \mathbf{z} - \mathbf{Y}\|_2^2,$$

where $\eta$ is a user-specified constant.

Let $(\boldsymbol{\beta}^k, \mathbf{z}^k, \theta^k)$ denote the $k$th iteration of the ADMM algorithm for $k = 0, 1, 2, \ldots$. The iterations for the standard ADMM algorithm are given by

$$\begin{aligned} \boldsymbol{\beta}^{k+1} &= \arg\min_{\boldsymbol{\beta}} \mathcal{L}_\eta(\boldsymbol{\beta}, \mathbf{z}^k, \theta^k) \\ \mathbf{z}^{k+1} &= \arg\min_{\mathbf{z}} \mathcal{L}_\eta(\boldsymbol{\beta}^{k+1}, \mathbf{z}, \theta^k) \\ \theta^{k+1} &= \theta^k - \eta(\mathbf{X}\boldsymbol{\beta}^{k+1} + \mathbf{z}^{k+1} - \mathbf{Y}), \end{aligned}$$

More specifically, the iterations are

$$\boldsymbol{\beta}^{k+1} = \arg\min_{\boldsymbol{\beta}} \lambda\sum_{j=1}^p w_j|\beta_j| - \langle\theta^k, \mathbf{X}\boldsymbol{\beta}\rangle + \frac{\eta}{2}\|\mathbf{X}\boldsymbol{\beta} + \mathbf{z}^k - \mathbf{Y}\|_2^2. \tag{6.14}$$

$$\mathbf{z}^{k+1} = \arg\min_{\mathbf{z}} \mathbb{Q}_\tau(\mathbf{z}) - \langle\theta^k, \mathbf{z}\rangle + \frac{\eta}{2}\|\mathbf{z} + \mathbf{X}\boldsymbol{\beta}^{k+1} - \mathbf{Y}\|_2^2. \tag{6.15}$$

The subproblem in (6.14) can be viewed as an adaptive Lasso linear regression problem (Zou, 2006) and it can be solved efficiently via the cyclic coordinate descent algorithm (Friedman, Hastie and Tibshirani, 2010). The subproblem in (6.15) actually has a closed form solution. This property directly addresses the computational difficulty caused by the nonsmoothness of the quantile regression check loss. Note that the update of $\mathbf{z}^{k+1}$ can be carried out component-wisely. For $i = 1, \ldots, n$, we have

$$z_i^{k+1} = \arg\min_{z_i} \rho_\tau(z_i) + \frac{n\eta}{2}\left[z_i - \left(Y_i - \mathbf{X}_i^T\boldsymbol{\beta} + \frac{1}{\eta}\theta_i^k\right)\right]^2. \tag{6.16}$$

Direct calculations show that

$$z_i^{k+1} = \text{Prox}_{\rho_\tau}\left[Y_i - \mathbf{X}_i^T\boldsymbol{\beta} + \frac{1}{\eta}\theta_i^k,\, n\eta\right],\ i = 1, \ldots, n, \tag{6.17}$$

where the operator Prox is defined as

$$\text{Prox}_{\rho_\tau}[\xi, \alpha] = \begin{cases} \xi - \frac{\tau}{\alpha}, & \text{if } \xi > \frac{\tau}{\alpha} \\ 0, & \text{if } \frac{\tau-1}{\alpha} \le \xi \le \frac{\tau}{\alpha} \\ \xi - \frac{\tau-1}{\alpha}, & \text{if } \xi < \frac{\tau-1}{\alpha}. \end{cases}$$

The above ADMM procedure is named *sparse coordinate descent ADMM* (*scdADMM*) in Gu *et al.* (2018). Gu *et al.* (2018) also proposed another variant of the ADMM algorithm called the proximal ADMM by using a "linearization" trick to modify the subproblem (6.14) such that the updated $\boldsymbol{\beta}^{k+1}$ also has a closed-form formula (Parikh and Boyd, 2013). Comparing the two ADMM algorithms, Gu *et al.*(2018) found that when $p$ is large scdADMM is much more efficient than the proximal ADMM, while the latter can be more efficient for smaller dimensions.

## 6.2 Penalized Composite Quantile Regression

Consider a linear regression model

$$Y = \sum_{j=1}^{p} X_j\beta_j^* + \varepsilon, \quad \varepsilon \sim (0, \sigma^2). \tag{6.18}$$

Our aim is to find high-efficiency robust estimators, having a better ability to adapt to the unknown distribution of $\varepsilon$. As shown in the previous section, quantile regression with any $\tau$ can deliver a root-n consistent estimator of $\boldsymbol{\beta}^*$ with asymptotic variance $\frac{\tau(1-\tau)}{f^2(b_\tau^*)}\boldsymbol{\Sigma}^{-1}$. This implies that there is no universal best choice of $\tau$ for the regression. Intuitively, one may try many levels of $\tau$ and then combine these quantile regression estimators. This motivates Zou and Yuan (2008) to propose the *composite quantile regression* (*CQR*) as follows:

$$(\widehat{b}_1, \cdots, \widehat{b}_K, \widehat{\boldsymbol{\beta}}^{\text{cqr}}) = \arg\min_{b_1, \cdots, b_k, \boldsymbol{\beta}} \sum_{k=1}^{K}\left\{\sum_{i=1}^{n}\rho_{\tau_k}(Y_i - b_k - \mathbf{X}_i^T\boldsymbol{\beta})\right\}, \tag{6.19}$$

where $0 < \tau_1 < \tau_2 < \ldots < \tau_K < 1$ are $K$ different quantile levels. A natural choice is $\tau_k = \frac{k}{K+1}$ for $k = 1, \ldots, K$.

Under the regularity conditions for establishing the asymptotic normality of a single quantile regression, the asymptotic normality of the CQR estimate can be established for finite dimension $p$ (Zou and Yuan, 2008).

**Theorem 6.3** *The limiting distribution of* $\sqrt{n}(\widehat{\boldsymbol{\beta}}^{\text{cqr}} - \boldsymbol{\beta}^*)$ *is* $N(0, \boldsymbol{\Sigma}_{\text{CQR}})$ *where*

$$\boldsymbol{\Sigma}_{\text{cqr}} = \boldsymbol{\Sigma}^{-1} \frac{\sum_{k,k'=1}^{K} \min(\tau_k, \tau_{k'})(1 - \max(\tau_k, \tau_{k'}))}{(\sum_{k=1}^{K} f(b^*_{\tau_k}))^2}.$$

Note that when $\sigma^2 < \infty$, the asymptotic variance of the least squares estimator is $\sigma^2\boldsymbol{\Sigma}^{-1}$. Therefore, the asymptotic relative efficiency (ARE) of the CQR with respect to the least squares is

$$\text{ARE}(K, f) = \frac{\sigma^2(\sum_{k=1}^{K} f(b^*_{\tau_k}))^2}{\sum_{k,k'=1}^{K} \min(\tau_k, \tau_{k'})(1 - \max(\tau_k, \tau_{k'}))}. \tag{6.20}$$

It is shown that when $K$ is large, there is a lower bound to the asymptotic relative efficiency.

**Theorem 6.4**

$$\lim_{K\to\infty} \frac{\sum_{k,k'=1}^{K} \min(\tau_k, \tau_{k'})(1 - \max(\tau_k, \tau_{k'}))}{(\sum_{k=1}^{K} f(b^*_{\tau_k}))^2} = \frac{1}{12(E_\varepsilon[f(\varepsilon)])^2},$$

*and*

$$\delta(f) \equiv \lim_{K\to\infty} \text{ARE}(K, f) = 12\sigma^2(E_\varepsilon[f(\varepsilon)])^2 \geq 0.864.$$

The proofs of Theorems 6.3 and 6.4 are good exercise (Exercise 6.5). It is recommended to use $K = 19$ in practice because ARE$(19, f)$ is very close to its large $K$ limit for many commonly used error distributions.

Table 6.2 lists the ARE of CQR for some error distributions. When the error is normal, CQR is almost as efficient as OLS. For other non-normal distributions, ARE is larger than one. In particular, the mixture case has ARE $g(r)$ where $r$ represents the percentage of "contamination" to the standard normal distribution. When $r \to 0$, $g(r) \approx \frac{3}{\pi}(1 + 1/r)^2 \to \infty$. The above discussions suggest that the CQR oracle estimator is a safe and potentially more efficient estimator than the OLS oracle estimator.

Under model (6.18) and further assuming $\boldsymbol{\beta}^*$ is sparse, let $\mathcal{S}$ denote the support of $\boldsymbol{\beta}^*$. The standard oracle estimator is $(\widehat{\boldsymbol{\beta}}_{\mathcal{S}}^{\text{ols}}, 0)$ where

$$\widehat{\boldsymbol{\beta}}_{\mathcal{S}}^{\text{ols}} = \arg\min_{\boldsymbol{\beta}_{\mathcal{S}}} \sum_{i=1}^{n} (Y_i - \sum_{j\in\mathcal{S}} X_{ij}\beta_j)^2.$$

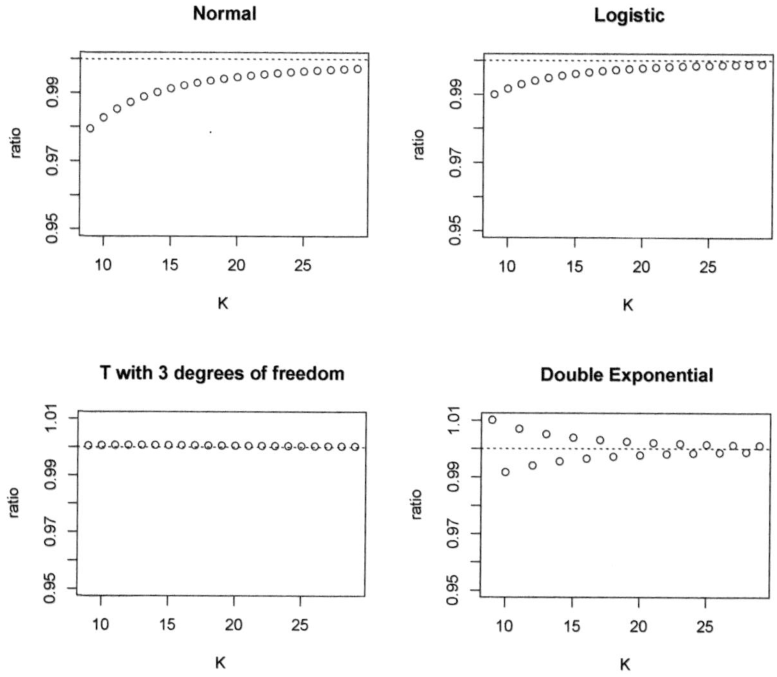


Figure 6.2: ARE$(K, f)$ of CQR as a function of $K$. The ratio is defined as $\frac{\text{ARE}(K,f)}{\delta(f)}$. When $K = 19$, the ratio is very close to 1 in all four examples. The example of double exponential distribution also shows that ARE$(K, f)$ can be larger than its limit.

Table 6.2: ARE of CQR for some error distributions.

| err. dist. | normal | $t_3$ | Laplace | Logistic | mixtures |
|---|---|---|---|---|---|
| ARE | 0.955 | 1.9 | 1.5 | 1.097 | $g(r)$ |

For mixtures normal $(1-r)N(0,1) + rN(0, r^6)$,
$g(r) = \frac{3}{\pi}\left((1-r)^2 + \frac{1}{r} + \frac{2\sqrt{2}r(1-r)}{\sqrt{1+r^6}}\right)^2 (1 - r + r^7)$.

The CQR oracle estimator is $(\widehat{\boldsymbol{\beta}}_{\mathcal{S}}^{\text{cqr}}, 0)$ defined as

$$\min_{b_1,\cdots,b_K,\boldsymbol{\beta}} \sum_{k=1}^{K} \left\{ \sum_{i=1}^{n} \rho_{\tau_k}(Y_i - b_k - \sum_{j\in\mathcal{S}} X_{ij}\beta_j) \right\}. \tag{6.21}$$

To approximate the oracle CQR estimator, we can use the folded concave penalty or the adaptive Lasso penalty. The penalized CQR estimator is defined as

$$\min_{b_1,\cdots,b_K,\boldsymbol{\beta}^{\text{pcqr}}} \sum_{k=1}^{K} \left\{ \frac{1}{n} \sum_{i=1}^{n} \rho_{\tau_k}(Y_i - b_k - \mathbf{X}_i^T \boldsymbol{\beta}) \right\} + \sum_{j=1}^{p} p_\lambda(|\beta_j|). \tag{6.22}$$

For the adaptive Lasso penalized CQR estimator, Zou and Yuan (2008) proved its asymptotic normality and weak oracle property for the fixed $p$ case. Bradic, Fan and Wang (2011) established the *biased oracle* property and asymptotic normality of the composite quasi-likelihood in a high-dimensional setting with weighted $L_1$-penalty. They also offered the optimal choice of weighting scheme. Gu and Zou (2019) established the strong oracle property of the SCAD penalized CQR under ultra-high dimensions. See also Gu (2017). The ADMM algorithm for penalized quantile regression introduced in the last section can be extended for penalized composite quantile regression by introducing new variables $\mathbf{z}_k = \mathbf{Y} - \mathbf{X}\boldsymbol{\beta} - b_k \mathbf{1}_n$ for $k = 1, \cdots, K$. It is a good exercise to extend the ADMM algorithm for penalized CQR (Exercise 6.6)

## 6.3 Variable Selection in Robust Regression

When the random error in the linear regression model follows a heavy-tail distribution, and/or there are potential outliers in the data, the *least absolute deviation* (LAD) regression is a natural alternative to the least squares method, which is not robust to outliers, and breaks down in the presence of infinite error variance. LAD can be viewed as a special case of quantile regression where $\tau = 0.5$. There are a large number of publications on robust regression in the literature. This section provides a brief introduction of robust regression and penalized Huber regression.

### 6.3.1 *Robust regression*

*Huber regression* based on Huber loss is another important technique for achieving robustness. Huber loss (Huber, 1964) can be viewed as a blend between the LAD loss and the squared error loss. The mathematical definition of Huber loss is given below

$$\ell_\delta(r) = \begin{cases} 2\delta|r| - \delta^2, & \text{if } |r| > \delta, \\ r^2, & \text{if } |r| \le \delta. \end{cases} \tag{6.23}$$

As $\delta$ varies from zero to $\infty$, $\ell_\delta(\cdot)$ changes from the LAD loss to the squared error loss. The Huber regression estimator is constructed via the following convex optimization problem:

$$\widehat{\boldsymbol{\beta}}^{\text{Huber}} = \arg\min_{\boldsymbol{\beta}} \frac{1}{n} \sum_{i=1}^{n} \ell_\delta(Y_i - \mathbf{X}_i^T \boldsymbol{\beta}). \tag{6.24}$$

The parameter $\delta$ is set to be $1.345\sigma$ such that when the error distribution is normal the Huber estimator is 95% efficient. In practice, $\sigma$ is estimated by $\widehat{s} = 1.483\text{median}\{|r_i|\}_{i=1}^n$, and $r_i$ is the OLS residual. Huber regression is computationally friendly owing to its smoothness. For example, one can directly apply a gradient-based algorithm to solve (6.24).

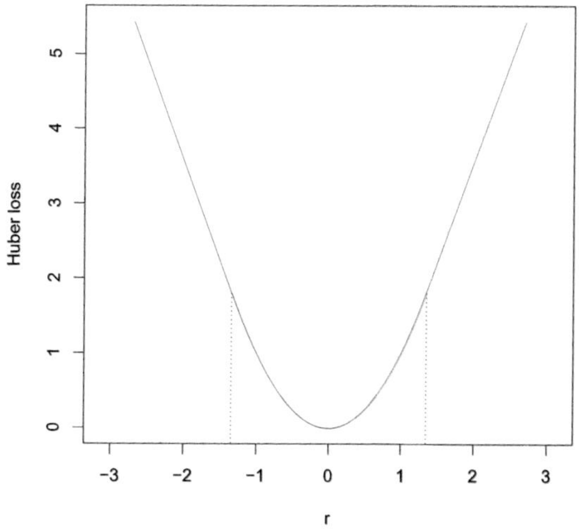


Figure 6.3: The Huber loss with $\delta = 1.345$.

Huber regression can be generalized to the so-called $M$ estimator (Huber, 1973) defined as

$$\widehat{\boldsymbol{\beta}}^{\text{M}} = \arg\min_{\boldsymbol{\beta}} \frac{1}{n}\sum_{i=1}^{n} \rho(Y_i - \mathbf{X}_i^T\boldsymbol{\beta}), \tag{6.25}$$

where $\rho(r)$ satisfies the following properties

(i). $\rho(\cdot)$ is convex,

(ii). $\rho(\cdot)$ is symmetric around zero, $\rho(0) = 0$,

(iii). $\rho(\cdot)$ is increasing for $t \in [0, \infty)$,

(iv). $\rho(r)$ is differentiable and let $\psi(r) = \rho'(r)$.

Suppose that the error distribution satisfies $\text{E}\{\psi(\varepsilon)\} = 0$, which obviously holds if the error distribution is symmetric around zero. The $M$ estimator is root-$n$ consistent and asymptotically normal with covariance $\frac{\text{E}[\psi^2(\varepsilon)]}{(\text{E}[\psi'])^2}\boldsymbol{\Sigma}^{-1}$, where $\boldsymbol{\Sigma} = \lim_{n\to\infty} n^{-1}\mathbf{X}^T\mathbf{X}$. We leave this as an exercise (Exercise 6.7).

Rousseeuw (1984) proposed the *least median of squares* regression (LMS) that directly minimizes the median of squared (or absolute) residuals. The

LMS estimator is defined as

$$\widehat{\boldsymbol{\beta}}^{\text{LMS}} = \arg\min_{\boldsymbol{\beta}} \text{median}\{(Y_i - \mathbf{X}_i^T \boldsymbol{\beta})^2\}_{i=1}^n. \tag{6.26}$$

One can also directly remove the large residuals (due to outliers) from the minimization criterion function. The *least trimmed squares* (LTS) solves the following problem:

$$\widehat{\boldsymbol{\beta}}^{\text{LTS}} = \arg\min_{\boldsymbol{\beta}} \sum_{i=1}^{K} r_{(i)}^2, \tag{6.27}$$

where $r_{(i)}$ is the ordered $i$th statistic of $\mathbf{r} = \mathbf{Y} - \mathbf{X}^T\boldsymbol{\beta}$. When $K = n$, then LTS reduced to OLS. If $K = 0.9n$, LTS removes the top 10% of the squared residuals.

LMS and LTS are computationally much more challenging than LAD and Huber regression. LMS is $n^{1/3}$ consistent (Rousseeuw, 1984) and LTS is $n^{1/2}$ consistent and asymptotic normal (Rousseeuw and Leroy, 1987). In order to improve efficiency, it is recommended to use LMS or LTS as the initial estimator and then do one-step $M$ estimation (Bickel, 1975) or compute a re-weighted least squares

$$\arg\min_{\boldsymbol{\beta}} \sum_{i=1}^{n} w_i (Y_i - \mathbf{X}_i^T \boldsymbol{\beta})^2$$

with

$$w_i = \begin{cases} |r_i|^{-1}, & \text{if } |r| \le 3\widehat{s} \\ 0, & \text{if } |r| > 3\widehat{s} \end{cases}$$

where $r_i$ is the residual from the initial estimator and $\widehat{s} = 1.483\text{median}_i|r_i|$ is a robust estimate of $\sigma$.

### 6.3.2 Variable selection in Huber regression

When $p$ is large, *penalized Huber regression* is a natural alternative to penalized least squares when the error distribution has heavy tails. Lambert-Lacroix and Zwald (2011) considered the adaptive Lasso penalized Huber regression. A more recent paper by Fan, Li and Wang (2017) suggested using a adaptive Huber loss in penalized regression. To use their notation, the *Huber loss* is defined as

$$\ell_\alpha(r) = \begin{cases} 2\alpha^{-1}|r| - \alpha^{-2}, & \text{if } |r| > \alpha^{-1} \\ r^2, & \text{if } |r| \le \alpha^{-1} \end{cases} \tag{6.28}$$

Comparing (6.23) and (6.28), we see $\alpha = \delta^{-1}$. The classical Huber regression uses a fixed $\delta$ or $\alpha$ (independent of $n$ and $p$). Fan, Li and Wang (20017) treated $\alpha$ as a tuning parameter and named $\ell_\alpha(\cdot)$ robust approximate quadratic loss or *adaptive Huber loss*. The motivation is to obtain a consistent estimate of the mean when the error distribution is not necessarily symmetric and to have an exponential concentration property. Consider the linear regression model

$$Y = \sum_{j=1}^{p} X_j \beta_j^* + \varepsilon,$$

with $\mathrm{E}\,\varepsilon = 0$ and $\mathrm{Var}(\varepsilon) = \sigma^2$. The *Lasso penalized Huber estimator* is given by

$$\widehat{\boldsymbol{\beta}}_\alpha = \arg\min_{\boldsymbol{\beta}} \frac{1}{n} \sum_{i=1}^{n} \ell_\alpha(Y_i - \mathbf{X}_i^T \boldsymbol{\beta}) + \lambda_n \sum_{j=1}^{p} |\beta_j|. \tag{6.29}$$

Define

$$\boldsymbol{\beta}_\alpha^* = \arg\min_{\boldsymbol{\beta}} \mathrm{E}[\ell_\alpha(Y - \mathbf{X}^T \boldsymbol{\beta})]. \tag{6.30}$$

In general $\boldsymbol{\beta}_\alpha^* \neq \boldsymbol{\beta}^*$. By the triangle inequality

$$\|\widehat{\boldsymbol{\beta}} - \boldsymbol{\beta}^*\| \le \|\widehat{\boldsymbol{\beta}}_\alpha - \boldsymbol{\beta}_\alpha^*\| + \|\boldsymbol{\beta}_\alpha^* - \boldsymbol{\beta}^*\|.$$

When $\alpha$ approaches zero, $\boldsymbol{\beta}_\alpha^*$ is expected to approach $\boldsymbol{\beta}^*$. Thus, without assuming the conditional error distribution given $\mathbf{X}$ is symmetric around zero, it is still possible for the Huber estimator to estimate $\boldsymbol{\beta}^*$ consistently.

More specifically, Fan, Li and Wang (2017) showed the following theorem.

**Theorem 6.5** *Under the following conditions:*

*(C1).* $\mathrm{E}\{\mathrm{E}(|\varepsilon|^k \mid \mathbf{X})\}^2 \le M_k < \infty$, *for some* $k \ge 2$.

*(C2).* $0 < \kappa_l \le \lambda_{\min}(\mathrm{E}[\mathbf{X}\mathbf{X}^T]) \le \lambda_{\max}(\mathrm{E}[\mathbf{X}\mathbf{X}^T]) \le \kappa_u < \infty$,

*(C3). For any* $\nu$, $\mathbf{X}^T \nu$ *is sub-Gaussian with parameter at most* $\kappa_0^2 \|\nu\|^2 m$, *i.e.,*

$$\mathrm{E}(\exp(t\mathbf{X}^T \nu)) \le \exp(t^2 \kappa_0^2 \|\nu\|^2 / 2), \quad \forall t$$

*there exists a universal positive constant* $C_1$, *such that*

$$\|\boldsymbol{\beta}_\alpha^* - \boldsymbol{\beta}^*\| \le C_1 \sqrt{\kappa_u} \kappa_l^{-1} (\kappa_0^k + \sqrt{M_k}) \alpha^{k-1}.$$

For the sparse least squares estimator, the assumption is that $\boldsymbol{\beta}^*$ belongs to an $\ell_q$ $(0 < q < 1)$ ball. Assume that

$$\sum_{j=1}^{p} |\beta_j^*|^q \le R_q / 2, \tag{6.31}$$

then

$$\begin{aligned}
\sum_{j=1}^{p} |\beta_{\alpha,j}^*|^q &\le \sum_{j=1}^{p} |\beta_{\alpha,j}^* - \beta_j^*|^q + \sum_{j=1}^{p} |\beta_j^*|^q \\
&\le p^{(2-q)/2} (\|\boldsymbol{\beta}_\alpha^* - \boldsymbol{\beta}^*\|^2)^{q/2} + \sum_{j=1}^{p} |\beta_j^*|^q \\
&\le O(\alpha^{2(k-1)}) + R_q / 2.
\end{aligned}$$

By Theorem 6.5, $\|\boldsymbol{\beta}_\alpha^* - \boldsymbol{\beta}^*\|^2 = O(\alpha^{2(k-1)})$. Thus, (6.31) implies

$$\sum_{j=1}^p |\beta_{\alpha,j}^*|^q \le R_q, \tag{6.32}$$

for $\alpha \le c\{R_q p^{-(2-q)/2}\}^{1/(q(k-1))}$. The following theorem is established in Fan, Li and Wang (2017).

**Theorem 6.6** *Define* $\kappa_\lambda = \frac{\lambda_n}{\sqrt{\log(p)/n}}$ *and choose* $c_l\sqrt{\log(p)/n} \le \alpha \le c_u\rho_2^{-1}$ *for some constants* $c_l, c_u, \rho_2$. *Under conditions (C1)–(C3), assume*

$$8\kappa_2\kappa_\lambda^{-q/2}\sqrt{R_q}\left(\frac{\log(p)}{n}\right)^{(1-q)/2} \le 1 \tag{6.33}$$

*for some universal positive constant* $\kappa_2$. *With probability at least* $1 - c_1\exp(-c_2 n)$,

$$\|\widehat{\boldsymbol{\beta}}_\alpha - \boldsymbol{\beta}_\alpha^*\| \le d_2\sqrt{R_q}\left(\frac{\log(p)}{n}\right)^{\frac{1}{2}-\frac{q}{4}}$$

*and*

$$\|\widehat{\boldsymbol{\beta}}_\alpha - \boldsymbol{\beta}^*\| \le d_1\alpha^{(k-1)} + d_2\sqrt{R_q}\left(\frac{\log(p)}{n}\right)^{\frac{1}{2}-\frac{q}{4}},$$

*where* $c_1, c_2, d_1, d_2$ *are positive constants.*

When the conditional distribution of $\varepsilon|\mathbf{X}$ is symmetric around zero, $\boldsymbol{\beta}^* = \boldsymbol{\beta}_\alpha^*$ for any $\alpha$. If the error distribution has heavy tails, one can use a large $\alpha$ to robustify the estimation. Theorem 6.6 implies that $\|\widehat{\boldsymbol{\beta}}_\alpha - \boldsymbol{\beta}^*\| \le d_2\sqrt{R_q}\left(\frac{\log(p)}{n}\right)^{\frac{1}{2}-\frac{q}{4}}$ which is the minimax rate (Raskutti, Wainwright and Yu, 2011) under the light tails.

If the conditional distribution of $\varepsilon|\mathbf{X}$ is asymmetric around zero, the Huber estimator still achieves the optimal rate provided that $\alpha \le \{d_1^{-1}d_2R_q[\log(p)/n]^{1-q/2}\}^{\frac{1}{2(k-1)}}$. Theorem 6.6 also requires $c_l\sqrt{\log(p)/n} \le \alpha \le c_u\rho_2^{-1}$. The assumption (6.33) guarantees the existence of such $\alpha$. The beauty of this theory is that it only requires the conditional error distribution to have a finite second moment, not the typical sub-Gaussian or other light tail assumptions.

## 6.4 Rank Regression and Its Variable Selection

Suppose that $\{\mathbf{X}_i, Y_i\}$, $i = 1, \cdots, n$ is a random sample from a linear regression model

$$Y = \alpha + \mathbf{X}^T\boldsymbol{\beta} + \varepsilon,$$

where the random error $\varepsilon$ is independent of $x$ and has a pdf $f(\cdot)$.

### 6.4.1 Rank regression

Denote $e_i = Y_i - \mathbf{X}_i^T\boldsymbol{\beta}$. Let $R_i(\boldsymbol{\beta})$ be the rank of $e_i$ among $e_1, \ldots, e_n$. Wilcoxon's *rank regression* (Jaeckel, 1972) is defined as

$$\widehat{\boldsymbol{\beta}}^{\text{Wilcoxon}} = \arg\min_{\boldsymbol{\beta}} \left( \frac{R_i(\boldsymbol{\beta})}{n+1} - \frac{1}{2} \right) (Y_i - \mathbf{X}_i^T\boldsymbol{\beta}). \tag{6.34}$$

This is the original definition of rank regression. It is intuitive, but it is difficult to carry out the corresponding minimization problem. Fortunately, it has an equivalent formulation

$$\widehat{\boldsymbol{\beta}}^{\text{Wilcoxon}} = \arg\min_{\boldsymbol{\beta}} \frac{1}{n^2} \sum_{i<j}^{n} |e_i - e_j|. \tag{6.35}$$

See, for example, Hettmansperger and McKean (1998). From (6.35), rank regression is robust to outliers along the direction of errors, and it is is one type of robust regression. It has been shown that it is also robust against the possible non-normal distributed error. (6.35) implies that the intercept $\alpha$ in the linear regression model cannot be estimated by rank regression. Since $\alpha = \mathrm{E}(y) - \boldsymbol{\beta}^T \mathrm{E}(\mathbf{X})$, a natural estimate of $\alpha$ is $\bar{y} - \bar{\mathbf{X}}^T\widehat{\boldsymbol{\beta}}^{\text{Wilcoxon}}$, where $\bar{y}$ and $\bar{\mathbf{X}}$ are the sample mean of $y$'s and $\mathbf{X}$'s respectively.

The rank regression defined in (6.35) is robust to only the direction of response rather than covariates. To achieve the robustness in both directions of the response and the covariates, we consider *weighted rank regression*

$$\widehat{\boldsymbol{\beta}}^{\text{Wilcoxon}} = \arg\min_{\boldsymbol{\beta}} \frac{1}{n^2} \sum_{i<j}^{n} b(\mathbf{X}_i, \mathbf{X}_j)|e_i - e_j|. \tag{6.36}$$

A non-uniform weight choice can be $b(\mathbf{X}_i, \mathbf{X}_j) = h(\mathbf{X}_i)h(\mathbf{X}_j)$ with

$$h(\mathbf{X}_i) = \min\{1, \frac{b_0}{(\mathbf{X}_i - \widehat{\boldsymbol{\mu}})^T\mathbf{S}^{-1}(\mathbf{X}_i - \widehat{\boldsymbol{\mu}})}\},$$

where $(\widehat{\boldsymbol{\mu}}, \mathbf{S})$ is the robust minimum volume ellipsoid estimator of the location and scatter (Rousseeuw and van Zomeren, 1990), and $b_0$ is the 95th percentile of $\chi^2_p$, where $p$ is the dimension of $\mathbf{X}$.

### 6.4.2 Penalized weighted rank regression

Similar to penalized quantile regression, *penalized weighted rank regression* can be defined to be

$$\widehat{\boldsymbol{\beta}}^{\text{prank}} = \arg\min_{\boldsymbol{\beta}} \frac{1}{n} \sum_{i<j}^{n} b(\mathbf{X}_i, \mathbf{X}_j)|e_i - e_j| + n \sum_{j=1}^{p} p_\lambda(|\beta_j|). \tag{6.37}$$

Penalized weighted rank regression inherits the merits of weighted rank regression. $\widehat{\boldsymbol{\beta}}^{\text{prank}}$ is robust against outliers in both directions of the response and covariates. When the error is normally distributed, $\widehat{\boldsymbol{\beta}}^{\text{prank}}$ is almost as efficient as the penalized least squares estimator. When the error is non-normal, the *relative efficiency* of $\widehat{\boldsymbol{\beta}}^{\text{prank}}$ with respect to the penalized least squares estimator has a lower bound of 0.864. Recall that the penalized CQR estimator also possesses this robust efficiency property. The loss function in (6.37) seems to be easier to handle than the composite quantile regression loss which is the sum of many nonsmooth check loss functions (see, for example, Wang and Li, 2009)

Wang and Li (2009) focused on the *SCAD penalized rank regression* and studied the theoretical properties of the resulting estimator. To state the theoretical properties of $\widehat{\boldsymbol{\beta}}^{\text{prank}}$, some notation and assumptions are needed. Let $\mathbf{W}$ be a $n \times n$ matrix with entries $w_{ij}$, $1 \le i, j \le n$ such that

$$w_{ij} = \begin{cases} -n^{-1} b(X_i, X_j), & \text{if } i \neq j \\ n^{-1} \sum_{k \neq i} b(X_i, X_k), & \text{if } i = j \end{cases}$$

Assume that $\frac{1}{n}\mathbf{X}^T\mathbf{X} \xrightarrow{P} \boldsymbol{\Sigma}$, $\frac{1}{n}\mathbf{X}^T\mathbf{W}\mathbf{X} \xrightarrow{P} \boldsymbol{\Sigma}$, $\frac{1}{n}\mathbf{X}^T\mathbf{W}^2\mathbf{X} \xrightarrow{P} \mathbf{V}$. Further assume the error density $f(\cdot)$ is absolutely continuous with finite Fisher information. Let $\mathcal{S} = \{j : \beta_j^* \neq 0\}$ and $s = |\mathcal{S}|$. Let $\lambda_n$ satisfy $\lambda_n \to 0, \sqrt{n}\lambda_n \to \infty$ as $n \to \infty$. For fixed and finite $p$, the SCAD penalized rank estimator must satisfy

$$P(\widehat{\boldsymbol{\beta}}_{\mathcal{S}^c}^{\text{prank}} = 0)$$

and

$$\sqrt{n}(\widehat{\boldsymbol{\beta}}_{\mathcal{S}}^{\text{prank}} - \boldsymbol{\beta}_{\mathcal{S}}^*) \xrightarrow{D} N(0, \tau^2 \boldsymbol{\Sigma}_{\mathcal{SS}}^{-1} \mathbf{V}_{\mathcal{SS}} \boldsymbol{\Sigma}_{\mathcal{SS}}^{-1})$$

with $\tau = (\sqrt{12} \int f^2(u) du)^{-1}$. This is due to Wang and Li (2009), and is left as an exercise (Exercise 6.9)

The case with $b(X_i, X_j) = 1$ is particularly important and interesting. In this case, $\mathbf{V}_{\mathcal{SS}} = \boldsymbol{\Sigma}_{\mathcal{SS}}$, so the asymptotic covariance becomes $\tau^2 \boldsymbol{\Sigma}_{\mathcal{SS}}^{-1}$. Then, the relative efficiency with respect to the SCAD penalized least squares is equal to $\text{ARE} = 12\sigma^2 [\int f^2(u) du]^2$, which is exactly the ARE of the CQR over least squares. We know $\text{ARE} \ge 0.864$ and can be often greater than one for non-normal errors.

## 6.5 Variable Selection for Survival Data

The penalized likelihood methodology introduced in Chapter 5 has been extended for variable selection in the *survival data analysis*. This section is devoted to variable selection for the most commonly-used models in the survival data analysis. Let us begin with some definition and notation used in the survival data analysis. Let $T$ and $C$ be the survival time and censoring time, respectively, and $\delta = I(T \le C)$ be the event indicator. The observed time is

$Z = \min\{T, C\}$. Let $\mathbf{X}$ be the $p$-dimensional covariate vector. It is assumed throughout this section that the censoring mechanism is noninformative. That is, given $\mathbf{X}$, $T$ and $C$ are conditionally independent. Define

$$f(t|\mathbf{x}), \quad S(t|\mathbf{x}) = P(T \geq t|\mathbf{x}) \quad \text{and} \quad h(t|\mathbf{x}) = \frac{f(t|\mathbf{x})}{S(t|\mathbf{x})}$$

to be the conditional density function, the conditional *survivor function* and the conditional *hazard function* of $T$ given $\mathbf{X} = \mathbf{x}$, respectively.

Suppose that the observed data $\{(\mathbf{X}_i, Z_i, \delta_i); i = 1, \ldots, n\}$ is a random sample from $(\mathbf{X}, Z, \delta)$; a complete likelihood of the data is given by

$$L = \prod_u f(Z_i|\mathbf{X}_i) \prod_c S(Z_i|\mathbf{X}_i) = \prod_u h(Z_i|\mathbf{X}_i) \prod_{i=1}^n S(Z_i|\mathbf{X}_i), \tag{6.38}$$

where the subscripts $c$ and $u$ denote the product of the censored and uncensored data, respectively. The likelihood function (6.38) is the foundation for developing statistical inference procedures in the survival data analysis. This section focuses on variable selection in proportional hazard models.

Consider *proportional hazards model*,

$$h(t|\mathbf{x}) = h_0(t) \exp(\mathbf{x}^T \boldsymbol{\beta}), \tag{6.39}$$

with the baseline hazard functions $h_0(t)$ and parameter $\boldsymbol{\beta}$. Let $t_1^0 < \cdots < t_N^0$ denote the ordered observed failure times. Let $(j)$ provide the label for the item falling at $t_j^0$ so that the covariates associated with the $N$ failures are $\mathbf{X}_{(1)}, \ldots, \mathbf{X}_{(N)}$. Let $R_j$ denote the risk set right before the time $t_j^0$ :

$$R_j = \{i : Z_i \geq t_j^0\}.$$

The likelihood in (6.38) becomes

$$L = \prod_{i=1}^N h_0(Z_{(i)}) \exp(\mathbf{X}_{(i)}^T \boldsymbol{\beta}) \prod_{i=1}^n \exp\{-H_0(Z_i) \exp(\mathbf{X}_i^T \boldsymbol{\beta})\}, \tag{6.40}$$

where $H_0(\cdot)$ is the cumulative baseline hazard function.

If the baseline hazard function has a parametric form, $h_0(\boldsymbol{\theta}, \cdot)$, then the corresponding penalized log-likelihood function is

$$\sum_{i=1}^N [\log\{h_0(\boldsymbol{\theta}, Z_{(i)})\} + \mathbf{X}_{(i)}^T \boldsymbol{\beta}] - \sum_{i=1}^n \{H_0(\boldsymbol{\theta}, Z_i) \exp(\mathbf{X}_i^T \boldsymbol{\beta})\} - n \sum_{j=1}^d p_\lambda(|\beta_j|). \tag{6.41}$$

Maximizing (6.41) with respect to $(\boldsymbol{\theta}, \boldsymbol{\beta})$ yields the maximum penalized likelihood estimator.

### 6.5.1 Partial likelihood

The baseline hazard function $h_0(t)$ in the Cox model is assumed to be unknown. Following *Breslow's idea*, consider the "least informative" nonparametric modeling for $H_0(\cdot)$, in which $H_0(t)$ has a possible jump $h_j$ at the observed failure time $t_j^0$. More precisely, let $H_0(t) = \sum_{j=1}^N h_j I(t_j^0 \le t)$. Then $H_0(Z_i) = \sum_{j=1}^N h_j I(i \in R_j)$. Thus, the logarithm of likelihood function in (6.40) becomes

$$\sum_{j=1}^N \{\log(h_j) + \mathbf{X}_{(j)}^T \boldsymbol{\beta}\} - \sum_{i=1}^n \left\{ \sum_{j=1}^N h_j I(i \in R_j) \exp(\mathbf{X}_i^T \boldsymbol{\beta}) \right\}. \tag{6.42}$$

Taking the derivative with respect to $h_j$ and setting it to be zero, we obtain that $\widehat{h}_j = \{\sum_{i \in R_j} \exp(\mathbf{X}_i^T \boldsymbol{\beta})\}^{-1}$. Substituting $\widehat{h}_j$ into (6.42), we obtain the logarithm of the partial likelihood function

$$\ell_p(\boldsymbol{\beta}) = \sum_{j=1}^N \left[ \mathbf{X}_{(j)}^T \boldsymbol{\beta} - \log \left\{ \sum_{i \in R_j} \exp(\mathbf{X}_i^T \boldsymbol{\beta}) \right\} \right], \tag{6.43}$$

after dropping a constant term "$-N$". The partial likelihood was first proposed by Cox (1975). Andersen and Gill (1982) established the asymptotical normality of the partial likelihood estimate by using the theory of counting process. Define $N_i(t) = I\{T_i \le t, T_i \le C_i\}$, and $Y_i(t) = I\{T_i \ge t, C_i \ge t\}$. Following Andersen and Gill (1982), the covariate $\mathbf{X}$ is allowed to be time-dependent, denoted by $\mathbf{X}(t)$, and assume that $T$ has a bounded support $[0, \tau]$. Without loss of generality, assume $\tau = 1$ through this section. One may extend the results to the interval $[0, \infty)$ using techniques developed in Andersen and Gill (1982). The following conditions, which are adapted from Fan and Li (2002), guarantee the asymptotic normality of the maximum partial likelihood estimates.

**Regularity Conditions for Partial Likelihood**

(A1) Assume that $T$ has a bounded support $[0, 1]$, and $\int_0^1 h_0(t)\, dt < \infty$.

(A2) The processes $\mathbf{X}(t)$ and $Y(t)$ are left-continuous with right limits, and

$$P\{Y(t) = 1 \quad \forall t \in [0, 1]\} > 0.$$

(A3) There exists a neighborhood $\mathcal{B}$ of $\boldsymbol{\beta}_0$, the true value of $\boldsymbol{\beta}$, such that

$$\mathrm{E} \sup_{t \in [0,1], \boldsymbol{\beta} \in \mathcal{B}} Y(t) \mathbf{X}(t)^T \mathbf{X}(t) \exp(\boldsymbol{\beta}^T \mathbf{X}(t)\} < \infty.$$

(A4) Define

$$\begin{aligned} s^{(0)}(\boldsymbol{\beta}, t) &= \mathrm{E}\, Y(t) \exp\{\boldsymbol{\beta}^T \mathbf{X}(t)\}, \\ s^{(1)}(\boldsymbol{\beta}, t) &= \mathrm{E}\, Y(t) \mathbf{X}(t) \exp\{\boldsymbol{\beta}^T \mathbf{X}(t)\}, \\ s^{(2)}(\boldsymbol{\beta}, t) &= \mathrm{E}\, Y(t) \mathbf{X}(t) \mathbf{X}(t)^T \exp\{\boldsymbol{\beta}^T \mathbf{X}(t)\}, \end{aligned}$$

where $s^{(0)}(\cdot,t)$, $s^{(1)}(\cdot,t)$ and $s^{(2)}(\cdot,t)$ are continuous of $\boldsymbol{\beta}\in\mathcal{B}$, uniformly in $t\in[0,1]$. $s^{(0)}$, $s^{(1)}$ and $s^{(2)}$ are bounded on $\mathcal{B}\times[0,1]$; $s^{(0)}$ is bounded away from zero on $\mathcal{B}\times[0,1]$. The matrix

$$I(\boldsymbol{\beta}_0)=\int_0^1 v(\boldsymbol{\beta}_0,t)s^{(0)}(\boldsymbol{\beta}_0,t)h_0(t)\,dt \tag{6.44}$$

is finite positive definite, where

$$v(\boldsymbol{\beta},t)=\frac{s^{(2)}}{s^{(0)}}-\left(\frac{s^{(1)}}{s^{(0)}}\right)\left(\frac{s^{(1)}}{s^{(0)}}\right)^T.$$

Under Conditions (A1)-(A4), the Barlett first and second identities are still valid. That is,

$$\mathrm{E}\left\{\frac{\partial\ell_p(\boldsymbol{\beta})}{\partial\boldsymbol{\beta}}\right\}=\mathbf{0},$$

$$\mathrm{E}\left\{\frac{\partial^2\ell_p(\boldsymbol{\beta})}{\partial\boldsymbol{\beta}\partial\boldsymbol{\beta}^T}\right\}=-\mathrm{E}\left(\left\{\frac{\partial\ell_p(\boldsymbol{\beta})}{\partial\boldsymbol{\beta}}\right\}\left\{\frac{\partial\ell_p(\boldsymbol{\beta})}{\partial\boldsymbol{\beta}}\right\}^T\right).$$

Thus, the maximum partial likelihood estimator is root $n$ consistent and efficient under some regularity conditions.

### 6.5.2 *Variable selection via penalized partial likelihood and its properties*

As a natural extension of penalized likelihood defined in Section 5.3, we may defined *penalized partial likelihood* as follows.

$$\ell_p(\boldsymbol{\beta})-n\sum_{j=1}^p p_\lambda(|\beta_j|). \tag{6.45}$$

Tibshirani (1997) proposed the penalized partial likelihood with the Lasso penalty in (6.44). Replacing Conditions (C1)—(C3) in Theorems 5.1 and 5.2 by Conditions (A1)-(A4), Fan and Li (2002) established the oracle property of the penalized partial likelihood estimator with the folded concave penalty under the setting of Section 5.7. Using the notation in Section 5.7, when $p$ is fixed and finite, if $\lambda_n\to 0$ and $\sqrt{n}\lambda_n\to\infty$ for the SCAD and MCP penalties, their corresponding penalized partial likelihood estimators enjoy the oracle property:

$$\widehat{\boldsymbol{\beta}}_2=0\quad\text{and}\quad\sqrt{n}(\widehat{\boldsymbol{\beta}}_1-\boldsymbol{\beta}_{10})\xrightarrow{D}N(\mathbf{0},\mathbf{I}_1(\boldsymbol{\beta}_{10})), \tag{6.46}$$

where $\mathbf{I}_1(\boldsymbol{\beta}_{10})=\mathbf{I}_1(\boldsymbol{\beta}_{10},\mathbf{0})$, the corresponding information matrix $\mathbf{I}(\boldsymbol{\beta}_0)$ defined in (6.44) knowing $\boldsymbol{\beta}_{20}=\mathbf{0}$.

The strategy for tuning parameter selection for penalized likelihood in Section 5.5 can be extended for the penalized partial likelihood. Li, Ren,

Yang and Yu (2018) studied the tuning parameter selection for the penalized partial likelihood. Interestingly they found that the partial likelihood function behaves differently from the likelihood function of independent and identically distributed samples. The result implies that Condition (C4) in Section 5.7 is invalid for the partial likelihood function.

Bradic, Fan and Jiang (2011) studied the asymptotic behavior of the penalized partial likelihood estimator in the presence of ultrahigh dimensional covariates by using the formulation of Fan and Lv (2011).

Bradic, Fan and Jiang (2011) first derive the appropriate necessary and sufficient conditions on the existence of the penalized partial likelihood estimate in a similar way to that in Fan and Lv (2011), and further studied the global optimality of the oracle estimator, which is defined as the maximizer of the penalized partial likelihood function with respect to $\boldsymbol{\beta}_1$ with $\boldsymbol{\beta}_2 = \mathbf{0}$.

To introduce some interesting results in Bradic, Fan and Jiang (2011), we need to introduce extra notation below. Denote $S_n^{(l)}(\boldsymbol{\beta}, t)$, $l = 0, 1$ and 2 to be the sample counterpart of $s_{(l)}(\boldsymbol{\beta}, t)$. For instance,

$$S_n^{(1)}(\boldsymbol{\beta}, t) = \frac{1}{n}\sum_{i=1}^{n} Y_i(t)\mathbf{X}_i(t)\exp\{\boldsymbol{\beta}^T\mathbf{X}(t)\}.$$

Denote by

$$\mathbf{e}(\boldsymbol{\beta}, t) = s^{(1)}(\boldsymbol{\beta}, t)/s^{(0)}(\boldsymbol{\beta}, t) \quad \text{and} \quad \mathrm{E}_n(\boldsymbol{\beta}, t) = S_n^{(1)}(\boldsymbol{\beta}, t)/S_n^{(0)}(\boldsymbol{\beta}, t).$$

Define

$$c_n = \sup_{t\in[0,1]} \|\mathrm{E}_n(\boldsymbol{\beta}_0, t) - \mathbf{e}(\boldsymbol{\beta}_0, t)\|_\infty \quad \text{and} \quad d_n = \sup_{t\in[0,1]} |S_n^{(0)}(\boldsymbol{\beta}_0, t) - s^{(0)}(\boldsymbol{\beta}_0, t)).$$

Denote by

$$r_{ij} = \int_0^1 \{X_{ij}(t) - e_j(\boldsymbol{\beta}_0, t)\}\, dM_i(t),$$

where $e_j(\boldsymbol{\beta}_0, t)$ is the $j$-th component of $\mathbf{e}(\boldsymbol{\beta}_0, t)$, and $M_i(t) = N_i(t) - H_i(t)$ is a martingale with compensator $H_i(t) = \int_0^t h(u|\mathbf{X}_i)\, du$. Let $\boldsymbol{\xi} = (\xi_1, \cdots, \xi_p)^T$ be the score vector of the partial likelihood function. That is,

$$\boldsymbol{\xi} = \sum_{i=1}^{n}\int_0^1 \{\mathbf{X}_i - \mathrm{E}_n(\boldsymbol{\beta}_0, t)\}\, dN_i(t).$$

**Theorem 6.7** *In addition to Assumptions (A1)—(A4), assume that the random sequences $c_n$ and $d_n$ are bounded almost surely, and*

$$\mathrm{E}\,|r_{ij}|^m \le m!M^{m-2}\sigma_j^2/2 \tag{6.47}$$

*for all $j$, where $M > 0$ is a constant, $m \ge 2$ and $\sigma_j^2 = \mathrm{Var}(r_{ij}) < \infty$. Then for any positive sequence $\{u_n\}$ bounded away from zero, there exist positive constants $c_0$ and $c_1$ such that*

$$P(|\xi_j| > \sqrt{n}u_n) \le c_0\exp(-c_1 u_n) \tag{6.48}$$

*uniformly over $j$, if $v_n = \max_j \sigma_j^2/u_n$ is bounded.*

This theorem is due to Bradic, Fan and Jiang (2011). Using this large deviation result, Bradic, Fan and Jiang (2011) established the strong oracle property of the penalized partial likelihood estimator under a set of regularity conditions. Namely, there exists a local maximizer $\widehat{\boldsymbol{\beta}}$ such that the probability of the event that $\boldsymbol{\beta}$ equals the oracle estimator tends to one at an exponential rate of the sample size $n$, under the set of conditions.

## 6.6 Theory of Folded-concave Penalized M-estimator

Let $\mathcal{D}_n$ be the collected sample with the sample size $n$ from a regression model involving parameter vector $\boldsymbol{\beta}$. Let $\ell_n(\boldsymbol{\beta}, \mathcal{D}_n)$ be a measure of the fit between $\boldsymbol{\beta}$ and the collected sample $\mathcal{D}_n$. The $\ell_n(\boldsymbol{\beta}, \mathcal{D}_n)$ is termed the empirical loss function in the literature. Of interest is to estimate $\boldsymbol{\beta}^*$, which is assumed to be the unique minimizer of the population risk. That is,

$$\boldsymbol{\beta}^* = \arg\min_{\boldsymbol{\beta}} \mathrm{E}\{\ell_n(\boldsymbol{\beta}, \mathcal{D}_n)\},$$

where the expectation is taken over the observed sample. Penalized M-estimator is referred to as the minimizer of the following function with respect to $\boldsymbol{\beta}$:

$$\ell_n(\boldsymbol{\beta}, \mathcal{D}_n) + p_\lambda(\boldsymbol{\beta}), \tag{6.49}$$

where $p_\lambda(\boldsymbol{\beta}) = \sum_{j=1}^p p_\lambda(|\beta_j|)$ and $p_\lambda(\cdot)$ is a penalty with a regularization parameter $\lambda$. For folded concave penalties such as the SCAD and the MCP, the minimizers of (6.49) typically are not unique. Wang, Liu and Zhang (2014) and Loh and Wainwright (2015) independently studied the properties of this regularized M-estimator. The empirical loss function $\ell_n(\boldsymbol{\beta}, \mathcal{D}_n)$ and the penalty $\sum_{j=1}^p p_\lambda(|\beta_j|)$ may be nonconvex in practice. Due to the potential nonconvexity, Loh and Wainwright (2015) includes a side constraint $g(\boldsymbol{\beta}) \le R$, where $g : R^p \to R_+$ is required to be a convex function satisfying the lower bound $g(\boldsymbol{\beta}) \ge \sum_{j=1}^p |\beta_j|$ for all $\boldsymbol{\beta} \in R^p$. See Section 6.6.3 for how to set up $g(\boldsymbol{\beta})$. Provided that the empirical loss and the penalty function are continuous, any feasible point for the optimization problem (6.50) below satisfies the constraint $\sum_{j=1}^p |\beta_j| \le R$:

$$\widehat{\boldsymbol{\beta}} = \arg\min_{\boldsymbol{\beta}} \ell_n(\boldsymbol{\beta}, \mathcal{D}_n) + p_\lambda(\boldsymbol{\beta}) \quad \text{subject to } g(\boldsymbol{\beta}) \le R. \tag{6.50}$$

The constrained minimization problem covers most optimization problems encountered in this book. For instance, the penalty function can be the $\ell_q$-penalty including $q = 1$, the SCAD penalty and the MCP penalty. The empirical loss function can be the least squares loss function, the Huber loss for robust regression, the check loss function for quantile regression, the loss function in rank regression, the negative log-likelihood for penalized likelihood studied in Chapter 5 and the negative logarithm partial likelihood for survival data studied in the last section. The empirical loss function can be the hinge loss in the support vector machine, and the loss function for graphical models. Thus, the theoretical formulation introduced in this section may be useful for various problems. Loh and Wainwright (2015) imposes conditions on the penalty term and the empirical loss function to achieve the generality of the penalized M-estimator.

### 6.6.1 *Conditions on penalty and restricted strong convexity*

To study statistical accuracy and computational accuracy, Loh and Wainwright (2015) considered penalties satisfying the following conditions.

**Condition P**:

(p1) The penalty function $p_\lambda$ satisfies that $p_\lambda(0) = 0$ and $p_\lambda(\theta) = p_\lambda(-\theta)$. For $\theta \in (0, \infty)$, $p_\lambda(\theta)$ is a nondecreasing function of $\theta$, while $p_\lambda(\theta)/\theta$ is a nonincreasing function of $\theta$.

(p2) The penalty function $p_\lambda(\theta)$ is differential for all $\theta \neq 0$ and subdifferentiable at $\theta = 0$ with $p'_\lambda(0+) = \lambda L$ for some positive constant $L$.

(p3) There exists $\mu > 0$ so that $p_{\lambda,\mu}(\theta) \widehat{=} p_\lambda(\theta) + \frac{\mu}{2}\theta^2$ is convex.

Condition (p1) is the same as Conditions (i) — (iii) in Loh and Wainwright (2015), and has been used in the literature (Zhang and Zhang, 2012). This condition shares the same spirit as Condition 1 in Fan and Lv (2011).

The SCAD and MCP penalties satisfy all conditions (p1)-(p3), while Condition (p2) rules out the $\ell_q$ penalty with $0 < q < 1$ (i.e. the bridge penalty), and the capped-$\ell_1$ penalty, which has points of non-differentiability over $(0, \infty)$. Condition (p3) is known as *weak convexity* (Vial, 1982, Chen and Gu, 2014), and is a type of curvature constraint that controls the level of nonconvexity of the penalty function. The capped-$\ell_1$ penalty does not satisfy this condition for any $\mu > 0$.

Under Condition P, $p_\lambda(\cdot)$ has some nice properties. It is shown in Lemma 4 of Loh and Wainwright (2015) that under Condition P, (a) $p_\lambda(\cdot)$ is $\lambda L$-Lipschitz, namely, $|p_\lambda(\theta_2) - p_\lambda(\theta_1)| \le \lambda L|\theta_2 - \theta_1|$ for any $\theta_1, \theta_2$; (b) all subgradients and derivatives of $p_\lambda(\cdot)$ are bounded by $\lambda L$; and (c)

$$\lambda L\|\boldsymbol{\beta}\|_1 \le p_\lambda(\boldsymbol{\beta}) + \frac{\mu}{2}\|\boldsymbol{\beta}\|_2^2. \tag{6.51}$$

This property enables us to take $g(\boldsymbol{\beta})$ in (6.50) to be $\lambda^{-1}\{p_\lambda(\boldsymbol{\beta}) + \frac{\mu}{2}\|\boldsymbol{\beta}\|_2^2\}$.

Let $\mathbf{v} \in R^p$, and define $\mathcal{A}$ to be the index set of the $k$ largest elements of $\mathbf{v}$ in magnitude. Suppose that $\xi > 0$ is such that $\xi p_\lambda(\mathbf{v}_\mathcal{A}) - p_\lambda(\mathbf{v}_{\mathcal{A}^c}) \ge 0$. It has been shown in Lemma 5 of Loh and Wainwright (2015) that

$$\xi p_\lambda(\mathbf{v}_\mathcal{A}) - p_\lambda(\mathbf{v}_{\mathcal{A}^c}) \le \lambda L(\xi\|\mathbf{v}_\mathcal{A}\|_1 - \|\mathbf{v}_{\mathcal{A}^c}\|_1). \tag{6.52}$$

Moreover, if $\boldsymbol{\beta}^* \in R^p$ is $k$-sparse (i.e., $\|\boldsymbol{\beta}^*\|_0 = k$), then for any vector $\boldsymbol{\beta} \in R^p$

such that $\xi p_\lambda(\boldsymbol\beta^*) - p_\lambda(\boldsymbol\beta) > 0$ and $\xi \ge 1$, we have

$$\xi p_\lambda(\boldsymbol\beta^*) - p_\lambda(\boldsymbol\beta) \le \lambda L(\xi\|\mathbf{v}_{\mathcal{A}}\|_1 - \|\mathbf{v}_{\mathcal{A}^c}\|_1), \tag{6.53}$$

where $\mathbf{v} = \boldsymbol\beta - \boldsymbol\beta^*$ and $\mathcal{A}$ is the index set of the $k$ largest elements of $\mathbf{v}$ in magnitude.

We next introduce *restricted strong convexity* (RSC) for the empirical loss function. In general, the empirical loss function is required to be differentiable, but not to be convex. To control the nonconvexity of the empirical loss function, the RSC, a weaker condition than convexity, is imposed on the empirical loss function. With slight abuse of the notation, denote $\ell_n(\boldsymbol\beta) = \ell_n(\boldsymbol\beta, \mathcal{D}_n)$.

To study the statistical accuracy of the penalized M-estimator, Loh and Wainwright (2015) imposes the following RSC condition. Let $\nabla h$ be the gradient or subgradient of $h$, if it exists. $\ell_n(\boldsymbol\beta)$ is said to satisfy the *RSC* condition if the following inequality holds.

$$\langle \nabla \ell_n(\boldsymbol\beta^* + \Delta) - \nabla \ell_n(\boldsymbol\beta^*), \Delta\rangle \ge \begin{cases} \alpha_1\|\Delta\|_2^2 - \tau_1 \frac{\log p}{n}\|\Delta\|_1^2, \forall \|\Delta\|_2 \le 1, \\ \alpha_2\|\Delta\|_2 - \tau_2 \frac{\log p}{n}\|\Delta\|_1, \forall \|\Delta\|_2 \ge 1, \end{cases} \tag{6.54}$$

where the $\alpha_j$s are strictly positive constants and the $\tau_j$'s are nonnegative constants. This RSC condition is used in Loh and Wainwright (2015), and is stronger than the RSC discussed in Negahban, Ravikumar, Wainwright and Yu (2012), which defines an RSC with respect to a fixed subset of index $\{1, \cdots, p\}$.

If $\ell_n(\boldsymbol\beta)$ is actually strongly convex, then the RSC condition holds with $\alpha_1 = \alpha_2 > 0$ and $\tau_1 = \tau_2 = 0$. However, for high dimensional data, it is typical that $p \gg n$, and hence the $\ell_n(\boldsymbol\beta)$ is not in general strongly convex or even convex. The RSC condition may still hold with strictly positive $(\alpha_j, \tau_j)$. The left-hand size in (6.54) is always nonnegative if $\ell_n(\boldsymbol\beta)$ is convex but not strongly convex. Thus (6.54) holds for $\|\Delta\|_1/\|\Delta\|_2 \ge \sqrt{\alpha_1/\tau_1}\sqrt{n/\log p}$ when $\|\Delta\|_2 \le 1$, and $\|\Delta\|_1/\|\Delta\|_2 \ge (\alpha_2/\tau_2)\sqrt{n/\log p}$ when $\|\Delta\|_2 \ge 1$. Therefore the RSC inequalities only enforce a type of strong convexity condition over a cone of the form $\{\|\Delta\|_1/\|\Delta\|_2 < c\sqrt{n/\log p}\}$.

### 6.6.2 *Statistical accuracy of penalized M-estimator with folded concave penalties*

Loh and Wainwright (2015) studies the error bound of any feasible solution $\widetilde{\boldsymbol\beta}$ that satisfies the *first-order necessary conditions* to be a local minimum of (6.50):

$$\langle \nabla \ell_n(\widetilde{\boldsymbol\beta}) + \nabla p_\lambda(\widetilde{\boldsymbol\beta}), \boldsymbol\beta - \widetilde{\boldsymbol\beta}\rangle \ge 0, \text{ for all possible } \boldsymbol\beta \in R^p. \tag{6.55}$$

When $\widetilde{\boldsymbol\beta}$ falls in the interior of the constraint set, the first-order necessary condition becomes the usual *zero-subgradient condition*:

$$\nabla \ell_n(\widetilde{\boldsymbol{\beta}}) + \nabla p_\lambda(\widetilde{\boldsymbol{\beta}}) = 0.$$

The $\widetilde{\boldsymbol{\beta}}$'s satisfying (6.55) are indeed the stationary points (Bertsekas, 1999). The set of stationary points also include interior local maxima and saddle points. One has to check the second-order necessary condition in order to distinguish these points.

Under Condition P and the RSC condition, the statistical accuracy of $\widetilde{\boldsymbol{\beta}}$ satisfying (6.55) is established in the following theorem, which is due to Loh and Wainwright (2015).

**Theorem 6.8** *Suppose that Condition P holds, $\ell_n(\boldsymbol{\beta})$ satisfies the RSC condition (6.54) with $\frac{3}{4}\mu < \alpha_1$. Consider any choice of $\lambda$ such that*

$$\frac{4}{L} \max\{\|\nabla \ell_n(\boldsymbol{\beta}^*)\|_\infty, \alpha_2 \sqrt{\log(p)/n}\} \le \lambda \le \frac{\alpha_2}{6RL}, \tag{6.56}$$

*and suppose that $n \ge (16R^2 \max(\tau_1^2, \tau_2^2)/\alpha_2) \log(p)$. Then any $\widetilde{\boldsymbol{\beta}}$ satisfying the first-order necessary condition (6.55) satisfies the error bounds*

$$\|\widetilde{\boldsymbol{\beta}} - \boldsymbol{\beta}^*\|_2 \le \frac{6\lambda L\sqrt{k}}{4\alpha_1 - 3\mu}, \quad \text{and} \quad \|\widetilde{\boldsymbol{\beta}} - \boldsymbol{\beta}^*\|_1 \le \frac{24\lambda L k}{4\alpha_1 - 3\mu}, \tag{6.57}$$

*and*

$$\langle \nabla \ell_n(\widetilde{\boldsymbol{\beta}}) - \nabla \ell_n(\boldsymbol{\beta}^*), \widetilde{\boldsymbol{\beta}} - \boldsymbol{\beta} \rangle \le \lambda^2 L^2 k \left( \frac{9}{4\alpha_1 - 3\mu} + \frac{27\mu}{(4\alpha_1 - 3\mu)^2} \right), \tag{6.58}$$

*where $k = \|\boldsymbol{\beta}^*\|_0$.*

The error bounds in (6.57) and (6.58) grow with $k$ and $\lambda$. To control the squared-$\ell_2$ error $\|\widetilde{\boldsymbol{\beta}} - \boldsymbol{\beta}^*\|_2^2$, we may choose $\lambda \propto \sqrt{\log(p)/n}$ and $R \propto \lambda^{-1}$ so that the condition (6.56) is satisfied for many statistical models, and hence the squared-$\ell_2$ error is of order $k \log(p)/n$, as expected.

For the penalized least squares estimator in linear regression model $\mathbf{Y} = \mathbf{X}\boldsymbol{\beta} + \boldsymbol{\varepsilon}$, $\ell_n(\boldsymbol{\beta}) = \frac{1}{2n}\|\mathbf{Y} - \mathbf{X}\boldsymbol{\beta}\|^2$. Thus,

$$\langle \nabla \ell_n(\widetilde{\boldsymbol{\beta}}) - \nabla \ell_n(\boldsymbol{\beta}^*), \widetilde{\boldsymbol{\beta}} - \boldsymbol{\beta} \rangle = \frac{1}{n}\|\mathbf{X}\widetilde{\boldsymbol{\beta}} - \mathbf{X}\boldsymbol{\beta}^*\|^2$$

which is the model error for the fixed-design setting. Using Theorem 6.8, we have the upper bound of the model error

$$\frac{1}{n}\|\mathbf{X}\widetilde{\boldsymbol{\beta}} - \mathbf{X}\boldsymbol{\beta}^*\|^2 \le \lambda^2 L^2 k \left( \frac{9}{4\alpha_1 - 3\mu} + \frac{27\mu}{(4\alpha_1 - 3\mu)^2} \right).$$

Under the assumptions of Theorem 6.8, any stationary points have error bounds in (6.57) and (6.58). Thus, these bounds are applied for not only local minima but also local maxima.

Results in Theorem 6.8 can be applied for various penalized M-estimators provided that the first order necessary condition (6.55) is valid. For example, Theorem 6.8 is directly applicable for the penalized likelihood estimate described in Chapter 5 and the penalized robust linear regression studied in Section 6.3. However, for the penalized quantile regression, penalized composite quantile regression and penalized rank regression, condition (6.55) may not be valid. One may have to modify this condition for the corresponding penalized M-estimator.

*Proof of Theorem 6.8.* Denote $\widetilde{\mathbf{v}} = \widetilde{\boldsymbol{\beta}} - \boldsymbol{\beta}^*$. We first show that $\|\widetilde{\mathbf{v}}\|_2 \le 1$. Otherwise, if $\|\widetilde{\mathbf{v}}\|_2 > 1$, then the RSC condition implies that

$$\langle \nabla \ell_n(\widetilde{\boldsymbol{\beta}}) - \nabla \ell_n(\boldsymbol{\beta}^*), \widetilde{\mathbf{v}}\rangle \ge \alpha_2 \|\widetilde{\mathbf{v}}\|_2 - \tau_2\sqrt{\log(p)/n}\|\widetilde{\mathbf{v}}\|_1. \tag{6.59}$$

Since $\boldsymbol{\beta}^*$ is a feasible solution, the first-order necessary condition holds for $\boldsymbol{\beta} = \boldsymbol{\beta}^*$. This together with (6.59) lead to

$$\langle -\nabla p_\lambda(\widetilde{\boldsymbol{\beta}}) - \nabla \ell_n(\boldsymbol{\beta}^*), \widetilde{\mathbf{v}}\rangle \ge \alpha_2 \|\widetilde{\mathbf{v}}\|_2 - \tau_2\sqrt{\log(p)/n}\|\widetilde{\mathbf{v}}\|_1. \tag{6.60}$$

Note that by Hölder's inequality and the triangle inequality, we have

$$\langle -\nabla p_\lambda(\widetilde{\boldsymbol{\beta}}) - \nabla \ell_n(\boldsymbol{\beta}^*), \widetilde{\mathbf{v}}\rangle \le \{\|\nabla p_\lambda(\widetilde{\boldsymbol{\beta}})\|_\infty + \|\nabla \ell_n(\boldsymbol{\beta}^*)\|_\infty\}\|\widetilde{\mathbf{v}}\|_1$$

Under Condition P, $\|\nabla p_\lambda(\widetilde{\boldsymbol{\beta}})\|_\infty \le \lambda L$. By (6.56), $\|\nabla \ell_n(\boldsymbol{\beta}^*)\|_\infty \le \frac{\lambda L}{2}$. Thus,

$$\langle -\nabla p_\lambda(\widetilde{\boldsymbol{\beta}}) - \nabla \ell_n(\boldsymbol{\beta}^*), \widetilde{\mathbf{v}}\rangle \le \frac{3\lambda L}{2}\|\widetilde{\mathbf{v}}\|_1.$$

Combining this bound with the bound in (6.60), it yields that

$$\|\widetilde{\mathbf{v}}\|_2 \le \frac{\|\widetilde{\mathbf{v}}\|_1}{\alpha_2}\left(\frac{3\lambda L}{2} + \tau_2\sqrt{\log(p)/n}\right) \le \frac{2R}{\alpha_2}\left(\frac{3\lambda L}{2} + \tau_2\sqrt{\log(p)/n}\right).$$

By using (6.56) and the assumed lower bound on the sample size, the right-hand side is at most 1, so $\|\widetilde{\mathbf{v}}\|_2 \le 1$.

Since $\|\widetilde{\mathbf{v}}\|_2 \le 1$, it follows by the RSC condition that

$$\langle \nabla \ell_n(\widetilde{\boldsymbol{\beta}}) - \nabla \ell_n(\boldsymbol{\beta}^*), \widetilde{\mathbf{v}}\rangle \ge \alpha_1 \|\widetilde{\mathbf{v}}\|_2^2 - \tau_1\frac{\log(p)}{n}\|\widetilde{\mathbf{v}}\|_1^2. \tag{6.61}$$

Since the function $p_{\lambda,\mu}(\boldsymbol{\beta}) = p_\lambda(\boldsymbol{\beta}) + \frac{\mu}{2}\|\boldsymbol{\beta}\|_2^2$ is convex under Condition P, it follows that

$$p_{\lambda,\mu}(\boldsymbol{\beta}^*) - p_{\lambda,\mu}(\widetilde{\boldsymbol{\beta}}) \ge \langle \nabla p_{\lambda,\mu}(\widetilde{\boldsymbol{\beta}}), \boldsymbol{\beta}^* - \widetilde{\boldsymbol{\beta}}\rangle = \langle \nabla p_\lambda(\widetilde{\boldsymbol{\beta}}) + \mu\widetilde{\boldsymbol{\beta}}, \boldsymbol{\beta}^* - \widetilde{\boldsymbol{\beta}}\rangle,$$

which implies that

$$\langle \nabla p_\lambda(\widetilde{\boldsymbol{\beta}}), \boldsymbol{\beta}^* - \widetilde{\boldsymbol{\beta}}\rangle \le p_\lambda(\boldsymbol{\beta}^*) - p_\lambda(\widetilde{\boldsymbol{\beta}}) + \frac{\mu}{2}\|\widetilde{\boldsymbol{\beta}} - \boldsymbol{\beta}^*\|^2. \tag{6.62}$$

Combining (6.61) with (6.55) and (6.62), we have

$$\alpha_1\|\widetilde{\mathbf{v}}\|_2^2 - \tau_1\frac{\log(p)}{n}\|\widetilde{\mathbf{v}}\|_1^2 \leq -\langle \nabla \ell_n(\boldsymbol{\beta}^*), \widetilde{\mathbf{v}}\rangle + p_\lambda(\boldsymbol{\beta}^*) - p_\lambda(\widetilde{\boldsymbol{\beta}}) + \frac{\mu}{2}\|\widetilde{\mathbf{v}}\|_2^2.$$

Rearranging and using Hölder's inequality, we then obtain

$$\begin{aligned} &(\alpha_1 - \frac{\mu}{2})\|\widetilde{\mathbf{v}}\|_2^2 \\ \leq\ & p_\lambda(\boldsymbol{\beta}^*) - p_\lambda(\widetilde{\boldsymbol{\beta}}) + \|\nabla \ell_n(\boldsymbol{\beta}^*)\|_\infty \cdot \|\widetilde{\mathbf{v}}\|_1 + \tau_1\frac{\log(p)}{n}\|\widetilde{\mathbf{v}}\|_1^2 \\ \leq\ & p_\lambda(\boldsymbol{\beta}^*) - p_\lambda(\widetilde{\boldsymbol{\beta}}) + (\|\nabla \ell_n(\boldsymbol{\beta}^*)\|_\infty + 4R\tau_1\frac{\log(p)}{n})\|\widetilde{\mathbf{v}}\|_1. \end{aligned} \tag{6.63}$$

By the assumption (6.56), it follows that

$$\|\nabla \ell_n(\boldsymbol{\beta}^*)\|_\infty + 4R\tau_1\frac{\log(p)}{n} \leq \frac{\lambda L}{4} + \alpha_2\sqrt{\log(p)/n} \leq \frac{\lambda L}{2}.$$

Combining this with (6.63) and (6.51), it follows by using the subadditivity of $p_\lambda(\cdot)$ that

$$\begin{aligned} (\alpha_1 - \frac{\mu}{2})\|\widetilde{\mathbf{v}}\|_2^2 &\leq p_\lambda(\boldsymbol{\beta}^*) - p_\lambda(\widetilde{\boldsymbol{\beta}}) + \frac{\lambda L}{2}\cdot\Big(\frac{p_\lambda(\widetilde{\mathbf{v}})}{\lambda L} + \frac{\mu}{2\lambda L}\|\widetilde{\mathbf{v}}\|_2^2\Big) \\ &\leq p_\lambda(\boldsymbol{\beta}^*) - p_\lambda(\widetilde{\boldsymbol{\beta}}) + \frac{p_\lambda(\boldsymbol{\beta}^*) + p_\lambda(\widetilde{\boldsymbol{\beta}})}{2} + \frac{\mu}{4}\|\widetilde{\mathbf{v}}\|_2^2, \end{aligned}$$

which implies that

$$0 \leq (\alpha_1 - \frac{3\mu}{4})\|\widetilde{\mathbf{v}}\|_2^2 \leq \frac{3}{2}p_\lambda(\boldsymbol{\beta}^*) - \frac{1}{2}p_\lambda(\widetilde{\boldsymbol{\beta}}). \tag{6.64}$$

In particular, we have $\xi p_\lambda(\boldsymbol{\beta}^*) - p_\lambda(\widetilde{\boldsymbol{\beta}}) \geq 0$ with $\xi = 3$. It follows by using (6.53) that

$$3p_\lambda(\boldsymbol{\beta}^*) - p_\lambda(\widetilde{\boldsymbol{\beta}}) \leq 3\lambda L\|\widetilde{\mathbf{v}}_{\mathcal{A}}\|_1 - \lambda L\|\widetilde{\mathbf{v}}_{\mathcal{A}^c}\|_1, \tag{6.65}$$

where $\mathcal{A}$ denotes the index set of the $k$ largest elements of $\widetilde{\boldsymbol{\beta}} - \boldsymbol{\beta}^*$ in magnitude. In particular, we have the cone condition

$$\|\widetilde{\mathbf{v}}_{\mathcal{A}^c}\|_1 \leq 3\|\widetilde{\mathbf{v}}_{\mathcal{A}}\|_1. \tag{6.66}$$

Substituting (6.66) into (6.65), we have

$$(2\alpha_1 - \frac{3\mu}{2})\|\widetilde{\mathbf{v}}\|_2^2 \leq 3\lambda L\|\widetilde{\mathbf{v}}_{\mathcal{A}}\|_1 - \lambda L\|\widetilde{\mathbf{v}}_{\mathcal{A}^c}\|_1 \leq 3\lambda L\|\widetilde{\mathbf{v}}_{\mathcal{A}}\|_1 \leq 3\lambda L\sqrt{k}\|\widetilde{\mathbf{v}}\|_2,$$

from which we can conclude that

$$\|\widetilde{\mathbf{v}}\|_2 \leq \frac{6\lambda L\sqrt{k}}{4\alpha_1 - 3\mu}$$

which is the desired $\ell_2$ bound. Using the cone inequality (6.66), it follows that $\|\widetilde{\mathbf{v}}\|_1 \le 4\|\widetilde{\mathbf{v}}_{\mathcal{A}}\|_1 \le 4\sqrt{k}\|\widetilde{\mathbf{v}}\|_2$. As a result

$$\|\widetilde{\mathbf{v}}\|_1 \le \frac{24\lambda Lk}{4\alpha_1 - 3\mu}$$

which is the $\ell_1$ bound in Theorem 6.8.

We next show (6.58). It follows by using (6.55) and (6.62) that

$$\begin{aligned} & \langle \nabla \ell_n(\widetilde{\boldsymbol{\beta}}) - \nabla \ell_n(\boldsymbol{\beta}^*), \widetilde{\mathbf{v}} \rangle \\ \le \; & -\langle p_\lambda(\widetilde{\boldsymbol{\beta}}) - \nabla \ell_n(\boldsymbol{\beta}^*), \widetilde{\mathbf{v}} \rangle \\ \le \; & p_\lambda(\boldsymbol{\beta}^*) - p_\lambda(\widetilde{\boldsymbol{\beta}}) + \frac{\mu}{2}\|\widetilde{\mathbf{v}}\|_2^2 + \|\nabla \ell_n(\boldsymbol{\beta}^*)\|_\infty \cdot \|\widetilde{\mathbf{v}}\|_1. \end{aligned} \tag{6.67}$$

By using (6.53) and (6.56), it follows that

$$\begin{aligned} & \|\nabla \ell_n(\boldsymbol{\beta}^*)\|_\infty \|\widetilde{\mathbf{v}}\|_1 \\ \le \; & \frac{\lambda L}{2}\Big(\frac{p_\lambda(\boldsymbol{\beta}^*) + p_\lambda(\widetilde{\boldsymbol{\beta}})}{\lambda L} + \frac{\mu}{2\lambda L}\|\widetilde{\mathbf{v}}\|_2^2\Big) \\ \le \; & \frac{p_\lambda(\boldsymbol{\beta}^*) + p_\lambda(\widetilde{\boldsymbol{\beta}})}{2} + \frac{\mu}{4}\|\widetilde{\mathbf{v}}\|_2^2. \end{aligned}$$

This combining with (6.67) yields

$$\begin{aligned} & \langle \nabla \ell_n(\widetilde{\boldsymbol{\beta}}) - \nabla \ell_n(\boldsymbol{\beta}^*), \widetilde{\mathbf{v}} \rangle \\ \le \; & \frac{3}{2}p_\lambda(\boldsymbol{\beta}^*) - \frac{1}{2}p_\lambda(\widetilde{\boldsymbol{\beta}}) + \frac{3\mu}{4}\|\widetilde{\mathbf{v}}\|_2^2 \\ \le \; & \frac{3\lambda L}{2}\|\widetilde{\mathbf{v}}_A\|_1 - \frac{\lambda L}{2}\|\widetilde{\mathbf{v}}_{A^c}\|_1 + \frac{3\mu}{4}\|\widetilde{\mathbf{v}}\|_2^2 \\ \le \; & \frac{3\lambda L\sqrt{k}}{2}\|\widetilde{\mathbf{v}}\|_2 + \frac{3\mu}{4}\|\widetilde{\mathbf{v}}\|_2^2. \end{aligned}$$

The proof is completed by plugging-in the $\ell_2$-bound. ■

### *6.6.3 Computational accuracy*

Wang, Liu and Zhang (2014) and Loh and Wainwright (2015) studied optimization problems related to penalized M-estimators. Loh and Wainwright (2015) introduced how to apply the *composite gradient descent* (Nesterov, 2007) to efficiently minimize (6.49), and show that it enjoys a linear rate of convergence under suitable conditions. Define

$$g_{\lambda,\mu}(\boldsymbol{\beta}) = \lambda^{-1}\{\sum_{j=1}^{p} p_\lambda(|\beta_j|) + \frac{\mu}{2}\|\boldsymbol{\beta}\|^2\},$$

which is possibly nonsmooth but convex under Condition P. Denote

$$\bar{\ell}_n(\boldsymbol{\beta}) = \ell_n(\boldsymbol{\beta}) - \frac{\mu}{2}\|\boldsymbol{\beta}\|_2^2,$$

which is a differentiable but possibly nonconvex function. We are interested in the computational accuracy of solution of (6.50) which can be written as

$$\widehat{\boldsymbol{\beta}} \in \arg\min_{\boldsymbol{\beta}} \bar{\ell}_n(\boldsymbol{\beta}) + \lambda g_{\lambda,\mu}(\boldsymbol{\beta}) \quad \text{subject to } g_{\lambda,\mu}(\boldsymbol{\beta}) \le R, \tag{6.68}$$

Here we represent the objective function into a sum of a differentiable but possibly nonconvex function and a possibly nonsmooth but convex penalty. This enables us to directly apply the composite gradient descent procedure of Nesterov (2007) to (6.68). The gradient descent procedure produces a sequence of iterates $\{\boldsymbol{\beta}^t : t = 0, \cdots, \infty\}$ via updates

$$\boldsymbol{\beta}^{t+1} \in \arg\min_{\{\boldsymbol{\beta}: g_{\lambda,\mu}(\boldsymbol{\beta}) \le R\}} \left\{ \frac{1}{2}\left\| \boldsymbol{\beta} - \left(\boldsymbol{\beta}^t - \frac{\nabla \bar{\ell}_n(\boldsymbol{\beta}^t)}{\eta}\right) \right\|_2^2 + \frac{\lambda}{\eta} g_{\lambda,\mu}(\boldsymbol{\beta}) \right\}, \tag{6.69}$$

where $\eta^{-1}$ is the step size. This minimization problem can be carried out in the following algorithm.

**Algorithm**

S1. First optimize the unconstrained program

$$\widehat{\boldsymbol{\beta}} \in \arg\min_{\boldsymbol{\beta}} \left\{ \frac{1}{2}\left\| \boldsymbol{\beta} - \left(\boldsymbol{\beta}^t - \frac{\nabla \bar{\ell}_n(\boldsymbol{\beta}^t)}{\eta}\right) \right\|_2^2 + \frac{\lambda}{\eta} g_{\lambda,\mu}(\boldsymbol{\beta}) \right\}. \tag{6.70}$$

which may have a closed-form solution for penalty functions such as the SCAD and MCP penalty.

S2. If $g_{\lambda,\mu}(\widehat{\boldsymbol{\beta}}) \le R$, define $\boldsymbol{\beta}^{t+1} = \widehat{\boldsymbol{\beta}}$.

S3. Otherwise, if $g_{\lambda,\mu}(\widehat{\boldsymbol{\beta}}) > R$, optimize the constrained program

$$\boldsymbol{\beta}^{t+1} \in \arg\min_{\{\boldsymbol{\beta}: g_{\lambda,\mu}(\boldsymbol{\beta}) \le R\}} \left\{ \frac{1}{2}\left\| \boldsymbol{\beta} - \left(\boldsymbol{\beta}^t - \frac{\nabla \bar{\ell}_n(\boldsymbol{\beta}^t)}{\eta}\right) \right\|_2^2 \right\}. \tag{6.71}$$

This algorithm is due to Loh and Wainwright (2015), whose Appendix C.1 provides a justification of this algorithm. It is of interest to derive the upper bound of $\|\boldsymbol{\beta}^t - \widehat{\boldsymbol{\beta}}\|^2$ for a given tolerance $\delta^2$.

Denote

$$T(\boldsymbol{\beta}_1, \boldsymbol{\beta}_2) = \ell_n(\boldsymbol{\beta}_1) - \ell_n(\boldsymbol{\beta}_2) - \langle \nabla \ell_n(\boldsymbol{\beta}_2), \boldsymbol{\beta}_1 - \boldsymbol{\beta}_2 \rangle, \tag{6.72}$$

which is the approximation error of $\ell_n(\boldsymbol{\beta}_1)$ by its linear Taylor's expansion at $\boldsymbol{\beta}_2$. Similarly to the definition of $\bar{\ell}_n(\boldsymbol{\beta})$, define

$$\bar{T}(\boldsymbol{\beta}_1, \boldsymbol{\beta}_2) = T(\boldsymbol{\beta}_1, \boldsymbol{\beta}_2) - \frac{\mu}{2}\|\boldsymbol{\beta}_1 - \boldsymbol{\beta}_2\|^2. \tag{6.73}$$

To study the difference between $\boldsymbol{\beta}^t$ and $\widehat{\boldsymbol{\beta}}$, Loh and Wainwright imposed another form of RSC:

$$T(\boldsymbol{\beta}_1, \boldsymbol{\beta}_2) \geq \begin{cases} \alpha_1 \|\boldsymbol{\beta}_1 - \boldsymbol{\beta}_2\|_2^2 - \tau_1 \frac{\log p}{n} \|\boldsymbol{\beta}_1 - \boldsymbol{\beta}_2\|_1^2, \forall \|\boldsymbol{\beta}_1 - \boldsymbol{\beta}_2\|_2 \leq 3, \\ \alpha_2 \|\boldsymbol{\beta}_1 - \boldsymbol{\beta}_2\|_2 - \tau_2 \frac{\log p}{n} \|\boldsymbol{\beta}_1 - \boldsymbol{\beta}_2\|_1, \forall \|\boldsymbol{\beta}_1 - \boldsymbol{\beta}_2\|_2 > 3 \end{cases} \tag{6.74}$$

for all vectors $\boldsymbol{\beta} \in B_2(3) \cap B_1(R)$, where $B_q(r)$ stands for the $\ell_q$-ball pf radius $r$ centered around 0. The form of RSC in (6.74) and the RSC in (6.54) have the same spirit but they are not identical because we are studying the computational accuracy rather than the statistical accuracy. In addition to the RSC, we need to impose the following Restricted SMoothness condition (*RSM*, Agarwal, Negahban and Wainwright, 2012):

$$T(\boldsymbol{\beta}_1, \boldsymbol{\beta}_2) \leq \alpha_3 \|\boldsymbol{\beta}_1 - \boldsymbol{\beta}_2\|_2^2 + \tau_3 \frac{\log(p)}{n} \|\boldsymbol{\beta}_1 - \boldsymbol{\beta}_2\|^2 \tag{6.75}$$

for all $\boldsymbol{\beta}_1$ and $\boldsymbol{\beta}_2$ lying in the solution region. In their theoretical analysis, Loh and Wainwright (2015) further impose the following assumptions.

**Assumptions.**

(A1) Assume that $2\alpha_1 > \mu$ for all $i$, and define $\alpha = \min\{\alpha_1, \alpha_2\}$ and $\tau = \max\{\tau_1, \tau_2, \tau_3\}$.

(A2) The population minimizer $\boldsymbol{\beta}^*$ is $k$-sparse (i.e., $\|\boldsymbol{\beta}^*\| = k$) and $\|\boldsymbol{\beta}\|_2 \leq 1$.

(A3) Assume the scaling $n > Ck\log(p)$ for a constant $C$ depending on the $\alpha_i$'s and $\tau_i$'s.

Denote

$$\varphi(n, p, k) + c\tau k \log(p)/\{n(2\alpha - \mu)\}$$

and

$$\kappa = \frac{1 - (2\alpha - \mu)/(8\eta) + \varphi(n, p, k)}{1 - \varphi(n, p, k)}. \tag{6.76}$$

Under Assumption (A3), $\kappa \in (0, 1)$ so it is a contraction factor. Loh and Wainwright (2015) showed that the squared optimization error will fall below $\delta^2$ within $T \sim \log(1/\delta^2)/\log(1/\kappa)$ iterations. More precisely, it guarantees $\delta$-accuracy for all iterations larger than

$$T^*(\delta) = \frac{2\log[\{\phi(\boldsymbol{\beta}^0) - \phi(\widehat{\boldsymbol{\beta}})\}/\delta^2]}{\log(1/\kappa)} + (1 + \frac{\log(2)}{\log(1/\kappa)}) \log\log(\lambda RL/\delta^2), \tag{6.77}$$

where $\phi(\boldsymbol{\beta}) = \ell_n(\boldsymbol{\beta}) + p_\lambda(\boldsymbol{\beta})$, the objective function to minimize.

Denote $\varepsilon_{stat}^2 = \|\widehat{\boldsymbol{\beta}} - \boldsymbol{\beta}^*\|_2^2$. Theorem 6.8 implies that the order of $\|\widehat{\boldsymbol{\beta}} - \boldsymbol{\beta}^*\|_2$ is $k\log(p)/n$ when $\lambda$ and $R$ are properly chosen. Thus, it makes sense to consider only the situation that $\delta$ shares the same order of $\varepsilon_{stat}$.

**Theorem 6.9** *Suppose that Condition P holds, Assumptions (A1)—(A3) are valid, and $\ell_n(\boldsymbol{\beta})$ satisfies the RSC condition (6.54) and RSM condition (6.75). Further assume that $\widehat{\boldsymbol{\beta}}$ is any global minimum of (6.68) with $\lambda$ chosen such that*

$$\frac{8}{L}\max\{\|\nabla \ell_n(\boldsymbol{\beta}^*)\|_\infty, c'\tau\sqrt{\log(p)/n}\} \le \lambda \le \frac{c''\alpha}{RL}.$$

*Suppose that $\mu < 2\alpha$. Then for any step size parameter $\eta \ge \max\{2\alpha_3 - \mu, \mu\}$ and tolerance $\delta^2 \ge \{c\varepsilon_{stat}^2/(1-\kappa)\}\{k\log(p)/n\}$, we have*

$$\|\boldsymbol{\beta}^t - \widehat{\boldsymbol{\beta}}\|_2^2 \le \frac{4}{2\alpha - \mu}(\delta^2 + \frac{\delta^4}{\tau} + c\tau\frac{k\log(p)}{n}\varepsilon_{stat}^2), \quad \forall t \ge T^*(\delta) \tag{6.78}$$

This theorem is due to Loh and Wainwright (2015).

For the optimal choice of $\delta \sim \{k\log(p)/n\}\varepsilon_{stat}$, the error bound in (6.78) takes the form

$$\frac{c\varepsilon_{stat}^2}{2\alpha - \mu}\cdot\frac{k\log(p)}{n}.$$

This implies that successive iterates of the composite gradient descent algorithm are guaranteed to converge to a region within statistical accuracy of the true global optimum $\widehat{\boldsymbol{\beta}}$. Under Assumption (A3), and $\lambda$ is chosen appropriately, Theorem 6.8 ensures that $\varepsilon_{stat} = O(\sqrt{k\log(p)/n})$ with probability tending to one. Using this fact together with Theorem 6.9, it follows that

$$\max\{\|\boldsymbol{\beta}^t - \widehat{\boldsymbol{\beta}}\|_2, \|\boldsymbol{\beta}^t - \boldsymbol{\beta}^*\|_2\} = O(\sqrt{k\log(p)/n}),$$

for all iterations $t \ge T(\varepsilon_{stat})$.

Wang, Liu and Zhang (2014) considered another definition of RSC. Under penalized least squares setting, Liu, Yao and Li (2016) compared these two different definitions of the RSC given in Wang, Liu and Zhang (2014) and Loh and Wainwright (2015), and derived a necessary condition of the RSC in their Lemma 4.1. They further conducted a "random RSC test" to see if the randomly generated sample instances can satisfy the RSC condition. Their simulation results clearly show that the RSC condition may not be satisfied when $n \ll p$ or the covariates are highly correlated. Readers are referred to Table 1 of Liu, Yao and Li (2016), and also Exercise 6.12. It needs some caution in using the RSC condition (6.75). Liu, Yao, Li and Ye (2017) studied the computational complexity, sparsity, statistical performance, and algorithm theory for local solution of penalized least squares with linear regression models.

## 6.7 Bibliographical Notes

A comprehensive review book of quantile regression is Koenker (2005). Lasso regularized quantile regression with fixed $p$ was first studied by Li and

Zhu (2008) and Wu and Liu (2009). For ultrahigh-dimensional $p$, Belloni and Chernozhukov (2011) derived an error bound for quantile regression with the Lasso-penalty, while Wang, Wu and Li (2012) studied regularized quantile regression with folded concave penalties. Fan, Fan and Barut (2014) studied penalized quantile regression with an adaptively weighted $\ell_1$ penalty to alleviate the bias problem and established the oracle property and asymptotic normality of their resulting estimator.

Zou and Yuan (2008) proposed composite quantile regression and further studied $\ell_1$-regularized composite quantile regression. The composite quantile regression was extended to the nonparametric regression model in Kai, Li and Zou (2010) and to semiparametric regression models in Kai, Li and Zou (2011). Li and Li (2016) further studied the composite quantile regression with Markov processes. In Gu's thesis (Gu (2017)) and Gu and Zou (2017), the efficient algorithm and the statistical theory are developed for penalized composite quantile regression in ultra-high dimensions.

Robust regression was first studied in Huber (1964). There are a huge number of publications on robust regression. Rousseeuw and Leroy (1987) is a popular reference for robust linear regression. Hettmansperger and McKean (1998) provides a comprehensive account on robust nonparametric statistical methods with focus on rank regression. When the dimension $p$ is fixed and finite, the $\ell_1$-penalized least absolute deviation (LAD) regression was first studied in Wang, Li and Jiang (2007), and penalized rank regression was proposed in Wang and Li (2009). Under the high dimensional setting, Wang (2013) studied $\ell_1$-penalized LAD regression and further showed that the estimator achieves near oracle risk performance. Fan, Li and Wang (2017) studied penalized robust regression with Huber loss under an ultrahigh dimensional setting.

Partial likelihood was proposed for analysis of survival data in Cox (1975). Tibshirani (1997) extended the Lasso method for survival data via $\ell - 1$- constrained partial likelihood. Fan and Li (2002) extended the penalized partial likelihood method for the Cox model and the Cox frailty model. Cai, Fan, Li and Zhou (2005) extended the penalized partial likelihood method for multivariate survival time. Zhang and Lu (2007) and Zou (2008) independently extended the adaptive Lasso for the Cox models and established its oracle property when $p$ is finite and fixed. Partial likelihood is closely related to profile likelihood. Fan and Li (2012) developed a variable selection procedure for a linear mixed model via penalized profile likelihood.

ADMM algorithms have found many successful applications in modern statistics and machine learning, such as compressed sensing (Yin, Osher, Goldfarb and Darbon, 2008; Goldstein and Osher 2009), sparse covariance matrix estimation (Xue, Ma and Zou, 2012), and sparse precision matrix estimation (Zhang and Zou, 2014). The readers are referred to Boyd, Parikh, Chu, Peleato and Eckstein (2011) for a good review paper on ADMM.

Wang, Liu and Zhang (2014) considered another definition of RSC. Liu, Yao and Li (2016) compared these two different definitions of the RSC under the penalized least squares setting. Liu, Yao, Li and Ye (2017) studied the computational complexity, sparsity, statistical performance, and algorithm theory for local solution of penalized least squares with linear regression models. Liu, Wang, Yao, Li and Ye (2019) modified the traditional sample average approximation scheme for stochastic programming for the penalized M-estimator with folded concave penalty with allowing the dimension $p$ to be much more than the sample size $n$.

## 6.8 Exercises

6.1 Develop numerical algorithms for quantile regression.

(a) Apply the strategy in Section 3.5.1 to derive a linear programming algorithm for quantile regression (6.3).

(b) Apply the idea of local quadratic approximation to the check loss function defined in Section 6.1 to derive an MM algorithm for quantile regression.

6.2 Use (6.4) to derive the asymptotic relative efficiency of the quantile regression to the least squares estimate. Further calculate the asymptotic relative efficiency when error distribution to be (1) zero-mean normal distribution, (2) t-distribution with degrees of freedom being 1, 2, and 3, (3) Laplace distribution.

6.3 Suppose that $p$, the dimension of $\mathbf{X}$, is finite.

(a) Establish the oracle property of the SCAD penalized quantile regression estimator defined in (6.11).

(b) Further prove Theorem 6.2.

6.4 Develop a numerical algorithm for penalized quantile regerssion.

(a) Derive the LARS algorithm for $\ell_1$ penalized quantile regression.

(b) Extend the coordinate descent algorithm in Section 3.5.6 for penalized quantile regression

6.5 Suppose that $p$, the dimension of $\mathbf{X}$, is finite. Prove Theorems 6.3 and 6.4.

6.6 Extend the scdADMM algorithm in Section 6.1.3 for the SCAD penalized CQR with LLA.

6.7 Suppose that $p$, the dimensional of $\mathbf{X}$ is finite.

(a) Show that the M-estimator defined in (6.25) is a root-$n$ consistent estimator.

(b) Establish the asymptotical normality of the M-estimator.

6.8 Let $\ell_\alpha(\cdot)$ be the Huber loss defined in (6.28). Define a penalized Huber regression for linear model as

$$\frac{1}{n}\sum_{i=1}^{n} \ell_\alpha(Y_i - \mathbf{X}^T)^T + \sum_{j=1}^{p} p_\lambda(|\beta_j|),$$

which is an extension of the $\ell_1$ penalized Huber regression defined in (6.29). Establish the oracle property of the penalized Huber regression with the penalty being the MCP and SCAD penalty when the dimension of $\mathbf{X}$ is finite and fixed.

6.9 Refer to Section 6.4

(a) Establish the oracle property of the SCAD penalized weighted rank regression estimator (6.37).

(b) Derive an ADMM algorithm for the $\ell_1$ penalized weighted rank regression.

(c) Extend the ADMM algorithm in (b) for the SCAD penalized weighted rank regression by applying LLA to the SCAD penalty with setting initial value to be the solution obtained in (b).

6.10 Establish the oracle property of the penalized partial likelihood estimator by proving (6.46) under regularity conditions (A1) — (A4) given in Section 6.5.

6.11 Refer to Section 6.6.

(a) Check whether the RSC condition (6.54) holds for penalized quantile regression and penalized Huber regression.

(b) Prove Theorem 6.9.

6.12 Design a simulation setting to check whether the RSC condition (6.74) holds with high probability. See Liu, Yao and Li (2016).

Chapter 7

# High Dimensional Inference

This chapter concerns statistical inference in the sense of interval estimation and hypothesis testing. The key difference from classical inference is that the parameters fall in high-dimensional sparse regions and regularized methods have been introduced to explore the sparsity. We have studied in previous chapters regularized estimation of high-dimensional (*HD*) objects such as the mean of the response vector, the coefficient vector and support set in linear regression, among others. In many cases, it has been shown that such regularized estimators have optimality properties such as rate minimaxity in various collections of unknown parameters classes. However, as these results concerns relatively large total error of HD quantities, they provide little information about low-dimensional (*LD*) features which could be of primary interest in common applications, such as a treatment effect. Indeed, these estimators introduce nonnegligible biases due to regularization and they need to be corrected first in order to have valid statistical inferences. In this section, we focus on statistical inferences of LD parameters with HD data.

A prototypical example of problems under consideration in this chapter is the efficient interval estimation of a preconceived effect in linear regression, namely, inferences on the regression coefficients for a few given covariates. In this and several related problems, our discussion would focus on a semi-low-dimensional (*semi-LD*) approach (Zhang, 2011) which is best described with the following model decomposition,

$$\text{HD model} = \text{LD component} + \text{HD component}, \tag{7.1}$$

where the LD component is of primary interest and the HD component is treated as a nuisance parameter. For example, if a coefficient $\beta_{j_0}$ represents the treatment effect of interest in a regression model $\mathbf{Y} = \mathbf{X}\boldsymbol{\beta} + \boldsymbol{\varepsilon}$, the LD component can be written as $\mathbf{X}_{j_0}\beta_{j_0}$ and the HD component $\mathbf{X}_{-j_0}\boldsymbol{\beta}_{-j_0}$. However, this decomposition is not necessarily unique or the most convenient in a semi-LD analysis.

The relationship between this semi-LD approach and regularized estimation of HD objects, such as the estimation of $\boldsymbol{\beta}$ or $\mathbf{X}\boldsymbol{\beta}$ in linear regression considered in Chapters 3 and 4, is parallel to the one between semi-parametric analysis and nonparametric (*NP*) estimation. In the semi-parametric approach

(Bickel, *et al.*, 1998) the statistical model is decomposed as

$$\text{model} = \text{parametric component} + \text{NP component}, \tag{7.2}$$

where the parametric component is of primary interest and the NP component is treated as a nuisance parameter. In both semi-LD and semi-parametric cases, efficient estimation of the parameter of interest requires estimation of the nuisance parameter at a reasonably fast rate. In this sense, the semi-parametric inference builds upon NP estimation theory whereas the semi-LD inference builds upon the theory of regularized estimation of HD objects. The main differences between the two models are the following. In the semi-parametric model (7.2), the data can be viewed as an independent and identically distributed sample from a fixed population as the sample size grows, and the NP component can be typically approximated by a member of a known subspace of moderate dimension depending on the sample size. However, in the semi-LD model (7.1), the dimension of the HD model and thus the population would change with the sample size and the HD component typically does not have a known approximating subspace. For example, when a regression coefficient vector $\boldsymbol{\beta}$ is assumed to be sparse in the sense of having a support $\mathcal{S}$ of moderate size, the set $\mathcal{S}$ is typically considered completely unknown. Nevertheless, in this chapter, the dependence of the model on sample size is suppressed throughout in both cases.

In Section 7.1, we will show how to apply the debiased technique to construct confidence intervals for low-dimensional regression coefficients and noise level. The techniques will be extended to generalized linear models in Section 7.2. Section 7.3 first gives the asymptotic efficiency theory for a general model and then applies it to the high-dimensional linear models and a partially linear model. Section 7.4 solves the inference and efficiency issues for estimating the partial correlation coefficient and elements of the precision matrix. More general solutions to statistical inference problems are given in Section 7.5. Section 7.3, 7.4 and 7.5 are technical, and these sections may be skipped without loss of continuity.

## 7.1 Inference in Linear Regression

This section focuses on statistical inference in the linear model

$$\mathbf{Y} = \mathbf{X}\boldsymbol{\beta} + \boldsymbol{\varepsilon} \tag{7.3}$$

with a design matrix $\mathbf{X} = (\mathbf{X}_1, \ldots, \mathbf{X}_p) \in I\!R^{n\times p}$, a sparse coefficient vector $\boldsymbol{\beta} = (\beta_1, \ldots, \beta_p)^T$ and a noise vector $\boldsymbol{\varepsilon}$ with $\mathrm{E}\,\boldsymbol{\varepsilon} = \mathbf{0}$ and $\mathrm{E}\,\boldsymbol{\varepsilon}\boldsymbol{\varepsilon}^T = \sigma^2\mathbf{I}_n$. In this section, we consider statistical inference for regression coefficients $\beta_j$ and the noise level $\sigma$. The penalized least-squares estimators for both quantities are biased in HD due to regularization. For inference about $\beta_j$, *debias* techniques are needed as $\mathbf{X}_j$ are typically correlated with other design variables in a non-trivial way. However, due to the orthogonality of the scores for the estimation of $\boldsymbol{\beta}$ and $\sigma$, efficient estimation of the noise level $\sigma$ can be achieved by joint estimation of $\boldsymbol{\beta}$ and $\sigma$ with regularization on $\boldsymbol{\beta}$ only. Unless otherwise stated, we assume in this section deterministic design $\mathbf{X}$ with normalized columns $\|\mathbf{X}_j\|_2^2 \leq n$ and Gaussian noise $\boldsymbol{\varepsilon} \sim N(\mathbf{0}, \sigma^2 \mathbf{I}_n)$.

### *7.1.1 Debias of regularized regression estimators*

For the linear model (7.3), a natural class of estimators for the regression coefficient $\beta_j$ is the linear estimators of form $\widehat{\beta}_j^* = \mathbf{w}_j^T \mathbf{Y}$ for a given weight vector $\mathbf{w}_j$ that depends only on $\mathbf{X}$. Then,

$$\mathrm{E}\,\widehat{\beta}_j^* = \mathbf{w}_j^T \mathbf{X}\boldsymbol{\beta},$$

where E denotes the conditional expectation given $\mathbf{X}$, if $\mathbf{X}$ are random, and the bias of the estimator $\widehat{\beta}_j^*$ is $\mathbf{w}_j^T \mathbf{X}\boldsymbol{\beta} - \beta_j$. Thus, for an initial estimator $\widehat{\boldsymbol{\beta}}^{(\text{init})}$, the bias corrected estimator is

$$\widehat{\beta}_j^* - (\mathbf{w}_j^T \mathbf{X} \widehat{\boldsymbol{\beta}}^{(\text{init})} - \widehat{\beta}_j^{(\text{init})}) = \widehat{\beta}_j^{(\text{init})} + \mathbf{w}_j^T \big(\mathbf{Y} - \mathbf{X}\widehat{\boldsymbol{\beta}}^{(\text{init})}\big).$$

An idea of Zhang and Zhang (2014) is to apply such a bias correction to regularized estimators $\widehat{\boldsymbol{\beta}}^{(\text{init})}$ such as those described in Chapter 3. This leads to a *de-biased estimator* of $\beta_j$ of the following form,

$$\widehat{\beta}_j = \widehat{\beta}_j^{(\text{init})} + \mathbf{w}_j^T \big(\mathbf{Y} - \mathbf{X}\widehat{\boldsymbol{\beta}}^{(\text{init})}\big) \tag{7.4}$$

for a suitable vector $\mathbf{w}_j$. As this estimator corrects the bias of the initial estimator through a one-dimensional linear projection of the residual $\mathbf{Y} - \mathbf{X}\widehat{\boldsymbol{\beta}}^{(\text{init})}$, it is called the low-dimensional projection estimator (*LDPE*).

By simple algebra, the estimation error of $\widehat{\beta}_j$ can be decomposed as

$$\widehat{\beta}_j - \beta_j = \mathbf{w}_j^T \boldsymbol{\varepsilon} + (\mathbf{X}^T \mathbf{w}_j - \boldsymbol{e}_j)^T \big(\widehat{\boldsymbol{\beta}}^{(\text{init})} - \boldsymbol{\beta}\big), \tag{7.5}$$

where $\boldsymbol{e}_j$ is the $p$-dimensional unit vector with 1 as the $j$-th element. The first term has expectation zero and the second term in (7.5), which is the one where the bias comes from, is bounded by

$$\|\mathbf{X}^T \mathbf{w}_j - \boldsymbol{e}_j\|_\infty \big\|\widehat{\boldsymbol{\beta}}^{(\text{init})} - \boldsymbol{\beta}\big\|_1 \tag{7.6}$$

The error is bounded by the products of two terms. They can be chosen to both converge to zero so that the error is of order $o(n^{-1/2})$, i.e. negligible. This will be formally stated in Theorem 7.1 below. Since the bias of the Lasso needs to be corrected, Theorem 7.1 is stated to facilitate such an application.

The performance of the *de-biased estimator* can be characterized by the following noise factor $\tau_j$ and bias factor $\eta_j$:

$$\tau_j = \|\mathbf{w}_j\|_2, \quad \eta_j = \|\mathbf{X}^T \mathbf{w}_j - \boldsymbol{e}_j\|_\infty / \tau_j. \tag{7.7}$$

The theorem below is a direct consequence of (7.5).

**Theorem 7.1** *Let $\widehat{\beta}_j$ be as in (7.4) with $\widehat{\boldsymbol{\beta}}^{(\text{init})}$ satisfying*

$$\|\widehat{\boldsymbol{\beta}}^{(\text{init})} - \boldsymbol{\beta}\|_1 = O_P\big(s^*\sigma\sqrt{(\log p)/n}\big) \tag{7.8}$$

*with a constant $s^*$. Let $\tau_j$ and $\eta_j$ be as in (7.7). Then, for all $J \subseteq \{1,\ldots,p\}$,*

$$\max_{j\in J}\left|\frac{\widehat{\beta}_j - \beta_j}{\tau_j\sigma} - \frac{\mathbf{w}_j^T\boldsymbol{\varepsilon}}{\tau_j\sigma}\right| = O_P(1)s^*\sqrt{(\log p)/n}\max_{j\in J}\eta_j. \tag{7.9}$$

*If $\mathbf{w}_j^T\boldsymbol{\varepsilon}/\|\mathbf{w}_j\|_2 \to N(0,1)$ and $(\eta_j \vee 1)s^*\sqrt{(\log p)/n} = o(1)$, then*

$$(\widehat{\beta}_j - \beta_j)/(\tau_j\widehat{\sigma}) \to N(0,1) \tag{7.10}$$

*for $\widehat{\sigma} = \|\mathbf{Y} - \mathbf{X}\widehat{\boldsymbol{\beta}}^{(\text{init})}\|_2/\sqrt{n}$ or any $\widehat{\sigma}$ satisfying $|\widehat{\sigma}/\sigma - 1| = o_P(1)$.*

Theorem 7.1 can be used to construct asymptotically normal estimators and confidence intervals for regression coefficient $\beta_j$. For example, by (7.10)

$$P\Big\{|\widehat{\beta}_j - \beta_j| \le 1.96\,\widehat{\sigma}\tau_j\Big\} \approx 0.95.$$

An essential element of (7.9) is the uniformity of the linearization in the set $J$, which facilitates simultaneous inference. We may use (7.9) to construct simultaneous confidence intervals for $\{\beta_j, j \in J\}$ and confidence intervals for linear combinations of $\beta_j$. For example, assuming $\boldsymbol{\varepsilon} \sim N(0,\mathbf{I})$ (this can be relaxed by using the central limit theorem), then

$$P\Big\{\big|(\widehat{\beta}_j - \widehat{\beta}_k) - (\beta_j - \beta_k)\big| \le 1.96\,\widehat{\sigma}\|\mathbf{w}_j - \mathbf{w}_k\|_2\Big\} \approx 0.95 \tag{7.11}$$

when $s^*\sqrt{(\log p)/n}\|(\mathbf{w}_j - \mathbf{w}_k)\mathbf{X}^T - (e_j - e_k)\|_\infty/\|\mathbf{w}_j - \mathbf{w}_k\|_2 = o(1)$.

Let $s^*$ be the capped-$\ell_1$ complex measure of $\boldsymbol{\beta}$, defined as

$$s^* = \sum_{j=1}^p \min\left(\frac{|\beta_j|}{\sigma\sqrt{(2/n)\log p}}, 1\right), \tag{7.12}$$

which is bounded by $\|\boldsymbol{\beta}\|_0$. As we have discussed in Chapter 4, when $\widehat{\boldsymbol{\beta}}^{(\text{init})}$ is a suitable regularized LSE, (7.8) holds under the sparsity condition $s^*\log p \ll n$ and proper conditions on the design matrix $\mathbf{X}$ and noise vector $\boldsymbol{\varepsilon}$. We will show in Section 7.3.2 that by suitable choices of $\mathbf{w}_j$, $\eta_j \le \sqrt{2\log p}$ is achievable for normalized Gaussian designs. Thus, Theorem 7.1 provides a valid inference for $\beta_j$ under the sparsity condition $s_*\log p \ll n^{1/2}$.

### *7.1.2 Choices of weights*

From (7.5), an ideal choice of $\mathbf{w}_j$ is such that $\mathbf{X}^T\mathbf{w}_j = \boldsymbol{e}_j$. In the low dimensional problem where $n \geq p$, this can be achieved by taking $\mathbf{w}_j = \mathbf{Z}_j/(\mathbf{Z}_j^T\mathbf{X}_j)$ where $\mathbf{Z}_j$ is the residual after regression of $\mathbf{X}_j$ against $\mathbf{X}_{-j}$, where $\mathbf{X}_{-j}$ is the $\mathbf{X}$ matrix without the $j^{th}$ column, namely,

$$\mathbf{Z}_j = (\mathbf{I} - \mathbf{X}_{-j}(\mathbf{X}_{-j}^T\mathbf{X}_{-j})^{-1}\mathbf{X}_{-j}^T)\mathbf{X}_j.$$

With this choice, the resulting estimator $\widehat{\beta}_j$ is simply the ordinary least-squares estimator and is unbiased.

For the high-dimensional problem where $p > n$, it is hard to have $\mathbf{X}^T\mathbf{w}_j = \boldsymbol{e}_j$, as we have more equations $p$ than the number of unknowns $n$ in $\mathbf{w}_j$. A natural relaxation is to use the penalized least-squares as proposed by Zhang and Zhang (2014), in which $\mathbf{Z}_j$ is now defined by the Lasso residual

$$\mathbf{Z}_j = \mathbf{X}_j - \mathbf{X}_{-j}\widehat{\gamma}^{(j)},$$

where $\widehat{\gamma}^{(j)}$ is the Lasso estimator:

$$\widehat{\gamma}^{(j)} = \widehat{\gamma}^{(j)}(\lambda_j) = \arg\min_{\gamma}\Big\{\|\mathbf{X}_j - \mathbf{X}_{-j}\gamma\|_2^2/(2n) + \lambda_j\|\gamma\|_1\Big\}. \tag{7.13}$$

By (3.32), we have

$$n^{-1}\|\mathbf{X}_{-j}\mathbf{Z}_j\|_\infty \leq \lambda_j.$$

Using this result, it is easy to see that the choice

$$\mathbf{w}_j = \frac{\mathbf{Z}_j}{\mathbf{Z}_j^T\mathbf{X}_j} \quad \text{and} \quad \widehat{\beta}_j = \widehat{\beta}_j^{(\text{init})} + \frac{\mathbf{Z}_j^T\big(\mathbf{Y} - \mathbf{X}\widehat{\boldsymbol{\beta}}^{(\text{init})}\big)}{\mathbf{Z}_j^T\mathbf{X}_j} \tag{7.14}$$

satisfies

$$\|\mathbf{X}^T\mathbf{w}_j - \mathbf{e}_j\|_\infty \leq n\lambda_j/(\mathbf{Z}_j^T\mathbf{X}_j). \tag{7.15}$$

The problem has since been considered and extended by many, including Bühlmann (2013), Belloni *et al.* (2014), van de Geer *et al.* (2014), Javanmard and Montanari (2014a, b), Chernozhukov (2015), Mitra and Zhang (2016) and Bloniarz *et al.* (2016), among many others. Battey, Fan, Liu, Lu, Zhu (2018) considered a distributed estimation of high-dimensional inferences, where the biases play a dominant role.

Various but similar choices of $\mathbf{w}_j$ have been considered. Zhang (2011) considered $\mathbf{w}_j = \mathbf{Z}_j/\|\mathbf{Z}_j\|_2^2$ as a special case of a general one-step correction (described in Section 7.5) and wrote the ideal $\mathbf{Z}_j$ as $\mathbf{Z}_j^o = \mathbf{X}\boldsymbol{\Omega}\boldsymbol{e}_j/\omega_{j,j}$ in random designs, where $\boldsymbol{\Omega} = (\omega_{j,k})_{p\times p}$ is the inverse of $\boldsymbol{\Sigma} = \mathrm{E}[\mathbf{X}^T\mathbf{X}/n]$ and $\mathrm{E}\,\|\mathbf{Z}_j^o\|_2^2 = n/\omega_{j,j}$. Now, let us specifically take $\mathbf{w}_j = \mathbf{X}\widehat{\boldsymbol{\Omega}}\mathbf{e}_j/n$ for a $p \times p$ matrix $\widehat{\boldsymbol{\Omega}}$ to be specified later. Then, by putting (7.4) as a vector, we have

$$\widehat{\boldsymbol{\beta}} = \widehat{\boldsymbol{\beta}}^{(\text{init})} + \widehat{\boldsymbol{\Omega}}\mathbf{X}^T\big(\mathbf{Y} - \mathbf{X}\widehat{\boldsymbol{\beta}}^{(\text{init})}\big)/n. \tag{7.16}$$

It follows that by (7.5) or simple algebra,

$$\widehat{\boldsymbol{\beta}} - \boldsymbol{\beta} = \widehat{\boldsymbol{\Omega}}\mathbf{X}^T\boldsymbol{\varepsilon} + (\widehat{\boldsymbol{\Omega}}\mathbf{X}^T\mathbf{X}/n - \mathbf{I}_p)(\widehat{\boldsymbol{\beta}}^{(\mathrm{init})} - \boldsymbol{\beta}). \tag{7.17}$$

In order for the second term to be small, $\widehat{\boldsymbol{\Omega}}$ should be chosen so that it is approximately the inversion of the matrix $\mathbf{X}^T\mathbf{X}/n$ or its limit $\boldsymbol{\Sigma}$.

An idea of van de Geer *et al.* (2014) and Javanmard and Montanari (2014a, b) is the global debiased estimation in the form (7.16). Thorough discussion on the estimation of the *precision matrix* $\boldsymbol{\Omega}$ can be found in Chapters 9 and 10. This typically requires that $\boldsymbol{\Omega}$ to be sparse (see Chapter 9), but this can be relaxed to a approximate factor structure (see Chapter 10). When this holds, Bradic, Fan and Zhu (2020) establish the asymptotic normality of $\boldsymbol{\beta}_j$ based on the fact that for random designs, $\boldsymbol{\beta} = \boldsymbol{\Omega}\,\mathrm{E}[\mathbf{X}Y]$, which is a sparse linear combination of $\mathrm{E}[\mathbf{X}Y]$ whose sample version has the asymptotic normality under the sparsity of $\boldsymbol{\Omega}$. This approach for constructing confidence intervals does not require the sparsity assumption of $\boldsymbol{\beta}$.

For the Lasso-based estimator (7.14),

$$\mathbf{X}_j^T\mathbf{Z}_j = \|\mathbf{Z}_j\|_2^2 + n\lambda_j\|\widehat{\gamma}^{(j)}\|_1$$

by the KKT condition (3.22). Thus, by (7.15), the noise and bias factors in (7.7) are given by

$$\tau_j = \tau_j(\lambda_j) = \frac{\|\mathbf{Z}_j\|_2}{\|\mathbf{Z}_j\|_2^2 + n\lambda_j\|\widehat{\gamma}^{(j)}\|_1}, \qquad \eta_j = \eta_j(\lambda_j) = \frac{n\lambda_j}{\|\mathbf{Z}_j\|_2}. \tag{7.18}$$

By (7.10), smaller $\tau_j$ would provide more efficiency for estimating $\widehat{\beta}_j$. However, a sufficiently small $\eta_j$ is needed to guarantee the validity of the inference and thus is a more crucial condition. Simulation experiments (Zhang and Zhang, 2014) have shown that after finding an initial $\tau_j$ with $\eta_j = \sqrt{2\log p}$, further minimizing $\eta_j$ with a somewhat larger $\tau_j$ would provide more accurate coverage probability for confidence intervals based on Theorem 7.1.

Instead of using the above Lasso-based $\mathbf{Z}_j$, we may optimize $\mathbf{w}_j$ by minimizing the noise factor $\tau_j$ for a pre-specified bias factor $\eta_j$ since $\tau_j$ is proportional to the asymptotic variance; see (7.10). As the $\mathbf{w}_j$ in (7.14) is scale free in $\mathbf{Z}_j$, we may set $\mathbf{Z}_j^T\mathbf{X}_j = n$, so that $\tau_j^2 = \|\mathbf{Z}_j\|_2^2/n^2$, $\eta_j = \|\mathbf{X}_{-j}^T\mathbf{Z}_j/n\|_\infty/\tau_j$ and the optimal solution is given by

$$\mathbf{Z}_j = \arg\min_{\mathbf{z}}\left\{\|\mathbf{z}\|_2^2 : \mathbf{X}_j^T\mathbf{z} = n, \|\mathbf{X}_{-j}^T\mathbf{z}/n\|_\infty \le \lambda_j\right\}, \tag{7.19}$$

subject to $\lambda_j = \eta_j\|\mathbf{Z}_j\|_2/n$. This can be solved recursively by computing $\mathbf{Z}_j$ with quadratic programming (7.19) given $\lambda_j$ and $\lambda_j = \eta_j\|\mathbf{Z}_j\|_2/n$ given $\mathbf{Z}_j$. The quadratic programming (7.19) was introduced in Zhang and Zhang (2014).

For the estimator in (7.16), since $\mathbf{w}_j = \mathbf{X}\widehat{\boldsymbol{\Omega}}\mathbf{e}_j/n$, we have

$$\tau_j^2 = \mathbf{b}_j^T\mathbf{X}^T\mathbf{X}\mathbf{b}_j/n^2,$$

where $\mathbf{b}_j = \widehat{\boldsymbol{\Omega}}\mathbf{e}_j$. Let us control

$$\|\mathbf{X}^T\mathbf{w}_j - \mathbf{e}_j\|_\infty = \|\mathbf{b}_j\mathbf{X}^T\mathbf{X}/n - \mathbf{e}_j\|_\infty \leq \lambda_j \tag{7.20}$$

for a sufficiently small $\lambda_j = o(1)$. We would take $\mathbf{b}_j$ to minimize $\tau_j^2$ subject to the above constraint (7.20). This slightly different form of quadratic programming was introduced in Javanmard and Montanari (2014a).

### 7.1.3 *Inference for the noise level*

The estimation of the noise level $\sigma$, which we study here, is an interesting problem in and of itself. Moreover, the theory of regularized least squares estimation typically requires the use of tuning parameters depending on the noise level. Section 3.7.2 gives a thorough treatment of the estimation of $\sigma$ via *refitted cross-validation* (Fan, Guo, and Hao, 2012). It can be used for constructing the confidence interval for $\sigma$.

We now consider the Lasso-based estimator. Recall the sparsity measure $s^*$ defined by (7.12). Suppose the coefficient vector $\boldsymbol{\beta}$ in (7.3) is sparse in the sense that $s^*(\log p)/n$ is sufficiently small. We have shown in Chapter 4 that under suitable regularity conditions on the design $\mathbf{X}$ and noise $\boldsymbol{\varepsilon}$ and for penalty level $\lambda = (\sigma/\eta)\sqrt{(2/n)\log p}$ with a constant $\eta \in (0,1)$, the Lasso

$$\widehat{\boldsymbol{\beta}} = \widehat{\boldsymbol{\beta}}(\lambda) = \arg\min_{\mathbf{b}} \left\{ \|\mathbf{Y} - \mathbf{X}\mathbf{b}\|_2^2/(2n) + \lambda\|\mathbf{b}\|_1 \right\} \tag{7.21}$$

is rate minimax in prediction and satisfies

$$\|\mathbf{X}\widehat{\boldsymbol{\beta}}(\lambda) - \mathbf{X}\boldsymbol{\beta}\|_2^2/n + \lambda\|\widehat{\boldsymbol{\beta}}(\lambda) - \boldsymbol{\beta}\|_1 \lesssim s^*\lambda^2. \tag{7.22}$$

A problem with (7.22) is that the penalty level $\lambda$ depends on the unknown $\sigma$. While the penalty level is often determined by cross-validation in practice, the theoretical guarantee of such schemes is unclear. Städler *et al.* (2010) proposed a penalized maximum likelihood method to estimate jointly the regression coefficients $\boldsymbol{\beta}$ and the noise level $\sigma$. Their method is scale-free but the resulting estimation of $\sigma$ is biased. A simple alternative approach is to use the joint loss (Antoniadis, 2010), which is the penalized Gaussian pseudo log-likelihood,

$$L_{\lambda_0}(\mathbf{b}, t) = \|\mathbf{Y} - \mathbf{X}\mathbf{b}\|_2^2/(2tn) + t/2 + \lambda_0\|\mathbf{b}\|_1, \quad t > 0, \tag{7.23}$$

where $\lambda_0 = \eta^{-1}\sqrt{(2/n)\log p}$ and to find the joint minimizer

$$\{\widehat{\boldsymbol{\beta}}, \widehat{\sigma}\} = \arg\min_{\mathbf{b},t} L_{\lambda_0}(\mathbf{b}, t), \tag{7.24}$$

which is termed *scaled Lasso* in Sun and Zhang (2012). It turns out that the $\widehat{\boldsymbol{\beta}}$ in (7.24) is identical to the square-root Lasso (easily be seen by optimizing $t$ first)

$$\widehat{\boldsymbol{\beta}} = \arg\min_{\mathbf{b}} \left\{ \|\mathbf{Y} - \mathbf{X}\mathbf{b}\|_2/n^{1/2} + \lambda_0\|\mathbf{b}\|_1 \right\}$$

(Belloni, Chernozhukov and Wang, 2011). Remark that the approach in Sun and Zhang (2010) for estimating $\sigma$ can be regarded as one-step implementation of (7.24): starting with $t = 1$, update iteratively $\mathbf{b}$ and then $t$ until convergence. Updating $\mathbf{b}$ given $t$ involves solving a Lasso problem and updating $\mathbf{t}$ involves computing the residual sum of squares of Lasso.

To study properties of (7.24), we consider the following condition on the design: For all $\delta_1, ..., \delta_p \in \{-1, 1\}$ and $1 \le j_0 < j_1 < \cdots < j_n \le p$,

$$\text{rank}\begin{pmatrix} \mathbf{X}_{j_0} & \mathbf{X}_{j_1} & \cdots & \mathbf{X}_{j_n} \\ \delta_{j_0} & \delta_{j_1} & \cdots & \delta_{j_n} \end{pmatrix}_{(n+1)\times(n+1)} = n + 1. \tag{7.25}$$

It is clear that the above condition holds for $\mathbf{X}$ almost everywhere in $I\!R^{n\times p}$.

**Theorem 7.2** *Let* $\lambda_0 = \eta_0^{-1}\sqrt{(2/n)\log(p)}$ *with certain* $\eta_0 \in (0, 1)$. *Let* $\{\widehat{\boldsymbol{\beta}}, \widehat{\sigma}\}$ *be as in* (7.24). *Suppose that for certain* $\epsilon_0 \in (0, 1)$, (7.22) *holds uniformly in* $\lambda \in [\sigma\lambda_0(1 - \epsilon_0), \sigma\lambda_0(1 + \epsilon_0)]$ *with probability* $1 + o(1)$ *and* $s^*\lambda_0^2 = o(1)$.
*(i) If* $\|\boldsymbol{\varepsilon}\|_2^2/(\sigma^2 n) = 1 + o_P(1)$, *then*

$$|\widehat{\sigma}/\sigma - 1| = o_P(1), \quad \|\mathbf{X}\widehat{\boldsymbol{\beta}} - \mathbf{X}\boldsymbol{\beta}\|_2^2/n + \sigma\lambda_0\|\widehat{\boldsymbol{\beta}} - \boldsymbol{\beta}\|_1 = O_P(\sigma^2 s^*\lambda_0^2).$$

*(ii) Let* $\widehat{\sigma}^o = \|\boldsymbol{\varepsilon}\|_2/\sqrt{n}$ *be an oracle estimate of* $\sigma$. *Suppose* (7.25) *holds. Then* $|\widehat{\sigma}/\widehat{\sigma}^o - 1| = O_P(s^*\lambda_0^2)$. *If* $s^*\lambda_0^2 = o(n^{-1/2})$ *and* $\boldsymbol{\varepsilon} \sim N(\mathbf{0}, \sigma^2\mathbf{I}_n)$, *then*

$$\sqrt{n}\big(\widehat{\sigma}/\sigma - 1\big) \to N(0, 1/2).$$

Theorem 7.2 (ii) can be used to construct confidence intervals for the noise level $\sigma$. For example, under the conditions of Theorem 7.2 (ii),

$$P\Big\{\big|\widehat{\sigma}/\sigma - 1\big| \le 1.96/\sqrt{2n}\Big\} \approx 0.95.$$

The proof of the theorem will be provided at the end of this section. Here is an outline of the proof. By (7.21) and (7.23),

$$L_{\lambda_0}(\widehat{\boldsymbol{\beta}}(t\lambda_0), t) = \inf_{\mathbf{b}} L_{\lambda_0}(\mathbf{b}, t).$$

Due to the joint convexity of $L_{\lambda_0}(\mathbf{b}, t)$, $L_{\lambda_0}(\widehat{\boldsymbol{\beta}}(t\lambda_0), t)$ is nondecreasing in $t \ge \widehat{\sigma}$ and non-increasing in $0 < t \le \widehat{\sigma}$. It turns out that under condition (7.25),

$$(\partial/\partial t)L_{\lambda_0}\big(\widehat{\boldsymbol{\beta}}(t\lambda_0), t\big) = 1/2 - \|\mathbf{Y} - \mathbf{X}\widehat{\boldsymbol{\beta}}(t\lambda_0)\|_2^2/(2nt^2). \tag{7.26}$$

By the risk bounds for the Lasso in (7.22), $\|\mathbf{Y} - \mathbf{X}\widehat{\boldsymbol{\beta}}(t\lambda_0)\|_2^2/n = \sigma^2(1 + o_P(1))$ when $t/\sigma$ is in $[1 - \epsilon, 1 + \epsilon]$ for sufficiently small $\epsilon > 0$. Thus, $\widehat{\sigma}/\sigma$ is within this small interval as the derivative in (7.26) is positive when $t/\sigma = 1 + \epsilon$ and negative when $t/\sigma = 1 - \epsilon$. Letting $\epsilon \to 0+$ would provide speed of convergence.

The proof of (7.26) involves the uniqueness of the Lasso and the differentiation of $\widehat{\boldsymbol{\beta}}(\lambda)$. Such problems has been considered in Osborne *et al.* (2000a,b), Efron *et al.* (2004), Zhang (2010) and Tibshirani (2013) among others. The first part of the following proposition is adapted from Bellec and Zhang (2018).

**Proposition 7.1** *Let $\widehat{\boldsymbol{\beta}}(\lambda)$ be the Lasso in (7.21). Suppose (7.25) holds.*
*(i) Let $B_\lambda = \{j : |\mathbf{X}_j^\top(\mathbf{Y} - \mathbf{X}\widehat{\boldsymbol{\beta}}(\lambda))| = \lambda n\}$. Then, $\mathbf{X}_{B_\lambda}$ has rank $|B_\lambda|$ and $\widehat{\boldsymbol{\beta}}(\lambda)$ is the unique solution to the optimization problem (7.21).*
*(ii) For each $j$, the Lasso path $\widehat{\beta}_j(t)$ is continuous and piecewise linear in $t$. Moreover, for every $\lambda > 0$, there exists $\epsilon > 0$ such that in the intervals $t \in [\lambda(1-\epsilon), \lambda)$ and $t \in (\lambda, \lambda(1+\epsilon)]$, the sign vector $\mathrm{sgn}(\widehat{\boldsymbol{\beta}}(t))$ does not change in $t$ and the derivative $\widehat{\boldsymbol{\beta}}'(t) = (\partial/\partial t)\widehat{\boldsymbol{\beta}}(t)$ exists and can be written as*

$$\widehat{\boldsymbol{\beta}}'_{\widehat{\mathcal{S}}_t}(t) = -\big(\mathbf{X}^T_{\widehat{\mathcal{S}}_t}\mathbf{X}_{\widehat{\mathcal{S}}_t}\big)^{-1}\mathrm{sgn}\big(\widehat{\boldsymbol{\beta}}_{\widehat{\mathcal{S}}_t}(t)\big), \quad \widehat{\boldsymbol{\beta}}'_{\widehat{\mathcal{S}}_t^c}(t) = \mathbf{0},$$

*where $\widehat{\mathcal{S}}_t = \{j : \widehat{\beta}_j(t) \neq 0\}$. Consequently, (7.26) holds for all $t > 0$.*

**Proof.** (i) Assume that $\mathbf{X}_{B_\lambda}$ has rank strictly less than $|B_\lambda|$. Then there must exist some $j \in B_\lambda$ and $A \subseteq B_\lambda \setminus \{j\}$ with $\mathbf{X}_j = \sum_{k\in A}\gamma_k\mathbf{X}_k$ and $\mathrm{rank}(\mathbf{X}_A) = \min(|A|, n)$. By the definition of $B_\lambda$

$$\lambda n\delta_j = \mathbf{X}_j^\top(\mathbf{Y} - \mathbf{X}\widehat{\boldsymbol{\beta}}(\lambda)) = \lambda n\sum_{k\in A}\gamma_k\delta_k$$

where $\delta_k = \mathbf{X}_k^\top(\mathbf{Y} - \mathbf{X}\widehat{\boldsymbol{\beta}}(\lambda))/(\lambda n) \in \{-1, 1\}$. This is impossible by (7.25). Hence $\mathbf{X}_{B_\lambda}$ has rank $|B_\lambda|$. For uniqueness, consider two Lasso solutions $\widehat{\boldsymbol{\beta}}(\lambda)$ and $\widehat{\mathbf{b}}$ of (7.21). It is easily seen that $\mathbf{X}\widehat{\boldsymbol{\beta}}(\lambda) = \mathbf{X}\widehat{\mathbf{b}}$ by the strict convexity of the squared loss in $\mathbf{X}\mathbf{b}$ in (7.21). Furthermore both $\widehat{\boldsymbol{\beta}}(\lambda)$ and $\widehat{\mathbf{b}}$ must be supported on $B_\lambda$. Hence $\widehat{\mathbf{b}} = \widehat{\boldsymbol{\beta}}(\lambda)$ because $\mathbf{X}_{B_\lambda}$ has rank $|B_\lambda|$.

(ii) Given $\lambda$ there exist $A \subseteq \{1, \ldots, p\}$, $\mathbf{s}_A \in \{-1, 1\}^A$ and $t_1 < t_2 < \cdots < \lambda$ with $t_m \to \lambda$ such that $\widehat{\mathcal{S}}_{t_m} = A$ and $\mathrm{sgn}(\widehat{\boldsymbol{\beta}}_A(t_m)) = \mathbf{s}_A$ for all $m \geq 1$. By (i), $\mathbf{X}_A$ is of full rank. Let $\widetilde{\boldsymbol{\beta}}(t)$ be given by $\widetilde{\boldsymbol{\beta}}_A(t) = (\mathbf{X}_A^T\mathbf{X}_A)^{-1}(\mathbf{X}^T\mathbf{Y}/n - t\mathbf{s}_A)$ and $\widetilde{\boldsymbol{\beta}}_{A^c}(t) = \mathbf{0}$. By the KKT conditions (see Section 3.2.5), $\widetilde{\boldsymbol{\beta}}(t_m) = \widehat{\boldsymbol{\beta}}(t_m)$ for all $m$. Moreover, as $\widetilde{\boldsymbol{\beta}}(t)$ is linear in $t$, the KKT condition holds for $\widetilde{\boldsymbol{\beta}}(t)$ for $t_m < t < t_{m+1}$ and in the limit $t_n \to \lambda$, so that $\widetilde{\boldsymbol{\beta}}(t) = \widehat{\boldsymbol{\beta}}(t)$ for all $t \in [t_1, \lambda]$. This gives the formula for the derivative of $\widehat{\boldsymbol{\beta}}(t)$ for $t \in [t_1, \lambda)$ and the left-continuity of $\widehat{\boldsymbol{\beta}}(t)$ at $\lambda$. Moreover, (7.26) holds for $t \in [t_1, \lambda)$ as

$$\begin{aligned}
& (\partial/\partial t)L_{\lambda_0}\big(\widehat{\boldsymbol{\beta}}(t\lambda_0), t\big) - 1/2 + \|\mathbf{Y} - \mathbf{X}\widehat{\boldsymbol{\beta}}(t\lambda_0)\|_2^2/(2nt^2). \\
= \; & \big\{(\partial/\partial t)\|\mathbf{Y} - \mathbf{X}\widehat{\boldsymbol{\beta}}(t\lambda_0)\|_2^2\big\}/(2nt) + \lambda_0(\partial/\partial t)\|\widehat{\boldsymbol{\beta}}(t\lambda_0)\|_1 \\
= \; & \big\{(\partial/\partial t)\widehat{\boldsymbol{\beta}}(t\lambda_0)\big\}\big\{(-\mathbf{X})^T(\mathbf{Y} - \mathbf{X}\widehat{\boldsymbol{\beta}}(t\lambda_0))/(nt) + \lambda_0\mathrm{sgn}(\widehat{\boldsymbol{\beta}}(t\lambda_0))\big\} \\
= \; & 0
\end{aligned}$$

due to the KKT conditions. The proof is complete as the same analysis applies to $t \in [\lambda, \lambda(1+\epsilon)]$ for a small $\epsilon > 0$, and the right-hand side of (7.26) is continuous in $t$. ■

**Proof of Theorem 7.2.** (i) It follows from the KKT conditions that

$$\big|\|\mathbf{Y} - \mathbf{X}\widehat{\boldsymbol{\beta}}(t\lambda_0)\|_2^2/n - (\widehat{\sigma}^o)^2\big|$$

$$
\begin{aligned}
&= \left|\left(\mathbf{X}\boldsymbol{\beta} - \mathbf{X}\widehat{\boldsymbol{\beta}}(t\lambda_0)\right)^T \left(2\mathbf{Y} - \mathbf{X}\widehat{\boldsymbol{\beta}}(t\lambda_0) - \mathbf{X}\boldsymbol{\beta}\right)/n\right| \\
&= \left|2\left(\boldsymbol{\beta} - \widehat{\boldsymbol{\beta}}(t\lambda_0)\right)^T \mathbf{X}^T\left(\mathbf{Y} - \mathbf{X}\widehat{\boldsymbol{\beta}}(t\lambda_0)\right)/n - \left\|\mathbf{X}\boldsymbol{\beta} - \mathbf{X}\widehat{\boldsymbol{\beta}}(t\lambda_0)\right\|_2^2/n\right| \\
&\leq 2t\lambda_0\|\boldsymbol{\beta} - \widehat{\boldsymbol{\beta}}(t\lambda_0)\|_1 + \left\|\mathbf{X}\boldsymbol{\beta} - \mathbf{X}\widehat{\boldsymbol{\beta}}(t\lambda_0)\right\|_2^2/n.
\end{aligned}
$$

As (7.22) holds uniformly in $t \in [\sigma(1-\epsilon_0), \sigma(1+\epsilon_0)]$ for some fixed $\epsilon_0 > 0$ depending on $\eta_0/\eta_2$ and $\eta_0/\eta_1$ only, it follows that

$$
\sup_{|t/\sigma - 1| \leq \epsilon_0} \left\{ \left| \|\mathbf{Y} - \mathbf{X}\widehat{\boldsymbol{\beta}}(t\lambda_0)\|_2^2/n - (\widehat{\sigma}^o)^2 \right| + t\lambda_0 \|\boldsymbol{\beta} - \widehat{\boldsymbol{\beta}}(t\lambda_0)\|_1 \right\} = O_P(1) s^*(\sigma\lambda_0)^2.
$$

Thus, $\sup_{|t/\sigma-1|\leq\epsilon_0} \left|L_{\lambda_0}\left(\widehat{\boldsymbol{\beta}}(t\lambda_0), t\right) - L_{\lambda_0}\left(\boldsymbol{\beta}, t\right)\right| = O_P(s^*(\sigma\lambda_0)^2)$. The consistency of $\widehat{\sigma}$ follows as $L_{\lambda_0}\left(\boldsymbol{\beta}, t\right)$ is strictly convex in $t$ and minimized at $t = \widehat{\sigma}^o$.

(ii) By (7.26) and the above calculation,

$$
\sup_{|t/\sigma - 1| \leq \epsilon_0} \left| (\partial/\partial t) L_{\lambda_0}\left(\widehat{\boldsymbol{\beta}}(t\lambda_0), t\right) - \left\{1/2 - (\widehat{\sigma}^o)^2/(2t^2)\right\} \right| = O_P(1) s^*(\sigma\lambda_0)^2.
$$

Thus, $(\partial/\partial t)L_{\lambda_0}\left(\widehat{\boldsymbol{\beta}}(t\lambda_0), t\right)$ is negative at some $t = t_1 < \widehat{\sigma}^o$ and positive at some $t = t_2 > \widehat{\sigma}^o$ with $(t_2 - t_1)/\widehat{\sigma}^o = O_P(1)s^*(\sigma\lambda_0)^2$. Because, $L_{\lambda_0}\left(\widehat{\boldsymbol{\beta}}(t\lambda_0), t\right)$ is nondecreasing in $t \geq \widehat{\sigma}$ and non-increasing in $0 < t \leq \widehat{\sigma}$, we have $|\widehat{\sigma}/\widehat{\sigma}^o - 1| = O_P(1)s^*(\sigma\lambda_0)^2$. The asymptotic normality follows from $n(\widehat{\sigma}^o/\sigma)^2 \sim \chi_n^2$. ■

## 7.2 Inference in Generalized Linear Models

We introduce generalized linear models and penalized likelihood methods in Chapter 5. In this section, we introduce several statistical inference procedures for high-dimensional generalized linear models. Let $Y$ be the response, and $\mathbf{X}$ be its associate fixed-design covariate vector. The generalized linear models with canonical link have the following probability density or mass function

$$
\exp\left(\frac{Y\mathbf{X}^T\boldsymbol{\beta} - b(\mathbf{X}^T\boldsymbol{\beta})}{\phi}\right) c(Y), \tag{7.27}
$$

where $\boldsymbol{\beta}$ is a $p$-dimensional vector of regression coefficients, and $\phi$ is some positive nuisance parameter. Throughout this section, it is assumed that $b(\cdot)$ is twice continuously differentiable with $b''(\cdot) > 0$. In this section, we will focus on model (7.27), although the statistical inference procedures introduced in this section can be applicable for a more general statistical setting such as the setting of the M-estimate.

Suppose that $Y_i$ is a sample from (7.27) along with covariates $\mathbf{X}_i$, $i = 1, \cdots, n$. Since we do not consider the over-dispersion issue, from now on, we suppress the scale parameter $\phi$, which equals 1 for the logistic regression and Poisson log-linear model, and error variance in the normal linear model. Denote by $\ell(\boldsymbol{\beta}) = \sum_{i=1}^{n} \{Y_i \mathbf{X}_i^T \boldsymbol{\beta} - b(\mathbf{X}_i^T \boldsymbol{\beta})\}$, and the score function

$$\mathbf{S}(\boldsymbol{\beta}) = \ell'(\boldsymbol{\beta}) = \mathbf{X}^T\{\mathbf{Y} - b'(\mathbf{X}\boldsymbol{\beta})\},$$

where $b'(\mathbf{X}\boldsymbol{\beta})$ is an $n$-dimensional vector with i-th element $b'(\mathbf{X}_i^T\boldsymbol{\beta})$.

### *7.2.1 Desparsified Lasso*

Here we give a slightly different motivation for (7.16) in van de Geer, Bühlmann, Ritov and Dezeure (2014). Consider linear regression model

$$\mathbf{Y} = \mathbf{X}\boldsymbol{\beta} + \boldsymbol{\varepsilon}, \tag{7.28}$$

where $\boldsymbol{\varepsilon}$ has zero mean and covariance matrix $\sigma^2 \mathbf{I}_n$. As defined in Chapter 3, the Lasso estimate is

$$\widehat{\boldsymbol{\beta}}_\lambda = \arg\min_{\mathbf{b}}\{\frac{1}{2n}\|\mathbf{Y} - \mathbf{X}\mathbf{b}\|_2 + \lambda\|\mathbf{b}\|_1\}.$$

The notation is slightly different from the last section to facilitate the presentation. From Theorem 3.1, its corresponding KKT conditions are

$$-\frac{1}{n}\mathbf{X}^T(\mathbf{Y} - \mathbf{X}\widehat{\boldsymbol{\beta}}_\lambda) + \lambda\widehat{\boldsymbol{\tau}} = 0$$

$$\|\widehat{\boldsymbol{\tau}}\|_\infty \le 1 \quad \text{and} \quad \widehat{\tau}_j = \mathrm{sign}(\widehat{\beta}_{\lambda,j}) \quad \text{if} \quad \widehat{\beta}_{\lambda,j} \neq 0.$$

Using the first equation, $\widehat{\boldsymbol{\beta}}_\lambda$ must satisfy that

$$\lambda\widehat{\boldsymbol{\tau}} = \frac{1}{n}\mathbf{X}^T(\mathbf{Y} - \mathbf{X}\widehat{\boldsymbol{\beta}}_\lambda).$$

Let $\boldsymbol{\beta}_0$ be the true value of $\boldsymbol{\beta}$. Then

$$\frac{1}{n}\mathbf{X}^T\mathbf{X}(\widehat{\boldsymbol{\beta}}_\lambda - \boldsymbol{\beta}_0) + \lambda\widehat{\boldsymbol{\tau}} = \frac{1}{n}\mathbf{X}^T\boldsymbol{\varepsilon}.$$

For high-dimensional settings, $\mathbf{X}^T\mathbf{X}$ becomes singular. Let $\widehat{\boldsymbol{\Theta}}$ be a reasonable approximation of the inverse of $\widehat{\boldsymbol{\Sigma}} = \frac{1}{n}\mathbf{X}^T\mathbf{X}$. Then

$$\widehat{\boldsymbol{\beta}}_\lambda + \widehat{\boldsymbol{\Theta}}(\lambda\widehat{\boldsymbol{\tau}}) = \boldsymbol{\beta}_0 + \widehat{\boldsymbol{\Theta}}\mathbf{X}^T\boldsymbol{\varepsilon}/n - \boldsymbol{\Delta}/\sqrt{n},$$

where $\boldsymbol{\Delta} = \sqrt{n}(\widehat{\boldsymbol{\Theta}}\widehat{\boldsymbol{\Sigma}} - \mathbf{I})(\widehat{\boldsymbol{\beta}}_\lambda - \boldsymbol{\beta}_0)$. As in (7.17), the product of these two factors in $\boldsymbol{\Delta}$ would make it negligible under certain sparsity assumptions. This leads to the use of the left-hand side above to define the estimator

$$\widehat{\boldsymbol{\beta}} = \widehat{\boldsymbol{\beta}}_\lambda + \widehat{\boldsymbol{\Theta}}\mathbf{X}^T(\mathbf{Y} - \mathbf{X}\widehat{\boldsymbol{\beta}}_\lambda)/n, \tag{7.29}$$

which is a special case of (7.16). van de Geer, *et al.* (2014) called (7.29) *desparsified Lasso* as they noticed that successful de-biasing of the Lasso, which provides the asymptotic normality of $\widehat{\beta}_j$ even when $\beta_j = 0$, must lead to desparcification of the Lasso. They suggest estimating $\widehat{\mathbf{\Theta}}$ by using the Lasso for nodewise regression, a regression techniques for estimating high-dimensional graphical models (Meinshausen and Bühlmann, 2006); see Section 9.4.

van de Geer,*et al.* (2014) further generalized (7.29) to the generalized linear model (7.27). Set $\widehat{\boldsymbol{\beta}}_\lambda$ to be the Lasso penalized likelihood estimate, and let $\widehat{\mathbf{\Sigma}} = -\frac{1}{n}\ell''(\widehat{\boldsymbol{\beta}}_\lambda)$, the Hessian matrix of $-\ell(\boldsymbol{\beta})$ at $\widehat{\boldsymbol{\beta}}_\lambda$, scaled by a factor of $1/n$, and $\widehat{\mathbf{\Theta}}$ be a reasonable approximation or an estimate of $\widehat{\mathbf{\Sigma}}$. van de Geer,*et al.* (2014) defined the desparsified Lasso as

$$\widehat{\boldsymbol{\beta}} = \widehat{\boldsymbol{\beta}}_\lambda + \widehat{\mathbf{\Theta}}\mathbf{S}(\widehat{\boldsymbol{\beta}}_\lambda)/n, \tag{7.30}$$

recalling $\mathbf{S}(\boldsymbol{\beta}) = \ell'(\boldsymbol{\beta})$. It is not hard to show that (7.30) coincides with (7.29) for linear models with normal errors.

For the generalized linear model (7.27), $-\ell''(\boldsymbol{\beta}) = \mathbf{X}^T\mathbf{W}(\boldsymbol{\beta})\mathbf{X}$, where $\mathbf{W}$ is a $n \times n$ diagonal matrix with elements $b''(\mathbf{X}_i^T\boldsymbol{\beta})$. Similar to the situation with linear models, we may apply the Lasso for nodewise regression on $\mathbf{W}^{1/2}(\widehat{\boldsymbol{\beta}}_\lambda)\mathbf{X}$ to construct $\widehat{\mathbf{\Theta}}$ as an estimate of the inverse of $\frac{1}{n}\mathbf{X}^T\mathbf{W}(\widehat{\boldsymbol{\beta}}_\lambda)\mathbf{X}$ (see Section 9.4). Then (7.30) leads to

$$\widehat{\boldsymbol{\beta}} = \widehat{\boldsymbol{\beta}}_\lambda + \widehat{\mathbf{\Theta}}\mathbf{X}^T\{\mathbf{Y} - b'(\mathbf{X}\widehat{\boldsymbol{\beta}}_\lambda)\}/n, \tag{7.31}$$

as $\mathbf{S}(\boldsymbol{\beta}) = \mathbf{X}^T\{\mathbf{Y} - b'(\mathbf{X}\boldsymbol{\beta})\}$ for the generalized linear model. Under a set of regularity conditions, the asymptotical normality of $\widehat{\boldsymbol{\beta}}$ has been established in van de Geer *et al.* (2014). Let $\widehat{\boldsymbol{\beta}}_1$ and $\boldsymbol{\beta}_{10}$ be a $p_1$-dimensional vector of $\widehat{\boldsymbol{\beta}}$ and $\boldsymbol{\beta}_0$, respectively. Then,

$$\sqrt{n}(\widehat{\boldsymbol{\beta}}_1 - \boldsymbol{\beta}_{10}) \xrightarrow{D} N(\mathbf{0}, \mathbf{F}_{11.2}^{-1}(\boldsymbol{\beta}_0)),$$

where $\mathbf{F}(\boldsymbol{\beta}) = n^{-1}\,\mathrm{E}\{-\ell''(\boldsymbol{\beta})\}$ is the Fisher information matrix, $\mathbf{F}_{kl}(\boldsymbol{\beta}))$, $k, l = 1, 2$ is its corresponding $2 \times 2$ partition with dimensionality $p_1$ and $p - p_1$, and

$$\mathbf{F}_{11.2}(\boldsymbol{\beta}) = \mathbf{F}_{11}(\boldsymbol{\beta}) - \mathbf{F}_{12}(\boldsymbol{\beta})\mathbf{F}_{22}^{-1}(\boldsymbol{\beta})\mathbf{F}_{21}(\boldsymbol{\beta}).$$

The result can be used to construct a confidence region for low dimensional parameter $\boldsymbol{\beta}_{10}$.

### *7.2.2 Decorrelated score estimator*

Motivated by the Rao score test, Ning and Liu (2017) proposed the *decorrelated score estimator*, which has similar form to (7.16) and (7.31). The decorrelated score estimator was motivated by the Rao score test, while the debiased Lasso and desparsified Lasso were proposed to construct inference intervals for low-dimensional parameters based on the Lasso estimator.

Partition the score function $\mathbf{S}(\boldsymbol{\beta})$ into $(\mathbf{S}_1(\boldsymbol{\beta})^T, \mathbf{S}_2(\boldsymbol{\beta})^T)^T$ according to the dimensions of $\boldsymbol{\beta}_1$ and $\boldsymbol{\beta}_2$. Define *decorrelated score function* for $\boldsymbol{\beta}_1$ as

$$\mathbf{S}_d(\boldsymbol{\beta}) = \mathbf{S}_1(\boldsymbol{\beta}) - \mathbf{F}_{12}(\boldsymbol{\beta})\mathbf{F}_{22}(\boldsymbol{\beta})^{-1}\mathbf{S}_2(\boldsymbol{\beta}). \tag{7.32}$$

Under certain regularity conditions, for the case with fixed dimensionality $p$, according to the traditional theory on MLE,

$$(1/\sqrt{n})\mathbf{S}(\boldsymbol{\beta}) \xrightarrow{D} N(\mathbf{0}, \mathbf{F}(\boldsymbol{\beta}))$$

in distribution. Thus, the well-known fact that

$$n^{-1/2}\mathbf{S}_d(\boldsymbol{\beta}) \xrightarrow{D} N(\mathbf{0}, \mathbf{F}_{11.2}(\boldsymbol{\beta})) \tag{7.33}$$

can be easily verified. Moreover, $\mathbf{S}_d(\boldsymbol{\beta})$ is asymptotically independent of $\mathbf{S}_2(\boldsymbol{\beta})$. Note that it follows from (7.33) that

$$n^{-1/2}\mathbf{S}_1(\boldsymbol{\beta}) \xrightarrow{D} N(\mathbf{0}, \mathbf{F}_{11}(\boldsymbol{\beta})),$$

but its asymptotic variance is bigger than that of $\mathbf{S}_d(\boldsymbol{\beta})$. The variance reduction is due to the use of correlation between $\mathbf{S}_1(\boldsymbol{\beta})$ and $\mathbf{S}_2(\boldsymbol{\beta})$.

The idea can be generalized to the high dimensional setting. Let $\widetilde{\boldsymbol{\beta}}$ be an estimator of $\boldsymbol{\beta}$ with a certain rate of convergence. Typically, $\widetilde{\boldsymbol{\beta}}$ is set to be a penalized likelihood estimate of $\boldsymbol{\beta}$. The *decorrelated score estimator* is the solution of the decorrelated score equation:

$$\mathbf{S}_d(\boldsymbol{\beta}_1, \widetilde{\boldsymbol{\beta}}_2) = \mathbf{S}_1(\boldsymbol{\beta}_1, \widetilde{\boldsymbol{\beta}}_2) - \widehat{\mathbf{A}}^T\mathbf{S}_2(\boldsymbol{\beta}_1, \widetilde{\boldsymbol{\beta}}_2) = 0, \tag{7.34}$$

where $\widehat{\mathbf{A}}$ is an estimate of $\mathbf{F}_{22}(\boldsymbol{\beta}_0)^{-1}\mathbf{F}_{21}(\boldsymbol{\beta}_0)$, denoted by $\mathbf{A}$; see (7.32). The estimation of $\mathbf{A}$ typically requires a regularization in high-dimension.

For normal linear models with fixed design, $\mathbf{F}(\boldsymbol{\beta}) = (\mathbf{X}^T\mathbf{X})/(n\sigma^2)$, and therefore $\mathbf{S}_d(\boldsymbol{\beta}) = \mathbf{X}_1^T\{\mathbf{I}_n - \mathbf{X}_2(\mathbf{X}_2^T\mathbf{X}_2)^{-1}\mathbf{X}_2^T\}(\mathbf{Y} - \mathbf{X}\boldsymbol{\beta})/\sigma^2$. Let $\mathbf{Z}_1 = \mathbf{X}_1 - \mathbf{X}_2\widehat{\boldsymbol{\Gamma}}$ be the residuals of penalized least-squares fitting $\mathbf{X}_1$ on $\mathbf{X}_2$ (for example, the Lasso estimator of every component in $\mathbf{X}_1$ regressed on $\mathbf{X}_2$). The decorrelated score method (7.34) reduces to solving the estimating equation

$$\mathbf{Z}_1^T(\mathbf{Y} - \mathbf{X}_1\boldsymbol{\beta}_1 - \mathbf{X}_2\widetilde{\boldsymbol{\beta}}_2) = \mathbf{0},$$

which becomes

$$\mathbf{Z}_1^T(\mathbf{Y} - \mathbf{Z}_1\boldsymbol{\beta}_1 - \mathbf{X}_2\widehat{\boldsymbol{\Gamma}}\boldsymbol{\beta}_1 - \mathbf{X}_2\widetilde{\boldsymbol{\beta}}_2) = \mathbf{0}$$

since $\mathbf{X}_1 = \mathbf{Z}_1 + \mathbf{X}_2\widehat{\boldsymbol{\Gamma}}$. As a result,

$$\widehat{\boldsymbol{\beta}}_1 = \widetilde{\boldsymbol{\beta}}_1 + (\mathbf{Z}_1^T\mathbf{Z}_1)^{-1}\mathbf{Z}_1^T(\mathbf{Y} - \mathbf{X}\widetilde{\boldsymbol{\beta}}) + (\mathbf{Z}_1^T\mathbf{Z}_1)^{-1}\mathbf{Z}_1^T\mathbf{X}_2\widehat{\boldsymbol{\Gamma}}(\widetilde{\boldsymbol{\beta}}_1 - \boldsymbol{\beta}_1). \tag{7.35}$$

Thus, if we set $\widetilde{\boldsymbol{\beta}}$ be the Lasso estimate, then the last term in the right-hand side of (7.35) is negligible, and (7.35) behaves similarly to (7.16) and (7.31).

Now back to the equation (7.34). To get an estimate of $\boldsymbol{\beta}_1$ based on the decorrelated score equation (7.34), we need to have $\widetilde{\boldsymbol{\beta}}$ and $\widehat{\mathbf{A}}$ first. In general, we may use penalized likelihood introduced in Chapter 5 to construct $\widetilde{\boldsymbol{\beta}}$. As to $\widehat{\mathbf{A}}$, note that $\mathbf{F}_{22}(\boldsymbol{\beta})\mathbf{A} = \mathbf{F}_{21}(\boldsymbol{\beta})$ by the definition of $\mathbf{A}$. This motivated

Ning and Liu (2017) to construct a Dantzig-type estimator for $\mathbf{A}$ by solving the following minimization problem

$$\widehat{\mathbf{A}} = \arg\min \|\mathbf{A}\| \quad \text{subject to} \quad \|\widehat{\mathbf{F}}_{21}(\widetilde{\boldsymbol{\beta}}) - \widehat{\mathbf{F}}_{22}(\widetilde{\boldsymbol{\beta}})\mathbf{A}\|_\infty < \lambda', \tag{7.36}$$

where $\lambda'$ is a regularization parameter in Dantzig selector, and $\widehat{\mathbf{F}}_{kl}(\widetilde{\boldsymbol{\beta}})$, $k, l = 1, 2$ consists of a partition of $\widehat{\mathbf{F}}(\widetilde{\boldsymbol{\beta}}) = n^{-1}\mathbf{X}^T\mathbf{W}(\widetilde{\boldsymbol{\beta}})\mathbf{X}$ according to the dimensions of $\boldsymbol{\beta}_1$ and $\boldsymbol{\beta}_2$.

The penalized least squares techniques introduced in Chapter 3 can also be used to construct an estimator of $\mathbf{A}$. Denote $\widetilde{\mathbf{X}} = \mathbf{W}^{1/2}(\widetilde{\boldsymbol{\beta}})\mathbf{X}$, and consider a linear model

$$\widetilde{\mathbf{X}}_1 = \widetilde{\mathbf{X}}_2\boldsymbol{\Gamma} + \widetilde{\mathbf{E}}, \tag{7.37}$$

where $\widetilde{\mathbf{E}}$ is a random error matrix. We may apply the penalized least squares method to (7.37) to obtain a sparse estimate $\widehat{\boldsymbol{\Gamma}}$, and then set $\widehat{\mathbf{A}} = \widehat{\boldsymbol{\Gamma}}$.

The procedure for obtaining a decorrelated score estimate can be summarized as the following steps.

**Step 1:** Obtain $\widetilde{\boldsymbol{\beta}}$ of $\boldsymbol{\beta}$ via a penalized likelihood approach.

**Step 2:** Obtain $\widehat{\mathbf{A}}$ by using regularization methods introduced in Chapter 3 such as (7.36) and (7.37).

**Step 3:** Solve the decorrelated score equation (7.34) to get an estimate of $\boldsymbol{\beta}_1$, denoted by $\widehat{\boldsymbol{\beta}}_1^{(\mathrm{ds})}$

It is typical that $\widetilde{\boldsymbol{\beta}}$ is a consistent estimate of $\boldsymbol{\beta}$ with a certain rate of convergence. Similar to the one-step estimate in the literature of robust statistics (Bickel, 1975), Ning and Liu (2017) further considered the one-step estimator for constructing the confidence interval of $\boldsymbol{\beta}_1$. The one-step decorrelated score estimator $\boldsymbol{\beta}_1^{(\mathrm{os})}$ is the solution of

$$\mathbf{S}_d(\widetilde{\boldsymbol{\beta}}_1, \widetilde{\boldsymbol{\beta}}_2) + n\widehat{\mathbf{F}}_{11.2}(\widetilde{\boldsymbol{\beta}})(\boldsymbol{\beta}_1 - \widetilde{\boldsymbol{\beta}}_1) = 0,$$

which is the first order Taylor expansion of (7.34) at $\boldsymbol{\beta}_1 = \widetilde{\boldsymbol{\beta}}_1$. This leads to

$$\widehat{\boldsymbol{\beta}}_1^{(\mathrm{os})} = \widetilde{\boldsymbol{\beta}}_1 - \widehat{\mathbf{F}}_{11.2}^{-1}(\widetilde{\boldsymbol{\beta}})\mathbf{S}_d(\widetilde{\boldsymbol{\beta}}_1, \widetilde{\boldsymbol{\beta}}_2)/n,$$

where $\widehat{\mathbf{F}}_{11.2}^{-1}(\widetilde{\boldsymbol{\beta}}) = \widehat{\mathbf{F}}_{11}(\widetilde{\boldsymbol{\beta}}) - \widehat{\mathbf{A}}^T\widehat{\mathbf{F}}_{21}(\widetilde{\boldsymbol{\beta}})$.

The properties of the decorrelated score estimator have been systematically studied in Ning and Liu (2017) under the following assumptions. Recall that $\boldsymbol{\beta}_0$ is the true value of $\boldsymbol{\beta}$.

**Assumption** 1. For some sequences $\eta_1(n)$ and $\eta_2(n)$ converging to 0 as $n \to \infty$, it holds

$$\|\widetilde{\boldsymbol{\beta}} - \boldsymbol{\beta}_0\|_1 = O_P(\eta_1(n)) \quad \text{and} \quad \|\widehat{\mathbf{A}} - \mathbf{A}\|_1 = O_P(\eta_2(n)).$$

**Assumption** 2. $\|n^{-1}\mathbf{S}(\boldsymbol{\beta}_0)\|_\infty = O_P(\sqrt{\log p/n})$. and $\|\widehat{\mathbf{F}}_{11.2}(\boldsymbol{\beta}_0) - \mathbf{F}_{11.2}(\boldsymbol{\beta}_0)\|_\infty = O_P(\sqrt{\log d/n})$, where $\widehat{\mathbf{F}}_{11.2}(\boldsymbol{\beta})$ is the sample counter part of $\mathbf{F}_{11.2}(\boldsymbol{\beta})$ based on the Hessian matrix of log-likelihood function.

**Assumption** 3. $n^{-1}\{\mathbf{S}_d(\widetilde{\boldsymbol{\beta}})-\mathbf{S}_d(\boldsymbol{\beta}_0)\}-\widehat{\mathbf{F}}_{11.2}(\boldsymbol{\beta}_0)(\widetilde{\boldsymbol{\beta}}-\boldsymbol{\beta}_0)=o_P(n^{-1/2})$ and $(\widehat{\mathbf{A}}-\mathbf{A})^T\{\mathbf{S}_2(\widetilde{\boldsymbol{\beta}})-\mathbf{S}_2(\boldsymbol{\beta}_0)\}/n=o_P(n^{-1/2})$

**Assumption** 4. The asymptotic normality holds for the decorrelated score function at $\boldsymbol{\beta}=\boldsymbol{\beta}_0$. That is, $\frac{1}{\sqrt{n}}\mathbf{S}_d(\boldsymbol{\beta}_0)\xrightarrow{D}N(\mathbf{0},\mathbf{F}_{11.2}(\boldsymbol{\beta}_0))$.

These assumptions are mild for generalized linear models. Under Assumptions 1–4 and model (7.27), the asymptotical normality of the one-step estimate can be established by using Theorem 3.2 of Ning and Liu (2017). If $\{\eta_1(n)+\eta_2(n)\}\sqrt{\log p}=o(1)$, $p_1$ is finite and fixed, $\widehat{\mathbf{F}}_{11.2}(\widetilde{\boldsymbol{\beta}})$ is a consistent estimate of $\mathbf{F}_{11.2}(\boldsymbol{\beta}_0)$, and $\mathbf{F}_{11.2}(\boldsymbol{\beta}_0)$ is positive definite, then under Assumptions 1–4, it follows that

$$\sqrt{n}(\widehat{\boldsymbol{\beta}}_1^{(\mathrm{os})}-\boldsymbol{\beta}_{10})\xrightarrow{D}N(\mathbf{0},\mathbf{F}_{11.2}^{-1}(\boldsymbol{\beta}_0)),$$

based on which we can construct the confidence region for $\boldsymbol{\beta}_1$. Note that in the classical setting in which the dimension $p$ is finite, the MLE $\widehat{\boldsymbol{\beta}}^{(\mathrm{mle})}$ of $\boldsymbol{\beta}$ has asymptotical covariance matrix $\mathbf{F}^{-1}(\boldsymbol{\beta}_0)$ and $\widehat{\boldsymbol{\beta}}_1^{(\mathrm{mle})}$ has asymptotical covariance matrix $\mathbf{F}_{11.2}^{-1}(\boldsymbol{\beta}_0)$. Thus, under the aforementioned assumptions, the one-step estimator $\widehat{\boldsymbol{\beta}}_1^{(\mathrm{os})}$ is asymptotically as efficient as $\widehat{\boldsymbol{\beta}}_1^{(\mathrm{mle})}$.

The decorrelated score function can be used to construct a Rao-score type test for hypothesis

$$H_0:\boldsymbol{\beta}_1=\mathbf{0}\quad\text{versus}\quad H_1:\boldsymbol{\beta}_1\neq\mathbf{0}. \tag{7.38}$$

Let $\breve{\boldsymbol{\beta}}=(\mathbf{0}^T,\widetilde{\boldsymbol{\beta}}_2^T)^T$, which sets the first component of $\widetilde{\boldsymbol{\beta}}$ to $\mathbf{0}$. Suppose that Assumption 3 is also valid with replacement of $\widetilde{\boldsymbol{\beta}}$ by $\breve{\boldsymbol{\beta}}$. If $\{\eta_1(n)+\eta_2(n)\}\sqrt{\log p}=o(1)$, then under Assumptions 1-4, it follows that

$$\frac{1}{\sqrt{n}}\mathbf{S}_d(\breve{\boldsymbol{\beta}})\xrightarrow{D}N(\mathbf{0},\mathbf{F}_{11.2}(\boldsymbol{\beta}_0))$$

in distribution. As a result, the Hotelling $T^2$ test statistic for (7.38)

$$T=n^{-1}\mathbf{S}_d(\breve{\boldsymbol{\beta}})^T\widehat{\mathbf{F}}_{11.2}^{-1}(\widetilde{\boldsymbol{\beta}})\mathbf{S}_d(\breve{\boldsymbol{\beta}}),$$

asymptotically follows a $\chi^2$-distribution with degrees of freedom $p_1$ under the null hypothesis. Thus, the p-value for the testing problem (7.38) can be computed by using the asymptotic null distribution.

### *7.2.3 Test of linear hypotheses*

In this section, we introduce the likelihood ratio-type test for the linear hypothesis

$$H_0:\mathbf{C}\boldsymbol{\beta}_{\mathcal{M}}=\mathbf{t}\quad\text{versus}\quad H_1:\mathbf{C}\boldsymbol{\beta}_{\mathcal{M}}\neq\mathbf{t} \tag{7.39}$$

in high-dimensional generalized linear models, where the index set $\mathcal{M}$ is known, and $\boldsymbol{\beta}_{\mathcal{M}}$ is a subvector of $\boldsymbol{\beta}$. The hypothesis in (7.38) is a special case of (7.39) with proper choice of $\mathbf{C}$ and setting $\mathbf{t} = \mathbf{0}$.

Following the likelihood ratio test, we will examine the difference between the likelihoods under $H = H_0 \cup H_1$ and $H_0$. If the likelihood under $H_0$ is significantly less than the likelihood under $H$, we reject $H_0$. To calculate the likelihoods under $H_0$ and $H_1$, we need to estimate $\boldsymbol{\beta}$ under $H_0$ and $H_1$. In general, we should not penalize parameters of primary interest in order to achieve power at local alternatives. For model (7.27), Shi, Song, Chen and Li (2019) define a *partial penalized likelihood* function

$$Q_n(\boldsymbol{\beta}, \lambda) = \ell(\boldsymbol{\beta}) - n \sum_{j \notin \mathcal{M}} p_\lambda(|\beta_j|)$$

for some penalty function $p_\lambda(\cdot)$ with a tuning parameter $\lambda$. Define partial penalized likelihood estimators as follows:

$$\widehat{\boldsymbol{\beta}}_{H_0} = \arg\max_{\boldsymbol{\beta}} Q_n(\boldsymbol{\beta}, \lambda_{n,0}) \text{ subject to } \mathbf{C}\boldsymbol{\beta}_{\mathcal{M}} = \mathbf{t}, \tag{7.40}$$

$$\widehat{\boldsymbol{\beta}}_{H} = \arg\max_{\boldsymbol{\beta}} Q_n(\boldsymbol{\beta}, \lambda_{n,a}), \tag{7.41}$$

which are the penalized likelihood estimators under $H_0$ and $H = H_0 \cup H_1$, respectively. Note that in (7.40) and (7.41), we do not penalize coefficients in $\boldsymbol{\beta}_{\mathcal{M}}$. This enables us to avoid imposing the minimal signal condition on elements of $\boldsymbol{\beta}_{\mathcal{M}}$. Thus, the corresponding likelihood ratio test has power at local alternatives. The ADMM algorithm introduced in Section 3.5.9 can be used to solve the optimization problem with linear constraints in (7.40). Other algorithms introduced in Chapters 3 and 5 may be used to solve the optimization problem in (7.41).

Define *partial penalized likelihood ratio test*,

$$T = 2\{\ell(\widehat{\boldsymbol{\beta}}_H) - \ell(\widehat{\boldsymbol{\beta}}_{H_0})\}/\widehat{\phi}, \tag{7.42}$$

where $\widehat{\phi}$ is a consistent estimator for the dispersion parameter $\phi$.

Suppose that the true coefficients $\boldsymbol{\beta}_0$ are sparse and satisfy $\mathbf{C}\boldsymbol{\beta}_{0,\mathcal{M}} - \mathbf{t} = \mathbf{h}_n$ for some sequence of vectors $\mathbf{h}_n \to \mathbf{0}$. When $\mathbf{h}_n = \mathbf{0}$, the null holds. Otherwise, the alternative holds. Let $\mathcal{S} = \{j \in \mathcal{M}^c : \beta_{0,j} \neq 0\}$ and $s = |\mathcal{S}|$. Denote $m = |\mathcal{M}|$ and $r$ to be the rank of $\mathbf{C}$, which is allowed to be diverging to $\infty$ as $n \to \infty$. Assume that $s + m = O(n^{1/3})$ and $\|\mathbf{h}_n\|_2 = O\left(\sqrt{\min(s+m-r, r)/n}\right)$, and the largest eigenvalue of $(\mathbf{C}\mathbf{C}^T)^{-1}$ is bounded. Under some regularity conditions on the penalty function, the fixed design matrix $\mathbf{X}$ and the distribution of $Y$, Shi, Song, Chen and Li (2019) established the rates of convergence and sparsity of $\widehat{\boldsymbol{\beta}}_{H_0}$ and $\widehat{\boldsymbol{\beta}}_H$. They further established asymptotic representations of non-vanished estimated coefficients. One may derive the asymptotic normality of $\widehat{\boldsymbol{\beta}}_{H_0}$ and $\widehat{\boldsymbol{\beta}}_H$ by using the asymptotical representations. Let $\chi^2(r, \gamma_n)$

be a chi-square random variable with $r$ degrees of freedom and noncentrality parameter $\gamma_n$ which is allowed to vary with $n$. Shi, *et al.* (2019) showed that

$$\sup_t |P\{T \le t\} - P\{\chi^2(r, \gamma_n) \le t\}| \to 0 \tag{7.43}$$

as $n \to \infty$, where $\gamma_n = n\mathbf{h}^T(\mathbf{C}\mathbf{\Sigma}_a\mathbf{C}^T)^{-1}\mathbf{h}/\phi$ with $\mathbf{\Sigma}_a$ being the asymptotical covariance matrix of $\widehat{\boldsymbol{\beta}}_{\mathcal{M}}$ under $H_1$. Under $H_0$, it follows from (7.43) that for any $0 < \alpha < 1$,

$$\lim_n P\{T > \chi^2_\alpha(r)\} = \alpha,$$

where $\chi^2_\alpha(r)$ is the $\alpha$-quantile of $\chi^2$-distribution with $r$ degrees of freedom at level $\alpha$.

Under $H_1 : \mathbf{C}\boldsymbol{\beta}_{\mathcal{M}} - \mathbf{t} = \mathbf{h}_n$, it follows from (7.43) that for any $0 < \alpha < 1$,

$$\lim_n |P\{T > \chi^2_\alpha(r)\} - P\{\chi^2(r, \gamma_n) > \chi^2_\alpha(r)\}| = 0.$$

Thus, the asymptotic power function of the partial penalized likelihood ratio test is

$$P\{\chi^2(r, \gamma_n) > \chi^2_\alpha(r)\}. \tag{7.44}$$

Based on the property of $\chi^2$-distribution (Ghosh, 1973), the asymptotic power function decreases as $r$ increases for a given $\gamma_n$. This is the same as that for the traditional likelihood ratio test. However, the asymptotic power function is unnecessary to be a monotonically increasing function of $r$ without given $\gamma_n$ since $\gamma_n$ can also grow with $r$.

### *7.2.4 Numerical comparison*

This section presents some numerical comparisons among the following test statistics: (1) the Wald test statistic based on the desparsified Lasso estimator (denoted by $T_W^D$) discussed in Section 7.2.1, (2) the decorrelated score test (denoted by $T_S^D$) defined in Section 7.2.2, and (3) the partial likelihood ratio test ($T$) given in Section 7.2.3.

The test statistic $T_W^D$ is computed via the R package `hdi` (Dezeure, Bühlmann, Meier and Meinshausen, 2015). We calculate $T_S^D$ according to Section 4.1 in Ning and Liu (2017) with $\widetilde{\boldsymbol{\beta}}$ being the SCAD estimate, and $\widehat{\mathbf{A}}$ being the Dantzig-type estimate. These penalized regressions are implemented via the R package `ncvreg` (Breheny and Huang, 2011). The tuning parameters are selected via 10-folded cross-validation.

Simulated data were generated from the normal linear model and logistic regression. For the normal linear model,

$$Y = 2X_1 - 2X_2 + hX_3 + \varepsilon, \tag{7.45}$$

where $\varepsilon \sim N(0, 1)$, and set $n = 100$ and $p = 200$.

Table 7.1: Rejection probabilities (%) of the three tests with standard errors in parenthesis (%). This table is adapted from Shi, *et al.* (2019)

| $h$ | $T$ | $T_W^D$ | $T_S^D$ |
|---|---|---|---|
| | Normal Linear Model | | |
| 0 | 5.67(0.94) | 6.50(1.01) | 2.67(0.66) |
| 0.1 | 13.67(1.40) | 3.67(0.77) | 8.17(1.12) |
| 0.2 | 39.17(1.99) | 9.67(1.21) | 24.67(1.76) |
| 0.4 | 91.50(1.14) | 51.33(2.04) | 80.50(1.62) |
| | Logistic Regression Model | | |
| 0 | 7.67(1.09) | 5.50(0.93) | 4.00(0.80) |
| 0.2 | 22.00(1.66) | 9.00(1.17) | 15.50(1.48) |
| 0.4 | 64.50(1.95) | 31.67(1.90) | 48.00(2.04) |
| 0.8 | 99.00(0.41) | 94.33(0.94) | 97.50(0.64) |

For the logistic regression model,

$$\text{logit}\{P(Y = 1|\mathbf{X})\} = 2X_1 - 2X_2 + hX_3, \tag{7.46}$$

where $\text{logit}(x) = \log\{x/(1-x)\}$, the logit link function, and set $n = 300$ and $p = 200$.

In (7.45) and (7.46), $\mathbf{X} \sim N(\mathbf{0}, \mathbf{\Sigma})$ and $h$ is a constant. The true value $\boldsymbol{\beta}_0 = (2, -2, h, \mathbf{0}_{p-3}^T)^T$ where $\mathbf{0}_q$ denotes a zero vector of length $q$. The covariance matrix $\mathbf{\Sigma}$ is of AR(1) covariance structure with the (i,j)-element $0.5^{|i-j|}$.

Consider testing the following hypotheses:

$$H_0 : \beta_3 = 0, \quad \text{versus} \quad H_1 : \beta_3 \neq 0,$$

which is equivalent to $h = 0$ in this example. We take $h = 0, 0.1, 0.2, 0.4$ for the normal linear model and $h = 0, 0.2, 0.4, 0.8$ for the logistic regression model to examine the Type I error rate and the power of the three tests. Table 7.1 displays the rejection probabilities of the three tests over 600 replications. It can be seen from Table 7.1 that all three tests perform well in terms of retaining the Type I error rate. It seems that $T_W^D$ is less powerful than the other two, and the partial penalized likelihood test performs the best in terms of power.

### 7.2.5 *An application*

In this section, we illustrate the partial penalized likelihood test by an empirical analysis of the European American single nucleotide polymorphisms (SNPs) data (Price, Patterson, Plenge, Weinblatt, Shadick and Reich, 2006). Material of this section was extracted from Shi, *et al.* (2019), in which the authors have conducted an empirical analysis of this data set to illustrate their partial penalized likelihood ratio test. This data set consists of 488 European American samples. We use the height phenotype (0/1, binary variable) of these European American samples as the response, and focus on finding variables that are associated with this phenotype among a set of 277 SNPs. The genotype for each SNP is a categorical variable, coded as 0/1/2. We removed the outlier individuals as in Price, *et al.* (2006). This gives us a total of 361 observations. There are approximately 2% missing values on SNPs for each subject on average. The R package `missForest` downloaded from CRAN was used to impute all the missing values.

To formulate the testing hypotheses, we adopt a data splitting procedure. First randomly sample 20% of data and perform a preliminary analysis based on this data subset. We independently fit 277 logistic regressions by maximizing the marginal likelihood with the response and each univariate covariate and obtain the p-values from each marginal model. There are a total of 13 out of 277 SNPs with p-values smaller than 0.05.

Based on the remaining 80% of the data set, we first test whether the regression coefficients of these 13 variables are zero or not. The p-value of the partial likelihood ratio test is $7.0 \times 10^{-3}$. So we reject the null hypothesis at level 0.05 and conclude that at least one of the 13 regression coefficients is not equal to zero.

Since each covariate $X_j$ is discrete and takes value on $\{0, 1, 2\}$, we can define the dummy variables $Z_{(j,1)} = I(X_j = 1)$, $Z_{(j,2)} = I(X_j = 2)$ and use these $Z_{(j,m)}$'s as covariates in the logistic regression model. This yields a total of 554 covariates. Denote $\beta_{j,m}$ to be the corresponding regression coefficient of $Z_{(j,m)}$. Testing the existence of dominant effects of all selected 13 SNPs can be formulated as the following hypothesis testing based on the remaining 80% of the samples:

$$H_0 : 2\beta_{j,1} = \beta_{j,2}, \quad \forall j \in \mathcal{M},$$

where $\mathcal{M}$ denotes the set of the 13 SNPs selected in the preliminary analysis. Under $H_0$, the effect of SNP $X_j$ is an additive effect since $\beta_{j,1}X_j = \beta_{j,1}Z_{(j,1)} + \beta_{j,2}Z_{(j,2)}$ for any $j \in \mathcal{M}$ under the null hypothesis. This corresponds to a lack of fit test regarding these important variables. The p-value of the partial likelihood ratio test is 0.086. Hence, we fail to reject $H_0$. This implies no existence of dominant effects of the 13 SNPs.

## 7.3 Asymptotic Efficiency*

We have studied in Section 7.1 asymptotically normal estimators and confidence intervals for the noise level $\sigma$ and regression coefficients $\beta_j$ in the linear regression model (7.3). Under the conditions of Theorem 7.2 (ii), the scaled Lasso estimator (7.24) for the noise level is clearly efficient as it is within $o_P(n^{-1/2})$ of the oracle efficient estimator $\widehat{\sigma}^o = \|\varepsilon\|_2/\sqrt{n}$ based on the knowledge of $\boldsymbol{\beta}$. While Theorem 7.1 provides the asymptotic normality of the LDPE (7.4), its asymptotic variance is implicit and its asymptotic efficiency does not directly follow. In this section, we prove the asymptotic efficiency of the LDPE for random design. However, we shall first discuss statistical efficiency and minimum Fisher information for the estimation of a general LD parameter and then apply these to the high-dimensional linear model and the partially linear model.

### 7.3.1 *Statistical efficiency and Fisher information*

Suppose we observe data points $\{\text{data}_i, 1 \le i \le n\}$ with the negative log-likelihood

$$\ell_{[n]}(\boldsymbol{\beta}) = \sum_{i=1}^{n} \ell_i(\boldsymbol{\beta}), \tag{7.47}$$

where $\ell_i(\boldsymbol{\beta}) = \ell_{i,n}(\boldsymbol{\beta}; \text{data}_i)$ is the negative log-likelihood of $\text{data}_i$ (possibly depending on $n$) and $\boldsymbol{\beta}$ represents the unknown parameter.

We assume that $\boldsymbol{\beta}$ belongs to a parameter space $\mathscr{B}_n$, and that $\mathscr{B}_n$ is a closed subset of a Hilbert space $\mathscr{H}_n$ equipped with an inner product $\langle \cdot, \cdot \rangle$ and the associated norm $\|\mathbf{u}\| = \langle \mathbf{u}, \mathbf{u} \rangle^{1/2}$. The dimension of $\text{data}_i$ and the spaces $\mathscr{B}_n$ and $\mathscr{H}_n$ are all allowed to depend on $n$. For simplicity, we treat $\boldsymbol{\beta}$ as a vector in $\mathbb{R}^p$, possibly with $p > n$, so that $\mathscr{H}_n = \mathbb{R}^p$ and $\langle \mathbf{u}, \mathbf{v} \rangle = \mathbf{u}^T \mathbf{v}$. The analysis in this subsection is also valid when $\boldsymbol{\beta}$ represents a general unknown object in a smooth manifold, for example an unknown smooth function.

For example, when $y_i$ belongs to an exponential family (see Section 5.1), $\text{data}_i = \{\mathbf{x}_i, y_i\}$ and

$$\ell_{[n]}(\boldsymbol{\beta}) = \sum_{i=1}^{n} \Big[ \psi_i\big(\langle T_i(\mathbf{x}_i), \boldsymbol{\beta} \rangle\big) - y_i \langle T_i(\mathbf{x}_i), \boldsymbol{\beta} \rangle \Big] \tag{7.48}$$

with observable $T_i(\mathbf{x}_i) \in \mathscr{H}_n$. In generalized linear regression with the canonical link, $\boldsymbol{\beta} \in \mathbb{R}^p$, $T_i(\mathbf{x}_i) = \mathbf{x}_i \in \mathbb{R}^p$, with $\psi_i(t) = t^2/2$ for linear regression. When $\boldsymbol{\beta}$ represents an unknown function $f$ in an *RKHS* $\mathscr{H}$, $\langle T_i(\mathbf{x}_i), f \rangle = f(\mathbf{x}_i) \in \mathbb{R}$ where $T(\mathbf{x}) \in \mathscr{H}$ is the reproducing kernel.

In the proportional hazards regression model (see Section 6.5), $\text{data}_i = \{t_i, R_i, \mathbf{x}_{i,j}, y_i\}$ and

$$\ell_{[n]}(\boldsymbol{\beta}) = \sum_{i=1}^{n} \Big\{ \log\Big( \sum_{j \in R_i} \exp\big[\langle T_{i,j}(\mathbf{x}_{i,j}), \boldsymbol{\beta} \rangle\big]\Big) - \langle T_{i,y_i}(\mathbf{x}_{i,y_i}), \boldsymbol{\beta} \rangle \Big\}, \tag{7.49}$$

where $t_i$ is the time of occurrence of the $i$-th event, $R_i$ is the set of subjects at risk at time $t_i-$, $\mathbf{x}_{i,j}$ is the covariate vector of the $j$-th subject at time $t_i$, and $y_i \in R_i$ indicates the subject to whom the event is observed at time $t_i$.

Consider the estimation of a real parameter $\theta = \theta(\boldsymbol{\beta})$ as a smooth function of $\boldsymbol{\beta}$. This serves two purposes. Firstly, it is of independent interest to consider the problem of statistical inference of a given LD parameter with HD data in the presence of an HD nuisance parameter vector. Secondly, statistical inference for a vector parameter can be carried out through its individual components.

Stein (1956) suggested characterizing the statistical efficiency of estimating the univariate parameter $\theta = \theta(\boldsymbol{\beta})$ at a point $\boldsymbol{\beta}$ by the minimum Fisher information in univariate sub-models (hardest subproblem). Given $\boldsymbol{\beta} \in \mathscr{B}_n$, the univariate sub-model, index by the direction $\mathbf{u}$ of its deviation from $\boldsymbol{\beta}$, can be written as

$$\{\mathbf{b}(\phi) : \|\mathbf{b}(\phi) - \boldsymbol{\beta} - \phi\mathbf{u}\| = o(\phi) : 0 \le \phi \le \epsilon^*_{\mathbf{u}}\} \subseteq \mathscr{B}_n$$

subject to the condition $\epsilon^*_{\mathbf{u}} \gg n^{-1/2}$. We denote by $\mathscr{U}_{n,m}$ certain collections of $m$ linearly independent directions $\mathbf{u} \in \mathscr{H}_n$ under consideration. While $\mathbf{b}(\phi)$ is not uniquely determined by $\mathbf{u}$ and $\phi$, for notational simplicity we use $\boldsymbol{\beta} + \phi\mathbf{u}$ to identify $\mathbf{b}(\phi)$ in such univariate sub-models with $\mathbf{u} \in \mathscr{U}_{n,m}$, even when $\boldsymbol{\beta} + \phi\mathbf{u} \notin \mathscr{B}_n$. In this convention, the univariate sub-model is written as $\{\boldsymbol{\beta}+\phi\mathbf{u}, 0 \le \phi \le \epsilon^*_{\mathbf{u}}\}$, $\mathbf{u} \in \mathscr{U}_{n,m}$ and any statement involving $\boldsymbol{\beta}+\phi\mathbf{u}$ must hold with $\boldsymbol{\beta}+\phi\mathbf{u}$ replaced by $\mathbf{b}(\phi)$ in all univariate sub-models under consideration.

Assume that $\theta(\boldsymbol{\beta})$ is differentiable with derivative

$$\mathbf{a}^o = \partial\theta(\boldsymbol{\beta})/\partial\boldsymbol{\beta} \in \mathscr{H}_{n,0} \setminus \{\mathbf{0}\} \tag{7.50}$$

in the sense of $\langle \mathbf{a}^o, h\mathbf{u}\rangle = (1+o(1))n^{1/2}\{\theta(\boldsymbol{\beta}+hn^{-1/2}\mathbf{u}) - \theta(\boldsymbol{\beta})\}, \mathbf{u} \in \mathscr{U}_{n,m}$ for every fixed $h \in I\!R$ and $m$, where $\mathscr{H}_{n,0}$ is a closed subspace containing all $\mathscr{U}_{n,m}$ under consideration. When $\boldsymbol{\beta} + hn^{-1/2}\mathbf{u} \notin \mathscr{B}_n$, this means $\langle \mathbf{a}^o, h\mathbf{u}\rangle = (1 + o(1))n^{1/2}\{\theta(\mathbf{b}(hn^{-1/2})) - \theta(\boldsymbol{\beta})\}, \mathbf{u} \in \mathscr{U}_{n,m}$. When the solution is not unique, $\mathbf{a}^o$ is defined as the one with the smallest $\|\mathbf{a}^o\|$. Here and in the sequel, the superscript $^o$ indicates that the quantity could be known in certain settings. For example, $\mathbf{a}^o$ is known when we estimate a linear functional $\theta = \langle \mathbf{a}^o, \boldsymbol{\beta}\rangle$.

Assume there exist certain negative scores $\dot{\ell}_i(\boldsymbol{\beta})$ as random elements in $\mathscr{H}_n$ such that $\langle \dot{\ell}_i(\boldsymbol{\beta}), \mathbf{u}\rangle = (\partial/\partial\phi)\ell(\boldsymbol{\beta} + \phi\mathbf{u})\big|_{\phi=0}$, $\mathrm{E}\langle \dot{\ell}_i(\boldsymbol{\beta}), \mathbf{u}\rangle = 0, 1 \le i \le n$, and

$$c_0 \le n^{-1}\textstyle\sum_{i=1}^n \mathrm{E}\langle \dot{\ell}_i(\boldsymbol{\beta}), \mathbf{u}\rangle^2 \le 1/c_0, \ \forall\, \mathbf{u} \in \mathscr{U}_{n,m},$$

for some fixed $c_0 \in (0,1)$. Define the Fisher information operator $\mathbf{F}$ at $\boldsymbol{\beta}$ by

$$\langle \mathbf{u}, \mathbf{F}\mathbf{v}\rangle = n^{-1}\textstyle\sum_{i=1}^n \mathrm{E}\langle \dot{\ell}_i(\boldsymbol{\beta}), \mathbf{u}\rangle\langle \dot{\ell}_i(\boldsymbol{\beta}), \mathbf{v}\rangle, \ \forall\ \mathbf{u}, \mathbf{v} \in \mathscr{H}_{n,0}, \tag{7.51}$$

as the bilinear extension from $\mathbf{u}, \mathbf{v} \in \mathscr{U}_{n,m}$ to $\mathscr{H}_{n,0}$, Assume the local asymptotic normality (*LAN*) condition (Le Cam, 1960) in the sense that for all fixed $\{h, c_1, \ldots, c_m\} \subset I\!R$

$$\begin{aligned}
&\sum_{j=1}^m c_j\Big(\sum_{i=1}^n \{\ell_i(\boldsymbol{\beta}) - \ell_i(\boldsymbol{\beta} + hn^{-1/2}\mathbf{u}_j)\}\Big) \\
= &\ -\sum_{i=1}^n \frac{h\langle \dot{\ell}_i(\boldsymbol{\beta}), \mathbf{u}\rangle}{n^{1/2}} - \sum_{j=1}^m c_j \frac{h^2\langle \mathbf{u}_j, \mathbf{F}\mathbf{u}_j\rangle}{2} + o_P(1) \qquad (7.52)\\
\approx &\ N\big(-\textstyle\sum_{j=1}^m c_j h^2\langle \mathbf{u}_j, \mathbf{F}\mathbf{u}_j\rangle/2, h^2\langle \mathbf{u}, \mathbf{F}\mathbf{u}\rangle\big)
\end{aligned}$$

when $\boldsymbol{\beta}$ is the true parameter under $P$, where $\{\mathbf{u}_1, \ldots, \mathbf{u}_m\} = \mathscr{U}_{n,m}$ and $\mathbf{u} =$

$\sum_{j=1}^m c_j \mathbf{u}_j$. Here $Z_n \approx N(\mu_n, \sigma_n^2)$ means the convergence of $(Z_n - \mu_n)/\sigma_n$ in distribution to $N(0,1)$. We note that (7.52) guarantees the convergence of the joint distribution of $n^{-1/2}\sum_{i=1}^n \langle \dot{\ell}_i(\boldsymbol{\beta}), \mathbf{u}\rangle, \mathbf{u} \in \mathscr{U}_{n,m}$. When $\boldsymbol{\beta} + hn^{-1/2}\mathbf{u}_j \notin \mathscr{B}_n$, the interpretation of (7.52) is its validity with $\boldsymbol{\beta} + hn^{-1/2}\mathbf{u}_j$ replaced by $\mathbf{b}_j(hn^{-1/2})$ in univariate sub-models satisfying $\|\mathbf{b}_j(\phi) - \boldsymbol{\beta} - \phi\mathbf{u}_j\| = o(\phi)$.

When $\boldsymbol{\beta} \in I\!R^p$ and $\ell_i(\boldsymbol{\beta})$ is smooth in $\boldsymbol{\beta}$, $\dot{\ell}_i(\boldsymbol{\beta}) = (\partial/\partial\boldsymbol{\beta}) \log \ell_i(\boldsymbol{\beta})$, $\mathbf{F}$ is the Hessian of the average expected negative log-likelihood

$$\mathbf{F} = (\partial/\partial\boldsymbol{\beta})^{\otimes 2}\,\mathrm{E}\left[\sum_{i=1}^n \ell_i(\boldsymbol{\beta})/n\right] \in I\!R^{p\times p},$$

and the classical definition of efficient estimation can be stated as

$$\mathrm{E}\left[g\big((n\mathbf{F})^{1/2}(\widehat{\boldsymbol{\beta}} - \boldsymbol{\beta})\big)\right] = (1+o(1))\,\mathrm{E}\left[g\big(N(\mathbf{0}, \mathbf{I}_p)\big)\right] + o(1)$$

for suitable classes of continuous functions $g$. However, when $\boldsymbol{\beta}$ lives in a large space, efficient estimation of the entire $\boldsymbol{\beta}$ may not be feasible for the given sample. Still, this does not exclude efficient estimation of the univariate parameter $\theta(\boldsymbol{\beta})$. Here and in the rest of this chapter, $\mathbf{F}$ is the average Fisher information per data point, and the dependence of $\mathbf{F}$ and related quantities on $\boldsymbol{\beta}$ and $n$ is suppressed.

When the sub-model is scaled to satisfy $\langle \mathbf{a}^o, \mathbf{u}\rangle = 1$, the parameter $\phi$ of the sub-model approximately agrees with a shifted parameter of interest,

$$\phi = \langle \mathbf{a}^o, \phi\mathbf{u}\rangle = (1+o(1))\big\{\theta(\boldsymbol{\beta} + \phi\mathbf{u}) - \theta(\boldsymbol{\beta})\big\},$$

for $\phi \asymp n^{-1/2}$. Thus, as the Fisher information for the estimation of $\theta(\boldsymbol{\beta} + \phi\mathbf{u})$ at $\phi = 0$ is $\langle \mathbf{u}, \mathbf{F}\mathbf{u}\rangle$ in the sub-model $\{\boldsymbol{\beta} + \phi\mathbf{u}, 0 \le \phi \le \epsilon_{\mathbf{u}}^*\}$, the *minimum Fisher information* for the estimation of $\theta(\boldsymbol{\beta})$ at $\boldsymbol{\beta}$ can be defined as

$$F_\theta = \liminf_{m\to\infty} \liminf_{n\to\infty} \min_{\mathbf{u}\in\mathscr{H}_{n,m}, \langle \mathbf{a}^o, \mathbf{u}\rangle = 1} \langle \mathbf{u}, \mathbf{F}\mathbf{u}\rangle, \tag{7.53}$$

where $\mathscr{H}_{n,m}$ is the linear span of $\mathscr{U}_{n,m}$. We note that the minimum in (7.53) is not taken over $\mathscr{U}_{n,m}$ but its span $\mathscr{H}_{n,m}$. As $\mathscr{U}_{n,m}$ is allowed to be optimized and to grow, the $F_\theta$ in (7.53) is often within an infinitesimal fraction of the Fisher information at the *least favorable direction* $\mathbf{u}^o$ in $\mathscr{H}_{n,0}$. With a slight abuse of notation, we write them as

$$F_\theta = \langle \mathbf{u}^o, \mathbf{F}\mathbf{u}^o\rangle, \quad \mathbf{u}^o = \mathop{\arg\min}_{\mathbf{u}\in\mathscr{H}_{n,0}, \langle \mathbf{a}^o, \mathbf{u}\rangle = 1} \langle \mathbf{u}, \mathbf{F}\mathbf{u}\rangle. \tag{7.54}$$

The $F_\theta$ in (7.54) becomes $F_{\theta,n,m}$ when $\mathscr{H}_{n,0}$ is replaced by $\mathscr{H}_{n,m}$, but the dependence on $m$ is not crucial as we are allowed to chose $\mathbf{u} \in \mathscr{H}_{n,m}$ with $\|\mathbf{F}^{1/2}(\mathbf{u} - \mathbf{u}^o)\|_2 \to 0$ in the limit process. More explicitly, we have

$$\mathbf{u}^o = \mathbf{F}^{-1}\mathbf{a}^o/\langle \mathbf{a}^o, \mathbf{F}^{-1}\mathbf{a}^o\rangle, \quad F_\theta = \langle \mathbf{u}^o, \mathbf{F}\mathbf{u}^o\rangle = 1/\langle \mathbf{a}^o, \mathbf{F}^{-1}\mathbf{a}^o\rangle, \tag{7.55}$$

where $\mathbf{F}^{-1}$ is the generalized inverse of $\mathbf{F}$ in $\mathscr{H}_n$.

The minimum Fisher information has the following interpretation through the local asymptotic theory (Le Cam, 1972, 1986; Hájek, 1972) even if the dimensions of the data and parameter space are allowed to change with $n$. In this theory, an estimator $\widehat{\theta}$ is regular at $\boldsymbol{\beta} \in \mathscr{B}_n$ if for all $\mathbf{u} \in \mathscr{U}_{n,m}$ and real $h$

$$\lim_{n\to\infty} \mathcal{L}_{\boldsymbol{\beta}+hn^{-1/2}\mathbf{u}}\Big(\sqrt{nF_\theta}\big\{\widehat{\theta} - \theta(\boldsymbol{\beta} + hn^{-1/2}\mathbf{u})\big\}\Big) = G \tag{7.56}$$

for a certain distribution $G$ not depending on $\mathscr{U}_{n,m}$, where $\mathcal{L}_{\boldsymbol{\beta}}(\xi)$ is the distribution of $\xi$ when $\boldsymbol{\beta}$ is the true unknown parameter. This regularity property requires that the confidence interval based on the estimator $\widehat{\theta}$ and limiting distribution $G$ has the same asymptotic coverage probability in $n^{-1/2}$-neighborhoods of $\boldsymbol{\beta}$ in the one-dimensional sub-models with $\mathbf{u} \in \mathscr{U}_{n,m}$. We state the statistical lower bound for regular estimation of $\theta$ as follows.

**Theorem 7.3** *Suppose the differentiability condition* (7.50) *and the LAN condition* (7.52) *hold. Suppose* $F_\theta > 0$ *in* (7.53).
*(i) For any regular estimator of* $\theta$*, the limiting distribution* $G$ *in* (7.56) *is of a random variable* $\xi$ *with decomposition* $\xi = \zeta_0 + \zeta_1$ *such that* $\zeta_0 \sim N(0,1)$ *and* $E\zeta_1\zeta_0 = 0$. *If in addition the differentiability and LAN conditions hold with* $\mathscr{U}_{n,m} = \{\mathbf{u}_1, \ldots, \mathbf{u}_m\}$ *replaced by* $\mathscr{U}_{n,m} \cup \{\mathbf{v}\}$ *for every* $\mathbf{v} = \sum_{j=1}^m b_j\mathbf{u}_j$ *with* $(b_1, \ldots, b_m)^T$ *in an open set of* $I\!R^m$*, then* $\zeta_1$ *is independent of* $\zeta_0$.
*(ii) An estimator* $\widehat{\theta}$ *is regular and attains the minimum asymptotic mean squared error among regular estimators if and only if it can be expressed in the following asymptotic linear form under the probability associated with the parameter* $\boldsymbol{\beta}$,

$$\widehat{\theta} = \theta - n^{-1}\textstyle\sum_{i=1}^n \langle \dot{\ell}_i(\boldsymbol{\beta}), \mathbf{u}^o\rangle / F_\theta + o_P\big((nF_\theta)^{-1/2}\big), \tag{7.57}$$

*with the minimum Fisher information* $F_\theta$ *and the direction* $\mathbf{u}^o$ *of the least favorable sub-model as in* (7.54). *In this case,* $\sqrt{nF_\theta}(\widehat{\theta} - \theta) \xrightarrow{D} N(0,1)$.

There is a vast literature on efficient estimation in parametric and semiparametric models. See Bickel *et al.* (1998), van der Vaart (2000) and references therein. Theorem 7.3 is called the convolution theorem (Hájek, 1970) as the limiting distribution is that of a sum of $\zeta_0 \sim N(0,1)$ and an additional noise variable independent of $\zeta_0$. The convolution theorem is typically stated under the condition that the least favorable direction can be directly approximated by the direction of the univariate sub-model, i.e. $F_\theta \approx F_{\theta,n,1}$ can be achieved with $m = 1$. However, more general versions were provided in Schick (1986), van der Vaart (2000, Theorem 25.20) and Zhang (2005, Theorem 6.1). The theorem justifies the following definition of statistical efficiency.

**Definition 7.1** *An estimator* $\widehat{\theta}$ *is asymptotically efficient for statistical inference about the parameter* $\theta = \theta(\boldsymbol{\beta})$ *at* $\boldsymbol{\beta}$ *if it is regular and attains the minimum asymptotic mean squared error among all regular estimators, or* *equivalently if it can be written as* (7.57) *under the probability associated with* $\boldsymbol{\beta}$. *In this case,* $[\widehat{L}, \widehat{U}]$ *is an efficient* $(1-\alpha)\times 100$ *percent confidence interval if*

$$\widehat{L} = \widehat{\theta} + (1+o_P(1))\frac{\Phi^{-1}(\alpha/2)}{(nF_\theta)^{1/2}}, \quad \widehat{U} = \widehat{\theta} - (1+o_P(1))\frac{\Phi^{-1}(\alpha/2)}{(nF_\theta)^{1/2}},$$

*where* $\Phi^{-1}$ *is the standard normal quantile function. The variable* $\langle \dot{\ell}_i(\boldsymbol{\beta}), \mathbf{u}^o\rangle / F_\theta$ *in* (7.57) *is called the efficient influence function of* $\text{data}_i$.

We prove Theorem 7.3 below and leave some details of the local asymptotic theory as an exercise.

**Proof of Theorem 7.3.** Let $\mathscr{U}_{n,m} = \{\mathbf{u}_1, \ldots, \mathbf{u}_m\}$ and

$$\mathbf{u}_0 = \arg\min_{\mathbf{u}} \Big\{ \langle \mathbf{u}, \mathbf{F}\mathbf{u}\rangle : \mathbf{u} \in \mathscr{H}_{n,m}, \langle \mathbf{a}^o, \mathbf{u}\rangle = 1 \Big\},$$

and $F_j = \langle \mathbf{u}_j, \mathbf{F}\mathbf{u}_j\rangle \in [c_0, 1/c_0]$. As we can always find $\mathscr{U}_{n,m}$ with $F_0/F_\theta = 1 + o(1)$ and $\|\mathbf{u}^o - \mathbf{u}_0\|_2 / \|\mathbf{u}\|_2 = o(1)$, it suffices to prove the theorem with $F_\theta$ replaced by $F_0 = \langle \mathbf{u}_0, \mathbf{F}\mathbf{u}_0\rangle = F_{\theta,n,m}$. As in (7.55), $\langle \mathbf{u}_0, \mathbf{F}\mathbf{u}_j\rangle = F_0\langle \mathbf{a}^o, \mathbf{u}_j\rangle$.

Taking subsequences if necessary we assume without loss of generality by the regularity of $\widehat{\theta}$ and the LAN condition (7.52) that

$$\left( \sqrt{n}(\widehat{\theta} - \theta), -\sum_{i=1}^n \frac{\langle \dot{\ell}_i(\boldsymbol{\beta}), \mathbf{u}_j\rangle}{n^{1/2}}, j = 0, \ldots, m \right)$$

converges in joint distribution to $(\xi', \xi_0, \xi_1, \ldots, \xi_m)$ with Gaussian $(\xi_0, \xi_1 \ldots, \xi_m)$, $\xi_0 \sim N(0, F_0)$, $\xi_j \sim N(0, F_j)$ and $\mathrm{E}\xi_0\xi_j = \langle \mathbf{u}_0, \mathbf{F}\mathbf{u}_j\rangle = F_0\langle \mathbf{a}^o, \mathbf{u}_j\rangle$, $j = 1, \ldots, m$.

By the regularity of $\widehat{\theta}$, the LAN condition (7.52) on the log-likelihood ratio, and the differentiability condition (7.50) on $\theta(\cdot)$, the characteristic function of $\xi = \sqrt{F_0}\xi' \sim G$ must satisfy that for $j = 1, \ldots, m$ and $C_j = \langle \mathbf{a}^o, \mathbf{u}_j\rangle$

$$\begin{aligned} \mathrm{E}\, e^{\sqrt{-1}t\xi'} &= \lim_{n\to\infty} \mathrm{E}\exp\Big[ \sqrt{-1}t\sqrt{n}\Big\{\widehat{\theta} - \theta\Big(\boldsymbol{\beta} + hn^{-1/2}\mathbf{u}_j\Big)\Big\} \\ &\qquad + \sum_{i=1}^n \Big\{ \ell_i(\boldsymbol{\beta}) - \ell_i(\boldsymbol{\beta} + hn^{-1/2}\mathbf{u}_j)\Big\}\Big] \\ &= \mathrm{E}\exp\Big[ \sqrt{-1}t(\xi' - hC_j) + h\xi_j - h^2F_j/2 \Big]. \end{aligned} \tag{7.58}$$

With $h = -\sqrt{-1}(C_j/F_j)t$ above and then $h = \sqrt{-1}(C_j/F_j)(-t+z)$, we have

$$\begin{aligned} \mathrm{E}\, e^{\sqrt{-1}t\xi'} &= \mathrm{E}\, e^{\sqrt{-1}t(\xi' - (C_j/F_j)\xi_j) - t^2C_j^2/(2F_j)} \\ &= \mathrm{E}\, e^{\sqrt{-1}\{t(\xi' - (C_j/F_j)\xi_j) + z(C_j/F_j)\xi_j\} + C_j^2\{t(-t+z) + (t-z)^2/2\}/F_j}, \end{aligned}$$

so that $\mathrm{E}\, e^{\sqrt{-1}\{t(\xi' - (C_j/F_j)\xi_j) + z(C_j/F_j)\xi_j\}} = \mathrm{E}\, e^{\sqrt{-1}t(\xi' - (C_j/F_j)\xi_j) - z^2C_j^2/(2F_j)}$.

Thus, $\xi' - (C_j/F_j)\xi_j$ is independent of $\xi_j \sim N(0, F_j)$ with $\mathrm{E}\xi_j\xi' = C_j = \langle \mathbf{a}^o, \mathbf{u}_j \rangle = \mathrm{E}\xi_j\xi_0/F_0$, $j = 1, \ldots, m$. It follows that $\xi' - \xi_0/F_0$ is orthogonal to $\xi_j$, including $\xi_0$. Consequently, the $L_2$ projection of $\xi = \sqrt{F_0}\xi'$ to $\zeta_0 = \xi_0/\sqrt{F_0} \sim N(0,1)$ is $\zeta_0$. When the conditions also holds for $\mathbf{v}$ in an open set in $\mathscr{H}_{n,m}$, the formula for the characteristic function also holds for all $\mathbf{v}$ in the entire $\mathscr{H}_{n,m}$ via analytic extension, so that $\xi' - \xi_0/F_0$ is independent of $\xi_0 \sim N(0, F_0)$ due to $C_0 = 1$. This completes of the proof of part (i).

If $\widehat{\theta}$ is regular and attains the Fisher information, then $\xi' = \xi_0/F_0$, so that $\widehat{\theta}$ has the asymptotic linear form given in part (ii). Conversely, if (7.57) holds, the second equality in (7.58) holds with $\xi' = \xi_0/F_0$ and its right-hand side becomes

$$\begin{aligned} & \mathrm{E}\exp\Big[\sqrt{-1}t\Big\{\xi_0/F_0 - hC_j\Big\} + h\xi_j - h^2F_j/2\Big] \\ = \ & \exp\Big[\mathrm{E}\big(\sqrt{-1}t\xi_0/F_0 + h\xi_j\big)^2/2 - \sqrt{-1}thC_j - h^2F_j/2\Big] \\ = \ & \exp\Big[\mathrm{E}\big(\sqrt{-1}t\xi_0/F_0\big)^2/2\Big] \end{aligned}$$

due to $\mathrm{E}\xi_j\xi_0/F_0 = C_j$. Thus, $\theta$ is regular. ∎

### 7.3.2 *Linear regression with random design*

We consider the efficient statistical inference of a linear functional

$$\theta = \theta(\boldsymbol{\beta}) = \langle \mathbf{a}^o, \boldsymbol{\beta} \rangle \tag{7.59}$$

for a general given $\mathbf{a}^o \in I\!\!R^p$ in the linear model (7.3) with random design. This includes the estimation of $\beta_j$ as a special case with $\mathbf{a}^o = \mathbf{e}_j$. Throughout this subsection, we assume Gaussian noise and random design with

$$\boldsymbol{\varepsilon} \sim N(\mathbf{0}, \sigma^2\mathbf{I}_n), \quad \|\mathbf{X}\mathbf{u}\|_2^2/n = (1 + o_P(1))\mathbf{u}^T\boldsymbol{\Sigma}\mathbf{u}, \tag{7.60}$$

for a suitable set $\mathscr{U}_n$ of $\mathbf{u} \subseteq I\!\!R^p$, where $\boldsymbol{\Sigma}$ is a positive-definite matrix. We provide below the Fisher information for the estimations of $\boldsymbol{\beta}$ and $\theta$.

**Proposition 7.2** *Suppose $\sigma$ is known in* (7.3) *and* (7.60) *holds for $\mathbf{u} \in \mathscr{U}_n$. If $\mathscr{U}_n = I\!\!R^p$, then $\mathbf{F} = \boldsymbol{\Sigma}/\sigma$ is the Fisher information for the estimation of $\boldsymbol{\beta}$ in the parameter space $I\!\!R^p$. Moreover, for the estimation of $\theta = \langle \mathbf{a}^o, \boldsymbol{\beta} \rangle$ in a restricted parameter space $\mathscr{B}_n$,*

$$\mathbf{u}^o = \frac{\boldsymbol{\Sigma}^{-1}\mathbf{a}^o}{\langle \mathbf{a}^o, \boldsymbol{\Sigma}^{-1}\mathbf{a}^o \rangle} \quad \textit{and} \quad F_\theta = \frac{\sigma^{-2}}{\langle \mathbf{a}^o, \boldsymbol{\Sigma}^{-1}\mathbf{a}^o \rangle} \tag{7.61}$$

*are respectively the direction of the least favorable sub-model and the minimum Fisher information at $\boldsymbol{\beta}$, as in* (7.54) *and* (7.55), *provided that $\mathbf{u}^o \in \mathscr{U}_n$ and $\boldsymbol{\beta} + \mathbf{u}^o h/\sqrt{n} \in \mathscr{B}_n$ for all $0 \le h \le b_n$ with certain $h_n \to \infty$.*

Proposition 7.2 is valid for both known and unknown distribution of the design matrix $\mathbf{X}$ as $\mathbf{X}$ is ancillary. It can be shown that Proposition 7.2 also holds in the case of unknown $\sigma$. The efficient estimation of $\sigma$ is provided in Theorem 7.2 (ii) as discussed at the very beginning of this section.

**Proof of Proposition 7.2.** As the differentiability condition (7.50) holds automatically, we first verify the LAN condition (7.52). By (7.60)

$$\sum_{i=1}^{n} \{\ell_i(\boldsymbol{\beta}) - \ell_i(\boldsymbol{\beta} + hn^{-1/2}\mathbf{u})\} = -\frac{h(\mathbf{X}\mathbf{u})^T\boldsymbol{\varepsilon}}{n^{1/2}\sigma^2} - \frac{h^2\|\mathbf{X}\mathbf{u}\|_2^2}{2n\sigma^2}$$

has the $N\big(-h^2\|\mathbf{X}\mathbf{u}\|_2^2/(2n\sigma^2), h^2\|\mathbf{X}\mathbf{u}\|_2^2/(n\sigma^2)\big)$ distribution given $\mathbf{X}$ and the approximate distribution $N\big(-h^2\mathbf{u}^T\mathbf{F}\mathbf{u}/2, h^2\mathbf{u}^T\mathbf{F}\mathbf{u}\big)$ with $\mathbf{F} = \boldsymbol{\Sigma}/\sigma^2$. Thus, (7.52) holds and Theorem 7.3 applies. Now consider the estimation of $\theta$ in the restricted parameter space $\mathscr{B}_n$. It still suffices to verify (7.52), but only for $\mathscr{U}_{n,m} = \{\mathbf{u}^o\}$. This condition holds when $\boldsymbol{\beta} + hn^{-1/2}\mathbf{u}^o \in \mathscr{B}_n$ for sufficiently large $n$ for each real number $h$ or equivalently for all $0 \le h \le h_n$ with $h_n \to \infty$. ∎

Now consider the normalized Gaussian design,

$$\mathbf{X} = \widetilde{\mathbf{X}}\overline{\mathbf{D}}^{-1/2}, \quad \overline{\mathbf{D}} = \text{diag}\left(\|\widetilde{\mathbf{X}}_1\|_2^2/n, \ldots, \|\widetilde{\mathbf{X}}_p\|_2^2/n\right), \tag{7.62}$$

where $\widetilde{\mathbf{X}} = (\widetilde{\mathbf{X}}_1, \ldots, \widetilde{\mathbf{X}}_p)$ is a random matrix with iid $N(0, \widetilde{\boldsymbol{\Sigma}})$ rows and a positive-definite population covariance matrix $\widetilde{\boldsymbol{\Sigma}}$. The regression model (7.3) has the interpretation

$$\mathbf{Y} = \mathbf{X}\boldsymbol{\beta} + \boldsymbol{\varepsilon} = \widetilde{\mathbf{X}}\boldsymbol{\beta}^{(\text{orig})} + \boldsymbol{\varepsilon}$$

where $\boldsymbol{\beta}^{(\text{orig})}$ can be viewed as the original coefficient vector and $\boldsymbol{\beta} = \overline{\mathbf{D}}^{1/2}\boldsymbol{\beta}^{(\text{orig})}$ the (random) normalized coefficient vector. The LDPE of $\beta_j^{(\text{orig})}$ is defined as

$$\widehat{\beta}_j^{(\text{orig})} = \overline{D}_j^{-1/2}\widehat{\beta}_j = (\sqrt{n}/\|\widetilde{\mathbf{X}}_j\|_2)\widehat{\beta}_j, \tag{7.63}$$

where $\widehat{\beta}_j$ is the LDPE of $\beta_j$ in (7.4), or more specifically in (7.14) or (7.14), with an initial estimator $\widehat{\boldsymbol{\beta}}^{(\text{init})}$ of $\boldsymbol{\beta}$.

Let $\boldsymbol{\Sigma}$ be the correlation matrix associated with $\widetilde{\boldsymbol{\Sigma}}$,

$$\boldsymbol{\Sigma} = \mathbf{D}^{-1/2}\widetilde{\boldsymbol{\Sigma}}\mathbf{D}^{-1/2} \text{ with } \mathbf{D} = \text{diag}(\widetilde{\boldsymbol{\Sigma}}) = \text{E}[\overline{\mathbf{D}}]. \tag{7.64}$$

We note that theoretical results on regularized estimation in HD regression are typically stated in terms of the normalized design, for example the sparsity condition on $\boldsymbol{\beta}$ and restricted eigenvalue condition on $\mathbf{X}^T\mathbf{X}/n$. Thus, for the Gaussian design, the population version of such conditions are stated in terms of $\mathbf{D}^{1/2}\boldsymbol{\beta}^{(\text{orig})}$ and $\boldsymbol{\Sigma}$.

For $s \in [0, p]$ consider the capped $\ell_1$ parameter space

$$\mathscr{B}_n(s) = \left\{ \mathbf{b} : \sum_{k=1}^{p} \min\left( \frac{|b_k|}{\sqrt{(2/n)\log p}}, 1 \right) \le s \right\}. \tag{7.65}$$

We recall that $\boldsymbol{\beta}$ is required to be sparse to guarantee condition (7.8) in Theorem 7.1. However, as $\boldsymbol{\beta}$ is random due to its dependence on $\overline{\mathbf{D}}$, we write the sparsity condition as $\mathbf{D}^{1/2}\boldsymbol{\beta}^{(\mathrm{orig})}/\sigma \in \mathscr{B}_n(s^*)$. The efficient estimation of the deterministic $\beta_j^{(\mathrm{orig})}$ and $D_j^{1/2}\beta_j^{(\mathrm{orig})}$ is equivalent to that of random $\beta_j$ given $\mathbf{X}$ via (7.9) due to the scaling invariance and convergence of $\|\widetilde{\mathbf{X}}_j\|_2^2/n$ to $D_j$.

Let $\mathbf{u}^{(j)} = \mathbf{\Sigma}^{-1}\boldsymbol{e}_j/(\boldsymbol{e}_j^T\mathbf{\Sigma}^{-1}\boldsymbol{e}_j)$ be the direction of the least favorable sub-model as defined in (7.61) with $\mathbf{a}^o = \boldsymbol{e}_j$ and the $\mathbf{\Sigma}$ in (7.64). This is for the estimation of $D_j^{1/2}\beta_j^{(\mathrm{orig})}$ as $\mathbf{\Sigma}$ is used. The corresponding direction of the least favorable sub-model for the estimation of $\beta_j$ is $\widetilde{\mathbf{D}}\mathbf{D}^{-1/2}\mathbf{u}^{(j)}$. Thus, the interpretation of the second part of condition (7.60) is

$$\left\|\mathbf{X}\widetilde{\mathbf{D}}\mathbf{D}^{-1/2}\mathbf{u}\right\|_2^2/n = \left\|\widetilde{\mathbf{X}}\mathbf{D}^{-1/2}\mathbf{u}\right\|_2^2/n = (1 + o_P(1))\mathbf{u}^T\mathbf{\Sigma}\mathbf{u}.$$

This holds automatically for the Gaussian design as $\left\|\widetilde{\mathbf{X}}\mathbf{D}^{-1/2}\mathbf{u}\right\|_2^2/\mathbf{u}^T\mathbf{\Sigma}\mathbf{u}$ has the chi-square distribution with $n$ degrees of freedom. In fact, for the Gaussian noise $\boldsymbol{\varepsilon}$ as in (7.60) and any random design $\widetilde{\mathbf{X}}$ satisfying the above condition,

$$F_j = \langle \mathbf{u}^{(j)}, \mathbf{F}\mathbf{u}^{(j)} \rangle = 1/\left\{\sigma^2(\boldsymbol{e}_j^T\mathbf{\Sigma}^{-1}\boldsymbol{e}_j)\right\} \tag{7.66}$$

is the minimum Fisher information for the estimation of both $\beta_j$ and $D_j^{1/2}\beta_j^{(\mathrm{orig})}$ by Proposition 7.2, provided that $\{\mathbf{D}^{1/2}\boldsymbol{\beta}^{(\mathrm{orig})} + h(nF_j)^{-1/2}\mathbf{u}^{(j)}, |h| \le h_n^*\}$ is contained in the parameter space for $\mathbf{D}^{1/2}\boldsymbol{\beta}^{(\mathrm{orig})}$ with some $h_n^* \to \infty$. Let

$$s_j = \sum_{k \neq j} \min\left\{ \frac{(\boldsymbol{e}_j^T\mathbf{\Sigma}^{-1}\boldsymbol{e}_j)^{1/2}|u_k^{(j)}|}{\sqrt{(2/n)\log p}}, 1 \right\} \tag{7.67}$$

for $\mathbf{u}^{(j)} = (u_1^{(j)}, \ldots, u_p^{(j)})^T$. When $\mathbf{D}^{1/2}\boldsymbol{\beta}^{(\mathrm{orig})}/\sigma \in \mathscr{B}_n(s^*)$, the parameter space $\sigma\mathscr{B}_n(s^* + s_j)$ indeed contains $\{\mathbf{D}^{1/2}\boldsymbol{\beta}^{(\mathrm{orig})} + h(nF_j)^{-1/2}\mathbf{u}^{(j)}, |h| \le h_n^*\}$ as the least favorable sub-model with $h_n^* \to \infty$.

For the estimation of $\beta_j^{(\mathrm{orig})}$, the Fisher information is $\widetilde{F}_j = F_j/D_j$ by Proposition 7.2 with $\mathbf{\Sigma}$ replaced by $\widetilde{\mathbf{\Sigma}}$, so that

$$\left(n\widetilde{F}_j\right)^{1/2}\left(\widehat{\beta}_j^{(\mathrm{orig})} - \beta_j^{(\mathrm{orig})}\right) = (1 + o_P(1))\sqrt{nF_j}(\widehat{\beta}_j - \beta_j) \tag{7.68}$$

in view of (7.63). The following theorem asserts that the above variables are approximately $N(0, 1)$ for the estimator in (7.4).

**Theorem 7.4** *Consider the linear model* (7.3) *with* $\boldsymbol{\varepsilon} \sim N(\mathbf{0}, \sigma^2 \mathbf{I}_n)$ *and normalized Gaussian design* (7.62) *with population correlation matrix* $\boldsymbol{\Sigma}$ *in* (7.64). *Suppose the minimum eigenvalue* $\phi_{\min}(\boldsymbol{\Sigma})$ *is no smaller than a certain constant* $c_* > 0$. *Let* $s_j$ *be as in* (7.67) *and* $\mathscr{B}_n(s)$ *be the parameter space in* (7.65). *Suppose* $\mathbf{D}^{1/2}\boldsymbol{\beta}^{(\mathrm{orig})}/\sigma \in \mathscr{B}_n(s^*)$ *as in* (7.12). *Let* $\eta_0 \in (0,1)$.
*(i) When* $\mathscr{B}_n((s^* + s_j) \wedge p)$ *is the parameter space of* $\mathbf{D}^{1/2}\boldsymbol{\beta}^{(\mathrm{orig})}$, $F_j$ *in* (7.66) *is the minimum Fisher information for the estimation of* $\beta_j$ *and* $D_j^{1/2}\beta_j^{(\mathrm{orig})}$, *and* $\widetilde{F}_j = F_j/D_j$ *is the minimum Fisher information for the estimation of* $\beta_j^{(\mathrm{orig})}$.
*(ii) Suppose* $s_j \ll n/\log p$ *in* (7.67). *Let* $\mathbf{w}_j = \mathbf{w}_j(\lambda_j)$, $\tau_j = \tau_j(\lambda_j)$ *and* $\eta_j = \eta_j(\lambda_j)$ *be as in* (7.13), (7.14) *and* (7.18) *with the largest* $\lambda_j$ *satisfying* $\eta_j(\lambda_j) \le \eta_0^{-1}\sqrt{2\log p}$. *Then,*

$$\sigma\tau_j(\lambda_j) = (1 + o_P(1))(nF_j)^{-1/2}. \tag{7.69}$$

*(iii) Suppose* $s_j \log p + (s^* \log p)^2 \ll n$. *Let* $\widehat{\beta}_j$ *be as in* (7.4) *with a certain* $\widehat{\boldsymbol{\beta}}^{(\mathrm{init})}$ *satisfying* (7.8) *and the* $\mathbf{w}_j$ *in (ii). Let* $\widehat{\sigma}$ *be as in* (7.10). *Then,*

$$(\widehat{\sigma}\tau_j)^{-1}(\widehat{\beta}_j - \beta_j) = \sqrt{nF_j}(\widehat{\beta}_j - \beta_j) + o_P(1) \to N(0,1), \tag{7.70}$$

*so that* $\widehat{\beta}_j$ *and* $\widehat{\beta}_j^{(\mathrm{orig})} = \overline{D}_j^{-1/2}\widehat{\beta}_j$ *are respectively asymptotically efficient estimators of* $\beta_j$ *and* $\beta_j^{(\mathrm{orig})}$.

Theorem 7.4 asserts that under the sparsity conditions $s_j \log p \ll n$ on $\mathbf{u}^{(j)}$ and $s^* \log p \ll n^{1/2}$ on $\mathbf{D}^{1/2}\boldsymbol{\beta}^{(\mathrm{orig})} \approx \overline{\mathbf{D}}^{1/2}\boldsymbol{\beta}^{(\mathrm{orig})} = \boldsymbol{\beta}$, the LDPE $\widehat{\beta}_j$ in (7.4) is an asymptotically normal and efficient estimator of $\beta_j$, $\widehat{\sigma}\tau_j = (1+o_P(1))/\sqrt{nF_j}$ is a consistent estimate of the standard error of $\widehat{\beta}_j$, and confidence intervals can be constructed based on (7.70). As a direct consequence, (7.63) provides asymptotically normal and efficient inference about $\beta_j^{(\mathrm{orig})}$ and approximate confidence intervals for $\beta_j^{(\mathrm{orig})}$ can be also constructed via (7.68).

The sparsity condition on the parameter space always hold when $s^* + s_j \ge p$, so that it is needed only when the parameter space $\sigma\mathscr{B}_n(s^* + s_j)$ is desired to reflect the sparsity condition imposed on the coefficients. For example, if we wish to impose the sparsity condition $s^* \log p \ll n^{1/2}$ on the entire parameter space, we would also need $s_j \log p \ll n^{1/2}$ for the asymptotic efficiency as stated in Theorem 7.4.

In Theorem 7.4, the weight vector $\mathbf{w}_j$ in (7.4) is constructed through the Lasso algorithm (7.13). Next, we prove that efficient statistical inference can be carried out without the sparsity condition $s_j \ll n/\log p$ when $\mathbf{w}_j$ is constructed through quadratic programming (7.19).

**Theorem 7.5** *Let* $\mathbf{X} \in I\!R^{n\times p}$ *be the normalized Gaussian design matrix* (7.62) *with* $\log p \lesssim n$, $p \to \infty$ *and a positive definite correlation matrix* $\boldsymbol{\Sigma}$ *in*

*(7.64). Let $F_j$ be as in (7.71). Then, with probability $1+o(1)$, the quadratic programming (7.19) is feasible under the constraint $\eta_j = n\lambda_j/\|\mathbf{Z}_j\|_2 \le \sqrt{2\log p}$, and the noise factor for the resulting $\mathbf{w}_j$ satisfies*

$$\sigma\tau_j \le (1+o_P(1))(nF_j)^{-1/2}. \tag{7.71}$$

*For this $\mathbf{w}_j$ and the $\widehat{\boldsymbol{\beta}}^{(\mathrm{init})}$ and $\widehat{\sigma}$ in Theorem 7.4 (iii), the $\widehat{\beta}_j$ in (7.4) satisfies*

$$(\widehat{\sigma}\tau_j)^{-1}(\widehat{\beta}_j - \beta_j) \xrightarrow{D} N(0,1) \tag{7.72}$$

*provided that $s^*\log p \ll n^{1/2}$. Consequently, (7.63) provides*

$$(\widehat{\sigma}\tau_j)^{-1}\overline{D}_j^{1/2}(\widehat{\beta}_j^{(\mathrm{orig})} - \beta_j^{(\mathrm{orig})}) \xrightarrow{D} N(0,1).$$

Theorem 7.5 requires a single sparsity condition $s^*\log p \ll n^{1/2}$ with the capped-$\ell_1$ complexity measure $s^*$ in (7.12). We will show in the next section that this condition is necessary for the estimation of an individual $\beta_j$ at the parametric $n^{-1/2}$ rate without imposing any additional condition on the correlation matrix $\mathbf{\Sigma}$ in the normalized Gaussian design.

We prove Theorems 7.4 and 7.5 in the rest of this subsection.

**Proof of Theorem 7.4.** Part (i) follows from Proposition 7.2 as we discussed above (7.66). Part (iii) follows from (7.69) and Theorem 7.1 as $\eta_j(\lambda_j) \le \eta_0^{-1}\sqrt{2\log p}$ by definition.

To prove (7.69), we assume without loss of generality $\mathrm{E}\,\|\widetilde{\mathbf{X}}_j\|_2^2/n = 1$ due to scale invariance, so that $\max_{1\le j\le p}|\overline{D}_j - 1| = O_P(\sqrt{(\log p)/n}) = o_P(1)$.

Write (7.13) as

$$\begin{aligned}\overline{D}_j^{1/2}\widehat{\gamma}^{(j)}(\lambda_j) &= \overline{D}_j^{1/2}\arg\min_{\boldsymbol{\gamma}}\left\{\|\mathbf{X}_j - \mathbf{X}_{-j}\boldsymbol{\gamma}\|_2^2/(2n) + \lambda_j\boldsymbol{\gamma}\right\}\\ &= \arg\min_{\mathbf{b}}\left\{\|\widetilde{\mathbf{X}}_j - \mathbf{X}_{-j}\mathbf{b}\|_2^2/(2n) + (\overline{D}_j^{1/2}\lambda_j)\|\mathbf{b}\|_1\right\}\end{aligned}$$

with $\mathbf{Z}_j = \mathbf{X}_j - \mathbf{X}_{-j}\widehat{\gamma}^{(j)}(\lambda_j) = \overline{D}_j^{-1/2}\{\widetilde{\mathbf{X}}_j - \mathbf{X}_{-j}\overline{D}_j^{1/2}\widehat{\gamma}^{(j)}(\lambda_j)\}$. As $\lambda_j$ is chosen to be the largest to satisfy $\eta_j = \eta_j(\lambda_j) \le \eta_0^{-1}\sqrt{2\log p}$ and (7.18) gives

$$\frac{\eta_j}{n^{1/2}} = \frac{\lambda_j}{\|\mathbf{Z}_j\|_2/n^{1/2}} = \frac{\overline{D}_j^{1/2}\lambda_j}{\|\widetilde{\mathbf{X}}_j - \mathbf{X}_{-j}\overline{D}_j^{1/2}\widehat{\gamma}^{(j)}(\lambda_j)\|_2/n^{1/2}},$$

(7.13) can be further written in the form of scaled Lasso (7.24) as

$$\begin{aligned}&\left\{\overline{D}_{j,j}^{1/2}\widehat{\gamma}^{(j)}(\lambda_j), \|\widetilde{\mathbf{X}}_j - \mathbf{X}_{-j}\overline{D}_{j,j}^{1/2}\widehat{\gamma}^{(j)}(\lambda_j)\|_2/n^{1/2}\right\}\\ =\;& \arg\min_{\mathbf{b},\sigma}\left\{\|\widetilde{\mathbf{X}}_j - \mathbf{X}_{-j}\mathbf{b}\|_2^2/(2nt) + t/2 + (\eta_j/\sqrt{n})\|\mathbf{b}\|_1\right\}.\end{aligned}$$

Moreover, $\eta_j = \eta_0^{-1}\sqrt{2\log p}$ if and only if $\widehat{\boldsymbol{\gamma}}^{(j)}(\lambda_j) \neq \mathbf{0}$, if and only if

$$\|\mathbf{X}_{-j}^T \widetilde{\mathbf{X}}_j/n\|_\infty \geq (\|\widetilde{\mathbf{X}}_j\|_2/n^{1/2})\eta_0^{-1}\sqrt{2\log p}.$$

Let $\boldsymbol{\gamma}^{(j)} = -\mathbf{u}_{-j}^{(j)} = -\boldsymbol{\Sigma}_{-j,-j}^{-1}\boldsymbol{\Sigma}_{-j,j}$ and $\mathbf{Z}_j^o = \widetilde{\mathbf{X}}_j - \widetilde{\mathbf{X}}_{-j}\boldsymbol{\gamma}^{(j)}$. Because $\mathbf{u}_j^{(j)} = 1$, $\mathbf{Z}_j^o = \mathbf{X}\mathbf{u}^{(j)}$, $\mathrm{E}\,\widetilde{\mathbf{X}}_{-j}^T\mathbf{Z}_j^o = \boldsymbol{\Sigma}_{-j,*}\mathbf{u}^{(j)} = \mathbf{0}$, and $\mathbf{Z}_j^o \sim N(\mathbf{0}, \mathbf{I}_n/(\boldsymbol{e}_j^T\boldsymbol{\Sigma}\boldsymbol{e}_j))$. The regression of $\widetilde{\mathbf{X}}_j$ against $\widetilde{\mathbf{X}}_{-j}$ and $\mathbf{X}_{-j}$ can be written as

$$\widetilde{\mathbf{X}}_j = \widetilde{\mathbf{X}}_{-j}\boldsymbol{\gamma}^{(j)} + \mathbf{Z}_j^o = \mathbf{X}_{-j}\overline{\mathbf{D}}_{-j,-j}^{1/2}\boldsymbol{\gamma}^{(j)} + \mathbf{Z}_j^o.$$

As $\phi_{\min}(\boldsymbol{\Sigma}_{-j,-j}) \geq \phi_{\min}(\boldsymbol{\Sigma}) \geq c_* > 0$, $\mathbf{X}_{-j}$ satisfies the restricted eigenvalue condition when $\overline{\mathbf{D}}_{-j,-j}^{1/2}\boldsymbol{\gamma}^{(j)}$ is capped-$\ell_1$ sparse in terms of

$$\widehat{s}_j = \sum_{k\neq j}\min\left\{\frac{\overline{D}_{k,k}^{1/2}|\gamma_k^{(j)}|}{(\boldsymbol{e}_j^T\boldsymbol{\Sigma}\boldsymbol{e}_j)^{-1/2}\sqrt{(2/n)\log p}}, 1\right\} = (1+o_P(1))s_j.$$

As $s_j = o_P(n/\log p)$ by assumption, Theorem 7.2 (i) yields

$$\frac{\|\mathbf{X}_{-j}\overline{D}_j^{1/2}\widehat{\boldsymbol{\gamma}}^{(j)}(\lambda_j) - \widetilde{\mathbf{X}}_{-j}\boldsymbol{\gamma}^{(j)}\|_2^2}{n/(\boldsymbol{e}_j^T\boldsymbol{\Sigma}\boldsymbol{e}_j)} + \frac{\|\overline{D}_j^{1/2}\widehat{\boldsymbol{\gamma}}^{(j)} - \overline{\mathbf{D}}_{-j,-j}^{1/2}\boldsymbol{\gamma}^{(j)}\|_1}{\sqrt{n/\log p}(\boldsymbol{e}_j^T\boldsymbol{\Sigma}\boldsymbol{e}_j)^{-1/2}} = o_P(1).$$

This and (7.18) imply

$$\begin{aligned}
\tau_j &= \frac{\|\mathbf{Z}_j\|_2}{\|\mathbf{Z}_j\|_2^2 + n\lambda_j\|\widehat{\boldsymbol{\gamma}}^{(j)}\|_1} \\
&= \frac{\|\widetilde{\mathbf{X}}_j - \mathbf{X}_{-j}\overline{D}_j^{1/2}\widehat{\boldsymbol{\gamma}}^{(j)}(\lambda_j)\|_2}{\overline{D}_j^{-1/2}\|\widetilde{\mathbf{X}}_j - \mathbf{X}_{-j}\overline{D}_j^{1/2}\widehat{\boldsymbol{\gamma}}^{(j)}(\lambda_j)\|_2^2 + \overline{D}_j^{1/2}n\lambda_j\|\widehat{\boldsymbol{\gamma}}^{(j)}\|_1} \\
&= \frac{(1+o_P(1))(\boldsymbol{e}_j^T\boldsymbol{\Sigma}\boldsymbol{e}_j)^{-1/2}}{(1+o_P(1))n^{1/2}/(\boldsymbol{e}_j^T\boldsymbol{\Sigma}\boldsymbol{e}_j) + (1+o_P(1))\eta_j(\boldsymbol{e}_j^T\boldsymbol{\Sigma}\boldsymbol{e}_j)^{-1/2}\|\boldsymbol{\gamma}^{(j)}\|_1} \\
&= \frac{1+o_P(1)}{n^{1/2}(\boldsymbol{e}_j^T\boldsymbol{\Sigma}\boldsymbol{e}_j)^{-1/2} + \eta_j\|\boldsymbol{\gamma}^{(j)}\|_1}
\end{aligned}$$

Moreover, as $\|\boldsymbol{\gamma}^{(j)}\|_2 \leq \{\phi_{\min}(\boldsymbol{\Sigma})\boldsymbol{e}_j^T\boldsymbol{\Sigma}^{-1}\boldsymbol{e}_j\}^{-1/2}$, $s_j = o_P(n/\log p)$ implies

$$\begin{aligned}
\eta_j\|\boldsymbol{\gamma}^{(j)}\|_1 &\leq \eta_j\{\|\boldsymbol{\gamma}^{(j)}\|_2 s_j^{1/2} + s_j(\boldsymbol{e}_j^T\boldsymbol{\Sigma}\boldsymbol{e}_j)^{-1/2}\sqrt{(2/n)\log p}\} \\
&= o_P(1)(\boldsymbol{e}_j^T\boldsymbol{\Sigma}\boldsymbol{e}_j)^{-1/2}\eta_j s_j^{1/2} \\
&= o_P(1)(\boldsymbol{e}_j^T\boldsymbol{\Sigma}\boldsymbol{e}_j)^{-1/2}n^{1/2}.
\end{aligned}$$

Consequently, $n^{1/2}\tau_j = (1+o(1))(\boldsymbol{e}_j^T\boldsymbol{\Sigma}\boldsymbol{e}_j)^{1/2} = (1+o(1))(\sigma^2 F_j)^{-1/2}$. This completes the proof of (7.69) and therefore the entire theorem. ■

**Proof of Theorem 7.5.** Let $\mathbf{u}^{(j)} = \boldsymbol{\Sigma}^{-1}\boldsymbol{e}_j/(\boldsymbol{e}_j^T\boldsymbol{\Sigma}^{-1}\boldsymbol{e}_j)$ and $\mathbf{Z}_j^o = \widetilde{\mathbf{X}}_j - \widetilde{\mathbf{X}}\mathbf{u}^{(j)}$. For $\mathbf{w}_j = \mathbf{Z}_j^o/(\mathbf{X}_j^T\mathbf{Z}_j^o)$, the noise and bias factors are given by

$$\tau_j^o = \frac{\|\mathbf{Z}_j^o\|_2}{|\mathbf{X}_j^T\mathbf{Z}_j^o|}, \quad \eta_j^o = \max_{k\neq j}\eta_{j,k}^o, \quad \eta_{j,k} = \frac{|\mathbf{X}_k^T\mathbf{Z}_j^o|}{\|\mathbf{Z}_j^o\|_2} = \frac{n^{1/2}|\widetilde{\mathbf{X}}_k^T\mathbf{Z}_j^o|}{\|\widetilde{\mathbf{X}}_k\|_2\|\mathbf{Z}_j^o\|_2}.$$

As (7.72) would follow from (7.71) with an application of Theorem 7.1, it suffices to prove $P\{\eta_j^o \leq \sqrt{2\log p}\} = 1 + o(1)$ and $\sqrt{n}\tau_j^o = (1+o(1))/(\sigma F_j^{1/2})$ due to the optimality of $\tau_j$.

Because $\widetilde{\mathbf{X}}_k$ and $\mathbf{Z}_j^o$ are independent Gaussian vectors each with iid entries, for $k \neq j$ $\eta_{j,k}^2/n$ has the beta$(1/2,(n-1)/2)$ distribution, so that

$$P\{\eta_{j,k}^2/n > x^2\} \leq 2c_n P\{N(0,1) > \sqrt{n-3/2}\sqrt{-\log(1-x^2)}\}, \qquad (7.73)$$

where $1 < c_n = \sqrt{2/(n-3/2)}\Gamma(n/2)\Gamma((n-1)/2) < \exp\big[1/(4n-6)^2\big]$. Thus,

$$P\{\eta_j^o > \sqrt{2\log p}\} \leq \frac{2pc_n\exp\big[-(n-3/2)(2/n)(\log p)/2\big]}{\sqrt{2\pi(n-3/2)(2/n)\log p}} = \frac{O(1)}{\sqrt{\log p}} \to 0.$$

as $\log p \lesssim n$ and $\log p \to \infty$. As $\mathrm{E}[\mathbf{X}_j^T\mathbf{Z}_j^o/n] = \mathrm{E}[\|\mathbf{Z}_j^o\|_2^2/n] = 1/(\boldsymbol{e}_j^T\boldsymbol{\Sigma}^{-1}\boldsymbol{e}_j)$, $\sqrt{n}\tau_j^o = (1+o(1))(\boldsymbol{e}_j^T\boldsymbol{\Sigma}^{-1}\boldsymbol{e}_j)^{1/2} = (1+o(1))\sigma/F_j^{1/2}$. ∎

### *7.3.3 Partial linear regression*

In this subsection, we illustrate the semi-parametric approach through partially linear regression with LD data. Suppose that $(W_i, X_i, Y_i), 1 \leq i \leq n$, is an independent and identically distributed sample from a population $(W, X, Y)$ satisfying the following partially linear model

$$Y = W\theta + f(X) + \varepsilon, \qquad (7.74)$$

where $\theta$ is an unknown real parameter of interest, $f$ is an unknown smooth function treated as the NP component in (7.2), and the noise $\varepsilon$ is independent of $(W, X)$ with $\mathrm{E}\,\varepsilon = 0$ and $\mathrm{E}\,\varepsilon^2 = \sigma^2$. The problem is inference and efficient estimation of $\theta$ and proper conditions for theoretical guarantee. The study of partially linear regression dates back to Engle *et al.* (1986), Rice (1986), Heckman (1988), Chen (1988) and Speckman (1988).

We assume $X$ is real for notational simplicity. The analysis in this subsection can be carried out for LD multivariate $X$ verbatim with proper change of notation, but NP estimation of $f(\cdot)$ is infeasible when the dimension of $X$ is high. Let $\mathbf{W} = (W_1, \ldots, W_n)^T$, $\mathbf{X} = (X_1, \ldots, X_n)^T$ and $\mathbf{Y} = (Y_1, \ldots, Y_n)^T$. Let $g(x) = \mathrm{E}[W|X = x]$ and $\boldsymbol{\psi} = \psi(\mathbf{X}) = \big(\psi(X_1), \ldots, \psi(X_n)\big)^T$ for any function $\psi$. In particular, $\mathbf{f} = (f(X_1), \cdots, f(X_n))^T$.

When $\theta$ is known and $f(\cdot)$ is smooth, the vector $\mathbf{f}$ is typically estimated by a linear smoother of the form $\mathbf{A}_n(\mathbf{Y} - \mathbf{W}\theta)$ with an $n \times n$ matrix $\mathbf{A}_n$ depending on $\mathbf{X}$ only, including the polynomial, spline, kernel and RKHS methods as discussed in Chapter 2. When $\theta$ is unknown, we estimate $\mathbf{f}$ by $\widetilde{\mathbf{f}} = \mathbf{A}_n(\mathbf{Y} - \mathbf{W}\widetilde{\theta})$ with an initial estimate of $\widetilde{\theta}$. As $\mathbf{Y} - \mathbf{W}\widetilde{\theta} - \widetilde{\mathbf{f}}$ is the residual, the de-biased estimator of $\theta$ can be then written as

$$\widehat{\theta} = \widetilde{\theta} + \frac{\mathbf{Z}^T(\mathbf{Y} - \mathbf{W}\widetilde{\theta} - \widetilde{\mathbf{f}})}{\mathbf{Z}^T(\mathbf{I}_n - \mathbf{A}_n)\mathbf{W}} = \frac{\mathbf{Z}^T(\mathbf{I}_n - \mathbf{A}_n)\mathbf{Y}}{\mathbf{Z}^T(\mathbf{I}_n - \mathbf{A}_n)\mathbf{W}}. \tag{7.75}$$

This is slightly different from (7.14) in HD linear regression as the linearity of $\widetilde{\mathbf{f}}$ allows complete removal of any impact of the choice of $\widetilde{\theta}$ on $\widehat{\theta}$. In fact,

$$\widehat{\theta} - \theta = \frac{\mathbf{Z}^T(\mathbf{I}_n - \mathbf{A}_n)\varepsilon}{\mathbf{Z}^T(\mathbf{I}_n - \mathbf{A}_n)\mathbf{W}} + \frac{\mathbf{Z}^T(\mathbf{I}_n - \mathbf{A}_n)\mathbf{f}}{\mathbf{Z}^T(\mathbf{I}_n - \mathbf{A}_n)\mathbf{W}}$$

by algebra, where $(\mathbf{I}_n - \mathbf{A}_n)\mathbf{f}$ is the bias in the estimation of $\mathbf{f}$. As $(\mathbf{I}_n - \mathbf{A}_n)\mathbf{f}$ depends on $\mathbf{X}$ only and $\mathbf{f} = f(\mathbf{X})$ with unknown $f(\cdot)$, when $\mathbf{g} = \mathrm{E}[\mathbf{W}|\mathbf{X}]$ is known it would be ideal to take $\mathbf{Z} = \mathbf{Z}^o$ in (7.75) where

$$\mathbf{Z}^o = \mathbf{W} - \mathbf{g} = (W_1 - \mathrm{E}[W_1|X_1], \ldots, W_n - \mathrm{E}[W_n|X_n])^T. \tag{7.76}$$

When $g(x)$ is an unknown smooth function of $x$, we may estimate $\mathbf{g}$ by a linear function of $\mathbf{W}$, e.g. $\mathbf{Z} = (\mathbf{I}_n - \mathbf{B}_n)\mathbf{W}$. For simplicity, we take $\mathbf{A}_n = \mathbf{B}_n = \mathbf{P}_{k_n}$ in the rest of this subsection where $\mathbf{P}_{k_n}$ is an orthogonal protection to the linear span of certain $k_n$ basis vectors. This leads to the de-biased estimator

$$\widehat{\theta} = \widetilde{\theta} + \frac{\mathbf{Z}^T(\mathbf{Y} - \mathbf{W}\widetilde{\theta} - \widetilde{\mathbf{f}})}{\mathbf{Z}^T\mathbf{W}} = \frac{\mathbf{Z}^T\mathbf{Y}}{\|\mathbf{Z}\|_2^2} = \frac{\mathbf{W}^T\mathbf{P}_{k_n}^{\perp}\mathbf{Y}}{\|\mathbf{P}_{k_n}^{\perp}\mathbf{W}\|_2^2} \tag{7.77}$$

with $\mathbf{Z} = \mathbf{W} - \mathbf{P}_{k_n}\mathbf{W} = \mathbf{P}_{k_n}^{\perp}\mathbf{W}$. See Speckman (1988). We shall study this estimator based on the error decomposition

$$\widehat{\theta} - \theta = \frac{\mathbf{W}^T\mathbf{P}_{k_n}^{\perp}\varepsilon}{\|\mathbf{P}_{k_n}^{\perp}\mathbf{W}\|_2^2} + \frac{(\mathbf{W} - \mathbf{g})^T\mathbf{P}_{k_n}^{\perp}\mathbf{f}}{\|\mathbf{P}_{k_n}^{\perp}\mathbf{W}\|_2^2} + \frac{\mathbf{g}^T\mathbf{P}_{k_n}^{\perp}\mathbf{f}}{\|\mathbf{P}_{k_n}^{\perp}\mathbf{W}\|_2^2}. \tag{7.78}$$

We note that $\mathbf{P}_{k_n}^{\perp}$ depends on $\mathbf{X}$ only, so that in the above decomposition $\mathrm{E}[\mathbf{W}^T\mathbf{P}_{k_n}^{\perp}\varepsilon] = \mathrm{E}[(\mathbf{W} - \mathbf{g})^T\mathbf{P}_{k_n}^{\perp}\mathbf{f}] = 0$, the variance of $(\mathbf{W} - \mathbf{g})^T\mathbf{P}_{k_n}^{\perp}\mathbf{f}$ is small when $\|\mathbf{P}_{k_n}^{\perp}\mathbf{f}\|_2$ is small, and the third term represents the bias. An analysis for other choices of $\mathbf{A}_n$ and $\mathbf{B}_n$ can be carried out along the same line.

The analysis of (7.77) is different from that of the LDPE (7.4) in HD regression due to the independence of $\{W_i - g(X_i), \varepsilon_i, i \le n\}$ given $\mathbf{X}$ in the partially linear model and the complete separation of the random noise as the first two terms in (7.78) and the bias as the third term in (7.78). As the variance of the random noise terms is stable as long as $k_n/n = o(1)$, some overfitting $\mathbf{f}$ or $\mathbf{g}$ does not prevent efficient estimation of $\theta$, and large $k_n$ can be taken to reduce the bias. This would not be feasible for (7.14) or (7.14) in the estimation of $\beta_j$ in linear regression as the correlations between the components of $\mathbf{X}(\widehat{\boldsymbol{\beta}}^{(\mathrm{init})} - \boldsymbol{\beta})$ and $\mathbf{X}_j - \mathbf{X}_{-j}\widehat{\gamma}^{(j)}$ are not explicit and the noise and bias terms are not separable in the analysis of the cross-product term $\mathbf{Z}^T\mathbf{X}(\boldsymbol{\beta} - \widehat{\boldsymbol{\beta}}^{(\mathrm{init})})$. Thus, overfitting $\boldsymbol{\beta}$ or $\gamma^{(j)}$ would produce poor theoretical results in our analysis.

The smoothness condition on $f$ can be stated in many ways. We consider here the following general notion of smoothness with an index $\alpha_f \ge 0$: There exist known basis vectors $\boldsymbol{\psi}_0, \ldots, \boldsymbol{\psi}_{n-1}$ in $I\!R^n$, depending on $\mathbf{X}$ only, such that

$$\mathrm{E}\left\|\mathbf{f} - \mathbf{P}_{k_n}\mathbf{f}\right\|_2^2 = o(n^{1-2\alpha_f}) \tag{7.79}$$

for certain integers $k_n \to \infty$ with $k_n/n \to 0$, where $\mathbf{P}_k$ is the orthogonal projection to the linear span of $\{\boldsymbol{\psi}_0, \ldots, \boldsymbol{\psi}_{k-1}\}$ in $I\!R^n$. For simplicity, we assume that $g(x) = \mathrm{E}[W|X = x]$ is also smooth with the same basis,

$$\mathrm{E}\left\|\mathbf{g} - \mathbf{P}_{k_n}\mathbf{g}\right\|_2^2 = o(n^{1-2\alpha_g}), \tag{7.80}$$

but with a different index $\alpha_g \ge 0$.

**Theorem 7.6** *Let $\{W, X, Y, \theta, f, \sigma\}$ be as in (7.74) with $\mathrm{E}\,\mathrm{Var}(W|X) \in (0, \infty)$. Suppose (7.79) and (7.80) hold with $\alpha_f + \alpha_g \ge 1/2$, and*

$$\mathrm{E}\left[\frac{1}{n}\sum_{i=1}^n \mathrm{Var}(W_i|X_i)(\mathbf{P}_{k_n}^{\perp}\mathbf{f})_i^2\right] = o(1). \tag{7.81}$$

*Let $\widehat{\theta}$ be the estimator in (7.77) and $\widehat{\sigma} = \|\mathbf{P}_{k_n}^{\perp}(\mathbf{Y} - \mathbf{W}\widehat{\theta})\|_2/\sqrt{n - k_n}$. Then,*

$$\left(\|\mathbf{P}_{k_n}^{\perp}\mathbf{W}\|_2/\widehat{\sigma}\right)(\widehat{\theta} - \theta) \xrightarrow{\mathrm{D}} N(0, 1), \tag{7.82}$$

*$\|\mathbf{P}_{k_n}^{\perp}\mathbf{W}\|_2^2/n = (1 + o_P(1))E\,\mathrm{Var}(W|X)$ and $\widehat{\sigma}/\sigma = 1 + o_P(1)$. Moreover, when $\varepsilon_i \sim N(0, \sigma^2)$ and $\alpha_f \le 1/2 + \alpha_g$, $F_\theta = \mathrm{E}\,\mathrm{Var}(W/\sigma|X)$ is the minimum Fisher information for the estimation of $\theta$ and $\widehat{\theta}$ is asymptotically linear and efficient,*

$$\sqrt{nF_\theta}(\widehat{\theta} - \theta) = \frac{(\mathbf{W} - \mathbf{g})^T\varepsilon}{\sigma\sqrt{n\,\mathrm{E}\,\mathrm{Var}(W|X)}} + o_P(1) \xrightarrow{\mathrm{D}} N(0, 1).$$

Condition (7.81) follows from (7.79) when $\mathrm{Var}(W|X = x)$ is uniformly bounded in $x$. As sufficient conditions for (7.79) and (7.80) can be derived in the same manner, we shall consider (7.79) only. The basis vectors $\boldsymbol{\psi}_j$ can be generated from the spline, Fourier, wavelet and other basis functions $\psi_j$ for fitting smooth functions. For simplicity, we assume further that the density of $X$ is supported in $[0, 1]$ and bounded from the above by a constant $C_0$ and consider the Fourier series $f(t) = \beta_0 + \sum_{j=1}^{\infty}\beta_j\sqrt{2}\cos(\pi j t)$ in $[0, 1]$. Let $f(t)$

be defined in $[-1,2]$ by the same Fourier series. For $\alpha \in (0,1]$ and $k \geq 1$

$$k^{2\alpha}\sum_{j=k}^{\infty}\beta_j^2 = k^{2\alpha}\int_0^1\Big(\sum_{j=k}^{\infty}\beta_j\sqrt{2}\cos(\pi jt)\Big)^2 dt \qquad (7.83)$$
$$\leq C_\alpha \sup_{0<\Delta\leq 1/(2k)}\int_{-1}^1\Big(\frac{f(t+\Delta)-f(t)}{\Delta^\alpha}\Big)^2 dt$$

with $C_\alpha = \max_{1/2\leq t\leq 1}(t^{2\alpha}/4)/\{1-\cos(\pi t)\} \leq (1-4^{-\alpha})^{-1}/4$. Thus, as

$$\mathrm{E}\left\|\mathbf{f}-\mathbf{P}_k\mathbf{f}\right\|_2^2 \leq n\,\mathrm{E}\left(\textstyle\sum_{j=k}^{\infty}\beta_j\psi_j(X)\right)^2 \leq nC_0\textstyle\sum_{j=k}^{\infty}\beta_j^2, \qquad (7.84)$$

(7.79) holds for $k_n = c_n n \to \infty$ with $c_n \to 0$ when

$$\sup_{0<\Delta\leq 1/(2k)}\int_{-1}^1\Big(\frac{f(t+\Delta)-f(t)}{\Delta^{\alpha_f}}\Big)^2 dt \ll c_n^{2\alpha_f}$$

We leave the verification of (7.83) and (7.84) as an exercise.

To derive the minimum Fisher information, we consider Gaussian noise $\varepsilon_i \sim N(0,\sigma^2)$ and functions $f$ satisfying (7.79). This condition on the NP component holds for $f-\phi g$ in the univariate model $\{\theta+\phi, f-\phi g : |\phi|\leq \epsilon_n^*\}$ when $(\epsilon_n^*)^2 E\|\mathbf{P}_{k_n}^{\perp}\mathbf{g}\|_2^2 = o(n^{1-2\alpha_f})$, which holds with $\epsilon_n^* \gg n^{-1/2}$ under (7.80) when $-2\alpha_g \leq 1-2\alpha_f$. In this case, the minimum Fisher information for the estimation of $\theta$ is no greater than

$$\left(d/d\phi\right)^2\mathrm{E}\left[Y-W(\theta+\phi)-\{f(X)-\phi g(X)\}\right]^2/(2\sigma^2) = \mathrm{E}\,\mathrm{Var}(W/\sigma|X).$$

**Proof of Theorem 7.6.** As $\mathbf{Z} = \mathbf{P}_{k_n}^{\perp}\mathbf{W}$ and $\mathbf{g} = \mathrm{E}[\mathbf{W}|\mathbf{X}]$,

$$\begin{aligned}\mathrm{E}\,\|\mathbf{Z}-(\mathbf{W}-\mathbf{g})\|_2^2 &= \mathrm{E}\|\mathbf{P}_{k_n}(\mathbf{W}-\mathbf{g})\|_2^2 + \mathrm{E}\|\mathbf{P}_{k_n}^{\perp}\mathbf{g}\|_2^2.\\ &= \sum_{i=1}^n \mathrm{E}\left[\mathrm{Var}(W_i|X_i)(\mathbf{P}_{k_n})_{i,i}^2\right] + o(n^{1-2\alpha_g}).\end{aligned}$$

As $(\mathbf{P}_{k_n})_{i,i}^2 \leq 1$, $\sum_{i=1}^n(\mathbf{P}_{k_n})_{i,i}^2 = k_n$ and $\mathrm{Var}(W_i|X_i)$ are uniformly integrable,

$$\mathrm{E}\,\|\mathbf{Z}-(\mathbf{W}-\mathbf{g})\|_2^2 = o(n) + O(k_n) + o(n^{1-2\alpha_g}) = o(n).$$

By the law of large numbers, $\|\mathbf{W}-\mathbf{g}\|_2^2/n = \mathrm{E}\,\mathrm{Var}(W|X) + o(1)$. Thus,

$$\frac{\|\mathbf{Z}\|_2^2/n}{\mathrm{E}\,\mathrm{Var}(W|X)} = 1 + o_P(1) \ \text{ and } \ \mathrm{E}\left[\left(\{\mathbf{Z}-(\mathbf{W}-\mathbf{g})\}^T\varepsilon\right)^2\right] \ll \sigma^2 n.$$

By (7.79) and (7.80), $|\mathbf{g}\mathbf{P}_{k_n}^{\perp}\mathbf{f}| \ll n^{1-\alpha_f-\alpha_g} \leq n^{-1/2}$. As $\mathbf{g} = \mathrm{E}[\mathbf{W}|\mathbf{X}]$,

$$\mathrm{E}\left((\mathbf{W}-\mathbf{g})\mathbf{P}_{k_n}^{\perp}\mathbf{f}\right)^2 = \sum_{i=1}^n \mathrm{E}\left[\mathrm{Var}(W_i|X_i)(\mathbf{P}_{k_n}^{\perp}\mathbf{f})_i^2\right] \ll n$$

by (7.81). It follows that $|\mathbf{Z}^T\mathbf{f}| \ll n^{1/2}$. Thus, by (7.78)

$$\|\mathbf{Z}\|_2(\widehat{\theta}-\theta) = \frac{(\mathbf{W}-\mathbf{g})^T\varepsilon}{\|\mathbf{Z}\|_2} + \frac{o_P(n^{1/2})}{\|\mathbf{Z}\|_2} = \frac{(\mathbf{W}-\mathbf{g})^T\varepsilon}{\{n\,\mathrm{E}\,\mathrm{Var}(W|X)\}^{1/2}} + o_P(1).$$

The right-hand side above converges to $N(0,\sigma^2)$ by the central limit theorem. Moreover, as $\|\mathbf{Z}\|_2^2/n = (1+o(1))\,\mathrm{E}\,\mathrm{Var}(W|X)$ and $\mathrm{E}\,\|\mathbf{P}_{k_n}\varepsilon\|_2^2 = \sigma^2 k_n \ll \sigma^2 n$,

$$\widehat{\sigma} = \frac{\|\mathbf{P}_{k_n}^{\perp}(\mathbf{Y}-\mathbf{W}\widehat{\theta})\|_2}{\sqrt{n-k_n}} = \frac{\|\mathbf{P}_{k_n}^{\perp}(\varepsilon+\mathbf{f})\|_2}{\sqrt{n-k_n}} + o_P(1) = \frac{\|\varepsilon\|_2}{\sqrt{n}} + o_P(1).$$

This completes the proof of the asymptotic normality (7.82) and the consistency statements $\|\mathbf{Z}\|_2^2/n = (1+o_P(1))\,\mathrm{E}\,\mathrm{Var}(W|X)$ and $\widehat{\sigma}/\sigma = 1+o_P(1)$.

When $\varepsilon_i \sim N(0,\sigma^2)$ and $\alpha_f \le 1/2+\alpha_g$, $\mathrm{E}\,\mathrm{Var}(W/\sigma|X)$ is an upper bound for the minimum Fisher information $F_\theta$. As $1/\mathrm{E}\,\mathrm{Var}(W/\sigma|X)$ is attained as the asymptotic variance of $\sqrt{n}(\widehat{\theta}-\theta)$, $F_\theta = \mathrm{E}\,\mathrm{Var}(W/\sigma|X)$ and the estimator is asymptotically efficient. ■

## 7.4 Gaussian Graphical Models

In a graphical model, the relationship between random variables are expressed in a graph in which vertices represent variables and edges represent conditional dependence of variable pairs given all other variables in the graph. Such a graphical model is Gaussian if the random variables in the graph are jointly Gaussian. Section 9.4 provides additional details.

Consider a Gaussian graphical model where vertices represents the elements of $(X_1,\ldots,X_p) \sim N(0,\mathbf{\Sigma})$. Suppose throughout this section that the covariance matrix $\mathbf{\Sigma}$ is of full rank. Let $\mathbf{\Omega} = (\omega_{j,k})_{p\times p} = \mathbf{\Sigma}^{-1}$ be the *precision matrix* associated with the covariance matrix $\mathbf{\Sigma}$. In the Gaussian graphical model, the strength of the edge between $X_j$ and $X_k$ is measured by the *partial correlation*

$$\rho_{j,k}^{(\mathrm{par})} = \mathrm{Corr}\Big(X_j, X_k \Big| X_i, i\in[p]\setminus\{j,k\}\Big) = \frac{-\omega_{j,k}}{\sqrt{\omega_{j,j}\omega_{k,k}}} \tag{7.85}$$

or the element $\omega_{j,k}$ of the precision matrix $\mathbf{\Omega}$. As $(X_j,X_k)$ is conditionally Gaussian, $\rho_{j,k}^{(\mathrm{par})}=0$ is equivalent to their conditional independence.

In this section, we study the statistical inference about $\rho_{j,k}^{(\mathrm{par})}$ and $\omega_{j,k}$, including asymptotic normality, efficient point and interval estimation, and sample size requirements. The result on the sample size requirement also implies the necessity of the sparsity condition $s^*\log p \ll n^{1/2}$ in Theorem 7.4 (iii). We will discuss feature screening via partial correlation in Subsection 8.6.3 and precision matrix estimation in Section 9.4.

### *7.4.1 Inference via penalized least squares*

Suppose a data matrix $\mathbf{X} \in I\!R^{n\times p}$ is observed with iid rows from the population $(X_1, \ldots, X_p) \sim N(0, \mathbf{\Sigma})$. Statistical inference about the partial correlation $\rho^{(\mathrm{par})}_{j,k}$ and individual elements $\omega_{j,k}$ of the precision matrix was considered in Sun and Zhang (2012) and Ren *et al.* (2015) among others. Their idea is to treat $\rho^{(\mathrm{par})}_{j,k}$ and $\omega_{j,k}$ as smooth functions of

$$\begin{aligned}\left(\mathbf{\Omega}_{\{j,k\},\{j,k\}}\right)^{-1} &= \begin{pmatrix}\omega_{j,j} & \omega_{j,k}\\ \omega_{k,j} & \omega_{k,k}\end{pmatrix}^{-1}\\ &= \mathrm{Cov}\left((X_j, X_k)^T, (X_j, X_k)\big|X_\ell, \ell \neq j, \ell \neq k\right) \qquad (7.86)\end{aligned}$$

and to estimate the above $2 \times 2$ conditional covariance matrix by the sample covariance of the residual vectors of the corresponding regularized regression of the $j$-th and $k$-th columns of $\mathbf{X}$ against other columns, e.g. running penalized least squares with $\mathbf{X}_j$ and $\mathbf{X}_k$ against $\mathbf{X}_{\{j,k\}^c}$.

The bivariate regression model can be written as

$$(X_j, X_k) = \left(X_\ell, \ell \neq j, \ell \neq k\right)\mathbf{\Gamma}^{\{j,k\}}_{\{j,k\}^c} + (Z_j^{\{j,k\}}, Z_k^{\{j,k\}})$$

with $\mathbf{\Gamma}^{\{j,k\}}_{\{j,k\}^c} = \mathbf{\Sigma}^{-1}_{\{j,k\}^c,\{j,k\}^c}\mathbf{\Sigma}_{\{j,k\}^c,\{j,k\}} = -\mathbf{\Omega}_{\{j,k\}^c,\{j,k\}}\mathbf{\Omega}^{-1}_{\{j,k\},\{j,k\}}$. The right-hand side of (7.86), or the covariance matrix of $(Z_j^{\{j,k\}}, Z_k^{\{j,k\}})$, is

$$\mathbf{\Sigma}_{\{j,k\},\{j,k\}} - \mathbf{\Sigma}_{\{j,k\},\{j,k\}^c}\mathbf{\Sigma}^{-1}_{\{j,k\}^c,\{j,k\}^c}\mathbf{\Sigma}_{\{j,k\}^c,\{j,k\}},$$

which implies the identity in (7.86) by the block inversion formula.

In the vector notation, the regression model with data can be written as

$$\mathbf{X}_j = \mathbf{X}_{\{j,k\}^c}\gamma^{(j),k}_{\{j,k\}^c} + \mathbf{Z}_j^{\{j,k\}} \qquad (7.87)$$

with $\mathbf{Z}_k^{\{j,k\}} = \mathbf{Z}_k^{\{k,j\}}$ and $\left(\gamma^{(j),k}_{\{j,k\}^c}, \gamma^{(k),j}_{\{j,k\}^c}\right) = \mathbf{\Gamma}^{\{j,k\}}_{\{j,k\}^c}$. We note that $\gamma^{(j),k}_\ell$ is the coefficient for the $\ell$-th variable in the regression of $\mathbf{X}_j$ against $\{j,k\}^c$, so that the $k$ in the superscript is needed. Let $\widehat{\gamma}^{(j),k}_{\{j,k\}^c}$ be penalized LSE of $\gamma^{(j),k}_{\{j,k\}^c}$ in the above regression model and $\widehat{\mathbf{Z}}_j^{\{j,k\}}$ be the residual

$$\widehat{\mathbf{Z}}_j^{\{j,k\}} = \mathbf{X}_j - \mathbf{X}_{\{j,k\}^c}\widehat{\gamma}^{(j),k}_{\{j,k\}^c} \qquad (7.88)$$

with $\widehat{\mathbf{Z}}_k^{\{j,k\}} = \widehat{\mathbf{Z}}_k^{(k,j)}$. The estimators of $\omega_{j,k}$ and $\rho^{(\mathrm{par})}_{j,k}$ are given by

$$\begin{pmatrix}\widehat{\omega}_{j,j} & \widehat{\omega}_{j,k}\\ \widehat{\omega}_{k,j} & \widehat{\omega}_{k,k}\end{pmatrix}^{-1} = \begin{pmatrix}\|\widehat{\mathbf{Z}}_j^{\{j,k\}}\|_2^2/n & (\widehat{\mathbf{Z}}_j^{\{j,k\}})^T(\widehat{\mathbf{Z}}_k^{\{j,k\}})/n\\ (\widehat{\mathbf{Z}}_k^{\{j,k\}})^T(\widehat{\mathbf{Z}}_j^{\{j,k\}})/n & \|\widehat{\mathbf{Z}}_k^{\{j,k\}}\|_2^2/n\end{pmatrix} \qquad (7.89)$$

as the $2 \times 2$ sample covariance matrix of the residuals, and

$$\widehat{\rho}^{(\mathrm{par})}_{j,k} = \frac{-\widehat{\omega}_{j,k}}{\sqrt{\widehat{\omega}_{j,j}\widehat{\omega}_{k,k}}} = \frac{(\widehat{\mathbf{Z}}_j^{\{j,k\}})^T\widehat{\mathbf{Z}}_k^{\{j,k\}}}{\|\widehat{\mathbf{Z}}_j^{\{j,k\}}\|_2\|\widehat{\mathbf{Z}}_k^{\{j,k\}}\|_2}, \quad 1 \leq j < k \leq p. \qquad (7.90)$$

While the above method involves $p(p-1)/2$ penalized LSE, its actual computational cost is much lower when $\mathbf{\Omega}$ is sparse. If we first carry out penalized LSE in the $p$ regression models

$$\mathbf{X}_j = \mathbf{X}_{-j}\boldsymbol{\beta}^{(j)}_{-j} + \varepsilon_j$$

with separable and possibly scaled penalties, we have $\widehat{\mathbf{Z}}^{\{j,k\}}_k = \mathbf{X}_j - \mathbf{X}_{-j}\widehat{\boldsymbol{\beta}}^{(j)}_{-j}$ when $\widehat{\beta}^{(j)}_k = 0$. Thus, as $\boldsymbol{\beta}^{(j)}_{-j} = -\mathbf{\Omega}_{j,-j}/\omega_{j,j}$, the computational cost of the above method is expected to be of no greater order than computing $\#\{(j,k) : \omega_{j,k} \neq 0\}$ penalized LSE by the bound on the size of the selected model in Subsection 4.3.3.

As (7.87) is a regression model with Gaussian error, Theorem 7.2 provides sufficient conditions under which $\|\widehat{\mathbf{Z}}^{\{j,k\}}_j\|_2/n^{1/2}$ is an asymptotically normal and efficient estimator of the noise level $\sigma^{\{j,k\}}_j = \big(\mathrm{E}\,\|\mathbf{Z}^{\{j,k\}}_j\|_2^2/n\big)^{1/2}$ in the regression model. Suppose for a certain $c_* > 0$ and $\mathbf{D} = \mathrm{diag}(\mathbf{\Sigma})$

$$\mathbf{u}^T\mathbf{D}^{-1/2}\mathbf{\Sigma}\mathbf{D}^{-1/2}\mathbf{u} \geq c_*\|\mathbf{u}\|_2^2,\ \forall \mathbf{u} \in I\!R^p \tag{7.91}$$

so that the restricted eigenvalue condition holds with overwhelming probability for (7.87). It turns out that the capped $\ell_1$ sparsity is bounded by

$$s^*_{j,k} = \sum_{\ell \in \{j,k\}^c} \min\left(1, \frac{|\rho^{(\mathrm{par})}_{\ell,j}| \vee |\rho^{(\mathrm{par})}_{\ell,k}|}{c_*\sqrt{(2/n)\log p}}\right) \leq s^* \ll n/\log p \tag{7.92}$$

in the regression model. Under conditions (7.91) and (7.92) and with scaled penalty level $\lambda_0 = \eta_0^{-1}\sqrt{(2/n)\log p}$, $\eta_0 \in (0,1)$, penalized LSE provides

$$\begin{aligned} &\|\overline{\mathbf{D}}^{-1/2}_{\{j,k\}}\mathbf{X}^T_{\{j,k\}}\widehat{\mathbf{Z}}^{\{j,k\}}_j\|_\infty/\big(n\sigma^{\{j,k\}}_j\big) \leq \lambda_0/\eta_0, \\ &\|\widehat{\mathbf{Z}}^{\{j,k\}}_j - \mathbf{Z}^{\{j,k\}}_j\|_2^2/\big\{n\big(\sigma^{\{j,k\}}_j\big)^2\big\} \leq C_0 s^*\lambda_0^2, \\ &\|\mathbf{D}^{1/2}_{\{j,k\}^c}\big(\widehat{\gamma}^{(j),k}_{\{j,k\}^c} - \gamma^{(j),k}_{\{j,k\}^c}\big)\|_1/\sigma^{\{j,k\}}_j \leq C_0 s^*\lambda_0, \end{aligned} \tag{7.93}$$

with overwhelming probability as normalized gradient, prediction error and $\ell_1$ estimation error bounds, where $\overline{\mathbf{D}} = \mathrm{diag}(\mathbf{X}^T\mathbf{X}/n)$. We note that (7.91), (7.92) and (7.93) are all stated for penalized LSE with column normalized data $\mathbf{X}\overline{\mathbf{D}}^{-1/2}$. With (7.93), an analysis of the off-diagonal element in (7.89) can be also carried out via the following error decomposition

$$\begin{aligned} &(\widehat{\mathbf{Z}}^{\{j,k\}}_j)^T(\widehat{\mathbf{Z}}^{\{j,k\}}_k) - (\mathbf{Z}^{\{j,k\}}_j)^T(\mathbf{Z}^{\{j,k\}}_k) \\ = \ &(\widehat{\mathbf{Z}}^{\{j,k\}}_j)^T(\widehat{\mathbf{Z}}^{\{j,k\}}_k - \mathbf{Z}^{\{j,k\}}_k) + (\mathbf{Z}^{\{j,k\}}_k)^T(\widehat{\mathbf{Z}}^{\{j,k\}}_j - \mathbf{Z}^{\{j,k\}}_j). \end{aligned}$$

This leads to the following theorem.

**Theorem 7.7** *Under conditions* (7.91) *and* (7.92), (7.93) *holds with large probability if* $\widehat{\mathbf{Z}}^{\{j,k\}}_j$ *in* (7.88) *is the residual vector of the scaled Lasso estimator* *with the column normalized data* $\mathbf{X}\overline{\mathbf{D}}^{-1/2}$ *in the linear model* (7.87). *Suppose* (7.92) *and* (7.93) *hold for all* $\{j,k\}$ *with any penalized LSE in* (7.88). *Then,*

$$P\left\{\max_{j,k}\left|\frac{(\widehat{\mathbf{Z}}_j^{\{j,k\}})^T\widehat{\mathbf{Z}}_k^{\{j,k\}}}{n\sigma_j^{\{j,k\}}\sigma_j^{\{j,k\}}} - \frac{(\mathbf{Z}_j^{\{j,k\}})^T\mathbf{Z}_k^{\{j,k\}}}{n\sigma_j^{\{j,k\}}\sigma_j^{\{j,k\}}}\right| > C_1 s^* \lambda_0^2\right\} = o(1), \quad (7.94)$$

*where* $\mathbf{Z}_j^{\{j,k\}}$ *is the noise vector in* (7.87), $\sigma_j^{\{j,k\}} = \left(E\|\mathbf{Z}_j^{\{j,k\}}\|_2^2/n\right)^{1/2}$, *and* $C_1$ *is a constant depending on* $c_*$ *only. Moreover, with probability* $1 + o(1)$,

$$\left(\max_{j\neq k}\frac{|\widehat{\rho}_{j,k}^{(\text{par})} - \rho_{j,k}^{(\text{par})}|}{\left(F_{j,k}^{(\text{par})}\right)^{-1/2}}\right) \vee \left(\max_{j,k}\frac{|\widehat{\omega}_{j,k} - \omega_{j,k}|}{F_{j,k}^{-1/2}}\right) \le \lambda_0(1 + C_1 s^* \lambda_0), \quad (7.95)$$

*where* $F_{j,k} = 1/(\omega_{j,k}^2 + \omega_{j,j}\omega_{k,k})$ *and* $F_{j,k}^{(\text{par})} = 1/\{1 - \omega_{j,k}^2/(\omega_{j,j}\omega_{k,k})\}^2$ *are the Fisher information for the estimation of* $\omega_{j,k}$ *and* $\rho_{j,k}^{(\text{par})}$ *respectively. If in addition* $s^* \log p \ll n^{1/2}$, *then*

$$\max_{1\le j\le k\le p}\sup_t\left|P\left\{\sqrt{nF_{j,k}}(\widehat{\omega}_{j,k} - \omega_{j,k}) \le t\right\} - \Phi(t)\right| = o(1)$$

*for the* $\widehat{\omega}_{j,k}$ *in* (7.89), *where* $\Phi(t)$ *is the* $N(0,1)$ *distribution function, and*

$$\max_{1\le j<k\le p}\sup_t\left|P\left\{\sqrt{nF_{j,k}^{(\text{par})}}(\widehat{\rho}_{j,k}^{(\text{par})} - \rho_{j,k}^{(\text{par})}) \le t\right\} - \Phi(t)\right| = o(1)$$

*for the* $\widehat{\rho}_{j,k}^{(\text{par})}$ *in* (7.90).

In addition to guaranteeing the restricted eigenvalue condition in regression models (7.87), condition (7.91) implies that $c_* \le 1$ and

$$1 \le \max_j |D_j\omega_{j,j}| \le 1/c_*, \quad \max_{j,k}|\rho_{j,k}^{(\text{par})}| \le \sqrt{1 - c_*}. \quad (7.96)$$

Under the sparsity condition $s^* \log p \ll n^{1/2}$, Theorem 7.7 provides the asymptotic efficiency of the estimator (7.89) for the elements $\omega_{j,k}$ of the precision matrix, including the diagonal elements $\omega_{j,j}$, and the asymptotic efficiency of the estimator (7.90) for the partial correlation. Under the weaker sparsity condition $s^* \log p \ll n$, it still provides the uniform asymptotic consistency of (7.89) and thus the consistency of plug-in estimators for the Fisher information $F_{j,k}$ and $F_{j,k}^{(\text{par})}$. Thus, Theorem 7.7 provides asymptotic efficiency of the confidence intervals given in the following corollary.

**Corollary 7.1** *Under conditions* (7.91) *and* (7.93) *and* $s^* \log p \ll n^{1/2}$,

$$\max_{1\le j\le k\le p}\sup_t\left|P\left\{\sqrt{n\widehat{F}_{j,k}}(\widehat{\omega}_{j,k} - \omega_{j,k}) \le t\right\} - \Phi(t)\right| = o(1)$$

*with* $\widehat{F}_{j,k} = 1/(\widehat{\omega}_{j,k}^2 + \widehat{\omega}_{j,j}\widehat{\omega}_{k,k})$, *and*

$$\max_{1\le j<k\le p}\sup_t\Big|P\Big\{\sqrt{n\widehat{F}_{j,k}^{(\mathrm{par})}}\big(\widehat{\rho}_{j,k}^{(\mathrm{par})} - \rho_{j,k}^{(\mathrm{par})}\big)\le t\Big\} - \Phi(t)\Big| = o(1)$$

*with* $\widehat{F}_{j,k}^{(\mathrm{par})} = 1/\{1-\widehat{\omega}_{j,k}^2/(\widehat{\omega}_{j,j}\widehat{\omega}_{k,k})\}^2$. *Moreover,* (7.95) *holds with* $F_{j,k}$ *and* $F_{j,k}^{(\mathrm{par})}$ *replaced by* $\widehat{F}_{j,k}$ *and* $\widehat{F}_{j,k}^{(\mathrm{par})}$ *respectively under the condition* $s^*\log p \le n$.

Based on (7.95) we may also construct an estimator for the entire partial correlation matrix $\boldsymbol{\rho}^{(\mathrm{par})} = \big(\rho_{j,k}^{(\mathrm{par})}\big)_{p\times p}$ by adaptive thresholding,

$$\widehat{\boldsymbol{\rho}}^{(\mathrm{par})} = \Big(I_{\{j=k\}} + I_{\{j\ne k\}}\mathrm{th}_{\lambda_1}\big(\widehat{\rho}_{j,k}^{(\mathrm{par})}\big)\Big)_{p\times p}, \tag{7.97}$$

where $\mathrm{th}_{\lambda_1}(t)$ is the soft- or hard-threshold estimator with threshold level $\lambda_1$ or any estimator in between, and the threshold level is given by $\lambda_1 = \lambda_1(s_1)$ with

$$s_1 = \min\Big\{s : \max_j \#\Big(k : |\widehat{\rho}_{j,k}^{(\mathrm{par})}| \ge \lambda_1(s)\Big) \le s\Big\}$$

where $\lambda_1(s) = \lambda_0(2 + C_1 s^*\lambda_0)$ with the $C_1$ in (7.94). For the estimation of the entire precision matrix $\boldsymbol{\Omega} = \boldsymbol{\Sigma}^{-1}$, we may use

$$\widehat{\boldsymbol{\Omega}} = \Big(\widehat{\omega}_{j,j}^{1/2}\mathrm{th}_{\lambda_1}\big(\widehat{\rho}_{j,k}^{(\mathrm{par})}\big)\widehat{\omega}_{k,k}^{1/2}\Big)_{p\times p}. \tag{7.98}$$

**Corollary 7.2** *Under conditions* (7.91) *and* (7.93) *and* $s^*\log p \ll n$,

$$P\Big\{\big\|\widehat{\boldsymbol{\rho}}^{(\mathrm{par})} - \boldsymbol{\rho}^{(\mathrm{par})}\big\|_1 \ge 2s^*\lambda_1(s^*)\Big\} = o(1)$$

*where* $\|(M_{j,k})_{p\times p}\|_1 = \max_j\sum_k |M_{j,k}|$ *is the* $\ell_1$ *operator norm, and*

$$P\Big\{\big\|\widehat{\boldsymbol{\Omega}} - \boldsymbol{\Omega}\big\|_1 \ge \sqrt{8}(s^*/c_*)\lambda_1(s^*)\Big\} = o(1).$$

Estimation of the precision matrix in the HD setting has been considered by many, including Meinshausen and Bühlmann (2006) for selection consistency and Yuan (2010), Cai (2011), and Sun and Zhang (2013) for estimation. The topic will be studied in Section 9.4 with detailed discussions of several approaches proposed in the literature. An interesting aspect of the above thresholding scheme is that it does not require an additional symmetrization step, and the error bounds in Corollary 7.2 does not require conditions on boundedness of $\|\boldsymbol{\Sigma}\|_S$ or $\|\boldsymbol{\Omega}\|_1$.

**Proof of Theorem 7.7.** The regression model (7.87) has normalized design $\mathbf{X}_{\{j,k\}^c}\overline{\mathbf{D}}_{\{j,k\}^c}^{-1/2}$, noise level $\sigma_j^{\{j,k\}}$ and normalized coefficients $\overline{D}_\ell^{1/2}\gamma_\ell^{(j),k}$.

We shall first prove that the capped-$\ell_1$ sparsity of the normalized coefficients satisfies

$$\sum_{\ell\in\{j,k\}^c} \min\left(1, \frac{|D_\ell^{1/2}\gamma_\ell^{(j),k}|}{\sigma_j^{\{j,k\}}\sqrt{(2/n)\log p}}\right) \le s^*_{j,k} \tag{7.99}$$

with the $s^*_{j,k}$ in (7.92) in this expression model. By algebra,

$$\mathbf{\Omega}^{-1}_{\{j,k\},\{j,k\}} = \begin{pmatrix} \omega_{j,j} & \omega_{j,k} \\ \omega_{k,j} & \omega_{k,k}\end{pmatrix}^{-1} = (\omega_{j,j}\omega_{k,k} - \omega^2_{j,k})^{-1}\begin{pmatrix} \omega_{k,k} & -\omega_{j,k} \\ -\omega_{k,j} & \omega_{j,j}\end{pmatrix}$$

is the covariance of the noise in the bivariate regression of $\mathbf{X}_{\{j,k\}}$ against $\mathbf{X}_{\{j,k\}^c}$, so that the variance of the elements of $\mathbf{Z}_j^{\{j,k\}}$ in (7.87) is $(\sigma_j^{\{j,k\}})^2 = \omega_{k,k}/(\omega_{j,j}\omega_{k,k} - \omega^2_{j,k})$. Moreover, as $\gamma^{(j),k}_{\{j,k\}^c}$ is the first column of $\mathbf{\Gamma}^{\{j,k\}}_{\{j,k\}^c} = -\mathbf{\Omega}_{\{j,k\}^c,\{j,k\}}\mathbf{\Omega}^{-1}_{\{j,k\},\{j,k\}}$, by algebra and (7.96),

$$\begin{aligned}
\left|D_\ell^{1/2}\gamma_\ell^{(j),k}/\sigma_j^{\{j,k\}}\right| &= \frac{D_\ell^{1/2}|\omega_{\ell,j}\omega_{k,k} - \omega_{\ell,k}\omega_{k,j}|}{\sqrt{\omega_{k,k}(\omega_{j,j}\omega_{k,k} - \omega^2_{j,k})}} \\
&= \sqrt{D_\ell\omega_{\ell,\ell}}\frac{|\rho^{(\text{par})}_{\ell,j} - \rho^{(\text{par})}_{\ell,k}\rho^{(\text{par})}_{k,j}|}{\sqrt{1-(\rho^{(\text{par})}_{j,k})^2}} \\
&\le |\rho^{(\text{par})}_{\ell,j} - \rho^{(\text{par})}_{\ell,k}\rho^{(\text{par})}_{k,j}|/c_*.
\end{aligned}$$

Thus, (7.99) holds with $s^*_{j,k} \le s^* \ll n/\log p$ by (7.92). Moreover, as the restricted eigenvalue condition holds with large probability for the normalized Gaussian design under condition (7.91), Theorem 7.2 provides (7.93).

We shall use the Gaussian concentration inequality in the following way. For any $n \times 2$ Gaussian matrix $(\mathbf{z}_1, \mathbf{z}_2)$ with i.i.d. rows, $N(0,1)$ entries and correlation $\rho$ between rows,

$$\begin{aligned}
\left|\mathbf{z}_1^T\mathbf{z}_2/n - \rho\right| &\le |\rho|\left\{\left(1 + (\varepsilon')_+/n^{1/2}\right)^2 - 1\right\} \\
&\quad + \sqrt{1-\rho^2}|\varepsilon''|\left(1 + (\varepsilon')_+/n^{1/2}\right)/n^{1/2}
\end{aligned} \tag{7.100}$$

for certain independent $N(0,1)$ random variables $\varepsilon_1$ and $\varepsilon_2$.

It follows from (7.93) and (7.100) that

$$\begin{aligned}
&\left|(\widehat{\mathbf{Z}}_j^{\{j,k\}})^T(\widehat{\mathbf{Z}}_k^{\{j,k\}}) - (\mathbf{Z}_j^{\{j,k\}})^T(\mathbf{Z}_k^{\{j,k\}})\right| \\
\le\ & \left|(\widehat{\mathbf{Z}}_j^{\{j,k\}})^T\mathbf{X}_{\{j,k\}^c}(\widehat{\gamma}^{(k),j}_{\{j,k\}^c} - \gamma^{(k),j}_{\{j,k\}^c})\right| \\
&+\left|(\mathbf{Z}_k^{\{j,k\}})^T\mathbf{X}_{\{j,k\}^c}(\widehat{\gamma}^{(j),k}_{\{j,k\}^c} - \gamma^{(j),k}_{\{j,k\}^c})\right| \\
\le\ & (n\sigma_j^{\{j,k\}}\lambda_0/\eta_0)\left\|\mathbf{D}^{1/2}_{\{j,k\}^c}(\widehat{\gamma}^{(k),j}_{\{j,k\}^c} - \gamma^{(k),j}_{\{j,k\}^c})\right\|_1 \\
&+\left\|(\mathbf{Z}_k^{\{j,k\}})^T\mathbf{X}_{\{j,k\}}\mathbf{D}^{-1/2}_{\{j,k\}^c}\right\|_\infty\left\|\mathbf{D}^{1/2}_{\{j,k\}^c}(\widehat{\gamma}^{(j),k}_{\{j,k\}^c} - \gamma^{(j),k}_{\{j,k\}^c})\right\|_1
\end{aligned}$$

$$\leq \left\{(n\lambda_0/\eta_0) + \max_j |\varepsilon_j''|(n^{1/2} + (\varepsilon_j')_+)\right\}(\sigma_j^{\{j,k\}}\sigma_j^{\{j,k\}})C_0 s^* \lambda_0.$$

Here the first inequality follows from the error decomposition above the statement of the theorem, the second follows from (7.93) and the third follows from (7.100) as $\mathbf{Z}_k^{\{j,k\}}/\sigma_k^{\{j,k\}}$ and $\mathbf{X}_\ell D_\ell^{-1/2}$ are independent standard Gaussian vectors for $\ell \in \{j,k\}^c$. This gives (7.94) as $P\{|\varepsilon_j'| \vee (\varepsilon_j'')_+ > t\} \leq 3e^{-t^2/2}$ and $\log p \ll n$.

It follows that the estimators (7.89) and (7.90) are within $o(n^{-1/2})$ from the sample covariance and correlation between standard Gaussian vectors $\mathbf{Z}_j^{\{j,k\}}/\sigma_j^{\{j,k\}}$ and $\mathbf{Z}_k^{\{j,k\}}/\sigma_k^{\{j,k\}}$. The asymptotic normality and efficiency then follow from the classical analysis in the case of $p = 2$, in view of (7.96). Moreover, it follows from (7.94) (7.96) and (7.100) that (7.95) holds with probability $1 + o(1)$ as the inequality holds when $\mathbf{Z}_j^{\{j,k\}}$ and $\mathbf{Z}_k^{\{j,k\}}$ are used to compute the estimators instead of their estimated version. ∎

**Proof of Corollary 7.2.** Let $S = \{(j,k) : |\rho_{j,k}^{(\text{par})}| \geq \lambda_0\}$. In the event (7.95), $|\widehat{\rho}_{j,k}^{(\text{par})}| \leq \lambda_0(2 + C_1 s^* \lambda_0) = \lambda_1(s^*)$ for $(j,k) \in S^c$, so that for $s \geq s^*$

$$\max_j \#\{k : |\widehat{\rho}_{j,k}^{(\text{par})}| \geq \lambda_1(s)\} \leq \max_j \#\{k : (j,k) \in S\} \leq s^*.$$

It follows that in the same event $s_1 \leq s^*$ and

$$\max_j \sum_{k=1}^p \left|\widehat{\rho}_{j,k}^{(\text{par})} - \rho_{j,k}^{(\text{par})}\right| \leq (s_1 + s^*)\lambda_1(s_1) \leq 2s^*\lambda_1(s^*).$$

As $\max_{j\leq p} \omega_{j,j} \leq 1/c_*$, the $\ell_1$ bound for $\widehat{\mathbf{\Omega}}$ follows with an application of (7.95) to the diagonal elements $\widehat{\omega}_{j,j}$. ∎

### *7.4.2 Sample size in regression and graphical models*

We have presented methods leading to regular and efficient statistical inference about LD parameters at the parametric $n^{-1/2}$ rate, including the noise level and individual coefficients in linear regression and the partial correlations and individual elements of the precision matrix in the Gaussian graphical model. However, compared with regularized estimation of HD quantities such as prediction and estimation of the entire coefficient vector or precision matrix, these results on efficient inference of LD quantities require the stronger sample size $n \gtrsim (s^* \log p)^2$, compared with the usual $n \gtrsim s^* \log p$. In this subsection we prove that this sample size requirement cannot be removed without further assumptions in the regression or Gaussian graphical model. These results were obtained in Ren *et al.* (2015) for Gaussian graphical models and in Cai and Guo (2017) for linear regression.

Consider the Gaussian graphical model where a data matrix $\mathbf{X}$ is observed with iid $N(\mathbf{0},\mathbf{\Sigma})$ rows. Consider the parameter class

$$\mathscr{G}_{s^*,c_*,c^*} = \Big\{\mathbf{\Sigma} : c_*\|\mathbf{u}\|_2^2 \le \mathbf{u}^T\mathbf{\Sigma}\mathbf{u} \le c^*\|\mathbf{u}\|_2^2\ \forall\mathbf{u}, \|\mathbf{\Omega}_{*,j}\|_0 \le s^*\ \forall j\Big\}, \quad (7.101)$$

where $\|\mathbf{\Omega}_{*,j}\|_0 = \#\{k : \omega_{k,j} \ne 0\}$ for the precision $\mathbf{\Omega} = \mathbf{\Sigma}^{-1}$. We shall prove that the minimax rate of convergence for the estimation of individual partial correlations $\rho_{j,k}^{(\text{par})}$ and elements $\omega_{j,k}$ of the precision matrix is actually $n^{-1/2} + s^*(\log p)/n$ when $s^*(\log p)/n$ is sufficiently small, so that the estimation at the $n^{-1/2}$ rate would be infeasible in this larger regime when $s^*(\log p)/n^{1/2}$ is large. Formally, we state our result in the following theorem.

**Theorem 7.8** *Let $P_{\mathbf{\Sigma}}$ be the probability under which $\mathbf{X} \in I\!R^{n\times p}$ has iid $N(\mathbf{0},\mathbf{\Sigma})$ rows. Let $r_{n,p,s^*} = \max\{n^{-1/2}, s^*(\log p)/n\}$, $c_* \in (0,1/2)$ and $c^* \in (2,\infty)$. For each $c_1 > 0$, there exists $c_2 \in (0,1)$ depending on $\{c_1,c_*,c^*\}$ only such that when $p \ge (s^*)^{2+c_1}$ and $1/c_1 \le s^* \le c_1 n/\log p$,*

$$\inf_{\delta_{j,k}(\mathbf{X})}\ \sup_{\mathbf{\Sigma}\in\mathscr{G}_{s^*,c_*,c^*}} P_{\mathbf{\Sigma}}\Big\{\big|\delta_{j,k}(\mathbf{X}) - \rho_{j,k}^{(\text{par})}\big| \ge c_2 r_{n,p,s^*}\Big\} \ge c_2 \quad (7.102)$$

*for all $1 \le j < k \le p$, and for all $1 \le j \le k \le p$*

$$\inf_{\delta_{j,k}(\mathbf{X})}\ \sup_{\mathbf{\Sigma}\in\mathscr{G}_{s^*,c_*,c^*}} P_{\mathbf{\Sigma}}\Big\{\big|\delta_{j,k}(\mathbf{X}) - \omega_{j,k}\big| \ge c_2 r_{n,p,s^*}\Big\} \ge c_2. \quad (7.103)$$

Theorem 7.8 shows that the estimation of the elements of a precision matrix is quite different from the estimation of individual covariances even when the precision matrix is assumed to be sparse. While the individual covariance $\sigma_{j,k}$ can always be estimated at the parametric rate in the Gaussian graphical model, the element $\omega_{j,k}$ in the precision matrix can be estimated at the rate $r_{n,p,s^*}$ at the best, and $r_{n,p,s^*} \gg n^{-1/2}$ when $s^*\log p \gg n^{1/2}$.

We note that as the capped-$\ell_1$ sparsity is more general than the hard sparsity, $\|\mathbf{b}\|_0 \le s^*$ implies $\mathbf{b} \in \mathscr{B}_n(s^*)$ for the parameter space in (7.65). Thus, the condition $\mathbf{\Sigma} \in \mathscr{G}_{s^*,c_*,c^*}$ is sufficient for Theorem 7.7 and its corollaries, and the lower bound in Theorem 7.8 is also valid in the more general setting.

Theorem 7.8 is proved by carefully constructing a parametric family of Gaussian densities $g(\mathbf{x}|\mathbf{w}) \sim N(\mathbf{0},\mathbf{\Sigma}(\mathbf{w}))$ in $I\!R^p$ satisfying the following three properties: (a) The covariance matrix belongs to the parameter class, $\mathbf{\Sigma}(\mathbf{w}) \in \mathscr{G}_{s^*,c_*,c^*}$; (b) There is a sufficiently large gap in parameter value between the cases of $\mathbf{w} = \mathbf{0}$ and $\mathbf{w} = \mathbf{1}$, in the sense that

$$\rho_{j,k}^{(\text{par})} = \begin{cases} \theta^{(0)}, & \mathbf{w} = \mathbf{0}, \\ \theta^{(1)}, & \mathbf{w} \ne \mathbf{0}, \end{cases} \quad (7.104)$$

for certain constants $\theta^{(0)}$ and $\theta^{(1)}$ satisfying $\big|\theta^{(1)} - \theta^{(0)}\big| \ge 2c_2 r_{n,p,s^*}$: (c) The two cases are not statistically separable in the sense that the sum of type-I and type-II errors for testing $H_0 : \mathbf{w} = \mathbf{0}$ against $H_1 : \mathbf{w} \neq \mathbf{0}$ is at least $2c_2$. We note that if we reject $H_0$ when $|\delta_{j,k}(\mathbf{X}) - \theta^{(1)}| < |\delta_{j,k}(\mathbf{X}) - \theta^{(0)}|$ and accept $H_0$ otherwise, the left-hand side of (7.102) would be no smaller than the minimum of the sum of the type-I and type-II errors. This is the so-called Le Cam's method of proving minimax lower bounds.

For the estimation of $\omega_{j,k}$, we use the same construction to prove (7.103) by replacing $\rho_{j,k}^{(\mathrm{par})}$ with $\omega_{j,k}$ in (7.104), possibly with different $\theta^{(0)}$ and $\theta^{(1)}$ satisfying $\left|\theta^{(1)} - \theta^{(0)}\right| \geq 2c_2 r_{n,p,s^*}$.

The same analysis also leads to parallel results for the estimation of the noise level and individual coefficients in linear regression as the Gaussian assumption provides the regression model

$$\mathbf{X}_j = \sum_{k \neq j} \mathbf{X}_k \beta_k^{(j)} + \boldsymbol{\varepsilon}_j \tag{7.105}$$

with $\boldsymbol{\varepsilon}_j \sim N(\mathbf{0}, \sigma_j^2 \mathbf{I}_n)$, where

$$\beta_k^{(j)} = -\omega_{j,k}/\omega_{j,j}, \quad \sigma_j^2 = 1/\omega_{j,j}. \tag{7.106}$$

Formally, we consider the linear model (7.3) with coefficient vector $\boldsymbol{\beta}$, design $\mathbf{X}$ with i.i.d. $N(\mathbf{0}, \boldsymbol{\Sigma})$ rows and $N(\mathbf{0}, \sigma^2 \mathbf{I}_n)$ noise $\boldsymbol{\varepsilon}$ with the triple $\{\boldsymbol{\beta}, \boldsymbol{\Sigma}, \sigma^2\}$ in the following parameter space

$$\mathscr{B}_{s^*,c_*,c^*} = \Big\{\{\boldsymbol{\beta}, \boldsymbol{\Sigma}, \sigma^2\} : \|\boldsymbol{\beta}\|_0 \leq s^*, \ \boldsymbol{\Sigma} \in \mathscr{G}_{s^*,c_*,c^*}, c_* \leq \sigma^2 \leq c^*\Big\}.$$

where $\mathscr{G}_{s^*,c_*,c^*}$ is as in (7.101).

**Theorem 7.9** *Let $P_{\boldsymbol{\beta},\boldsymbol{\Sigma},\sigma^2}$ be the probability under which $\boldsymbol{\beta}$ is the coefficient vector, $\mathbf{X} \in I\!R^{n\times p}$ has iid $N(\mathbf{0}, \boldsymbol{\Sigma})$ rows and $\boldsymbol{\varepsilon} \sim N(\mathbf{0}, \sigma^2\mathbf{I}_n)$ in the regression model* (7.3). *Let $r_{n,p,s^*} = \max\{n^{-1/2}, s^*(\log p)/n\}$, $c_* \in (0, 1/2)$ and $c^* \in (2, \infty)$. For each $c_1 > 0$, there exists $c_2 \in (0, 1)$ depending on $\{c_1, c_*, c^*\}$ only such that when $p \geq (s^*)^{2+c_1}$ and $1/c_1 \leq s^* \leq c_1 n/\log p$,*

$$\inf_{\delta_j(\mathbf{X})} \sup_{\{\boldsymbol{\beta},\boldsymbol{\Sigma},\sigma^2\} \in \mathscr{B}_{s^*,c_*,c^*}} P_{\boldsymbol{\beta},\boldsymbol{\Sigma},\sigma^2}\Big\{\left|\delta_j(\mathbf{X}) - \beta_j\right| \geq c_2 r_{n,p,s^*}\Big\} \geq c_2 \tag{7.107}$$

*for all $1 \leq j \leq p$, and*

$$\inf_{\delta(\mathbf{X})} \sup_{\{\boldsymbol{\beta},\boldsymbol{\Sigma},\sigma^2\} \in \mathscr{B}_{s^*,c_*,c^*}} P_{\boldsymbol{\beta},\boldsymbol{\Sigma},\sigma^2}\Big\{\left|\delta(\mathbf{X}) - \sigma\right| \geq c_2 r_{n,p,s^*}\Big\} \geq c_2. \tag{7.108}$$

The construction in the proof of Theorem 7.8 is also applicable to Theorem 7.9 via (7.105) by putting $\beta_k^{(j)} = -\omega_{j,k}/\omega_{j,j}$ or $\sigma_j^2 = 1/\omega_{j,j}$ in (7.104).

**Proof of Theorems 7.8 and 7.9.** Consider $(j, k) = (1, 2)$ without loss of generality. As the $n^{-1/2}$ rate follows from the parametric estimation theory in the LD setting or with the additional knowledge of $\omega_{j,k}$ for $3 \le j \le k \le p$, it suffices to consider the case of $s^* \log p > n^{1/2}$.

We construct the density family $g(\mathbf{x}|\mathbf{w}) \sim N(\mathbf{0}, \mathbf{\Sigma}(\mathbf{w}))$ as follows. Let $S = \{1, 2\}$ and define

$$\mathbf{\Sigma}^{(0)} = \mathbf{\Sigma}(\mathbf{0}) = \begin{pmatrix} b^2+\sigma^2 & b & \mathbf{0} \\ b & 1 & \mathbf{0} \\ \mathbf{0} & \mathbf{0} & \mathbf{I}_{S^c \times S^c} \end{pmatrix} = \begin{pmatrix} \mathbf{\Sigma}_{S,S} & \mathbf{0} \\ \mathbf{0} & \mathbf{I}_{S^c \times S^c} \end{pmatrix}.$$

For $\mathbf{w} = \mathbf{0}$, we define $g(\mathbf{x}|\mathbf{w})$ as the joint density of $N(\mathbf{0}, \mathbf{\Sigma}^{(0)})$ in $I\!R^p$. The two eigenvalues of $\mathbf{\Sigma}_{S,S}$ are given by

$$\lambda_{\pm} = \{(1+b^2+\sigma^2) \pm \sqrt{(1+b^2+\sigma^2)^2 - 4\sigma^2}\}/2,$$

so that for sufficiently small $|b|$ and $b^2 + \sigma^2 \approx 1$, $c_* \le \min(\lambda_-, 1)$ and $\max(\lambda_+, 1) \le c^*$, due to $c_* < 1/2$ and $c^* > 2$. As the inverse of $\mathbf{\Sigma}^{(0)}$ is

$$\mathbf{\Omega}^{(0)} = \begin{pmatrix} \mathbf{\Sigma}_{S,S}^{-1} & \mathbf{0} \\ \mathbf{0} & \mathbf{I}_{S^c \times S^c} \end{pmatrix} = \begin{pmatrix} 1/\sigma^2 & -b/\sigma^2 & \mathbf{0} \\ -b/\sigma^2 & 1+b^2/\sigma^2 & \mathbf{0} \\ \mathbf{0} & \mathbf{0} & \mathbf{I}_{S^c \times S^c} \end{pmatrix} \tag{7.109}$$

we have $\|\mathbf{\Omega}_{j,*}\|_0 \le 2$, so that $\mathbf{\Sigma} \in \mathscr{G}_{s^*, c_*, c^*}$ as in (7.101).

For the alternative, we define

$$\mathbf{\Sigma}^{(1)} = \begin{pmatrix} \mathbf{\Sigma}_{S,S} & (\rho,1)^T \mathbf{w}^T \\ \mathbf{w}(\rho,1) & \mathbf{I}_{S^c \times S^c} \end{pmatrix} \tag{7.110}$$

with $\mathbf{w} \in I\!R^{p-2}$. As the operator norm of the difference is give by

$$\left\| \mathbf{\Sigma}^{(1)} - \mathbf{\Sigma}^{(0)} \right\|_{op} = \left\| \begin{pmatrix} \mathbf{0} & (\rho,1)^T \mathbf{w}^T \\ \mathbf{w}(\rho,1) & \mathbf{0} \end{pmatrix} \right\|_{op} = 2\sqrt{1+\rho^2}\|\mathbf{w}\|_2,$$

the eigenvalues of $\mathbf{\Sigma}^{(1)}$ are bounded in between $(\lambda_- \wedge 1) - 2\sqrt{1+\rho^2}\|\mathbf{w}\|_2$ and $(\lambda_- \vee 1) + 2\sqrt{1+\rho^2}\|\mathbf{w}\|_2$. We choose fixed $\{\sigma^2, |b|, |\rho|, \|\mathbf{w}\|_2\}$ such that these upper and lower bounds are both in $[c_*, c^*]$. By the block inversion formula, the inverse of $\mathbf{\Sigma}^{(1)}$ is given by

$$\mathbf{\Omega}^{(1)} = \begin{pmatrix} \mathbf{\Omega}_{S,S}^{(1)} & -\mathbf{\Omega}_{S,S}^{(1)}(\rho,1)^T \mathbf{w}^T \\ -\mathbf{w}(\rho,1)\mathbf{\Omega}_{S,S}^{(1)} & \mathbf{I}_{S^c \times S^c} + \mathbf{w}(\rho,1)\mathbf{\Omega}_{S,S}^{(1)}(\rho,1)^T \mathbf{w}^T \end{pmatrix}$$

with $\mathbf{\Omega}_{S,S}^{(1)} = \left(\mathbf{\Sigma}_{S,S} - (\rho,1)^T(\rho,1)\|\mathbf{w}\|_2^2\right)^{-1}$. Thus, as the column space of $\mathbf{\Omega}^{(1)}$ is contained in the span of $\{(a_1, a_2, \mathbf{w}^T)^T, a_1 \in I\!R, a_2 \in I\!R\}$, or equivalently

$$\max_j \left\|\mathbf{\Omega}_{*,j}^{(1)}\right\|_0 \le 2 + \|\mathbf{w}\|_0.$$

We consider $\mathbf{w} \in \{0, a\}^{p-2}$ with $\|\mathbf{w}\|_0 = s^* - 2$ and $a = \tau_0\sqrt{(\log p)/n}$ with a constant $\tau_0 > 0$, so that $\max_j \left\|\mathbf{\Omega}^{(1)}_{*,j}\right\|_0 \le s^*$ and $\|\mathbf{w}\|_2^2 = a^2(s^*-2)$ depends on $(n,p,s_*)$ only as desired. Consequently, $\mathbf{\Sigma}^{(1)} \in \mathscr{G}_{s^*,c_*,c^*}$ for all such $\mathbf{w}$.

Next, we show that for $(j,k)=(1,2)$, the partial correlation, element of the precision matrix, regression coefficient and noise level are all separated by a gap of the order $\|\mathbf{w}\|_2^2 = \tau_0^2(s^*-2)(\log p)/n$. For this purpose, we explicitly derive

$$\begin{aligned}
\mathbf{\Omega}^{(1)}_{S,S} &= \begin{pmatrix} \omega^{(1)}_{1,1} & \omega^{(1)}_{1,2} \\ \omega^{(1)}_{2,1} & \omega^{(1)}_{2,2} \end{pmatrix} \\
&= \left(\mathbf{\Sigma}_{S,S} - (\rho,1)^T\mathbf{w}^T\mathbf{w}(\rho,1)\right)^{-1} \\
&= \begin{pmatrix} \sigma^2+b^2-\rho^2\|\mathbf{w}\|_2^2 & b-\rho\|\mathbf{w}\|_2^2 \\ b-\rho\|\mathbf{w}\|_2^2 & 1-\|\mathbf{w}\|_2^2 \end{pmatrix}^{-1} \\
&= \frac{\begin{pmatrix} 1-\|\mathbf{w}\|_2^2 & -b+\rho\|\mathbf{w}\|_2^2 \\ -b+\rho\|\mathbf{w}\|_2^2 & \sigma^2+b^2-\rho^2\|\mathbf{w}\|_2^2 \end{pmatrix}}{(\sigma^2+b^2-\rho^2\|\mathbf{w}\|_2^2)(1-\|\mathbf{w}\|_2^2)-(b-\rho\|\mathbf{w}\|_2^2)^2}.
\end{aligned} \tag{7.111}$$

We note that the denominator is approximately $\sigma^2$ as $\|\mathbf{w}\|_2^2$ is small. This gives

$$\begin{aligned}
\rho^{(\mathrm{par})}_{1,2} &= \frac{b-\rho\|\mathbf{w}\|_2^2}{\{1-\|\mathbf{w}\|_2^2\}^{1/2}\{\sigma^2+b^2-\rho^2\|\mathbf{w}\|_2^2\}^{1/2}} \\
\omega_{1,2} &= \frac{-b+\rho\|\mathbf{w}\|_2^2}{(\sigma^2+b^2-\rho^2\|\mathbf{w}\|_2^2)(1-\|\mathbf{w}\|_2^2)-(b-\rho\|\mathbf{w}\|_2^2)^2}. \\
\beta^{(1)}_2 &= (b-\rho\|\mathbf{w}\|_2^2)/\{1-\|\mathbf{w}\|_2^2\} \\
\sigma_1^2 &= \sigma^2+b^2-\rho^2\|\mathbf{w}\|_2^2-(b-\rho\|\mathbf{w}\|_2^2)^2/(1-\|\mathbf{w}\|_2^2) \\
&= \sigma^2-(b-\rho)^2\|\mathbf{w}\|_2^2/(1-\|\mathbf{w}\|_2^2).
\end{aligned} \tag{7.112}$$

We note that (7.112) also gives the values of these parameters under $H_0$ by setting $\|\mathbf{w}\|_2^2=0$, in view of (7.109). This proves (7.104) for $\rho^{(\mathrm{par})}_{1,2}$ and with $\rho^{(\mathrm{par})}_{1,2}$ replaced by all other parameters in (7.112).

We have proved $g(\mathbf{x}|\mathbf{w})$ has properties (a) and (b) as discussed in the paragraph with (7.104). It remains to prove property (c), that is to find a nonzero lower bound for the sum of Type I and Type II errors in testing $H_0 : \mathbf{w} = \mathbf{0}$ against $H_1 : \mathbf{w} \ne \mathbf{0}$ with signal strength $\|\mathbf{w}\|_2^2 \asymp s^*(\log p)/n$.

Recall that $g(\mathbf{x}|\mathbf{w})$ is the joint density of one copy of $N(\mathbf{0}, \mathbf{\Sigma}(\mathbf{w}))$. Let $f(\cdot|\mathbf{w})$ be the joint density of $n$-copies of $N(\mathbf{0}, \mathbf{\Sigma}(\mathbf{w}))$. Let $f_k = \int f(\cdot|\mathbf{w})\pi_k(d\mathbf{w})$ with $\pi_0$ a prior putting the unit mass at $\mathbf{w}=\mathbf{0}$ and $\pi_1$ a prior putting the entire mass in the set of $\mathbf{w} \in \{0,a\}^{p-2}$ satisfying $\|\mathbf{w}\|_0 = s^*-2$. We notice that the value of the parameters is still given by (7.112) and (7.109) under $f_1$ and $f_0$ respectively.

To apply Le Cam's method, we utilize the fact that the minimum of the sum of type-I and type-II errors for testing $H_0 : f_0$ against $H_1 : f_1$, achieved by rejecting $H_0$ when $f_1 > f_0$, is given by

$$\int_{f_1>f_0} f_0 + \int_{f_1<f_0} f_1 = 1 - \frac{1}{2}\int |f_1 - f_0|,$$

and the $L_1$ distance is bounded by the chi-square distance, $\big(\int |f_1 - f_0|\big)^2 \le \int |f_1/f_0 - 1|^2 f_0 = \int f_1^2/f_0 - 1$. Thus, it suffices to prove

$$\chi^2(f_0, f_1) = \int f_1^2/f_0 - 1 = o(1). \tag{7.113}$$

The advantage of the chi-square distance is that it factorizes when $f_k$ are the joint densities of $n$-copies of observations from $g_k$, $k = 1, 2$,

$$\begin{aligned}\chi^2(f_0, f_1) &= \int\int\Big[\int_{I\!R^{n\times p}} \frac{f(\cdot|\mathbf{w})f(\cdot|\widetilde{\mathbf{w}})}{f(\cdot|\mathbf{0})} - 1\Big]\pi_1(d\mathbf{w})\pi_1(d\widetilde{\mathbf{w}}) \\ &= \int\int\Big[\int_{I\!R^{p}} \frac{g(\cdot|\mathbf{w})g(\cdot|\widetilde{\mathbf{w}})}{g(\cdot|\mathbf{0})}\Big]^n \pi_1(d\mathbf{w})\pi_1(d\widetilde{\mathbf{w}}) - 1.\end{aligned}$$

As $g(\cdot|\mathbf{w})$ is jointly Gaussian given $\mathbf{w}$,

$$\begin{aligned}&\int_{I\!R^p} \frac{g(\mathbf{x}|\mathbf{w})g(\mathbf{x}|\widetilde{\mathbf{w}})}{g(\mathbf{x}|\mathbf{0})}d\mathbf{x} \\ =& \int_{I\!R^p} \frac{\exp\big[-\operatorname{trace}\big((\mathbf{x}\mathbf{x}^T/2)\big(\mathbf{\Sigma}^{-1}(\mathbf{w}) + \mathbf{\Sigma}^{-1}(\widetilde{\mathbf{w}}) - \mathbf{\Sigma}^{-1}(\mathbf{0})\big)\big)\big]}{\big\{\det(\mathbf{\Sigma}(\mathbf{w})\mathbf{\Sigma}(\widetilde{\mathbf{w}}))/\det(\mathbf{\Sigma}(\mathbf{0}))\big\}^{1/2}}d\mathbf{x} \\ =& \left\{\frac{\det\big(\mathbf{\Sigma}^{-1}(\mathbf{w}) + \mathbf{\Sigma}^{-1}(\widetilde{\mathbf{w}}) - \mathbf{\Sigma}^{-1}(\mathbf{0})\big)}{\det(\mathbf{\Sigma}(\mathbf{w})\mathbf{\Sigma}(\widetilde{\mathbf{w}}))/\det(\mathbf{\Sigma}(\mathbf{0}))}\right\}^{1/2} \\ =& \det\Big(\mathbf{I} - \mathbf{\Omega}^{(0)}\big(\mathbf{\Sigma}(\mathbf{w}) - \mathbf{\Sigma}^{(0)}\big)\mathbf{\Omega}^{(0)}\big(\mathbf{\Sigma}(\widetilde{\mathbf{w}}) - \mathbf{\Sigma}^{(0)}\big)\Big)^{-1/2}.\end{aligned}$$

It follows from (7.109) and (7.110) that

$$\begin{aligned}\mathbf{\Omega}^{(0)}\big(\mathbf{\Sigma}(\mathbf{w}) - \mathbf{\Sigma}^{(0)}\big) &= \begin{pmatrix}\mathbf{\Sigma}_{S,S}^{-1} & \mathbf{0} \\ \mathbf{0} & \mathbf{I}_{S^c\times S^c}\end{pmatrix}\begin{pmatrix}\mathbf{0} & (\rho,1)^T\mathbf{w}^T \\ \mathbf{w}(\rho,1) & \mathbf{0}\end{pmatrix} \\ &= \begin{pmatrix}\mathbf{0} & \mathbf{\Sigma}_{S,S}^{-1}(\rho,1)^T\mathbf{w}^T \\ \mathbf{w}(\rho,1) & \mathbf{0}\end{pmatrix},\end{aligned}$$

so that

$$\begin{aligned}&\mathbf{\Omega}^{(0)}\big(\mathbf{\Sigma}(\mathbf{w}) - \mathbf{\Sigma}^{(0)}\big)\mathbf{\Omega}^{(0)}\big(\mathbf{\Sigma}(\widetilde{\mathbf{w}}) - \mathbf{\Sigma}^{(0)}\big) \\ =& \begin{pmatrix}\mathbf{0} & \mathbf{\Sigma}_{S,S}^{-1}(\rho,1)^T\mathbf{w}^T \\ \mathbf{w}(\rho,1) & \mathbf{0}\end{pmatrix}\begin{pmatrix}\mathbf{0} & \mathbf{\Sigma}_{S,S}^{-1}(\rho,1)^T\widetilde{\mathbf{w}}^T \\ \widetilde{\mathbf{w}}(\rho,1) & \mathbf{0}\end{pmatrix} \\ =& \begin{pmatrix}\mathbf{\Sigma}_{S,S}^{-1}(\rho,1)^T\|\mathbf{w}\|_2^2(\rho,1) & \mathbf{0} \\ \mathbf{0} & \mathbf{w}(\rho,1)\mathbf{\Sigma}_{S,S}^{-1}(\rho,1)^T\widetilde{\mathbf{w}}^T\end{pmatrix}.\end{aligned}$$

The two diagonal blocks above are both of rank 1 with the identical eigenvalue

$$\begin{aligned} & \operatorname{trace}\Big(\boldsymbol{\Sigma}_{S,S}^{-1}(\rho,1)^T\widetilde{\mathbf{w}}^T\mathbf{w}(\rho,1)\Big) \\ = \ & \widetilde{\mathbf{w}}^T\mathbf{w}(\rho,1)\begin{pmatrix} 1/\sigma^2 & -b/\sigma^2 \\ -b/\sigma^2 & 1+b^2/\sigma^2 \end{pmatrix}\begin{pmatrix}\rho \\ 1\end{pmatrix} \\ = \ & \widetilde{\mathbf{w}}^T\mathbf{w}\big(\rho^2/\sigma^2 - 2\rho b/\sigma^2 + 1 + b^2/\sigma^2\big) \\ = \ & \widetilde{\mathbf{w}}^T\mathbf{w}\big(1+(\rho-b)^2/\sigma^2\big) \end{aligned}$$

in view of the formula of $\boldsymbol{\Sigma}_{S,S}^{-1}$ from (7.109). Thus,

$$\begin{aligned} & \det\Big(\mathbf{I} - \boldsymbol{\Omega}^{(0)}\big(\boldsymbol{\Sigma}^{(1)} - \boldsymbol{\Sigma}^{(0)}\big)\boldsymbol{\Omega}^{(0)}\big(\widetilde{\boldsymbol{\Sigma}}^{(1)} - \boldsymbol{\Sigma}^{(0)}\big)\Big) \\ = \ & \Big(1 - \widetilde{\mathbf{w}}^T\mathbf{w}\big(1+(\rho-b)^2/\sigma^2\big)\Big)^2. \end{aligned}$$

Let $a = \tau_0\sqrt{(\log p)/n}$ with a small $\tau_0 > 0$, $p_- = p-2$ and $s_- = s^* - 2$. Let $\pi_1$ be the prior under which $\mathbf{w}$ is uniformly distributed among the $\binom{p_-}{s_-}$ choices of $\mathbf{w}$ satisfying $\mathbf{w} \in \{0,a\}^{p_-}$ and $\|\mathbf{w}\|_0 = s_-$. As $\widetilde{\mathbf{w}}$ is an independent copy of $\mathbf{w}$, $J = \widetilde{\mathbf{w}}^T\mathbf{w}/a^2$ has the Hypergeometric$(p_-, s_-, s_-)$ distribution. Thus,

$$\widetilde{\mathbf{w}}^T\mathbf{w}\big(1+(\rho-b)^2/\sigma^2\big) = a^2 J\big(1+(\rho-b)^2/\sigma^2\big) = \tau_1 J(\log p)/(2n)$$

with $\tau_1 = 2\tau_0^2\big(1+(\rho-b)^2/\sigma^2\big)$. As $J(\log p)/n \le s_-(\log p)/n \le c_1$ is small, the above calculation yields

$$\begin{aligned} \chi^2(f_0,f_1)+1 &= \mathrm{E}\Big(1 - \widetilde{\mathbf{w}}^T\mathbf{w}\big(1+(\rho-b)^2/\sigma^2\big)\Big)^{-n} \\ &= \mathrm{E}\Big(1 - \tau_1 J(\log p)/(2n)\Big)^{-n} \\ &\le \mathrm{E}\exp\big(\tau_1 J\log p\big) \\ &\le \exp\Big({s_-}^2 p^{\tau_1}/(p_- - s_-)\Big) \\ &= 1 + o(1) \end{aligned}$$

when $\tau_0 > 0$ is chosen to satisfy $2/(1-\tau_1) < 2 + c_1$, as $(s^*)^{2+c_1} \le p$. Here in the last inequality above we use the following bound for the moment generating function of the Hypergeometric$(p_-, s_-, s_-)$ distribution,

$$\begin{aligned} \mathrm{E}\, e^{t\tau} &= \binom{p_-}{s_-}^{-1}\sum_{j=0}^{s_-}\binom{s_-}{j}\binom{p_- - s_-}{s_- - j}e^{tj} \\ &\le \sum_{j=0}^{s_-}\binom{s_-}{j}\frac{(s_- e^t)^j (p_- - s_-)^{s_- - j}}{(p_- - s_-)^{s_-}} \\ &= \left(1 + \frac{s_- e^t}{p_- - s_-}\right)^{s_-} \\ &\le \exp\Big({s_-}^2 e^t/(p_- - s_-)\Big). \end{aligned}$$

This completes the proof of (7.113) and thus that of the entire theorem. ∎

## 7.5 General Solutions*

In this section, we describe some general solutions in the semi-LD approach for asymptotically normal estimation and efficient inference about a real smooth function $\theta = \theta(\boldsymbol{\beta})$ based on a general loss function

$$\ell_{[n]}(\boldsymbol{\beta}) = \textstyle\sum_{i=1}^{n} \ell_i(\boldsymbol{\beta}), \tag{7.114}$$

where $\ell_i(\boldsymbol{\beta}) = \ell_{i,n}(\boldsymbol{\beta}; \text{data}_i)$ is the loss function with $\text{data}_i = \text{data}_{i,n}$, $\boldsymbol{\beta} \in \mathscr{B}_n$ is the unknown parameter, and the parameter space $\mathscr{B}_n$ is a subset of a Euclidean or Hilbert space $\mathscr{H}_n$ equipped with an inner product $\langle \cdot, \cdot \rangle$ and the corresponding norm $\|\mathbf{u}\| = \langle \mathbf{u}, \mathbf{u} \rangle^{1/2}$. We note that negative log-likelihood functions can be treated as loss functions. See (7.47). As in (7.48), we assume that $\theta = \theta(\mathbf{b})$ is smooth in $\mathscr{B}_n$ with the Fréchet derivative

$$\mathbf{a}^o = \partial \theta(\boldsymbol{\beta}) / \partial \boldsymbol{\beta} \in \mathscr{H}_{n,0} \setminus \{\mathbf{0}\} \tag{7.115}$$

for a suitable subspace $\mathscr{H}_{n,0}$ of $\mathscr{H}_n$.

Let $\mathbf{F} = \mathbf{F}_n$ be the Hessian of $\mathrm{E}\, \ell_{[n]}(\mathbf{b})$ at $\mathbf{b} = \boldsymbol{\beta}$. Assume that $\mathbf{F}$ is positive definite in $\mathscr{H}_{n,0}$ and denote by $\mathbf{F}^{-1}$ its inverse in the subspace. Let

$$\mathbf{u}^o = \mathbf{F}^{-1}\mathbf{a}^o / \langle \mathbf{a}^o, \mathbf{F}^{-1}\mathbf{a}^o \rangle, \quad F_\theta = \langle \mathbf{u}^o, \mathbf{F}\mathbf{u}^o \rangle = 1/\langle \mathbf{a}^o, \mathbf{F}^{-1}\mathbf{a}^o \rangle, \tag{7.116}$$

be respectively the least favorable direction of deviation from $\boldsymbol{\beta}$ and the minimum Fisher information for the estimation of $\theta$ in the vicinity of the true $\boldsymbol{\beta}$ as in (7.54). We assume that $\boldsymbol{\beta} + \phi \mathbf{u}^o \in \mathscr{B}_n$ for all $|\phi| \le \epsilon_n^*$ and certain $\epsilon_n^* \gg n^{-1/2}$. Our analysis is also valid when $\boldsymbol{\beta} + \phi \mathbf{u}^o \notin \mathscr{B}_n$ but $\mathbf{b}^o(\theta) \in \mathscr{B}_n$ with $\|\mathbf{b}^o(\theta) - \boldsymbol{\beta} - \phi \mathbf{u}^o\| = o(\phi)$. See (7.52) and (7.55) and related discussion in Section 7.1 where the weaker condition is sufficient for the derivation of the lower bound for the asymptotic variance in the $n^{-1/2}$ neighborhood.

### *7.5.1 Local semi-LD decomposition*

Suppose we have the knowledge that $\boldsymbol{\beta}$ is in a relatively small neighborhood of a certain $\mathbf{b}_0$, so that it suffices to estimate the difference $\theta(\boldsymbol{\beta}) - \theta(\mathbf{b}_0)$. Consider a local decomposition of the difference $\boldsymbol{\beta} - \mathbf{b}_0$ of the form (7.1),

$$\boldsymbol{\beta} - \mathbf{b}_0 = \mathbf{u}^o \phi_0 + \boldsymbol{\nu}, \quad \langle \mathbf{a}^o, \boldsymbol{\nu} \rangle = 0, \tag{7.117}$$

with $\mathbf{u}^o \phi_0$ as the LD component and a nuisance parameter $\boldsymbol{\nu}$ as the HD component, where $\mathbf{u}^o$ is as in (7.116). As $\langle \mathbf{a}^o, \mathbf{u}^o \rangle = 1$ by (7.116),

$$\theta(\boldsymbol{\beta}) - \theta(\mathbf{b}_0) \approx \langle \mathbf{a}^o, \mathbf{u}^o \phi_0 + \boldsymbol{\nu} \rangle = \phi_0$$

by (7.115), so that $\phi_0$ is approximately the parameter value of interest. The decomposition (7.117) is uniquely determined by

$$\phi_0 = \langle \mathbf{a}^o, \boldsymbol{\beta} - \mathbf{b}_0 \rangle \ \text{ and } \ \boldsymbol{\nu} = (\boldsymbol{\beta} - \mathbf{b}_0) - \mathbf{u}^o \langle \mathbf{a}^o, \boldsymbol{\beta} - \mathbf{b}_0 \rangle.$$

Let $\dot{\ell}_i(\mathbf{b}) = (\partial/\partial\mathbf{b})\ell_i(\mathbf{b})$. Similar to (7.52), assume

$$n^{-1/2}\sum_{i=1}^{n}\langle\dot{\ell}_i(\boldsymbol{\beta}),\mathbf{u}^o\rangle \approx N\big(0,V_\theta\big) \tag{7.118}$$

for a certain $V_\theta \in (0,\infty)$, and that

$$\begin{aligned} & n^{-1/2}\textstyle\sum_{i=1}^{n}\big\langle\dot{\ell}_i(\mathbf{b}_0+\phi\mathbf{u}^o),\mathbf{u}^o\big\rangle \\ = \;& n^{-1/2}\textstyle\sum_{i=1}^{n}\big\langle\dot{\ell}_i(\boldsymbol{\beta}),\mathbf{u}^o\big\rangle + n^{1/2}\big\langle\mathbf{b}_0+\phi\mathbf{u}^o-\boldsymbol{\beta},\mathbf{F}\mathbf{u}^o\big\rangle + o_P(1) \\ = \;& n^{-1/2}\textstyle\sum_{i=1}^{n}\big\langle\dot{\ell}_i(\boldsymbol{\beta}),\mathbf{u}^o\big\rangle + n^{1/2}(\phi-\phi_0)F_\theta + o_P(1) \end{aligned} \tag{7.119}$$

for each $\phi \in [\phi_0-\epsilon_n^*,\phi_0+\epsilon_n^*]$, with $\epsilon_n^* \gg n^{-1/2}$. The last step above is due to $\mathbf{b}_0+\phi\mathbf{u}^o-\boldsymbol{\beta} = -\boldsymbol{\nu}+(\phi-\phi_0)\mathbf{u}^o$, $F_\theta = \langle\mathbf{u}^o,\mathbf{F}\mathbf{u}^o\rangle$ and $\langle\boldsymbol{\nu},\mathbf{F}\mathbf{u}^o\rangle = \langle\boldsymbol{\nu},\mathbf{a}^o\rangle F_\theta = 0$. This suggests the estimation of $\phi_0$ by minimizing the univariate loss function in the least favorable direction,

$$\widehat{\phi} = \arg\min_{\phi}\sum_{i=1}^{n}\ell_i(\mathbf{b}_0+\phi\mathbf{u}^o). \tag{7.120}$$

As $\theta(\boldsymbol{\beta})-\theta(\mathbf{b}_0)\approx\phi_0$, (7.120) estimates the original parameter of interest by

$$\begin{aligned} \widehat{\theta}(\mathbf{b}_0;\mathbf{u}^o) &= \theta(\mathbf{b}_0)+\arg\min_{\phi}\textstyle\sum_{i=1}^{n}\ell_i\big(\mathbf{b}_0+\mathbf{u}^o\phi\big) \\ &= \arg\min_{t}\textstyle\sum_{i=1}^{n}\ell_i\big(\mathbf{b}_0+\mathbf{u}^o\{t-\theta(\mathbf{b}_0)\}\big). \end{aligned} \tag{7.121}$$

Zhang (2011) calls the above estimator LDPE as it can be viewed as an extension of (7.14) from linear regression to the general setting. When (7.118) and (7.119) hold,

$$\sqrt{n}(\widehat{\phi}-\phi_0) = \sum_{i=1}^{n}\frac{\big\langle\dot{\ell}_i(\boldsymbol{\beta}),\mathbf{u}^o\big\rangle}{-F_\theta n^{1/2}} + o_P(1) \approx N(0,V_\theta/F_\theta^2) \tag{7.122}$$

for at least one local solution $\widehat{\phi}$ in $[\phi_0-\epsilon_n^*,\phi_0+\epsilon_n^*]$. As $\widehat{\theta}(\mathbf{b}_0;\mathbf{u}^o)-\theta = \widehat{\phi}-\phi_0-(\theta(\boldsymbol{\beta})-\theta(\mathbf{b}_0)-\langle\mathbf{a}^o,\boldsymbol{\beta}-\mathbf{b}_0\rangle)$, the asymptotic normality of $\widehat{\theta}(\mathbf{b}_0;\mathbf{u}^o)$ follows once

$$\theta(\mathbf{b}_0)-\theta(\boldsymbol{\beta})-\langle\mathbf{a}^o,\mathbf{b}_0-\boldsymbol{\beta}\rangle = o(n^{-1/2}). \tag{7.123}$$

The estimator (7.121) requires the knowledge of an approximation $\mathbf{b}_0$ of the entire unknown $\boldsymbol{\beta}$ and the direction $\mathbf{u}^o$ of the least favorable sub-model. These quantities can be replaced by suitable estimates $\widehat{\boldsymbol{\beta}}\approx\boldsymbol{\beta}$ and $\widehat{\mathbf{u}}\approx\mathbf{u}^o$, with the option $\widehat{\mathbf{u}}=\mathbf{u}^o$ when $\mathbf{u}^o$ is known. In this case (7.121) becomes

$$\widehat{\theta} = \widehat{\theta}\big(\widehat{\boldsymbol{\beta}};\widehat{\mathbf{u}}\big) = \arg\min_{t}\textstyle\sum_{i=1}^{n}\ell_i\big(\widehat{\boldsymbol{\beta}}+\widehat{\mathbf{u}}\{t-\theta(\widehat{\boldsymbol{\beta}})\}\big). \tag{7.124}$$

For example, with the $\ell_{[n]}(\mathbf{b})$ in (7.114), we may use

$$\widehat{\boldsymbol{\beta}} = \text{ a regularized minimizer of } \ell_{[n]}(\mathbf{b}), \tag{7.125}$$

$\widehat{\mathbf{a}} = (\partial/\partial \mathbf{b})\theta(\mathbf{b})\big|_{\mathbf{b}=\widehat{\boldsymbol{\beta}}}$, $\widehat{\mathbf{F}}$ as the Hessian of $\ell_{[n]}(\mathbf{b})$ at $\mathbf{b} = \widehat{\boldsymbol{\beta}}$, and

$$\widehat{\mathbf{u}} = \text{ a regularized minimizer of } \langle \mathbf{u}, \widehat{\mathbf{F}}\mathbf{u} \rangle \text{ subject to } \langle \widehat{\mathbf{a}}, \mathbf{u} \rangle = 1. \quad (7.126)$$

Moreover, the asymptotic variance $V_\theta / F_\theta^2$ in (7.122) can be estimated via

$$\widehat{F}_\theta = \langle \widehat{\mathbf{u}}, \widehat{\mathbf{F}}\widehat{\mathbf{u}} \rangle, \quad \widehat{V}_\theta = \frac{1}{n}\sum_{i=1}^{n} \langle \dot{\ell}(\widehat{\boldsymbol{\beta}}), \widehat{\mathbf{u}} \rangle^2. \quad (7.127)$$

However, while (7.123) follows from the quarter rate convergence $\|\mathbf{b}_0 - \boldsymbol{\beta}\| = o(n^{-1/4})$ when the spectrum norm of the Hessian of $\theta(\mathbf{b})$ is bounded in a neighborhood of $\boldsymbol{\beta}$, (7.119) is often nontrivial in semi-LD settings, especially with estimated $\mathbf{b}_0$ and $\mathbf{u}^o$.

### *7.5.2 Data swap*

Under suitable regularity conditions, the asymptotic efficiency of (7.124) and the consistency of (7.127) can be verified in many ways, although the analysis is nontrivial. When $\widehat{\boldsymbol{\beta}}$ and $\widehat{\mathbf{u}}$ are available through domain knowledge or estimated from historical data, we may treat them as deterministic. When $(\widehat{\boldsymbol{\beta}}, \widehat{\mathbf{u}})$ and (7.124) are based on the same data, the analysis is typically more complicated. To avoid such technicality, a data swap scheme can be used to obtain the desired theoretical results. The idea of data swap dates back at least to Schick (1986) and Klaassen (1987) in their studies of efficient semi-parametric estimation.

An $m$-fold data swap scheme can be described as follows. Let $I_1, \dots I_m$ form a partition of $[n] = \{1, \dots, n\}$ and $I_k^c = [n] \setminus I_k$. Typically, the data are equally divided between different folds $\lfloor n/m \rfloor \le |I_k| \le \lceil n/m \rceil$. In the data swap scheme the data points with label $i \in I_k^c$ are used to construct estimators of nuisance parameters with the loss function for the data points with label $i \in I_k$, $k = 1, \dots, m$, and then the loss function for all the data points is used to estimate the parameter of primary interest. Formally, letting $\widehat{\boldsymbol{\beta}}^{(k)}$ and $\widehat{\mathbf{u}}^{(k)}$ be respectively estimates of $\boldsymbol{\beta}$ and $\mathbf{u}^o$ based on $\{\text{data}_i, i \in I_k^c\}$, similar to (7.121) the data-swapped LDPE can be defined as

$$\widehat{\theta} = \arg\min_t \textstyle\sum_{k=1}^m \sum_{i \in I_k} \ell_i\Big(\widehat{\boldsymbol{\beta}}^{(k)} + \widehat{\mathbf{u}}^{(k)}\big(t - \theta(\widehat{\boldsymbol{\beta}}^{(k)})\big)\Big). \quad (7.128)$$

For example, $\widehat{\boldsymbol{\beta}}^{(k)}$ and $\widehat{\mathbf{u}}^{(k)}$ can be constructed respectively as in (7.125) and (7.126) with $\ell_{[n]}(\mathbf{b})$ replaced by $\sum_{i \in I_k^c} \ell_i(\mathbf{b})$. As $\theta$ is a function of $\boldsymbol{\beta}$, we may view $\widehat{\boldsymbol{\beta}}^{(k)}$ as initial estimates of $\boldsymbol{\beta}$ and $\mathbf{u}^{(k)}$ as estimates of the nuisance parameter $\mathbf{u}^o$, but we treat them both as estimates of nuisance parameters as the conceptual distinction has no bearing in the performance and analysis of (7.128) in the estimation of the parameter of primary interest.

In this data swap scheme, the full sample is used in the univariate optimization in (7.128) to retain full efficiency in the estimation of primary parameter

$\theta = \theta(\boldsymbol{\beta})$, while the nuisance parameters are estimated using subsamples. An analytical benefit of such a data swap is that when a certain version of (7.119) can be obtained for $\sum_{i\in I_k}\langle \dot{\ell}_i(\widehat{\boldsymbol{\beta}}^{(k)} + \widehat{\mathbf{u}}^{(k)}(t - \theta(\widehat{\boldsymbol{\beta}}^{(k)}))), \widehat{\mathbf{u}}^{(k)}\rangle$ as a perturbed linear function of $t$ conditionally on $\widehat{\boldsymbol{\beta}}^{(k)}$ and $\widehat{\mathbf{u}}^{(k)}$, their sum is also a perturbed linear function of $t$.

We note that when the data swap is not deployed or partially deployed to estimate one of $\boldsymbol{\beta}$ or $\mathbf{u}^o$, (7.128) is still valid as it allows common $\widehat{\boldsymbol{\beta}}^{(k)} = \widehat{\boldsymbol{\beta}}$ and/or $\widehat{\mathbf{u}}^{(k)} = \widehat{\mathbf{u}}$. In particular, (7.124) is a special case of (7.128).

In the rest of this subsection, we study theoretical properties of (7.128). Let

$$\xi_k(t, \mathbf{b}, \mathbf{u}) = \textstyle\sum_{i\in I_k}\langle \dot{\ell}_i(\mathbf{b} + \mathbf{u}\{t + \theta(\boldsymbol{\beta}) - \theta(\mathbf{b})\}), \mathbf{u}\rangle.$$

Suppose that for certain nonnegative constants $\{C_{j,n}, \delta_{j,n}, \alpha_1\}$,

$$\|\mathbf{F}^{1/2}(\widehat{\boldsymbol{\beta}}^{(k)} - \boldsymbol{\beta})\| \le \delta_{1,n}, \quad \|\mathbf{F}^{1/2}(\widehat{\mathbf{u}}^{(k)} - \mathbf{u}^o)\| \le F_\theta^{1/2}\delta_{2,n}, \tag{7.129}$$

with probability $1 + o(1)$, and for deterministic $\mathbf{h}$ and $\mathbf{u}$ in $\mathscr{H}_n$

$$\big|\theta(\mathbf{h} + \boldsymbol{\beta}) - \theta(\boldsymbol{\beta}) - \langle \mathbf{a}^o, \mathbf{h}\rangle\big| \le C_{1,n}\|\mathbf{h}\|^{1+\alpha_1}, \tag{7.130}$$

$$\mathrm{Var}\big(|I_k|^{-1/2}\textstyle\sum_{i\in I_k}\langle \dot{\ell}_i(\mathbf{h} + \boldsymbol{\beta}) - \dot{\ell}_i(\boldsymbol{\beta}), \mathbf{u}\rangle\big) \le \delta_{3,n}^2\|\mathbf{F}^{1/2}\mathbf{u}\|^2, \tag{7.131}$$

$$\big||I_k|^{-1}\textstyle\sum_{i\in I_k}E\langle \dot{\ell}_i(\mathbf{h} + \boldsymbol{\beta}), \mathbf{u}\rangle - \langle \mathbf{F}\mathbf{h}, \mathbf{u}\rangle\big| \le C_{2,n}\|\mathbf{F}^{1/2}\mathbf{h}\|^2\|\mathbf{F}^{1/2}\mathbf{u}\|, \tag{7.132}$$

when $\|\mathbf{F}^{1/2}\mathbf{h}\| \le 4\delta_{1,n}$ and $\|\mathbf{F}^{1/2}(\mathbf{u} - \mathbf{u}^o)\| \le F_\theta^{1/2}\delta_{2,n}$. We recall that $\|\mathbf{v}\|$ is the inner-product norm in $\mathscr{H}_n$ and $\|\mathbf{v}\| = \|\mathbf{v}\|_2$ when $\mathscr{H}_n = I\!R^p$.

**Theorem 7.10** *(i) Let $\mathbf{u}^o$ and $F_\theta$ be as in (7.116). Suppose (7.118) holds and*

$$\max_{k\le m}\Big|\xi_k(t, \widehat{\boldsymbol{\beta}}^{(k)}, \widehat{\mathbf{u}}^{(k)}) - \xi_k(0, \boldsymbol{\beta}, \mathbf{u}^o)\} - t|I_k|F_\theta\Big| = o_P(n^{1/2}). \tag{7.133}$$

*for $|t| \le \epsilon_n^*$ and certain $\epsilon_n^* \gg n^{-1/2}$. Then,*

$$\sqrt{n}(\widehat{\theta} - \theta) = \sum_{i=1}^n \frac{\langle \dot{\ell}_i(\boldsymbol{\beta}), \mathbf{u}^o\rangle}{-F_\theta n^{1/2}} + o_P(1) \approx N(0, V_\theta/F_\theta^2) \tag{7.134}$$

*for a certain local minimizer $\widehat{\theta}$ of (7.128) in the interval $[\theta - \epsilon_n^*, \theta + \epsilon_n^*]$. If $\ell_i(\mathbf{b})$ is convex in $\mathbf{b}$, then (7.134) holds for all solutions $\widehat{\theta}$. If $\ell_{[n]}(\mathbf{b})$ is the negative log-likelihood, then $V_\theta = F_\theta$ and $\widehat{\theta}$ is asymptotically efficient.*
*(ii) Suppose* data$_i$ *are independent and a data-swap scheme is used to estimate $\widehat{\boldsymbol{\beta}}^{(k)}$ and $\widehat{\mathbf{u}}^{(k)}$. Suppose conditions (7.129), (7.130), (7.131) and (7.132) hold. Then, (7.133) holds when* $\max_{1\le j\le 3}\delta_{j,n} = o(1)$ *and*

$$\delta_{1,n}\delta_{2,n} + F_\theta^{1/2}C_{1,n}\delta_{1,n}^{1+\alpha_1} + C_{2,n}\delta_{1,n}^2 \ll n^{-1/2} \le \delta_{1,n}. \tag{7.135}$$

Under conditions (7.129), (7.131) and (7.132), a certain data-swapped version of (7.127), say $\widehat{F}_\theta$ and $\widehat{V}_\theta$, provides consistent estimates of $F_\theta$ and $V_\theta$. Thus, confidence intervals for $\theta$ can be constructed based on

$$\widehat{F}_\theta\big(n/\widehat{V}_\theta\big)^{1/2}(\widehat{\theta}-\theta) \xrightarrow{D} N(0,1)$$

by Theorem 7.10 (ii). We leave this as an exercise.

Theorem 7.10 (i) applies to both (7.124) and (7.128) as both (7.128) and (7.133) allow data-dependent common $\widehat{\boldsymbol{\beta}}^{(k)}$ and $\widehat{\mathbf{u}}^{(k)}$ across different subsamples $I_k$. Theorem 7.10 (ii) verifies the key condition (7.133) in the data-swap case from error bounds for $\widehat{\boldsymbol{\beta}}^{(k)}$ and $\widehat{\mathbf{u}}^{(k)}$ in the combination of (7.129) and (7.135) and mild additional conditions on the smoothness of $\theta(\mathbf{b})$ and the mean and variance of $\dot{\ell}_i(\mathbf{b})-\dot{\ell}_i(\boldsymbol{\beta})$. We note that Theorem 7.10 (ii) imposes no rate requirements on $\delta_{3,n}$ other than $\max_{1\le j\le 3}\delta_{j,n}=o(1)$.

Other than (7.135), the conditions imposed in Theorem 7.10 (ii) are relatively easy to verify. Condition (7.130) holds with $\alpha_1=1$ when the Hessian of $\ell(\mathbf{b})$ is bounded by $2C_{1,n}$ in the specified neighborhood of $\boldsymbol{\beta}$ and with $C_{n,1}=0$ when $\theta(\mathbf{b})$ is linear. Condition (7.131) can be viewed as an $L_2$ Lipschitz condition on $\langle\dot{\ell}(\mathbf{b})-\mathrm{E}\,\dot{\ell}(\mathbf{b}),\mathbf{u}\rangle$ with $\delta_{3,n}\lesssim\delta_{1,n}$. Condition (7.132) can be viewed as a third order smoothness condition on the mean $\mathrm{E}\,\dot{\ell}_i(\mathbf{b})$ in a small neighborhood of $\boldsymbol{\beta}$, and $C_{2,n}=0$ when the loss function $\ell_i(\mathbf{b})$ is quadratic in $\mathbf{b}$.

In the generalized linear model (7.48) with i.i.d. $T_i(\mathbf{x}_i)=\mathbf{x}_i\in I\!R^p$,

$$\dot{\ell}_i(\mathbf{b})=\dot{\psi}_i\big(\langle\mathbf{x}_i,\mathbf{b}\rangle\big)\mathbf{x}_i-y_i\mathbf{x}_i,\quad \mathbf{F}=n^{-1}\textstyle\sum_{i=1}^n\mathrm{E}\big[\ddot{\psi}_i\big(\langle\mathbf{x}_i,\boldsymbol{\beta}\rangle\big)\mathbf{x}_i\mathbf{x}_i^T\big],$$

with $\dot{\psi}_i(t)=(d/dt)\psi_i(t)$ and $\ddot{\psi}_i(t)=(d/dt)^2\psi_i(t)$. Thus, (7.131) holds with

$$\delta_{3,n}^2=(4\delta_{1,n})^2\max_{(\mathbf{h},\mathbf{u})\in\mathscr{S}}\left|\int_0^1\frac{\sum_{i\in I_k}\mathrm{Var}\big[\ddot{\psi}_i\big(\langle\mathbf{x}_i,\boldsymbol{\beta}+t\mathbf{h}\rangle\big)(\mathbf{x}_i^T\mathbf{h})(\mathbf{x}_i^T\mathbf{u})\big]}{|I_k|\,\|\mathbf{F}^{1/2}\mathbf{h}\|_2^2\|\mathbf{F}^{1/2}\mathbf{u}\|_2^2}dt\right|$$

where $\mathscr{S}=\{(\mathbf{h},\mathbf{u}):0<\|\mathbf{F}^{1/2}\mathbf{h}\|_2\le 4\delta_{1,n},\mathbf{F}\mathbf{u}\neq 0\}$, and (7.132) holds with

$$C_{2,n}=\max_{(\mathbf{h},\mathbf{u})\in\mathscr{S}}\left|\int_0^1\frac{\sum_{i\in I_k}\mathrm{E}\big[\dddot{\psi}_i\big(\langle\mathbf{x}_i,\boldsymbol{\beta}+t\mathbf{h}\rangle\big)(\mathbf{x}_i^T\mathbf{h})^2(\mathbf{x}_i^T\mathbf{u})\big]}{|I_k|\,\|\mathbf{F}^{1/2}\mathbf{h}\|_2^2\|\mathbf{F}^{1/2}\mathbf{u}\|_2}dt\right|$$

where $\dddot{\psi}_i(t)=(d/dt)^3\psi_i(t)$. Thus, as $\|\mathbf{F}\mathbf{u}\|_2^2=\mathrm{E}\sum_{i=1}^n\mathrm{E}\langle\mathbf{x}_i,\mathbf{u}\rangle^2/n$, (7.132) and (7.131) are respective weighted third and fourth moment conditions on the design vectors $\mathbf{x}_i$. In logistic regression, $\ddot{\psi}(t)=e^t/(1+e^t)^2$ and $\dddot{\psi}(t)=e^t(1-e^t)/(1+e^t)^3$ are both bounded by $1/4$.

For the estimation of linear functionals in linear regression, $C_{1,n}=C_{2,n}=0$, so that Theorem 7.10 (ii) requires just one condition $\delta_{1,n}\delta_{2,n}\ll n^{-1/2}$ due to an application of the Cauchy-Schwarz inequality to the cross-product remainder term. In the HD setting of Theorems 7.1 and 7.4, an $\ell_1$-$\ell_\infty$ split of the cross-product remainder term leads to the condition $s^*\log p\ll n^{1/2}$, while the condition $\delta_{1,n}\delta_{2,n} \ll n^{-1/2}$ is equivalent to $\sqrt{s_j s^*}\log p \ll n^{1/2}$ due to an application of Cauchy-Schwarz on the cross-product remainder term.

It is also interesting to compare Theorem 7.10 (ii) with Theorem 7.6. Both theorems are proved using Cauchy-Schwarz to bound the main cross-product remainder term. As the optimal rates for the estimations of $f$ and $g$ are respectively $\delta_{1,n} = n^{-\alpha_f/(1+2\alpha_f)}$ and $\delta_{2,n} = n^{-\alpha_g/(1+2\alpha_g)}$, Theorem 7.10 (ii) would require $\alpha_f/(1+2\alpha_f) + \alpha_g/(1+2\alpha_g) > 1/2$ for efficient estimation. However, as the main cross-product remainder term $\mathbf{g}\mathbf{P}_{k_n}^{\perp}\mathbf{f}$ involves only the bias in Theorem 7.6, larger $k_n$ is taken so that the bias is of smaller order than the variability in the estimation of nuisance parameters $\mathbf{f}$ and $\mathbf{g}$, resulting in the weaker assumption $\alpha_f + \alpha_g > 1/2$ for the efficient inference of the linear functional $\theta$.

**Proof of Theorem 7.10.** (i) It follows from (7.128) and the definition of $\xi_k$ that $\widehat{\theta}$ is a local solution of (7.128) if the random function of $t$

$$\textstyle\sum_{k=1}^{m}\xi_k\big(t, \widehat{\boldsymbol{\beta}}^{(k)}, \widehat{\mathbf{u}}^{(k)}\big)$$

crosses sign from the negative side to the positive side at $t = \widehat{\theta} - \theta$. By (7.133), this random function is approximately linear in the interval $[\theta - \epsilon_n^*, \theta + \epsilon_n^*]$ with $\epsilon_n^* \gg n^{-1/2}$, so that (7.134) holds as in (7.122) and (7.123).

(ii) Taking a smaller $\epsilon_n^*$ if necessary, assume $n^{-1/2} \ll \epsilon_n^* \le \delta_{1,n}$. Let

$$\delta_{4,n} = (2+\delta_{2,n})\delta_{1,n} + (1+\delta_{2,n})(\epsilon_n^* + F_\theta^{1/2} C_{1,n}\delta_{1,n}^{1+\alpha_1}).$$

By (7.135) $2\delta_{1,n} \le \delta_{4,n} \le 4\delta_{1,n}$. Let $\mathbf{h}^{(k)} = \widehat{\boldsymbol{\beta}}^{(k)} - \boldsymbol{\beta} + \widehat{\mathbf{u}}^{(k)}\{t + \theta(\boldsymbol{\beta}) - \theta(\widehat{\boldsymbol{\beta}}^{(k)})\}$. Let $\mathrm{E}_k$ be the conditional expectation given $\{\text{data}_i, i \in I_k^c\}$. The main task is to bound the bias of $\widehat{\theta}$ by proving

$$\Big| |I_k|^{-1}\textstyle\sum_{i\in I_k}\mathrm{E}_k\langle \dot{\ell}_i(\boldsymbol{\beta}+\mathbf{h}^{(k)}), \widehat{\mathbf{u}}^{(k)}\rangle - tF_\theta\Big| = o_P(n^{-1/2}).$$

Consider the case where the events in (7.129) both happen. We still need to prove $\|\mathbf{F}^{1/2}\mathbf{h}^{(k)}\| \le 4\delta_{1,n}$ for the application of (7.130), (7.131) and (7.132) to $\mathbf{h} = \mathbf{h}^{(k)}$ and $\mathbf{u} = \widehat{\mathbf{u}}^{(k)}$. As (7.130) is applicable to $\mathbf{h} = \widehat{\boldsymbol{\beta}}^{(k)} - \boldsymbol{\beta}$ by (7.129),

$$\big|\theta(\widehat{\boldsymbol{\beta}}^{(k)}) - \theta(\boldsymbol{\beta})\big| \le \|\mathbf{F}^{-1/2}\mathbf{a}^o\|_2\delta_{1,n} + C_{1,n}\delta_{1,n}^{1+\alpha_1} = F_\theta^{-1/2}\delta_{1,n} + C_{1,n}\delta_{1,n}^{1+\alpha_1}.$$

As $\big\|\mathbf{F}^{1/2}\widehat{\mathbf{u}}^{(k)}\big\|_2 \le \|\mathbf{F}^{1/2}\mathbf{u}^o\|_2 + F_\theta^{1/2}\delta_{2,n} = (1+\delta_{2,n})F_\theta^{1/2}$, for $|t| \le \epsilon_n^*$

$$\big\|\mathbf{F}^{1/2}\mathbf{h}^{(k)}\big\|_2 \le \delta_{1,n} + \big\|\mathbf{F}^{1/2}\widehat{\mathbf{u}}^{(k)}\big\|_2\big(|t| + F_\theta^{-1/2}\delta_{1,n} + C_{1,n}\delta_{1,n}^{1+\alpha_1}\big) \le \delta_{4,n}.$$

Thus, the regularity conditions can be applied to $\mathbf{h} = \mathbf{h}^{(k)}$. By (7.132),

$$\Big||I_k|^{-1}\textstyle\sum_{i\in I_k}\mathrm{E}_k\langle \dot{\ell}_i(\boldsymbol{\beta}+\mathbf{h}^{(k)}), \widehat{\mathbf{u}}^{(k)}\rangle - \langle \mathbf{F}\mathbf{h}^{(k)}, \widehat{\mathbf{u}}^{(k)}\rangle\Big| \le C_{2,n}\delta_{4,n}^2(1+\delta_{2,n})F_\theta^{1/2}.$$

Let $\widetilde{\mathbf{h}}^{(k)} = \widehat{\boldsymbol{\beta}}^{(k)} - \boldsymbol{\beta} + \mathbf{u}^o\{t + \theta(\boldsymbol{\beta}) - \theta(\widehat{\boldsymbol{\beta}}^{(k)})\}$. As in the proof of $\|\mathbf{F}^{1/2}\mathbf{h}^{(k)}\|_2 \le \delta_{4,n}$,

$$\begin{aligned} & \left|\langle \mathbf{F}\mathbf{h}^{(k)}, \widehat{\mathbf{u}}^{(k)}\rangle - \langle \mathbf{F}\widetilde{\mathbf{h}}^{(k)}, \mathbf{u}^o\rangle\right| \\ \le\ & \sqrt{F_\theta}\delta_{4,n}\delta_{2,n} + \left|\langle \mathbf{F}\mathbf{h}^{(k)} - \mathbf{F}\widetilde{\mathbf{h}}^{(k)}, \mathbf{u}^o\rangle\right| \\ =\ & \sqrt{F_\theta}\delta_{4,n}\delta_{2,n} + \left|\langle \mathbf{u}^{(k)} - \mathbf{u}^o, \mathbf{F}\mathbf{u}^o\rangle\{t + \theta(\boldsymbol{\beta}) - \theta(\widehat{\boldsymbol{\beta}}^{(k)})\}\right| \\ \le\ & 2\sqrt{F_\theta}\delta_{2,n}\delta_{4,n}. \end{aligned}$$

Moreover,

$$\begin{aligned} \left|\langle \mathbf{F}\widetilde{\mathbf{h}}^{(k)}, \mathbf{u}^o\rangle - tF_\theta\right| &= \left|\langle \widehat{\boldsymbol{\beta}}^{(k)} - \boldsymbol{\beta} + \mathbf{u}^o\{\theta(\boldsymbol{\beta}) - \theta(\widehat{\boldsymbol{\beta}}^{(k)})\}, \mathbf{F}\mathbf{u}^o\rangle\right| \\ &= F_\theta\left|\langle \widehat{\boldsymbol{\beta}}^{(k)} - \boldsymbol{\beta}, \mathbf{a}^o\rangle + \theta(\boldsymbol{\beta}) - \theta(\widehat{\boldsymbol{\beta}}^{(k)})\right| \\ &\le F_\theta C_{1,n}\delta_{1,n}^{1+\alpha_1}. \end{aligned}$$

Putting the above inequalities together, we find that

$$\begin{aligned} & \left| |I_k|^{-1}\textstyle\sum_{i\in I_k} \mathrm{E}_k\langle \dot{\ell}_i(\boldsymbol{\beta} + \mathbf{h}^{(k)}), \widehat{\mathbf{u}}^{(k)}\rangle - tF_\theta \right| \\ \le\ & C_{2,n}\delta_{4,n}^2(1 + \delta_{2,n})F_\theta^{1/2} + 2\sqrt{F_\theta}\delta_{2,n}\delta_{4,n} + F_\theta C_{1,n}\delta_{1,n}^{1+\alpha_1} \\ \lesssim\ & F_\theta^{1/2}\{C_{2,n}\delta_{1,n}^2 + \delta_{2,n}\delta_{1,n} + F_\theta^{1/2}C_{1,n}\delta_{1,n}^{1+\alpha_1}\} \\ \ll\ & F_\theta^{1/2}n^{-1/2}. \end{aligned}$$

Finally, as $\xi_k(t, \mathbf{b}, \mathbf{u})$ is a sum of independent variables, (7.131) implies

$$\mathrm{Var}(\xi_k(t, \mathbf{b}, \mathbf{u}))\big|_{\mathbf{b}=\boldsymbol{\beta}+\mathbf{h}^{(k)}, \mathbf{u}=\widehat{\mathbf{u}}^{(k)}} \le \delta_{3,n}(1 + \delta_{2,n})^2 F_\theta |I_k|.$$

Consequently, the above bounds for the conditional mean and variance yields

$$\left|\xi_k(t, \widehat{\boldsymbol{\beta}}^{(k)}, \widehat{\mathbf{u}}^{(k)}) - |I_k|tF_\theta\right| = o_P(F_\theta^{1/2}n^{-1/2})|I_k| + O_P(F_\theta|I_k|)^{1/2}\delta_{3,n}.$$

This completes the proof of (7.133). □

### *7.5.3 Gradient approximation*

A variation of (7.124) is its gradient approximation

$$\widehat{\theta} = \theta(\widehat{\boldsymbol{\beta}}) - n^{-1}\textstyle\sum_{i=1}^n \langle \dot{\ell}_i(\widehat{\boldsymbol{\beta}}), \widehat{\mathbf{v}}\rangle, \tag{7.136}$$

where $\widehat{\mathbf{v}}$ are estimates of

$$\mathbf{v}^o = \mathbf{u}^o / F_\theta = \mathbf{F}^{-1}\mathbf{a}^o. \tag{7.137}$$

Similar to (7.128), the data-swapped version of (7.136) is

$$\widehat{\theta} = n^{-1}\textstyle\sum_{k=1}^m \sum_{i\in I_k}\{\theta(\widehat{\boldsymbol{\beta}}^{(k)}) - \langle \dot{\ell}_i(\widehat{\boldsymbol{\beta}}^{(k)}), \widehat{\mathbf{v}}^{(k)}\rangle\}, \tag{7.138}$$

with $\widehat{\boldsymbol{\beta}}^{(k)}$ and $\widehat{\mathbf{v}}^{(k)}$ based on $\{\text{data}_i, i \in I_k^c\}$. Again, (7.136) can be viewed as a special case of (7.138) with $\widehat{\boldsymbol{\beta}}^{(k)}$ and $\widehat{\mathbf{v}}^{(k)}$ not dependent on $k$.

If the parameter of interest is linear, $\theta(\boldsymbol{\beta}) = \langle \mathbf{a}^o, \boldsymbol{\beta}\rangle$, and $\mathbf{v}^o = \mathbf{F}^{-1}\mathbf{a}^o$ is known and taken as $\widehat{\mathbf{v}}^{(k)}$, (7.138) can be written as

$$\widehat{\langle \mathbf{a}^o, \boldsymbol{\beta}\rangle} = n^{-1}\sum_{k=1}^m \sum_{i\in I_k}\big\{\langle \mathbf{a}^o, \widehat{\boldsymbol{\beta}}^{(k)}\rangle - \langle \mathbf{F}^{-1}\dot{\ell}_i(\widehat{\boldsymbol{\beta}}^{(k)}), \mathbf{a}^o\rangle\big\},$$

or symbolically as $\widehat{\boldsymbol{\beta}} = n^{-1}\sum_{k=1}^m \sum_{i\in I_k}\big\{\widehat{\boldsymbol{\beta}}^{(k)} - \mathbf{F}^{-1}\dot{\ell}_i(\widehat{\boldsymbol{\beta}}^{(k)})\big\}$. This leads to

$$\widehat{\boldsymbol{\beta}} = \sum_{k=1}^m (|I_k|/n)\big\{\widehat{\boldsymbol{\beta}}^{(k)} - \big(|I_k|\widehat{\mathbf{F}}^{(k)}\big)^{-1}\sum_{i\in I_k}\dot{\ell}_i(\widehat{\boldsymbol{\beta}}^{(k)})\big\}, \qquad (7.139)$$

where $\widehat{\mathbf{F}}^{(k)} = \mathbf{F}$ if $\mathbf{F}$ is known and invertible and otherwise estimated. In a data swap scheme, $\widehat{\boldsymbol{\beta}}^{(k)}$ and $\widehat{\mathbf{F}}^{(k)}$ are estimated based on $\{\text{data}_i, i \in I_k^c\}$. If data-swap is not deployed, (7.139) becomes $\widehat{\boldsymbol{\beta}} = \widehat{\boldsymbol{\beta}}^{(\text{init})} - \big(n\widehat{\mathbf{F}}\big)^{-1}\sum_{i=1}^n \dot{\ell}_i(\widehat{\boldsymbol{\beta}}^{(\text{init})})$ and (7.138) becomes

$$\widehat{\theta} = \theta(\widehat{\boldsymbol{\beta}}^{(\text{init})}) + \big\langle \widehat{\mathbf{a}}, \widehat{\boldsymbol{\beta}} - \widehat{\boldsymbol{\beta}}^{(\text{init})}\big\rangle. \qquad (7.140)$$

These formulas for $\widehat{\boldsymbol{\beta}}$ provide a global perspective about the estimator. However, for statistical inference about a specific $\theta(\boldsymbol{\beta})$, the knowledge, estimation or the computation of the entire $\mathbf{F}$ is typically not necessary.

To state the counter part of Theorem 7.10 for (7.138), we define

$$\widetilde{\xi}_k(\mathbf{b}, \mathbf{v}) = \sum_{i\in I_k}\big\langle \dot{\ell}_i(\mathbf{b}), \mathbf{v}\big\rangle.$$

**Theorem 7.11** *(i) Let $\widehat{\theta}$ be as in (7.138). Suppose (7.118) holds,*

$$\Big|\sum_{k=1}^m \big\{\widetilde{\xi}_k(\widehat{\boldsymbol{\beta}}^{(k)}, \widehat{\mathbf{v}}^{(k)}) - \widetilde{\xi}_k(\boldsymbol{\beta}, \mathbf{v}^o) - |I_k|\langle \mathbf{a}^o, \widehat{\boldsymbol{\beta}}^{(k)} - \boldsymbol{\beta}\rangle\big\}\Big| = o_P\big(n/F_\theta\big)^{1/2} \qquad (7.141)$$

*and (7.130) holds with $C_{1,n}\delta_{1,n}^{1+\alpha_1} \ll (nF_\theta)^{-1/2}$. Then, (7.134) holds, and $\widehat{\theta}$ is asymptotically efficient when $V_\theta = F_\theta$.*

*(ii) Suppose* $\text{data}_i$ *are iid and the data swap is deployed with both $\widehat{\boldsymbol{\beta}}^{(k)}$ and $\widehat{\mathbf{v}}^{(k)}$ being functions of $\{\text{data}_i, i \in I_k^c\}$, $k \le m$. Suppose that conditions (7.129), (7.131) and (7.132) hold with $\widehat{\mathbf{u}}^{(k)} = \widehat{\mathbf{v}}^{(k)}/F_\theta$. Then, (7.141) holds when*

$$\delta_{1,n}\delta_{2,n} + C_{2,n}\delta_{1,n}^2 \ll n^{-1/2} \quad \textit{and} \quad \delta_{3,n} = o(1).$$

As we have discussed below Theorem 7.10, Theorem 7.11 (i) is applicable with or without data swap. Theorem 7.11 (ii) provides a way of checking the key condition (7.141) in the data-swap case based on the mean and variance of $\dot{\ell}_i(\mathbf{b}) - \dot{\ell}_i(\boldsymbol{\beta})$ provided convergence rates for $\widehat{\boldsymbol{\beta}}^{(k)}$ and $\mathbf{v}^{(k)}$. Moreover, consistent estimates of $F_\theta$ and $V_\theta$, and therefore confidence intervals for $\theta$, can be constructed based on data-swapped $\widehat{\boldsymbol{\beta}}^{(k)}$ and $\mathbf{v}^{(k)}$ as an exercise (Exercise 7.16). Compared with Theorem 7.10, one does not need to deal with local minima to apply Theorem 7.11. See also below Theorem 7.10 for discussion of conditions (7.130), (7.131) and (7.132).

**Proof of Theorem 7.11.** (i) By (7.138), (7.141), and (7.130),

$$\begin{aligned}
& \big|\widehat{\theta} - \theta(\boldsymbol{\beta}) + n^{-1}\textstyle\sum_{k=1}^m \widetilde{\xi}_k(\boldsymbol{\beta}, \mathbf{v}^o)\big| \\
= \; & \big|n^{-1}\textstyle\sum_{k=1}^m \big\{|I_k|\theta(\widehat{\boldsymbol{\beta}}^{(k)}) - |I_k|\theta(\boldsymbol{\beta}) - \widetilde{\xi}_k(\widehat{\boldsymbol{\beta}}^{(k)}, \widehat{\mathbf{v}}^{(k)}) + \widetilde{\xi}_k(\boldsymbol{\beta}, \mathbf{v}^o)\big\}\big| \\
\le \; & o_P(1)\big(nF_\theta\big)^{-1/2} + O_P(1)C_{1,n}\delta_{1,n}^{1+\alpha_1} \\
= \; & o_P(1)\big(nF_\theta\big)^{-1/2}.
\end{aligned}$$

By (7.137) and (7.118), $\sqrt{F_\theta/n}\sum_{k=1}^m \widetilde{\xi}_k(\boldsymbol{\beta}, \mathbf{v}^o) \approx N(0, V_\theta)$.

(ii) Let $\mathbf{h} = \widehat{\boldsymbol{\beta}}^{(k)} - \boldsymbol{\beta}$. Similar to the proof of Theorem 7.10 (ii), we have

$$\begin{aligned}
& \widetilde{\xi}_k(\widehat{\boldsymbol{\beta}}^{(k)}, \widehat{\mathbf{v}}^{(k)}) - \widetilde{\xi}_k(\boldsymbol{\beta}, \mathbf{v}^o) - |I_k|\langle \mathbf{a}^o, \widehat{\boldsymbol{\beta}}^{(k)} - \boldsymbol{\beta}\rangle \\
= \; & \sum_{i\in I_k} \big\{\langle \dot{\ell}_i(\widehat{\boldsymbol{\beta}}^{(k)}) - \dot{\ell}_i(\boldsymbol{\beta}^o), \widehat{\mathbf{v}}^{(k)}\rangle - \langle \mathbf{a}^o, \widehat{\boldsymbol{\beta}}^{(k)} - \boldsymbol{\beta}\rangle\big\} \\
= \; & O_P\big(n^{1/2}\delta_{3,n}\big\|\mathbf{F}^{1/2}\widehat{\mathbf{v}}^{(k)}\big\|_2 + nC_{2,n}\|\mathbf{F}^{1/2}\mathbf{h}\|_2^2\big\|\mathbf{F}^{1/2}\widehat{\mathbf{v}}^{(k)}\big\|_2\big) \\
& + |I_k|\big\langle \mathbf{F}\widehat{\mathbf{v}}^{(k)} - \mathbf{a}^o, \widehat{\boldsymbol{\beta}}^{(k)} - \boldsymbol{\beta}\big\rangle \\
= \; & O_P(1)\big\{\delta_{3,n}(n/F_\theta)^{1/2} + C_{2,n}\delta_{1,n}^2 n/F_\theta^{1/2} + n\delta_{1,n}\delta_{2,n}/F_\theta^{1/2}\big\}.
\end{aligned}$$

Thus, (7.141) holds when $\delta_{3,n} = o(1)$ and $C_{2,n}\delta_{1,n}^2 + \delta_{1,n}\delta_{2,n} \ll n^{-1/2}$. □

## 7.6 Bibliographical Notes

Statistical inference with HD data can be carried out in many ways. As discussed in Chapters 3 and 4, LD inference at the parametric rate can be carried out after consistent variable selection in regression and graphical models among others. However, selection consistency requires a uniform level of signal strength. Leeb and Potscher (2006) proved that it is impossible to consistently estimate the sampling distribution of estimation errors after model selection if small perturbations of the parameter are allowed. This essentially asserts that estimation after model selection cannot be regular in the classical sense, essentially as a consequence of super efficiency when the selection is consistent (Theorem 7.3). The focus of this chapter is the semi-LD approach of regular statistical inference of LD parameters with HD data (Zhang, 2011), which is very much inspired from the semi-parametric literature (Bickel *et al.*, 1998). The idea is to correct the bias of regularized estimators. In linear regression, this can be carried out by LD projections in a suitable direction (Zhang and Zhang 2014). Alternatively, inference after model selection can be carried out conditionally on the selected model (Lockhart *et al.*, 2014; Lee, 2016) or by multiplicity adjustment over all candidate models (Berk *et al.*, 2013). In the presence of small perturbation of parameters, the target parameter of such inference is dependent on the selected models. Another interesting topic which we do not cover is bootstrap. Important contributions in bootstrap inference with HD data include Chatterjee and Lahiri (2011, 2013), Chernozhukov (2013, 2017), Dezeure *et al.* (2017),

The materials in Section 7.1 is mostly based on the results of Sun and Zhang (2012) and Zhang and Zhang (2014), but inference in linear regression has been considered by many, including Bühlmann (2013), Belloni *et al.* (2014), van de Geer *et al.* (2014), Javanmard and Montanari (2014a, b) and Chernozhukov (2015) among many others. The materials in Subsections 7.3.1 and 7.5.1 are adapted from Zhang (2011) as an extension of the semi-parametric inference to HD data. The one-step approximation dates back at least to Le Cam (1969) and Bickel (1975) and the data-swap to Schick (1986). Subsection 7.4.1 is adapted from Ren *et al.* (2015), and the analysis of the sample size requirement in Subsection 7.4.2 is adapted from Cai and Guo (2017).

Desparsified Lasso was proposed in van de Geer, Bühlmann, Ritov and Dezeure (2014) for generalized linear models and M-estimate. It is essentially the same as the debiased Lasso (7.14) in the linear model and similar to the low-dimensional projection estimator (7.121) in the generalized linear model. The decorrelated score estimator was proposed in Ning and Liu (2017). The one-step decorrelated score estimator is similar to the low-dimensional projection estimator and the desparsified Lasso, although the decorrelated score estimate was motivated from Rao's score test. Decorrelated score tests have been developed for survival data analysis in Fang, Ning and Liu (2017) and for high-dimensional longitudinal data in Fang, Ning and Li (2020). An estimation and valid inference method for low-dimensional regression coefficients in high-dimensional generalized linear models via recursive online-score estimation was developed in Shi, Song, Lu and Li (2020).

Under model (7.27), Fan and Peng (2004) developed the penalized likelihood ratio test for $H_0 : \mathbf{C}\boldsymbol{\beta}_S = \mathbf{0}$, where $\boldsymbol{\beta}_S$ is the vector consisting of all nonzero elements of $\boldsymbol{\beta}_0$ when $p = o(n^{1/5})$. Shi, Song, Chen and Li (2019) developed the partial penalized likelihood test for hypothesis in (7.39) under $\log p = O(n^a)$ for some $0 < a < 1$. The authors also developed the partial Wald test and partial Rao score test for (7.39). Under certain conditions, the authors further shows that the limiting distribution of the partial likelihood ratio test, the partial penalized Wald test and the partial penalized score test have the same limiting distribution. They also derived the power function under local alternatives.

## 7.7 Exercises

7.1 Suppose (7.25) holds and $\boldsymbol{\varepsilon} \sim N(\mathbf{0}, \sigma^2 \mathbf{I}_n)$ in (7.3).

(a) For fixed $\lambda > 0$. Prove that the KKT condition holds strictly and almost surely for the Lasso (7.21) in the sense of $P\{|\mathbf{X}_j^T(\mathbf{Y} - \mathbf{X}\widehat{\boldsymbol\beta}(\lambda))/n| < \lambda \; \forall \widehat{\beta}_j = 0\} = 1$.

(b) The Lasso path (7.21) has the "one-at-a-time" property in the sense of

$$P\big\{\#\{j : \widehat{\beta}_j(\lambda+) \neq \widehat{\beta}_j(\lambda-)\} \leq 1 \; \forall \lambda > 0\big\} = 1.$$

7.2 Suppose (7.25) holds and $\boldsymbol\varepsilon \sim N(\mathbf{0}, \sigma^2\mathbf{I}_n)$ in (7.3). Let $\widehat{\boldsymbol\beta}$ be the Lasso (7.21) with fixed $\lambda > 0$ and $\widehat{\mathcal{S}} = \{j : \widehat{\beta}_j \neq 0\}$. Let $\nabla$ denote the differentiation with respect to $\mathbf{Y}$. Prove that with probability one,

$$(\nabla\widehat{\boldsymbol\beta})_{\widehat{\mathcal{S}}} = \big(\mathbf{X}_{\widehat{\mathcal{S}}}^T\mathbf{X}_{\widehat{\mathcal{S}}}\big)^{-1}\mathbf{X}_{\widehat{\mathcal{S}}}^T, \quad (\nabla\widehat{\boldsymbol\beta})_{\widehat{\mathcal{S}}^c} = \mathbf{0},$$

and $\nabla(\mathbf{X}\widehat{\boldsymbol\beta}) = \mathbf{P}_{\widehat{\mathcal{S}}} = \mathbf{X}_{\widehat{\mathcal{S}}}\big(\mathbf{X}_{\widehat{\mathcal{S}}}^T\mathbf{X}_{\widehat{\mathcal{S}}}\big)^{-1}\mathbf{X}_{\widehat{\mathcal{S}}}^T$ is the orthogonal projection to the linear space generated by the selected design vectors $\{\mathbf{X}_j : \widehat{\beta}_j \neq 0\}$.

7.3 Prove Theorem 7.1 and (7.11).

7.4 Prove that in (7.18), both $\|\mathbf{Z}_j\|_2 = \|\mathbf{X}_j - \mathbf{X}_{-j}\widehat{\gamma}^{(j)}(\lambda_j)\|_2$ and $\eta_j(\lambda_j)$ are non-decreasing functions of $\lambda_j$.

7.5 Extend the ADMM algorithm in Section 3.5.9 for the optimization problem in (7.40) under model (7.39)

(a) For linear models, derive an ADMM algorithm for (7.40). Further generate from model (7.45), and test $H_0 : \beta_1 + \beta_2 + \beta_3 = 0$ with $h = 0, 0.1, 0.2$ and 0.4.

(b) For logistic regression models, derive an ADMM algorithm for (7.40). Further generate from model (7.46), and test $H_0 : \beta 1 + \beta_2 + \beta_3 = 0$ with $h = 0, 0.2, 0.4$ and 0.8.

7.6 Study the asymptotical properties of $\widehat{\boldsymbol\beta}_{H_0}$ and $\widehat{\boldsymbol\beta}_H$ defined in (7.40) and (7.41), respectively.

(a) First establish the rate of convergence of these two estimators.

(b) Prove that these estimators possess the sparsity property.

(c) What are the asymptotical distributions of these two estimators?

7.7 Let $f_{n,i}(x;\phi) = \exp\{-\ell_i(\boldsymbol\beta + \phi\mathbf{u})\}$ be probability density functions on the real line, with negative log-likelihood $\ell_i(\boldsymbol\beta + \phi\mathbf{u})$, and $s_{n,i} = -\langle \dot{\ell}_i(\boldsymbol\beta), \mathbf{u}\rangle = (d/d\phi)f_{n,i}(x;\phi)\big|_{\phi=0}$. Suppose data$_i$ are iid variables from $f_{n,i}(x;\phi)$ and

$$\int\Big(n^{1/2}\{f_{n,i}^{1/2}(x; hn^{-1/2}) - f_{n,i}^{1/2}(x;0)\} - (h/2)s_{n,i}(x)f_{n,i}^{1/2}(x;0)\Big)^2 dx = o(h^2)$$

for all real $h$. Verify the LAN condition (7.52) under the Lindeberg condition.

7.8 Prove that Proposition 7.2 also holds when $\sigma$ is unknown.

7.9 Provide details for the proof of (7.73) for the normalized Gaussian design by deriving the distribution of $\eta_{j,k}$ and its tail probability bound. Hint: Take the transformation $y = \sqrt{-(n-3/2)\log(1-t)}$ in the computation of the integral $\int_{x^2}^1 t^{-1/2}(1-t)^{n/2-3/2}dt$.

7.10 Prove that the asymptotic normality in Theorem 7.6 still holds when the independence of $\varepsilon$ and $(W, X)$ in (7.74) is weakened to $\mathrm{E}[\varepsilon|X] = 0$, $\mathrm{E}[\varepsilon(W - g(X))|X] = 0$, $\mathrm{E}\{\varepsilon(W - g(X))\}^2 = \sigma^2 \,\mathrm{E}\,\mathrm{Var}(W|X)$ and

$$\mathrm{E}\left[\frac{1}{n}\sum_{i=1}^{n}\mathrm{Var}(\varepsilon_i|X_i)\{(\mathbf{P}_{k_n}^{\perp}\mathbf{g})_i^2 + (\mathbf{P}_{k_n}(\mathbf{W}-\mathbf{g}))_i^2\}\right] = o(1).$$

7.11 Suppose $f(t) = \beta_0 + \sum_{j=1}^{\infty}\beta_j\sqrt{2}\cos(\pi jt)$ for all real $t$. Prove that $\int_0^1 f^2(t)dt = \sum_{j=0}^{\infty}\beta_j^2$ and that (7.83) and (7.84) hold.

7.12 Consider the Gaussian graphical model.

(a) Prove that $F_{j,k} = 1/(\omega_{j,j}\omega_{k,k} + \omega_{j,k}^2)$ is the minimum Fisher information for the estimation of $\omega_{j,k}$.

(b) Prove that $F_{j,k}^{(\mathrm{par})} = 1/\{1 - (\rho_{j,k}^{(\mathrm{par})})^2\}^2$ is the minimum Fisher information for the estimation of $\omega_{j,k}$.

7.13 Let $\widehat{\rho}_n$ be the sample correlation based on iid observations $(X_i, Y_i)$, $1 \le i \le n$, with correlation $\rho$,

(a) Prove that $\sqrt{n}(\widehat{\rho}_n - \rho)/(1-\rho^2)$ converges in distribution to $N(0,1)$ when $(X_1, Y_1)$ is bivariate Gaussian.

(b) Without assuming normality, under what moment conditions is the limiting distribution in part (i) still valid?

7.14 Prove that (7.96) holds under condition (7.91).

7.15 Suppose that the observation $X_n$ has density $f_{0,n}$ against a measure $\nu_n$ under $P_0$ and $f_{1,n}$ under $P_1$. Prove that the minimum of the sum of type-I and type-II errors for testing $H_0 : f_{0,n}$ against $H_1 : f_{1,n}$ is $1 - \int |f_{1,0} - f_{0,n}|\nu_n(dx)/2$, and that the $L_1$ distance converges to 0 iff the Hellinger distance $\int (f_{1,n}^{1/2} - f_{0,n}^{1/2})^2$ converges to 0.

7.16 Suppose (7.129), (7.131) and (7.132) hold. Prove that

$$(\delta_n n)^{-1}\textstyle\sum_{k=1}^{n}\sum_{i\in I_k}\big\langle \dot{\ell}_i(2\delta_n\mathbf{u}^{(k)} + \widehat{\boldsymbol{\beta}}^{(k)}) - \dot{\ell}_i(\delta_n\mathbf{u}^{(k)} + \widehat{\boldsymbol{\beta}}^{(k)}), \mathbf{u}^{(k)}\big\rangle$$

is a consistent estimate of $F_\theta$ when $n^{-1/2} \ll \delta_n/F_\theta^{1/2} \le \delta_{1,n}$. If in addition the data points are iid, prove that

$$n^{-1}\sum_{k=1}^{n}\sum_{i\in I_k}\langle \dot{\ell}_i(\widehat{\boldsymbol{\beta}}^{(k)}), \mathbf{u}^{(k)}\rangle^2 = V_\theta + o_P(1).$$

Chapter 8

# Feature Screening

In previous chapters we introduced variable selection methods for selecting the correct submodel. We now present feature screening methods in which the goal is to discard as many noise features as possible and at the same time to retain all important features. Let $Y$ be the response variable, and $\mathbf{X} = (X_1, \cdots, X_p)^T$ consists of the $p$-dimensional predictors, from which we obtain a set of random sample $\{\mathbf{X}_i, Y_i\}_{i=1}^n$. The dimensionality $p$ can be ultrahigh (i.e. the dimensionality $p$ grows at the exponential rate of the sample $n$). Even though the variable selection methods introduced in the previous chapters can be used to identify the important variables, the algorithms used for optimization can still be very expensive when the dimension is extremely high. In practice, we may consider naturally a two-stage approach: variable screening followed by variable selection. We first use feature screening methods to reduce ultrahigh dimensionality $p$ to a moderate scale $d_n$ with $d_n \le n$, and then apply the existing variable selection algorithm to select the correct submodel from the remaining variables. This idea can also be employed iteratively. If all important variables are retained in the dimension reduction stage, then the two-stage approach is much more economical.

Throughout this chapter, we adopt the following notation. Denote $\mathbf{Y} = (Y_1, \ldots, Y_n)^T$ and $\mathbf{X} = (\mathbf{X}_1, \cdots, \mathbf{X}_n)^T$, the $n \times p$ design matrix. Let $\mathbf{X}_j$ be the $j$-th column of $\mathbf{X}$. Thus, $\mathbf{X} = (\mathbf{X}_1, \ldots, \mathbf{X}_p)$. We slightly abuse notation $\mathbf{X}$ as well as $\mathbf{X}_i$ and $\mathbf{X}_j$, but their meanings are clear in the context. Let $\varepsilon$ be a general random error and $\boldsymbol{\varepsilon} = (\varepsilon_1, \ldots, \varepsilon_n)^T$ with $\varepsilon_i$ being a random errors. Let $\mathcal{M}_*$ stand for the true model with the size $s = |\mathcal{M}_*|$, and $\widehat{\mathcal{M}}$ be the selected model with the size $d = |\widehat{\mathcal{M}}|$. The definitions of $\mathcal{M}_*$ and $\widehat{\mathcal{M}}$ may be different for different models and contexts.

## 8.1 Correlation Screening

For linear regression model (2.2), its matrix form is

$$\mathbf{Y} = \mathbf{X}\boldsymbol{\beta} + \boldsymbol{\varepsilon}. \tag{8.1}$$

When $p > n$, $\mathbf{X}^T\mathbf{X}$ is singular, and hence the least squares estimator of $\boldsymbol{\beta}$ is not well defined. In such a situation, ridge regression is particularly useful.

The *ridge regression* estimator for model (8.1) is given by

$$\widehat{\boldsymbol{\beta}}_\lambda = (\mathbf{X}^T\mathbf{X} + \lambda \mathbf{I}_p)^{-1}\mathbf{X}^T\mathbf{Y},$$

where $\lambda$ is a ridge parameter. As discussed in Chapter 3, the ridge regression estimator is the solution of penalized least squares with $L_2$ penalty for the linear model. When $\lambda \to 0$, then $\widehat{\boldsymbol{\beta}}_\lambda$ tends to the least squares estimator if $\mathbf{X}$ is full rank, while $\lambda\widehat{\boldsymbol{\beta}}_\lambda$ tends to $\mathbf{X}^T\mathbf{Y}$ as $\lambda \to \infty$. This implies that $\widehat{\boldsymbol{\beta}}_\lambda \propto \mathbf{X}^T\mathbf{Y}$ when $\lambda \to \infty$.

Suppose that all covariates and the response are marginally standardized so that their sample means and variances equal 0 and 1, respectively. Then $\frac{1}{n}\mathbf{X}^T\mathbf{Y}$ becomes the vector consisting of the sample version of Pearson correlations between the response and individual covariate. This motivates people to use the Pearson correlation as a marginal utility for feature screening. Specifically, we first marginally standardize $\mathbf{X}_j$ and $\mathbf{Y}$, and then evaluate

$$\omega_j = \frac{1}{n}\mathbf{X}_j^T\mathbf{Y}, \quad \text{for } j = 1, 2, \ldots, p, \tag{8.2}$$

which is the sample correlation between the $j$th predictor and the response variable.

Intuitively, the higher correlation between $X_j$ and $Y$ is, the more important $X_j$ is. Fan and Lv (2008) proposed to rank the importance of the individual predictor $X_j$ according to $|\omega_j|$, and developed a feature screening procedure based on the Pearson correlation as follows. For a pre-specified proportion $\gamma \in (0, 1)$, the $[\gamma n]$ top ranked predictors are selected to obtain the submodel

$$\widehat{\mathcal{M}}_\gamma = \{1 \le j \le p : |\omega_j| \text{ is among the first } [\gamma n] \text{ largest of all}\}, \tag{8.3}$$

where $[\gamma n]$ denotes the integer part of $\gamma n$. It reduces the ultrahigh dimensionality down to a relatively moderate scale $[\gamma n]$, i.e. the size of $\widehat{\mathcal{M}}_\gamma$, and then a penalized variable selection method is applied to the submodel $\widehat{\mathcal{M}}_\gamma$. We have to set the value $\gamma$ to carry the screening procedure. In practice, one may take the value of $[\gamma n]$ to be $[n/\log(n)]$.

For binary response (coding $Y$ as $\pm 1$), the marginal correlation (8.2) is equivalent to the two-sample $t$-test for comparing the difference of means in the $j^{th}$ variable. It has been successfully applied in Ke, Kelly and Xiu (2019) to select words that are associated with positive returns and negative returns in financial markets; see Figure 1.3.

### *8.1.1 Sure screening property*

For model (8.1), define $\mathcal{M}_* = \{1 \le j \le p : \beta_j \ne 0\}$ with the size $s = |\mathcal{M}_*|$. Under the sparsity assumption, the size $s$ should be much smaller than $n$. The success of the feature screening procedure relies on whether it is valid that $\mathcal{M}_* \subset \widehat{\mathcal{M}}_\gamma$ with probability tending to one.

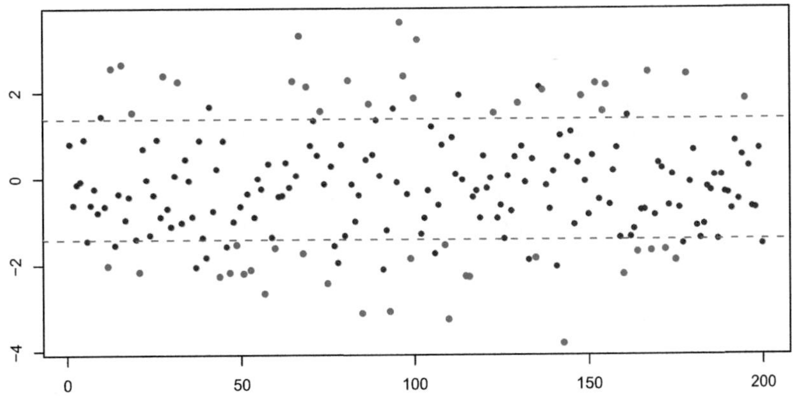


Figure 8.1: Illustration of sure screening. Presented are the estimated coefficients $\{\widehat{\beta}_j\}_{j=1}^p$ (red dots are from $\mathcal{M}_*$ and blue dots from $\mathcal{M}_*^c$). A threshold is applied to screen variables ($|\widehat{\beta}_j| > 1.5$). Sure screening means that all red dots jump over bars. Blue dots exceeding the bars are false selections.

A variable screen or selection method is said to have a *sure screening* property if

$$P\{\mathcal{M}_* \subset \widehat{\mathcal{M}}_\gamma\} \to 1$$

Figure 8.1 illustrates the idea of sure screening. Fan and Lv (2008) proved that such a sure screening property holds with an overwhelming probability under the following technical conditions:

(A1) The covariate vector follows an elliptical contoured distribution (Fang, Kotz and Ng, 1990) and has the concentration property (See Fan and Lv (2008) for details about the concentration property.)

(A2) The variance of the response variable is finite. For some $\kappa \geq 0$ and $c_1, c_2 > 0$, $\min_{j \in \mathcal{M}_*} |\beta_j| \geq c_1 n^{-\kappa}$ and $\min_{j \in \mathcal{M}_*} |\text{cov}(\beta_j^{-1} Y, X_j)| \geq c_2$.

(A3) For some $\tau \geq 0$ and $c_3 > 0$, the largest eigenvalue of $\mathbf{\Sigma} = \text{cov}(\mathbf{X})$ satisfies $\lambda_{\max}(\mathbf{\Sigma}) \leq c_3 n^\tau$, and $\varepsilon \sim \mathcal{N}(0, \sigma^2)$ for some $\sigma > 0$.

(A4) There exists a $\xi \in (0, 1 - 2\kappa)$ such that $\log p = O(n^\xi)$, and $p > n$.

**Theorem 8.1** *Under Conditions (A1)-(A4), if* $2\kappa + \tau < 1$, *then there exists some* $0 < \theta < 1 - 2\kappa - \tau$ *such that when* $\gamma \sim cn^{-\theta}$ *with* $c > 0$. *Then for some* $C > 0$,

$$P(\mathcal{M}_* \subset \widehat{\mathcal{M}}_\gamma) = 1 - O(\exp(-Cn^{1-2\kappa}/\log n)) \to 1, \tag{8.4}$$

*as* $n \to \infty$.

This theorem is due to Fan and Lv (2008). It actually has two parts. Condition (A2) is the key to make the *sure screening property* (8.4) hold [even without condition (A3)]. Due to this property, the marginal (independence) screening procedure defined by (8.3) is referred to as *sure independence screening* (*SIS*) in the literature. It ensures that with probability tending to one, the submodel selected by SIS would not miss any truly important predictor, and hence the false negative rate can be controlled. Note that no condition is imposed on $\mathbf{\Sigma}$ yet. This is an advantage of the independence screening procedure. However, having the sure screening property does not tell the whole story. For example, taking $\widehat{\mathcal{M}} = \{1, \cdots, p\}$, i.e. selecting all variables, also has the sure screening property. That is where condition (A3) comes in to the picture. Under such a condition, the selected model size $|\widehat{\mathcal{M}}| = O_P(n^{1-\theta})$ is also well controlled. Therefore, the false positive rate is no larger than $O_P(n^{1-\theta}/p)$, which is small when $p > n$.

The sure screening property is essential for implementation of all screening procedures in practice, since any post-screening variable selection method introduced in the previous chapters is based on the screened submodels. It is worth pointing out that Conditions (A1)-(A4) certainly are not the weakest conditions to establish this property. See, for example, Li, Liu and Lou (2017) for a different set of conditions that ensure the sure screening property. Conditions (A1)-(A4) are only used to facilitate the technical proofs from a theoretical point of view. Although these conditions are sometimes difficult to check in practice, the numerical studies in Fan and Lv (2008) demonstrated that the SIS can effectively shrink the ultrahigh dimension $p$ down to a relatively large scale $O(n^{1-\theta})$ for some $\theta > 0$ and still can contain all important predictors into the submodel $\widehat{\mathcal{M}}_\gamma$ with probability approaching one as $n$ tends to $\infty$.

### *8.1.2 Connection to multiple comparison*

To illustrate the connection between the SIS and multiple comparison methods, let us consider the two-sample testing problem on the normal mean. Suppose that a random sample with size $n$ was collected from two groups: $\mathbf{X}_i$, $i = 1, \cdots, n_1$ from the disease group, and $\mathbf{X}_i$, $i = n_1+1, \cdots, n_1+n_2 = n$ from the normal group. Of interest is to identify which variables $X_j$s associate with the status of disease. For example, $\mathbf{X}$ consists of gene expression information, and we are interested in which genetic markers are associated with the disease. Such problems are typically formulated as identifying $X_j$ with different means in different groups.

For simplicity, assume that $\mathbf{X}_i \sim N(\boldsymbol{\mu}_1, \mathbf{\Sigma})$ for observations from the disease group, and $\mathbf{X}_i \sim N(\boldsymbol{\mu}_2, \mathbf{\Sigma})$ for observations from the control group. Hotelling $T^2$ (Anderson, 2003) may be used for testing $H_0 : \boldsymbol{\mu}_1 = \boldsymbol{\mu}_2$, but it does not serve for the purpose of identifying which variables associate with the status of disease. In the literature of multiple comparison, people have used the two-sample $t$-test to identify variables associated with the disease status.

Denote $\bar{x}_{1j}$ and $\bar{x}_{2j}$ to be the sample mean for disease and normal groups, respectively. The two-sample $t$-test for $H_{0j}: \mu_{1j} = \mu_{2j}$ can be written as

$$t_j = \frac{\sqrt{n_1 n_2/n}(\bar{X}_{1j} - \bar{X}_{2j})}{\sqrt{\{(n_1-1)s_{1j}^2 + (n_2-1)s_{2j}^2)\}/(n-2)}},$$

where $s_{1j}^2$ and $s_{2j}^2$ are the sample variance for disease and normal groups, respectively. Methods in multiple tests are to derive a critical value or proper thresholding for all $t$-values based on the *Bonferroni correction* or *false discovery rate* method of Benjamini and Hockberg (1995)(see Section 11.2 for further details) .

By marginally standardizing the data, assume with loss of generality that $s_{1j}^2 = s_{2j}^2 = 1$ (by marginally standardizing). Then

$$t_j = \sqrt{n_1 n_2/n}(\bar{X}_{1j} - \bar{X}_{2j})$$

As a result, ranking $P$-values of the two-sample t-test is equivalent to ranking $|t_j|$, which is proportional to $|\bar{X}_{1j} - \bar{X}_{2j}|$. To see its connection with the Pearson correlation under linear model framework, define $Y_i = (1/n_1)\sqrt{n_1 n_2/n}$ if the $i$-th subject comes from the disease group, and $Y_i = -(1/n_2)\sqrt{n_1 n_2/n}$ if the $i$-th subject comes from the normal group. Then $t_j = \mathbf{X}_j^T\mathbf{Y}$. As a result, multiple comparison methods indeed rank the importance of variables in the two sample problem by using the sample Pearson correlation.

Fan and Fan (2008) proposed the $t$-test statistic for the two-sample mean problem as a marginal utility for feature screening. Feature screening based on $\omega_j$ is distinguished from the multiple test methods in that the screening method aims to rank the importance of predictors rather than directly judge whether an individual variable is significant. Thus, further fine-tuning analysis based on the selected variables from the screening step is necessary. Note that it is not involved in any numerical optimization to evaluate $\omega_j$ or $\mathbf{X}^T\mathbf{Y}$. Thus, feature screening based on the Pearson correlation can be easily carried out at low computational and memory burden. In addition to the computational advantage, Fan and Fan (2008) showed that this feature screening procedure also enjoys a nice theoretical property. Mai and Zou (2013) further proposed a feature screening procedures for the two-sample problem based on the Kolmogorov's goodness-of-fit test statistic, and established its sure screening property. See Section 8.5.3 for more details.

### 8.1.3 *Iterative SIS*

The screening procedure may fail when some key conditions are not valid. For example, when a predictor $X_j$ is jointly correlated but marginally uncorrelated with the response, then $\text{cov}(Y, X_j) = 0$. As a result, Condition (A2) is not valid. In such situations, the SIS may fail to select this important variable. On the other hand, the SIS tends to select the unimportant predictor which is jointly uncorrelated but highly marginally correlated with the response. To refine the screening performance, Fan and Lv (2008) provided an *iterative SIS procedure* (*ISIS*) by iteratively replacing the response with the residual obtained from the regression of the response on selected covariates in the previous step. Section 8.3 below will introduce iterative feature screening procedures under a general parametric model framework, to better control the false negative rate in the finite sample case than the one-step screening procedures. Forward regression is another way to cope with the situation with some individual covariates being jointly correlated but marginally uncorrelated with the response. Wang (2009) studied the property of forward regression with ultrahigh dimensional predictors. To achieve the sure screening property, the forward regression requires that there exist two positive constants $0 < \tau_1 < \tau_2 < \infty$ such that $\tau_1 < \lambda_{\min}(\mathrm{cov}(\mathbf{X})) \leq \lambda_{\max}(\mathrm{cov}(\mathbf{X})) < \tau_2$. See Section 8.6.1 for more details. Recently, Zhang, Jiang and Lan (2019) show that such a procedure has a sure screening property, without imposing conditions on marginal correlation which gives a theoretical justification of ISIS. Their results also include forward regression as a specific example.

## 8.2 Generalized and Rank Correlation Screening

The SIS procedure in the previous section performs well for the linear regression model with ultrahigh dimensional predictors. It is well known that the Pearson correlation is used to measure linear dependence. However, it is very difficult to specify a regression structure for the ultrahigh dimensional data. If the linear model is misspecified, SIS may fail, since the Pearson correlation can only capture the linear relationship between each predictor and the response. Thus, SIS is most likely to miss some nonlinearly important predictors. In the presence of nonlinearity, one may try to make a transformation such as the Box-Cox transformation on the covariates. This motivates us to consider the Pearson correlation between transformed covariates and the response as the marginal utility.

To capture both linearity and nonlinearity, Hall and Miller (2009) defined the *generalized correlation* between the $j$th predictor $X_j$ and $Y$ to be,

$$\rho_g(X_j, Y) = \sup_{h\in\mathcal{H}} \frac{\mathrm{cov}\{h(X_j), Y\}}{\sqrt{\mathrm{var}\{h(X_j)\}\mathrm{var}(Y)}}, \quad \text{for each } j = 1, \ldots, p, \tag{8.5}$$

where $\mathcal{H}$ is a class of functions including all linear functions. For instance, it is a class of polynomial functions up to a given degree. Remark that if $\mathcal{H}$ is a class of all linear functions, $\rho_g(X_j, Y)$ is the absolute value of the Pearson correlation between $X_j$ and $Y$. Therefore, $\rho_g(X, Y)$ is considered as a generalization of the conventional Pearson correlation. Suppose that $\{(X_{ij}, Y_i), i = 1, 2, \ldots, n\}$ is a random sample from the population $(X_j, Y)$.

The generalized correlation $\rho_g(X_j, Y)$ can be estimated by

$$\widehat{\rho}_g(X_j, Y) = \sup_{h\in\mathcal{H}} \frac{\sum_{i=1}^n \{h(X_{ij}) - \bar{h}_j\}(Y_i - \overline{Y})}{\sqrt{\sum_{i=1}^n \{h(X_{ij}) - \bar{h}_j\}^2 \sum_{i=1}^n (Y_i - \overline{Y})^2}}, \tag{8.6}$$

where $\bar{h}_j = n^{-1}\sum_{i=1}^n h(X_{ij})$ and $\overline{Y} = n^{-1}\sum_{i=1}^n Y_i$. In practice, $\mathcal{H}$ can be defined as a set of polynomial functions or cubic splines (Hall and Miller, 2009).

The above generalized correlation is able to characterize both linear and nonlinear relationships between two random variables. Thus, Hall and Miller (2009) proposed using the generalized correlation $\rho_g(X_j, Y)$ as a marginal screening utility and ranking all predictors based on the magnitude of estimated generalized correlation $\widehat{\rho}_g(X_j, Y)$. In other words, select predictors whose generalized correlation exceeds a given threshold. Hall and Miller (2009) introduced a bootstrap method to determine the threshold. Let $r(j)$ be the ranking of the $j$th predictor $X_j$; that is, $X_j$ has the $r(j)$ largest empirical generalized correlation. Let $r^*(j)$ be the ranking of the $j$th predictor $X_j$ using the bootstrapped sample. Then, a nominal $(1-\alpha)$-level two-sided prediction interval of the ranking, $[\widehat{r}_-(j), \widehat{r}_+(j)]$, is computed based on the bootstrapped $r^*(j)$'s. For example, if $\alpha = 0.05$, based on 1000 bootstrap samples, $\widehat{r}_+(j)$ is the $975^{th}$ largest rank of variable $X_j$. Hall and Miller (2009) recommended a criterion to regard the predictor $X_j$ as influential if $\widehat{r}_+(j) < \frac{1}{2}p$ or some smaller fraction of $p$, such as $\frac{1}{4}p$. Therefore, the proposed generalized correlation ranking reduces the ultrahigh $p$ down to the size of the selected model $\widehat{\mathcal{M}}_k = \{j : \widehat{r}_+(j) < kp\}$, where $0 < k < 1/2$ is a constant multiplier to control the size of the selected model $\widehat{\mathcal{M}}_k$. Although the generalized correlation ranking can detect both linear and nonlinear features in the ultrahigh dimensional problems, how to choose an optimal transformation $h(\cdot)$ remains an open issue and the associated sure screening property needs to be justified.

As an alternative to making transformations on predictors, one may make a transformation on the response and define a correlation between the transformed response and a covariate. A general transformation regression model is defined to be

$$H(Y_i) = \mathbf{X}_i^T\boldsymbol{\beta} + \varepsilon_i. \tag{8.7}$$

Li, Peng, Zhang and Zhu (2012) proposed the rank correlation as a measure of the importance of each predictor by imposing an assumption on strict monotonicity on $H(\cdot)$ in model (8.7). Instead of using the sample Pearson correlation defined in Section 8.1.1, Li, Peng, Zhang and Zhu (2012) proposed the marginal *rank correlation*

$$\omega_j = \frac{1}{n(n-1)}\sum_{i\neq\ell}^n I(X_{ij} < X_{\ell j})I(Y_i < Y_\ell) - \frac{1}{4}, \tag{8.8}$$

to measure the importance of the $j$th predictor $X_j$. Note that the marginal rank correlation equals a quarter of the *Kendall* $\tau$ *correlation* between the response and the $j$th predictor. According to the magnitudes of all $\omega_j$'s, the feature screening procedure based on the rank correlation selects a submodel

$$\widehat{\mathcal{M}}_{\gamma_n} = \{1 \le j \le p : |\omega_j| > \gamma_n\},$$

where $\gamma_n$ is the predefined threshold value.

Because the Pearson correlation is not robust against heavy-tailed distributions, outliers or influence points, the rank correlation can be considered as an alternative way to robustly measure the relationship between two random variables. Li, Peng, Zhang and Zhu (2012) referred to the rank correlation based feature screening procedure as a *robust rank correlation screening* (*RRCS*). From the definition of the marginal rank correlation, it is robust against heavy-tailed distributions and invariant under monotonic transformation, which implies that there is no need to estimate the transformation $H(\cdot)$. This may save a lot of computational cost and is another advantage of RRCS over the feature screening procedure based on generalized correlation.

For model (8.7) with $H(\cdot)$ being an unspecified strictly increasing function, Li, Peng, Zhang and Zhu (2012) defined the true model as $\mathcal{M}_* = \{1 \le j \le p : \beta_j \neq 0\}$ with the size $s = |\mathcal{M}_*|$. To establish the sure screening property of the RRCS, Li, *et al.* (2012) imposes the following *marginally symmetric conditions* for model (8.7):

**(M1)** Denote $\Delta H(Y) = H(Y_1) - H(Y_2)$, where $H(\cdot)$ is the link function in (8.7), and $\Delta X_k = X_{1k} - X_{2k}$. The conditional distribution $F_{\Delta H(Y)|\Delta X_k}(t)$ is symmetric about zero when $k \in \mathcal{M}_*^c$

**(M2)** Denote $\Delta\varepsilon_k = \Delta H(Y) - \rho_k^* \Delta X_k$. Assume that the conditional distribution $F_{\Delta\varepsilon_k|\Delta X_k}(t) = \pi_{0k} F_0(t, \sigma_0^2|\Delta X_k) + (1-\pi_{0k}) F_1(t, \sigma_1^2|\Delta X_k)$ follows a symmetric finite mixture distribution where $F_0(t, \sigma_0^2|\Delta X_k)$ follows a symmetric unimodal distribution with the conditional variance $\sigma_0^2$ related to $\Delta X_k$ and $F_1(t, \sigma_1^2|\Delta X_k)$ is a symmetric distribution function with the conditional variance $\sigma_1^2$ related to $\Delta X_k$ when $k \in \mathcal{M}_*$. $\pi_{0k} \ge \pi^*$, where $\pi^*$ is a given positive constant in $(0, 1]$ for any $\Delta X_k$ and any $k \in \mathcal{M}_*$.

**Theorem 8.2** *Under Conditions (M1) and (M2), further assume that (a1)* $p = O(\exp(n^\delta))$ *for some* $\delta \in (0,1)$ *satisfying* $\delta + 2\kappa < 1$ *for any* $\kappa \in (0, 0.5)$, *(a2) assume* $\min_{j \in \mathcal{M}_*} \mathrm{E}\,|X_j|$ *is a positive constant free of* $p$, *and (a3) the error and covariates in model (8.7) are independent. Then there exist positive constants* $c_1, c_2 > 0$ *such that*

$$P\left(\max_{1 \le j \le p} |\omega_j - \mathrm{E}(\omega_j)| \ge c_1 n^{-\kappa}\right) \le p \exp(-c_2 n^{1-2\kappa}).$$

*Furthermore, by taking* $\gamma_n = c_3 n^{-\kappa}$ *for some* $c_3 > 0$ *and assuming* $|\rho_j| > c n^{-\kappa}$ *for* $j \in \mathcal{M}_*$ *and some* $c > 0$, *where* $\rho_j$ *is the Pearson correlation between the response and the* $j$*th predictor.*

$$P\left(\mathcal{M}_* \subseteq \widehat{\mathcal{M}}_{\gamma_n}\right) \ge 1 - 2s \exp(-c_2 n^{1-2\kappa}) \to 1, \quad \text{as } n \to \infty. \tag{8.9}$$

This theorem is due to Li, Peng, Zhang and Zhu (2012) and establishes the sure screening property of the RRCS procedure. The authors further demonstrated that the RRCS procedure is robust to outliers and influence points in the observations.

## 8.3 Feature Screening for Parametric Models

Consider fitting a simple linear regression model

$$Y = \beta_{j0} + X_j\beta_{j1} + \varepsilon_j^*, \tag{8.10}$$

where $\varepsilon_j^*$ is a random error with $\mathrm{E}(\varepsilon_j^*|X_j) = 0$. The discussion in Section 8.1.1 actually implies that we may use the magnitude of the least squares estimate $\widehat{\beta}_{j1}$ or the residual sum of squares $\mathrm{RSS}_j$ to rank the importance of predictors. Specifically, let $\mathbf{Y}$ and $\mathbf{X}_j$ both be marginally standardized so that their sample means equals 0 and sample variance equal 1. Thus, $\widehat{\beta}_{j1} = n^{-1}\mathbf{X}_j^T\mathbf{Y}$, which equals $\omega_j$ defined in (8.2), is used to measure the importance of $X_j$. Furthermore, note that

$$\mathrm{RSS}_j \stackrel{\text{def}}{=} \|\mathbf{Y} - \widehat{\beta}_{j0} - \mathbf{X}_j\widehat{\beta}_{j1}\|^2 = \|\mathbf{Y}\|^2 - n\|\widehat{\beta}_{j1}\|^2,$$

as $\mathbf{X}_j$ is marginally standardized (i.e, $\|\mathbf{X}_j\|^2 = n$), and the least squares estimate $\widehat{\beta}_{j0} = \bar{Y} = 0$. Thus, a greater magnitude of $\widehat{\beta}_{j1}$, or equivalently a smaller $\mathrm{RSS}_j$, results in a higher rank of the $j$th predictor $X_j$. The same rationale can be used to develop the marginal utilities for a much broader class of models, with a different estimate of $\beta_{j1}$ or a more general definition of loss function than RSS, such as the negative likelihood function for the generalized linear models. In fact, the negative likelihood function is a monotonic function of RSS in the Gaussian linear models.

### *8.3.1 Generalized linear models*

Suppose that $\{\mathbf{X}_i, Y_i\}_{i=1}^n$ are a random sample from a generalized linear model introduced in Chapter 5. Denote by

$$\ell(Y_i, \beta_0 + \mathbf{X}_i^T\boldsymbol{\beta}) = Y_i(\beta_0 + \mathbf{X}_i^T\boldsymbol{\beta}) - b(\beta_0 + \mathbf{X}_i^T\boldsymbol{\beta})$$

be the negative logarithm of (quasi)-likelihood function of $Y_i$ given $\mathbf{X}_i$ using the canonical link function(dispersion parameter is taken as $\phi = 1$ without loss of generality), defined in Chapter 5 for the $i$-th observation $\{\mathbf{X}_i, Y_i\}$. Note that the minimizer of the negative log-likelihood $\sum_{i=1}^n \ell(\beta_0 + \mathbf{X}_i^T\boldsymbol{\beta}, Y_i)$ is not well defined when $p > n$.

Parallel to the least squares estimate for model (8.10), assume that each predictor is standardized to have mean zero and standard deviation one, and define the maximum marginal likelihood estimator (MMLE) $\widehat{\boldsymbol{\beta}}_j^M$ for the $j$th predictor $X_j$ as

$$\widehat{\boldsymbol{\beta}}_j^M = (\widehat{\beta}_{j0}^M, \widehat{\beta}_{j1}^M) = \arg\min_{\beta_{j0},\beta_{j1}} \sum_{i=1}^n \ell(Y_i, \beta_{j0} + \beta_{j1}X_{ij}), \tag{8.11}$$

Similar to the marginal least squares estimate for model (8.10), it is reasonable to consider the magnitude of $\widehat{\beta}_{j1}^M$ as a marginal utility to rank the importance of $X_j$ and select a submodel for a given prespecified threshold $\gamma_n$,

$$\widehat{\mathcal{M}}_{\gamma_n} = \{1 \le j \le p : |\widehat{\beta}_{j1}^M| \ge \gamma_n\}. \tag{8.12}$$

This was proposed for feature screening in the generalized linear model by Fan and Song (2010). To establish the theoretical properties of MMLE, Fan and Song (2010) defined the population version of the marginal likelihood maximizer as

$$\boldsymbol{\beta}_j^M = (\beta_{j0}^M, \beta_{j1}^M) = \arg\min_{\beta_{j0},\beta_{j1}} \mathrm{E}\,\ell(Y, \beta_{j0} + \beta_{j1}X_j).$$

They first showed that the marginal regression parameters $\beta_{j1}^M = 0$ if and only if $\mathrm{cov}(Y, X_j) = 0$ for $j = 1, \ldots, p$. Thus, when the important variables are correlated with the response, $\beta_{j1}^M \ne 0$. Define the true model as $\mathcal{M}_* = \{1 \le j \le p : \beta_j \ne 0\}$ with the size $s = |\mathcal{M}_*|$. Fan and Song (2010) showed that under some conditions if

$$|\mathrm{cov}(Y, X_j)| \ge c_1 n^{-\kappa} \quad \text{for } j \in \mathcal{M}_* \text{ and some } c_1 > 0,$$

then

$$\min_{j\in\mathcal{M}_*} |\beta_{j1}^M| \ge c_2 n^{-\kappa}, \quad \text{for some } c_2, \kappa > 0.$$

Thus, the marginal signals $\beta_{j1}^M$'s are stronger than the stochastic noise provided that $X_j$'s are marginally correlated with $Y$. Under some technical assumptions, Fan and Song (2010) proved that the MMLEs are uniformly convergent to the population values and established the sure screening property of the MMLE screening procedure. That is, if $n^{1-2\kappa}/(k_n^2 K_n^2) \to \infty$, where $k_n = b'(K_n B + B) + m_0 K_n^\alpha / s_0$, $B$ is the upper bound of the true value of $\beta_j^M$, $m_0$ and $s_0$ are two positive constants that are related to the model, and $K_n$ is any sequence tending to $\infty$ that satisfies the above condition, then for any $c_3 > 0$, there exists some $c_4 > 0$ such that

$$\begin{aligned} &P\left(\max_{1\le j\le p} |\widehat{\beta}_{j1}^M - \beta_{j1}^M| \ge c_3 n^{-\kappa}\right) \\ \le\ & p\left\{\exp(-c_4 n^{1-2\kappa}/(k_n K_n)^2) + n m_1 \exp(-m_0 K_n^\alpha)\right\}. \end{aligned}$$

This inequality is obtained by the *concentration inequality* for maximum likelihood established in Fan and Song (2010). Therefore, if the right-hand side converges to zero ($p$ can still be ultrahigh dimensional), then the maximum marginal estimation error is bounded by $c_3 n^{-\kappa}$ with probability tending to one. In addition, assume that $|\text{cov}(Y, X_j)| \geq c_1 n^{-\kappa}$ for $j \in \mathcal{M}_*$ and take $\gamma_n = c_5 n^{-\kappa}$ with $c_5 \leq c_2/2$, the following inequality holds,

$$P\left(\mathcal{M}_* \subseteq \widehat{\mathcal{M}}_{\gamma_n}\right)$$
$$\geq \; 1 - s\left\{\exp(-c_4 n^{1-2\kappa}/(k_n K_n)^2) + n m_1 \exp(-m_0 K_n^{\alpha})\right\} \to 1 \quad (8.13)$$

as $n \to \infty$. It further implies that the MMLE can handle the ultrahigh dimensionality as high as $\log p = o(n^{1-2\kappa})$ for the logistic model with bounded predictors, and $\log p = o(n^{(1-2\kappa)/4})$ for the linear model without the joint normality assumption.

Fan and Song (2010) further studied the size of the selected model $\widehat{\mathcal{M}}_{\gamma_n}$ so as to understand the false positive rate of the marginal screening method. Under some regularity conditions, they showed that with probability approaching one, $|\widehat{\mathcal{M}}_{\gamma_n}| = O\{n^{2\kappa}\lambda_{\max}(\boldsymbol{\Sigma})\}$, where the constant $\kappa$ determines how large the threshold $\gamma_n$ is, and $\lambda_{\max}(\boldsymbol{\Sigma})$ is the maximum eigenvalue of the covariance matrix $\boldsymbol{\Sigma}$ of predictors $\mathbf{x}$, which controls how correlated the predictors are. If $\lambda_{\max}(\boldsymbol{\Sigma}) = O(n^{\tau})$, the size of $\widehat{\mathcal{M}}_{\gamma_n}$ has the order $O(n^{2\kappa+\tau})$.

The iterative feature screening procedure significantly improves the simple marginal screening in that it can select weaker but still significant predictors and delete inactive predictors which are spuriously marginally correlated with the response. However, Xu and Chen (2014) claimed that the gain of the iterative procedure is apparently built on higher computational cost and increased complexity. To this end, they proposed the *sparse MLE* (SMLE) method for the generalized linear models and demonstrated that the SMLE retains the virtues of the iterative procedure in a conceptually simpler and computationally cheaper manner. SMLE will be introduced in Section 8.6.2 in details.

### 8.3.2 *A unified strategy for parametric feature screening*

As discussed in the beginning of this section, one may use the RSS of marginal simple linear regression as a marginal utility for feature screening. The RSS criterion can be extended to a general parametric model framework.

Suppose that we are interested in exploring the relationship between $\mathbf{X}$ and $Y$. A general parametric framework is to minimize an objective function

$$Q(\beta_0, \boldsymbol{\beta}) = \sum_{i=1}^{n} L(Y_i, \beta_0 + \boldsymbol{\beta}^T \mathbf{X}_i), \quad (8.14)$$

where $L(\cdot, \cdot)$ is a loss function. By taking different loss functions, many commonly-used statistical frameworks can be unified under (8.14). Let us provide a few examples under this model framework.

1. **Linear models**. Let $L(Y_i, \beta_0 + \boldsymbol{\beta}^T \mathbf{X}_i) = (Y_i - \beta_0 - \boldsymbol{\beta}^T \mathbf{X}_i)^2$. This leads to the least squares method for a linear regression model.

2. **Generalized linear models.** Let $L(Y_i, \beta_0 + \boldsymbol{\beta}^T \mathbf{X}_i) = Y_i(\beta_0 + \mathbf{X}_i^T \boldsymbol{\beta}) - b(\beta_0 + \mathbf{X}_i^T \boldsymbol{\beta})$ be the negative logarithm of (quasi)-likelihood function of $Y_i$ given $\mathbf{X}_i$ using the canonical link function(dispersion parameter is taken as $\phi = 1$ without loss of generality), defined in Chapter 5. This leads to the likelihood-based method for the generalized linear model, which includes normal linear regression models, the logistic regression model and Poisson log-linear models as special cases.
3. **Quantile linear regression.** For a given $\tau \in (0, 1)$, define $\rho_\tau(u) = u\{\alpha - I(u < 0)\}$, the check loss function used in the quantile regression (see Section 6.1), where $I(A)$ is the indicator function of a set $A$. Taking $L(Y_i, \beta_0 + \boldsymbol{\beta}^T \mathbf{X}_i) = \rho_\tau(Y_i - \beta_0 - \boldsymbol{\beta}^T \mathbf{X}_i)$ leads to the quantile regression. In particular, $\tau = 1/2$ corresponds to the median regression or the least absolute deviation method. As discussed in Section 6.1, quantile regression instead of the least squares method is particularly useful in the presence of heteroscedastic errors.
4. **Robust linear regression.** As discussed in Section 6.3, robust linear regression has been proved to be more suitable than the least squares method in the presence of outliers or heavy-tailed errors. The commonly-used loss function includes the $L_1$-loss: $\rho_1(u) = |u|$, the Huber loss function: $\rho_2(u) = (1/2)|u|^2 I(|u| \le \delta) + \delta\{|u| - (1/2)\delta\} I(|u| > \delta)$ (Huber, 1964). Setting $L(Y_i, \beta_0 + \boldsymbol{\beta}^T \mathbf{X}_i) = \rho_k(Y_i - \beta_0 + \boldsymbol{\beta}^T \mathbf{X}_i)$, $k = 1$, or 2 leads to a robust linear regression.
5. **Classification.** In machine learning, the response variable typically is set to be the input class label such as $Y = \{-1, +1\}$ as in the classification method in the statistical literature. The hinge loss function is defined to be $h(u) = \{(1 - u) + |1 - u|\}/2$ (Vapnik, 1995). Taking $L(Y_i, \beta_0 + \boldsymbol{\beta}^T \mathbf{X}_i) = h(Y_i - \beta_0 - \boldsymbol{\beta}^T \mathbf{X}_i)$ in (8.14) leads to a classification rule.

Similar to $\text{RSS}_j$ defined in the beginning of this section, a natural marginal utility of the $j$th predictor is

$$L_j = \min_{\beta_{j0}, \beta_{j1}} \frac{1}{n} \sum_{i=1}^{n} L(Y_i, \beta_{j0} + X_{ij}\beta_{j1}).$$

According to the definition, the smaller the value $L_j$, the more important the corresponding $j$th predictor. This was first proposed in Fan, Samworth and Wu (2009), in which numerical studies clearly demonstrated the potential of this marginal utility.

As mentioned in the end of Section 8.1.3, the marginal feature screening methodology may fail if the predictor is marginally uncorrelated but jointly related with the response, or jointly uncorrelated with the response but has higher marginal correlation than some important features. Fan, Samworth and Wu (2009) proposed an iterative screening and selection procedure (Iterative SIS) under the general statistical framework (8.14). The proposed iterative procedure consists of the following steps. We illustrate the idea in Figure 8.2.

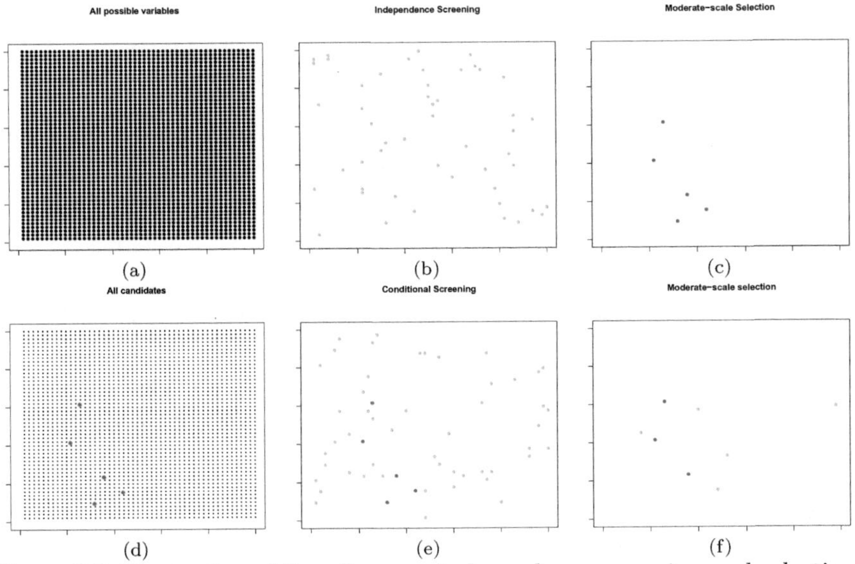


Figure 8.2: Illustration of iterative sure independence screening and selection. Starting from all variables (each dot in (a) represents a variable), apply an independence screening to select variables $\widehat{\mathcal{A}}_1$ (green dots in (b)). Now select variables in (b) by applying a Lasso or SCAD and obtain the results (red variables $\widehat{\mathcal{M}}$) in (c). Conditioning on red variables $\widehat{\mathcal{M}}$ in (d) (the same as those in (c)), we recruit variables with additional contributions via conditional sure independence screening and obtain green variables $\widehat{\mathcal{A}}_2$ in (e). Among all variables $\widehat{\mathcal{M}} \bigcup \mathcal{A}_2$in (e), apply Lasso or SCAD to further select the variables. This results in (f), which deletes some of the red variables and some of the green variables.

**S1.** Compute the vector of marginal utilities $(L_1, \ldots, L_p)$ (Figure 8.2(a)) and select the set $\widehat{\mathcal{A}}_1 = \{1 \leq j \leq p : L_j \text{ is among the first } k_1 \text{ smallest of all}\}$ (green dots in Figure 8.2(b)). Then apply a penalized method to the model with index set $\widehat{\mathcal{A}}_1$, such as the Lasso (Tibshirani, 1996) and SCAD (Fan and Li, 2001), to select a subset $\widehat{\mathcal{M}}$ (red dots in Figure 8.2(c)).

**S2.** Compute, for each $j \in \{1, \ldots, p\}/\widehat{\mathcal{M}}$, the conditional marginal utility of variable $X_j$:

$$L_j^{(2)} = \min_{\beta_0, \boldsymbol{\beta}_{\widehat{\mathcal{M}}}, \beta_j} \frac{1}{n} \sum_{i=1}^{n} L(Y_i, \beta_0 + \mathbf{X}_{i,\widehat{\mathcal{M}}}^T \boldsymbol{\beta}_{\widehat{\mathcal{M}}} + X_{ij}\beta_j),$$

where $\mathbf{X}_{i,\widehat{\mathcal{M}}}$ denotes the sub-vector of $\mathbf{X}_i$ consisting of those elements in $\widehat{\mathcal{M}}$ (Figure 8.2(d)). $L_j^{(2)}$ can be considered as the additional contribution of the $j$th predictor given the existence of predictors in $\widehat{\mathcal{M}}$. Then, select, by screening, the set (green dots in Figure 8.2)

$$\widehat{\mathcal{A}}_2 = \{j \in \{1, \ldots, p\}/\widehat{\mathcal{M}} : L_j^{(2)} \text{ is among the first } k_2 \text{ smallest of all}\}.$$

**S3.** Employ a penalized (pseudo-)likelihood method such as the Lasso and SCAD on the combined set $\widehat{\mathcal{M}} \bigcup \widehat{\mathcal{A}}_2$ (green and red dots in Figure 8.2(e)) to select an updated selected set $\widehat{\mathcal{M}}$. That is, use the penalized likelihood to obtain

$$\widehat{\boldsymbol{\beta}}_2 = \operatorname*{arg\,min}_{\beta_0, \boldsymbol{\beta}_{\widehat{\mathcal{M}}}, \boldsymbol{\beta}_{\widehat{\mathcal{A}}_2}} \frac{1}{n} \sum_{i=1}^{n} L(Y_i, \beta_0 + \mathbf{X}_{i,\widehat{\mathcal{M}}}^T \boldsymbol{\beta}_{\widehat{\mathcal{M}}} + \mathbf{X}_{i,\widehat{\mathcal{A}}_2}^T \boldsymbol{\beta}_{\widehat{\mathcal{A}}_2}) + \sum_{j \in \widehat{\mathcal{M}} \bigcup \widehat{\mathcal{A}}_2} p_\lambda(|\beta_j|),$$

where $p_\lambda(\cdot)$ is a penalty function such as Lasso or SCAD. Thus the indices set of $\widehat{\boldsymbol{\beta}}_2$ that are none-zero yield a new updated set $\widehat{\mathcal{M}}$ (green and red dots in Figure 8.2(f)).

**S4.** Repeat S2 and S3 until $|\widehat{\mathcal{M}}| \leq d$, where $d$ is the prescribed number and $d \leq n$. The indices set $\widehat{\mathcal{M}}$ is the final selected submodel.

This iterative procedure extends the Iterative SIS (Fan and Lv, 2008), without explicit definition of residuals, to a general parametric framework. It also allows the procedure to delete predictors from the previously selected set. Zhang, Jiang and Lan (2019) show that such a screening and selection procedure has a sure screening property, without imposing conditions on marginal regression coefficients and hence justifying the use of iterations.

In addition, to reduce the false selection rates, Fan, Samworth and Wu (2009) further introduced a variation of their iterative feature screening procedure. One can partition the sample into two halves at random and apply the previous iterative procedure separately to each half to obtain the two submodels, denoted by $\widehat{\mathcal{M}}^{(1)}$ and $\widehat{\mathcal{M}}^{(2)}$. By the sure screening property, both sets may satisfy $P(\mathcal{M}_* \subseteq \widehat{\mathcal{M}}^{(h)}) \to 1$, as $n \to \infty$, for $h = 1, 2$, where $\mathcal{M}_*$ is the true model set. Then, the intersection $\widehat{\mathcal{M}} = \widehat{\mathcal{M}}^{(1)} \bigcap \widehat{\mathcal{M}}^{(2)}$ can be considered the final estimated set of $\mathcal{M}_*$. This estimate also satisfies that $P(\mathcal{M}_* \subseteq \widehat{\mathcal{M}}) \to 1$. This variant can effectively reduce the false selection rates in the feature screening stage.

### *8.3.3 Conditional sure independence screening*

In many applications, one can know a priori that a set $\mathcal{C}$ of variables are related to the response $Y$ and would like to recruit additional variables to explain or predict $Y$ via independence screening. This naturally leads to *conditional sure independence screening (CSIS)*. For each variable $j \in \{1, \ldots, p\}/\mathcal{C}$, the conditional marginal utility (indeed loss) of variable $X_j$ is

$$L_j = \min_{\beta_0, \boldsymbol{\beta}_{\mathcal{C}}, \beta_j} \frac{1}{n} \sum_{i=1}^{n} L(Y_i, \beta_0 + \mathbf{X}_{i,\mathcal{C}}^T \boldsymbol{\beta}_{\mathcal{C}} + X_{ij}\beta_j),$$

Now rank these variables by using $\{L_j, j \notin \mathcal{C}^c\}$: the smaller the better and choose top-K variables.

Barut, Fan, and Verhasselt (2016) investigated the properties of such a conditional SIS. It gives the conditions under which that sure screening property can be obtained. In particular, they showed that CSIS overcomes several drawbacks of the SIS mentioned in Section 8.1 and helps reduce the number of false selections when the covariates are highly correlated. Note that CSIS is one of step (S2) in the Iterative SIS. Therefore, the study there shed some light on understanding the Iterative SIS.

## 8.4 Nonparametric Screening

In practice, parametric models such as linear and generalized linear models may lead to model mis-specification. Further, it is often that little prior information about parametric models is available at the initial stage of modeling. Thus, nonparametric models become very useful in the absence of a priori information about the model structure and may be used to enhance the overall modelling flexibility. Therefore, the feature screening techniques for the nonparametric models naturally drew great attention from researchers.

### *8.4.1 Additive models*

Fan, Feng and Song (2011) proposed a *nonparametric independence screening (NIS)* for the ultrahigh dimensional *additive model*

$$Y = \sum_{j=1}^{p} m_j(X_j) + \varepsilon, \tag{8.15}$$

where $\{m_j(X_j)\}_{j=1}^p$ have mean 0 for identifiability. An intuitive population level marginal screening utility is $\mathrm{E}(f_j^2(X_j))$, where $f_j(X_j) = \mathrm{E}(Y|X_j)$ is the projection of $Y$ onto $X_j$. Spline regression can be used to estimate the marginal regression function $\mathrm{E}(Y|X_j)$. Let $\mathbf{B}_j(x) = \{B_{j1}(x), \ldots, B_{jd_n}(x)\}^T$ be a normalized B-spline basis, and we approximate $f_j(x)$ by $\boldsymbol{\beta}_j^T \mathbf{B}_j(x)$, a linear combination of $B_{jk}(x)$'s. Suppose that $\{(\mathbf{X}_i, Y_i), i = 1, \ldots, n\}$ is a random sample from the additive model (8.15). Then $\boldsymbol{\beta}_j$ can be estimated by its least squares estimate

$$\widehat{\boldsymbol{\beta}}_j = \operatorname*{argmin}_{\boldsymbol{\beta}_j \in \mathbb{R}^{d_n}} \sum_{i=1}^{n} (Y_i - \boldsymbol{\beta}_j^T \mathbf{B}_j(X_{ij}))^2.$$

As a result,

$$\widehat{f}_{nj}(x) = \widehat{\boldsymbol{\beta}}_j^T \mathbf{B}_j(x), \quad 1 \le j \le p, \tag{8.16}$$

Thus, the screened model index set is

$$\widehat{\mathcal{M}}_\nu = \{1 \le j \le p : \|\widehat{f}_{nj}\|_n^2 \ge \nu_n\}, \tag{8.17}$$

for some predefined threshold value $\nu_n$, with $\|\widehat{f}_{nj}\|_n^2 = n^{-1}\sum_{i=1}^n \widehat{f}_{nj}(X_{ij})^2$. In practice, $\nu_n$ can be determined by the random permutation idea (Zhao and Li, 2012), i.e., $\nu_n$ is taken as the $q$th quantile ($q$ is given and close to 1) of $\|\widehat{f}_{nj}^*\|_n^2$, where $\widehat{f}_{nj}^*$ is estimated in the same fashion as above, but based on the random permuted decouple $\{(\mathbf{X}_{\pi(i)}, Y_i)\}_{i=1}^n$, and $\{\pi(1),\dots,\pi(n)\}$ is a random permutation of the index $\{1,\dots,n\}$. The rationale is that for the permuted data $\{(\mathbf{X}_{\pi(i)}, Y_i)\}_{i=1}^n$, $\mathbf{X}$ and $Y$ are not associated and $\|\widehat{f}_{nj}^*\|_n^2$ provides the baseline value for such a no-association setting.

Fan, Feng and Song (2011) established the sure screening property of NIS of including the true model $\mathcal{M}_* = \{j : \mathrm{E}\, m_j(X_j)^2 > 0\}$ in the following theorem.

**Theorem 8.3** *Suppose that the following conditions are valid: (b1) The rth derivative of $f_j$ is Lipschitz of order $\alpha$ for some $r > 0$, $\alpha \in (0, 1]$ and $s = r + \alpha > 0.5$; (b2) the marginal density function of $X_j$ is bounded away from 0 and infinity; (b3)* $\min_{j\in\mathcal{M}_*} \mathrm{E}\{f_j^2(X_j)\} \ge c_1 d_n n^{-2\kappa}$, $0 < \kappa < s/(2s+1)$ *and* $c_1 > 0$; *(b4) the sup norm $\|m\|_\infty$ is bounded, the number of spline basis $d_n$ satisfies $d_n^{2s-1} \le c_2 n^{-2\kappa}$ for some $c_2 > 0$; and the i.i.d. random error $\varepsilon_i$ satisfies the sub-exponential tail probability: for any $B_1 > 0$, $\mathrm{E}\{\exp(B_2|\varepsilon_i|)|\mathbf{X}_i\} < B_2$ for some $B_2 > 0$. It follows that*

$$P\left(\mathcal{M}_* \subset \widehat{\mathcal{M}}_\nu\right) \to 1 \tag{8.18}$$

*for* $p = \exp\{n^{1-4\kappa} d_n^{-3} + n d_n^{-3}\}$.

In addition, if $\mathrm{var}(Y) = O(1)$, then the size of the selected model $|\widehat{\mathcal{M}}_\nu|$ is bounded by the polynomial order of $n$ and the false selection rate is under control.

Fan, Feng, Song (2011) further refined the NIS by two strategies. The first strategy consists of the iterative NIS along with the post-screening penalized method for additive models (ISIS-penGAM) – first conduct the regular NIS with a data-driven threshold; then apply penGAM (Meier, Geer and Bühlmann, 2009) to further select components based on the screened model; re-apply the NIS with the response $Y$ replaced by the residual from the regression using selected components; repeat the process until convergence. The second strategy is the greedy INIS – in each NIS step, a predefined small model size $p_0$, often taken to be 1 in practice, is used instead of the data-driven cutoff $\nu_n$. The false positive rate can be better controlled by this method, especially when the covariates are highly correlated or conditionally correlated.

### *8.4.2 Varying coefficient models*

*Varying coefficient models* are another class of popular nonparametric regression models in the literature and defined by

$$Y = \sum_{j=1}^{p} \beta_j(U)X_j + \varepsilon, \tag{8.19}$$

where $\beta_j(U)$'s are coefficient functions, depending on the observed (exposure) variable $U$ such as age. An intercept can be included by setting $X_1 \equiv 1$. Fan, Ma and Dai (2014) extended the NIS introduced in the last section to varying coefficient models, and Song, Yi and Zou (2014) further explored this methodology to the longitudinal data. From a different perspective, Liu, Li and Wu (2014) proposed a conditional correlation screening procedure based on the kernel regression approach.

Fan, Ma and Dai (2014) started from the marginal regression problem for each $X_j$, $j = 1, \ldots, p$, i.e., finding $a_j(u)$ and $b_j(u)$ to minimize $\mathrm{E}\{(Y - a_j(U) - b_j(U)X_j)^2|U = u\}$. Thus, the marginal contribution of $X_j$ for predicting $Y$ can be described as

$$u_j = \mathrm{E}\{a_j(U) + b_j(U)X_j\}^2 - \mathrm{E}\{a_0(U)^2\},$$

where $a_0(U) = \mathrm{E}(Y|U)$.

Suppose that $\{(U_i, \mathbf{X}_i, Y_i), i = 1, \ldots, n\}$ is a random sample from (8.19). Similar to the B-spine regression, we may obtain an estimate of $a_j(u), b_j(u)$ and $a_0(u)$ through the basis expansion. Specifically, we fit the marginal model, using variables $U$ and $X_j$,

$$Y = \boldsymbol{\eta}_j^T \mathbf{B}(U) + \boldsymbol{\theta}_j^T \mathbf{B}(U)X_j + \varepsilon_j,$$

where $\mathbf{B}(u) = \{B_1(u), \ldots, B_{d_n}(u)\}^T$ consists of normalized B-spline bases and $\boldsymbol{\eta}_j$ and $\boldsymbol{\theta}_j$ are the coefficients for the functions $a_j(\cdot)$ and $b_j(\cdot)$, respectively. This leads to the following least-squares: find $\boldsymbol{\eta}_j$ and $\boldsymbol{\theta}_j$ to minimize

$$\sum_{i=1}^{n} \Big( Y_i - \boldsymbol{\eta}_j^T \mathbf{B}(U_i) - \boldsymbol{\theta}_j^T \mathbf{B}(U_i)X_{ij} \Big)^2.$$

This least-squares problem admits an analytic solution. Let

$$\mathbf{Q}_{nj} = \begin{pmatrix} \mathbf{B}(U_1)^T, & X_{1j}\mathbf{B}(U_1)^T \\ \vdots & \vdots \\ \mathbf{B}(U_n)^T, & X_{1j}\mathbf{B}(U_n)^T \end{pmatrix}_{n \times 2d_n}$$

and $\mathbf{Y} = (Y_1, \cdots, Y_n)^T$. Then,

$$(\widehat{\boldsymbol{\eta}}_j^T, \widehat{\boldsymbol{\theta}}_j^T)^T = (\mathbf{Q}_{nj}^T \mathbf{Q}_{nj})^{-1} \mathbf{Q}_{nj}^T \mathbf{Y}.$$

Thus the marginal coefficient functions $a_j(\cdot)$ and $b_j(\cdot)$ can be estimated as

$$\widehat{a}_j(u) = \mathbf{B}(u)^T \widehat{\boldsymbol{\eta}}_j, \quad \text{and} \quad \widehat{b}_j(u) = \mathbf{B}(u)^T \widehat{\boldsymbol{\theta}}_j.$$

We can estimate $a_0(\cdot)$ similarly (though this baseline function is not used in comparison of marginal utility).

The sample marginal utility for screening is

$$\widehat{\mu}_{nj} = \|\widehat{a}_j(u) + \widehat{b}_j(u)X_j\|_n^2 - \|\widehat{a}_0(u)\|_n^2, \tag{8.20}$$

where $\|\cdot\|_n^2$ is the sample average, defined in the same fashion as in the last section. The selected model is then $\widehat{\mathcal{M}}_\tau = \{1 \le j \le p : \widehat{u}_{nj} \ge \tau_n\}$ for the pre-specified data-driven threshold $\tau_n$. Note that the baseline estimator $\|\widehat{a}_0(u)\|_n^2$ is not needed for ranking the marginal utility.

Fan, Ma and Dai (2014) established the sure screening property $P(\mathcal{M}_* \subset \widehat{\mathcal{M}}_\tau) \to 1$ for $p = o(\exp\{n^{1-4\kappa}d_n^{-3}\})$ with $\kappa > 0$ under the regularization conditions listed in the previous section with an additional bounded constraint imposed on $U$, and assuming both $X_j$ and $\mathrm{E}(Y|\mathbf{X}, U)$, as well as the random error $\varepsilon$ follow distributions with the sub-exponential tails. Furthermore, the iterative screening and greedy iterative screening can be conducted in a manner similar to INIS.

Song, Yi and Zou (2014) proposed a feature screening procedure for varying coefficient models (8.19) under the setting of longitudinal data $\{(U_{il}, Y_{il}, \mathbf{X}_{il}), i = 1, \ldots, n,\ l = 1, \ldots, n_i\}$, where $\mathbf{X}_{il} = (X_{il1}, \ldots, X_{ilp})^T$. Instead of using $U_j$ for the marginal utility, they proposed the following marginal utility:

$$\Omega_j = \frac{1}{|\mathcal{U}|}\int_{\mathcal{U}} b_j^2(u)du, \tag{8.21}$$

where $\mathcal{U}$ is the bounded support of the covariate $U$, and $b_j(u)$ is the minimizer of the marginal objective function $Q_j = \mathrm{E}\{Y - X_j b_j(u)\}^2$. To estimate $\Omega_j$, the predictor-specific B-spline basis $\mathbf{B}_j(u) = \{B_{j1}(u), \ldots, B_{jd_j}(u)\}^T$ is used, thus $\widehat{b}_{nj} = \mathbf{B}_j(u)^T\widehat{\gamma}_j$, where $\widehat{\gamma}_j = (\gamma_{j1}, \ldots, \gamma_{jd_j})^T$ is the minimizer of the sample objective function

$$\widehat{Q}_{nj} = \sum_{i=1}^n w_i \sum_{l=1}^{n_i}\{Y_{il} - \sum_{k=1}^{d_j} \mathbf{x}_{il}B_{jk}(u_{il})\gamma_{jk}\}^2,$$

with the weight $w_i = 1$ or $1/n_i$. Specifically,

$$\widehat{\gamma}_j = \left(\sum_{i=1}^n \mathbf{B}_{ji}\mathcal{X}_{ji}\mathbf{W}_i\mathcal{X}_{ji}\mathbf{B}_{ji}^T\right)^{-1} \sum_{i=1}^n \mathbf{B}_{ji}\mathcal{X}_{ji}\mathbf{W}_i\mathbf{Y}_i,$$

where $\mathcal{X}_{ji} = \mathrm{diag}(X_{i1j}, \ldots, X_{in_ij})_{n_i \times n_i}$, $\mathbf{W}_i = \mathrm{diag}(w_i, \ldots, w_i)_{n_i \times n_i}$,

$$\mathbf{B}_{ji} = \begin{pmatrix} B_{j1}(u_{i1}), & \cdots, & B_{j1}(u_{in_i}) \\ \vdots & \cdots & \vdots \\ B_{jd_j}(u_{i1}), & \cdots, & B_{jd_j}(u_{in_i}) \end{pmatrix}_{d_j \times n_i}, \text{ and } \mathbf{Y}_i = \begin{pmatrix} Y_{i1} \\ \vdots \\ Y_{in_i} \end{pmatrix}_{n_i \times 1}.$$

Therefore, a natural estimate of $\Omega_j$ is

$$\widehat{\Omega}_j = \widehat{\gamma}_j^T \frac{1}{|\mathcal{U}|} \int_{\mathcal{U}} \{\mathbf{B}_j(u)\mathbf{B}_j(u)^T du\} \widehat{\gamma}_j. \tag{8.22}$$

The integral in (8.22) does not depend on data, and can be easily evaluated by numerical integration since $u$ is univariate. As a result, we select the submodel $\widehat{\mathcal{M}}_\kappa$ with a pre-specified cutoff $\kappa_n$: $\widehat{\mathcal{M}}_\kappa = \{1 \le j \le p : \widehat{\Omega}_j \ge \kappa_n\}$. Song, Yi and Zou (2014) also established the sure screening property of their procedure under the ultrahigh dimension setting with a set of regularity conditions.

From a different point of view, Liu, Li and Wu (2014) proposed another sure independent screening procedure for the ultrahigh dimensional varying coefficient model (8.19) based on *conditional correlation learning* (CC-SIS). Since (8.19) is indeed a linear model when conditioning on $U$, the conditional correlation between each predictor $X_j$ and $Y$ given $U$ is defined similarly as the Pearson correlation:

$$\rho(X_j, Y|U) = \frac{\text{cov}(X_j, Y|U)}{\sqrt{\text{cov}(X_j, X_j|U)\text{cov}(Y, Y|U)}},$$

where $\text{cov}(X_j, Y|U) = \text{E}(X_jY|U) - \text{E}(X_j|U)\,\text{E}(Y|U)$, and the population level marginal utility to evaluate the importance of $X_j$ is $\rho_{j0}^* = \text{E}\{\rho^2(X_j, Y|U)\}$. To estimate $\rho_{j0}^*$, or equivalently, the conditional means $\text{E}(Y|U)$, $\text{E}(Y^2|U)$, $\text{E}(X_j|U)$, $\text{E}(X_j^2|U)$ and $\text{E}(X_jY|U)$, with sample $\{(U_i, \mathbf{X}_i, Y_i), i = 1, \ldots, n\}$, the kernel regression is adopted, for instance,

$$\widehat{\text{E}}(Y|U = u) = \frac{\sum_{i=1}^n K_h(U_i - u)Y_i}{\sum_{i=1}^n K_h(U_i - u)},$$

where $K_h(t) = h^{-1}K(t/h)$ is a rescaled kernel function. Therefore, all the plug-in estimates for the conditional means and conditional correlations $\widehat{\rho}^2(X_j, Y|U_i)$'s are obtained, and thus also that of $\rho_{j0}^*$:

$$\widehat{\rho}_j^* = \frac{1}{n}\sum_{i=1}^n \widehat{\rho}^2(X_j, Y|U_i). \tag{8.23}$$

And the screened model is defined as

$$\widehat{\mathcal{M}} = \{j : 1 \le j \le p,\ \widehat{\rho}_j^* \text{ ranks among the first } d\},$$

where the size $d = [n^{4/5}/\log(n^{4/5})]$ follows the hard threshold by Fan and Lv (2008) but modified for the nonparametric regression, where $[a]$ is the integer part of $a$.

The sure screening property of CC-SIS has been established under some regularity conditions: (c1) the density function of $u$ has continuous second-order derivative; (c2) the kernel function $K(\cdot)$ is bounded uniformly over its finite support; (c3) $X_jY$, $Y^2$, $X_j^2$ satisfy the sub-exponential tail probability uniformly; (c4) all the conditional means, along with their first- and second-order derivatives are finite and the conditional variances are bounded away from 0. See Liu, Li and Wu (2014) for details.

### *8.4.3 Heterogeneous nonparametric models*

As is known, nonparametric quantile regression is useful to analyze the heterogeneous data, by separately studying different conditional quantiles of the response given the predictors. As to the ultrahigh-dimensional data, He, Wang and Hong (2013) proposed an adaptive nonparametric screening approach based on this quantile regression methodology.

At any quantile $\tau \in (0,1)$, the true sparse model is defined as

$$\mathcal{M}_\tau = \{1 \le j \le p : \ Q_\tau(Y|\mathbf{X}) \text{ functionally depends on } X_j\},$$

where $Q_\tau(Y|\mathbf{X})$ is the $\tau$th conditional quantile of $Y$ given $\mathbf{X} = (X_1, \ldots, X_p)^T$. Thus, their population screening criteria can be defined as

$$q_{\tau j} = \mathrm{E}\{Q_\tau(Y|X_j) - Q_\tau(Y)\}^2, \tag{8.24}$$

with $Q_\tau(Y|X_j)$, the $\tau$th conditional quantile of $Y$ given $X_j$, and $Q_\tau(Y)$, the unconditioned quantile of $Y$. Thus $Y$ and $X_j$ are independent if and only if $q_{\tau j} = 0$ for every $\tau \in (0,1)$. Notice that

$$Q_\tau(Y|X_j) = \mathrm{argmin}_f \, \mathrm{E}[\rho_\tau(Y - f(X_j))|X_j],$$

where $\rho_\tau(z) = z[\tau - I(z < 0)]$, the check loss function defined in Section 6.1.

To estimate $q_{\tau j}$ based on the sample $\{(\mathbf{X}_i, Y_i), i = 1, \ldots, n\}$, one may consider the B-spline approximation based on a set of basis functions $\mathbf{B}(x) = \{B_1(x), \ldots, B_{d_n}(x)\}^T$. That is, one considers $\widehat{Q}_\tau(Y|x) = \mathbf{B}(x)^T \widehat{\mathbf{\Gamma}}_j$, where

$$\widehat{\mathbf{\Gamma}} = \mathrm{argmin}_{\mathbf{\Gamma}} \sum_{i=1}^n \rho_\tau(Y_i - \mathbf{B}(X_{ij})^T \mathbf{\Gamma}).$$

Furthermore, $Q_\tau(Y)$ in (8.24) might be estimated by $F_{Y,n}^{-1}(\tau)$, the $\tau$th sample quantile function based on $Y_1, \ldots, Y_n$. Therefore, $q_{\tau j}$ is estimated by

$$\widehat{q}_{\tau j} = \frac{1}{n} \sum_{i=1}^n \left\{\mathbf{B}(X_{ij})^T \widehat{\mathbf{\Gamma}}_j - F_{Y,n}^{-1}(\tau)\right\}^2, \tag{8.25}$$

and the selected submodel $\widehat{\mathcal{M}}_\gamma = \{1 \le j \le p : \widehat{q}_{\tau j} \ge \gamma_n\}$ for some $\gamma_n > 0$.

To guarantee the sure screening property and control the false selection rate, the $r$th derivative of $Q_\tau(Y|X_j)$ is required to satisfy the Lipschitz condition of order $c$, where $r + c > 0.5$; the active predictors need to have strong enough marginal signals; the conditional density function of $Y$ given $\mathbf{X}$ is locally bounded away from 0 and infinity – this relaxes the sub-exponential tail probability condition in literature; however, global upper and lower bounds are needed for the marginal density functions of $X_j$'s; and an additional restriction on $d_n$, the number of basis functions is applied.

## 8.5 Model-free Feature Screening

Feature screening procedures described in Sections 8.1-8.4 were developed for a class of specific models. In high-dimensional modeling, it may be very challenging in specifying the model structure on the regression function without priori information. Zhu, Li, Li and Zhu (2011) advocated model-free feature screening procedures for ultrahigh dimensional data. We introduce several feature screening procedures without imposing a specific model structure on regression function in this section.

### *8.5.1 Sure independent ranking screening procedure*

Let $Y$ be the response variable with support $\Psi_y$. Here $Y$ can be both univariate and multivariate. Let $\mathbf{X} = (X_1, \cdots, X_p)^T$ be a covariate vector. Zhu, Li, Li and Zhu (2011) developed the notion of active predictors and inactive predictors without specifying a regression model. Consider the conditional distribution function of $Y$ given $\mathbf{X}$, denoted by $F(y \mid \mathbf{X}) = P(Y < y \mid \mathbf{X})$. Define the true model

$$\mathcal{M}_* = \{k : F(y \mid \mathbf{X}) \text{ functionally depends on } X_k \text{ for some } y \in \Psi_y\},$$

If $k \in \mathcal{M}_*$, $X_k$ is referred to as an active predictor, otherwise it is referred to as an inactive predictor. Again $|\mathcal{M}_*| = s$. Let $\mathbf{X}_{\mathcal{M}_*}$, an $s \times 1$ vector, consist of all $X_k$ with $k \in \mathcal{M}_*$. Similarly, let $\mathbf{X}_{\mathcal{M}_*^c}$, a $(p-s) \times 1$ vector, consist of all inactive predictors $X_k$ with $k \in \mathcal{M}_*^c$.

Zhu, Li, Li and Zhu (2011) considered a general model framework under which a unified screening approach was developed. Specifically, they considered that $F(y \mid \mathbf{X})$ depends on $\mathbf{X}$ only through $\mathbf{B}^T\mathbf{X}_{\mathcal{M}_*}$ for some $p_1 \times K$ constant matrix $\mathbf{B}$. In other words, assume that

$$F(y \mid \mathbf{X}) = F_0(y \mid \mathbf{B}^T\mathbf{X}_{\mathcal{M}_*}), \tag{8.26}$$

where $F_0(\cdot \mid \mathbf{B}^T\mathbf{X}_{\mathcal{M}_*})$ is an unknown distribution function for a given $\mathbf{B}^T\mathbf{X}_{\mathcal{M}_*}$. The model parameter $\mathbf{B}$ may not be identifiable. However, the identifiability of $\mathbf{B}$ is of no concern here because our primary goal is to identify active variables rather than to estimate $\mathbf{B}$ itself. Moreover, the model-free screening procedure does not require an explicit estimation of $\mathbf{B}$.

Many existing models with either continuous or discrete response are special examples of (8.26). Moreover, $K$ is as small as just one, two or three in frequently encountered examples. For instance, many existing regression models can be written in the following form:

$$h(Y) = f_1(\boldsymbol{\beta}_1^T\mathbf{X}_{\mathcal{M}_*}) + \boldsymbol{\beta}_2^T\mathbf{X}_{\mathcal{M}_*} + f_2(\boldsymbol{\beta}_3^T\mathbf{X}_{\mathcal{M}_*})\varepsilon, \tag{8.27}$$

for a continuous response and

$$g\{\mathrm{E}(Y|\mathbf{X})\} = f_1(\boldsymbol{\beta}_1^T\mathbf{X}_{\mathcal{M}_*}) + \boldsymbol{\beta}_2^T\mathbf{X}_{\mathcal{M}_*} \tag{8.28}$$

for a binary or count response by extending the framework of the generalized linear model to semiparametric regression, where $h(\cdot)$ is a monotonic function, $g(\cdot)$ is a link function, $f_2(\cdot)$ is a nonnegative function, $\boldsymbol{\beta}_1, \boldsymbol{\beta}_2$ and $\boldsymbol{\beta}_3$ are unknown coefficients, and it is assumed that $\varepsilon$ is independent of $\mathbf{X}$. Here $h(\cdot), g(\cdot), f_1(\cdot)$ and $f_2(\cdot)$ may be either known or unknown. Clearly, model (8.27) is a special case of (8.26) if $\mathbf{B}$ is chosen to be a basis of the column space spanned by $\boldsymbol{\beta}_1, \boldsymbol{\beta}_2$ and $\boldsymbol{\beta}_3$ for model (8.27) and by $\boldsymbol{\beta}_1$ and $\boldsymbol{\beta}_2$ for model (8.28). Meanwhile, it is seen that model (8.27) with $h(Y) = Y$ includes the following special cases: the linear regression model, the partially linear model (Härdle, Liang and Gao, 2000) and the single-index model (Härdle, Hall and Ichimura, 1993), and the generalized partially linear single-index model (Carroll, Fan, Gijbels and Wand, 1997) is a special of model (8.28). Model (8.27) also includes the transformation regression model for a general transformation $h(Y)$. As a consequence, a feature screening approach developed under (8.26) offers a unified approach that works for a wide range of existing models.

As before, assume that $\mathrm{E}(X_k) = 0$ and $\mathrm{Var}(X_k) = 1$ for $k = 1, \ldots, p$. By the law of iterated expectations, it follows that

$$\mathrm{E}[\mathbf{X}\,\mathrm{E}\{\mathbf{1}(Y < y) \mid \mathbf{X}\}] = \mathrm{cov}\{\mathbf{X}, \mathbf{1}(Y < y)\}.$$

To avoid specifying a structure for regression function, Zhu, *et al.*(2011) considered the correlation between $X_j$ and $\mathbf{1}(Y < y)$ to measure the importance of the $j$-th predictor instead of the correlation between $X_j$ and $Y$. Specifically, define $\mathbf{M}(y) = \mathrm{E}\{\mathbf{X}F(y \mid \mathbf{X})\}$, and let $\Omega_k(y)$ be the $k$-th element of $\mathbf{M}(y)$. Zhu, *et al.*(2011) proposed to use

$$\omega_k = \mathrm{E}\left\{\Omega_k^2(Y)\right\}, \quad k = 1, \ldots, p. \tag{8.29}$$

as the marginal utility for feature screening. Intuitively, if $X_k$ and $Y$ are independent, $\Omega_k(y) = 0$ for any $y \in \Psi_y$ and $\omega_k = 0$. On the other hand, if $X_k$ and $Y$ are related, then there exists some $y \in \Psi_y$ such that $\Omega_k(y) \neq 0$, and hence $\omega_k$ must be positive. This is the motivation to employ the sample estimate of $\omega_k$ to rank all the predictors.

Suppose that $\{(\mathbf{X}_i, Y_i), i = 1, \cdots, n\}$ is a random sample from $\{\mathbf{X}, Y\}$. For ease of presentation, we assume that the sample predictors are all marginally standardized. For any given $y$, the sample moment estimator of $\mathbf{M}(y)$ is

$$\widehat{\mathbf{M}}(y) = n^{-1}\sum_{i=1}^{n} \mathbf{X}_i I(Y_i < y).$$

Consequently, a natural estimator for $\omega_k$ is

$$\widehat{\omega}_k = \frac{1}{n}\sum_{j=1}^{n} \widehat{\Omega}_k^2(Y_j) = \frac{1}{n}\sum_{j=1}^{n}\left\{\frac{1}{n}\sum_{i=1}^{n} X_{ik} I(Y_i < Y_j)\right\}^2, \quad k = 1, \ldots, p,$$

where $\widehat{\Omega}_k(y)$ denotes the $k$-th element of $\widehat{\mathbf{M}}(y)$, and $X_{ik}$ denotes the $k$-th element of $\mathbf{X}_i$. Zhu, *et al.*, (2011) proposed ranking all the candidate predictors $X_k, k = 1, \ldots, p$, according to $\widehat{\omega}_k$ from the largest to smallest, and then selecting the top ones as the active predictors.

Motivated by the following fact, Zhu, *et al.* (2011) established the consistency in ranking property for their procedure. If $K = 1$ and $\mathbf{X} \sim N_p(\mathbf{0}, \sigma^2 I_p)$ with unknown $\sigma^2$ in model (8.26), it follows by a direct calculation that

$$\mathbf{M}(y) = \mathrm{E}\{\mathbf{X}F_0(y \mid \mathbf{b}^T\mathbf{X})\} = c(y)\mathbf{b},$$

where $c(y) = \|\mathbf{b}\|^{-1} \int_{-\infty}^{\infty} vF_0(y \mid v\|\mathbf{b}\|)\phi(v; 0, \sigma^2)\, dv$ with $\phi(v; 0, \sigma^2)$ being the density function of $N(0, \sigma^2)$ at $v$. Thus, if $\mathrm{E}\{c^2(Y)\} > 0$, then

$$\max_{k \in \mathcal{M}_*^c} \omega_k < \min_{k \in \mathcal{M}_*} \omega_k, \tag{8.30}$$

and $\omega_k = 0$ if and only if $k \in \mathcal{M}_*^c$. This implies that the quantity $\omega_k$ may be used for feature screening in this setting. Under much more general conditions than the normality assumption, Zhu, *et al.* (2011) showed that (8.30) holds. The authors further established a concentration inequality for $\widehat{\omega}_k$ and, $\max_{k \in \mathcal{M}_*^c} \widehat{\omega}_k < \min_{k \in \mathcal{M}_*} \widehat{\omega}_k$, which is referred to as the property of consistency in ranking (CIR). Due to this property, they referred to their procedure as the *sure independent ranking screening* (*SIRS*) procedure. They studied several issues related to implementation of the SIRS method. Since the SIRS possesses the CIR property, the authors further developed a soft cutoff value for $\widehat{\omega}_j$'s to determine which indices should be included in $\widehat{\mathcal{M}_*}$ by extending the idea of introducing auxiliary variables for thresholding proposed by Luo, Stefanski and Boos (2006). Zhu, *et al.* (2011) empirically demonstrated that the combination of the soft cutoff and hard cutoff by setting $d = [n/\log(n)]$ works quite well in their simulation studies. Lin, Sun and Zhu (2013) proposed an improved version of the SIRS procedure for a setting in which the relationship between the response and an individual predictor is symmetric.

### *8.5.2 Feature screening via distance correlation*

The SIRS procedure was proposed for multi-index models that include many commonly-used models in order to be a model-free screening procedure. Another strategy to achieve the model-free quality is to employ the measure of independence to efficiently detect linearity and nonlinearity between predictors and the response variable, and construct feature screening procedures for ultrahigh dimensional data. Li, Zhong and Zhu (2012) proposed a SIS procedure based on the distance correlation (Székely, Rizzo and Bakirov, 2007). Unlike the Pearson correlation coefficient, rank correlation coefficient and generalized correlation coefficient, which all are defined for two random variables, the distance covariance is defined for two random vectors which are allowed to have different dimensions.

The *distance correlation* between two random vectors $\mathbf{U} \in R^{q_1}$ and $\mathbf{V} \in$

$R^{q_2}$ is defined by

$$\mathrm{dcov}^2(\mathbf{U},\mathbf{V}) = \int_{R^{q_1+q_2}} \|\phi_{\mathbf{U},\mathbf{V}}(\mathbf{t},\mathbf{s}) - \phi_{\mathbf{U}}(\mathbf{t})\phi_{\mathbf{V}}(\mathbf{s})\|^2 w(\mathbf{t},\mathbf{s})\, d\mathbf{t}\, ds, \tag{8.31}$$

where $\phi_{\mathbf{U}}(\mathbf{t})$ and $\phi_{\mathbf{V}}(\mathbf{s})$ are the marginal characteristic functions of $\mathbf{U}$ and $\mathbf{V}$, respectively, $\phi_{\mathbf{U},\mathbf{V}}(\mathbf{t},\mathbf{s})$ is the joint characteristic function of $\mathbf{U}$ and $\mathbf{V}$, and

$$w(\mathbf{t},\mathbf{s}) = \left\{c_{q_1}c_{q_2}\|\mathbf{t}\|_{q_1}^{1+q_1}\|\mathbf{s}\|_{q_2}^{1+q_2}\right\}^{-1}$$

with $c_d = \pi^{(1+d)/2}/\Gamma\{(1+d)/2\}$ (This choice is to facilitate the calculation of the integral). Here $\|\phi\|^2 = \phi\bar{\phi}$ for a complex-valued function $\phi$ with $\bar{\phi}$ being the conjugate of $\phi$. From the definition (8.31), $\mathrm{dcov}^2(\mathbf{U},\mathbf{V}) = 0$ if and only if $\mathbf{U}$ and $\mathbf{V}$ are independent. Székely, Rizzo and Bakirov (2007) proved that

$$\mathrm{dcov}^2(\mathbf{U},\mathbf{V}) = S_1 + S_2 - 2S_3,$$

where

$$\begin{aligned} S_1 &= \mathrm{E}\left(\|\mathbf{U}-\widetilde{\mathbf{U}}\|\|\mathbf{V}-\widetilde{\mathbf{V}}\|\right), \qquad S_2 = \mathrm{E}\left(\|\mathbf{U}-\widetilde{\mathbf{V}}\|\right)E(\|\mathbf{U}-\widetilde{\mathbf{V}}\|) \\ S_3 &= \mathrm{E}\left\{E(\|\mathbf{U}-\widetilde{\mathbf{U}}\||\mathbf{U})E(\|\mathbf{V}-\widetilde{\mathbf{V}}\||\ \mathbf{V})\right\}, \end{aligned}$$

and $(\widetilde{\mathbf{U}},\widetilde{\mathbf{V}})$ is an independent copy of $(\mathbf{U},\mathbf{V})$. Thus, the distance covariance between $\mathbf{U}$ and $\mathbf{V}$ can be estimated by substituting in its sample counterpart. Specifically, the distance covariance between $\mathbf{U}$ and $\mathbf{V}$ is estimated by

$$\widehat{\mathrm{dcov}}^2(\mathbf{U},\mathbf{V}) = \widehat{S}_1 + \widehat{S}_2 - 2\widehat{S}_3$$

with

$$\begin{aligned} \widehat{S}_1 &= \frac{1}{n^2}\sum_{i=1}^n\sum_{j=1}^n \|\mathbf{U}_i - \mathbf{U}_j\|\|\mathbf{V}_i - \mathbf{V}_j\|, \\ \widehat{S}_2 &= \frac{1}{n^2}\sum_{i=1}^n\sum_{j=1}^n \|\mathbf{U}_i - \mathbf{U}_j\|\frac{1}{n^2}\sum_{i=1}^n\sum_{j=1}^n \|\mathbf{V}_i - \mathbf{V}_j\|, \text{ and} \\ \widehat{S}_3 &= \frac{1}{n^3}\sum_{i=1}^n\sum_{j=1}^n\sum_{l=1}^n \|\mathbf{U}_i - \mathbf{U}_l\|\|\mathbf{V}_j - \mathbf{V}_l\|. \end{aligned}$$

based on a random sample $\{(\mathbf{U}_i,\mathbf{V}_i), i = 1,\cdots,n\}$ from the population $(\mathbf{U},\mathbf{V})$.

Accordingly, the distance correlation between $\mathbf{U}$ and $\mathbf{V}$ is defined by

$$\mathrm{dcorr}(\mathbf{U},\mathbf{V}) = \mathrm{dcov}(\mathbf{U},\mathbf{V})/\sqrt{\mathrm{dcov}(\mathbf{U},\mathbf{V})\mathrm{dcov}(\mathbf{V},\mathbf{V})},$$

and estimated by plugging in the corresponding estimate of distance covariances. As mentioned above, $\mathrm{dcov}^2(\mathbf{U},\mathbf{V}) = 0$ if and only if $\mathbf{U}$ and $\mathbf{V}$ are independent. Furthermore, $\text{dcorr}(U, V)$ is strictly increasing in the absolute value of the Pearson correlation between two univariate normal random variables $U$ and $V$.

Let $\mathbf{Y} = (Y_1, \cdots, Y_q)^T$ be the response vector and $\mathbf{X} = (X_1, \ldots, X_p)^T$ be the predictor vector. Motivated by the appealing properties of distance correlation, Li, *et al.* (2012) proposed a SIS procedure to rank the importance of the $k$th predictor $X_k$ for the response using its distance correlation $\widehat{\omega}_j = \widehat{\text{dcorr}}^2(X_j, \mathbf{Y})$. Then, a set of important predictors with large $\widehat{\omega}_j$ is selected and denoted by $\widehat{\mathcal{M}} = \{j : \widehat{\omega}_j \geq cn^{-\kappa}, \text{ for } 1 \leq j \leq p\}$.

Suppose that both $\mathbf{X}$ and $\mathbf{Y}$ satisfy the sub-exponential tail probability uniformly in $p$. That is, $\sup_p \max_{1\leq j \leq p} \text{E}\left\{\exp(sX_j^2)\right\} < \infty$, and $\text{E}\{\exp(s\mathbf{Y}^T\mathbf{Y})\} < \infty$, for $s > 0$. Li, *et al.* (2012) proved that $\widehat{\omega}_j$ is consistent to $\omega_j$ uniformly in $p$,

$$
\begin{aligned}
& P\left(\max_{1\leq j\leq p} |\widehat{\omega}_j - \omega_j| \geq cn^{-\kappa}\right) \\
\leq \;& O\left(p\left[\exp\left\{-c_1 n^{1-2(\kappa+\gamma)}\right\} + n\exp\left(-c_2 n^{\gamma}\right)\right]\right), \qquad (8.32)
\end{aligned}
$$

for any $0 < \gamma < 1/2 - \kappa$, some constants $c_1 > 0$ and $c_2 > 0$. If the minimum distance correlation of active predictors is further assumed to satisfy $\min_{j\in\mathcal{M}_*} \omega_j \geq 2cn^{-\kappa}$, for some constants $c > 0$ and $0 \leq \kappa < 1/2$, as $n \to \infty$,

$$
\begin{aligned}
& P\left(\mathcal{M}_* \subseteq \widehat{\mathcal{M}}\right) \\
\geq \;& 1 - O\left(s\left[\exp\left\{-c_1 n^{1-2(\kappa+\gamma)}\right\} + n\exp\left(-c_2 n^{\gamma}\right)\right]\right) \to 1, \quad (8.33)
\end{aligned}
$$

where $s$ is the size of $\mathcal{M}_*$, the indices set of the active predictors is defined by

$$\mathcal{M}_* = \{1 \leq j \leq p : F(\mathbf{Y} \mid \mathbf{X}) \text{ functionally depends on } X_j \text{ for some } \mathbf{Y} \in \Psi_y\},$$

without specifying a regression model. The result in (8.32) shows that the feature screening procedure enjoys the sure screening property without assuming any regression function of $\mathbf{Y}$ on $\mathbf{X}$, and therefore it has been referred to as distance correlation based SIS (DC-SIS for short). Clearly, the DC-SIS provides a unified alternative to existing model-based sure screening procedures. If $(U, V)$ follows a bivariate normal distribution, then $\text{dcorr}(U, V)$ is a strictly monotone increasing function of the Pearson correlation between $U$ and $V$. Thus, the DC-SIS is asymptotically equivalent to the SIS proposed in Fan and Lv (2008) for normal linear models. The DC-SIS is model-free since it does not impose a specific model structure on the relationship between the response and the predictors. From the definition of the distance correlation, the DC-SIS can be directly employed for screening grouped variables, and it can be directly utilized for ultrahigh dimensional data with multivariate responses. Feature screening for multivariate responses and/or grouped predictors is of great interest in pathway analysis (Chen, Paul, Prentice and Wang, 2011).

As discussed in Section 8.1.3, there exist some predictors that are marginally uncorrelated with or independent of, but jointly are not independent of the response. Similar to marginal screening procedures introduced in Sections 8.1 and 8.2, the SIRS and DC-SIS both will fail to select such predictors. Thus, an iterative screening procedure may be considered, since both SIRS and DC-SIS do not rely on the regression function (i.e., the conditional mean of the response given the predictor). Thus, the iterative procedure introduced in Section 8.3 cannot be applied for SIRS and the DC-SIS because it may be difficult to obtain the residuals. Motivated by the idea of partial residuals, one may apply the SIRS and DC-SIS to the residual of the predictors on the selected predictor and the response. This strategy was used to construct an iterative procedure for the SIRS by Zhu, *et al.* (2011). Zhong and Zhu (2015) applied this strategy to the DC-SIS and developed a general iterative procedure for the DC-SIS, which consists of the three following steps.

**T1.** Apply the DC-SIS for $\mathbf{Y}$ and $\mathbf{X}$ and select $p_1$ predictors, which are denoted by $\mathbf{X}_{\mathcal{M}_1} = \{X_1^{(1)}, \ldots, X_{p_1}^{(1)}\}$, where $p_1 < d$, where $d$ is a user-specified model size (for example, $d = [n/\log n]$). Let $\widehat{\mathcal{M}} = \mathcal{M}_1$.

**T2.** Denote by $\mathbf{X}_1$ and $\mathbf{X}_1^c$ the corresponding design matrix of variables in $\widehat{\mathcal{M}}$ and $\widehat{\mathcal{M}}^c$, respectively. Define $\mathbf{X}_{new} = \left\{\mathbf{I}_n - \mathbf{X}_1\left(\mathbf{X}_1^T\mathbf{X}_1\right)^{-1}\mathbf{X}_1^T\right\}\mathbf{X}_1^c$. Apply the DC-SIS for $\mathbf{Y}$ and $\mathbf{X}_{new}$ and select $p_2$ predictors $\mathbf{X}_{\mathcal{M}_2} = \{X_1^{(2)}, \ldots, X_{p_2}^{(2)}\}$. Update $\widehat{\mathcal{M}} = \widehat{\mathcal{M}} \cup \mathcal{M}_2$.

**T3.** Repeat Step T2 until the total number of selected predictors reaches $d$. The final model we select is $\widehat{\mathcal{M}}$.

### *8.5.3 Feature screening for high-dimensional categorial data*

Analysis of high-dimensional categorical data is common in scientific researches. For example, it is of interest to identify which genes are associated with a certain types of tumors. The types of tumors are categorical and may be binary if researchers are concerned with only case and control. This is a typical example of categorical responses. Genetic markers such as SNPs are categorical covariates. Classification and discriminant analysis are useful for analysis of categorical response data. Traditional methods of classification and discriminant analysis may break down when the dimensionality is extremely large. Even when the covariance matrix is known, Bickel and Levina (2004) showed that the Fisher discriminant analysis performs poorly in a minimax sense due to the diverging spectra. Fan and Fan (2008) demonstrated further that almost all linear discriminants can perform as poorly as random guessing. To this end, it is important to choose a subset of important features for high dimensional classification and discriminant analysis. High-dimensional classification is discussed in Chapter 12 in detail. This section introduces feature screening procedures for ultrahigh dimensional categorical data.

Let $Y$ be a categorical response with $K$ classes $\{Y_1, Y_2, \ldots, Y_K\}$. If an individual covariate $X_j$ is associated with the response $Y$, then $\mu_{jk} = \mathrm{E}(X_j|Y = y_k)$ are likely different from the population mean $\mu_j = \mathrm{E}(X_j)$. Thus, it is intuitive to use the test statistic for the multi-sample mean problem as a marginal utility for feature screening. Based on this idea, Tibshirani, Hastie, Narasimhan and Chu (2002) propose a nearest shrunken centroid approach to cancer class prediction from gene expression data. Fan and Fan (2008) proposed using the two-sample $t$-statistic as a marginal utility for feature screening in high-dimensional binary classification. The authors further showed that the $t$-statistic does not miss any of the important features with probability one under some technical conditions.

Although variable screening based on the two-sample $t$-statistic (Fan and Fan, 2008) performs generally well in the high dimensional classification problems, it may break down for heavy-tailed distributions or data with outliers, and it is not model-free. To overcome these drawbacks, Mai and Zou (2013) proposed a new feature screening method for binary classification based on the *Kolmogorov-Smirnov statistic*. For ease of notation, relabel $Y = +1, -1$ to be the class label. Let $F_{1j}(x)$ and $F_{2j}(x)$ denote the conditional cumulative distribution function of $X_j$ given $Y = +1, -1$, respectively. Thus, if $X_j$ and $Y$ are independent, then $F_{1j}(x) \equiv F_{2j}(x)$. Based on this observation, the nonparametric test of independence may be used to construct a marginal utility for feature screening. Mai and Zou (2013) proposed the following Kolmogorov-Smirnov statistic to be a marginal utility for feature screening;

$$\omega_j = \sup_{x \in \mathbb{R}} |F_{1j}(x) - F_{2j}(x)|, \tag{8.34}$$

which can be estimated by $\omega_{nj} = \sup_{x\in\mathbb{R}} |\widehat{F}_{1j}(x) - \widehat{F}_{2j}(x)|$, where $\widehat{F}_{1j}(x)$ and $\widehat{F}_{2j}(x)$ are the corresponding empirical conditional cumulative distribution functions. Mai and Zou (2013) named this feature screening method as the Kolmogorov filter which sets the selected subset as

$$\widehat{\mathcal{M}} = \{1 \leq j \leq p: \ K_{nj} \text{ ranks among the first } d_n \text{ largest of all}\}. \tag{8.35}$$

Under the conditions that $\min_{j\in\mathcal{M}_*} \{K_j\} - \max_{j\in\mathcal{M}_*^c} \{K_j\} > (\log p/n)^{1/2}$ and $|\mathcal{M}_*| < d_n$, Mai and Zou (2013) showed that the sure screening property of the Kolmogorov filter holds with probability approaching one. That is, $P(\mathcal{M}_* \subseteq \widehat{\mathcal{M}}) \to 1$ as $n \to \infty$.

The Kolmogorov filter is model-free and robust to heavy-tailed distributions of predictors and the presence of potential outliers. However, it is limited to binary classification. Mai and Zou (2015) further developed the *fused Kolmogorov filter* for model-free feature screening with multicategory response, continuous response and counts response.

Cui, Li and Zhong (2014) proposed a sure independence screening using the mean-variance index for ultrahigh dimensional discriminant analysis. It not only retains the advantages of the Kolmogorov filter, but also allows the categorical response to have a diverging number of classes in the order of $O(n^\kappa)$ with some $\kappa \geq 0$.

To emphasize the number of classes being allowed to diverge as the sample size $n$ grows, we use $K_n$ for $K$. Denote by $F_j(x) = P(X_j \leq x)$ the unconditional distribution function of the $j$th feature $X_j$ and $F_{jk}(x) = P(X_j \leq x|Y = y_k)$ the conditional distribution function of $X_j$ given $Y = y_k$. If $X_j$ and $Y$ are statistically independent, then $F_{jk}(x) = F_j(x)$ for any $k$ and $x$. This motivates one to consider the following marginal utility

$$\mathrm{MV}(X_j|Y) = \mathrm{E}_{X_j}[\mathrm{Var}_Y(F(X_j|Y))]. \tag{8.36}$$

This index is named as the mean variance index of $X_j$ given $Y$. Cui, Li and Zhong (2014) showed that

$$\mathrm{MV}(X_j|Y) = \sum_{k=1}^{K_n} p_k \int [F_{jk}(x) - F_j(x)]^2 dF_j(x),$$

which is a weighted average of Cramér-von Mises distances between the conditional distribution function of $X_j$ given $Y = y_k$ and the unconditional distribution function of $X_j$, where $p_k = P(Y = k)$. The authors further showed that $\mathrm{MV}(X_j|Y) = 0$ if and only if $X_j$ and $Y$ are statistically independent. Thus, the proposal of Cui, Li and Zhong (2014) is basically to employ a test of the independence statistic as a marginal utility of feature screening for ultrahigh-dimensional categorical data.

Let $\{(X_{ij}, Y_i) : 1 \leq i \leq n\}$ be a random sample of size $n$ from the population $(X_j, Y)$. Then, $\mathrm{MV}(X_j|Y)$ can be estimated by its sample counterpart

$$\widehat{\mathrm{MV}}(X_j|Y) = \frac{1}{n} \sum_{k=1}^{K_n} \sum_{i=1}^{n} \widehat{p}_k [\widehat{F}_{jk}(X_{ij}) - \widehat{F}_j(X_{ij})]^2, \tag{8.37}$$

where $\widehat{p}_k = n^{-1} \sum_{i=1}^{n} I\{Y_i = y_k\}$, $\widehat{F}_{jk}(x) = n^{-1} \sum_{i=1}^{n} I\{X_{ij} \leq x, Y_i = y_k\}/\widehat{p}_k$, and $\widehat{F}_j(x) = n^{-1} \sum_{i=1}^{n} I\{X_{ij} \leq x\}$.

Without specifying any classification model, Cui, Li and Zhong (2014) defined the important feature subset by

$$\mathcal{M}_* = \{j : F(y|\mathbf{x}) \text{ functionally depends on } X_j \text{ for some } y = y_k\}.$$

$\widehat{\mathrm{MV}}(X_j|Y)$ is used to rank the importance of all features and the subset $\widehat{\mathcal{M}} = \{j : \widehat{\mathrm{MV}}(X_j|Y) \geq cn^{-\tau}, \text{ for } 1 \leq j \leq p\}$ is selected. This procedure is referred to as the *MV-based sure independence screening* (*MV-SIS*). Assume that (1) for some positive constants $c_1$ and $c_2$,

$$c_1/K_n \leq \min_{1 \leq k \leq K_n} p_k \leq \max_{1 \leq k \leq K_n} p_k \leq c_2/K_n,$$

and $K_n = O(n^\kappa)$ for $\kappa \geq 0$; (2) for some positive constants $c > 0$ and $0 \leq$

$\tau < 1/2$, $\min_{j\in\mathcal{M}_*} \text{MV}(X_j|Y) \geq 2cn^{-\tau}$. Under Conditions (1) and (2), Cui, Li and Zhong (2014) proved that

$$P\Big(\mathcal{M}_* \subseteq \widehat{\mathcal{M}}\Big) \geq 1 - O\left(s\exp\{-bn^{1-(2\tau+\kappa)} + (1+\kappa)\log n\}\right), \qquad (8.38)$$

where $b$ is a positive constant and $s = |\mathcal{M}_*|$. As $n \to \infty$, $P\Big(\mathcal{M}_* \subseteq \widehat{\mathcal{M}}\Big) \to 1$ for the NP-dimensionality problem $\log p = O(n^\alpha)$, where $\alpha < 1 - 2\tau - \kappa$ with $0 \leq \tau < 1/2$ and $0 \leq \kappa < 1 - 2\tau$. Thus, the sure screening property of the MV-SIS holds.

It is worth pointing out that the MV-SIS is directly applicable for the setting in which the response is continuous, but the feature variables are categorical (e.g., the covariates are SNP). To implement the MV-SIS, one can simply use $\text{MV}(Y|X_j)$ as the marginal utility for feature screening. When both response and feature variables are categorical, it is not difficult to use a test of independence statistic as marginal utility for feature screening. Huang, Li and Wang (2014) employed the Pearson $\chi^2$-test statistic for independence as a marginal utility for feature screening. They further established the sure screening procedure of their screening procedure under mild conditions.

## 8.6 Screening and Selection

In previous sections, we focus on marginal screening procedures, which enjoy the sure screening property under certain conditions. Comparing with variable selection procedures introduced in Chapters 3 - 6, the marginal screening procedures can be implemented with very low computational cost. To perform well, they typically require quite strong conditions. For instance, these marginal screening procedures may fail in the presence of highly correlated covariates or marginally independent but jointly dependent covariates. See Fan and Lv (2008) for detailed discussions on these issues. Although iterative SIS or its similar strategy may provide a remedy in such situations, it would be more natural to consider joint screening procedures that require screening and selection.

### 8.6.1 *Feature screening via forward regression*

Wang (2009) proposed using the classical *forward regression method*, which recruits one best variable at a time, for feature screening in the linear regression model (8.1). Given a subset $\mathcal{S}$ that is already recruited, the next step is to find one variable $X_{\widehat{j}}$ so that $(\mathbf{X}_\mathcal{S}, X_{\widehat{j}}\}$ has the smallest RSS among all models $\{(\mathbf{X}_\mathcal{S}, X_j), j \notin \mathcal{S}\}$. Specifically, starting with the initial set $\mathcal{S}^{(0)} = \emptyset$, and for $k = 1, \cdots, n$, let $a_k \in \mathcal{S}^{(k-1)c}$, the complement of $\mathcal{S}^{(k-1)}$, be the index so that the covariate $X_{a_k}$ along with all predictors in $\mathcal{S}^{(k-1)}$ achieves the minimum of the residual sum of squares over $j \in \mathcal{S}^{(k-1)c}$, and set $S^{(k)} = \mathcal{S}^{(k-1)} \cup \{a_k\}$ accordingly. Wang (2009) further suggested using the *extended BIC* (Chen and Chen, 2008) to determine the size of the active predictor set. The extended

BIC score for a model $\mathcal{M}$ is defined as

$$\text{BIC}(\mathcal{M}) = \log \widehat{\sigma}^2_{\mathcal{M}} + n^{-1}|\mathcal{M}|\{\log(n) + 2\log(p)\},$$

where $\widehat{\sigma}^2_{\mathcal{M}}$ is the mean squared error under model $\mathcal{M}$, and $|\mathcal{M}|$ stands for the model size of model $\mathcal{M}$. Choose

$$\widehat{m} = \min_{0\le k\le n} \text{BIC}(\mathcal{S}^{(k)})$$

and set $\widehat{\mathcal{S}} = \mathcal{S}^{(\widehat{m})}$. Wang (2009) imposed the following conditions to establish the sure screening property. Recall $\mathcal{M}_* = \{j : \beta_j \ne 0\}$, the index set of active predictors, defined in Section 8.1.1.

(B1) Assume that both $\mathbf{X}$ and $\varepsilon$ in model (8.1) follow normal distributions.

(B2) There exist two positive constants $0 < \tau_{\min} < \tau_{\max} < \infty$ such that $\tau_{\min} < \lambda_{\min}(\mathbf{\Sigma}) < \tau_{\max}(\mathbf{\Sigma}) < \tau_{\max}$, where $\mathbf{\Sigma} = \text{E}(\mathbf{X}\mathbf{X}^T)$.

(B3) Assume that $\|\boldsymbol{\beta}\|_2 \le C_{\boldsymbol{\beta}}$ for some constant $C_{\boldsymbol{\beta}}$ and $\min_{1\le j\le p}|\beta_j| \ge v_{\boldsymbol{\beta}} n^{-\xi_{\min}}$ for some positive constants $v_{\boldsymbol{\beta}}$ and $\xi_{\min} > 0$.

(B4) There exists constants $\xi$, $\xi_0$ and $v$ such that $\log(p) \le vn^{\xi}$, $|\mathcal{M}_*| \le vn^{\xi_0}$, and $\xi + 6\xi_0 + 12\xi_{\min} < 1$.

**Theorem 8.4** *Under model (8.1) and Conditions (B1)-(B4), it follows that*

$$P(\mathcal{M}_* \subset \widehat{\mathcal{S}}) \to 1, \quad \text{as } n \to \infty.$$

### *8.6.2 Sparse maximum likelihood estimate*

Suppose that a random sample was collected from an ultrahigh dimensional generalized linear model with the canonical link, and the log-likelihood function is

$$\ell_n(\boldsymbol{\beta}) = \sum_{i=1}^{n} \left\{(\mathbf{X}_i^T\boldsymbol{\beta})Y_i - b(\mathbf{X}_i^T\boldsymbol{\beta})\right\}. \tag{8.39}$$

In genetic study, we may encounter such a situation in which we have resources to examine only $k$ genetic markers out of a huge number of potential genetic markers. Thus, it requires the feature screening stage to choose $k$ covariates from $p$ potential covariates, in which $k \ll p$. Xu and Chen (2014) formulates this problem as maximum likelihood estimation with $\ell_0$ constraint. They define the *sparse maximum likelihood estimate* (SMLE) of $\boldsymbol{\beta}$ to be

$$\widehat{\boldsymbol{\beta}}_{[k]} = \arg\max_{\boldsymbol{\beta}} \ell_n(\boldsymbol{\beta}), \quad \text{subject to } \|\boldsymbol{\beta}\|_0 \le k, \tag{8.40}$$

where $\|\cdot\|_0$ denotes the $\ell_0$-norm. This problem of finding the best subset of size

$k$ finds its applications also in guarding the spurious correlation in statistical machine learning (Fan and Zhou, 2016; Fan, Shao, and Zhou, 2018) who also independently used the following algorithm (Section 2.3, Fan and Zhou, 2016).

Let $\widehat{\mathcal{M}} = \{1 \le j \le p : \widehat{\boldsymbol{\beta}}_{[k]j} \neq 0\}$ correspond to nonzero entries of $\widehat{\boldsymbol{\beta}}_{[k]}$. For a given number $k$, the SMLE method can be considered as a joint-likelihood-supported feature screening procedure. Of course, the maximization problem in (8.40) cannot be solved when $p$ is large. To tackle the computation issue, Xu and Chen (2014) suggested a quadratic approximation to the log-likelihood function. Let $S_n(\boldsymbol{\beta}) = \ell_n'(\boldsymbol{\beta})$ be the score function. For a given $\boldsymbol{\beta}$, approximate $\ell_n(\boldsymbol{\gamma})$ by the following isotropic quadratic function

$$h_n(\boldsymbol{\gamma};\boldsymbol{\beta}) = \ell_n(\boldsymbol{\beta}) + (\boldsymbol{\gamma}-\boldsymbol{\beta})^T S_n(\boldsymbol{\beta}) - (u/2)\|\boldsymbol{\gamma}-\boldsymbol{\beta}\|_2^2, \tag{8.41}$$

for some scaling parameter $u > 0$. The first two terms in (8.41) come from the first-order Taylor's expansion of $\ell_n(\boldsymbol{\beta})$, while the third term is viewed as an $\ell_2$-penalty to avoid $\boldsymbol{\gamma}$ too far away from $\boldsymbol{\beta}$. The approximation (8.41) coincides with (3.77) in the iterative shrinkage-thresholding algorithm introduced in Section 3.5.7 Instead of maximizing (8.40) directly, we iteratively update $\boldsymbol{\beta}^{(t)}$ by

$$\boldsymbol{\beta}^{(t+1)} = \arg\max_{\boldsymbol{\gamma}} h_n(\boldsymbol{\gamma};\boldsymbol{\beta}^{(t)}), \quad \text{subject to} \quad \|\boldsymbol{\gamma}\|_0 \le k,$$

for a given $\boldsymbol{\beta}^{(t)}$ in the $t$-th step. Due to the form of $h_n(\boldsymbol{\gamma};\boldsymbol{\beta})$, the solution $\boldsymbol{\beta}^{(t+1)}$ is a hard thresholding rule and has an explicit form. Let

$$\mathbf{g} = (g_1, \cdots, g_p) = \boldsymbol{\beta}^{(t)} + u^{-1}\mathbf{X}^T\left\{\mathbf{Y} - b'(\mathbf{X}\boldsymbol{\beta}^{(t)})\right\},$$

and $r_k$ be the $k$th largest value of $\{|g_1|, \cdots, |g_p|\}$. Then

$$\beta_j^{(t+1)} = g_j I(|g_j| > r_k)$$

which is a hard-thresholding rule. Thus, Xu and Chen (2014) referred to this algorithm as the *iterative hard thresholding* (IHT) algorithm. Notice that

$$h(\boldsymbol{\beta};\boldsymbol{\beta}^{(t)}) = \ell_n(\boldsymbol{\beta}) \quad \text{and} \quad h(\boldsymbol{\beta};\boldsymbol{\beta}^{(t)}) \ge \ell_n(\boldsymbol{\beta})$$

for any $\boldsymbol{\beta}$ with proper chosen $u$. These two properties make the IHT algorithm a specific case of the MM algorithm and lead to the ascent property of the IHT algorithm. That is, under some conditions, $\ell_n(\boldsymbol{\beta}^{(t+1)}) \ge \ell_n(\boldsymbol{\beta}^{(t)})$. Therefore, one can start with an initial $\boldsymbol{\beta}^{(0)}$ and carry out the above iteration until $\|\boldsymbol{\beta}^{(t+1)} - \boldsymbol{\beta}^{(t)}\|_2$ falls below some tolerance level.

To see why SMLE can be used for feature screening, let us examine the case of the normal linear regression model, in which $b'(\theta) = \theta$. Thus, for any $\boldsymbol{\beta}^{(t)}$, $\mathbf{g} = (1-u^{-1})\boldsymbol{\beta}^{(t)} + u^{-1}\mathbf{X}^T\mathbf{Y}$. As a result, if $u = 1$, $g_j$ is proportional to the sample correlation between $Y$ and $X_j$. Clearly, the SMLE with $u = 1$ is a marginal screening procedure for normal linear models.

Xu and Chen (2014) further demonstrated that the SMLE procedure enjoys the sure screening property in Theorem 8.5 under the following conditions. Recall the notation $\mathcal{M}_* = \{1 \leq j \leq p : \beta_j \neq 0\}$ the true model, where $\boldsymbol{\beta}^* = (\beta_1^*, \cdots, \beta_p^*)^T$ is the true value of $\boldsymbol{\beta}$. For an index set $\mathcal{S} \subset \{1, \cdots, p\}$, $\boldsymbol{\beta}_\mathcal{S} = \{\beta_j, j \in \mathcal{S}\}$ and $\boldsymbol{\beta}_\mathcal{S}^* = \{\beta_j^*, j \in \mathcal{S}\}$.

**(C1)** $\log p = O(n^\alpha)$ for some $0 \leq \alpha < 1$.

**(C2)** There exists $w_1, w_2 > 0$ and some nonnegative constants $\tau_1, \tau_2$ such that

$$\min_{j \in \mathcal{M}_*} |\beta_j^*| \geq w_1 n^{-\tau_1} \text{ and } |\mathcal{M}_*| < k \leq w_2 n^{\tau_2}.$$

**(C3)** There exist constants $c_1 > 0$, $\delta_1 > 0$, such that for sufficiently large $n$, $\lambda_{\min}\{n^{-1} H_n(\boldsymbol{\beta}_\mathcal{S})\} \geq c_1$ for $\boldsymbol{\beta}_\mathcal{S} \in \{\boldsymbol{\beta}_\mathcal{S} : \|\boldsymbol{\beta}_\mathcal{S} - \boldsymbol{\beta}_\mathcal{S}^*\|_2 \leq \delta_1\}$ and $\mathcal{S} \in \{\mathcal{M} : \mathcal{M}_* \subset \mathcal{M}; \|\mathcal{M}\|_0 \leq 2k\}$, where $H_n(\boldsymbol{\beta}_\mathcal{S})$ is the Hessian matrix of $\ell_n(\boldsymbol{\beta}_\mathcal{M})$

**(C4)** There exist positive constants $c_2$ and $c_3$, such that all covariates are bounded by $c_2$, and when n is sufficiently large,

$$\max_{1 \leq j \leq p} \max_{1 \leq i \leq n} \frac{X_{ij}^2}{\sum_{i=1}^n X_{ij} b''(\mathbf{X}_i^T \boldsymbol{\beta}^*)} \leq c_3 n^{-1}.$$

**Theorem 8.5** *Under model (8.39) and Conditions (C1)-(C4) with* $\tau_1 + \tau_2 < (1-\alpha)/2$, *let* $\widehat{\mathcal{M}}$ *be the model selected (8.40). It follows that*

$$P(\mathcal{M}_* \subseteq \widehat{\mathcal{M}}) \to 1,$$

*as* $n \to \infty$.

### *8.6.3 Feature screening via partial correlation*

Based on the partial correlation, Bühlmann, Kalisch and Maathuis (2010) proposed a variable selection method for the linear model (8.1) with normal response and predictors. To extend this method to the elliptical distribution, Li, Liu and Lou (2017) first studied the limiting distributions for the sample partial correlation under the elliptical distribution assumption, and further developed feature screening and variable selection for ultra-high dimensional linear models.

A $p$-dimensional random vector $\mathbf{X}$ is said to have an *elliptical distribution* if its characteristic function $\mathrm{E}\exp(i\mathbf{t}^T\mathbf{X})$ can be written as $\exp(i\mathbf{t}^T\boldsymbol{\mu})\phi(\mathbf{t}^T\boldsymbol{\Sigma}\mathbf{t})$ for some characteristic generator $\phi(\cdot)$. We write $\mathbf{X} \sim \mathrm{EC}_p(\boldsymbol{\mu}, \boldsymbol{\Sigma}, \phi)$. Fang, Kotz and Ng (1990) gives a comprehensive study on elliptical distribution theory. Many multivariate distributions such as multivariate normal distribution and multivariate t-distribution are special cases of elliptical distributions.

For an index set $\mathcal{S} \subseteq \{1, 2, \cdots, p\}$, define $\mathbf{X}_\mathcal{S} = \{X_j : j \in \mathcal{S}\}$ to be a subset of covariates with index set $\mathcal{S}$. The *partial correlation* between two random variables $X_j$ and $Y$ given a set of controlling variables $\mathbf{X}_\mathcal{S}$, denoted by $\rho(Y, X_j|\mathbf{X}_\mathcal{S})$, is defined as the correlation between the residuals $r_{X_j,\mathbf{X}_\mathcal{S}}$ and $r_{Y,\mathbf{X}_\mathcal{S}}$ from the linear regression of $X_j$ on $\mathbf{X}_\mathcal{S}$ and that of $Y$ on $\mathbf{X}_\mathcal{S}$, respectively.

The corresponding sample partial correlation between $Y$ and $X_j$ given $\mathbf{X}_\mathcal{S}$ is denoted by $\widehat{\rho}(Y, X_j|\mathbf{X}_\mathcal{S})$. Suppose that $(\mathbf{X}_1^T, Y_1), \cdots, (\mathbf{X}_n^T, Y_n)$ are a random samples from an elliptical distribution $\mathrm{EC}_{p+1}(\boldsymbol{\mu}, \boldsymbol{\Sigma}, \phi)$ with all finite fourth moments and kurtosis $\kappa$. For any $j = 1, \cdots, p$, and $\mathcal{S} \subseteq \{j\}^c$ with cardinality $|\mathcal{S}| = o(\sqrt{n})$, if there exists a positive constant $\delta_0$ such that $\lambda_{\min}\{\mathrm{cov}(\mathbf{X}_\mathcal{S})\} > \delta_0$, then

$$\sqrt{n}\,\{\widehat{\rho}(Y, X_j|\mathbf{X}_\mathcal{S})\} - \rho(Y, X_j|\mathbf{X}_\mathcal{S})\} \to N\Big(0, (1+\kappa)\{1 - \rho^2(Y, X_j|\mathbf{X}_\mathcal{S})\}^2\Big). \tag{8.42}$$

See Theorem 1 of Li, Liu and Lou (2017). Further apply the Fisher Z-transformation on $\widehat{\rho}(Y, X_j|\mathbf{X}_\mathcal{S})$ and $\rho(Y, X_j|\mathbf{X}_\mathcal{S})$:

$$\widehat{Z}(Y, X_j|\mathbf{X}_\mathcal{S}) = \frac{1}{2}\log\left\{\frac{1 + \widehat{\rho}(Y, X_j|\mathbf{X}_\mathcal{S})}{1 - \widehat{\rho}(Y, X_j|\mathbf{X}_\mathcal{S})}\right\}, \tag{8.43}$$

$$Z(Y, X_j|\mathbf{X}_\mathcal{S}) = \frac{1}{2}\log\left\{\frac{1 + \rho(Y, X_j|\mathbf{X}_\mathcal{S})}{1 - \rho(Y, X_j|\mathbf{X}_\mathcal{S})}\right\}. \tag{8.44}$$

Then, it follows by the delta method and (8.42) that

$$\sqrt{n}\left\{\widehat{Z}(Y, X_j|\mathbf{X}_\mathcal{S}) - Z(Y, X_j|\mathbf{X}_\mathcal{S})\right\} \to N(0, 1+\kappa). \tag{8.45}$$

The asymptotic variance of $\widehat{\rho}(Y, X_j|\mathbf{X}_\mathcal{S})$ in (8.42) involves $\rho(Y, X_j|\mathbf{X}_\mathcal{S})$, while the asymptotic variance of $\widehat{Z}(Y, X_j|\mathbf{X}_\mathcal{S})$ does not. Thus, it is easier to derive the selection threshold for $\widehat{Z}(Y, X_j|\mathbf{X}_\mathcal{S})$ than for $\widehat{\rho}(Y, X_j|\mathbf{X}_\mathcal{S})$.

Let $\emptyset$ be the empty set, and $\widehat{\rho}(Y, X_j|\mathbf{X}_\emptyset)$ and $\rho(Y, X_j|\mathbf{X}_\emptyset)$ stand for $\widehat{\rho}(Y, X_j)$ and $\rho(Y, X_j)$, respectively. Then (8.42) is also valid for $\mathcal{S} = \emptyset$:

$$\sqrt{n}\,\{\widehat{\rho}(Y, X_j) - \rho(Y, X_j)\} \to N\Big(0, (1+\kappa)\{1 - \rho^2(Y, X_j)\}^2\Big), \tag{8.46}$$

which was shown in Theorem 5.1.6 of Muirhead (1982).

The limiting distributions of partial correlation and sample correlation are given in (8.42) and (8.46), respectively. They provide insights into the impact of the normality assumption on the variable selection procedure proposed by Bülhmann, *et al.* (2010) through the marginal kurtosis under ellipticity assumption. This enables one to modify their algorithm by taking into account the marginal kurtosis to ensure its good performance.

To use partial correlation for variable selection, Bühlmann, Kalisch and Maathuis (2010) imposed the partial faithful condition which implies that

$$\rho(Y, X_j|\mathbf{X}_\mathcal{S}) \neq 0 \text{ for all } \mathcal{S} \subseteq \{j\}^c \text{ if and only if } \beta_j \neq 0. \tag{8.47}$$

if the covariance matrix of $\mathbf{X}$ is positive definite. That is, $X_j$ is important (or $\beta_j \neq 0$) if and only if the partial correlations between $Y$ and $X_j$ given all subsets $\mathcal{S}$ contained in $\{j\}^c$ are not zero. Extending the PC-simple algorithm proposed by Bühlmann, Kalisch and Maathuis (2010), Li, Liu and Lou (2017) proposed to identify active predictors by iteratively testing the series of hypotheses

$$H_0: \ \rho(Y, X_j|\mathbf{X}_\mathcal{S}) = 0 \text{ for } |\mathcal{S}| = 0, 1, \ldots, \widehat{m}_{reach},$$

where $\widehat{m}_{reach} = \min\{m : |\widehat{\mathcal{A}}^{[m]}| \leq m\}$, and $\widehat{\mathcal{A}}^{[m]}$ is the chosen model index set in the $m$th step with cardinality $|\widehat{\mathcal{A}}^{[m]}|$. Based on the limiting distribution (8.45), the rejection region at level $\alpha$ is $|\widehat{Z}(Y, X_j|\mathbf{X}_\mathcal{S})| > \sqrt{1+\widehat{\kappa}}\, z_{\alpha/2}/\sqrt{n}$ with $\widehat{\kappa}$ being a consistent estimate of $\kappa$, where $z_\alpha$ is the critical value of $N(0,1)$ at level $\alpha$. In practice, the factor $\sqrt{n}$ in the rejection region is replaced by $\sqrt{n-1-|\mathcal{S}|}$ due to the loss of degrees of freedom used in the calculation of residuals. Therefore, an equivalent form of the rejection region with small sample correction is

$$|\widehat{\rho}(Y, X_j|\mathbf{X}_\mathcal{S})| > T(\alpha, n, \widehat{\kappa}, |\mathcal{S}|), \tag{8.48}$$

where

$$T(\alpha, n, \widehat{\kappa}, |\mathcal{S}|) = \frac{\exp\left\{2\sqrt{1+\widehat{\kappa}}\Phi^{-1}(1-\alpha/2)/\sqrt{n-1-|\mathcal{S}|}\right\} - 1}{\exp\left\{2\sqrt{1+\widehat{\kappa}}\Phi^{-1}(1-\alpha/2)/\sqrt{n-1-|\mathcal{S}|}\right\} + 1} \tag{8.49}$$

with $\kappa$ being estimated by its sample counterpart:

$$\widehat{\kappa} = \frac{1}{p}\sum_{j=1}^{p}\left\{\frac{\frac{1}{n}\sum_{i=1}^n (X_{ij} - \bar{X}_j)^4}{3\{\frac{1}{n}\sum_{i=1}^n (X_{ij} - \bar{X}_j)^2\}^2} - 1\right\}, \tag{8.50}$$

where $\bar{X}_j$ is the sample mean of the $j$-the element of $\mathbf{X}$ and $X_{ij}$ is the $j$-th element of $\mathbf{X}_i$. In practice, the sample partial correlations can be computed recursively: for any $k \in \mathcal{S}$,

$$\widehat{\rho}(Y, X_j|\mathbf{X}_\mathcal{S}) = \frac{\widehat{\rho}(Y, X_j|\mathbf{X}_{\mathcal{S}\backslash\{k\}}) - \widehat{\rho}(Y, X_k|\mathbf{X}_{\mathcal{S}\backslash\{k\}})\widehat{\rho}(X_j, X_k|\mathbf{X}_{\mathcal{S}\backslash\{k\}})}{[\{1-\widehat{\rho}(Y, X_k|\mathbf{X}_{\mathcal{S}\backslash\{k\}})^2\}\{1-\widehat{\rho}(X_j, X_k|\mathbf{X}_{\mathcal{S}\backslash\{k\}})^2\}]^{1/2}}. \tag{8.51}$$

A thresholded partial correlation variable selection can be summarized in the following algorithm, which was proposed by Li, Liu and Lou (2017).

**Algorithm for TPC variable selection.**

**S1.** Set $m = 1$ and $\mathcal{S} = \emptyset$, obtain the marginally estimated active set by

$$\widehat{\mathcal{A}}^{[1]} = \{j = 1, \cdots, p : |\widehat{\rho}(Y, X_j)| > T(\alpha, n, \widehat{\kappa}, 0)\}.$$

**S2.** Based on $\widehat{\mathcal{A}}^{[m-1]}$, construct the $m$th step estimated active set by

$$\begin{aligned}\widehat{\mathcal{A}}^{[m]} &= \{j \in \widehat{\mathcal{A}}^{[m-1]} : |\widehat{\rho}(Y, X_j|\mathbf{X}_\mathcal{S})| > T(\alpha, n, \widehat{\kappa}, |\mathcal{S}|), \\ &\qquad \forall \mathcal{S} \subseteq \widehat{\mathcal{A}}^{[m-1]}\backslash\{j\}, |\mathcal{S}| = m-1\}.\end{aligned}$$

**S3.** Repeat S2 until $m = \widehat{m}_{reach}$.

This algorithm results in a sequence of estimated active sets

$$\widehat{\mathcal{A}}^{[1]} \supseteq \widehat{\mathcal{A}}^{[2]} \supseteq \ldots \widehat{\mathcal{A}}^{[m]} \supseteq \ldots \supseteq \widehat{\mathcal{A}}^{[\widehat{m}_{reach}]}.$$

Since $\kappa = 0$ for normal distributions, the TPC is indeed the PC-simple algorithm proposed by Bühlmann, *et al.* (2010) under the normality assumption. Thus, (8.45) clearly shows that the PC-simple algorithm tends to over-fit (under-fit) the models under those distributions where the kurtosis is larger (smaller) than the normal kurtosis 0. As suggested by Bühlmann, *et al.* (2010), one further applies the ordinary least squares approach to estimate the coefficients of predictors in $\widehat{\mathcal{A}}^{[\widehat{m}_{reach}]}$ after running the TPC variable selection algorithm.

Li, Liu and Lou (2017) imposed the following regularity conditions to establish the asymptotic theory of the TPC. These regularity conditions may not be the weakest ones.

**(D1)** The joint distribution of $(\mathbf{X}^T, Y)$ satisfies the partial faithfulness condition (8.47).

**(D2)** $(\mathbf{X}^T, Y)$ follows $\text{EC}_{p+1}(\boldsymbol{\mu}, \boldsymbol{\Sigma}, \phi)$ with $\boldsymbol{\Sigma} > 0$. Furthermore, there exists $s_0 > 0$, such that for all $0 < s < s_0$,

$$\text{E}\{\exp(sY^2)\} < \infty, \quad \max_{1\le j\le p} \text{E}\{\exp(sX_jY)\} < \infty,$$
$$\text{and} \quad \max_{1\le j,k\le p} \text{E}\{\exp(sX_jX_k)\} < \infty.$$

**(D3)** There exists $\delta > -1$, such that the kurtosis satisfies $\kappa > \delta > -1$.

**(D4)** For some $c_n = O(n^{-d})$, $0 < d < 1/2$, the partial correlations $\rho(Y, X_j|\mathbf{X}_{\mathcal{S}})$ satisfy $\inf\Big\{|\rho(Y, X_j|\mathbf{X}_{\mathcal{S}})| : 1 \le j \le p, \mathcal{S} \subseteq \{j\}^c|\mathcal{S}| \le d_0,$ $\rho(Y, X_j|\mathbf{X}_{\mathcal{S}}) \neq 0\Big\} \ge c_n$.

**(D5)** The partial correlations $\rho(Y, X_j|\mathbf{X}_{\mathcal{S}})$ and $\rho(X_j, X_k|\mathbf{X}_{\mathcal{S}})$ satisfy:
i). $\sup\{|\rho(Y, X_j|\mathbf{X}_{\mathcal{S}})| : 1 \le j \le p, \mathcal{S} \subseteq \{j\}^c, |\mathcal{S}| \le d_0\} \le \tau < 1$,
ii). $\sup\{|\rho(X_j, X_k|\mathbf{X}_{\mathcal{S}})| : 1 \le j \neq k \le p, \mathcal{S} \subseteq \{j,k\}^c, |\mathcal{S}| \le d_0\} \le \tau < 1$.

The partial faithfulness condition in (D1) ensures the validity of the TPC method as a variable selection criterion. The assumption on elliptical distribution in (D2) is crucial when deriving the asymptotic distribution of the sample partial correlation. Condition (D3) puts a mild condition on the kurtosis, and is used to control Type I and II errors. The lower bound of the partial correlations in (D4) is used to control Type II errors for the tests. This condition has the same spirit as that of the penalty-based methods which requires the non-zero coefficients to be bounded away from 0. The upper bound of partial correlations in the condition i) of (D5) is used to control Type I error, and the condition ii) of (D5) imposes a fixed upper bound on the population partial correlations between the covariates, which excludes the perfect collinearity between the covariates.

Based on the above regularity conditions, the following consistency property has been established. First we consider the model selection consistency of the final estimated active set by TPC. Since the TPC depends on the significance level $\alpha = \alpha_n$, we rewrite the final chosen model to be $\widehat{\mathcal{A}}_n(\alpha_n)$.

**Theorem 8.6** *Consider linear model (8.1). Under Conditions (D1)-(D5), there exists a sequence $\alpha_n \to 0$ and a positive constant $C$, such that if $d_0$ is fixed, then for $p = o(\exp(n^\xi))$, $0 < \xi < 1/5$, the estimated active set can be identified with the following rate*

$$P\{\widehat{\mathcal{A}}_n(\alpha_n) = \mathcal{A}\} \quad \geq \quad 1 - O\{\exp(-n^\nu/C)\}, \tag{8.52}$$

*where $\xi < \nu < 1/5$; and if $d_0 = O(n^b)$, $0 < b < 1/5$, then for $p = o(\exp(n^\xi))$, $0 < \xi < 1/5 - b$, (8.52) still holds, with $\xi + b < \nu < 1/5$.*

This theorem is due to Li, Liu and Lou (2017). Theorem 8.6 implies that the TPC method including the original PC-simple algorithm enjoys the model selection consistency property when dimensionality increases at an exponential rate of the sample size. Bülhmann, *et al.* (2010) suggested that $\alpha_n = 2\{1 - \Phi(c_n\sqrt{n/(1+\kappa)}/2)\}$.

Bühlmann, *et al.* (2010) established results in Theorem 8.6 for the normal linear model with $p_n$ increasing at a polynomial rate of $n$. In other words, Theorem 8.6 relaxes normality assumptions on model error and dimensionality from the polynomial rate to the exponential rate.

Since the first step in the TPC variable selection algorithm is indeed correlation learning, it is essentially equivalent to the SIS procedure described in Section 8.1.1. Thus, it is of interest to study under what conditions, the estimated active set from the first step of the TPC, denoted by $\widehat{\mathcal{A}}_n^{[1]}(\alpha_n)$, possesses the sure screening property. To this end, we impose the following conditions on the population marginal correlations:

**(D6)** $\inf\{|\rho_n(Y, X_j)| : j = 1, \cdots, \mathrm{p}, \rho_n(y, X_j) \neq 0\} \quad \geq \quad c_n$, where $c_n = O(n^{-d})$, and $0 < d < 1/2$.

**(D7)** $\sup\{|\rho_n(Y, X_j)| : j = 1, \cdots, p_n,\} \leq \tau < 1$.

**Theorem 8.7** *Consider linear model (8.1) and assume (D1)-(D3), (D6) and (D7). For $p = O(\exp(n^\xi))$, where $0 < \xi < 1$, there exists a sequence $\alpha_n \to 0$ such that $P\{\widehat{\mathcal{A}}_n^{[1]} \supseteq \mathcal{A}\} \geq 1 - O\{\exp(-n^\nu/C^*)\}$, where $C^*$ is a positive constant and $\xi < \nu < 1/5$.*

This theorem confirms the sure screening property of the marginal screening procedure based on the Pearson correlation under a different set of regularity conditions given in Section 8.1.1.

## 8.7 Refitted Cross-Validation

The impact of spurious correlation on residual variance estimation was demonstrated in Section 1.3.3 (see Figure 1.9 there). The refitted cross-validation (RCV) method was briefly introduced in Section 3.7 for estimating the error variance in ultrahigh dimensional linear regression. It is effective in reducing the impact of spurious correlated variables and hence a good companion with sure screening methods, where many variables can be spurious. As an effective tool to get rid of the effect of spurious correlated variables, the RCV method can be used in many regression models beyond linear models. This section provides additional insights.

### *8.7.1 RCV algorithm*

Consider a general regression model

$$Y = m(\mathbf{X}) + \varepsilon, \tag{8.53}$$

where $\varepsilon$ is a random error with $\mathrm{E}(\varepsilon|\mathbf{X}) = 0$ and $\mathrm{Var}(\varepsilon|\mathbf{X}) = \sigma^2$. Of interest is to estimate $\sigma^2$. In the presence of ultrahigh dimensional covariates, we typically assume that only a small number of covariates are active. In other words, $m(\mathbf{x})$ depends on $\mathbf{x}$ only through a low dimensional subvector of $\mathbf{x}$. Thus, model selection techniques are used to identify the small subset of active predictors. Due to ultra-high dimensionality, the selected model very likely includes some predictors that are spuriously correlated with the response. As a result, the mean squared errors of the selected model leads to an underestimate of $\sigma^2$. Thus, the estimation of $\sigma^2$ indeed is challenging.

The *RCV method* is to randomly split the entire sample into two data subsets, denoted by $\mathcal{D}_1$ and $\mathcal{D}_2$ respectively. We use $\mathcal{D}_1$ for model selection, and then refit the selected model to $\mathcal{D}_2$. This leads to an estimate of $\sigma^2$, denoted by $\widehat{\sigma}_1^2$. Intuitively, the selected model based on $\mathcal{D}_1$ may include some spuriously correlated predictors, which likely are not spuriously correlated predictors any more in $\mathcal{D}_2$ due to randomly splitting, and will play minor roles in estimating $\sigma^2$ based on $\mathcal{D}_2$. Thus, $\widehat{\sigma}_1^2$ may be a good estimate of $\sigma^2$. Unlike the cross-validation method, both $\mathcal{D}_1$ and $\mathcal{D}_2$ play an important role, and their sample size cannot be too small since we need to ensure enough sample size in $\mathcal{D}_1$ to select a good model and ensure enough sample size in $\mathcal{D}_2$ to achieve estimation accuracy.

Similarly, we may use $\mathcal{D}_2$ for model selection, and then use $\mathcal{D}_1$ for error variance estimation. This leads to another estimate of $\sigma^2$, denoted by $\widehat{\sigma}_2^2$. Then the *RCV estimate* of $\sigma^2$ is the average of $\widehat{\sigma}_1^2$ and $\widehat{\sigma}_2^2$:

$$\widehat{\sigma}_{\mathrm{RCV}}^2 = (\widehat{\sigma}_1^2 + \widehat{\sigma}_2^2)/2. \tag{8.54}$$

Similar to (3.103), one may consider a weighted average of $\widehat{\sigma}_1^2$ and $\widehat{\sigma}_2^2$ as an estimate of $\sigma^2$. Figure 8.3 depicts the algorithm for estimating $\sigma^2$ by using the RCV method for general regression settings.

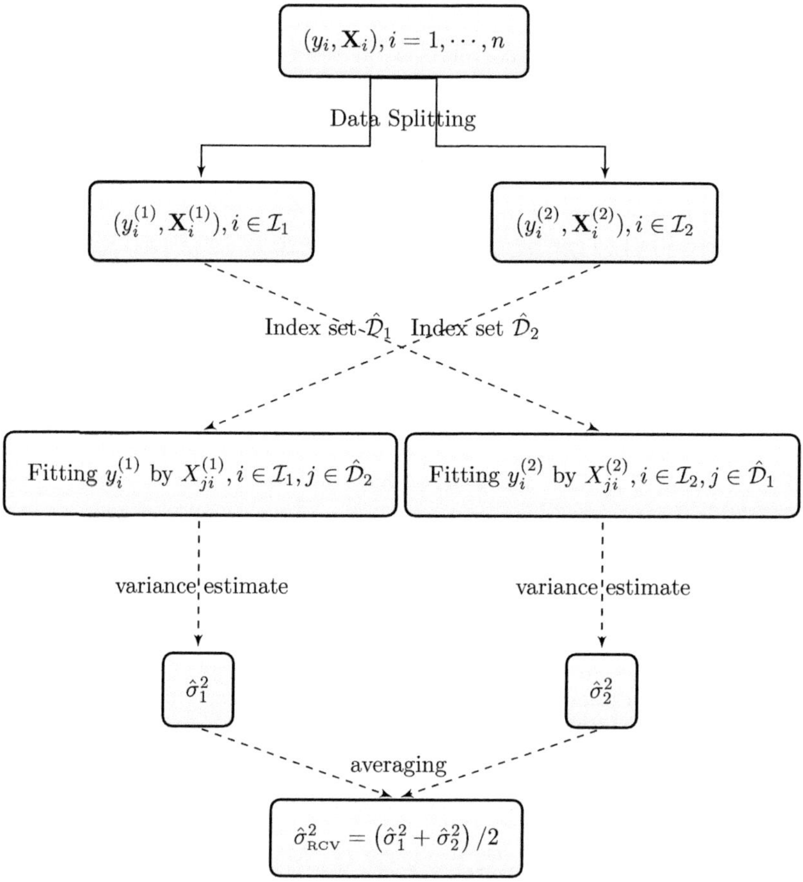


Figure 8.3: RCV algorithm for estimating $\sigma^2$. Adapted from Chen, Fan and Li (2018).

### *8.7.2 RCV in linear models*

Consider linear regression model

$$\mathbf{Y} = \mathbf{X}\boldsymbol{\beta} + \boldsymbol{\varepsilon}.$$

Let $\mathbf{X}_M$ be the submatrix containing the columns of X that are indexed by $M$ and denote $\mathbf{P}_M = \mathbf{X}_M(\mathbf{X}_M^T\mathbf{X}_M)^{-1}\mathbf{X}_M^T$, the corresponding projection matrix of $\mathbf{X}_M$.

A naive two-stage procedure consists of (a) selecting a model $\widehat{M}$ by variable selection or feature screening procedures introduced in Chapter 3 and Sections 8.1-8.6 and (b) then estimate $\sigma^2$ by the mean squared errors based on model $\widehat{M}$:

$$\widehat{\sigma}^2_{\widehat{M}} = \mathbf{Y}^T(\mathbf{I}_n - \mathbf{P}_{\widehat{M}})\mathbf{Y}/(n - |\widehat{M}|).$$

Denote $\widehat{\gamma}_n^2 = \boldsymbol{\varepsilon}^T\mathbf{P}_{\widehat{M}}\boldsymbol{\varepsilon}/\boldsymbol{\varepsilon}^T\boldsymbol{\varepsilon}$. Thus, $\widehat{\sigma}^2_{\widehat{M}}$ can be rewritten as

$$\widehat{\sigma}_n^2 = (1 - \widehat{\gamma}_n^2)\boldsymbol{\varepsilon}^T\boldsymbol{\varepsilon}/(n - |\widehat{M}|).$$

Fan, Guo and Hao (2012) showed that under certain regularization conditions, $\widehat{\sigma}_n^2/(1-\widehat{\gamma}_n^2) \to \sigma^2$ in probability, and

$$\{\widehat{\sigma}_n^2/(1-\widehat{\gamma}_n^2) - \sigma^2\}/\sqrt{n} \to N(0, \mathrm{Var}(\varepsilon^2)) \tag{8.55}$$

in distribution. Thus, if $\widehat{\gamma}_n^2$ does not tend to 0 in probability, the naive estimator $\widehat{\sigma}^2_{\widehat{M}}$ is not consistent. Fan, Guo and Hao (2012) further showed that under certain conditions and if $\log(p/n) = O(1)$, $\widehat{\gamma}_n^2 = O_P\{|\widehat{M}|\log(p)/n\}$. This rate is often sharp. This implies that $\widehat{\gamma}_n^2$ may converge to a positive constant with a non-negligible probability and explains the reason why the naive estimator underestimates the noise level.

The RCV method for the linear model is to randomly split $(\mathbf{X}, \mathbf{Y})$ into two even data sets $(\mathbf{X}_1, \mathbf{Y}_1)$ and $(\mathbf{X}_2, \mathbf{Y}_2)$, based on which, we can obtain two selected models $\widehat{M}_1$ and $\widehat{M}_2$ by a variable selection such as the Lasso and the SCAD or a feature screening procedure such as SIS, respectively. Then we fit data subset $(\mathbf{X}_2, \mathbf{Y}_2)$ to model $\widehat{M}_1$ and calculate its mean squared errors $\widehat{\sigma}_1^2$, and fit data subset $(\mathbf{X}_1, \mathbf{Y}_1)$ to model $\widehat{M}_2$ to obtain its mean squared errors $\widehat{\sigma}_2^2$. More specifically,

$$\widehat{\sigma}_k^2 = \mathbf{Y}_{3-k}^T(\mathbf{I}_{n/2} - \mathbf{P}_{3-k})\mathbf{Y}_{3-k}/(n/2 - |\widehat{M}_k|)$$

for $k = 1$ and 2, where $\mathbf{P}_{3-k}$ is the projection matrix associated with model $\widehat{M}_k$ based on data $(\mathbf{X}_{3-k}, \mathbf{Y}_{3-k})$. As defined in (8.54), $\widehat{\sigma}^2_{\mathrm{RCV}}$ is the average of $\widehat{\sigma}_1^2$ and $\widehat{\sigma}_2^2$. Fan, Guo and Hao (2012) studied the theoretical properties of $\widehat{\sigma}^2_{\mathrm{RCV}}$ in the following theorem.

**Theorem 8.8** *Suppose that (a) the errors $\varepsilon_i$, $i = 1, \cdots, n$ are independent and identically distributed with zero mean and finite variance $\sigma^2$ and independent of the design matrix $\mathbf{X}$; and (b) there is a constant $\lambda_0 > 0$ and $b_n$ such that $b_n/n \to 0$ such that $P\{\phi_{\min}(b_n) \geq \lambda_0\} = 1$ for all $n$, where $\phi_{\min}(m) = \min_{M:|M|\leq m}\{\lambda_{\min}(n^{-1}\mathbf{X}_M^T\mathbf{X}_M)\}$. Further assume that $E(\varepsilon_1^4) < \infty$ and with probability tending to one, the selected models $\widehat{\mathbf{M}}_1$ and $\widehat{\mathbf{M}}_2$ include the true model $\mathbf{M}_* = \{j : \beta_j \neq 0\}$ with $\max\{|\widehat{\mathbf{M}}_1|, |\widehat{\mathbf{M}}_2|\} \leq b_n$. Then*

$$(\widehat{\sigma}^2_{RCV} - \sigma^2)/\sqrt{n} \to N(0, \mathrm{Var}(\varepsilon^2)) \tag{8.56}$$

*in distribution.*

Assumption (a) and the assumption on the finite fourth moment of $\varepsilon$ are natural. Assumption (b) implies that the variables selected in Stage 1 are not highly correlated. In the presence of highly correlated predictors in the selected variables, ridge regression or penalization methods can be applied in the refitted stage. Moreover, robust regression techniques can be utilized in the refitted state in the presence of heavy-tailed error distribution.

This theorem requires the model selection procedure used for obtaining $\widehat{M}_1$ and $\widehat{M}_2$ possesses a sure screening property rather than selection consistency. As shown in Theorem 8.1, the SIS procedure enjoys the sure screening property and can be used for obtaining $\widehat{M}_1$ and $\widehat{M}_2$.

This theorem implies that under certain conditions, the RCV estimate $\widehat{\sigma}^2_{\text{RCV}}$ and the oracle estimator $\sigma^2_O = \|\varepsilon\|^2/n$ share the same asymptotic normal distribution. It is of interest to compare the RCV estimator with error variance estimators $\widehat{\sigma}^2_L$, the mean squared errors of the Lasso estimate, and $\widehat{\sigma}^2_{\text{SCAD}}$, the mean squared errors of the SCAD estimate. Fan, Guo and Hao (2012) established the asymptotic normality of $\widehat{\sigma}^2_L$ and $\widehat{\sigma}^2_{\text{SCAD}}$. Under extra strong assumptions in additions to the assumptions in Theorem 8.8, $\widehat{\sigma}^2_L$ and $\widehat{\sigma}^2_{\text{SCAD}}$ may have the asymptotic normal distribution in (8.55). Simulation results in Fan, *et al.* (2012) imply that the impacts of different variable selection and feature screening procedures such as SIS and Lasso on the performance of RCV are negligible. Both $\widehat{\sigma}^2_L$ and $\widehat{\sigma}^2_{\text{SCAD}}$ may lead to an underestimate.

As mentioned in Fan, Guo and Hao (2012), there are two natural extensions of the aforementioned RCV techniques.

The first natural extension is to use a *K-fold data splitting* technique rather than twofold splitting. That is, one can randomly split the data into $K$ groups and select the model with all groups except one, which is used to estimate the variance with refitting. This may improve the sure screening probability with this $K$-fold method since more data are utilized in the first stage. However, there are only $[n/K]$ data points in the second stage for refitting. This means that the number of variables that are selected in the first stage should be much less than $[n/K]$. This makes the ability of sure screening difficult in the first stage. For this reason, the twofold RCV is recommended.

The second extension is to repeatedly and randomly split data. There are many ways to split the data randomly. Hence, many RCV variance estimators can be obtained. We may take the average of the resulting estimators. This reduces the influence of the randomness in the data splitting, while this may significantly increase the computational cost.

### *8.7.3 RCV in nonparametric regression*

The idea of the RCV method is applicable for nonparametric and semi-parametric regression models. This section focuses on high dimensional additive models introduced in Section 8.4.1. Chen, Fan and Li (2018) found that the spurious correlation may be much more severe in ultrahigh dimensional nonparametric regression models than in the linear model.

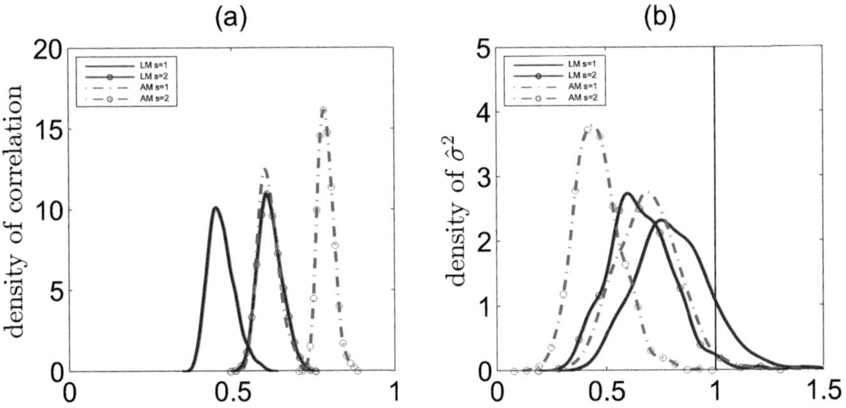


Figure 8.4: Distributions of the maximum "linear" and "nonparametric" spurious correlations for $s = 1$ and $s = 2$ (left panel, $n = 50$ and $p = 1000$) and their consequences on the estimating of noise variances (right panel). The legend 'LM' stands for linear model, and 'AM' stands for additive model, i.e., a nonparametric model. Adapted from Chen, Fan and Li (2018).

To illustrate the spurious correlation in the nonparametric regression setting, we generate a random sample $(\mathbf{X}_i^T, Y_i)$, $i = 1, \cdots, n = 50$ from $N_{p+1}(0, I_{p+1})$. As in Section 1.3.3, we compute the maximum "linear" spurious correlation $r_L = \max_{1 \le j \le p} |\widehat{\mathrm{corr}}(X_j, Y)|$ and the maximum "nonparametric" spurious correlation $r_N = \max_{1 \le j \le p} |\widehat{\mathrm{corr}}(\widehat{f}_j(X_j), Y)|$, where $\widehat{f}_j(X_j)$ is the best cubic spline fit of variable $X_j$ to the response $Y$, using 3 equally spaced knots in the range of the variable $X_j$ which create $d_n = 6$ B-spline bases for $X_j$. The concept of the maximum spurious "linear" and spurious "nonparametric" (additive) correlations can easily be extended to $s$ variables, which are the correlation between the response and fitted values using the best subset of $s$-variables. Based on 500 simulated data sets, Figure 8.4 depicts the results, which show the big increase of spurious correlations from linear to nonparametric fit. As a result, the noise variance is significantly underestimated.

Similar to Section 8.4.1, we can approximate $m_j(x) = \sum_{k=1}^{d_j} \gamma_{jk} B_{jk}(x)$, where $B_{jk}(\cdot)$, $k = 1, \cdots, d_j$ is a set of B-spline bases with knots depending on $j$. For ease of notation and without loss of generality, assume that all $d_j$'s are the same and are denoted by $d_n$. Replacing $m_j(X_j)$ by its spline approximation, we have

$$Y \approx \sum_{j=1}^{p} \sum_{k=1}^{d_n} \gamma_{jk} B_{jk}(X_j) + \varepsilon,$$

which can be viewed as an extended linear model. The strategy of the RCV method for linear models can be directly used for the ultrahigh dimensional additive model. Note that the dimension for this extended linear model is $d_n p$.

A naive estimate $\widehat{\sigma}^2$ can be described as follows. We first apply a feature screening procedure such as NIS and DC-SIS to the whole data set, and fit the data to the corresponding selected spline regression model, denoted by $\mathcal{D}^*$ the indices of all true predictors and $\widehat{\mathcal{D}}$ the indices of the selected predictors respectively, satisfying the sure screening property $\mathcal{D}^* \subset \widehat{\mathcal{D}}$. Then, we minimize

$$\sum_{i=1}^{n} \Big\{Y_i - \sum_{j\in\widehat{\mathcal{D}}} \sum_{k=1}^{d_n} \gamma_{jk} B_{jk}(X_{ij})\Big\}^2. \tag{8.57}$$

with respect to $\gamma$. Denote by $\widehat{\gamma}_{jk}$ the resulting least squares estimate. Then, the nonparametric residual variance estimator is

$$\widehat{\sigma}^2_{\widehat{\mathcal{D}}} = \frac{1}{n - |\widehat{\mathcal{D}}| \cdot d_n} \sum_{i=1}^{n} \Big\{Y_i - \sum_{j\in\widehat{\mathcal{D}}} \sum_{k=1}^{d_n} \widehat{\gamma}_{jk} B_{jk}(X_{ij})\Big\}^2.$$

Here we have implicitly assumed that the choice of $\widehat{\mathcal{D}}$ and $d_n$ is such that $n \geq |\widehat{\mathcal{D}}| \cdot d_n$. Let $\mathbf{P}_{\widehat{\mathcal{D}}}$ be the corresponding projection matrix of model (8.57) with the entire samples, denoted by $\widehat{\gamma}^2_n = \boldsymbol{\varepsilon}^T \mathbf{P}_{\widehat{\mathcal{D}}} \boldsymbol{\varepsilon} / \boldsymbol{\varepsilon}^T \boldsymbol{\varepsilon}$. Under a certain regularity condition, Chen, Fan and Li (2018) show that (8.55) is still valid, and

$$\widehat{\gamma}^2_n = O\Big(\big(\frac{2}{1-\delta}\big)^{|\widehat{\mathcal{D}}|} \frac{d_n \log(p d_n)}{n}\Big), \quad \text{for some } \delta \in (0,1). \tag{8.58}$$

These results imply that $\widehat{\sigma}^2_{\widehat{\mathcal{D}}}$ may significantly underestimates $\sigma^2$ due to spurious correlation.

The RCV method is to randomly split the random samples into two even data subsets denoted by $\mathcal{D}_1$ and $\mathcal{D}_2$. We apply a feature screening procedure (e.g., DC-SIS or NIS) for each set, and obtain two index sets of selected $x$-variables, denoted by $\widehat{\mathcal{D}}_1$ and $\widehat{\mathcal{D}}_2$. Both of them retain all important predictors. The refitted cross-validation procedure consists of three steps. In the first step, we fit data in $\mathcal{D}_l$ to the selected additive model $\widehat{\mathcal{D}}_{3-l}$ for $l = 1$ and 2 by the least squares method. These results in two least squares estimates $\widehat{\gamma}^{(3-l)}$ based on $\mathcal{D}_l$, respectively. In the second step, we calculate the mean squared errors for each fit:

$$\widehat{\sigma}^2_l = \frac{1}{n/2 - |\widehat{\mathcal{D}}_{3-l}| \cdot d_n} \sum_{i\in\mathcal{D}_l} \Big\{Y_i - \sum_{j\in\widehat{\mathcal{D}}_{3-l}} \sum_{k=1}^{d_n} \widehat{\gamma}^{(3-l)}_{jk} B_{jk}(X_{ij})\Big\}^2$$

for $l = 1$ and 2. As defined in (8.54), $\widehat{\sigma}^2_{\text{RCV}}$ is the average of $\widehat{\sigma}^2_1$ and $\widehat{\sigma}^2_2$. Under certain regularity conditions, Chen, Fan and Li (2018) showed that the asymptotic normality of $\widehat{\sigma}^2_{\text{RCV}}$ in (8.56) still holds.

## 8.8 An Illustration

This section is devoted to an illustration of the feature screening procedures and RCV method via an empirical analysis of a supermarket data set (Wang, 2009). This illustration is adapted from Chen, Fan and Li (2018). This supermarket data set consists of a total of $n = 464$ daily records of the number of customers ($Y_i$) and the sale amounts of $p = 6,398$ products, denoted as $X_{i1}, \cdots, X_{ip}$, which are used as predictors. For data confidentiality, both the response and predictors are marginally standardized so that they have zero sample mean and unit sample variance. We consider an additive model

$$Y_i = m_1(X_1) + \cdots + m_p(X_p) + \varepsilon_i,$$

where $\varepsilon_i$ is a random error with $\mathrm{E}(\varepsilon_i) = 0$ and $\mathrm{Var}(\varepsilon_i|\mathbf{X}_i) = \sigma^2$.

Table 8.1: Error variance estimate for market data. Adapted from Chen, Fan and Li (2018)

| $\widehat{s}$ | 40 | 35 | 30 | 28 | 25 |
|---|---|---|---|---|---|
| Naive | 0.0866 | 0.0872 | 0.0910 | 0.0938 | 0.0990 |
| RCV | 0.1245 | 0.1104 | 0.1277 | 0.1340 | 0.1271 |

In this example, we use DS-SIS for feature screening in the model selection stage since it is a model-free feature screening procedure and possesses the sure screening property. To illustrate the impact of the selected model size $\widehat{s}$ on the error variance estimation, we let $\widehat{s} = 25, 30, 35$ and 40. Since the sample size $n = 464$, the suggested threshold $\widehat{s} = [n^{4/5}/\log(n^{4/5})] = 28$ for nonparametric screening by Liu, Li and Wu (2014) is also considered. For these five model sizes, we obtain the naive error variance estimate and the RCV error variance estimate introduced in Section 8.7.3. Table 8.1 depicts the resulting error variance estimates. Table 8.1 clearly implies that the RCV estimate of error variance is stable with different choices of $\widehat{s}$, while the naive error variance estimate reduces as $\widehat{s}$ increases. With $\widehat{s} = 28$, the naive error variance estimate equals 0.0938, while the RCV error variance estimate equals 0.1340, an 43% increase of the estimated value when the spurious correlation is reduced.

Regarding the selected models with $\widehat{s} = 28$ predictors as a correct model and ignoring the approximation errors (if any) due to B-spline, we apply the Wald's $\chi^2$-test for the hypothesis whether $(\gamma_{j1}, \cdots, \gamma_{jd_j})^T$ equals zero, namely whether the $j^{th}$ variable is active when the rest of the variables are included. The Wald's $\chi^2$ statistics may provide us a rough picture of whether $X_j$ is active or not. The Wald's $\chi^2$-test with the naive error variance estimate concludes 12 significant predictors at significant level 0.05, while the Wald's $\chi^2$-test with the RCV error variance estimate concludes seven significant predictors at the same significance level. Figure 8.5 depicts the Q-Q plot of values of the $\chi^2$-test statistic of those insignificant predictors identified by the Wald's test.

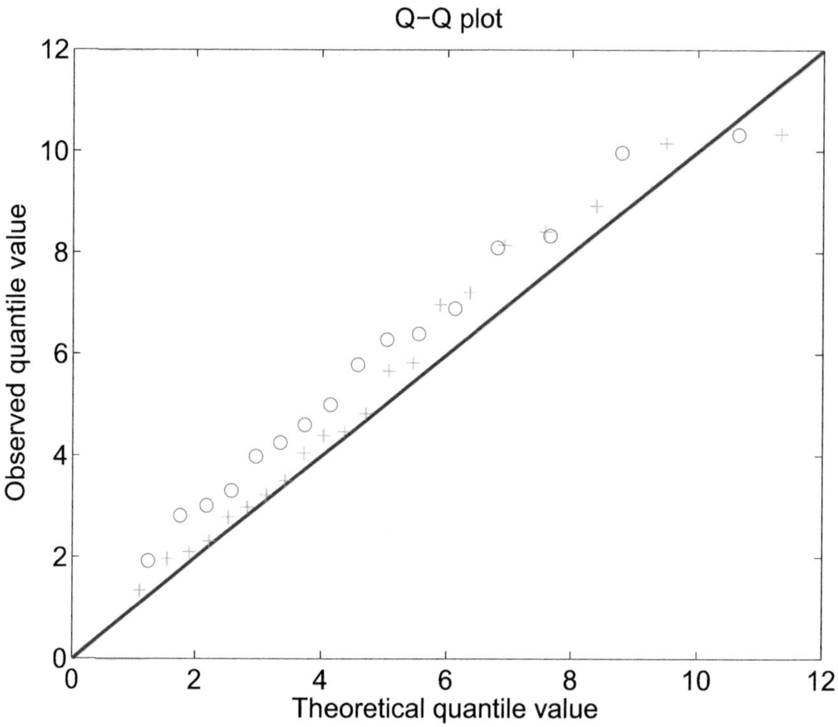


Figure 8.5: Quantile-quantile plot of the $\chi^2$-test values. "o" stands for $\chi^2$-test using the naive error variance estimate. "+" stands for $\chi^2$-test using the RCV error variance estimate. Taken from Chen, Fan and Li (2018).

Figure 8.5 clearly shows that the $\chi^2$-test values using the naive error variance estimate systematically deviate from the 45-degree line. This implies that the naive method results in an underestimate of error variance, while the RCV method results in a good estimate of error variance.

The Wald's test at level 0.05 affirms that seven predictors, $X_{11}$, $X_{139}$, $X_3$, $X_{39}$, $X_6$, $X_{62}$ and $X_{42}$, are significant. We refit the data with the additive model with these 7 predictors. The corresponding mean squared errors is 0.1207, which is close to the $\widehat{\sigma}^2_{\text{RCV}} = 0.1340$. Note that $\sigma^2$ is the minimum possible prediction error. It provides a benchmark for other methods to compare with and is achievable when modeling bias and estimation errors are negligible.

To see how the above selected variables perform in terms of prediction, we further use the leave-one-out cross-validation (CV) and five-fold CV to estimate the mean squared prediction errors (MSPE). The leave-one-out CV yields MSPE=0.1414, and the average of the MSPE obtained from five-fold CV based on 400 randomly splitting data yields is 0.1488 with the 2.5th

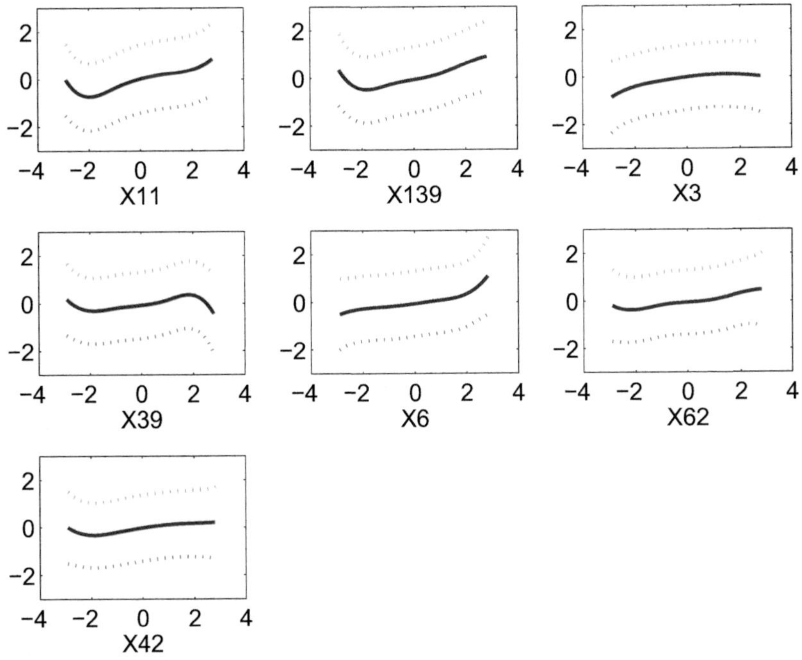


Figure 8.6: Estimated functions based on 7 variables selected from 28 variables that survive DC-SIS screening by the $\chi^2$-test with the RCV error variance estimator. Taken from Chen, Fan and Li (2018).

percentile and 97.5th percentile being 0.1411 and 0.1626, respectively. The MSPE is slightly greater than $\widehat{\sigma}^2_{\text{RCV}}$. This is expected as the uncertainty of parameter estimation has not been accounted for. This bias can be corrected with the theory of linear regression analysis.

For a linear regression model $\mathbf{Y} = \mathbf{X}^T\boldsymbol{\beta} + \varepsilon$, let $\widehat{\boldsymbol{\beta}}$ be the least squares estimate of $\boldsymbol{\beta}$. The linear predictor $\widehat{Y} = \mathbf{x}^T\widehat{\boldsymbol{\beta}}$ has prediction error at a new observation $\{\mathbf{x}_*, Y_*\}$: $\mathrm{E}\{(Y_* - \mathbf{x}_*^T\widehat{\boldsymbol{\beta}})^2|\mathbf{X}, \mathbf{x}^*\} = \sigma^2(1 + \mathbf{x}_*^T(\mathbf{X}^T\mathbf{X})^{-1}\mathbf{x}_*)$, where $\sigma^2$ is the error variance. This explains why the MSPE is slightly greater than $\widehat{\sigma}^2_{\text{RCV}}$. To further gauge the accuracy of the RCV estimate of $\sigma^2$, define the weighted prediction error $|y_* - \mathbf{x}_*^T\widehat{\boldsymbol{\beta}}|/\sqrt{1 + \mathbf{x}_*^T(\mathbf{X}^T\mathbf{X})^{-1}\mathbf{x}_*}$. Then the leave-one-out method leads to the mean squared weighted predictor error (MSWPE) 0.1289 and the average of the five-fold CV based on 400 randomly splitting data yields MSWPE 0.1305 with the 2.5th percentile and 97.5th percentile being 0.1254 and 0.1366, respectively. These results imply (a) the seven selected variables achieves the benchmark prediction; (b) the modeling biases using the additive models of these seven variables are negligible; (c) $\widehat{\sigma}^2_{\text{RCV}}$ provides a very good estimate for $\sigma^2$.

Their estimated functions $\widehat{m}_j(x_j)$ are depicted in Figure 8.6, from which

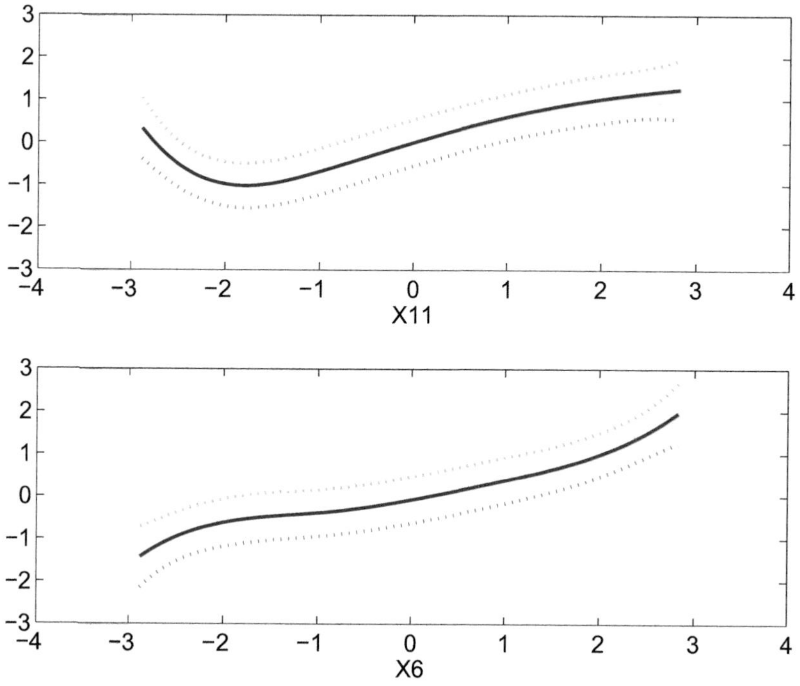


Figure 8.7: Estimated functions based on 2 variables selected from 28 variables that survive DC-SIS screening by the $\chi^2$-test with the RCV error variance estimator and the Bonferroni adjustment. Taken from Chen, Fan and Li (2018).

it seems that all predictors shown in Figure 8.6 are not significant since zero crosses the entire confidence interval. This can be due to the fact that we have used too many variables which increases the variance of the estimate.

We further employ the Wald's test with Bonferroni correction for 28 null hypotheses. This leads to only two significant predictors, $X_{11}$ and $X_6$, at level 0.05. We refit the data with the two selected predictors. Figure 8.7 depicts the plot of $\widehat{m}_{11}(x_{11})$ and $\widehat{m}_6(x_6)$.

## 8.9 Bibliographical Notes

Fan and Lv (2008) first proposed the concept of SIS for linear regression models. Hall and Miller (2009) proposed generalized correlation for feature screening and proposed using bootstrap to quantify the uncertainty of feature rankings. Li, Peng, Zhang and Zhu (2012) proposed a rank correlation based feature screening procedure to achieve the robustness. Fan and Fan (2008) apply the idea of SIS for binary classification problems based on a two-sample $t$-statistic. Pan, Wang and Li (2016) proposed a feature screening procedure for multiclass linear discriminant analysis based on partial residuals. Fan, Samworth and Wu (2009) developed a feature screening procedure under a general parametric framework. Fan and Song (2010) proposed using the marginal likelihood estimate for feature screening in generalized linear models. Fan, Feng and Wu (2012) extended the idea of Fan and Song (2010) to the partial likelihood setting for survival data. Zhao and Li (2012) proposed using the marginal $t$-statistic derived from marginal partial likelihood estimate for feature screening in high dimensional survival data. Xu, Zhu and Li (2014) developed a feature screening procedure for longitudinal data based on the generalized estimating equation method.

There are many publications on feature screening for nonparametric and semiparametric models. Fan, Feng and Song (2011) extended the SIS for high dimensional additive models by using spline regression techniques. Liu, Li and Wu (2014) developed a SIS procedure for varying coefficient models based on conditional Pearson correlation. Fan, Ma and Dai (2014) further used spline regression techniques for high dimensional varying coefficient models. Song, Yi and Zou (2014) and Chu, Li and Reimherr (2016) developed two feature screening procedures for varying coefficient models with longitudinal data. Cheng, Honda, Li and Peng (2014) developed an automatic procedure for finding a sparse semi-varying coefficient models in the longitudinal data analysis. Extending the forward regression for ultrahigh dimensional linear model in Wang (2009), Cheng, Honda and Zhang (2016) proposed using forward regression for feature screening in ultrahigh dimensional varying coefficient models and studied its theoretical property. Li, Ke and Zhang (2015) proposed using penalized likelihood methods for structure specification in high-dimensional generalized semi-varying coefficient models. Chu, Li, Liu and Reimherr (2020) developed a feature screening procedure for generalized varying coefficient mixed effect models.

Model-free feature screening was first proposed in Zhu, Li, Li and Zhu (2011) for multi-index regression models. Based on Zhu et al, (2011), Lin, Sun and Zhu (2013) proposed a nonparametric feature screening for multi-index models. He, Wang and Hong (2013) proposed a model-free feature screening procedure using quantile regression techniques. Ma, Li and Tsai (2017) further developed partial quantile correlation for feature screening. Li, Zhong and Zhu (2012) introduced ideas of a test of independence for feature screening and proposed a feature screening procedure based on distance correlation. Mai and Zou (2013) proposed using the Kolmogorov statistic for a test of independence in the literature of goodness of fit as a marginal utility for feature screening in high dimensional binary classification. Mai and Zou (2015) further developed the fused Kolmogorov filter for model-free feature screening with multicategory, counts and continuous response variables. Cui, Li and Zhong (2014) developed a statistic to measure the independence between a continuous random variable and a categorical random variable and further using this statistic as a marginal utility for feature screening in discriminant analysis. Huang, Li and Wang (2014) employed the Pearson $\chi^2$-test statistic for independence as a marginal utility for feature screening when the response and covariates are categorical. Wang, Liu, Li and Li (2017) developed a model-free conditional independence feature screening procedure and established the sure screening property and consistency in ranking property under a set of regularity conditions. Zhou, Zhu, Xu and Li (2019) developed a new metric named cumulative divergence (CD), and further proposed a CD-based forward screening procedure, which is also a model-free forward screening procedure.

The idea of using data splitting for improving statistical inference may go back to Cox (1975). Wasserman and Roeder (2009) suggested using data splitting in high dimensional variable selection to achieve valid statistical inference after model selection. Fan, Guo and Hao (2012) proposed refitted cross-validation for error variance estimation. Unlike the procedure in Wasserman and Roeder (2009), which uses only parts of data for statistical inference, the refitted cross-validation method uses all data for estimation of error variance. Chen, Fan and Li (2018) developed the refitted cross-validation method for high dimensional additive models.

## 8.10 Exercises

8.1 Construct an example of a linear regression model, in which for some $j$, $X_j$ is jointly dependent but marginally independent with the response $Y$.

8.2 Based on the relation between the Pearson correlation and the Kendall $\tau$ correlation under normality assumption, show that the screening procedures based on the Pearson correlation and the rank correlation are equivalent for ultrahigh dimensional normal linear models.

8.3 For ultrahigh dimensional linear regression models, show that under some conditions, the screening procedure in (8.12) is equivalent to the SIS based on the Pearson correlation proposed in Section 8.1. Further show the marginal utility $L_j$ defined in Section 8.3.2 is also equivalent to the SIS based on the Pearson correlation for ultrahigh dimensional linear models.

8.4 Let $Y$ be the response along with a continuous covariate $U$ and covariate vector $\mathbf{X} = (X_1, \cdots, X_p)^T$. Partial linear model is defined by

$$Y = \beta_0(U) + \beta_1 X_1 + \cdots + \beta_p X_p + \epsilon,$$

where $\beta_0(\cdot)$ is the unknown baseline function. Construct a sure independent screening procedure by applying related techniques to screening procedures for varying coefficient models studied in Section 8.4.2.

8.5 Construct sure screening procedures in Section 8.5.3 for the settings in which the response $Y$ is continuous, while $\mathbf{X}$ consists of categorical covariates such as SNPs in a genome wise association study.

8.6 Refer to Section 8.6.2. Show that $h(\boldsymbol{\beta};\boldsymbol{\beta}^{(t)}) = \ell_n(\boldsymbol{\beta})$ and $h(\boldsymbol{\beta};\boldsymbol{\beta}^{(t)}) \geq \ell_n(\boldsymbol{\beta})$ for any $\boldsymbol{\beta}$ with a proper chosen $u$. Further establish the ascent property of the IHT algorithm.

8.7 Prove Theorem 8.5.

8.8 A cardiomyopathy microarray data set can be downloaded from this book website. This data set was once analyzed by Segal, Dahlquist and Conklin (2003), Hall and Miller (2009) and Li, Zhong and Zhu (2012). The data set consists of $n = 30$ samples, each of which has $p = 6319$ gene expressions. The goal is to identify the most influential genes for overexpression of a G protein-coupled receptor (Ro1) in mice. The response Y is the Ro1 expression level, and the predictors $X_k$s are other gene expression levels.

(a) Apply the Pearson correlation marginal screening procedure introduced in Section 8.1 to this data set.

(b) Apply the generalized correlation marginal screening procedure introduced in Section 8.2 to this data set.

(c) Apply the DC-SIS procedure introduced in Section 8.5.2 to this data.

(d) Comment upon your findings in (a), (b) and (c).

8.9 The data set 'RatEyeExpression.txt' consists of microarray data of tissue harvesting from the eyes of a 120 twelve-week-old male rat. The microarrays are used to analyze the RNA from the eyes of these rats. There are 18,976 probes for each rat in this data set, which was used by Scheetz, et al. (2006) and Huang, Ma and Zhang (2008), and re-analyzed by Li, Zhong and Zhu (2012) for illustrating their iterative DC-SIS. The data can be downloaded from this book's website. Following Huang, Ma and Zhang (2008), take the response variable to be TRIM32, which is one of the selected 18,976 probes and corresponds to the row with probe.name "1389163_at".

(a) Select 3,000 probes with the largest variances from the 18,975 probes.

(b) Apply the DC-SIS in Section 8.5.2 to the regression model with the response TRIM32 and the selected 3,000 covariates.

(c) Apply the iterative DC-SIS in Section 8.5.2 to the regression model with the response TRIM32 and the selected 3,000 covariates

(d) Compare your findings in (b) and (c).

Chapter 9

# Covariance Regularization and Graphical Models

Covariance and precision matrix estimation and inference pervade every aspect of statistics and machine learning. They appear in the Fisher's discriminant and optimal portfolio choice, in Gaussian graphical models where edge represents conditional dependence, and in statistical inference such as Hotelling $T^2$-test and false discovery rate controls. They have also widely been applied to various disciplines such as economics, finance, genomics, genetics, psychology, and computational social sciences.

Covariance learning, consisting of unknown elements of square order of dimensionality, is ultra-high dimension in nature even when the dimension of data is moderately high. It is necessary to impose some structure assumptions in covariance learning such as sparsity or bandability. Factor models offer another useful approach to covariance learning from high-dimensional data. We can infer the latent factors that drive the dependence of observed variables. These latent factors can then be used to predict the outcome of responses, reducing the dimensionality. They can also be used to adjust the dependence in model selection, achieving better model selection consistency and reducing spurious correlation. They can further be used to adjust the dependence in large scale inference, achieving better false discovery controls and enhancing high statistical power. These applications will be outlined in Chapter 11.

This chapter introduces theory and methods for estimating large covariance and precision matrices. We will also introduce robust covariance inputs to be further regularized. In Chapter 10, we will focus on estimating a covariance matrix induced by factor models. We will show there how to extract latent factors, factor loadings and idiosyncratic components. They will be further applied to high-dimensional statistics and statistical machine learning problems in Chapter 11.

## 9.1 Basic Facts about Matrices

We here collect some basic facts and introduce some notations. They will be used throughout this chapter.

For a symmetric $m \times m$ matrix $\mathbf{A}$, we have its *eigen-decomposition* as follows

$$\mathbf{A} = \mathbf{V}\boldsymbol{\Lambda}\mathbf{V}^T = \sum_{i=1}^{m} \lambda_i \mathbf{v}_i \mathbf{v}_i^T, \tag{9.1}$$

where $\lambda_i$ is the eigenvalue of $\mathbf{A}$ and $\mathbf{v}_i$ is its associated eigenvector. For a $m \times n$ matrix, it admits the *singular value decomposition*

$$\mathbf{A} = \mathbf{U}\boldsymbol{\Lambda}\mathbf{V}^T = \sum_{i=1}^{\min(m,n)} \lambda_i \mathbf{u}_i \mathbf{v}_i^T \tag{9.2}$$

where $\mathbf{U}$ is an $m \times m$ unitary (othornormal) matrix, $\boldsymbol{\Lambda}$ is an $m \times n$ rectangular diagonal matrix, and $\mathbf{V}$ is an $n \times n$ unitary matrix. The diagonal entries $\lambda_i$ of $\boldsymbol{\Lambda}$ are called the *singular values* of $\mathbf{A}$. The columns of $\mathbf{U}$ and the columns of $\mathbf{V}$ are known as the *left-singular* vectors and *right-singular* vectors of $\mathbf{A}$, respectively:

- the left-singular vectors of $\mathbf{A}$ are orthonormal eigenvectors of $\mathbf{A}\mathbf{A}^T$;
- the right-singular vectors of $\mathbf{A}$ are orthonormal eigenvectors of $\mathbf{A}^T\mathbf{A}$;
- the non-vanishing singular values of $\mathbf{A}$ are the square roots of the non-vanishing eigenvalues of both $\mathbf{A}^T\mathbf{A}$ and $\mathbf{A}\mathbf{A}^T$.

We will assume the singular values are ordered in decreasing order and use $\lambda_i(\mathbf{A})$ to denote the $i^{th}$ largest singular value, $\lambda_{\max}(\mathbf{A}) = \lambda_1(\mathbf{A})$ the largest singular value, and $\lambda_{\min}(\mathbf{A}) = \lambda_m(\mathbf{A})$ the minimum eigenvalue.

**Definition 9.1** *A norm of an $n \times m$ matrix satisfies*

1. $\|\mathbf{A}\| \geq 0$ *and* $\|\mathbf{A}\| = 0$ *if and only if* $\mathbf{A} = 0$;
2. $\|\alpha\mathbf{A}\| = |\alpha|\|\mathbf{A}\|$;
3. $\|\mathbf{A} + \mathbf{B}\| \leq \|\mathbf{A}\| + \|\mathbf{B}\|$;
4. $\|\mathbf{A}\mathbf{B}\| \leq \|\mathbf{A}\|\|\mathbf{B}\|$, *if* $\mathbf{A}\mathbf{B}$ *is properly defined.*

A common method to define a matrix norm is the *induced norm*: $\|\mathbf{A}\| = \max_{\|\mathbf{x}\|=1} \|\mathbf{A}\mathbf{x}\|$. In particular, $\|\mathbf{A}\|_{p,q} = \max_{\|\mathbf{x}\|_p=1} \|\mathbf{A}\mathbf{x}\|_q$, which satisfies, by definition,

$$\|\mathbf{A}\mathbf{x}\|_q \leq \|\mathbf{A}\|_{p,q} \, \|\mathbf{x}\|_p.$$

More specifically, when $p = q$, it becomes the matrix $L_p$ norm:

$$\|\mathbf{A}\|_p = \|\mathbf{A}\|_{p,p} = \max_{\|\mathbf{x}\|_p=1} \|\mathbf{A}\mathbf{x}\|_p.$$

It includes, as its specific cases,

1. Operator norm: $\|\mathbf{A}\|_2 = \lambda_{\max}(\mathbf{A}^T\mathbf{A})^{1/2}$, which is also written as $\|\mathbf{A}\|_{op}$ or simply $\|\mathbf{A}\|$;
2. *$L_1$-norm*: $\|\mathbf{A}\|_1 = \max_j \sum_{i=1}^m |a_{ij}|$, maximum $L_1$-norm of columns;
3. *$L_\infty$-norm*: $\|\mathbf{A}\|_\infty = \max_i \sum_{j=1}^n |a_{ij}|$, maximum $L_1$-norm of rows.

**Definition 9.2** *The Frobenius norm is defined as*

$$\|\mathbf{A}\|_F^2 = \mathrm{tr}(\mathbf{A}^T\mathbf{A}) = \sum_{i=1}^{m}\sum_{j=1}^{n} a_{i,j}^2 = \lambda_1^2 + \cdots + \lambda_{\min(m,n)}^2,$$

*and the nuclear norm is given by*

$$\|\mathbf{A}\|_* = |\lambda_1| + \cdots + |\lambda_{\min(m,n)}|,$$

*where* $\{\lambda_i\}$ *are singular values defined in* (9.2).

Operator norm is hard to manipulate, whereas $\|\mathbf{A}\|_\infty$ and $\|\mathbf{A}\|_1$ are easier. We have the following inequalities.

**Proposition 9.1** *For any* $m \times n$ *matrix* $\mathbf{A}$*, we have*

1. $\|\mathbf{A}\|_2^2 \le \|\mathbf{A}\|_\infty \|\mathbf{A}\|_1$ *and* $\|\mathbf{A}\|_2 \le \|\mathbf{A}\|_1$ *if* $\mathbf{A}$ *is symmetric.*
2. $\|\mathbf{A}\|_{\max} \le \|\mathbf{A}\|_2 \le (mn)^{1/2}\|\mathbf{A}\|_{\max}$, $\quad \|\mathbf{A}\|_{\max} = \max_{ij}|a_{ij}|$
3. $\|\mathbf{A}\|_2 \le \|\mathbf{A}\|_F \le \sqrt{r}\|\mathbf{A}\|_2$, $\quad r = rank(\mathbf{A})$;
4. $\|\mathbf{A}\|_F \le \|\mathbf{A}\|_* \le \sqrt{r}\|\mathbf{A}\|_F$, $\quad r = rank(\mathbf{A})$;
5. $n^{-1/2}\|\mathbf{A}\|_\infty \le \|\mathbf{A}\|_2 \le m^{1/2}\|\mathbf{A}\|_\infty$ *and* $m^{-1/2}\|\mathbf{A}\|_1 \le \|\mathbf{A}\|_2 \le n^{1/2}\|\mathbf{A}\|_1$

Here are two basic results about *matrix perturbation*. For two $m \times m$ symmetric matrices $\mathbf{A}$ and $\widehat{\mathbf{A}}$, let $\lambda_1 \ge \cdots \ge \lambda_m$ and $\widehat{\lambda}_1 \ge \cdots \ge \widehat{\lambda}_m$ be their eigenvalues with associated eigenvectors $\mathbf{v}_1, \cdots, \mathbf{v}_m$ and $\widehat{\mathbf{v}}_1, \cdots, \widehat{\mathbf{v}}_m$, respectively. For two given integers $r$ and $s$, write $d = s - r + 1$,

$$\mathbf{V} = (\mathbf{v}_r, \mathbf{v}_{r+1}, \cdots, \mathbf{v}_s) \quad \text{and} \quad \widehat{\mathbf{V}} = (\widehat{\mathbf{v}}_r, \widehat{\mathbf{v}}_{r+1}, \cdots, \widehat{\mathbf{v}}_s).$$

and let $\sigma_1 \ge \cdots \ge \sigma_d$ be the singular values of $\widehat{\mathbf{V}}^T\mathbf{V}$. The $d$ *principal angles* between their column spaces $\mathbf{V}$ and $\widehat{\mathbf{V}}$ are defined as $\{\theta_j\}_{j=1}^d$, where $\theta_j = \arccos\sigma_j$. Let

$$\sin\mathbf{\Theta}(\widehat{\mathbf{V}}, \mathbf{V}) = \mathrm{diag}(\sin\theta_1, \cdots, \sin\theta_d)$$

be the $d \times d$ diagonal matrix. Note that the Frobenius norm of two projection matrices

$$\begin{aligned}\|\widehat{\mathbf{V}}\widehat{\mathbf{V}}^T - \mathbf{V}\mathbf{V}^T\|_F^2 &= 2d - 2\|\widehat{\mathbf{V}}^T\mathbf{V}\|_F^2 = 2\sum_{i=1}^d (1-\sigma_i^2) = 2\sum_{i=1}^d \sin^2(\theta_i) \\ &= 2\|\sin\mathbf{\Theta}(\widehat{\mathbf{V}}, \mathbf{V})\|_F^2. \end{aligned} \tag{9.3}$$

**Proposition 9.2** *For two* $m \times m$ *symmetric matrices* $\mathbf{A}$ *and* $\widehat{\mathbf{A}}$*, we have*

- *Weyl Theorem:* $\lambda_{\min}(\widehat{\mathbf{A}} - \mathbf{A}) \le \widehat{\lambda}_i - \lambda_i \le \lambda_{\max}(\widehat{\mathbf{A}} - \mathbf{A})$. *Consequently,* $|\widehat{\lambda}_i - \lambda_i| \le \|\widehat{\mathbf{A}} - \mathbf{A}\|$.

- *Davis-Kahan sin θ Theorem: With* $\widehat{\lambda}_0 = \infty$, $\widehat{\lambda}_{m+1} = -\infty$ *and* $\delta = \min\{|\widehat{\lambda} - \lambda| : \lambda \in [\lambda_s, \lambda_r], \widehat{\lambda} \in (-\infty, \widehat{\lambda}_{s-1}] \cup [\widehat{\lambda}_{r+1}, \infty)\}$ *denoting the eigen gap, we have*
$$\|\sin \mathbf{\Theta}(\widehat{\mathbf{V}}, \mathbf{V})\|_F \le \|\widehat{\mathbf{A}} - \mathbf{A}\|_F/\delta$$
*The result also holds for the operator norm or any other norm that is invariant under an orthogonal transformation.*

The above result is due to Davis and Kahan (1970). Specifically, by taking $r = s = i$, we have

$$\sin\Theta(\widehat{\mathbf{v}}_i, \mathbf{v}_i) \le \frac{\|\widehat{\mathbf{A}} - \mathbf{A}\|_2}{\min(|\widehat{\lambda}_{i-1} - \lambda_i|, |\lambda_i - \widehat{\lambda}_{i+1}|)}.$$

Note that

$$\begin{aligned} \|\widehat{\mathbf{v}}_i - \mathbf{v}_i\|^2 &= 2 - 2\widehat{\mathbf{v}}_i^T \mathbf{v}_i \\ &= 2[1 - \cos\Theta(\widehat{\mathbf{v}}_i, \mathbf{v}_i)] \\ &\le 2[1 - \cos^2\Theta(\widehat{\mathbf{v}}_i, \mathbf{v}_i)] \\ &= 2\sin^2\Theta(\widehat{\mathbf{v}}_i, \mathbf{v}_i). \end{aligned}$$

Combining the last two results, we have

$$\|\widehat{\mathbf{v}}_i - \mathbf{v}_i\| \le \frac{\sqrt{2}\|\widehat{\mathbf{A}} - \mathbf{A}\|_2}{\min(|\widehat{\lambda}_{i-1} - \lambda_i|, |\lambda_i - \widehat{\lambda}_{i+1}|)}.$$

The *eigen gap* $\delta$ involves both sets of eigenvalues. In statistical applications, it typically replaces $\widehat{\lambda}_i's$ by its $\lambda_i$'s using Weyl's theorem. Yu, Wang and Samworth (2014) give a simple calculation and obtain

$$\|\sin \mathbf{\Theta}(\widehat{\mathbf{V}}, \mathbf{V})\|_F \le \frac{2\max(d^{1/2}\|\widehat{\mathbf{A}} - \mathbf{A}\|_2, \|\widehat{\mathbf{A}} - \mathbf{A}\|_F)}{\min(\lambda_{r-1} - \lambda_r, \lambda_s - \lambda_{s+1})}, \tag{9.4}$$

and in particular

$$\|\widehat{\mathbf{v}}_i - \mathbf{v}_i\| \le \frac{2\|\widehat{\mathbf{A}} - \mathbf{A}\|_2}{\min(|\lambda_{i-1} - \lambda_i|, |\lambda_i - \lambda_{i+1}|)}. \tag{9.5}$$

We conclude this section by introducing *Sherman-Morrison-Woodbury Formula* and block matrix inversion. If $\mathbf{\Sigma} = \mathbf{B}\mathbf{\Sigma}_f\mathbf{B}^T + \mathbf{\Sigma}_u$, then

$$\mathbf{\Sigma}^{-1} = \mathbf{\Sigma}_u^{-1} - \mathbf{\Sigma}_u^{-1}\mathbf{B}[\mathbf{\Sigma}_f^{-1} + \mathbf{B}^T\mathbf{\Sigma}_u^{-1}\mathbf{B}]^{-1}\mathbf{B}^T\mathbf{\Sigma}_u^{-1}. \tag{9.6}$$

Let $\mathbf{S} = \mathbf{A} - \mathbf{B}\mathbf{D}^{-1}\mathbf{C}$ be the Schur complement of $\mathbf{D}$. Then

$$\begin{pmatrix} \mathbf{A} & \mathbf{B} \\ \mathbf{C} & \mathbf{D} \end{pmatrix}^{-1} = \begin{pmatrix} \mathbf{S}^{-1} & -\mathbf{S}^{-1}\mathbf{B}\mathbf{D}^{-1} \\ -\mathbf{D}^{-1}\mathbf{C}\mathbf{S}^{-1} & \mathbf{D}^{-1} + \mathbf{D}^{-1}\mathbf{C}\mathbf{S}^{-1}\mathbf{B}\mathbf{D}^{-1} \end{pmatrix}.$$

## 9.2 Sparse Covariance Matrix Estimation

To estimate high-dimensional covariance and precision matrices, we need to impose some covariance structure. Conditional sparsity was imposed in Fan, Fan, and Lv (2008) for estimating a high-dimensional covariance matrix and their applications to portfolio selection and risk estimation. This *factor model* induces a low-rank plus sparse covariance structure (Wright, Ganesh, Rao, Peng, and Ma 2009; Candés, Li, Ma and Wright, 2011), which includes (unconditional) sparsity structure imposed by Bickel and Levina (2008b) as a specific example. Let us begin with the sparse covariance matrix estimation.

### 9.2.1 Covariance regularization by thresholding and banding

Let us first consider the Gaussian model where we observed data $\{\mathbf{X}_i\}_{i=1}^n$ from $p$-dimensional Gaussian distribution $N(\boldsymbol{\mu}, \boldsymbol{\Sigma})$. Natural estimators for $\boldsymbol{\mu}$ and $\boldsymbol{\Sigma}$ are the sample mean vector and sample covariance matrix

$$\widehat{\boldsymbol{\mu}} = \frac{1}{n}\sum_{i=1}^n \mathbf{X}_i, \qquad \mathbf{S} = \frac{1}{n}\sum_{i=1}^n (\mathbf{X}_i - \widehat{\boldsymbol{\mu}})(\mathbf{X}_i - \widehat{\boldsymbol{\mu}})^T. \tag{9.7}$$

They are the maximum likelihood estimator when $p \le n-1$. However, the sample covariance is degenerate when $p \ge n$. Since we estimate $O(p^2)$ elements[1], the *noise accumulation* will make the sample covariance matrix differs substantially from the population one. See Figure 9.1. It shows that the sample covariance matrix is not a good estimator, without any regularization. This is well documented in the *random matrix* theory (Bai and Silverstein, 2009).

To make the problem estimable in high-dimension, we need to impose a structure on the *covariance matrix*. The simplest structure is the sparsity of the covariance $\boldsymbol{\Sigma}$. This leads to a penalized likelihood method. The log-likelihood for the observed Gaussian data is

$$-\frac{n}{2}\log(2\pi) - \frac{n}{2}\log|\boldsymbol{\Sigma}| - \frac{1}{2}\sum_{i=1}^n (\mathbf{X}_i - \boldsymbol{\mu})^T \boldsymbol{\Sigma}^{-1}(\mathbf{X}_i - \boldsymbol{\mu}).$$

Substitution $\boldsymbol{\mu}$ by its MLE $\bar{\mathbf{X}}$, expressing $(\mathbf{X}_i - \bar{\mathbf{X}})\boldsymbol{\Sigma}^{-1}(\mathbf{X}_i - \bar{\mathbf{X}})^T$ as $\mathrm{tr}(\boldsymbol{\Sigma}^{-1}(\mathbf{X}_i - \bar{\mathbf{X}})(\mathbf{X}_i - \bar{\mathbf{X}})^T)$ where $\mathrm{tr}(\mathbf{A})$ is the trace of a matrix $\mathbf{A}$, we obtain the negative profile log-likelihood

$$Q(\boldsymbol{\Sigma}) = -\log|\boldsymbol{\Sigma}^{-1}| + \mathrm{tr}(\boldsymbol{\Sigma}^{-1}\mathbf{S}), \tag{9.8}$$

after ignoring the constant term $-\frac{n}{2}\log(2\pi)$ and dividing the result by $-n/2$. Thus, the penalized likelihood estimator is to minimize

$$-\log|\boldsymbol{\Sigma}^{-1}| + \mathrm{tr}(\boldsymbol{\Sigma}^{-1}\mathbf{S}) + \sum_{i \ne j} p_{\lambda_{ij}}(|\sigma_{ij}|). \tag{9.9}$$

[1] $1,000 \times 1,000$ matrix involves over half a million elements, whereas $10,000 \times 10,000$ contains over 50 million elements!

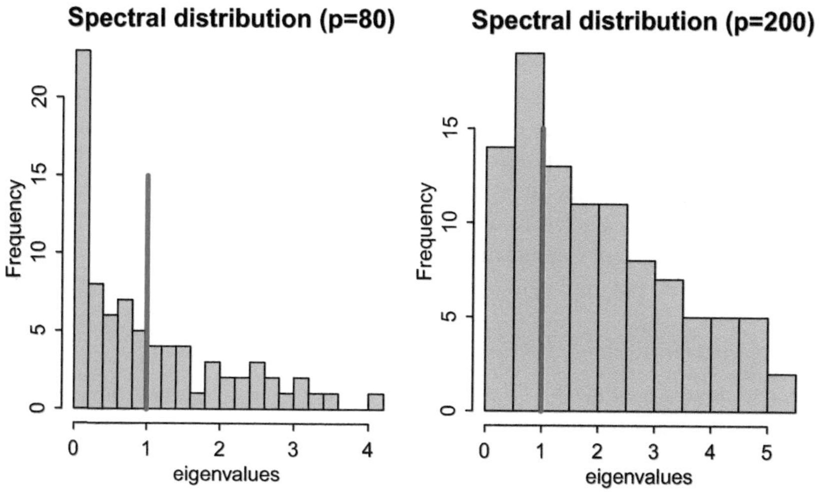


Figure 9.1: Empirical distributions of the eigenvalues of the data $\{\mathbf{X}_i\}_{i=1}^n$ simulated from $N(0, I_p)$ for $n = 100$ and $p = 80$ (left panel) and $p = 200$ (right panel). For the case $p = 200$, there are 100 empirical eigenvalues that are zero, which are not plotted in the figure. Note that the theoretical eigenvalues are 1 for the identity matrix $\mathbf{I}_p$. Theoretical and empirical spectral distributions are very different when $n$ and $p$ are of the same order.

Here, we do not penalize on the diagonal elements, since they are not supposed to be sparse.

The *penalized likelihood estimator* is unfortunately hard to compute because (9.9) is a non-convex problem even if the penalty function is convex. Bickel and Levina (2008b) explored the sparsity by a simple thresholding operator:

$$\widehat{\mathbf{\Sigma}}_\lambda = \left(\widehat{\sigma}_{i,j}\mathbb{I}(|\widehat{\sigma}_{i,j}| \geq \lambda)\right) \tag{9.10}$$

for a given *threshold parameter* $\lambda$, where $\widehat{\sigma}_{ij} = s_{ij}$. In other words, we kill small entries of the sample covariance matrix, whose value is smaller than $\lambda$. Again, in (9.10), we do not apply the thresholding to the diagonal elements. The thresholding estimator (9.10) can be regarded as a penalized least-squares estimator: Minimize with respect to $\sigma_{ij}$

$$\sum_{i,j}(\sigma_{ij} - s_{ij})^2 + \sum_{i \neq j} p_\lambda(|\sigma_{ij}|) \tag{9.11}$$

for the hard thresholding penalty function (3.15). Allowing other penalty functions in Sections 3.2.1 and 3.2.2 leads to *generalized thresholding* rules (Antoniadis and Fan, 2001; Rothman, Levina and Zhu, 2009).

**Definition 9.3** *The function $\tau_\lambda(\cdot)$ is called a generalized thresholding function, if it satisfies* *a)* $|\tau_\lambda(z)| \le a|y|$ *for all* $z$, $y$ *that satisfy* $|z-y| \le \lambda/2$ *and some* $a > 0$;

*b)* $|\tau_\lambda(z) - z| \le \lambda$, *for all* $z \in \mathbb{R}$.

Note that by taking $y = 0$, we have $\tau_\lambda(z) = 0$ for $|z| \le \lambda/2$ and hence it is a thresholding function. It is satisfied for the soft-thresholding function by taking $a = 1$ and the hard-thresholding function by taking $a = 2$. It is also satisfied by other penalty functions in Sections 3.2.2.

The simple thresholding (9.10) does not even take the varying scales of the covariance matrix into account. One way to account for this is to threshold on $t$-type statistics. For example, we can define the *adaptive thresholding* estimator (Cai and Liu, 2011) as

$$\widehat{\mathbf{\Sigma}}_\lambda = \Big(\widehat{\sigma}_{i,j}\mathbb{I}(|\widehat{\sigma}_{i,j}|/\mathrm{SE}(\widehat{\sigma}_{i,j}) \ge \lambda)\Big),$$

where $\mathrm{SE}(\widehat{\sigma}_{i,j})$ is the estimated standard error of $\widehat{\sigma}_{i,j}$.

A simpler method to take the scale into account is to apply thresholding on the correlation matrix (Fan, Liao and Mincheva, 2013). Let $\widehat{\mathbf{\Psi}}_\lambda$ be the thresholded correlation matrix with thresholding parameter $\lambda \in [0, 1]$. Then,

$$\widehat{\mathbf{\Sigma}}^*_\lambda = \mathrm{diag}(\widehat{\mathbf{\Sigma}})^{1/2}\widehat{\mathbf{\Psi}}_\lambda\,\mathrm{diag}(\widehat{\mathbf{\Sigma}})^{1/2} \tag{9.12}$$

is an estimator of the covariance matrix. In particular, when $\lambda = 0$, it is the sample covariance matrix, whereas when $\lambda = 1$, it is the diagonal matrix, consisting of the sample variances. The estimator (9.12) is equivalent to applying the *entry dependent thresholding* $\lambda_{ij} = \sqrt{\widehat{\sigma}_{i,i}\widehat{\sigma}_{j,j}}\;\lambda$ to the sample covariance matrix $\widehat{\mathbf{\Sigma}}$. This leads to the more general estimator

$$\widehat{\mathbf{\Sigma}}^\tau_\lambda = \Big(\tau_{\lambda_{ij}}(\widehat{\sigma}_{ij})\Big) \equiv \Big(\widehat{\sigma}^\tau_{ij}\Big), \qquad \lambda_{ij} = \lambda\sqrt{\widehat{\sigma}_{i,i}\widehat{\sigma}_{j,j}\frac{\log p}{n}}. \tag{9.13}$$

where $\tau_{\lambda_{ij}}(\cdot)$ is a generalized thresholding function, $(\widehat{\sigma}_{ij})$ is a generic estimator of covariance matrix $\mathbf{\Sigma}$, and $\sqrt{(\log p)/n}$ is its uniform rate of convergence, which can be more generally denoted by $a_n$ (see Remark 9.1). In (9.13), we use $\lambda\sqrt{\log p/n}$ to denote the level of thresholding to facilitate future presentations.

An advantage of thresholding is that it avoids estimating small elements so that noises do not accumulate as much. The decision of whether an element should be estimated is much easier than the attempt to estimate it accurately. Thresholding creates biases. Since the matrix $\mathbf{\Sigma}$ is sparse, the biases are also controlled. This is demonstrated in the following section.

Bickel and Levina (2008a) considered the estimation of a large covariance matrix with banded structure in which $\sigma_{ij}$ is small when $|i-j|$ is large. This occurs for time series measurements where observed data are weakly correlated when they are far apart. They also appear in the spatial statistics in which correlation decays with the distance where measurements are taken. They proposed a simple banded estimator as

$$\mathbf{B}_k(\widehat{\mathbf{\Sigma}}) = \Big(\widehat{\sigma}_{ij}\mathbb{I}(|i-j| \le k)\Big)$$

and derived the convergence rate. Cai, Zhang, and Zhou (2010) established the minimax optimal rates of convergence for estimating large banded matrices. See Bien, Bunea, and Xiao (2016) for convex banding, Yu and Bien (2017) for varying banding, and Bien (2018) for graph-guided banding. Li and Zou (2016) propose two Stein's Unbiased Risk Estimation (SURE) criteria for selecting the banding parameter $k$ and show that one SURE criterion is risk optimal and the other SURE criterion is selection consistent if the true covariance matrix is banded.

### *9.2.2 Asymptotic properties*

To study the estimation error in the operator norm, we note, by Proposition 9.1, that for any general thresholding estimator (9.13),

$$\|\widehat{\mathbf{\Sigma}}_\lambda^\mathcal{T} - \mathbf{\Sigma}\|_2 \le \max_i \sum_{j=1}^p |\widehat{\sigma}_{ij}^\mathcal{T} - \sigma_{ij}|. \tag{9.14}$$

Thus, we need to control only the $L_1$-norm in each row, whose error depends on the effective number of parameters, i.e. the maximum number of nonzero elements in each row

$$m_{p,0} = \max_{i \le p} \sum_{j=1}^p \mathbb{I}\{\sigma_{ij} \ne 0\}.$$

A generalization of the exact sparsity measure $m_0$ is the approximate sparsity measure

$$m_{p,q} = \max_{i \le p} \sum_{j=1}^p |\sigma_{ij}|^q, \qquad q < 1. \tag{9.15}$$

Note that $q = 0$ corresponds to exact sparsity $m_{p,0}$. Bickel and Levina (2008b) considered the parameter (covariance) space

$$\{\mathbf{\Sigma} \succeq 0 : \sigma_{ii} \le C, \ \sum_{j=1}^p |\sigma_{ij}|^q \le m_p\},$$

with generalized sparsity controlled at a given level $m_p$. As a result, by using the Cauchy-Schwartz inequality $|\sigma_{ij}|^2 \le \sigma_{ii}\sigma_{jj}$, we have

$$\|\mathbf{\Sigma}\|_2 \le \max_i \sum_j |\sigma_{ij}| \le \max_i \sum_j (\sigma_{ii}\sigma_{jj})^{(1-q)/2} |\sigma_{ij}|^q \le C^{1-q} m_p.$$

This leads to the generalization of the parameter space to (Cai and Liu, 2011)

$$\mathcal{C}_q(m_p) = \Big\{\mathbf{\Sigma} : \ \max_i \sum_j (\sigma_{ii}\sigma_{jj})^{(1-q)/2} |\sigma_{ij}|^q \le C^{1-q} m_p\Big\}. \tag{9.16}$$

The following result is very general. It encompasses those of Bickel and Levina (2008b) and Cai and Liu (2011).

**Theorem 9.1** *Suppose that the pilot estimator $\widehat{\mathbf{\Sigma}}$ of $\mathbf{\Sigma}$ satisfies*

$$\sup_{\mathbf{\Sigma}\in \mathcal{C}_q(m_p)} P\big(\|\widehat{\mathbf{\Sigma}} - \mathbf{\Sigma}\|_{\max} > C_0\sqrt{(\log p)/n}\big) \le \epsilon_{n,p}, \tag{9.17}$$

*where $C_0$ is a positive constant. If $\log p = o(n)$ and $\min_{i\le p}\sigma_{ii} = \gamma > 0$, there exists a positive constant $C_1$ such that for sufficiently large $n$ and constant $\lambda$, we have*

$$\sup_{\mathbf{\Sigma}\in \mathcal{C}_q(m_p)} P\Big\{\|\widehat{\mathbf{\Sigma}}_\lambda^{\mathcal{T}} - \mathbf{\Sigma}\|_2 > C_1 m_p\Big(\frac{\log p}{n}\Big)^{(1-q)/2}\Big\} \le 3\epsilon_{n,p}.$$

*A similar result also holds uniformly over $\mathcal{C}_q(m_p)$ under the Frobenius norm:*

$$p^{-1}\|\widehat{\mathbf{\Sigma}}_\lambda^{\mathcal{T}} - \mathbf{\Sigma}\|_F^2 = O_P\left(m_p\Big(\frac{\log p}{n}\Big)^{1-q/2}\right), \tag{9.18}$$

*if $\max_{i\le p}\sigma_{ii} \le C_2$. In addition, if $\|\mathbf{\Sigma}^{-1}\|$ is bounded from below, then*

$$\left\|\left(\widehat{\mathbf{\Sigma}}_\lambda^{\mathcal{T}}\right)^{-1} - \mathbf{\Sigma}^{-1}\right\| = O_P\left(m_p\Big(\frac{\log p}{n}\Big)^{(1-q)/2}\right).$$

**Remark 9.1** The last part of Theorem 9.1 follows easily from (Exercise 9.1)

$$\|\mathbf{A}^{-1} - \mathbf{B}^{-1}\| \le \|\mathbf{A} - \mathbf{B}\|\|\mathbf{B}^{-1}\|(\|\mathbf{B}^{-1}\| + \|\mathbf{A}^{-1} - \mathbf{B}^{-1}\|),$$

for any matrix $\mathbf{A}$ and $\mathbf{B}$. The result of Theorem 11.1 holds more generally by replacing $\sqrt{(\log p)/n}$ with any rate $a_n \to 0$. In this case, the condition (9.17) and the threshholding are modified as

$$\sup_{\mathbf{\Sigma}\in \mathcal{C}_q(m_p)} P\big(\|\widehat{\mathbf{\Sigma}} - \mathbf{\Sigma}\|_{\max} > C_0 a_n\big) \le \epsilon_{n,p}, \qquad \lambda_{ij} = \lambda a_n\sqrt{\widehat{\sigma}_{i,i}\widehat{\sigma}_{j,j}}.$$

The results of Theorem 9.1 hold with replacing $\frac{\log p}{n}$ by $a_n^2$, using identically the same proof.

The above Theorem and its remark are a slight generalization of that in Avella-Medina, Battey, Fan, and Li (2018), whose proof is simplified and relegated to Section 9.6.1. It shows that as long as the pilot estimator uniformly converges at a certain rate and the problem is sparse enough, we have the convergence of the estimated covariance matrix. In particular, when $q = 0$, the convergence rate is $m_p\sqrt{\log p/n}$. This matches with the upper bound (9.14): effectively we estimate $m_p$ parameters, each with rate $\sqrt{\log p/n}$. Cai and Zhou (2011) showed the rate given in Theorem 9.1 is minimax optimal.

If the data $\{\mathbf{X}_i\}$ are drawn independently from $N(\boldsymbol{\mu}, \mathbf{\Sigma})$, and $\lambda =$

$M(n^{-1}\log p)^{1/2}$ for large $M$, Bickel and Levina (2008b) verified the condition (9.17):

$$\|\widehat{\mathbf{\Sigma}} - \mathbf{\Sigma}\|_{\max} = O_P\Big(\sqrt{(\log p)/n}\Big), \tag{9.19}$$

uniformly over the parameter space $\mathcal{C}_q(m_p)$. Here, we use (9.19) to denote the meaning of (9.17). The first result in Theorem 9.1 will also be more informally written as

$$\|\widehat{\mathbf{\Sigma}}_{\lambda}^{\mathcal{T}} - \mathbf{\Sigma}\|_2 = O_P\left(m_p\Big(\frac{\log p}{n}\Big)^{(1-q)/2}\right). \tag{9.20}$$

Condition (9.19) holds more generally for the data drawn from *sub-Gaussian* distributions (4.20) using the concentration inequality (Theorem 3.2). The sample covariance matrix involves the product of two random variables. If they are both sub-Gaussian, their product is sub-exponential and a concentration inequality for sub-exponential distribution is needed. The calculation is somewhat tedious. Instead, we appeal to the following theorem, which can be found in Vershynin (2010). See also Tropp (2015).

**Theorem 9.2** *Let* $\mathbf{\Sigma}$ *be the covariance matrix of* $\mathbf{X}_i$. *Assume that* $\{\mathbf{\Sigma}^{-\frac{1}{2}}\mathbf{X}_i\}_{i=1}^n$ *are i.i.d. sub-Gaussian random vectors, and denote by* $\kappa = \sup_{\|\mathbf{v}\|_2=1}\|\mathbf{v}^T\mathbf{X}_i\|_{\psi_2}$, *where* $\|X\|_{\psi_2} = \inf\{s > 0 : \quad E\exp(X^2/s^2) \le 2\}$ *is the Orlicz norm. Then for any* $t \ge 0$, *there exist constants* $C$ *and* $c$ *only depending on* $\kappa$ *such that*

$$P\Big(\|\mathbf{S} - \mathbf{\Sigma}\|_2 \ge \max(\delta, \delta^2)\|\mathbf{\Sigma}\|_2\Big) \le 2\exp(-ct^2),$$

*where* $\delta = C\sqrt{p/n} + t/\sqrt{n}$, *and* $\mathbf{S}$ *is the sample covariance matrix* (9.7).

The above bound depends on the ambient dimension $p$. It can be replaced by the *effective rank* of $\mathbf{\Sigma}$, defined as $\mathrm{tr}(\mathbf{\Sigma})/\|\mathbf{\Sigma}\|_2$. See Koltchinskii and Lounici (2017) for such a refined result. In particular, if we take $p = 2$ in Theorem 9.2, we obtain a concentration inequality for the sample covariance, which is stated in Theorem 9.3.

**Theorem 9.3** *Let* $s_{ij}$ *be the* $(i,j)$*-element of the sample covariance matrix* $\mathbf{S}$ *in* (9.7) *and* $\mathbf{\Sigma}_{ij}$ *be the* $2\times 2$ *covariance matrix of* $(X_i, X_j)^T$. *Under conditions of Theorem 9.2, if* $\{\sigma_{ii}\}$ *are bounded, there exists a sufficiently large constant* $M$ *such that when* $|t| \ge M$, *we have, for all* $i$ *and* $j$,

$$P(|s_{ij} - \sigma_{ij}| \ge 2(\sigma_{ii} + \sigma_{jj})t/\sqrt{n}) \le 2\exp(-ct^2), \qquad \text{when } t < \sqrt{n}/2.$$

*If in addition* $\log p = o(n)$, *we have*

$$\|\mathbf{S} - \mathbf{\Sigma}\|_{\max} = O_P\Big(\sqrt{\frac{\log p}{n}}\Big).$$

This theorem can be used to verify condition (9.17). Hence (9.18) and (9.20) hold for thresholded estimator (9.13), when data are drawn from a sub-Gaussian distribution.

### *9.2.3 Nearest positive definite matrices*

All regularized estimators are symmetric, but not necessarily semipositive definite. This calls for the projection of a generic symmetric matrix $\widehat{\mathbf{\Sigma}}_\lambda$ into the space of semipositive definite matrices.

The simplest method is probably the $L_2$-projection. As in (9.2), let us decompose

$$\widehat{\mathbf{\Sigma}}_\lambda = \mathbf{V}\,\mathrm{diag}(\lambda_1, \cdots, \lambda_p)\mathbf{V}^T, \tag{9.21}$$

and compute its $L_2$*-projected matrix*

$$\widehat{\mathbf{\Sigma}}_\lambda^+ = \mathbf{V}\,\mathrm{diag}(\lambda_1^+, \cdots, \lambda_p^+)\mathbf{V}^T,$$

where $\lambda_j^+ = \max(\lambda_j, 0)$ is the positive part of $\lambda_j$. This $L_2$-projection is the solution to the least-squares problems: Minimize, with respect to $\mathbf{X} \succeq 0$ the problem:

$$\|\widehat{\mathbf{\Sigma}}_\lambda - \mathbf{X}\|_F^2, \quad \text{s.t.} \quad \mathbf{X} \succeq 0.$$

Another simple method is to compute

$$\widehat{\mathbf{\Sigma}}_\lambda^+ = (\widehat{\mathbf{\Sigma}}_\lambda + \lambda_{\min}^- \mathbf{I}_p)/(1 + \lambda_{\min}^-).$$

where $\lambda_{\min}^-$ is the negative part of the minimum eigenvalue of $\widehat{\mathbf{\Sigma}}_\lambda$, which equals $-\lambda_{\min}$ when $\lambda_{\min} < 0$ and 0 otherwise. Both projections do not alter eigenvectors, using the same matrix $\mathbf{V}$ as the original decomposition (9.21). When applied to the correlation matrix, the second method still yields a correlation matrix.

The above projection gives a simple and analytic solution. However, it is not necessarily optimal. For example, Qi and Sun (2006) introduced an algorithm for computing the *nearest correlation matrix*: For a given symmetric matrix $\mathbf{A}$, find a correlation matrix $\mathbf{R}$ to minimize

$$\|\mathbf{A} - \mathbf{R}\|_F^2, \quad \text{s.t.} \quad \mathbf{R} \succeq 0, \mathrm{diag}(\mathbf{R}) = \mathbf{I}_p. \tag{9.22}$$

To apply the algorithm (9.22) to our thresholding estimator $\widehat{\mathbf{\Sigma}}_\lambda$, we compute the standardized input

$$\mathbf{A} = \mathrm{diag}(\widehat{\mathbf{\Sigma}}_\lambda)^{-1/2}\widehat{\mathbf{\Sigma}}_\lambda\,\mathrm{diag}(\widehat{\mathbf{\Sigma}}_\lambda)^{-1/2}$$

first and then apply (9.22) to obtain $\widehat{\mathbf{R}}_\lambda$ and transform back to the covariance matrix as

$$\widehat{\mathbf{\Sigma}}_\lambda^* = \mathrm{diag}(\widehat{\mathbf{\Sigma}}_\lambda)^{1/2}\widehat{\mathbf{R}}_\lambda\,\mathrm{diag}(\widehat{\mathbf{\Sigma}}_\lambda)^{1/2}.$$

Of course, if the *thresholding* is applied to the correlation matrix as in (9.12), we can then take the input $\mathbf{A} = \widehat{\mathbf{\Psi}}_\lambda$. After the nearest correlation matrix projection, we transform the estimate back.

There is a R-package called *nearPD* that computes (9.22). The function is also called `nearPD`.

As a generalization of (9.22), one can also solve the following optimization problem

$$\|\mathbf{A} - \mathbf{R}\|_F^2, \quad \text{s.t.} \quad \lambda_{\min}(\mathbf{R}) \geq \delta, \text{diag}(\mathbf{R}) = \mathbf{I}_p$$

for a given $\delta \geq 0$. Namely, we wish to find a correlation matrix $\mathbf{R}$ with minimum eigenvalue $\delta$ such that it is closest to $\mathbf{A}$. When $\delta = 0$, it reduces to problem (9.22).

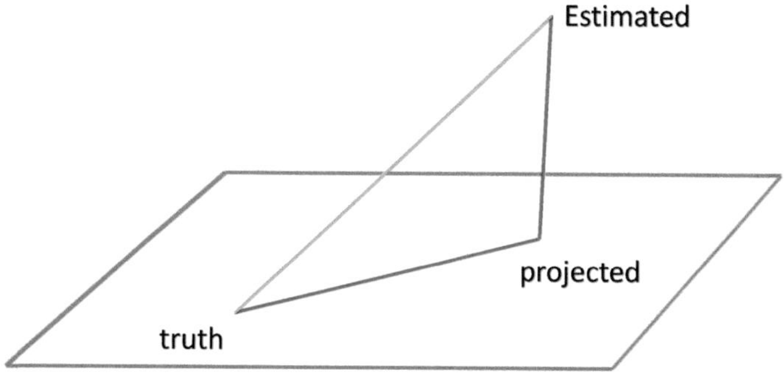


Figure 9.2: The projected matrix cannot be twice worse than the unprojected one by using the triangular inequality. In the case of a rotationally invariant norm, the projected matrix is indeed closer to the truth.

More generally, we can cast the following projection problem:

$$\min \|\widehat{\mathbf{\Sigma}}_\lambda - \mathbf{X}\|, \quad \text{s.t.} \quad \mathbf{X} \in \mathcal{S} \tag{9.23}$$

where $\mathcal{S}$ is a set of matrices such that $\mathbf{\Sigma} \in \mathcal{S}$ is a feasible solution. Let $\mathbf{S}_\lambda$ be a solution. Then, we have

$$\|\mathbf{S}_\lambda - \mathbf{\Sigma}\| \leq \|\widehat{\mathbf{\Sigma}}_\lambda - \mathbf{\Sigma}\| + \|\widehat{\mathbf{\Sigma}}_\lambda - \mathbf{S}_\lambda\| \leq 2\|\widehat{\mathbf{\Sigma}}_\lambda - \mathbf{\Sigma}\|.$$

Figure 9.2 gives an illustration. In other words, the projection cannot make the estimator much worse than the original one. For example, if we take the elementwise $L_\infty$-norm $\|\mathbf{A}\| = \|\mathbf{A}\|_{\max}$, then the projected estimator $\mathbf{S}_\lambda$ can not be twice as worse as the original $\widehat{\mathbf{\Sigma}}_\lambda$ in terms of elementwise $L_\infty$-norm. This minimization problem can be solved by using *cvx* solver in MATLAB (Grant and Boyd, 2014) for the constraint $\lambda_{\min}(\mathbf{X}) \geq \delta$.

Another way of enforcing semi-positive definiteness is to impose a constraint on the covariance regularization problem. For example, in minimizing (9.11), we can impose the constraint $\lambda_{min}(\mathbf{\Sigma}) \geq \tau$. Xue, Ma and Zou (2012) propose this approach for enforcing the positive definiteness in thresholding covariance estimator:

$$\arg\min_{\mathbf{\Sigma} \succ \tau \mathbf{I}_p} \frac{1}{2}\|\mathbf{\Sigma} - \mathbf{S}\|_F^2 + \lambda \sum_{i \neq j} |\sigma_{ij}|,$$

and the above eigenvalue constrained sparse covariance estimator can be solved efficiently via an alternating direction method of multiplier algorithm. A similar estimator was considered by Liu, Wang, and Zhao (2014).

## 9.3 Robust Covariance Inputs

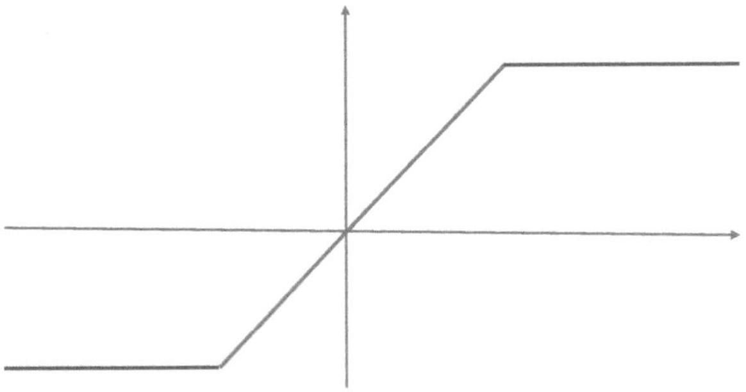

Figure 9.3: Winsorization or truncation of data.

The sub-Gaussian assumption is too strong for high-dimensional statistics. Requiring tens of thousands of variables to have all light tails is a mathematical fiction rather than reality. This is documented in Fan, Wang and Zhu (2016). Catoni (2012), Fan, Wang, Zhu (2016) and Fan, Li, and Wang (2017) tackle the problem by first considering estimation of a univariate mean $\mu$ from a sequence of i.i.d random variables $Y_1, \cdots, Y_n$ with variance $\sigma^2$. In this case, the sample mean $\bar{Y}$ provides an estimator but without exponential concentration. Indeed, by the Markov inequality, we have

$$P(|\bar{Y} - \mu| \geq t\sigma/\sqrt{n}) \leq t^{-2},$$

which is tight in general and has a Cauchy tail (in terms of $t$). On the other hand, if we truncate the data $\widetilde{Y}_i = \operatorname{sign}(Y_i)\min(|Y_i|, \tau)$ with $\tau \asymp \sigma\sqrt{n}$ (see Figure 9.3) and compute the mean of the truncated data (usually called Winsorized data), then we have (Fan, Wang and Zhu, 2016)

$$P\Big(\big|\frac{1}{n}\sum_{i=1}^{n}\widetilde{Y}_i - \mu\big| \geq t\frac{\sigma}{\sqrt{n}}\Big) \leq 2\exp(-ct^2), \tag{9.24}$$

for a universal constant $c > 0$ (see also Theorem 3.2). In other words, the mean of truncated data with only a finite second moment behaves very much the same as the sample mean from the Gaussian data: both estimators have Gaussian tails (in terms of $t$). This sub-Gaussian concentration is fundamental in high-dimensional statistics as the sample mean is computed tens of thousands or even millions of times, and with sub-Gaussianity, the maximum error will not accumulate to be too large.

As an example, estimating the high-dimensional covariance matrix $\mathbf{\Sigma} = (\sigma_{ij})$ involves $O(p^2)$ univariate mean estimation, since the covariance can be expressed as $\sigma_{ij} = \mathrm{E}(X_i X_j) - \mathrm{E}(X_i)\,\mathrm{E}(X_j)$. Estimating each component by the truncated mean yields an *elementwise robust covariance* estimator

$$\widehat{\mathbf{\Sigma}}_E = \left(\widehat{\mathrm{E}(X_i X_j)}^a - \widehat{\mathrm{E}\,X_i}^a \widehat{\mathrm{E}\,X_j}^a\right), \tag{9.25}$$

where $a$ is the method of estimating the univariate mean. When $a = T$, we refer to the truncated sample mean as in (9.24) and when $a = H$, we refer to the adaptive *Huber estimator* in Theorem 3.2. For example, for the Winsorized estimator ($a = T$), we have

$$\widehat{\mathrm{E}(X_i X_j)}^T = n^{-1} \sum_{k=1}^{n} \mathrm{sign}(X_{ki} X_{kj}) \min(|X_{ki} X_{kj}|, \tau)$$

and $\widehat{\mathrm{E}(X_i)}^T = n^{-1} \sum_{k=1}^{n} \mathrm{sign}(X_{ki}) \min(|X_{ki}|, \tau)$ in (9.25). We would expect this method performs similarly to the sample covariance matrix based on the truncated data:

$$\widetilde{\mathbf{\Sigma}}_T = \frac{1}{n} \sum_{i=1}^{n} \widetilde{\mathbf{X}}_i \widetilde{\mathbf{X}}_i^T - \bar{\mathbf{X}}_T \bar{\mathbf{X}}_T^T, \qquad \text{with } \bar{\mathbf{X}}_T = \frac{1}{n} \sum_{i=1}^{n} \widetilde{\mathbf{X}}_i, \tag{9.26}$$

where $\widetilde{\mathbf{X}}_i$ is the elementwise truncated data at level $\tau$.

Assuming the fourth moment is bounded (as the covariance itself involves second moments), by using the union bound and the above concentration inequality (9.24), we can easily obtain (see the proof in Section 9.6.3)

$$\|\widehat{\mathbf{\Sigma}}_E - \mathbf{\Sigma}\|_{\max} = O_P\left(\sqrt{\frac{\log p}{n}}\right), \tag{9.27}$$

and condition (9.17) is met. In other words, with robustification, when the data have merely bounded fourth moments, we can achieve the same estimation rate as the sample covariance matrix under the Gaussian data, including the regularization result in Theorem 9.1. Formally, we have

**Theorem 9.4** *Let* $u = \max_{i,j} \mathrm{SD}(X_i X_j)$ *and* $v = \max_i \mathrm{SD}(\mathrm{X_i})$ *and* $\sigma = \max(u, v)$ *be bounded. Assume that* $\log p = o(n)$.

1. *For the elementwise truncated estimator* $\widehat{\mathbf{\Sigma}}_T$*, when* $\tau \asymp \sigma\sqrt{n}$*, we have*

$$P\left(\|\widehat{\mathbf{\Sigma}}_T - \mathbf{\Sigma}\|_{\max} \geq \sqrt{\frac{a \log p}{c' n}}\right) \leq 4p^{2-a}$$

*for any* $a > 0$ *and a constant* $c' > 0$.

2. *For the elementwise adaptive Huber estimator* $\widehat{\mathbf{\Sigma}}_H$, *when* $\tau = \sqrt{n/(a\log p)}\sigma$, *we have*

$$P\Big(\|\widehat{\mathbf{\Sigma}}_H - \mathbf{\Sigma}\|_{\max} \geq \sqrt{\frac{a\log p}{c'n}}\Big) \leq 4p^{2-a/16},$$

*for any* $a > 0$ *and a constant* $c' > 0$.

Instead of using elementwise truncation, we can also apply the global truncation. Note that

$$\mathbf{\Sigma} = \frac{1}{2}\,\mathrm{E}(\mathbf{X}_i - \mathbf{X}_j)(\mathbf{X}_i - \mathbf{X}_j)^T = \frac{1}{2}\,\mathrm{E}\,\|\mathbf{X}_i - \mathbf{X}_j\|^2 \frac{(\mathbf{X}_i - \mathbf{X}_j)(\mathbf{X}_i - \mathbf{X}_j)^T}{\|\mathbf{X}_i - \mathbf{X}_j\|^2},$$

which avoids the use of mean. Truncating the norm $\|\mathbf{X}_i - \mathbf{X}_j\|^2$ at $\tau$ leads to the *robust U-covariance*

$$\begin{aligned}\widehat{\mathbf{\Sigma}}_U &= \frac{1}{2\binom{n}{2}}\sum_{i\neq j}\min(\|\mathbf{X}_i - \mathbf{X}_j\|^2, \tau)\frac{(\mathbf{X}_i - \mathbf{X}_j)(\mathbf{X}_i - \mathbf{X}_j)^T}{\|\mathbf{X}_i - \mathbf{X}_j\|^2} \\ &= \frac{1}{2\binom{n}{2}}\sum_{i\neq j}\min\Big(1, \frac{\tau}{\|\mathbf{X}_i - \mathbf{X}_j\|^2}\Big)(\mathbf{X}_i - \mathbf{X}_j)(\mathbf{X}_i - \mathbf{X}_j)^T, \quad (9.28)\end{aligned}$$

which does not involve estimation of the mean due to symmetrization of $\mathbf{X}_i - \mathbf{X}_j$. Fan, Ke, Sun and Zhou (2018+) prove the follow result.

**Theorem 9.5** *Let* $v^2 = \frac{1}{2}\big\|\,\mathrm{E}\{(\mathbf{X}-\boldsymbol{\mu})(\mathbf{X}-\boldsymbol{\mu})^T\}^2 + \mathrm{tr}(\mathbf{\Sigma})\mathbf{\Sigma} + 2\mathbf{\Sigma}^2\big\|_2$. *For any* $t > 0$, *when* $\tau \geq vn^{1/2}/(2t)$, *we have*

$$P\{\|\widehat{\mathbf{\Sigma}}_U - \mathbf{\Sigma}\|_2 \geq 4vtn^{-1/2}\} \leq 2p\exp(-t^2).$$

As in the proof of Theorem 9.3, the result in Theorem 9.5 also implies the sharp result in elementwise $L_\infty$-norm $\|\cdot\|_{\max}$, namely condition (9.17).

Before the introduction of the robust U-covariance matrix estimator, Fan, Wang and Zhu (2016) and Minsker (2016) independently propose shrinkage variants of the sample covariance using fourth moments. Specifically, when $E\mathbf{X} = 0$, they propose the following estimator:

$$\widehat{\mathbf{\Sigma}}_s = \frac{1}{n}\sum_{i=1}^{n}\widetilde{\mathbf{X}}_i\widetilde{\mathbf{X}}_i^T, \qquad \widetilde{\mathbf{X}}_i := (\|\mathbf{X}_i\|_4 \wedge \tau)\mathbf{X}_i/\|\mathbf{X}_i\|_4. \quad (9.29)$$

It admits the sub-Gaussian behavior under the spectral norm, as long as the fourth moments of $\mathbf{X}$ are finite.

For an elliptical family of distributions, Liu, Han, Yuan, Lafferty, and Wasserman (2012) and Xue and Zou (2012) used rank correlation such as the *Kendall* $\tau$ *correlation* (9.46) or Spearman correlation and (9.49) (see Section 9.5) to obtain a robust estimation of the correlation matrix. After estimating the correlation matrix, we can use Catoni's (2012) method to obtain robust estimation of individual variances. Such constructions also possess the uniform elementwise convergence as in Theorem 9.4.

## 9.4 Sparse Precision Matrix and Graphical Models

The inverse of a covariance matrix, also called a *precision matrix*, is prominently featured in many statistical problems. Examples include Fisher discriminant, Hotelling $T^2$-test, and the Markowitz optimal portfolio allocation vector. It plays an important role in Gaussian graphical models (Lauritzen, 1996, Whittaker, 2009). This section considers estimation of the sparse precision matrix and its applications to graphical models.

### *9.4.1 Gaussian graphical models*

Let $\mathbf{X} \sim N(\boldsymbol{\mu}, \boldsymbol{\Sigma})$ and $\boldsymbol{\Omega} = \boldsymbol{\Sigma}^{-1}$. Then, it follows that the conditional distribution of a block of elements given the other block of variables is also normally distributed. In particular, if we partition

$$\mathbf{X} = \begin{pmatrix} \mathbf{X}_1 \\ \mathbf{X}_2 \end{pmatrix}, \quad \boldsymbol{\Sigma} = \begin{pmatrix} \boldsymbol{\Sigma}_{11} & \boldsymbol{\Sigma}_{12} \\ \boldsymbol{\Sigma}_{21} & \boldsymbol{\Sigma}_{22} \end{pmatrix} \quad \text{and} \quad \boldsymbol{\Omega} = \begin{pmatrix} \boldsymbol{\Omega}_{11} & \boldsymbol{\Omega}_{12} \\ \boldsymbol{\Omega}_{21} & \boldsymbol{\Omega}_{22} \end{pmatrix}$$

then the conditional distribution of

$$(\mathbf{X}_1|\mathbf{X}_2) \sim N\big(\boldsymbol{\mu}_1 + \boldsymbol{\Sigma}_{12}\boldsymbol{\Sigma}_{22}^{-1}(\mathbf{X}_2 - \boldsymbol{\mu}_2), \boldsymbol{\Sigma}_{11.2}\big), \tag{9.30}$$

where $\boldsymbol{\Sigma}_{11.2} = \boldsymbol{\Sigma}_{11} - \boldsymbol{\Sigma}_{12}\boldsymbol{\Sigma}_{22}^{-}\boldsymbol{\Sigma}_{21}$ is the *Schur completement* of $\boldsymbol{\Sigma}_{22}$.

Let $\boldsymbol{\Omega} = \boldsymbol{\Sigma}^{-1}$ be the precision matrix, and $\boldsymbol{\Omega}_{11}, \boldsymbol{\Omega}_{12}, \boldsymbol{\Omega}_{21}$ and $\boldsymbol{\Omega}_{22}$ be its partition in the same way as $\boldsymbol{\Sigma}$. Then, it follows from the block matrix inversion formula at the end of Section 9.1, we have $\boldsymbol{\Omega}_{11} = \boldsymbol{\Sigma}_{11.2}^{-1}$. In particular, when $\mathbf{X}_1 = (X_1, X_2)^T$ and $\mathbf{X}_2 = (X_3, \cdots, X_p)^T$, we can compute $\boldsymbol{\Sigma}_{11.2}^{-1}$ analytically for the $2 \times 2$ matrix $\boldsymbol{\Sigma}_{11.2}$. It then follows easily that

$$\omega_{12} = 0 \Longleftrightarrow \text{the } (1,2) \text{ element of } \boldsymbol{\Sigma}_{11.2} \text{ is zero.}$$

In other words, $\omega_{12} = 0$ if and only if $X_1$ and $X_2$ are independent given $X_3, \cdots, X_p$. We summarize the result as follows.

**Proposition 9.3** *For a Gaussian random vector* $\mathbf{X}$ *with precision matrix* $\boldsymbol{\Omega} = (\omega_{ij})$, $\omega_{ij} = 0$ *if and only if* $X_i$ *and* $X_j$ *are conditionally independent, given the rest of the other random variables.*

This result will be extended to the Gaussian copula in Section 9.5 and serves as the basis of the *graphical model*. The latter summarizes the conditional dependence *network* among the random variables $\{X_j\}_{j=1}^p$ via a graph. The vertices of the graph represent variables $X_1, \cdots, X_p$ and the edges represent the conditional dependence. The edge between $i$ and $j$ exists if and only if $\omega_{ij} \neq 0$. Figure 9.4 gives such an example for $p = 6$.

$$\boldsymbol{\Omega} = \begin{bmatrix} 1 & 0.2 & 0.2 & 0.2 & 0 & 0 \\ 0.2 & 1 & 0 & 0 & 0 & 0 \\ 0.2 & 0 & 1 & 0 & 0.2 & 0.2 \\ 0.2 & 0 & 0 & 1 & 0.2 & 0.2 \\ 0 & 0 & 0.2 & 0.2 & 1 & 0 \\ 0 & 0 & 0.2 & 0.2 & 0 & 1 \end{bmatrix}$$

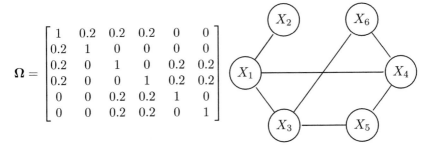

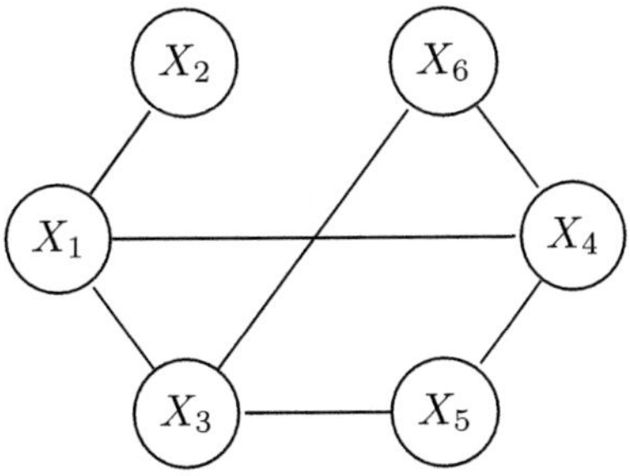


Figure 9.4: The left panel is a precision matrix and the right panel is its graphical representation. Note that only the zero pattern of the precision matrix is used in the graphical representation.

### *9.4.2 Penalized likelihood and M-estimation*

For given data from a Gaussian distribution, its likelihood is given by (9.8). By (9.8), reparameterizing $\boldsymbol{\Sigma}$ by $\boldsymbol{\Omega} = (\omega_{ij})$ and adding the penalty term to the likelihood leads to the following penalized likelihood:

$$-\log|\boldsymbol{\Omega}| + \mathrm{tr}(\boldsymbol{\Omega}\mathbf{S}) + \sum_{i\neq j} p_{\lambda_{ij}}(|\omega_{ij}|). \tag{9.31}$$

The optimization problem (9.31) is challenging due to the singularity of the penalty function and the positive definite constraint on $\boldsymbol{\Omega}$. When the penalty is the Lasso penalty, Li and Gui (2006) used a *threshold gradient descent* procedure to approximately solve (9.31). Their method was motivated by the close connection between $\epsilon$-boosting and Lasso. Various algorithms have been proposed to solve (9.31) with the Lasso penalty. Yuan and Lin (2007) used the maxdet algorithm but that algorithm is very slow for high-dimensional data. A more efficient clockwise descent algorithm was proposed in Banerjee, El Ghaoui and d'Aspremont (2008) and Friedman, Hastie and Tibshirani (2008). Duchi *et al.* (2008) proposed a projected gradient method and Lu (2009) proposed a method by applying Nesterov's smooth optimization technique. Scheinberg *et al.* (2010) developed an alternating direction method of multiplier for solving the Lasso penalized Gaussian likelihood estimator. Fan, Feng and Wu (2009) studied (9.31) with the adaptive Lasso penalty. Rothman, Bickel, Levina and Zhu (2008) and Lam and Fan (2009) derived the asymptotic properties of the estimator. They showed that

$$\|\widehat{\boldsymbol{\Omega}}_{\mathrm{MLE}} - \boldsymbol{\Omega}\|_F = O_p\left(|\boldsymbol{\Omega}^*|\sqrt{\frac{\log p}{n}}\right),$$

and the consistency of sparsity selection for folded concave penalty function, where $|\boldsymbol{\Omega}^*|$ is the number of non-zero elements of true $\boldsymbol{\Omega}^*$. Under an irrepresentable condition, Ravikumar *et al.* (2011) established the selection consistency and rates of convergence of the Lasso penalized Gaussian likelihood estimator.

Zhang and Zou (2014) introduced an *empirical loss minimization* framework to the precision matrix estimation problem:

$$\arg\min_{\boldsymbol{\Omega}\succ 0} L(\boldsymbol{\Omega}, \mathbf{S}) + \lambda \sum_{i\neq j} |\omega_{ij}| \tag{9.32}$$

where $L$ is a proper loss function that satisfies two conditions:

(i) $L(\boldsymbol{\Omega}, \boldsymbol{\Sigma})$ is a smooth convex function of $\boldsymbol{\Omega}$;

(ii) the unique minimizer of $L(\boldsymbol{\Omega}, \boldsymbol{\Sigma}^*)$ occurs at the true $\boldsymbol{\Omega}^* = (\boldsymbol{\Sigma}^*)^{-1}$.

Condition (i) is required for computational reasons and condition (ii) is needed to justify (9.32). In (9.31) the negative log-Gaussian likelihood is identical to the loss function $L_G(\boldsymbol{\Omega}, \boldsymbol{\Sigma}) = \langle \boldsymbol{\Omega}, \boldsymbol{\Sigma}\rangle - \log\det(\boldsymbol{\Omega})$. Zhang and Zou (2014) further proposed a new loss function named *D-Trace loss*:

$$L_D(\boldsymbol{\Omega}, \boldsymbol{\Sigma}) = \frac{1}{2}\operatorname{tr}(\boldsymbol{\Omega}^2\boldsymbol{\Sigma}) - \operatorname{tr}(\boldsymbol{\Omega}). \tag{9.33}$$

One can easily verify that $L_D(\boldsymbol{\Omega}, \boldsymbol{\Sigma})$ satisfies conditions (i) and (ii). In particular, the D-Trace loss has a constant Hessian matrix of the form $(\boldsymbol{\Sigma}\otimes\mathbf{I}+\mathbf{I}\otimes\boldsymbol{\Sigma})/2$, where $\otimes$ denotes the *Kronecker product*. For the empirical D-Trace loss $L_D(\boldsymbol{\Omega}, \mathbf{S})$ its Hessian matrix is $(\mathbf{S}\otimes\mathbf{I}+\mathbf{I}\otimes\mathbf{S})/2$. With the D-Trace loss, (9.32) is solved very efficiently by an alternating direction method of multiplier. Selection consistency and rates of convergence of the *Lasso penalized D-Trace estimator* are established under an irrepresentable condition. It is interesting to see that (9.32) provides a good estimator of $\boldsymbol{\Sigma}^{-1}$ without the normality assumption on the distribution of $\mathbf{X}$. The penalized D-Trace estimator can even outperform the penalized Gaussian likelihood estimator when the data distribution is Gaussian. See Zhang and Zou (2014) for numerical examples.

### *9.4.3 Penalized least-squares*

The aforementioned methods directly estimate the precision matrix $\boldsymbol{\Omega}$. If we relax the positive definiteness condition for $\boldsymbol{\Omega}$, a computational easy alternative is to estimate $\boldsymbol{\Omega}$ via column-by-column methods. Without loss of generality, we assume $\boldsymbol{\mu} = 0$ for simplicity of presentation.

For a given $j$, let $\mathbf{X}_{-j}$ be the subvector of $\mathbf{X}$ without using $X_j$. Correspondingly, let $\boldsymbol{\Sigma}_j$ be the $(p-1)\times(p-1)$ matrix with $j^{th}$ row and $j^{th}$ column deleted, and $\boldsymbol{\sigma}_j$ be the $j^{th}$ column of $\boldsymbol{\Sigma}$ with the $j^{th}$ element deleted. For example, when $j = 1$, the introduced notations are in the partition:

$$\boldsymbol{\Sigma} = \begin{pmatrix} \sigma_{11} & \boldsymbol{\sigma}_1^T \\ \boldsymbol{\sigma}_1 & \boldsymbol{\Sigma}_1 \end{pmatrix}.$$

Set $\tau_j = \sigma_{jj} - \boldsymbol{\sigma}_j^T \boldsymbol{\Sigma}_j^{-1} \boldsymbol{\sigma}_j$. By (9.30), we have

$$(X_j | \mathbf{X}_{-j}) \sim N(\mu_j + \boldsymbol{\beta}_j^T \mathbf{X}_{-j}, \tau_j), \qquad \boldsymbol{\beta}_j = \boldsymbol{\Sigma}_j^{-1} \boldsymbol{\sigma}_j. \tag{9.34}$$

Using the similar notations to those for $\boldsymbol{\Sigma}$, we define similarly $\boldsymbol{\Omega}_j$ and $\boldsymbol{\omega}_j$ based on $\boldsymbol{\Omega}$. By the block inversion formula at the end of Section 9.1, we have, for example,

$$\boldsymbol{\Omega} = \boldsymbol{\Sigma}^{-1} = \begin{pmatrix} \tau_1^{-1} & \tau_1^{-1} \boldsymbol{\sigma}_1^T \boldsymbol{\Sigma}_1^{-1} \\ \tau_1^{-1} \boldsymbol{\Sigma}_1^{-1} \boldsymbol{\sigma}_1 & \text{xyz} \end{pmatrix},$$

where "xyz" is a quantity that does not matter in our calculation. Therefore, $\omega_{11} = \tau_1^{-1}$ and $\boldsymbol{\omega}_1 = \boldsymbol{\beta}_1 / \tau_1$ with $\boldsymbol{\beta}_1$ given by (9.34). Note that $\boldsymbol{\beta}_1$ is the regression coefficient of $X_1$ given $\mathbf{X}_{-1}$ and $\tau_1$ is its residual variance. The sparsity of $\boldsymbol{\beta}_1$ inherits from that of $\boldsymbol{\omega}_1$. Therefore, the first column can be recovered from the sparse regression of $X_1$ given the rest of the variables.

The above derivation holds for every $j$. It also holds for non-normal distributions, as all the properties are the results of linear algebra, rather than those of the normal distributions. We summarize the result as follows.

**Proposition 9.4** *Let $\alpha_j^*$ and $\boldsymbol{\beta}_j^*$ be the solution to the least-squares problem:*

$$\min \mathrm{E}(X_j - \alpha_j - \boldsymbol{\beta}_j^T \mathbf{X}_{-j})^2 \qquad \text{and} \qquad \tau_j^* = \min \mathrm{E}(X_j - \alpha_j - \boldsymbol{\beta}_j^T \mathbf{X}_{-j})^2.$$

*Then, the true parameters $\boldsymbol{\Omega}^*$ are given column-by-column as follows:*

$$\omega_{jj}^* = 1/\tau_j^*, \qquad \boldsymbol{\omega}_j^* = \boldsymbol{\beta}_j^* / \tau_j^*.$$

Proposition 9.4 shows the relation between the matrix inversion and the least-squares regression. It allows us to estimate the precision matrix $\boldsymbol{\Omega}$ column-by-column. It prompts us to estimate the $j^{th}$ column via the penalized least-squares

$$\sum_{i=1}^{n} (X_{ij} - \alpha - \boldsymbol{\beta}^T \mathbf{X}_{i,-j})^2 + \sum_{k=1}^{p-1} p_\lambda(|\beta_k|). \tag{9.35}$$

This is the key idea behind Meinshausen and Bühlmann (2006) who considered the $L_1$-penalty. Note that each regression has different residual variance $\tau_j$ and therefore the penalty parameter should depend on $j$. To make this less dependence sensitive, Liu and Wang (2017) propose to use the *square-root Lasso* of Belloni, Chernozhukov, and Wang (2011), which minimizes

$$\left\{ \sum_{i=1}^{n} (X_{ij} - \alpha - \boldsymbol{\beta}^T \mathbf{X}_{i,-j})^2 \right\}^{1/2} + \sum_{k=1}^{p-1} p_\lambda(|\beta_k|).$$

Instead of using the Lasso, one can use the Dantzig selector:

$$\widehat{\boldsymbol{\beta}}_j = \operatorname*{arg\,min}_{\boldsymbol{\beta}_j \in \mathbb{R}^{p-1}} \|\boldsymbol{\beta}_j\|_1 \quad \text{subject to} \quad \|\mathbf{s}_j - \mathbf{S}_j \boldsymbol{\beta}_j\|_\infty \le \gamma_j, \tag{9.36}$$

for a regularization parameter $\gamma_j$, where $\mathbf{S}_j$ and $\mathbf{s}_j$ is the partition of the sample covariance matrix $\mathbf{S}$ in a similar manner to that of $\mathbf{\Sigma}$. This method is proposed by Yuan (2010).

Sun and Zhang (2013) propose to estimate $\boldsymbol{\beta}_j$ and $\tau_j$ by a *scaled-Lasso*: Find $\widehat{\mathbf{b}}_j$ and $\widehat{\sigma}_j$ to minimize (to be explained below)

$$\frac{\mathbf{b}^T\mathbf{S}\mathbf{b}}{2\sigma} + \frac{\sigma}{2} + \lambda \sum_{k=1}^{p} s_{kk} |b_k| \text{ subject to } b_j = -1. \tag{9.37}$$

Once $\widehat{\mathbf{b}}_j$ and $\widehat{\sigma}_j$ are obtained, we get $\widehat{\boldsymbol{\beta}}_j = (\widehat{b}_1, \ldots, \widehat{b}_{j-1}, \widehat{b}_{j+1}, \ldots, \widehat{b}_p)^T$ and $\widehat{\tau}_j = \widehat{\sigma}_j^2$. Thus, we obtain a column-by-column estimator using Proposition 9.4. They provide the spectral-norm rate of convergence of the obtained precision matrix estimator.

Note that $\mathbf{b}^T\mathbf{S}\mathbf{b}$ in (9.37) is a compact way of writing the quadratic loss in (9.35), after minimizing with respect $\alpha_j$ (Exercise 9.9), also called *profile least-squares*. It also shows innately the relationship between covariance estimation and least-squares regression. Indeed,

$$\mathrm{E}(Y - \mathbf{X}^T\boldsymbol{\beta})^2 = (-1, \boldsymbol{\beta}^T)^T \mathbf{\Sigma}^* (-1, \boldsymbol{\beta}^T), \qquad \mathbf{\Sigma}^* = \mathrm{Var}((Y, \mathbf{X}^T)^T). \tag{9.38}$$

We can replace $\mathbf{\Sigma}^*$ by any of its robust covariance estimators in Section 9.3. This yields a robust least-squares estimator, requiring only bounded fourth moments in high-dimensional problems. In such a kind of least-squares problem, it is essential to use a semi-positive definite covariance estimate. Projections in Section 9.2.3 are essential before plugging it into (9.38).

The column-by-column estimator $\widehat{\mathbf{\Omega}}$ is not necessarily symmetric. A simple fix is to do projection:

$$\widehat{\mathbf{\Omega}}_s = \arg\min_{\mathbf{\Omega}} \|\mathbf{\Omega} - \widehat{\mathbf{\Omega}}\|_* \quad \text{s.t. } \mathbf{\Omega} = \mathbf{\Omega}^T,$$

where $\|\cdot\|_*$ can be the operator, Frobenius, or elementwise max norm. For both the Frobenius and elementwise max norms, the solution is

$$\widehat{\mathbf{\Omega}}_s = \frac{1}{2}\left(\widehat{\mathbf{\Omega}} + \widehat{\mathbf{\Omega}}^T\right).$$

Another way of symmetrization is to perform a shrinkage and symmetrization simultaneously: $\widehat{\mathbf{\Omega}}_s = (\widehat{\omega}_{ij}^s)$ with

$$\widehat{\omega}_{ij}^s = \widehat{\omega}_{ji}^s = \widehat{\omega}_{ij}\mathbb{I}(|\widehat{\omega}_{ij}| \le |\widehat{\omega}_{ji}|) + \widehat{\omega}_{ji}\mathbb{I}(|\widehat{\omega}_{ij}| > |\widehat{\omega}_{ji}|). \tag{9.39}$$

This was suggested by Cai, Liu and Luo (2011) for their CLIME estimator so that it has the property

$$\|\widehat{\mathbf{\Omega}}_s\|_1 \le \|\widehat{\mathbf{\Omega}}\|_1. \tag{9.40}$$

As an illustration, we use gene expression arrays from Arabidopsis thaliana in Wille *et al.* (2004) and construct the genomic network that involves 39

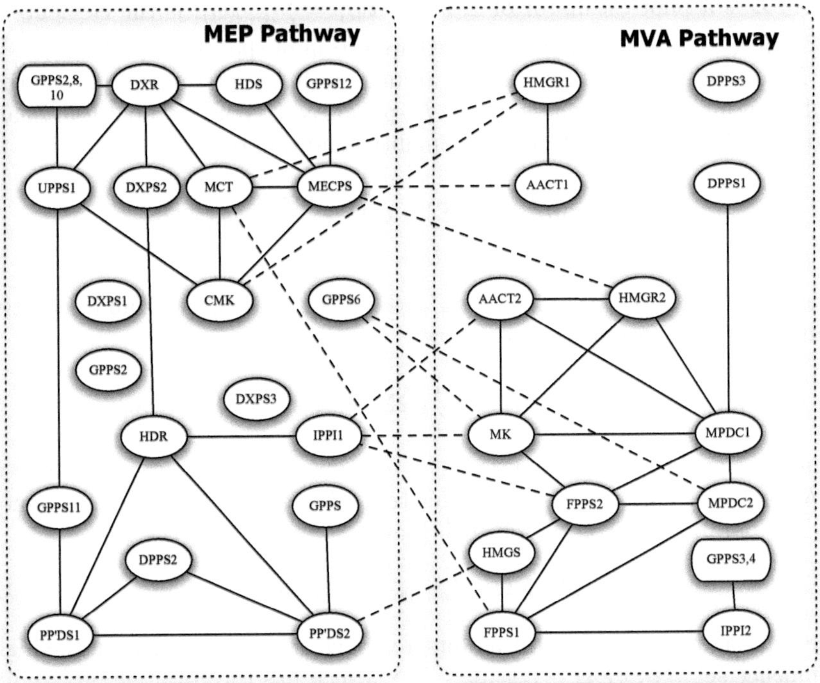


Figure 9.5: The estimated gene networks of the Arabadopsis data. The within-pathway links are shown in solid lines and between-pathway links are presented by dashed lines. Adapted from Liu and Wang (2017).

genes in two isoprenoid metabolic pathways: 16 from the mevalonate (MVA) pathway are located in the cytoplasm, 18 from the plastidial (MEP) pathway are located in the chloroplast, and 5 are located in the mitochondria. While the MVA and MEP pathways generally operate independently, crosstalks can happen too. Figure 9.5 gives the conditional dependence graph using TIGER in Liu and Wang (2017). The graph has 44 edges and suggests that the connections from genes AACT1 and HMGR2 to gene MECPS indicate a primary sources of the crosstalk between the MEP and MVA pathways.

### *9.4.4 CLIME and its adaptive version*

For a given covariance input $\widetilde{\boldsymbol{\Sigma}}$, the constrained $\ell_1$-minimization for inverse matrix estimation (CLIME), introduced by Cai, Liu, and Luo (2011), solves the following problem:

$$\min \|\boldsymbol{\Omega}\|_{\ell_1}, \qquad \text{subject to } \|\widetilde{\boldsymbol{\Sigma}}\boldsymbol{\Omega} - \mathbf{I}_p\|_{\max} \leq \lambda_n, \tag{9.41}$$

where $\widetilde{\mathbf{\Sigma}}$ is a pilot estimator of $\mathbf{\Sigma}$ and $\|\mathbf{\Omega}\|_{\ell_1} = \sum_{i,j} |\omega_{ij}|$ is the elementwise $\ell_1$-norm. It is used to facilitate the computation of the solution $\widehat{\mathbf{\Omega}}$.

Following the Dantzig-selector argument, it is interesting to see that CLIME is more closely related to the $L_1$-penalized D-Trace estimator in Zhang and Zou (2014). The first order optimality condition of

$$\arg\min_{\mathbf{\Omega}} \frac{1}{2}\operatorname{tr}(\mathbf{\Omega}^2\widetilde{\mathbf{\Sigma}}) - \operatorname{tr}\mathbf{\Omega} + \sum_{i,j} \lambda_n |\omega_{ij}|$$

is $\frac{1}{2}(\mathbf{\Omega}\widetilde{\mathbf{\Sigma}} + \widetilde{\mathbf{\Sigma}}\mathbf{\Omega}) - \mathbf{I}_p = \lambda_n \mathbf{Z}$, where $Z_{ij} \in [-1, 1]$. This suggests an alternative Dantzig-selector formulation as follows:

$$\min \|\mathbf{\Omega}\|_{\ell_1}, \qquad \text{subject to } \|\frac{1}{2}(\mathbf{\Omega}\widetilde{\mathbf{\Sigma}} + \widetilde{\mathbf{\Sigma}}\mathbf{\Omega}) - \mathbf{I}_p\|_{\max} \leq \lambda_n$$

We call the above formulation the *symmetrized CLIME* because it is easy to show that its solution is always symmetric, while the CLIME solution is not.

The parameter $\lambda_n$ is chosen such that

$$P\{\|\widetilde{\mathbf{\Sigma}}\mathbf{\Omega} - \mathbf{I}_p\|_{\max} \leq \lambda_n\} \to 1. \tag{9.42}$$

The Dantzig-type of estimator admits a very clear interpretation: in the high-confidence set $\{\mathbf{\Omega} : \|\widetilde{\mathbf{\Sigma}}\mathbf{\Omega} - \mathbf{I}_p\|_{\max} \leq \lambda_n\}$, which has the coverage probability tending to one, let us find the sparsest solution gauged by the $\ell_1$-norm.

Let $\mathbf{e}_j$ be the unit vector with 1 at the $j^{th}$ position. Then, the problem (9.41) can be written as

$$\min \sum_{j=1}^{p} \|\boldsymbol{\omega}_j\|_1, \qquad \text{subject to } \max_j \|\widetilde{\mathbf{\Sigma}}\boldsymbol{\omega}_j - \mathbf{e}_j\|_\infty \leq \lambda_n, \tag{9.43}$$

It can be casted as $p$ separate optimization problems: find $\widehat{\boldsymbol{\omega}}_j$ to solve

$$\min \|\boldsymbol{\omega}_j\|_1, \qquad \text{subject to } \|\widetilde{\mathbf{\Sigma}}\boldsymbol{\omega}_j - \mathbf{e}_j\|_\infty \leq \lambda_n. \tag{9.44}$$

This problem can be solved by a linear program. Let $\widetilde{\boldsymbol{\sigma}}_k$ be the $k^{th}$ column of $\widetilde{\mathbf{\Sigma}}$. Write $\boldsymbol{\omega} = \boldsymbol{\omega}_j$ (dropping the index $j$ in $\boldsymbol{\omega}_j$) and $x_i = |\omega_i|$, where $\omega_i$ is the $i^{th}$ component of $\boldsymbol{\omega}$. Then, the problem (9.44) can be expressed as

$$\begin{aligned}
\min \textstyle\sum_{i=1}^{p} x_i \quad \text{s.t.} \quad & x_i \geq -\omega_i, & i = 1, \cdots, p \\
& x_i \leq \omega_i, & i = 1, \cdots, p \\
& \widetilde{\boldsymbol{\sigma}}_k^T \boldsymbol{\omega} - I(k=j) \geq -\lambda_n, & k = 1, \cdots, p \\
& \widetilde{\boldsymbol{\sigma}}_k^T \boldsymbol{\omega} - I(k=j) \leq \lambda_n, & k = 1, \cdots, p.
\end{aligned}$$

The first two constants are on the fact that $x_i = |\omega_i|$ and the last two constraints are on $\|\widetilde{\mathbf{\Sigma}}\boldsymbol{\omega}_j - \mathbf{e}_j\|_\infty \leq \lambda_n$. Thus, we have $2p$ variables $\mathbf{x}$ and $\boldsymbol{\omega}$ and $4p$ linear inequality constraints.

After obtaining the column-by column estimator, we need to symmetrize the estimator as in (9.39), resulting in $\widehat{\mathbf{\Omega}}_s$. Now consider a generalized sparsity measure that is similar to (9.16), namely,

$$\mathbf{\Omega}^* \in \mathcal{C}_q^*(m_p) = \{\mathbf{\Omega} \succ 0 : \|\mathbf{\Omega}\|_1 \le D_n, \quad \sum_{j=1}^p |\omega_{ij}|^q \le m_p\}.$$

Using the sample covariance as the initial estimator, Cai, Liu and Luo (2011) established the rate of convergence for the CLIME estimator. Inspecting the proof there, the results hold more generally for any pilot estimator that satisfies certain conditions, which we state below.

**Theorem 9.6** *If we take $\lambda_n \ge \|\mathbf{\Omega}^*\|_1 \|\widetilde{\mathbf{\Sigma}} - \mathbf{\Sigma}^*\|_{\max}$, then uniformly over true precision matrix $\mathbf{\Omega}^* \in \mathcal{C}_q^*(m_p)$,*

$$\begin{aligned} \|\widehat{\mathbf{\Omega}}_s - \mathbf{\Omega}^*\|_{\max} &\le 4\|\mathbf{\Omega}^*\|_1 \lambda_n \\ \|\widehat{\mathbf{\Omega}}_s - \mathbf{\Omega}^*\|_2 &\le 12 m_p (4\|\mathbf{\Omega}^*\|_1 \lambda_n)^{1-q}, \\ \frac{1}{p}\|\widehat{\mathbf{\Omega}}_s - \mathbf{\Omega}^*\|_F^2 &\le 12 m_p (4\|\mathbf{\Omega}^*\|_1 \lambda_n)^{2-q}. \end{aligned}$$

*In particular, if the pilot estimator $\widetilde{\mathbf{\Sigma}}$ satisfies*

$$\sup_{\mathbf{\Sigma} \in \mathcal{C}_q^*(m_p)} P\big(\|\widetilde{\mathbf{\Sigma}} - \mathbf{\Sigma}\|_{\max} > C_0 \sqrt{(\log p)/n}\big) \le \epsilon_{n,p} \to 0, \tag{9.45}$$

*with probability at least $1 - \varepsilon_{n,p}$, by taking $\lambda_n = C_0 D_n \sqrt{(\log p)/n}$, we have*

$$\begin{aligned} \|\widehat{\mathbf{\Omega}}_s - \mathbf{\Omega}^*\|_{\max} &\le \left(\frac{16 C_0^2 D_n^4 \log p}{n}\right)^{1/2}, \\ \|\widehat{\mathbf{\Omega}}_s - \mathbf{\Omega}^*\|_2 &\le 12 m_p \left(\frac{16 C_0^2 D_n^4 \log p}{n}\right)^{1/2-q/2}, \\ \frac{1}{p}\|\widehat{\mathbf{\Omega}}_s - \mathbf{\Omega}^*\|_F^2 &\le 12 m_p \left(\frac{16 C_0^2 D_n^4 \log p}{n}\right)^{1-q/2}. \end{aligned}$$

The results in Theorem 9.6 indicate that any pilot estimator that satisfies (9.45) can be used for CLIME regularization. In particular, the robust covariance inputs in Section 9.3 are applicable.

The CLIME above uses a constant regularization parameter $\lambda_n$, which is not ideal. Cai, Liu and Zhou (2016) introduce an adaptive version, called *ACLIME*. It is based on the observation that the asymptotic variance of $(\mathbf{S}\mathbf{\Omega} - \mathbf{I}_p)_{ij}$ in (9.41) is heteroscedastic and depends on $\sigma_{ii}\omega_{jj}$ for the sample covariance matrix $\mathbf{S}$. Therefore, the regularization parameter should be of form

$$|(\mathbf{S}\mathbf{\Omega} - \mathbf{I}_p)_{ij}| \le \lambda_n \sqrt{\sigma_{ii}\omega_{jj}},$$

which depends on unknown quantities $\sigma_{ii}$ and $\omega_{jj}$. These unknown quantities need to be estimated first.

Fan, Xue and Zou (2014) show that the CLIME estimator can be used as a good initial estimator for computing the SCAD penalized Gaussian likelihood estimator such that the resulting estimator enjoys the *strong oracle property*. Recall that the SCAD penalized Gaussian likelihood estimator of $\mathbf{\Omega}$ is defined as

$$-\log|\mathbf{\Omega}| + \mathrm{tr}(\mathbf{\Omega}\mathbf{S}) + \sum_{i\neq j} p_{\lambda}^{\mathrm{scad}}(|\omega_{ij}|).$$

Let $\mathcal{A}$ denote the support of the true $\mathbf{\Omega}^*$. The oracle estimator of $\mathbf{\Omega}^*$ is defined as

$$\widehat{\mathbf{\Omega}}^{\mathrm{oracle}} = \arg\min_{\mathbf{\Omega}_{\mathcal{A}^c}=0} \; -\log\det(\mathbf{\Omega}) + \mathrm{tr}(\mathbf{\Omega}\widehat{\mathbf{S}}).$$

Let $\mathbf{\Omega}^{(0)}$ be the CLIME estimator and compute the following weighted $\ell_1$ penalized Gaussian likelihood estimator for $k = 1, 2$

$$\mathbf{\Omega}^{(k)} = \arg\min_{\mathbf{\Omega}} -\log|\mathbf{\Omega}| + \mathrm{tr}(\mathbf{\Omega}\mathbf{S}) + \sum_{i\neq j} p'^{\mathrm{scad}}_{\lambda}(|\omega_{ij}^{(k-1)}|)|\omega_{ij}|.$$

Fan, Xue and Zou (2014) show that with an overwhelming probability $\mathbf{\Omega}^{(2)}$ equals the oracle estimator. Simulation examples in Fan, Xue and Zou (2014) also show that $\mathbf{\Omega}^{(2)}$ performs better than the Lasso penalized Gaussian likelihood estimator and CLIME.

Avella-Medina, Battey, Fan and Li (2018) generalize the procedure to a general pilot estimator of covariance, which includes robust covariance inputs in Section 9.3. It adds a projection step to the original ACLIME. The method consists of the following 4 steps. The first step is to make sure that the covariance input $\mathbf{\Sigma}^+$ is positive definite. The second step is a preliminarily regularized estimator to get a consistent estimator for $\omega_{jj}$. It can be a non-adaptive version too. The output $\omega_{jj}^{(1)}$ is to make the mathematical argument valid. The third step is the implementation of the adaptive CLIME that takes into account of the aforementioned heteroscedacity. The last step is merely a symmetrization.

1. Given a covariance input $\widetilde{\mathbf{\Sigma}}$, let

   $$\mathbf{\Sigma}^+ = \arg\min_{\mathbf{\Sigma}\succeq\epsilon\mathbf{I}_p} \|\mathbf{\Sigma} - \widetilde{\mathbf{\Sigma}}\|_{\max}$$

   for an small $\varepsilon > 0$. This can be easily implemented in MATLAB using the cvx solver, as mentioned in Section 9.2.3. We then estimate $\sigma_{ii}$ by $\sigma_{ii}^+$, the $i^{th}$ diagonal element of the projected matrix $\mathbf{\Sigma}^+$.

2. Get an estimate of $\{\omega_{jj}\}$ by using CLIME with an adaptive regularization parameter via solving

   $$\boldsymbol{\omega}_j^\dagger = \arg\min_{\boldsymbol{\omega}_j}\Big\{\|\boldsymbol{\omega}_j\|_1 : |(\mathbf{\Sigma}^+\boldsymbol{\omega}_j - \mathbf{e}_j)_i| \le \delta_n \max(\sigma_{ii}^+, \sigma_{jj}^+)\omega_{jj},\ \omega_{jj} > 0\Big\}.$$

This is still a linear program. Output

$$\omega_{jj}^{(1)} = \omega_{jj}^{\dagger} I\Big\{\sigma_{jj}^{+} \le \sqrt{n/\log p}\Big\} + \sqrt{(\log p)/n} I\Big\{\sigma_{jj}^{+} > \sqrt{n/\log p}\Big\}.$$

3. Run the CLIME with adaptive tuning parameter

$$\boldsymbol{\omega}_j^{(2)} = \arg\min_{\boldsymbol{\omega}_j}\Big\{\|\boldsymbol{\omega}_j\|_1 : \big|(\boldsymbol{\Sigma}^{+}\boldsymbol{\omega}_j - \mathbf{e}_j)_i\big| \le \lambda_n\sqrt{\sigma_{ii}^{+}\omega_{jj}^{(1)}}, 1 \le i \le p\Big\}.$$

4. The final estimator is constructed as the symmetrization of the last estimator:

$$\widehat{\boldsymbol{\Omega}} = (\widehat{\omega}_{ij}) \ \text{ where } \widehat{\omega}_{ij} = \widehat{\omega}_{ji} = \omega_{ij}^{(2)} 1\big\{|\omega_{ij}^{(2)}| \le |\omega_{ji}^{(2)}|\big\} + \omega_{ji}^{(2)} 1\big\{|\omega_{ij}^{(2)}| > |\omega_{ji}^{(2)}|\big\}.$$

Avella-Medina *et al.* (2018) derive the asymptotic property for the ACLIME estimator with a general covariance input that satisfies (9.17). The results and the proofs are similar to those in Theorem 9.6.

Avella-Medina *et al.* (2018) conduct extensive simulation studies and show the big benefit of using robust covariance estimators as a pilot estimator, for both estimating the covariance matrix and the precision matrix. In particular, they used the elementwise adaptive Huber estimator in their simulation studies.

To assess whether their proposed robust estimator can improve inference on gene regulatory networks, they used a microarray dataset in the study by Huang *et al.* (2011) on the inflammation process of cardiovascular disease. Huang *et al.* (2011) found that the Toll-like receptor signaling (TLR) pathway plays a key role in the inflammation process. Their study involves $n = 48$ patients and data is available on the Gene Expression Omnibus via the accession name "GSE20129". Avella-Medina *et al.* (2018) considered 95 genes from the TLR pathway and another 62 genes from the PPAR signaling pathway which is known to be unrelated to cardiovascular disease. We would expect that a good method discovers more connections for genes within each of their pathways but not across them. In this study, $n = 48$ and $p = 95 + 62$. To see whether a robust procedure gives some extra merits, they used both the adaptive Huber estimator and the sample covariance matrix as pilot estimators. We will call them robust ACLIME or "RACLIME" and ACLIME, respectively.

They first chose the tuning parameters that deliver the top 100 connections for each method. The selection results are presented in the left panel in Table 9.1 and also displayed graphically in Figure 9.6. The robust ACLIME identifies a greater number of connections within pathways and fewer connections across pathways.

They additionally took the same tuning parameter in the ACLIME optimization problem for each procedure. The results are summarized in the right panel of Table 9.1. The robust procedure detects fewer connections, but yields higher percentage (64.2% by robust ACLIME) of within-pathway connections (in comparison with 55% by ACLIME).

Table 9.1: Summary of regulatory network graphs: Number of connections detected by using robust ACLIME and regular ACLIME, with two methods of choosing regularization parameters. Adapted from Avella-Medina *et al.* (2018).

| Top 100 connections | | | | Equal tuning parameters | | | |
|---|---|---|---|---|---|---|---|
| | within | between | total | | within | between | total |
| RACLIME | 60 | 40 | 100 | robust ACLIME | 27 | 15 | 42 |
| ACLIME | 55 | 45 | 100 | ACLIME | 55 | 45 | 100 |

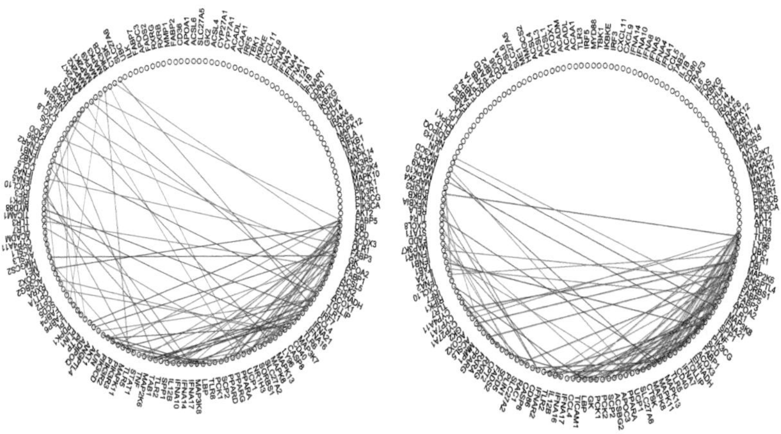

Figure 9.6: Connections estimated by ACLIME using the sample covariance pilot estimator (left) and the adaptive Huber type pilot estimator (right). Blue lines are within pathway connections and red lines are between pathway connections. Taken from Avella-Medina *et al.* (2018).

## 9.5 Latent Gaussian Graphical Models

Graphical models are useful representations of the dependence structure among multiple random variables. Yet, the normality assumption in the last two sections is restrictive. This assumption has been relaxed to the Gaussian *copula* model by Xue and Zou (2012) and Liu, Han, Yuan, Lafferty, and Wasserman (2012). It has been further extended by Fan, Liu, Ning, and Zou (2017) to binary data and the mix of continuous and binary types of data. In particular, the method allows one to infer about the latent graphical model based only on binary data. This section follows the development in Fan *et al.* (2017).

**Definition 9.4** *A random vector* $\mathbf{X} = (X_1, ..., X_p)^T$ *is said to follow a Gaussian copula distribution, if there exists strictly increasing transformations* $\mathbf{g} = (g_j)_{j=1}^p$ *such that* $\mathbf{g}(\mathbf{X}) = (g_1(X_1), ..., g_p(X_p))^T \sim N(0, \mathbf{\Sigma})$ *with* $\sigma_{jj} = 1$

*for any* $1 \leq j \leq p$. *Write* $\mathbf{X} \sim NPN(0, \mathbf{\Sigma}, \mathbf{g})$, *also called a nonpara-normal model.*

In the above definition, we assume without loss of generality that $\mathbf{\Sigma}$ is a correlation matrix. This is due to the identifiability: the scale of functions $\mathbf{g}$ can also be taken such that $\mathbf{\Sigma}$ is a correlation matrix. Let $F_j(\cdot)$ be the cumulative distribution function of a continuous random variable $X_j$. Then, $F_j(X_j)$ follows the standard uniform distribution and $\Phi^{-1}(F_j(X_j)) \sim N(0,1)$, where $\Phi(\cdot)$ is the cdf of the standard normal distribution. Therefore,

$$g_j(\cdot) = \Phi^{-1}(F_j(\cdot))$$

and can easily be estimated by substitution of its empirical distribution $\widehat{F}_{n,j}(\cdot)$ or its smoothed version. Therefore, $\mathbf{\Omega}$ can be estimated as in the previous two sections using the transformed data $\{\widehat{\mathbf{g}}(\mathbf{X}_i)\}_{i=1}^n$.

It follows from Proposition 9.3 that $g_i(X_i)$ and $g_j(X_j)$ given the rest of the transformed variables are independent if and only if $\omega_{ij} = 0$. Since the functions in $\mathbf{g}$ are strictly increasing, this is equivalent to the statement that $X_i$ and $X_j$ are independent given the rest of the untransformed data.

**Proposition 9.5** *If* $\mathbf{X} \sim NPN(0, \mathbf{\Sigma}, \mathbf{g})$, *then* $X_i$ *and* $X_j$ *are independent given the rest of the variables in* $\mathbf{X}$ *if and only if* $\omega_{ij} = 0$. *In addition,* $g_j(x) = \Phi^{-1}(F_j(x))$.

There are several ways to model binary data. One way is through the *Ising model* (Ising,1925). Ravikumar, Wainwright and Lafferty (2010) proposed the neighborhood Lasso logistic regression estimator for estimating the sparse Ising model. The Lasso and SCAD penalized *composite likelihood* estimators were considered in Höfling and Tibshirani (2009) and Xue, Zou and Cai (2012), respectively. Another way is to dichotomize the continuous variables. This leads us to the following definition for the data generation process for the mix of binary and continuous type of data. In particular, it includes all binary data as a specific example.

**Definition 9.5** *Let* $\mathbf{X} = (\mathbf{X}_1^T, \mathbf{X}_2^T)^T$, *where* $\mathbf{X}_1$ *represents* $d_1$*-dimensional binary variables and* $\mathbf{X}_2$ *represents* $d_2$*-dimensional continuous variables. The random vector* $\mathbf{X}$ *is said to follow a latent Gaussian copula model, if there exists a* $d_1$*-dimensional random vector* $\mathbf{Z}_1 = (Z_1, \cdots, Z_{p_1})^T$ *such that* $\mathbf{Z} = (\mathbf{Z}_1, \mathbf{X}_2) \sim NPN(0, \mathbf{\Sigma}, \mathbf{g})$ *and*

$$X_j = \mathbb{I}(Z_j < C_j) \quad \text{for all } j = 1, \ldots, p_1,$$

*where* $\mathbb{I}(\cdot)$ *is the indicator function and* $\mathbf{C} = (C_1, \cdots, C_{p_1})$ *is a vector of constants. Denote by* $\mathbf{X} \sim LNPN(0, \mathbf{\Sigma}, \mathbf{g}, \mathbf{C})$, *where* $\mathbf{\Sigma}$ *is the latent correlation matrix.*

Based on the observed data of mixed types, we are interested in estimating $\boldsymbol{\Omega}$ and hence the conditional dependence graph of latent variables $\mathbf{Z}$. To obtain this, we need a covariance input that satisfies (9.17) so that the methods introduced in the previous two sections can be used. For continuous variables, its covariance can be estimated by using the estimated transformations (Liu *et al.* , 2012; Xue and Zou, 2012). For binary data and binary-continuous data, how to estimate their covariance matrices is unclear.

Fan, Liu, Ning and Zou (2017) provide a unified method to estimate the covariance matrix for mixed data, including continuous data, using the *Kendall $\tau$ correlation*

$$\widehat{\tau}_{jk} = \frac{2}{n(n-1)} \sum_{1 \le i < i' \le n} \mathrm{sgn}\Big\{ (X_{ij} - X_{i'j})(X_{ik} - X_{i'k}) \Big\}. \tag{9.46}$$

It estimates the population quantity

$$\begin{aligned} \tau_{jk} &= \mathrm{E}\,\mathrm{sgn}\Big\{ (X_{ij} - X_{i'j})(X_{ik} - X_{i'k}) \Big\} \\ &= \mathrm{E}\,\mathrm{sgn}\,(X_{ij} - X_{i'j})\,\mathrm{sgn}\,(X_{ik} - X_{i'k}) \end{aligned} \tag{9.47}$$

using $\mathrm{sgn}(ab) = \mathrm{sgn}(a)\mathrm{sgn}(b)$.

For binary data $a$ and $b$, using $\mathrm{sgn}(a-b) = a - b$, it follows from (9.47) that

$$\begin{aligned} \tau_{jk} &= \mathrm{E}\,(X_{ij} - X_{i'j})(X_{ik} - X_{i'k}) \\ &= 2\,\mathrm{E}(X_{ij}X_{ik}) - 2\,\mathrm{E}(X_{ij})\,\mathrm{E}(X_{ik}) \end{aligned}$$

Let $\Delta_j = g_j(C_j)$ and $\Phi_2(x_1, x_2, \rho)$ be the cumulative distribution function of the bivariate normal distribution with mean 0 and correlation $\rho$. Then, since $g_j(\cdot)$ is strictly increasing and $g_j(Z_{ij}) \sim N(0,1)$, we have

$$\begin{aligned} \tau_{jk} &= 2P\{g_j(Z_{ij}) < \Delta_j, g_k(Z_{ik}) < \Delta_k\} \\ &\quad -2P\{g_j(Z_{ij}) < \Delta_j\} \cdot P\{g_k(Z_{ik}) < \Delta_k) \\ &= 2\Big\{\Phi_2(\Delta_j, \Delta_k, \sigma_{jk}) - \Phi(\Delta_j)\Phi(\Delta_k)\Big\}. \end{aligned} \tag{9.48}$$

Note that $P(X_j = 1) = \Phi(\Delta_j)$. This leads to a natural estimator

$$\widehat{\Delta}_j = \Phi^{-1}(\widehat{p}_j) \quad \text{with} \quad \widehat{p}_j = n^{-1} \sum_{i=1}^{n} \mathbb{I}(X_{ij} = 1).$$

Note that $\Phi_2(x_1, x_2, \rho)$ is a strictly increasing function of $\rho$. Hence, we can estimate $\sigma_{ij}$ via the inversion of (9.48).

For continuous data, note that

$$\mathrm{sgn}\,(X_{ij} - X_{i'j}) = \mathrm{sgn}(U_j), \qquad U_j = \{g_j(X_{ij}) - g_j(X_{i'j})\}/\sqrt{2}$$

and $(U_j, U_k)$ follow the bivariate normal distribution with mean 0 and correlation $\rho$. Thus, by (9.47), we have

$$\tau_{ij} = 2\{P(U_j < 0, U_k < 0) - P(U_j > 0, U_k < 0)\} = 2\sin^{-1}(\sigma_{jk})/\pi \quad (9.49)$$

or

$$\sigma_{jk} = \sin\left(\frac{\pi}{2}\tau_{ij}\right).$$

The latter was shown in Kendall (1948) that spells out the relationship between Kendall's $\tau$-correlation and Pearson's correlation. This forms the basis of the rank based covariance matrix estimation in Liu et al. (2012) and Xue and Zou (2012). Note that the relation (9.49) holds for elliptical distributions or more generally the transformations of elliptical distributions (Liu, Han, and Zhang, 2012; Han and Liu, 2017). We leave this as an exercise.

Now, for the binary and continuous blocks of the covariance matrix, similar calculation shows (Exercise 9.13) that

$$\tau_{jk} = 4\Phi_2(\Delta_j, 0, \sigma_{ij}/\sqrt{2}) - 2\Phi(\Delta_j), \quad (9.50)$$

where we assume that $X_j$ is binary and $X_k$ is continuous.

**Proposition 9.6** *Suppose the mixed data* $\{\mathbf{X}_i\}_{i=1}^n$ *are drawn from* $\mathbf{X} \sim$ *LNPN*$(0, \mathbf{\Sigma}, \mathbf{g}, \mathbf{C})$. *Then, the relation between Kendall's* $\tau$ *correlation and Pearson's correlation are given by* $\tau_{ij} = G(\sigma_{ij})$, *where the function* $G(\cdot)$ *is given respectively in* (9.48), (9.49) *and* (9.50), *for binary, continuous and binary-continuous blocks.*

Now a natural initial covariance estimator is $\widetilde{\mathbf{\Sigma}}$ in which $\widetilde{\sigma}_{jk} = G^{-1}(\widehat{\tau}_{jk})$ for $j \neq k$ and $\widetilde{\sigma}_{jj} = 1$. It is easy to show that $G^{-1}(x)$ is a Lipschitz function of $x$, for $|x| \leq c$, for any constant $c < 1$. Therefore, the concentration inequality of $\widetilde{\sigma}_{jk}$ follows from that of $\widehat{\tau}_{ij}$, which is established in Fan, Liu, Ning and Zou (2017). The basic idea in establishing the concentration inequality is that $\Delta_j$ and $\tau_{ij}$ can be uniformly estimated by using the Hoeffding inequality for $\widehat{p}_j$ and the $U$-statistics $\widehat{\tau}_{ij}$. As a result, they showed that the pilot estimator satisfies condition (9.17):

$$\|\widetilde{\mathbf{\Sigma}} - \mathbf{\Sigma}\|_{\max} = O_p\left(\sqrt{\frac{\log p}{n}}\right).$$

With this pilot covariance estimator, we can now estimate the precision matrix using the methods in the previous two sections and construct the corresponding conditional graph.

## 9.6 Technical Proofs

### 9.6.1 *Proof of Theorem 9.1*

By (9.14), we need to bound individual $L_1$-error. Consider the event $\mathsf{E}_1 = \{|\widehat{\sigma}_{ij} - \sigma_{ij}| \leq \lambda_{ij}/2, \forall i, j\}$. For this event, we have

$$\left|\tau_{\lambda_{ij}}(\widehat{\sigma}_{ij}) - \sigma_{ij}\right| \leq \left|\tau_{\lambda_{ij}}(\widehat{\sigma}_{ij}) - \widehat{\sigma}_{ij}\right| + \left|\widehat{\sigma}_{ij} - \sigma_{ij}\right| \leq 1.5\lambda_{ij},$$

by using the property of the generalized thresholding function (see Definition 9.3b)). Thus, by considering the cases $|\sigma_{ij}| \geq \lambda_{ij}$ and $|\sigma_{ij}| < \lambda_{ij}$ separately and using property a) of the generalized thresholding function for the second case, we have that for the event $\mathsf{E}_1$,

$$\begin{aligned} \left|\tau_{\lambda_{ij}}(\widehat{\sigma}_{ij}) - \sigma_{ij}\right| &\leq 1.5\lambda_{ij}1\{|\sigma_{ij}| \geq \lambda_{ij}\} + (1+a)|\sigma_{ij}|1\{|\sigma_{ij}| < \lambda_{ij}\} \\ &\leq 1.5|\sigma_{ij}|^q \lambda_{ij}^{1-q} + (1+a)|\sigma_{ij}|^q \lambda_{ij}^{1-q} \\ &= (2.5+a)|\sigma_{ij}|^q \lambda_{ij}^{1-q}. \end{aligned} \tag{9.51}$$

Hence, we have

$$\sum_{j=1}^p \left|\tau_{\lambda_{ij}}(\widehat{\sigma}_{ij}) - \sigma_{ij}\right| \leq (2.5+a)\sum_{j=1}^p \lambda_{ij}^{1-q}|\sigma_{ij}|^q,$$

which is bounded, for the event $\mathsf{E}_2 = \{\widehat{\sigma}_{ii}\widehat{\sigma}_{jj} \leq 2\sigma_{ii}\sigma_{jj}, \forall i, j\}$, further by

$$(2.5+a)(2\lambda)^{1-q}\left(\frac{\log p}{n}\right)^{(1-q)/2}\sum_{j=1}^p (\sigma_{ii}\sigma_{jj})^{(1-q)/2}|\sigma_{ij}|^q,$$

by using the definition of (9.13). Hence, for the parameter space $\mathcal{C}_q(m_p)$, we have by using (9.14) that

$$\|\widehat{\mathbf{\Sigma}}_\lambda^\tau - \widehat{\mathbf{\Sigma}}\|_2 \leq \max_i \sum_{j=1}^p |\widehat{\sigma}_{ij}^\tau - \sigma_{ij}| \leq C_1 m_p \left(\frac{\log p}{n}\right)^{(1-q)/2},$$

for a constant $C_1 = (2.5+a)(2C\lambda)^{1-q}$.

It remains to lower bound the probability of the event $\mathcal{E}_n = \{\mathsf{E}_1 \cap \mathsf{E}_2\}$. Notice that

$$\widehat{\sigma}_{ii}\widehat{\sigma}_{jj} = \sigma_{ii}\sigma_{jj} + (\widehat{\sigma}_{ii} - \sigma_{ii})\widehat{\sigma}_{jj} + (\widehat{\sigma}_{jj} - \sigma_{jj})\widehat{\sigma}_{ii} - (\widehat{\sigma}_{ii} - \sigma_{ii})(\widehat{\sigma}_{jj} - \sigma_{jj}). \tag{9.52}$$

Since the last three terms tend to zero and $\sigma_{ii} \geq \gamma$, it therefore follows from (9.17) that, for large $n$

$$P\Big(\widehat{\sigma}_{ii}\widehat{\sigma}_{jj} \geq \gamma^2/2, \forall i, j\Big) \geq 1 - \epsilon_{n,p} \quad \text{and} \quad P(\mathsf{E}_2) \geq 1 - \epsilon_{n,p}.$$

Consequently,

$$\begin{aligned} P(\mathsf{E}_1^c) &= P\left\{\max_{i,j}\frac{|\widehat{\sigma}_{ij}-\sigma_{ij}|}{(\widehat{\sigma}_{ii}\widehat{\sigma}_{jj})^{1/2}} > \frac{\lambda}{2}\sqrt{\frac{\log p}{n}}\right\} \\ &\le P\left\{\max_{i,j}|\widehat{\sigma}_{ij}-\sigma_{ij}| \ge \frac{\lambda}{2}\sqrt{\frac{\gamma^2\log p}{2n}}\right\} + \epsilon_{n,p} \\ &\le 2\epsilon_{n,p}, \end{aligned}$$

where the last one follows from (9.17) when $\lambda$ is sufficiently large. Hence, we have

$$P(\mathcal{E}_n^c) \le P(\mathsf{E}_1^c) + P(\mathsf{E}_2^c) \le 3\epsilon_{n,p}.$$

This completes the proof of the first result.

The proof of the second result follows a similar calculation. Using the same calculation as in (9.51), replacing $q$ there by $q/2$, we have for the event $\mathsf{E}_1$

$$\left|\tau_{\lambda_{ij}}(\widehat{\sigma}_{ij}) - \sigma_{ij}\right|^2 \le (2.5+a)^2|\sigma_{ij}|^q\lambda_{ij}^{2-q}.$$

Hence, for the event $\mathsf{E}_2$, using $\sigma_{ii} \le C_2$, we have

$$\sum_{j=1}^p \left|\tau_{\lambda_{ij}}(\widehat{\sigma}_{ij}) - \sigma_{ij}\right|^2 \le C_3 m_p \left(\frac{\log p}{n}\right)^{1-q/2},$$

where $C_3 = (2.5+a)^2(2\lambda)^{2-q}C_2$. This completes the proof.

### *9.6.2 Proof of Theorem 9.3*

We plan to use Theorem 9.2 to prove this theorem. Consider a $2\times 2$ submatrix $\mathbf{S}_{ij}$ of $\mathbf{S}$ that involves $s_{ij}$. Then, the corresponding $\kappa_{ij}$ is bounded by $\kappa$ by definition. Similarly, the corresponding

$$\|\mathbf{\Sigma}_{ij}\|_2 \le \operatorname{tr}(\mathbf{\Sigma}_{ij}) \le \sigma_{ii} + \sigma_{jj} \equiv d_{ij},$$

which is bounded by assumption. Hence, the constant $c_{ij}$ is lower bounded by some $c > 0$ and the constant $C_{ij}$ is upper bounded by a constant $C$. Thus, for sufficiently large $t$ that is bounded by $\sqrt{n}/2$,

$$\delta = C\sqrt{2/n} + t/\sqrt{n} \le 2t/\sqrt{n} \le 1.$$

Now applying Theorem 9.2 to this $2\times 2$ submatrix $\mathbf{S}_{ij}$, we obtain

$$P(|s_{ij}-\sigma_{ij}| \ge 2d_{ij}t/\sqrt{n}) \le P(\|\mathbf{S}_{ij} - \mathbf{\Sigma}_{ij}\|_2 \ge \|\mathbf{\Sigma}_{ij}\|_2\delta) \le 2\exp(-ct^2).$$

Now, take $t = \sqrt{a\log p} = o(\sqrt{n})$, we obtain

$$P\left(|s_{ij}-\sigma_{ij}| \ge 2d_{ij}\sqrt{\frac{a\log p}{n}}\right) \le 2p^{-ac}.$$

By the union bound, we have

$$P\Big(\|\mathbf{S} - \mathbf{\Sigma}\|_{\max} \geq 2\max d_{ij}\sqrt{\frac{a\log p}{n}}\Big) \leq 2p^{2-ac},$$

which tends to zero by taking $a$ is a sufficiently large constant. This completes the proof.

*9.6.3 Proof of Theorem 9.4*

By using (9.24), for the truncated estimator, with $t = \sqrt{(a/c)\log p}$, we have

$$P\Big(|\widehat{\mathrm{E}(X_iX_j)}^T - \mathrm{E}(X_iX_j)| \geq \sqrt{\frac{a\sigma^2\log p}{cn}}\Big) \leq 2\exp(-a\log p) = 2p^{-a},$$

and similarly

$$P\Big(|\widehat{\mathrm{E}(X_i)}^T - \mathrm{E}(X_i)| \geq \sqrt{\frac{a\sigma^2\log p}{cn}}\Big) \leq 2\exp(-a\log p) = 2p^{-a}.$$

Therefore, by the union bound, we have $P(\mathsf{E}_1^c) \leq 2p^{2-a}$, where

$$\mathsf{E}_1 = \Big(\max_{i,j}|\widehat{\mathrm{E}(X_iX_j)}^T - \mathrm{E}(X_iX_j)| \leq \sqrt{\frac{a\sigma^2\log p}{cn}}\Big)$$

and $P(\mathsf{E}_2^c) \leq 2p^{1-a}$, where

$$\mathsf{E}_2 = \Big(\max_{i}|\widehat{\mathrm{E}(X_i)}^T - \mathrm{E}(X_i)| \leq \sqrt{\frac{a\sigma^2\log p}{cn}}\Big).$$

Using the decomposition (9.52) and $\log p = o(n)$, when $n$ is sufficiently large, for the event $\mathsf{E}_2$, we have

$$|\widehat{\mathrm{E}(X_i)}^T\widehat{\mathrm{E}(X_j)}^T - \mathrm{E}(X_i)\,\mathrm{E}(X_j)| \leq 3\max_i \mathrm{E}\,|X_i|\sqrt{\frac{a\sigma^2\log p}{cn}}.$$

Therefore, for the event $\mathsf{E}_1 \cap \mathsf{E}_2$, we have

$$|\widehat{\mathrm{E}(X_iX_j)}^T - \widehat{\mathrm{E}(X_i)}^T\widehat{\mathrm{E}(X_j)}^T - \sigma_{ij}| \leq (1 + 3\max_i \mathrm{E}\,|X_i|)\sqrt{\frac{a\sigma^2\log p}{cn}}.$$

The result now follows from the fact that

$$P(\mathsf{E}_1^c \cup \mathsf{E}_2^c) \leq 2p^{2-a} + 2p^{1-a} \leq 4p^{2-a}$$

by taking $c' = c/(1 + 3\max_i \mathrm{E}\,|X_i|)^2$.

The same proof applies to the elementwise adaptive Huber estimator by using the concentration inequality in Theorem 3.2 d). For example, with $t = \sqrt{a \log p}$, we have $\tau = \sqrt{n/(a \log p)}\sigma$ and

$$P\Big(\big|\widehat{\mathrm{E}(X_iX_j)}^H - \mathrm{E}(X_iX_j)\big| \geq \sqrt{\frac{a\sigma^2 \log p}{n}}\Big) \leq 2p^{-a/16}.$$

The rest follows virtually the same line of the proof.

### 9.6.4 *Proof of Theorem 9.6*

Let $\mathbf{\Omega}^*$ and $\mathbf{\Sigma}^*$ be the true precision and covariance matrices, respectively.

(i) Let us first establish the result under $\|\cdot\|_{\max}$ norm. Using $\|\mathbf{AB}\|_{\max} \leq \|\mathbf{A}\|_{\max}\|\mathbf{B}\|_1$, we have

$$\|\widetilde{\mathbf{\Sigma}}\mathbf{\Omega}^* - \mathbf{I}_p\|_{\max} = \|(\widetilde{\mathbf{\Sigma}} - \mathbf{\Sigma}^*)\mathbf{\Omega}^*\|_{\max} \leq \|\widetilde{\mathbf{\Sigma}} - \mathbf{\Sigma}^*\|_{\max}\|\mathbf{\Omega}^*\|_1 \leq \lambda_n.$$

This shows that $\mathbf{\Omega}^*$ is in the feasible set of (9.41) or equivalently (9.43). Let $\boldsymbol{\omega}_j^*$ and $\widehat{\boldsymbol{\omega}}_j$ be respectively the $j^{th}$ column of $\mathbf{\Omega}^*$ and $\widehat{\mathbf{\Omega}}$. Since $\boldsymbol{\omega}_j^*$ is a feasible solution to the problem (9.43), we have $\|\widehat{\boldsymbol{\omega}}_j\|_1 \leq \|\boldsymbol{\omega}_j^*\|_1$ by the equivalent optimization problem (9.44). Hence, $\|\widehat{\mathbf{\Omega}}\|_1 \leq \|\mathbf{\Omega}^*\|_1$. By (9.40), we have

$$\|\widehat{\mathbf{\Omega}}_s\|_1 \leq \|\mathbf{\Omega}^*\|_1. \tag{9.53}$$

The triangular inequality entails that

$$\|\widetilde{\mathbf{\Sigma}}(\widehat{\mathbf{\Omega}}_s - \mathbf{\Omega}^*)\|_{\max} \leq \|\widetilde{\mathbf{\Sigma}}\widehat{\mathbf{\Omega}}_s - \mathbf{I}_p\|_{\max} + \|\widetilde{\mathbf{\Sigma}}\mathbf{\Omega}^* - \mathbf{I}_p\|_{\max} \leq 2\lambda_n,$$

where we used the fact that $\widehat{\mathbf{\Omega}}_s$ satisfies the constraint in (9.41). The inequality (9.53) implies that $\|\widehat{\mathbf{\Omega}}_s - \mathbf{\Omega}^*\|_1 \leq 2\|\mathbf{\Omega}^*\|_1$ and it follows from the last result that

$$\begin{aligned}\|\mathbf{\Sigma}^*(\widehat{\mathbf{\Omega}}_s - \mathbf{\Omega}^*)\|_{\max} &\leq \|\widetilde{\mathbf{\Sigma}}(\widehat{\mathbf{\Omega}}_s - \mathbf{\Omega}^*)\|_{\max} + \|(\widetilde{\mathbf{\Sigma}} - \mathbf{\Sigma}^*)(\widehat{\mathbf{\Omega}}_s - \mathbf{\Omega}^*)\|_{\max}\\ &\leq 2\lambda_n + \|\widehat{\mathbf{\Omega}}_s - \mathbf{\Omega}^*\|_1\|\widetilde{\mathbf{\Sigma}} - \mathbf{\Sigma}^*\|_{\max} \leq 4\lambda_n.\end{aligned}$$

Therefore, we conclude that

$$\|\widehat{\mathbf{\Omega}}_s - \mathbf{\Omega}^*\|_{\max} \leq \|\mathbf{\Omega}^*\mathbf{\Sigma}^*(\widehat{\mathbf{\Omega}}_s - \mathbf{\Omega}^*)\|_{\max} \leq 4\|\mathbf{\Omega}^*\|_1\lambda_n,$$

where the last inequality uses again from $\|\mathbf{AB}\|_{\max} \leq \|\mathbf{A}\|_{\max}\|\mathbf{B}\|_1$.

(ii) We next consider the loss under the operator norm. Recall that $\|\widehat{\mathbf{\Omega}}_s - \mathbf{\Omega}^*\|_2 \leq \|\widehat{\mathbf{\Omega}}_s - \mathbf{\Omega}^*\|_1$. Thus, it suffices to bound the matrix $L_1$-norm. Let

$$\widehat{\mathbf{\Omega}}_1 = (\widehat{\omega}_{ij}^s I(|\widehat{\omega}_{ij}^s| > 2a_n)) \quad \text{and} \quad \widehat{\mathbf{\Omega}}_2 = \widehat{\mathbf{\Omega}}_s - \widehat{\mathbf{\Omega}}_1,$$

where $a_n = \|\widehat{\mathbf{\Omega}}_s - \mathbf{\Omega}^*\|_{\max}$. Let $\widehat{\boldsymbol{\omega}}_j^1$, $\widehat{\boldsymbol{\omega}}_j^2$ and $\widehat{\boldsymbol{\omega}}_j^s$ be the $j^{th}$ column of $\widehat{\mathbf{\Omega}}_1$, $\widehat{\mathbf{\Omega}}_2$, $\widehat{\mathbf{\Omega}}_s$, respectively. Then, by (9.40), we have

$$\|\widehat{\boldsymbol{\omega}}_j^1\|_1 + \|\widehat{\boldsymbol{\omega}}_j^2\|_1 = \|\widehat{\boldsymbol{\omega}}_j^s\|_1 \leq \|\widehat{\boldsymbol{\omega}}_j\|_1 \leq \|\boldsymbol{\omega}_j^*\|_1.$$

On the other hand,

$$\|\widehat{\boldsymbol{\omega}}_j^1\|_1 \geq \|\boldsymbol{\omega}_j^*\|_1 - \|\widehat{\boldsymbol{\omega}}_j^1 - \boldsymbol{\omega}_j^*\|_1.$$

The combination of the last two displays leads to

$$\|\widehat{\boldsymbol{\omega}}_j^2\|_1 \leq \|\widehat{\boldsymbol{\omega}}_j^1 - \boldsymbol{\omega}_j^*\|_1.$$

In other words, $\|\widehat{\boldsymbol{\Omega}}_2\|_1 \leq \|\widehat{\boldsymbol{\Omega}}_1 - \boldsymbol{\Omega}^*\|_1$. Hence,

$$\|\widehat{\boldsymbol{\Omega}}_s - \boldsymbol{\Omega}^*\|_1 \leq \|\widehat{\boldsymbol{\Omega}}_1 - \boldsymbol{\Omega}^*\|_1 + \|\widehat{\boldsymbol{\Omega}}_2\|_1 \leq 2\|\widehat{\boldsymbol{\Omega}}_1 - \boldsymbol{\Omega}^*\|_1.$$

It remains to bound the estimation error $\|\widehat{\boldsymbol{\Omega}}_1 - \boldsymbol{\Omega}^*\|_1$. This is indeed a hard-thresholding estimator $\widehat{\boldsymbol{\Omega}}_1$ for the sparse matrix $\boldsymbol{\Omega}^*$. The results then follow from the proof of Theorem 9.1, which requires only sparsity and uniform consistency of the pilot estimator. More specifically, note that for the hard-thresholding function, we have $a = 2$. Adopting (9.51) to the current setting, we have $\lambda_{ij} = 2a_n$ there. Set $d_{ij} = |\widehat{\omega}_{ij}^s \mathbb{I}(|\widehat{\omega}_{ij}^s| \geq 2a_n) - \omega_{ij}^*|$. It follows from property b) of Definition 9.3 with $a = 2$ that

$$d_{ij} \leq |\widehat{\omega}_{ij}^s \mathbb{I}(|\widehat{\omega}_{ij}^s| \geq 2a_n) - \widehat{\omega}_{ij}^s| + |\widehat{\omega}_{ij}^s - \omega_{ij}^*| \leq 3a_n.$$

Now, separating the bounds for two cases with $\mathbb{I}\{|\omega_{ij}^*| \geq a_n\}$ and $\mathbb{I}\{|\omega_{ij}^*| < a_n\}$, using the property a) of Definition 9.3 for the second case for the hard-thresholding function, we have

$$\begin{aligned} d_{ij} &\leq 3a_n \mathbb{I}\{|\omega_{ij}^*| \geq a_n\} + 3|\omega_{ij}^*| \mathbb{I}\{|\omega_{ij}^*| < a_n\} \\ &\leq 3|\omega_{ij}^*|^q a_n^{1-q} + 3|\omega_{ij}^*|^q a_n^{1-q} \\ &= 6|\omega_{ij}^*|^q a_n^{1-q}. \end{aligned}$$

Hence,

$$\|\widehat{\boldsymbol{\omega}}_j^1 - \boldsymbol{\omega}_j^*\|_1 \leq 6a_n^{1-q} \sum_{i=1}^{p} |\omega_{ij}^*|^q \leq 6a_n^{1-q} m_p,$$

using $\boldsymbol{\Omega}^* \in \mathcal{C}_q^*(m_p)$, namely, $\|\widehat{\boldsymbol{\Omega}}_1 - \boldsymbol{\Omega}^*\|_1 \leq 6a_n^{1-q} m_p$. Therefore,

$$\|\widehat{\boldsymbol{\Omega}}_s - \boldsymbol{\Omega}^*\|_1 \leq 12a_n^{1-q} m_p.$$

(iii) The result for the Frobenius norm follows immediately from that

$$\|\mathbf{A}\|_F^2 = \sum_{i,j} a_{ij}^2 \leq \|\mathbf{A}\|_{\max} \sum_{i=1}^{p} \sum_{j=1}^{p} |a_{ij}| \leq p \|\mathbf{A}\|_{\max} \, \|\mathbf{A}\|_1$$

and the results in (i) and (ii).

For the event $(\|\widetilde{\boldsymbol{\Sigma}} - \boldsymbol{\Sigma}\|_{\max} > C_0\sqrt{(\log p)/n})$, the second part of the result follows directly from the choice of $\lambda_n$ and the fact $\|\boldsymbol{\Omega}\|_1 \leq D_n$. This completes the proof. ∎

## 9.7 Bibliographical Notes

Wu and Pourahmadi (2003) gave a nonparametric estimation for large covariance matrices of longitudinal data. Estimation of a high-dimensional covariance matrix was independently studied by Bickel and Levina (2008a,b) for banded and sparse convariance and Fan, Fan, and Lv (2008) for low-rank plus diagonal matrix. Levina, Rothman and Zhu (2008) and Lam and Fan (2009) investigated various maximum likelihood estimators with different sparse structures. Xue, Ma and Zou (2012) derived an efficient convex program for estimating a sparse covariance matrix under the positive definite constraint. Xue and Zou (2014) showed an adaptive minimax optimal estimator of sparse correlation matrices of a semiparametric Gaussian copula model. The optimality of large covariance estimation was investigated by Cai, Zhang and Zhou (2010), Cai and Liu (2011), Cai and Zhou (2011, 2012a, b), Rigollet and Tsybakov (2012), Cai, Ren, and Zhou (2013), among others. Gao, Ma, Ren, and Zhou (2015) investigated optimal rates in sparse canonical correlation analysis. Cai and Yuan (2012) studied an adaptive covariance matrix estimation through block thresholding. Bien, Bunea, and Xiao (2016) proposed convex banding, Yu and Bien (2017) considered a varying banded matrix estimation problem, and Bien (2018) introduced a generalized notation of sparsity and proposed graph-guided banding of the covariance matrix. Xue and Zou (2014) proposed a ranked-based banding estimator for estimating the bandable correlation matrices. Li and Zou (2016) proposed a Stein's Unbiased Risk Estimation (SURE) criterion for selecting the banding parameter in the banded covariance estimator. Additional references can be found in the book by Pourahmadi (2013).

There is a large literature on estimation of the precision matrix and its applications to graphical models. An early study on the high-dimensional graphical model is Meinshausen and Bühlmann (2006). Penalized likelihood is a popular approach to this problem. See for example Banerjee, El Ghaoui and d'Aspremont (2008), Yuan and Lin (2007), Friedman, Hastie and Tibshirani (2008), Levina, Rothman and Zhu (2008) and Lam and Fan (2009), Fan, Feng and Wu (2009), Yuan (2010), Shen, Pan and Zhu (2012). Ravikumar, Wainwright, Raskutti, and Yu (2011) proposed estimating the precision matrix by the penalized log-determinant divergence. Zhang and Zou (2014) proposed the penalized D-trace estimator for sparse precision matrix estimation. Cai, Liu and Luo (2011) and Cai, Liu and Zhou (2016) proposed a constrained $\ell_1$ minimization approach and its adaptive version to sparse precision matrix estimation. Liu and Wang (2017) proposed a tuning-insensitive approach. Ren, Sun, Zhang and Zhou (2015) established the asymptotic normality for entries of the precision matrix and obtained optimal rates of convergence for general sparsity. The results were further extended to the situation where heterogeneity should be adjusted first before building a graphical model in Chen, Ren, Zhao, and Zhou (2016).

The literature has been further expanded into robust estimation based on regularized rank-based approaches (Liu, Han, Yuan, Lafferty, and Wasserman, 2012; Xue and Zou, 2012). The related work includes, for instance, Mitra and Zhang (2014), Wegkamp and Zhao (2016), Han and Liu (2017), Fan, Liu, and Wang (2018). It requires elliptical distributions on the data generating process. Such an assumption can be removed by the truncation and adaptive Huber estimator; see Avella-Medina, Battey, Fan, and Li (2018).

## 9.8 Exercises

9.1(a) Suppose that $\mathbf{A}$ and $\mathbf{B}$ are two invertible and compatible matrices. Show that for any matrix norm

$$\|\mathbf{A}^{-1} - \mathbf{B}^{-1}\| \leq \|\mathbf{A} - \mathbf{B}\| \|\mathbf{B}^{-1}\| (\|\mathbf{B}^{-1}\| + \|\mathbf{A}^{-1} - \mathbf{B}^{-1}\|).$$

*Hint*: Use $\mathbf{A}^{-1} - \mathbf{B}^{-1} = \mathbf{A}^{-1}(\mathbf{B} - \mathbf{A})\mathbf{B}^{-1}$.

(b) Suppose that the operator norms $\|\mathbf{\Sigma}\|$ and $\|\mathbf{\Sigma}^{-1}\|$ are bounded. If there is an estimate $\widehat{\mathbf{\Sigma}}$ such that $\|\widehat{\mathbf{\Sigma}} - \mathbf{\Sigma}\| = O_P(a_n)$ for a sequence $a_n \to 0$, show that $\widehat{\mathbf{\Sigma}}^{-1}$ is invertible with probability tending to 1 and that

$$\|\widehat{\mathbf{\Sigma}}^{-1} - \mathbf{\Sigma}^{-1}\| = O_P(a_n).$$

9.2 Suppose $\mathbf{\Sigma}$ is a $p \times p$ covariance matrix of a random vector $\mathbf{X} = (X_1, X_2, \cdots, X_p)^T$.

(a) Prove the second part of Theorem 9.1: If $\max_{1 \leq i \leq p} \sigma_{ii} \leq C_2$, uniformly over $\mathcal{C}_q(m_p)$

$$p^{-1} \|\widehat{\mathbf{\Sigma}}_\lambda^\mathcal{T} - \mathbf{\Sigma}\|_F^2 = O_\mathbb{P}\left( m_p \left( \frac{\log p}{n} \right)^{1-q/2} \right).$$

You may use the notation and results proved in the first part of Theorem 9.1.

(b) Assume that $\mathrm{E}(X_i) = 0$ for all $i$, and $\sigma = \max_{i,j} \mathrm{SD}(X_i X_j)$ is bounded. Consider the elementwise adaptive Huber estimator $\widehat{\mathbf{\Sigma}}_H = (\widehat{\sigma}_{ij}^\tau)$ where $\widehat{\sigma}_{ij}^\tau$ is the adaptive Huber estimator of the mean $\mathrm{E}(X_i X_j) = \sigma_{ij}$ based on the data $\{X_{ki} X_{kj}\}_{k=1}^n$ with parameter $\tau$. Show that when $\tau = \sqrt{n/(a \log p)}\sigma$, we have

$$P\left( \|\widehat{\mathbf{\Sigma}}_H - \mathbf{\Sigma}\|_{\max} \geq \sqrt{\frac{a\sigma^2 \log p}{n}} \right) \leq 4p^{2-a/16},$$

for any constant $a > 0$. **Hint**: Use Theorem 3.2(d) of the book.

9.3 Deduce Theorem 9.3 from Theorem 9.2.
*Hint*: Consider a $2 \times 2$ submatrix that involves $s_{ij}$ and apply Theorem 9.2. The solution is furnished in Section 9.6, but it is a good practice to carry out the proof on your own.

9.4 Complete the proof of the second part of Theorem 9.4.

9.5 Use Theorem 9.5 to show that

$$\|\widehat{\mathbf{\Sigma}}_U - \mathbf{\Sigma}\|_{\max} = O_P(\sqrt{(\log p)/n}).$$

9.6 Suppose that the covariance input $\widehat{\mathbf{\Sigma}}_\lambda$ satisfies $P\{\|\widehat{\mathbf{\Sigma}}_\lambda - \mathbf{\Sigma}\|_{\max} \geq a_n\} \leq \varepsilon_n$. Show that the projected estimator $\mathbf{S}_\lambda$ that minimizes (9.23) with $\|\cdot\|_{\max}$-norm satisfies

$$P\{\|\mathbf{S}_\lambda - \mathbf{\Sigma}\|_{\max} \geq 2a_n\} \leq \varepsilon_n.$$

9.7 Let $\mathbf{X} \sim N(\boldsymbol{\mu}, \mathbf{\Sigma})$ and $\mathbf{\Omega} = \mathbf{\Sigma}^{-1}$ be the precision matrix. Let $\omega_{ij}$ be the $(i,j)^{th}$ element of $\mathbf{\Omega}$. Show that $X_i$ and $X_j$ are independent given the rest of the variables in $\mathbf{X}$ if and only if $\omega_{ij} = 0$.

9.8 This exercise intends to give a least-squares interpretation of (9.30).

(a) Without the normality assumption, find the best linear prediction of $\mathbf{X}_1$ by $\mathbf{a} + \mathbf{B}\mathbf{X}_2$. Namely, find $\mathbf{a}$ and $\mathbf{B}$ to minimize $\mathrm{E}\,\|\mathbf{X}_1 - \mathbf{a} - \mathbf{B}\mathbf{X}_2\|^2$. Let $\mathbf{a}^*$ and $\mathbf{B}^*$ be the solution. Show that

$$\mathbf{B}^* = \mathrm{cov}(\mathbf{X}_2, \mathbf{X}_1)\,\mathrm{Var}(\mathbf{X}_2)^{-1}, \qquad \mathbf{a}^* = \mathrm{E}\,\mathbf{X}_1 - \mathbf{B}^*\,\mathrm{E}\,\mathbf{X}_2,$$

where $\mathrm{cov}(\mathbf{X}_2, \mathbf{X}_1) = \mathrm{E}(\mathbf{X}_2 - \mathrm{E}\,\mathbf{X}_2)(\mathbf{X}_1 - \mathrm{E}\,\mathbf{X}_1)^T$ and $\mathrm{Var}(\mathbf{X}_2) = \mathrm{E}(\mathbf{X}_2 - \mathrm{E}\,\mathbf{X}_2)(\mathbf{X}_2 - \mathrm{E}\,\mathbf{X}_2)^T$.

(b) Let $\mathbf{U} = \mathbf{X}_1 - \mathbf{a}^* - \mathbf{B}^*\mathbf{X}_2$ be the residuals. Show that $\mathrm{E}\,\mathbf{U} = 0$ and the covariance between $\mathbf{X}_2$ and $\mathbf{U}$ is zero:

$$\mathrm{E}(\mathbf{X}_2\mathbf{U}^T) = 0.$$

(c) Show that

$$\mathrm{Var}(\mathbf{U}) = \mathrm{E}\,\mathbf{U}\mathbf{U}^T = \mathbf{\Sigma}_{11.2}.$$

9.9 Let $\mathbf{S}$ be the sample covariance matrix of data $\{\mathbf{X}_i\}_{i=1}^n$. Show that

$$\min_{\alpha_j} n^{-1} \sum_{i=1}^{n} (X_{ij} - \alpha_j - \boldsymbol{\beta}_j^T \mathbf{X}_{i,-j})^2 = \boldsymbol{\beta}^T \mathbf{S} \boldsymbol{\beta}$$

subject to the constraint that $\beta_j = -1$.

9.10 Prove Proposition 9.4.

9.11 Consider the optimization problem:

$$\widehat{\mathbf{\Omega}}_s = \arg\min_{\mathbf{\Omega}} \|\mathbf{\Omega} - \widehat{\mathbf{\Omega}}\|_* \quad \text{s.t. } \mathbf{\Omega} = \mathbf{\Omega}^T.$$

For both the Frobenius norm $\|\cdot\|_* = \|\cdot\|_F$ and elementwise max norm $\|\cdot\|_* = \|\cdot\|_{\max}$, show that the solution is

$$\widehat{\mathbf{\Omega}}_s = \frac{1}{2}\left(\widehat{\mathbf{\Omega}} + \widehat{\mathbf{\Omega}}^T\right).$$

9.12 Let $F(\cdot)$ be the cumulative distribution function of a continuous random variable $X$.

(a) If $g(X) \sim N(0,1)$, show that $g(\cdot) = \Phi^{-1}(F(\cdot))$, where $\Phi(\cdot)$ is the cumulative distribution function of the standard normal distribution.

(b) Let $\widehat{F}_n(x) = n^{-1}\sum_{i=1}^n \mathbb{I}(X_i \le x)$ be the empirical distribution based on a random sample $\{X_i\}_{i=1}^n$ and $\widehat{g}_n(x) = \Phi^{-1}(\widehat{F}_n(x))$. Show that

$$\sqrt{n}\,(\widehat{g}_n(x) - g(x)) \stackrel{L}{\longrightarrow} N(0,\sigma^2), \quad \text{with} \quad \sigma^2 = \frac{F(x)(1-F(x))}{\phi(\Phi^{-1}(F(x)))^2},$$

where $\phi(\cdot) = \Phi'(\cdot)$ is the density of the standard normal distribution.

9.13 We use the notation introduced in Section 9.5.

(a) Show that the function $\Phi_2(x_1, x_2, \rho)$ is strictly increasing in $\rho$.

(b) Prove (9.49).

(c) Prove (9.50).

9.14 This is a simplified version of the next exercise.

(a) Prove the result in Exercise 9.15(b) for continuous data, rather than the mixed data.

(b) Prove the result in Exercise 9.15(b) for binary data, rather than the mixed data.

9.15 Let the function $G$ be defined in Proposition 9.6.

(a) Show that $G^{-1}(x)$ satisfies

$$|G^{-1}(x) - G^{-1}(y)| \le D|x-y|, \qquad \text{for } |x| \le c, |y| \le c,$$

for a sufficiently large constant $D$, where $0 < c < 1$.

(b) Let $\widetilde{\sigma}_{jk} = G^{-1}(\widehat{\tau}_{jk})$. Prove that

$$\|\widetilde{\mathbf{\Sigma}} - \mathbf{\Sigma}\|_{\max} = O_p\left(\sqrt{\frac{\log p}{n}}\right).$$

9.16 Download the mice protein expression data from UCI Machine Learning Repository:

`https://archive.ics.uci.edu/ml/datasets/Mice+Protein+Expression.`

There are in total 1080 measurements. Use attributes 2–78 (the expression levels of 77 proteins), as the input variables for large covariance estimation. We clean the missing values by dropping 6 columns "BAD_N", "BCL2_N", "pCFOS_N", "H3AcK18_N", "EGR1_N", "H3MeK4_N" and then removing all rows with at least 1 missing value. For your convenience, you can also download directly the clean input data from the book website. You should have an input data of size $1047 \times 71$.

(a) Compute the correlation-thresholded covariance estimate with $\lambda = 4$. Report the entries of zeros and the minimum and maximum eigenvalues of the resulting estimate. Compare them with those from the input covariance matrix.

(**Hint**: the entry-dependence thresholds for the covariance matrix are $\lambda_{ij} = \lambda\sqrt{\widehat{\sigma}_{ii}, \widehat{\sigma}_{jj}\frac{\log p}{n}}$.)

(b) Estimate the precision matrix by CLIME. (Hint: you can use the package "`flare`".)

(c) Construct the protein network by using the estimated sparsity pattern of the precision matrix. (Hint: you can use the package "`graph`" or "`qgraph`")

(d) Use the winsorized data to repeat part (a), where the winsorized parameter is taken to be 2.5 standard deviation.

Chapter 10

# Covariance Learning and Factor Models

High-dimensional measurements are often strongly dependent, as they usually measure similar quantities under similar environments. Returns of financial assets are strongly correlated due to market conditions. Gene expressions are frequently stimulated by cytokines or regulated by biological processes. Factor models are frequently used to model the dependence among high-dimensional measurements. They also induce a low-rank plus sparse covariance structure among the measured variables, which are referred to as *factor-driven covariance* and *idiosyncratic covariance*, respectively.

This chapter introduces the theory and methods for estimating a high-dimensional covariance matrix under factor models. The theory and methods for estimating both the covariance matrices and the idiosyncratic covariance will be presented. We will also show how to estimate the latent factors and their associated *factor loadings* and how well they can be estimated. The methodology is summarized in Section 10.2, where any robust covariance inputs can be utilized and the methods for choosing the number of factors are also presented. *Principal component analysis* plays a critical role in unveiling latent factors as well as estimating a structured covariance matrix. We will defer the applications of factor models to variable selection, large-scale inferences, prediction, as well as other statistical machine learning problems to Chapter 11.

## 10.1 Principal Component Analysis

*Principal component analysis* (*PCA*) is a popular data processing and dimension reduction technique. It has been widely used to extract latent factors for high dimensional statistical learning and inferences. Other applications include handwritten zip code classification (Hastie, Tibshirani and Friedman, 2009), human face recognition (Hancock, Burton and Bruce, 1996), gene expression data analysis (Misra *et al.* 2002, Hastie *et al.* 2000), to name a few.

### *10.1.1 Introduction to PCA*

PCA can be mathematically described as a set of orthogonal linear transformations of the original variables $\mathbf{X}$ such that the transformed variables have the greatest variance. Imagine that we have $p$ measurements on an individual's ability and we wish to combine these measurements linearly into an overall score $Z_1 = \sum_{j=1}^p \xi_{1j} X_j = \boldsymbol{\xi}_1^T \mathbf{X}$. We wish such a combination to differentiate as much as possible individual abilities, namely makes the individual scores vary as large as possible subject to the normalization $\|\boldsymbol{\xi}\| = 1$. This leads to choosing $\boldsymbol{\xi}_1$ to maximize the variance of $Z_1$, i.e.,

$$\boldsymbol{\xi}_1 = \text{argmax}_{\boldsymbol{\xi}} \boldsymbol{\xi}^T \boldsymbol{\Sigma} \boldsymbol{\xi} \quad \text{subject to } \|\boldsymbol{\xi}\| = 1.$$

Now suppose that we wish to find another metric (a linear combination of the measurements), whose information is unrelated to what we already have, $Z_1$, that summarizes other aspects of the individual's ability. It is again to find the linear combination to maximize the variance subject to the normalization and uncorrelatedness constraint. In general, the principal components can be defined sequentially. Given the first $k$ principal components $Z_l = \boldsymbol{\xi}_l^T \mathbf{X}$ for $1 \le l \le k$, the $(k+1)$th principal component is defined as $Z_{k+1} = \boldsymbol{\xi}_{k+1}^T \mathbf{X}$ with

$$\begin{aligned} \boldsymbol{\xi}_{k+1} &= \text{argmax}_{\boldsymbol{\xi}} \boldsymbol{\xi}^T \boldsymbol{\Sigma} \boldsymbol{\xi} \\ \text{subject to} \quad & \|\boldsymbol{\xi}\| = 1 \quad \text{and} \quad \boldsymbol{\xi}^T \boldsymbol{\xi}_l = 0, \ \forall 1 \le l \le k. \end{aligned} \tag{10.1}$$

In other words, it is defined as the projection direction that maximizes the variance subject to the orthogonality constraints. Note that the sign of *principal components* can not be determined as $\pm \boldsymbol{\xi}_{k+1}$ are both the solutions to (10.1). It will soon be seen that the orthogonality constraints here are equivalent to the uncorrelatedness constraints. Figure 10.1 illustrates the two principal components based on the sample covariance of a random sample depicted in the figure.

Assume that $\boldsymbol{\Sigma}$ has distinct eigenvalues. Then, $\boldsymbol{\xi}_1$ is the first eigenvector of $\boldsymbol{\Sigma}$ and $\lambda_1$ is the variance of $Z_1$. By induction, it is easy to see that the $k$th principal component $\boldsymbol{\xi}_k$ is the $k$th eigenvector of $\boldsymbol{\Sigma}$. Therefore, PCA can be found via spectral decomposition of the covariance matrix $\boldsymbol{\Sigma}$:

$$\boldsymbol{\Sigma} = \sum_{j=1}^p \lambda_j \boldsymbol{\xi}_j \boldsymbol{\xi}_j^T = \boldsymbol{\Gamma} \boldsymbol{\Lambda} \boldsymbol{\Gamma}^T, \tag{10.2}$$

where $\boldsymbol{\Gamma} = (\boldsymbol{\xi}_1, \cdots, \boldsymbol{\xi}_p)$ and $\boldsymbol{\Lambda} = \text{diag}(\lambda_1, \cdots, \lambda_p)$. Then, the $k$th principal component is given by $Z_k = \boldsymbol{\xi}_k^T \mathbf{X}$, in which $\boldsymbol{\xi}_k$ will be called the *loadings* of $Z_k$. Note that

$$\text{cov}(Z_j, Z_k) = \boldsymbol{\xi}_j^T \boldsymbol{\Sigma} \boldsymbol{\xi}_k = \lambda_k \boldsymbol{\xi}_j^T \boldsymbol{\xi}_k = \begin{cases} 0 & \text{if } j \neq k \\ \lambda_k & \text{if } j = k. \end{cases}$$

Hence, the orthogonality constraints and uncorrelatedness constraints are equivalent.

In practice, one replaces $\boldsymbol{\Sigma}$ by an estimate such as the sample covariance

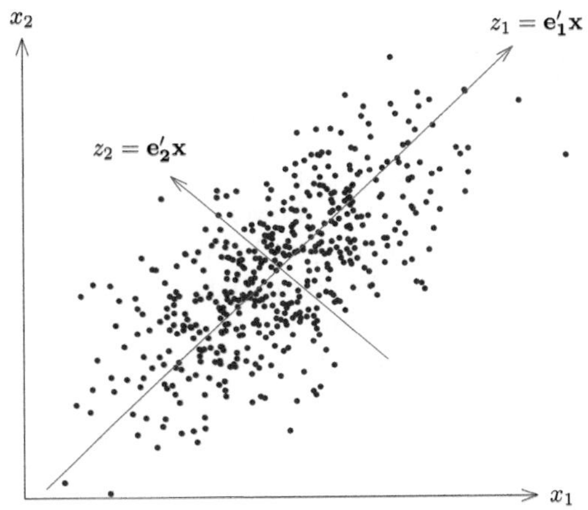


Figure 10.1: An illustration of PCA on a simulated dataset with two variables. $z_1$ is the first principal component and $z_2$ is the second one.

matrix. As an example, let us use the sample covariance matrix $\mathbf{S} = n^{-1}\widetilde{\mathbf{X}}\widetilde{\mathbf{X}}^T$, where $\widetilde{\mathbf{X}}$ is the $n \times p$ data matrix whose column means have been removed. One may replace $n^{-1}$ by $(n-1)^{-1}$ to achieve an unbiased estimate of $\mathbf{\Sigma}$, but it does not matter in PCA. Do the *singular value decomposition* (SVD) for the data matrix $\widetilde{\mathbf{X}}$

$$\widetilde{\mathbf{X}} = \mathbf{UDV}^T$$

where $\mathbf{D}$ is a diagonal matrix with elements $d_1, \cdots, d_p$ in a descending order, and $\mathbf{U}$ and $\mathbf{V}$ are is $n \times p$ and $p \times p$ orthonormal matrices, respectively. Then,

$$\mathbf{S} = n^{-1}\widetilde{\mathbf{X}}^T\widetilde{\mathbf{X}} = n^{-1}\mathbf{VD}^2\mathbf{V}^T.$$

Hence, the columns of $\mathbf{V}$ are the eigenvectors of $\mathbf{S}$. Let $\mathbf{v}_k$ be its $k^{th}$ column. The $k^{th}$ (estimated) principal component $z_k = \mathbf{v}_k^T\mathbf{X}$ with realizations $\{z_{k,i} = \mathbf{v}_k^T\mathbf{x}_i\}_{i=1}^n$. Put this as a vector $\mathbf{z}_k$. By $\widetilde{\mathbf{X}}\mathbf{V} = \mathbf{UD}$, we have $\mathbf{z}_k = d_k\mathbf{u}_k$, where $\mathbf{u}_k$ is the $k$th column of $\mathbf{U}$.

### *10.1.2 Power method*

In addition to computing the PCA by the SVD, the *power method* (also known as *power iteration*) is very simple to use. Given a diagonalizable matrix $\mathbf{\Sigma}$ in (10.2), its leading eigenvector can be computed iteratively via

$$\boldsymbol{\xi}_1^{(t)} = \frac{\mathbf{\Sigma}\boldsymbol{\xi}_1^{(t-1)}}{\|\mathbf{\Sigma}\boldsymbol{\xi}_1^{(t-1)}\|}, \quad t = 1, \cdots. \tag{10.3}$$

It requires that the initial value $\boldsymbol{\xi}_1^{(0)}$ is not orthogonal to the true eigenvector $\boldsymbol{\xi}_1$. Note that

$$\boldsymbol{\xi}_1^{(t)} = \frac{\boldsymbol{\Sigma}^t \boldsymbol{\xi}_1^{(0)}}{\|\boldsymbol{\Sigma}^t \boldsymbol{\xi}_1^{(0)}\|} = \frac{\boldsymbol{\Gamma}\,\mathrm{diag}(1, (\lambda_2/\lambda_1)^t, \cdots, (\lambda_p/\lambda_1)^t)\boldsymbol{\Gamma}^T \boldsymbol{\xi}_1^{(0)}}{\|\boldsymbol{\Gamma}\,\mathrm{diag}(1, (\lambda_2/\lambda_1)^t, \cdots, (\lambda_p/\lambda_1)^t \boldsymbol{\Gamma}^T) \boldsymbol{\xi}_1^{(0)}\|},$$

by using (10.2). Thus, the power suppresses smaller eigenvalues to zero with geometric convergence rate $(\lambda_2/\lambda_1)^t$. Power iteration is a very simple algorithm, but it may converge slowly if $\lambda_2 \approx \lambda_1$. With the computed eigenvector, its corresponding eigenvalue can be computed by using $\lambda_1 = \boldsymbol{\xi}_1^T \boldsymbol{\Sigma} \boldsymbol{\xi}_1$.

Now the second eigenvector can be computed by applying the power method to the matrix $\boldsymbol{\Sigma} - \lambda_1 \boldsymbol{\xi}_1 \boldsymbol{\xi}_1^T$. It is easy to see from (10.2), $\boldsymbol{\xi}_2$ will be correctly computed so long as $\lambda_3/\lambda_2 < 1$. In general, after computing the first $k$ eigenvalues and eigenvectors, we apply the power method to the matrix

$$\boldsymbol{\Sigma} - \sum_{j=1}^{k} \lambda_j \boldsymbol{\xi}_j \boldsymbol{\xi}_j^T$$

to obtain $\boldsymbol{\xi}_{k+1}$. After that, the eigenvalue $\lambda_{k+1}$ can be computed via $\lambda_{k+1} = \boldsymbol{\xi}_{k+1}^T \boldsymbol{\Sigma} \boldsymbol{\xi}_{k+1}$.

## 10.2 Factor Models and Structured Covariance Learning

Sparse covariance matrix assumption is not reasonable in many applications. For example, financial returns depend on common equity market risks; housing prices depend on economic health and locality; gene expressions can be stimulated by cytokines; fMRI data can have local correlation among their measurements. Due to the presence of common factors, it is unrealistic to assume that many outcomes are weakly correlated: in fact, the covariance matrix is dense due to the presence of the common factor (see Example 10.2 below). A natural extension is the conditional sparsity, namely, conditioning on the common factors, the covariance matrix of the outcome variables is sparse. This induces a factor model.

The factor model is one of the most useful tools for understanding the common dependence among multivariate outputs. It has broad applications in statistics, finance, economics, psychology, sociology, genomics, and computational biology, among others, where the dependence among multivariate measurements is strong. The following gives a formal definition. Recall the notation $[p] = \{1, \cdots, p\}$ for the index set.

**Definition 10.1** *Suppose that we have p-dimensional vector* $\mathbf{X}$ *of measurements, whose dependence is driven by* $K$ *factors* $\mathbf{f}$. *The factor model assumes*

$$\begin{aligned} \mathbf{X} &= \mathbf{a} + \mathbf{B}\mathbf{f} + \mathbf{u}, \qquad \mathrm{E}\,\mathbf{u} = 0, \quad \mathrm{cov}(\mathbf{f}, \mathbf{u}) = 0 \quad or \\ X_j &= a_j + \mathbf{b}_j^T \mathbf{f} + u_j = a_j + b_{j1} f_1 + \cdots + b_{jK} f_K + u_j. \end{aligned} \tag{10.4}$$

*for each component* $j = 1, \cdots, p$. *Here,* $\mathbf{u}$ *is the idiosyncratic component that is uncorrelated with the common factors* $\mathbf{f}$, $\mathbf{B} = (\mathbf{b}_1, \cdots, \mathbf{b}_p)^T$ *are the* $p \times$

*$K$ factor loading matrix with $b_{jk}$ indicating how much the $j^{th}$ outcome $X_j$ depends on the $k^{th}$ factor $f_k$, and* $\mathbf{a}$ *is an intercept term. When* $\mathbf{\Sigma}_u = \mathrm{Var}(\mathbf{u})$ *is diagonal, it is referred to as a strict factor model; when* $\mathbf{\Sigma}_u$ *is sparse, it is called an approximate factor model.*

The factor model induces a covariance structure for $\mathbf{\Sigma} = \mathrm{Var}(\mathbf{X})$:

$$\mathbf{\Sigma} = \mathbf{B}\mathbf{\Sigma}_f\mathbf{B}^T + \mathbf{\Sigma}_u, \qquad \mathbf{\Sigma}_f = \mathrm{Var}(\mathbf{f}), \quad \mathbf{\Sigma}_u = \mathrm{Var}(\mathbf{u}). \tag{10.5}$$

It is frequently assumed that $\mathbf{\Sigma}_u$ is sparse, since common dependence has already been taken out. This condition is referred to as the *conditional sparsity* at the beginning of this section. It is indeed the case when $\mathbf{f}$ and $\mathbf{u}$ are independent so that $\mathbf{\Sigma}_u$ is the conditional variance of $\mathbf{X}$ given $\mathbf{f}$.

### 10.2.1 Factor model and high-dimensional PCA

Suppose that we have observed $n$ data points $\{\mathbf{X}_i\}_{i=1}^n$ from the factor model (10.4):

$$\mathbf{X}_i = \mathbf{a} + \mathbf{B}\mathbf{f}_i + \mathbf{u}_i, \quad \text{or} \quad X_{ij} = a_j + \mathbf{b}_j^T\mathbf{f}_i + u_{ij}, \tag{10.6}$$

for each variable $j = 1, \cdots, p$. In many applications, the realized factors $\{\mathbf{f}_i\}_{i=1}^n$ are unknown and we need to learn them from the data. Intuitively, the larger the $p$, the better we can learn about the common factors $\{\mathbf{f}_i\}$. However, the loading matrix and the latent factors are not identifiable as we can write the model (10.6) as

$$\mathbf{X}_i = \mathbf{a} + (\mathbf{B}\mathbf{H})(\mathbf{H}^{-1}\mathbf{f}_i) + \mathbf{u}_i$$

for any invertible matrix $\mathbf{H}$. Thus, we can take $\mathbf{H}$ to make $\mathrm{Var}(\mathbf{H}^{-1}\mathbf{f}_i) = \mathbf{I}_p$. There are many such matrices $\mathbf{H}$ and we can even take the one such that the columns of $(\mathbf{B}\mathbf{H})$ are orthogonal. In addition, $\mathrm{E}\mathbf{X} = \mathbf{a} + \mathbf{B}\mathrm{E}\mathbf{f}$. Therefore, $\mathbf{a}$ and $\mathrm{E}\mathbf{f}$ can not be identified and to make them identifiable, we assume $\mathrm{E}\mathbf{f} = 0$ so that $\mathbf{a} = \mathrm{E}\mathbf{X}$. This leads to the identifiability condition as follows.

Identifiability Condition: $\mathbf{B}^T\mathbf{B}$ is diagonal, $\mathrm{E}\mathbf{f} = 0$ and $\mathrm{cov}(\mathbf{f}) = \mathbf{I}_p$.

Under the identifiability condition, we have the covariance structure

$$\mathbf{\Sigma} = \mathbf{B}\mathbf{B}^T + \mathbf{\Sigma}_u. \tag{10.7}$$

This matrix admits a low-rank + sparse structure, with the first matrix being of rank $K$. This leads naturally to the following penalized least-squares problem: minimize with respect to $\mathbf{\Theta}$ and $\mathbf{\Gamma}$

$$\|\mathbf{S} - \mathbf{\Theta} - \mathbf{\Gamma}\|_F^2 + \lambda\|\mathbf{\Theta}\|_* + \lambda\nu\sum_{i\neq j}|\gamma_{ij}| \tag{10.8}$$

where $\mathbf{S}$ is the sample covariance matrix and $\|\mathbf{\Theta}\|_*$ is the nuclear norm (Definition 9.2) that encourages the low-rankness and the second penalty encourages sparseness of $\mathbf{\Gamma} = (\gamma_{ij})$. The above formulation is also known as *Robust PCA* (Wright *et al.*, 2009; Candés *et al.*, 2011).

How can we differentiate the low-rank and sparse components? The analogy is that real numbers can be uniquely decomposed into integers + fractions. The separation between the first and the second part in (10.7) is the spikeness of the eigenvalues (Wright, *et al.*, 2009; Candés *et al.*, 2011; Agarwal, Negahban, and Wainwright 2012; Fan, Liao and Mincheva, 2013). It is assumed that the smallest non-zero eigenvalues for $\mathbf{B}\mathbf{B}^T$ are much bigger than the largest eigenvalue of $\mathbf{\Sigma}_u$ so that the eigenspace spanned by the nonzero eigenvalues of $\mathbf{B}\mathbf{B}^T$, which is the same as that spanned by the column space of $\mathbf{B}$, can be identified approximately as those of $\mathbf{\Sigma}$. This leads to the following condition. Let $\Omega(1)$ denote a quantity that is bounded away from 0 and $\infty$.

Pervasiveness assumption: The first $K$ eigenvalues of $\mathbf{B}\mathbf{B}^T$ have order $\Omega(p)$, whereas $\|\mathbf{\Sigma}_u\|_2 = O(1)$.

This condition is not the weakest possible (see Onatski, 2012), but is often imposed for convenience. It implies that the common factors affect a large fraction of the measurements in $\mathbf{X}$. If the factor loadings $\{\mathbf{b}_j\}_{j=1}^p$ are regarded as a realization from a non-degenerative $K$-dimensional population, then by the law of large numbers, we have

$$p^{-1}\sum_{j=1}^{p}\mathbf{b}_j\mathbf{b}_j^T \rightarrow \mathrm{E}\,\mathbf{b}\mathbf{b}^T,$$

which has non-vanishing eigenvalues. Thus, by the Weyl theorem (Proposition 9.2), the eigenvalues of $\mathbf{B}^T\mathbf{B}$ are of order $p$ and so are the non-vanishing eigenvalues of $\mathbf{B}\mathbf{B}^T$, as they are the same. The second condition in the pervasive assumption fulfills easily by the sparsity condition; see the equation just before (9.16).

Note that the non-vanishing eigenvalues of $\mathbf{B}\mathbf{B}^T$ and their associated eigenvectors are

$$\lambda_j(\mathbf{B}^T\mathbf{B}) = \|\widetilde{\mathbf{b}}_j\|^2, \qquad \xi_j(\mathbf{B}^T\mathbf{B}) = \widetilde{\mathbf{b}}_j/\|\widetilde{\mathbf{b}}_j\|, \tag{10.9}$$

where $\widetilde{\mathbf{b}}_j$ is the $j^{th}$ column of $\mathbf{B}$, ordered decreasingly in $\|\widetilde{\mathbf{b}}_j\|$ (Exercise 10.1). Applying Weyl's theorem and the Davis-Kahan sin $\theta$ theorem (Proposition 9.2), we have

**Proposition 10.1** *Let $\lambda_j = \lambda_j(\mathbf{\Sigma})$ and $\boldsymbol{\xi}_j = \boldsymbol{\xi}_j(\mathbf{\Sigma})$. Under the identifiability and pervasive condition, we have*

$$|\lambda_j - \|\widetilde{\mathbf{b}}_j\|_2^2| \le \|\mathbf{\Sigma}_u\|_2, \ \textit{for } j \le K, \qquad |\lambda_j| \le \|\mathbf{\Sigma}_u\|_2, \ \textit{for } j > K.$$

*Furthermore,*

$$\|\boldsymbol{\xi}_j - \widetilde{\mathbf{b}}_j/\|\widetilde{\mathbf{b}}_j\|_2\|_2 = O(p^{-1}\|\mathbf{\Sigma}_u\|_2), \qquad \textit{for all } j \le K.$$

We leave the proof of this proposition as an exercise (Exercise 10.2). The first part shows that the eigenvalues $\lambda_j = \Omega(p)$ for $j \le K$ and $\lambda_j = O(1)$ for $j > K$. Thus, the eigenvalues are indeed well separated. In addition, the second part shows that the principal component direction $\boldsymbol{\xi}_j$ approximates well the normalized $j^{th}$ column of the matrix $\mathbf{B}$, for $j \le K$. In other words

$$|\lambda_j/\|\widetilde{\mathbf{b}}_j\|_2^2 - 1| = O(p^{-1}), \qquad \widetilde{\mathbf{b}}_j = \lambda_j^{1/2}\boldsymbol{\xi}_j + O(p^{-1/2}), \tag{10.10}$$

for $j \le K$, where we used

$$\|\widetilde{\mathbf{b}}_j\| - \lambda_j^{1/2} = \frac{\|\widetilde{\mathbf{b}}_j\|^2 - \lambda_j}{\|\widetilde{\mathbf{b}}_j\| + \lambda_j^{1/2}} = O(p^{-1/2}).$$

In addition, by (10.6) and the orthogonality of matrix $\mathbf{B}$, we have

$$\widetilde{\mathbf{b}}_j^T(\mathbf{X}_i - \mathbf{a}) = \|\widetilde{\mathbf{b}}_j\|_2^2 \; f_{ij} + \widetilde{\mathbf{b}}_j^T\mathbf{u}_i$$

The last component is the weighted average of weakly correlated noises and is often negligible. Therefore, by using (10.10), we have

$$f_{ij} \approx \widetilde{\mathbf{b}}_j^T(\mathbf{X}_j - \mathbf{a})/\|\widetilde{\mathbf{b}}_j\|_2^2$$

(see Exercise 10.3) or in the matrix notation, by using (10.10),

$$\mathbf{f}_i \approx \mathrm{diag}(\lambda_1, \cdots, \lambda_K)^{-1}\mathbf{B}^T(\mathbf{X}_i - \mathbf{a}), \tag{10.11}$$

where $\mathbf{a} = E\mathbf{X}$ can be estimated by, for example, the sample mean. This demonstrates that factor models and PCA are approximately the same for high-dimensional problems with the pervasiveness assumption, though such a connection is much weaker for the finite dimensional problem (Jolliffe, 1986).

**Example 10.1** *Consider the specific example in which observed random vector $\mathbf{X}$ has the covariance matrix $\boldsymbol{\Sigma} = (1-\rho^2)\mathbf{I}_p + \rho^2\mathbf{1}\mathbf{1}^T$. This is an equi-correlation matrix. It can be decomposed as*

$$\mathbf{X} = \rho f \mathbf{1} + \sqrt{1-\rho^2}\mathbf{u}, \qquad \mathrm{Var}(\mathbf{u}) = \mathbf{I}_p$$

*In this case, the factor-loading $\mathbf{b} = \rho\mathbf{1}$. Averaging out both sides, the latent factor $f$ can be estimated through*

$$p^{-1}\sum_{j=1}^{p} X_j = \rho f + \sqrt{1-\rho^2}\, p^{-1}\sum_{j=1}^{p} u_j \approx \rho f.$$

*Applying this to each individual $i$, we estimate*

$$\widehat{f}_i = \rho^{-1}p^{-1}\sum_{j=1}^{p} X_{ij},$$

*which has error $\sqrt{1-\rho^2}\, p^{-1}\sum_{j=1}^{p} u_{ij} = O_p(p^{-1/2})$. This shows that the realized latent factors can be consistently estimated.*

### *10.2.2 Extracting latent factors and POET*

The connections between the principal component analysis and factor analysis in the previous section lead naturally to the following procedure based on the empirical data.

a) Obtain an estimator $\widehat{\boldsymbol{\mu}}$ and $\widehat{\boldsymbol{\Sigma}}$ of $\boldsymbol{\mu}$ and $\boldsymbol{\Sigma}$, e.g., the sample mean and covariance matrix or their robust versions.

b) Compute the eigen-decomposition of

$$\widehat{\boldsymbol{\Sigma}} = \sum_{j=1}^{p} \widehat{\lambda}_j \widehat{\boldsymbol{\xi}}_j \widehat{\boldsymbol{\xi}}_j^T. \tag{10.12}$$

Let $\{\widehat{\lambda}_k\}_{k=1}^K$ be the top $K$ eigenvalues and $\{\widehat{\boldsymbol{\xi}}_k\}_{k=1}^K$ be their corresponding eigenvectors.

c) Obtain the estimators for the factor loading matrix

$$\widehat{\mathbf{B}} = (\widehat{\lambda}_1^{1/2}\widehat{\boldsymbol{\xi}}_1, \cdots, \widehat{\lambda}_K^{1/2}\widehat{\boldsymbol{\xi}}_K) \tag{10.13}$$

and the latent factors

$$\widehat{\mathbf{f}}_i = \mathrm{diag}(\widehat{\lambda}_1, \cdots, \widehat{\lambda}_K)^{-1}\widehat{\mathbf{B}}^T(\mathbf{X}_i - \widehat{\boldsymbol{\mu}}), \tag{10.14}$$

namely, $\widehat{\mathbf{B}}$ consists of the top-$K$ rescaled eigenvectors of $\widehat{\boldsymbol{\Sigma}}$ and $\widehat{\mathbf{f}}_i$ is just the rescaled projection of $\mathbf{X}_i - \widehat{\boldsymbol{\mu}}$ onto the space spanned by the eigenvectors.

d) Obtain a pilot estimator for $\boldsymbol{\Sigma}_u$ by $\widehat{\boldsymbol{\Sigma}}_u = \sum_{j=K+1}^p \widehat{\lambda}_j \widehat{\boldsymbol{\xi}}_j \widehat{\boldsymbol{\xi}}_j^T$ and its regularized estimator $\widehat{\boldsymbol{\Sigma}}_{u,\lambda}^{\mathcal{T}}$ by using the correlation thresholding estimator (9.13).

e) Estimate the covariance matrix $\boldsymbol{\Sigma} = \mathrm{Var}(\mathbf{X})$ by

$$\widehat{\boldsymbol{\Sigma}}_\lambda = \widehat{\mathbf{B}}\widehat{\mathbf{B}}^T + \widehat{\boldsymbol{\Sigma}}_{u,\lambda}^{\mathcal{T}} = \sum_{j=1}^{K} \widehat{\lambda}_j \widehat{\boldsymbol{\xi}}_j \widehat{\boldsymbol{\xi}}_j^T + \widehat{\boldsymbol{\Sigma}}_{u,\lambda}^{\mathcal{T}}. \tag{10.15}$$

The use of the high-dimensional PCA to estimate latent factors has been studied by a number of authors. In particular, Bai (2003) proves the asymptotic normalities of estimated latent factors, estimated loading matrix and their products. The covariance estimator (10.15) is introduced in Fan, Liao and Mincheva (2013) and is referred to as the Principal Orthogonal ComplEment Thresholding (*POET*) estimator. They demonstrate that POET is insensitive to the over estimation on the number of factors $K$.

When a part of the factors is known such as the Fama-French 3-factor models, we can subtract these factors' effect out by running the regression as in Section 10.3.1 below to obtain the covariance matrix $\widehat{\boldsymbol{\Sigma}}_u$ of residuals $\{\mathbf{u}_i\}_{i=1}^n$ therein. Now apply the PCA above further to $\widehat{\boldsymbol{\Sigma}}_u$ to extract extra latent factors. Note that these extra latent factors are orthogonal (uncorrelated) with the initial factors and augmented the original observed factors.

The above procedure reveals that

$$\widehat{f}_{ij} = \widehat{\lambda}_j^{-1/2} \widehat{\boldsymbol{\xi}}_j^T (\mathbf{X}_i - \widehat{\boldsymbol{\mu}}) \quad \text{so that} \quad \widehat{f}_{ij} - \overline{\widehat{f}}_j = \widehat{\lambda}_j^{-1/2} \widehat{\boldsymbol{\xi}}_j^T (\mathbf{X}_i - \overline{\mathbf{X}}).$$

Therefore, the sample covariance between the estimated $j^{th}$ and $k^{th}$ factor is

$$\widehat{\lambda}_j^{-1/2} \widehat{\lambda}_k^{-1/2} \widehat{\boldsymbol{\xi}}_j^T \mathbf{S} \widehat{\boldsymbol{\xi}}_k,$$

where $\mathbf{S}$ is the sample covariance matrix defined by (9.7). It does not depend on the input $\widehat{\boldsymbol{\mu}}$. When the sample covariance is used as the initial estimator, these estimated factors $\{\widehat{f}_{ij}\}_{i=1}^n$ and $\{\widehat{f}_{ik}\}_{i=1}^n$ are uncorrelated for $j \neq k$ with sample variance 1. In general, they should be approximately uncorrelated with the sample variance approximately 1.

In many applications, it is typically assumed that the mean of each variable has been removed so that intercepts $\mathbf{a}$ can be set to zero. When the initial covariance matrix is the sample covariance matrix (9.7), the solution admits the least-squares interpretation. From (10.6), it is natural to find $\mathbf{B}$ and $\mathbf{F} = (\mathbf{f}_1, \cdots, \mathbf{f}_n)^T$ to minimize

$$\sum_{i=1}^n \|\mathbf{X}_i - \mathbf{B}\mathbf{f}_i\|^2 = \|\mathbf{X} - \mathbf{F}\mathbf{B}^T\|_F^2 \tag{10.16}$$

subject to $n^{-1}\sum_{i=1}^n \mathbf{f}_i\mathbf{f}_i^T = \mathbf{I}_K$ and $\mathbf{B}^T\mathbf{B}$ diagonal (the identifiability condition). For given $\mathbf{B}$, the solution for the least-squares problem is

$$\widehat{\mathbf{f}}_i = \text{diag}(\mathbf{B}^T\mathbf{B})^{-1}\mathbf{B}^T\mathbf{X}_i$$

or in the matrix form $\widehat{\mathbf{F}} = \mathbf{XBD}$ where $\mathbf{X}$ is a $n \times p$ data matrix and $\mathbf{D} = \text{diag}(\mathbf{B}^T\mathbf{B})^{-1}$. Substituting it in (10.16), the objective function now becomes

$$\|\mathbf{X}(\mathbf{I}_p - \mathbf{BDB}^T)\|_F^2 = \text{tr}[(\mathbf{I}_p - \mathbf{BDB}^T)\mathbf{X}^T\mathbf{X}].$$

The solution of this problem is the same as the one given in step c) above, namely, $\{\|\widetilde{\mathbf{b}}_j\|^2\}_{j=1}^K$ are the top $K$ eigenvalues of the sample covariance matrix $n^{-1}\mathbf{X}^T\mathbf{X}$ and the columns $\{\widetilde{\mathbf{b}}_j/\|\widetilde{\mathbf{b}}_j\|\}_{j=1}^K$ are their associated eigenvectors $\{\widehat{\boldsymbol{\xi}}_j\}_{j=1}^K$ (Exercise 10.5). This gives a least-squares interpretation.

We can derive similarly that for given $\mathbf{F}$, since the columns of the design matrix $\mathbf{F}$ are orthonormal, the least-squares estimator is $\widehat{\mathbf{B}} = n^{-1}\mathbf{X}^T\mathbf{F}$. Substituting it in, the objective function now becomes

$$\|\mathbf{X} - \mathbf{F}\mathbf{F}^T\mathbf{X}/n\|_F^2 = \text{tr}[(\mathbf{I}_n - \mathbf{F}\mathbf{F}^T/n)\mathbf{X}\mathbf{X}^T].$$

The solution to this problem is that the columns of $\widehat{\mathbf{F}}/\sqrt{n}$ are the top $K$ eigenvectors of the $n \times n$ matrix $\mathbf{XX}^T$ and $\widehat{\mathbf{B}} = n^{-1}\mathbf{X}^T\widehat{\mathbf{F}}$ (Stock and Watson, 2002). This provides an alternative formula for computing $\widehat{\mathbf{B}}$ and $\widehat{\mathbf{F}}$. It involves computing top $K$ eigenvalues and their associated eigenvectors for a $n \times n$ matrix $\mathbf{XX}^T$, instead of a $p \times p$ sample covariance matrix $\mathbf{X}^T\mathbf{X}$. If $p \gg n$, this method is much faster to compute. We summarize the result as the following proposition. We leave the second result as an exercise (Exercise 10.6).

**Proposition 10.2** *The solution to the least-squares problem* (10.16) *is*

$$\widehat{\mathbf{F}} = \sqrt{n} \times \text{top } K \text{ eigenvectors of } \mathbf{X}\mathbf{X}^T \quad \text{and} \quad \widehat{\mathbf{B}} = n^{-1}\mathbf{X}^T\widehat{\mathbf{F}}.$$

*This is an alternative formula for estimating latent factors based on the sample covariance matrix* $\mathbf{S}$, *namely, it is the same as* (10.13) *and* (10.14) *with* $\widehat{\boldsymbol{\Sigma}} = \mathbf{S}$, *when* $\bar{\mathbf{X}} = 0$ *(the demeaned data).*

The advantages of the robust covariance inputs have been demonstrated by Fan, Liu and Wang (2018) and Fan, Wang and Zhong (2019) for different covariance inputs. As an illustration, we simulate data from the factor model with $(\mathbf{f}_i, \mathbf{u}_i)$ jointly following a multivariate t-distribution with degrees of freedom $\nu$ whose covariance matrix is given by $\text{diag}(\mathbf{I}_K, 5\mathbf{I}_p)$ and factor loadings from $N(\mathbf{0}, \mathbf{I}_K)$. Larger $\nu$ corresponds to a lighter tail and $\nu = \infty$ corresponds to a multivariate normal distribution. Note that the observed data is generated as $\mathbf{X}_i = \mathbf{B}\mathbf{f}_i + \mathbf{u}_i$, which requires us to have the matrix $\mathbf{B}$ first and then jointly drawing $(\mathbf{f}_i, \mathbf{u}_i)$ in order to compute $\mathbf{X}_i$. Each row of the loading matrix $\mathbf{B}$ is independently sampled from a standard normal distribution. The true covariance is $\boldsymbol{\Sigma} = \mathbf{B}\mathbf{B}^T + 5\mathbf{I}_p$ in this factor model.

We vary $p$ from 200 to 900 with sample size $n = p/2$, and fixed number of factors $K = 3$. The thresholding parameter in (9.13) was set to $\lambda = 2$. For data simulated from heavy distribution ($\nu = 3$), moderately heavy distribution ($\nu = 5$) and the light distribution ($\nu = \infty$, normal), we compared the performance of the following four methods

(1) the sample covariance;

(2) the adaptive Huber's robust estimator;

(3) the marginal Kendall's tau estimator; [marginal estimated variances are taken from (2)]

(4) the spatial Kendall's tau estimator.

For each method, we compute the estimation errors $\|\widehat{\boldsymbol{\Sigma}}_u^{\mathcal{T}} - \boldsymbol{\Sigma}_u\|$, $\|(\widehat{\boldsymbol{\Sigma}}^{\mathcal{T}})^{-1} - \boldsymbol{\Sigma}^{-1}\|$ and $\|\widehat{\boldsymbol{\Sigma}}^{\mathcal{T}} - \boldsymbol{\Sigma}\|_{\Sigma}$ (see the definition after Example 10.2) and report its ratio with respect to method (1), the sample covariance method. The results are depicted in Figure 10.2 based on 100 simulations. For the first two settings, all robust methods outperform the sample covariance; more so for $\nu = 3$. For the third setting (normal distribution), method (1) outperforms, but the price that we pay is relatively small.

### *10.2.3 Methods for selecting number of factors*

In applications, we need to choose the number of factors $K$ before estimating the loading matrix, factors, and so on. The number $K$ can be usually estimated from the eigenvalues of the pilot covariance estimator $\widehat{\boldsymbol{\Sigma}}$. Classical methods include likelihood ratio tests (Bartlett, 1950), the scree plot (Cattell,

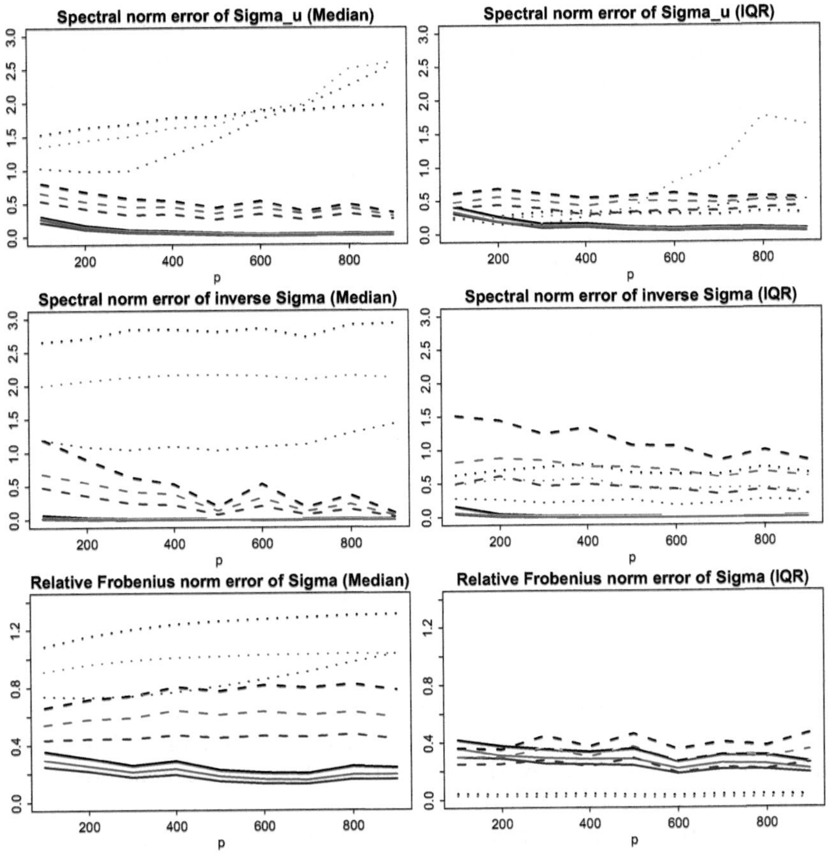


Figure 10.2: Error ratios of robust estimates against varying dimensions. Blue lines represent errors of Method (2) over Method (1) under different norms; black lines errors of Method (3) over Method (1); red lines errors of Method (4) over Method (1). The median errors and their IQR's (interquartile range) over 100 simulations are reported. Taken from Fan, Wang, Zhong (2019).

1966), among others. The latter basically examines the percentage of variance explained by the top $k$ eigenvectors, defined as

$$\widehat{p}_k = \frac{\sum_{j=1}^{k} \lambda_j(\widehat{\boldsymbol{\Sigma}})}{\sum_{j=1}^{p} \lambda_j(\widehat{\boldsymbol{\Sigma}})}.$$

One chooses such a $k$ so that $\widehat{p}_k$ fails to increase noticeably.

Let $\widehat{\mathbf{R}} = \mathrm{diag}(\widehat{\boldsymbol{\Sigma}})^{-1/2}\widehat{\boldsymbol{\Sigma}}\,\mathrm{diag}(\widehat{\boldsymbol{\Sigma}})^{-1/2}$ be the correlation matrix with corresponding eigenvalues $\{\lambda_j(\widehat{\mathbf{R}})\}$. A classical approach is to choose

$$\widehat{K}_1 = \#\{j : \lambda_j(\widehat{\mathbf{R}}) > 1\}$$

as the number of factors. See, for example, Guttman (1954) and Johnson and Wichern (2007, page 491). When $p$ is comparable with $n$, an appropriate modification is the following *Adjusted Eigenvalues Thresholding* (ACT) estimator by Fan, Guo and Zheng, 2020)

$$\widehat{K}_2 = \#\{j : \lambda_j^C(\widehat{\mathbf{R}}) > 1 + \sqrt{p/n}\},$$

where $\lambda_j^C(\widehat{\mathbf{R}})$ the bias-corrected estimator of the $j^{th}$ largest eigenvalue. This is a tuning-free and scale-invariant method.

Here, we introduce a few more recent methods: the first one is based on the eigenvalue ratio, the second one based on eigenvalue differences, and the third one based on information criteria. They are in the order of progressive complexity. As noted before, for estimating covariance matrix $\mathbf{\Sigma}$, POET estimator $\widehat{\mathbf{\Sigma}}_\lambda$ is insensitive to the overestimation of $K$, but is sensitive to the underestimation of $K$.

For a pre-determined $k_{\max}$, the *eigenvalue ratio estimator* is

$$\widehat{K}_3 = \operatorname*{argmax}_{j \le k_{\max}} \frac{\lambda_j(\widehat{\mathbf{\Sigma}})}{\lambda_{j+1}(\widehat{\mathbf{\Sigma}})}, \tag{10.17}$$

where $k_{\max}$ is a predetermined parameter. This method was introduced by Luo, Wang, and Tsai (2009), Lam and Yao (2012) and Ahn and Horenstein (2013). Intuitively, when the signal eigenvalues are well separated from the rest of the eigenvalues, the ratio at $k = K$ should be the largest. Under some conditions, the consistency of this estimator, which does not involve tuning parameters, has been established.

Onatski (2010) proposed to use the differences of consecutive eigenvalues. For a given $\delta > 0$ and pre-determined integer $k_{\max}$, define

$$\widehat{K}_4(\delta) = \max\{j \le k_{\max} : \lambda_j(\widehat{\mathbf{\Sigma}}) - \lambda_{j+1}(\widehat{\mathbf{\Sigma}}) \ge \delta\}.$$

Using a result on the empirical distribution of eigenvalues from random matrix theory, Onatski (2010) proved the consistency of $\widehat{K}_2(\delta)$ under the pervasiveness assumption. The paper also proposed a data-driven way to determine $\delta$ from the empirical eigenvalue distribution of the sample covariance matrix.

A third possibility is to use an information criterion. Define

$$V(k) = \frac{1}{np} \min_{\mathbf{B} \in I\!R^{p\times k}, \mathbf{F} \in I\!R^{n \times k}} \|\mathbf{X} - \mathbf{1}_n \overline{\mathbf{X}}^T - \mathbf{F}\mathbf{B}^T\|_F^2 = p^{-1} \sum_{j>k} \lambda_j(\mathbf{S}),$$

where the second equality is well known (Exercise 10.7). For a given $k$, $V(k)$ is interpreted as the sum of squared residuals, which measures how well $k$ factors fit the data. A natural estimator $\widehat{K}_3$ is to find the best $k \le k_{\max}$ such that the following penalized version of $V(k)$ is minimized (Bai and Ng, 2002):

$$PC(k) = V(k) + k\,\widehat{\sigma}^2 g(n,p), \qquad \text{where} \quad g(n,p) := \frac{n+p}{np} \log\left(\frac{np}{n+p}\right),$$

and $\widehat{\sigma}^2$ is any consistent estimate of $(np)^{-1}\sum_{i=1}^n\sum_{j=1}^d \mathrm{E}\, u_{ji}^2$. Consistency results are established under more general choices of $g(n,p)$.

## 10.3 Covariance and Precision Learning with Known Factors

In some applications, factors are observable such as the famous *Fama-French* 3 factors or 5 factors models in finance (Fama and French, 1992, 2015). In this case, the available data are $\{(\mathbf{f}_i, \mathbf{X}_i)\}_{i=1}^n$ and the factor model is really a *multivariate linear model* (10.6). This model also sheds lights on the accuracy of covariance learning when the factors are unobservable.

In other applications, factors can be partially known. In this case, we can run the regression first to obtain the residuals. Then, apply the methods in Section 10.2.2 to an estimate of the residual covariance matrix. This gives new estimated factors and the regularized estimation of the covariance matrix. They can be further used for prediction and portfolio theory. See Chapter 11.

### *10.3.1 Factor model with observable factors*

Running the multiple regression to the model (10.6), we obtain the least-squares estimator for the regression coefficients in each component:

$$(\widehat{a}_j, \widehat{\mathbf{b}}_j^T)^T = \operatorname*{arg\,min}_{a_j, \mathbf{b}_j} \sum_{i=1}^n (X_{ij} - a_j - \mathbf{b}_j^T \mathbf{f}_i)^2.$$

This yields an estimator of the loading matrix $\widehat{\mathbf{B}} = (\widehat{\mathbf{b}}_1^T, \cdots, \widehat{\mathbf{b}}_p^T)^T$. Let

$$\widehat{\mathbf{u}}_i = (\widehat{u}_{i1}, \cdots, \widehat{u}_{ip})^T \quad \text{with} \quad \widehat{u}_{ij} = X_{ij} - \widehat{a}_j - \widehat{\mathbf{b}}_j^T \mathbf{f}_i$$

be the residual of model (10.6). Denote by $\widehat{\boldsymbol{\Sigma}}_u$ the sample covariance of the residual vectors $\{\widehat{\mathbf{u}}_i\}_{i=1}^n$.

Since $\boldsymbol{\Sigma}_u$ is sparse, we can regularize the estimator $\widehat{\boldsymbol{\Sigma}}_u$ by the correlation thresholding (9.13), resulting in $\widehat{\boldsymbol{\Sigma}}_{u,\lambda}$. This leads to the regularized estimation of covariance matrix by

$$\widehat{\boldsymbol{\Sigma}}_\lambda = \widehat{\mathbf{B}}\widehat{\boldsymbol{\Sigma}}_f\widehat{\mathbf{B}}^T + \widehat{\boldsymbol{\Sigma}}_{u,\lambda}, \tag{10.18}$$

where $\widehat{\boldsymbol{\Sigma}}_f$ is the sample covariance matrix of $\{\mathbf{f}_i\}_{i=1}^n$. In particular, when $\lambda = 0$, $\widehat{\boldsymbol{\Sigma}}_{u,\lambda} = \widehat{\boldsymbol{\Sigma}}_u$, the sample covariance matrix of $\{\widehat{\mathbf{u}}_i\}_{i=1}^n$, and $\widehat{\boldsymbol{\Sigma}}_\lambda$ reduces to the sample covariance matrix of $\{\mathbf{X}_i\}_{i=1}^n$. When $\lambda = \sqrt{n/\log p}$, which is equivalent to applying thresholding 1 to the correlation matrix so that off-diagonals are set to zero, $\widehat{\boldsymbol{\Sigma}}_{u,\lambda} = \operatorname{diag}(\widehat{\boldsymbol{\Sigma}}_u)$, and our estimator becomes

$$\widehat{\boldsymbol{\Sigma}} = \widehat{\mathbf{B}}\widehat{\boldsymbol{\Sigma}}_f\widehat{\mathbf{B}}^T + \operatorname{diag}(\widehat{\boldsymbol{\Sigma}}_u). \tag{10.19}$$

This estimator, proposed and studied in Fan, Fan, Lv (2008), is always positive definite and is suitable for the *strict factor model*.

Let $m_{u,p}$ be the sparsity measure (9.15) applied to $\mathbf{\Sigma}_u$ and $r_n$ be the uniform rate for the residuals in $L_2$:

$$\max_{j\in[p]} n^{-1}\sum_{i=1}^{n}(\widehat{u}_{ij}-u_{ij})^2 = O_P(r_n). \tag{10.20}$$

Under some regularity conditions, Fan, Liao and Mincheva (2011) showed that

$$\|\widehat{\mathbf{\Sigma}}_u - \mathbf{\Sigma}_u\|_2 = O_P(m_{u,p}\omega_n^{1-q}), \quad \|\widehat{\mathbf{\Sigma}}_u^{-1} - \mathbf{\Sigma}_u^{-1}\|_2 = O_P(m_{u,p}\omega_n^{1-q}),$$

where $\omega_n = K\sqrt{\frac{\log p}{n}} + r_n$. The results can be obtained by checking the conditions of pilot estimator $\widehat{\mathbf{\Sigma}}_u$ (see Theorem 9.1). For the sub-Gaussian case, it can be shown that $r_n = O(\sqrt{\frac{\log p}{n}})$ so that the rates are the same as those given in Theorem 9.1.

What are the benefits of using the regularized estimator (10.18) or (10.19)? Compared with the sample covariance matrix, Fan, Fan and Lv (2008) showed that the factor-model based estimator has a better rate for estimating $\widehat{\mathbf{\Sigma}}^{-1}$ and the same rate for estimating $\mathbf{\Sigma}$. This is also verified by the extensive studies therein. The following simple example, given in Fan, Fan, Lv (2008), sheds some light upon this.

**Example 10.2** *Consider the specific case $K = 1$ with the known loading $\mathbf{B} = \mathbf{1}_p$ and $\mathbf{\Sigma}_u = \mathbf{I}_p$, where $\mathbf{1}_p$ is a p-dimensional vector with all elements 1. Then $\mathbf{\Sigma} = \sigma_f^2\mathbf{1}_p\mathbf{1}_p' + \mathbf{I}_p$, and we only need to estimate $\sigma_f^2 \equiv \mathrm{Var}(f)$ using, say, the sample variance $\widehat{\sigma}_f^2 = \frac{1}{n}\sum_{i=1}^n(f_i - \bar{f})^2$. Then, it follows that*

$$\|\widehat{\mathbf{\Sigma}} - \mathbf{\Sigma}\|_2 = p|\widehat{\sigma}_f^2 - \sigma_f^2|.$$

*Therefore, by the central limit theorem, $\frac{\sqrt{n}}{p}\|\widehat{\mathbf{\Sigma}} - \mathbf{\Sigma}\|_2$ is asymptotically normal. Hence $\|\widehat{\mathbf{\Sigma}} - \mathbf{\Sigma}\|_2 = O_p(p/\sqrt{n})$ even for such a toy model, which is not consistent when $p \asymp \sqrt{n}$.*

*To provide additional insights, the eigenvalues for this toy model are $(1+\sigma_f^2)p, 1, \cdots, 1$. They are very spiked. The largest eigenvalues can at best be estimated only at the rate $O_p(p/\sqrt{n})$. On the other hand, $\mathbf{\Sigma}^{-1}$ has eigenvalues $(1+\sigma_f^2)^{-1}/p, 1, \cdots, 1$, which are much easier to estimate, turning the spike ones into an easily estimated one.*

The above example inspires Fan, Fan and Lv (2008) to consider the relative losses based on the discrepancies between the matrix $\mathbf{\Sigma}^{-1/2}\widehat{\mathbf{\Sigma}}\mathbf{\Sigma}^{-1/2}$ and $\mathbf{I}_p$. Letting $\|\mathbf{A}\|_\Sigma = p^{-1/2}\|\mathbf{\Sigma}^{-1/2}\mathbf{A}\mathbf{\Sigma}^{-1/2}\|_F$, they define the quadratic loss

$$\|\widehat{\mathbf{\Sigma}} - \mathbf{\Sigma}\|_\Sigma = p^{-1/2}\|\mathbf{\Sigma}^{-1/2}\widehat{\mathbf{\Sigma}}\mathbf{\Sigma}^{-1/2} - \mathbf{I}_p\|_F$$

and the entropy loss (James and Stein, 1961; see also (9.8))

$$\mathrm{tr}(\widehat{\mathbf{\Sigma}}\mathbf{\Sigma}^{-1}) - \log|\widehat{\mathbf{\Sigma}}\mathbf{\Sigma}^{-1}| - p.$$

These two metrics are highly related, but the former is much easier to manipulate mathematically. To see this, let $\widetilde{\lambda}_1, \cdots, \widetilde{\lambda}_p$ be the eigenvalues of the matrix $\mathbf{\Sigma}^{-1/2}\widehat{\mathbf{\Sigma}}\mathbf{\Sigma}^{-1/2}$. Then

$$\|\mathbf{\Sigma}^{-1/2}\widehat{\mathbf{\Sigma}}\mathbf{\Sigma}^{-1/2} - \mathbf{I}_p\|_F^2 = \sum_{j=1}^{p}(\widetilde{\lambda}_j - 1)^2,$$

whereas the entropy loss can be written as

$$\sum_{j=1}^{p}(\widetilde{\lambda}_j - 1 - \log\widetilde{\lambda}_j) \approx \frac{1}{2}\sum_{j=1}^{p}(\widetilde{\lambda}_j - 1)^2,$$

by Taylor expansion at $\widetilde{\lambda}_j = 1$. Therefore, they are approximately the same (modulus a constant factor) when $\widehat{\mathbf{\Sigma}}$ is close to $\mathbf{\Sigma}$.

The discrepancy of $\mathbf{\Sigma}^{-1/2}\widehat{\mathbf{\Sigma}}\mathbf{\Sigma}^{-1/2}$ and $\mathbf{I}_p$ is also related to the relative estimation errors of *portfolio risks*. Thinking of $\mathbf{X}$ as a vector of returns of $p$ assets, $\mathbf{w}^T\mathbf{X}$ is the return of the portfolio with the allocation vector $\mathbf{w}$, whose portfolio variance is given by $\text{Var}(\mathbf{w}^T\mathbf{X}) = \mathbf{w}^T\mathbf{\Sigma}\mathbf{w}$. The relative estimation error of the portfolio variance is bounded by

$$\sup_{\mathbf{w}} |\mathbf{w}^T\widehat{\mathbf{\Sigma}}\mathbf{w}/\mathbf{w}^T\mathbf{\Sigma}\mathbf{w} - 1| = \|\mathbf{\Sigma}^{-1/2}\widehat{\mathbf{\Sigma}}\mathbf{\Sigma}^{-1/2} - \mathbf{I}_p\|_2,$$

another measure of relative matrix estimation error. Note that the absolute error is bounded by

$$|\mathbf{w}^T\widehat{\mathbf{\Sigma}}\mathbf{w} - \mathbf{w}^T\mathbf{\Sigma}\mathbf{w}| \le \|\widehat{\mathbf{\Sigma}} - \mathbf{\Sigma}\|_{\max}\|\mathbf{w}\|_1^2.$$

In comparison with the sample covariance, Fan, Fan, Lv (2008) showed that there are substantial gains in estimating quantities that involve $\mathbf{\Sigma}^{-1}$ using regularized estimator (10.19), whereas there are no gains for estimating the quantities that use $\mathbf{\Sigma}$ directly. Fan, Liao and Mincheva (2011) established the rates of convergence for a regularized estimator (10.18).

### *10.3.2 Robust initial estimation of covariance matrix*

Fan, Wang and Zhong (2019) generalized the theory and methods for robust versions of the covariance estimators when factors are observable. Let $\mathbf{Z}_i = (\mathbf{X}_i^T, \mathbf{f}_i^T)^T$. Then, its covariance admits the form (Exercise 10.10)

$$\mathbf{\Sigma}_z = \begin{pmatrix} \mathbf{B}\mathbf{\Sigma}_f\mathbf{B}^T + \mathbf{\Sigma}_u & \mathbf{B}\mathbf{\Sigma}_f \\ \mathbf{\Sigma}_f\mathbf{B}^T & \mathbf{\Sigma}_f \end{pmatrix} =: \begin{pmatrix} \mathbf{\Sigma}_{11} & \mathbf{\Sigma}_{12} \\ \mathbf{\Sigma}_{21} & \mathbf{\Sigma}_{22} \end{pmatrix}.$$

Using the estimator $\widehat{\mathbf{\Sigma}}_z$, we estimate $\mathbf{B}\mathbf{\Sigma}_f\mathbf{B}^T$ through the identity

$$\mathbf{B}\mathbf{\Sigma}_f\mathbf{B}^T = \mathbf{\Sigma}_{12}\mathbf{\Sigma}_{22}^{-1}\mathbf{\Sigma}_{21}$$

and obtain

$$\widehat{\mathbf{\Sigma}}_u = \widehat{\mathbf{\Sigma}}_{11} - \widehat{\mathbf{\Sigma}}_{12}\widehat{\mathbf{\Sigma}}_{22}^{-1}\widehat{\mathbf{\Sigma}}_{21}.$$

Note that each element of the $p \times p$ matrix $\mathbf{\Sigma}_{12}\mathbf{\Sigma}_{22}^{-1}\mathbf{\Sigma}_{21}$ involves summation of $K^2$ elements and each element is estimated with rate $\sqrt{\log p/n}$. Therefore, we would expect that

$$\|\widehat{\mathbf{\Sigma}}_u - \mathbf{\Sigma}_u\|_{\max} = O_p(\omega_n), \qquad \omega_n = K^2\sqrt{\frac{\log p}{n}}.$$

Here, we allow $K$ to slowly grow with $n$. Apply the correlation thresholding (9.13) to obtain $\widehat{\mathbf{\Sigma}}_u^{\mathcal{T}}$ and produce the final estimator for $\mathbf{\Sigma}$:

$$\widehat{\mathbf{\Sigma}}^{\mathcal{T}} = \widehat{\mathbf{\Sigma}}_{12}\widehat{\mathbf{\Sigma}}_{22}^{-1}\widehat{\mathbf{\Sigma}}_{21} + \widehat{\mathbf{\Sigma}}_u^{\mathcal{T}}.$$

As noted in Remark 9.1, the thresholding level should now be $\lambda\omega_n\sqrt{\widehat{\sigma}_{u,ii}\widehat{\sigma}_{u,jj}}$, with aforementioned $\omega_n$.

**Theorem 10.1** *Suppose that the minimum and largest eigenvalues of* $\mathbf{\Sigma}_u \in C_q(m_p)$ *and* $\mathbf{\Sigma}_f$ *are bounded from above and below and that* $\|\mathbf{B}\|_{\max}$ *is bounded and* $\lambda_K(\mathbf{\Sigma}) > cp$ *for some* $c > 0$*. If the initial estimator satisfies*

$$\|\widehat{\mathbf{\Sigma}}_z - \mathbf{\Sigma}_z\|_\infty = O_P(\sqrt{\log p/n}) \tag{10.21}$$

*and* $Km_p\omega_n^{1-q} \to 0$ *with* $\omega_n = K^2\sqrt{\log p/n}$*, then we have*

$$\|\widehat{\mathbf{\Sigma}}_u^{\mathcal{T}} - \mathbf{\Sigma}_u\|_2 = \|(\widehat{\mathbf{\Sigma}}_u^{\mathcal{T}})^{-1} - \mathbf{\Sigma}_u^{-1}\|_2 = O_P\Big(m_p\omega_n^{1-q}\Big), \tag{10.22}$$

*and*

$$\begin{aligned} \|\widehat{\mathbf{\Sigma}}^{\mathcal{T}} - \mathbf{\Sigma}\|_{\max} &= O_P\Big(\omega_n\Big), \\ \|\widehat{\mathbf{\Sigma}}^{\mathcal{T}} - \mathbf{\Sigma}\|_{\Sigma} &= O_P\Big(\frac{K\sqrt{p}\log p}{n} + m_p\omega_n^{1-q}\Big), \\ \|(\widehat{\mathbf{\Sigma}}^{\mathcal{T}})^{-1} - \mathbf{\Sigma}^{-1}\|_2 &= O_P\Big(K^2 m_p\omega_n^{1-q}\Big). \end{aligned} \tag{10.23}$$

The above results are applicable to the robust covariance estimators in Section 9.3, since they satisfy the elementwise convergence condition (10.21). They are also applicable to the sample covariance matrix, when data $\mathbf{Z}_i$ admits sub-Gaussian tails. In particular, when $K$ is bounded, we have $\omega_n = \sqrt{\log p/n}$. If, in addition, $q = 0$, then the rates now become

$$\|\widehat{\mathbf{\Sigma}}_u^{\mathcal{T}} - \mathbf{\Sigma}_u\|_2 = \|(\widehat{\mathbf{\Sigma}}_u^{\mathcal{T}})^{-1} - \mathbf{\Sigma}_u^{-1}\|_2 = O_P\Big(m_p\sqrt{\frac{\log p}{n}}\Big),$$

and

$$\begin{aligned} \|\widehat{\mathbf{\Sigma}}^{\mathcal{T}} - \mathbf{\Sigma}\|_{\max} &= O_P\Big(\sqrt{\tfrac{\log p}{n}}\Big), \\ \|\widehat{\mathbf{\Sigma}}^{\mathcal{T}} - \mathbf{\Sigma}\|_{\Sigma} &= O_P\Big(\tfrac{\sqrt{p}\log p}{n} + m_p\sqrt{\tfrac{\log p}{n}}\Big), \\ \|(\widehat{\mathbf{\Sigma}}^{\mathcal{T}})^{-1} - \mathbf{\Sigma}^{-1}\|_2 &= O_P\Big(m_p\sqrt{\tfrac{\log p}{n}}\Big). \end{aligned}$$

To provide some numerical insights, Fan, Liao and Mincheva (2011) simulate data from a three-factor model, whose parameters are calibrated from the Fama-French model, using the returns of 30 industrial portfolios and 3 Fama-French factors from Jan $1^{st}$, 2009 to Dec $31^{st}$, 2010 ($n$=500). More specifically,

1. Calculate the least-squares estimator $\widehat{\mathbf{B}}$ based on the regression of the returns of the 30 portfolios on the Fama-French factors. Then, compute the sample mean vector $\boldsymbol{\mu}_B$ and sample covariance matrix $\boldsymbol{\Sigma}_B$ of the rows of $\mathbf{B}$. The results are depicted in Table 10.1. We then simulate $\{\mathbf{b}_i\}_{i=1}^p$ from the trivariate normal $N(\boldsymbol{\mu}_B, \boldsymbol{\Sigma}_B)$.

Table 10.1: Mean and covariance matrix used to generate $\mathbf{b}$

| $\boldsymbol{\mu}_B$ | $\boldsymbol{\Sigma}_B$ | | |
|---|---|---|---|
| 1.0641 | 0.0475 | 0.0218 | 0.0488 |
| 0.1233 | 0.0218 | 0.0945 | 0.0215 |
| -0.0119 | 0.0488 | 0.0215 | 0.1261 |

2. For $j = 1, \cdots, 30$, let $\widehat{\sigma}_j$ denote the standard deviation of the residuals of the regression fit to the $j$th portfolio returns, using the Fama-French 3 factors. These standard deviations have mean $\bar{\sigma} = 0.6055$ and standard deviations $\sigma_{SD} = 0.2621$ with $\min(\widehat{\sigma}_j) = 0.3533$ and $\max(\widehat{\sigma}_j) = 1.5222$. In the simulation, for each given dimensionality $p$, we generate $\sigma_1, \cdots, \sigma_p$ independently from the Gamma distribution $G(\alpha, \beta)$, with mean $\alpha\beta$ and standard deviation $\sqrt{\alpha}\beta$ selected to match $\bar{\sigma} = 0.6055$ and $\sigma_{SD} = 0.2621$. This yields $\alpha = 5.6840$ and $\beta = 0.1503$. Further, we create a loop that only accepts the value of $\sigma_i$ if it is between $[0.3533, 1.5222]$ to ensure the side constraints. Set $\mathbf{D} = \mathrm{diag}\{\sigma_1^2, ..., \sigma_p^2\}$. Simulate a sparse vector $\mathbf{s} = (s_1, \cdots, s_p)'$ by drawing $s_i \sim N(0,1)$ with probability $\frac{0.2}{\sqrt{p}\log p}$, and $s_i = 0$ otherwise. Create a sparse covariance matrix $\boldsymbol{\Sigma}_u = \mathbf{D} + \mathbf{s}\mathbf{s}' - \mathrm{diag}\{s_1^2, ..., s_p^2\}$. Create a loop to generate $\boldsymbol{\Sigma}_u$ multiple times until it is positive definite.
3. Compute the sample mean $\boldsymbol{\mu}_f$ and sample covariance matrix $\boldsymbol{\Sigma}_f$ of the observed Fama-French factors and fit the vector autoregressive (VAR(1)) model $\mathbf{f}_i = \boldsymbol{\mu} + \boldsymbol{\Phi}\mathbf{f}_{i-1} + \boldsymbol{\varepsilon}_i$ ($i \in [500]$) to the Fama-French 3 factors to obtain $3 \times 3$ matrix $\boldsymbol{\Phi}$ and covariance matrix $\boldsymbol{\Sigma}_\epsilon$. These fitted values are summarized in Table 10.2. In the simulation, we draw the factors $\mathbf{f}_i$ from the VAR(1) model with $\boldsymbol{\varepsilon}_i$'s drawn independently from $N_3(0, \boldsymbol{\Sigma}_\epsilon)$.

Based on the 200 simulations, Fan, Liao and Mincheva (2011) computed the estimation errors $\|\widehat{\boldsymbol{\Sigma}}_\lambda - \boldsymbol{\Sigma}\|_{\Sigma}$, $\|\widehat{\boldsymbol{\Sigma}}_\lambda - \boldsymbol{\Sigma}\|_{\max}$ and $\|\widehat{\boldsymbol{\Sigma}}_\lambda^{-1} - \boldsymbol{\Sigma}^{-1}\|_2$. The results for the sample covariance are also included for comparison and are depicted in Figure 10.3 for $p$ ranging from 20 to 600.

First of all, the estimation error grows with dimensionality $p$. The regularized covariance matrix outperforms the sample covariance matrix whenever $\boldsymbol{\Sigma}^{-1}$ is involved. In terms of estimating $\boldsymbol{\Sigma}$, they are approximately the same.

Table 10.2: Parameters for generating observed factors.

| $\boldsymbol{\mu}_f$ | $\boldsymbol{\Sigma}_f$ | | | $\boldsymbol{\Phi}$ | | |
|---|---|---|---|---|---|---|
| 0.1074 | 2.2540 | 0.2735 | 0.9197 | -0.1149 | 0.0024 | 0.0776 |
| 0.0357 | 0.2735 | 0.3767 | 0.0430 | 0.0016 | -0.0162 | 0.0387 |
| 0.0033 | 0.9197 | 0.0430 | 0.6822 | -0.0399 | 0.0218 | 0.0351 |

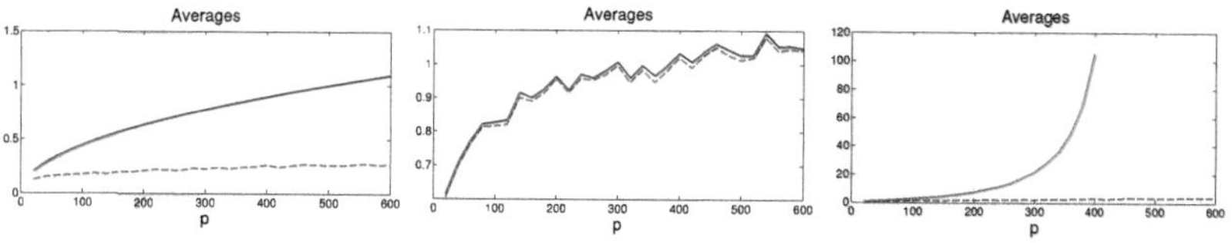


Figure 10.3: The averages of estimation errors $\|\widehat{\boldsymbol{\Sigma}}_\lambda - \boldsymbol{\Sigma}\|_\Sigma$, $\|\widehat{\boldsymbol{\Sigma}}_\lambda - \boldsymbol{\Sigma}\|_{\max}$ and $\|\widehat{\boldsymbol{\Sigma}}_\lambda^{-1} - \boldsymbol{\Sigma}^{-1}\|_2$, based on 200 simulations (red). The results for the sample covariance matrix are also included for comparison (blue). Adapted from Fan, Liao and Mincheva (2011)

This is consistent with the theory and phenomenon observed in Fan, Fan and Lv (2008). For estimating $\boldsymbol{\Sigma}^{-1}$, when $p$ is large, the sample covariance gets nearly singular and performs much worse.

## 10.4 Augmented Factor Models and Projected PCA

So far, we have treated $\mathbf{B}$ and $\mathbf{f}_i$ as completely unknown and infer them from the data by using PCA. However, in many applications, we do have side information available. For example, the factor loadings of the $j^{th}$ stock on the market risk factors should depend on the firm's attributes such as the size, value, momentum and volatility and we wish to use the augmented information in extracting latent factors and their associated loading matrix. Fan, Liao, and Wang (2016) assumed that factor loadings of the $j^{th}$ firm can be partially explained by the additional covariate vector $\mathbf{W}_j$ and modeled it through the semiparametric model

$$b_{jk} = g_k(\mathbf{W}_j) + \gamma_{jk}, \qquad \mathrm{E}(\gamma_{jk}|\mathbf{W}_j) = 0, \tag{10.24}$$

where $g_k(\mathbf{W}_j)$ is the part of factor loading of the $j^{th}$ firm on the $k^{th}$ factor that can be explained by its covariates. This is a *random-effect model*, thinking of coefficients $\{b_{jk}\}$ as realizations from model (10.24). Using again the matrix notation (10.6), the observed data can now be expressed as

$$\mathbf{X}_i = (\mathbf{G}(\mathbf{W}) + \boldsymbol{\Gamma})\mathbf{f}_i + \mathbf{u}_i, \tag{10.25}$$

where the $(j,k)^{th}$ element of $\mathbf{G}(\mathbf{W})$ and $\mathbf{\Gamma}$ are respectively $g_k(\mathbf{W}_j)$ and $\gamma_{jk}$.

Similarly, in many applications, we do have some knowledge about the latent factors $\mathbf{f}_i$. For example, while it is not known whether the Fama-French factors $\mathbf{W}_i$ are the true factors, they can at least explain a part of the true factors, namely,

$$\mathbf{f}_i = \mathbf{g}(\mathbf{W}_i) + \boldsymbol{\gamma}_i, \qquad \text{with} \quad \mathrm{E}(\boldsymbol{\gamma}_i|\mathbf{W}_i) = 0. \tag{10.26}$$

Thus, the Fama-French factors are used as the augmented information $\{\mathbf{W}_i\}$. Using the matrix notation (10.6), the observed data can be expressed as

$$\mathbf{X}_i = \mathbf{B}[\mathbf{g}(\mathbf{W}_i) + \boldsymbol{\gamma}_i] + \mathbf{u}_i.$$

Using (10.26), assuming $\mathbf{u}_i$ is *exogenous* in the sense that $\mathrm{E}(\mathbf{u}_i|\mathbf{W}_i) = 0$, we have

$$\underbrace{\mathrm{E}(\mathbf{X}_i|\mathbf{W}_i)}_{\widetilde{\mathbf{X}}_i} = \mathbf{B}\,\underbrace{\mathbf{g}(\mathbf{W}_i)}_{\widetilde{\mathbf{f}}_i}. \tag{10.27}$$

This is a noiseless factor model with "new data" $\widetilde{\mathbf{X}}_i$ and common factor $\widetilde{\mathbf{f}}_i$. The matrix $\mathbf{B}$ can be estimated with better precision. Indeed, it can be solved from the equation (10.27) if $n \geq K$.

The *projected PCA* was first introduced by Fan, Liao and Wang (2016) for model (10.25), but it is probably easier to understand for model (10.26), which was considered by Fan, Ke and Liao (2019+). Regress each component of $\mathbf{X}_i$ on $\mathbf{W}_i$ to obtain the fitted value $\widehat{\mathbf{X}}_i$. This can be linear regression or nonparametric *additive model* (see below) or *kernel machine*, using the data $\{(\mathbf{W}_i, X_{ij})\}_{i=1}^n$ for each variable $X_j$, $j = 1, \cdots, p$. Run PCA on the projected data $\{\widehat{\mathbf{X}}_i\}_{i=1}^n$ to obtain an estimate of $\mathbf{B}$ and $\widetilde{\mathbf{f}}_i = \mathbf{g}(\mathbf{W}_i)$ as shown in (10.27). Denote them by $\widehat{\mathbf{B}}$ and $\widehat{\mathbf{g}}(\mathbf{W}_i)$, respectively. With this more precisely estimated $\widehat{\mathbf{B}}$, we can now estimate the latent factor $\widehat{\mathbf{f}}_i$ using (10.14) and estimate $\boldsymbol{\gamma}_i$ by $\widehat{\boldsymbol{\gamma}}_i = \widehat{\mathbf{f}}_i - \widehat{\mathbf{g}}(\mathbf{W}_i)$. We can even compute the percentage of $\mathbf{f}_i$ that is explained by $\mathbf{W}_i$ as

$$1 - \sum_{i=1}^n \|\widehat{\boldsymbol{\gamma}}_i\|^2 / \sum_{i=1}^n \|\widehat{\mathbf{f}}_i\|^2.$$

We now return to model (10.24). It can be expressed as

$$X_{ij} = (\mathbf{g}(\mathbf{W}_j) + \boldsymbol{\gamma}_j)^T \mathbf{f}_i + u_{ij},$$

where $\mathbf{g}(\mathbf{W}_j)$ and $\boldsymbol{\gamma}_j$ are the $j^{th}$ row of $\mathbf{G}(\mathbf{W})$ and $\mathbf{\Gamma}$ in (10.25), respectively. Suppose that $u_{ij}$ is exogenous to $\mathbf{W}_j$. Then, by (10.24), we have

$$\underbrace{\mathrm{E}(X_{ij}|\mathbf{W}_j)}_{\widetilde{X}_{ij}} = \underbrace{\mathbf{g}(\mathbf{W}_j)}_{\widetilde{\mathbf{b}}_j}{}^T \mathbf{f}_i.$$

This is again a noiseless factor model with the "new data" $\widetilde{X}_{ij}$ and the same latent factor $\mathbf{f}_i$. The regression is now conducted across $j$, the firms. To see why $\mathbf{f}_i$ is not altered in the above operation, let us consider the sample version. Let $\mathbf{P}$ be the $p \times p$ projection matrix, corresponding to regressing $\{X_{ij}\}_{j=1}^p$ on $\{\mathbf{W}_j\}_{j=1}^p$ (see the end of this section for an example of $\mathbf{P}$) for each given $i$. Now, multiplying $\mathbf{P}$ to (10.25), we have

$$\mathbf{P}\mathbf{X}_i = \{\mathbf{P}(\mathbf{G}(\mathbf{W})) + \mathbf{P}\boldsymbol{\Gamma}\}\mathbf{f}_i + \mathbf{P}\mathbf{u}_i \approx \mathbf{P}(\mathbf{G}(\mathbf{W}))\mathbf{f}_i, \tag{10.28}$$

using

$$\mathbf{P}\boldsymbol{\Gamma} \approx 0 \quad \text{and} \quad \mathbf{P}\mathbf{u}_i \approx 0. \tag{10.29}$$

This is an approximate noiseless factor model and the latent factors $\mathbf{f}_i$ can be better estimated. As noted at the end of Section 10.2.2, $\widehat{\mathbf{F}}/\sqrt{n}$ are the $K$ eigenvectors of the $n \times n$ matrix $\mathbf{P}\mathbf{X}(\mathbf{P}\mathbf{X})^T = \mathbf{X}\mathbf{P}\mathbf{X}^T$ and $\widehat{\mathbf{B}} = n^{-1}\mathbf{X}^T\widehat{\mathbf{F}}$ using a better estimated factor $\widehat{\mathbf{F}}$.

The above discussion reveals that as long as the projection matrix $\mathbf{P}$ can smooth out the noises $\mathbf{u}_i$ and $\{\gamma_{jk}\}_{j=1}^p$, namely, satisfying (10.24), the above method is valid. To see the benefit of projection, Fan, Liao, and Wang (2016) considered the following simple example to illustrate the intuition.

**Example 10.3** *Consider the specific case where $K = 1$ so that the model (10.24) reduces to*

$$X_{ij} = (g(W_j) + \gamma_j)f_i + u_{ij}.$$

*The projection matrix should now be the local averages to estimate $g$. Assume that $g(\cdot)$ is so smooth that it is in fact a constant $\beta > 0$ so that the model is now simplified to*

$$X_{ij} = (\beta + \gamma_j)f_i + u_{ij}.$$

*In this case, the local smoothing now becomes the global average over $j$, which yields*

$$\bar{X}_i = (\beta + \bar{\gamma})f_i + \bar{u}_i \approx \beta f_i,$$

*where $\bar{X}_i$, $\bar{\gamma}$ and $\bar{u}_i$ are the averages over $j$. Hence, an estimator for the factor is $\widehat{f}_i = \bar{X}_i/\widehat{\beta}$. Using the normalization $n^{-1}\sum_{i=1}^n \widehat{f}_i^2 = 1$, we have*

$$\widehat{\beta} = \left(n^{-1}\sum_{i=1}^n \bar{X}_i^2\right)^{1/2}, \qquad \widehat{f}_i = \bar{X}_i/\widehat{\beta}.$$

*Thus, the rate of convergence for $\widehat{f}_i$ is $O(1/\sqrt{p})$, regardless of the sample size $n$, as $\bar{X}_i$ is the average over $p$ random variables.*

We now show how to construct the projection matrix $\mathbf{P}$ for model (10.24); similar methods apply to model (10.26). A common model for multivariate function $g_k(\mathbf{W}_j)$ is the nonparametric *additive model* and each function is approximated by a sieve basis, namely,

$$g_k(\mathbf{W}_j) = \sum_{l=1}^L g_{kl}(W_{jl}), \qquad g_{kl}(W_{jl}) \approx \sum_{m=1}^M a_{m,kl}\phi_m(W_{jl}),$$

where $L$ is the number of components in $\mathbf{W}_i$, $\{\phi_k(\cdot)\}_{m=1}^M$ is the sieve basis (the same for each $kl$ for simplicity of presentation). Thus,

$$g_k(\mathbf{W}_j) \approx \mathbf{\Phi}(\mathbf{W}_j)^T \mathbf{a}_k,$$

where $\mathbf{a}_k$ is the $LM$-dimensional regression coefficients and

$$\boldsymbol{\phi}(\mathbf{W}_j) = (\phi_1(W_{j1}), \cdots, \phi_M(W_{j1}), \cdots, \phi_1(W_{jL}), \cdots, \phi_M(W_{jL})^T.$$

Set $\mathbf{\Phi}(\mathbf{W}) = (\boldsymbol{\phi}(\mathbf{W}_1), \cdots, \boldsymbol{\phi}(\mathbf{W}_p))^T$. Then,

$$\mathbf{P} = \mathbf{\Phi}(\mathbf{W})(\mathbf{\Phi}(\mathbf{W})^T\mathbf{\Phi}(\mathbf{W}))^{-1}\mathbf{\Phi}(\mathbf{W})^T,$$

is the projection of the matrix of regression of any response variable $\{Z_j\}_{j=1}^p$ on $\{\mathbf{W}_j\}_{j=1}^p$ through the additive sieve basis:

$$Z_j = \sum_{l=1}^{L}\sum_{m=1}^{M} a_{ml}\phi_m(W_{jl}) + \varepsilon_j, \qquad j = 1, \cdots, p,$$

namely the fitted value $\widehat{\mathbf{Z}} = \mathbf{PZ}$. See Section 2.5. Clearly, the projection matrix $\mathbf{P}$ satisfies (10.29). In our application, $Z_j = X_{ij}$ $(j = 1, \cdots, p)$ for each fixed $i$.

In summary, as mentioned before, we extract the latent factors $\widehat{\mathbf{F}}$ by taking the top $K$ eigenvectors of the matrix $\mathbf{XPX}^T$ multiplied by $\sqrt{n}$ and $\widehat{\mathbf{B}} = n^{-1}\mathbf{X}^T\widehat{\mathbf{F}}$.

## 10.5 Asymptotic Properties

This section gives some properties for the estimated parameters in the factor model given in Section 10.2.2. In particular, we present the asymptotic properties for the factor loading matrix $\mathbf{B}$, estimated latent factors $\{\mathbf{f}_i\}_{i=1}^n$, as well as POET estimators $\widehat{\mathbf{\Sigma}}_\lambda^T$ and $\widehat{\mathbf{\Sigma}}_\lambda$. These properties were established by Fan, Liao and Mincheva (2013). We extend it further to cover robust inputs. The asymptotic normalities of these quantities were established by Bai (2003) for the sample covariance matrix as an input, with possible weak serious correlation.

### 10.5.1 *Properties for estimating loading matrix*

As in (10.13), $\widehat{\mathbf{B}}$ consists of two parts. Let $\widetilde{\lambda}_j = \|\widetilde{\mathbf{b}}_j\|^2$ and $\widetilde{\boldsymbol{\xi}}_j = \widetilde{\mathbf{b}}_j/\|\widetilde{\mathbf{b}}_j\|$. Then, by Proposition 10.1 and the pervasiveness assumption, we have $\widetilde{\lambda}_j = \Omega(p)$ if $\|\mathbf{\Sigma}_u\| = o(p)$. Thus, $\|\widetilde{\boldsymbol{\xi}}_j\|_{\max} = \Omega(p^{-1/2})$ as long as $\|\mathbf{B}\|_{\max}$ is bounded. In other words, there are no spiked components in the eigenvectors $\{\widetilde{\boldsymbol{\xi}}_j\}_{j=1}^K$. By Proposition 10.1 again, there are also no spiked components in the eigenvectors $\{\boldsymbol{\xi}_j\}_{j=1}^K$.

Following Fan, Liu and Wang (2018), we make $\{\widehat{\lambda}_j\}_{j=1}^K$ and $\{\widehat{\boldsymbol{\xi}}_j\}_{j=1}^K$ modulus to facilitate the flexibility of various applications. In others, $\{\widehat{\lambda}_j\}_{j=1}^K$ and $\{\widehat{\boldsymbol{\xi}}_j\}_{j=1}^K$ do not need to be the eigenvalues and their associated eigenvectors of $\widehat{\boldsymbol{\Sigma}}$. They require only initial pilot estimators $\widehat{\boldsymbol{\Sigma}}, \widehat{\boldsymbol{\Lambda}}, \widehat{\boldsymbol{\Gamma}}$ for covariance matrix $\boldsymbol{\Sigma}$, its leading eigenvalues $\boldsymbol{\Lambda} = \mathrm{diag}(\lambda_1, \ldots, \lambda_K)$ and their corresponding leading eigenvectors $\boldsymbol{\Gamma}_{p\times K} = (\boldsymbol{\xi}_1, \ldots, \boldsymbol{\xi}_K)$ to satisfy

$$\begin{aligned} &\|\widehat{\boldsymbol{\Sigma}} - \boldsymbol{\Sigma}\|_{\max} = O_P(\sqrt{\log p/n})\,, \\ &\|(\widehat{\boldsymbol{\Lambda}} - \boldsymbol{\Lambda})\boldsymbol{\Lambda}^{-1}\|_{\max} = O_P(\sqrt{\log p/n})\,, \\ &\|\widehat{\boldsymbol{\Gamma}} - \boldsymbol{\Gamma}\|_{\max} = O_P(\sqrt{\log p/(np)})\,. \end{aligned} \tag{10.30}$$

These estimators can be constructed separately from different methods or even sources of data. We estimate $\widehat{\mathbf{B}}$ as in (10.13) and $\widehat{\mathbf{f}}_i$ as in (10.14) with $\widehat{\boldsymbol{\xi}}_j$ now denoting the $j^{th}$ column of $\widehat{\boldsymbol{\Gamma}}$. We then compute $\widehat{\boldsymbol{\Sigma}}_u = \widehat{\boldsymbol{\Sigma}} - \widehat{\boldsymbol{\Gamma}}\widehat{\boldsymbol{\Lambda}}\widehat{\boldsymbol{\Gamma}}^T$, and construct the correlation thresholded estimator $\widehat{\boldsymbol{\Sigma}}_u^{\mathcal{T}}$ and $\widehat{\boldsymbol{\Sigma}}^{\mathcal{T}}$ as in (10.15). Here, we drop the dependence on the constant $\lambda$ in (10.15) for simplicity and replace the rate $\sqrt{\log p/n}$ there by

$$w_n = K^2(\sqrt{\log p/n} + 1/\sqrt{p}) \tag{10.31}$$

to reflect the price that needs to be paid for learning the $K$ latent factors. The term $1/\sqrt{p}$ reflects the bias of approximating factor loadings by PCA (see (10.10)) and the term $\sqrt{\log p/n}$ reflects the stochastic error in the estimation of PCA.

**Theorem 10.2** *Suppose that the top $K$ eigenvalues are distinguished and the pervasiveness condition holds. In addition $\|\mathbf{B}\|_{\max}$, $\|\boldsymbol{\Sigma}_u\|_2$ are bounded. If (10.30) holds, then we have*

$$\begin{aligned} &\|\widehat{\boldsymbol{\Gamma}}\widehat{\boldsymbol{\Lambda}}\widehat{\boldsymbol{\Gamma}}^T - \mathbf{B}\mathbf{B}^T\|_{\max} = O_P(w_n), \\ &\|\widehat{\mathbf{B}} - \mathbf{B}\|_{\max} = O_p\Big(\sqrt{\tfrac{\log p}{n}} + \tfrac{1}{\sqrt{p}}\Big). \end{aligned} \tag{10.32}$$

Fan, Liao and Mincheva (2013) showed that for the sample covariance matrix under the sub-Gaussian assumption, condition (10.30) holds even for weakly dependent data. Therefore, the conclusion of Theorem 10.2 holds. Fan, Liu and Wang (2018) proved that condition (10.30) holds for the marginal Kendall's $\tau$ estimator (9.46) and the spatial Kendall's $\tau$ estimator,

$$\frac{2}{n(n-1)} \sum_{i<i'} \frac{(\mathbf{x}_i - \mathbf{x}_{i'})(\mathbf{x}_i - \mathbf{x}_{i'})'}{\|\mathbf{x}_i - \mathbf{x}_{i'}\|_2^2}\,. \tag{10.33}$$

They also showed that (10.30) is also sufficient for the CLIME estimator of the sparse $\boldsymbol{\Sigma}_u^{-1}$, which is needed for constructing the sparse graphical model of $\mathbf{u}$. Fan, Wang and Zhong (2019) proved the condition (10.30) holds for the elementwise adaptive Huber estimator in Section 9.3. An $\ell_\infty$ eigenvector perturbation bound was introduced therein in order to establish the desired result.

### 10.5.2 Properties for estimating covariance matrices

We now state the properties for estimating $\mathbf{\Sigma}_u$ and $\mathbf{\Sigma}$. The results here are a slight extension of those in Fan, Liao and Mincheva (2013) and Fan, Liu and Wang (2018).

**Theorem 10.3** *Under condition of Theorem 10.2, if* $\mathbf{\Sigma}_u \in \mathcal{C}_q(m_p)$, $m_p w_n^{1-q} = o(1)$ *and* $\|\mathbf{\Sigma}_u^{-1}\|_2 = O(1)$, *then we have for estimating* $\mathbf{\Sigma}_u$

$$\begin{aligned}
\|\widehat{\mathbf{\Sigma}}_u^{\mathcal{T}} - \mathbf{\Sigma}_u\|_{\max} &= O_P\Big(w_n\Big), \\
\|\widehat{\mathbf{\Sigma}}_u^{\mathcal{T}} - \mathbf{\Sigma}_u\|_2 &= O_P\Big(m_p w_n^{1-q}\Big) \\
\|(\widehat{\mathbf{\Sigma}}_u^{\mathcal{T}})^{-1} - \mathbf{\Sigma}_u^{-1}\|_2 &= O_P\Big(m_p w_n^{1-q}\Big),
\end{aligned} \tag{10.34}$$

*and for estimating* $\mathbf{\Sigma}$

$$\begin{aligned}
\|\widehat{\mathbf{\Sigma}}^{\mathcal{T}} - \mathbf{\Sigma}\|_{\max} &= O_P\Big(w_n\Big), \\
\|\widehat{\mathbf{\Sigma}}^{\mathcal{T}} - \mathbf{\Sigma}\|_{\Sigma} &= O_P\Big(\frac{K^{3/2}p^{1/2}\log p}{n} + m_p w_n^{1-q} + K w_n/p\Big), \\
\|(\widehat{\mathbf{\Sigma}}^{\mathcal{T}})^{-1} - \mathbf{\Sigma}^{-1}\|_2 &= O_P\Big(K^2 m_p w_n^{1-q}\Big).
\end{aligned} \tag{10.35}$$

**Remark 10.1** The condition $\|\mathbf{\Sigma}_u^{-1}\|_2 = O(1)$ is used only to establish the third result in (10.34) and the third result in (10.35). It was not used anywhere else. The third term in the second rate in (10.35) is dominated by the second term if $K = O(p/m_p)$. In establishing the second rate in (10.35), we used the bound

$$\|\widehat{\mathbf{\Gamma}} - \mathbf{\Gamma}\|^2 \le Kp\|\widehat{\mathbf{\Gamma}} - \mathbf{\Gamma}\|_{\max}^2$$

so that we can use the third condition in (10.30). This is usually too crude when $K$ depends on $n$.

When $K$ is finite, Theorem 10.3 shows that in comparison with the case where the factors are observable (Theorem 10.1), we need to pay an additional price of $1/\sqrt{p}$ for learning the latent factors. This price is negligible when $p \gg n \log n$.

*10.5.3 Properties for estimating realized latent factors*

We now derive the properties for estimating the latent factor by (10.14).

**Theorem 10.4** *Under condition of Theorem 10.2, if* $\|\widehat{\boldsymbol{\mu}} - \boldsymbol{\mu}\| = O_P(\sqrt{p \log p/n})$*, then we have for estimating the realized latent factors*

$$\begin{aligned}
n^{-1}\sum_{i=1}^{n} \|\widehat{\mathbf{f}}_i - \mathbf{f}_i\| &= O_P\Big(K/\sqrt{p} + K\sqrt{\log p/n}\Big), \\
n^{-1}\sum_{i=1}^{n} \|\widehat{\mathbf{f}}_i - \mathbf{f}_i\|^2 &= O_P\Big(K^2/p + K^2 \log p/n\Big), \\
\max_{i\le n} \|\widehat{\mathbf{f}}_i - \mathbf{f}_i\| &= O_P\Big(\sqrt{K/p}(\sqrt{\log p/n} + 1/\sqrt{p}) \max_{i\le n} \|\mathbf{X}_i - \boldsymbol{\mu}\| \\
&\quad + p^{-1} \max_{i\le n} \|\mathbf{B}^T\mathbf{u}_i\|\Big).
\end{aligned} \tag{10.36}$$

Note that $E\|\mathbf{X}_i - \boldsymbol{\mu}\|^2 = \mathrm{tr}(\boldsymbol{\Sigma}) = O(Kp)$ as we have $K$ spiked eigenvalues of order $O(p)$ and similarly $E\|\mathbf{B}^T\mathbf{u}_i\|^2 = O(pK^2)$ (see (10.54) and (10.55)). Therefore, we expect

$$\begin{aligned}
\max\nolimits_{i\le n} \|\mathbf{X}_i - \boldsymbol{\mu}\| &= O_p(\sqrt{pK \log n}), \\
\max\nolimits_{i\le n} \|\mathbf{B}^T\mathbf{u}_i\| &= O_p(\sqrt{pK^2 \log n}),
\end{aligned} \tag{10.37}$$

if the random variables $\|\mathbf{X}_i - \boldsymbol{\mu}\|$ and $\|\mathbf{B}^T\mathbf{u}_i\|$ are sub-Gaussian. Consequently,

$$\max_{i\le n} \|\widehat{\mathbf{f}}_i - \mathbf{f}_i\| = O_P\Big(K\sqrt{\log n/p} + K\sqrt{(\log p)(\log n)/n}\Big). \tag{10.38}$$

**Corollary 10.1** *Under the conditions of Theorem 10.4 and* (10.37)*, we have*

$$\max_{i\le n, j\le p} |\widehat{\mathbf{b}}_j^T\widehat{\mathbf{f}}_i - \mathbf{b}_j^T\mathbf{f}_i| = O_P\Big(K^{3/2}\sqrt{\log n/p} + K^{3/2}\sqrt{(\log p)(\log n)/n}\Big),$$

*provided* $\max_{i\le n}\|\mathbf{f}_i\| = O_P(\sqrt{K \log n})$.

**Proof.** It is easy to see that

$$|\widehat{\mathbf{b}}_j^T\widehat{\mathbf{f}}_i - \mathbf{b}_j^T\mathbf{f}_i| \le \|\widehat{\mathbf{b}}_j\|\|\widehat{\mathbf{f}}_i - \mathbf{f}_i\| + \|\widehat{\mathbf{b}}_j - \mathbf{b}_j\|\|\mathbf{f}_i\|.$$

Using (10.32) and $\|\mathbf{B}\|_{\max} = O(1)$, the first term is of order

$$O_P\left(K^{1/2}\Big(K\sqrt{\log n/p} + K\sqrt{(\log p)(\log n)/n}\Big)\right),$$

by (10.38). The second term is bounded by

$$O_p\Big(\sqrt{\frac{\log p}{n}} + \frac{1}{\sqrt{p}}\Big) \max_{i\le n}\|\mathbf{f}_i\| = O_P\Big(\sqrt{K}\Big[\sqrt{\frac{\log p}{n}} + \frac{1}{\sqrt{p}}\Big]\sqrt{K \log n}\Big).$$

Combining the last two terms, the conclusion follows.

*10.5.4 Properties for estimating idiosyncratic components*

By (10.6), a natural estimator for $\mathbf{u}_i$ is

$$\widehat{\mathbf{u}}_i = \mathbf{X}_i - \widehat{\boldsymbol{\mu}} - \widehat{\mathbf{B}}\widehat{\mathbf{f}}_i. \tag{10.39}$$

By definition, we have

$$\widehat{u}_{ij} - u_{ij} = \mu_i - \widehat{\mu}_i + \mathbf{b}_j^T \mathbf{f}_i - \widehat{\mathbf{b}}_j^T \widehat{\mathbf{f}}_i.$$

By Corrollary 10.1, we have

**Corollary 10.2** *Under the conditions of Corollary 10.1, we have*

$$\max_{i \le n} \|\widehat{\mathbf{u}}_i - \mathbf{u}_i\|_{\max} = O_P\Big(K^{3/2}\sqrt{\log n/p} + K^{3/2}\sqrt{(\log p)(\log n)/n}\Big).$$

Note that if we are interested in the $L_2$-error, the condition $\max_{j \le n} \|\mathbf{f}_j\| = O_P(\sqrt{K \log n})$ can be dropped. To see this, note that

$$\|\widehat{\mathbf{u}}_i - \mathbf{u}_i\| \le \|\widehat{\boldsymbol{\mu}} - \boldsymbol{\mu}\| + \|\widehat{\mathbf{B}}\widehat{\mathbf{f}}_i - \mathbf{B}\mathbf{f}_i\|.$$

The second term is further bounded by

$$\|\widehat{\mathbf{B}}\|\|\widehat{\mathbf{f}}_i - \mathbf{f}_i\| + \|\widehat{\mathbf{B}} - \mathbf{B}\|\|\mathbf{f}_i\|.$$

Therefore,

$$n^{-1}\sum_{i=1}^{n}\|\widehat{\mathbf{u}}_i - \mathbf{u}_i\|^2 \le 3\Big(\|\widehat{\boldsymbol{\mu}} - \boldsymbol{\mu}\|^2 + \|\widehat{\mathbf{B}}\|^2 n^{-1}\sum_{i=1}^{n}\|\widehat{\mathbf{f}}_i - \mathbf{f}_i\|^2 + \|\widehat{\mathbf{B}} - \mathbf{B}\|^2 n^{-1}\sum_{i=1}^{n}\|\mathbf{f}_i\|^2\Big).$$

This depends only on the average $n^{-1}\sum_{i=1}^{n}\|\mathbf{f}_i\|^2$, rather than the maximum. By using Theorems 10.2 and 10.4, we conclude (Exercise 10.11) that

$$(np)^{-1}\sum_{i=1}^{n}\|\widehat{\mathbf{u}}_i - \mathbf{u}_i\|^2 = O_P(K^3/p + K^3 \log p/n). \tag{10.40}$$

## 10.6 Technical Proofs

### *10.6.1 Proof of Theorem 10.1*

We first give two lemmas. The technique in the proof of Lemma 10.1 will be repeatedly used in the proof of Theorem 10.1. Therefore, we furnish a bit more details. Throughout this section, we use $\|\cdot\|$ to denote $\|\cdot\|_2$.

**Lemma 10.1** *For any compatible matrices* $\mathbf{A}$ *and* $\mathbf{B}$*, we have*

$$\|\mathbf{A}\mathbf{B}\|_\Sigma \le p^{-1/2}\min(\|\mathbf{A}\|_F\|\mathbf{B}\|, \|\mathbf{A}\|\|\mathbf{B}\|_F)\ \|\boldsymbol{\Sigma}^{-1}\|.$$

*Further, for any three compatible matrices* $\mathbf{A}$, $\mathbf{B}$ *and* $\mathbf{C}$*, we have*

$$\|\mathbf{A}\mathbf{B}\mathbf{C}\|_F \le \min(\|\mathbf{A}\|\|\mathbf{B}\|\|\mathbf{C}\|_F, \|\mathbf{A}\|\|\mathbf{B}\|_F\|\mathbf{C}\|, \|\mathbf{A}\|_F\|\mathbf{B}\|\|\mathbf{C}\|).$$

**Proof.** Note that for any symmetric matrix $\mathbf{D}$, $\|\mathbf{D}\|\mathbf{I} - \mathbf{D} \succeq 0$ and hence

$$\operatorname{tr}(\mathbf{E}^T(\|\mathbf{D}\|\mathbf{I} - \mathbf{D})\mathbf{E}) \geq 0 \quad \text{or} \quad \operatorname{tr}(\mathbf{E}^T\mathbf{D}\mathbf{E}) \leq \|\mathbf{D}\| \operatorname{tr}(\mathbf{E}^T\mathbf{E}). \tag{10.41}$$

Let $\tau_n = p\|\mathbf{AB}\|_\Sigma^2$. Then, by (10.41), we have

$$\begin{aligned} \tau_n &= \operatorname{tr}(\boldsymbol{\Sigma}^{-1/2}(\mathbf{AB})^T\boldsymbol{\Sigma}^{-1}(\mathbf{AB})\boldsymbol{\Sigma}^{-1/2}) \\ &\leq \|\boldsymbol{\Sigma}^{-1}\| \operatorname{tr}(\boldsymbol{\Sigma}^{-1/2}(\mathbf{AB})^T(\mathbf{AB})\boldsymbol{\Sigma}^{-1/2}). \end{aligned}$$

Now using the property that $\operatorname{tr}(\mathbf{CD}) = \operatorname{tr}(\mathbf{DC})$, we have

$$\tau_n \leq \|\boldsymbol{\Sigma}^{-1}\| \operatorname{tr}((\mathbf{AB})\boldsymbol{\Sigma}^{-1}(\mathbf{AB})^T) \leq \|\boldsymbol{\Sigma}^{-1}\|^2 \operatorname{tr}(\mathbf{ABB}^T\mathbf{A}^T),$$

using again (10.41). Finally, using (10.41) one more time, we obtain

$$\tau_n \leq \|\boldsymbol{\Sigma}^{-1}\|^2\|\mathbf{BB}^T\|\|\mathbf{A}\|_F^2 = \|\boldsymbol{\Sigma}^{-1}\|^2\|\mathbf{B}\|^2\|\mathbf{A}\|_F^2.$$

Similarly, by using $\operatorname{tr}(\mathbf{ABB}^T\mathbf{A}^T) = \operatorname{tr}(\mathbf{B}^T\mathbf{A}^T\mathbf{AB})$ and using (10.41), we obtain

$$\tau_n \leq \|\boldsymbol{\Sigma}^{-1}\|^2\|\mathbf{A}\|^2\|\mathbf{B}\|_F^2.$$

The first conclusion follows by combining the last two results.

To prove the second conclusion, using (10.41), we have

$$\|\mathbf{ABC}\|_F^2 = \operatorname{tr}(\mathbf{C}^T\mathbf{B}^T\mathbf{A}^T\mathbf{ABC}) \leq \|\mathbf{A}\|^2 \operatorname{tr}(\mathbf{C}^T\mathbf{B}^T\mathbf{BC}) \leq \|\mathbf{A}\|^2\|\mathbf{B}\|^2 \operatorname{tr}(\mathbf{C}^T\mathbf{C}).$$

On the other hand, by using (10.41) again,

$$\|\mathbf{ABC}\|_F^2 \leq \|\mathbf{A}\|^2 \operatorname{tr}(\mathbf{BCC}^T\mathbf{B}^T) \leq \|\mathbf{A}\|^2\|\mathbf{C}\|^2 \operatorname{tr}(\mathbf{BB}^T).$$

Shuffling the order of $\mathbf{A}$, $\mathbf{B}$ and $\mathbf{C}$ in the calculation of trace, we can also get

$$\|\mathbf{ABC}\|_F^2 = \operatorname{tr}(\mathbf{ABCC}^T\mathbf{B}^T\mathbf{A}^T) \leq \|\mathbf{C}\|^T\|\mathbf{B}\|^2 \operatorname{tr}(\mathbf{AA}^T).$$

The result follows from the combination of the above three inequalities. ■

**Lemma 10.2** *Let* $\boldsymbol{\Sigma} = \mathbf{B}^T\boldsymbol{\Sigma}_f\mathbf{B} + \boldsymbol{\Sigma}_u$. *Then, under the assumption of Theorem 10.1, we have*

$$\|\mathbf{B}^T\boldsymbol{\Sigma}^{-1}\mathbf{B}\|^2 \leq \|\boldsymbol{\Sigma}_f^{-1}\| = O(1).$$

**Proof.** By the Sherman-Morrison-Woodbury formula (9.6), we have

$$\begin{aligned} \mathbf{B}^T\boldsymbol{\Sigma}^{-1}\mathbf{B} &= \mathbf{B}^T\boldsymbol{\Sigma}_u^{-1}\mathbf{B} - \mathbf{B}^T\boldsymbol{\Sigma}_u^{-1}\mathbf{B}\left[\boldsymbol{\Sigma}_f^{-1} + \mathbf{B}^T\boldsymbol{\Sigma}_u^{-1}\mathbf{B}\right]^{-1}\mathbf{B}^T\boldsymbol{\Sigma}_u^{-1}\mathbf{B} \\ &= \mathbf{B}^T\boldsymbol{\Sigma}_u^{-1}\mathbf{B} - \mathbf{B}^T\boldsymbol{\Sigma}_u^{-1}\mathbf{B}\left[\boldsymbol{\Sigma}_f^{-1} + \mathbf{B}^T\boldsymbol{\Sigma}_u^{-1}\mathbf{B}\right]^{-1} \\ &\quad \times \{(\boldsymbol{\Sigma}_f^{-1} + \mathbf{B}^T\boldsymbol{\Sigma}_u^{-1}\mathbf{B}) - \boldsymbol{\Sigma}_f\}. \end{aligned}$$

By multiplying the two terms in the curly bracket, we obtain

$$\begin{aligned}\mathbf{B}^T\boldsymbol{\Sigma}^{-1}\mathbf{B} &= \mathbf{B}^T\boldsymbol{\Sigma}_u^{-1}\mathbf{B}\left[\boldsymbol{\Sigma}_f^{-1}+\mathbf{B}^T\boldsymbol{\Sigma}_u^{-1}\mathbf{B}\right]^{-1}\boldsymbol{\Sigma}_f^{-1}\\ &= \{(\boldsymbol{\Sigma}_f^{-1}+\mathbf{B}^T\boldsymbol{\Sigma}_u^{-1}\mathbf{B})-\boldsymbol{\Sigma}_f^{-1}\}\left[\boldsymbol{\Sigma}_f^{-1}+\mathbf{B}^T\boldsymbol{\Sigma}_u^{-1}\mathbf{B}\right]^{-1}\boldsymbol{\Sigma}_f^{-1}\\ &= \boldsymbol{\Sigma}_f^{-1}-\boldsymbol{\Sigma}_f^{-1}\left[\boldsymbol{\Sigma}_f^{-1}+\mathbf{B}^T\boldsymbol{\Sigma}_u^{-1}\mathbf{B}\right]^{-1}\boldsymbol{\Sigma}_f^{-1}\\ &\le \boldsymbol{\Sigma}_f^{-1},\end{aligned}$$

where the third inequality follows by multiplying the two terms in the curly bracket out. Therefore, the conclusion follows. ■

**Proof of Theorem 10.1**. (i) By Theorem 9.1, to establish (10.22), we need to verify the condition $\|\widehat{\boldsymbol{\Sigma}}_u-\boldsymbol{\Sigma}_u\|_{\max}=O_P(\omega_n)$. Since $\|\widehat{\boldsymbol{\Sigma}}_z-\boldsymbol{\Sigma}_z\|_{\max}=O_P(\sqrt{\log p/n})$, all sub-blocks $\boldsymbol{\Sigma}_{11}$, $\boldsymbol{\Sigma}_{12}$ and $\boldsymbol{\Sigma}_{22}$ are estimated with the same uniform rate of convergence. Therefore, we need only to verify the rate for estimating $\boldsymbol{\Sigma}_{12}\boldsymbol{\Sigma}_{22}^{-1}\boldsymbol{\Sigma}_{21}$. Since each element of this $p\times p$ matrix involves only summations $K^2$ elements, it follows that

$$\|\widehat{\boldsymbol{\Sigma}}_{12}\widehat{\boldsymbol{\Sigma}}_{22}^{-1}\widehat{\boldsymbol{\Sigma}}_{21}-\boldsymbol{\Sigma}_{12}\boldsymbol{\Sigma}_{22}^{-1}\boldsymbol{\Sigma}_{21}\|_{\max}=O_P(\omega_n). \tag{10.42}$$

This establishes (10.22).

(ii) First rate in (10.23). We next establish the rate for $\|\widehat{\boldsymbol{\Sigma}}^{\mathcal{T}}-\boldsymbol{\Sigma}\|_{\max}$. First of all

$$\|\widehat{\boldsymbol{\Sigma}}_u^{\mathcal{T}}-\boldsymbol{\Sigma}_u\|_{\max}\le\|\widehat{\boldsymbol{\Sigma}}_u^{\mathcal{T}}-\widehat{\boldsymbol{\Sigma}}_u\|_{\max}+\|\widehat{\boldsymbol{\Sigma}}_u-\boldsymbol{\Sigma}_u\|_{\max}.$$

The former is bounded by $\max\lambda_{ij}$, which has the same rate as $\|\widehat{\boldsymbol{\Sigma}}_u-\boldsymbol{\Sigma}_u\|_{\max}=O_P(\omega_n)$. Thus, $\|\widehat{\boldsymbol{\Sigma}}_u^{\mathcal{T}}-\boldsymbol{\Sigma}_u\|_{\max}=O_P(\omega_n)$. This and (10.42) entail that

$$\|\widehat{\boldsymbol{\Sigma}}^{\mathcal{T}}-\boldsymbol{\Sigma}\|_{\max}\le\|\widehat{\boldsymbol{\Sigma}}_{12}\widehat{\boldsymbol{\Sigma}}_{22}^{-1}\widehat{\boldsymbol{\Sigma}}_{21}-\boldsymbol{\Sigma}_{12}\boldsymbol{\Sigma}_{22}^{-1}\boldsymbol{\Sigma}_{21}\|_{\max}+\|\widehat{\boldsymbol{\Sigma}}_u^{\mathcal{T}}-\boldsymbol{\Sigma}_u\|_{\max}=O_p(\omega_n).$$

(iii) Second rate in (10.23). We next deal with the relative Frobenius norm convergence. Decompose

$$\begin{aligned}\|\widehat{\boldsymbol{\Sigma}}^{\mathcal{T}}-\boldsymbol{\Sigma}\|_{\Sigma} &\le \|\widehat{\boldsymbol{\Sigma}}_{12}\widehat{\boldsymbol{\Sigma}}_{22}^{-1}\widehat{\boldsymbol{\Sigma}}_{12}^{T}-\boldsymbol{\Sigma}_{12}\boldsymbol{\Sigma}_{22}^{-1}\boldsymbol{\Sigma}_{12}^{T}\|_{\Sigma}+\|\widehat{\boldsymbol{\Sigma}}_u^{\mathcal{T}}-\boldsymbol{\Sigma}_u\|_{\Sigma}\\ &\le \|(\widehat{\boldsymbol{\Sigma}}_{12}-\boldsymbol{\Sigma}_{12})\widehat{\boldsymbol{\Sigma}}_{22}^{-1}(\widehat{\boldsymbol{\Sigma}}_{12}-\boldsymbol{\Sigma}_{12})^T\|_{\Sigma}+2\|(\widehat{\boldsymbol{\Sigma}}_{12}-\boldsymbol{\Sigma}_{12})\widehat{\boldsymbol{\Sigma}}_{22}^{-1}\boldsymbol{\Sigma}_{12}^{T}\|_{\Sigma}\\ &\quad+\|\boldsymbol{\Sigma}_{12}(\widehat{\boldsymbol{\Sigma}}_{22}^{-1}-\boldsymbol{\Sigma}_{22}^{-1})\boldsymbol{\Sigma}_{12}^{T}\|_{\Sigma}+\|\widehat{\boldsymbol{\Sigma}}_u^{\mathcal{T}}-\boldsymbol{\Sigma}_u\|_{\Sigma}\\ &=:\ \Delta_1+2\Delta_2+\Delta_3+\Delta_4\,.\end{aligned} \tag{10.43}$$

We bound the four terms one by one.

The last term is the easiest. Using Lemma 10.1 and (10.22), we have

$$\Delta_4 \le p^{-1/2}\|\widehat{\mathbf{\Sigma}}_u^{\mathcal{T}} - \mathbf{\Sigma}_u\|_F\|\mathbf{\Sigma}^{-1}\| = O_P(\|\widehat{\mathbf{\Sigma}}_u^{\mathcal{T}} - \mathbf{\Sigma}_u\|) = O_P(m_p\omega_n^{1-q}).$$

Here we used $\|\mathbf{\Sigma}^{-1}\| \le \|\mathbf{\Sigma}_u^{-1}\|$, which is bounded by assumption.

The bound for $\Delta_1$ uses the obvious inequality

$$\|\widehat{\mathbf{\Sigma}}_{12} - \mathbf{\Sigma}_{12}\|_F \le \sqrt{Kp}\|\widehat{\mathbf{\Sigma}}_{12} - \mathbf{\Sigma}_{12}\|_{\max} = O_P(\sqrt{Kp\log p/n}) \tag{10.44}$$

and the fact that $\|\widehat{\mathbf{\Sigma}}_{22}^{-1}\|$ and $\|\mathbf{\Sigma}^{-1}\|$ are $O_P(1)$. By Lemma 10.1, we have

$$\begin{aligned}\Delta_1 &\le p^{-1/2}\|\widehat{\mathbf{\Sigma}}_{12} - \mathbf{\Sigma}_{12}\|_F\|\widehat{\mathbf{\Sigma}}_{12} - \mathbf{\Sigma}_{12}\|\|\widehat{\mathbf{\Sigma}}_{22}^{-1}\|\|\mathbf{\Sigma}^{-1}\| \\ &\le p^{-1/2}\|\widehat{\mathbf{\Sigma}}_{12} - \mathbf{\Sigma}_{12}\|_F^2\|\widehat{\mathbf{\Sigma}}_{22}^{-1}\|\|\mathbf{\Sigma}^{-1}\| \\ &= O_P\Big(\frac{K\sqrt{p}\log p}{n}\Big).\end{aligned}$$

We next bound the second term. Using the definition and the argument as in Lemma 10.1, it is easy to see that

$$\begin{aligned}\Delta_2^2 &= p^{-1}\,\mathrm{tr}(\mathbf{\Sigma}^{-1/2}(\widehat{\mathbf{\Sigma}}_{12} - \mathbf{\Sigma}_{12})\widehat{\mathbf{\Sigma}}_{22}^{-1}\mathbf{\Sigma}_{12}^T\mathbf{\Sigma}^{-1}\mathbf{\Sigma}_{12}\widehat{\mathbf{\Sigma}}_{22}^{-1}(\widehat{\mathbf{\Sigma}}_{12} - \mathbf{\Sigma}_{12})^T\mathbf{\Sigma}^{-1/2}) \\ &\le p^{-1}\,\mathrm{tr}(\mathbf{\Sigma}^{-1/2}(\widehat{\mathbf{\Sigma}}_{12} - \mathbf{\Sigma}_{12})(\widehat{\mathbf{\Sigma}}_{12} - \mathbf{\Sigma}_{12})^T\mathbf{\Sigma}^{-1/2})\;\|\widehat{\mathbf{\Sigma}}_{22}^{-1}\mathbf{\Sigma}_{12}^T\mathbf{\Sigma}^{-1}\mathbf{\Sigma}_{12}\widehat{\mathbf{\Sigma}}_{22}^{-1}\| \\ &\le p^{-1}\|\widehat{\mathbf{\Sigma}}_{12} - \mathbf{\Sigma}_{12}\|_F^2\|\mathbf{\Sigma}^{-1}\|\|\widehat{\mathbf{\Sigma}}_{22}^{-1}\|^2\|\mathbf{\Sigma}_{21}^T\mathbf{\Sigma}^{-1}\mathbf{\Sigma}_{12}\|.\end{aligned}$$

Using (10.44), the first four products are of order $O_P(K\log p/n)$. Recalling that $\mathbf{B} = \mathbf{\Sigma}_{12}\mathbf{\Sigma}_{22}^{-1}$, we have

$$\|\mathbf{\Sigma}_{12}^T\mathbf{\Sigma}^{-1}\mathbf{\Sigma}_{12}\| \le \|\mathbf{B}^T\mathbf{\Sigma}^{-1}\mathbf{B}\|\|\mathbf{\Sigma}_{22}\|^2,$$

which is bounded by Lemma 10.2. Hence $\Delta_2 = O_P(\sqrt{K\log p/n})$.

We next bound $\Delta_3$. Again, using the argument as in Lemma 10.1, we have

$$\Delta_3 \le p^{-1/2}\|\widehat{\mathbf{\Sigma}}_{22}^{-1} - \mathbf{\Sigma}_{22}^{-1}\|_F\|\mathbf{\Sigma}_{12}^T\mathbf{\Sigma}^{-1}\mathbf{\Sigma}_{12}\|.$$

The second factor is bounded as shown above and the first factor is

$$\|\mathbf{\Sigma}_{22}^{-1}(\widehat{\mathbf{\Sigma}}_{22} - \mathbf{\Sigma}_{22})\mathbf{\Sigma}_{22}^{-1}\|_F \le \|\mathbf{\Sigma}_{22}^{-1}\|^2\|\widehat{\mathbf{\Sigma}}_{22} - \mathbf{\Sigma}_{22}\|_F = O_P(K\sqrt{\log p/n}),$$

by Lemma 10.1. Namely, $\Delta_3 = O_P(Kp^{-1/2}\sqrt{\log p/n})$.

Combining results above, by (10.43), we conclude that

$$\|\widehat{\mathbf{\Sigma}}^{\mathcal{T}} - \mathbf{\Sigma}\|_{\mathbf{\Sigma}} = O_P(K\sqrt{p}\log p/n + m_p\omega_n^{1-q}).$$

(iv) Third rate in (10.23). We now establish the rate of convergence for $\|(\widehat{\mathbf{\Sigma}}^{\mathcal{T}})^{-1} - \mathbf{\Sigma}^{-1}\|$. By the Sherman-Morrison-Woodbury formula (9.6), we have

$$\mathbf{\Sigma}^{-1} = \mathbf{\Sigma}_u^{-1} - \mathbf{D}\mathbf{A}^{-1}\mathbf{D}^T,$$

where $\mathbf{A} = \mathbf{\Sigma}_{22} + \mathbf{\Sigma}_{12}^T \mathbf{\Sigma}_u^{-1} \mathbf{\Sigma}_{21}$ and $\mathbf{D} = \mathbf{\Sigma}_u^{-1} \mathbf{\Sigma}_{12}$. Denote by the substitution estimators of $\mathbf{A}$ and $\mathbf{D}$, $\widehat{\mathbf{A}}$ and $\widehat{\mathbf{D}}$ respectively. Then, we have the following bound similar to (10.43):

$$\begin{aligned}
\|(\widehat{\mathbf{\Sigma}}^{\mathcal{T}})^{-1} - \mathbf{\Sigma}^{-1}\| &\le \|\widehat{\mathbf{D}}\widehat{\mathbf{A}}^{-1}\widehat{\mathbf{D}}^T - \mathbf{D}\mathbf{A}^{-1}\mathbf{D}^T\| + \|(\widehat{\mathbf{\Sigma}}_u^{\mathcal{T}})^{-1} - \mathbf{\Sigma}_u^{-1}\| \\
&\le \|(\widehat{\mathbf{D}} - \mathbf{D})\widehat{\mathbf{A}}^{-1}(\widehat{\mathbf{D}} - \mathbf{D})^T\| + 2\|(\widehat{\mathbf{D}} - \mathbf{D})\widehat{\mathbf{A}}^{-1}\mathbf{D}^T\| \\
&\quad + \|\mathbf{D}(\widehat{\mathbf{A}}^{-1} - \mathbf{A}^{-1})\mathbf{D}^T\| + \|(\widehat{\mathbf{\Sigma}}_u^{\mathcal{T}})^{-1} - \mathbf{\Sigma}_u^{-1}\| \\
&=: \widetilde{\Delta}_1 + 2\widetilde{\Delta}_2 + \widetilde{\Delta}_3 + \widetilde{\Delta}_4 .
\end{aligned}$$

From (10.22), $\widetilde{\Delta}_4 = O_P(m_p \omega_n^{1-q})$. For the remaining terms, we need to find the rates for $\|\widehat{\mathbf{D}} - \mathbf{D}\|$, $\|\widehat{\mathbf{A}}^{-1}\|$, $\|\mathbf{D}\|$ and $\|\widehat{\mathbf{A}}^{-1} - \mathbf{A}^{-1}\|$ separately. Note that

$$\|\mathbf{\Sigma}_{12}\| = \|\mathbf{B}\mathbf{\Sigma}_{22}\| \le \|\mathbf{B}\|\|\mathbf{\Sigma}_{22}\| \le \sqrt{Kp}\|\mathbf{B}\|_{\max}\|\mathbf{\Sigma}_{22}\| = O_P(\sqrt{Kp})$$

and hence $\|\mathbf{D}\| = O_P(\sqrt{Kp})$. By (10.44) and (10.22), we have

$$\|\widehat{\mathbf{D}} - \mathbf{D}\| \le \|(\widehat{\mathbf{\Sigma}}_u^{\mathcal{T}})^{-1}\|\|\widehat{\mathbf{\Sigma}}_{12} - \mathbf{\Sigma}_{12}\| + \|\mathbf{\Sigma}_{12}\|\|(\widehat{\mathbf{\Sigma}}_u^{\mathcal{T}})^{-1} - \mathbf{\Sigma}_u^{-1}\| = O_P(\sqrt{Kp}m_p\omega_n^{1-q}).$$

In addition, it is not hard to show

$$\|\widehat{\mathbf{A}} - \mathbf{A}\| = O_p(\|\mathbf{\Sigma}_{12}\|^2\|(\widehat{\mathbf{\Sigma}}_u)^{-1} - \mathbf{\Sigma}_u\|) = O_P(Kpm_p\omega_n^{1-q}).$$

Note that

$$\lambda_{\min}(\mathbf{A}) \ge \lambda_{\min}(\mathbf{\Sigma}_{12}^T\mathbf{\Sigma}_u^{-1}\mathbf{\Sigma}_{21}) \ge \lambda_{\min}(\mathbf{\Sigma}_u^{-1})\lambda_{\min}(\mathbf{\Sigma}_{22})\lambda_K(\mathbf{B}\mathbf{\Sigma}_{22}\mathbf{B}^T),$$

and by Weyl's inequality,

$$\lambda_K(\mathbf{B}\mathbf{\Sigma}_{22}\mathbf{B}^T) \ge \lambda_K(\mathbf{\Sigma}) - \|\mathbf{\Sigma}_u\| \ge cp.$$

Therefore, $\|\mathbf{A}^{-1}\| = O_P(p^{-1})$ and

$$\|\widehat{\mathbf{A}}^{-1} - \mathbf{A}^{-1}\| \le \|\mathbf{A}^{-1}\|\|\widehat{\mathbf{A}}^{-1}\|\|\widehat{\mathbf{A}} - \mathbf{A}\|$$

implies

$$\|\widehat{\mathbf{A}}^{-1} - \mathbf{A}^{-1}\| = O_P(Kp^{-1}m_p\omega_n^{1-q}),$$

and $\|\widehat{\mathbf{A}}^{-1}\| = O_P(p^{-1})$. Incorporating the above rates together, we conclude

$$\begin{aligned}
\widetilde{\Delta}_1 &= O_P(p^{-1}\|\widehat{\mathbf{D}} - \mathbf{D}\|^2) = O_P(Km_p^2\omega_n^{2(1-q)}), \\
\widetilde{\Delta}_2 &= O_P(\sqrt{K/p}\|\widehat{\mathbf{D}} - \mathbf{D}\|) = O_P(Km_p\omega_n^{1-q}), \\
\widetilde{\Delta}_3 &= O_P(Kp\|\widehat{\mathbf{A}}^{-1} - \mathbf{A}^{-1}\|) = O_P(K^2m_p\omega_n^{1-q}).
\end{aligned}$$

Combining rates for $\widetilde{\Delta}_i, i = 1, 2, 3, 4$, we complete the proof.

*10.6.2 Proof of Theorem 10.2*

To obtain (10.32), we need to bound separately

$$\Delta_1 := \|\widehat{\mathbf{\Gamma}}\widehat{\mathbf{\Lambda}}\widehat{\mathbf{\Gamma}}^T - \mathbf{\Gamma}\mathbf{\Lambda}\mathbf{\Gamma}^T\|_{\max} \quad \text{and} \quad \Delta_2 := \|\mathbf{B}\mathbf{B}^T - \mathbf{\Gamma}\mathbf{\Lambda}\mathbf{\Gamma}^T\|_{\max}.$$

Recall $\mathbf{B} = (\widetilde{\mathbf{b}}_1, \ldots, \widetilde{\mathbf{b}}_K)$ and $\mathbf{B}\mathbf{B}^T = \widetilde{\mathbf{\Gamma}}\widetilde{\mathbf{\Lambda}}\widetilde{\mathbf{\Gamma}}^T$, where $\widetilde{\mathbf{\Lambda}} = \mathrm{diag}(\|\widetilde{\mathbf{b}}_1\|^2, \ldots, \|\widetilde{\mathbf{b}}_K\|^2)$ and the $j^{th}$ column of $\widetilde{\mathbf{\Gamma}}$ is $\widetilde{\mathbf{b}}_j/\|\widetilde{\mathbf{b}}_j\|$. By assumption (10.30) and Proposition 10.1, we have for some constant $c > 0$

$$\begin{array}{lll} \lambda_j \geq cp, & \widehat{\lambda}_j \geq cp + o_P(1), & \widetilde{\lambda}_j \geq cp, \\ \lambda_j = O(p), & \widehat{\lambda}_j = O_P(p), & \widetilde{\lambda}_j = O(p), \end{array} \quad \text{for } j \leq K. \tag{10.45}$$

Furthermore, by Proposition 10.1, we have

$$|\lambda_j - \widetilde{\lambda}_j| \leq \|\mathbf{\Sigma}_u\|, \qquad \|\widetilde{\mathbf{\Gamma}} - \mathbf{\Gamma}\|_{\max} = O(1/p). \tag{10.46}$$

As remarked at the beginning of Section 10.5.1, $\|\widetilde{\mathbf{\Gamma}}\|_{\max} = O(1/\sqrt{p})$. This together with (10.46) and assumption (10.30) entail

$$\|\mathbf{\Gamma}\|_{\max} = O(1/\sqrt{p}) \quad \text{and} \quad \|\widehat{\mathbf{\Gamma}}\|_{\max} = O_P(1/\sqrt{p}). \tag{10.47}$$

By definition, it is easy to verify for any $p \times K$ matrix $\mathbf{A}$ and $q \times K$ matrix $\mathbf{B}$,

$$\|\mathbf{A}\mathbf{B}^T\|_{\max} = \max_{i,j} |\sum_{k=1}^{K} a_{ik} b_{jk}| \leq K\|\mathbf{A}\|_{\max}\|\mathbf{B}\|_{\max}.$$

Using this and the decomposition

$$\begin{aligned} \widehat{\mathbf{\Gamma}}\widehat{\mathbf{\Lambda}}\widehat{\mathbf{\Gamma}}^T - \mathbf{\Gamma}\mathbf{\Lambda}\mathbf{\Gamma}^T &= \widehat{\mathbf{\Gamma}}(\widehat{\mathbf{\Lambda}} - \mathbf{\Lambda})\widehat{\mathbf{\Gamma}}^T + (\widehat{\mathbf{\Gamma}} - \mathbf{\Gamma})\mathbf{\Lambda}(\widehat{\mathbf{\Gamma}} - \mathbf{\Gamma})^T \\ &\quad + \mathbf{\Gamma}\mathbf{\Lambda}(\widehat{\mathbf{\Gamma}} - \mathbf{\Gamma})^T + (\widehat{\mathbf{\Gamma}} - \mathbf{\Gamma})\mathbf{\Lambda}\mathbf{\Gamma}^T, \end{aligned} \tag{10.48}$$

we have

$$\begin{aligned} \Delta_1 &\leq \|\widehat{\mathbf{\Gamma}}(\widehat{\mathbf{\Lambda}} - \mathbf{\Lambda})\widehat{\mathbf{\Gamma}}^T\|_{\max} + \|(\widehat{\mathbf{\Gamma}} - \mathbf{\Gamma})\mathbf{\Lambda}(\widehat{\mathbf{\Gamma}} - \mathbf{\Gamma})^T\|_{\max} + 2\|\mathbf{\Gamma}\mathbf{\Lambda}(\widehat{\mathbf{\Gamma}} - \mathbf{\Gamma})^T\|_{\max} \\ &\leq K^2\|\widehat{\mathbf{\Gamma}}\|_{\max}^2\|\widehat{\mathbf{\Lambda}} - \mathbf{\Lambda}\|_{\max} + K^2\|\widehat{\mathbf{\Gamma}} - \mathbf{\Gamma}\|_{\max}^2\|\mathbf{\Lambda}\|_{\max} \\ &\quad + 2K^2\|\mathbf{\Gamma}\|_{\max}\|\mathbf{\Lambda}\|_{\max}\|\widehat{\mathbf{\Gamma}} - \mathbf{\Gamma}\|_{\max} \end{aligned}$$

By using assumption (10.30) and (10.47), we have

$$\begin{aligned} \Delta_1 &= O_P\Big(K^2(p^{-1}p\sqrt{\log p/n} + \log p/(np)p + p^{-1/2}p\sqrt{\log p/(np)})\Big) \\ &= O_P(K^2\sqrt{\log p/n}). \end{aligned}$$

Similarly, by (10.46) and (10.47), we have

$$\begin{aligned}\Delta_2 &= \|\widetilde{\mathbf{\Gamma}}\widetilde{\mathbf{\Lambda}}\widetilde{\mathbf{\Gamma}}^T - \mathbf{\Gamma}\mathbf{\Lambda}\mathbf{\Gamma}^T\|_{\max} \\ &\le \|\widetilde{\mathbf{\Gamma}}(\widetilde{\mathbf{\Lambda}} - \mathbf{\Lambda})\widetilde{\mathbf{\Gamma}}^T\|_{\max} + \|(\widetilde{\mathbf{\Gamma}} - \mathbf{\Gamma})\mathbf{\Lambda}(\widetilde{\mathbf{\Gamma}} - \mathbf{\Gamma})^T\|_{\max} + 2\|\mathbf{\Gamma}\mathbf{\Lambda}(\widetilde{\mathbf{\Gamma}} - \mathbf{\Gamma})^T\|_{\max} \\ &\le K^2(\|\widetilde{\mathbf{\Gamma}}\|_{\max}^2\|\widetilde{\mathbf{\Lambda}} - \mathbf{\Lambda}\|_{\max} + \|\mathbf{\Lambda}\|_{\max}\|\|\widetilde{\mathbf{\Gamma}} - \mathbf{\Gamma}\|_{\max}^2 \\ &\quad + 2\|\mathbf{\Gamma}\|_{\max}\|\mathbf{\Lambda}\|_{\max}\|\widetilde{\mathbf{\Gamma}} - \mathbf{\Gamma}\|_{\max}).\end{aligned}$$

Substituting the rates in (10.46) into the above bound, we have

$$\Delta_2 = O\Big(K^2(p^{-1} + p/p^2 + p^{-1/2}p/p)\Big) = O(K^2/\sqrt{p}).$$

Combining the rates of $\Delta_1$ and $\Delta_2$, we prove the first rate of (10.32).

Second rate in (10.32). The idea of this part of the proof follows very similarly to that of the first part. For simplicity of notation, let

$$\widehat{a}_j = \widehat{\lambda}_j^{1/2}, \quad \widetilde{a}_j = \widetilde{\lambda}_j^{1/2}, \quad a_j = \lambda_j^{1/2}.$$

Recall $\widehat{\boldsymbol{\xi}}_j$ denotes the $j^{th}$ column of $\widehat{\mathbf{\Gamma}}$. Then, the $j^{th}$ column of $\widehat{\mathbf{B}} - \mathbf{B}$ is

$$\|\widehat{a}_j\widehat{\boldsymbol{\xi}}_j - \widetilde{a}_j\widetilde{\boldsymbol{\xi}}_j\|_{\max} = \|\widehat{a}_j\widehat{\boldsymbol{\xi}}_j - a_j\boldsymbol{\xi}_j\|_{\max} + \|a_j\boldsymbol{\xi}_j - \widetilde{a}_j\widetilde{\boldsymbol{\xi}}_j\|_{\max} = \Delta_3 + \Delta_4. \quad (10.49)$$

Using the same decomposition as (10.48), we have

$$\begin{aligned}\Delta_3 &\le |\widehat{a}_j - a_j|\|\widehat{\boldsymbol{\xi}}_j\|_{\max} + a_j\|\widehat{\boldsymbol{\xi}}_j - \boldsymbol{\xi}_j\|_{\max} \\ &= O_p\Big(\sqrt{p\log p/n}/\sqrt{p} + \sqrt{p}\sqrt{\log p/(np)}\Big) = O_P\Big(\sqrt{\frac{\log p}{n}}\Big).\end{aligned}$$

Similarly, we have

$$\Delta_4 = O\Big(\frac{\|\mathbf{\Sigma}_u\|}{\sqrt{p}}/p + \sqrt{p}\|\mathbf{\Sigma}_u\|/p\Big) = O(1/\sqrt{p}).$$

This proves the second part of (10.32).

### *10.6.3 Proof of Theorem 10.3*

(i) To obtain the rates of convergence in (10.34), by Theorem 9.1, it suffices to prove $\|\widehat{\mathbf{\Sigma}}_u - \mathbf{\Sigma}_u\|_{\max} = O_P(w_n)$. According to (10.30) and (10.32), we have

$$\|\widehat{\mathbf{\Sigma}}_u - \mathbf{\Sigma}_u\|_{\max} = \|\widehat{\mathbf{\Sigma}} - \mathbf{\Sigma} - (\widehat{\mathbf{\Gamma}}\widehat{\mathbf{\Lambda}}\widehat{\mathbf{\Gamma}}^T - \mathbf{B}\mathbf{B}^T)\|_{\max} = O_p(w_n).$$

Hence, (10.34) follows by Theorem 9.1.

(ii) First rate in (10.35): Note that

$$\|\widehat{\mathbf{\Sigma}}_u^{\mathcal{T}} - \mathbf{\Sigma}_u\|_{\max} \le \|\widehat{\mathbf{\Sigma}}_u^{\mathcal{T}} - \widehat{\mathbf{\Sigma}}_u\|_{\max} + \|\widehat{\mathbf{\Sigma}}_u - \mathbf{\Sigma}_u\|_{\max} = O_P(w_n)$$

since the order of the adaptive thresholding is chosen to be the same order as $w_n$. It then follows from Theorem 10.2 that

$$\|\widehat{\boldsymbol{\Sigma}}^{\mathcal{T}} - \boldsymbol{\Sigma}\|_{\max} \le \|\widehat{\boldsymbol{\Gamma}}\widehat{\boldsymbol{\Lambda}}\widehat{\boldsymbol{\Gamma}} - \mathbf{B}\mathbf{B}^T\|_{\max} + \|\widehat{\boldsymbol{\Sigma}}_u^{\mathcal{T}} - \boldsymbol{\Sigma}_u\|_{\max} = O_P(w_n).$$

(iii) Second rate in (10.35): Let

$$\boldsymbol{\Sigma} = \boldsymbol{\Gamma}_p \boldsymbol{\Lambda}_p \boldsymbol{\Gamma}_p^T, \qquad \boldsymbol{\Gamma}_p = (\boldsymbol{\Gamma}, \boldsymbol{\Gamma}_2), \qquad \boldsymbol{\Lambda}_p = \mathrm{diag}(\boldsymbol{\Lambda}, \boldsymbol{\Theta}).$$

be the spectral decomposition of $\boldsymbol{\Sigma}$. Note that

$$\begin{aligned}\|\widehat{\boldsymbol{\Sigma}}^{\mathcal{T}} - \boldsymbol{\Sigma}\|_{\Sigma} \le & p^{-1/2}\left\|\boldsymbol{\Sigma}^{-\frac{1}{2}}(\widehat{\boldsymbol{\Gamma}}\widehat{\boldsymbol{\Lambda}}\widehat{\boldsymbol{\Gamma}}^T - \mathbf{B}\mathbf{B}^T)\boldsymbol{\Sigma}^{-\frac{1}{2}}\right\|_F \\ & + p^{-1/2}\|\boldsymbol{\Sigma}^{-\frac{1}{2}}(\widehat{\boldsymbol{\Sigma}}_u^{\mathcal{T}} - \boldsymbol{\Sigma}_u)\boldsymbol{\Sigma}^{-\frac{1}{2}}\|_F \equiv \Delta_L + \Delta_S.\end{aligned} \tag{10.50}$$

The second term can easily be bounded as, by Lemma 10.1,

$$\Delta_S \le p^{-1/2}\|\boldsymbol{\Sigma}^{-1}\|\|\widehat{\boldsymbol{\Sigma}}_u^{\mathcal{T}} - \boldsymbol{\Sigma}_u\|_F \le C\|\widehat{\boldsymbol{\Sigma}}_u^{\mathcal{T}} - \boldsymbol{\Sigma}_u\| = O_P(m_p w_n^{1-q}), \tag{10.51}$$

where the last inequality follows from Proposition 9.1(3). We now deal with the first term. It is easy to see that

$$\begin{aligned}\Delta_L = & p^{-1/2}\left\|\begin{pmatrix}\boldsymbol{\Lambda}^{-\frac{1}{2}}\boldsymbol{\Gamma}^T \\ \boldsymbol{\Theta}^{-\frac{1}{2}}\boldsymbol{\Gamma}_2^T\end{pmatrix}(\widehat{\boldsymbol{\Gamma}}\widehat{\boldsymbol{\Lambda}}\widehat{\boldsymbol{\Gamma}}^T - \mathbf{B}\mathbf{B}^T)\begin{pmatrix}\boldsymbol{\Gamma}\boldsymbol{\Lambda}^{-\frac{1}{2}} & \boldsymbol{\Gamma}_2\boldsymbol{\Theta}^{-\frac{1}{2}}\end{pmatrix}\right\|_F \\ & \le \Delta_{L1} + \Delta_{L2} + \sqrt{2}\Delta_{L3},\end{aligned} \tag{10.52}$$

where

$$\begin{aligned}\Delta_{L1} &= \|\boldsymbol{\Lambda}^{-\frac{1}{2}}\boldsymbol{\Gamma}^T(\widehat{\boldsymbol{\Gamma}}\widehat{\boldsymbol{\Lambda}}\widehat{\boldsymbol{\Gamma}}^T - \mathbf{B}\mathbf{B}^T)\boldsymbol{\Gamma}\boldsymbol{\Lambda}^{-\frac{1}{2}}\|_F/\sqrt{p}, \\ \Delta_{L2} &= \|\boldsymbol{\Theta}^{-\frac{1}{2}}\boldsymbol{\Gamma}_2^T(\widehat{\boldsymbol{\Gamma}}\widehat{\boldsymbol{\Lambda}}\widehat{\boldsymbol{\Gamma}}^T - \mathbf{B}\mathbf{B}^T)\boldsymbol{\Gamma}_2\boldsymbol{\Theta}^{-\frac{1}{2}}\|_F/\sqrt{p}, \\ \Delta_{L3} &= \|\boldsymbol{\Lambda}^{-\frac{1}{2}}\boldsymbol{\Gamma}^T(\widehat{\boldsymbol{\Gamma}}\widehat{\boldsymbol{\Lambda}}\widehat{\boldsymbol{\Gamma}}^T - \mathbf{B}\mathbf{B}^T)\boldsymbol{\Gamma}_2\boldsymbol{\Theta}^{-\frac{1}{2}}\|_F/\sqrt{p}.\end{aligned}$$

We now deal with each of the terms separately. By Lemma 10.1, we have

$$\Delta_{L1} \le p^{-1/2}\|\boldsymbol{\Lambda}^{-1}\|\|\boldsymbol{\Gamma}\|^2\|\widehat{\boldsymbol{\Gamma}}\widehat{\boldsymbol{\Lambda}}\widehat{\boldsymbol{\Gamma}}^T - \mathbf{B}\mathbf{B}^T\|_F = O_P(p^{-3/2}\|\widehat{\boldsymbol{\Gamma}}\widehat{\boldsymbol{\Lambda}}\widehat{\boldsymbol{\Gamma}}^T - \mathbf{B}\mathbf{B}^T\|_F).$$

Using Proposition 9.1(3) and (2), we have

$$\|\widehat{\boldsymbol{\Gamma}}\widehat{\boldsymbol{\Lambda}}\widehat{\boldsymbol{\Gamma}}^T - \mathbf{B}\mathbf{B}^T\|_F \le K^{1/2}\|\widehat{\boldsymbol{\Gamma}}\widehat{\boldsymbol{\Lambda}}\widehat{\boldsymbol{\Gamma}}^T - \mathbf{B}\mathbf{B}^T\| \le K\sqrt{p}\|\widehat{\boldsymbol{\Gamma}}\widehat{\boldsymbol{\Lambda}}\widehat{\boldsymbol{\Gamma}}^T - \mathbf{B}\mathbf{B}^T\|_{\max}.$$

Therefore, by (10.32), we conclude that

$$\Delta_{L1} = O(Kp^{-1}w_n).$$

By the triangular inequality, $\Delta_{L2}$ is bounded by

$$p^{-1/2}\left(\|\boldsymbol{\Theta}^{-\frac{1}{2}}\boldsymbol{\Gamma}_2^T\widehat{\boldsymbol{\Gamma}}\widehat{\boldsymbol{\Lambda}}\widehat{\boldsymbol{\Gamma}}^T\boldsymbol{\Gamma}_2\boldsymbol{\Theta}^{-\frac{1}{2}}\|_F + \|\boldsymbol{\Theta}^{-\frac{1}{2}}\boldsymbol{\Gamma}_2^T\widetilde{\boldsymbol{\Gamma}}\widetilde{\boldsymbol{\Lambda}}\widetilde{\boldsymbol{\Gamma}}^T\boldsymbol{\Gamma}_2\boldsymbol{\Theta}^{-\frac{1}{2}}\|_F\right) =: \Delta_{L2}^{(1)} + \Delta_{L2}^{(2)}.$$

By Lemma 10.1 and $\mathbf{\Gamma}_2^T\mathbf{\Gamma} = 0$, we have

$$\begin{aligned}\Delta_{L2}^{(1)} &\le p^{-1/2}\|\mathbf{\Theta}^{-1}\|\|\mathbf{\Gamma}_2^T\widehat{\mathbf{\Gamma}}\widehat{\mathbf{\Lambda}}\widehat{\mathbf{\Gamma}}^T\mathbf{\Gamma}_2\|_F \\ &\le p^{-1/2}\|\mathbf{\Theta}^{-1}\|\|\mathbf{\Gamma}_2^T\widehat{\mathbf{\Gamma}}\|^2\|\widehat{\mathbf{\Lambda}}\|_F \\ &= O_P(\sqrt{pK}\|\mathbf{\Gamma}_2^T(\widehat{\mathbf{\Gamma}} - \mathbf{\Gamma})\|^2).\end{aligned}$$

By Proposition 9.1(2), we have

$$\|\mathbf{\Gamma}_2^T(\widehat{\mathbf{\Gamma}} - \mathbf{\Gamma})\| \le \|\widehat{\mathbf{\Gamma}} - \mathbf{\Gamma}\| \le \sqrt{pK}\|\widehat{\mathbf{\Gamma}} - \mathbf{\Gamma}\|_{\max}$$

Therefore,

$$\Delta_{L2}^{(1)} = O_P(K^{3/2}p^{1/2}\log p/n),$$

since $\|\widehat{\mathbf{\Gamma}} - \mathbf{\Gamma}\|_{\max} = O_P(\sqrt{\log p/(np)})$.

Using similar arguments, by Lemma 10.1, we have

$$\Delta_{L2}^{(2)} \le p^{-1/2}\|\mathbf{\Theta}^{-1}\|\|\widetilde{\mathbf{\Lambda}}\|_F\|\mathbf{\Gamma}_2^T\widetilde{\mathbf{\Gamma}}\|^2 = O_P(\sqrt{pK}\|\mathbf{\Gamma}_2^T\widetilde{\mathbf{\Gamma}}\|^2).$$

By the $\sin\theta$ theorem (Proposition 10.1), we have

$$\|\mathbf{\Gamma}_2^T\widetilde{\mathbf{\Gamma}}\| = \|\mathbf{\Gamma}_2^T(\widetilde{\mathbf{\Gamma}} - \mathbf{\Gamma})\| \le \|\mathbf{\Gamma} - \widetilde{\mathbf{\Gamma}}\| = O(\|\mathbf{\Sigma}_u\|/p).$$

Hence, $\Delta_{L2}^{(2)} = O(K^{1/2}/p^{3/2})$. Combining the bounds, we have

$$\Delta_{L2} = O_P(K^{3/2}p^{1/2}\log p/n + K^{1/2}/p^{3/2}).$$

By similar analysis, $\Delta_{L3}$ is dominated by $\Delta_{L1}$ and $\Delta_{L2}$. Combining the terms $\Delta_{L1}$, $\Delta_{L2}$, $\Delta_{L3}$ and $\Delta_S$ together, we complete the proof for the relative Frobenius norm.

(iv) Third rate in (10.35): We now turn to the analysis of the spectral norm error of the inverse covariance matrix. By the Sherman-Morrison-Woodbury formula (9.6), we have

$$\|(\widehat{\mathbf{\Sigma}}^{\tau})^{-1} - \mathbf{\Sigma}^{-1}\|_2 \le \|(\widehat{\mathbf{\Sigma}}_u^{\tau})^{-1} - \mathbf{\Sigma}_u^{-1}\| + \Delta,$$

where with $\widehat{\mathbf{B}} = \widehat{\mathbf{\Gamma}}\widehat{\mathbf{\Lambda}}^{\frac{1}{2}}$

$$\Delta = \|(\widehat{\mathbf{\Sigma}}_u^{\tau})^{-1}\widehat{\mathbf{B}}(\mathbf{I}_K + \widehat{\mathbf{J}})^{-1}\widehat{\mathbf{B}}^T(\widehat{\mathbf{\Sigma}}_u^{\tau})^{-1} - \mathbf{\Sigma}_u^{-1}\mathbf{B}(\mathbf{I}_K + \mathbf{J})^{-1}\mathbf{B}^T\mathbf{\Sigma}_u^{-1}\|,$$

with $\widehat{\mathbf{J}} = \widehat{\mathbf{B}}^T(\widehat{\mathbf{\Sigma}}_u^{\tau})^{-1}\widehat{\mathbf{B}}$ and $\mathbf{J} = \mathbf{B}^T\mathbf{\Sigma}_u^{-1}\mathbf{B}$. The right-hand side can be bounded by the sum of the following five terms:

$$\begin{aligned}T_1 &= \|((\widehat{\mathbf{\Sigma}}_u^{\tau})^{-1} - \mathbf{\Sigma}_u^{-1})\widehat{\mathbf{B}}(\mathbf{I}_K + \widehat{\mathbf{J}})^{-1}\widehat{\mathbf{B}}^T\widehat{\mathbf{\Sigma}}_u^{-1}\|, \\ T_2 &= \|\mathbf{\Sigma}_u^{-1}(\widehat{\mathbf{B}} - \mathbf{B})(\mathbf{I}_K + \widehat{\mathbf{J}})^{-1}\widehat{\mathbf{B}}^T\widehat{\mathbf{\Sigma}}_u^{-1}\|, \\ T_3 &= \|\mathbf{\Sigma}_u^{-1}\mathbf{B}[(\mathbf{I}_K + \widehat{\mathbf{J}})^{-1} - (\mathbf{I}_K + \mathbf{J})^{-1}]\widehat{\mathbf{B}}^T\widehat{\mathbf{\Sigma}}_u^{-1}\|. \\ T_4 &= \|\mathbf{\Sigma}_u^{-1}\mathbf{B}(\mathbf{I}_K + \mathbf{J})^{-1}(\widehat{\mathbf{B}} - \mathbf{B})^T\widehat{\mathbf{\Sigma}}_u^{-1}\|. \\ T_5 &= \|\mathbf{\Sigma}_u^{-1}\mathbf{B}(\mathbf{I}_K + \mathbf{J})^{-1}\mathbf{B}^T(\widehat{\mathbf{\Sigma}}_u^{-1} - \mathbf{\Sigma}_u^{-1})\|.\end{aligned}$$

Note that $\mathbf{I}_K + \widehat{\mathbf{J}} \geq \lambda_{\min}(\widehat{\boldsymbol{\Sigma}}_u^{-1})\widehat{\mathbf{B}}^T\widehat{\mathbf{B}}$. Hence,

$$\|(\mathbf{I}_K + \widehat{\mathbf{J}})^{-1}\| = O_P(1/p).$$

Thus, the first term is bounded by

$$T_1 \leq \lambda_{\min}(\widehat{\boldsymbol{\Sigma}}_u^{-1})^{-1}\|\widehat{\mathbf{B}}(\widehat{\mathbf{B}}^T\widehat{\mathbf{B}})^{-1}\widehat{\mathbf{B}}^T\|\|(\widehat{\boldsymbol{\Sigma}}_u^{\tau})^{-1} - \boldsymbol{\Sigma}_u^{-1}\|\|(\widehat{\boldsymbol{\Sigma}}_u^{\tau})^{-1}\| = O_P(m_p w_n^{1-q}),$$

where we used the fact that the operator norm of any projection matrix of 1. The second term is bounded by

$$T_2 = O_p(\sqrt{Kp}\|\widehat{\mathbf{B}} - \mathbf{B}\|_{\max} p^{-1}\sqrt{Kp}\|\widehat{\mathbf{B}}\|_{\max}) = O_p(w_n).$$

To bound the third term, note that

$$\begin{aligned}
\|(\mathbf{I}_K + \widehat{\mathbf{J}})^{-1} - (\mathbf{I}_K + \mathbf{J})^{-1}\| &\leq \|(\mathbf{I}_K + \widehat{\mathbf{J}})^{-1}\|\|\widehat{\mathbf{J}} - \mathbf{J}\|\|(\mathbf{I}_K + \mathbf{J})^{-1}\| \\
&= O_P(p^{-2}\|\mathbf{B}\|^2\|\|(\widehat{\boldsymbol{\Sigma}}_u^{\tau})^{-1} - \boldsymbol{\Sigma}_u^{-1}\|).
\end{aligned}$$

The third term is of order

$$T_3 = O_P(p^{-2}\|\mathbf{B}\|^4 m_p \omega_p^{1-q}) = O_P(K^2 m_p \omega_p^{1-q}),$$

by using $\|\mathbf{B}\| \leq \sqrt{Kp}\|\mathbf{B}\|_{\max}$ (Proposition 9.1 (2)). Following the same arguments, it is easy to see that $T_4 = O_P(T_2)$ and $T_5 = O_P(T_1)$.

With the combination of the above results, we conclude that $\Delta = O_P(K^2 m_p w_n^{1-q})$ and that $\|(\widehat{\boldsymbol{\Sigma}}^{\tau})^{-1} - \boldsymbol{\Sigma}^{-1}\|_2 = O_P(K^2 m_p w_n^{1-q})$. This completes the proof.

### *10.6.4 Proof of Theorem 10.4*

Let $\boldsymbol{\mu} = \mathrm{E}\,\mathbf{X}_i$. First of all, by (10.6),

$$\begin{aligned}
\mathbf{f}_i &= (\mathbf{B}^T\mathbf{B})^{-1}\mathbf{B}^T(\mathbf{X}_i - \boldsymbol{\mu} - \mathbf{u}_i) \\
&= \widetilde{\boldsymbol{\Lambda}}^{-1}\mathbf{B}^T(\mathbf{X}_i - \boldsymbol{\mu}) - \widetilde{\boldsymbol{\Lambda}}^{-1}\mathbf{B}^T\mathbf{u}_i.
\end{aligned}$$

Let $f_i^* = \boldsymbol{\Lambda}^{-1}\mathbf{B}^T(\mathbf{X}_i - \boldsymbol{\mu})$. Then,

$$\|\widehat{\mathbf{f}}_i - \mathbf{f}_i\| \leq \|\widehat{\mathbf{f}}_i - \mathbf{f}_i^*\| + \|\mathbf{f}_i^* - \mathbf{f}_i\|.$$

We deal with these two terms separately.

Let us deal with the first term. Observe that

$$\begin{aligned}
\|\widehat{\mathbf{f}}_i - \mathbf{f}_i^*\| &= \|(\widehat{\boldsymbol{\Lambda}}^{-1} - \boldsymbol{\Lambda}^{-1})\widehat{\mathbf{B}}^T(\mathbf{X}_i - \widehat{\boldsymbol{\mu}})\| + \|\boldsymbol{\Lambda}^{-1}(\widehat{\mathbf{B}} - \mathbf{B})(\mathbf{X}_i - \widehat{\boldsymbol{\mu}})\| \\
&\quad + \|\boldsymbol{\Lambda}^{-1}\mathbf{B}(\widehat{\boldsymbol{\mu}} - \boldsymbol{\mu})\| \\
&\leq \|\widehat{\boldsymbol{\Lambda}}^{-1} - \boldsymbol{\Lambda}^{-1}\|\|\widehat{\mathbf{B}}^T\|\|\mathbf{X}_i - \widehat{\boldsymbol{\mu}}\| + \|\boldsymbol{\Lambda}^{-1}\|\|\widehat{\mathbf{B}} - \mathbf{B}\|\|\mathbf{X}_i - \widehat{\boldsymbol{\mu}}\| \\
&\quad + \|\boldsymbol{\Lambda}^{-1}\|\|\mathbf{B}\|\|\widehat{\boldsymbol{\mu}} - \boldsymbol{\mu}\|. \qquad (10.53)
\end{aligned}$$

By the assumptions and (10.32), we have

$$\|\widehat{\mathbf{\Lambda}}^{-1} - \mathbf{\Lambda}^{-1}\| \|\widehat{\mathbf{B}}^T\| = O_P(p^{-1}\sqrt{\log p/n}\sqrt{pK}) = O_P(\sqrt{K/p}\sqrt{\log p/n}),$$

and by Theorem 10.2,

$$\|\mathbf{\Lambda}^{-1}\| \|\widehat{\mathbf{B}} - \mathbf{B}\| = O_P\Big(p^{-1}\sqrt{Kp}(\sqrt{\log p/n} + 1/\sqrt{p})\Big).$$

Thus, the first term in (10.53) is always demonstrated by the second term. Similarly, by the assumption, the third term is bounded by

$$\|\mathbf{\Lambda}^{-1}\| \|\mathbf{B}\| \|\widehat{\boldsymbol{\mu}} - \boldsymbol{\mu}\| = O_p(p^{-1}\sqrt{pK}\sqrt{p\log p/n}) = O_P(\sqrt{K\log p/n}).$$

Let $U_n = n^{-1}\sum_{i=1}^n \|\mathbf{X}_i - \widehat{\boldsymbol{\mu}}\|$. Then, it follows from (10.53) that

$$n^{-1}\sum_{i=1}^n \|\widehat{\mathbf{f}}_i - \mathbf{f}_i^*\| = O_P(\sqrt{K/p}(\sqrt{\log p/n} + 1/\sqrt{p})U_n + \sqrt{K\log p/n}).$$

Now, let us deal with $U_n$. Notice that

$$\mathrm{E}\,\|\mathbf{X}_i - \boldsymbol{\mu}\| \le (E\|\mathbf{X}_i - \boldsymbol{\mu}\|^2)^{1/2} = (\mathrm{tr}(\mathbf{\Sigma}))^{1/2} = O(\sqrt{pK}). \tag{10.54}$$

Therefore, by the law of large numbers, we have

$$U_n \le n^{-1}\sum_{i=1}^n \|\mathbf{X}_i - \boldsymbol{\mu}\| + \|\widehat{\boldsymbol{\mu}} - \boldsymbol{\mu}\| = O_P(\sqrt{pK} + \sqrt{p\log p/n}).$$

Thus, we conclude that

$$n^{-1}\sum_{i=1}^n \|\widehat{\mathbf{f}}_i - \mathbf{f}_i^*\| \le O_P\Big(K(\sqrt{\log p/n} + 1/\sqrt{p})\Big).$$

We now deal with the second term.

$$\|\mathbf{f}_i^* - \mathbf{f}_i\| \le \|\widetilde{\mathbf{\Lambda}}^{-1} - \mathbf{\Lambda}^{-1}\| \|\mathbf{B}^T\| \|\mathbf{X}_i - \boldsymbol{\mu}\| + \|\widetilde{\mathbf{\Lambda}}^{-1}\| \|\mathbf{B}^T\mathbf{u}_i\|.$$

Thus, by (10.54), we have

$$n^{-1}\sum_{i=1}^n \|\mathbf{f}_i^* - \mathbf{f}_i\| = O_P(p^{-2}pK + p^{-1}E\|\mathbf{B}^T\mathbf{u}_i\|).$$

It remains to calculate

$$(\mathrm{E}\,\|\mathbf{B}^T\mathbf{u}_i\|)^2 \le \mathrm{E}\,\|\mathbf{B}^T\mathbf{u}_i\|^2 = \mathrm{tr}(\mathbf{B}^T\mathbf{\Sigma}_u\mathbf{B}) \le \|\mathbf{\Sigma}_u\| \,\mathrm{tr}(\mathbf{B}^T\mathbf{B}) = O(pK^2). \tag{10.55}$$

Combinations of these two terms yield the desired results.

Inspecting the above proof, the second result holds using basically the same proof. The third result also holds by using identically the same calculation. Here, we note that the maximum is no smaller than average. Therefore, negligible terms in the previous arguments continue to be negligible. This completes the proof.

## 10.7 Bibliographical Notes

PCA was first proposed by Pearson in 1901 and later by Hotelling in 1933. A popular classical book on PCA is Jolliffe (1986). Hastie and Stuetzle (1989) developed principal curves and principal surfaces. Kernel PCA in a reproducing kernel Hilbert space was discussed in Schölkopf, Smola and Müller (1999). Johnstone and Lu (2009) considered a rank-1 spiked covariance model and showed the standard PCA produces an inconsistent estimate to the leading eigenvector of the true covariance matrix when $p/n \to c > 0$ and the first principal component is not sufficiently spiked. A large amount of literature contributed to the asymptotic analysis of PCA under the ultra-high dimensional regime, including Paul (2007), Johnstone and Lu (2009), Shen, Shen, Zhu and Marron (2016), and Wang and Fan (2017), among others.

High-dimensional factor models have been extensively studied in econometrics and statistics literature. The primary focus of econometrics literature is on estimating latent factors and the factor loading matrix, whereas in statistics literature the focus is on estimating large covariance and precision matrices of observed variables and idiosyncratic components. Stock and Watson (2002a, b) and Boivin and Ng (2006) estimated latent factors and applied them to forecast macroeconomics variables. Bai (2003) derived asymptotic properties for estimated factors, factor loadings, among others, which were extended further to the maximum likelihood estimator in Bai and Li (2012). Onatski (2012) estimated factors under weak factor models. Bai and Liao (2016) investigated the penalized likelihood method for the approximate factor model. Connor and Linton (2007) and Connor, Matthias and Oliver (2012) consider a semiparametric factor model. See also related work by Park, Mammen, Härdle and Borzk (2009).

The dynamic factor model and *generalized dynamic model* were considered and studied in Molenaar (1985), Forni, Hallin, Lippi, and Reichlin (2000, 2005). As shown in Forni and Lippi (2001), the dynamic factor model does not really impose a restriction on the data-generating process, rather it is a decomposition into the common component and idiosyncratic error. The related literature includes, for example, Forni *et al.*(2004), Hallin and Liška (2007, 2011), Forni *et al.*(2015), and references therein.

The separations between the common factors and idiosyncratic components are carried out by the low-rank plus sparsity decomposition. It corresponds to the identifiability issue of statistical problems. See, for example, Candés and Recht (2009), Wright, Ganesh, Rao, Peng, and Ma (2009), Candés, Li, Ma and Wright (2011), Negahban and Wainwright (2011), Agarwal, Negahban, and Wainwright (2012). Statistical estimation under such a model has been investigated by Fan, Liao and Mincheva (2011, 2013), Cai, Ma and Wu (2013), Koltchinskii, Lounici and Tsybakov (2011), Ma (2013), among others.

## 10.8 Exercises

10.1 Let $\mathbf{B}$ be a $p \times K$ matrix, whose columns $\widetilde{\mathbf{b}}_j$ are orthogonal. If $\{\|\widetilde{\mathbf{b}}_j\|\}_{j=1}^K$ are ordered in a non-increasing manner, show

$$\lambda_j(\mathbf{B}^T\mathbf{B}) = \|\widetilde{\mathbf{b}}_j\|^2, \qquad \xi_j(\mathbf{B}^T\mathbf{B}) = \widetilde{\mathbf{b}}_j/\|\widetilde{\mathbf{b}}_j\|$$

for $j = 1, \cdots K$. What is $\lambda_{K+1}(\mathbf{B}^T\mathbf{B})$?

10.2 Prove Proposition 10.1.

10.3 Consider the linear combination $\mathbf{b}^T\mathbf{u}$. Show that

$$\text{Var}(\mathbf{b}^T\mathbf{u}) \leq \|\mathbf{\Sigma}_u\|_1\|\mathbf{b}\|^2.$$

Referring to the notation in (10.10), deduce from this that

$$\widetilde{\mathbf{b}}_j^T\mathbf{u}/\|\widetilde{\mathbf{b}}_j^T\|^2 = O_p(\sqrt{\|\mathbf{\Sigma}_u\|_1/p}),$$

which tends to zero if $\|\mathbf{\Sigma}_u\|_1 = o(p)$ (sparse).

10.4 Consider the one-factor model:

$$\mathbf{X}_i = \mathbf{b}f_i + \mathbf{u}_i, \qquad \text{Var}(\mathbf{u}) = \mathbf{I}_p.$$

Show that

(a) The largest eigenvalue of $\mathbf{\Sigma} = \text{Var}(\mathbf{X})$ is $(1 + \|\mathbf{b}\|^2)$ with the associated eigenvector $\mathbf{b}/\|\mathbf{b}\|$.

(b) Let $\widehat{f}_i = \mathbf{b}^T\mathbf{X}_i/\|\mathbf{b}\|^2$. If $\mathbf{u}_i \sim N(0, \mathbf{I}_p)$, show that

$$\max_{i \leq n}(\widehat{f}_i - f_i)^2 = O_P(\log n\|\mathbf{b}\|^{-2}).$$

10.5 Suppose that $\widehat{\mathbf{\Sigma}} = \sum_{j=1}^p \widehat{\lambda}_j\widehat{\boldsymbol{\xi}}_j\widehat{\boldsymbol{\xi}}_j$. Let $\mathbf{B}$ be a $p \times K$ matrix and $\mathbf{P} = \mathbf{B}(\mathbf{B}^T\mathbf{B})^{-1}\mathbf{B}^T$ be the project matrix into the column space spanned by $\mathbf{B}$.

(a) Show

$$\text{tr}[(\mathbf{I}_p - \mathbf{B}(\mathbf{B}^T\mathbf{B})^{-1}\mathbf{B}^T)\widehat{\mathbf{\Sigma}}].$$

is minimized at $\mathbf{B}_0$ = linear space spanned by $\widehat{\boldsymbol{\xi}}_1, \cdots, \widehat{\boldsymbol{\xi}}_K$.

(b) The minimum value of the above minimization problem is $\lambda_{K+1} + \cdots + \lambda_p$.

(c) Prove that $\widehat{\mathbf{B}} = (\widehat{\lambda}_1^{1/2}\widehat{\boldsymbol{\xi}}_1, \cdots, \widehat{\lambda}_K^{1/2}\widehat{\boldsymbol{\xi}}_K)$ is a solution to problem (a).

10.6 Show that the solution given in Proposition 10.2 is the same as (10.13) and (10.14) with $\widehat{\mathbf{\Sigma}} = \mathbf{S}$, when $\bar{\mathbf{X}} = 0$.

10.7 Prove the following

$$V(k) = \frac{1}{np} \min_{\mathbf{B}\in I\!R^{p\times k}, \mathbf{F}\in I\!R^{n\times k}} \|\mathbf{X} - \mathbf{1}_n\bar{\mathbf{X}}^T - \mathbf{F}\mathbf{B}^T\|_F^2 = p^{-1}\sum_{j>k}\lambda_j(\mathbf{S}),$$

where $\mathbf{S}$ is the sample covariance. Note that you can use the results in Exercise 10.5 to solve this problem.

10.8 Assume that the factor model (10.6) holds. Note that the least-squares solution is to minimize $\sum_{i=1}^n \|\mathbf{X}_i - \mathbf{a} - \mathbf{B}\mathbf{f}_i\|^2$ with respect to $\mathbf{a}$ and $\mathbf{B}$ with observable factors $\{\mathbf{f}_i\}_{i=1}^n$. This is equivalent to the componentwise minimization outlined below (10.6).

(a) For given $\mathbf{B}$, show that the least-squares estimator for $\mathbf{a}$ is $\widehat{\mathbf{a}} = \bar{\mathbf{X}} - \mathbf{B}\bar{\mathbf{f}}$, where $\bar{\mathbf{X}} = n^{-1}\sum_{i=1}^n \mathbf{X}_i$ and $\bar{\mathbf{f}} = n^{-1}\sum_{i=1}^n \mathbf{f}_i$.

(b) Show that the least-squares estimator for $\mathbf{B}$ is given by

$$\begin{aligned}\widehat{\mathbf{B}} &= n^{-1}\sum_{i=1}^n (\mathbf{X}_i - \bar{\mathbf{X}})(\mathbf{f}_i - \bar{\mathbf{f}})^T\, \mathbf{S}_f^{-1} \\ &= \mathbf{B} + n^{-1}\sum_{i=1}^n \mathbf{u}_i(\mathbf{f}_i - \bar{\mathbf{f}})^T\, \mathbf{S}_f^{-1},\end{aligned}$$

where $\mathbf{S}_f^{-1}$ is the sample covariance matrix of $\{\mathbf{f}_i\}_{i=1}^n$.

(c) Show that

$$\widehat{\mathbf{u}}_i - \mathbf{u}_i = -n^{-1}\sum_{j=1}^n \mathbf{u}_j(\mathbf{f}_j - \bar{\mathbf{f}})^T\mathbf{S}_f^{-1}(\mathbf{f}_i - \bar{\mathbf{f}}) - \bar{\mathbf{u}}.$$

10.9 Continuing with the previous problem, assume further that $\mathbf{f}_i$ and $\mathbf{u}_i$ are independent.

(a) Conditional on $\{\mathbf{f}_i\}$, show that

$$\mathrm{E}\left[(\widehat{\mathbf{B}} - \mathbf{B})(\widehat{\mathbf{B}} - \mathbf{B})^T | \{\mathbf{f}_i\}\right] = n^{-1}\,\mathrm{tr}(\mathbf{S}_f^{-1})\boldsymbol{\Sigma}_u.$$

(b) Show that the conditional mean-square error

$$\mathrm{E}\left[\|\widehat{\mathbf{u}}_i - \mathbf{u}_i\|^2 | \{\mathbf{f}_i\}\right] = n^{-1}\,\mathrm{tr}(\boldsymbol{\Sigma}_u)\left[(\mathbf{f}_i - \bar{\mathbf{f}})^T\mathbf{S}_f^{-1}(\mathbf{f}_i - \bar{\mathbf{f}}) + 1\right].$$

10.10 Let $\mathbf{Z} = (\mathbf{X}^T, \mathbf{f}^T)^T$, where $\mathbf{X}$ follows the factor model (10.6).

(a) Show that the covariance matrix of $\mathbf{Z}$ is given by

$$\boldsymbol{\Sigma}_z = \begin{pmatrix} \mathbf{B}\boldsymbol{\Sigma}_f\mathbf{B}^T + \boldsymbol{\Sigma}_u & \mathbf{B}\boldsymbol{\Sigma}_f \\ \boldsymbol{\Sigma}_f\mathbf{B}^T & \boldsymbol{\Sigma}_f \end{pmatrix} := \begin{pmatrix} \boldsymbol{\Sigma}_{11} & \boldsymbol{\Sigma}_{12} \\ \boldsymbol{\Sigma}_{21} & \boldsymbol{\Sigma}_{22} \end{pmatrix}.$$

(b) Prove that the covariance of the idiosyncratic component is given by

$$\mathbf{\Sigma}_u = \mathbf{\Sigma}_{11} - \mathbf{\Sigma}_{12}\mathbf{\Sigma}_{22}^{-1}\mathbf{\Sigma}_{21}.$$

(c) If $\widehat{\mathbf{\Sigma}}_z$ is obtained by the sample covariance matrix of the data $\mathbf{Z}_i = (\mathbf{X}_i^T, \mathbf{f}_i)^T$, show that $\widehat{\mathbf{\Sigma}}_u = \widehat{\mathbf{\Sigma}}_{11} - \widehat{\mathbf{\Sigma}}_{12}\widehat{\mathbf{\Sigma}}_{22}^{-1}\widehat{\mathbf{\Sigma}}_{21}$ is the same as the sample covariance matrix of $\{\widehat{\mathbf{u}}_i\}_{i=1}^n$, where $\widehat{\mathbf{u}}_i$ is the residual vector based on the least-squares fit, as in Exercise 10.15.

10.11 Prove (10.40).

10.12 Let us consider the 128 macroeconomic time series from Jan. 1959 to Dec. 2018, which can be downloaded from the book website. We extract the data from Jan. 1960 to Oct. 2018 (in total 706 months) and remove the feature named "sasdate" and the features with missing entries. We use $\mathbf{M}$ to denote the preprocessed data matrix.

(a) Plot the time series of the variable "HWI" (Help-Wanted Index). Given this, can you explain why people usually use a very small fraction of the historical data to train their models for forecasting?

(b) Extract the first 120 months (i.e. the first 120 rows of $\mathbf{M}$), and standardize the data such that all the variables have zero mean and unit variance.

i. For the standardized data, draw a scree plot of the 20 leading eigenvalues of the sample covariance matrix. Use the eigen-ratio method with $k_{\max} = 10$ to determine the number of factors $K$. Compare the result with that using the eigenvalue thresholding $\{j : \lambda_j(R) > 1 + p/n\}$. Here, we do not do eigenvalue adjustments to facilitate the implementations for both methods.

ii. Use the package `POET` to estimate the covariance matrix. Report the maximum and minimum eigenvalues of the estimated matrix. Print the sub-covariance matrix of the first 3 variables.

(c) The column "UNRATE" of $\mathbf{M}$ corresponds to monthly changes of the unemployment rate. We will predict the future "UNRATE" using current values of the other macroeconomic variables.

i. Pair each month's "UNRATE" with the other macroeconomic variables in the previous month (to do so, you need to drop the "UNRATE" in the first month and other variables in the last month). Let $\{(\mathbf{X}_t, Y_t)\}_{t=1}^N$ denote the derived pairs of covariates and responses.

ii. We next train the model once a year using the past 120 months' data and use the model to forecast the next 12 months' unemployment rates. More precisely, for each month $t \in \{121 + 12m : m = 0, 1, \cdots 40\}$, we want to forecast the next 12 monthly "UNRATE"s $\{Y_{t+i}\}_{i=0}^{11}$ based on $\{\mathbf{X}_{t+i}\}_{i=0}^{11}$, using the past 120 months' data $\{(\mathbf{X}_{t-i}, Y_{t-i})\}_{i=1}^{120}$ for training. Implement the FarmSelect method by following the steps: standardize the covariates; fit a factor model; fit Lasso on augmented covariates; output the predicted values $\{Y_{t+i}^{\text{farm}}\}_{i=0}^{11}$ for $\{Y_{t+i}\}_{i=0}^{11}$. Also, compute the baseline predictions $\bar{Y}_{t+i} = \frac{1}{120}\sum_{j=1}^{120} Y_{t-j}$ for all $i = 0, 1, \cdots, 11$.

iii. Report the out-of-sample $R^2$ using

$$R^2 = 1 - \frac{\sum_{t=121}^{612}(Y_t - Y_t^{\text{farm}})^2}{\sum_{t=121}^{612}(Y_t - \bar{Y}_t)^2}.$$

(d) Run principal component regression (PCR) with 5 components for the same task as Part (c) and report the out-of-sample $R^2$.

(e) Treat $\{\mathbf{X}_i\}_{i=1}^{60}$ and $\{\mathbf{X}_{60+i}\}_{i=1}^{120}$ as independent and identically distributed samples from two distributions and let $\boldsymbol{\mu}_1, \boldsymbol{\mu}_2$ denote their expectations. Use the R package `FarmTest` to test

$$H_0: \ \mu_{1j} = \mu_{2j}, \ \forall j \qquad \text{v.s.} \qquad H_1: \ \mu_{1j} \neq \mu_{2j} \text{ for some } j.$$

Print the indices of the rejected hypotheses.

Chapter 11

# Applications of Factor Models and PCA

As mentioned in the introduction of Chapter 10, the dependence of high-dimensional measurements is often driven by several common factors: different variables have different dependence on these common factors. This leads to the factor model (10.4):

$$\begin{aligned} \mathbf{X} &= \mathbf{a} + \mathbf{B}\mathbf{f} + \mathbf{u}, \qquad \operatorname{E}\mathbf{u} = 0, \quad \operatorname{cov}(\mathbf{f}, \mathbf{u}) = 0 \quad \text{or} \\ X_j &= a_j + \mathbf{b}_j^T \mathbf{f} + u_j = a_j + b_{j1} f_1 + \cdots + b_{jK} f_K + u_j \end{aligned} \tag{11.1}$$

for each component $j = 1, \cdots, p$. These factors $\mathbf{f}$ can be consistently estimated by using the principal component analysis as shown in the previous chapter.

This chapter applies the principal component analysis and factor models to solve several statistics and machine learning problems. These include factor adjustments in high-dimensional model selection and multiple testing and high-dimensional regression using augmented factor models. Principal component analysis is a form of *spectral learning*. We also discuss other applications of spectral learning in problems such as matrix completion, community detection, and item ranking, and applications that serve as initial values for non-convex optimization problems such as mixture models. Detailed treatments of these require a whole book. Rather, we focus only on some methodological powers of spectral learning to give the reader some idea of its importance to statistical machine learning. Indeed, the role of PCA in statistical machine learning is very similar to regression analysis in statistics.

## 11.1 Factor-adjusted Regularized Model Selection

Model selection is critical in high dimensional regression analysis. Numerous methods for model selection have been proposed in the past two decades, including Lasso, SCAD, the elastic net, the Dantzig selector, among others. However, these regularization methods work only when the covariates satisfy certain regularity conditions such as the irrepresentable condition (3.38). When covariates admit a factor structure (11.1), such a condition breaks down and the model selection consistency can not be granted.

### *11.1.1 Importance of factor adjustments*

As a simple illustrative example, we simulate 100 random samples of size $n = 100$ from the $p = 250$ dimensional sparse regression model

$$Y = \mathbf{X}^T\boldsymbol{\beta} + \varepsilon, \qquad \boldsymbol{\beta} = (\overbrace{3, \cdots, 3}^{10}, \overbrace{0, \cdots, 0}^{p-10})^T, \qquad \varepsilon \sim N(0, 0.3^2), \qquad (11.2)$$

where $\mathbf{X} \sim N(0, \boldsymbol{\Sigma})$ and $\boldsymbol{\Sigma}$ is a compound symmetry correlation matrix with equal correlation $\rho$. We apply Lasso to each sample and record the average model size and the model selection consistency rate. The results are depicted in Figure 11.1. The model selection consistency rate is low for Lasso when $\rho \geq 0.2$, due to over estimation of model size.

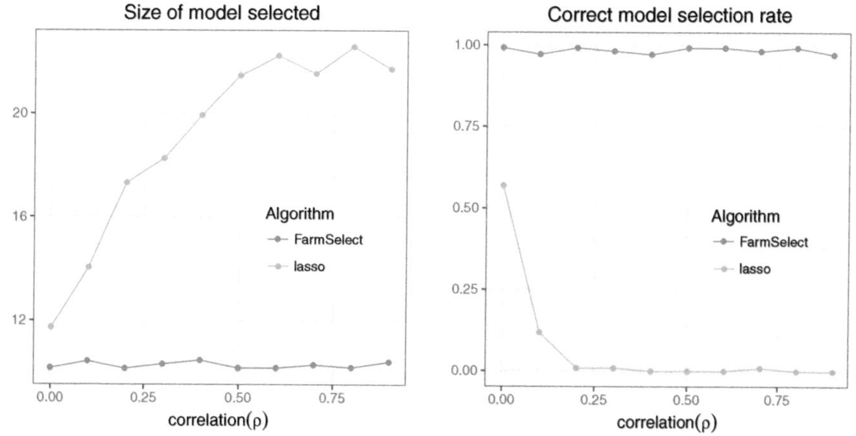


Figure 11.1: Comparison of model selection consistency rates by Lasso with (FarmSelect) and without (Lasso) factor adjustments.

The failure of model selection consistency is due to high correlation among the original variables $\mathbf{X}$ in $p$ dimensions. If we use the factor model (11.1), the regression model (11.2) can now be written as

$$Y = \alpha + \mathbf{u}^T\boldsymbol{\beta} + \boldsymbol{\gamma}^T\mathbf{f} + \varepsilon, \qquad (11.3)$$

where $\alpha = \mathbf{a}^T\boldsymbol{\beta}$ and $\boldsymbol{\gamma} = \mathbf{B}^T\boldsymbol{\beta}$. If $\mathbf{u}$ and $\mathbf{f}$ were observable, we obtain a $(p + 1 + K)$-dimensional regression problem if $\boldsymbol{\gamma}$ and $\alpha$ are regarded as free parameters. In this lifted space, the covariates $\mathbf{f}$ and $\mathbf{u}$ are now weakly correlated (indeed, they are independent under specific model (11.2)) and we can now apply a regularized method to fit model (11.3). Note that the high-dimensional coefficient vector $\boldsymbol{\beta}$ in (11.3) is the same as that in (11.2) and is hence sparse.

In practice, $\mathbf{u}$ and $\mathbf{f}$ are not observable. However, they can be consistently estimated by PCA as shown in Section 10.5.4. Regarding the estimated latent factors $\{\mathbf{f}_i\}_{i=1}^n$ and the idiosyncratic components $\{\mathbf{u}_i\}$ as the covariates, we can now apply a regularized model selection method to the regression problem (11.3). The resulting method is called a *Factor-Adjusted Regularized Model* selector or *FarmSelect* for short. It was introduced by Kneip and Sarda (2011) and expanded and developed by Fan, Ke and Wang (2020). Figure 11.1 shows the effectiveness of FarmSelect: The model selection consistency rate is nearly 100%, regardless of $\rho$.

Finally, we would like to remark in representation (11.3), the parameter $\boldsymbol{\gamma} = \mathbf{B}^T\boldsymbol{\beta}$ is not free ($\mathbf{B}$ is given by the factor model). How can we assure that the least-squares problem (11.3) with free $\boldsymbol{\gamma}$ would yield a solution $\boldsymbol{\gamma} = \mathbf{B}^T\boldsymbol{\beta}$? This requires the condition

$$\mathrm{E}\,\mathbf{u}\varepsilon = 0 \quad \text{and} \quad \mathrm{E}\,\mathbf{f}\varepsilon = 0. \tag{11.4}$$

See Exercise 11.1. This is somewhat stronger than the original condition that $\mathrm{E}\,\mathbf{X}\varepsilon = 0$ and holds easily as $\mathbf{X}$ is exogenous to $\varepsilon$.

### *11.1.2 FarmSelect*

The lifting via factor adjustments is used to weaken the dependence of covariates. It is insensitive to the over-estimate on the number of factors, as the expended factor models include the original factor models as a specific case and hence the dependence of new covariates has also been weakened. The idea of factor adjustments is more generally applicable to a family of sparse linear model

$$Y_i = g(\mathbf{X}_i^T\boldsymbol{\beta}, \varepsilon_i), \quad \text{with} \quad \mathbf{X}_i = \mathbf{a} + \mathbf{B}\mathbf{f}_i + \mathbf{u}_i \tag{11.5}$$

following a factor model, where $g(\cdot,\cdot)$ is a known function. This model encompasses the *generalized linear models* in Chapter 5 and *proportional hazards model* in Chapter 6.

Suppose that we fit model (11.5) by using the loss function $L(Y_i, \mathbf{X}_i^T\boldsymbol{\beta})$, which can be a negative log-likelihood. This can be the log-likelihood function or $M$-estimation. Instead of regularizing directly via

$$\sum_{i=1}^n L(Y_i, \mathbf{X}_i^T\boldsymbol{\beta}) + \sum_{j=1}^p p_\lambda(|\beta_j|), \tag{11.6}$$

where $p_\lambda(\cdot)$ is a folded concave penalty such as Lasso, SCAD and MCP with a regularization parameter $\lambda$, the Factor-Adjusted Regularized (or Robust when so implemented) Model selection (*FarmSelect*) consists of two steps:

- **Factor analysis**. Obtain the estimates of $\widehat{\mathbf{f}}_i$ and $\widehat{\mathbf{u}}_i$ in model (11.5) via PCA.
- **Augmented regularization**. Find $\alpha$, $\boldsymbol{\beta}$ and $\boldsymbol{\gamma}$ to minimize

$$\sum_{i=1}^n L(Y_i, \alpha + \widehat{\mathbf{u}}_i^T\boldsymbol{\beta} + \widehat{\mathbf{f}}_i^T\boldsymbol{\gamma}) + \sum_{j=1}^p p_\lambda(|\beta_j|). \tag{11.7}$$

The resulting estimator is denoted as $\widehat{\boldsymbol{\beta}}$. The procedure is implemented in R package `FarmSelect`. Note that the $\boldsymbol{\beta}$ in augmented model (11.7) is the same as that in model (11.5).

As an illustration, we consider the sparse linear model (11.2) again with $p = 500$ and $\boldsymbol{\beta} = (\beta_1, \cdots, \beta_{10}, \mathbf{0}_{p-10})^\top$, where $\{\beta_j\}_{j=1}^{10}$ are drawn uniformly at random from $[2, 5]$. Instead of taking the equi-correlation covariance matrix, the correlation structure is now calibrated to that for the excess monthly returns of the constituents of the S&P 500 Index between 1980 and 2012. Figure 11.2 reports the model selection consistency rates for $n$ ranges 50 to 150. The effectiveness of factor adjustments is evidenced.

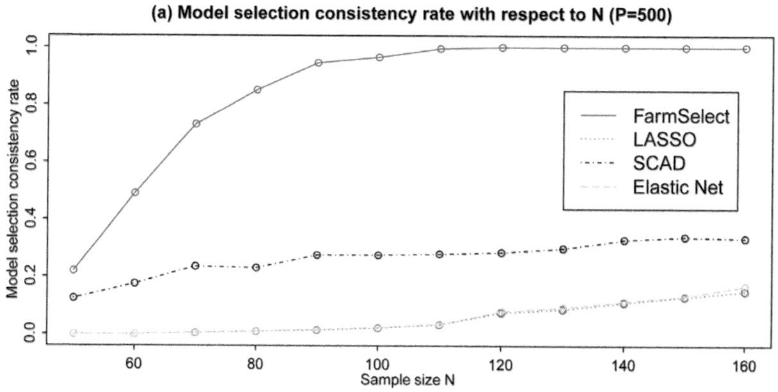


Figure 11.2: FarmSelect is compared with Lasso, SCAD, Elastic Net in terms of selection consistency rates. Taken from Fan, Ke and Wang (2020).

As pointed out in Fan, Ke and Wang (2020), factor adjustments can also be applied to variable screening. This can be done via an application of an independent screening rule to variables $\mathbf{u}$ or by the conditional sure independent screening rule, conditioned on the variables $\mathbf{f}$. See Chapter 8 for details.

### *11.1.3 Application to forecasting bond risk premia*

As an empirical application, we take the example in Fan, Ke, Wang (2019+) that predicts the U.S. government bond risk premia by a large panel of macroeconomic variables mentioned in Section 1.1.4. The response variable is the monthly data of U.S. bond risk premia with maturity of 2 to 5 years between January 1980 and December 2015, consisting of 432 data points. The bond risk premia is calculated as the one year return of an $n$ years maturity bond excessing the one year maturity bond yield as the risk-free rate. The covariates are 131 monthly U.S. disaggregated macroeconomic variables in the

FRED-MD database[1]. These macroeconomic variables are driven by economic fundamentals and are highly correlated. To wit, we apply principal component analysis to these macroeconomic variables and obtain the scree plot of the top 20 principal components in Figure 11.3. It shows the first principal component alone explains more than 60% of the total variance. In addition, the first 5 principal components together explain more than 90% of the total variance.

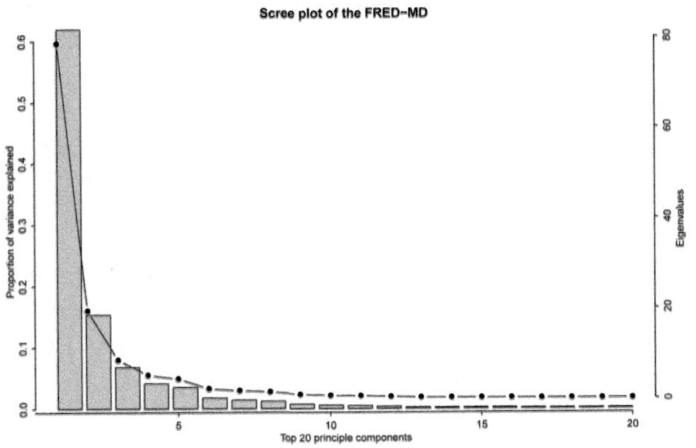


Figure 11.3: Eigenvalues (dotted line) and proportions of variance (bar) explained by the top 20 principal components for the macroeconomic data. Taken from Fan, Ke and Wang (2020).

We forecast one month bond risk premia using a rolling window of size 120 months. Within each window, the risk premia is predicted by a sparse linear regression of dimensionality 131. The performance is evaluated by the out-of-sample $R^2$, defined as

$$R^2 = 1 - \frac{\sum_{t=121}^{432}(Y_t - \widehat{Y}_t)^2}{\sum_{t=121}^{432}(Y_t - \bar{Y}_t)^2},$$

where $y_t$ is the response variable at time $t$, $\widehat{Y}_t$ is the predicted $Y_t$ using the previous 120 months data, and $\bar{Y}_t$ is the sample mean of the previous 120 months responses $(Y_{t-120}, \ldots, Y_{t-1})$, representing a naive baseline predictor.

We compare FarmSelect with Lasso in terms of prediction and model size. We include also the *principal component regression* (PCR), which uses $\mathbf{f}$ as predictors, in the prediction competition. The FarmSelect is implemented by the R package `FarmSelect` with default settings. To be specific, the loss function is $L_1$, the number of factors is estimated by the eigen-ratio method and the regularized parameter is selected by multi-fold cross validation. The Lasso

[1]The FRED-MD is a monthly economic database updated by the Federal Reserve Bank of St. Louis which is publicly available at http://research.stlouisfed.org/econ/mccracken/sel/

method is implemented by the `glmnet` R package. The PCR method is implemented by the `pls` package in R. The number of principal components is chosen as 8 as suggested in Ludvigson and Ng (2009).

Table 11.1: Out of sample $R^2$ and average selected model size by FarmSelect, Lasso, and PCR.

| Maturity of Bond | Out of sample $R^2$ | | | Average model size | |
|---|---|---|---|---|---|
| | FarmSelect | Lasso | PCR | FarmSelect | Lasso |
| 2 Years | 0.2586 | 0.2295 | 0.2012 | 8.80 | 22.72 |
| 3 Years | 0.2295 | 0.2166 | 0.1854 | 8.92 | 21.40 |
| 4 Years | 0.2137 | 0.1801 | 0.1639 | 9.03 | 20.74 |
| 5 Years | 0.2004 | 0.1723 | 0.1463 | 9.21 | 20.21 |

Table 11.1, taken from Fan, Ke and Wang (2019+), reports the results. It is clear that FarmSelect outperforms Lasso and PCR, in terms of out-of-sample $R^2$, even though Lasso uses many more variables. Note that PCR always uses 8 variables, which are about the same as FarmSelect.

### *11.1.4 Application to a neuroblastoma data*

Neuroblastoma is a malignant tumor of neural crest origin that is the most frequent solid extracranial neoplasia in children. It is responsible for about 15% of all pediatric cancer deaths. Oberthuer *et al.*(2006) analyzed the German Neuroblastoma Trials NB90-NB2004 (diagnosed between 1989 and 2004) and developed a classifier based on gene expressions. For 246 neuroblastoma patients, aged from 0 to 296 months (median 15 months), gene expressions over 10,707 probe sites were measured. We took 3-year event-free survival information of the patients (56 positive and 190 negative) as the binary response and modeled the response as a high dimensional sparse logistic regression model of gene expressions.

First of all, due to co-expressions of genes, the covariates are correlated. This is evidenced by the scree plot in Figure 11.4. The figure and the numerical results here are taken from an earlier version of Fan, Ke and Wang (2020). The scree plot shows the top ten principal components together can explain more than 50% of the total variance. The eigen ratio method suggests a $\widehat{K} = 4$ factor model. Therefore, FarmSelect should be advantageous in selecting important genes.

FarmSelect selects a model with 17 genes. In contrast, Lasso, SCAD and Elastic Net (with $\lambda_1 = \lambda_2 \equiv \lambda$) select much larger models, including 40, 34 and 86 genes respectively. These methods over-fit the model as they ignore the dependence among covariates. The over fitted models make it harder to interpret the molecular mechanisms and biological processes.

To assess the performance of each model selection method, we apply a

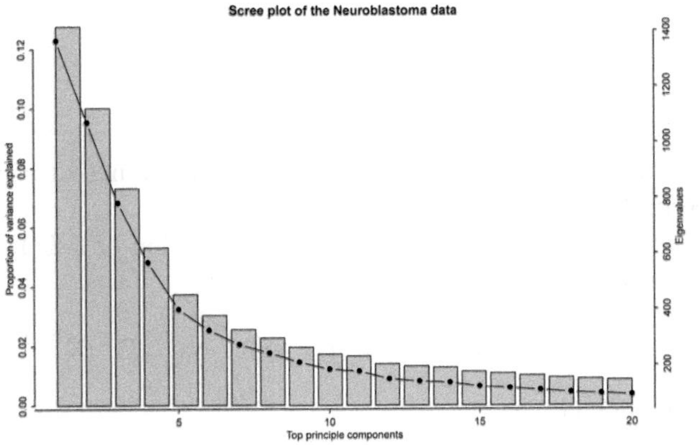


Figure 11.4: Eigenvalues (dotted line) and proportional of variance (bar) explained by the top 20 principal components for the neuroblastoma data.

bootstrap based out-of-sample prediction as follows. For each replication, we randomly draw 200 observations as the training set and leave the remaining 46 data points as the testing set. We use the training set to select and fit a model. Then, for each observation in the testing set, we use the fitted model to calculate the conditional probabilities and to classify outcomes. We record the selected model size and the correct prediction rate (No. of correct predictions/46). By repeating this procedure over 2,000 replications, we report the average model sizes and average correct prediction rates in Table 11.2. It shows that FarmSelect has the smallest average model size and the highest correct prediction rate. Lasso, SCAD and Elastic Net tend to select much larger models and result in lower prediction rates. In Table 11.2 we also report the out-of-sample correct prediction rate using the first 17 variables that entered the solution path of each model selection method. As a result, the correct prediction rates of Lasso, SCAD and Elastic Net decrease further. This indicates that the Lasso, SCAD and Elastic Net are likely to select overfitted models.

Table 11.2: Bootstrapping model selection performance of neuroblastoma data

| Bootstrap sample average | Model selection methods | | | |
|---|---|---|---|---|
| | FarmSelect | Lasso | SCAD | Elastic Net |
| Model size | **17.6** | 46.2 | 34.0 | 90.0 |
| Correct prediction rate | **0.813** | 0.807 | 0.809 | 0.790 |
| Prediction performance with first 17 variables entering the solution path | | | | |
| | FarmSelect | Lasso | SCAD | Elastic Net |
| Correct prediction rate | **0.813** | 0.733 | 0.764 | 0.705 |

### 11.1.5 Asymptotic theory for FarmSelect

To derive the asymptotic theory for FarmSelect, we need to solve the following three problems:

1. Like problem (11.3), the unconstrained minimization of $\mathrm{E}\, L(Y, \alpha + \mathbf{u}^T\boldsymbol{\beta} + \mathbf{f}^T\boldsymbol{\gamma})$ with respect to $\alpha$, $\boldsymbol{\beta}$ and $\boldsymbol{\gamma}$ is obtained at

$$\alpha_0 = \mathbf{a}^T\boldsymbol{\beta}^*, \quad \boldsymbol{\beta}_0 = \boldsymbol{\beta}^* \quad \text{and} \quad \boldsymbol{\gamma}_0 = \mathbf{B}^T\boldsymbol{\beta}^* \tag{11.8}$$

under model (11.1).

2. Let $\boldsymbol{\theta} = (\alpha, \boldsymbol{\beta}^T, \boldsymbol{\gamma}^T)^T$. The error bounds $\|\widehat{\boldsymbol{\theta}} - \boldsymbol{\theta}^*\|_\ell$ for $\ell = 1, 2, \infty$ and the sign consistency $\mathrm{sgn}(\widehat{\boldsymbol{\beta}}) = \mathrm{sgn}(\boldsymbol{\beta}^*)$ should be available for the penalized M-estimator (11.7), when $\{(\mathbf{u}_i^T, \mathbf{f}_i)^T\}_{i=1}^n$ were observable.
3. The estimation errors in $\{(\widehat{\mathbf{u}}_i^T, \widehat{\mathbf{f}}_i)^T\}_{i=1}^n$ are negligible for the penalized M-estimator (11.7).

Similar to condition (11.4), Problem 1 can be resolved by imposing a condition

$$\mathrm{E}\, \nabla L(Y, \mathbf{X}^T\boldsymbol{\beta}^*)\mathbf{u} = 0 \quad \text{and} \quad \mathrm{E}\, \nabla L(Y, \mathbf{X}^T\boldsymbol{\beta}^*)\mathbf{f} = 0, \tag{11.9}$$

where $\nabla L(s,t) = \frac{\partial L(s,t)}{\partial t}$. See Exercise 11.2. Problem 2 is solved in Fan, Ke and Wang (2020) for strong convex loss function under the $L_1$-penalty. Their derivations are related to the results on the error bounds and sign consistency for the $L_1$-penalized M-estimator (Lee, Sun and Taylor, 2015). Problem 3 is handled by restricting the likelihood function to the generalized linear model, along with the consistency of estimated latent factors and idiosyncratic components in Sections 10.5.3 and 10.5.4. In particular, the representable conditions are now imposed based on $\mathbf{u}$ and $\mathbf{f}$, which is much easier to satisfy. We refer to Fan, Ke and Wang (2019+) for more careful statements and technical proofs.

Finally, we would also to note that in the regression (11.7), only the linear space spanned by $\{\widehat{\mathbf{f}}_i\}_{i=1}^n$ matters. Therefore, we only require the consistent estimation of the eigen-space spanned by $\{\widehat{\mathbf{f}}_i\}_{i=1}^n$, rather than consistency of each individual factors.

## 11.2 Factor-adjusted Robust Multiple Testing

Sparse linear regression attempts to find a set of variables that explain jointly the response variable. In many high-dimensional problems, we are also asking about individual effects of treatments such as which genes expressed differently within different cells (e.g. tumor vs. normal control), which mutual fund managers have positive $\alpha$-skills, namely, risk-adjusted excess returns. These questions result in high-dimensional multiple testing problems that have been popularly studied in statistics. For an overview, see the monograph by Efron (2012) and reference therein.

Due to co-expression of genes or herding effect in fund management as well as other latent factors, the test statistics are often dependent. This results in a correlated test whose false discovery rate is hard to control (Efron, 2007, 2010; Fan, Han, Gu, 2012). Factor adjustments are a powerful tool for this purpose.

### *11.2.1 False discovery rate control*

Suppose that we have $n$ measurements from the model

$$\mathbf{X}_i = \boldsymbol{\mu} + \boldsymbol{\varepsilon}_i, \qquad \mathrm{E}\,\boldsymbol{\varepsilon}_i = 0, \qquad i = 1, \cdots, n. \tag{11.10}$$

Here $\mathbf{X}_i$ can be in one case the $p$-dimensional vector for the logarithms of gene expression ratios between the tumor and normal cells for individual $i$ or in another case the risk adjusted returns. It can also be the risk-adjusted mutual fund returns over the $i^{th}$ period. Our interest is to test multiple times the hypotheses

$$H_{0j} : \mu_j = 0 \quad \text{versus} \quad H_{1j} : \mu_j \neq 0, \quad \text{for } j = 1, \cdots, p. \tag{11.11}$$

Let $T_j$ be a generic test statistic for $H_{0j}$ with rejection region $|T_j| \geq z$ for a given critical value $z > 0$. Here, we implicitly assumed that the test statistic $T_j$ has the same null distribution; otherwise the critical value $z$ should be $z_j$. The numbers of total discoveries $R(z)$ and false discoveries $V(z)$ are the number of rejections and the number of false rejections, defined as

$$R(z) = \#\{j : |T_j| \geq z\} \quad \text{and} \quad V(z) = \#\{j : |T_j| \geq z,\ \mu_j = 0\}. \tag{11.12}$$

Note that $R(z)$ is observable, which is really the total number of rejected hypotheses, while $V(z)$ needs to be estimated, depending on the unknown set

$$\mathcal{S}_0 = \{j \in [p] : \mu_j = 0\},$$

which is also called *true null*. The *false discovery proportion* and *false discovery rate* are defined respectively as

$$\mathrm{FDP}(z) = V(z)/R(z) \quad \text{and} \quad \mathrm{FDR}(z) = \mathrm{E}(\mathrm{FDP}(z)), \tag{11.13}$$

with convention $0/0 = 0$. Our goal is to control the false discovery proportion or false discovery rate. The FDP is more relevant for the current data. But the difference is small when the test statistics are independent and $p$ is large. Indeed, by the law of large numbers, under mild conditions,

$$\mathrm{FDP}(z) \approx \mathrm{E}\,V(z)/\mathrm{E}\,R(z) \approx \mathrm{FDR}(z).$$

However, the difference can be substantial when the observed data are dependent, as can be seen in Example 11.1.

Let $P_j$ be the P-value for the $j^{th}$ hypothesis and $P_{(1)} \leq \cdots P_{(p)}$ be the ordered p-values. The significance of the $j^{th}$ hypothesis (or the importance of the $j^{th}$ variable) is sorted by their p-values: the smaller the more importance. To control the false discovery rate at a prescribed level $\alpha$, Benjamini and Hochberg (1995) propose to choose

$$\widehat{k} = \max\{j : P_{(j)} \leq \frac{j}{p}\alpha\} \tag{11.14}$$

and reject all null hypotheses $H_{0,(j)}$ for $j = 1, \cdots, \widehat{k}$. They prove such a procedure controls FDR at level $\alpha$, under the independence assumption of P-values. The procedure can also be implemented by defining the *Benjamini-Hochberg adjusted p-value*:

$$\text{adjusted P-value for } (j)^{th} \text{ hypothesis} = \min_{k \geq j} P_{(k)} \frac{p}{k} \tag{11.15}$$

where the minimum is taken to maintain the monotonic property, and rejects the null hypotheses with adjusted P-value less than $\alpha$. The function `p.adjust` in R implements the procedure.

To better understand the above procedure, let us express the testing procedure in terms of P-values: reject $H_{0j}$ when $P_j \leq t$ for a given significant level $t$. Then, the total number of rejections and total number of false discoveries are respectively

$$r(t) = \sum_{j=1}^{p} I(P_j < t) \quad \text{and} \quad v(t) = \sum_{j \in \mathcal{S}_0} I(P_j < t) \tag{11.16}$$

and the false rejection proportion of the test is $\text{FDP}(t) = \frac{v(t)}{r(t)}$. Note that under the null hypothesis, $P_j$ is uniformly distributed. If the test statistics $\{T_j\}$ are independent, then the sequence $\{I(P_j < t)\}$ are the i.i.d. Bernoulli random variable with the probability of success $t$. Thus, $v(t) \approx p_0 t$, where $p_0 = |\mathcal{S}_0|$ is the number of true nulls. This is unknown, but can be upper bounded by $p$. Using this bound, we have an estimate

$$\widehat{\text{FDP}}(t) = \frac{pt}{r(t)} \tag{11.17}$$

The Benjamini-Hockberg method takes $t = P_{(\widehat{k})}$ with $\widehat{k}$ given by (11.14). Consequently,

$$\widehat{\text{FDP}}(P_{(\widehat{k})}) = \frac{pP_{(\widehat{k})}}{\widehat{k}} \leq \alpha,$$

by (11.14). This explains why the Benjamini-Hockberg approach controls the FDR at level $\alpha$.

As mentioned above, estimating $\pi_0 = p_0/p$ by its upper bound 1 will control the FDP correctly but it can be too crude. Let us now briefly describe the method for estimating $\pi_0$ by Storey (2002). Now the histogram of P-values for the p hypotheses is available to us (e.g. Figure 11.5), which is a mixture of those from true nulls $\mathcal{S}_0$ and false nulls $\mathcal{S}_0^c$. Assume P-values for false nulls (significant hypotheses) are all small so that P-values exceeding a given threshold $\eta \in (0,1)$ are all from the true nulls (e.g. $\eta = 0.5$). Under this assumption, $\pi_0(1-\eta)$ should be approximately the same as the percentage of P-values that exceed $\eta$. This leads to the estimator:

$$\widehat{\pi}_0(\eta) = \frac{1}{(1-\eta)p} \sum_{j=1}^{p} I(P_j > \eta). \tag{11.18}$$

for a given $\eta = 0.5$, say. Figure 11.7, taken from Fan, Wang, Zhong, Zhu, (2019+), illustrates the idea.

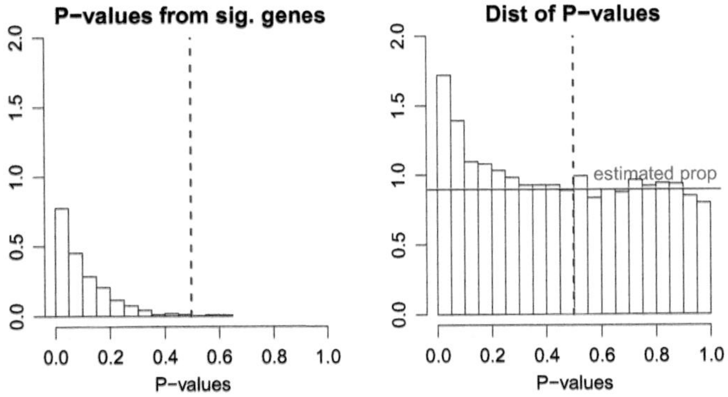


Figure 11.5: Estimation of proportion of true nulls. The observed P-values (right panel) consist of those from significant variables, which are usually small, and those from insignificant variables, which are uniformly distributed. Assuming the P-values for significant variables are mostly less than $\eta$ (taken to be 0.5 in this illustration, left panel), the contributions of observed P-values $> \eta$ are mostly from true nulls and this yields a natural estimator (11.20), which is the average height of the histogram with P-values $> \eta$ (red line). Note that the histograms above the red line estimates the distributions of P-values from the significant variables in the left panel.

### *11.2.2 Multiple testing under dependence measurements*

Due to co-expressions of genes and latent factors, the gene expressions in the vector $\mathbf{X}_i$ are dependent. Similarly, due to the "herding effect" and unadjusted latent factors, the fund returns within the $i^{th}$ period are correlated. This calls for further modeling the dependence in $\varepsilon_i$ in (11.10).

A common technique for modeling dependence is the factor model. Modeling $\varepsilon_i$ by a factor model, by (11.10), we have

$$\mathbf{X}_i = \boldsymbol{\mu} + \underbrace{\mathbf{B}\mathbf{f}_i + \mathbf{u}_i}_{\varepsilon_i}, \qquad \mathrm{E}\,\mathbf{f}_i = 0, \mathrm{E}\,\mathbf{u}_i = 0, \mathrm{E}\,\mathbf{f}_i\mathbf{u}_i^T = 0, \tag{11.19}$$

for all $i = 1, \cdots, n$. In the presence of strong dependence, the FDR and FDP can now be very different, making the interpretation harder. The following example is adapted from Fan and Han (2017).

**Example 11.1** *To gain insight on how the dependence impacts on the number of false discoveries, consider model* (11.19) *with only one factor* $f_i \sim N(0,1)$ *and* $\mathbf{B} = \rho\mathbf{1}$ *and* $\mathbf{u}_i \sim N(0, (1-\rho^2)\mathbf{I}_p)$*, namely,*

$$\mathbf{X}_i = \boldsymbol{\mu} + \rho f_i \mathbf{1} + \mathbf{u}_i.$$

*Assume in addition that* $f_i$ *and* $\mathbf{u}_i$ *are independent. The sample-average test statistics for problem* (11.11) *admit*

$$\mathbf{Z} \equiv \sqrt{n}\bar{\mathbf{X}} = \sqrt{n}\boldsymbol{\mu} + \rho W \mathbf{1} + \sqrt{1-\rho^2}\mathbf{U},$$

*where* $W = \sqrt{n}\bar{f} \sim N(0,1)$ *is independent of* $\mathbf{U} = \sqrt{n/(1-\rho^2)}\bar{\mathbf{u}} \sim N(0, \mathbf{I}_p)$*. Let* $p_0 = |\mathcal{S}_0|$ *be the size of true nulls. Then, the number of false discoveries*

$$V(z) = \sum_{j\in\mathcal{S}_0} I(|Z_j| > z) = \sum_{j\in\mathcal{S}_0}\Big[I\big(U_j > a(z-\rho W)\big) + I\big(U_j < a(-z-\rho W)\big)\Big],$$

*where* $a = (1-\rho^2)^{-1/2}$*. By using the law of large numbers, conditioning on* $W$*, we have as* $p_0 \to \infty$ *that*

$$p_0^{-1}V(z) = \Phi\left(\frac{-z+\rho W}{\sqrt{1-\rho^2}}\right) + \Phi\left(\frac{-z-\rho W}{\sqrt{1-\rho^2}}\right) + o_p(1),$$

*Therefore, the number of false discoveries depends on the realization of* $W$*. When* $\rho = 0$*,* $p_0^{-1}V(z) \approx 2\Phi(-z)$*. In this case, the FDP and FDR are approximately the same. To quantify the dependence on the realization of* $W$*, let* $p_0 = 1000$ *and* $z = 2.236$ *(the* $99^{th}$ *percentile of the normal distribution) and* $\rho = 0.8$ *so that*

$$p_0^{-1}V(t) \approx [\Phi((-2.236 + 0.8W)/0.6) + \Phi((-2.236 - 0.8W)/0.6)].$$

*When* $W = -3, -2, -1, 0$*, the values of* $p_0^{-1}V(z)$ *are approximately 0.608, 0.145, 0.008 and 0, respectively, which depend heavily on the realization of* $W$*. This is in contrast with the independence case in which* $p_0^{-1}V(z)$ *is always approximately 0.02.*

### 11.2.3 *Power of factor adjustments*

Assume for a moment that $\mathbf{B}$ and $\mathbf{f}_i$ are known in (11.19). It is then very natural to base the test on the adjusted data

$$\underbrace{\mathbf{X}_i - \mathbf{B}\mathbf{f}_i}_{\mathbf{Y}_i} = \boldsymbol{\mu} + \mathbf{u}_i. \tag{11.20}$$

This has two advantages: the factor-adjusted data has weaker dependence since the idiosyncratic component $\mathbf{u}_i$ is assumed to be so. This makes FDP and FDR approximately the same. In other words, it is easier to control type I errors. More importantly, $\mathbf{Y}_i$ has less variance than $\mathbf{X}_i$. In other words, the factor adjustments reduce the noise without suppressing the signals. Therefore, the tests based on $\{\mathbf{Y}_i\}$ also increase the power.

To see the above intuition, let us consider the following simulated example.

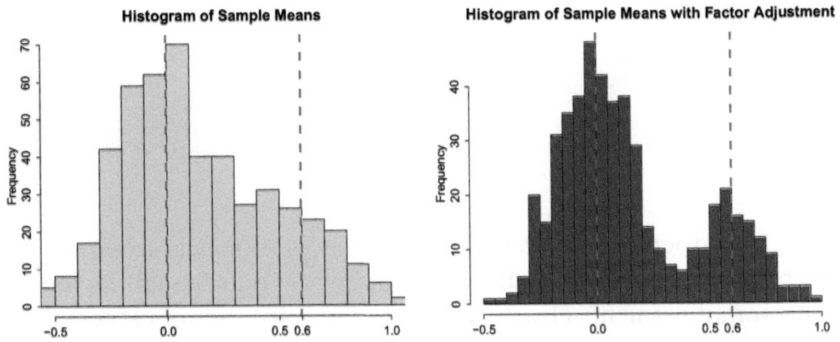


Figure 11.6: Histograms for the sample means from a synthetic three-factor models without (left panel) and with (right panel) factor adjustments. Factor adjustments clearly decrease the noise in the test statistics.

**Example 11.2** *We simulate a random sample of size* $n = 100$ *from a three-factor model* (11.19), *in which*

$$\mathbf{f}_i \sim N(0, \mathbf{I}_3), \quad \mathbf{B} = (b_{jk}) \text{ with } b_{jk} \sim_{iid} \text{Unif}(-1, 1), \quad \mathbf{u}_i \sim t_3(0, \mathbf{I}_p)$$

*with* $p = 100$. *We take* $\mu_j = 0.6$ *for* $j \le p/4$ *and 0, otherwise. Based on the data, we compute the sample average* $\bar{\mathbf{X}}$. *This results in a vector of* $p$ *test statistics, whose histogram is presented in the left panel of Figure 11.6. The bimodality of the distribution is blurred by large stochastic errors in the estimates. On the other hand, the histogram (right panel of Figure 11.6) for the sample averages based on the factor-adjusted data* $\{\mathbf{X}_i - \mathbf{B}\mathbf{f}_i\}_{i=1}^n$ *reveals clearly the bimodality, thanks to the decreased noise in the data. Hence the power of the testing is enhanced. At the same time, the factor-adjusted data are now uncorrelated due to the removal of the common factors. If the errors* $\mathbf{u}$ *were generated from* $N(0, 3\mathbf{I}_p)$, *then the factor-adjusted data were independent and FDR and FDP are approximately the same.*

*We generated data from* $t_3(0, \mathbf{I}_p)$ *to demonstrated further the power of robustification. The distribution has* $3 - \delta$ *moment for any* $\delta > 0$, *but is not sub-Gaussian. Therefore, we employ a robust estimation of means via the adaptive Huber estimator in Theorem 3.2. The resulting histograms of estimated means are depicted in Figure 11.7. It demonstrates that the robust procedure helps further reduce the noise in the estimation: the bimodality is more clearly revealed.*

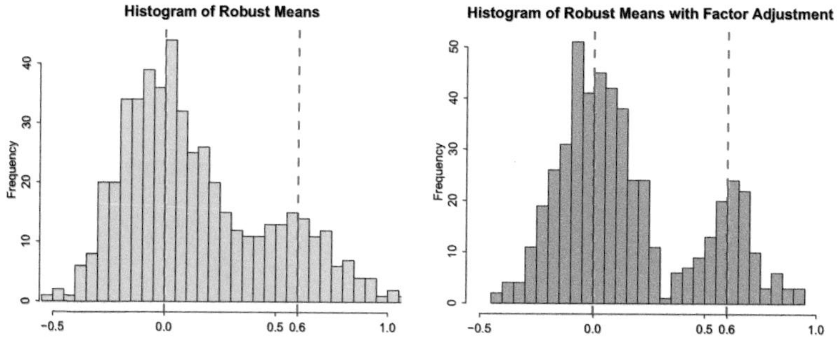


Figure 11.7: Histograms for the robust estimates of means from a synthetic three-factor models without (left panel) and with (right panel) factor adjustments. Factor adjustments clearly decreases the noise and robustness enhances the bimodality further.

### *11.2.4 FarmTest*

The factor-adjusted test essentially applies a multiple testing procedure to the factor-adjusted data $\mathbf{Y}_i$ in (11.20). This requires first to learn $\mathbf{B}\mathbf{f}_i$ via PCA, apply the $t$-test to $\mathbf{Y}_i$, and then control FDR by the Benjamini-Hochberg procedure. In other words, the existing software is fed by using the factor-adjusted data $\{\mathbf{Y}_i\}_{i=1}^n$ rather than the original data $\{\mathbf{X}_i\}_{i=1}^n$.

The Factor Adjusted Robust Multiple test (*FarmTest*), introduced by Fan, Ke, Sun and Zhou (2019+), is a robust implementation of the above basic idea. Instead of using the sample mean, we use the adaptive Huber estimator to estimate $\mu_j$ for $j = 1, \cdots, p$. With a robustification parameter $\tau_j > 0$, we consider the following $M$-estimator of $\mu_j$:

$$\widehat{\mu}_j = \arg\min_{\theta} \sum_{i=1}^{n} \rho_{\tau_j}(X_{ij} - \theta),$$

where $\rho_\tau(u)$ is the Huber loss given in Theorem 3.2. It was then shown in Fan, Ke, Sun, Zhou (2018) that

$$\sqrt{n}\,(\widehat{\mu}_j - \mu_j - \mathbf{b}_j^T\bar{\mathbf{f}}) \to N(0, \sigma_{u,j}) \text{ uniformly in } j = 1, \cdots, p,$$

where $\mathbf{b}_j^T$ is the $j^{th}$ row of $\mathbf{B}$ and $\bar{\mathbf{f}} = n^{-1}\sum_{i=1}^n \mathbf{f}_i$, and

$$\sigma_{u,j} = \mathrm{Var}(u_j) = \mathrm{Var}(X_{ij}) - \|\mathbf{b}_j\|_2^2. \tag{11.21}$$

By (11.19), unknown $\bar{\mathbf{f}}$ can be estimated robustly from the regression problem

$$\bar{X}_j = \mu_j + \mathbf{b}_j^T\bar{\mathbf{f}} + \bar{u}_j, \qquad j = 1, \cdots, p, \tag{11.22}$$

if $\{\mathbf{b}_j\}$ are given, by regarding sparse $\{\mu_j\}_{j=1}^p$ as outliers. Thus, given a robust estimate $\widehat{\mathbf{B}}$, we obtain the robust estimate of $\widehat{\mathbf{f}}$ for $\bar{\mathbf{f}}$ by using (11.22) and use the factor adjusted statistic

$$T_j = \sqrt{n}(\widehat{\mu}_j - \widehat{\mathbf{b}}_j^T\widehat{\mathbf{f}})/\widehat{\sigma}_{u,j},$$

where $\widehat{\sigma}_{u,j}$ is a robust estimator of $\sigma_{u,j}$. This test statistic is simply a robust version of the $z$-test based on the factor-adjusted data $\{\mathbf{Y}_i\}$.

We expect that the factor-adjusted test statistic $T_j$ is approximately normally distributed. Hence, by the law of large numbers, the number of false discoveries

$$V(z) = \sum_{j\in\mathcal{S}_0} I(|T_j| \geq z) \approx 2p_0\Phi(-z).$$

This is due to the fact that we assume that the correlation matrix of the idiosyncratic component vector $\mathbf{u}_i$ is sparse so that $\{T_j\}$ are nearly independent. Therefore, by definition (11.13), we have $\mathrm{FDP}(z) \approx \mathrm{FDP}^A(z)$, where

$$\mathrm{FDP}^A(z) = 2\pi_0 p\Phi(-z)/R(z), \tag{11.23}$$

where $\pi_0 = p_0/p$. Note that $\mathrm{FDP}^A(z)$ is nearly known, since only $\pi_0$ is unknown. In many applications, true discoveries are sparse so that $\pi_0 \approx 1$. It can also be consistently estimated by (11.18).

We now summarize the FarmTest of Fan, Ke, Sun, Zhou (2019+). The inputs include $\{\mathbf{X}_i\}_{i=1}^n$, a generic robust covariance matrix estimator $\widehat{\mathbf{\Sigma}} \in \mathbb{R}^{p\times p}$ from the data, a pre-specified level $\alpha \in (0,1)$ for FDP control, the number of factors $K$, and the robustification parameters $\gamma$ and $\{\tau_j\}_{j=1}^p$. Many of those parameters can be simplified in the implementation. For example, $K$ can be estimated by the methods in Section 10.2.3, and overestimating $K$ has little impact on final outputs. Similarly, the robustification parameters $\{\tau_j\}$ can be selected via a five-fold cross-validation with their theoretically optimal orders taken into account.

FarmTest consists of the following three steps.

a) Compute the eigen-decomposition of $\widehat{\mathbf{\Sigma}}$, set $\{\widehat{\lambda}_j\}_{j=1}^K$ to be its top $K$ eigenvalues in descending order, and $\{\widehat{\mathbf{v}}_j\}_{j=1}^K$ to be their corresponding eigenvectors. Let $\widehat{\mathbf{B}} = (\widetilde{\lambda}_1^{1/2}\widehat{\mathbf{v}}_1, \ldots, \widetilde{\lambda}_K^{1/2}\widehat{\mathbf{v}}_K) \in \mathbb{R}^{p\times K}$ where $\widetilde{\lambda}_j = \max\{\widehat{\lambda}_j, 0\}$, and denote its rows by $\{\widehat{\mathbf{b}}_j\}_{j=1}^p$.

b) Let $\bar{x}_j = \frac{1}{n}\sum_{i=1}^n x_{ij}$ for $j = 1, \cdots, p$ and $\widehat{\mathbf{f}} = \mathrm{argmax}_{\mathbf{f}\in I\!R^K} \sum_{j=1}^p \ell_\gamma(\bar{x}_j - \widehat{\mathbf{b}}_j^\top \mathbf{f})$, which is an implementation of (11.22) by regarding sparse $\{\mu_j\}$ as "outliers". Construct factor-adjusted test statistics

$$T_j = \sqrt{n/\widehat{\sigma}_{u,jj}}(\widehat{\mu}_j - \widehat{\mathbf{b}}_j^T \widehat{\mathbf{f}}) \quad \text{for } j = 1, \cdots, p, \qquad (11.24)$$

where $\widehat{\sigma}_{u,jj} = \widehat{\theta}_j - \widehat{\mu}_j^2 - \|\widehat{\mathbf{b}}_j\|_2^2$, $\widehat{\theta}_j = \arg\min_{\theta \geq \widehat{\mu}_j^2 + \|\widehat{\mathbf{b}}_j\|_2^2} \ell_{\tau_j}(x_{ij}^2 - \theta)$. This estimate of variance is based on (11.21).

c) Calculate the critical value $z_\alpha = \inf\{z \geq 0 : \mathrm{FDP}^{\mathrm{A}}(z) \leq \alpha\}$, where $\mathrm{FDP}^{\mathrm{A}}(z) = 2\widehat{\pi}_0 p \Phi(-z)/R(z)$ (see (11.23)), and reject $H_{0j}$ whenever $|T_j| \geq z_\alpha$.

Step c) of FarmTest is similar to controlling the FDR by the Benjamini-Hockberg procedure, as noted in Section 11.2.1. Even though many heuristics are used in the introduction of the FarmTest procedure, the validity of the procedure has been rigorously proved by Fan, Ke, Sun, and Zhou (2019+). The proofs are sophisticated and we refer readers to the paper for details. The R software package `FarmTest` implements the above procedure.

In the implementation, Fan, Ke, Sun and Zhou (2019+) propose to use the robust U-covariance (9.28) or elementwise robust covariance (9.25) as their robust covariance matrix estimator $\widehat{\boldsymbol{\Sigma}}$. They provide extensive simulations to demonstrate the advantages of the FarmTest. They showed that FarmTest controls accurately the false discovery rate and has higher power in improving missed discovery rates, in comparison to the methods without factor adjustments. They also show convincingly that when the error distribution $\mathbf{u}_i$ has light tails, FarmTest performs about the same as its non-robust implementation; whereas when $\mathbf{u}_i$ has heavy tails, FarmTest outperforms.

### *11.2.5 Application to neuroblastoma data*

We now revisit the neuroblastoma data used in Section 11.1.4. The figures and the results are taken from Fan, Ke, Sun and Zhou (2019+). Instead of examining the joint effect via the sparse logistic regression, we now investigate the marginal effect: which genes expressed statistically differently in the neuroblastoma from the normal controls. The dependence of gene expression has been demonstrated in Figure 11.4. Indeed, the top 4 principal components (PCs) explain 42.6% and 33.3% of the total variance for the positive and negative groups, respectively.

Another stylized feature of high-dimensional data is that many variables have heavy tails. To wit, Figure 11.8 depicts the histograms of the excess kurtosis of the gene expressions for both positive (event-free survivals) and negative groups. For the positive group, the left panel of Figure 11.8 shows that 6518 gene expressions have positive excess kurtosis and 420 of them have kurtosis greater than 6. In other words, about 4% of gene expressions are

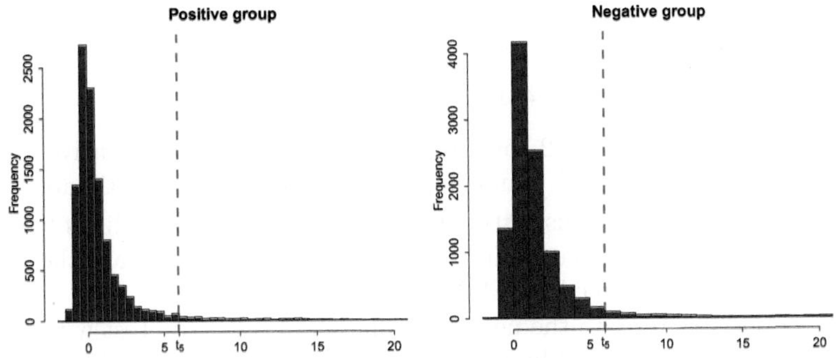


Figure 11.8: Histograms of excess kurtosises for the gene expressions in positive and negative groups of the neuroblastoma data. The vertical dashed line shows the excess kurtosis (equals 6) of $t_5$-distribution.

severely heavy tailed as their tails are fatter than the $t$-distribution with 5 degrees of freedom. Similarly, in the negative group, 9341 gene expressions exhibit positive excess kurtosis with 671 of them have kurtosis greater than 6. Such a heavy-tailed feature indicates the necessity of using robust methods to estimate the mean and covariance of the data.

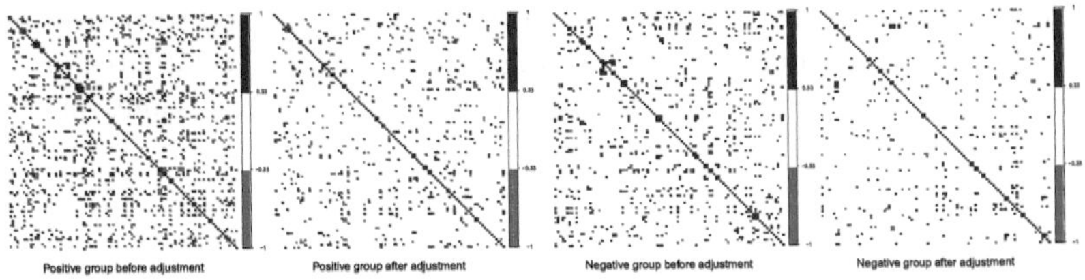


Figure 11.9: Correlations among the first 100 genes before and after factor adjustments. The blue and red pixels represent the pairs of gene expressions whose correlations are greater than 1/3 or less than $-1/3$, respectively.

The effectiveness of the factor adjustments is evidenced in Figure 11.9. Therefore, for each group, we plot the correlation matrices of the first 100 gene expressions before and after adjusting the top 4 PCs. It shows convincingly that the correlations are significantly weakened after the factor adjustments. More specifically, the number of pairs of genes with absolute correlation bigger than 1/3 drops from 1452 to 666 for the positive group and from 848 to 414 for the negative group.

The two-sample version of FarmTest uses the following two-sample $t$-type statistics:

$$T_j = \frac{(\widehat{\mu}_{1j} - \widehat{\mathbf{b}}_{1j}^{\mathrm{T}}\widehat{\mathbf{f}}_1) - (\widehat{\mu}_{2j} - \widehat{\mathbf{b}}_{2j}^{\mathrm{T}}\widehat{\mathbf{f}}_2)}{(\widehat{\sigma}_{1u,jj}/56 + \widehat{\sigma}_{2u,jj}/190)^{1/2}}, \quad j = 1, \ldots, 10707, \tag{11.25}$$

where the subscripts 1 and 2 correspond to the positive and negative groups, respectively. Specifically, $\widehat{\mu}_{1j}$ and $\widehat{\mu}_{2j}$ are the robust mean estimators, and $\widehat{\mathbf{b}}_{1j}$, $\widehat{\mathbf{b}}_{2j}$, $\widehat{\mathbf{f}}_1$ and $\widehat{\mathbf{f}}_2$ are robust estimators of the factors and loadings based on a robust covariance estimator. In addition, $\widehat{\sigma}_{1u,jj}$ and $\widehat{\sigma}_{2u,jj}$ are the robust variance estimators defined in (11.24).

We apply four tests to the neuroblastoma data set: FarmTest with covariance inputs (9.25) (denoted by FARM-H) and (9.28) (denoted FARM-U), the non-robust factor-adjusted version using sample mean and sample variance (denoted as FAM) and the naive method without factor adjustments and robustification. At level $\alpha = 0.01$, FARM-H and FARM-U methods identify, respectively, 3912 and 3855 probes with different gene expressions, among which 3762 probes are identical. This shows an approximately 97% similarity in the two methods. In contrast, FAM and the naive method discover 3509 and 3236 probes, respectively. Clearly, the more powerful method leads to more discoveries.

## 11.3 Factor Augmented Regression Methods

Principal component regression (PCR) is very useful for dimensionality reduction in high-dimensional regression problems. Factor models provide new prospects for understanding why such a method is useful for statistical modeling. Such factor models are usually augmented by other covariates, leading to *augmented factor models.* This section introduces Factor Augmented Regression Methods for Prediction (*FarmPredict*)

### 11.3.1 Principal component regression

Principal component analysis is very frequently used in compressing high-dimensional variables into a couple of principal components. These principal components can also be used for regression analysis, achieving the dimensionality reduction.

Why are the principal components, rather than other components, used as regressors? Factor models offer a new modeling prospective. The basic assumption is that there are latent factors that drive the independent variables and the response variables simultaneously, as is schematically demonstrated in Figure 11.10. Therefore, we extract the latent factors from the independent variables and use the estimated latent factors as the predictors for the response variables. The latent factors are frequently estimated by principal components. Since principal components are used as the predictors in the regression for $y$, it is called *principal component regression* (PCR).

To understand the last paragraph better, suppose that we have data

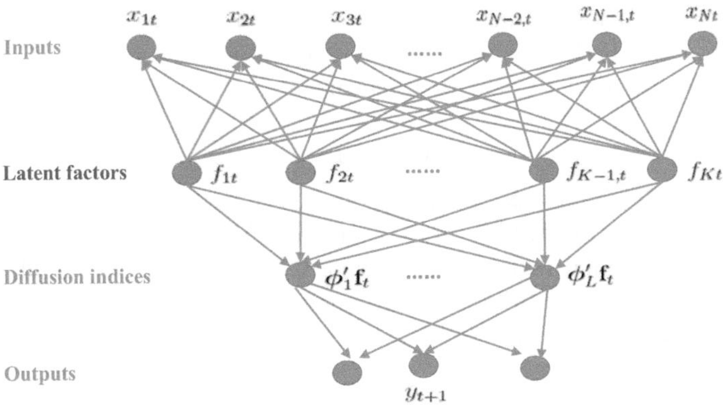


Figure 11.10: Illustration of principal components regression. Latent factors $f_1, \cdots, f_K$ drive both the input variables $x_1, \cdots, x_N$ and output variables $y_1, \cdots, y_q$ at time or subject $t$. The analysis is then to extract these latent factors via PCA and use these estimated factors $\widehat{f}_1, \cdots, \widehat{f}_K$ as predictors, leading to the principal component regression. An example of PCR is to construct multiple indices $\{\boldsymbol{\phi}_j^T \mathbf{f}_t\}$ of factors to predict $\mathbf{Y}$; see (11.34).

$\{(\mathbf{X}_i, Y_i)\}_{i=1}^n$ and $\mathbf{X}_i$ and $Y_i$ are both driven by the latent factors via

$$\begin{cases} \mathbf{X}_i &= \mathbf{a} + \mathbf{B}\mathbf{f}_i + \mathbf{u}_i, \\ Y_i &= g(\mathbf{f}_i) + \varepsilon_i, \end{cases} \tag{11.26}$$

where $g(\cdot)$ is a regression function. The model is also applicable in time series prediction, in which $i$ indexes time $t$. Under model (11.26), we use data $\{\mathbf{X}_i\}_{i=1}^n$ to estimate $\{\mathbf{f}_i\}_{i=1}^n$ via PCA, resulting in $\{\widehat{\mathbf{f}}_i\}_{i=1}^n$. We then fit the regression

$$Y_i = g(\widehat{\mathbf{f}}_i) + \varepsilon_i, \qquad i = 1, \cdots, n. \tag{11.27}$$

PCR corresponds to multiple regression in model (11.27):

$$Y_i = \alpha + \boldsymbol{\beta}^T \widehat{\mathbf{f}}_i + \varepsilon_i, \qquad i = 1, \cdots, n. \tag{11.28}$$

Due to the identifiability constraints of latent factors, we often normalize $\{\mathbf{f}_i\}_{i=1}^n$ to have sample mean zero and sample variance identity. Under such a condition, the least-squares estimate admits the sample formula:

$$\widehat{\alpha} = n^{-1} \sum_{i=1}^n Y_i, \qquad \widehat{\boldsymbol{\beta}} = n^{-1} \sum_{i=1}^n (Y_i - \bar{Y}) \widehat{\mathbf{f}}_i. \tag{11.29}$$

where $\widehat{f}_{ij}$ is the $j^{th}$ component of $\widehat{\mathbf{f}}_i$, and $K$ is the number of factors. Paul, Bair, Hastie and Tibshirani (2008) employed "*preconditioning*" (*correlation screening*) to reduce the dimensionality in learning latent factors.

PCR is traditionally motivated differently from ours. For simplicity, we assume that both the means of the covariates and the response are zero so that we do not need to deal with the mean. Write the linear model in the matrix form

$$\mathbf{Y} = \mathbf{X}\boldsymbol{\beta} + \boldsymbol{\varepsilon}.$$

Let $\mathbf{X} = \mathbf{U}\boldsymbol{\Delta}\mathbf{V}^T$ where $\boldsymbol{\Delta} = \mathrm{diag}(\delta_1, \cdots, \delta_p)$ is the diagonal matrix, consisting of its singular values, $\mathbf{U} = (\mathbf{u}_1, \cdots, \mathbf{u}_p)$ and $\mathbf{V} = (\mathbf{v}_1, \cdots, \mathbf{v}_p)$ are the $n \times p$ and $p \times p$ orthogonal matrices, consisting of left- and right- singular vectors of matrix $\mathbf{X}$. Then, the principal components based on the sample covariance matrix $n^{-1}\mathbf{X}^T\mathbf{X}$ are simply

$$\widehat{\mathbf{f}}_j = \mathbf{X}\mathbf{v}_j/\delta_j, \qquad j = 1, \cdots, p,$$

where $\mathbf{v}_j$ is the principal direction. Then, the PCR can be regarded as a constrained least-squares problem:

$$\min_{\boldsymbol{\beta}} \|\mathbf{Y} - \mathbf{X}\boldsymbol{\beta}\|^2, \qquad \text{s.t. } \boldsymbol{\beta}^T\mathbf{v}_j = 0, \qquad \text{for } j > K. \tag{11.30}$$

The constraint can be regarded as regularization to reduce the complexity of $\boldsymbol{\beta}$.

PCR also admits the optimality in the following sense. Let $\mathbf{L}$ be the $p \times (p-K)$ full rank matrix. Consider the generalization of the constrained least-squares problem (11.30)

$$\min_{\boldsymbol{\beta}} \|\mathbf{Y} - \mathbf{X}\boldsymbol{\beta}\|^2, \qquad \text{s.t. } \mathbf{L}^T\boldsymbol{\beta} = 0. \tag{11.31}$$

Let $\widehat{\boldsymbol{\beta}}_L$ be the solution and MSE($\mathbf{L}$) be the mean square error for estimating the true parameter vector $\boldsymbol{\beta}^*$. The optimal choice of $\mathbf{L}$ that minimizes the MSE is the one given by (11.30), or PCR (Park, 1981).

### *11.3.2 Augmented principal component regression*

In many applications, we have more than one type of data. For example, in the macroeconomic data given in Section 11.1.3, in addition to 131 disaggregated macroeconomic data, we have 8 aggregated macroeconomics variables; see Section 11.3.3 for addition details. Similarly, in the neuroblastoma data in Section 11.1.4, in additional to the microarray gene expression data, we have demographical and clinical information such as age, weight, and social economic status, among others. Clearly these two types of variables can not be aggregated via PCA.

Assume that we have additional covariates $\mathbf{W}_i$ so that the observed data are now $\{(\mathbf{W}_i, \mathbf{X}_i, Y_i)\}_{i=1}^n$. We assume that $\mathbf{W}_i$ is low-dimensional; otherwise, we will use the estimated common factors from $\mathbf{W}_i$. The *augmented factor model* is to extract the latent factors $\{\mathbf{f}_i\}$ from $\{\mathbf{X}_i\}$ and run the regression using both the estimated latent factors and the augmented variables:

$$Y_i = g(\widehat{\mathbf{f}}_i, \mathbf{W}_i) + \varepsilon_i, \qquad i = 1, \cdots, n. \tag{11.32}$$

We will refer this class of prediction method to the Factor-Augmented Regression Model for Prediction or *FarmPredict* for short.

The regression problem in (11.32) can be a multiple regression model:

$$Y_i = \alpha + \boldsymbol{\beta}^T \widehat{\mathbf{f}}_i + \boldsymbol{\gamma}^T \mathbf{W}_i + \varepsilon_i, \qquad i = 1, \cdots, n. \tag{11.33}$$

This results in the *augmented principal component regression.* The regression function in (11.32) can also be estimated by using nonparametric function estimation such as the kernel and spline methods in Chapter 2 or deep neural network models in Chapter 14. It can also be estimated by using a *multi-index model* as is schematically shown in Figure 11.10. There, we attempt to extract several linear combinations of estimated factors, also called a diffusion index by Stock and Watson (2002b), to predict the response $Y$. See Fan, Xue and Yao (2017) for a study on this approach, which imposes the model:

$$Y_i = g(\boldsymbol{\phi}_1^T \mathbf{X}_i^*, \cdots, \boldsymbol{\phi}_L^T \mathbf{X}_i^*) + \varepsilon_i \tag{11.34}$$

where $\mathbf{X}_i^* = (\mathbf{f}_i^T, \mathbf{W}_i^T)^T$ is now the expanded predictors. In particular, when there is no $\mathbf{W}_i$ component, the model reduces to the case depicted in Figure 11.10.

Since the study of the multi-index model by Li (1991, 1992), there are a lot of developments in this area on the estimation of the diffusion indices. See Cook (2007, 2009) and the references therein.

### *11.3.3 Application to forecast bond risk premia*

We now revisit the macroeconomic data presented in Section 11.1.3. In addition to the 131 disaggregated macroeconomic variables, we have 8 aggregated macroeconomic variables:

$W_1$ = Linear combination of five forward rates;

$W_2$ = GDP;

$W_3$ = Real category development index (CDI);

$W_4$ = Non-agriculture employment;

$W_5$ = Real industrial production;

$W_6$ = Real manufacturing and trade sales;

$W_7$ = Real personal income less transfer;

$W_8$ = Consumer price index (CPI)

These variables $\{\mathbf{W}_t\}$ can be combined with the 5 principal components $\{\widehat{\mathbf{f}}_t\}$ extracted from the 131 macroeconomic variables to predict the bond risk premia. This yields the PCR based on covariates $\mathbf{f}_t$ and the augmented PCR based on $(\mathbf{W}_t^T, \mathbf{f}_t^T)$. The out-of-sample $R^2$ for these two models is presented in the last block of Table 11.3 under the heading of PCA. The augmented PCR improves somewhat the out-of-sample $R^2$, but not by a lot.

Another usage of the augmented variables $\mathbf{W}_t$ is to blend them in extracting the latent factors via projected PCA introduced in Section 10.4. It can be

Table 11.3: Out-of-sample $R^2$ (%) for linear prediction of bond risk premia. Adapted from Fan, Ke, and Liao (2018).

| predictors | PPCA | | | | PCA | | | |
|---|---|---|---|---|---|---|---|---|
| | Maturity(Year) | | | | Maturity(Year) | | | |
| | 2 | 3 | 4 | 5 | 2 | 3 | 4 | 5 |
| $\mathbf{f}_t$ | 38.0 | 32.7 | 25.6 | 22.9 | 23.0 | 20.7 | 16.8 | 16.5 |
| $(\mathbf{f}_t^T, \mathbf{W}_t^T)^T$ | 37.7 | 32.4 | 25.4 | 22.7 | 23.9 | 21.4 | 17.4 | 17.5 |

implemented robustly (labeled PPCA). After extracting the latent factors, we apply the multi-index model (11.34) to predict the bond risk premia. The diffusion indices coefficient $\{\phi_j\}_{j=1}^L$ is estimated via the slice inverse regression (Li, 1991) with the number of diffusion index $L$ chosen by the eigen-ratio method. The choice is usually $L = 2$ or 3. The nonparametric function $g$ in (11.34) is fitted by using the additive model $g(x_1, \cdots, x_L) = g_1(x_1) + \cdots + g_L(x_L)$ via the local linear fit. The prediction results are shown in the first block of Table 11.3. It shows that extracting latent factors via projected PCA is far more effective than the augmented PCR. We also implemented a non-robust version of PPCA and obtained similar but somewhat weaker results than those by its robust implementation.

Table 11.4: Out-of-sample $R^2$ (%) by the multi-index model for the bond risk premia. Adapted from Fan, Ke, and Liao (2018).

| Predictors | PPCA | | | | PCA | | | |
|---|---|---|---|---|---|---|---|---|
| | Maturity(Year) | | | | Maturity(Year) | | | |
| | 2 | 3 | 4 | 5 | 2 | 3 | 4 | 5 |
| $\mathbf{f}_t$ | 44.6 | 43.0 | 38.8 | 37.3 | 30.1 | 25.5 | 23.2 | 21.3 |
| $(\mathbf{f}_t^T, \mathbf{W}_t^T)^T$ | 41.5 | 38.7 | 35.2 | 33.8 | 30.8 | 26.3 | 24.6 | 22.0 |

Fan, Ke, and Liao (2018) also applied the multi-index model to the PCA extracted factors along with the augmented variables. The out-of-sample $R^2$ is presented in the second block of Table 11.4. It is substantially better than the linear predictor but is not as effective as the factors extracted by PPCA.

## 11.4 Applications to Statistical Machine Learning

Principal component analysis has been widely used in statistical machine learning and applied mathematics. The applications range from dimensionality reduction such as factor analysis, manifold learning, and multidimensional scaling to clustering analysis such as image segmentation and community detection. Other applications include item ranking, matrix completion, $Z_2$-synchronization, and blind convolution and initialization for nonconvex optimization problems such as mixture models and phase retrieval. The role of PCA in statistical machine learning is very similar to that of regression analysis in statistics. It requires a whole book in order to give a comprehensive account on this subject. Rather, we only highlight some applications to give readers an idea of how PCA can be used to solve seemingly complicated statistical machine learning and applied mathematics problems.

In the applications so far, PCA is mainly applied to the sample covariance matrix or robust estimates of the covariance matrix to extract latent factors. In the applications below, the PCA is applied to a class of *Wigner matrices*, which are random symmetric (or Hermitian) matrices with independent elements in the upper diagonal matrix. It is often assumed that the random elements in the Wigner matrices have mean zero and finite second moment. In contrast, the elements in the sample covariance matrix are dependent. Factor modeling and principal components are closely related. We can view PCA as a form of spectral learning which is applied to the sample covariance matrix or robust estimates of the covariance matrix to extract latent factors.

### *11.4.1 Community detection*

*Community detection* is basically a clustering problem based on *network* data. It has diverse applications in social networks, citation networks, genomic networks, economics and financial networks, among others. The observed network data are represented by a graph in which nodes represent members in the communities and edges indicate the links between two members. The data can be summarized as an *adjacency matrix* $\mathbf{A} = (a_{ij})$, where the element $a_{ij} = 0$ or 1 depends on whether or not there is a link between the $i^{th}$ and $j^{th}$ node.

The simplest probabilistic model is probably the *stochastic block model*, introduced in Holland and Leinhardt (1981), Holland, Laskey, and Leinhardt (1983) and Wang and Wong (1987). For an overview of recent developments, see Abbe (2018).

**Definition 11.1** *Suppose that we have $n$ vertices, labeled $\{1, \cdots, n\}$, that can be partitioned into $K$ disjoint communities $\mathcal{C}_1, \cdots, \mathcal{C}_K$. For two vertices $k \in \mathcal{C}_i$ and $l \in \mathcal{C}_j$, a stochastic block model assumes that they are connected with probability $p_{ij}$, independent of other connections. In other words, the adjacency matrix $\mathbf{A} = (A_{kl})$ is regarded as a realization of a sequence of independent Bernoulli random variables with*

$$P(A_{kl} = 1) = p_{ij}, \qquad \text{for } k \in \mathcal{C}_i, l \in \mathcal{C}_j. \tag{11.35}$$

*The $K \times K$ symmetric matrix $\mathbf{P} = (p_{i,j})$ is referred to as the edge probability matrix.*

In the above stochastic block model, the edge probability $p_{ij}$ describes the connectivity between the two communities $\mathcal{C}_i$ and $\mathcal{C}_j$. Within a community $\mathcal{C}_i$, the connection probability is $p_{ii}$. The simplest model is the one in which $p_{ij} = p$ for all $i$ and $j$, and is referred to as the *Erdös-Rényi graph* or model (Erdös and Rényi, 1960). In other words, the partition is irrelevant, corresponding to a degenerate case, resulting in a *random graph*. Denote such a model as $G(n, p)$.

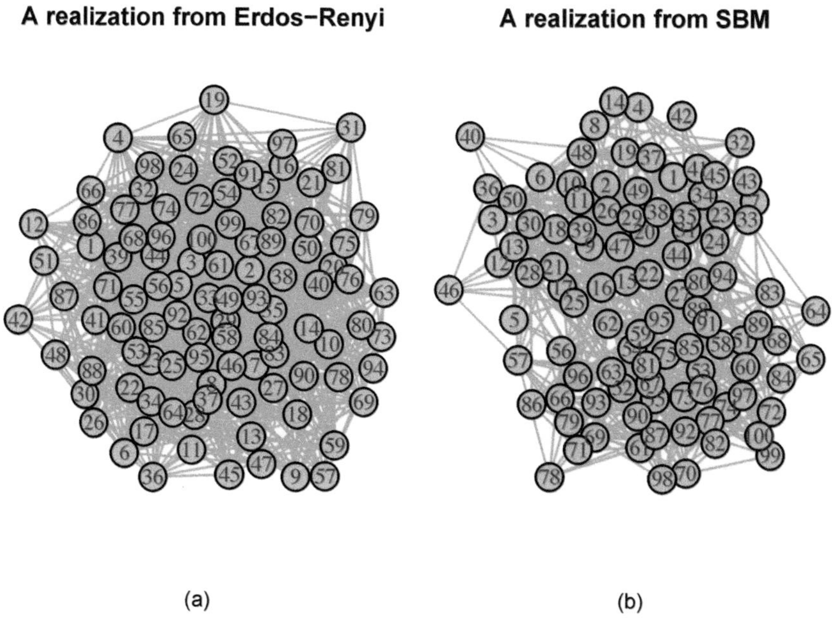


Figure 11.11: Simulated network data from (a) Erdös-Rényi graph and (b) stochastic block model with $n = 100$, $p = 5\log(n)/n$, and $q = p/4$. For SMB model, there were two communities, with the first $n/2$ members from group 1 and the second $n/2$ members from group 2; they are hard to differentiate because $p$ and $q$ are quite close.

A *planted partition model* corresponds to the case where $p_{ii} = p$ for all $i$ and $p_{ij} = q$ for all $i \neq j$ and $p \neq q$. In other words, the probability of within-community connections is $p$ and the probability of between-community connections is $q$. When $r = 2$, this corresponds to the two community case. Figure 11.11 gives a realization from the Erdös-Rényi model and the stochastic block model. The simulations and plots are generated from the functions `sample_gnp` and `sample_sbm` in the R package `igraph` [`install.package("igraph"); library(igraph)`].

Given an observed adjacency matrix $\mathbf{A} = (a_{ij})$, statistical tasks include detecting whether the graph has a latent structure and recovering the latent partition of the communities. We naturally appeal to the maximum likelihood method. Let $C(i)$ be the unknown community to which node $i$ belongs. Then the likelihood function of observing $\{a_{ij}\}_{i>j}$ is simply the multiplication of the Bernoulli likelihood:

$$\prod_{i>j} p_{C(i)C(j)}^{a_{ij}} (1 - p_{C(i)C(j)})^{1-a_{ij}},$$

where the unknown parameters are $\{C(i)\}_{i=1}^K$ and $\mathbf{P} = (p_{ij}) \in R^{K \times K}$. This is an NP-hard computation problem. Therefore, various relaxed algorithms including *semidefinite programs* and *spectral methods* have been proposed; see Abbe (2018).

The spectral method is based on the *method of moments*. Let $\mathbf{\Gamma}$ be an $n \times K$ membership matrix, whose $i^{th}$ row indicates the membership of node $i$: It has an element 1 at the location $C(i)$ and zero otherwise. The matrix $\mathbf{\Gamma}$ classifies each node into a community, with rows indicating the communities. Then, it is easy to see that $\mathrm{E}\, A_{ij} = p_{C(i)C(j)}$ for $i \neq j$ and

$$\mathrm{E}\,\mathbf{A} = \mathbf{\Gamma P \Gamma}^T, \tag{11.36}$$

except the diagonal elements, which are zero on the left (assuming no self-loop).[2] It follows immediately that the membership matrix $\mathbf{\Gamma}$ falls in the eigen-space spanned by the first $K$ eigenvectors of $\mathrm{E}\,\mathbf{A}$. Indeed, the column space of $\mathbf{\Gamma}$ is the same as the eigen-space spanned by the top $K$ eigenvalues of $\mathrm{E}\,\mathbf{A}$, if $\mathbf{P}$ is not degenerate. See Exercise 11.9.

The above discussion reveals that the column space of $\mathbf{\Gamma}$ can be estimated via its empirical counterpart, namely the eigen-space spanned by the top $K$ eigenvalues (in absolute value) of $\mathbf{A}$. To find a consistent estimate of $\mathbf{\Gamma}$, an unknown rotation is needed. This can be challenging to find. Instead of rotating, the second step is typically replaced by a clustering algorithm such as the $K$-means algorithm (see Section 13.1.1). This is due to the fact that if two members are in the same community, their corresponding rows of the eigenvectors associated with the top $K$ eigenvalues are the same, by using simple permutation arguments. See Example 11.3. Therefore, there are $K$ distinct rows, which can be found by using the $K$-means algorithm. Spectral clustering is also discussed in Chapter 13. We summarize the algorithm for *spectral clustering* as in Algorithm 11.1.

**Example 11.3** *As a specific example, let us consider the planted partition model with two communities. For simplicity, we assume that we have two groups of equal size $n/2$. The first group is in the index set $J$ and the second group is in the index set $J^c$. Then,*

$$\mathrm{E}(\mathbf{A}) = \begin{pmatrix} p\mathbf{1}_{J,J} & q\mathbf{1}_{J,J^c} \\ q\mathbf{1}_{J^c,J} & p\mathbf{1}_{J^c,J^c} \end{pmatrix}$$

[2]More rigorously, $\mathrm{E}\,\mathbf{A} = \mathbf{\Gamma P \Gamma}^T - \mathrm{diag}(\mathbf{\Gamma P \Gamma}^T)$. Since $\|\mathrm{diag}(\mathbf{\Gamma P \Gamma}^T)\| \leq \max_{k\in[K]} p_{kk}$, this diagonal matrix is negligible in the analysis of the top $K$ eigenvalues of $\mathrm{E}\,\mathbf{A}$ under some mild conditions. Putting heuristically, $\mathrm{E}\,\mathbf{A} \approx \mathbf{\Gamma P \Gamma}^T$. We will ignore this issue throughout this section.

**Algorithm 11.1** Spectral clustering for community detection

1. From the adjacency matrix $\mathbf{A}$, get the $n \times K$ matrix $\widehat{\mathbf{\Gamma}}$, consisting of the eigenvectors of the top $K$ absolute eigenvalues;
2. Treat each of the $n$ rows in $\widehat{\mathbf{\Gamma}}$ as the input data, run a cluster algorithm such as $k$-means clustering to group the data points into $K$ clusters. This puts $n$ members into $K$ communities. (Normalizing each rows to unit norm is recommended and is needed for more general situations)

*has two non-vanishing eigenvalues. They are* $\lambda_1^* = \frac{n(p+q)}{2}$ *and* $\lambda_2^* = \frac{n(p-q)}{2}$, *with corresponding eigenvectors*

$$\mathbf{u}_1^* = \frac{1}{\sqrt{n}}\mathbf{1}, \ \mathbf{u}_2^* = \frac{1}{\sqrt{n}}(\mathbf{1}_J - \mathbf{1}_{J^c}),$$

*In this case, the sign of the second eigenvector identifies memberships. Our spectral method is particularly simple. Get the second eigenvector* $\mathbf{u}_2$ *from the observed adjacency matrix* $\mathbf{A}$ *and use* $\mathrm{sgn}(\mathbf{u}_2)$ *to classify each membership. Abbe, Fan, Wang and Zhong (2019+) show such a simple method is optimal in exact and nearly exact recovery cases.*

We now apply the spectral method in Example 11.3 to the simulated data given in Figure 11.11. The results are depicted in Figure 11.12. From the screeplot in the left panel of Figure 11.12, we can see that the data from the Erdös-Rényi model has only one distinguished eigenvalue, whereas the data from the SBM model has two distinguished eigenvalues. This corresponds to the fact that $\mathrm{E}\,\mathbf{A}$ is rank one for the Erdös-Rényi model and rank two for the SBM model. The right panel of Figure 11.12 demonstrates the effectiveness of the spectral method. Using the signs of the empirical second eigenvector as a classifier, we recover correctly the communities for almost all members based on the data generated from SBM. In contrast, as it should be, we only identify 50% correctly for the data generated from Erdös-Rényi model.

There are a number of variations on the applications of the spectral method. For each node, let $d_i = \sum_{j \neq i} a_{ij}$ be the *degree of node* $i$, which is the number of edges that node $i$ has and $\mathbf{D} = \mathrm{diag}(d_1, \cdots, d_n)$ be the degree matrix. The *Laplacian matrix*, *symmetric normalized Laplacian* and *random walk normalized Laplacian* are defined as

$$\mathbf{L} = \mathbf{D} - \mathbf{A}, \qquad \mathbf{L}^{\mathrm{sym}} = \mathbf{I} - \mathbf{D}^{-1/2}\mathbf{A}\mathbf{D}^{-1/2}, \qquad \mathbf{L}^{\mathrm{rw}} = \mathbf{D}^{-1}\mathbf{L}. \quad (11.37)$$

In step 1 of Algorithm 11.1, the adjacency matrix $\mathbf{A}$ can be replaced by one of the Laplacian matrices in (11.37). In step 2, the row of $\widehat{\mathbf{\Gamma}}$ should be normalized to have the unit norm before running the $K$-mean algorithm. See Qin and Rohe (2013) and the arguments at the end of this subsection.

There are various extensions of the stochastic block models to make it more

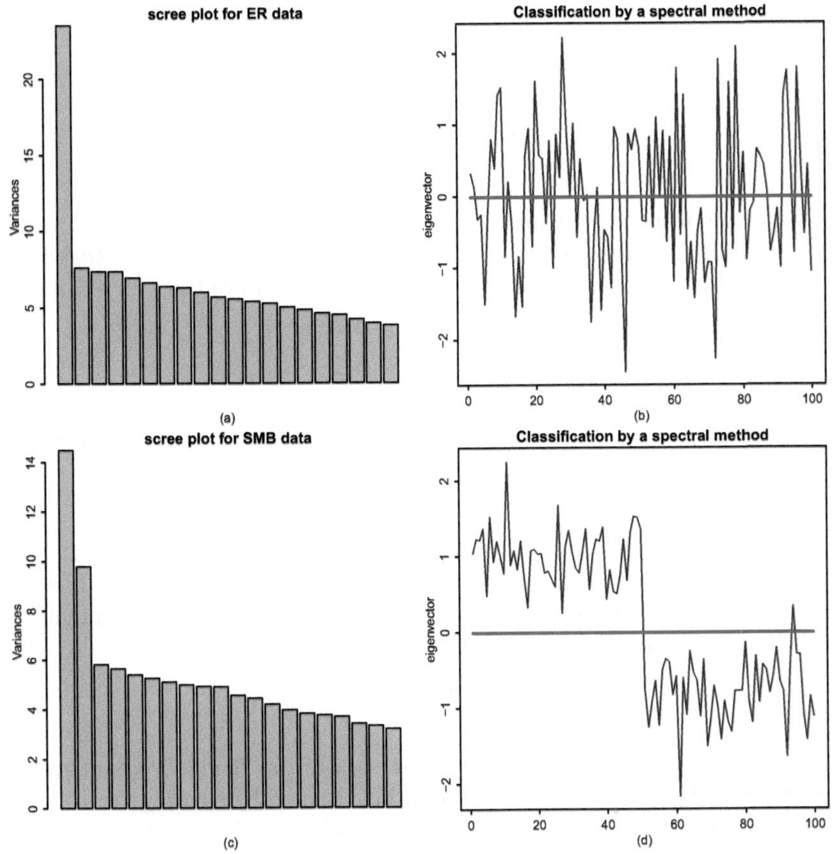


Figure 11.12: Spectral analysis for the simulated network data given in Figure 11.11. Top panel is for the data generated from the Erdös-Rényi model and the bottom is for the data generated from the stochastic block model. The left panel is the scree plot based on $\{|\lambda_i|\}_{i=1}^{20}$ and the right panel is $\sqrt{n}\mathbf{u}_2$.

realistic. *Mixed membership models* have been introduced by Airoldi, Blei, Fienberg and Xing (2008) to accommodate members having multiple communities. Karrer and Newman (2011) introduced a degree corrected stochastic block model to account for varying degrees of connectivity among members in each community. We describe these models herewith to give readers some insights. We continue to use the notations introduced in Definition 11.1.

**Definition 11.2** *For each node $i = 1, \cdots, n$, let $\boldsymbol{\pi}_i$ be a $K$-dimensional probability vector that describes the mixed membership, with $\pi_i(k) = P(i \in \mathcal{C}_k)$ for $k \in [K]$ and $\theta_i > 0$ representing the degree of heterogeneity in affinity of node $i$. Under the degree corrected mixed membership model, we assume that*

$$P(A_{kl} = 1 | k \in \mathcal{C}_i, l \in \mathcal{C}_j) = \theta_k \theta_l p_{ij}. \tag{11.38}$$

*Here we implicitly assume that $0 \le \theta_k \theta_l p_{ij} \le 1$ for all $i, j, k$ and $l$. Consequently, the probability that there is an edge between nodes $k$ and $l$ is*

$$P(A_{kl} = 1) = \theta_k \theta_l \sum_{i=1}^{K} \sum_{j=1}^{K} \pi_k(i) \pi_l(j) p_{ij} \tag{11.39}$$

Note that when $\theta_i = 1$ and $\boldsymbol{\pi}_i$ is an indicator vector for all $i = 1, \cdots, n$, model (11.39) reduces to the stochastic block model (11.35). If $\{\theta_i\}$ varies but $\boldsymbol{\pi}_i$ is an indicator vector for all $i = 1, \cdots, n$, we have the *degree-corrected stochastic block model*:

$$P(A_{kl} = 1) = \theta_k \theta_l p_{ij}, \qquad k \in \mathcal{C}_i, l \in \mathcal{C}_j. \tag{11.40}$$

Return to the general model (11.39). Let $\boldsymbol{\Theta} = \mathrm{diag}(\theta_1, \cdots, \theta_n)$. Then,

$$\mathrm{E}\,\mathbf{A} = \boldsymbol{\Theta}\boldsymbol{\Pi}\mathbf{P}\boldsymbol{\Pi}^T\boldsymbol{\Theta}, \tag{11.41}$$

where $\boldsymbol{\Pi}$ is an $n \times K$ matrix, with $\boldsymbol{\pi}_i^T$ as its $i^{th}$ row. Thus, $\boldsymbol{\Theta}\boldsymbol{\Pi}$ falls in the space spanned by the eigenvectors of the top $K$ eigenvalues of $\mathrm{E}\,\mathbf{A}$. They can be estimated by their empirical counterparts. A simple way to eliminate the degree heterogeneity $\boldsymbol{\Theta}$ is the ratios of eigenvectors

$$Y_{ik} = \widehat{\gamma}_{ik} / \widehat{\gamma}_{i1}, \qquad k = 2, \cdots, K,$$

where $\widehat{\gamma}_{ik}$ is the $(i, k)$ element of $\widehat{\boldsymbol{\Gamma}}$, the matrix of top $K$ eigenvectors given in Algorithm 11.1. The community can now be detected by running the $K$-means algorithm on the rows $\{(Y_{i2}, \cdots, Y_{iK})\}_{i=1}^n$. This is *SCORE* by Jin (2015). Another way to eliminate the degree heterogeneity is through normalization of rows of $\widehat{\boldsymbol{\Gamma}}$ to the unit norm before running the $K$-mean algorithm.

To see why degree heterogeneity can be removed by the above two methods, let us consider the population version: $\mathrm{E}\,\mathbf{A}$ or a symmetric normalized Laplacian version: $(\mathrm{E}\,\mathbf{D})^{-1/2}\,\mathrm{E}\,\mathbf{A}(\mathrm{E}\,\mathbf{D})^{-1/2}$ (Note that $\mathbf{L} = \mathbf{D}^{1/2}\mathbf{L}^{\mathrm{sym}}\mathbf{D}^{1/2}$ falls also in this category). Whatever the version is used, the population matrix is of form $\widetilde{\boldsymbol{\Theta}}\boldsymbol{\Pi}\mathbf{P}\boldsymbol{\Pi}^T\widetilde{\boldsymbol{\Theta}}$, where $\widetilde{\boldsymbol{\Theta}}$ is a diagonal matrix. Let $\boldsymbol{\Gamma}$ be the matrix that comprises of the top $K$ eigenvectors of $\widetilde{\boldsymbol{\Theta}}\boldsymbol{\Pi}\mathbf{P}\boldsymbol{\Pi}^T\widetilde{\boldsymbol{\Theta}}$. Then, the column space spanned by $\boldsymbol{\Gamma}$ is the same as that spanned by $\widetilde{\boldsymbol{\Theta}}\boldsymbol{\Pi}$, assuming $\mathbf{P}$ is not degenerate. Therefore, there exists a nondegenerate $K \times K$ matrix $\mathbf{V} = (\mathbf{v}_1, \cdots, \mathbf{v}_K)$ such that

$$\boldsymbol{\Gamma} = \widetilde{\boldsymbol{\Theta}}\boldsymbol{\Pi}\mathbf{V} = (\widetilde{\boldsymbol{\theta}} \circ \boldsymbol{\Pi}\mathbf{v}_1, \cdots, \widetilde{\boldsymbol{\theta}} \circ \boldsymbol{\Pi}\mathbf{v}_K),$$

where $\widetilde{\boldsymbol{\theta}} = (\widetilde{\theta}_1, \cdots, \widetilde{\theta}_n)^T$ are the diagonal elements of $\widetilde{\boldsymbol{\Theta}}$ and $\circ$ is the Hadamard product (componentwise product). Let $\boldsymbol{\gamma}_j$ be the $j^{th}$ column of $\boldsymbol{\Gamma}$. Then, the ratios of eigenvectors $\{\boldsymbol{\gamma}_j / \boldsymbol{\gamma}_1\}_{j=2}^K$ (componentwise division) will eliminate the parameter vector $\widetilde{\boldsymbol{\theta}}$, resulting in $\{\boldsymbol{\Pi}\mathbf{v}_j / \boldsymbol{\Pi}\mathbf{v}_1\}_{j=2}^K$. In addition, if $\boldsymbol{\Pi}$ has at most $m$ distinct rows, so are the ratios of eigenvectors.

Similarly, letting $\{\mathbf{u}_i^T\}_{i=1}^n$ be the rows of $\boldsymbol{\Pi}\mathbf{V}$. Then, $\mathbf{u}_i^T = \boldsymbol{\pi}_i^T\mathbf{V}$ and the $i^{th}$ row of $\mathbf{\Gamma}$ is $\widetilde{\theta}_i \mathbf{u}_i^T$. If $\{\boldsymbol{\pi}_i\}_{i=1}^n$ have at most $m$ distinct vectors, so do $\{\mathbf{u}_i\}_{i=1}^n$. Within a cluster, there might be different $\widetilde{\theta}_i$ but rows of $\mathbf{\Gamma}$ point the same direction $\mathbf{u}_i = \mathbf{V}^T \boldsymbol{\pi}_i$. Therefore, the normalization to the unit norm puts these rows at the same point so that the $K$-means algorithm can be applied.

Membership estimation by *vertex hunting* can be found in the paper by Jin, Ke and Luo (2017). Statistical inference on membership profiles $\boldsymbol{\pi}_i$ is given in Fan, Fan, Han and Lv (2019).

### *11.4.2 Topic model*

A *topic model* is used to classify a collection of documents into "topics" (see Figure 1.3). Suppose that we are given a collection of $n$ documents and a bag of $p$ words. Note that one of these words can be "none of the above words". Then, we can count the frequencies of occurrence of $p$ words in each document and obtain a $p \times n$ observation matrix $\mathbf{D} = (\mathbf{d}_1, \cdots, \mathbf{d}_n)$, called the *text corpus* matrix, where $\mathbf{d}_j$ is the observed frequencies of *bag of words* in document $j$. Suppose that there are $K$ topics and the bag of words appears in topic $k$ with a probability vector $\mathbf{p}_k, k \in [K]$, with $p_{ik}$ indicating the probability of word $i$ in topic $k$. Like the mixed membership in the previous section, we assume that document $j$ is a mixture of $K$ topics with probability vector $\boldsymbol{\pi}_j = (\pi_j(1), \cdots, \pi_j(K))^T$, i.e. document $j$ puts weight $\pi_j(k)$ on topic $k$. Therefore, the probability of observing the $i^{th}$ word in document $j$ is

$$d_{ij}^* = \sum_{k=1}^{K} \pi_j(k) p_{ik}.$$

Let $n_j$ be the length of the $j^{th}$ document. We assume that

$$n_j \mathbf{d}_j \sim \text{Multinomial}(n_j, \mathbf{d}_j^*), \qquad j \in [n],$$

where $\mathbf{d}_j^* = (d_{1j}^*, \cdots, d_{pj}^*)^T$. This is the probabilistic *latent semantic indexing* model introduced by Hofmann (1999).

Let $\mathbf{\Pi}$ be the $n \times K$ matrix with the $j^{th}$ row $\boldsymbol{\pi}_j^T$,

$$\mathbf{D}^* = (\mathbf{d}_1^*, \cdots, \mathbf{d}_n^*), \quad \text{and} \quad \mathbf{P} = (\mathbf{p}_1, \cdots, \mathbf{p}_K).$$

Then, the observed data matrix $\mathbf{D}$ is related to unknown parameters through

$$\mathrm{E}\,\mathbf{D} = \mathbf{D}^* = \mathbf{P}\mathbf{\Pi}^T,$$

which is of rank at most $K$. Let

$$\mathbf{D}^* = \mathbf{L}^* \mathbf{\Lambda}^* (\mathbf{R}^*)^T$$

be the *singular value decomposition* (*SVD*) of $\mathbf{D}^*$, where $\mathbf{L}^*$ and $\mathbf{R}^*$ are respectively the left and right singular vectors of $\mathbf{D}^*$ and $\mathbf{\Lambda}^*$ is the $K \times K$ diagonal matrix, consisting of $K$ non-vanishing singular values. Then, the column space spanned by $\mathbf{P}$ is the same as that spanned by $\mathbf{L}^*$:

$$\mathbf{P} = \mathbf{L}^*\mathbf{U}, \qquad \text{for some } \mathbf{U} \in \Re^{K\times K},$$

or $\mathbf{P}$ is identifiable by $\mathbf{L}^*$ up to a right affine matrix $\mathbf{U}$. Similarly,

$$\mathbf{\Pi} = \mathbf{R}^*\mathbf{V}, \qquad \text{for some } \mathbf{V} \in \Re^{K\times K}.$$

The unknown matrix $\mathbf{P}$ and $\mathbf{\Pi}$ can be estimated by a spectral method. Let $\mathbf{D} = \mathbf{L\Lambda R}$ be the SVD of the text corpus. Then $\mathbf{L}$ is an estimate of $\mathbf{P}$ up to a right $K \times K$ matrix $\mathbf{U}$. This can be further identified by *anchor words*, a concept that is similar to the pure nodes in the mixed membership model. As explained at the end of Section 11.4.1, we can normalize the matrix $\mathbf{D}$ first, followed by post-SVD normalization. The $K$-means algorithm can be conducted by using different rows $\mathbf{L}$ for grouping words or $\mathbf{R}$ for clustering documents. We refer to Ke and Wang (2019) for details.

A popular Bayesian approach to topic modeling is the *Latent Dirichlet Allocation*, introduced by Blei, Ng and Jordan (2003), in which a Dirichlet prior on $\{\boldsymbol{\pi}_i\}$ is imposed. The parameters are then estimated by a variational EM algorithm (Hoffman, Blei, Wang, and Paisley, 2013).

### *11.4.3 Matrix completion*

A motivating example is the Netflix problem in movie-rating: Customer $i$ rates movie $j$ if she has watched the movie; otherwise the $(i, j)$-entry of the ratings matrix is missing. Since Netflix has millions of customers and tens of thousands of movies, most entries of the ratings matrix are missing. The task is to predict the remaining entries in order to make good recommendations to customers on what to watch next, namely to complete the matrix. Similar problems appear in ratings and recommendations for books, music and CDs, which are the basis for *collaborative filtering* in the *recommender systems*. Another example of *matrix completion* is the word-document matrix: The frequencies of words used in a collection of documents can be represented as a matrix and the task is to classify the documents.

Let $\mathbf{\Theta}$ be the $n_1 \times n_2$ true preference matrix that we would like to have: the preference score of the $i^{th}$ customer on the $j^{th}$ item. Instead, we observe only $\mathbf{X}$ on a small subset $\mathbf{\Omega}$ of $\{1, \cdots, n_1\} \times \{1, \cdots, n_2\}$, which can possibly further be corrupted with noise (or uncertainties in ratings) so that the observed data are

$$X_{ij} = \theta_{ij} + \varepsilon_{ij}, \qquad \text{for } (i, j) \in \mathbf{\Omega}. \tag{11.42}$$

The problem as stated is under-determined even when there are no noises. What makes the solution feasible is the low-rank assumption of $\mathbf{\Theta}$. This assumption leads to the *factor model* interpretation. Take the movie rating as an example. Each movie has a number of features (factors) and different users have different loadings on these features plus some idiosyncratic components of personal tastes. This leads to the factor model (10.6) and low rank matrix $\mathbf{\Theta}$ in (11.42). The main difference here is that we only observe a small fraction of the data in the factor model.

A common assumption in the sampling is that $\mathbf{\Omega}$ is randomly chosen in some sense. It is often assumed that the $(i, j)$-item is observed with probability $p$, independently of other entries. In other words, data are missing at random. Let $\{I_{ij}\}$ be the i.i.d. Bernoulli random variables with probability of success $p$. We model the observed entries as

$$\mathbf{\Omega} = \{(i, j) : I_{ij} = 1\}. \tag{11.43}$$

A natural solution to problem (11.42) is the penalized least-squares: Find $\widehat{\mathbf{\Theta}}$ to minimize

$$\sum_{(i,j)\in\mathbf{\Omega}} (X_{ij} - \theta_{ij})^2 + \lambda\|\mathbf{\Theta}\|_*, \tag{11.44}$$

where the nuclear norm penalty is a relaxation of the low rank constraint. Candés and Recht (2009) show that in the noiseless situation, solving the relaxed problem yields the same solution as the nonconvex rank constrained optimization problem with high probability and is optimal. Chen, Chen, Fan, Ma, and Yan (2019) showed further such an optimality continues to hold for a noisy setting and Chen, Fan, Ma, and Yan (2019) refined further the optimality result and derived a method for constructing confidence intervals.

We now offer a quick solution from the spectral point of view. First of all, let us create a complete data matrix via the *inverse probability weighting*:

$$\mathbf{Y} = (X_{ij} I_{(i,j)\in\mathbf{\Omega}}/\widehat{p}) \tag{11.45}$$

where $\widehat{p}$ is the proportion of the observed entries. When there are sufficiently many entries, $\widehat{p}$ is an accurate estimate of $p$. Ignoring the error in estimating $p$, we then have

$$\mathrm{E}\,\mathbf{Y} = \mathbf{\Theta},$$

a low-rank matrix. This suggests a spectral method as follows. Let us assume that the rank of $\mathbf{\Theta}$ is known to be $K$. As in (9.2), let

$$\mathbf{Y} = \sum_{i=1}^{\min(n_1,n_2)} \lambda_i \mathbf{u}_i \mathbf{v}_i^T$$

be its singular value decomposition. Then, a spectral estimator is

$$\widehat{\mathbf{\Theta}} = \sum_{i=1}^{K} \lambda_i \mathbf{u}_i \mathbf{v}_i^T. \tag{11.46}$$

The maximum entry-wise estimation error of such a procedure was established by Abbe, Fan, Wang and Zhong (2019+). They show that the rate is optimal, as it can recover the Frobenius bound (Keshavan et al., 2010) up to a log factor.

### *11.4.4 Item ranking*

Item ranking is a classical subject, dated back at least to the 17th century. Humans have a better ability to express preferences between two items than among many items. Thus, one of important tasks is to rank the top $K$-items based on many pairwise comparisons. Such a ranking problem finds its applications in web searchs such as Page rank (Dwork, 2001), recommendation systems (Baltrunas, Makcinskas and Ricci (2010)), sports competitions (Masse, 1997)), among others.

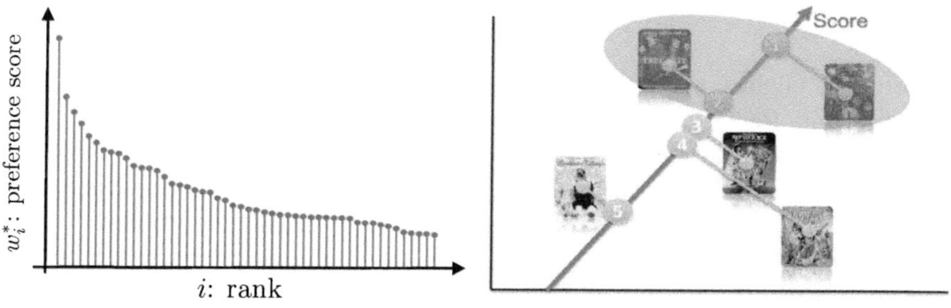


Figure 11.13: One way to rank top $K$ items is to first estimate the latent scores $\{w_i^*\}_{i=1}^n$ and then rank them (left panel). This yields the actual rank of items (right panel)

To accomplish the above complicated task, we need a statistical model. We assume that there is a positive latent score $w_i^*$ that is associated with item $i$ so that

$$P(\text{item } j \text{ beats item } i) = \frac{w_j^*}{w_i^* + w_j^*} \tag{11.47}$$

for any pairs $(i, j)$. Such a model is referred to as the Bradley-Terry-Luce model (Bradley and Terry, 1952; Luce, 1959). Thus, our task is to get an estimate of the latent score vector $\mathbf{w}^* = (w_1^*, \cdots, w_n^*)^T$ so that we can rank the items according to the preference scores. The idea is depicted in Figure 11.13. Note that $\mathbf{w}^*$ can only be identified up to a constant multiple so that we will assume that it is a probability vector. Hunter (2004) discussed an MM algorithm for fitting the Bradley-Terry-Luce model.

We do not assume that we have all pairwise comparisons available. In fact, there are only a small fraction of pairs that have ever competed. The pairwise comparisons form a graph $\mathcal{G}$, in which items are vertices and edges indicate whether the two items have been compared. See Figure 11.14. Let $\mathcal{E}$ be the edge set of the graph $\mathcal{G}$. For two items $(i, j) \in \mathcal{E}$, we have obtained $L_{ij}$

independent comparisons:

$$Y_{ij}^{(l)} \sim_{\text{ind.}} \begin{cases} 1, & \text{with prob. } \frac{w_j^*}{w_i^*+w_j^*} \\ 0, & \text{else} \end{cases} \qquad 1 \le l \le L_{ij}. \tag{11.48}$$

Let $\widehat{p}_{ij}$ be the proportion that the $j^{th}$ item beats the $i^{th}$ one. For theoretical studies, we may assume that $\mathcal{G}$ is a realization of the Erdös-Rényi graph.

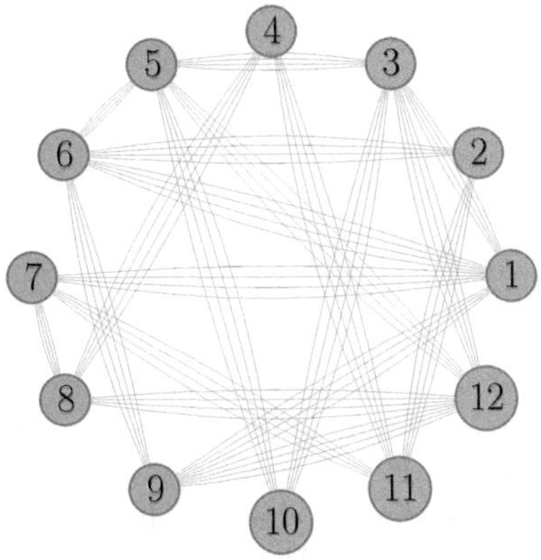


Figure 11.14: Comparison graph for top $K$ ranking. Each pair in the graph has matched $L_{ij}$ times, with proportion $\widehat{p}_{ij}$ that the $j^{th}$ item beats the $i^{th}$ item.

A simple and naive ranking is to compute the overall proportion $\widehat{p}_i$ of winning that item $i$ has ever competed. Then, rank the items according to $\{\widehat{p}_i\}_{i=1}^n$. This method ignores the ranks of the competitors item $i$ compared with. With a probabilistic model, a natural method is the maximum likelihood method. Conditioning on $\mathcal{G}$, the likelihood of the observed data is the product of those of Bernoulli's:

$$L(\mathbf{w}) = \prod_{(i,j)\in\mathcal{E}} \left(\frac{w_j^*}{w_i^* + w_j^*}\right)^{L_{ij}\widehat{p}_{ij}} \left(\frac{w_i^*}{w_i^* + w_j^*}\right)^{L_{ij}(1-\widehat{p}_{ij})}.$$

The log-likelihood is given by

$$\ell(\mathbf{w}) = \sum_{(i,j)\in\mathcal{E}} L_{ij}\widehat{p}_{ij} \log\left(\frac{w_j}{w_i + w_j}\right) + L_{ij}(1-\widehat{p}_{ij}) \log\left(\frac{w_i}{w_i + w_j}\right). \tag{11.49}$$

Parametrizing $\theta_i = \log w_i$ yields a logistic type of regression and the log-likelihood function is convex. To stabilize the high-dimensional likelihood, one can also regularize the likelihood function by using

$$-\ell(\exp(\boldsymbol{\theta})) + \lambda \sum_{i=1}^n \theta_i^2,$$

which is given by

$$\sum_{(i,j)\in\mathcal{E}} \{L_{ij}\widehat{p}_{ij}\theta_j + L_{ij}(1-\widehat{p}_{ij})\theta_i - L_{ij}\log(\exp(\theta_i)+\exp(\theta_j))\} + \lambda\sum_{i=1}^n \theta_i^2. \quad (11.50)$$

Now, let us consider the spectral method, which is also referred to as *rank centrality*. Without loss of generality, we assume that $\mathbf{w}^*$ has been normalized to a probability vector. The idea is to create a *Markov chain* on the graph $\mathcal{G}$ such that $\mathbf{w}^*$ is its *invariant distribution*, also called *stationary distribution*. Let us define the transition matrix $\mathbf{P}^*$ as

$$p^*_{i,j} = \begin{cases} \frac{1}{d}\frac{w_j^*}{w_i^*+w_j^*}, & \text{if } (i,j)\in\mathcal{E}, \\ 1-\frac{1}{d}\sum_{k:(i,k)\in\mathcal{E}}\frac{w_k^*}{w_i^*+w_k^*}, & \text{if } i=j, \\ 0, & \text{otherwise,} \end{cases} \quad (11.51)$$

where $\mathcal{E}$ is the set of edges of the graph $\mathcal{G}$ and $d$ is a parameter. When $d$ is sufficiently large (e.g. the maximum of the degrees) so that the diagonal entries are non-negative, the matrix $\mathbf{P}^*$ is a transition probability matrix: each row has a sum equals to 1. This Markov chain prefers to walk to the state with a higher score. It is easy to verify that $\mathbf{P}$ is a transition matrix with the detailed balance

$$w_i^* P^*_{i,j} = w_j^* P^*_{j,i}, \qquad \forall\,(i,j).$$

Therefore,

$$\sum_{i=1}^n w_i^* P^*_{i,j} = w_j^* \sum_{i=1}^n P^*_{j,i} = w_j^*$$

or put in the matrix notation

$$\mathbf{w}^{*T}\mathbf{P}^* = \mathbf{w}^{*T}. \quad (11.52)$$

In other words, $\mathbf{w}^*$ is the leading left eigenvector of $\mathbf{P}^*$ with eigenvalue 1. It is also the stationary distribution of the Markov chain with transition matrix $\mathbf{P}^*$.

The spectral method replaces the unknown transition matrix by its empirical counterpart $\widehat{\mathbf{P}}$, whose $(i,j)$ element is given by

$$\widehat{P}_{i,j} = \begin{cases} \frac{1}{d}\widehat{p}_{i,j}, & \text{if } (i,j)\in\mathcal{E}, \\ 1-\frac{1}{d}\sum_{k:(i,k)\in\mathcal{E}}\widehat{p}_{i,k}, & \text{if } i=j, \\ 0, & \text{otherwise,} \end{cases} \quad (11.53)$$

If $d > d_{\max}$, the maximum degree of the graph, then $\widehat{\mathbf{P}}$ is a transition matrix (Exercise 11.13). In practice, we take $d = 2d_{\max}$.

Negahban, Oh and Shah (2017) establish the $L_2$-rate of convergence for the spectral estimator $\widehat{\mathbf{w}}$. Chen, Fan, Ma, and Wang (2019) improve their

**Algorithm 11.2** Spectral method for top $K$-ranking (Rank Centrality)

1. Input the comparison graph $\mathcal{G}$, sufficient statistics $\{\widehat{p}_{ij}, (i,j) \in \mathcal{E}\}$, and the normalization factor $d$.
2. Define the probability transition matrix $\widehat{\mathbf{P}} = [\widehat{P}_{i,j}]_{1\le i,j\le n}$ as in (11.53).
3. Compute the leading left eigenvector $\widehat{\mathbf{w}}$ of $\widehat{\mathbf{P}}$.
4. Output the $K$ items that correspond to the $K$ largest entries of $\widehat{\mathbf{w}}$.

results and establish further the maximum entrywise estimation error. As a result, they were able to obtain the selection consistency for both the spectral method and the regularized MLE. They show that the sampling complexity achieves the information low bound given by Chen and Suh (2015). Additional references can be found in the papers by Negahban, Oh and Shah (2017) and Chen, Fan, Ma, and Wang (2019). In particular, the latter give a comprehensive review on the recent development on this topic (see Section 2.7 there).

### *11.4.5 Gaussian mixture models*

As an illustration of the applications of spectral methods to nonconvex optimization, let us consider the Gaussian mixture model:

$$\mathbf{X} \sim w_1 N(\boldsymbol{\mu}_1, \sigma_1^2\mathbf{I}_p) + \cdots + w_K N(\boldsymbol{\mu}_K, \sigma_K^2\mathbf{I}_p) \tag{11.54}$$

with unknown parameters $\{w_k, \boldsymbol{\mu}_k, \sigma_k^2\}_{k=1}^K$. The likelihood function based on a random sample $\{\mathbf{X}_i\}_{i=1}^n$ from model (11.54) can easily be found (Exercise 11.14), but the function is non-convex, which imposes computational challenges to find the MLE.

PCA allows us to find a good initial value for the nonconvex optimization problem. Indeed, the classical result of Bickel (1975) shows that for finite dimensional parametric problems, if we start from a root-n consistent estimator, the one-step Newton-Raphson estimator will be statistically efficient. In other words, *optimization errors* (the distance to a global optimum) will be of smaller order than *statistical errors* (standard deviations of estimators). Further iterations will only improve the optimization errors, as is more precisely stated in Robison (1988).

The method of moments is very simple in finding root-n consistent estimators. For the mixture model (11.54), the first two moments (Exercise 11.15) are given by

$$\mathrm{E}\,\mathbf{X} = \sum_{i=1}^K w_k \boldsymbol{\mu}_k, \qquad \mathrm{E}[\mathbf{X}\otimes\mathbf{X}] = \sum_{i=1}^K w_k \boldsymbol{\mu}_k \otimes \boldsymbol{\mu}_k + \sigma_w^2 \mathbf{I}_p \tag{11.55}$$

where $\otimes$ denote the Kronecker product and $\sigma_w^2 = \sum_{i=1}^K w_k \sigma_k^2$, the weighted average variances. Therefore, the column space spanned by $\{\boldsymbol{\mu}_k\}_{k=1}^p$ is the same as the eigen-space spanned by the top $K$ principal components, and $\sigma_w^2$ is the minimum eigenvalue of $\mathrm{E}[\mathbf{X} \otimes \mathbf{X}]$.

To further determine rotation, Hsu and Kakade (2013) introduce the third order *tensor* method. Suppose that $\{\boldsymbol{\mu}_k\}_{k=1}^K$ are linearly independent and $w_k > 0$ for all $k \in [K]$ so that the problem does not degenerate. Then $\sigma_w^2$ is the minimum eigenvalue of $\boldsymbol{\Sigma} = \mathrm{cov}(\mathbf{X})$; this will be shown at the end of this section. Let $\mathbf{v}$ be any eigenvector of $\boldsymbol{\Sigma}$ that is associated with the eigenvalue $\sigma_w^2$ as shown at the end of this section. Define the following quantities:

$$\mathbf{M}_1 = \mathrm{E}[(\mathbf{v}^T(\mathbf{X} - E\mathbf{X}))^2\mathbf{X}] \in R^p,$$
$$\mathbf{M}_2 = \mathrm{E}[\mathbf{X} \otimes \mathbf{X}] - \sigma_w^2 \cdot \mathbf{I}_p \in R^{p\times p},$$
$$\mathbf{M}_3 = \mathrm{E}[\mathbf{X} \otimes \mathbf{X} \otimes \mathbf{X}] - \sum_{j=1}^p \sum_{\mathrm{cyc}} \mathbf{M}_1 \otimes \mathbf{e}_j \otimes \mathbf{e}_j \in R^{p\times p\times p}.$$

where we use the notation

$$\sum_{\mathrm{cyc}} \mathbf{a} \otimes \mathbf{b} \otimes \mathbf{c} := \mathbf{a} \otimes \mathbf{b} \otimes \mathbf{c} + \mathbf{b} \otimes \mathbf{c} \otimes \mathbf{a} + \mathbf{c} \otimes \mathbf{a} \otimes \mathbf{b}$$

and $\mathbf{e}_j$ is the unit vector with $j^{th}$ position 1. Then Hsu and Kakade (2013) show that

$$\mathbf{M}_1 = \sum_{k=1}^K w_k \sigma_k^2 \boldsymbol{\mu}_k, \quad \mathbf{M}_2 = \sum_{k=1}^K w_k \boldsymbol{\mu}_k \otimes \boldsymbol{\mu}_k,$$
$$\mathbf{M}_3 = \sum_{k=1}^K w_k \boldsymbol{\mu}_k \otimes \boldsymbol{\mu}_k \otimes \boldsymbol{\mu}_k. \tag{11.56}$$

With $\{\mathbf{M}_i\}_{i=1}^3$ replaced by their empirical versions, the remaining task is to solve for all the parameters of interest via (11.56). Hsu and Kakade (2013) and Anandkumar, Ge, Hsu, Kakade and Telgarsky (2014) propose a fast method called the *robust tensor power* method to compute the estimators. The idea is to orthogonalize $\{\boldsymbol{\mu}_k\}$ in $\mathbf{M}_3$ by using the matrix $\mathbf{M}_2$ so that the power method can be used in computing the tensor decomposition.

Let $\mathbf{M}_2 = \mathbf{U}\mathbf{D}\mathbf{U}^T$ be the spectral decomposition of $\mathbf{M}_2$, where $\mathbf{D}$ is the $K \times K$ non-vanishing eigenvalues of $\mathbf{M}_2$. Set

$$\mathbf{W} = \mathbf{U}\mathbf{D}^{-1/2} \in R^{p\times K} \quad \text{and} \quad \widetilde{\boldsymbol{\mu}}_k = \sqrt{w_k}\mathbf{W}^T\boldsymbol{\mu}_k \in R^K. \tag{11.57}$$

Note that $\mathbf{W}$ is really a generalized inverse of $\mathbf{M}_2^{1/2}$. Then, $\mathbf{W}^T\mathbf{M}_2\mathbf{W} = \mathbf{I}_K$, which implies by (11.56) that

$$\mathbf{W}^T\mathbf{M}_2\mathbf{W} = \sum_{i=1}^K w_k \mathbf{W}^T \boldsymbol{\mu}_k \boldsymbol{\mu}_k^T \mathbf{W} = \sum_{i=1}^K \widetilde{\boldsymbol{\mu}}_k \widetilde{\boldsymbol{\mu}}_k^T = \mathbf{I}_K.$$

Thus, $\{\widetilde{\boldsymbol{\mu}}_k\}_{k=1}^K$ are orthonormal and

$$\boldsymbol{\mu}_k = \mathbf{U}\mathbf{D}^{1/2}\widetilde{\boldsymbol{\mu}}_k / \sqrt{w_k}. \tag{11.58}$$

Note that the quadratic form $\mathbf{W}^T\mathbf{M}_2\mathbf{W}$ can be denoted as $\mathbf{M}_2(\mathbf{W},\mathbf{W})$, which is defined as

$$\mathbf{M}_2(\mathbf{W},\mathbf{W}) = \sum_{k=1}^{K} w_k(\mathbf{W}^T\boldsymbol{\mu}_k)^{\otimes 2}$$

where $\mathbf{a}^{\otimes 2} = \mathbf{a}\otimes\mathbf{a}$. We can define similarly

$$\begin{aligned}\widetilde{\mathbf{M}}_3 &:= \mathbf{M}_3(\mathbf{W},\mathbf{W},\mathbf{W}) := \sum_{k=1}^{K} w_k(\mathbf{W}^T\boldsymbol{\mu}_k)^{\otimes 3} \\ &= \sum_{k=1}^{K}\frac{1}{\sqrt{\omega_k}}\widetilde{\boldsymbol{\mu}}_k^{\otimes 3} \in I\!R^{K\times K\times K}, \qquad (11.59)\end{aligned}$$

where $\mathbf{a}^{\otimes 3} = \mathbf{a}\otimes\mathbf{a}\otimes\mathbf{a}$. Therefore, $\widetilde{\mathbf{M}}_3$ admits an orthogonal tensor decomposition that is similar to the spectral decomposition for a symmetric matrix and the decomposition can be rapidly computed by the power method. It then can be verified that

$$\widetilde{\mathbf{M}}_3 = \mathrm{E}[(\mathbf{W}^T\mathbf{X})^{\otimes 3}] - \sum_{j=1}^{p}\sum_{\mathrm{cyc}}(\mathbf{W}^T\mathbf{M}_1)\otimes(\mathbf{W}^T\mathbf{e}_j)\otimes(\mathbf{W}^\top\mathbf{e}_j). \qquad (11.60)$$

which is merely a rotated version of (11.56).

The idea of estimation is now very clear. First of all, using the method of moments, we get an estimator of $\widetilde{\mathbf{M}}_3$ by (11.60). Using the tensor power method (Anandkumar et al., 2014), we find estimators for the orthogonal tensor decomposition $\{\widetilde{\boldsymbol{\mu}}_k\}$ and its associated eigenvalue $\{1/\sqrt{w_k}\}_{1=1}^{K}$ in (11.59). The good piece of the news is that we operate in the $K$-dimensional rather than the original $p$-dimensional space. We omit the details. Using (11.57) and (11.58), we can get an estimator for $\boldsymbol{\mu}_k$. We summarize the idea in Algorithm 11.3.

Note that in Algorithm 11.3, the sample covariance can be replaced by the empirical second moment matrix $n^{-1}\sum_{i=1}^{n}\mathbf{X}_i\mathbf{X}_i^T$. To see this, by (11.55) and (11.56),

$$\mathrm{E}\,\mathbf{X}\otimes\mathbf{X} = \mathbf{M}_2 + \sigma_w^2\mathbf{I}_p = \mathbf{U}\mathbf{D}\mathbf{U}^T + \sigma_w^2\mathbf{I}_p$$

Thus, the top $K$ eigenvalues are $\{d_1+\sigma_w^2,\cdots,d_K+\sigma_w^2\}$ and the remaining $(p-K)$ eigenvalues are $\sigma_w^2$. The top $K$ eigenvectors are in the linear span of $\{\boldsymbol{\mu}_k\}_{k=1}^{K}$. Let $\mathbf{v}$ be an eigenvector that corresponds to the minimum eigenvalue of $\mathrm{E}[\mathbf{X}\otimes\mathbf{X}]$. Then, $\boldsymbol{\mu}_k^T\mathbf{v} = 0$ for all $k\in[K]$. Using the population covariance

$$\boldsymbol{\Sigma} = \mathrm{E}[\mathbf{X}\otimes\mathbf{X}] - (\mathrm{E}\,\mathbf{X})\otimes(\mathrm{E}\,\mathbf{X}),$$

we have

$$\boldsymbol{\Sigma}\mathbf{v} = \sigma_w^2\mathbf{v} - (\mathrm{E}\,\mathbf{X})(\mathrm{E}\,\mathbf{X})^T\mathbf{v} = \sigma_w^2\mathbf{v}.$$

**Algorithm 11.3** The *tensor power method* for estimating parameters in the normal mixtures

1. Calculate the sample covariance matrix, its minimum eigenvalue $\widehat{\sigma}_w^2$ and its associated eigenvector $\widehat{\mathbf{v}}$.
2. Derive the estimators $\widehat{\mathbf{M}}_1$ and $\widehat{\mathbf{M}}_2$ by using the empirical moments of $\mathbf{X}$, $\widehat{\mathbf{v}}$ and $\widehat{\sigma}_w^2$.
3. Calculate the spectral decomposition $\widehat{\mathbf{M}}_2 = \widehat{\mathbf{U}}\widehat{\mathbf{D}}\widehat{\mathbf{U}}^T$. Let $\widehat{\mathbf{W}} = \widehat{\mathbf{U}}\widehat{\mathbf{D}}^{-1/2}$. Construct an estimator of $\widetilde{\mathbf{M}}_3$, denoted by $\widehat{\mathbf{M}}_3$, based on (11.60) by substituting the empirical moments of $\widehat{\mathbf{W}}^T\mathbf{X}$, $\widehat{\mathbf{W}}$ and $\widehat{\mathbf{M}}_1$. Apply the robust tensor power method in Anandkumar et al. (2014) to $\widehat{\mathbf{M}}_3$ and obtain $\{\widehat{\widetilde{\boldsymbol{\mu}}}_k\}_{k=1}^K$ and $\{\widehat{w}_k\}_{k=1}^K$.
4. Compute $\widehat{\boldsymbol{\mu}}_k = \widehat{\mathbf{U}}\widehat{\mathbf{D}}^{1/2}\widehat{\widetilde{\boldsymbol{\mu}}}_k/\sqrt{\widehat{w}_k}$. Solve the linear equations $\widehat{\mathbf{M}}_1 = \sum_{k=1}^{K} \widehat{w}_k\widehat{\sigma}_k^2\widehat{\boldsymbol{\mu}}_k$ for $\{\widehat{\sigma}_k^2\}_{k=1}^K$.

Therefore, $\sigma_w^2$ is an eigenvalue and $\mathbf{v}$ is an eigenvector $\boldsymbol{\Sigma}$. We can show further $\sigma_w^2$ is indeed the minimum eigenvalue.

The above gives us an idea of how the spectral methods can be used to obtain an initial estimator. Anandkumar *et al.*(2014) used tensor methods to solve a number of statistical machine learning problems. In addition to the aforementioned mixtures of spherical Gaussians with heterogeneous variances, they include *hidden Markov models* and *latent Dirichlet allocation*. Sedghi, Janzamin and Anandkumar (2016) apply tensor methods to learning mixtures of generalized linear models.

## 11.5 Bibliographical Notes

There is a huge literature on controlling false discoveries for independent test statistics. After the seminal work of Benjamini and Hochberg (1995) on controlling false discovery rates (FDR), the field of large-scale multiple testing received a lot of attention. Important work in this area includes Storey (2002); Genovese and Wasserman (2004), Lehmann and Romano(2005), Lehmann, Romano and Shaffer (2005), among others. The estimation of the proportion of nulls has been studied by, for example, Storey (2002), Langaas and Lindqvist (2005), Meinshausen and Rice (2006), Jin and Cai (2007) and Jin (2008), among others.

Although Benjamini and Yekutieli (2001), Sarkar (2002), Storey, Taylor and Siegmund (2004), Clarke and Hall (2009), Blanchard and Roquain(2009) and Liu and Shao (2014) showed that the Benjamini and Hochberg and its related procedures continue to provide valid control of FDR under positive regression dependence on subsets or weak dependence, they will still suffer from efficiency loss without considering the actual dependence information.

Efron (2007, 2010) pioneered the work in the field and noted that correlation must be accounted for in deciding which null hypotheses are significant because the accuracy of FDR techniques will be compromised in high correlation situations. Leek and Storey (2008), Friguet, Kloareg and Causeur (2009), Fan, Han and Gu(2012), Desai and Storey (2012) and Fan and Han (2017) proposed different methods to control FDR or FDP under factor dependence models. Other related studies on controlling FDR under dependence include Owen (2005), Sun and Cai (2009), Schwartzman and Lin (2011), Wang, Zhao, Hastie and Owen (2017), among others.

There is a surge of interest in community detection in statistical machine learning literature. We only brief mention some of these. Since the introduction of stochastic block models by Holland, Laskey and Leinhardt (1983), there have been various extensions of this model. See Abbe (2018) for an overview. In addition to the references in Section 11.4.1, Bickel and Chen (2009) gave a nonparametric view of network models. Other related work on nonparametric estimation of graphons includes Wolfe and Olhede (2013), Olhede and Wolfe (2014), Gao, Lu and Zhou (2015). Bickel, Chen and Levina (2011) employed the method of moments. Amini *et al.*(2013) studied pseudo-likelihood methods for network models. Rohe, Chatterjee and Yu (2011) and Lei and Rinaldo (2015) investigated the properties of spectral clustering for the stochastic block model. Zhao, Levina and Zhu (2012) established consistency of community detection under degree-corrected stochastic block models. Jin (2015) proposed fast community detection by score and Jin, Ke and Luo (2017) proposed a simplex hunting approach. Abbe, Bandeira and Hall (2016) investigated the exact recovery problem. Inferences on network models have been studied by Bickel and Sarkar (2016) and Lei (2016), Fan, Fan, Han and Lv(2019a,b).

There has been a surge of interest in the matrix completion problem over the last decade. Candés and Recht (2009) and Recht (2011) studied the exact recovery via convex optimization. Candés and Plan (2010) investigated matrix completion with noisy data. Candés and Tao (2010) studied the optimality of the matrix completion. Cai, Candés and Shen (2010) proposed a singular value thresholding algorithm for matrix completion. Keshavan, Montanari and Oh(2010) investigated the sampling complexity of matrix completion. Eriksson, Balzano and Nowak (2011) studied high-rank matrix completion. Negahban and Wainwright (2011, 2012) derived the statistical error of the penalized M-estimator and showed that it is optimal up to logarithmic factors. Fan, Wang and Zhu (2016) proposed a robust procedure for low-rank matrix recovery, including the matrix completion as a specific example. Cai, Cai and Zhang (2016) and Fan and Kim (2019) applied structured matrix completion to genomic data integration and volatility matrix estimation for high-frequency financial data.

## 11.6 Exercises

11.1 Suppose that we have linear model $Y = \mathbf{X}^T\boldsymbol{\beta}^* + \varepsilon$ with $\mathbf{X}$ following model (11.1). Here, we use $\boldsymbol{\beta}^*$ to indicate the true parameter of the model. Consider the least squares solution to the problem (11.3) without constraints:

$$\min_{\alpha,\gamma,\boldsymbol{\beta}} \mathrm{E}(Y - \alpha - \mathbf{u}^T\boldsymbol{\beta} - \gamma^T\mathbf{f})^2.$$

Let $\alpha_0$, $\boldsymbol{\beta}_0$, and $\gamma_0$ be the solution. Show that under condition (11.4), the solution to the above least-squares problem is obtained at

$$\alpha_0 = \mathbf{a}^T\boldsymbol{\beta}^*, \quad \boldsymbol{\beta}_0 = \boldsymbol{\beta}^*, \quad \text{and} \quad \gamma = \mathbf{B}^T\boldsymbol{\beta}^*.$$

11.2 Suppose that the population parameter $\boldsymbol{\beta}^*$ is the unique solution to $\mathrm{E}\,\nabla L(Y, \mathbf{X}^T\boldsymbol{\beta})\mathbf{X} = 0$ where $\nabla L(s,t) = \frac{\partial L(s,t)}{\partial t}$. Let $\mathbf{W} = (1, \mathbf{u}^T, \mathbf{f}^T)^T$ and assume that $\mathrm{E}\,\nabla L(Y, \mathbf{W}^T\boldsymbol{\theta})\mathbf{W} = 0$ has a unique solution. Then, under condition (11.9), the unconstrained minimization to the population problem $\mathrm{E}\,L(Y, \alpha + \mathbf{u}^T\boldsymbol{\beta} + \mathbf{f}^T\gamma)$ has a unique solution (11.8).

11.3 Let us consider the macroeconomic time series in the Federal Reserve Bank of St. Louis

http://research.stlouisfed.org/econ/mccracken/sel/

from January 1963 to December of the last year. Let us take the unemployment rate as the $Y$ variable and the remaining as $\mathbf{X}$ variables. Use `FarmSelect` and rolling windows of the past 120, 180, and 240 months to predict the monthly change of unemployment rates. Compute the out-of-sample $R^2$ and produce a table that is similar to Table 11.1. Also report the top 10 most frequently selected variables by FarmSelect in the rolling windows. Note that several variables such as "Industrial Production Index", "Real Personal Consumption", "Real M2 Money Stock", "Consumer Price Index", "S&P 500 index", etc, are non-stationary and they are better replaced by computing log-returns: namely the difference of the logarithm of time series.

11.4 Let us dichotomize the monthly changes of unemployment rate as 0 and 1, depending upon whether the changes are non-negative or positive. Run the similar analysis, using FarmSelect for logistic regression, as the previous exercise and report the percent of correct prediction using rolling windows of the past 120, 180, and 240 months. Report also the top 10 variables selected most frequently by FarmSelect in the rolling windows.

11.5 As a generalization of Example 11.1, we consider the one-factor model

$$X_{ij} = \mu_j + b_j f_i + a_j u_{ij}, \qquad i = 1, \cdots, n, j = 1, \cdots, p.$$

Consider again the sample average test statistics $Z_j = \sqrt{n}\bar{X}_{\cdot j}$ for the testing problem (11.11). Show that under some mild conditions,

$$p_0^{-1}V(z) = p_0^{-1}\sum_{j\in\mathcal{S}_0}[\Phi(a_j(z+b_jW)) + \Phi(a_j(z-b_jW))] + o_p(1).$$

11.6 Analyze the neuroblastoma data again and reproduce the results in Section 11.2.5 using `FarmTest` and other functions in the R software package. Produce in particular

(a) the scree plots and the number of factors selected in both positive and negative groups;

(b) the correlation coefficient matrix (similar plot to Figure 11.9) for the last 100 genes before and after factor adjustments;

(c) the top 30 significant genes with and without factor adjustments.

11.7 Prove (11.29) and (11.30).

11.8 Use the data in Exercise 11.3 and run principal component regression and augmented principal component regression to forecast the changes of unemployment rates using a rolling windows of 120 months, 180 months, and 240 months, and report the out-of-sample $R^2$ results similar to the PCA block in Table 11.3.

11.9 Let us verify (11.36) using a specific example with three communities.

(a) Assume that communities 1, 2 and 3 consist of nodes $\{1,\cdots,m\}$, $\{m+1,\cdots,2m\}$, and $\{2m+1,\cdots,3m\}$. Verify (11.36)

(b) Show that the column space of $\mathbf{\Gamma}$ is the eigen-space spanned by the top 3 eigen-vectors of the matrix $\mathrm{E}\,\mathbf{A}$, if $\mathbf{P}$ is not degenerate.

(c) Verify the eigenvalues and eigenvectors given in Example 11.3

11.10 For the Erdös-Rényi graph $G(n,p)$,

(a) What is the probability that a node is isolated?

(b) What is the limit of the probability as $n\to\infty$ if $p=c/n$.

(c) Show that the graph has no isolated node with probability tending to one if $p=c(\log n)/n$ for $c>1$.

11.11 Simulate a data set from the model given in Exercise 9 with $n=150$ ($m=50$), diagonal elements of $\mathbf{P}$ matrix $p=5(\log n)/n$, and off-diagonal elements $q=p/10$. Give the resulting graph, the scree plot of absolute eigenvalues of the adjacency matrix, and the results of spectral analysis using R function `kmeans`.

11.12 Following the notation in Section 11.4.3, let $Z_{ij} = X_{ij}I_{(i,j)\in\Omega}/p$. Show that $\mathrm{E}\,Z_{ij} = \mathrm{E}\,X_{ij} = \theta_{ij}$ under the Bernoulli sampling scheme.

11.13 Show that the matrix (11.53) is a transition matrix if $d > d_{\max}$.

11.14 Suppose that we have a random sample $\{\mathbf{X}_i\}_{i=1}^n$ from the Gaussian mixture model (11.54).

(a) Write down the likelihood function.

(b) Simulate a random sample of size $n = 1000$ from the model (11.54) with $p = 1$, $K = 3$, $\sigma_k = 1$ and $w_k = 1/3$ for all $k$, and $\mu_1 = -1$, $\mu_2 = 1$ and $\mu_3 = 10$. Plot the resulting likelihood function as a function of $(\mu_1, \mu_2)$ and a function of $(\mu_2, \mu_3)$. Are they convex?

11.15 Prove (11.55).

# Chapter 12

# Supervised Learning

*Supervised learning*, also referred to as *classification*, is one of the fundamental problems in statistical machine learning. Suppose that we have a random sample of size $n$ from some unknown distribution of $(\mathbf{X}, Y)$, where $Y$ denotes the class label and $\mathbf{X}$ represents a $p$-dimensional predictor. Let $\mathcal{C}$ be the set of all possible class labels. A classification rule $\delta$ is a mapping from $\mathbf{X}$ to $\mathcal{C}$ such that a label $\delta(\mathbf{X})$ is assigned to the data point $\mathbf{X}$. Throughout this chapter we will use the standard 0-1 loss to measure the classification performance. Then the misclassification error of $\delta$ is $R(\delta) = \Pr(Y \neq \delta(\mathbf{X}))$. It can be fairly easy to show that the smallest *classification error* is achieved by the *Bayes rule* defined as $\text{argmax}_{c_i \in \mathcal{C}} \Pr(Y = c_i|\mathbf{X})$. The misclassification error rate of the Bayes rule is called the *Bayes error*. In theory, the learning goal is to approximate the Bayes rule as closely as possible, given a finite amount of data.

There is a huge amount of work on classification algorithms in the literature. We have to be selective in order to cover this big topic in one chapter. We first describe some popular classical classifiers including linear discriminant analysis, logistic regression, kernel density classifier and nearest neighbor classifier. We then introduce more modern classifiers such as random forests, boosting and the support vector machine. Last, we discuss various sparse classifiers with high-dimensional features. *Deep learning* classifiers will be introduced in Chapter 14.

## 12.1 Model-based Classifiers

### *12.1.1 Linear and quadratic discriminant analysis*

*Linear discriminant analysis* (*LDA*) and *quadratic discriminant analysis* (*QDA*) are probably the oldest model-based methods for classification. Nevertheless, they are still being routinely used in real applications. In fact, LDA and QDA can be as accurate as much fancier classifiers tested on many benchmark data sets (Michie, Spiegelhalter and Taylor, 1994; Hand, 2006).

LDA and QDA models start with the assumption that the conditional distribution of the predictor given the class label is a multivariate normal distribution. Define $\pi_k = \Pr(Y = c_k)$, $f_k(\mathbf{x})$ as the density function of $\mathbf{X}$ conditioned on $y = c_k$ and $f(\mathbf{x})$ as the marginal density function of $\mathbf{X}$. Assume that $(\mathbf{X}|y = c_k) \sim N(\boldsymbol{\mu}_k, \boldsymbol{\Sigma}_k)$, namely,

$$f_k(\mathbf{x}) = \frac{1}{\sqrt{2\pi|\boldsymbol{\Sigma}_k|}} \exp\left(-\frac{1}{2}(\mathbf{x} - \boldsymbol{\mu}_k)^T \boldsymbol{\Sigma}_k^{-1}(\mathbf{x} - \boldsymbol{\mu}_k)\right). \qquad (12.1)$$

With the above modelling assumption, the Bayes rule can be easily derived by using the Bayes formula:

$$\Pr(Y = c_k|\mathbf{X} = \mathbf{x}) = \frac{\pi_k f_k(\mathbf{x})}{f(\mathbf{x})},$$

where $f(\mathbf{x}) = \sum_k \pi_k f_k(\mathbf{x})$. Under (12.1), the Bayes rule can be written as

$$\arg\max_k \delta_k^{\text{qda}}(\mathbf{x})$$

where

$$\delta_k^{\text{qda}}(\mathbf{x}) = \log \pi_k - \frac{1}{2}\log|\boldsymbol{\Sigma}_k| - \frac{1}{2}(\mathbf{x} - \boldsymbol{\mu}_k)^T \boldsymbol{\Sigma}_k^{-1}(\mathbf{x} - \boldsymbol{\mu}_k). \qquad (12.2)$$

We can see that the Bayes rule is a quadratic function of $\mathbf{x}$. This classification rule is very intuitive: except for some constants to reflect prior knowledge, the QDA classifies a point $\mathbf{x}$ according to its *Mahalanobis distance* to the centroids $\boldsymbol{\mu}_k$, defined by

$$(\mathbf{x} - \boldsymbol{\mu}_k)^T \boldsymbol{\Sigma}_k^{-1}(\mathbf{x} - \boldsymbol{\mu}_k).$$

When $\pi_k$ and $\boldsymbol{\Sigma}_k$ are independent of $k$, this is exactly the *nearest centroid* classification. Figure 12.1 gives an illustration when $\boldsymbol{\Sigma}_1 = \boldsymbol{\Sigma}_2$ for binary classification.

Given the parametric form of the Bayes rule, QDA substitutes the parameters in (12.2) with the following estimates:

$$\begin{aligned}
\widehat{\pi}_k &= \frac{n_k}{n}, \\
\widehat{\boldsymbol{\mu}}_k &= \frac{1}{n_k}\sum_{Y_i = c_k} \mathbf{X}_i, \\
\widehat{\boldsymbol{\Sigma}}_k &= \frac{1}{n_k - 1}\sum_{Y_i = c_k} (\mathbf{X}_i - \widehat{\boldsymbol{\mu}}_k)(\mathbf{X}_i - \widehat{\boldsymbol{\mu}}_k)^T,
\end{aligned}$$

where $n_k$ is the number of observations in class $c_k$. Thus the QDA rule is

$$\text{argmax}_k\{\log \widehat{\pi}_k - \frac{1}{2}\log|\widehat{\boldsymbol{\Sigma}}_k| - \frac{1}{2}(\mathbf{x} - \widehat{\boldsymbol{\mu}}_k)^T \widehat{\boldsymbol{\Sigma}}_k^{-1}(\mathbf{x} - \widehat{\boldsymbol{\mu}}_k)\}.$$

When data are not necessarily from normal distributions, Fan, Ke, Liu and Xia (2015) propose to use quadratic rule $Q(\mathbf{X}) = \mathbf{X}^T\boldsymbol{\Omega}\mathbf{X} - 2\boldsymbol{\delta}^T\mathbf{X}$ that

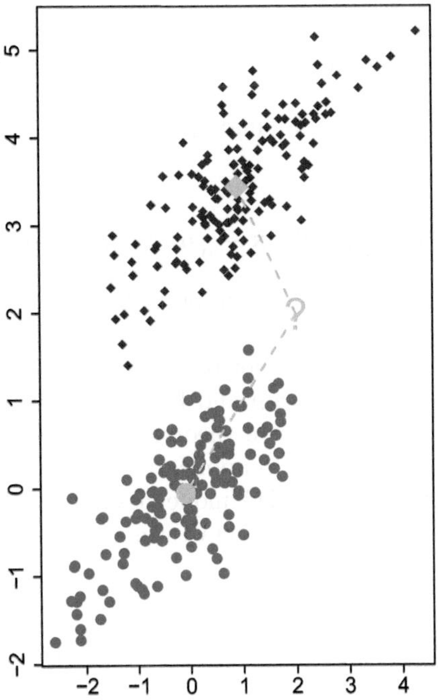


Figure 12.1: Quadratic classification basically classifies a point, indicated by a question mark, to a class with the closest distance to the centroid of the class.

optimizes the *Raleigh quotient* for binary classification: Find $\boldsymbol{\Omega}$ and $\boldsymbol{\delta}$ to maximize

$$R(\boldsymbol{\Omega}, \boldsymbol{\delta}) = \frac{\left\{\mathrm{E}[Q(\mathbf{X})|Y=0] - \mathrm{E}[Q(\mathbf{X})|Y=1]\right\}^2}{\pi \mathrm{Var}[Q(\mathbf{X})|Y=0] + (1-\pi)\mathrm{Var}[Q(\mathbf{X})|Y=1]},$$

where $\pi$ is the prior proportion for class 0. This makes two classes as far as possible using the quadratic classifier $Q(\mathbf{X})$, after normalizing by its variance. The variance $\mathrm{Var}[Q(\mathbf{X})]$ depends on all the fourth cross-moments of $\mathbf{X}$ and hence there are too many (of order $O(p^4)$) to estimate when $p$ is moderately large. To circumvent this issue, they assume the elliptical distribution of $\mathbf{X}$ and need only to estimate one additional kurtosis parameter on top of the first and second moments. The resulting procedure is called QUADRO (quadratic dimension reduction via Rayleigh optimization) that classifies a point $\mathbf{x}$ to class 0 if $Q(\mathbf{x}) < c$, in which $c$ is chosen to minimizes the misclassification error. See Fan, *et al.* (2015) for additional details.

LDA uses the additional homogeneous covariance assumption

$$\boldsymbol{\Sigma}_k = \boldsymbol{\Sigma} \quad \text{for all } c_k \in \mathcal{C}. \tag{12.3}$$

As a result, the quadratic term $\frac{1}{2}\mathbf{x}^T\widehat{\boldsymbol{\Sigma}}_k^{-1}\mathbf{x}$ and the $\log|\widehat{\boldsymbol{\Sigma}}_k|$ are independent of $k$. Hence, the Bayes rule has a simpler form:

$$\arg\max_k \delta_k^{\text{lda}}(\mathbf{x})$$

where

$$\delta_k^{\text{lda}}(\mathbf{x}) = \boldsymbol{\mu}_k^T\boldsymbol{\Sigma}^{-1}\mathbf{x} - \frac{1}{2}\boldsymbol{\mu}_k^T\boldsymbol{\Sigma}^{-1}\boldsymbol{\mu}_k + \log\pi_k. \tag{12.4}$$

Now the Bayes rule is a linear function of $\mathbf{x}$ and hence a linear discriminant. It is also referred to as *Fisher's discriminant analysis.* Note that Fisher's original derivation was geometric, not based on the probabilistic LDA model. See Chapter 4 of Hastie, Tibshirani and Friedman (2009) for the discussion. The LDA model is best viewed as a theoretical way to prove the optimality of Fisher's discriminant analysis. In applications, it is observed that LDA can perform quite well although the LDA model assumptions are clearly violated.

Let $K$ be the number of classes. LDA estimates $\boldsymbol{\Sigma}$ with the pooled sample covariance:

$$\widehat{\boldsymbol{\Sigma}} \quad = \quad \frac{1}{\sum_{k=1}^K (n_k - 1)} \sum_{k=1}^K (n_k - 1)\widehat{\boldsymbol{\Sigma}}_k.$$

The LDA rule is

$$\text{argmax}_k\{\widehat{\boldsymbol{\mu}}_k^T\widehat{\boldsymbol{\Sigma}}^{-1}\mathbf{x} - \frac{1}{2}\widehat{\boldsymbol{\mu}}_k^T\widehat{\boldsymbol{\Sigma}}^{-1}\widehat{\boldsymbol{\mu}}_k + \log\widehat{\pi}_k\}.$$

In particular, if $\mathcal{C} = \{1, 2\}$, then the Bayes rule classifies an observation to Class 2 if and only if

$$(\mathbf{x} - \widehat{\boldsymbol{\mu}}_a)^T\widehat{\boldsymbol{\Sigma}}^{-1}(\widehat{\boldsymbol{\mu}}_2 - \widehat{\boldsymbol{\mu}}_1) + \log\frac{\widehat{\pi}_2}{\widehat{\pi}_1} > 0. \tag{12.5}$$

where $\boldsymbol{\mu}_a = (\widehat{\boldsymbol{\mu}}_1 + \widehat{\boldsymbol{\mu}}_2)/2$.

When $p$ is moderately large, $\widehat{\boldsymbol{\Sigma}}$ can not be accurately estimated for moderate sample size, and the computation of $(\widehat{\boldsymbol{\Sigma}})^{-1}$ can be very unstable. If $p \geq n$, $\widehat{\boldsymbol{\Sigma}}$ is not even full rank. Friedman (1989) suggested *regularized discriminant analysis* (*RDA*) which uses the following shrinkage estimators:

$$\widehat{\boldsymbol{\Sigma}}^{\text{rda}}(\gamma) = \gamma\widehat{\boldsymbol{\Sigma}} + (1-\gamma)\frac{\text{tr}(\widehat{\boldsymbol{\Sigma}})}{p}\widehat{\mathbf{I}}, \quad 0 \leq \gamma \leq 1,$$

is shrunken toward an identity matrix (up to a scalar) as $\gamma \to 0$ and

$$\widehat{\boldsymbol{\Sigma}}_k^{\text{rda}}(\alpha, \gamma) = \alpha\widehat{\boldsymbol{\Sigma}}_k + (1-\alpha)\widehat{\boldsymbol{\Sigma}}^{\text{rda}}(\gamma), \tag{12.6}$$

which is shrunken toward a common covariance matrix as $\alpha \to 0$. In practice,

$(\alpha, \gamma)$ are chosen from the data by cross-validation. Chapter 9 gives a comprehensive account of robust high-dimensional covariance matrix estimation and regularization. They can be applied herewith.

Let us briefly discuss the misclassification rate, which is the *testing error* rate when the testing sample size is infinity, for linear classifiers. For simplicity, let us consider the binary classification and assume that random variables representing two classes $\mathcal{C}_1$ and $\mathcal{C}_2$ follow $p$-variate normal distributions $N(\boldsymbol{\mu}_1, \boldsymbol{\Sigma})$ and $N(\boldsymbol{\mu}_2, \boldsymbol{\Sigma})$, respectively. For a given coefficient vector $\mathbf{w}$, consider a linear classifier

$$\delta(\mathbf{X}) = I(\mathbf{w}^T(\mathbf{X} - \boldsymbol{\mu}_a) > 0) + 1, \qquad \boldsymbol{\mu}_a = (\boldsymbol{\mu}_1 + \boldsymbol{\mu}_2)/2 \tag{12.7}$$

that classifies $\mathbf{X}$ to class 2 if $\mathbf{w}^T(\mathbf{X} - \boldsymbol{\mu}_a) > 0$. Note that the classifier coefficient $\mathbf{w}$ is only identifiable up to a positive constant. It can easily be computed (see Exercise 12.2) that the misclassification rate is

$$R(\mathbf{w}) \equiv P(\delta(\mathbf{X}) \neq Y) = 1 - \Phi\Big(\frac{\mathbf{w}^T \boldsymbol{\mu}_d}{(\mathbf{w}^T \boldsymbol{\Sigma} \mathbf{w})^{1/2}}\Big), \qquad \boldsymbol{\mu}_d = (\boldsymbol{\mu}_2 - \boldsymbol{\mu}_1)/2. \tag{12.8}$$

Furthermore, the optimal coefficient $\mathbf{w}_{opt}$ that minimizes the misclassification rate is given by $\mathbf{w}_{opt} = \boldsymbol{\Sigma}^{-1}\boldsymbol{\mu}_d$, the Fisher classifier (see Exercise 12.2). Compare with (12.5).

### *12.1.2 Logistic regression*

We discuss the logistic regression model for binary classification. The multi-class version of logistic regression is often referred to as multinomial logistic regression (see Chapter 5). Let $\mathcal{C} = \{1, 0\}$. For notation convenience, write

$$p(\mathbf{x}) = \Pr(Y = 1|\mathbf{X} = \mathbf{x}), \quad \text{and} \quad 1 - p(\mathbf{x}) = \Pr(Y = 0|\mathbf{X} = \mathbf{x}).$$

The log-odds function $f(\mathbf{x})$ is defined as $\log(p(\mathbf{x})/(1 - p(\mathbf{x}))$. The linear logistic regression model assumes that

$$\log(p(\mathbf{x})/(1 - p(\mathbf{x})) = \beta_0 + \beta_1 x_1 + \cdots + \beta_p x_p = \beta_0 + \mathbf{x}^T\boldsymbol{\beta}, \tag{12.9}$$

or equivalently

$$p(\mathbf{x}) = \frac{\exp(\beta_0 + \beta_1 x_1 + \cdots + \beta_p x_p)}{1 + \exp(\beta_0 + \beta_1 x_1 + \cdots + \beta_p x_p)}.$$

The model parameter, $(\beta_0, \boldsymbol{\beta})$, is estimated by maximum likelihood

$$(\widehat{\beta}_0, \widehat{\boldsymbol{\beta}}) = \operatorname{argmax}_{\beta_0, \boldsymbol{\beta}} \ell(\beta_0, \boldsymbol{\beta}).$$

where

$$\ell(\beta_0, \boldsymbol{\beta}) = \sum_{i=1}^n Y_i(\beta_0 + \mathbf{X}_i^T\boldsymbol{\beta}) - \sum_{i=1}^n \log\left(1 + \exp(\beta_0 + \mathbf{X}_i^T\boldsymbol{\beta})\right). \tag{12.10}$$

The logistic regression classifier is $\widehat{y} = I(\widehat{\beta}_0 + \mathbf{x}^T\widehat{\boldsymbol{\beta}} > 0)$, which is equivalent to classifying $\mathbf{x}$ as class 1 if its estimated probability is over 0.5. Note that logistic regression is also applicable to the transformed features as in Section 2.5.

The linear logistic regression model is an important special case of GLIM discussed in Chapter 5. The general algorithm for fitting GLIM can be used to compute the linear logistic regression. For sake of completeness, we briefly mention the classical *iterative re-weighted least squares* (*IRLS*) algorithm. For notation convenience, let us expand the predictor matrix $\mathbf{X}^*$ by adding the constant vector of 1. Likewise, write $\boldsymbol{\beta}^* = (\beta_0, \boldsymbol{\beta})$. Then we have

$$\nabla\ell(\boldsymbol{\beta}^*) = \sum_{i=1}^{n} \mathbf{X}_i^*(Y_i - p(\mathbf{X}_i^*)),$$

$$\nabla^2\ell(\boldsymbol{\beta}^*) = -\sum_{i=1}^{n} \mathbf{X}_i^*(\mathbf{X}_i^*)^T p(\mathbf{X}_i^*)(1 - p(\mathbf{X}_i^*)).$$

The Newton-Raphson update is

$$\boldsymbol{\beta}^*(\text{new}) \leftarrow \boldsymbol{\beta}^* - [\nabla^2\ell(\boldsymbol{\beta}^*)]^{-1}\nabla\ell(\boldsymbol{\beta}^*) \mid_{\boldsymbol{\beta}^*} . \tag{12.11}$$

Equation (12.11) has a very nice connection to the weighted least squares. Define the "working response"

$$z_i = (\mathbf{X}_i^*)^T\boldsymbol{\beta}^* + \frac{Y_i - p(\mathbf{X}_i^*)}{w_i},$$

where $w_i = p(\mathbf{X}_i^*)(1 - p(\mathbf{X}_i^*))$ is called the "working weights". Then simple algebra shows that the updating formula in (12.11) is equivalent to

$$(\beta_0^{\text{new}}, \boldsymbol{\beta}^{\text{new}}) \leftarrow \arg\min_{\beta_0, \boldsymbol{\beta}} \sum_{i=1}^{n} w_i(z_i - \beta_0 - \mathbf{X}_i^T\boldsymbol{\beta})^2. \tag{12.12}$$

Based on (12.12) the Newton-Raphson update is the solution to a weighted least squares problem. Therefore, the entire algorithm can be viewed as iterative re-weighted least squares.

The logistic regression model has an intimate connection to the linear discriminant analysis model. By the Bayes theorem we compute the log-odds under the LDA model as follows:

$$\mathit{logodds}(\mathbf{x}) = \log(\frac{\pi_1}{\pi_0}) - \frac{1}{2}(\boldsymbol{\mu}_1 + \boldsymbol{\mu}_0)^T\boldsymbol{\Sigma}^{-1}(\boldsymbol{\mu}_1 - \boldsymbol{\mu}_0) + \mathbf{x}^T\boldsymbol{\Sigma}^{-1}(\boldsymbol{\mu}_1 - \boldsymbol{\mu}_0).$$

This corresponds to the linear logistic regression model with $\beta_0 = \log(\pi_1/\pi_0) - \frac{1}{2}(\boldsymbol{\mu}_1 + \boldsymbol{\mu}_0)^T\boldsymbol{\Sigma}^{-1}(\boldsymbol{\mu}_1 - \boldsymbol{\mu}_0)$ and $\boldsymbol{\beta} = \boldsymbol{\Sigma}^{-1}(\boldsymbol{\mu}_1 - \boldsymbol{\mu}_0)$, though the logistic regression uses a different estimate of $\boldsymbol{\beta}$. The difference between these two methods lies in the marginal distribution of $\mathbf{X}$. The LDA model explicitly assumes that the marginal distribution of $\mathbf{X}$ is a mixture of two normal distributions with common covariance. In contrast, the logistic regression does not specify the marginal distribution of $\mathbf{X}$ at all. As a result, the logistic regression classifier is more robust than LDA but can be less efficient than LDA when the LDA model is correct. Efron (1975) showed that the loss of efficiency is about 30 percent.

## 12.2 Kernel Density Classifiers and Naive Bayes

The kernel density classifier uses a nonparametric kernel density estimator to estimate the density function of $\mathbf{X}$ given $y$ and then plugs the density estimator into the Bayes formula to estimate the Bayes rule. Let $f_k(\mathbf{x})$ denote the density function of $\mathbf{X}$ conditioning on $Y = c_k$ and $\pi_k = \Pr(Y = c_k)$. The Bayes theorem says that

$$\Pr(Y = c_k|\mathbf{X} = \mathbf{x}) = \frac{f_k(\mathbf{x})\pi_k}{\sum_k f_k(\mathbf{x})\pi_k}.$$

Then we can write the Bayes rule as

$$\operatorname{argmax}_{c_k \in \mathcal{C}} \pi_k f_k(\mathbf{x}). \tag{12.13}$$

We estimate $\pi_k$ by $\widehat{\pi}_k = \frac{n_k}{n}$ where $n_k$ is the number of observations in class $c_k$. Suppose that $\widehat{f}_k(\mathbf{x})$ is a reasonably good estimator of $f_k(\mathbf{x})$. Then we could approximate the Bayes rule by

$$\operatorname{argmax}_{c_k \in \mathcal{C}} \widehat{\pi}_k \widehat{f}_k(\mathbf{x}). \tag{12.14}$$

To carry out the idea in (12.14), we often use the *kernel density estimator* (*KDE*) (Parzen, 1962; Epanechnikov, 1969; Silverman, 1998). Let $\mathbf{X}_1, \ldots, \mathbf{X}_n$ be a random sample from a $p$-dimensional distribution with density $f(\mathbf{x})$. Then a kernel density estimator of $f(\mathbf{x})$ is given by

$$\widehat{f}(\mathbf{x}) = \frac{1}{n}\sum_{i=1}^{n} h^{-p} K((\mathbf{X}_i - \mathbf{x})/h), \tag{12.15}$$

where $K$ is a non-negative density function, called a *kernel function* so that $\widehat{f}(\mathbf{x})$ is a legitimate density function. For example, a commonly-used kernel function is the standard Gaussian density. The parameter $h$ is called *bandwidth*.

KDE simply distributes the point pass $\mathbf{X}_i$ by a smoothed function $h^{-p}K((\mathbf{X}_i - \mathbf{x})/h)$ for a small $h$. See Figure 12.2 for the case with $p = 1$. Typically, the kernel function is fixed and the bandwidth $h$ is chosen to trade off biases and variances. There are many papers on data-driven choices of the bandwidth for KDE (Sheather and Jones, 1991; Jones, Marron and Sheather, 1996). Basically, the optimal $h$ depends on the kernel function and the underlying density function.

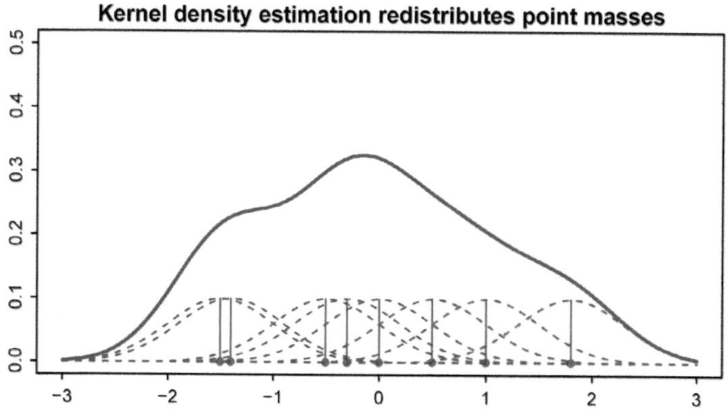


Figure 12.2: Illustration of kernel density estimation. KDE redistributes the point mass (red dots, with mass $1/n$) to a smooth density $\frac{1}{nh}\phi((x - x_i)/h)$ indicated by a dash curve, where $\phi(\cdot)$ is the standard normal distribution. KDE, shown as the solid curve, is the summation of those dashed curves

Consider the univariate density estimation problem and define the optimal bandwidth as the one minimizing the *mean integrated squared error*

$$\text{MISE}(h) = \text{E}\int(\widehat{f}(x) - f(x))^2 dx.$$

Here and hereafter, we use $\int$ to denote the integration over the entire line or more generally space. Asymptotically, for the univariate density estimation, if $\int xK(x)dx = 0$, MISE$(h)$ has the expansion

$$\text{MISE}(h) = \frac{\int K(x)^2 dx}{nh} + \frac{h^4}{4}\left(\int x^2 K(x)dx\right)^2 \left(\int (f''(x))^2 dx\right) + o(\frac{1}{nh} + h^4).$$

See Exercise 12.5 for a more general result. Hence the asymptotically optimal $h$ is given by

$$h^{\text{opt}} = n^{-1/5}\left(\frac{\int K(x)^2 dx}{(\int x^2 K(x)dx)^2(\int (f''(x))^2 dx)}\right)^{1/5}. \tag{12.16}$$

In particular, when $K$ is the Gaussian kernel and $f$ is the normal density, we have

$$h = 1.06\sigma n^{-1/5},$$

where $\sigma$ is the standard deviation of the normal density. Substituting $\sigma$ by the sample standard deviation yields a very simple rule of thumb choice of bandwidth (Silverman, 1998)

For estimating a multivariate density, typically, but not always, the kernel function is radially symmetric and unimodal and the KDE is now expressed as

$$\widehat{f}(\mathbf{x}) = \frac{1}{nh^p} \sum_{i=1}^{n} K(\|\mathbf{X}_i - \mathbf{x}\|_2/h),$$

where $K(\|\mathbf{x}\|)$ is a $p$-dimensional density function. For example, we can use the density function of the standard normal distribution and then the KDE is written as

$$\widehat{f}(\mathbf{x}) = \frac{1}{n(2\pi)^{p/2}h^p} \sum_{i=1}^{n} \exp\left(-\frac{\|\mathbf{X}_i - \mathbf{x}\|_2^2}{2h^2}\right), \tag{12.17}$$

for a given bandwidth $h$. Another commonly used kernel smoothing function is the product kernel: $K(\mathbf{x}) = \prod_{j=1}^{p} K_j(x_j)$ with $K_j \geq 0$ and $\int K_j(t)dt = 1$. Then we have

$$\widehat{f}(\mathbf{x}) = \frac{1}{n} \sum_{i=1}^{n} \prod_{j=1}^{p} h_j^{-1} K_j((\mathbf{X}_j - \mathbf{x})/h_j), \tag{12.18}$$

where bandwidths $\{h_j\}_{j=1}^{p}$ are given, but allowed to be different. Note that using the product kernel does not imply or assume the independence of predictors.

Now let us go back to equation (12.14). For each sub-sample of class $k$, we apply the kernel density estimator to construct $\widehat{f_k}(\mathbf{x})$ and then use the argmax rule to predict the class label at $\mathbf{x}$. The kernel density classifier is quite straightforward both conceptually and computationally, but it is not recommended when the dimension is 3 or higher, due to the *"curse-of-dimensionality"*.

The *naive Bayes classifier* can be viewed as a much simplified kernel density classifier when the dimension is relatively large. The basic idea is very simple. Assume that given the class label the features are conditionally independent. That is,

$$f_k(\mathbf{x}) = \prod_{j=1}^{p} f_{jk}(x_j) \quad \text{for all } k. \tag{12.19}$$

The above independence assumption drastically simplifies the problem of density estimation. Instead of estimating a $p$-dimensional density function, we now estimate $p$ univariate density functions. Combining (12.14) and (12.19) yields the naive Bayes classifier

$$\text{argmax}_{c_k \in \mathcal{C}} \widehat{\pi}_k \prod_{j=1}^{p} \widehat{f}_{jk}(x_j), \tag{12.20}$$

where $\widehat{f}_{jk}(x_j)$ is the univariate KDE for variable $X_j$ based on the $k^{th}$ class of data.

If $X_j$ is continuous, we can use the kernel density estimator for $\widehat{f}_{jk}(x_j)$.

For example, if we use the Gaussian kernel with bandwidth, applying the univariate KDE (12.17) to the $j^{th}$ variable in class $k$, we have

$$\widehat{f}_{jk}(x_j) = \frac{1}{n_k h} \sum_{i \in \mathcal{C}_k} \frac{1}{\sqrt{2\pi}} \exp\left( -\frac{(x_{ij} - x_j)^2}{2h^2} \right), \tag{12.21}$$

where $\mathcal{C}_k = \{i : Y_i = k\}$ is the labels for the data from class $k$. If $X_j$ is discrete then simply let $\widehat{f}_{jk}(x_j)$ be the histogam estimate, which computes the relative frequencies in each category. Thus, the naive Bayes classifier can naturally handle both continuous and discrete predictors.

Although the conditional independence assumption is very convenient, it is rather naive and too optimistic to be remotely true in reality. Hence one might wonder if the naive Bayes classifier is practically useful at all. Surprisingly, naive Bayes classifiers have worked very well in many complex real-world applications such as text classification (McCallum and Nigam, 1998). A possible explanation is that although individual class density estimates ($\prod_{j=1}^p \widehat{f}_{jk}(x_j)$) are poor estimators for the joint conditional density of the predictor vector, they might be good enough to separate the most probable class from the rest. Another important use of the naive Bayes rule is to create augmented features. See Section 12.8.1.

**Example 12.1** (Comparison of misclassification rates between the Naive Bayes and Bayes classifiers, taken from Fan, Feng, Tong, 2012) *To see the loss of the naive Bayes classifier, let us consider the binary classification from the normal populations with equal probability: The random variables representing two classes $C_1$ and $C_2$ follows p-variate normal distributions $N(\boldsymbol{\mu}_1, \boldsymbol{\Sigma})$ and $N(\boldsymbol{\mu}_2, \boldsymbol{\Sigma})$, respectively. The Bayes classifier uses the direction $\mathbf{w}_{opt} = \boldsymbol{\Sigma}^{-1}(\boldsymbol{\mu}_2 - \boldsymbol{\mu}_1)$, whereas the naive Bayes classifier is the one that ignores the correlation in $\boldsymbol{\Sigma}$, resulting in the classifier that uses the coefficients*

$$\mathbf{w}_{\rm NB} = \operatorname{diag}(\boldsymbol{\Sigma})^{-1}(\boldsymbol{\mu}_2 - \boldsymbol{\mu}_1),$$

*Clearly, $\mathbf{w}_{\rm NB}$ is easier to estimate than $\mathbf{w}_{opt}$ in high-dimension, but it is less efficient. To see this, by (12.8), the misclassification rate for the naive Bayes classifier is*

$$1 - \Phi(\Gamma_p^{1/2}), \qquad \Gamma_p = (\boldsymbol{\mu}_d^T \operatorname{diag}(\boldsymbol{\Sigma})^{-1} \boldsymbol{\mu}_d)^2 / \boldsymbol{\mu}_d^T \boldsymbol{\Sigma} \boldsymbol{\mu}_d$$

*and the Bayes classifier has the misclassification rate*

$$1 - \Phi(\Delta_p^{1/2}), \qquad \Delta_p = \boldsymbol{\mu}_d^T \boldsymbol{\Sigma}^{-1} \boldsymbol{\mu}_d.$$

*Without loss of generality, assume that each variable has been marginally standardized by its standard deviation so that $\boldsymbol{\Sigma}$ is a correlation matrix:* $\operatorname{diag}(\boldsymbol{\Sigma}) = \mathbf{I}_p$. *Let us decompose the covariance matrix as*

$$\boldsymbol{\Sigma} = \lambda_1 \boldsymbol{\xi}_1 \boldsymbol{\xi}_1^T + \cdots + \lambda_1 \boldsymbol{\xi}_p \boldsymbol{\xi}_p^T$$

*where* $\{\lambda_i\}_{i=1}^p$ *and* $\{\boldsymbol{\xi}_i\}_{i=1}^p$ *are respectively the eigenvalues and eigenvectors of the matrix* $\boldsymbol{\Sigma}$*. Similarly, decompose*

$$\boldsymbol{\mu}_d = a_1\boldsymbol{\xi}_1 + \cdots + a_p\boldsymbol{\xi}_p,$$

*where* $\{a_i\}_{i=1}^p$ *are the coefficients of* $\boldsymbol{\mu}_d$ *in this new orthonormal basis. Then, we have*

$$\Delta_p = \sum_{j=1}^p a_j^2/\lambda_j, \qquad \Gamma_p = \Big(\sum_{j=1}^p a_j^2\Big)^2 \Big/ \sum_{j=1}^p \lambda_j a_j^2.$$

*The relative efficiency of the Fisher discriminant over naive Bayes is characterized by* $\Delta_p/\Gamma_p$*. By the Cauchy-Schwartz inequality,*

$$\Delta_p/\Gamma_p \geq 1.$$

*The naive Bayes method performs as well as the Fisher discriminant only when* $\boldsymbol{\mu}_d$ *is an eigenvector of* $\boldsymbol{\Sigma}$*, i.e., the coefficients* $\{a_i\}_{i=1}^p$ *are spiked.*

*In general,* $\Delta_p/\Gamma_p$ *can be far greater than one. Since* $\boldsymbol{\Sigma}$ *is the correlation matrix,* $\sum_{j=1}^p \lambda_j = \mathrm{tr}(\boldsymbol{\Sigma}) = p$*. If* $\boldsymbol{\mu}_d$ *is equally loaded on* $\boldsymbol{\xi}_j$*, then the ratio*

$$\Delta_p/\Gamma_p = p^{-2}\sum_{j=1}^p \lambda_j \sum_{j=1}^p \lambda_j^{-1} = p^{-1}\sum_{j=1}^p \lambda_j^{-1}.$$

*More generally, if* $\{a_j\}_{j=1}^p$ *are realizations from a distribution with the second moment* $\sigma^2$*, then by the law of large numbers,*

$$\sum_{j=1}^p a_j^2/\lambda_j \approx \sigma^2 \sum_{j=1}^p 1/\lambda_j, \quad p^{-1}\sum_{j=1}^p a_j^2 \approx \sigma^2, \quad \sum_{j=1}^p \lambda_j a_j^2 \approx \sigma^2 \sum_{j=1}^p \lambda_j.$$

*Hence,*

$$\Delta_p/\Gamma_p \approx p^{-1}\sum_{j=1}^p \lambda_j^{-1}.$$

*Suppose that half of the eigenvalues of* $\boldsymbol{\Sigma}$ *are* $c$ *and the other half are* $2-c$*. Then, the right-hand side of the ratio is* $(c^{-1}+(2-c)^{-1})/2$*. When the condition number is 10, say, this ratio is about 3. For example, when* $\Gamma_p^{1/2} = 0.5$*, we have* $1-\Phi(0.5) = 30.9\%$ *and* $1-\Phi(3*0.5) = 6.7\%$ *error rates respectively for the naive Bayes and Fisher discriminant.*

## 12.3 Nearest Neighbor Classifiers

The nearest neighbor classifier is a localized classification algorithm in the predictor space. Let $d(\mathbf{x}, \mathbf{x}')$ be the distance metric between two points in the predictor space. Some commonly used distance functions include

$$d(\mathbf{x}, \mathbf{x}') = \|\mathbf{x} - \mathbf{x}'\|_q = (\sum_{j=1}^{p} |x_j - x_j'|^q)^{1/q},$$

with $q = 2, 1, 0$. When $q = 0$, $d(\mathbf{x}, \mathbf{x}') = \sum_{j=1}^{p} I(x_j \neq x_j')$ which is called the *Hamming distance*. Given any predictor vector $\mathbf{x}$, let $d_i = d(\mathbf{X}_i, \mathbf{x})$ for $1 \leq i \leq n$. Define the $k$-nearest neighborhood of $\mathbf{x}$ as the $k$ data points that are closest to $\mathbf{x}$, namely,

$$\mathcal{N}_k(\mathbf{x}) = \{\mathbf{X}_i : d(\mathbf{X}_i, \mathbf{x}) \leq d_{(k)}\} \tag{12.22}$$

where $d_{(k)}$ is the $k$th smallest value of $(d_1, \ldots, d_n)$. Then the nearest neighbor classifier predicts the class label according to the simple majority vote:

$$\text{argmax}_{c_j \in C} \sum_{i \in \mathcal{N}_k(\mathbf{x})} I(Y_i = c_j). \tag{12.23}$$

The parameter $k$ is critically important for the performance of the nearest neighbor classifier. Just like other local smoothers, a smaller $k$ yields a more jumpy classification boundary. On the other hand, using a larger $k$ can cause large bias. In practice, it is recommended to use some data-driven tuning methods such as cross-validation or bootstrap to choose the optimal $k$. Hall, Park and Samworth (2008) discussed the theory for choosing the optimal $k$.

The nearest neighbor classifier can be computationally intensive, especially when $n$ is large. Over the years many fast algorithms for searching nearest neighbors have been proposed in the literature (Fukunaga and Narendra, 1975; Friedman, Baskett and Shustek, 1975; Djouadi and Bouktache, 1997; Arya *et al.*, 1998). Due to its local learning nature, the nearest neighbor classifier suffers severely from the "curse of dimensionality". In the high dimension case the choice of the distance function becomes more critically important. The reason is that when the dimension is high even the 1-nearest neighbor of $\mathbf{x}$ is not close enough to bear enough similarity to $\mathbf{x}$ in order to help make prediction at $\mathbf{x}$. Of course, if the data lie on a lower dimensional manifold, we might be able to construct a more meaningful distance metric to make the nearest neighbor classifier successful. Hastie and Tibshirani (1996) proposed a locally adaptive form of nearest neighbor classification in which a local linear discriminant analysis is used to estimate an effective metric for computing neighborhoods.

## 12.4 Classification Trees and Ensemble Classifiers

### *12.4.1 Classification trees*

There are several popular classification tree methods in the literature, including CART (Breiman Friedman, Olsen and Stone, 1984), C4.5 (Quinlan, 1993) and GUIDE (Loh., 2009). See also Zhang and Singer (1999). The earliest tree-based algorithm in the literature was introduced in Morgan and Sonquist (1963). A tree-structured classifier is typically constructed by recursive partitioning algorithms in the predictor space. The process of *recursive partitioning* can be graphically presented by a *decision tree* in which the *terminals* correspond to the final partitioned regions. The classifier makes a constant prediction within each region by using a simple majority vote. Let $R_1, \dots, R_S$ be the final partitioned regions. Given a query point $\mathbf{x}$, the decision tree puts $\mathbf{x}$ into a region, say $R(\mathbf{x})$, then the predicted label is the majority in that region:

$$\widehat{y} = \text{argmax}_{c_k \in C} \sum_{\mathbf{X}_i \in R(\mathbf{x})} I(Y_i = c_k). \tag{12.24}$$

Binary splits are often used in the recursive partitioning process. One may also consider multiway splits. In general, binary splits are preferred for two main reasons. First, we want to have a reasonably sized sub-sample in each node but doing multiway splits can fragment the training data too quickly. Second, if a multiway split is really needed, it can be done via a series of binary splits. In what follows we discuss the greedy algorithm used to construct the binary tree in *CART*.

CART uses simple splits such that the final partitions consist of "rectangles". To split each node, CART uses a splitting variable $X_j$ and produces the pair of half-planes

$$R_1(j,t) = \{X : X_j \leq t\} \quad \text{and} \quad R_2(j,t) = \{X : X_j > t\}, \tag{12.25}$$

if $X_j$ is a continuous variable. The greedy algorithm seeks for the optimal pair $j$ and $t$ such that the node impurity is reduced by the greatest amount. For a categorical predictor with $s$ possible levels, there are $2^{s-1} - 1$ ways to partition. If $s$ is large, for example, the variable is zip code, then the exclusive search is computationally prohibitive. For binary classification $y = 0, 1$, CART uses a very neat trick to overcome the computation difficulty. We can order the unordered categorical variable levels based on the proportion falling in outcome class 1 and then split this categorical variable as if it were ordered. The greedy partition method is applied to each of the two resulting regions. If a region has only a few observations (say, less than 5), then no splitting is needed and the region becomes a terminal.

The *impurity function* of each node can be either the Gini index or the cross-entropy. Let $R$ be the node to be split into two regions. Define

$$p_k = \frac{1}{|R|} \sum_{\mathbf{X}_i \in R} I(Y_i = c_k), \quad 1 \leq k \leq p.$$

The Gini index of $R$ is defined as

$$\mathrm{GI}(R) = \sum_k p_k(1 - p_k), \tag{12.26}$$

and the cross-entropy of $R$ is defined as

$$\mathrm{CE}(R) = -\sum_k p_k \log(p_k). \tag{12.27}$$

Using one of these two measures, the optimal pair of $j$ and $t$ is defined as

$$(j,t)^{\mathrm{opt}} = \arg\min_{j,t} \left[ \frac{|R_1(j,t)|}{|R|} \mathrm{GI}(R_1(j,t)) + \frac{|R_2(j,t)|}{|R|} \mathrm{GI}(R_2(j,t)) \right],$$

where $R_1(j,t)$ and $R_2(j,t)$ are the half planes defined before, or

$$(j,t)^{\mathrm{opt}} = \arg\min_{j,t} \left[ \frac{|R_1(j,t)|}{|R|} \mathrm{CE}(R_1(j,t)) + \frac{|R_2(j,t)|}{|R|} \mathrm{CE}(R_2(j,t)) \right].$$

CART uses the above described recursive partitioning method to grow the tree until every terminal reaches the lower bound of minimum number of observations. However, CART does not stop there. It further uses cost-complexity pruning to prune the fully grown tree to a smaller one. The basic idea is to assign a positive "penalty term" $\alpha$ to each terminal. The chosen tree should minimize the sum of the impurity and penalty over all terminals. CART uses cross-validation to select the penalty constant $\alpha$. Figure 12.3 shows a fitted classification tree for the South African heart disease data (Hastie, Tibshirani and Friedman, 2009).

Instead of using the simple split $X_j \leq t$, one can also consider using a linear split function defined as $\sum_j a_j X_j \leq t$ where the coefficients $a_j$ and the split point $t$ are jointly optimized by minimizing the impurity function of the two sub-regions. Although this strategy could improve the prediction ability of the tree, CART does not use this strategy. Note that the biggest advantage of a tree classifier lies in its interpretability. Using the linear combination of variables weakens the easy interpretability of the simple univariate split. Another reason is that there is little room to improve the classification accuracy of a single classification tree, because a single tree constructed by recursive partitioning and greedy search can be highly variable. Often a small change in the training data could greatly distort the tree: the series of splits could be very different. Fortunately, the damage due to instability of a single tree

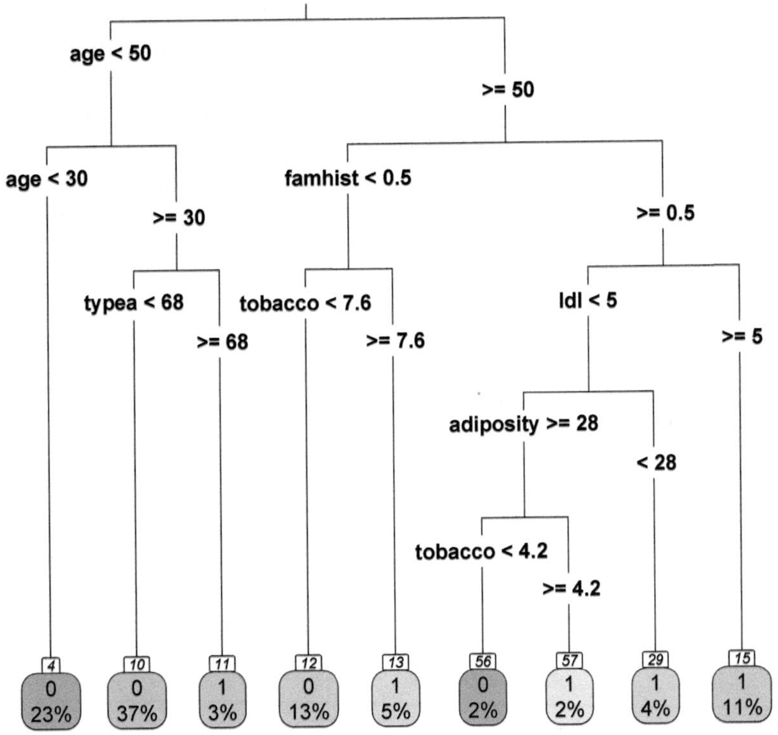


Figure 12.3: CART on South African heart disease data. The fitted classification tree has nine terminal nodes. The label of a terminal node is the predicted label for any future data falling into this region. "1"="Yes, Disease", "0"="No". The plot also shows the percentage of training data in each terminal node.

can be mitigated if we combine multiple trees to form a final classification by using ensemble learning techniques such as *bagging* (Section 12.4.2) or *random forest* (Section 12.4.3).

### *12.4.2 Bagging*

**B**ootstrap **agg**regat**ing** (Bagging) was proposed by Breiman (1996a) as a general technique for improving unstable predictive algorithms. Since classification trees are notorious for their instability, Bagging decision trees is particularly attractive.

Bagging is straightforward to implement. Let $Z_n = \{(Y_i, \mathbf{X}_i), 1 \leq i \leq n\}$

denote the training data. Consider a learning algorithm that takes the training data as its input and produces a prediction outcome $g(Z_n)$. Let us draw a *bootstrap* (Efron, 1979; Efron and Tibshirani, 1993) sample $Z^{(*b)} = \{(y_i^{(*b)}, \mathbf{x}_i^{(*b)}), 1 \leq i \leq n\}$ by sampling from the training data with replacement. Apply the sample learning algorithm to the bootstrap sample to produce a bootstrap prediction outcome $g(Z_n^{(*b)})$. Repeat the procedure for $b = 1, 2, \ldots, B$. In Breiman (1996a) $B = 50$ was used in all numeric examples, but there is no particular reason for this choice. If the learning algorithm is fast to compute then one can certainly use a larger $B$ (say $B = 500$).

For regression problems, the Bagging estimator is defined as

$$\widehat{y}^{\text{bag}} = \frac{1}{B}\sum_{b=1}^{B} g(Z_n^{(*b)}). \tag{12.28}$$

For classification problems, $g(Z_n)$ is the predicted class label and the Bagging estimator is defined as the outcome of the majority vote:

$$\widehat{y}^{\text{bag}} = \text{argmax}_{c_j \in \mathcal{C}} \sum_{b=1}^{B} I(g(Z_n^{(*b)}) = c_j). \tag{12.29}$$

Figure 12.4 illustrates the basic idea of Bagging.

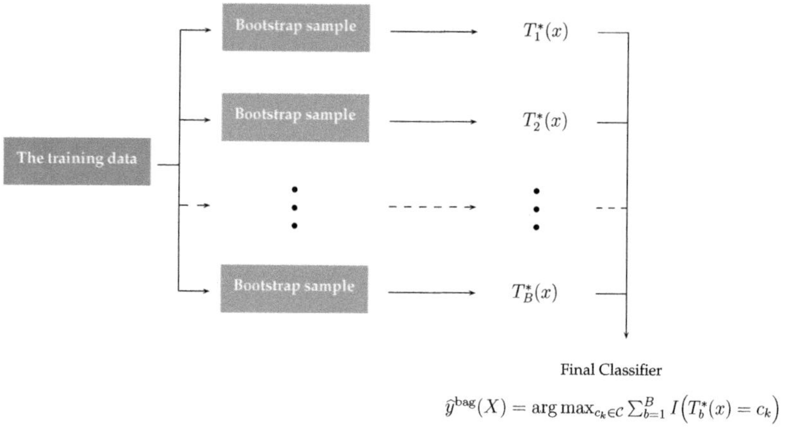


Figure 12.4: An illustration of bagging in classification by taking the majority-voted outcomes.

Unstable learning algorithms are very sensitive to a small change in the training data, resulting in high variability in the prediction outcome. Instability was first systematically studied in Breiman (1996b) where he showed that neural network, CART and subset selection are unstable, while nearest neighbor methods are stable. Breiman's experiments (Breiman, 1996a) showed that

Bagging can greatly reduced the misclassification error of CART on a variety of simulated and real data sets. The reduction in the test misclassification rates ranges from 20% to 47%. However, on the same data sets Bagging does not help the nearest neighbor classifier.

Several papers have been devoted to theoretical understanding of Bagging. Friedman and Hall (2000) assumed that $g(Z_n)$ is a smooth nonlinear function which can be decomposed into a linear part and higher orders. They showed, with some heuristic arguments, that Bagging reduces variance for the higher order terms but leaves the linear part unaffected. Bühlmann and Yu (2002) studied the effect of Bagging on nonsmooth and unstable predictors. They showed that Bagging improves the first order dominant variance and mean squared error asymptotic terms, by as much as a factor of 3. Buja and Stuetzle (2006) showed that Bagging $U$-statistics often but not always decreases variance, whereas it always increases bias.

### *12.4.3 Random forests*

*Random forests* is an ensemble learning classifier invented by Breiman (Breiman, 2001). It is claimed to be one of the most accurate "general purpose" classifiers (Breiman, 2001). Note that Bagging uses a randomized method, namely bootstrap, to generate a collection of classification trees. Random forests put another level of randomization on top of Bagging by using a randomly selected a subset of predictors to construct each classification tree. The idea of using randomly selected features in classification trees was previously used in Ho (1998) and Amit and Geman (1997). By using randomly selected variables in Bagging, Breiman (2001) aimed to reduce the correlations between the trees, which can further enhance the variance reduction capability of Bagging. Figure 12.5 shows how to grow a random forest. Note that although it looks very much like Figure 12.4 for Bagging, a key and fundamental difference is that a randomized tree-growing algorithm is applied to get the tree classifier based on each bootstrap sample. Algorithm 12.1 covers the details of random forests.

Random forests has only two tuning parameters. Breiman (2001) suggested the default values: for regression $m = \lfloor p/3 \rfloor$ and $n_{\min} = 5$; for classification $m = \lfloor \sqrt{p} \rfloor$ and $n_{\min} = 1$. Random forests uses *out-of-bag* (OOB) samples to estimate its generalization error. For each observation $(\mathbf{X}_i, Y_i)$, compute its OOB error by averaging those trees whose bootstrap samples do not include $(\mathbf{X}_i, Y_i)$. The average OOB error of all training data is monitored as a function of $B$. Once the OOB curve stabilizes, no more bootstrapping tree is needed. Unlike many nonlinear learning algorithms, random forests does not need to use cross-validation to determine the value of $B$.

Breiman (2001) also suggested using random linear combinations of predictors in random forests. Before each split, $m$ variables are randomly selected and $F$ linear coefficient vectors are generated as well, with each element of

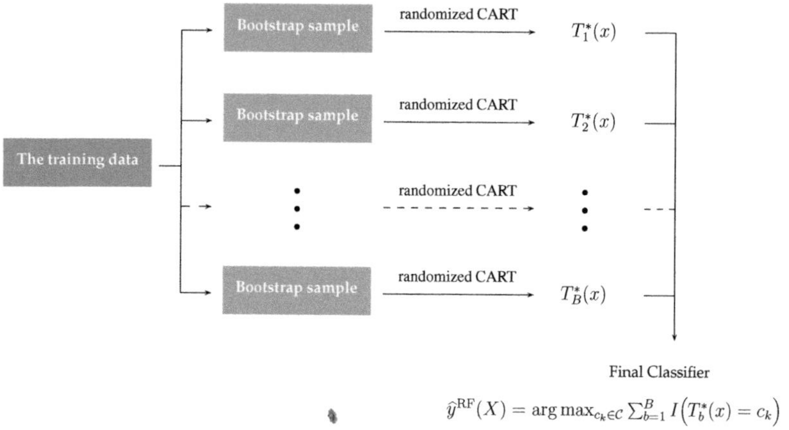


Figure 12.5: Schematic of random forests.

---

**Algorithm 12.1** Random forests

---

- Generate B bootstrap samples $Z_n^{(*b)}$, $b = 1, \ldots, B$.
- Apply a recursive partition method to each bootstrap sample to build each tree. Before splitting each node, randomly select $m$ variables out of the total of $p$ predictors. Then use the greedy search method as that used in CART to split the node into two sub-regions. A node becomes a terminal when the minimum node size $n_{\min}$ is reached or there is only one class in the node.
- Save the ensemble of classification trees $T_b^*$, $b = 1, \ldots, B$.
- Given a query $\mathbf{X} = \mathbf{x}$, let $T_b^*(\mathbf{x})$ be the prediction of the $b$th tree. Then the random forest prediction for classification is given by

$$\hat{y}^{\mathrm{RF}}(\mathbf{x}) = \mathrm{argmax}_{c_k \in C} \sum_{b=1}^{B} I(T_b^*(\mathbf{x}) = c_k).$$

For regression problems, the random forest prediction is $\frac{1}{B}\sum_{b=1}^{B} T_b^*(\mathbf{x})$.

---

coefficients being uniform random numbers on $[-1, 1]$. Then a greedy search is used to find the best split. Breiman (2001) named this method Forest-RC and tried using $F = 1$ and $F = \lfloor \log_2(p) \rfloor + 1$ in his numeric study.

Breiman (2001) gave some theoretical analysis of random forests. He showed that as $B$ increases the generalization error of random forests always converges. Lin and Jeon (2006) draw a nice connection between random forests and an adaptive nearest neighbor classifier. They introduced a concept of potential nearest neighbors (PNN) and showed that random forests can be seen as adaptively weighted PNN methods.

### *12.4.4 Boosting*

Similar to Bagging and random forests, *boosting* is a powerful ensemble learning idea. Unlike Bagging and random forests, boosting sequentially finds a "committee" of base learners and then makes a collective decision by using a weighted linear combination of all base learners. The first successful and popular boosting algorithm is *AdaBoost* (Freund and Shapire, 1997). Figure 12.6 shows the schematic of AdaBoost.

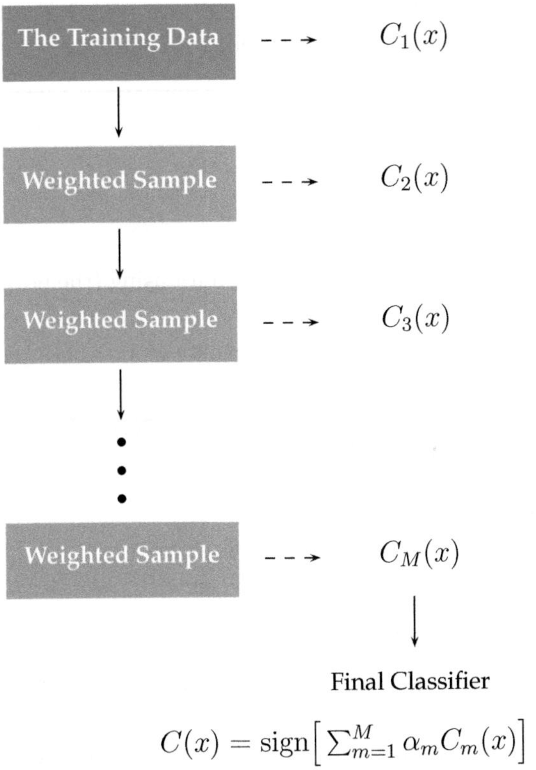


Figure 12.6: Schematic illustration of AdaBoost. Each weak learner is trained sequentially on weighted training data. The final classification is based on a weighted combination of all weak learners. Details are given in Algorithm 12.2.

The details of AdaBoost are shown in Algorithm 12.2. The current classifier is denoted by $C_m(\mathbf{x})$ whose value is either 1 or $-1$. We assume that $C_m(\mathbf{x})$ is better than random guessing, which implies that $\text{err}^{(m)} < 0.5$ and hence $\alpha_m > 0$. Note that after fitting $C_m(\mathbf{x})$ to the weighted training data, the weights get updated in 2(d). For those incorrectly classified observations, their weights get inflated by a factor $e^{\alpha_m}$, while the weights for the correctly classified observations get shrunken by a factor $e^{-\alpha_m}$. Thus, this re-weighting encourages the next classifier $C_{m+1}(\mathbf{x})$ to focus more on the observations misclassified by $C_m(\mathbf{x})$.

Bagging and random forests are almost perfect for doing parallel computing, because their base learners are completely independent of each other in terms of computation. This is not true in AdaBoost (or other boosting algorithms), because the base learners are sequentially built. When implementing Algorithm 12.2 in practice, each base classifier $C_m(\mathbf{x})$ is typically a classification tree. In fact, AdaBoost with classification trees is said to be the "best off-the-shelf classifier in the world" (Breiman, 1998).

**Algorithm 12.2** AdaBoost

1. Initialize the observation weights $w_i = 1/n,\ i = 1, 2, \ldots, n$. Code the class label with 1 and $-1$.
2. For $m = 1, \cdots, M$:
   (a) Fit a classifier $C_m(\mathbf{x})$ to the training data using (the normalized) weights $w_i$, aiming to minimize the weighted error $\sum_{i=1}^n w_i I(Y_i \neq C_m(\mathbf{X}_i))$.
   (b) Compute the weighted misclassification error
   $$\mathrm{err}^{(m)} = \sum_{i=1}^n w_i I\left(Y_i \neq C_m(\mathbf{X}_i)\right) / \sum_{i=1}^n w_i.$$
   (c) Compute
   $$\alpha_m = \log \frac{1 - \mathrm{err}^{(m)}}{\mathrm{err}^{(m)}}.$$
   (d) Update the weights by
   $$w_i \leftarrow w_i \cdot \exp\left(\alpha_m \cdot I\left(Y_i \neq C_m(\mathbf{X}_i)\right)\right),\ i = 1, 2, \ldots, n.$$
3. Output the final classified label as $\mathrm{sign}\left(\sum_{m=1}^M \alpha_m C_m(\mathbf{x})\right)$.

There are many papers devoted to theoretical understanding of AdaBoost. Schapire, Freund, Bartlett and Lee (1998) suggested a *margin-maximization* explanation. They showed that if the training data are separable the boosting iterations increase the $\ell_1$ margin. This interesting result offers a geometric view of boosting and relates boosting to the support vector machine: the latter directly maximizes the $\ell_2$ margin in the case of separable data (See Section 12.5). Breiman (1999) pointed out that AdaBoost can be viewed as a gradient descent algorithm in function space. This viewpoint was enthusiastically embraced by many researchers (Mason, Baxter, Bartlett and Frean, 1999; Friedman, Hastie and Tibshirani, 2000; Bühlmann and Hothorn, 2007). In particular, Friedman, Hastie and Tibshirani (2000) provided a nice statistical interpretation of AdaBoost by showing that AdaBoost is equivalent to a *forward additive modeling* procedure that tries to minimize an exponential loss. Here "additive" refers to the fact that the final boosting classifier is an additive combination of many base learners.

Consider a generic classification rule $\text{sign}(f(\mathbf{x}))$. The exponential loss function is defined as $\exp(-Yf(\mathbf{x}))$ and the empirical exponential loss is

$$L(f) = \frac{1}{n}\sum_{i=1}^{n} \exp(-Y_i f(\mathbf{X}_i)).$$

Friedman, Hastie and Tibshirani (2000) justified the exponential loss by considering its population minimizer, i.e., the minimizer of $L(f)$ when $n = \infty$. In that ideal case, $L(f)$ becomes $\text{E}[\exp(-Yf(\mathbf{X}))]$. Observe that

$$\text{E}[\exp(-Yf(\mathbf{X}))] = \text{E}_{\mathbf{X}}[\Pr(Y=1|\mathbf{X})e^{-f(\mathbf{X})} + \Pr(Y=-1|\mathbf{X})e^{f(\mathbf{X})}]. \quad (12.30)$$

It is easy to show (Exercise 12.8) that the minimizer of $\text{E}[\exp(-Yf(\mathbf{X}))]$ with respect to $f(\cdot)$ is

$$f^*(\mathbf{x}) = \frac{1}{2}\log\left(\frac{\Pr(Y=1|\mathbf{X}=\mathbf{x})}{\Pr(Y=-1|\mathbf{X}=\mathbf{x})}\right). \quad (12.31)$$

Moreover, the Bayes rule can be written as $\text{sign}(f^*(\mathbf{x}))$. The above arguments suggest that by minimizing the empirical exponential loss we can approximate the *Bayes classifier*.

Now let us try to minimize the empirical exponential loss in a function space defined as $\{f(\mathbf{x}) : f(\mathbf{x}) = \sum_{m=1}^{M} \beta_m C_m(\mathbf{x})\}$, where $\beta_m > 0$ and $C_m(\mathbf{x})$ is a classification function whose value is either 1 or $-1$. We conduct the optimization by using an iterative greedy search algorithm. Set the initial value to be $f^{(0)}(\mathbf{x}) = 0$. Given the current $f(\mathbf{x}) = f^{(m-1)}(\mathbf{x})$, we update $f(\mathbf{x})$ to $f^{(m)}(\mathbf{x}) = f^{(m-1)}(\mathbf{x}) + \beta^{(m)} C_m(\mathbf{x})$, where

$$\begin{aligned}(\beta^{(m)}, C_m(\mathbf{x})) &= \arg\min_{\beta, C(\mathbf{x})} \frac{1}{n}\sum_{i=1}^{n} \exp\left\{-Y_i[f^{(m-1)}(\mathbf{X}_i) + \beta C(\mathbf{X}_i)]\right\} \\ &= \arg\min_{\beta, C(\mathbf{x})} \sum_{i=1}^{n} w_i^{(m)} \exp\{-Y_i \beta C(\mathbf{X}_i)\},\end{aligned} \quad (12.32)$$

where

$$w_i^{(m)} = \frac{1}{n}\exp(-Y_i f^{(m-1)}(\mathbf{X}_i)). \quad (12.33)$$

Note that with $f^{(0)}(\mathbf{x}) = 0$ the initial weights equal $\frac{1}{n}$. Fixing $\beta > 0$ in (12.32), the solution for $C_m(\mathbf{x})$ is

$$\begin{aligned} C_m(\mathbf{x}) &= \arg\min_{C(\mathbf{x})} \sum_{i=1}^{n} [w_i^{(m)} e^{-\beta} I(Y_i = C(\mathbf{X}_i)) + w_i^{(m)} e^{\beta} I(Y_i \neq C(\mathbf{X}_i))] \\ &= \arg\min_{C(\mathbf{x})} \sum_{i=1}^{n} w_i^{(m)} I(Y_i \neq C(\mathbf{X}_i)), \end{aligned}$$

after ignoring the constant factors that are not related to the classifier $C(\mathbf{x})$. Thus, $C_m(\mathbf{x})$ is the classification function minimizing the weighted error rate on training data. Given $C_m(\mathbf{x})$, $\beta^{(m)}$ is given by

$$\beta^{(m)} = \arg\min_{\beta} \sum_{i=1}^{n} [e^{-\beta} w_i^{(m)} I(Y_i = C_m(\mathbf{X}_i)) + e^{\beta} w_i^{(m)} I(Y_i \neq C_m(\mathbf{X}_i))]. \tag{12.34}$$

The solution to (12.34) is

$$\beta^{(m)} = \frac{1}{2} \log \left( \frac{\sum_{i=1}^{n} w_i^{(m)} I(Y_i = C_m(\mathbf{X}_i))}{\sum_{i=1}^{n} w_i^{(m)} I(Y_i \neq C_m(\mathbf{X}_i))} \right). \tag{12.35}$$

Using notation in Algorithm 12.2, we can write

$$\beta^{(m)} = \frac{1}{2} \alpha_m.$$

Then we have

$$f^{(m)}(\mathbf{x}) = f^{(m-1)}(\mathbf{x}) + \frac{1}{2} \alpha_m C_m(\mathbf{x}).$$

By (12.33) the weights for finding $C^{(m+1)}(\mathbf{x})$ become

$$w_i^{(m+1)} = \frac{1}{n} \exp(-Y_i f^{(m)}(\mathbf{X}_i)) = w_i^{(m)} \exp\left\{-\frac{1}{2} \alpha_m Y_i C_m(\mathbf{X}_i)\right\}.$$

Using $Y_i C_m(\mathbf{X}_i) = 1 - 2I(Y_i \neq C_m(\mathbf{X}_i))$, we write

$$w_i^{(m+1)} = w_i^{(m)} e^{-\alpha_m/2} \exp\left\{\alpha_m I(Y_i \neq C_m(\mathbf{X}_i))\right\}. \tag{12.36}$$

We can drop the constant factor $e^{-\alpha_m/2}$ (which is independent of $i$) when using $w^{(m+1)}$ to get $C^{(m+1)}(\mathbf{x})$. So (12.36) is equivalent to 2(d) in Algorithm 12.2. Suppose that we repeat this greedy search procedure $M$ times. Then we have

$$f^{(M)}(\mathbf{x}) = \sum_{m=1}^{M} \frac{1}{2} \alpha_m C_m(\mathbf{x}).$$

The factor $\frac{1}{2}$ can be dropped when the sign of $f^{(M)}(\mathbf{x})$ is concerned. This is exactly the final classifier in Algorithm 12.2.

**Algorithm 12.3** MART

1. Compute the constant function $f_0(\mathbf{x}) = \arg\min_\gamma \sum_{i=1}^n L(Y_i, \gamma)$. Set $\nu < 0.01$ as a small positive constant.
2. For $m = 1, \cdots, M$:
   (a) For $i = 1, \ldots, n$, compute $g_{im}$.
   $$g_{im} = -\left[\frac{\partial L(Y_i, f(\mathbf{X}_i))}{\partial f(\mathbf{X}_i)}\right]_{f(\mathbf{X}_i)=f_{m-1}(\mathbf{X}_i)}$$
   (b) Treat $g_{im}$ $1 \le i \le n$ as a response variable. Use $\mathbf{X}$ as input variables to fit a regression tree. Denote the terminal regions as $R_{jm}$, $j = 1, \ldots, J_m$.
   (c) For $j = 1, \ldots, J_m$, compute
   $$\gamma_{jm} = \arg\min_\gamma \sum_{\mathbf{X}_i \in R_{jm}} L(Y_i, f_{m-1}(\mathbf{X}_i) + \gamma)$$
   (d) Update $f_m(\mathbf{x}) = f_{m-1}(\mathbf{x}) + \nu(\sum_{j=1}^{J_m} \gamma_{jm} \cdot I(\mathbf{x} \in R_{jm}))$.
3. Output $f(\mathbf{x}) = f_M(\mathbf{x})$.

Friedman (2001) further developed the forward stage-wise modeling view of AdaBoost into a generic *gradient tree-boosting* algorithm named "Multiple Additive Regression Trees", or *MART*. Algorithm 12.3 summarizes the details of MART. As its name suggests, the base learners in MART are regression trees. The constant $\nu$ in Algorithm 12.3 is a shrinkage factor, another regularizer of MART used to slow down the forward stage-wise modeling procedure. The loss function $L(y, f(\mathbf{x}))$ is not limited to the exponential loss in AdaBoost. For example, it can be the negative binomial log-likelihood in nonparametric logistic regression. It is straightforward to derive a *multicategory boosting algorithm* too: we simply use the negative multinomial log-likelihood as the loss function in MART.

A very popular implementation of MART and gradient boosting is XGBoost by Chen and Guestrin (2016).

## 12.5 Support Vector Machines

The *support vector machine* (SVM) has proven to be one of the most successful classification tools (Vapnik, 1996).

### *12.5.1 The standard support vector machine*

The *SVM* uses 1 and $-1$ to code the class label in binary classification.

When the training data can be perfectly separated by a hyperplane, the SVM has a nice geometric interpretation: it finds the hyperplane that maximizes the so-called *margin*. A hyperplane is defined by $\{\mathbf{x} : \boldsymbol{\beta}_0 + \mathbf{x}^T\boldsymbol{\beta} = 0\}$, where $\boldsymbol{\beta}$ is a unit vector ($\|\boldsymbol{\beta}\|_2 = 1$). If the training data are linearly separable, then there exists a hyperplane such that

$$Y_i(\boldsymbol{\beta}_0 + \mathbf{X}_i^T\boldsymbol{\beta}) > 0, \quad \forall i,$$

which means the hyperplane is a perfect classifier. The margin is defined as the smallest distance from the training data to the hyperplane. Thus, the SVM problem is formulated as

$$\max_{\boldsymbol{\beta},\boldsymbol{\beta}_0,\|\boldsymbol{\beta}\|_2=1} \quad C,$$
$$\text{subject to} \quad Y_i(\beta_0 + \mathbf{X}_i^T\boldsymbol{\beta}) \geq C \quad i = 1, \ldots, n,$$

Figure 12.7 depicts the margin maximization idea of the SVM.

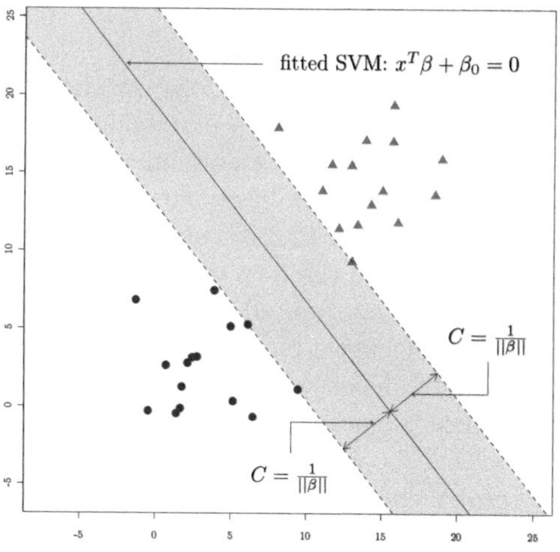


Figure 12.7: Geometry of the support vector machine. The data points are separable. The solid line is the fitted SVM boundary. The shaded region between two dotted lines has width $2C = 2/\|\boldsymbol{\beta}\|_2$. The fitted SVM maximizes this width among all separation hyperplanes.

In applications the training data are usually not linearly separable. Then the SVM allows some training data to be on the wrong side of the hyperplane.

The general SVM problem is defined by

$$\max_{\boldsymbol{\beta},\beta_0,\|\boldsymbol{\beta}\|_2=1} \quad C, \tag{12.37}$$

$$\text{subject to} \quad Y_i(\beta_0 + \mathbf{X}_i^T\boldsymbol{\beta}) \geq C(1-\xi_i), \quad i = 1,\ldots,n, \tag{12.38}$$

$$\xi_i \geq 0, \sum \xi_i \leq B, \tag{12.39}$$

where $\xi_i, \xi_i \geq 0$ are *slack variables*, and $B$ is a pre-specified positive number which can be regarded as a *tuning parameter*.

The SVM also has an equivalent *loss + penalty* formulation. Let $\boldsymbol{\gamma} = \boldsymbol{\beta}/C$ and $\gamma_0 = \beta_0/C$. Then, $C = 1/\|\boldsymbol{\gamma}\|_2$ and constraint (12.38) now becomes $\xi_i \geq 1 - Y_i(\gamma_0 + \mathbf{X}_i^T\boldsymbol{\gamma})$. This and the non-negative constraints in (12.39) are equivalent to

$$\xi_i \geq [1 - Y_i(\gamma_0 + \mathbf{X}_i^T\boldsymbol{\gamma})]_+,$$

where $z_+$ denotes the positive part of a real valued number $z$. Clearly, the optimal choice of $\xi_i$ is obtained at the boundary

$$\xi_i = [1 - Y_i(\gamma_0 + \mathbf{X}_i^T\boldsymbol{\gamma})]_+$$

and the constraint (12.39) now reduces to

$$\sum_{i=1}^n [1 - Y_i(\gamma_0 + \mathbf{X}_i^T\boldsymbol{\gamma})]_+ \leq B.$$

With this constraint, we wish to maximize $1/\|\boldsymbol{\gamma}\|_2$ or equivalently minimize $\|\boldsymbol{\gamma}\|^2$. By using the Lagrange multiplier method, we are minimizing

$$\|\boldsymbol{\gamma}\|^2 + \lambda_1 \sum_{i=1}^n [1 - Y_i(\gamma_0 + \mathbf{X}_i^T\boldsymbol{\gamma})]_+$$

for a multipler $\lambda_1$ or equivalently

$$\min_{\beta_0,\boldsymbol{\beta}} \frac{1}{n}\sum_{i=1}^n \left[1 - Y_i(\beta_0 + \mathbf{X}_i^T\boldsymbol{\beta})\right]_+ + \lambda\|\boldsymbol{\beta}\|_2^2, \tag{12.40}$$

by writing $\gamma_1 = \frac{1}{n\lambda}$ and $\beta$ as $\gamma$. This shows that the problem (12.37)–(12.39) is equivalent to (12.40) and $\lambda$ has a one-to-one correspondence to $B$ in (12.39). The loss function $(1-t)_+$ is called the *hinge loss*.

From (12.40) we can directly obtain the so-called kernel SVM that uses nonlinear decision rules. Consider the *Hilbert space* $\mathcal{H}_K$ with a *reproducing kernel* $K(\mathbf{x}, \mathbf{x}')$. The kernel SVM is defined by

$$\min_{f\in\mathcal{H}_K} \frac{1}{n}\sum_{i=1}^n [1 - Y_i f(\mathbf{X}_i)]_+ + \lambda\|f\|_{\mathcal{H}_K}^2. \tag{12.41}$$

The theory of *reproducing kernel Hilbert space* (Wahba, 1990) guarantees that the solution to (12.41) has a finite dimensional expression in terms of the *reproducing kernel* function. That is,

$$\widehat{f}(\mathbf{x}) = \widehat{\beta}_0 + \sum_{i=1}^{n} \widehat{\beta}_i K(\mathbf{x}, \mathbf{X}_i). \tag{12.42}$$

and $(\widehat{\beta}_0, \widehat{\boldsymbol{\beta}})$ is the solution to

$$\min_{\beta_0, \boldsymbol{\beta}} \frac{1}{n} \sum_{i=1}^{n} \left[ 1 - Y_i(\beta_0 + \sum_{l=1}^{n} \beta_l K(\mathbf{X}_i, \mathbf{X}_l)) \right]_+ + \lambda \boldsymbol{\beta}^T \mathbf{K} \boldsymbol{\beta}, \tag{12.43}$$

where $\mathbf{K}_{i,l} = K(\mathbf{X}_i, \mathbf{X}_l), 1 \leq i, l \leq n$. See also Section 2.6.5.

In practice, the Gaussian *radial basis function* (RBF) kernel is perhaps the most popular kernel function. The Gaussian RBF kernel function is

$$K(\mathbf{x}_1, \mathbf{x}_2) = \exp(-\gamma \|\mathbf{x}_1 - \mathbf{x}_2\|_2^2).$$

where the parameter $\gamma$ controls the nonlinearity of the fitted classification boundary. For a given $\lambda$, the larger the $\gamma$ the higher degree of nonlinearity. Thus, $\gamma$ serves as an important regularizer like $\lambda$. It is wise to jointly select $\lambda, \gamma$ by using data-driven tuning methods such as cross-validation. Another popular kernel function is the polynomial kernel defined as

$$K(\mathbf{x}_1, \mathbf{x}_2) = (\mathbf{x}_1^T \mathbf{x}_2 + 1)^d.$$

where $d$ is the degree parameter.

### *12.5.2 Generalizations of SVMs*

In light of (12.41), Lin (2002) provided a statistical explanation of the support vector machine. Note that $\frac{1}{n} \sum_{i=1}^{n} [1 - Y_i f(\mathbf{X}_i))]_+$ is the empirical version of the expected hinge loss $\mathrm{E}\left([1 - Y f(\mathbf{X}))]_+\right)$, whose the minimizer is exactly the Bayes rule, i.e.,

$$\arg\min_f \mathrm{E}\left([1 - Y f(\mathbf{X})]_+\right) = \mathrm{sign}(p_+(\mathbf{x}) - p_-(\mathbf{x})). \tag{12.44}$$

where $p_+(\mathbf{x}) = \Pr(Y = 1|\mathbf{X} = \mathbf{x})$ and $p_-(\mathbf{x}) = \Pr(Y = -1|\mathbf{X} = \mathbf{x})$. This statistical viewpoint suggests that the SVM directly approximates the Bayes rule without directly estimating the conditional class probability.

The statistical explanation of the SVM naturally motivates us to ask whether other loss functions can be used in (12.41) to replace the hinge loss and produce a good classifier. The answer is positive. Let $\phi(t)$ denote a general loss function. The corresponding risk is $\mathrm{E}\left(\phi(Y f(\mathbf{X}))\right)$ and the conditional risk is $\mathrm{E}\left[\phi(Yf(\mathbf{X})) \mid \mathbf{X}\right]$. Because $Y = 1, -1$ with probability $p_+(\mathbf{X})$ and $p_-(\mathbf{X})$, respectively, we can write

$$\mathrm{E}\left[\phi(Yf(\mathbf{X})) \mid \mathbf{X} = \mathbf{x}\right] = \phi(f(\mathbf{x}))p_+(\mathbf{x}) + \phi(-f(\mathbf{x}))p_-(\mathbf{x}). \tag{12.45}$$

Given $\mathbf{x}$, define

$$f^*(\mathbf{x}) = \arg\min_f \phi(f)p_+(\mathbf{x}) + \phi(-f)p_-(\mathbf{x}). \tag{12.46}$$

Then $f^*$ is the population minimizer of $\phi$-risk $\mathrm{E}\left(\phi(Yf(\mathbf{X}))\right)$. We say $\phi$ is *Fisher consistent* if

$$\mathrm{sign}(f^*(\mathbf{x})) = \mathrm{sign}(p_+(\mathbf{x}) - p_-(\mathbf{x})) \quad \text{for any} \quad \mathbf{x}. \tag{12.47}$$

The hinge loss is Fisher consistent and so is the exponential loss discussed for AdaBoost in Section 12.4.4.

Lin (2004) and Bartlett, Jordan and McAuliffe (2006) offered the next two Theorems that allow us to easily verify whether a given loss function is Fisher consistent without computing the population minimizer explicitly.

**Theorem 12.1** *Consider a function $\phi$ satisfying the following assumptions:*

*A1.* $\phi(t) < \phi(-t)$, $\forall t > 0$.

*A2.* $\phi'(0) \neq 0$ *exists.*

*A3. the conditional $\phi$ risk* $\mathrm{E}[\phi(Yf(\mathbf{X})|\mathbf{X})]$ *has a global minimizer.*

*Then the loss function $\phi$ is Fisher consistent.*

**Theorem 12.2** *Let $\phi$ be a convex loss function and if it is differentiable at* 0 *with* $\phi'(0) < 0$, *then $\phi$ is Fisher-consistent.*

For binary data with coding $Y = \pm 1$, the conditional log-likelihood is

$$\log P(Y = -1|\mathbf{X}) = -\log(1 + \exp(f(\mathbf{X}))) = -\log(1 + \exp(-Yf(\mathbf{X})))$$

since $Y = -1$ and

$$\log P(Y = 1|\mathbf{X}) = \log \frac{\exp(f(\mathbf{X}))}{1 + \exp(f(\mathbf{X}))} = -\log(1 + \exp(-Yf(\mathbf{X})))$$

as $Y = 1$. Thus, the negative log-likelihood is $\log(1 + \exp(-Yf(\mathbf{X})))$ and the logistic regression loss function

$$\phi^{\mathrm{logit}}(t) = \log(1 + e^{-t}).$$

By Theorem 12.2 it is easy to verify that the logistic regression loss is Fisher consistent. Indeed, the population minimizer can be shown to be

$$f^*(\mathbf{x}) = \log\left(\frac{p_+(\mathbf{x})}{p_-(\mathbf{x})}\right).$$

Another popular Fisher consistent loss is the Huberized hinge loss (Wang, Zhu and Zou, 2008) defined as

$$\phi^{\text{Huber}}(t) = \begin{cases} 0, & t > 1 \\ (1-t)^2/2\delta, & 1-\delta < t \le 1 \\ 1-t-\delta/2, & t \le 1-\delta, \end{cases} \tag{12.48}$$

where $\delta > 0$ is a pre-specific constant. Substituting the Huberized hinge loss into (12.49) yields the Huberized SVM. Figure 12.8 shows a graphical comparison of the logistic regression loss, the hinge loss and the Huberized hinge loss.

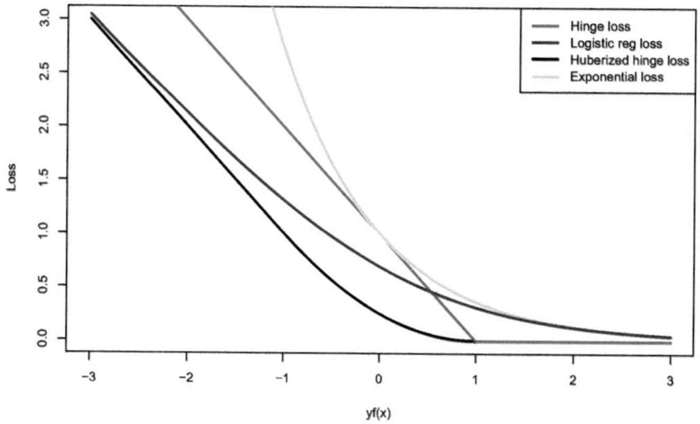


Figure 12.8: Three popular large margin loss functions: the hinge loss, the logistic regression loss and the Huberized hinge loss.

Given a Fisher consistent loss function, its empirical loss is $\frac{1}{n}\sum_{i=1}^{n} \phi(Y_i f(\mathbf{X}_i))$. Then a general large margin classifier is defined by

$$\min_{f \in \mathcal{H}_K} \frac{1}{n}\sum_{i=1}^{n} \phi(Y_i f(\mathbf{X}_i)) + \lambda \|f\|^2_{\mathcal{H}_K}. \tag{12.49}$$

Similarly, the theory of reproducing kernel Hilbert spaces guarantees that the solution to (12.49) has a finite dimensional expression $\widehat{f}(\mathbf{x}) = \beta_0 + \sum_{i=1}^{n} \beta_i K(\mathbf{x}, \mathbf{X}_i)$ and $\widehat{\beta}_0, \widehat{\boldsymbol{\beta}}$ are the solutions to the following finite dimensional optimization problem

$$\min_{\beta_0, \boldsymbol{\beta}} \frac{1}{n}\sum_{i=1}^{n} \phi\left( Y_i(\beta_0 + \sum_{l=1}^{n} \beta_l K(\mathbf{X}_i, \mathbf{X}_l)) \right) + \lambda \boldsymbol{\beta}^T \mathbf{K} \boldsymbol{\beta}. \tag{12.50}$$

For example, with the logistic regression loss, (12.49) yields the *kernel logistic regression classifier*

$$\min_{f \in \mathcal{H}_K} \frac{1}{n} \sum_{i=1}^{n} \phi^{\text{logit}}(Y_i f(\mathbf{X}_i)) + \lambda \|f\|^2_{\mathcal{H}_K}. \tag{12.51}$$

The kernel logistic regression classifier offers a natural estimator of the conditional class probability

$$\widehat{p}_+(\mathbf{x}) = \frac{\exp(\widehat{f}(\mathbf{x}))}{1+\exp(\widehat{f}(\mathbf{x}))}, \quad \widehat{p}_-(\mathbf{x}) = \frac{1}{1+\exp(\widehat{f}(\mathbf{x}))}. \tag{12.52}$$

Two other interesting examples of the general formulation (12.49) are *$\psi$-learning* (Shen, Tseng, Zhang and Wong, 2003) and *distance-weighted discrimination* (Marron, Todd and Ahn, 2007; Wang and Zou, 2018).

## 12.6 Sparse Classifiers via Penalized Empirical Loss

### *12.6.1 The importance of sparsity under high-dimensionality*

In high-dimensional classification problems the number of covariates is much bigger than the sample size. For example, consider Golub's leukemia data (Golub *et al.*,1999) where there are 47 patients with acute lymphoblastic leukemia and 25 patients with acute myeloid leukemia and for each patient 7129 gene expressions are measured. Modern classifiers such as random forests and kernel support vector machines are engineered to solve the traditional classification problems where the dimension is small. One should be cautious when applying these methods to high-dimensional problems. The reason is that these classifiers treat all covariates equally in their algorithm, but high-dimensional datasets tend to have many unimportant variables. If we do not try to differentiate important and unimportant features, we would compromise classification accuracy due to *noise accumulation* (see Section 1.3.2) and the resulting classifier does not have good interpretability, a very important goal in scientific studies.

We conduct two experiments to demonstrate the negative impact of unimportant variables on classification. In the first example we apply random forests to the Email Spam dataset (Hastie, Tibshirani and Friedman, 2009) that consists of 4601 email messages of which about 40% are spam emails. There are also 57 predictors such as frequencies of certain words, total number of capital letters, etc. This dataset is publicly available at the UCI Machine Learning Repository (Bache and Lichman, 2013). The goal is to classify spams from real emails. We randomly split the dataset into a training set of 400 observations and a testing set of 4201 observations. We used R package `randomForest` to fit a random forest classifier to the training data and computed its misclassification error on the testing data ($\approx 7.25\%$). Next, we add

443 standard normal random variables into the predictor set, which are clearly unimportant features for this email classification. When random forests are applied to this augmented data with 500 predictors, which consists of at least 443 noise features, its classification error increases significantly as shown in Figure 12.9. In the same figure we also present another implementation of random forests called $t$-test-screening-random forests (T-RF). In the $t$-test screening step (Fan and Fan (2008), see Chapter 8.5.3) we conduct the two-sample $t$ test on each predictor and pick the top 67 (which is approximately $\sqrt{n/\log(n)}, n = 400$) with the largest absolute $t$-test value. The $t$-test screening helps remove most of the noise variables but still kept a few. Consequently, T-RF performs significantly better than the naive RF but slightly worse than RF using the original 57 predictors.

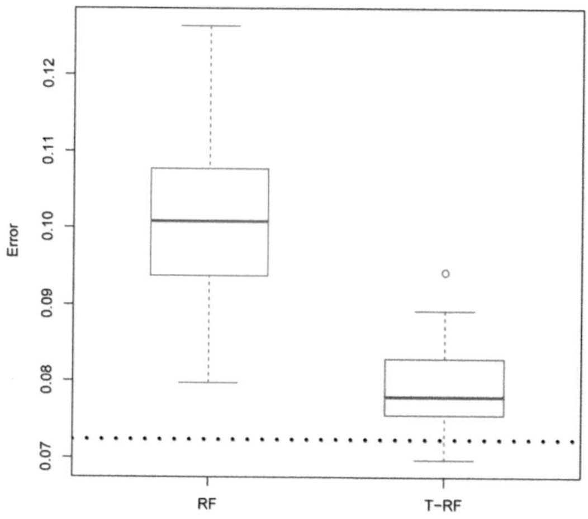


Figure 12.9: Email spam data with noise features. We generated 100 augmented data and fitted 100 naive RF (red) and T-RF (blue). Their median error rates are 10% and 7.8%, respectively. The dotted line corresponds to RF using the original 57 predictors.

The second example is on the SVM. For the experiment we consider a simulation data model (Zhu, Rosset, Hastie and Robert, 2003; Zou, 2007). As shown in Figure 12.10, one class (blue triangle) has two independent standard normal predictors and the other class (red circle) also has two independent standard normal predictors $(x_1, x_2)$ but conditioned on $4.5 \leq x_1^2 + x_2^2 \leq 8$. Both classes have the same prior probability. We further add $q$ independent standard normal variables to the predictor set. Figure 12.10 suggests that a linear classifier will not perform well. We use an enlarged dictionary of predictors $D = \{\sqrt{2}X_j, \sqrt{2}X_jX_k, X_j^2, j, k = 1, 2, \ldots, q+2\}$ to build the linear

SVM classifier. We fit three different SVM classifiers to the training data of 100 observations and compute their misclassification errors on an independent test set of 20000 observations. When $q$ is increased from 2 to 8, the number of working predictors jumps from 14 to 65. The standard SVM uses all predictors. Figure 12.11 shows that the performance of the standard SVM degrades quickly as noise variables cumulate. In Figure 12.11 the $\ell_1$ SVM is much more resistant to the presence of noise variables. The reason is that the $\ell_1$ SVM is able to conduct meaningful feature selection, hence its performance is much less influenced by these added noise variables.

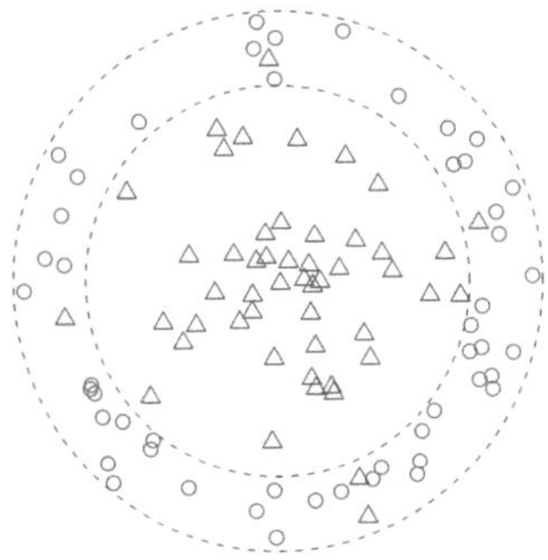

Figure 12.10: A sample drawn from the simulation model. The data are projected onto the plane of the first two predictors.

### *12.6.2 Sparse support vector machines*

To motivate the $\ell_1$ SVM, let us consider the *loss* + *penalty* formulation of the standard SVM where

$$\min_{\beta_0,\boldsymbol{\beta}} \frac{1}{n}\sum_{i=1}^{n}\left[1 - Y_i(\beta_0 + \mathbf{X}_i^T\boldsymbol{\beta})\right]_+ + \lambda\|\boldsymbol{\beta}\|_2^2. \tag{12.53}$$

Because of the $\ell_2$ penalty the standard SVM uses all variables in its classification, i.e., the solution has no zero elements. Ideally, we should discard all unimportant features and only use the important features in the model. To this end, we can replace the $\ell_2$ penalty with another penalty that is good at regularizing variability and also encourages sparsity. The $\ell_1$ penalty (or Lasso penalty) is such a good choice. The $\ell_1$ SVM is defined as follows:

$$\min_{\beta_0,\boldsymbol{\beta}} \frac{1}{n}\sum_{i=1}^{n}\left[1 - Y_i(\beta_0 + \mathbf{X}_i^T\boldsymbol{\beta})\right]_+ + \lambda\|\boldsymbol{\beta}\|_1. \tag{12.54}$$

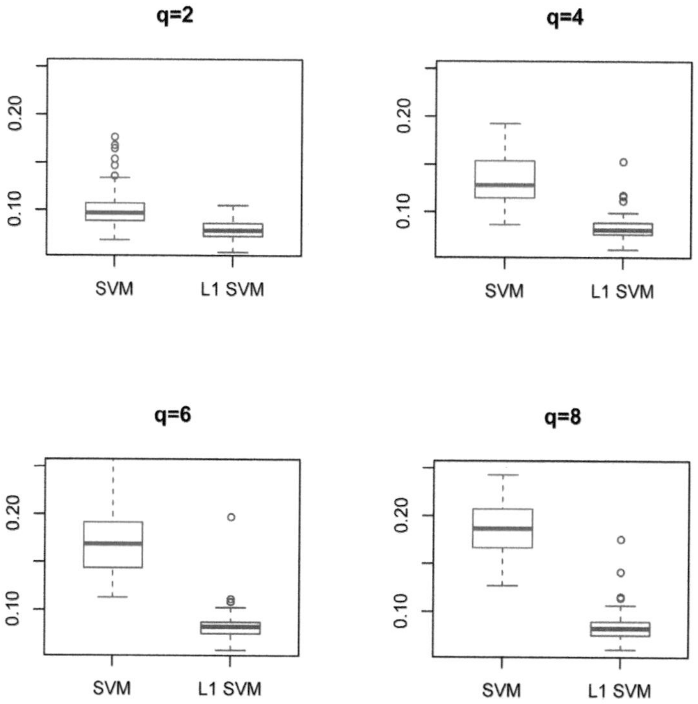


Figure 12.11: Comparisons of misclassification rates with and without feature selection. The simulation results are based on 100 runs. Shown in the figure are the boxplots of misclassification errors. The predictor size $p$ grows quadratically with the number of added noise variables $q$. The $\ell_1$ SVM uses the Lasso $\ell_1$ penalty to perform automatic variable selection.

The $\ell_1$ SVM was discussed in Bradley and Mangasarian (1998). Zhu *et al.* (2003) derived a path-following algorithm for computing the entire solution paths of the $\ell_1$ SVM. Their algorithm is a generalization of the least angle regression algorithm by Efron, Hastie, Johnstone and Tibshirani (2004) for solving the Lasso penalized least squares. Statistical risk properties inherit from those in Section 5.9.

One can also employ concave penalty functions to achieve sparsity in the SVM. For example, Zhang, Ahn, Lin and Park (2006) applied the SCAD penalized SVM to do gene selection in cancer classification.

### *12.6.3 Sparse large margin classifiers*

We can generalize the $\ell_1$ SVM to a whole class of sparse *large margin classifiers*. Let $\phi(\cdot)$ be a Fisher-consistent margin loss function and consider a general sparse-induced penalty function $p_\lambda(\cdot)$. A sparse large margin classifier is defined as $\text{sign}(\widehat{\beta}_0 + \mathbf{x}^T\widehat{\boldsymbol{\beta}})$, where $(\widehat{\beta}_0, \widehat{\boldsymbol{\beta}})$ is obtained by solving the following minimization problem:

$$\min_{\beta_0,\boldsymbol{\beta}} \frac{1}{n}\sum_{i=1}^{n} \phi\left(Y_i(\beta_0 + \mathbf{X}_i^T\boldsymbol{\beta})\right) + \sum_j p_\lambda(|\beta_j|). \tag{12.55}$$

When $\phi$ is the hinge loss and $p_\lambda(|\beta|) = \lambda|\beta|$, the above formulation yields the $\ell_1$ SVM. When $\phi$ is chosen to be $\phi^{\text{logit}}(t) = \log(1 + e^{-t})$, (12.55) becomes the sparse penalized logistic regression. Theoretical properties of the $\ell_1$ and folded-concave penalized logistic regression have been presented in Chapter 5. Friedman, Hastie and Tibshirani (2010) derived a very efficient algorithm to compute the Elastic Net penalized logistic regression. Their algorithm is a combination of Newton-Raphson and coordinate descent. First, we follow the Newton-Raphson algorithm to get a quadratic approximation of the logistic regression loss. Then, a very fast coordinate descent algorithm can be used to find the minimizer of the approximate objective function. The final solution is obtained by iterating between the Newton-Raphson step and the coordinate descent step. Their algorithm has been implemented in the R package `glmnet`.

The *Huberized SVM* is another popular large margin classifier which uses a Huberized hinge loss (Wang, Zhu and Zou, 2008). The Huberized hinge loss is convex and differentiable everywhere. However, the algorithm used for `glmnet` does not work for the $\ell_1$ penalized Huberized SVM, because the Huberized hinge loss does not have a second order derivative. Yang and Zou (2013) developed a unified *coordinate-majorization-descent* (CMD) algorithm for fitting a whole class of sparse large margin classifiers, as long as the loss function satisfies a *quadratic majorization condition.* Yang and Zou's algorithm is implemented in the R package `gcdnet`.

A margin loss function is said to satisfy the quadratic majorization condition if

$$\phi(t + a) \leq \phi(t) + \phi'(t)a + \frac{M}{2}a^2 \quad \forall t, a. \tag{12.56}$$

It is easy to check that $M = \frac{1}{4}$ for the logistic regression loss and $M = \frac{2}{\delta}$ for the Huberized hinge loss. We now explain the CMD algorithm for Elastic Net penalized $\phi$ classifier under (12.56). The optimization problem is

$$\min_{\beta_0,\boldsymbol{\beta}} \frac{1}{n}\sum_{i=1}^{n} \phi\left(Y_i(\beta_0 + \mathbf{X}_i^T\boldsymbol{\beta})\right) + \sum_j p_{\lambda_1,\lambda_2}(|\beta_j|).$$

where

$$p_{\lambda_1,\lambda_2}(\boldsymbol{\beta}) = \sum_{j=1}^{p} p_{\lambda_1,\lambda_2}(\beta_j) = \sum_{j=1}^{p}\left(\lambda_1|\beta_j| + \frac{\lambda_2}{2}\beta_j^2\right).$$

The Elastic Net outperforms the Lasso when there are strongly correlated variables in the predictor set, which is common in high-dimensional data. Without loss of generality, assume the input data are standardized: $\frac{1}{n}\sum_{i=1}^{n} X_{ij} = 0$,

$\frac{1}{n}\sum_{i=1}^{n} X_{ij}^2 = 1$, for $j = 1, \ldots, p$. Define the current margin $r_i = Y_i(\widetilde{\beta}_0 + \mathbf{X}_i^T \widetilde{\boldsymbol{\beta}})$ and

$$F(\beta_j|\widetilde{\beta}_0, \widetilde{\boldsymbol{\beta}}) = \frac{1}{n}\sum_{i=1}^{n} \phi(r_i + Y_i X_{ij}(\beta_j - \widetilde{\beta}_j)) + p_{\lambda_1,\lambda_2}(|\beta_j|).$$

The standard coordinate descent algorithm sequentially updates $\widetilde{\beta}_j$ by

$$\widetilde{\beta}_j = \arg\min_{\beta_j} F(\beta_j|\widetilde{\beta}_0, \widetilde{\boldsymbol{\beta}}) \tag{12.57}$$

till convergence is declared. The major difficulty is that (12.57) does not have a closed form solution, since the coordinate descent algorithm usually needs to be used for solving such a convex problem tens of thousands (or more) times. Therefore, it is crucial that each coordinate update should remain simple. The CMD algorithm overcomes that computational obstacle by integrating the Majorization-Minimization principle (Hunter and Lange, 2000) into the coordinate descent framework. By (12.56), we construct a majorization function of $F(\beta_j|\widetilde{\beta}_0, \widetilde{\boldsymbol{\beta}})$, as follows:

$$\begin{aligned} Q(\beta_j|\widetilde{\beta}_0, \widetilde{\boldsymbol{\beta}}) &= n^{-1}\sum_{i=1}^{n} \phi(r_i) + n^{-1}(\sum_{i=1}^{n} \phi'(r_i) Y_i X_{ij})(\beta_j - \widetilde{\beta}_j) \\ &\quad + \frac{M}{2}(\beta_j - \widetilde{\beta}_j)^2 + p_{\lambda_1,\lambda_2}(|\beta_j|). \end{aligned} \tag{12.58}$$

CMD sets $\widetilde{\beta}_j = \widehat{\beta}_j^{\mathrm{C}}$ as the new estimate. It is easy to see that

$$\widehat{\beta}_j^{\mathrm{C}} = (M + \lambda_2)^{-1}\, S\left(M\widetilde{\beta}_j - n^{-1}\sum_{i=1}^{n} \phi'(r_i) Y_i X_{ij}, \lambda_1\right), \tag{12.59}$$

where $S(z,t) = (|z| - t)_+ \mathrm{sign}(z)$ is the soft-thresholding function. The intercept $\beta_0$ can be updated similarly by

$$\widehat{\beta}_0^{\mathrm{C}} = \widetilde{\beta}_0 - \frac{1}{Mn}\sum_{i=1}^{n} \phi'(r_i) Y_i. \tag{12.60}$$

Algorithm 12.4 summarizes the above CMD algorithm.

## 12.7 Sparse Discriminant Analysis

Despite its simplicity, LDA has been proven useful in many applications with conventional data where sample size is larger than the number of covariates. The simplicity of LDA is actually considered as its most attractive merit. Because of that, many researchers have tried to apply LDA to high-dimensional classification problems. The direct application of LDA has an obvious technical issue: when the number of covariates exceeds the sample size, the estimator of the common within-class covariance matrix, $\widehat{\boldsymbol{\Sigma}}$, becomes

**Algorithm 12.4** The CMD algorithm for Elastic Net penalized large margin classifiers under the quadratic majorization condition (12.56) .

1. Initialize $(\widetilde{\beta}_0, \widetilde{\boldsymbol{\beta}})$.
2. Iterate 2(a)-2(b) until convergence:

2(a) Cyclic coordinate descent: for $j = 1, 2, \ldots, p$,

- (2.a.1) Compute $r_i = y_i(\widetilde{\beta}_0 + \mathbf{x}_i^{\mathsf{T}}\widetilde{\boldsymbol{\beta}})$.
- (2.a.2) Compute (12.59).
- (2.a.3) Set $\widetilde{\beta}_j = \widehat{\beta}_j^{\mathrm{C}}$.

2(b) Update the intercept term

- (2.b.1) Re-compute $r_i = y_i(\widetilde{\beta}_0 + \mathbf{x}_i^{\mathsf{T}}\widetilde{\boldsymbol{\beta}})$.
- (2.b.2) Compute (12.60).
- (2.b.3) Set $\widetilde{\beta}_0 = \widehat{\beta}_0^{\mathrm{C}}$.

---

singular, and hence does not have a proper inverse. To avoid the singularity issue, an ad hoc solution is replacing $\widehat{\boldsymbol{\Sigma}}^{-1}$ with the generalized inverse $\widehat{\boldsymbol{\Sigma}}^{-}$, but this practice is far from being satisfactory. A better solution is Friedman's RDA where we can replace $\widehat{\boldsymbol{\Sigma}}^{-1}$ by $(\widehat{\boldsymbol{\Sigma}} + \alpha\mathbf{I})^{-1}$ for some data-driven positive $\alpha$. RDA only provides a way to bypass the singularity issue, but it does not handle high dimensionality nicely. In high-dimensional classification problems the fundamental challenge is that we cannot afford to estimate the full model with a much smaller sample size. When the dimension exceeds the sample size, Bickel and Levina (2004) showed that a naive Bayes version of LDA (in which $\widehat{\boldsymbol{\Sigma}}$ is a diagonal matrix) can perform much better than the usual LDA.

However, it is interesting to point out that the difficulty of estimating $\boldsymbol{\Sigma}$ is not the root for the failure of LDA for high-dimensional classification problems. To illustrate this point, consider a very simple LDA model with $\boldsymbol{\Sigma}$ being the identity matrix. Suppose that we know that $\boldsymbol{\Sigma}$ is the identity matrix and thus need not to estimate $\boldsymbol{\Sigma}$ at all. In this much simplified situation, we do not have the difficulty of estimating a large matrix $\boldsymbol{\Sigma}$ or its inverse. However, Fan and Fan (2008) showed that the noise accumulation in estimating the mean vectors may degrade the LDA classifier to no better than random guessing. This striking result reveals the fundamental role of feature selection in high-dimensional classification and suggests that a successful high-dimensional generalization of LDA relies on a proper handling of sparsity.

We will introduce several methods that generalize LDA to high-dimensional classification by performing good feature selection. The pioneering methods are based on the *independent rule*. Two most important examples include *nearest shrunken centroids classifier* (NSC) (Tibshirani, Hastie, Narasimhan and Chu, 2002) and *Features Annealed Independent Rule* (FAIR)

(Fan and Fan, 2008). An *independent rule* completely ignores the correlation between features and performs feature selection by simply comparing the mean vectors. They are straightforward, computationally efficient and often have surprisingly good performance in applications. However, ignoring correlation leads to biased feature selection, which in turns implies that the independent rule methods may not be theoretically optimal. There are other approaches to sparse discriminant analysis. An incomplete list includes Wu *et al.* (2008), Witten and Tibshirani (2011), Cai and Liu (2011), Clemmensen *et al.* (2012), Mai, Yuan and Zou (2012) and Fan, Feng and Tong (2012). Rigorous theories have been established for *direct sparse discriminant analysis* (Mai, Yuan and Zou, 2012), *Linear Programming Discriminant* (Cai and Liu, 2011) and Regularized Optimal Affine Discriminant (Fan, Feng and Tong, 2012). These methods are generally more reliable than the sparse classifiers based on independence rules.

### 12.7.1 Nearest shrunken centroids classifier

The *nearest shrunken centroids classifier* (*NSC*) is a sparse version of the so-called independence rule. The independence rule aims to bypass the difficulty in estimating $\mathbf{\Sigma}$ and $\mathbf{\Sigma}^{-1}$ by using a very simple (but biased) estimator: it only takes the diagonals of the usual sample estimator of $\mathbf{\Sigma}$. In other words, an independent rule ignores the correlation structure between variables. In this sense, the independent rule is like a parametric naive Bayes classifier. Bickel and Levina (2004) showed an interesting result that the independent rule can be better than the usual LDA when dimension is high.

The nearest shrunken centroids classifier (NSC) proposed by Tibshirani *et al.* (2002) further uses shrinkage estimators of the mean vectors. Let $n_k$ be the number of training sample for class $k$ for $k \in [K]$,

$$\widehat{\mu}_{kj} = n_k^{-1} \sum_{Y_i = c_k} X_{ij} \quad \text{and} \quad \widehat{\mu}_j = n^{-1} \sum_{i=1}^{n} X_{ij}, \quad \text{for } j = 1, \cdots, p$$

be the sample mean for the $j^{th}$ variable in class $C_k$ and the overall mean for the $j^{th}$ feature, respectively. Define $t$-type of statistic

$$d_{kj} = \frac{\widehat{\mu}_{kj} - \widehat{\mu}_j}{m_k(s_j + s_0)}, \tag{12.61}$$

where $m_k = \sqrt{1/n_k - 1/n}$ and $s_j$ is the pooled variance given by

$$s_j^2 = \frac{1}{n-K} \sum_{k=1}^{K} \sum_{Y_i = c_k} (X_{ij} - \widehat{\mu}_{kj})^2.$$

The number $s_0$ is a positive constant across all features, added to stabilize the algorithm numerically. Tibshirani *et al.* (2002) recommended using $s_0$

as the median value of $\{s_j\}_{j=1}^p$ in practice. NSC shrinks $d_{jk}$ to $d'_{jk}$ by soft-thresholding

$$d'_{kj} = \text{sign}(d_{jk})(|d_{jk}| - \Delta)_+, \tag{12.62}$$

where $\Delta$ is a positive constant that determines the amount of shrinkage. It is recommended that $\Delta$ is chosen by cross-validation. Then the *shrunken centroids* are defined as

$$\widehat{\mu}'_{kj} = \widehat{\mu}_j + m_k(s_j + s_0)d'_{jk}, \tag{12.63}$$

which shrinks the centroid $\widehat{\mu}_{jk}$ towards the common one $\widehat{\mu}_j$. Define the discriminant score for class $\mathcal{C}_k$ as

$$\delta_k(\mathbf{x}) = \sum_{j=1}^p \frac{(x_j - \widehat{\mu}'_{kj})^2}{(s_j + s_0)^2} - 2\log\widehat{\pi}_k, \tag{12.64}$$

and the NSC classification rule is $\arg\min_k \delta_k(\mathbf{x})$ for each given $\mathbf{x}$.

The shrunken centroids provide variable selection in NSC. Given $\Delta$, let $\mathcal{N} = \{j : d'_{kj} = 0 \;\forall\; k\}$. For $j \in \mathcal{N}$, the corresponding shrunken centroid mean $\widehat{\mu}'_{kj}$ equals $\widehat{\mu}_j$ for all $k$. Then discriminant score can be written as

$$\delta_k(\mathbf{x}) = \sum_{j\notin\mathcal{N}} \frac{(x_j - \widehat{\mu}'_{kj})^2}{(s_j + s_0)^2} + \sum_{j\in\mathcal{N}} \frac{(x_j - \widehat{\mu}_j)^2}{(s_j + s_0)^2} - 2\log\widehat{\pi}_k.$$

Since the second term is independent of $k$, NSC classification rule can be written as

$$\arg\min_k \sum_{j\notin\mathcal{N}} \frac{(x_j - \widehat{\mu}'_{kj})^2}{(s_j + s_0)^2} - 2\log\widehat{\pi}_k.$$

Therefore, $\mathcal{N}$ represents the set of discarded features.

### *12.7.2 Features annealed independent rule*

*Features annealed independent rule* (*FAIR*) (Fan and Fan ,2008) inherits the idea behind NSC that only the features with distinct means should be responsible for classification, except that it uses hard-thresholding rather than soft-thresholding for feature selection. The main focus in FAIR is on the theoretical understanding of sparse independent rules. For clarity in theories, FAIR focuses on binary classification.

For the $j$th variable, FAIR computes its modified two-sample $t$-statistic

$$t_j = \frac{\widehat{\mu}_{1j} - \widehat{\mu}_{2j}}{\sqrt{s_{1j}^2 + s_{2j}^2}},$$

where $s_{1j}^2$ and $s_{2j}^2$ are with-in group sample variance defined by

$$s_{kj}^2 = (n_k - 1)^{-1} \sum_{Y_i = c_k} (X_{ij} - \widehat{\mu}_{kj})^2, \qquad k = 1, 2, \quad j = 1, \cdots, p.$$

For a given threshold $\Delta > 0$, FAIR chooses features with $|t_j| \geq \Delta$ and applies the naive Bayes rule (LDA (12.4) with correlation ignored; see also Example 12.1) to the remaining features. Specifically, it classifies an observation to Class 2 if and only if

$$\sum_{j=1}^{p} s_j^{-1}(\widehat{\mu}_{2j} - \widehat{\mu}_{ij})(x_j - \frac{\widehat{\mu}_{1j} + \widehat{\mu}_{2j}}{2})I(|t_j| > \Delta) > 0, \tag{12.65}$$

where $s_j$ is the pooled sample variance for the $j^{th}$ feature defined in the previous section. Under suitable regularity conditions, FAIR is proven to be capable of separating the features with distinct means from those without, even when the equal-variance and the normality assumptions are not satisfied. Specifically, Fan and Fan (2008) proved the following results.

**Theorem 12.3** *Let $\mu_{kj} = \operatorname{E} X_{kj}$ be the population mean of the $j^{th}$ variable in class $k$. Define $\mathcal{S} = \{j : \mu_{1j} \neq \mu_{2j}\}$ and $s$ is the cardinality of the set $\mathcal{S}$. If the following conditions are satisfied:*

1. *there exist constants $\nu_1, \nu_2, M_1, M_2$, such that $\operatorname{E}|X_{kj} - \mu_{kj}|^m \leq m!M_1^{m-2}\nu_1/2$ and $\operatorname{E}|(X_{kj} - \mu_{kj})^2 - \operatorname{Var}(X_{kj})|^m \leq m!M_2\nu_2/2$;*
2. *$\sigma_{k,jj}$ are bounded away from 0;*
3. *there exists $0 < \gamma < 1/3$ and a sequence $b_n \to 0$ such that $\log(p - s) = o(n^{\gamma})$ and $\log s = o(n^{1/2-\gamma}b_n)$;*

*then for $\Delta \sim cn^{\gamma/2}$, where $c$ is a positive constant, we have*

$$\Pr(\min_{j\in\mathcal{S}} |t_j| \geq \Delta, \ \text{and} \ \max_{j\notin\mathcal{S}} |t_j| < \Delta) \to 1.$$

The result shows that set $\mathcal{S}$ can be consistently estimated by using the two-sample $t$-statistics. Fan and Fan (2008) also provided a theoretical solution for choosing the threshold $\Delta$. First, FAIR lets $\Delta = t^*_{(m)}$, where $t^*_{(m)}$ represents the $m$th order statistic of $|t_j|$. Then FAIR chooses $m$ by minimizing the upper bound of the misclassification error rate. Under the normality assumption, the misclassification error rate of FAIR is bounded above by

$$1 - \Phi\left(\frac{\frac{n_1n_2}{pn}(\boldsymbol{\mu}_2 - \boldsymbol{\mu}_1)^T\mathbf{D}^{-1}(\boldsymbol{\mu}_2 - \boldsymbol{\mu}_1)(1 + o_P(1)) + \sqrt{\frac{p}{nn_1n_2}}(n_1 - n_2)}{2\sqrt{\lambda_{\max}}\{1 + \frac{n_1n_2}{pn}(\boldsymbol{\mu}_2 - \boldsymbol{\mu}_1)^T\mathbf{D}^{-1}(\boldsymbol{\mu}_2 - \boldsymbol{\mu}_1)(1 + o_P(1))\}^{1/2}}\right),$$

where $\mathbf{D} = \operatorname{diag}(\boldsymbol{\Sigma})$ and $\lambda_{\max}$ is the largest eigenvalue of $\mathbf{R} = \mathbf{D}^{-\frac{1}{2}}\boldsymbol{\Sigma}\mathbf{D}^{-\frac{1}{2}}$. Without loss of generality, assume that $|t_j|$ is decreasing in $j$. We choose top $m$ features to minimize the upper bound of the misclassification error (replacing $p$ in the above expression by $m$). Such an $m$ is given by, after substituting of estimates,

$$m = \arg\max_m \frac{1}{\widehat{\lambda}^m_{\max}} \frac{n[\sum_{j=1}^m t_j^2 + m(n_1 - n_2)/n]^2}{mn_1n_2 + n_1n_2\sum_{j=1}^m t_j^2},$$

where $\widehat{\lambda}^m_{\max}$ is the largest eigenvalue of the upper-left $m \times m$ block of the sample estimate of $\mathbf{R}$.

### 12.7.3 Selection bias of sparse independence rules

Independence rules such as NSC and FAIR completely ignore the correlation between variables. Although this practice is computationally convenient, one would question how this affects the classification accuracy and the feature selection. Consider the binary LDA model where the Bayes rule is a linear classifier with the classification coefficient vector $\boldsymbol{\beta}^{\text{Bayes}} = \boldsymbol{\Sigma}^{-1}(\boldsymbol{\mu}_2 - \boldsymbol{\mu}_1)$. Define $\mathcal{D} = \{j : \boldsymbol{\beta}^{\text{Bayes}}_j \neq 0\}$ as the discriminative set, which should be the target of sparse LDA methods. The oracle LDA rule is the LDA using only variables in $\mathcal{D}$. As long as the variables in $\mathcal{D}$ are correlated, it is easy to see that any sparse independence rule will have an estimation bias even when it correctly selects the set $\mathcal{D}$. In the following discussion we show that there is actually a selection bias as well.

By definition, sparse independence rules such as NSC and FAIR select important variables by checking their mean differences. Thus, sparse independence rules target the so-called signal set $\mathcal{S} = \{j : \mu_{1j} \neq \mu_{2j}\}$. The following theorem given in Mai, Yuan and Zou (2012) shows that $\mathcal{D}$ and $\mathcal{S}$ need not be identical, and hence sparse independence rules may select a wrong set of variables.

**Theorem 12.4** *Let us partition*

$$\boldsymbol{\Sigma} = \begin{pmatrix} \boldsymbol{\Sigma}_{\mathcal{S},\mathcal{S}} & \boldsymbol{\Sigma}_{\mathcal{S},\mathcal{S}^c} \\ \boldsymbol{\Sigma}_{\mathcal{S}^c,\mathcal{S}} & \boldsymbol{\Sigma}_{\mathcal{S}^c,\mathcal{S}^c} \end{pmatrix} \text{ and } \boldsymbol{\Sigma} = \begin{pmatrix} \boldsymbol{\Sigma}_{\mathcal{D},\mathcal{D}} & \boldsymbol{\Sigma}_{\mathcal{D},\mathcal{D}^c} \\ \boldsymbol{\Sigma}_{\mathcal{D}^c,\mathcal{D}} & \boldsymbol{\Sigma}_{\mathcal{D}^c,\mathcal{D}^c} \end{pmatrix}.$$

*1. If and only if $\boldsymbol{\Sigma}_{\mathcal{S}^c,\mathcal{S}}\boldsymbol{\Sigma}^{-1}_{\mathcal{S},\mathcal{S}}(\boldsymbol{\mu}_{2,\mathcal{S}} - \boldsymbol{\mu}_{1,\mathcal{S}}) = 0$, we have $\mathcal{D} \subseteq \mathcal{S}$;*

*2. If and only if $\boldsymbol{\mu}_{2,\mathcal{D}^c} = \boldsymbol{\mu}_{1,\mathcal{D}^c}$ or $\boldsymbol{\Sigma}_{\mathcal{D}^c,\mathcal{D}}\boldsymbol{\Sigma}^{-1}_{\mathcal{D},\mathcal{D}}(\boldsymbol{\mu}_{2,\mathcal{D}} - \boldsymbol{\mu}_{1,\mathcal{D}}) = 0$, we have $\mathcal{S} \subseteq \mathcal{D}$.*

By Theorem 12.4, we can easily construct examples in which $\mathcal{D}$ and $\mathcal{S}$ are not identical. Consider an LDA model with $p = 25$, $\boldsymbol{\mu}_1 = 0_p$, $\sigma_{ii} = 1$ and $\sigma_{ij} = 0.5, i \neq j$. If $\boldsymbol{\mu}_2 = (\mathbf{1}_5^T, \mathbf{0}_{20}^T)^T$, then $\mathcal{S} = \{1, 2, 3, 4, 5\}$ and $\mathcal{D} = \{j : j = 1, \ldots, 25\}$, since $\boldsymbol{\beta}^{\text{Bayes}} = (1.62 \times \mathbf{1}_5^T, -0.38 \times \mathbf{1}_{20}^T)^T$. On the other hand, if we let $\boldsymbol{\mu}_2 = (3 \times \mathbf{1}_5^T, 2.5 \times \mathbf{1}_5^T)^T$, then $\mathcal{S} = \{1, \ldots, 25\}$ but $\mathcal{D} = \{1, 2, 3, 4, 5\}$, because $\boldsymbol{\beta}^{\text{Bayes}} = (\mathbf{1}_5^T, \mathbf{0}_{20}^T)^T$. These two examples show that ignoring correlation can yield inconsistent variable selection.

The discussion here strongly suggests that when developing a sparse high-dimensional extension of LDA, we should respect the correlation structure.

### *12.7.4 Regularized optimal affine discriminant*

For the Gaussian data with common variance, the linear classifier (12.7) has misclassification rate (12.8). Since the misclassification rate is scale-invariant, the optimal coefficient vector $\boldsymbol{\beta}$ can be found via the optimization problem

$$\min_{\boldsymbol{\beta}} \boldsymbol{\beta}^T \boldsymbol{\Sigma} \boldsymbol{\beta}, \quad \text{s.t. } \boldsymbol{\beta}^T \boldsymbol{\mu}_d = 1.$$

Fan, Feng and Tong (2012) proposed Regularized Optimal Affine Discriminant (*ROAD*) as a sparse LDA method for binary classification that optimizes the misclassification rate:

$$\widehat{\boldsymbol{\beta}}^{\text{ROAD}} = \arg\min_{\boldsymbol{\beta}} \boldsymbol{\beta}^T \widehat{\boldsymbol{\Sigma}} \boldsymbol{\beta}, \text{ s.t. } \boldsymbol{\beta}^T \widehat{\boldsymbol{\mu}}_d = 1 \text{ and } \sum_{j=1}^{p} |\beta_j| \le \tau, \tag{12.66}$$

where $\widehat{\boldsymbol{\mu}}_d = (\widehat{\boldsymbol{\mu}}_2 - \widehat{\boldsymbol{\mu}}_1)/2$. Here, the additional $\ell_1$-constraint is used to encourage the sparsity. ROAD uses the following approximation to (12.66):

$$\widehat{\boldsymbol{\beta}}_{\lambda,\gamma} = \arg\min \frac{1}{2} \boldsymbol{\beta}^T \widehat{\boldsymbol{\Sigma}} \boldsymbol{\beta} + \lambda \|\boldsymbol{\beta}\|_1 + \gamma(\boldsymbol{\beta}^T \widehat{\boldsymbol{\mu}}_d - 1)^2, \tag{12.67}$$

where $\gamma$ is a sufficiently large constant that enforces the equality constraint. Then coordinate descent is used to solve (12.67).

To simplify the theoretical discussion, assume $\Pr(Y=1) = \Pr(Y=2) = 1/2$. Define

$$\begin{aligned} \boldsymbol{\beta}_\tau^{(1)} &= \arg\min_{\boldsymbol{\beta}} \boldsymbol{\beta}^T \boldsymbol{\Sigma} \boldsymbol{\beta}, & \text{s.t. } \boldsymbol{\beta}^T \boldsymbol{\mu}_d = 1 \text{ and } \|\boldsymbol{\beta}\|_1 \le \tau; \\ \boldsymbol{\beta}_\lambda^{(2)} &= \arg\min_{\boldsymbol{\beta}} \boldsymbol{\beta}^T \boldsymbol{\Sigma} \boldsymbol{\beta} + \lambda \|\boldsymbol{\beta}\|_1, & \text{s.t. } \boldsymbol{\beta}^T \boldsymbol{\mu}_d = 1. \end{aligned}$$

Note that $\boldsymbol{\beta}_\tau^{(1)}$ or $\boldsymbol{\beta}_\lambda^{(2)}$ can be viewed as the ROAD solution on the population level for problems (12.66) and (12.67). In particular, $\boldsymbol{\beta}_\infty^{(1)} = \boldsymbol{\beta}_0^{(2)} = \dfrac{\boldsymbol{\Sigma}^{-1}\boldsymbol{\mu}_d}{\boldsymbol{\mu}_d^T \boldsymbol{\Sigma}^{-1} \boldsymbol{\mu}_d}$ is the *Fisher discriminant*.

Fan, Feng and Tong (2012) proved the following two theorems. The first theorem says that so long as $\lambda$ and $s = \|\boldsymbol{\beta}_0^{(2)}\|_0$ are small, the regularized $\boldsymbol{\beta}_\lambda^{(2)}$ is close to the Fisher discriminant at the population level.

**Theorem 12.5** *Let $s = \|\boldsymbol{\beta}_0^{(2)}\|_0$, the number of non-zero elements of the Fisher discriminant. Then, we have*

$$\|\boldsymbol{\beta}_\lambda^{(2)} - \boldsymbol{\beta}_0^{(2)}\|_2 \le \frac{\lambda\sqrt{s}}{\lambda_{\min}(\boldsymbol{\Sigma})},$$

*where $\lambda_{\min}(\boldsymbol{\Sigma})$ is the smallest eigenvalue of $\boldsymbol{\Sigma}$.*

Recall the misclassification error rate $R(\mathbf{w})$ given by (12.8). To show that the misclassification rate of the classifier using

$$\widehat{\boldsymbol{\beta}}_\tau^{(1)} = \arg\min_{\boldsymbol{\beta}} \boldsymbol{\beta}^T \widehat{\boldsymbol{\Sigma}} \boldsymbol{\beta}, \quad \text{s.t. } \|\boldsymbol{\beta}\|_1 \le \tau,\ \boldsymbol{\beta}^T \widehat{\boldsymbol{\mu}}_d = 1$$

is close to that using $\boldsymbol{\beta}_\tau^{(1)}$, we define an intermediate quantity

$$\widehat{\boldsymbol{\beta}}_\tau^{(1)*} = \arg\min_{\boldsymbol{\beta}} \boldsymbol{\beta}^T \boldsymbol{\Sigma} \boldsymbol{\beta}, \quad \text{s.t. } \|\boldsymbol{\beta}\|_1 \le \tau,\ \boldsymbol{\beta}^T \widehat{\boldsymbol{\mu}}_d = 1.$$

The following theorem shows the difference in the misclassification rate.

**Theorem 12.6** *Let* $s_\tau = \|\boldsymbol{\beta}_\tau^{(1)}\|_0$, $\widehat{s}_\tau^* = \|\widehat{\boldsymbol{\beta}}_\tau^{(1)*}\|_0$ *and* $\widehat{s}_\tau = \|\widehat{\boldsymbol{\beta}}_\tau^{(1)}\|_0$. *Assume that* $\lambda_{\min}(\boldsymbol{\Sigma}) = \sigma_0^2 > 0$, $\|\widehat{\boldsymbol{\Sigma}} - \boldsymbol{\Sigma}\|_\infty = O_p(a_n)$ *and* $\|\widehat{\boldsymbol{\mu}}_d - \boldsymbol{\mu}_d\|_\infty = O_p(a_n)$ *for a sequence* $a_n \to 0$. *Then*

$$R(\widehat{\boldsymbol{\beta}}_\tau^{(1)}) - R(\boldsymbol{\beta}_\tau^{(1)}) = O_p(d_n),$$

*where* $b_n = a_n(\tau^2 \vee s_\tau \vee \widehat{s}_\tau^*)$ *and* $d_n = b_n \vee (a_n \widehat{s}_\tau)$.

There are many robust ways to construct estimated means and covariance matrices with uniform convergence. We refer to Chapter 9 for these.

### *12.7.5 Linear programming discriminant*

Cai and Liu (2011) suggested a *linear programming discriminant* (*LPD*) for a sparse LDA. It begins with the assumption that the Bayes classification vector $\boldsymbol{\beta}^{\text{Bayes}} = \boldsymbol{\Sigma}^{-1}(\boldsymbol{\mu}_2 - \boldsymbol{\mu}_1)$ is sparse and the observation that it satisfies the linear equation $\boldsymbol{\Sigma}\boldsymbol{\beta} = \boldsymbol{\mu}_2 - \boldsymbol{\mu}_1$. Then LPD estimates $\boldsymbol{\beta}^{\text{Bayes}}$ by using a *Dantzig selector* formulation:

$$\widehat{\boldsymbol{\beta}} = \arg\min_{\boldsymbol{\beta}} \|\boldsymbol{\beta}\|_1, \text{ s.t. } \|\widehat{\boldsymbol{\Sigma}}\boldsymbol{\beta} - (\widehat{\boldsymbol{\mu}}_2 - \widehat{\boldsymbol{\mu}}_1)\|_\infty \le \lambda_n. \tag{12.68}$$

LPD can be solved by using the primal-dual interior-point method. Under the equal class probability assumption $\Pr(Y = 1) = \Pr(Y = 2) = 1/2$, LPD classifies an observation $\mathbf{x}$ to class 2 if $\widehat{\boldsymbol{\beta}}^T(\mathbf{x} - \widehat{\boldsymbol{\mu}}_a) > 0$, where $\boldsymbol{\mu}_a = (\widehat{\boldsymbol{\mu}}_2 + \widehat{\boldsymbol{\mu}}_1)/2$.

Cai and Liu (2011) showed that the misclassification error of LPD can approach the Bayes error under the LDA model as $n, p \to \infty$. We use the notation in Example 12.1. Recall the Bayes error there is given by $R_{\text{Bayes}} = 1 - \Phi(\Delta_p)$. It is not hard to show that the LPD classifier, conditioned on the given data, has the misclassification error [see Exercise 12.1(c)]

$$R_n = 1 - 0.5\Phi\left(-\frac{(\widehat{\boldsymbol{\mu}}_a - \boldsymbol{\mu}_1)^T\widehat{\boldsymbol{\beta}}}{\widehat{\boldsymbol{\beta}}^T\boldsymbol{\Sigma}\widehat{\boldsymbol{\beta}}}\right) - 0.5\Phi\left(\frac{(\widehat{\boldsymbol{\mu}}_a - \boldsymbol{\mu}_2)^T\widehat{\boldsymbol{\beta}}}{\widehat{\boldsymbol{\beta}}^T\boldsymbol{\Sigma}\widehat{\boldsymbol{\beta}}}\right),$$

where $\widehat{\boldsymbol{\beta}}$ is given by (12.68). It reduces to $R(\widehat{\boldsymbol{\beta}})$ with $R(\cdot)$ given by (12.8) if $\widehat{\boldsymbol{\mu}}_a = \boldsymbol{\mu}_a$. The difference lies in that $R_n$ takes into account the estimation error of $\boldsymbol{\mu}_a$.

**Theorem 12.7** *Suppose $n_1 \asymp n_2, \log p = o(n)$, $\max \sigma_{ii} \le c_0$ and $\Delta_p > c_1$, and $\lambda_n = c_2\sqrt{\Delta_p \log p/n}$, for some positive constants $c_0$ and $c_1$, and sufficiently large constant $c_2$. Then, under the model assumption in Example 12.1, we have*

1. *if*

$$\|\boldsymbol{\beta}^{\text{Bayes}}\|_1/\Delta_p^{1/2} + \|\boldsymbol{\beta}^{\text{Bayes}}\|_1^2/\Delta_p^2 = o\left(\sqrt{\frac{n}{\log p}}\right),$$

*then the misclassification error rate of LPD satisfies*

$$R_n - R_{\text{Bayes}} \to 0$$

*in probability as $n, p \to \infty$;*

2. *if*

$$\|\boldsymbol{\beta}^{\text{Bayes}}\|_1\Delta_p^{1/2} + \|\boldsymbol{\beta}^{\text{Bayes}}\|_1^2 = o\left(\sqrt{\frac{n}{\log p}}\right),$$

*then*

$$\frac{R_n}{R_{\text{Bayes}}} - 1 = O\left((\|\boldsymbol{\beta}^{\text{Bayes}}\|_1\Delta^{1/2} + \|\boldsymbol{\beta}^{\text{Bayes}}\|_1^2)\sqrt{\frac{n}{\log p}}\right)$$

*with probability greater than $1 - O(p^{-1})$.*

### 12.7.6 *Direct sparse discriminant analysis*

Mai, Yuan and Zou (2012) proposed *direct sparse discriminant analysis* (*DSDA*) by utilizing a least squares representation of LDA in the binary case. Let $\widehat{\boldsymbol{\beta}}^{\text{Bayes}}$ be the LDA estimator of the Bayes rule $\boldsymbol{\beta}^{\text{Bayes}} = \boldsymbol{\Sigma}^{-1}(\boldsymbol{\mu}_2 - \boldsymbol{\mu}_1)$. Suppose that the class label is coded by two distinct numeric values $y = -n/n_1$ for class 1 and $y = n/n_2$ for class 2, where $n = n_1 + n_2$. Define

$$(\widehat{\beta}_0^{\text{ols}}, \widehat{\boldsymbol{\beta}}^{\text{ols}}) = \arg\min_{(\beta_0, \boldsymbol{\beta})} \sum_{i=1}^n (Y_i - \beta_0 - \mathbf{X}_i^T\boldsymbol{\beta})^2.$$

It can be shown that $\widehat{\boldsymbol{\beta}}^{\text{ols}} = c\widehat{\boldsymbol{\beta}}^{\text{Bayes}}$ for some positive constant $c$ (see Exercise 12.3), when $p < n$. This means that the sample LDA classification direction can be exactly recovered by doing the least squares computation. Note that the classification direction vector is all we need in binary classification. However, this result does not hold in high-dimensional cases because the

LDA is not well defined. We can consider a penalized least squares formulation to produce a properly modified sparse LDA method. DSDA is defined by

$$\left(\widehat{\beta}_0, \widehat{\boldsymbol{\beta}}^{\text{DSDA}}\right) = \arg\min_{(\beta_0,\boldsymbol{\beta})} n^{-1}\sum_{i=1}^{n}(Y_i - \beta_0 - \mathbf{X}_i^T\boldsymbol{\beta})^2 + p_\lambda(\boldsymbol{\beta}), \tag{12.69}$$

where $p_\lambda(\cdot)$ is a sparse penalty function, such as Lasso and SCAD penalty. We use only the direction $\widehat{\boldsymbol{\beta}}^{\text{DSDA}}$, not the intercept $\widehat{\beta}_0$ to construct the classifier. Mai, Yuan and Zou (2012) show that the optimal choice of the intercept is

$$\widetilde{\beta}_0 = -\widehat{\boldsymbol{\mu}}_a^T\widehat{\boldsymbol{\beta}}^{\text{DSDA}} + \frac{(\widehat{\boldsymbol{\beta}}^{\text{DSDA}})^T\widehat{\boldsymbol{\Sigma}}\widehat{\boldsymbol{\beta}}^{\text{DSDA}}}{\{(\widehat{\boldsymbol{\mu}}_2 - \widehat{\boldsymbol{\mu}}_1)^T\widehat{\boldsymbol{\beta}}^{\text{DSDA}}\}}\log(\frac{n_2}{n_1}), \tag{12.70}$$

among classifiers that predict the class label 2 when $\mathbf{x}^T\widehat{\boldsymbol{\beta}}^{\text{DSDA}} + \beta_0 > 0$, where $\widehat{\boldsymbol{\mu}}_a = (\widehat{\boldsymbol{\mu}}_1 + \widehat{\boldsymbol{\mu}}_2)/2$. In other words, DSDA classifies $\mathbf{x}$ as class 2 when

$$\mathbf{x}^T\widehat{\boldsymbol{\beta}}^{\text{DSDA}} + \widetilde{\beta}_0 > 0.$$

When $n_1 = n_2$, it reduces to $\mathbf{x}^T(\widehat{\boldsymbol{\beta}}^{\text{DSDA}} - \widehat{\boldsymbol{\mu}}_a) > 0$.

DSDA is computationally very efficient, because we can take advantage of the fast algorithms for computing the penalized least squares. Theoretical results have been established for DSDA with the Lasso and SCAD penalties. Note that although the estimator is obtained via penalized least squares, we no longer have a sparse linear regression model as the truth for the theoretical analysis, which is different from settings considered in Chapters 3 and 4.

Recall that $\boldsymbol{\beta}^{\text{Bayes}} = \boldsymbol{\Sigma}^{-1}(\boldsymbol{\mu}_2 - \boldsymbol{\mu}_1)$ represents the Bayes classifier coefficient vector. Recall that $\mathcal{D} = \{j : \beta_j^{\text{Bayes}} \neq 0\}$ and let $s = \|\mathcal{D}\|_0$. We use $\boldsymbol{\Sigma}$ to represent the covariance matrix of the predictors and partition $\boldsymbol{\Sigma}$ as

$$\boldsymbol{\Sigma} = \begin{pmatrix} \boldsymbol{\Sigma}_{\mathcal{D}\mathcal{D}} & \boldsymbol{\Sigma}_{\mathcal{D}\mathcal{D}^c} \\ \boldsymbol{\Sigma}_{\mathcal{D}^c\mathcal{D}} & \boldsymbol{\Sigma}_{\mathcal{D}^c\mathcal{D}^c} \end{pmatrix}.$$

Denote $\boldsymbol{\beta}^* = (\boldsymbol{\Sigma}_{\mathcal{D}\mathcal{D}})^{-1}(\boldsymbol{\mu}_{2\mathcal{D}} - \boldsymbol{\mu}_{1\mathcal{D}})$ and define $\widetilde{\boldsymbol{\beta}}^{\text{Bayes}}$ by letting $\widetilde{\boldsymbol{\beta}}_{\mathcal{D}}^{\text{Bayes}} = \boldsymbol{\beta}^*$ and $\widetilde{\boldsymbol{\beta}}_{\mathcal{D}^c}^{\text{Bayes}} = 0$ and define

$$\kappa = \|\boldsymbol{\Sigma}_{\mathcal{D}^c\mathcal{D}}(\boldsymbol{\Sigma}_{\mathcal{D}\mathcal{D}})^{-1}\|_\infty, \quad \varphi = \|(\boldsymbol{\Sigma}_{\mathcal{D}\mathcal{D}})^{-1}\|_\infty, \quad \Delta = \|(\boldsymbol{\mu}_2 - \boldsymbol{\mu}_1)_{\mathcal{D}}\|_\infty.$$

**Theorem 12.8** *The quantities $\widetilde{\boldsymbol{\beta}}^{\text{Bayes}}$ and $\boldsymbol{\beta}^{\text{Bayes}}$ are equivalent in the sense that $\widetilde{\boldsymbol{\beta}}^{\text{Bayes}} = c\boldsymbol{\beta}^{\text{Bayes}}$ for some positive constant $c$ and the Bayes classifier can be written as assigning $\mathbf{X}$ to class 2 if*

$$\{\mathbf{X}-(\boldsymbol{\mu}_1+\boldsymbol{\mu}_2)/2\}^T\widetilde{\boldsymbol{\beta}}^{\text{Bayes}} + (\widetilde{\boldsymbol{\beta}}^{\text{Bayes}})^T\boldsymbol{\Sigma}\widetilde{\boldsymbol{\beta}}^{\text{Bayes}}\{(\boldsymbol{\mu}_2-\boldsymbol{\mu}_1)^T\widetilde{\boldsymbol{\beta}}^{\text{Bayes}}\}^{-1}\log\frac{\pi_2}{\pi_1} > 0.$$

According to the above theorem it suffices to show that DSDA can consistently recover the support of $\widetilde{\boldsymbol{\beta}}^{\text{Bayes}}$ and estimate $\boldsymbol{\beta}^*$. Non-asymptotic analysis was given in Mai, Yuan and Zou (2012). Here we only present the corresponding asymptotic results to highlight the main points of the theory for DSDA.

We assume the following regularity conditions:

(1). $n, p \to \infty$ and $\log(ps)s^2/n \to 0$;
(2). $\min_{j\in A} |\beta_j^*| \gg \{\log(ps)s^2/n\}^{1/2}$.
(3). $\Delta, \kappa, \varphi$ are constants.

**Theorem 12.9** *Let $\widehat{\boldsymbol{\beta}}^{\text{lasso}}$ be the DSDA estimator with Lasso penalty and $\widehat{\mathcal{D}}^{\text{lasso}} = \{j : \widehat{\beta}_j^{\text{lasso}} \neq 0\}$. If we choose the Lasso penalty parameter $\lambda = \lambda_n$ such that $\min_{j\in\mathcal{D}} |\beta_j^*| \gg \lambda_n \gg \{\log(ps)s^2/n\}^{1/2}$, and further assume $\kappa = \|\boldsymbol{\Sigma}_{\mathcal{D}^c\mathcal{D}}(\boldsymbol{\Sigma}_{\mathcal{D}\mathcal{D}})^{-1}\|_\infty < 1$, then with probability tending to 1, $\widehat{\mathcal{D}}^{\text{lasso}} = \mathcal{D}$ and $\|\widehat{\boldsymbol{\beta}}_{\mathcal{D}}^{\text{lasso}} - \boldsymbol{\beta}^*\|_\infty \leq 4\varphi\lambda_n$.*

### *12.7.7 Solution path equivalence between ROAD and DSDA*

Mai and Zou (2014) proved an interesting equivalence result for ROAD and DSDA. For the sake of presentation, we use the following definition for ROAD:

$$\widehat{\boldsymbol{\beta}}_\lambda^{\text{ROAD}} = \arg\min_{\boldsymbol{\beta}} \frac{n-2}{n}\boldsymbol{\beta}^T\widehat{\boldsymbol{\Sigma}}\boldsymbol{\beta} + \lambda\|\boldsymbol{\beta}\|_1, \quad \text{s.t. } (\widehat{\boldsymbol{\mu}}_2 - \widehat{\boldsymbol{\mu}}_1)^T\boldsymbol{\beta} = 1. \tag{12.71}$$

**Theorem 12.10** *Define $c(\lambda) = (\widehat{\boldsymbol{\mu}}_2 - \widehat{\boldsymbol{\mu}}_1)^T\widehat{\boldsymbol{\beta}}^{\text{DSDA}}$ and $\widetilde{\boldsymbol{\beta}}_\lambda^{\text{DSDA}} = \widehat{\boldsymbol{\beta}}^{\text{DSDA}}/c(\lambda)$. Then*

$$\widetilde{\boldsymbol{\beta}}_\lambda^{\text{DSDA}} = \widehat{\boldsymbol{\beta}}_{\widetilde{\lambda}}^{\text{ROAD}},$$

*where $\widetilde{\lambda} = \frac{\lambda}{n|c(\lambda)|}$.*

Theorem 12.10 is a finite sample result that holds for all data sets. It says that ROAD and DSDA have the same solution paths. We use the prostate cancer dataset (Singh *et al.*, 2002) to provide a numerical demonstration of the path equivalence between ROAD and DSDA. This dataset contains the expression levels of 6033 genes, measured on 50 normal tissues and 52 prostate cancer tumors. We normalized the predictors such that each predictor has zero sample mean and unit sample variance. We took a fine grid of $\lambda$ values and computed the corresponding $\widehat{\boldsymbol{\beta}}^{\text{DSDA}}$. For each $\lambda$ we computed $\widetilde{\lambda}$ according to Theorem 12.10 and then $\boldsymbol{\beta}_{\widetilde{\lambda}}^{\text{ROAD}}$. Figure 12.12 shows that ROAD and DSDA have identical solution paths.

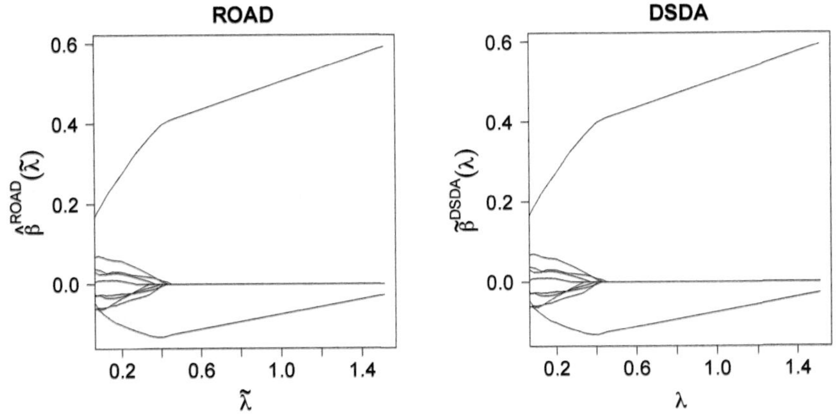


Figure 12.12: Solution paths of ROAD and DSDA on the prostate cancer data. We have computed 6033 coefficient curves but only show 10 curves here for ease of presentation.

## 12.8 Feature Augmention and Sparse Additive Classifiers

### *12.8.1 Feature augmentation*

Fan, Feng, Jiang and Tong (2016) proposed a simple idea to create nonlinear features. The main idea is motivated by the naive Bayes model and logistic regression. It creates nonlinear features by using the marginal log-likelihood ratio. Let $f_{1j}, f_{2j}$ denote the density of $X_j$ in class 1 and 2. Then the naive Bayes classifier will classify $\mathbf{x}$ as class 2 if $\sum_j \log(\frac{f_{2j}(x_j)}{f_{1j}(x_j)}) > 0$. It can be easily implemented by plugging in the kernel density estimators for $f_{2j}(x_j)$ and $f_{1j}(x_j)$. Fan *et al.* (2016) propose to treat $z_j = \log(\frac{f_{2j}(x_j)}{f_{1j}(x_j)})$ as a derived feature, which is the Bayes classifier if only the jth feature is used. These *augmented features* can be applied to any classifiers introduced so far.

As an application, Fan *et al.*(2016) proposed *feature augmentation via nonparametrics and selection* (*FANS*) to incorporate an additive structure in sparse classification. By coding class 2 as 1 and class 1 as 0, they consider a linear logistic regression model for $(\mathbf{Z}, Y)$ with

$$\log\left(\frac{P(Y=1|\mathbf{Z}=\mathbf{z})}{1-P(Y=1|\mathbf{Z}=\mathbf{z})}\right) = a + \beta_1 z_1 + \cdots + \beta_p z_p.$$

The *naive Bayes* classifier corresponds to setting coefficients $a = 0$, $\beta_1 = \cdots = \beta_p = 1$ and is hence improved by this logistic regression. When $p$ is large, we can fit a sparse logistic regression model to $\{(\widehat{\mathbf{z}}_i, Y_i)\}_{i=1}^n$ where the

$j^{th}$ component of $\widehat{\mathbf{z}}_i$ is given by

$$\widehat{z}_{ij} = \log(\frac{\widehat{f}_{2j}(x_j)}{\widehat{f}_{1j}(x_j)}), \quad j = 1, \cdots, p$$

and $\widehat{f}_{1j}$ and $\widehat{f}_{2j}$ are the kernel density estimators based on the data in classes 1 and 2. The kernel density estimator can be computed by using `density` in R. A typical choice of kernel is the Gaussian kernel and a simple rule of thumb for selecting bandwidth is $h = 1.06sn^{-1/5}$ under the Gaussian kernel, where $s$ is the standard deviation of the data.

To avoid using the same data to fit the kernel density and the penalized logistic regression, Fan *et al.* (2016) proposed a data-split method to implement FANS. See Algorithm 12.5.

**Algorithm 12.5** FANS: feature augmentation via nonparametrics and selection

1. Randomly split the data $D = (\mathbf{X}_i, Y_i)_{i=1}^n$ into two parts $\mathcal{D}_1$ and $\mathcal{D}_2$.
2. On $\mathcal{D}_1$, apply kernel density estimation to obtain $\widehat{f}_{1j}(x_j)$ and $\widehat{f}_{2j}(x_j)$ for $j = 1, \cdots, p$.
3. On $\mathcal{D}_2$, compute augmented features $\widehat{Z}_{ij} = \log(\frac{\widehat{f}_{2j}(x_{ij})}{\widehat{f}_{1j}(x_{ij})})$ for $i \in \mathcal{D}_2$ and $j = 1, \cdots, p$.
4. On $\mathcal{D}_2$, fit a sparse penalized linear logistic regression model to $(\widehat{\mathbf{z}}_i, Y_i)$, $i \in \mathcal{D}_2$. Let the estimated coefficients be $(\widehat{a}, \widehat{\boldsymbol{\beta}})$.
5. Given any new $\mathbf{x}^{\text{new}}$, compute $z_j^{\text{new}} = \log(\frac{\widehat{f}_{2j}(x_j^{\text{new}})}{\widehat{f}_{1j}(x_j^{\text{new}})})$ and the estimated conditional probability $\widehat{p}(\mathbf{x}) = \frac{e^{\widehat{a}+\widehat{\boldsymbol{\beta}}^T z^{\text{new}}}}{1+e^{\widehat{a}+\widehat{\boldsymbol{\beta}}^T z^{\text{new}}}}$ for $\Pr(Y = 1|\mathbf{X} = \mathbf{x}^{\text{new}})$.
6. Repeat steps (1)–(5) $L$ times and denote the estimated conditional probabilities as $\widehat{p}_1, \ldots, \widehat{p}_L$. The classified label is 2 if $\frac{1}{L}\sum_{l=1}^L \widehat{p}_l > 0.5$.

A variant of FANS named FANS2 also includes the original $\mathbf{X}$ variables in step 4 and step 5 of Algorithm 12.5, which amounts to fitting the sparse logistic regression model

$$\log\left(\frac{P(Y = 1|\mathbf{X} = \mathbf{x})}{1 - P(Y = 1|\mathbf{X} = \mathbf{x})}\right) = a + \beta_1 Z_1 + \cdots + \beta_p Z_p + \alpha_1 X_1 + \cdots + \alpha_p X_p$$

in Step 4 by penalized logistic regression. Thus, the original features $\{X_j\}_{j=1}^p$ are augmented by $\{Z_j\}_{j=1}^p$. The original features and *augmented features* are then selected by penalized logistic regression. Figure 12.13, compares the misclassification rates for several classifiers on the Email Spam data (Hastie, Tibshirani and Friedman, 2009) with different training/test split ratio. It is clear that FANS2 performs very nicely.

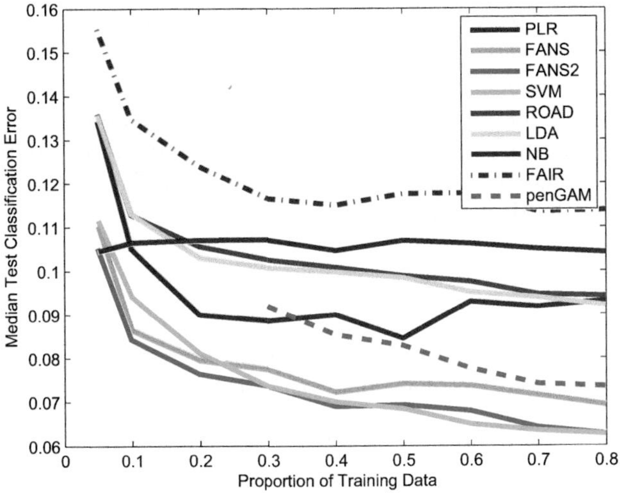


Figure 12.13: E-mail spam data: median test classification error using various proportions of the data as training sets for several classification methods. Taken from Fan, Feng and Tong (2016).

### *12.8.2 Penalized additive logistic regression*

In the generalized additive model for logistic regression, the log-odds function is modelled as

$$f(\mathbf{x}) = \beta_0 + \sum_{j=1}^{p} f_j(X_j)$$

and hence $P(Y = 1|\mathbf{X} = \mathbf{x}) = \frac{e^{f(\mathbf{x})}}{1+e^{f(\mathbf{x})}}$. Furthermore, each univariate function $f_j$ can be modelled by a linear combination of basis functions such as B-splines:

$$f_j(X_j) = \sum_{m=1}^{M_j} \beta_{jm} B_m(X_j).$$

We use $\pm 1$ to code each label and adopt the margin-based logistic regression loss $\phi(t) = \log(1 + e^{-t})$. Meier, van de Geer, and Bühlmann (2008) considered the penalized additive logistic regression estimator

$$\arg\min_{\boldsymbol{\beta}} \frac{1}{n} \sum_{i=1}^{n} \log(1 + e^{-Y_i f(\mathbf{X}_i)}) + \lambda \sum_{j=1}^{p} \sqrt{M_j} \|\boldsymbol{\beta}^{(j)}\|_2 \qquad (12.72)$$

where $\|\boldsymbol{\beta}^{(j)}\|_2 = \sqrt{\sum_m^{M_j} \beta_{jm}^2}$ is the *group-Lasso* penalty (Antoniadis and Fan,

2001; Yuan and Lin, 2006). Obviously, we can use other Fisher-consistent loss functions in (12.72) to get a different sparse additive classifier.

Meier *et al.* (2008) extended the blockwise descent algorithm to solve the group-Lasso penalized logistic regression. A much faster *blockwise-majorization-descent* (*BMD*) was proposed by Yang and Zou (2015) for solving a whole class of group-Lasso learning problems, including the group-Lasso penalized logistic regression and the Huberized SVM. The BMD algorithm is implemented in the R package `gglasso`.

### *12.8.3 Semiparametric sparse discriminant analysis*

The *semiparametric linear discriminant analysis* (*SeLDA*) model (Lin and Jeon, 2003) follows the basic idea from the famous *Box-Cox model*. It assumes that transformed input variables follow the LDA model. Specifically, we have

$$(h_1(X_1), \cdots, h_p(X_p)) \mid Y \sim N(\boldsymbol{\mu}_Y, \boldsymbol{\Sigma}), \tag{12.73}$$

where $h = (h_1, \cdots, h_p)$ is a set of strictly monotone univariate transformations. It is clear that each transformation function $h$ is only unique up to location and scale shifts. For identifiability we assume that $n_2 \geq n_1, \boldsymbol{\mu}_2 = 0$, $\sigma_{jj} = 1, 1 \leq j \leq p$ so that $\boldsymbol{\Sigma}$ is a correlation matrix. The Bayes rule under the SeLDA model is

$$Y^{\text{Bayes}} = I\left((h(\mathbf{X}) - \boldsymbol{\mu}_a)^T \boldsymbol{\beta}^{\text{Bayes}} + \log\frac{\pi_2}{\pi_1}\right) + 1,$$

where $\boldsymbol{\beta}^{\text{Bayes}} = \boldsymbol{\Sigma}^{-1}(\boldsymbol{\mu}_2 - \boldsymbol{\mu}_1)$. A variable $X_j$ is important if and only if $\beta_j^{\text{Bayes}} \neq 0$. For variable selection, we only need to estimate the support of $\boldsymbol{\beta}^{\text{Bayes}}$, but for classification we need to estimate both $\boldsymbol{\beta}^{\text{Bayes}}$ and the transformation function $h_j$, $j = 1, \ldots, p$. Mai and Zou (2015) established the high-dimensional estimation method and theory of SeLDA.

We first estimate the transformation functions. Note that for any continuous univariate random variable $X$, we have

$$\Phi^{-1} \circ F(X) \sim N(0, 1), \tag{12.74}$$

where $F$ is the *cumulative probability function* (*CDF*) of $X$ and $\Phi$ is the CDF of the standard normal distribution. Based on this fact and the assumption (12.74) that $h_j(X_j) \sim N(0, 1)$, we have

$$h_j = \Phi^{-1} \circ F_{2j}, \tag{12.75}$$

where $F_{2j}$ is the conditional CDF of $X_j$ given $Y = 2$. We can have a similar expression based on $F_{1j}$ and we defer this discussion. Let $\widetilde{F}_{2j}$ be the empirical conditional CDF. We need some modification to $\widetilde{F}_{2j}$ before substituting it into (12.75). Define a *winsorization* estimator as

$$\widehat{F}_{2j}^{a,b}(x) = \begin{cases} b & \text{if } \widetilde{F}_{2j}(x) > b; \\ \widetilde{F}_{2j}(x) & \text{if } a \leq \widetilde{F}_{2j}(x) \leq b; \\ a & \text{if } \widetilde{F}_{2j}(x) < a. \end{cases} \tag{12.76}$$

Then our estimator of $h_j$ is

$$\widehat{h}_j(\cdot) = \Phi^{-1} \circ \widehat{F}_{2j}^{a,b}(\cdot). \tag{12.77}$$

Lin and Jeon (2003) suggested choosing $a = 0.025$ and $b = 0.975$. However, it is unclear whether their choice of Winsorization parameters $(a, b)$ is still valid when $p$ is much larger than $n$. In order to handle the high-dimensional SeLDA model, Mai and Zou (2015) proposed using

$$(a, b) = (a_n, b_n) = (\frac{1}{n_2^2}, 1 - \frac{1}{n_2^2}). \tag{12.78}$$

In the above discussion we only use data in class 2 to estimate $\widehat{h}_j$. One can also use data in class 1 to estimate $\widehat{h}_j$. To better use the data, we can consider a pooled estimator of $h_j$ defined as

$$\widehat{h}_j^{(\text{pool})} = \widehat{\pi}_2 \widehat{h}_j^{(2)} + \widehat{\pi}_1 \widehat{h}_j^{(1)}, \tag{12.79}$$

where $\widehat{h}_j^{(2)}$ and $\widehat{h}_j^{(1)}$ represent the estimator in (12.77) based solely on the class 2 or the class 1, respectively.

After estimating the transformation functions, one can apply any good sparse LDA method to the transformed data $(\widehat{h}(\mathbf{X}), Y)$. For example, Mai and Zou (2015) studied SeSDA by following DSDA:

$$\widehat{\boldsymbol{\beta}} = \arg\min_{\boldsymbol{\beta}} n^{-1} \sum_{i=1}^{n} (Y_i - \beta_0 - \widehat{h}(\mathbf{x}_i)^T \boldsymbol{\beta})^2 + \sum_{j=1}^{p} p_\lambda(|\beta_j|), \tag{12.80}$$

where $Y_i = -1$ or $Y_i = 1$ depends on whether $\mathbf{X}_i$ is in class 1 or 2.

Define $\boldsymbol{\beta}^\star = \boldsymbol{\Sigma}^{-1}(\boldsymbol{\mu}_2 - \boldsymbol{\mu}_1)$, where $\boldsymbol{\Sigma}$ is the covariance matrix of $(h_1(\mathbf{X}_1), \ldots, h_p(\mathbf{X}_p))$. Recall that $\boldsymbol{\beta}^\star$ is equal to $c\boldsymbol{\Sigma}^{-1}(\boldsymbol{\mu}_2 - \boldsymbol{\mu}_1) = c\boldsymbol{\beta}^{\text{Bayes}}$ for some positive constant (Mai, Zou and Yuan, 2012). Define

$$\mathcal{D} = \{j : \beta_j^{\text{Bayes}} \neq 0\} = \{j : \beta_j^\star \neq 0\}.$$

Let $s$ be the cardinality of $\mathcal{D}$. We partition $\boldsymbol{\Sigma}$ as

$$\boldsymbol{\Sigma} = \begin{pmatrix} \boldsymbol{\Sigma}_{\mathcal{D}\mathcal{D}} & \boldsymbol{\Sigma}_{\mathcal{D}\mathcal{D}^c} \\ \boldsymbol{\Sigma}_{\mathcal{D}^c\mathcal{D}} & \boldsymbol{\Sigma}_{\mathcal{D}^c\mathcal{D}^c} \end{pmatrix}.$$

**Theorem 12.11** *Suppose that we use the Lasso penalty in (12.80). Assume the irrepresentable condition* $\kappa = \|\boldsymbol{\Sigma}_{\mathcal{D}^c\mathcal{D}}(\boldsymbol{\Sigma}_{\mathcal{D}\mathcal{D}})^{-1}\|_\infty < 1$ *and*

*(C1).* $n, p \to \infty$ *and* $s^2 \log(ps)/n^{\frac{1}{3}-\rho} \to 0$, *for some* $\rho$ *in* $(0, 1/3)$;

*(C2).* $\min_{j \in A} |\beta_j| \gg \max\left\{sn^{-1/4}, \sqrt{\log(ps)s^2/n^{\frac{1}{3}-\rho}}\right\}$ *for some* $\rho$ *in* $(0, 1/3)$.

*Let* $\widehat{\mathcal{D}} = \{j : \widehat{\beta}_j \neq 0\}$. *If we choose* $\lambda = \lambda_n$ *such that*

$$\min_{j \in A} |\beta_j| \gg \lambda_n \gg \sqrt{\log(ps)s^2/n^{\frac{1}{3}-\rho}},$$

*then we have*

$$P(\widehat{\mathcal{D}} = \mathcal{D}) \to 1 \quad \text{and} \quad P\left(\|\widehat{\boldsymbol{\beta}}_{\mathcal{D}} - \boldsymbol{\beta}_{\mathcal{D}}\|_\infty \leq 4\varphi\lambda_n\right) \to 1,$$

*where* $\varphi = \|(\boldsymbol{\Sigma}_{\mathcal{D}\mathcal{D}})^{-1}\|_\infty$ *Moreover,* $R_n - R \to 0$ *in probability, where* $R = \Pr(Y \neq sign(h(\mathbf{X})^T\boldsymbol{\beta}^* + \beta_0))$ *and* $R_n = \Pr(Y \neq sign(\widehat{h}(\mathbf{X})^T\widehat{\boldsymbol{\beta}} + \widehat{\beta}_0))$ *are respectively the Bayes error rate and the error rate for the estimated Bayes classifier.*

## 12.9 Bibliographical Notes

Efficient computation of the Elastic Net penalized multinomial regression was discussed in Friedman, Hastie and Tibshirani (2010). Wang and Shen (2006) extended the $\ell_1$ penalized SVM to the multicategory case. Clemmensen *et al.* (2011) proposed a sparse discriminant analysis method based on sparse penalized optimal scoring. Witten and Tibshirani (2011) proposed a penalized Fisher's linear discriminant based on a nonconvex formulation. A multiclass extension of DSDA was proposed by Mai, Yang and Zou (2019) and is implemented in R package `msda`. Wang and Zou (2016) proposed sparse distance-weighted discrimination. Hao, Dong and Fan (2015) propose a simple and novel method to rotate the observed variables so that the resulting Fisher discriminant coefficients are sparse. QUADRO proposed by Fan *et al.* (2015) is a high-dimensional generalization of quadratic discriminant analysis. Fan *et al.* (2015) proposed IIS-SQDA for high-dimensional classification under a QDA model where a new feature/interaction screen technique is used to reduce the dimensionality. Most of works on classification consider the usual classification paradigm in which the goal is to minimize the sum (or weighted sum) of type I and type II misclassification errors. Zhao, Feng, Wang and Tong (2016) studied the high-dimensional Neyman-Pearson classification problem where the goal is to minimize the type II misclassification error when the type I misclassification error is controlled by a pre-specified level. See also Tong, Feng and Li (2018).

## 12.10 Exercises

12.1 Suppose that random variables representing two classes $\mathcal{C}_1$ and $\mathcal{C}_2$ follow $p$-variate normal distributions $N(\boldsymbol{\mu}_1, \boldsymbol{\Sigma})$ and $N(\boldsymbol{\mu}_2, \boldsymbol{\Sigma})$, respectively. Consider the linear classifier $\delta(\mathbf{X}) = I(\mathbf{w}^T(\mathbf{X} - \boldsymbol{\mu}) > 0) + 1$ which classifies $\mathbf{X}$ to class 2 when $\mathbf{w}^T(\mathbf{X} - \boldsymbol{\mu}) > 0$ for a given parameter $\boldsymbol{\mu}$.

(a) The misclassification rate of classifying a data point from class $C_2$ is $P_2(\mathbf{w}^T(\mathbf{X}-\boldsymbol{\mu}) \le 0)$ where $P_2$ is the probability distribution under $N(\boldsymbol{\mu}_2, \boldsymbol{\Sigma})$. Show that this misclassification rate is given by

$$1 - \Phi\Big(\frac{\mathbf{w}^T(\boldsymbol{\mu}_2 - \boldsymbol{\mu})}{(\mathbf{w}^T\boldsymbol{\Sigma}\mathbf{w})^{1/2}}\Big).$$

(b) The misclassification rate of classifying a data point from class $C_1$ is $P_1(\mathbf{w}^T(\mathbf{X}-\boldsymbol{\mu}) > 0)$ where $P_1$ is the probability distribution under $N(\boldsymbol{\mu}_1, \boldsymbol{\Sigma})$. Show that this misclassification rate is given by

$$1 - \Phi\Big(\frac{\mathbf{w}^T(\boldsymbol{\mu} - \boldsymbol{\mu}_1)}{(\mathbf{w}^T\boldsymbol{\Sigma}\mathbf{w})^{1/2}}\Big).$$

(c) Let the prior probability $\pi = P(Y=2)$. Show that the misclassification rate

$$P(\delta(\mathbf{X}) \ne Y) = 1 - \pi\Phi\Big(\frac{\mathbf{w}^T(\boldsymbol{\mu}_2 - \boldsymbol{\mu})}{(\mathbf{w}^T\boldsymbol{\Sigma}\mathbf{w})^{1/2}}\Big) - (1-\pi)\Phi\Big(\frac{\mathbf{w}^T(\boldsymbol{\mu} - \boldsymbol{\mu}_1)}{(\mathbf{w}^T\boldsymbol{\Sigma}\mathbf{w})^{1/2}}\Big)$$

12.2 Let us revisit the problem in the previous exercise and consider the linear classifier $\delta(\mathbf{X}) = I(\mathbf{w}^T(\mathbf{X} - \boldsymbol{\mu}_a) > 0) + 1$, where $\boldsymbol{\mu}_a = (\boldsymbol{\mu}_1 + \boldsymbol{\mu}_2)/2$.

(a) Show that the misclassification rate

$$P(\delta(\mathbf{X}) \ne Y) = 1 - \Phi\Big(\frac{\mathbf{w}^T\boldsymbol{\mu}_d}{(\mathbf{w}^T\boldsymbol{\Sigma}\mathbf{w})^{1/2}}\Big), \qquad \boldsymbol{\mu}_d = (\boldsymbol{\mu}_2 - \boldsymbol{\mu}_1)/2.$$

(b) The optimal weight that minimizes the misclassification rate is $\mathbf{w}_{opt} = c\boldsymbol{\Sigma}^{-1}\boldsymbol{\mu}_d$, the Fisher classifier, for any constant $c > 0$.

12.3 Let $\widehat{\boldsymbol{\beta}}^{\text{lda}}$ be the LDA rule coefficient on a dataset $(\mathbf{X}_i, Y_i)_{i=1}^n$ with $Y_i = 0$ or 1. On the same dataset we consider an ordinary least squares problem:

$$(\widehat{\beta}_0, \widehat{\boldsymbol{\beta}}^{\text{ols}}) = \arg\min_{\beta_0, \boldsymbol{\beta}} \sum_{i=1}^n (Y_i - \beta_0 - \mathbf{X}_i^T\boldsymbol{\beta})^2.$$

Prove that $\widehat{\boldsymbol{\beta}}^{\text{lda}}$ and $\widehat{\boldsymbol{\beta}}^{\text{ols}}$ have the same direction.

12.4 Consider the kernel density estimator (12.15), where $h \to 0$.

(a) Show that $\mathrm{E}\,\widehat{f}(\mathbf{x}) = \int K(\mathbf{Y})f(\mathbf{x} + h\mathbf{Y})d\mathbf{Y}$.

(b) If $\int Y_i K(\mathbf{Y})d\mathbf{Y} = 0$ for all $i$ and $f''(\cdot)$ is continuous at point $\mathbf{x}$, then

$$\mathrm{E}\,\widehat{f}(\mathbf{x}) = f(\mathbf{x}) + \frac{h^2}{2}\int \mathbf{Y}^T f''(\mathbf{x})\mathbf{Y}K(\mathbf{Y})d\mathbf{Y} + o(h^2).$$

If further $\int Y_i y_j K(\mathbf{Y}) d\mathbf{Y} = 0$ for all $i \neq j$, show that the bias

$$\mathrm{E}\,\widehat{f}(x) - f(\mathbf{x}) = \frac{h^2}{2} \sum_{j=1}^{p} \frac{\partial^2}{\partial x_j^2} f(\mathbf{x}) \int y_j^2 K(\mathbf{Y}) d\mathbf{Y} + o(h^2)$$

**Hint**: Use Taylor expansion of $f(\mathbf{x} + h\mathbf{Y})$. You may assume that the support of $K(\cdot)$ is bounded so that the limit and integral are exchangeable.

(c) Show that the variance

$$\mathrm{Var}(\widehat{f}(\mathbf{x})) = \frac{1}{nh^p} f(\mathbf{x}) \int K^2(\mathbf{Y}) d\mathbf{Y}(1 + o(1)).$$

12.5 Using the results in Exercise 12.4,

(a) show that the mean-square errors

$$\begin{aligned} \mathrm{E}[\widehat{f}(\mathbf{x}) - f(\mathbf{x})]^2 &= \frac{h^4}{4}[\int \mathbf{Y}^T f''(\mathbf{x}) \mathbf{Y} K(\mathbf{Y}) d\mathbf{Y}]^2 + \frac{1}{nh^p} f(\mathbf{x}) \int K^2(\mathbf{Y}) d\mathbf{Y} \\ &\quad + o\Big(h^4 + \frac{1}{nh^p}\Big); \end{aligned}$$

(b) find the bandwidth $h$ to optimize the main term of the mean-square error;

(c) compute the optimal mean-square error.

12.6(a) Find the bandwidth $h$ to minimize the main term of the integrated mean-square error given in Exercise 12.5.

(b) Give the expression for the minimum integrated mean-square error in (a).

(c) Show that the optimal bandwidth is given by (12.16) when $p = 1$.

12.7 Write a program to implement Algorithm 12.2. Apply it to the Parkinsons dataset from the UCI Machine Learning Repository. The link is https://archive.ics.uci.edu/ml/datasets/parkinsons.

12.8 Show (12.31) is the minimizer of (12.30).

12.9 Derive equation (12.40) from (12.37)–(12.39).

12.10 Demonstrate (12.44), namely

$$\arg\min_f \mathrm{E}\left([1 - Y f(\mathbf{X})]_+\right) = \mathrm{sign}(p_+(\mathbf{x}) - p_-(\mathbf{x})),$$

where $p_+(\mathbf{x}) = \Pr(Y = 1 | X = \mathbf{x})$ and $p_-(\mathbf{x}) = \Pr(Y = -1 | X = \mathbf{x})$.

12.11 Prove Theorem 12.2.

12.12 Let $\phi(t)$ be the Huberized hinge loss in (12.48).
 (a) Prove that (12.56) holds and find the constant $M$.
 (b) Derive (12.58) from (12.57).

12.13 Prove Theorem 12.4.

12.14 Prove Theorem 12.10.

12.15 Download the malaria data (Ockenhouse *et al.*, 2006) from `http://www.ncbi.nlm.nih.gov/sites/GDSbrowser?acc=GDS2362`. The predictors are the expression levels of 22283 genes.
 (a) The 71 samples are split with a roughly 1:1 ratio to form training and testing sets. Fit $\ell_1$ penalized logistic regression and $\ell_1$ penalized DSDA based on the training set and then use the testing set to report the test error. Repeat the process 100 times.
 The 2059th gene (`IRF1`) is scientifically known to be important. It is the first identified interferon regulatory transcription factor. For each method, report the frequency of selecting `IRF1`.
 (b) If you were given the information that the 2059th gene is scientifically important, how could you modify your analysis? Carry out the new analysis and compare the classification and selection results to those from (a).

12.16 Let $Y$ be a random variable, taking values in $\{1, -1\}$. Let $p_+(\mathbf{x}) = P(Y = 1|\mathbf{X} = \mathbf{x})$ and $p_-(\mathbf{x}) = P(Y = -1|\mathbf{X} = \mathbf{x})$.
 (a) Show that

$$\arg\min_f \mathrm{E}\left([1 - Yf(\mathbf{X})]_+\right) = \mathrm{sign}(p_+(\mathbf{x}) - p_-(\mathbf{x})),$$

 i.e. the function $\mathrm{sign}(p_+(\mathbf{x}) - p_-(\mathbf{x}))$ achieves the minimum of $\mathrm{E}\left([1 - Yf(\mathbf{X})]_+\right)$.
 (b) Show that

$$\arg\min_f \mathrm{E}\log(1 + e^{-Yf(\mathbf{X})}) = \log\left(\frac{p_+(\mathbf{x})}{p_-(\mathbf{x})}\right).$$

 (c) Let $\phi$ be a convex loss function and if it is differentiable at 0 with $\phi'(0) < 0$, then $\phi$ is Fisher-consistent, namely

$$\mathrm{sign}(f^*(\mathbf{x})) = \mathrm{sign}(p_+(\mathbf{x}) - p_-(\mathbf{x})), \quad \text{where} \quad f^* = \arg\min_f \mathrm{E}\left(\phi(Yf(\mathbf{X}))\right).$$

12.17 Consider the M-step in the EM algorithm for the Gaussian mixture model, which minimizes

$$\operatorname*{argmax}_{\{\pi_k, \boldsymbol{\mu}_k, \boldsymbol{\Sigma}_k\}_{k=1}^K} \sum_{i=1}^n \sum_{k=1}^K [\log \phi(\mathbf{X}_i; \boldsymbol{\mu}_k, \boldsymbol{\Sigma}_k) + \log \pi_k] \widetilde{p}_{ik}.$$

Show that the solution is given by

$$\begin{aligned}
\pi_k^{(t+1)} &= \frac{1}{n}\sum_{i=1}^{n}\widetilde{p}_{ik}, \\
\boldsymbol{\mu}_k^{(t+1)} &= \frac{\sum_{i=1}^{n}\widetilde{p}_{ik}\mathbf{X}_i}{\sum_{i=1}^{n}\widetilde{p}_{ik}}, \\
\boldsymbol{\Sigma}_k^{(t+1)} &= \frac{\sum_{i=1}^{n}\widetilde{p}_{ik}(\mathbf{X}_i-\boldsymbol{\mu}_k^{(t+1)})(\mathbf{X}_i-\boldsymbol{\mu}_k^{(t+1)})^T}{\sum_{i=1}^{n}\widetilde{p}_{ik}}.
\end{aligned} \tag{12.81}$$

12.18 The Email Spam dataset consists of 4601 email messages of which about 40% are spam emails. There are also 57 predictors such as frequencies of certain words, total number of capital letters, etc. This dataset is publicly available at the UCI Machine Learning Repository (Bache and Lichman, 2013): `https://archive.ics.uci.edu/ml/datasets/Spambase`.
The goal is to distinguish spams from real emails. Let us randomly split the dataset into a training set of 1000 observations and a testing set of 3601 observations (set the random seed 525 beforehand). To compute the augmented features, use the R command `density` and its default bandwidth. Report the misclassfication error rates on the testing set by using

(a) penalized logistic regression;

(b) penalized logistic regression with augmented features;

(c) linear support vector machine with augmented features;

(d) kernel support vector machine with Gaussian kernel;

(e) CART;

(f) random forest.

12.19 Let us consider the Zillow data again in Exercise 2.9. We want to predict the house prices with all features except the first three columns ("*(empty)*", "*id*", "*date*"). Set a random seed by *set.seed*(525). Report out-of-sample $R^2$ using CART and Random Forest.
(**Hint**: you can use the package **rpart** for CART and the package **randomForest** for the random forest algorithm. *randomForest* cannot handle a categorical variable with more than 32 categories. Please use One-Hot encoding for *zipcode*. To speed up computation or to tune model parameters (optimal), you can use the package **ranger** in which the algorithm is implemented in C++.)

# Chapter 13

# Unsupervised Learning

Cluster analysis in statistics is a learning task in which class labels are unobservable. The aim is to group data into several clusters such that each cluster is considered as a homogeneous subpopulation. It has numerous applications in many fields, such as sociology, psychology, business, information retrieval, linguistics, bioinformatics, computer graphics, and so on. Principal component analysis (*PCA*) is a widely used dimension reduction technique in multivariate data analysis where we construct a small number of new features via linear transformations from the original dataset such that there is little information loss. The basic idea of PCA was introduced in Section 10.1. Both clustering and PCA involve no response variable. The two are important examples of the so-called *unsupervised learning* which is a big topic in machine learning to be covered in a single chapter. This chapter will be devoted to cluster analysis and PCA and their high-dimensional extensions.

## 13.1 Cluster Analysis

Let $\{\mathbf{x}_i\}_{i=1}^n$ be the $n$ observed data of $p$-dimensional vector. We expect to see that the data within each cluster is similar. In order to quantify the similarity, we define a *dissimilarity measure* $d_j(x_{ij}, x_{lj})$ for the $j$th variable, which measures distance between two variables, and define the overall dissimilarity measure between $\mathbf{x}_i$ and $\mathbf{x}_l$ as

$$D(\mathbf{x}_i, \mathbf{x}_l) = \sum_{j=1}^{p} w_j d_j(x_{ij}, x_{lj}),$$

where $w_j$ is a weight assigned to the $j$th variable. A simple choice is to set $w_j = 1$, but the optimal choice may well depend on specific applications. For the continuous variables, the most common choice for $d$ is the squared distance $d_j(x_{ij}, x_{lj}) = (x_{ij} - x_{lj})^2$. More generally, it is also natural to use $d_j(x_{ij}, x_{lj}) = h(|x_{ij} - x_{lj}|)$ with an increasing function $h$. If the variable is not continuous, another distance, such as the *Hamming distance* $d_j(x_{ij}, x_{lj}) = I(x_{ij} \neq x_{lj})$, is more appropriate.

Suppose that we want to separate the data into $K$ clusters. We discuss ways to determine the value of $K$ in Section 13.2. For now, $K$ is given. A clustering algorithm will assign each observation to one and only one cluster. Denote by $\mathcal{C}(i)$ the cluster label assigned to the $i$th observation by the clustering algorithm. Then, we define the *within-cluster* "distance" as the sum of all distances within each cluster, which is given by

$$W(C) = \frac{1}{2}\sum_{k=1}^{K} \sum_{i:\mathcal{C}(i)=k} \sum_{l:\mathcal{C}(l)=k} D(\mathbf{x}_i, \mathbf{x}_l). \tag{13.1}$$

Here, one half is used because $D(\mathbf{x}_i, \mathbf{x}_l)$ and $D(\mathbf{x}_l, \mathbf{x}_i)$ are both counted in the above expression. Likewise, the between-cluster "distance" is

$$B(C) = \frac{1}{2}\sum_{k=1}^{K} \sum_{i:\mathcal{C}(i)=k} \sum_{l:\mathcal{C}(l)\neq k} D(\mathbf{x}_i, \mathbf{x}_l). \tag{13.2}$$

Note that $W(C) + B(C) = \frac{1}{2}\sum_{i=1}^{n}\sum_{l=1}^{n} D(\mathbf{x}_i, \mathbf{x}_l)$. Thus minimizing the within-cluster "distance" is equivalent to maximizing the between-cluster "distance".

### 13.1.1 K-means clustering

*K-means clustering*, also referred to as the *Lloyd algorithm* (Lloyd, 1957), is a simple but popular clustering algorithm, especially when all variables are quantitative. It iteratively updates the centers of the clusters and the cluster memberships. It aims to solve the following optimization problem

$$(\mathcal{C}^{\text{opt}}, \boldsymbol{\mu}_1^*, \ldots, \boldsymbol{\mu}_K^*) = \underset{\mathcal{C}, \boldsymbol{\mu}_1, \ldots, \boldsymbol{\mu}_K}{\arg\min} \sum_{i:\mathcal{C}(i)=k} \|\mathbf{x}_i - \boldsymbol{\mu}_k\|^2, \tag{13.3}$$

Namely, find the centroids and memberships that minimize the sum of the distance of each data point to the centroid of its group.

Problem (13.3) is naturally solved by an alternating minimization algorithm. Given the clustering results $\{\mathcal{C}_k\}_{k=1}^K$, the optimal $\boldsymbol{\mu}_k$ is just

$$\widehat{\boldsymbol{\mu}}_k = \frac{1}{n_k} \sum_{i:\mathcal{C}(i)=k} \mathbf{x}_i, \qquad n_k = |\mathcal{C}_k|,$$

the gravity center of the data in the $k$th cluster. Given estimated centroids $\{\boldsymbol{\mu}_k\}_{k=1}^K$, the optimal clustering for the $i^{th}$ data is the one that is closest to its centroid:

$$\mathcal{C}(i) = \arg\min_k \|\mathbf{x}_i - \boldsymbol{\mu}_k\|^2. \tag{13.4}$$

Algorithm 13.1 furnishes the details.

It is easy to see that the target function is non-increasing in each iteration and therefore the algorithm will converge in terms of the object function. The convergence is very fast and so is the computation. However, the

**Algorithm 13.1** K-means clustering

1. Randomly choose $K$ centroids: $\boldsymbol{\mu}_1, \ldots, \boldsymbol{\mu}_K$. A typical choice is to pick $K$ points at random from the data $\{\mathbf{x}_i\}_{i=1}^n$.
2. Cluster each observation $\mathbf{x}_i$ to the cluster $k_i^*$: $k_i^* = \arg\min_k \|\mathbf{x}_i - \boldsymbol{\mu}_k\|^2$
3. Replace the centroids $\boldsymbol{\mu}_k$ within the sample mean with the $k$th cluster.
4. Iterate steps 2 and 3 until the clustering results do not change.
5. Try multiple initial values, pick the solution with the best objective value.

estimated controids and clusters, if converging, will converge only to local minima. Indeed, the K-means clustering algorithm depends sensitively on the initial values. Hence, it is often recommended to try multiple random centroids as initialization and then report the best cluster outcome, the one with the minimum object value. Figure 13.1 displays a clustering result for a simulated data set, in which the true membership is indicated by numbers $\{1, 2, 3\}$ and the clustering results indicated as {red, blue and black}. Mislabeling can be seen, but the results are quite satisfactory.

### *13.1.2 Hierarchical clustering*

Hierarchical clustering is another popular clustering algorithm in practice. It uses a recursive procedure to sequentially build up a hierarchical structure to segment data into different clusters. There are two ways to search for the hierarchy: *agglomerative* and *divisive*, which correspond to greedy bottom-up and top-down algorithms. Agglomerative algorithms start at the bottom by treating each observation as a cluster. Then, we recursively merge a selected pair of clusters into a bigger cluster. The selected pair of clusters has the smallest dissimilarity measure among all possible pairs. In other words, each time we merge two most similar clusters, until there is only one cluster left. In case of tied dissimilarity, all tied pairs may be joined at the same time. On the other hand, divisive algorithms first cluster all observations into a single cluster and then recursively split one of the current clusters into two smaller clusters such that the two sub-clusters have the largest dissimilarity measure. By design, hierarchical clustering can generate a *dendrogram* to show the clustering procedure, which makes it a popular tool for cluster analysis in some applications such as clustering genes in computational biology.

Figure 13.2 displays such a dendrogram based on the same simulated dataset used in Figures 13.1, using agglomerative hierarchical clustering. The height represents the dissimilarities where two clusters are merged. Cutting the dendrogram at height 4.5 gives three clusters, represented by red lines; whereas cutting the dendrogram at height 4 would give 4 clusters. The bottom nodes can show the data index or names; however, in this simulated data, we show the true class labels so that misclassifications can be inspected.

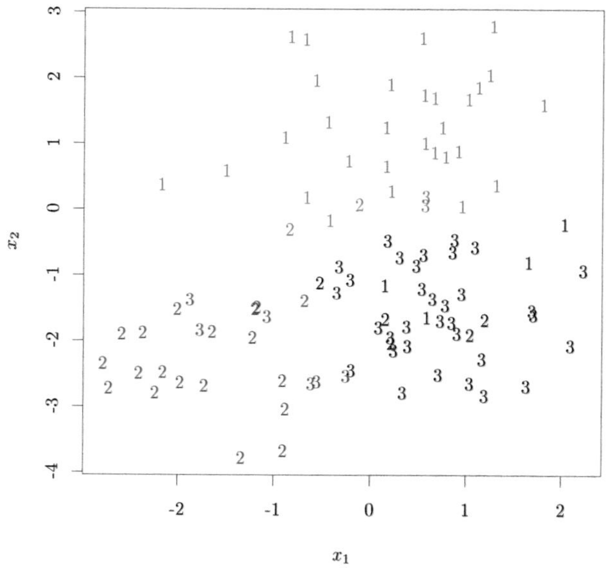


Figure 13.1: K-means clustering on a simulated dataset with two variables. We set $K = 3$ in Algorithm 13.1. The discovered clusters are colored red, blue and black. The true class labels {1,2,3} are shown for comparison of misclassification and are unknown in practice (only points are seen).

The dissimilarity measure between two clusters is the key metric in hierarchical clustering. There are three commonly used statistics: *single linkage*, *complete linkage* and *average linkage*. Let $\mathcal{G}_1, \mathcal{G}_2$ denote two clusters. Single linkage is defined as

$$d_S(\mathcal{G}_1, \mathcal{G}_2) = \min_{\mathbf{x}_1 \in \mathcal{G}_1} \min_{\mathbf{x}_2 \in \mathcal{G}_2} D(\mathbf{x}_1, \mathbf{x}_2),$$

where $D$ is the distance measure between two observations such as the squared distance. Complete linkage is defined as

$$d_C(\mathcal{G}_1, \mathcal{G}_2) = \max_{\mathbf{x}_1 \in \mathcal{G}_1} \max_{\mathbf{x}_2 \in \mathcal{G}_2} D(\mathbf{x}_1, \mathbf{x}_2),$$

and the average linkage is defined as

$$d_A(\mathcal{G}_1, \mathcal{G}_2) = \frac{\sum_{\mathbf{x}_1 \in \mathcal{G}_1} \sum_{\mathbf{x}_2 \in \mathcal{G}_2} D(\mathbf{x}_1, \mathbf{x}_2)}{|\mathcal{G}_1| \cdot |\mathcal{G}_2|},$$

where $|G|$ is the number of observations in $G$. Using one of these dissimilarity measures, we can run the hierarchical cluster algorithms.

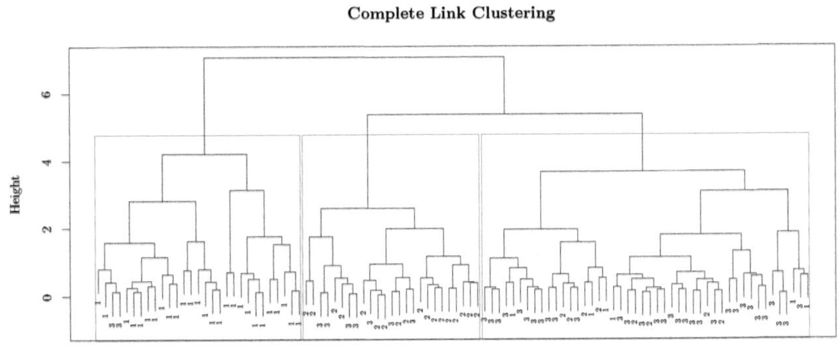


Figure 13.2: A dendrogram for hierarchical clustering on a simulated dataset with two variables. The height measures the dissimilarity. The true class labels $\{1, 2, 3\}$ are indicated at the bottom so that we can see the misclassifications. In practice, they are unknown and only indices (or names) of individual data are labeled at the bottom.

### *13.1.3 Model-based clustering*

The definition of cluster analysis naturally implies that we can take a probabilistic approach to clustering. We assume the observations are independent and identically distributed according to a *finite mixture model* with $K$ components where the probability density function is written as

$$f(\mathbf{x}) = \sum_{k=1}^{K} \pi_k f_k(\mathbf{x}; \boldsymbol{\theta}_k). \tag{13.5}$$

It is often assumed that each $f_k(\cdot; \boldsymbol{\theta}_k)$ represents a relatively simple distribution with unknown parameter vector $\boldsymbol{\theta}_k$, but this is not required in principle. Clustering based on the model (13.5) is called *model-based clustering*. Nice review papers on model-based clustering include Fraley and Raftery (2002) and Städler, Bühlmann, van de Geer (2010).

The mixture model has a simple probabilistic interpretation and relates cluster analysis to multiclass classification. Imagine a data generating process in which we first draw a random sample $\xi$, the class label, from a multinomial distribution with probability parameters $\{\pi_k\}_{k=1}^{K}$, and after seeing the value of $\xi$ (say $\xi = k$) we draw a random sample $\mathbf{X}$ from the $k$th distribution with density $f_k(\cdot; \boldsymbol{\theta}_k)$. This process will generate a class label and variables $\mathbf{X}$. However, the class label is not actually observed. What we observed is realizations of $\mathbf{X}$, which has the mixture distribution given by (13.5). From this perspective, clustering is like classification without class labels.

The conditional distribution of $\xi$ given $\mathbf{X}$ is multinomial with the class probability

$$P(\xi = k \mid \mathbf{X} = \mathbf{x}) = \frac{\pi_k f_k(\mathbf{x})}{\sum_{l=1}^{K} \pi_l f_l(\mathbf{x})}.$$

The Bayes rule classifies $\mathbf{X} = \mathbf{x}$ into a class with the highest conditional probability:

$$\mathcal{C}_{\text{Bayes}}(\mathbf{x}) = \text{argmax}_{k \in [K]} P(\xi = k \mid \mathbf{X} = \mathbf{x}).$$

If we can estimate such probabilities, then a natural clustering algorithm follows the Bayes rule.

A popular model-based clustering algorithm is based on *Gaussian mixtures* where each $f_k$ is assumed to be the density function of a $p$-dimensional Gaussian distribution with mean $\boldsymbol{\mu}_k$ and covariance $\boldsymbol{\Sigma}_k$, whose density function is denoted by $\phi(\cdot; \boldsymbol{\mu}_k, \boldsymbol{\Sigma}_k)$. Then, the log-likelihood function of observed data $\{\mathbf{x}_i\}_{i=1}^n$ is

$$\ell(\boldsymbol{\theta}) = \sum_{i=1}^{n} \log(\sum_{k=1}^{K} \pi_k \phi(\mathbf{x}_i; \boldsymbol{\mu}_k, \boldsymbol{\Sigma}_k)), \tag{13.6}$$

where $\boldsymbol{\theta}$ represents all unknown parameters in the model. This is a nonconvex optimization problem.

One possible way to compute the *maximum likelihood estimator* is a classical application of the *expectation-maximization* (*EM*) algorithm (Dempster, Laird and Rubin, 1977). Technically, it can be regarded as a special case of the (MM) algorithm (Hunter and Lange, 2000) in Section 3.5.3, but the EM algorithm has a special place in statistics and machine learning. Introduce $n$ independent multinomial random variable $\xi_i$ such that $P(\xi_i = k) = \pi_k$ and $(\mathbf{X}_i \mid \xi_i = k) \sim N(\boldsymbol{\mu}_k, \boldsymbol{\Sigma}_k)$. Consider $\{\xi_i\}_1^n$ as the *"missing data"* in the EM algorithm. The full log-likelihood of the *"complete data"* $\{(\xi_i, \mathbf{x}_i)\}_{i=1}^n$ is

$$\ell_{\text{full}}(\boldsymbol{\theta}) = \sum_{i=1}^{n} \sum_{k=1}^{K} [\log(\phi(\mathbf{x}_i; \boldsymbol{\mu}_k, \boldsymbol{\Sigma}_k) + \log \pi_k] I(\xi_i = k). \tag{13.7}$$

Let $\boldsymbol{\theta}^{(t)}$ be the EM estimate at the $t$th iteration. The EM algorithm iteratively updates the estimator via an E-step and an M-step. The E-step computes $\mathrm{E}_{\boldsymbol{\theta}^{(t)}}(\ell_{\text{full}}(\boldsymbol{\theta}) \mid \mathbf{X}_1, \ldots, \mathbf{X}_n)$. This yields

$$\begin{aligned} &\mathrm{E}_{\boldsymbol{\theta}^{(t)}}(\ell_{\text{full}}(\boldsymbol{\theta}) \mid \mathbf{X}_1 = \mathbf{x}_1, \ldots, \mathbf{X}_n = \mathbf{x}_n) \\ = &\sum_{i=1}^{n} \sum_{k=1}^{K} [\log(\phi(\mathbf{x}_i; \boldsymbol{\mu}_k, \boldsymbol{\Sigma}_k) + \log \pi_k] \widetilde{p}_{ik} \end{aligned} \tag{13.8}$$

with

$$\widetilde{p}_{ik} = P(\xi_i = k \mid \mathbf{X}_i = \mathbf{x}_i) = \frac{\pi_k^{(t)} \phi(\mathbf{x}_i; \boldsymbol{\mu}_k^{(t)}, \boldsymbol{\Sigma}_k^{(t)})}{\sum_{l=1}^{K} \pi_l^{(t)} \phi(\mathbf{x}_i; \boldsymbol{\mu}_l^{(t)}, \boldsymbol{\Sigma}_l^{(t)})}. \tag{13.9}$$

In the M-step we update the current estimate by maximizing the function in (13.8) with respect to $\boldsymbol{\theta} = \{\pi_k, \boldsymbol{\mu}_k, \boldsymbol{\Sigma}_k\}_{k=1}^K$. That is,

$$\boldsymbol{\theta}^{(t+1)} = \underset{\{\pi_k, \boldsymbol{\mu}_k, \boldsymbol{\Sigma}_k\}_{k=1}^K}{\operatorname{argmax}} \sum_{i=1}^n \sum_{k=1}^K [\log(\phi(\mathbf{x}_i; \boldsymbol{\mu}_k, \boldsymbol{\Sigma}_k) + \log \pi_k]\widetilde{p}_{ik}.$$

It is straightforward to show (Exercise 13.4) that $\boldsymbol{\theta}^{(t+1)}$ is unique and given by

$$\begin{aligned} \pi_k^{(t+1)} &= \frac{1}{n}\sum_{i=1}^n \widetilde{p}_{ik}, \\ \boldsymbol{\mu}_k^{(t+1)} &= \frac{\sum_{i=1}^n \widetilde{p}_{ik}\mathbf{x}_i}{\sum_{i=1}^n \widetilde{p}_{ik}}, \\ \boldsymbol{\Sigma}_k^{(t+1)} &= \frac{\sum_{i=1}^n \widetilde{p}_{ik}(\mathbf{x}_i - \boldsymbol{\mu}^{(t+1)})(\mathbf{x}_i - \boldsymbol{\mu}^{(t+1)})^T}{\sum_{i=1}^n \widetilde{p}_{ik}}. \end{aligned} \tag{13.10}$$

Denote the outcome of the EM algorithm as $\widehat{\boldsymbol{\theta}}$. The EM algorithm provides only a local maximizer of the likelihood function and depends on the initial values. As such, it is recommended to try multiple random initial values and choose the one with the largest $\ell(\widehat{\boldsymbol{\theta}})$ value. Plugging $\widehat{\boldsymbol{\theta}}$ into (13.9) yields

$$\widehat{p}_{ik} = P(\xi_i = k \mid \mathbf{X}_i = \mathbf{x}_i) = \frac{\widehat{\pi}_k \phi(\mathbf{x}_i; \widehat{\boldsymbol{\mu}}_k, \widehat{\boldsymbol{\Sigma}}_k)}{\sum_{l=1}^K \widehat{\pi}_l \phi(\mathbf{x}_i; \widehat{\boldsymbol{\mu}}_l, \widehat{\boldsymbol{\Sigma}}_l)}. \tag{13.11}$$

This probability can be interpreted as the maximum likelihood estimate of the probability of drawing the $i$th observation from the $k$th Gaussian distribution. Then we assign $\mathbf{x}_i$ into cluster $\mathcal{C}(i)$ where

$$\mathcal{C}(i) = \operatorname{argmax}_k \widehat{p}_{ik}.$$

Note that the denominator in (13.11) does not need to be computed as it is independent of $k$.

---

**Algorithm 13.2** Clustering based on mixture of Gaussians

---

1. Randomly choose initial values $\boldsymbol{\theta}^{(0)}$.
2. Repeat the following for $t = 0, 1, 2, \ldots$ until convergence
   - (2.a) Compute (13.9)
   - (2.b) Update the estimate by (13.10)
3. Assign $\mathbf{x}_i$ into cluster $\mathcal{C}(i)$ where $\mathcal{C}(i) = \operatorname{argmax}_k \widetilde{p}_{ik}$.

---

It is interesting to note that a limiting special case of Algorithm 13.2 is equivalent to K-means clustering. Suppose that we fix $\boldsymbol{\Sigma}_k$ to be $\sigma^2\mathbf{I}$ in

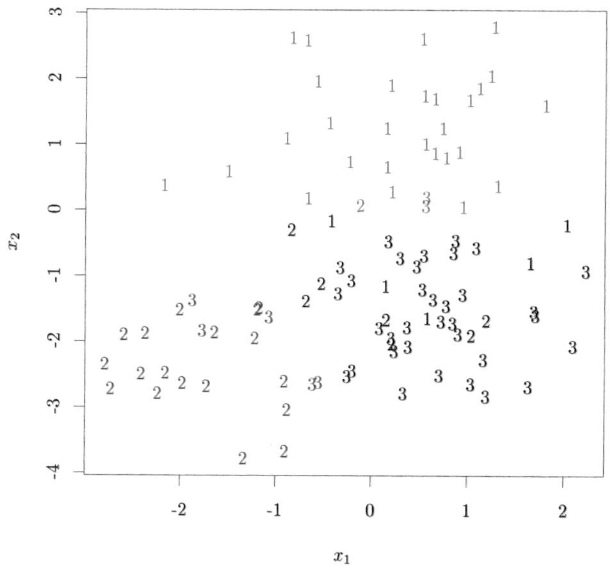


Figure 13.3: The results of model-based clustering on a simulated dataset with two variables and a class label. The numbers (1,2,3) indicate the class label. We set $K = 3$ in Algorithm 13.2. The fitted clusters are colored by red, blue and black.

the Gaussian mixture model and $\sigma^2$ is a known constant. The step (2.a) in Algorithm 13.2 becomes

$$\widetilde{p}_{ik} = \frac{\pi_k^{(t)}}{\sum_{l=1}^K \pi_l^{(t)} w_{ik}(l;\sigma^2)}$$

with

$$w_{ik}(l;\sigma^2) = \exp(-\frac{1}{2\sigma^2}(\|\mathbf{x}_i - \boldsymbol{\mu}_l^{(t)}\|^2 - \|\mathbf{x}_i - \boldsymbol{\mu}_k^{(t)}\|^2)).$$

In step (2.b) we do not need to update the covariance estimate since it is given. Note that $\lim_{\sigma^2 \to 0} w_{ik}(l;\sigma^2)$ is one of three possible values: 0, 1 and $\infty$ depending on the sign of $\|\mathbf{x}_i - \boldsymbol{\mu}_l^{(t)}\|^2 - \|\mathbf{x}_i - \boldsymbol{\mu}_k^{(t)}\|^2$. Assume no ties in K-means clustering,

$$\lim_{\sigma^2 \to 0} \widetilde{p}_{ik} = I(k = k_i^*)$$

where $k_i^* = \arg\min_k \|\mathbf{x}_i - \boldsymbol{\mu}_k^{(t)}\|^2$ as defined in (13.4). Then, $\boldsymbol{\mu}_k^{(t+1)}$ in step (2.b) is exactly the vector of the updated centroids in K-means clustering.

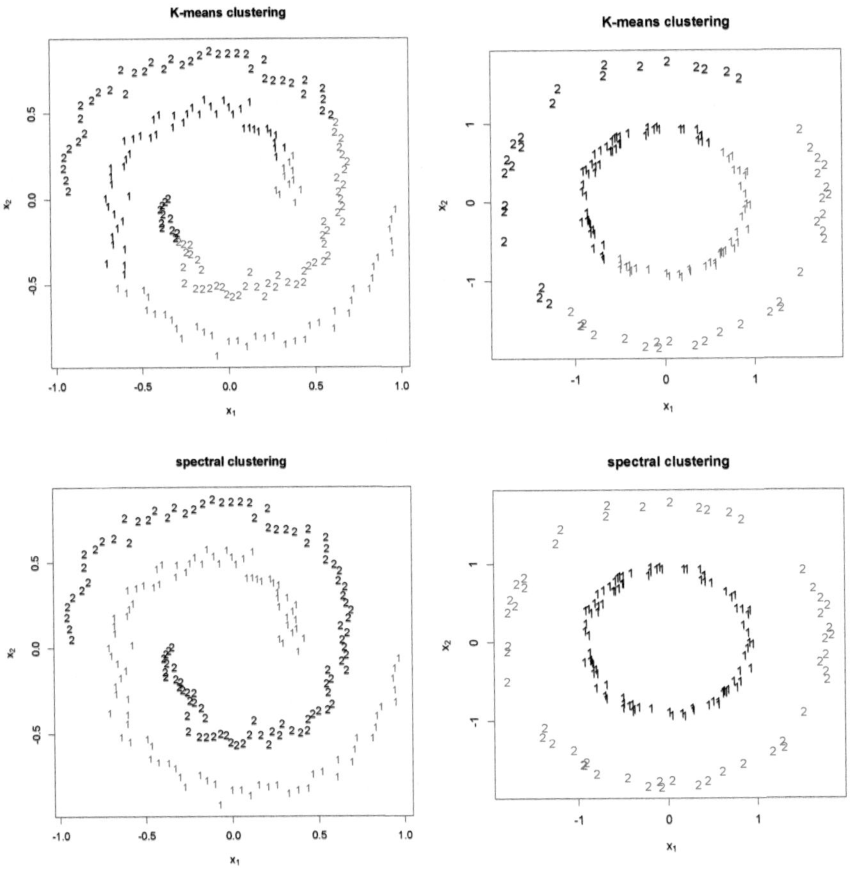


Figure 13.4: Two simulated datasets. The left panel is the spiral data with 100 observations in each cluster. The right panel is the two rings data with 92 observations in cluster 1 and 52 observations in cluster 2. Red and black are used to indicate the estimated cluster membership by a spectral clustering. It is very clear that K-means fails badly on both data sets, but spectral clustering does a perfect job.

### *13.1.4 Spectral clustering*

Spectral clustering is a relatively more modern clustering algorithm with a close link to spectral graph partition. It is easy to implement and often can have better performance when the more traditional algorithms such as K-means perform poorly. Figure 13.4 displays two such examples.

The definition starts with an un-directed similarity graph denoted by $G = (V, E)$, where each of the $n$ observations is a vertex and the edge between $\mathbf{x}_i, \mathbf{x}_j$ is weighted by a non-negative value $w_{ij}$. If $w_{ij} = 0$ then the vertices

$i, j$ are not connected. The weights are required to be symmetric: $w_{ij} = w_{ji}$. The matrix $\mathbf{W} = (w_{ij}) 1 \le i, j \le n$ is called the *adjacency matrix*. If nonzero weights are equal to constant then the graph is unweighted. For a weighted graph, we usually let $w_{ij} = s_{ij}$ (for $w_{ij} \neq 0$) where $s_{ij}$ is the similarity between $\mathbf{x}_i$ and $\mathbf{x}_j$ such as $s_{ij} = \exp(-\|\mathbf{x}_i - \mathbf{x}_j\|^2/\sigma^2)$. For the unweighted graph, the adjacency matrix consists of only two values: 0 or 1, indicating whether vertices are connected or not.

The degree of a vertex $v_i$ $1 \le i \le n$ is

$$d_i = \sum_j w_{ij}.$$

Then define the *degree matrix* $\mathbf{D} = \text{diag}(d_1, \cdots, d_n)$. The *graph Laplacian matrix* is defined as

$$\mathbf{L} = \mathbf{D} - \mathbf{W}, \tag{13.12}$$

It can be easily checked that

$$\boldsymbol{\alpha}^T \mathbf{L} \boldsymbol{\alpha} = \frac{1}{2} \sum_{i=1,j=1}^{n} w_{ij}(\alpha_i - \alpha_j)^2, \tag{13.13}$$

which shows that $\mathbf{L}$ is semi-definite positive and has an eigenvalue 0 (The corresponding eigenvector is $\boldsymbol{\alpha} = \mathbf{1}/\sqrt{n}$; see Exercise 13.5). A random walk *normalized graph Laplacian matrix* is defined as

$$\widetilde{\mathbf{L}} = \mathbf{I} - \mathbf{D}^{-1}\mathbf{W}, \tag{13.14}$$

or a *symmetric normalized graph Laplacian matrix* is defined as

$$\widetilde{\mathbf{L}}^{\text{sym}} = \mathbf{I} - \mathbf{D}^{-1/2}\mathbf{W}\mathbf{D}^{-1/2}. \tag{13.15}$$

If the graph has $K$ connected subgraphs and there is no connection between any two subgraphs, then the graph Laplacian matrix is block diagononal after suitable permutation. The $k^{th}$ block diagonal matrix is a graph Laplacian matrix for that subgraph $(V_k, E_k)$ and hence has a zero eigenvalue with eigenvector $\mathbf{1}_{V_k}/\sqrt{n_k}$, where $n_k = |V_k|$ is the number of vertices on the $k^{th}$ subgraph. Thus, the original Laplacian matrix has eigenvalue zero with multiplicity $K$ and the eigenspace of zero eigenvalue is spanned by the indicator vectors of the connected subgraphs. When the graph has weakly connected subgraphs that have strong connections within them, we expect to see very small eigenvalues. Spectral clustering takes advantage of this graph property. Shi and Malik (2000) proposed a popular *normalized cut* algorithm which amounts to using the eigenvectors of $\widetilde{\mathbf{L}}$ in (13.14). Ng, Jordan and Weiss (2002) proposed a normalized spectral clustering algorithm by considering the eigenvectors of $\widetilde{\mathbf{L}}^{\text{sym}}$ in (13.15). The details are presented in Algorithm 13.3. Note that the minimum eigenvalue of $\widetilde{\mathbf{L}}^{\text{sym}}$ is zero with associated eigenvector proportional

**Algorithm 13.3** Spectral clustering; Ng, Jordan and Weiss (2001)

1. Compute the first $K$ eigenvectors of $\mathbf{I} - \widetilde{\mathbf{L}}^{\text{sym}} = \mathbf{D}^{-1/2}\mathbf{W}\mathbf{D}^{-1/2}$. Denote them by $\mathbf{z}_1, \ldots, \mathbf{z}_K$.
2. Construct an $n \times K$ data matrix $\mathbf{Z}$ with $\mathbf{z}_1, \ldots, \mathbf{z}_K$ as columns.
3. Normalize each row of $\mathbf{Z}$ by $Z_{ij} \leftarrow \frac{z_{ij}}{(\sum_{k=1}^{K} z_{ik}^2)^{1/2}}$.
4. Treat the $i$-th row of $\mathbf{Z}$ as the $i$th observation, $1 \le i \le n$. Apply K-means clustering to cluster these $n$ observations into $K$ clusters.

to $\mathbf{D}^{1/2}\mathbf{1}/\|\mathbf{D}^{1/2}\mathbf{1}\|$ (see Exercise 13.6), which is an eigenvector associated with the largest eigenvalue of the matrix $\mathbf{I} - \widetilde{\mathbf{L}}^{\text{sym}}$.

There are several commonly-used similarity graphs, which give rises to several adjacency matrices. We can use a fully connected graph where all vertices are connected and each edge has a positive weight. For example, $w_{ij} = \exp(-\|\mathbf{x}_i - \mathbf{x}_j\|^2/\sigma^2)$ where $\sigma$ is a positive parameter (similar to the bandwidth parameter in local smoothing). We can also consider the *$k$-nearest-neighbor graph* where $v_i$ and $v_j$ are connected if $v_j$ is among the $k$-nearest neighbors of $v_i$ or $v_i$ is among the $k$-nearest neighbors of $v_j$. In the *mutual $k$-nearest-neighbor graph* $v_i$ and $v_j$ are connected if and only if $v_j$ is among the $k$-nearest neighbors of $v_i$ and $v_i$ is among the $k$-nearest neighbors of $v_j$. Then, the nonzero weight equals the similarity of the two vertices. In the $\delta$-neighbor graph $v_i, v_j$ are connected if their distance is less than $\delta$. The nonzero weight is usually a constant. In applications we need to select the graph parameter such as $\sigma$, $k$ and $\delta$.

## 13.2 Data-driven Choices of the Number of Clusters

In some applications we may know the number of clusters beforehand. Frequently, we also use a data-driven method to assist selecting the number of clusters. A number of methods have been proposed to choose the number of clusters. In the following discussion, we use $K^*$ to denote the chosen 'optimal' number of clusters in each method.

For the model-based clustering algorithm, because there is a natural definition of the right number of clusters based on the underlying probabilistic finite mixture model, it is natural to use the *Bayesian information criterion* to select $K^*$ (Fraley and Raftery, 1998). Given a finite mixture model with $K$ components $\mathcal{M}_K$, let $S_{\mathcal{M}_K}$ denote the number of free parameters to be estimated. The BIC method chooses $K^*$ as follows:

$$K^*_{\text{BIC}} = \text{argmax}_K\{2\text{log-likelihood}(\mathcal{M}_K) - \log(n)S_{\mathcal{M}_K}\}. \tag{13.16}$$

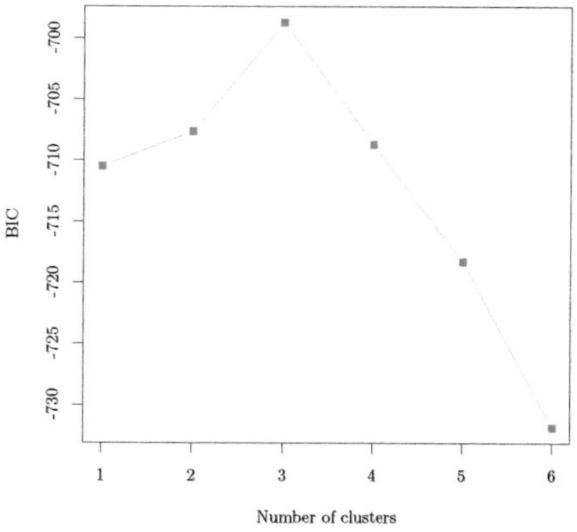


Figure 13.5: An illustration of using BIC to determine the number of clusters in model-based clustering. The data are simulated from a mixture model with three components.

Under the squared Euclidean distance, define

$$\begin{aligned}\mathrm{WGSS}(K) &= \sum_{k=1}^{K}\sum_{C(i)=k}\|\mathbf{x}_i - \bar{\boldsymbol{\mu}}_k\|^2,\\ \mathrm{BGSS}(K) &= \sum_{i=1}^{n}\|\mathbf{x}_i - \bar{\boldsymbol{\mu}}\|^2 - \mathrm{WGSS}(K)\end{aligned}$$

where $\bar{\boldsymbol{\mu}}_k = \frac{1}{n_k}\sum_{C(i)=k}\mathbf{x}_i$ and $\bar{\boldsymbol{\mu}} = \frac{1}{n}\sum_{i=1}^{n}\mathbf{x}_i$. We can plot WGSS($K$) against $K$ and hope to see a clear 'elbow'. There are a number of statistics proposed to make this eyeballing approach more formal. Caliński and Harabasz (1974) proposed using

$$K^*_{\mathrm{CH}} = \mathrm{argmax}_K \frac{\mathrm{BGSS}(K)/(K-1)}{\mathrm{WGSS}(K)/(n-K)} \tag{13.17}$$

called CH index, a $F$-type of test statistic. Hartigan (1975) proposed the statistic $\mathrm{H}_K = (\frac{\mathrm{WGSS}(K)}{\mathrm{WGSS}(K+1)} - 1)/(n-K-1)$ and suggested that a new cluster to be added if $\mathrm{H}_K \geq 10$. Hence, his optimal $K^*$ is

$$K^*_{\mathrm{H}} = \min_K\{\mathrm{H}_{K-1} \geq 10 \text{ and } \mathrm{H}_K < 10\}. \tag{13.18}$$

Krzanowski and Lai (1985) defined

$$\mathrm{DIFF}(K) = (K-1)^{2/p}\mathrm{WGSS}(K-1) - K^{2/p}\mathrm{WGSS}(K)$$

and suggested using

$$K^*_{\text{KL}} = \text{argmax}_K \frac{|\text{DIFF}(K)|}{|\text{DIFF}(K+1)|}. \tag{13.19}$$

Kaufman and Rousseeuw (1990) proposed the *silhouette statistic* for measuring how well each data point is clustered. The average silhouette statistics can be used to select the optimal number of clusters. Consider the $i$th observation $\mathbf{x}_i$ with the cluster label $\mathcal{C}(i)$. Define $a(i)$ as the average distance between $\mathbf{x}_i$ and other observations in the cluster $\mathcal{C}(i)$. Given any other cluster $\mathcal{C}'$, compute the average distance between $\mathbf{x}_i$ and the observations in $\mathcal{C}'$. Define $b(i)$ as the smallest average distance over all $\mathcal{C}' \neq \mathcal{C}(i)$. We can view $b(i)$ as the average distance between $\mathbf{x}_i$ and its nearest neighbor cluster. The silhouette statistic is defined as

$$\text{s}(i) = \frac{b(i) - a(i)}{\max(b(i), a(i))},$$

and

$$K^*_{\text{s}} = \text{argmax}_K \frac{1}{n} \sum_{i=1}^{n} s(i). \tag{13.20}$$

Tibshirani, Walther and Hastie (2001) proposed a *gap statistic* for choosing the number of clusters. Let $W_K = \sum_{k=1}^{K} \frac{1}{2n_k} \sum_{i,l \in \mathcal{C}_k} D(\mathbf{x}_i, \mathbf{x}_l)$. If $D$ is the squared Euclidean distance, $W_K$ is the pooled within-cluster sum of squares around the cluster means used in the CH index, because

$$\frac{1}{2n_k} \sum_{i,l \in \mathcal{C}_k} (\mathbf{x}_i - \mathbf{x}_l)^2 = \sum_{i \in \mathcal{C}_k} (\mathbf{x}_i - \boldsymbol{\mu}_k)^2$$

and hence

$$W_k = \text{WGSS}(K) = \sum_{k=1}^{K} \sum_{C(i)=k} \|\mathbf{x}_i - \bar{\boldsymbol{\mu}}_k\|^2.$$

The gap statistic is defined as

$$\text{Gap}_K = \text{E}^*_n(\log W_K) - \log W_K,$$

where the expectation is taken with respect to a reference distribution with sample size $n$ and has roughly the same distribution shape and range of the observed data. Let $\text{sd}_K$ be the standard deviation of $\log W_K$ with respect to the reference distribution. Then the chosen $K^*$ is defined as follows:

$$K^*_{\text{Gap}} = \inf_K \{\text{Gap}_K \geq \text{Gap}_{K+1} - \text{sd}_{K+1}\}. \tag{13.21}$$

Tibshirani *et al.* (2001) suggested two reference distributions. The first one is uniform over the range of the observed values. The second one uses the uniform distribution over the range of the principal components in order to take into account the shape of the data. Once the reference distribution is decided, the gap statistic and its standard deviation can be directly computed by Monte Carlo.

Sugar and James (2003) proposed a *jump statistic* to choose $K^*$. Suppose that a clustering algorithm produces $K$ clusters with cluster centers $\mathbf{c}_1, \ldots, \mathbf{c}_K$. Define

$$\widehat{d}_K = \frac{1}{p} \sum_{i=1}^{n} \min_{k \in [K]} (\mathbf{x}_i - \mathbf{c}_k)^T \mathbf{\Gamma}^{-1} (\mathbf{x}_i - \mathbf{c}_k), \quad \widehat{d}_0 = 0 \tag{13.22}$$

where $\mathbf{\Gamma}$ is assumed to be the same within-cluster covariance matrix across different clusters and $p$ is the dimensionality of the data. Sugar and James (2003) recommended using the identity matrix to replace $\mathbf{\Gamma}$ in (13.22). Choose the number of cluster

$$K^*_{\text{Jump}} = \text{argmax}_K J_k \quad \text{with} \quad J_K = \widehat{d}_K^{-a} - \widehat{d}_{K-1}^{-a} \tag{13.23}$$

for some $a > 0$, called the jump statistic. Sugar and James (2003) suggested to use $a = p/2$.

## 13.3 Variable Selection in Clustering

When dimension $p$ is high, it is more desirable to select a subset of variables that have high clustering power. Recall that the dissimilarity between two observations is defined as $D(\mathbf{x}_i, \mathbf{x}_l) = \sum_{j=1}^{p} w_j d_j(x_{ij}, x_{lj})$, where $w_j$ is a weight assigned to the $j$th variable. Assume that there is a subset of variables $\mathcal{A}$ that should be used in clustering and the rest of the variables are pure noise variables. Then, we set $w_j = 1$ for $j \in \mathcal{A}$ and $w_j = 0$ if $j \in \mathcal{A}^c$. The difficulty is that we do not know $\mathcal{A}$ and we need to do cluster analysis and variable selection simultaneously.

### 13.3.1 *Sparse clustering*

Witten and Tibshirani (2011) proposed a general sparse clustering framework. Recall the within-cluster and between-cluster dissimilarities defined in (13.1) and (13.2) and

$$W(\mathcal{C}) = \frac{1}{2} \sum_{i} \sum_{l} D(\mathbf{x}_i, \mathbf{x}_l) - B(\mathcal{C}). \tag{13.24}$$

A clustering algorithm can be derived by maximizing $B(\mathcal{C})$. In the case of equal weights, we have $D(\mathbf{x}_i, \mathbf{x}_l) = \sum_{j=1}^{p} d_j(x_{ij}, x_{lj})$ and we can rewrite $B(\mathcal{C})$ as

$$B(\mathcal{C}) = \sum_{j=1}^{p} g_j(\mathcal{C}) \tag{13.25}$$

where $g_j(\mathcal{C}) = \frac{1}{2}\sum_{k=1}^{K}\sum_{\mathcal{C}(i)=k}\sum_{\mathcal{C}(l)\neq k} d_j(x_{ij}, x_{lj})$ is the between-cluster dissimilarity contributed by the $j^{th}$ variable. To choose the variables for *sparse clustering*, let us consider the following constrained optimization problem:

$$\text{argmax}_{\mathcal{C},\mathbf{w}} \sum_{j=1}^{p} w_j g_j(\mathcal{C}) \tag{13.26}$$

$$\text{subject to} \qquad w_j \geq 0, \quad \sum_{j=1}^{p} w_j \leq s, \quad \sum_{j=1}^{p} w_j^2 \leq 1.$$

An alternating maximizing algorithm was proposed to solve (13.26). Given the weights $\mathbf{w}$, finding the best $\mathcal{C}$ can be done via a usual clustering algorithm such as the K-means algorithm. Given $\mathcal{C}$, let $a_j = g_j(\mathcal{C})$. The best weights are

$$\widehat{\mathbf{w}} \quad = \quad \text{argmax}_{w} \sum_{j=1}^{p} w_j a_j \tag{13.27}$$

$$\text{subject to} \qquad w_j \geq 0, \quad \sum_{j=1}^{p} w_j \leq s, \quad \sum_{j=1}^{p} w_j^2 \leq 1.$$

For the solution to (13.27) we need some notation. Denote by $z_+$ the positive part of a real number $z$. For a vector $\mathbf{z}$, let $\mathbf{z}_+$ be a vector with the $j$th element being $(z_j)_+$. Let $S(x,t) = \text{sign}(x)(|x| - t)_+$ be the soft-thresholding operator. For a vector $\mathbf{z}$, denote by $S(\mathbf{z}, t)$ a vector with the $j$th element being $S(z_j, t)$.

**Lemma 13.1** *Consider the following optimization problem*

$$\boldsymbol{\beta}^* = \text{argmax}_{\boldsymbol{\beta}} \boldsymbol{\beta}^T \mathbf{x} \quad \text{subject to} \quad \|\boldsymbol{\beta}\| \leq 1, \|\boldsymbol{\beta}\|_1 \leq s.$$

*The solution is*

$$\boldsymbol{\beta}^* = \frac{S(\mathbf{x}, \boldsymbol{\Delta})}{\|S(\mathbf{x}, \boldsymbol{\Delta})\|},$$

*with* $\boldsymbol{\Delta} = 0$ *if the corresponding* $\boldsymbol{\beta}^*$ *satisfies* $\|\boldsymbol{\beta}^*\|_1 \leq s$*; otherwise* $\Delta > 0$ *is chosen to satisfy* $\|\boldsymbol{\beta}^*\|_1 = s$.

We leave the proof as an exercise (Exercise 13.7). In (13.27) it is obvious that $\widehat{w}_j = 0$ if $a_j \leq 0$. By Lemma 13.1 we have

$$\widehat{\mathbf{w}} = \frac{S(\mathbf{a}_+, \Delta)}{\|S(\mathbf{a}_+, \Delta)\|}, \tag{13.28}$$

where $\Delta = 0$ if the corresponding $\widehat{w}$ satisfies $\sum_{j=1}^{p} \widehat{w}_j \leq s$, otherwise $\Delta > 0$ is chosen to satisfy $\sum_{j=1}^{p} \widehat{w}_j = s$.

As an example of the general sparse clustering framework, consider the

**Algorithm 13.4** Sparse K-means clustering

1. Start with $w_j^{(0)} = 1/\sqrt{p}$ for $j = 1, 2, \ldots, p$ and let $\mathcal{C}^{(0)}$ be the standard K-means clustering results.
2. For $m = 1, 2, 3, \ldots$ iterate (2.a)-(2.c) until convergence

(2.a) Compute $a_j = \frac{1}{2}\sum_{k=1}^{K}\sum_{\mathcal{C}^{(m-1)}(i)=k}\sum_{\mathcal{C}^{(m-1)}(l)\neq k}(x_{ij} - x_{lj})^2$

(2.b) Calculate $\mathbf{w}^{(m)}$ by using (13.28) and let $\mathcal{A} = \{j : w_j^{(m)} > 0\}$.

(2.c) Let $x_{ij}^* = \sqrt{w_j^{(m)}}x_{ij}$ for $j \in \mathcal{A}$. Do K-means clustering on $(x_{ij}^*)_{1\leq i\leq n, j\in\mathcal{A}}$, which gives $\mathcal{C}^{(m)}$.

squared distance $d_j(x_{ij}, x_{lj}) = (x_{ij} - x_{lj})^2$. We need to work out the clustering algorithm for maximizing $\sum_{j,w_j>0} w_j g_j(\mathbf{x}_j, \mathcal{C})$, since only features with positive weight have clustering power. Note that

$$\sum_{j:w_j>0} w_j g_j(\mathbf{x}_j, \mathcal{C}) = \sum_{j:w_j>0} g_j^*(\mathcal{C}),$$

where $g_j^*(\mathcal{C})$ is the same as $g_j(\mathcal{C})$ but uses the transformed data $x_{ij}^* = \sqrt{w_j}x_{ij}$. This is exactly the clustering problem with transformed data $\{\mathbf{x}_j^*\}$. If we use the K-means clustering to estimate $\mathcal{C}$, then the iterative procedure produces a *sparse K-means clustering* algorithm as in Algorithm 13.4.

### 13.3.2 Sparse model-based clustering

Sparse penalization techniques can be naturally used in model-based clustering and variable selection, resulting in a *sparse model-based clustering* algorithm. Pan and Shen (2007) considered the Lasso penalized mixture of Gaussians with a common diagonal covariance matrix. In their approach, variables are centered to have zero mean before clustering. The log-likelihood function is

$$\ell(\boldsymbol{\theta}) = \sum_{i=1}^{n} \log(\sum_{k=1}^{K} \pi_k \phi(\mathbf{x}_i; \boldsymbol{\mu}_k, \boldsymbol{\Sigma})),$$

where $\boldsymbol{\Sigma}$ is a diagonal matrix with elements $\sigma_j^2$, $j = 1, \ldots, p$. They considered the Lasso penalized MLE:

$$\operatorname{argmax}\Big\{\sum_{i=1}^{n} \log\Big(\sum_{k=1}^{K} \pi_k \phi(\mathbf{x}_i; \boldsymbol{\mu}_k, \boldsymbol{\Sigma})\Big) - \lambda \sum_j \sum_k |\mu_{kj}|\Big\}. \tag{13.29}$$

One can use other sparse penalties such as SCAD (Fan and Li, 2001) .

The EM algorithm was used to compute the penalized MLE. Let $\boldsymbol{\theta}^{(t)}$ be the estimate at the $t$th iteration. The E-step calculations remain the same. For the M-step, we maximize

$$\boldsymbol{\theta}^{(t+1)} = \mathrm{argmax}_{\{\pi_k, \boldsymbol{\mu}_k\}_1^K, \boldsymbol{\Sigma}}\{\sum_{i=1}^{n}\sum_{k=1}^{K}[\log(\phi(\mathbf{x}_i;\boldsymbol{\mu}_k,\boldsymbol{\Sigma})+\log\pi_k]\widetilde{p}_{ik} - \lambda\sum_{j}\sum_{k}|\mu_{kj}|\},$$

with

$$\widetilde{p}_{ik} = \frac{\pi_k^{(t)}\phi(\mathbf{x}_i;\boldsymbol{\mu}_k^{(t)},\boldsymbol{\Sigma}^{(t)})}{\sum_{l=1}^{K}\pi_l^{(t)}\phi(\mathbf{x}_i;\boldsymbol{\mu}_l^{(t)},\boldsymbol{\Sigma}^{(t)})}. \tag{13.30}$$

It is easy to see that $\pi_k^{(t+1)} = \frac{1}{n}\sum_{i=1}^{n}\widetilde{p}_{ik}$ but $(\boldsymbol{\mu}_k^{(t+1)})_{k=1}^K$ and $\boldsymbol{\Sigma}^{(t+1)}$ do not have a closed form solution. Given $\boldsymbol{\mu}_k$, the optimal $\boldsymbol{\Sigma}$ is given by

$$\sigma_j^{2,(t+1)} = \frac{\sum_{k=1}^{K}\sum_{i=1}^{n}\widetilde{p}_{ik}(x_{ij}-\mu_j)^2}{n} \quad 1 \le j \le p. \tag{13.31}$$

Given $\boldsymbol{\Sigma}$, owing to the diagonal structure of $\boldsymbol{\Sigma}$ and the Lasso penalty, we have a soft-thresholding operator for updating $\boldsymbol{\mu}$

$$\mu_{kj}^{(t+1)} = S(\widetilde{\mu}_{kj}, \gamma_j) \tag{13.32}$$

where $\gamma_j = \lambda\frac{\sigma_j^{2,(t+1)}}{\sum_{i=1}^{n}\widetilde{p}_{ik}}$ and $\widetilde{\mu}_k$ is the standard mean update without the penalization term

$$\widetilde{\mu}_k = \frac{\sum_{i=1}^{n}\widetilde{p}_{ik}\mathbf{x}_i}{\sum_{i=1}^{n}\widetilde{p}_{ik}}.$$

We can iterate between (13.31) and (13.32) to compute $(\boldsymbol{\mu}_k^{(t+1)})_{k=1}^K$ and $\boldsymbol{\Sigma}^{(t+1)}$. We can also simply do one iteration and the overall EM algorithm becomes the generalized EM algorithm, keeping the crucial ascent property of the EM algorithm.

The purpose of using the Lasso penalty is to set many estimated $\mu_{kj}$ to be zero, thus realizing variable selection. A closer inspection of the clustering algorithm shows that under Pan and Shen's model, the $j$th variable appears in the clustering probability (13.11) if and only if $\mu_{1j} = \mu_{2j} = \ldots = \mu_{Kj}$. Because the overall mean is zero (after centering variables), we would prefer to have $\mu_{1j} = \mu_{2j} = \ldots = \mu_{Kj} = 0$ for many variables. To this end, a *group penalization* method is more appropriate. The corresponding penalized MLE is defined as

$$\mathrm{argmax}\Big\{\sum_{i=1}^{n}\log\Big(\sum_{k=1}^{K}\pi_k\phi(\mathbf{x}_i;\boldsymbol{\mu}_k,\boldsymbol{\Sigma})\Big) - \sum_{j}p_\lambda\Big(\Big(\sum_{k}\mu_{kj}^2\Big)^{1/2}\Big)\Big\}.$$

The same EM algorithm can be applied and we only need to modify (13.32). For example, consider using the group Lasso penalty (Yuan and Lin, 2006)

$$p_\lambda\Big(\Big(\sum_{k}\mu_{kj}^2\Big)^{1/2}\Big) = \lambda\Big(\sum_{k}\mu_{kj}^2\Big)^{1/2},$$

and the mean updating formula becomes

$$\mu_{kj} = \widetilde{\mu}_{kj} \left( 1 - \frac{\gamma_j}{(\sum_k |\widetilde{\mu}_{kj}|^2)^{1/2}} \right)_+ .$$

Algorithm 13.5 summarizes the penalized MLE method for variable selection and clustering.

---

**Algorithm 13.5** Penalized model-based clustering with variable selection

---

1. Randomly choose initial values $\boldsymbol{\theta}^{(0)}$.
2. Repeat the following for $t = 0, 1, 2, \ldots$ until convergence:
   - (2.a) Using (13.30) to compute $\widetilde{p}_{ik}$.
   - (2.b) $\pi_k^{(t+1)} = \frac{1}{n} \sum_{i=1}^n \widetilde{p}_{ik}$ and $\widetilde{\mu}_k = \frac{\sum_{i=1}^n \widetilde{p}_{ik} \mathbf{x}_i}{\sum_{i=1}^n \widetilde{p}_{ik}}$.
   - (2.c) For $j = 1, \ldots, p$,

$$\sigma_j^{2,(t+1)} = \frac{\sum_{k=1}^K \sum_{i=1}^n \widetilde{p}_{ik} (x_{ij} - \mu_j^{(t)})^2}{n}$$

   then let $\gamma_j = \lambda \frac{\sigma_j^{2,(t+1)}}{\sum_{i=1}^n \widetilde{p}_{ik}}$ and the mean update is

$$\mu_{kj}^{(t+1)} = S(\widetilde{\mu}_{kj}, \gamma_j) \quad \text{if using the Lasso penalty}$$

$$\text{or } \mu_{kj} = \widetilde{\mu}_{kj} \left( 1 - \frac{\gamma_j}{(\sum_k |\widetilde{\mu}_{kj}|^2)^{1/2}} \right)_+ \quad \text{if using the group Lasso penalty.}$$

3. Assign $X_i$ into cluster $C(i) = \text{argmax}_k \widetilde{p}_{ik}$.

---

### *13.3.3 Sparse mixture of experts model*

An important goal of using big data is to find heterogeneous subpopulations that require different treatments in data rich environments. The big data gives us hope to identify rare subpopulations that require special treatments. A reasonable model for this effort is the following *mixture of experts model*. Given observed data $\mathbf{x}$ of an individual, with certain probability $\pi_k(\mathbf{x}; \boldsymbol{\alpha}_k)$ that the individual belongs to subpopulation $k$. In this homogeneous subpopulation, the covariates $\mathbf{x}$ have an impact on the response $y$ according to the conditional density $f_k(y|\mathbf{x}; \boldsymbol{\beta}_k)$ with $\boldsymbol{\beta}_k$ as an unknown parameter vector. Therefore, the observed data $\{(\mathbf{x}_i, y_i)\}_{i=1}^n$ are as realizations from the following mixture model

$$f(y|\mathbf{x}) = \sum_k \pi_k(\mathbf{x}; \boldsymbol{\alpha}_k) f_k(y|\mathbf{x}; \boldsymbol{\beta}_k). \tag{13.33}$$

This is schematically summarized in Figure 13.6. The response in each subpopulation is allowed to associate with different sets of variables, genes, and biomarkers, and so is the probability $\pi_k(\mathbf{x}; \boldsymbol{\alpha}_k)$ in classification. We may parametrize the mixture probability by logistic regression model $\log(\pi_k(\mathbf{x})/\pi_K(\mathbf{x})) = \mathbf{x}^T\boldsymbol{\alpha}_k$, in which $K$ is the total number of the subpopulation and $\mathbf{x}$ includes the intercept term. We will write them as $\pi_k(\mathbf{x}; \boldsymbol{\alpha})$ with $\boldsymbol{\alpha} = (\boldsymbol{\alpha}_1, \cdots, \boldsymbol{\alpha}_{K-1})$. We can further parametrize the conditional density in the regression model as $f_k(y; \mathbf{x}^T\boldsymbol{\beta}_k, \phi_k)$. This conditional density can be $N(\mathbf{x}^T\boldsymbol{\beta}_k, \sigma_k^2)$ for the continuous response, or the generalized linear model (5.1) with a canonical link such as logistic regression for the binary response.

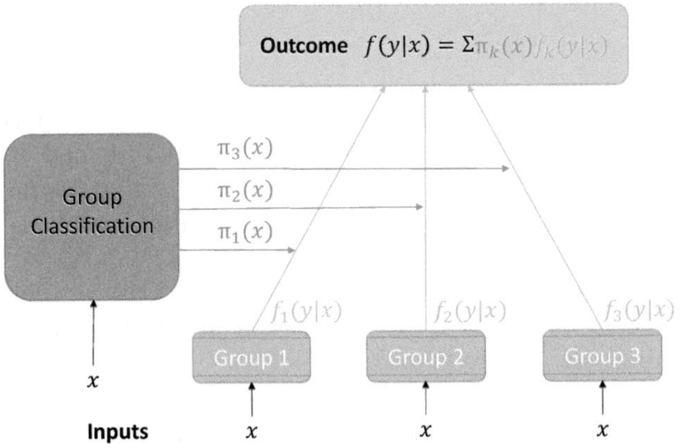


Figure 13.6: Each individual with covariate $\mathbf{x}$ is classified with probability $\pi_k(\mathbf{x})$ into subpopulation $k$. Within homogeneous subpopulation $k$, the response $Y$ is associated the covariate $\mathbf{x}$ according to the conditional law $f_k(y|\mathbf{x})$

Model (13.33) models simultaneously the covariates' effect on the group classifications and on the response $Y$. It is also called the *mixture of experts* (MOE) in statistical machine learning (Jacobs, Jordan, Nowlan, and Hinton, 1991; Jordan and Jacobs, 1994; Jiang and Tanner, 1999).

Despite their flexibility and popularity in applications, limited studies have been conducted on the problems of variable selection and parameter estimation in MOE models (13.33), when the number of features $p$ is comparable with or much larger than the sample size $n$. The scalable statistical methods and algorithms are in imminent need.

To take up the challenges of variable selection in model (13.33), we assume further that the parameters $\boldsymbol{\theta} = \{\boldsymbol{\alpha}_k, \boldsymbol{\beta}_k\}_{k=1}^K$ are sparse. By (13.33), the conditional log-likelihood for the problem is

$$\sum_{i=1}^{n} \log\Big\{\sum_{k=1}^{K} \pi_k(\mathbf{x}_i;\boldsymbol{\alpha}) f_k(y_i;\mathbf{x}_i^T\boldsymbol{\beta}_k)\Big\} \equiv \ell_n(\boldsymbol{\theta}). \tag{13.34}$$

To select important features and explore the sparsity of the parameters, we use the penalized likelihood to estimate the parameters, which amounts to minimizing

$$-\ell_n(\boldsymbol{\theta}) + p_n(\boldsymbol{\theta}), \quad \text{with} \quad p_n(\boldsymbol{\theta}) = p_1(\boldsymbol{\alpha}) + p_2(\boldsymbol{\beta}), \tag{13.35}$$

where $\ell_n(\boldsymbol{\theta})$ is the conditional log-likelihood and

$$p_1(\boldsymbol{\alpha}) = \sum_{k=1}^{K-1}\sum_{l=1}^{p} p_n(|\alpha_{kl}|;\lambda_1) \quad \text{and} \quad p_2(\boldsymbol{\beta}) = \sum_{k=1}^{K}\sum_{l=1}^{p} p_n(|\beta_{kl}|;\lambda_2),$$

and $p_n(x;\lambda)$ is a penalty function with regularization parameter $\lambda$ such as Lasso $p_n(x;\lambda) = \lambda|x|$ and SCAD. A widely used algorithm for this kind of mixture problem is the EM algorithm (Dempster, Laird, and Rubin, 1977 and McLachlan and Peel, 2000). The challenge here is that we need to solve penalized high-dimensional regression and classification problems in the M-step, which does not admit analytical solutions.

Let $Z_{ij}$ be the indicator, indicating whether the $i^{th}$ subject belongs to the $j^{th}$ group ($j = 1, \cdots, K$). Then, the penalized (complete) log-likelihood when the missing labels $\{Z_{ij}\}$ were known is

$$-\sum_{i=1}^{n}\sum_{j=1}^{n} Z_{ij}\Big\{\log \pi_k(\mathbf{x}_i;\boldsymbol{\alpha}) + \log f_k(y_i|\mathbf{x}_i^T;\boldsymbol{\beta}_k)\Big\} + p_n(\boldsymbol{\theta}). \tag{13.36}$$

Given the current estimate of the parameters $\boldsymbol{\theta}^{(m)}$, the E-step can be computed analytically:

$$\tau_{ij}^{(m)} \equiv \mathrm{E}(Z_{ij}|\boldsymbol{\theta}^{(m)}, \mathbf{x}_i, y_i) = \frac{\pi_j(\mathbf{x}_i, \boldsymbol{\alpha}^{(m)}) f_j(y_i, \mathbf{x}_i^T\boldsymbol{\beta}^{(m)})}{\sum_{k=1}^{K} \pi_k(\mathbf{x}_i, \boldsymbol{\alpha}^{(m)}) f_k(y_i, \mathbf{x}_i; \boldsymbol{\beta}^{(m)})}. \tag{13.37}$$

In the M-step, we replace the unknown $Z_{ij}$ by $\tau_{ij}^{(m)}$ in (13.36). Because of the separable penalty function in $\boldsymbol{\alpha}$ and $\boldsymbol{\beta}$ in (13.36), our minimization in the M-step reduces to solve both classification and regression separately:

$$\min_{\alpha}\Big\{-\sum_{i=1}^{n}\sum_{j=1}^{n} \tau_{ij}^{(m)} \log \pi_k(\mathbf{x}_i;\boldsymbol{\alpha}) + p_1(\boldsymbol{\alpha})\Big\}, \tag{13.38}$$

and

$$\min_{\boldsymbol{\beta}}\Big\{-\sum_{i=1}^{n}\sum_{j=1}^{n} \tau_{ij}^{(m)} \log f_k(y_i;\mathbf{x}_i;\boldsymbol{\beta}_k) + p_2(\boldsymbol{\beta})\Big\}, \tag{13.39}$$

where $p_1(\boldsymbol{\alpha})$ and $p_2(\boldsymbol{\beta})$ are defined in (13.35). The EM algorithm is to iterate between (13.37), (13.38) and (13.39).

## 13.4 An Introduction to High Dimensional PCA

PCA is a useful tool for summarizing high-dimensional data. However, it has an obvious drawback: each principal component uses all variables. This not only requires high signal-to-noise ratio as in Chapter 10, but also makes principal components hard to interpret. Rotation techniques are commonly used to help practitioners interpret principal components (Jolliffe, 1995). Sparse PCA is a method that not only achieves the dimensionality reduction but also selects a subset of variables to present the principal components. Here, sparse PCA means that the loadings of the principal components are sparse. Before introducing proposals for sparse PCA, we should mention that the simple thresholding of the loadings can lead to sparse PCA (Cadima and Jolliffe, 1995), though the method can be improved.

### *13.4.1 Inconsistency of the regular PCA*

The regular PCA uses the eigen decomposition of the sample covariance matrix. When the dimension $p$ is high, the sample covariance matrix is no longer a good estimator of the true covariance matrix and hence the regular PCA may have poor performance when its corresponding eigenvalues are not large enough. In fact, the inconsistency phenomenon of PCA was first observed in the unsupervised learning theory literature in physics; see, for example, Biehl and Mietzner (1994) and Watkin and Nadal (1994). A series of papers in statistics (Baik and Silverstein, 2006; Paul, 2007; Nadler, 2008; Johnstone and Lu, 2009) proved the inconsistency of the regular PCA when estimating the leading principal eigenvectors in the high-dimensional setting when eigenvalues are not spiked enough.

Let us consider the following simple one-factor model:

$$\mathbf{X}_i = a_i \mathbf{v}_1 + \sigma \mathbf{Z}_i, \quad i = 1, \dots, n \tag{13.40}$$

where $a_i \sim N(0,1)$ are iid Gaussian random variables, and $\mathbf{Z}_i \sim N(0, \mathbf{I}_p)$ are independent noise vectors. Let $\widehat{\mathbf{v}}_1$ be the first eigenvector of the sample covariance matrix and $R(\widehat{\mathbf{v}}_1, \mathbf{v}_1) = \cos\alpha(\widehat{\mathbf{v}}_1, \mathbf{v}_1)$ be the cosine of the angle between $\widehat{\mathbf{v}}_1$ and $\mathbf{v}_1$. Assume $c = \lim_{n\to\infty} p/n$. Let $\omega = \lim_{n\to\infty} \|\mathbf{v}_1\|^2/\sigma^2$ be the limiting signal-to-noise ratio. Johnstone and Lu (2009) proved that

$$P\left(\lim_{n\to\infty} R^2(\widehat{\mathbf{v}}_1, \mathbf{v}_1) = R^2_\infty(\omega, c)\right) = 1 \tag{13.41}$$

where

$$R^2_\infty(\omega, c) = (\omega^2 - c)_+/(\omega^2 + c\omega).$$

Note that $R^2_\infty(\omega, c) < 1$ if and only if $c > 0$. Thus, $\widehat{\mathbf{v}}_1$ is a consistent estimate of $\mathbf{v}_1$ if and only if $c = 0$, which also implies the inconsistency of the regular PCA when $c > 0$.

Note that the covariance matrix implied by the factor model (13.40) is

$$\mathbf{\Sigma} = \mathbf{v}_1 \mathbf{v}_1 + \boldsymbol{\sigma}^2 \mathbf{I}_p \tag{13.42}$$

so that the eigenvalues are

$$\|\mathbf{v}_1\|^2 + \sigma^2, \underbrace{\sigma^2, \cdots, \sigma^2}_{p-1}.$$

The difference here from what was discussed in Chapter 10 is that the signal does not assume to converge to $\infty$. If we assume $\|\mathbf{v}_1\| \to \infty$ while keeping $\sigma$ fixed, then $\omega \to \infty$ and the consistency can be seen from (13.41). The eigenvalues in Chapter 10 are stronger than what is needed here because we assume higher dimensionality and give the rates of convergence.

The phenomenon holds more generally for the spike covariance model. Consider a rank-$k$ spike covariance model in which the $n \times p$ data matrix $\mathbf{X}$ is generated as follows:

$$\mathbf{X} = \mathbf{U}\mathbf{D}\mathbf{V}^T + \mathbf{Z} \tag{13.43}$$

where $\mathbf{U}$ is the $n \times k$ random effects matrix with i.i.d. $N(0,1)$ entries, $\mathbf{D} = \text{diag}(\lambda_1^{1/2}, \cdots, \lambda_k^{1/2})$ is a diagonal matrix with order $\lambda_1 \geq \cdots \geq \lambda_k > 0$, $\mathbf{V}$ is an orthonormal matrix, $\mathbf{Z}$ is a random matrix with i.i.d. $N(0, \sigma^2)$ entries, and $\mathbf{U}$ and $\mathbf{Z}$ are independent. Writing in the vector form, let $\mathbf{V} = (\mathbf{v}_1, \cdots, \mathbf{v}_k)$. Then, the $i^{th}$ observation (the transpose of the $i^{th}$ row of $\mathbf{X}$) can be written as

$$\mathbf{X}_i = \mathbf{B}\mathbf{U}_i + \mathbf{Z}_i, \qquad \mathbf{B} = (\lambda_1^{1/2}\mathbf{v}_1, \cdots, \lambda_k^{1/2}\mathbf{v}_k), \tag{13.44}$$

where $\mathbf{U}_i \sim N(0, \mathbf{I}_k)$ and $\mathbf{Z}_i \sim N(0, \sigma^2 \mathbf{I}_p)$ are the transposes of the $i^{th}$ column of $\mathbf{U}$ and $\mathbf{Z}$, respectively. This is a specific case of the factor model discussed in Chapter 10. When $k = 1$, (13.43) reduces to (13.40).

Denote by $\boldsymbol{\Sigma}$ the covariance matrix of $\mathbf{X}_i$. By (13.44) we have

$$\boldsymbol{\Sigma} = \mathbf{V}\boldsymbol{\Lambda}\mathbf{V}^T + \sigma^2 \mathbf{I}_p, \tag{13.45}$$

where $\boldsymbol{\Lambda} = \text{diag}(\lambda_1, \cdots, \lambda_k)$. For the multiple-factor case, by (13.45), the top $k$ eigenvectors are $\mathbf{v}_1, \cdots, \mathbf{v}_k$ and we consider estimating the subspace spanned by these top $k$ eigenvectors. Let $\widehat{\mathbf{V}}$ be an estimator of $\mathbf{V}$ and $\widehat{\mathbf{V}}^T\widehat{\mathbf{V}} = \mathbf{I}_k$, and the loss function be $L(\widehat{\mathbf{V}}, \mathbf{V}) = \|\widehat{\mathbf{V}}\widehat{\mathbf{V}}^T - \mathbf{V}\mathbf{V}^T\|_F^2$. The inconsistency of the regular PCA was proved in Paul (2007) and Nadler (2008).

### *13.4.2 Consistency under sparse eigenvector model*

A positive message from Johnstone and Lu (2009) is that it is possible to obtain a consistent estimator when the eigenvector is sparse. They considered the one-factor model (13.40) assuming $\mathbf{v}_1$ is sparse. Therefore, we select variables first and apply the PCA to a subset of selected variables to reduce the dimensionality. Under model (13.40), by (13.42), the nonzero elements of the eigenvector $\mathbf{v}_1/\|\mathbf{v}_1\|$ (the principal components) lie only on the coordinates whose corresponding variables $X_j$'s have higher variance than $\sigma^2$. This leads to the following simple screening procedure.

Let $\widehat{\sigma}_j^2$ be the sample variance of $\{X_{ij}\}_{i=1}^n$. The selected subset is defined as

$$\widehat{\mathcal{A}} = \{j : \widehat{\sigma}_j^2 \geq \widehat{\sigma}^2(1 + \gamma_n)\}, \tag{13.46}$$

where $\gamma_n = \gamma\sqrt{n^{-1}\log n}$ and $\gamma > \sqrt{12}$ are constants and $\widehat{\sigma}^2$ is the estimated noise variance given by

$$\widehat{\sigma}^2 = \text{median}(\widehat{\sigma}_j^2).$$

Let $\widehat{\mathbf{V}}(\mathcal{A})$ be the PCA outcome based on variables in the subset $\mathcal{A}$. Johnstone and Lu (2009) proved that $\widehat{\mathbf{V}}(\mathcal{A})$ is consistent.

Ma (2013) proposed an iterative QR thresholding method for estimating the principal subspace with sparse eigenvectors and proved its consistency. His method is based on the QR algorithm for eigen-analysis (Golub and Van Loan,1996). Suppose $\mathbf{\Sigma}$ is a $p \times p$ covariance matrix, and we want to compute its leading eigenspace of dimension $k$. Starting with a $p \times k$ orthonormal matrix $\mathbf{Q}^{(0)}$, the QR algorithm generates a sequence of $p \times k$ orthonormal matrices $\mathbf{Q}^{(t)}$ by alternating the following two steps till convergence:

(1). Multiplication: $\mathbf{G}^{(t)} = \mathbf{\Sigma}\mathbf{Q}^{(t-1)}$;

(2). QR factorization: $\mathbf{Q}^{(t)}\mathbf{R}^{(t)} = \mathbf{G}^{(t)}$.

Ma (2013) modified the iterative QR procedure by including an thresholding step. His method proceeds as follows: Given the the sample covariance matrix $\mathbf{S}$ and the initial estimate $\mathbf{Q}^{(0)}$, iteratively run the following algorithm:

(1). Multiplication: $\mathbf{G}^{(t)} = \mathbf{S}\mathbf{Q}^{(t-1)}$;

(2). Thresholding: $g_{ij}^{(t)} = S(g_{ij}^{(t)}, \Delta_{nj})$, where $g_{ij}^{(t)}$ is the $(i, j)$ element of $\mathbf{G}^{(t)}$ and we use the same notation to denote the thresholded result;

(3). QR factorization: $\mathbf{Q}^{(t)}\mathbf{R}^{(t)} = \mathbf{G}^{(t)}$,

where $S(a, \Delta)$ denotes a thresholding operator on $a$ with threshold $\Delta > 0$. For example, $S$ can be the soft thresholding operator $\text{sgn}(a)(|a| - \Delta)_+$ or the hard thresholding $aI(|a| \geq \Delta)$. To initialize the iterative thresholding QR method, Ma (2013) used the estimator from Johnstone and Lu (2009). Specifically, in (13.46) use

$$\gamma_n = \gamma\sqrt{n^{-1}\log(\max(p, n))}$$

to select $\mathcal{A}$ and $\widehat{\mathbf{V}}(\mathcal{A}) = [\widehat{v}_1, \ldots, \widehat{v}_{|\mathcal{A}|}]$, then $\mathbf{Q}^{(0)} = [\widehat{v}_1, \ldots, \widehat{v}_k]$. The thresholding parameter $\Delta$ is set to be

$$\Delta_{nj} = \alpha\sqrt{n^{-1}\log(\max(p, n))\max(s_j^{\mathcal{A}}, 1)},$$

where $s_j^{\mathcal{A}}$ is the $j$th largest eigenvalue of $\mathbf{S}_{\mathcal{A}\mathcal{A}}$. In practice, $\gamma, \alpha$ are the user specified constants.

The minimax rate for estimating the rank-k spiked covariance model was established in Cai, Ma and Wu (2013) who considered the following parameter space for $\mathbf{\Sigma}$:

$$\begin{aligned}&\Theta(s, p, k, \lambda)\\ &= \{\mathbf{\Sigma} = \mathbf{V}\mathbf{\Lambda}\mathbf{V}^T + \sigma^2\mathbf{I}_p : \kappa\lambda \geq \lambda_1 \geq \cdots \geq \lambda_k \geq \lambda > 0, \mathbf{V}^T\mathbf{V} = \mathbf{I}_k, \|\mathbf{V}\|_w \leq s\}\end{aligned}$$

where $\kappa > 1$, $\mathbf{\Lambda} = \text{diag}(\lambda_1, \cdots, \lambda_k)$, and $\|\mathbf{V}\|_w = \max_{j\in[p]} j\|\mathbf{V}_{(j)*}\|_0$ is the weak $\ell_0$ radius of $\mathbf{V}$ and $\mathbf{V}_{(j)*}$ is the row of $\mathbf{V}$ with the $j^{th}$ largest $\ell_0$-norm. Note that the union of the column supports of $\mathbf{V}$ is of size at most $s$. The minimax risk bound is

$$\inf_{\widehat{\mathbf{V}}} \sup_{\mathbf{\Sigma}\in\Theta(s,p,k,\lambda)} \text{E}[L(\mathbf{V}, \widehat{\mathbf{V}})] \asymp \left[\frac{\lambda/\sigma^2+1}{n(\lambda/\sigma^2)^2}\left(k(s-k) + s\log\frac{ep}{s}\right)\right] \wedge 1.$$

The result depends on the sparsity $s$ rather than the total number of elements. The benefits of exploring sparsity is self evident.

## 13.5 Sparse Principal Component Analysis

The benefit of sparse PCA can be seen from the idealized model (13.43)–(13.45) in Section 13.4. However, these models are utilized to facilitate theoretical analysis and we need to introduce ideas that are suitable for more general models.

In this section, we assume that the mean of each variable in the $n \times p$ data matrix $\mathbf{X}$ whose column means have been removed. We use $\mathbf{x}_i$ to denote the $i^{th}$ observed data, namely, the $i^{th}$ row of the data.

### 13.5.1 Sparse PCA

A direct way to select variables is by adding an $\ell_1$ constraint to the loadings. This approach was considered by Jolliffe and Uddin (2003) where they proposed *SCoTLASS* to derive the sparse principal components one by one. SCoTLASS maximizes

$$\boldsymbol{\xi}_k^T(\mathbf{X}^T\mathbf{X})\boldsymbol{\xi}_k, \quad \text{s.t. } \boldsymbol{\xi}_k^T\boldsymbol{\xi}_k = 1 \quad \text{and} \quad \boldsymbol{\xi}_l^T\boldsymbol{\xi}_k = 0, \quad l < k; \qquad (13.47)$$

and the extra constraints

$$\|\boldsymbol{\xi}_k\|_1 \leq t$$

for some tuning parameter $t$. When $t$ is small enough, some elements of $\boldsymbol{\xi}_k$ will be zero. By its construction, SCoTLASS has high computational cost. Moreover, the examples in Jolliffe and Uddin (2003) suggest the obtained loadings by SCoTLASS are not sparse enough when requiring a high percentage of explained variance.

Zou, Hastie and Tibshirani (2006) proposed a sparse principal component (*SPCA*) algorithm based on a generalization of the *closest linear manifold approximation* view of PCA. Let $\mathbf{V}_k = [\mathbf{v}_1, \cdots, \mathbf{v}_k]$, where $\mathbf{v}_j$ is the $j^{th}$ principal component for $j \leq k$. Then, it is a $p \times k$ orthonormal matrix. Let $\mathcal{S}_k$ denote the column space spanned by $\{\mathbf{v}_j\}_{j=1}^k$ The projection operator to $\mathcal{S}_k$ is $\mathbf{P}_k = \mathbf{V}_k\mathbf{V}_k^T$. Viewing the projected data as an approximation of the original dataset, one way to define the best projection is by minimizing the total $\ell_2$

approximation error

$$\min_{\mathbf{V}_k} \sum_{i=1}^{n} \|\mathbf{x}_i - \mathbf{V}_k \mathbf{V}_k^T \mathbf{x}_i\|^2. \tag{13.48}$$

Observe that the objective function in (13.48) can be written as $\mathrm{tr}(\mathbf{S}(\mathbf{I}-\mathbf{P}_k))$, where $\mathbf{S} = n^{-1}\sum_{i=1}^{n} \mathbf{x}_i \mathbf{x}_i^T$ is the sample covariance matrix. Hence, the optimal $\mathbf{V}_k$ is defined as

$$\max_{\mathbf{V}_k} \mathrm{tr}(\mathbf{V}_k^T \mathbf{S} \mathbf{V}_k) \quad \text{subject to } \mathbf{V}_k^T \mathbf{V}_k = \mathbf{I}. \tag{13.49}$$

The solution gives the desired subspace $\mathcal{S}_k$, although the optimal $\mathbf{V}_k$ is only unique up to an orthonormal transformation. This view motivates us to consider the sparse PCA problem as the problem of finding a sparse representation of $\mathcal{S}_k$. If we directly add a sparse penalty on $\mathbf{V}_k$ in (13.48), we need to deal with the following optimization problem

$$\min_{\mathbf{V}_k:\mathbf{V}_k^T\mathbf{V}_k=\mathbf{I}} \sum_{i=1}^{n} \|\mathbf{x}_i - \mathbf{V}_k \mathbf{V}_k^T \mathbf{x}_i\|^2 + \lambda \sum_{j=1}^{k} \|\mathbf{v}_j\|_1. \tag{13.50}$$

However, it is not easy to handle both the $\ell_1$ penalty and the orthonormal constraint simultaneously. To overcome this drawback, Zou, Hastie and Tibshirani (2006) generalized the closest linear manifold approximation formulation to a more relaxed form that allows for an efficient computation of sparse loadings. Their sparse PCA formulation is motivated by the following theorem.

**Theorem 13.1** *Let* $\mathbf{A}$ *and* $\mathbf{B}$ *be* $p \times k$ *matrices. Let* $\widetilde{d}_1, \ldots, \widetilde{d}_k$ *be the first* $k$ *eigenvalues of* $\mathbf{X}^T\mathbf{X}$. *For any* $\lambda > 0$, *let* $\widehat{\mathbf{A}}$ *and* $\widehat{\mathbf{B}}$ *be the solution to the optimization problem*

$$\min_{\mathbf{A},\mathbf{B}} \sum_{i=1}^{n} \|\mathbf{x}_i - \mathbf{A}\mathbf{B}^T \mathbf{x}_i\|^2 + \lambda\, \mathrm{tr}(\mathbf{B}\mathbf{B}^T) \tag{13.51}$$
$$\textit{subject to} \quad \mathbf{A}^T\mathbf{A} = \mathbf{I}$$

*Then* $\widehat{\mathbf{B}} = \mathbf{V}_k \mathbf{I}_\lambda \mathbf{O}$ *where* $\mathbf{O}$ *is any* $k \times k$ *orthonormal matrix and* $\mathbf{I}_\lambda$ *is a diagonal matrix whose diagonal elements are* $\widetilde{d}_1/(\widetilde{d}_1+\lambda), \ldots, \widetilde{d}_k/(\widetilde{d}_k+\lambda)$.

**Proof of Theorem 13.1.** Let $\mathbf{a}_j$ be the $j$th column of $\mathbf{A}$ and $\mathbf{X}$ be the data matrix whose $i^{th}$ row is $\mathbf{x}_i^T$. By $\mathbf{A}^T\mathbf{A} = \mathbf{I}$, we have

$$\begin{aligned} &\sum_{i=1}^{n} \|\mathbf{x}_i - \mathbf{A}\mathbf{B}^T \mathbf{x}_i\|^2 + \lambda\, \mathrm{tr}(\mathbf{B}\mathbf{B}^T) \\ =\ & \mathrm{tr}(\mathbf{X}^T\mathbf{X}) - 2\,\mathrm{tr}(\mathbf{A}^T\mathbf{X}^T\mathbf{X}\mathbf{B}) + \mathrm{tr}(\mathbf{B}^T(\mathbf{X}^T\mathbf{X}+\lambda)\mathbf{B}). \end{aligned} \tag{13.52}$$

The last two terms have the following vector form:

$$\sum_{j=1}^{k} \left\{ \mathbf{b}_j^T (\mathbf{X}^T\mathbf{X} + \lambda\mathbf{I})\mathbf{b}_j - 2\mathbf{a}_j^T \mathbf{X}^T\mathbf{X}\mathbf{b}_j \right\}.$$

Thus given $\mathbf{A}$, the optimal $\mathbf{B}$ is $\mathbf{b}_j = \left(\mathbf{X}^T\mathbf{X} + \lambda\mathbf{I}\right)^{-1}\mathbf{X}^T\mathbf{X}\mathbf{a}_j$ for $j = 1, \ldots, k$; or equivalently

$$\mathbf{B} = \left(\mathbf{X}^T\mathbf{X} + \lambda\mathbf{I}\right)^{-1}\mathbf{X}^T\mathbf{X}\mathbf{A}. \tag{13.53}$$

Substituting (13.53) into (13.52), we have

$$\begin{aligned} \widehat{\mathbf{A}} &= \operatorname{argmax}_{\mathbf{A}} \operatorname{tr} \mathbf{A}^T\mathbf{X}^T\mathbf{X}(\mathbf{X}^T\mathbf{X} + \lambda\mathbf{I})^{-1}\mathbf{X}^T\mathbf{X}\mathbf{A} \\ & \quad \text{subject to} \quad \mathbf{A}^T\mathbf{A} = \mathbf{I}. \end{aligned}$$

By (13.49), the solution is that $\mathbf{A}$ consists of the first $k$ principal components of matrix $\mathbf{X}^T\mathbf{X}(\mathbf{X}^T\mathbf{X} + \lambda\mathbf{I})^{-1}\mathbf{X}^T\mathbf{X}$. Let us now find them. Do the spectral decomposition

$$\mathbf{X}^T\mathbf{X} = \mathbf{V}\widetilde{\mathbf{D}}\mathbf{V}^T$$

where $\widetilde{\mathbf{D}} = \operatorname{diag}(\widetilde{d}_1, \cdots, \widetilde{d}_p)$ is the diagonal elements, consisting of eigenvalues and $\mathbf{V} = (\mathbf{v}_1, \cdots, \mathbf{v}_p)$ is an orthonormal matrix, consisting of eigenvectors. It is then easy to see

$$\mathbf{X}^T\mathbf{X}(\mathbf{X}^T\mathbf{X} + \lambda\mathbf{I})^{-1}\mathbf{X}^T\mathbf{X} = \mathbf{V}\widetilde{\mathbf{D}}(\widetilde{\mathbf{D}} + \lambda\mathbf{I})^{-1}\widetilde{\mathbf{D}}\mathbf{V}^T.$$

Note that $\widetilde{\mathbf{D}}(\widetilde{\mathbf{D}} + \lambda\mathbf{I})^{-1}\widetilde{\mathbf{D}}$ is also a diagonal matrix. Hence, $\mathbf{A}$ equals $\mathbf{V}_k$ up to an orthonormal transformation. Write $\mathbf{A} = \mathbf{V}_k\mathbf{O}$. Substituting this into (13.53), we have $\widehat{\mathbf{B}} = \mathbf{V}_k\mathbf{I}_\lambda\mathbf{O}$. This completes the proof. ∎

Based on the proof of Theorem 13.1, we see that the parameter $\lambda$ is only needed when $\mathbf{X}^T\mathbf{X}$ is not invertible. For the $n > p$ case, we can use $\lambda = 0$ and $\mathbf{I}_\lambda$ becomes the identity matrix. For the $p > n$ case, we can use a very small $\lambda$ such that $\mathbf{I}_\lambda$ is almost an identity matrix. This can be done in practice because we can compute $\widetilde{d}_1, \ldots, \widetilde{d}_k$. Note that the operator norm of $\widehat{\mathbf{B}} - \mathbf{V}_k\mathbf{O}$ is upper bounded by $\lambda/(\widetilde{d}_k + \lambda)$. Therefore, it is proper to view $\widehat{\mathbf{B}}$ as $\mathbf{V}_k\mathbf{O}$, a set of orthonormal basis of $\mathcal{S}_k$.

With the formulation in Theorem 13.1, we now add the $\ell_1$ penalty to $\mathbf{B}$ in (13.51) in order to encourage zero loadings. This leads to the sparse PCA criterion (Zou, Hastie and Tibshirani, 2006) defined as

$$\begin{aligned} (\widehat{\mathbf{A}}, \widehat{\mathbf{B}}) &= \arg\min \sum_{i=1}^{n} \|\mathbf{x}_i - \mathbf{A}\mathbf{B}^T\mathbf{x}_i\|^2 + \lambda \sum_{j=1}^{k} \|\mathbf{b}_j\|^2 + \sum_{j=1}^{k} \lambda_{1,j}\|\mathbf{b}_j\|_1 \\ & \quad \text{subject to} \quad \mathbf{A}^T\mathbf{A} = \mathbf{I}. \end{aligned} \tag{13.54}$$

The normalized columns of $\widehat{\mathbf{B}}$ are taken as the loadings of the first $k$ sparse principal components. In the formulation (13.54), the drawback of simultaneous orthonormal constraints and $\ell_1$ constraints is avoided. This leads to the following alternating minimization algorithm for solving (13.54).

We first consider the optimal $\mathbf{B}$ solution given $\mathbf{A}$. Note that

$$\sum_{i=1}^{n} \|\mathbf{x}_i - \mathbf{A}\mathbf{B}^T\mathbf{x}_i\|^2 = \|\mathbf{X} - \mathbf{X}\mathbf{B}\mathbf{A}^T\|_{\mathrm{F}}^2, \tag{13.55}$$

where $\|\cdot\|_\mathrm{F}$ denotes the matrix Frobenius norm. Let $\mathbf{A}_\perp$ be any orthonormal matrix such that $[\mathbf{A}; \mathbf{A}_\perp]$ is $p \times p$ orthonormal. Then, we have

$$\begin{aligned}\|\mathbf{X} - \mathbf{X}\mathbf{B}\mathbf{A}^T\|_\mathrm{F}^2 &= \|\mathbf{X}\mathbf{A}_\perp\|_\mathrm{F}^2 + \|\mathbf{X}\mathbf{A} - \mathbf{X}\mathbf{B}\|_\mathrm{F}^2 \\ &= \|\mathbf{X}\mathbf{A}_\perp\|_\mathrm{F}^2 + \sum_{j=1}^{k} \|\mathbf{X}\mathbf{a}_j - \mathbf{X}\mathbf{b}_j\|^2.\end{aligned}$$

Define working response $\widetilde{\mathbf{Y}}_j = \mathbf{X}\mathbf{a}_j$. Then the optimal $\mathbf{B}$ must satisfy

$$\widehat{\mathbf{b}}_j = \arg\min_{\mathbf{b}_j} \|\widetilde{\mathbf{Y}}_j - \mathbf{X}\mathbf{b}_j\|^2 + \lambda\|\mathbf{b}_j\|^2 + \lambda_{1,j}\|\mathbf{b}_j\|_1 \qquad (13.56)$$

for $j = 1, \ldots, k$. The optimization problem (13.56) is an *Elastic Net* penalized least squares (Zou and Hastie, 2005).

Next, we consider the optimal $\mathbf{A}$ solution given $\mathbf{B}$. By (13.55), the best $\mathbf{A}$ must be

$$\arg\min_{\mathbf{A}} \|\mathbf{X} - \mathbf{X}\mathbf{B}\mathbf{A}^T\|_\mathrm{F}^2 \quad \text{subject to} \quad \mathbf{A}^T\mathbf{A} = \mathbf{I}. \qquad (13.57)$$

The solution has an explicit form given by the *reduced rank Procrustes rotation* theorem in Zou, Hastie and Tibshirani (2006). Explicitly, the solution is $\widehat{\mathbf{A}} = \mathbf{U}\mathbf{V}^T$ where $\mathbf{U}$ and $\mathbf{V}$ are the matrices in the SVD of $\mathbf{X}^T\mathbf{X}\mathbf{B} = \mathbf{U}\mathbf{D}\mathbf{V}^T$. This is justified by the following theorem and the resulting algorithm is summarized in Algorithm 13.6.

**Theorem 13.2** *Let $\mathbf{M}_{n\times p}$ and $\mathbf{N}_{n\times k}$ be two matrices. Let the SVD of $\mathbf{M}^T\mathbf{N}$ be $\mathbf{U}\mathbf{D}\mathbf{V}^T$. Then, the solution to the constrained minimization problem*

$$\arg\min_{\mathbf{A}} \|\mathbf{M} - \mathbf{N}\mathbf{A}^T\|_\mathrm{F}^2 \quad \textit{subject to} \quad \mathbf{A}^T\mathbf{A} = \mathbf{I}.$$

*is $\mathbf{U}\mathbf{V}^T$.*

We leave the proof of this theorem to an exercise (Exercise 13.10).

There are numerous successful applications of sparse PCA in the scientific literature. See, for example, Sjöstrand *et al.* (2007), Gravuer *et al.* (2008), Baden *et al.* (2016). R code for sparse PCA is in the package `elasticnet`. The Matlab implementation of sparse PCA is available in the toolbox `SpaSM` maintained by Karl Sjöstrand. The url is http://www2.imm.dtu.dk/projects/spasm/.

### *13.5.2 An iterative SVD thresholding approach*

PCA can be computed via the *singular value decomposition* (SVD) of the data matrix. Thus, it is natural to consider a sparse PCA algorithm based on

**Algorithm 13.6** Sparse PCA

1. Let $\mathbf{A}$ start at $\mathbf{V}_k$, the loadings of the first $k$ ordinary principal components.
2. Write $\mathbf{A} = [\mathbf{a}_1, \cdots, \mathbf{a}_k]$. For $j = 1, \ldots, k$, let $\widetilde{\mathbf{Y}}_j = \mathbf{X}\mathbf{a}_j$ and solve

$$\widehat{\mathbf{b}}_j = \arg\min_{\mathbf{b}_j} \|\widetilde{\mathbf{Y}}_j - \mathbf{X}\mathbf{b}_j\|^2 + \lambda\|\mathbf{b}_j\|^2 + \lambda_{1,j}\|\mathbf{b}_j\|_1$$

3. Write $\mathbf{B} = [\widehat{\mathbf{b}}_1, \cdots, \widehat{\mathbf{b}}_k]$ and compute the SVD of $\mathbf{X}^T\mathbf{X}\mathbf{B} = \mathbf{U}\mathbf{D}\mathbf{V}^T$, then update $\mathbf{A} = \mathbf{U}\mathbf{V}^T$.
4. Repeat steps 2–3, until convergence.
5. Output: The sparse PCA is $\widehat{\mathbf{v}}_j = \frac{\mathbf{b}_j}{\|\mathbf{b}_j\|}$, $j = 1, \ldots, k$.

the sparse modification of the SVD of $\mathbf{X}$. This idea was explored in Shen and Huang (2008) and Witten, Tibshirani and Hastie (2009).

Let the SVD of $\mathbf{X}$ be $\mathbf{X} = \mathbf{U}\mathbf{D}\mathbf{V}^T$. Consider the first principal component. We know the loading vector is $\mathbf{v}_1$, the first column of $\mathbf{V}$. It is a well known result that SVD is related to the best low rank approximation of $\mathbf{X}$ (Eckart and Young, 1936). Specifically, let $\widetilde{\mathbf{u}}$ be an $n$-dimensional vector with unit norm and $\widetilde{\mathbf{v}}$ be a $p$-dimensional vector. The best rank one approximation is defined as

$$\min_{\widetilde{\mathbf{u}},\widetilde{\mathbf{v}}} \|\mathbf{X} - \widetilde{\mathbf{u}}\widetilde{\mathbf{v}}^T\|_{\mathrm{F}}^2 \quad \text{subject to} \quad \|\widetilde{\mathbf{u}}\| = 1, \tag{13.58}$$

and the solution is $\widetilde{\mathbf{u}} = \mathbf{u}_1$ and $\widetilde{\mathbf{v}} = d_1\mathbf{v}_1$ where $d_1$ is the first singular value.

Based on (13.58), Shen and Huang (2008) proposed the following optimization problem

$$(\widehat{\mathbf{u}}, \widehat{\mathbf{v}}) = \arg\min_{\mathbf{u},\mathbf{v}} \|\mathbf{X} - \mathbf{u}\mathbf{v}^T\|_{\mathrm{F}}^2 + \lambda\|\mathbf{v}\|_1 \quad \text{subject to} \quad \|\mathbf{u}\| = 1, \tag{13.59}$$

and the sparse loading vector is taken to be the normalized $\widehat{\mathbf{v}}$, namely, $\frac{\widehat{\mathbf{v}}}{\|\widehat{\mathbf{v}}\|}$. In (13.59) the Lasso penalty can be replaced with another sparse penalty such as SCAD (Fan and Li, 2001). An alternating minimization algorithm can be used to solve (13.59). Note that given $\mathbf{v}$, the optimal $\mathbf{u}$ is given by a rank one Procrustes rotation. By Theorem 13.2, the optimal $\mathbf{u}$ is the normalized $\mathbf{X}\mathbf{v}$: $\mathbf{u} = \mathbf{X}\mathbf{v}/\|\mathbf{X}\mathbf{v}\|$. Given $\mathbf{u}$, the optimal $\mathbf{v}$ is

$$\arg\min_{\mathbf{v}} -2\,\mathrm{tr}(\mathbf{X}^T\mathbf{u}\mathbf{v}^T) + \|\mathbf{v}\|^2 + \lambda\|\mathbf{v}\|_1$$

and the solution is given by a vector soft-thresholding operator:

$$\mathbf{v} = S(\mathbf{X}^T\mathbf{u}, \lambda/2),$$

where the soft-thresholding operator is applied to each component of the vector. See Algorithm 13.7.

Shen and Huang (2008) applied Algorithm 13.7 sequentially to compute

**Algorithm 13.7** An iterative thresholding algorithm for the first sparse principal component

1. Let $\mathbf{u}$ start at the first column of the $\mathbf{U}$ matrix from the SVD of $\mathbf{X} = \mathbf{UDV}^T$.
2. Update $\mathbf{v} = S(\mathbf{X}^T\mathbf{u}, \lambda/2)$
3. Update $\mathbf{u} = \frac{\mathbf{v}}{\|\mathbf{v}\|}$.
4. Repeat steps 2–3, until convergence.
5. Output: the loading for the sparse PCA is $\frac{\mathbf{v}}{\|\mathbf{v}\|}$.

the rest of the sparse principal components. After computing the first $k$ sparse principal components, let $\mathbf{X}_{(k+1)} = \mathbf{X} - \sum_l^k \mathbf{u}_l \mathbf{v}_l^T$ and $(\mathbf{u}_{(k+1)}, \mathbf{v}_{(k+1)})$ be the outcome of using Algorithm 13.7 on $\mathbf{X}_{(k+1)}$. The normalized $\mathbf{v}_{(k+1)}$ is the loading vector of the $(k+1)$th sparse principal component. The $\lambda$ parameter is allowed to differ for different principal components.

### *13.5.3 A penalized matrix decomposition approach*

Witten, Tibshirani and Hastie (2009) proposed a *penalized matrix decomposition* (PMD) criterion as follows:

$$\begin{aligned}(\widehat{\mathbf{u}}, \widehat{\mathbf{v}}, \widehat{d}) &= \arg\min_{\mathbf{u},\mathbf{v},d} \|\mathbf{X} - d\mathbf{u}\mathbf{v}^T\|_{\mathrm{F}}^2 \\ \text{subject to} &\quad \|\mathbf{u}\| = 1, \|\mathbf{u}\|_1 \le c_1; \quad \|\mathbf{v}\| = 1, \|\mathbf{v}\|_1 \le c_2.\end{aligned} \tag{13.60}$$

By straightforward calculation, it can be shown that (13.60) is equivalent to the following optimization problem

$$\begin{aligned}(\widehat{\mathbf{u}}, \widehat{\mathbf{v}}) &= \text{argmax}_{\mathbf{u},\mathbf{v}} \mathbf{u}^T\mathbf{X}\mathbf{v} \\ \text{subject to} &\quad \|\mathbf{u}\| = 1, \|\mathbf{u}\|_1 \le c_1; \quad \|\mathbf{v}\| = 1, \|\mathbf{v}\|_1 \le c_2.\end{aligned} \tag{13.61}$$

and $\widehat{d} = \widehat{\mathbf{u}}^T\mathbf{X}\widehat{\mathbf{v}}$.

They also used an alternating minimization algorithm to compute (13.61). Given $\mathbf{v}$, we update $\mathbf{u}$ by solving

$$\max_{\mathbf{u}} \mathbf{u}^T\mathbf{X}\mathbf{v} \quad \text{subject to} \quad \|\mathbf{u}\| = 1, \|\mathbf{u}\|_1 \le c_1. \tag{13.62}$$

Given $\mathbf{u}$, we update $\mathbf{v}$ by solving

$$\max_{\mathbf{v}} \mathbf{u}^T\mathbf{X}\mathbf{v} \quad \text{subject to} \quad \|\mathbf{v}\| = 1, \|\mathbf{v}\|_1 \le c_2. \tag{13.63}$$

It is easy to see that the equality constraint $\|\mathbf{u}\| = 1, \|\mathbf{v}\| = 1$ in (13.62) and (13.63) can be replaced with the inequality constraint $\|\mathbf{u}\| \le 1, \|\mathbf{v}\| \le 1$ and the solutions remain the same. The solutions to these problems are analytically

**Algorithm 13.8** PMD algorithm for the first sparse principal component

1. Initialize $\mathbf{v}$ by a vector with unity norm.
2. Update $\mathbf{u}$ by $\frac{S(\mathbf{X}\mathbf{v},\Delta_{c_1})}{\|S(\mathbf{X}\mathbf{v},\Delta_{c_1})\|}$, where $\Delta_{c_1}$ is given in Lemma 13.1
3. Update $\mathbf{v}$ by $\frac{S(\mathbf{X}^T\mathbf{u},\Delta_{c_2})}{\|S(\mathbf{X}^T\mathbf{u},\Delta_{c_2})\|}$.
4. Repeat steps 2–3, until convergence.
5. $d = \mathbf{u}^T\mathbf{X}\mathbf{v}$.

given in Lemma 13.1 in Section 13.3.1. We summarize the computation steps in Algorithm 13.8.

PMD can be repeatedly used to derive the rest of the sparse principal components. Suppose that we have computed $(\mathbf{u}_j, d_j, \mathbf{v}_j)$ for $j = 1, \ldots, k$, then we let $\mathbf{X}_{(k+1)} = \mathbf{X} - \sum_{j=1}^{k} d_j \mathbf{u}_j \mathbf{v}_j^T$ and apply Algorithm 13.8 to $\mathbf{X}_{(k+1)}$ to compute $(\mathbf{u}_{(k+1)}, d_{(k+1)}, \mathbf{v}_{(k+1)})$.

### *13.5.4 A semidefinite programming approach*

We introduce some necessary notation first. For a given matrix $\mathbf{M}$, we use $\text{Card}(\mathbf{M})$ to denote the number of nonzero elements of $\mathbf{M}$ and $|\mathbf{M}|$ to mean a new matrix with each element of $\mathbf{M}$ replaced by its absolute value. Let $\mathbf{1}_p$ be the $p$-dimensional vector with all elements 1.

Consider the first $k$-sparse principal component with at most $k$ nonzero loadings. A natural definition of the optimal $k$-sparse loading vector is

$$\begin{aligned} & \text{argmax}_{\boldsymbol{\alpha}} \boldsymbol{\alpha}^T \widehat{\boldsymbol{\Sigma}} \boldsymbol{\alpha} \\ \text{subject to} \quad & \|\boldsymbol{\alpha}\| = 1, \quad \text{Card}(\boldsymbol{\alpha}) \leq k. \end{aligned} \tag{13.64}$$

When $k = p$, the above definition reduces the loadings of the first principal component. However, (13.64) is nonconvex and computationally difficult, especially when $p$ is large. Convex relaxation is a standard technique used in optimization to handle difficult nonconvex optimization problems. d'Aspremont, Ghaoui, Jordan and Lanckriet (2007) developed a *convex relaxation* of (13.64), which is expressed as a *semidefinite programming* problem.

Let $\mathbf{P} = \boldsymbol{\alpha}\boldsymbol{\alpha}^T$. Then, $\boldsymbol{\alpha}^T \widehat{\boldsymbol{\Sigma}} \boldsymbol{\alpha} = \text{tr}(\widehat{\boldsymbol{\Sigma}}\mathbf{P})$. The norm-1 constraint on $\boldsymbol{\alpha}$ leads to a linear equality constraint on $\mathbf{P}$: $\text{tr}(\mathbf{P}) = 1$. Moreover, the cardinality constraint $\|\boldsymbol{\alpha}\|_0 \leq k$ implies $\text{Card}(\mathbf{P}) \leq k^2$. Hence, we consider the following relaxed optimization problem of $\mathbf{P}$:

$$\begin{aligned} & \text{argmax}_{\mathbf{P}} \, \text{tr}(\widehat{\boldsymbol{\Sigma}}\mathbf{P}) \\ \text{subject to} \quad & \text{tr}\,\mathbf{P} = 1, \quad \text{Card}(\mathbf{P}) \leq k^2, \\ & \mathbf{P} \succeq 0 \quad \text{and} \quad \text{rank}(\mathbf{P}) = 1. \end{aligned} \tag{13.65}$$

The above formulation in (13.65) is still nonconvex and difficult to handle due to the cardinality constraint and the rank one constraint. By definition, $\mathbf{P}$ is symmetric and $\mathbf{P}^2 = \mathbf{P}$. Observe that

$$\|\mathbf{P}\|_F^2 = \mathrm{tr}(\mathbf{P}^T\mathbf{P}) = \mathrm{tr}(\mathbf{P}) = 1.$$

By the Cauchy-Schwartz inequality,

$$\mathbf{1}_p^T|\mathbf{P}|\mathbf{1}_p \le \sqrt{\mathrm{Card}(\mathbf{P})\|\mathbf{P}\|_{\mathrm{F}}^2} \le k. \tag{13.66}$$

Therefore, d'Aspremont, Ghaoui, Jordan and Lanckriet (2007) suggested to relax the cardinality constraint in (13.65) to a linear inequality constraint $\mathbf{1}_p^T|\mathbf{P}|\mathbf{1}_p \le k$. Furthermore, they dropped the rank one constraint and ended up with a convex optimization problem of $\mathbf{P}$

$$\begin{aligned} & \mathrm{argmax}_{\mathbf{P}}\, \mathrm{tr}(\widehat{\mathbf{\Sigma}}\mathbf{P}) \\ \text{subject to} \quad & \mathrm{tr}(\mathbf{P}) = 1, \quad \mathbf{1}_p^T|\mathbf{P}|\mathbf{1}_p \le k, \mathbf{P} \succeq 0\,. \end{aligned} \tag{13.67}$$

The above is recognized as a semidefinite programming problem and can be solved by Matlab software `SDPT3` (Toh, Todd and Tutuncu,1999) and `CVX` (Grant and Boyd, 2013).

The computational complexity of (13.67) is $O(p^3)$ which can be a huge problem for large $p$. The semidefinite programming method only solves for $\mathbf{P}$ but not $\boldsymbol{\alpha}$. To compute the loading vector $\boldsymbol{\alpha}$, d'Aspremont *et al.* (2007) recommended truncating $\mathbf{P}$ and retaining only the dominant (sparse) eigenvector of $\mathbf{P}$. For the second sparse principal component, it is suggested to replace $\widehat{\mathbf{\Sigma}}$ with $\widehat{\mathbf{\Sigma}} - (\boldsymbol{\alpha}^T\widehat{\mathbf{\Sigma}}\boldsymbol{\alpha})\boldsymbol{\alpha}\boldsymbol{\alpha}^T$ in (13.67). The same procedure can be repeated to compute the rest of the sparse principal components.

### *13.5.5 A generalized power method*

A direct formulation of the $\ell_1$ penalized sparse principal component is

$$\begin{aligned} & \mathrm{argmax}_{\|\boldsymbol{\alpha}\|=1}\boldsymbol{\alpha}^T\mathbf{X}^T\mathbf{X}\boldsymbol{\alpha} \\ \text{subject to} \quad & \|\boldsymbol{\alpha}\|_1 \le t. \end{aligned} \tag{13.68}$$

Equivalently, we can solve

$$\begin{aligned} & \mathrm{argmax}_{\|\boldsymbol{\alpha}\|=1}\sqrt{\boldsymbol{\alpha}^T\mathbf{X}^T\mathbf{X}\boldsymbol{\alpha}} \\ \text{subject to} \quad & \|\boldsymbol{\alpha}\|_1 \le t. \end{aligned} \tag{13.69}$$

Journée, Nesterov, Richtárik and Sepulchre (2010) considered the Lagrangian form of (13.64)

$$\mathrm{argmax}_{\|\boldsymbol{\alpha}\|=1}\sqrt{\boldsymbol{\alpha}^T\mathbf{X}^T\mathbf{X}\boldsymbol{\alpha}} - \lambda\|\boldsymbol{\alpha}\|_1. \tag{13.70}$$

They offered a *generalized power method* for solving (13.70). Their idea takes advantage of this simple fact: let $\widetilde{\mathbf{u}} = \text{argmax}_{\|\mathbf{u}\|=1} \mathbf{u}^T\mathbf{z}$, then $\widetilde{\mathbf{u}} = \mathbf{z}/\|\mathbf{z}\|$ and $\widetilde{\mathbf{u}}^T\mathbf{z} = \|\mathbf{z}\|$. Thus, an equivalent formulation of (13.69) is

$$\begin{aligned} (\mathbf{u}^*, \boldsymbol{\alpha}^*) &= \text{argmax}_{\mathbf{u},\boldsymbol{\alpha}} \mathbf{u}^T\mathbf{X}\boldsymbol{\alpha} - \lambda\|\boldsymbol{\alpha}\|_1 \\ \text{subject to} & \quad \|\mathbf{u}\| = 1, \ \|\boldsymbol{\alpha}\| = 1. \end{aligned} \tag{13.71}$$

Notice that the formulation (13.71) is the Lagrangian form of the PMD formulation (13.61) without imposing the $\ell_1$ constraint on $\mathbf{u}$. PMD used an alternating algorithm which in general finds a local optimum. The generalized power method uses a different algorithm.

For any $\mathbf{u}$, the optimal $\boldsymbol{\alpha}$ and $\mathbf{X}^T\mathbf{u}$ must share the same sign for each component. Thus, the objective function in (13.71) can be expressed as

$$\sum_{j=1}^{p} (|\mathbf{X}^T\mathbf{u}|_j - \lambda)|\alpha_j|.$$

Let $\mathbf{z}$ be a vector with its $j$th element being $z_j = |\alpha_j|$. Then the optimal $\mathbf{z}$ can be found by the following program:

$$\begin{aligned} \mathbf{z}^* &= \text{argmax}_{\mathbf{z}} \sum_{j=1}^{p} (|\mathbf{X}^T\mathbf{u}|_j - \lambda) z_j \\ \text{subject to} & \quad z_j \geq 0, \ \sum_{j=1}^{p} z_j^2 = 1. \end{aligned} \tag{13.72}$$

When $|\mathbf{X}^T\mathbf{u}|_j - \lambda \leq 0$, $z_j^* = 0$. By the Cauchy-Schwartz inequality, it is easy to see that the solution to (13.72) is

$$z_j^* = \frac{(|\mathbf{X}^T\mathbf{u}|_j - \lambda)_+}{\sqrt{\sum_{j=1}^{p}(|\mathbf{X}^T\mathbf{u}|_j - \lambda)_+^2}}. \tag{13.73}$$

Since $\boldsymbol{\alpha}^* = \text{sgn}(\mathbf{X}^T\mathbf{u})|\mathbf{z}^*|$, we have

$$\boldsymbol{\alpha}^* = \frac{S(\mathbf{X}^T\mathbf{u}, \lambda)}{\|S(\mathbf{X}^T\mathbf{u}, \lambda)\|}, \tag{13.74}$$

where $S$ is the soft-thresholding operator. Plugging (13.73) back into the objective function in (13.72), we obtain a new optimization criterion of $\mathbf{u}$:

$$\mathbf{u}^* = \text{argmax}_{\mathbf{u}:\|\mathbf{u}\|=1} \sqrt{\sum_{j=1}^{p}(|\mathbf{X}^T\mathbf{u}|_j - \lambda)_+^2}, \tag{13.75}$$

or equivalently

$$\mathbf{u}^* = \text{argmax}_{\mathbf{u}:\|\mathbf{u}\|\leq 1} \sum_{j=1}^{p}(|\mathbf{X}^T\mathbf{u}|_j - \lambda)_+^2. \tag{13.76}$$

Solving $\mathbf{u}^*$ is a $n$-dimensional optimization problem, although the original formulation (13.67) is a $p$-dimensional optimization problem. Moreover, the objective function in (13.76) is differentiable and convex, and the constraint set is compact and convex. Journée, Nesterov, Richtárik and Sepulchre (2010) used an efficient gradient method to compute $\mathbf{u}^*$, and $\boldsymbol{\alpha}^* = \frac{S(\mathbf{X}^T\mathbf{u}^*,\lambda)}{\|S(\mathbf{X}^T\mathbf{u}^*,\lambda)\|}$.

## 13.6 Bibliographical Notes

Two popular classical books on clustering are Hartigan(1975) and Kaufman and Rousseeuw (1990). The K-medoids algorithm was first proposed in Kaufman and Rousseeuw (1987) and it is also known as the Partitioning Around Medoids (PAM) algorithm. Raftery's group hosts a website for research on model-based clustering

https://www.stat.washington.edu/raftery/Research/mbc.html.

The EM algorithm was first introduced in Dempster, Laird and Rubin (1977). It is frequently used for fitting mixture models. See also, for example, Foss et al. (2016), McLachlan and Peel (2000), Khalili and Chen (2007), Khalili, Chen and Lin(2011), and references therein for additional studies on the mixture models.

Two well-cited papers on spectral clustering are Shi and Malik (2000) and Ng, Jordan and Weiss (2002). von Luxburg, Belkin and Bousquet (2008) established the consistency of spectral clustering by showing that the large sample limit of $\widetilde{\mathbf{L}}$ is an operator on the space of continuous functions. Their analysis also showed why the normalized spectral clustering is preferred over the un-normalized version. Friedman and Meulman (2004) proposed *COSA* for clustering with a subset of important variables. Witten and Tibshirani (2010) noticed that COSA does not really select variables and hence proposed their general sparse clustering framework. They also derived a sparse hierarchical clustering algorithm, in addition to a sparse K-means clustering algorithm. By using different sparse penalty functions, various sparse model-based clustering algorithms can be derived. See, for example, Wang and Zhu (2008), Xie, Pan and Shen (2008) and Guo, Levina, Michailidis and Zhu (2010). Jin and Wang (2016) proposed *IF-PCA* for high-dimensional clustering, where marginal testing is used to discard many useless features and then K-means clustering is done in the first few principal components of the reduced dataset.

There has been extensive literature on sparse principal components, which enhances the convergence of the principal components in high dimensional space. See Zou, Hastie, and Tibshirani (2006), d'Aspremont, El Ghaoui, Jordan, and Lanckriet (2007), d'Aspremont, Bach and El Ghaoui (2008), Shen and Huang (2008), Johnstone and Lu (2009), Amini and Wainwright (2009), Witten, Tibshirani, and Hastie (2009), Journée, Nesterov, Richtárik and Sepulchre (2010) Ma (2013), and Berthet and Rigollet (2013). Amini and Wainwright (2009) studied the support recovery property of the semidefinite programming approach of d'Aspremont *et al.* (2007) under the $k$-sparse assumption for the leading eigenvector in the rank-1 spiked covariance model.

Berthet and Rigollet (2013) considered a high-dimensional covariance testing problem in which the null hypothesis is that the covariance matrix is an identity matrix and the alternative is a $k$-sparse rank-1 spiked covariance model with noise variance one. They studied a minimax optimal test based on the $k$ sparse largest eigenvalue of the empirical covariance matrix. Birnbaum, Johnstone, Nadler and Paul (2013) established the minimax lower bounds for the individual eigenvectors under a rank-r spiked covariance model. Cai, Ma and Wu (2015) established the minimax rates under the spectral norm for sparse spiked covariance matrices and the optimal rate for the rank detection boundary. Considering a row sparsity assumption, Vu and Lei (2013) established the minimax rates for estimating the principal subspace without the spiked covariance model. Sparse factor analysis was first proposed in Choi, Oehlert and Zou (2010).

## 13.7 Exercises

13.1 Suppose that $p(\xi = k) = \pi_k(x)$. Show that the conditional distribution of $\xi$ given $\mathbf{X} = \mathbf{x}$ is multinomial with the class probability

$$P(\xi = k \mid \mathbf{X} = \mathbf{x}) = \frac{\pi_k(\mathbf{x}) f_k(\mathbf{x})}{\sum_{l=1}^{K} \pi_l(\mathbf{x}) f_l(\mathbf{x})}.$$

13.2 Download the mice protein expression data from the UCI Machine Learning Repository. The link is

`https://archive.ics.uci.edu/ml/datasets/Mice+Protein+Expression.`

Use attributes 2–78 (the expression levels of 77 proteins) as the input variables for clustering analysis. Attribute 82 shows the true class labels (there are 8 classes). Apply k-means clustering and spectral clustering to this dataset with $k = 8$. Compare the clustering result to the ground truth.

13.3 Download the seeds dataset from the UCI Machine Learning Repository. The link is https://archive.ics.uci.edu/ml/datasets/seeds Apply model-based clustering to the seeds dataset. Use BIC to select $K$.

13.4 Download the mice protein expression data from the UCI Machine Learning Repository:

`https://archive.ics.uci.edu/ml/datasets/Mice+Protein+Expression.`

(a) There are in total 1080 examples. Remove the examples with missing values. Report the total number of examples afterwards. Use attributes 2–78 (the expression levels of 77 proteins) as the input variables for clustering analysis. Attribute 82 shows the true class labels (there are 8 classes).

(b) Apply k-means clustering, spectral clustering and agglomerative hierarchical clustering to this dataset with the number of clusters $k = 8$. Compare the clustering results to the ground truth.

13.5 Verify (13.10).

13.6 Verify (13.13) and shows that $\mathbf{L}$ has an eigenvalue 0 with corresponding eigenvector $\boldsymbol{\alpha} = \mathbf{1}/\sqrt{n}$.

13.7 Show that the matrix $\widetilde{\mathbf{L}}^{\text{sym}}$ has eigenvalue zero with eigenvector $\mathbf{D}^{1/2}\mathbf{1}/c$, where c is the normalization constant $c = \|\mathbf{D}^{1/2}\mathbf{1}\|$.

13.8 Prove Lemma 13.1.

13.9 Verify (13.73).

13.10 Prove Theorem 13.2.

13.11 Let $G$ be an undirected graph with non-negative weights. Show that the multiplicity $K$ of the eigenvalue 0 of the graph Laplacian equals the number of connected components in the graph and that eigenspace of eigenvalue zero is spanned by the indicator vectors of the connected components.

13.12 Consider the following data generating model:

$$\begin{aligned} V_1 &\sim N(0, 290), & V_2 &\sim N(0, 300) \\ V_3 &= -0.3V_1 + 0.925V_2 + \epsilon, & \epsilon &\sim N(0, 1) \\ V_1, V_2 &\text{ and } \epsilon & &\text{are independent.} \end{aligned}$$

Ten observable variables are simulated as follows:

$$\begin{aligned} X_i &= V_1 + \epsilon_i^1, & \epsilon_i^1 &\sim N(0,1), & i &= 1,2,3,4, \\ X_i &= V_2 + \epsilon_i^2, & \epsilon_i^2 &\sim N(0,1), & i &= 5,6,7,8, \\ X_i &= V_3 + \epsilon_i^3, & \epsilon_i^3 &\sim N(0,1), & i &= 9,10, \\ \{\epsilon_i^j\} &\text{ are independent}, & j &= 1,2,3 & i &= 1,\cdots,10. \end{aligned}$$

Let $\boldsymbol{\Sigma}$ be the population covariance matrix of $\mathbf{X} = (X_1, \cdots, X_{10})$.

(a) The first two oracle sparse principal components are defined as

$$\boldsymbol{\alpha}_1 = \text{argmax}_{\boldsymbol{\alpha}} \boldsymbol{\alpha}^T \boldsymbol{\Sigma} \boldsymbol{\alpha} \quad \text{subject to } \|\boldsymbol{\alpha}_1\| = 1, \|\boldsymbol{\alpha}_1\|_0 = 4$$

$$\boldsymbol{\alpha}_2 = \text{argmax}_{\boldsymbol{\alpha}} \boldsymbol{\alpha}^T \boldsymbol{\Sigma} \boldsymbol{\alpha} \quad \text{subject to } \|\boldsymbol{\alpha}_2\| = 1, \boldsymbol{\alpha}_1^T \boldsymbol{\alpha}_2 = 0, \|\boldsymbol{\alpha}_2\|_0 = 4.$$

Derive $\boldsymbol{\alpha}_1$ and $\boldsymbol{\alpha}_2$.

(b) Apply a sparse PCA algorithm to $\boldsymbol{\Sigma}$. Do you recover the oracle sparse principal components?

Chapter 14

# An Introduction to Deep Learning

Deep learning or deep neural networks has achieved tremendous success in recent years. In simple words, *deep learning* uses many compositions of linear transformations followed by a nonlinear gating to approximate high-dimensional functions. The family of such functions is very flexible so that they can approximate most of target functions very well. While *neural networks* have a long history, recent advances have greatly improved their performance in computer vision, natural language processing, machine translations, among others, where the information set $\mathbf{x}$ is given but highly complex such as images, texts and voices and the signal-to-noise ratio is high.

What makes deep learning so successful nowadays? The arrivals of big data allows us to reduce variance in the deep neural networks and modern computing architects and powers permit us to use deeper networks to better approximate high-dimensional functions and hence reduces the biases. In other words, deep learning is a great family of scalable nonparametric methods that achieve great bias and variance trade-off for high-dimensional function estimation when sample size is very large.

From the statistical and scientific perspective, it is natural to ask: What is deep learning? What are the new characteristics of deep learning, compared with classical methods? What are the theoretical foundations of deep learning? To answer these questions, we introduce common neural network models (e.g., convolutional neural nets, recurrent neural nets, generative adversarial nets) and training techniques (e.g., stochastic gradient descent, dropout, batch normalization) from a statistical point of view. Along the way, we highlight new characteristics of deep learning (including depth and over-parametrization) and explain their practical and theoretical benefits. While a complete understanding of deep learning remains elusive, we hope that our introduction provides readers with a quick idea on the subject, which is advancing rapidly.

This chapter is adapted from the recent survey paper on this subject by Fan, Ma and Zhong (2019). A lot of the materials are taken from that paper. We thank particularly Mr. Cong Ma and Yiqiao Zhong (2019) for their tremendous help in writing this chapter.

## 14.1 Rise of Deep Learning

Modern machine learning and statistics deal with the problem of *learning from data*: given a training dataset $\{(y_i, \mathbf{x}_i)\}_{1 \le i \le n}$ where $\mathbf{x}_i \in \mathbb{R}^d$ is the input and $y_i \in \mathbb{R}$ is the output[1], one seeks a function $f : \mathbb{R}^d \mapsto \mathbb{R}$ from a certain function class $\mathcal{F}$ that has good prediction performance on test data. This problem is of fundamental significance and finds applications in numerous scenarios. For instance, in image recognition, the input $\mathbf{x}$ (reps. the output $y$) corresponds to the raw image (reps. its category) and the goal is to find a mapping $f(\cdot)$ that can classify future images accurately. Decades of research efforts in statistics and machine learning have been devoted to developing methods to find $f(\cdot)$ efficiently with provable guarantees. Prominent examples include linear classifiers (e.g., linear / logistic regression, linear discriminant analysis), kernel methods (e.g., support vector machines), tree-based methods (e.g., decision trees, random forests), nonparametric regression (e.g., nearest neighbors, local kernel smoothing), etc. See Chapter 12 for an in-depth discussion of the methods. Roughly speaking, each aforementioned method corresponds to a different function class $\mathcal{F}$ from which the final classifier $f(\cdot)$ is chosen.

Deep learning (LeCun, Bengio, Hinton, 2015), in its abstract form, proposes the following *compositional* function class:

$$\{f(\mathbf{x};\boldsymbol{\theta}) = \mathbf{W}_L \boldsymbol{\sigma}_L(\mathbf{W}_{L-1} \cdots \boldsymbol{\sigma}_1(\mathbf{W}_1\mathbf{x} + \mathbf{b}_1) + \mathbf{b}_{L-1}) + b_L\}, \tag{14.1}$$

where the parameters are $\boldsymbol{\theta} = \{\mathbf{W}_1, \ldots, \mathbf{W}_L, \mathbf{b}_1, \cdots, \mathbf{b}_L\}$, consisting of matrices $\{\mathbf{W}_\ell\}$ and vectors $\{\mathbf{b}_\ell\}$ with appropriate sizes. Here, for each $1 \le l \le L$, $\boldsymbol{\sigma}_\ell(\cdot)$ is some given nonlinear functions, called *activation functions*, applied to each component of inputs in the previous layer. In other words, starting from $\mathbf{x}^{(0)} = \mathbf{x}$, recursively compute

$$\mathbf{x}^{(\ell)} = \boldsymbol{\sigma}_\ell(\mathbf{W}_\ell \mathbf{x}^{(\ell-1)} + \mathbf{b}_\ell), \quad \text{and} \quad f(\mathbf{x};\boldsymbol{\theta}) = \mathbf{x}^{(L)}.$$

Though simple, deep learning has made significant progress towards addressing the problem of learning from data over the past decade. Specifically, it has performed close to or better than humans in various important tasks in artificial intelligence, including image recognition (He, Zhang, Ren and Sun, 2016), game playing (Silver, *et al.*, 2017), and machine translation (Wu, *et al.*, 2016). Owing to its great promise, the impact of deep learning is also growing rapidly in areas beyond artificial intelligence; examples include statistics (Bauer and Kohler, 2017; Liang, 2017; Romano, Sesia and Candés, 2018; Gao, Liu, Yao and Zhu, 2018; Schmidt-Hieber, 2020), applied mathematics (E,

[1]When the label $y$ is given, this problem is often known as *supervised learning*. We mainly focus on this paradigm throughout this chapter and remark sparingly on its counterpart, *unsupervised learning*, where $y$ is not given.

Table 14.1: Winning models for the ILSVRC image classification challenge over time.

| Model | Year | # Layers | # Params | Top-5 error |
|---|---|---|---|---|
| Shallow | $< 2012$ | — | — | $> 25\%$ |
| AlexNet | 2012 | 8 | 61M | 16.4% |
| VGG19 | 2014 | 19 | 144M | 7.3% |
| GoogleNet | 2014 | 22 | 7M | 6.7% |
| ResNet-152 | 2015 | 152 | 60M | 3.6% |

Han and Jentzen, 2017; Chen, Rubanova, Bettencourt and Duvenaud, 2018), clinical research (De Fauw *et al.*, 2018), etc.

To get a better idea of the success of deep learning, let us take the ImageNet Challenge (also known as *ILSVRC*, Russakovsky, *et al.*, 2015) as an example. In the classification task, one is given a training dataset consisting of 1.2 million color images with 1000 categories, and the goal is to classify images based on the input pixels. The performance of a classifier is then evaluated on a test dataset of 100 thousand images, and in the end the top-5 error[2] is reported. Table 14.1 highlights a few popular models and their corresponding performance. As can be seen, starting from 2012, the deep learning model *AlexNet* outperforms, by quite a wide margin, shallow models[3] (the first row) that fit linear models / tree-based models on handcrafted features, which generates a surge of interest in deep learning research. In addition, the improvements are further realized as we use deeper networks.

It is widely acknowledged that two indispensable factors contribute to the success of deep learning, namely (1) huge datasets that often contain millions of samples and (2) immense computing power resulting from clusters of graphics processing units (GPUs). Admittedly, these resources are only recently available: the latter allows us to train larger neural networks which reduces biases and the former enables variance reduction. However, these two alone are not sufficient to explain the tremendous success of deep learning. Several "dreadful" characteristics such as *over-parametrization* and *nonconvexity* will emerge along with our introduction of different deep learning models and training methods.

There has been a very rich literature on deep learning and its frontier is still being pushed on a daily basis. Therefore it is impossible to cover every aspect of deep learning in such a small chapter. We will then first introduce basic deep

[2]The algorithm makes an error if the true label is not contained in the 5 predictions made by the algorithm.

[3]Shallow models are typically feature extraction followed by linear/nonlinear SVMs, which were used in the ILSVRC classification challenge before 2012.

learning models including feed-forward neural nets, convolutional neural nets and recurrent neural nets in Sections 14.2 and 14.3. Section 14.4 introduces autoencoders and generative adversarial networks. Section 14.5 is devoted to training algorithms and their ability of driving the training error small. Section 14.6 uses an experiment on images to demonstrate the superiority of deep learning models. In the end, we remark in passing on the theoretical questions in the research of deep learning.

## 14.2 Feed-forward Neural Networks

Before delving into the vanilla feed-forward neural nets, let us set up necessary notation for the rest of this section. We focus primarily on *classification* problems, as regression problems can be addressed similarly. Given the training dataset $\{(y_i, \mathbf{x}_i)\}_{1\le i\le n}$ where $y_i \in [K] \triangleq \{1, 2, \ldots, K\}$ and $\mathbf{x}_i \in \mathbb{R}^d$ are independent across $i \in [n]$, supervised learning aims at finding a (possibly random) function $\widehat{f}(\mathbf{x})$ that predicts the outcome $y$ for a new input $\mathbf{x}$, assuming $(y, \mathbf{x})$ follows the same distribution as $(y_i, \mathbf{x}_i)$. In the terminology of machine learning, the input $\mathbf{x}_i$ is often called the *feature*, the output $y_i$ called the *label*, and the pair $(y_i, \mathbf{x}_i)$ is an *example*. The function $\widehat{f}$ is called the *classifier*, and estimation of $\widehat{f}$ is *training* or *learning*. The performance of $\widehat{f}$ is evaluated through the prediction error $\mathbb{P}(y \neq \widehat{f}(\mathbf{x}))$, which can often be estimated from a separate test dataset.

As with classical statistical estimation, for each $k \in [K]$, a classifier approximates the conditional probability $\mathbb{P}(y = k|\mathbf{x})$ using a function $f_k(\mathbf{x}; \boldsymbol{\theta}_k)$ parametrized by $\boldsymbol{\theta}_k$. Then the category with the highest probability is predicted. Thus, learning is essentially estimating the parameters $\boldsymbol{\theta}_k$. In statistics, one of the most popular methods is *multinomial logistic regression*, which stipulates a specific form for the functions

$$f_k(\mathbf{x}; \boldsymbol{\theta}_k) = \frac{\exp(\mathbf{x}^\top \boldsymbol{\beta}_k + \alpha_k)}{\sum_{j=1}^K \exp(\mathbf{x}^\top \boldsymbol{\beta}_j + \alpha_j)}.$$

It is clear that multinomial logistic regression induces linear decision boundaries in $\mathbb{R}^d$, and hence it is restrictive in modeling nonlinear dependency between $y$ and $\mathbf{x}$. The deep neural networks we introduce below provide a flexible framework for modeling nonlinearity in a fairly general way.

### *14.2.1 Model setup*

From the high level, deep neural networks (DNNs) use composition of a series of simple nonlinear functions to model nonlinearity

$$\mathbf{h}^{(L)} = \mathbf{g}^{(L)} \circ \mathbf{g}^{(L-1)} \circ \ldots \circ \mathbf{g}^{(1)}(\mathbf{x}),$$

where $\circ$ denotes the composition of two functions and $L$ is the number of hidden layers, and is usually called *depth* of a NN model. Letting $\mathbf{h}^{(0)} \triangleq \mathbf{x}$, one can recursively define $\mathbf{h}^{(\ell)} = \mathbf{g}^{(\ell)}(\mathbf{h}^{(\ell-1)})$ for all $\ell = 1, 2, \ldots, L$. The *feed-forward neural networks*, also called the *multilayer perceptrons* (MLPs), are neural nets with a specific choice of $\mathbf{g}^{(\ell)}$: for $\ell = 1, \ldots, L$, define

$$\mathbf{h}^{(\ell)} = \mathbf{g}^{(\ell)}(\mathbf{h}^{(\ell-1)}) \triangleq \boldsymbol{\sigma}(\mathbf{W}^{(\ell)}\mathbf{h}^{(\ell-1)} + \mathbf{b}^{(\ell)}), \tag{14.2}$$

where $\mathbf{W}^{(\ell)}$ and $\mathbf{b}^{(\ell)}$ are the weight matrix and the bias / intercept, respectively, associated with the $l$-th layer, and $\boldsymbol{\sigma}(\cdot)$ is usually a simple given (known) nonlinear function called the *activation function*. In words, in each layer $\ell$, the input vector $\mathbf{h}^{(\ell-1)}$ goes through an affine transformation first and then passes through a fixed nonlinear function $\boldsymbol{\sigma}(\cdot)$. See Figure 14.1 for an illustration of a simple MLP with two hidden layers. The activation function $\boldsymbol{\sigma}(\cdot)$ is usually applied element-wise, and a popular choice is the ReLU (*Rectified Linear Unit*) function:

$$\sigma(z) = \max\{z, 0\}. \tag{14.3}$$

Other choices of activation functions include leaky ReLU, tanh function (Maas, Hannun and Ng, 2013) and the classical sigmoid function $(1 + e^{-z})^{-1}$, which is less popular now.

It is worthwhile noting that the ReLU activation function is a crucial element in most deep learning models. Its popularity is largely due to the fact that its derivative is either 0 or 1, which makes training more efficient. See Section 14.5 for discussion on numerical stability.

Given an output $\mathbf{h}^{(L)}$ from the final hidden layer and a label $y$, we can define a loss function to minimize. A common loss function for classification problems is the multinomial logistic loss. Using the terminology of deep learning, we say that $\mathbf{h}^{(L)}$ goes through an affine transformation and then the *soft-max* function:

$$f_k(\mathbf{x}; \boldsymbol{\theta}) \triangleq \frac{\exp(z_k)}{\sum_k \exp(z_k)}, \quad \forall\, k \in [K], \quad \text{where } \mathbf{z} = \mathbf{W}^{(L+1)}\mathbf{h}^{(L)} + \mathbf{b}^{(L+1)} \in \mathbb{R}^K.$$

Then the loss is defined to be the cross-entropy between the label $y$ (in the form of an indicator vector) and the score vector $(f_1(\mathbf{x}; \boldsymbol{\theta}), \ldots, f_K(\mathbf{x}; \boldsymbol{\theta}))^\top$, which is exactly the negative log-likelihood of the multinomial logistic regression model:

$$\mathcal{L}(\mathbf{f}(\mathbf{x}; \boldsymbol{\theta}), y) = -\sum_{k=1}^{K} \mathbf{1}\{y = k\} \log f_k(\mathbf{x}; \boldsymbol{\theta}), \tag{14.4}$$

where $\boldsymbol{\theta} \triangleq \{\mathbf{W}^{(\ell)}, \mathbf{b}^{(\ell)} : 1 \le \ell \le L + 1\}$. As a final remark, the number of parameters scales with both the depth $L$ and the width (i.e., the dimensionality of $\mathbf{W}^{(\ell)}$), and hence it can be very large for deep neural nets.

### *14.2.2 Back-propagation in computational graphs*

Training neural networks follows the *empirical risk minimization* paradigm

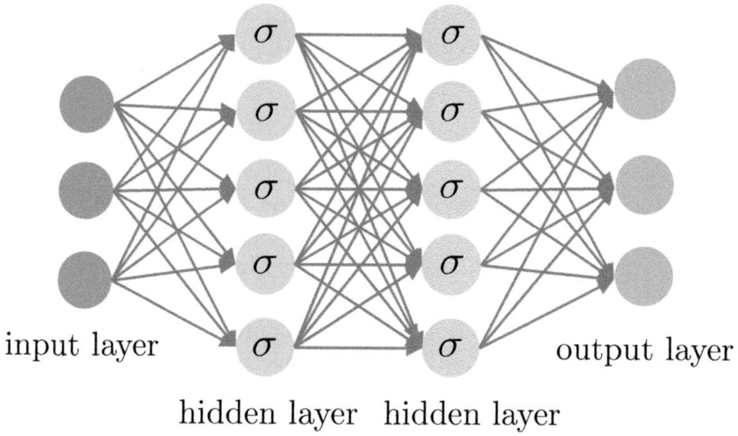


Figure 14.1: A feed-forward neural network with an input layer, two hidden layers and an output layer. The input layer represents raw features $\{\mathbf{x}_i\}_{1\le i\le n}$. Both hidden layers compute an affine transform (a.k.a. indices) of the input and then apply an element-wise activation function $\boldsymbol{\sigma}(\cdot)$. Finally, the output returns a linear transform followed by the softmax activation (resp. simply a linear transform) of the hidden layers for the classification (resp. regression) problem. Taken from Fan, Ma and Zhong (2019).

that minimizes the loss (e.g., (14.4)) over all the training data. This minimization is usually done via *stochastic gradient descent* (SGD). In a way similar to gradient descent, SGD starts from a certain initial value $\boldsymbol{\theta}^0$ and then iteratively updates the parameters $\boldsymbol{\theta}^t$ by moving it in the direction of the negative gradient. The difference is that, in each update, a small subsample $\mathcal{B} \subset [n]$ called a *mini-batch*—which is typically of size 32–512—is randomly drawn and the gradient calculation is only on $\mathcal{B}$ instead of the full batch $[n]$. This saves considerably the computational cost in calculation of gradient. By the law of large numbers, this stochastic gradient should be close to the full sample one, albeit with some random fluctuations. A pass of the whole training set is called an *epoch*. Usually, after several or tens of epochs, the error on a validation set levels off and training is complete. See Section 14.5 for more details and variants on training algorithms.

The key to the above training procedure, namely SGD, is the calculation of the gradient $\nabla \ell_{\mathcal{B}}(\boldsymbol{\theta})$, where

$$\ell_{\mathcal{B}}(\boldsymbol{\theta}) \triangleq |\mathcal{B}|^{-1} \sum_{i\in\mathcal{B}} \mathcal{L}(\mathbf{f}(\mathbf{x}_i;\boldsymbol{\theta}), y_i). \tag{14.5}$$

The computation savings come from the fact that the size of $\mathcal{B}$ is much smaller than $n$. Gradient computation, however, is in general nontrivial for complex models, and it is susceptible to numerical instability for a model with large depth. Here, we introduce an efficient approach, namely *back-propagation*, for computing gradients in neural networks.

Back-propagation (Rumelhart, Hinton and Williams, 1985) is a direct application of the chain rule for calculation derivatives of composition functions in networks. As the name suggests, the calculation is performed in a backward fashion: one first computes $\partial \ell_{\mathcal{B}}/\partial \mathbf{h}^{(L)}$, then $\partial \ell_{\mathcal{B}}/\partial \mathbf{h}^{(L-1)}$, ..., and finally $\partial \ell_{\mathcal{B}}/\partial \mathbf{h}^{(1)}$. For example, in the case of the ReLU activation function[4], by (14.2), we have the following recursive / backward relation

$$\frac{\partial \ell_{\mathcal{B}}}{\partial \mathbf{h}^{(\ell-1)}} = \frac{\partial \mathbf{h}^{(\ell)}}{\partial \mathbf{h}^{(\ell-1)}} \cdot \frac{\partial \ell_{\mathcal{B}}}{\partial \mathbf{h}^{(\ell)}} = (\mathbf{W}^{(\ell)})^{\top} \mathsf{diag}\left(\mathbf{1}\{\mathbf{W}^{(\ell)}\mathbf{h}^{(\ell-1)} + \mathbf{b}^{(\ell)} \geq \mathbf{0}\}\right) \frac{\partial \ell_{\mathcal{B}}}{\partial \mathbf{h}^{(\ell)}} \tag{14.6}$$

where $\mathsf{diag}(\cdot)$ denotes a diagonal matrix with elements given by the argument and the indicator is the derivative of the ReLU function. Note that the calculation of $\partial \ell_{\mathcal{B}}/\partial \mathbf{h}^{(\ell-1)}$ depends on $\partial \ell_{\mathcal{B}}/\partial \mathbf{h}^{(\ell)}$, which is the partial derivatives from the next layer. In this way, the derivatives are "back-propagated" from the last layer to the first layer. These derivatives $\{\partial \ell_{\mathcal{B}}/\partial \mathbf{h}^{(\ell)}\}$ are then used to update the parameters. For instance, the gradient update for $\mathbf{W}^{(\ell)}$ is given by

$$\mathbf{W}^{(\ell)} \leftarrow \mathbf{W}^{(\ell)} - \eta \frac{\partial \ell_{\mathcal{B}}}{\partial \mathbf{W}^{(\ell)}}, \quad \text{where} \quad \frac{\partial \ell_{\mathcal{B}}}{\partial W^{(\ell)}_{jm}} = \frac{\partial \ell_{\mathcal{B}}}{\partial h^{(\ell)}_{j}} \cdot \sigma' \cdot h^{(\ell-1)}_{m}, \tag{14.7}$$

where $\sigma' = 1$ if the $j$-th element of $\mathbf{W}^{(\ell)}\mathbf{h}^{(\ell-1)} + \mathbf{b}^{(\ell)}$ is nonnegative, and $\sigma' = 0$ otherwise. The step size $\eta > 0$, also called the *learning rate*, controls how much parameters are changed in a single update. The derivative in (14.7) follows from the definition (14.2) and the factor in the derivative can be computed through the back-propagation.

A more general way to think about neural network models and training is to consider *computational graphs*. Computational graphs are directed acyclic graphs that represent functional relations between variables. They are very convenient and flexible to represent function composition, and moreover, they also allow an efficient way of computing gradients. Consider an MLP with a single hidden layer and an $\ell_2$ regularization:

$$\ell^{\lambda}_{\mathcal{B}}(\boldsymbol{\theta}) = \ell_{\mathcal{B}}(\boldsymbol{\theta}) + r_{\lambda}(\boldsymbol{\theta}) = \ell_{\mathcal{B}}(\boldsymbol{\theta}) + \lambda\Big(\sum_{j,j'} (W^{(1)}_{j,j'})^2 + \sum_{j,j'} (W^{(2)}_{j,j'})^2\Big), \tag{14.8}$$

where $\ell_{\mathcal{B}}(\boldsymbol{\theta})$ is the same as (14.5), and $\lambda \geq 0$ is a tuning parameter. A similar example is considered in Goodfellow, Bengio, and Courville (2016). The corresponding computational graph is shown in Figure 14.2. Each node represents a function (inside a circle), which is associated with an output of that function (outside a circle). For example, we view the term $\ell_{\mathcal{B}}(\boldsymbol{\theta})$ as a result of 4 compositions: first the input data $\mathbf{x}$ multiplies the weight matrix $\mathbf{W}^{(1)}$ resulting

[4]The issue of non-differentiability at the origin is often ignored in implementation.

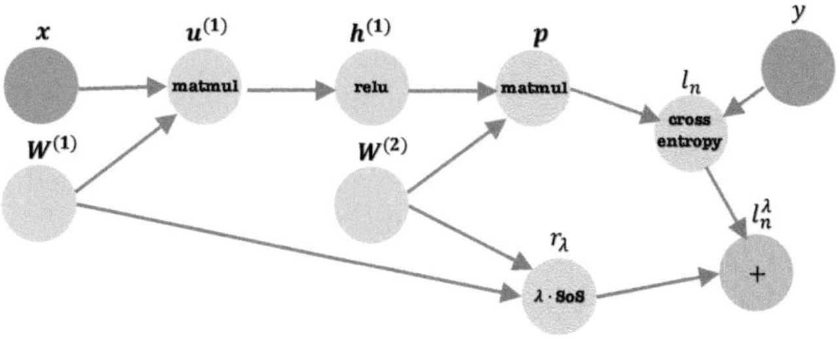


Figure 14.2: The computational graph illustrates the loss (14.8). For simplicity, we omit the bias terms. Symbols inside nodes represent functions, and symbols outside nodes represent function outputs (vectors/scalars). `matmul` is matrix multiplication, `relu` is the ReLU activation, `cross entropy` is the cross entropy loss, and `SoS` is the sum of squares. From Fan, Ma and Zhong (2019).

in $\mathbf{u}^{(1)}$, then it goes through the ReLU activation function `relu` resulting in $\mathbf{h}^{(1)}$, then it multiplies another weight matrix $\mathbf{W}^{(2)}$ leading to $\mathbf{p}$, and finally it produces the cross-entropy with label $y$ as in (14.4). The regularization term is incorporated in the graph similarly.

A forward pass is complete when all nodes are evaluated starting from the input $\mathbf{x}$. A backward pass then calculates the gradients of $\ell_{\mathcal{B}}^{\lambda}$ with respect to all other nodes in the reverse direction. Due to the chain rule, the gradient calculation for a variable (say, $\partial \ell_{\mathcal{B}}/\partial \mathbf{u}^{(1)}$) is simple: it only depends on the gradient value of the variables ($\partial \ell_{\mathcal{B}}/\partial \mathbf{h}$) the current node points to, and the function derivative evaluated at the current variable value ($\boldsymbol{\sigma}'(\mathbf{u}^{(1)})$). Thus, in each iteration, a computation graph only needs to (1) calculate and store the function evaluations at each node in the forward pass, and then (2) calculate all derivatives in the backward pass.

Back-propagation in computational graphs forms the foundations of popular deep learning programming softwares, including TensorFlow (Abadi *et al.*, 2015) and PyTorch (Paszke, *et al.*2017), which allows more efficient building and training of complex neural net models.

## 14.3 Popular Models

Moving beyond vanilla feed-forward neural networks, we introduce two other popular deep learning models, namely, the convolutional neural networks (CNNs) and the recurrent neural networks (RNNs). One important characteristic shared by the two models is *weight sharing*, that is some model parameters are identical across locations in CNNs or across time in RNNs. This is related to the notion of translational invariance in CNNs and stationarity in RNNs. At the end of this section, we introduce a modular thinking for constructing more flexible neural nets.

### *14.3.1 Convolutional neural networks*

The convolutional neural network (CNN) (LeCun, Bottou, Bengio and Haffner, 1998; Fukushima and Miyake, 1982) is a special type of feed-forward neural networks that is tailored for image processing. More generally, it is suitable for analyzing data with salient spatial structures. In this subsection, we focus on image classification using CNNs, where the raw input (image pixels) and features of each hidden layer are represented by a 3D tensor $\mathbf{X} \in \mathbb{R}^{d_1 \times d_2 \times d_3}$. Here, the first two dimensions $d_1, d_2$ of $\mathbf{X}$ indicate spatial coordinates of an image while the third $d_3$ indicates the number of channels. For instance, $d_3$ is 3 for the raw inputs due to the red, green and blue channels, and $d_3$ can be much larger (say, 256) for hidden layers. Each channel is also called a *feature map*, because each feature map is specialized to detect the same feature at different locations of the input, which we will soon explain. We now introduce two building blocks of CNNs, namely the convolutional layer (linear transforms) and the pooling layer (nonlinear gating).

1. *Convolutional layer (CONV).* A convolutional layer has the same functionality as described in (14.2), where the input feature $\mathbf{X} \in \mathbb{R}^{d_1 \times d_2 \times d_3}$ goes through an affine transformation first and then an element-wise nonlinear activation. The difference lies in the specific form of the affine transformation. A convolutional layer uses a number of *filters* to extract local features from the previous input. More precisely, each filter is represented by a 3D tensor $\mathbf{F}_k \in \mathbb{R}^{w \times w \times d_3}$ $(1 \le k \le \widetilde{d}_3)$, where $w$ is the size of the filter (typically 3 or 5) and $\widetilde{d}_3$ denotes the total number of filters. Note that the third dimension $d_3$ of $\mathbf{F}_k$ is equal to that of the input feature $\mathbf{X}$. For this reason, one usually says that the filter has size $w \times w$, while suppressing the third dimension $d_3$. Each filter $\mathbf{F}_k$ then convolves with the input feature $\mathbf{X}$ to obtain one single feature map $\mathbf{O}^k \in \mathbb{R}^{(d_1 - w+1) \times (d_1 - w+1)}$, where[5]

$$O_{ij}^k = \left\langle [\mathbf{X}]_{ij}, \mathbf{F}_k \right\rangle = \sum_{i'=1}^{w} \sum_{j'=1}^{w} \sum_{l=1}^{d_3} [\mathbf{X}]_{i+i'-1, j+j'-1, l} [\mathbf{F}_k]_{i', j', l}, \tag{14.9}$$

which is a locally (since $w$ is small) weighted average of $[\mathbf{X}]_{ij}$ using weights $\mathbf{F}_k$. Here $[\mathbf{X}]_{ij} \in \mathbb{R}^{w \times w \times d_3}$ is a small "patch" of $\mathbf{X}$ starting at location $(i, j)$. See Figure 14.3 for an illustration of the convolution operation. If we view the 3D tensors $[\mathbf{X}]_{ij}$ and $\mathbf{F}_k$ as vectors, then each filter essentially computes their inner product with a part of $\mathbf{X}$ indexed by $i, j$ (which can be also viewed as convolution, as its name suggests). We then pack the resulted feature maps $\{\mathbf{O}^k\}$ into a 3D tensor $\mathbf{O}$ with size $(d_1 - w+1) \times (d_1 - w+1) \times \widetilde{d}_3$, where

$$[\mathbf{O}]_{ijk} = [\mathbf{O}^k]_{ij}.$$

Note that $\widetilde{d}_3$ can be larger than $d_3$ and the parameters to be optimized from the data are the elements in $\{\mathbf{F}_k\}_{k=1}^{\widetilde{d}_3}$.

[5]To simplify notation, we omit the bias/intercept term associated with each filter.

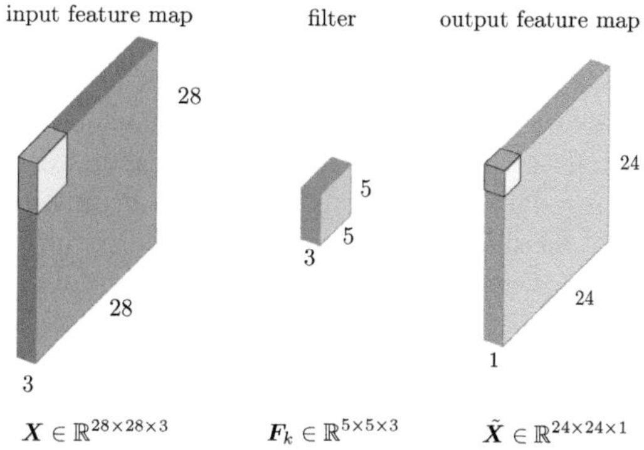


Figure 14.3: $\mathbf{X} \in \mathbb{R}^{28\times 28\times 3}$ represents the input feature consisting of $28 \times 28$ spatial coordinates in a total number of 3 channels / feature maps. $\mathbf{F}_k \in \mathbb{R}^{5\times 5\times 3}$ denotes the $k$-th filter with size $5 \times 5$. The third dimension 3 of the filter automatically matches the number 3 of channels in the previous input. Every 3D patch of $\mathbf{X}$ gets convolved with the filter $\mathbf{F}_k$ and this as a whole results in a single output feature map $\widetilde{X}_{:,:,k}$ with size $24 \times 24 \times 1$. Stacking the outputs of all the filters $\{\mathbf{F}_k\}_{1\leq k\leq K}$ will lead to the output feature with size $24 \times 24 \times K$. Taken from Fan, Ma and Zhong (2019).

The outputs of convolutional layers are then followed by nonlinear activation functions. In the ReLU case, we have

$$\widetilde{X}_{ijk} = \sigma(O_{ijk}), \qquad \forall i \in [d_1 - w + 1], j \in [d_2 - w + 1], k \in [\widetilde{d}_3]. \quad (14.10)$$

The convolution operation (14.9) and the ReLU activation (14.10) work together to extract features $\widetilde{\mathbf{X}}$ from the input $\mathbf{X}$, which functions in a way similar to the feedforward neural net (14.2). More specifically than feedforward neural nets, the filters $\mathbf{F}_k$ are shared across all locations $(i, j)$. A patch $[\mathbf{X}]_{ij}$ of an input responds strongly (that is, producing a large value) to a filter $\mathbf{F}_k$ if they are positively correlated. Therefore intuitively, each filter $\mathbf{F}_k$ serves to extract features similar to $\mathbf{F}_k$.

As a side note, after the convolution (14.9), the spatial size $d_1 \times d_2$ of the input $\mathbf{X}$ shrinks to $(d_1 - w + 1) \times (d_2 - w + 1)$ of $\widetilde{\mathbf{X}}$. However one may want the spatial size unchanged. This can be achieved via *padding*, where one appends zeros to the margins of the input $\mathbf{X}$ to enlarge the spatial size to $(d_1 + w - 1) \times (d_2 + w - 1)$. In addition, a *stride* in the convolutional layer determines the gap $i' - i$ and $j' - j$ between two patches $\mathbf{X}_{ij}$ and $\mathbf{X}_{i'j'}$: in (14.9) the stride is 1, and a larger stride would lead to feature maps with smaller sizes.

2. *Pooling layer (POOL).* A pooling layer aggregates the information of nearby features into a single one. This downsampling operation reduces the size of the features for subsequent layers and saves computation. One common

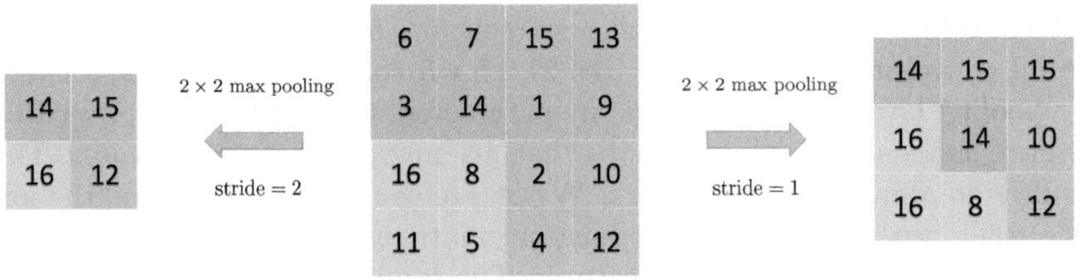


Figure 14.4: A 2 × 2 max pooling layer extracts the maximum of 2 by 2 neighboring pixels / features across the spatial dimension. Taken from Fan, Ma and Zhong (2019).

form of the pooling layer is composed of the 2 × 2 max-pooling filter. It computes $\max\{X_{i,j,k}, X_{i+1,j,k}, X_{i,j+1,k}, X_{i+1,j+1,k}\}$, that is, the maximum of the 2 × 2 neighborhood in the spatial coordinates; see Figure 14.4 for an illustration. Note that the pooling operation is done separately for each feature map $k$. As a consequence, a 2 × 2 max-pooling filter acting on $\mathbf{X} \in \mathbb{R}^{d_1 \times d_2 \times d_3}$ will result in an output of size $d_1/2 \times d_2/2 \times d_3$. In addition, the pooling layer does not involve any parameters to optimize. Pooling layers serve to reduce redundancy since a small neighborhood around a location $(i, j)$ in a feature map is likely to contain the same information.

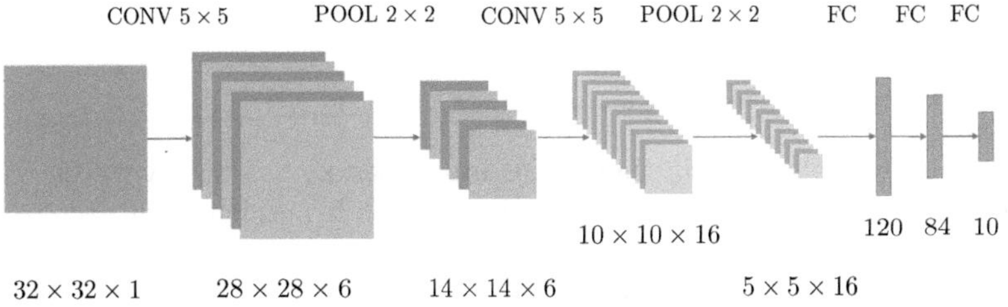


Figure 14.5: LeNet is composed of an input layer, two convolutional layers (with $\widetilde{d}_3 = 6$ and $\widetilde{d}_3 = 16$), two pooling layers (stride = 2), and three fully-connected layers. Both convolutions are valid and use filters with size 5 × 5. In addition, the two pooling layers use 2 × 2 average pooling. Taken from Fan, Ma and Zhong (2019).

In addition, we also use fully-connected layers as building blocks, which we have already seen in Section 14.2. Each fully-connected layer treats input tensor $\mathbf{X}$ as a vector $\text{Vec}(\mathbf{X})$, and computes $\widetilde{\mathbf{X}} = \boldsymbol{\sigma}(\mathbf{W}\text{Vec}(\mathbf{X}))$. A fully-connected layer does not use weight sharing and is often used in the last few layers of a CNN. As an example, Figure 14.5 depicts the well-known LeNet 5 (LeCun, Bottou, Bengio and Haffner, 1998), which is composed of two sets of CONV-POOL layers and three fully-connected layers. To help us better understand the underlying statistical model, let us count the number of parameters in LeNet 5. First, we have six 5 × 5 convolution filters in the first convolution layer and another sixteen $5 \times 5$ convolution filters in the second convolution layer. This gives $21 * 5 * 5$ convolution parameters. The two pooling layers do not involve any parameters to be optimized. After this, we have sixteen $5 \times 5$ feature maps that are mapped into 120 features. This requires an affine transform matrix $\mathbf{W}_1$ of size $(5*5*16) \times 120$ to do the job. After that, we still need affine transform matrices $\mathbf{W}_2$ of size $120 \times 84$ and $\mathbf{W}_3$ of size $84 \times 10$ to map the features. The total number of parameters is

$$6*5*5+16*5*5+5*5*16*120+120*84+84*10 = 59470$$

The final 10 features are used to classify 10 digits.

### *14.3.2 Recurrent neural networks*

Recurrent neural nets (RNNs) are another family of powerful models, which are designed to process time series data and other sequence data. RNNs have successful applications in speech recognition (Sak, Senior and Beaufays, 2014), machine translation (Wu, Schuster, *et al.*, 2016) genome sequencing (Cao, *et al.*, 2018), among others. The structure of an RNN naturally forms a computational graph, and can be easily combined with other structures such as CNNs to build large computational graph models for complex tasks. Here we introduce vanilla RNNs and improved variants such as long short-term memory (LSTM).

#### *14.3.2.1 Vanilla RNNs*

Suppose we have general time series inputs $\mathbf{x}_1, \mathbf{x}_2, \ldots, \mathbf{x}_T$. A vanilla RNN models the "hidden state" at time $t$ by a vector $\mathbf{h}_t$, which is subject to the recursive formula

$$\mathbf{h}_t = \mathbf{f}_{\boldsymbol{\theta}}(\mathbf{h}_{t-1}, \mathbf{x}_t). \tag{14.11}$$

Here, $f_{\boldsymbol{\theta}}$ is generally a nonlinear function parametrized by $\boldsymbol{\theta}$. Concretely, a vanilla RNN with one hidden layer has the following form[6]

$$\begin{aligned}
\mathbf{h}_t &= \mathbf{tanh}\left(\mathbf{W}_{hh}\mathbf{h}_{t-1} + \mathbf{W}_{xh}\mathbf{x}_t + \mathbf{b_h}\right), \qquad \text{where } \tanh(a) = \tfrac{e^{2a}-1}{e^{2a}+1}, \\
\mathbf{z}_t &= \boldsymbol{\sigma}\left(\mathbf{W}_{hy}\mathbf{h}_t + \mathbf{b_z}\right),
\end{aligned}$$

where $\mathbf{W}_{hh}, \mathbf{W}_{xh}, \mathbf{W}_{hy}$ are trainable weight matrices, $\mathbf{b_h}, \mathbf{b_z}$ are trainable bias vectors, and $\mathbf{z}_t$ is the output at time $t$. Like many classical time series models, those parameters are shared across time. Note that in different applications, we may have different input/output settings (cf. Figure 14.6). Examples include

- **One-to-many:** a single input with multiple outputs; see Figure 14.6(a). A typical application is image captioning, where the input is an image and outputs are a series of words.

[6]Similar to the activation function $\boldsymbol{\sigma}(\cdot)$, the function $\mathbf{tanh}(\cdot)$ means element-wise operations.

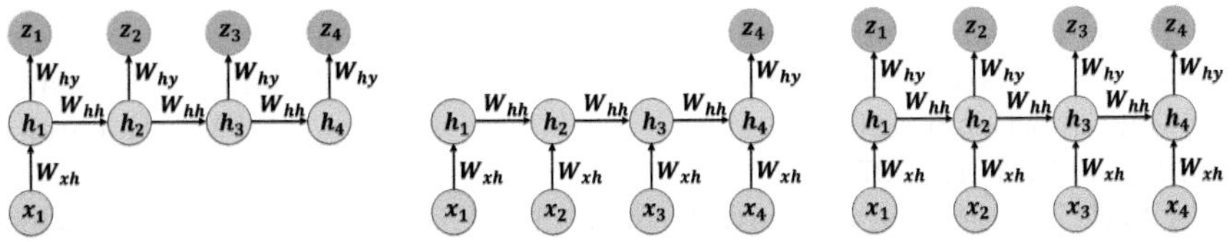


Figure 14.6: Vanilla RNNs with different inputs/outputs settings. (a) has one input but multiple outputs; (b) has multiple inputs but one output; (c) has multiple inputs and outputs. Note that the parameters are shared across time steps. Taken from Fan, Ma and Zhong (2019).

- **Many-to-one:** multiple inputs with a single output; see Figure 14.6(b). One application is text sentiment classification, where the input is a series of words in a sentence and the output is a label (e.g., positive vs. negative).
- **Many-to-many:** multiple inputs and outputs; see Figure 14.6(c). This is adopted in machine translation, where inputs are words of a source language (say Chinese) and outputs are words of a target language (say English).

As is the case with feed-forward neural nets, we minimize a loss function using back-propagation, where the loss is typically

$$\ell_{\mathcal{T}}(\boldsymbol{\theta}) = \sum_{t \in \mathcal{T}} \mathcal{L}(y_t, \mathbf{z}_t) = -\sum_{t \in \mathcal{T}} \sum_{k=1}^{K} I(y_t = k) \log \left( \frac{\exp([\mathbf{z}_t]_k)}{\sum_k \exp([\mathbf{z}_t]_k)} \right),$$

where $K$ is the number of categories for classification (e.g., size of the vocabulary in machine translation), and $\mathcal{T} \subset [T]$ is the length of the output sequence. During the training, the gradients $\partial \ell_{\mathcal{T}} / \partial \mathbf{h}_t$ are computed in the reverse time order (from $T$ to $t$). For this reason, the training process is often called *back-propagation through time*.

One notable drawback of vanilla RNNs is that, they have difficulty in capturing long-range dependencies in sequence data when the length of the sequence is large. This is sometimes due to the phenomenon of *exploding / vanishing gradients.* Take Figure 14.6(c) as an example. Computing $\partial \ell_{\mathcal{T}} / \partial \mathbf{h}_1$ involves the product $\prod_{t=1}^{3} (\partial \mathbf{h}_{t+1} / \partial \mathbf{h}_t)$ by the chain rule. However, if the sequence is long, the product will be the multiplication of many Jacobian matrices, which usually results in exponentially large or small singular values. To alleviate this issue, in practice, the forward pass and backward pass are implemented in a shorter sliding window $\{t_1, t_1 + 1, \ldots, t_2\}$, instead of the full sequence $\{1, 2, \ldots, T\}$. Though effective in some cases, this technique alone does not fully address the issue of long-term dependency.

#### *14.3.2.2 GRUs and LSTM*

There are two improved variants that alleviate the above issue: *gated recurrent units* (GRUs) (Cho, Van Merriënboer, Gulcehre, Bahdanau, Bougares, Schwenk and Bengio, 2014) and *long short-term memory* (LSTM) (Hochreiter and Schmidhuber, 1997).

- A **GRU** refines the recursive formula (14.11) by introducing *gates*, which are vectors of the same length as $\mathbf{h}_t$. The gates, which take values in $[0, 1]$ elementwise, multiply with $\mathbf{h}_{t-1}$ elementwise and determine how much they keep the old hidden states.
- An **LSTM** similarly uses gates in the recursive formula. In addition to $\mathbf{h}_t$, an LSTM maintains a *cell state*, which takes values in $\mathbb{R}$ elementwise and are analogous to counters.

Here we only discuss LSTM in detail. Denote by $\odot$ the element-wise multiplication. We have a recursive formula replacing (14.11):

$$\begin{pmatrix} \boldsymbol{i}_t \\ \mathbf{f}_t \\ \mathbf{o}_t \\ \mathbf{g}_t \end{pmatrix} = \begin{pmatrix} \sigma \\ \sigma \\ \sigma \\ \mathbf{tanh} \end{pmatrix} \mathbf{W} \begin{pmatrix} \mathbf{h}_{t-1} \\ \mathbf{x}_t \\ 1 \end{pmatrix},$$

$$\mathbf{c}_t = \mathbf{f}_t \odot \mathbf{c}_{t-1} + \boldsymbol{i}_t \odot \mathbf{g}_t,$$

$$\mathbf{h}_t = \mathbf{o}_t \odot \mathbf{tanh}(\mathbf{c}_t),$$

where $\mathbf{W}$ is a big weight matrix with appropriate dimensions. The cell state vector $\mathbf{c}_t$ carries information of the sequence (e.g., singular/plural form in a sentence). The forget gate $\mathbf{f}_t$ determines how much the values of $\mathbf{c}_{t-1}$ are kept for time $t$, the input gate $\boldsymbol{i}_t$ controls the amount of update to the cell state, and the output gate $\mathbf{o}_t$ gives how much $\mathbf{c}_t$ reveals to $\mathbf{h}_t$. Ideally, the elements of these gates have nearly binary values. For example, an element of $\mathbf{f}_t$ being close to 1 may suggest the presence of a feature in the sequence data. Similar to the skip connections in residual nets, the cell state $\mathbf{c}_t$ has an additive recursive formula, which helps back-propagation and thus captures long-range dependencies.

#### *14.3.2.3 Multilayer RNNs*

Multilayer RNNs are a generalization of the one-hidden-layer RNN discussed above. Figure 14.7 shows a vanilla RNN with two hidden layers. In place of (14.11), the recursive formula for an RNN with $L$ hidden layers now reads

$$\mathbf{h}_t^\ell = \mathbf{tanh}\left[\mathbf{W}^\ell \begin{pmatrix} \mathbf{h}_t^{\ell-1} \\ \mathbf{h}_{t-1}^\ell \\ 1 \end{pmatrix}\right], \quad \text{for all}\, \ell \in [L], \qquad \mathbf{h}_t^0 \triangleq \mathbf{x}_t.$$

Note that a multilayer RNN has two dimensions: the sequence length $T$ and depth $L$. Two special cases are the feed-forward neural nets (where $T = 1$)

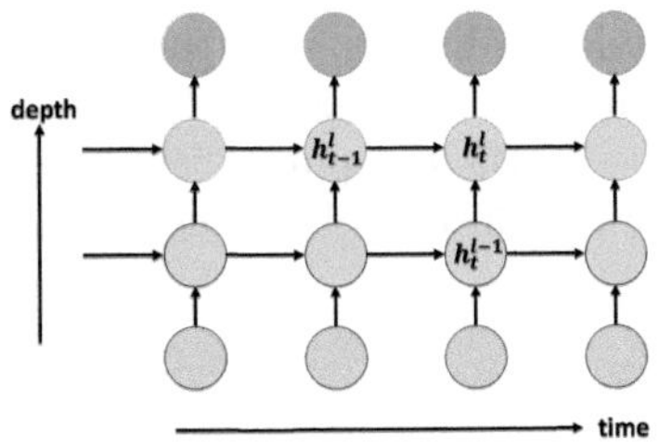


Figure 14.7: A vanilla RNN with two hidden layers. Higher-level hidden states $\mathbf{h}_t^\ell$ are determined by the old states $\mathbf{h}_{t-1}^\ell$ and lower-level hidden states $\mathbf{h}_t^{\ell-1}$. Multilayer RNNs generalize both feed-forward neural nets and one-hidden-layer RNNs. Taken from Fan, Ma and Zhong (2019).

introduced in Section 14.2, and RNNs with one hidden layer (where $L = 1$). Multilayer RNNs usually do not have very large depth (e.g., 2–5), since $T$ is already very large.

Finally, we remark that CNNs, RNNs, and other neural nets can be easily combined to tackle tasks that involve different sources of input data. For example, in image captioning, the images are first processed through a CNN, and then the high-level features are fed into an RNN as inputs. Theses neural nets combined together form a large computational graph, so they can be trained using back-propagation. This generic training method provides much flexibility in various applications.

### *14.3.3 Modules*

Deep neural nets are essentially composition of many nonlinear functions. A component function may be designed to have specific properties in a given task, and it can itself result from composing a few simpler functions. In LSTM, we have seen that the building block consists of several intermediate variables, including cell states and forget gates that can capture long-term dependency and alleviate numerical issues.

This leads to the idea of designing *modules* for building more complex neural net models. Desirable modules usually have low computational costs, alleviate numerical issues in training, and lead to good statistical accuracy. Since modules and the resulting neural net models form computational graphs, training follows the same principle briefly described in Section 14.2.

Here, we use the examples of *Inception* and *skip connections* to illustrate the ideas behind modules. Figure 14.8(a) is an example of "Inception" modules used in GoogleNet (Szegedy *et al.*, 2015). As before, all the convolutional layers are followed by the ReLU activation function. The concatenation of information from filters with different sizes gives the model great flexibility to capture spatial information. Note that $1 \times 1$ filters are $1 \times 1 \times d_3$ tensor (where $d_3$ is the number of feature maps), so its convolutional operation does not interact with other spatial coordinates, only serving to aggregate information from different feature maps at the same coordinate. This reduces the number of parameters and speeds up the computation. Similar ideas appear in other work (Lin, Chen and Yan, 2013; Iandola, *et al.*, 2016).

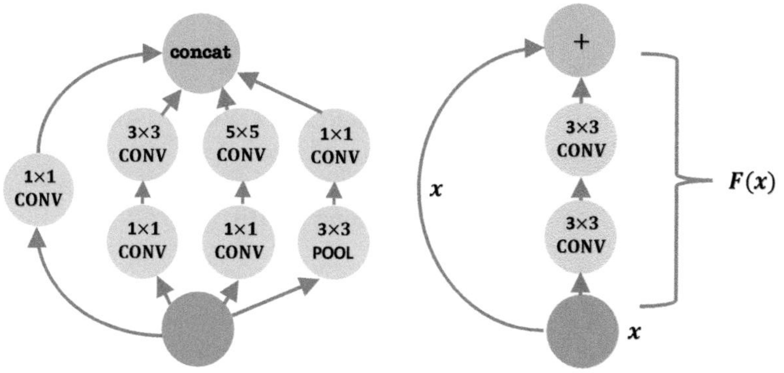


(a) "Inception" module (b) Skip connections

Figure 14.8: (a) The "Inception" module from GoogleNet. `Concat` means combining all features maps into a tensor. (b) Skip connections are added every two layers in ResNets. Taken from Fan, Ma and Zhong (2019).

Another module, usually called *skip connections*, is widely used to alleviate numerical issues in very deep neural nets, with additional benefits in optimization efficiency and statistical accuracy. Training very deep neural nets is generally more difficult, but the introduction of skip connections in *residual networks* (He, Zhang, Ren and Sun, 2016a,b) has greatly eased the task.

The high level idea of skip connections is to add an identity map to an existing nonlinear function. Let $\mathbf{F}(\mathbf{x})$ be an arbitrary nonlinear function represented by a (fragment of) neural net, then the idea of skip connections is simply replacing $\mathbf{F}(\mathbf{x})$ with $\mathbf{x}+\mathbf{F}(\mathbf{x})$. Figure 14.8(b) shows a well-known structure from residual networks (He, Zhang, Ren and Sun, 2016a, b)—for every two layers, an identity map is added:

$$\mathbf{x} \longmapsto \boldsymbol{\sigma}(\mathbf{x} + \mathbf{F}(\mathbf{x})) = \boldsymbol{\sigma}(\mathbf{x} + \mathbf{W}'\boldsymbol{\sigma}(\mathbf{W}\mathbf{x} + \mathbf{b}) + \mathbf{b}'), \tag{14.12}$$

where $\mathbf{x}$ can be hidden nodes from any layer and $\mathbf{W}, \mathbf{W}', \mathbf{b}, \mathbf{b}'$ are corresponding parameters. By repeating (namely composing) this structure throughout all layers, He, Zhang, Ren and Sun(2016a, b) are able to train neural nets with hundreds of layers easily, which overcomes well-observed training difficulties in deep neural nets. Moreover, deep residual networks also improve statistical accuracy, as the classification error on the ImageNet challenge was reduced by 46% from 2014 to 2015. As a side note, skip connections can be used flexibly. They are not restricted to the form in (14.12), and can be used between any pair of layers $\ell, \ell'$ (Huang, Liu, Van Der Maaten and Weinberger, 2017).

## 14.4 Deep Unsupervised Learning

This section shows how deep neural network functions (14.2) can be applied to *unsupervised learning*. Again, they are critical to approximating high-dimensional nonlinear functions.

In *supervised learning*, given a labelled training set $\{(y_i, \mathbf{x}_i)\}$, we focus on discriminative models, which essentially represents $\mathbb{P}(y \mid \mathbf{x})$ by a deep neural net $f(\mathbf{x}; \boldsymbol{\theta})$ with parameters $\boldsymbol{\theta}$. Unsupervised learning, in contrast, aims at extracting *information* from *unlabeled* data $\{\mathbf{x}_i\}$, where the labels $\{y_i\}$ are absent. In regard to this information, it can be a low-dimensional embedding of the data $\{\mathbf{x}_i\}$ or a generative model with latent variables to approximate the distribution $\mathbb{P}_{\mathbf{X}}(\mathbf{x})$. To achieve these goals, we introduce two popular unsupervised deep leaning models, namely, autoencoders and generative adversarial networks (GANs). The first one can be viewed as a dimension reduction technique, and the second as a density estimation method. Both problems require estimating high-dimensional nonlinear functions and DNNs are the key elements for both of these two models.

### *14.4.1 Autoencoders*

Recall that in dimension reduction, the goal is to reduce the dimensionality of the data and at the same time preserve its salient features. In particular, in principal component analysis (PCA), the goal is to embed the data $\{\mathbf{x}_i\}_{1 \le i \le n}$ into a low-dimensional space via a linear function $\mathbf{f}$ such that maximum variance can be explained. Equivalently, we want to find linear functions $\mathbf{f} : \mathbb{R}^d \to \mathbb{R}^k$ and $\mathbf{g} : \mathbb{R}^k \to \mathbb{R}^d$ $(k \le d)$ such that the difference between $\mathbf{x}_i$ and $\mathbf{g}(\mathbf{f}(\mathbf{x}_i))$ is minimized. Formally, we let

$$\mathbf{f}(\mathbf{x}) = \mathbf{W}_f \mathbf{x} \triangleq \mathbf{h} \text{ and } \mathbf{g}(\mathbf{h}) = \mathbf{W}_g \mathbf{h}, \quad \text{where } \mathbf{W}_f \in \mathbb{R}^{k \times d} \text{ and } \mathbf{W}_g \in \mathbb{R}^{d \times k}.$$

Here, for simplicity, we assume that the intercept/bias terms for $\mathbf{f}$ and $\mathbf{g}$ are zero. Then, PCA amounts to minimizing the quadratic loss function

$$\text{minimize}_{\mathbf{W}_f, \mathbf{W}_g} \quad \frac{1}{n} \sum_{i=1}^{n} \|\mathbf{x}_i - \mathbf{W}_f \mathbf{W}_g \mathbf{x}_i\|_2^2 . \tag{14.13}$$

It is the same as minimizing $\|\mathbf{X} - \mathbf{W}\mathbf{X}\|_{\mathrm{F}}^2$ subject to $\mathrm{rank}(\mathbf{W}) \le k$, where $\mathbf{X} \in \mathbb{R}^{p \times n}$ is the design matrix. The solution is given by the singular value decomposition of $\mathbf{X}$ (Thm. 2.4.8 in Golub and Van Loan, 2013), which is exactly what PCA does. It turns out that PCA is a special case of *autoencoders*, which is often known as the *undercomplete linear autoencoder*.

More broadly, autoencoders are neural network models for (nonlinear) dimension reduction, which generalize PCA. An autoencoder has two key components, namely, the *encoder* function $\mathbf{f}(\cdot)$, which maps the input $\mathbf{x} \in \mathbb{R}^d$ to a hidden code/representation $\mathbf{h} \triangleq \mathbf{f}(\mathbf{x}) \in \mathbb{R}^k$, and the *decoder* function $\mathbf{g}(\cdot)$, which maps the hidden representation $\mathbf{h}$ back to a point $\mathbf{g}(\mathbf{h}) \in \mathbb{R}^d$. Both functions can be multilayer neural network functions (14.2). See Figure 14.9 for an

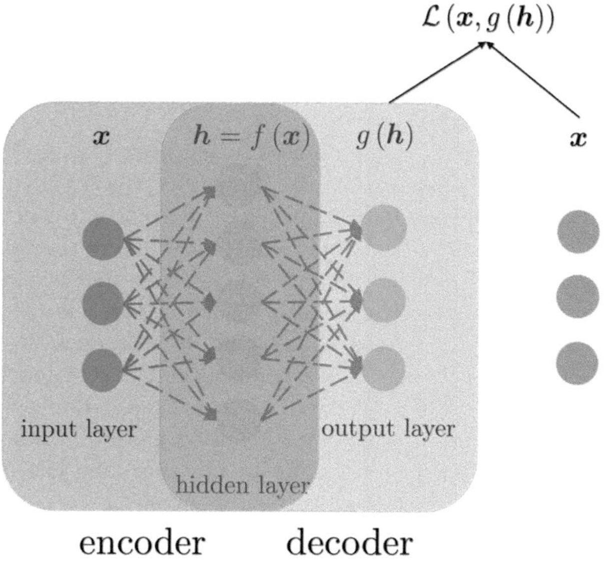


Figure 14.9: First an input $\mathbf{x}$ goes through the encoder $\mathbf{f}(\cdot)$, and we obtain its hidden representation $\mathbf{h} = \mathbf{f}(\mathbf{x})$. Then, we use the decoder $\mathbf{g}(\cdot)$ to get $\mathbf{g}(\mathbf{h})$ as a reconstruction of $\mathbf{x}$. Finally, the loss is determined from the difference between the original input $\mathbf{x}$ and its reconstruction $\mathbf{g}(\mathbf{f}(\mathbf{x}))$. Taken from Fan, Ma and Zhong (2019).

illustration of autoencoders. Let $\mathcal{L}(\mathbf{x}_1, \mathbf{x}_2)$ be a loss function that measures the difference between $\mathbf{x}_1$ and $\mathbf{x}_2$ in $\mathbb{R}^d$. Similar to PCA, an autoencoder is used to find the encoder $\mathbf{f}(\cdot)$ and decoder $\mathbf{g}(\cdot)$ such that $\mathcal{L}(\mathbf{x}, \mathbf{g}(\mathbf{f}(\mathbf{x})))$ is as small as possible. Mathematically, this amounts to solving the following minimization problem

$$\text{minimize}_{\mathbf{f},\mathbf{g}} \quad \frac{1}{n}\sum_{i=1}^{n} \mathcal{L}\left(\mathbf{x}_i, \mathbf{g}\left(\mathbf{f}\left(\mathbf{x}_i\right)\right)\right). \tag{14.14}$$

One needs to make structural assumptions on the functions $\mathbf{f}$ and $\mathbf{g}$ in order to find useful representations of the data, which leads to different types of autoencoders. Indeed, if no assumption is made, choosing $\mathbf{f}$ and $\mathbf{g}$ to be identity functions clearly minimizes the above optimization problem. To avoid this trivial solution, one natural way is to require that the encoder $f$ maps the data onto a space with a smaller dimension, i.e., $k < d$. This is the *undercomplete autoencoder* that includes PCA as a special case. One natural application of the neural network models is to replace $\mathbf{f}$ and $\mathbf{g}$ by neural network functions (14.2) and optimize the parameters via minimizing (14.14).

Just as PCA has a probabilistic model underpinning, namely factor models, autoencoders (14.14) can also be cast in the language of latent variable models. Let the latent variables $\mathbf{h}$ be $\mathcal{N}(\mathbf{0}, \mathbf{I}_k)$, and suppose the conditional probability $p_{\boldsymbol{\theta}}(\mathbf{x}|\mathbf{h})$ is associated with a decoder neural network with parameters $\boldsymbol{\theta}$; for example, $\mathbf{x} = \mathbf{g}_{\boldsymbol{\theta}}(\mathbf{h}) + \sigma \cdot \mathcal{N}(\mathbf{0}, \mathbf{I}_d)$. To find the MLE for $\boldsymbol{\theta}$, one difficulty is to compute $p_{\boldsymbol{\theta}}(\mathbf{x})$, which involves a complicated integral after expanding this probability in term of $\mathbf{h}$. The idea of *variational autoencoders* is to use an encoder neural network $f_{\boldsymbol{\phi}}$ to approximate the posterior $p_{\boldsymbol{\theta}}(\mathbf{h}|\mathbf{x})$. See Doersch (2014) for details of derivation and examples. The final optimization formulation of variational autoencoders involves a loss term that represents the reconstruction error, and a Kullback-Leibler divergence that serves as a regularizer.

There are other structured autoencoders which add desired properties to the model such as sparsity or robustness, mainly through regularization terms. Below we present two other common types of autoencoders.

- *Sparse autoencoders.* One may believe that the dimension $k$ of the hidden code $\mathbf{h}_i = \mathbf{f}(\mathbf{x}_i)$ is larger than the input dimension $d$, and that $\mathbf{h}_i$ admits a sparse representation. As with Lasso (Tibshirani, 1996) or SCAD (Fan and Li, 2001), one may add a regularization term to the reconstruction loss $\mathcal{L}$ in (14.14) to encourage sparsity (Poultney *et al.*, 2007). A sparse autoencoder solves

$$\min_{\mathbf{f},\mathbf{g}} \underbrace{\frac{1}{n}\sum_{i=1}^{n} \mathcal{L}\left(\mathbf{x}_i, \mathbf{g}\left(\mathbf{h}_i\right)\right)}_{\text{loss}} + \underbrace{\lambda \left\|\mathbf{h}_i\right\|_1}_{\text{regularizer}} \quad \text{with} \quad \mathbf{h}_i = \mathbf{f}\left(\mathbf{x}_i\right), \text{ for all } i \in [n].$$

  This is similar to *dictionary learning*, where one aims at finding a sparse representation of input data on an overcomplete basis. Due to the imposed sparsity, the model can potentially learn useful features of the data.

- *Denoising autoencoders.* One may hope that the model is robust to noise in the data: even if the input data $\mathbf{x}_i$ are corrupted by small noise $\boldsymbol{\xi}_i$ or miss some components (the noise level or the missing probability is typically small), an ideal autoencoder should faithfully recover the original data. A denoising autoencoder (Vincent, Larochelle, Bengio and Manzagol, 2008) achieves this robustness by explicitly building a noisy data $\widetilde{\mathbf{x}}_i = \mathbf{x}_i + \boldsymbol{\xi}_i$ as the new input, and then solves an optimization problem similar to (14.14) where $\mathcal{L}\left(\mathbf{x}_i, \mathbf{g}\left(\mathbf{h}_i\right)\right)$ is replaced by $\mathcal{L}\left(\mathbf{x}_i, \mathbf{g}\left(\mathbf{f}(\widetilde{\mathbf{x}}_i)\right)\right)$. A denoising autoencoder encourages the encoder/decoder to be stable in the neighborhood of an input, which is generally a good statistical property. An alternative way could be constraining $f$ and $g$ in the optimization problem, but that would be very difficult to optimize. Instead, sampling by adding small perturbations in the input provides a simple implementation. We shall see similar ideas in Section 14.5.3.3.

### *14.4.2 Generative adversarial networks*

Given unlabeled data $\{\mathbf{x}_i\}_{1\le i\le n}$, density estimation aims to estimate the underlying probability density function $\mathbb{P}_{\mathbf{X}}$ from which the data is generated. Both parametric and nonparametric estimators (Silverman, 1998) have been proposed and studied under various assumptions on the underlying distribution. Different from these classical density estimators, where the density function is explicitly defined in relatively low dimension, *generative adversarial networks* (GANs) (Goodfellow *et al.*, 2014) can be categorized as an *implicit* density estimator in much higher dimension such as the images of birds, or natural sceneries or handwritings. The reasons are twofold: (1) GANs put more emphasis on sampling from the distribution $\mathbb{P}_{\mathbf{X}}$ than estimation; (2) GANs define the density estimation implicitly through a source distribution $\mathbb{P}_{\mathbf{Z}}$ and a generator function $g(\cdot)$, which is usually a deep neural network. This shows once more how deep net functions (14.2) are powerful in approximating high-dimensional nonlinear functions.

We introduce GANs from the perspective of sampling from $\mathbb{P}_{\mathbf{X}}$ and later we will generalize the vanilla GANs using their relation to density estimators.

#### *14.4.2.1 Sampling view of GANs*

Suppose the data $\{\mathbf{x}_i\}_{1\le i\le n}$ at hand are all real images, and we want to generate *new* natural images. Unlike bootstrap which simply memorizes the old images, GAN would like to learn how to generate images from the distribution of the data $\{\mathbf{x}_i\}_{1\le i\le n}$ that were generated from. With this goal in mind, GAN models a *zero-sum* game between two players, namely, the generator $\mathcal{G}$ and the discriminator $\mathcal{D}$. The generator $\mathcal{G}$ tries to generate fake images akin to the true images $\{\mathbf{x}_i\}_{1\le i\le n}$ while the discriminator $\mathcal{D}$ aims at differentiating the fake ones from the true ones. Intuitively, one hopes to learn a generator $\mathcal{G}$ to generate images where the *best* discriminator $\mathcal{D}$ cannot distinguish. Therefore the payoff is higher for the generator $\mathcal{G}$ if the probability of the discriminator $\mathcal{D}$ getting wrong is higher, and correspondingly the payoff for the discriminator correlates positively with its ability to tell wrong from truth.

Recall that any continuous distributions in $\mathbb{R}^d$ can be represented as the distribution of $\mathbf{g}(\mathbf{Z})$ for a given function $\mathbf{g}$ and a random vector $\mathbf{Z} \in \mathbb{R}^d$ such as the standard multivariate Gaussian or uniform random vector. Usually, the distributions of real images have structures so that we can choose $\mathbf{Z}$ in a much lower-dimensional space to approximate the true distribution (dimension reduction). A natural candidate to then model $\mathbf{g}(\cdot)$ is by the deep neural network function (14.2). This leads naturally to the following procedure.

Mathematically, the generator $\mathcal{G}$ consists of two components, an source distribution $\mathbb{P}_{\mathbf{Z}}$ (usually a standard multivariate Gaussian distribution with hundreds of dimensions) and a function $\mathbf{g}(\cdot)$ which maps a sample $\mathbf{z}$ from $\mathbb{P}_{\mathbf{Z}}$ to

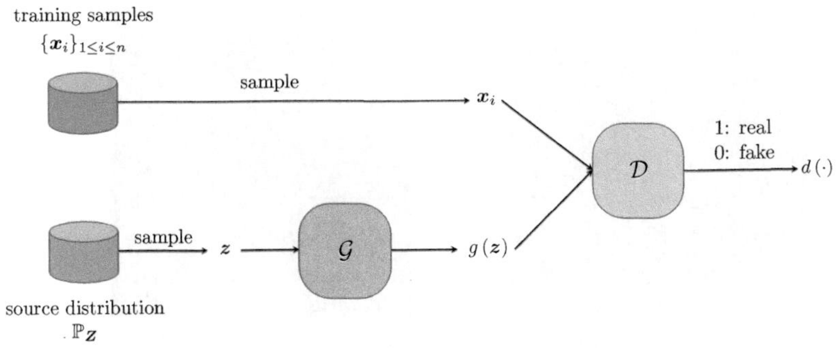


Figure 14.10: GANs consist of two components, a generator $\mathcal{G}$ which generates fake samples and a discriminator $\mathcal{D}$ which differentiates the true ones from the fake ones.

a point $\mathbf{g}(\mathbf{z})$ living in the same space as $\mathbf{x}$. For generating images, $\mathbf{g}(\mathbf{z})$ would be a 3D tensor. Here $\mathbf{g}(\mathbf{z})$ is the fake sample generated from $\mathcal{G}$. Similarly the discriminator $\mathcal{D}$ is composed of one function which takes an image $\mathbf{x}$ (real or fake) and return a number $d(\mathbf{x}) \in [0, 1]$, the probability of $\mathbf{x}$ being a real sample from $\mathbb{P}_{\mathbf{X}}$ or not. Oftentimes, both the generating function $\mathbf{g}(\cdot)$ and the discriminating function $d(\cdot)$ are realized by deep neural networks, e.g., CNNs introduced in Section 14.3.1. See Figure 14.10 for an illustration for GANs. Denote $\boldsymbol{\theta}_{\mathcal{G}}$ and $\boldsymbol{\theta}_{\mathcal{D}}$ the parameters in $\mathbf{g}(\cdot)$ and $d(\cdot)$, respectively. Then GAN tries to solve the following *min-max* problem:

$$\min_{\boldsymbol{\theta}_{\mathcal{G}}} \max_{\boldsymbol{\theta}_{\mathcal{D}}} \quad \mathrm{E}_{\mathbf{x}\sim\mathbb{P}_{\mathbf{X}}}\left[\log\left(d\left(\mathbf{x}\right)\right)\right] + \mathrm{E}_{\mathbf{z}\sim\mathbb{P}_{\mathbf{Z}}}\left[\log\left(1 - d\left(\mathbf{g}\left(\mathbf{z}\right)\right)\right)\right]. \tag{14.15}$$

Recall that $d(\mathbf{x})$ models the belief / probability that the discriminator thinks that $\mathbf{x}$ is a true sample. Fix the parameters $\boldsymbol{\theta}_{\mathcal{G}}$ and hence the generator $\mathcal{G}$ and consider the inner maximization problem. We can see that the goal of the discriminator is to maximize its ability of differentiation. Similarly, if we fix $\boldsymbol{\theta}_{\mathcal{D}}$ (and hence the discriminator), the generator tries to generate more realistic images $\mathbf{g}(\mathbf{z})$ to fool the discriminator.

Figure 14.11 shows the result of training a GAN using the MNIST dataset, which consists of 60000 images of handwritten digits. Most of the generated images are very similar to the real images and are recognizable by humans.

### *14.4.2.2 Minimum distance view of GANs*

Let us now take a density-estimation view of GANs. Fixing the source distribution $\mathbb{P}_{\mathbf{Z}}$, any generator $\mathcal{G}$ induces a distribution $\mathbb{P}_{\mathcal{G}}$ over the space of images. Removing the restrictions on $d(\cdot)$, one can then rewrite (14.15) as

$$\min_{\mathbb{P}_{\mathcal{G}}} \max_{d(\cdot)} \quad \mathrm{E}_{\mathbf{x}\sim\mathbb{P}_{\mathbf{X}}}\left[\log\left(d\left(\mathbf{x}\right)\right)\right] + \mathrm{E}_{\mathbf{x}\sim\mathbb{P}_{\mathcal{G}}}\left[\log\left(1 - d\left(\mathbf{x}\right)\right)\right]. \tag{14.16}$$

Observe that the inner maximization problem is solved by the likelihood ra-

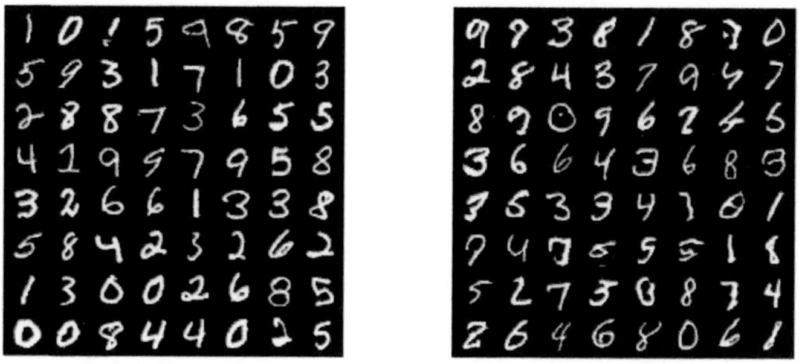

(a) MNIST real images (b) Fake images generated by GAN

Figure 14.11: (a) Sampled images from the MNIST handwritten digits dataset. (b) Sampled images generated by the generator $\mathcal{G}$, which is obtained by training a GAN for 25 epochs.

tio, i.e.

$$d^*(\mathbf{x}) = \frac{\mathbb{P}_{\mathbf{X}}(\mathbf{x})}{\mathbb{P}_{\mathbf{X}}(\mathbf{x}) + \mathbb{P}_{\mathcal{G}}(\mathbf{x})}.$$

As a result, (14.16) can be simplified as

$$\min_{\mathbb{P}_{\mathcal{G}}} \quad \mathrm{JS}(\mathbb{P}_{\mathbf{X}} \parallel \mathbb{P}_{\mathcal{G}}), \tag{14.17}$$

where $\mathrm{JS}(\cdot\|\cdot)$ denotes the Jensen–Shannon divergence between two distributions

$$\mathrm{JS}(\mathbb{P}_{\mathbf{X}} \| \mathbb{P}_{\mathcal{G}}) = \frac{1}{2}\mathrm{KL}\big(\mathbb{P}_{\mathbf{X}} \parallel \tfrac{\mathbb{P}_{\mathbf{X}}+\mathbb{P}_{\mathcal{G}}}{2}\big) + \frac{1}{2}\mathrm{KL}\big(\mathbb{P}_{\mathcal{G}} \parallel \tfrac{\mathbb{P}_{\mathbf{X}}+\mathbb{P}_{\mathcal{G}}}{2}\big).$$

In other words, the vanilla GAN (14.15) seeks a density $\mathbb{P}_{\mathcal{G}}$ that is closest to $\mathbb{P}_{\mathbf{X}}$ in terms of the Jensen–Shannon divergence. This view allows us to generalize GANs to other variants, by changing the distance metric. Examples include f-GAN (Nowozin, Cseke and Tomioka, 2016), Wasserstein GAN (W-GAN) (Arjovsky, Chintala and Bottou, 2017), MMD GAN (Li, Swersky and Zemel, 2015), etc. We single out the Wasserstein GAN (W-GAN) to introduce due to its popularity. As the name suggests, it minimizes the Wasserstein distance between $\mathbb{P}_{\mathbf{X}}$ and $\mathbb{P}_{\mathcal{G}}$:

$$\min_{\boldsymbol{\theta}_{\mathcal{G}}} \quad \mathrm{WS}(\mathbb{P}_{\mathbf{X}} \| \mathbb{P}_{\mathcal{G}}) = \min_{\boldsymbol{\theta}_{\mathcal{G}}} \sup_{f:f \text{ 1-Lipschitz}} \mathrm{E}_{\mathbf{x}\sim\mathbb{P}_{\mathbf{X}}}[f(\mathbf{x})] - \mathrm{E}_{\mathbf{x}\sim\mathbb{P}_{\mathcal{G}}}[f(\mathbf{x})], \tag{14.18}$$

where $f(\cdot)$ is taken over all Lipschitz functions with coefficient 1. Comparing W-GAN (14.18) with the original formulation of GAN (14.15), one finds that the Lipschitz function $f$ in (14.18) corresponds to the discriminator $\mathcal{D}$ in (14.15) in the sense that they share similar objectives to differentiate the true distribution $\mathbb{P}_{\mathbf{X}}$ from the fake one $\mathbb{P}_{\mathcal{G}}$. In the end, we would like to mention that GANs are more difficult to train than supervised deep learning models such as CNNs (Salimans *et al.*, 2016). Apart from the training difficulty, how to evaluate GANs objectively and effectively is an ongoing research.

## 14.5 Training deep neural nets

In this section, we introduce standard methods, namely *stochastic gradient descent* (SGD) and its variants, to train deep neural networks. As with many statistical machine learning tasks, training DNNs follows the *empirical risk minimization* (ERM) paradigm which solves the following optimization problem

$$\text{minimize}_{\boldsymbol{\theta}\in\mathbb{R}^p} \qquad \ell_n(\boldsymbol{\theta}) \triangleq \frac{1}{n}\sum_{i=1}^{n} \mathcal{L}\left(f\left(\mathbf{x}_i;\boldsymbol{\theta}\right), y_i\right). \tag{14.19}$$

Here $\mathcal{L}(f(\mathbf{x}_i;\boldsymbol{\theta}), y_i)$ measures the discrepancy between the prediction $f(\mathbf{x}_i;\boldsymbol{\theta})$ of the neural network and the true label $y_i$. Correspondingly, denote by $\ell(\boldsymbol{\theta}) \triangleq \mathrm{E}_{(\mathbf{x},y)\sim\mathcal{D}}[\mathcal{L}(f(\mathbf{x};\boldsymbol{\theta}), y)]$ the theoretical *testing error* (with infinity testing samples), where $\mathcal{D}$ is the joint distribution over $(y, \mathbf{x})$. Solving ERM (14.19) for deep neural nets faces various challenges that roughly fall into the following three categories.

- *Scalability and nonconvexity.* Both the sample size $n$ and the number of parameters $p$ can be huge for modern deep learning applications, as we have seen in Table 14.1. Many optimization algorithms are not practical due to the computational costs and memory constraints. What is worse, the empirical loss function $\ell_n(\boldsymbol{\theta})$ in deep learning is often nonconvex. It is *a priori* not clear whether an optimization algorithm can drive the empirical loss (14.19) small.
- *Numerical stability.* With a large number of layers in DNNs, the magnitudes of the hidden nodes can be drastically different, which may result in the "exploding gradients" or "vanishing gradients" issue during the training process. This is because the recursive relations across layers often lead to exponentially increasing / decreasing values in both forward passes and backward passes.
- *Generalization performance.* Our ultimate goal is to find a parameter $\widehat{\boldsymbol{\theta}}$ such that the out-of-sample error $\ell(\widehat{\boldsymbol{\theta}})$ is small. However, in the over-parametrized regime where $p$ is much larger than $n$, the underlying neural network has the potential to fit the training data perfectly while performing poorly on the test data. To avoid this overfitting issue, proper regularization, whether explicit or implicit, is needed in the training process for the neural nets to generalize.

In the following three subsections, we discuss practical solutions / proposals to address these challenges.

### *14.5.1 Stochastic gradient descent*

Stochastic gradient descent (SGD) (Robbins and Monro, 1951) is by far the most popular optimization algorithm to solve ERM (14.19) for large-scale problems. It has the following simple update rule:

$$\boldsymbol{\theta}^{t+1} = \boldsymbol{\theta}^t - \eta_t \mathbf{G}(\boldsymbol{\theta}^t) \qquad \text{with} \qquad \mathbf{G}\left(\boldsymbol{\theta}^t\right) = \nabla \mathcal{L}\left(f\left(\mathbf{x}_{i_t}; \boldsymbol{\theta}^t\right), y_{i_t}\right) \tag{14.20}$$

for $t = 0, 1, 2, \ldots$, where $\eta_t > 0$ is the *step size* (or *learning rate*), $\boldsymbol{\theta}^0 \in \mathbb{R}^p$ is an initial point and $i_t$ is chosen randomly from $\{1, 2, \cdots, n\}$. It is easy to verify that $\mathbf{G}(\boldsymbol{\theta}^t)$ is an unbiased estimate of $\nabla \ell_n(\boldsymbol{\theta}^t)$, despite only one data point being used. There are two variants of SGD:

- *One-pass SGD.* Each $i_t$ is drawn *without* replacement from $\{1, 2, \cdots, n\}$. In this setting, the stochastic gradient $\mathbf{G}(\boldsymbol{\theta}^t)$ is an unbiased estimate of $\nabla \ell(\boldsymbol{\theta}^t)$, the gradient of the *population* loss and one-pass SGD is regarded as a method to minimize $\ell(\boldsymbol{\theta})$.
- *Multi-pass SGD.* Each $i_t$ is drawn *with* replacement from $\{1, 2, \cdots, n\}$ independently. Under this rule, the stochastic gradient $\mathbf{G}(\boldsymbol{\theta}^t)$ is an unbiased estimate of $\nabla \ell_n(\boldsymbol{\theta}^t)$, the gradient of the *empirical* loss and multi-pass SGD can be viewed as an algorithm to minimize $\ell_n(\boldsymbol{\theta})$.

This division is merely for theoretical purposes and practitioners often strike a balance between these two approaches: we often run SGD for a number of *epochs*, where in each epoch we follow the rule for one-pass SGD. The advantage of SGD is clear: compared with gradient descent, which goes over the entire dataset in every update, SGD uses a single example in each update and hence is considerably more efficient in terms of both computation and memory (especially in the first few iterations).

There are certainly other challenges for vanilla SGD to train deep neural nets: (1) training algorithms are often implemented in GPUs, and therefore it is important to tailor the algorithm to such computing infrastructure, (2) the vanilla SGD might converge very slowly for deep neural networks, albeit providing good theoretical guarantees for well-behaved problems, and (3) the learning rates $\{\eta_t\}$ can be difficult to tune in practice. To address the aforementioned challenges, three important variants of SGD, namely *mini-batch SGD*, *momentum-based SGD*, and *SGD with adaptive learning rates* are introduced.

#### *14.5.1.1 Mini-batch SGD*

Modern computational infrastructures (e.g., GPUs) can evaluate the gradient on a number (say 64) of examples as efficiently as evaluating a gradient on a single example. To utilize this advantage, mini-batch SGD with batch size $K \geq 1$ forms the stochastic gradient through $K$ random samples:

$$\boldsymbol{\theta}^{t+1} = \boldsymbol{\theta}^t - \eta_t \mathbf{G}(\boldsymbol{\theta}^t) \quad \text{with} \quad \mathbf{G}(\boldsymbol{\theta}^t) = \frac{1}{K}\sum_{k=1}^{K} \nabla \mathcal{L}\big(f(\mathbf{x}_{i_t^k}; \boldsymbol{\theta}^t), y_{i_t^k}\big), \tag{14.21}$$

where for each $k \in [K]$, $i_t^k$ is sampled uniformly from $\{1, 2, \cdots, n\}$. Mini-batch SGD, which is an "interpolation" between gradient descent and stochastic gradient descent, achieves the best of both worlds: (1) using $1 \ll K \ll n$ samples to estimate the gradient, one effectively reduces the variance and hence accelerates the convergence, and (2) by taking the batch size $K$ appropriately (say 64 or 128), the stochastic gradient $\mathbf{G}(\boldsymbol{\theta}^t)$ can be efficiently computed using the matrix computation toolboxes on GPUs.

#### *14.5.1.2 Momentum-based SGD*

While mini-batch SGD forms the foundation of training neural networks, it can sometimes be slow to converge due to its oscillation behavior (Sutskever, Martens, Dahl and Hinton, 2013). The optimization community has long investigated how to accelerate the convergence of gradient descent, which results in a beautiful technique called *momentum methods* (Polyak, 1964; Nesterov, 1983). Similar to gradient descent with moment, *momentum-based SGD*, instead of moving the iterate $\boldsymbol{\theta}^t$ in the direction of the current stochastic gradient $\mathbf{G}(\boldsymbol{\theta}^t)$, smoothes the past (stochastic) gradients $\{\mathbf{G}(\boldsymbol{\theta}^t)\}$, via exponential smoothing, to stabilize the update directions. Starting from $\mathbf{v}^0 = \mathbf{G}(\boldsymbol{\theta}^0)$, $t = 0, 1, \cdots$, update the current estimates

$$\boldsymbol{\theta}^{t+1} = \boldsymbol{\theta}^t - \eta_t \mathbf{v}^t$$

and the gradient direction $\mathbf{v}^t \in \mathbb{R}^p$ by the exponential smoothing with a smoothing parameter $\rho \in [0, 1)$

$$\mathbf{v}^{t+1} = \rho \mathbf{v}^t + (1 - \rho)\mathbf{G}(\boldsymbol{\theta}^{t+1}). \tag{14.22}$$

A typical choice of $\rho$ is 0.9. Notice that $\rho = 0$ recovers the mini-batch SGD (14.21), where no past information of gradients is used. A simple unrolling of $\mathbf{v}^t$ reveals that $\mathbf{v}^t$ is actually an exponentially weighted average of the past gradients, i.e., $\mathbf{v}^t = (1 - \rho)\sum_{j=0}^{t} \rho^{t-j}\mathbf{G}(\boldsymbol{\theta}^j)$. Compared with vanilla mini-batch SGD, the inclusion of the momentum "smoothes" the oscillation direction and accumulates the persistent descent direction. We want to emphasize that theoretical justifications of momentum in the *stochastic* setting is not fully understood (Kidambi, Netrapalli, Jain and Kakade, 2018; Jain, Kakade, Kidambi, Netrapalli and Sidford, 2017).

#### *14.5.1.3 SGD with adaptive learning rates*

In optimization, *preconditioning* is often used to accelerate first-order optimization algorithms. In principle, one can apply this to SGD, which yields the following update rule:

$$\boldsymbol{\theta}^{t+1} = \boldsymbol{\theta}^t - \eta_t \mathbf{P}_t^{-1} \mathbf{G}(\boldsymbol{\theta}^t) \tag{14.23}$$

with $\mathbf{P}_t \in \mathbb{R}^{p \times p}$ being a preconditioner at the $t$-th step. Newton's method can be viewed as one type of preconditioning where $\mathbf{P}_t = \nabla^2 \ell_n(\boldsymbol{\theta}^t)$. The advantages of preconditioning are two-fold: first, a good preconditioner reduces the condition number by changing the local geometry to be more homogeneous, which is amenable to fast convergence; second, a good preconditioner frees practitioners from laborious tuning of the step sizes, as is the case with Newton's method. *AdaGrad*, an adaptive gradient method proposed by Duchi, Hazan and Singer (2011), builds a preconditioner $\mathbf{P}_t$ based on information of the past gradients:

$$\mathbf{P}_t = \left\{ \mathsf{diag}\Big( \sum_{j=0}^{t} \mathbf{G}\left(\boldsymbol{\theta}^t\right) \mathbf{G}\left(\boldsymbol{\theta}^t\right)^\top \Big) \right\}^{1/2}. \tag{14.24}$$

Since we only require the diagonal part, this preconditioner (and its inverse) can be efficiently computed in practice. In addition, investigating (14.23) and (14.24), one can see that AdaGrad adapts to the importance of each coordinate of the parameters by setting smaller learning rates for frequent features, whereas it sets larger learning rates for those infrequent ones. In practice, one adds a small quantity $\delta > 0$ (say $10^{-8}$) to the diagonal entries to avoid singularity (numerical underflow). A notable drawback of AdaGrad is that the effective learning rate vanishes quickly along the learning process. This is because the historical sum of the gradients can only increase with time. RMSProp (Hinton, Srivastava and Swersky, 2012) is a popular remedy for this problem which incorporates the idea of exponential averaging:

$$\mathbf{P}_t = \left\{ \mathsf{diag}\Big( \rho \mathbf{P}_{t-1} + (1-\rho) \mathbf{G}\left(\boldsymbol{\theta}^t\right) \mathbf{G}\left(\boldsymbol{\theta}^t\right)^\top \Big) \right\}^{1/2}. \tag{14.25}$$

Again, the decaying parameter $\rho$ is usually set to be 0.9. Later, Adam (Kingma and Bai, 2014; Reddi, Kale and Kumar, 2019) combines the momentum method and adaptive learning rate and becomes the default training algorithms in many deep learning applications.

### *14.5.2 Easing numerical instability*

For very deep neural networks or RNNs with long dependencies, training difficulties often arise when the values of nodes have different magnitudes or when the gradients "vanish" or "explode" during back-propagation. Here we discuss three partial solutions to alleviate this problem.

#### *14.5.2.1 ReLU activation function*

One useful characteristic of the ReLU function is that its derivative is either 0 or 1, and the derivative remains 1 even for a large input. This is in sharp contrast with the standard sigmoid function $(1+e^{-t})^{-1}$ which results in a very small derivative when inputs have large magnitude. The consequence of small derivatives across many layers is that gradients tend to be "killed", which means that gradients become approximately zero in deep nets.

The popularity of the ReLU activation function and its variants (e.g., leaky ReLU) is largely attributable to the above reason. It has been well observed that the ReLU activation function has superior training performance over the sigmoid function (Krizhevsky, Sutskever and Hinton, 2012; Maas, Hannun and Ng, 2013).

#### *14.5.2.2 Skip connections*

We have introduced skip connections in Section 14.3.3. Why are skip connections helpful for reducing numerical instability? This structure does not introduce a larger function space, since the identity map can be also represented with ReLU activations: $\mathbf{x} = \boldsymbol{\sigma}(\mathbf{x}) - \boldsymbol{\sigma}(-\mathbf{x})$.

One explanation is that skip connections bring ease to the training process. Suppose that we have a general nonlinear function $\mathbf{F}(\mathbf{x}_\ell; \boldsymbol{\theta}_\ell)$. With a skip connection, we represent the map as $\mathbf{x}_{\ell+1} = \mathbf{x}_\ell + \mathbf{F}(\mathbf{x}_\ell; \boldsymbol{\theta}_\ell)$ instead. Now the gradient $\partial \mathbf{x}_{\ell+1}/\partial \mathbf{x}_\ell$ becomes

$$\frac{\partial \mathbf{x}_{\ell+1}}{\partial \mathbf{x}_\ell} = \mathbf{I} + \frac{\partial \mathbf{F}(\mathbf{x}_\ell; \boldsymbol{\theta}_\ell)}{\partial \mathbf{x}_\ell} \quad \text{instead of} \quad \frac{\partial \mathbf{F}(\mathbf{x}_\ell; \boldsymbol{\theta}_\ell)}{\partial \mathbf{x}_\ell}, \tag{14.26}$$

where $\mathbf{I}$ is an identity matrix. By the chain rule, a gradient update requires computing products of many components, e.g., $\frac{\partial \mathbf{x}_L}{\partial \mathbf{x}_1} = \prod_{\ell=1}^{L-1} \frac{\partial \mathbf{x}_{\ell+1}}{\partial \mathbf{x}_\ell}$, so it is desirable to keep the spectra (singular values) of each component $\frac{\partial \mathbf{x}_{\ell+1}}{\partial \mathbf{x}_\ell}$ close to 1. In neural nets, with skip connections, this is easily achieved if the parameters have small values; otherwise, this may not be achievable even with careful initialization and tuning. Notably, training neural nets with hundreds of layers is possible with the help of skip connections.

#### *14.5.2.3 Batch normalization*

Recall that in regression analysis, one often standardizes the design matrix so that the features have zero mean and unit variance. Batch normalization extends this standardization procedure from the input layer to all the hidden layers. Mathematically, fix a mini-batch of input data $\{(\mathbf{x}_i, y_i)\}_{i\in\mathcal{B}}$, where $\mathcal{B} \subset [n]$. Let $\mathbf{h}_i^{(\ell)}$ be the feature of the $i$-th example in the $\ell$-th layer ($\ell = 0$ corresponds to the input $\mathbf{x}_i$). The batch normalization layer computes the normalized version of $\mathbf{h}_i^{(\ell)}$ via the following steps:

$$\boldsymbol{\mu} \triangleq \frac{1}{|\mathcal{B}|}\sum_{i\in\mathcal{B}} \mathbf{h}_i^{(\ell)}, \qquad \boldsymbol{\sigma}^2 \triangleq \frac{1}{|\mathcal{B}|}\sum_{i\in\mathcal{B}} \left(\mathbf{h}_i^{(\ell)} - \boldsymbol{\mu}\right)^2 \quad \text{and} \quad \mathbf{h}_{i,\text{norm}}^{(l)} \triangleq \frac{\mathbf{h}_i^{(\ell)} - \boldsymbol{\mu}}{\boldsymbol{\sigma}}.$$

Here all the operations are element-wise. In other words, batch normalization computes the z-score for each feature over the mini-batch $\mathcal{B}$ and uses that as inputs to subsequent layers. To make it more versatile, a typical batch normalization layer has two additional learnable parameters $\boldsymbol{\gamma}^{(\ell)}$ and $\boldsymbol{\beta}^{(\ell)}$ such that

$$\mathbf{h}_{i,\text{new}}^{(l)} = \boldsymbol{\gamma}^{(l)} \odot \mathbf{h}_{i,\text{norm}}^{(l)} + \boldsymbol{\beta}^{(l)}.$$

Again $\odot$ denotes the element-wise multiplication. As can be seen, $\boldsymbol{\gamma}^{(\ell)}$ and $\boldsymbol{\beta}^{(\ell)}$ set the new feature $\mathbf{h}_{i,\text{new}}^{(l)}$ to have mean $\boldsymbol{\beta}^{(\ell)}$ and standard deviation $\boldsymbol{\gamma}^{(\ell)}$. The introduction of batch normalization makes the training of neural networks much easier and smoother. More importantly, it allows the neural nets to perform well over a large family of hyper-parameters including the number of layers, the number of hidden units, etc. At test time, the batch normalization layer needs more care. For brevity we omit the details and refer to Ioffe and Szegedy (2015).

### *14.5.3 Regularization techniques*

So far we have focused on training techniques to drive the empirical loss (14.19) to a small value efficiently. Here we proceed to discuss common practice to improve the generalization power of trained neural nets.

#### *14.5.3.1 Weight decay*

One natural regularization idea is to add an $\ell_2$ penalty to the loss function. This regularization technique is known as the weight decay in deep learning. We have seen one example in (14.8). For general deep neural nets, the loss to optimize is $\ell_n^\lambda(\boldsymbol{\theta}) = \ell_n(\boldsymbol{\theta}) + r_\lambda(\boldsymbol{\theta})$ where

$$r_\lambda(\boldsymbol{\theta}) = \lambda \sum_{\ell=1}^{L} \sum_{j,j'} \left[W_{j,j'}^{(\ell)}\right]^2.$$

Note that the bias (intercept) terms are not penalized. If $\ell_n(\boldsymbol{\theta})$ is a least squares loss, then regularization with weight decay gives precisely ridge regression. The penalty $r_\lambda(\boldsymbol{\theta})$ is a smooth function and thus it can also be implemented efficiently with back-propagation.

#### *14.5.3.2 Dropout*

Dropout, introduced by Hinton *et al.*(2012), prevents overfitting by randomly dropping out subsets of features during training. It shares a similar spirit to the bagging and random forests introduced in Sections 12.4.2 and Section 12.4.3. Take the $l$-th layer of the feed-forward neural network as an example. Instead of propagating all the features in $\mathbf{h}^{(\ell)}$ for later computations, dropout randomly omits some of its entries by

$$\mathbf{h}_{\text{drop}}^{(\ell)} = \mathbf{h}^{(\ell)} \odot \mathsf{mask}^{\ell},$$

where $\odot$ denotes element-wise multiplication as before, and $\mathsf{mask}^\ell$ is a vector of Bernoulli variables with success probability $p$. It is sometimes useful to rescale the features $\mathbf{h}^{(\ell)}_{\text{inv drop}} = \mathbf{h}^{(\ell)}_{\text{drop}}/p$, which is called *inverted dropout*. During training, $\mathsf{mask}^\ell$ are i.i.d. vectors across mini-batches and layers. However, when testing on fresh samples, dropout is disabled and the original features $\mathbf{h}^{(\ell)}$ are used to compute the output label $y$. It has been nicely shown by Wager, Wang and Liang (2013) that for generalized linear models, dropout serves as adaptive regularization. In the simplest case of linear regression, it is equivalent to $\ell_2$ regularization. Another possible way to understand the regularization effect of dropout is through the lens of bagging (Goodfellow, Bengio and Courville, 2016) in Section 12.4.2. Since different mini-batches have different masks, dropout can be viewed as training a large ensemble of classifiers at the same time, with a further constraint that the parameters are shared. Theoretical justification remains elusive.

#### *14.5.3.3 Data augmentation*

Data augmentation is a technique of enlarging the dataset when we have knowledge about the invariance structure of data. It implicitly increases the sample size and usually regularizes the model effectively. For example, in image classification, we have strong prior knowledge about what invariance properties a good classifier should possess. The label of an image should not be affected by translation, rotation, flipping, and even crops of the image. Hence one can augment the dataset by randomly translating, rotating and cropping the images in the original dataset.

Formally, during training we want to minimize the loss $\ell_n(\boldsymbol{\theta}) = \sum_i \mathcal{L}(f(\mathbf{x}_i;\boldsymbol{\theta}), y_i)$ w.r.t. parameters $\boldsymbol{\theta}$, and we know a priori that certain transformation $T \in \mathcal{T}$ where $T : \mathbb{R}^d \to \mathbb{R}^d$ (e.g., affine transformation) should not change the category / label of a training sample. In principle, if computation costs were not a consideration, we could convert this knowledge to a constraint $f_{\boldsymbol{\theta}}(T\mathbf{x}_i) = f_{\boldsymbol{\theta}}(\mathbf{x}_i), \forall T \in \mathcal{T}$ in the minimization formulation. Instead of solving a constrained optimization problem, *data augmentation* enlarges the training dataset by sampling $T \in \mathcal{T}$ and generating new data $\{(T\mathbf{x}_i, y_i)\}$. In this sense, data augmentation induces invariance properties through sampling, which results in a much bigger dataset than the original one.

## 14.6 Example: Image Classification

We now demonstrate the power of deep neural nets on the CIFAR-10 dataset, which is a popular dataset for image classification. The dataset consists of 50,000 training images and 10,000 test images, all of which are of size $32 \times 32 \times 3$ (32 pixels in two dimensions, and 3 color channels). Each image shows a common object (e.g., planes, cars) and belongs to one of 10 categories. These categories contain the same number of images in the training and test sets. Figure 14.12 shows three sample images for each category.

The classification task aims at finding a classifier based on the pixels'

Figure 14.12: Some sample images from CIFAR-10. The name of each category is shown at the bottom.

inputs. We tested both deep learning and, as a comparison, classical methods in supervised learning. If we simply treat an image as a vector of length $32 \times 32 \times 3$ and use the methods from Chapter 12, the results are not impressive. For example, we found that the multinomial logistic regression and the nearest neighborhood method both yield a misclassification error of more than 60% on the test set. If unsupervised learning (e.g., clustering) is used first to construct features, it is reported (Coates, Ng and Lee, 2011) that 20%–40% errors can be achieved. Results are summarized in Table 14.2.

Table 14.2: A summary of test errors of different methods.

| Method | NN | logistic reg. | unsuperv. + SVM | MLP-3 | VGG-19 |
|---|---|---|---|---|---|
| Error | 68.8% | 60.0% | 20% – 40% | 42.5% | 10.5% |

In contrast, deep learning is very effective for this kind of task if an appropriate model is chosen. We found that VGG-19 is able to attain 10% error on the test set. VGG-19 (Simonyan and Zisserman, 2014) is a well-known 19-layer convolutional neural net. It uses a stack of convolutional layers (all with $3 \times 3$ kernels) and pooling layers to map the pixel inputs into a 512-dimensional feature vector, which is then followed by a multinomial logistic regression. Note that the general-purpose model MLP does not perform well, as it does not utilize the spatial structure of images. An MLP with 3 hidden layers, named as MLP-3, yields an error of more than 40%.

Two deep neural nets are trained, VGG-19 and MLP-3, on a single GPU (nVIDIA K20) using PyTorch. Both nets are trained for 150 epochs using SGD with momentum 0.9 and weight decay $5 \times 10^{-4}$. The learning rate is set to be 0.1 for epoch 0–49, 0.01 for epoch 50–99, and 0.001 for epoch 100–149. We used data augmentation by randomly cropping and flipping the images. Training took a few hours to complete. Figure 14.13 shows the training/test errors of the two nets during the training process. Two noticeable drops of errors occur at epoch 50 and 100, where the learning rate is reduced.

There is an interesting phenomenon about the generalization error of neu-

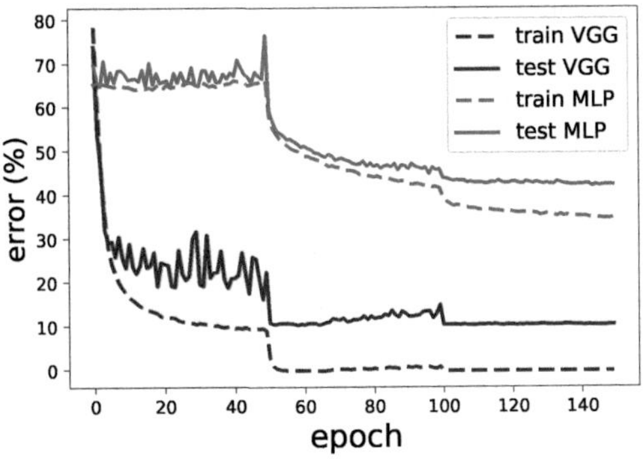


Figure 14.13: Training/test errors during the training process and as a function of the number of epoches. Two noticeable drops of errors occur at epoch 50 and 100, where the learning rate is reduced.

ral nets. Despite the fact that the number of parameters (millions) is much larger than the sample size, the generalization error (namely the difference between the training error and the test error) is not huge. This phenomenon is persistent across different deep learning experiments. It seems to be at odds with classical statistical theory: with a huge model, normally we should expect severe overfitting, but this is not the case here in terms of prediction error. This discrepancy is currently not well understood and has stirred continued statistical research.

## 14.7 Additional Examples using FensorFlow and $R$

This section illustrates applications of deep neural nets using three datasets and three different models. We have no intension to tune the networks to obtain the best predictive results, which is time-consuming. Rather, we use this opportunity to introduce the R software package `keras` that calls *tensorflow* written in `Python`. To install the R-package *keras*, type

```
> install.packages("keras")              #install the package
> library(keras)                         #use the package
> install_keras()                   #needed only for the first time
```

The third line is needed for the first use of `keras`. It will install the related *Python* files for $R$. Additional instructions can be found in

`https://tensorflow.rstudio.com/guide/keras/`

**Example 14.1** *Reuters Newswire data consist of 11228 short newswires and their topics (46 topics) as published in 1986. This dataset is frequently used as a testbed for text classification and analysis. It is included in the R package*

Keras. *We will use 20% of the data for testing. Below is how the data are extracted. See*

https://tensorflow.rstudio.com/guide/keras/examples/reuters_mlp/

```
library(keras)
max_words = 1000      #most frequent 1000 words will be used as x
reuters <- dataset_reuters(num_words = max_words, test_split = 0.2)
                                            ## 20% used as testing.
x_train <- reuters$train$x       #extract the training data
y_train <- reuters$train$y       #extract response (topic of newswires)
x_test <- reuters$test$x
y_test <- reuters$test$y
x_train[[1]];                    #take a peek on the first news data
[1]   1   2   2   8  43  10 447   5  25 207 270   5   2 111  16 369 186  90
[19]  67   7  89   5  19 102   6  19 124  15  90  67  84  22 482  26   7  48
[37]   4  49   8 864  39 209 154   6 151   6  83  11  15  22 155  11  15   7
[55]  48   9   2   2 504   6 258   6 272  11  15  22 134  44  11  15  16   8
[73] 197   2  90  67  52  29 209  30  32 132   6 109  15  17  12
##show the vocabulary in the top 1000 words of the first training article
> y_train[1:20]                    #a peek of topics for first 20 articles
 [1]  3  4  3  4  4  4  4  3  3 16  3  3  4  4 19  8 16  3  3 21
      #shape the data into a matrix format
 tokenizer <- text_tokenizer(num_words = max_words)
 x_train <- sequences_to_matrix(tokenizer, x_train, mode = 'binary')
 x_test  <- sequences_to_matrix(tokenizer, x_test, mode = 'binary')
 dim(x_train)                      #take a peek of data dim
[1] 8982 1000
 dim(x_test)
[1] 2246 1000
> x_train[1:3, 1:10]               #take a peek of the actual data
     [,1] [,2] [,3] [,4] [,5] [,6] [,7] [,8] [,9] [,10]
[1,]    0    1    1    0    1    1    1    1    1     1
[2,]    0    1    1    0    1    1    0    1    1     1
[3,]    0    1    1    0    1    1    0    1    0     1
y_train = to_categorical(y_train, 46)  #put response as categorical data
y_test  = to_categorical(y_test,  46)
y_train[1:3,]                     #a peek of response for first 3 articles after coding
     [,1] [,2] [,3] [,4] [,5] [,6] [,7] [,8] [,9] [,10] [,11] [,12] [,13] [,14]
[1,]    0    0    0    1    0    0    0    0    0     0     0     0     0     0
[2,]    0    0    0    0    1    0    0    0    0     0     0     0     0     0
[3,]    0    0    0    1    0    0    0    0    0     0     0     0     0     0
     [,15] [,16] [,17] [,18] [,19] [,20] [,21] [,22] [,23] [,24] [,25] [,26] [,27]
[1,]     0     0     0     0     0     0     0     0     0     0     0     0     0
[2,]     0     0     0     0     0     0     0     0     0     0     0     0     0
[3,]     0     0     0     0     0     0     0     0     0     0     0     0     0
     [,28] [,29] [,30] [,31] [,32] [,33] [,34] [,35] [,36] [,37] [,38] [,39] [,40]
[1,]     0     0     0     0     0     0     0     0     0     0     0     0     0
[2,]     0     0     0     0     0     0     0     0     0     0     0     0     0
[3,]     0     0     0     0     0     0     0     0     0     0     0     0     0
     [,41] [,42] [,43] [,44] [,45] [,46]
[1,]     0     0     0     0     0     0
[2,]     0     0     0     0     0     0
[3,]     0     0     0     0     0     0
```

We are now ready to analyze the data. Let us first employ the feed *forward neural network* (FNN) also called *Multi Layer Perceptron*(MLP). See

https://tensorflow.rstudio.com/guide/keras/examples/reuters_mlp/

We use one hidden layer with 512 hidden nodes, mapped from the 1000 feature space (1000 words). This needs $1001 * 512$ parameters (1 for bias in each hidden node). The output layer has 46 outputs (topics), consisting of $513 * 46$ parameters. The R-code and the results are given below.

```
library(keras)
model <- keras_model_sequential()   #creates an instance of a model for an FNN
```

```
model %>%
    layer_dense(units = 512, input_shape = c(1000)) %>%
    layer_activation(activation = 'relu') %>%
    layer_dropout(rate = 0.5) %>%
    layer_dense(units = 46) %>%
    layer_activation(activation = 'softmax')
    ###One hidden layer: 1000*512 with 1001*512 parameters
    ### Output layer:  512*46 with 513*46 parameters

> model         #take a peek of the model used

Layer (type)                        Output Shape                    Param #
===========================================================================
dense (Dense)                       (None, 512)                     512512
---------------------------------------------------------------------------
activation (Activation)             (None, 512)                     0
---------------------------------------------------------------------------
dropout (Dropout)                   (None, 512)                     0
---------------------------------------------------------------------------
dense_1 (Dense)                     (None, 46)                      23598
---------------------------------------------------------------------------
activation_1 (Activation)           (None, 46)                      0
===========================================================================
Total params: 536,110
Trainable params: 536,110
Non-trainable params: 0

model %>% compile(
    loss = 'categorical_crossentropy', optimizer = 'sgd',
    metrics = c('accuracy'))  #training model options

history <- model %>% fit(x_train, y_train, batch_size = 32,
    epochs = 50, verbose = 1, validation_split = 0.1 )
            #The actual computation of training, using 10 fold CV

model %>% evaluate( x_test, y_test, batch_size = batch_size,
            verbose = 1 )     #evaluation on the testing results
$loss
[1] 0.9397183

$accuracy
[1] 0.7894034
```

The training loss function and goodness of fits are presented in Figure 14.14. The testing result is 78.94%. This is a good achievement, when taking into account that there are 46 topics and we only use the dichotomized predictors.

**Example 14.2** *The Mixed National Institute of Standards and Technology (MNIST for short) data consist of 70000 handwritten digits (*$28 \times 28$ *grey images): 60K for training and 10K for testing. It has been popularly used as a benchmark data set for testing machine learning algorithms. See*

https://tensorflow.rstudio.com/guide/keras/examples/mnist_cnn/

*and Figures 14.11. For this image input, we employ CNN. Let us first get the training and testing data.*

```
library(keras)
mnist <- dataset_mnist()
x_train <- mnist$train$x
y_train <- mnist$train$y
x_test <- mnist$test$x
```

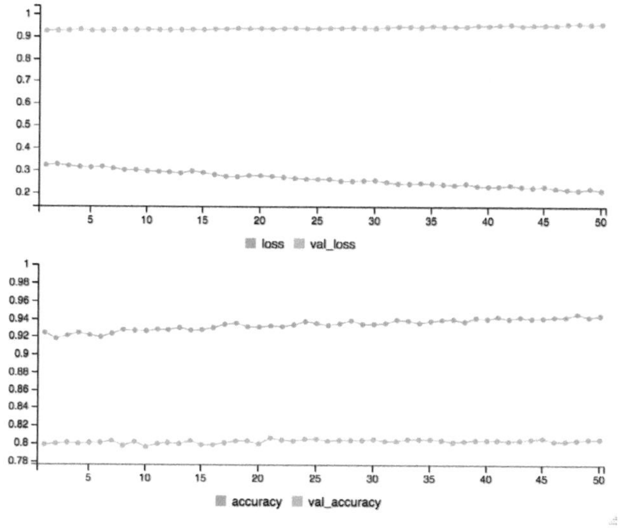


Figure 14.14: Training errors (cross-entropy loss or accuracy) during the training process as a function of the number of epochs.

```
y_test <- mnist$test$y
      ### reshape the data #######
dim(x_train) <- c(60000, 28, 28, 1)       #  add a channel dimension
x_train <- x_train / 255                  #  normalize grey level to [0,1]
dim(x_test) <- c(10000, 28, 28, 1)
x_test <- x_test / 255
y_train <- to_categorical(y_train)        #  convert into categorical
y_test <- to_categorical(y_test)
```

Let us now build a convolution network. See the code below. The first layer of convolution uses 32 of 3×3 filters and the `relu` activation function, followed by maximum pooling of size 2×2. Each convolution of size 3×3 has 9 parameters and one bias parameter. This yields $32 * 10 = 320$ parameters from the input layer to the first hidden layer of convolutions. It transforms an input image of $28 \times 28$ into 32 images of $26 * 26$. The maximum pooling does not involve any parameters, but reduces the images to 32 images of 13*13, as shown in the output of `model` below. The second convolution layer uses 64 3×3 filters and the `relu` activation function. The number of parameters from the first maximum pooling layer to the second layer of convolution is $32*64*9+64 = 18496$, where 64 are the parameters for biases. See the output of `model` below. After dropout and maximum pooling, we get 64 images of $5 \times 5$. These 64 images are further convolved by using $3 \times 3$ filters. This results in 64 images of 3×3. The number of parameters used is $64*64*9+64 = 36,928$. This output layer has $64 * 3 * 3 = 576$ pixels (variables). These 576 variables are used to fit the multinomial logistic regression with 10 classes.

```
#########Building the Convolution Model  #################
```

Bertsekas., D. P. (1999). *Nonlinear Programming.* Athena Scientific, Belmont, MA, 1999.

Bickel, P.J. (1975). One-step Huber estimates in the linear model. *Jour. Ameri. Statist. Assoc.*, **70**, 428–434.

Bickel, P.J. (1983). Minimax estimation of a normal mean subject to doing well at a point. In *Recent Advances in Statistics* (M.H. Rizvi, J.S. Rustagi, and D. Siegmund, eds), 511–528. Academic Press, New York.

Bickel, P. J. (2008). Discussion on the paper by Fan and Lv. *Jour. Roy. Statist. Soc. B*, **70**, 883–884.

Bickel, P.J., and Chen, A. (2009). A nonparametric view of network models and Newman-Girvan and other modularities. *Proc. Natl. Acad. Sci.*, **106**, 21068–21073.

Bickel, P.J., Chen, A., and Levina, E. (2011). The method of moments and degree distributions for network models. *Ann. Statist.*, **39**, 2280–2301.

Bickel, P. J. and Levina, E. (2004). Some theory for Fisher's linear discriminant function, naive Bayes, and some alternatives when there are many more variables than observations. *Bernoulli*, **10**, 989–1010.

Bickel, P. J., and Levina, E. (2008a). Regularized estimation of large covariance matrices. *Ann. Statist.*, **36**, 199–227.

Bickel, P. J. and Levina, E. (2008b). Covariance regularization by thresholding. *Ann. Statist.*, **36**, 2577–2604.

Bickel, P. J. and Sarkar, P. (2016). Hypothesis testing for automated community detection in networks. *Jour. Roy. Statist. Soc. B*, **78**, 253-273.

Bickel, P.J., Ritov, Y. and Tsybakov, A. (2009). Simultaneous analysis of Lasso and Dantzig selector. *Ann. Statist.*, **37**, 1705–1732.

Bien, J. (2019). Graph-guided banding of the covariance matrix. *Jour. Ameri. Statist. Assoc.*, **114**, 782–792.

Bien, J., Bunea, F., and Xiao, L. (2016) Convex banding of the covariance matrix. *Jour. Ameri. Statist. Assoc.*, **111**, 834–845.

Birnbaum, A., Johnstone, I.M., Nadler, B. and Paul, D. (2013). Minimax bounds for sparse PCA with noisy high-dimensional data. *Ann. Statist.*, **41**, 1055–1081.

Blanchard, G. and Roquain, E. (2009). Adaptive false discovery rate control under independence and dependence. *Jour. Mach. Learn. Res.*, **10**, 2837–2871.

Blei, D. M., Ng, A. Y., and Jordan, M. I. (2003). Latent dirichlet allocation. *Jour. Mach. Learn. Res.*, **3**, 993-1022.

Bloniarz, A., Liu, H., Zhang, C.-H., Sekhon, J. S., and Yu, B. (2016). Lasso adjustments of treatment effect estimates in randomized experiments. *Proc. Natl. Acad. Sci.*, **113**, 7383–7390.

Bogdan, M., van den Berg, E., Sabatti, C., Su, W. and Candès, E. J.

(2015). Slope adaptive variable selection via convex optimization. *Ann. Appl. Statist.*, **9**, 1103 – 1140.

Boivin, J. and Ng, S.(2006). Are more data always better for factor analysis? *Jour. Econ.*, **132**, 169–194.

Boucheron, S., Lugosi, G. and Massart, P. (2013). *Concentration Inequalities: A Nonasymptotic Theory of Independence.* Oxford University Press.

Box, G. and Cox, D. R. (1964). An analysis of transformations. *Jour. Roy. Statist. Soc. B*, **26**, 211–252.

Boyd, S., and Vandenberghe, L. (2004). *Convex Optimization.* Cambridge University Press, Cambridge, UK.

Boyd, S., Parikh, N., Chu, E., Peleato, B., and Eckstein, J. (2011). Distributed optimization and statistical learning via the alternating direction method of multipliers. *Foundations and Trends in Machine Learning*, **3**, 1–122.

Bradley, P. and Mangasarian, O. (1998). Feature selection via concave minimization and support vector machines. *Proceedings of the Fifteenth International Conference (ICML'98)*, Citeseer, 82–90.

Bradley, R. A. and Terry, M. E. (1952). Rank analysis of incomplete block designs: I. The method of paired comparisons. *Biometrika*, **39**, 324-345.

Bradic, J., Fan, J. and Jiang, J. (2011). Regularization for Cox's proportional hazards model with NP-dimensionality. *Ann. Statist.*, **39**, 3092-3120.

Bradic, J., Fan, J. and Wang, W. (2011). Penalized composite quasi-likelihood for ultrahigh-dimensional variable selection. *Jour. Roy. Statist. Soc. B*, **73**, 325–349.

Bradic, J., Fan, J., and Zhu, Y. (2020). Testability of high-dimensional linear models with non-sparse structures. *Annals of Statistics*, to appear.

Breheny, P. and Huang, J. (2011). Coordinate descent algorithms for nonconvex penalized regression, with applications to biological feature selection. *Ann. Appl. Statist.*, **5**, 232 - 253.

Breiman, L. (1995). Better subset regression using the nonnegative garrote. *Technometrics*, **37**, 373–384.

Breiman, L. (1996a). Bagging predictors. *Machine Learning*, **24**, 123–140.

Breiman, L. (1996b). Heuristics of instability and stabilization in model selection, *Ann. Statist.*, **24**, 2350–2383.

Breiman, L. (1998). Arcing classifiers (with discussion). *Ann. Statist.*, **26**, 801–849.

Breiman, L. (1999). Prediction games and arcing algorithms. *Neural Computation*, **11**, 1493–1517.

Breiman, L. (2001). Random forests. *Machine Learning*, **45**, 5–32.

Breiman, L., Friedman, J., Olsen, R. and Stone, C. (1984). *Classification and Regression Trees.* Wadsworth, Belmont, CA.

Brownlees, C.T., and Engle, R.B. (2017). SRISK: a conditional capital shortfall measure of systemic risk. *The Review of Financial Studies*, **30**, 48 - 79.

Bühlmann, P. (2013). Statistical significance in high-dimensional linear models. *Bernoulli*, **19**, 1212–1242.

Bühlmann, P and Hothorn, T. (2007). Boosting algorithms: regularization, prediction and model fitting. *Statist. Sci.*, 22, 477–505.

Bühlmann, P., Kalisch, M., and Maathuis, M. (2010). Variable selection in high-dimensional linear models: partially faithful distributions and the PC-simple algorithm. *Biometrika*, **97**, 261–278.

Bühlmann, P. and van de Geer, S. (2011). *Statistics for High-Dimensional Data: Methods, Theory and Applications.* Springer.

Bühlmann, P. and Yu, B. (2002). Analyzing Bagging. *Ann. Statist.*, **30**, 927–961.

Buja, A. and Stuetzle, W. (2006). Observations on Bagging. *Statist. Sinica*, **16**, 323–352.

Bunea, F., Tsybakov, A. and Wegkamp, M. H. (2007). Sparsity oracle inequalities for the lasso. *Electron. Jour. Stat.*, **1**, 169–194.

Buu, A. Johnson, N.J., Li, R. and Tan, X. (2011). New variable selection methods for zero-inflated count data with applications to the substance abuse field. *Stat. in Med.*, **30**, 2326–2340.

Cadima, J. and Jolliffe, I.T. (1995). Loadings and correlations in the interpretation of principal components. *Jour. Appl. Stat.*, **22**, 203–214.

Cai, J., Fan, J., Li, R. and Zhou, H. (2005). Variable selection for multivariate failure time data. *Biometrika*, **92**, 303–316.

Cai, J.-F.; Candés, E.J. and Shen, Z. (2010). A singular value thresholding algorithm for matrix completion. *SIAM Jour. Optim.*, **20**, 1956-1982.

Cai, T., Cai, T. T., and Zhang, A. (2016). Structured matrix completion with applications to genomic data integration. *Jour. Ameri. Statist. Assoc.*, **111**, 621–633.

Cai, T. and Guo, Z. (2017). Confidence intervals for high-dimensional linear regression: Minimax rates and adaptivity. *Ann. Statist.*, **45**, 615–646.

Cai, T. and Liu, W. (2011). A direct estimation approach to sparse linear discriminant analysis. *Jour. Ameri. Statist. Assoc.*, **106**, 1566–1577.

Cai, T., Liu, W. and Luo, X. (2011). A constrained $\ell_1$ minimization approach to sparse precision matrix estimation. *Jour. Ameri. Statist. Assoc.*, **106**, 594–607.

Cai, T., Liu, W. and Zhou, H. (2016). Estimating sparse precision matrix: Optimal rates of convergence and adaptive estimation. *Ann. Statist.*, **44**, 455–488.

Cai, T., Ma, Z. and Wu, Y. (2013). Sparse PCA: Optimal rates and adaptive estimation. *Ann. Statist.*, **41**, 3074–3110.

Cai, T., Ma, Z. and Wu, Y. (2015). Optimal estimation and rank detection for sparse spiked covariance matrices. *Probab. Theory and Related Fields*, **161**, 781–815.

Cai, T., Ren, Z., and Zhou, H. H. (2013). Optimal rates of convergence for estimating Toeplitz covariance matrices. *Probab. Theory Related Fields*, **156**, 101-143.

Cai, T., Wang, L. and Xu, G. (2010). Shifting inequality and recovery of sparse signals. *IEEE Trans. Signal Process.*, **58**, 1300–1308.

Cai, T.T. and Yuan, M. (2012). Adaptive covariance matrix estimation through block thresholding. *Ann. Statist.*, **40**, 2014–2042.

Cai, T.T. and Zhang, A. (2013). Sharp rip bound for sparse signal and low-rank matrix recovery. *Applied and Computational Harmonic Analysis*, **35**, 74–93.

Cai, T.T., Zhang, C.-H., and Zhou, H. (2010). Optimal rates of convergence for covariance matrix estimation *Ann. Statist.*, **38**, 2118–2144.

Cai, T.T., and Zhou, H. (2012a). Optimal rates of convergence for sparse covariance matrix estimation. *Ann. Statist.*, **40**, 2389–2420.

Cai, T.T. and Zhou, H. (2012b). Minimax estimation of large covariance matrices under $\ell_1$ norm (with discussion). *Statist. Sinica*, **22** 1319–1378.

Cai, T. T. and Zhou, W. X. (2016). Matrix completion via max-norm constrained optimization. *Electron. Jour. Stat.* , **10**, 1493–1525.

Caliński, T. and Harabasz, J. (1974). A dendrite method for cluster analysis. *Comm. Statist. Theory Methods* , **3**, 1–27.

Candés, E. J., Fan, Y., Janson, L. and Lv, J. (2018). Panning for gold: Model-X knockoffs for high-dimensional controlled variable selection. *Jour. Roy. Statist. Soc. B*, **80**, 551–577.

Candés, E. J., Li, X., Ma, Y., and Wright, J. (2011). Robust principal component analysis? *Jour. ACM*, **58**, 1–37

Candés, E. J. and Plan, Y. (2009). Near-ideal model selection by $\ell_1$ minimization. *Ann. Statist.*, **37**, 2145–2177.

Candés, E.J. and Plan, Y. (2010). Matrix completion with noise. *Proceedings of the IEEE.*, **98**, 925–936.

Candés, E. J. and Recht, B. (2009). Exact matrix completion via convex optimization. *Foundations of Computational Mathematics*, **9**, 717–772.

Candés, E. J. and Tao, T. (2005). Decoding by linear programming. *IEEE Trans. on Information Theory*, **51**, 4203–4215.

Candés, E. J. and Tao, T. (2007). The Dantzig selector: statistical estimation when $p$ is much larger than $n$ (with discussion). *Ann. Statist.*, **35**, 2313–2404.

Candés, E.J. and Tao, T. (2010). The power of convex relaxation: near-optimal matrix completion. *IEEE Trans. Inform. Theory*, **56**, 2053–2080.

Cao, C., Liu, F., Tan, H., Song, D., Shu, W., Li, W., Zhou, Y., Bo, X. and Xie, Z. (2018). Deep learning and its applications in biomedicine. *Genomics, Proteomics & Bioinformatics*, **16**, 17–32.

Carroll, R. J., Fan, J., Gijbels, I. and Wand, M. P. (1997). Generalized partially linear single-index models. *Jour. Ameri. Statist. Assoc.*, **92**, 477–489.

Catoni, O. (2012). Challenging the empirical mean and empirical variance: a deviation study. *Annales de l'Institut Henri Poincaré, Probabilités et Statistiques*, **48**, 1148–1185.

Cattell, R. B. (1966). The scree test for the number of factors. *Multivar. Behavioral Res.*, **1**, 245–276.

Chatterjee, A. and Lahiri, S. (2011). Bootstrapping lasso estimators. *Jour. Ameri. Statist. Assoc.*, **106**, 608–625.

Chen, H. (1988). Convergence rates for parametric components in a partly linear model. *Ann. Statist.*, **16**, 136–146.

Chen, J., and Chen, Z. (2008). Extended Bayesian information criteria for model selection with large model spaces. *Biometrika*, **95**, 759–771.

Chen, K. and Lei, J. (2018). Network cross-validation for determining the number of communities in network data. *Jour. Ameri. Statist. Assoc.*, **113**, 241–251.

Chen, L. S., Paul, D., Prentice, R. L. and Wang, P. (2011). A regularized Hotelling's $T^2$ test for pathway analysis in proteomic studies. *Jour. Ameri. Statist. Assoc.*, **106**, 1345–1360.

Chen, L. and Gu, Y. (2014). The convergence guarantees of a non-convex approach for sparse recovery. *IEEE Trans. Signal Process.*, **62**, 3754 – 3767.

Chen, M., Ren, Z., Zhao, H., and Zhou, H. (2016). Asymptotically normal and efficient estimation of covariate-adjusted Gaussian graphical model. *Jour. Ameri. Statist. Assoc.*, **111**, 394-406.

Chen, S., Donoho, D.L. and Sanders, M. A. (1998). Atomic decomposition by basis pursuit. *SIAM Jour. Sci. Comput.*, **20**, 33–61.

Chen, T. and Guestrin, C. (2016). XGBoost: A scalable tree boosting system. *KDD 2016.*

Chen, T., Rubanova, Y., Bettencourt, J. and Duvenaud, D. (2018). Neural ordinary differential equations. *arXiv preprint arXiv:1806.07366.*

Chen, X. and Xie, M. G. (2014). A split-and-conquer approach for analysis of extraordinarily large data. *Statist. Sinica*, **24**, 1655–1684.

Chen, Y., Chi, Y., Fan, J., Ma, C., and Yang, Y. (2019). Noisy matrix completion: Understanding statistical errors of convex relaxation via nonconvex optimization. *arXiv preprint arXiv:1902.07698.*

Chen, Y., Fan, J., Ma, C., and Wang, K. (2019). Spectral method and regularized MLE are both optimal for Top-$K$ ranking. *Ann. Statist.*, **47**, 2204-2235.

Chen, Y., Fan, J., Ma, C., and Yan, Y. (2019). Inference and uncertainty quantification for noisy matrix completion. *Proc. Natl. Acad. Sci. USA*, **116**, 22931-2293.

Chen, Y. and Suh, C. (2015). Spectral MLE: Top-K rank aggregation from pairwise comparisons. In *International Conference on Machine Learning*, 371–380

Chen, Z., Fan, J. and Li, R. (2018). Error variance estimation in ultrahigh dimensional additive models. *Jour. Ameri. Statist. Assoc.*. **113**, 315–327.

Cheng, M. Y., Honda, T., Li, J. and Peng, H. (2014) . Nonparametric independence screening and structure identification for ultra-high dimensional longitudinal data. *Ann. Statist.*, **42**, 1819–1849.

Cheng, M. Y., Honda, T. and Zhang, J.-T. (2016). Forward variable selection for sparse ultra-high dimensional varying coefficient models. *Jour. Ameri. Statist. Assoc.*., **111**, 1209–1221.

Chernozhukov, V., Chetverikov, D., and Kato, K. (2013). Gaussian approximations and multiplier bootstrap for maxima of sums of high-dimensional random vectors. *Ann. Statist.*, **41**, 2786–2819.

Chizat, L. and Bach, F. (2018). On the global convergence of gradient descent for over-parameterized models using optimal transport. In *Advances in Neural Information Processing Systems*, 3040–3050.

Cho, K., Van Merrianboer, B., Gulcehre, C., Bahdanau, D., Bougares, F., Schwenk, H. and Bengio, Y. (2014). Learning phrase representations using RNN encoder decoder for statistical machine translation. *arXiv preprint arXiv:1406.1078.*

Choi, J., Zou, H. and Oehlert, G. (2010). A penalized maximum likelihood approach to sparse factor analysis. *Statistics and Its Interface*, **3**, 429–436.

Chu, W., Li, R., Liu, J. and Reimherr, M. (2020). Feature screening for generalized varying coefficient mixed effect models with application to obesity GWAS. *Ann. Appl. Statist.* To appear.

Chu, W., Li, R. and Reimherr, M. (2016). Feature screening for time-varying coefficient models with ultrahigh dimensional longitudinal data. *Ann. Appl. Statist.*, **10**, 596 - 617.

Clarke, S. and Hall, P. (2009). Robustness of multiple testing procedures against dependence. *Ann. Statist.* **37**, 332–358.

Clemmensen, L., Hastie, T., Witten, D. and Ersboll, B. (2011). Sparse discriminant analysis. *Technometrics*, **53**, 406–413.

Coates, A., Ng, A. and Lee, H. (2011). An analysis of single-layer networks in unsupervised feature learning. In *Proceedings of the Fourteenth International Conference on Artificial Intelligence and Statistics*, 215–223.

Connor, G., Matthias, H. and Oliver, L. (2012). Efficient semiparametric estimation of the fama-french model and extensions. *Econometrica*, **80**, 713–754.

Connor, G. and Linton, O. (2007). Semiparametric estimation of a characteristic-based factor model of stock returns. *Jour. of Empir. Fin.*, **14**, 694–717.

Cook, R.D. (2007). Fisher lecture: Dimension reduction in regression (with discussion). *Statist. Sci.*, **22**, 1–26.

Cook, R.D. (2009). *Regression Graphics: Ideas for Studying Regressions through Graphics*, Vol. 482, John Wiley & Sons.

Cox, D. R. (1975a). Partial likelihood. *Biometrika*, **62**, 269–276.

Cox, D. R. (1975b). A note on data-splitting for the evaluation of significance levels. *Biometrika*, **62**, 441 – 444.

Craven, P. and Wahba, G. (1979). Smoothing noisy data with spline functions: estimating the correct degree of smoothing by the method of generalized cross-validation. *Numer. Math.*, **31**, 377–403.

Cui, H., Li, R. and Zhong, W. (2015). Model-free feature screening for ultrahigh dimensional discriminant analysis. *Jour. Ameri. Statist. Assoc.,* **110,** 630–641.

d'Aspremont, A., El Ghaoui, L., Jordan, M. I., and Lanckriet, G. R. (2007). A direct formulation for sparse PCA using semidefinite programming. *SIAM Review*, **49**, 434–448.

d'Aspremont, A., Bach, F. and El Ghaoui, L. (2008). Optimal solution for sparse principal component analysis. *Jour. Mach. Learn. Res.*, **9**, 1269–1294.

Daubechies, I., Defrise, M., De Mol, C. (2004). An iterative thresholding algorithm for linear inverse problems with a sparsity constraint. *Comm. Pure Appl. Math.*, **57**, 1413–1457.

Davis, C. and Kahan, W. (1970). The rotation of eigenvectors by a perturbation III. *SIAM Jour. Numer. Anal.* , **7**, 1–46.

Dempster, A.P., Laird, N.M. and Rubin, D.B. (1977). Maximum likelihood from incomplete data via the EM algorithm. *Jour. Roy. Statist. Soc. B*, **39**, 1–38.

Fauw, J., Ledsam, J.R., Romera-Paredes, B. et al. (2018). Clinically applicable deep learning for diagnosis and referral in retinal disease. *Nature Medicine*, **24**, 1342–1350.

Dempster, A. P., Laird, N.M. and Rubin, D. B. (1977). Maximum likelihood from incomplete data via the EM algorithm (with discussion). *Jour. Roy. Statist. Soc. B*, **39**, 1–38.

Desai, K.H. and Storey, J.D. (2012). Cross-dimensional inference of dependent high-dimensional data. *Jour. Ameri. Statist. Assoc.*, **107**, 135–151.

Devroye, L., Lerasle, M., Lugosi, G. and Oliveira, R.I. (2016). Sub-Gaussian mean estimators. *Ann. Statist.*, **44**, 2695–2725.

Dezeure, R., Bühlmann, P., Meier, L. and Meinshausen, N. (2015). High-dimensional inference: Confidence intervals, p-values and R-software hdi. *Statist. Sci.*, **30**, 533– 558.

Djouadi, A. and Bouktache, E. (1997). A fast algorithm for the nearest-neighbor classifier. *IEEE Trans. on Pattern Analysis and Machine Intelligence*, **19**, 277–282.

Doersch, C. (2016). Tutorial on variational autoencoders. *arXiv preprint arXiv:1606.05908.*

Douglas, J. and Rachford, H. H. (1956). On the numerical solution of the heat conduction problem in 2 and 3 space variables. *Transactions of the American Mathematical Society*, **82**, 421–439.

Donoho, D. L. (2000). High-dimensional data analysis: The curses and blessings of dimensionality. *Aide-Memoire of a Lecture at AMS Conference on Math Challenges of the 21st Century.*

Donoho, D.L. and Johnstone, I.M. (1994). Ideal spatial adaptation by wavelet shrinkage. *Biometrika*, **81**, 425–455.

Du, S., Lee, J., Li, H, Wang, L., and Zhai, X. (2018). Gradient descent finds global minima of deep neural networks. *arXiv preprint arXiv:1811.03804.*

Duchi, J. Gould, S. and Koller, D. (2008). Projected subgradient methods for learning sparse Gaussians. In *Proceedings of the Twenty-fourth Conference on Uncertainty in AI (UAI)*, 145–152.

Duchi, J., Hazan, E. and Singer, Y. (2011). Adaptive subgradient methods for online learning and stochastic optimization. *Jour. Mach. Learn. Res.*, **12**, 2121–2159.

Dwork, C., Kumar, R., Naor, M. and Sivakumar, D. (2001). Rank aggregation methods for the web. In *International Conference on World Wide Web*, 613–622.

E, W., Han, J. and Jentzen, A. (2017). Deep learning-based numerical methods for high-dimensional parabolic partial differential equations and backward stochastic differential equations. *Commun. Math. Stat.*, **5**, 349–380.

Eckart, C. and Young, G. (1936). The approximation of one matrix by another of lower rank. *Psychometrika*, **1**, 211–218.

Eckstein, J. and Bertsekas, D. P. (1992). On the Douglas-Rachford splitting method and the proximal point algorithm for maximal monotone operators. *Math. Program.*, **55**, 293–318.

Efron, B. (1975). The efficiency of logistic regression compared to normal discriminant analysis. *Jour. Ameri. Statist. Assoc.*, **70**, 892–898.

Efron, B. (1979). Bootstrap methods: Another look at the jackknife. *Ann. Statist.*, **7**, 1–26.

Efron, B. (1986). How biased is the apparent error rate of a prediction rule? *Jour. Ameri. Statist. Assoc.*, **81**, 461–470.

Efron, B. (2004). The estimation of prediction error: Covariance penalties and crossvalidation (with discussion). *Jour. Amer. Statist. Assoc.* **99** 619–642.

Efron, B. (2007). Correlation and large-scale simultaneous significance testing. *Jour. Ameri. Statist. Assoc.*, **102**, 93–103.

Efron, B. (2010a). *Large-scale Inference: Empirical Bayes Methods for Estimation, Testing, and Prediction.* IMS Monographs Vol 1. Cambridge University Press, Cambridge.

Efron, B. (2010b). Correlated z-values and the accuracy of large-scale statistical estimates. *Jour. Ameri. Statist. Assoc.*, **105**, 1042–1055.

Efron, B., Hastie, T., Johnstone, I. and Tibshirani, R. (2004). Least angle regression (with discussions), *Ann. Statist.*, **32**, 407–499.

Efron, B. and Tibshirani, R. (1993). *An Introduction to the Bootstrap*, Chapman and Hall/CRC.

El Karoui, N. and d'Aspremont, A. (2010). Second order accurate distributed eigenvector computation for extremely large matrices. *Electron. Jour. Stat.*, **4**, 1345–1385.

Elter, M., Schulz-Wendtland, R., and Wittenberg, T. (2007). The prediction of breast cancer biopsy outcomes using two CAD approaches that both emphasize an intelligible decision process. *Medical Physics*, **34**, 4164–4172.

Engle, R. F., Granger, C. W. J., Rice, J., and Weiss, A. (1986). Semiparametric estimates of the relation between weather and electricity sales. *Jour. Amer. Statist. Assoc.*, **81**, 310–320.

Epanechnikov, V. (1969). Non-parametric estimation of a multivariate probability density. *Theory Probab. Appl.*, **14**, 153–158.

Erdös, P. and Rényi, A. (1960). On the evolution of random graphs. *Publ. Math. Inst. Hung. Acad. Sci.*, **5**, 17–60.

Eriksson, B., Balzano, L. and Nowak, R. (2011). High-rank matrix completion and subspace clustering with missing data. *arXiv preprint, arXiv:1112.5629*

Fama, E. and French, K. (1992). The cross-section of expected stock returns. *Jour. Fin.*, **47**, 427–465.

Fama, E. and French, K. (2015). A five-factor asset pricing model. *Jour. of Fin. Econ.*, **116**, 1–22.

Fan, J. (1997). Comments on "Wavelets in statistics: A review," by A. Antoniadis. *J. Italian Statist. Assoc.*, **6**, 131–138.

Fan, J. (2014). Features of big data and sparsest solution in high confidence set. In *Past, Present and Future of Statistical Science* (X, Lin, C. Genest, D. L. Banks, G. Molenberghs, D. W. Scott, J.-L. Wang, Eds.), Chapman & Hall, New York, 507–523.

Fan, J. and Fan, Y. (2008). High dimensional classification using features annealed independence rules. *Ann. Statist.*, **36**, 2605–2637.

Fan, J., Fan, Y. and Barut, E. (2014). Adaptive robust variable selection. *Ann. Statist.*, **42**, 324–351.

Fan, J., Fan, Y., Han, X., and Lv, J. (2019a). Asymptotic theory of eigenvectors for large random matrices. *arXiv preprint arXiv1902.06846*

Fan, J., Fan, Y., Han, X. and Lv, J. (2019b). SIMPLE: statistical inference on membership profiles in large network. *arXiv preprint arXiv:1910.01734v1*

Fan, J., Fan, Y. and Lv, J. (2008). High dimensional covariance matrix estimation using a factor model. *Jour. Econ.*, **147**, 186–197.

Fan, J., Feng, Y., Jiang, J. and Tong, X., (2016). Feature Augmentation via Nonparametrics and Selection (FANS) in high-dimensional classification, *Jour. Ameri. Statist. Assoc.*, **111**, 275–287.

Fan, J., Feng, Y. and Song, R. (2011). Nonparametric independence screening in sparse ultra-high dimensional additive models. *Jour. Ameri. Statist. Assoc.*, **106**, 544–557

Fan, J., Feng, Y. and Tong, X. (2012). A ROAD to classification in high dimensional space. *Jour. Roy. Statist. Soc. B*, **74**, 745–771.

Fan, J., Feng, Y. and Wu, Y. (2009). Network exploration via the adaptive LASSO and SCAD penalties. *Ann. Appl. Statist.*, **3**, 521–541.

Fan, J., Feng, Y. and Wu, Y. (2010). Ultrahigh dimensional variable selection for Cox's proportional hazards model. *IMS Collection*, **6**, 70–86.

Fan, J. and Gijbels, I. (1996). *Local Polynomial Modelling and its Applications.* 341pp. Chapman and Hall, London.

Fan, J., Guo, J. and Zheng, S. (2020). Estimating number of factors by adjusted eigenvalues thresholding. *Jour. Ameri. Statist. Assoc.*, to appear.

Fan, J., Guo, S. and Hao, N. (2012). Variance estimation using refitted cross-validation in ultrahigh dimensional regression. *Jour. Roy. Statist. Soc. B*, **74**, 37–65.

Fan, J., Han, F., and Liu, H. (2014). Challenges of big data analysis. *National Science Review*, **1**, 293–314.

Fan, J. and Han, X. (2017). Estimation of false discovery proportion with unknown dependence. *Jour. Roy. Statist. Soc. B*, **79**, 1143–1164.

Fan, J., Han, X., and Gu, W.(2012). Estimating false discovery proportion under arbitrary covariance dependence (with discussion). *Jour. Ameri. Statist. Assoc.*, **107**, 1019–1048.

Fan, J., Ke, Y., and Wang, K. (2020). Factor-adjusted regularized model selection. *Jour. Econ.*, **216**, 71 - 85.

Fan, J., Ke, Y., Sun, Q., and Zhou, W.X. (2019). FarmTest: Factor-adjusted robust multiple testing with false discovery control. *Jour. Ameri. Statist. Assoc.*, **114**, 1880 - 1893

Fan, J., Ke, Z., Liu, H. and Xia, L. (2015). QUADRO: a supervised dimension reduction method via Rayleigh quotient optimization. *Ann. Statist.*, **43**, 1493–1534.

Fan, J. and Kim, D. (2019). Structured volatility matrix estimation for non-synchronized high-frequency financial data. *Jour. Econ.*, **209**, 61-78.

Fan, J., Li, Q., and Wang, Y. (2017). Estimation of high-dimensional mean regression in absence of symmetry and light-tail assumptions. *Jour. Roy. Statist. Soc. B*, **79**, 247–265.

Fan, J. and Li, R. (2001). Variable selection via nonconcave penalized likelihood and its oracle properties. *Jour. Ameri. Statist. Assoc.*, **96**, 1348–1360.

Fan, J. and Li, R. (2002). Variable selection for Cox's proportional hazards model and frailty model. *Ann. Statist.*, **30**, 74–99.

Fan, J. and Li, R. (2006). Statistical challenges with high-dimensionality: feature selection in knowledge discovery. *Proceedings of International Congress of Mathematicians* (M. Sanz-Solé, J. Soria, J.L. Varona, J. Verdera, eds.), Vol. III, 595–622.

Fan, J., Li, Q., and Wang, Y. (2017). Estimation of high-dimensional mean regression in absence of symmetry and light-tail assumptions. *Jour. Roy. Statist. Soc. B*, **79**, 247–265.

Fan, J., Li, Y. and Yu, K. (2012). Vast volatility matrix estimation using high frequency data for portfolio selection. *Jour. Ameri. Statist. Assoc.*, **107**, 412–428.

Fan, J. and Liao, Y. (2014). Endogeneity in ultrahigh dimension. *Ann. Statist.*, **42**, 872–917.

Fan, J., Liao, Y., and Mincheva, M. (2011). High dimensional covariance matrix estimation in approximate factor models. *Ann. Statist.*, **39**, 3320–3356.

Fan, J., Liao, Y., and Mincheva, M. (2013). Large covariance estimation by thresholding principal orthogonal complements (with discussion). *Jour. Roy. Statist. Soc. B*, **75**, 603–680.

Fan, J., Liu, H., Ning, Y., and Zou, H. (2017). High dimensional semiparametric latent graphical model for mixed data. *Jour. Roy. Statist. Soc. B*, **79**, 405–421.

Fan, J., Liu, H., Sun, Q., and Zhang, T. (2018). I-LAMM: Simultaneous control of algorithmic complexity and statistical error. *Ann. Statist.*, **46**, 814–841.

Fan, J., Liu, H., and Wang, W. (2018). Large covariance estimation through elliptical factor models. *Ann. Statist.*, **46**, 1383–1414.

Fan, J. and Lv, J. (2008). Sure independence screening for ultrahigh dimensional feature space (with discussion), *Jour. Roy. Statist. Soc. B*, **70**, 849–911.

Fan, J. and Lv, J. (2010). A selective overview of variable selection in high dimensional feature space (Invited review article). *Statist. Sinica*, **20**, 101–148.

Fan, J. and Lv, J. (2011). Non-concave penalized likelihood with NP-Dimensionality. *IEEE Trans. Inform. Theory*, **57**, 5467–5484.

Fan, J., Lv, J., and Qi, L. (2011). Sparse high-dimensional models in economics. *Ann. Rev. Econ.*, **3**, 291–317.

Fan, J. and Peng, H. (2004). On non-concave penalized likelihood with diverging number of parameters. *Ann. Statist.*, **32**, 928–961.

Fan, J., Ma, C., and Zhong, Y. (2019). A selective overview of deep learning. *arXiv preprint, arXiv:1904.05526.*

Fan, J., Ma, Y. and Dai, W. (2014). Nonparametric independence screening in sparse ultra-high dimensional varying coefficient models, *Jour. Ameri. Statist. Assoc.,* **109**, 1270–1284.

Fan, J., Samworth, R., and Wu, Y. (2009). Ultrahigh dimensional variable selection: beyond the linear model. *Jour. Mach. Learn. Res.*, **10**, 1829–1853.

Fan, J., Shao, Q., and Zhou, W. (2018). Are discoveries spurious? Distributions of maximum spurious correlations and their applications. *Ann. Statist.*, **46**, 989 - 1017.

Fan, J. and Song, R. (2010). Sure independence screening in generalized linear models with NP-dimensionality. *Ann. Statist.*, **38**, 3567–3604.

Fan, J., Wang, D., Wang, K., and Zhu, Z. (2019). Distributed estimation of principal eigenspaces. *Ann. Statist.*, **47**, 3009-3031.

Fan, J., Wang, K., Zhong, Y., and Zhu, Z. (2019+). Robust high dimensional factor models with applications to statistical machine learning. *Statist. Sci.*, invited.

Fan, J., Wang, W. and Zhong, Y. (2019). Robust covariance estimation for approximate factor models. *Jour. Econ.*, **208**, 5-22.

Fan, J., Wang, W., and Zhu, Z. W. (2016). A shrinkage principle for heavy-tailed data: High-dimensional robust low-rank matrix recovery. *arXiv preprint, arXiv:1603.08315.*

Fan, J., Xue, L., and Yao, J. (2017). Sufficient forecasting using factor models. *Jour, of Econ.*, **201**, 292–306.

Fan, J., Xue, L., and Zou, H. (2014). Strong oracle optimality of folded concave penalized estimation. *Ann. Statist.*, **42**, 819-849.

Fan, J. and Zhang, W. (2008). Statistical methods with varying coefficient models. *Stat. Interface*, **1**, 179–195.

Fan, J. and Zhou, W. (2016). Guarding from spurious discoveries in high dimension. *Jour. Mach. Learn. Res.*, **17** (203), 1–34.

Fan, Y., Kong, Y., Li, D. and Zheng, Z. (2015). Innovated interaction screening for high-dimensional nonlinear classification. *Ann. Statist.*, **43**, 1243–1272.

Fan, Y. and Li, R. (2012). Variable selection in linear mixed effects models. *Ann. Statist.*, **40**, 2043–2068.

Fan, Y. and Tang, C. Y. (2013). Tuning parameter selection in high dimensional penalized likelihood. *Jour. Roy. Statist. Soc. B*, **75**, 531–552.

Fang, K.T., Kotz, S., and Ng, K. W. (1990). *Symmetric Multivariate and Related Distributions*, Chapman and Hall, New York, NY.

Fang, X. E., Ning, Y. and Li, R. (2020). Test of significance for high-dimensional longitudinal data. *Ann. Statist.*, To appear.

Fano, R. (1961). *Transmission of Information: a Statistical Theory of Communications.* M.I.T. Press.

Feng, L. and Zhang, C.-H. (2019). Sorted concave penalized regression. *Ann. Statist.*, **47**, 3069 - 3098.

Ferńandez-Delgado, M., Cernadas, E., Barro, S. and Amorim, D. (2014). Do we need hundreds of classifiers to solve real world classification problems? *Jour. Mach. Learn. Res.*, **15**, 3133–3181.

Forni, M., Hallin, M., Lippi, M., and Reichlin, L. (2000). The generalized dynamic-factor model: Identification and estimation. *Rev. Econ. Statist.*, **82**, 540–554.

Forni, M., Hallin, M., Lippi, M., and Reichlin, L. (2004). The generalized dynamic factor model consistency and rates. *Jour. Econ.*, **119**, 231–255.

Forni, M., Hallin, M., Lippi, M., and Reichlin, L. (2005). The generalized dynamic factor model. *Jour. Ameri. Statist. Assoc.*, **100**, 830–840.

Forni, M., Hallin, M., Lippi, M., and Zaffaroni, P. (2015). Dynamic factor models with infinite-dimensional factor spaces: One-sided representations. *Jour. Econ.*, **185**, 359–371.

Forni, M. and Lippi, M. (2001). The generalized dynamic factor model: representation theory. *Econometric Theory*, **17**, 1113–1141.

Foss, A., Markatou, M., Ray, B. and Heching, A. (2016). A semi-parametric method for clustering mixed data. *Machine Learning*, **105**, 419–458.

Foster, D. P., and George, E. I. (1994). The risk inflation criterion for multiple regression. *Ann. Statist.*, **22**, 1947–1975.

Fraley, C. and Raftery, A. (2002). Model-based clustering, discriminant analysis, and density estimation. *Jour. Ameri. Statist. Assoc.*, **97**, 611–631.

Frank, I.E. and Friedman, J. H. (1993). A statistical view of some chemometrics regression tools. *Technometrics*, **35**, 109–148.

Freund, Y. and Schapire, R. (1997). A decision-theoretic generalization of on-line learning and an spplication to boosting. *Journal of Computer and System Sciences*, **55**, 119–139.

Friedman, J. (1989). Regularized discriminant analysis. *Jour. Ameri. Statist. Assoc.*, **84**, 165–175.

Friedman, J. (2001). Greedy function approximation: a gradient boosting machine. *Ann. Statist.*, **29**, 1189–1232.

Friedman, J., Baskett, F. and Shustek, L. (1975). An algorithm for finding nearest neighbors. *IEEE Trans. Comput.*, **24**, 1000–1006.

Friedman, J. and Hall, P. (2000). On Bagging and nonlinear estimation. *Technical report, Stanford University, Department of Statistics.*

Friedman, J., Hastie, T., Höfling, H., and Tibshirani, R. (2007). Pathwise coordinate optimization. *Ann. Appl. Statist.*, **1**, 302–332.

Friedman, J., Hastie, T. and Tibshirani, R. (2000). Additive logistic regression: a statistical view of boosting (with discussion). *Ann. Statist.*, **28**, 337–407.

Friedman, J., Hastie, T. and Tibshirani, R. (2008). Sparse inverse covariance estimation with the graphical LASSO. *Biostatistics*, **9**, 432–441.

Friedman, J., Hastie, T. and Tibshirani, R. (2010). Regularization paths for generalized linear models via coordinate descent. *Journal of Statistical Software*, **33**, 1–22.

Friedman, J. and Meulman, J. (2004). Clustering objects on subsets of attributes. *Jour. Roy. Statist. Soc. B*, **66**(4), 815–849.

Friguet, G., Kloareg, M. and Causeur, D. (2009). A factor model approach to multiple testing under dependence. *Jour. Amier. Statist. Assoc.*, **104**, 1406–1415.

Fu, W. J. (1998). Penalized regression: the bridge versus the LASSO, *Jour. Comput. Graph. Statist.* , **7**, 397–416.

Fukunaga, K. and Narendra, P. (1975). A branch and bound algorithm for computing k-nearest neighbors. *IEEE Trans. Computers*, **24**, 750–753.

Fukushima, K. and Miyake, S. (1982). Neocognitron: A self-organizing neural network model for a mechanism of visual pattern recognition. In *Competition and Cooperation in Neural Nets*, 267–285. Springer.

Gao, C., Liu, J., Yao, Y. and Zhu, W. (2018). Robust estimation and generative adversarial nets. *arXiv preprint arXiv:1810.02030.*

Gao, C., Ma, Z., Ren, Z., and Zhou, H. (2015). Minimax estimation in sparse canonical correlation analysis. *Ann. Statist.*, **43**, 2168–2197.

Garber, D., Shamir, O. and Srebro, N. (2017). Communication-efficient algorithms for distributed stochastic principal component analysis. *arXiv preprint, arXiv:1702.08169.*

Genovese, C. and Wasserman, L. (2004). A stochastic process approach to false discovery control. *Ann. Statist.*, **32**, 1035–1061.

George, E. I. and Foster, D. P. (2000). Calibration and empirical Bayes variable selection. *Biometrika*, **87** 731–747.

George, E. I. and McCulloch, R. E. (1993). Variable selection via Gibbs sampling. *Jour. Amer. Statist. Assoc.*, **88**, 881–889.

Ghosh, B. K. (1973). Some monotonicity theorems for $\chi^2$, $F$ and $t$ distributions with application. *Jour. Roy. Statist. Soc. B*, **79**, 1415-1437.

Goldstein, T. and Osher, S. (2009). The split Bregman method for $\ell_1$-regularized problems. *SIAM J. Imaging Sci.*, **2**, 323 - 343.

Golowich, N., Rakhlin, A., and Shamir, O. (2017). Size-independent sample complexity of neural networks. *arXiv preprint arXiv:1712.06541.*

Golub, G., Heath, M. and Wahba, G. (1979). Generalized cross-validation as a method for choosing a good ridge parameter. *Technometrics*, **21**, 215–224.

Golub, G. H. and Van Loan, C. F. (2013). *Matrix Computations* (4 Ed.). Baltimore: Johns Hopkins University Press.

Golub, T., Slonim, D., Tamayo, P., Huard, C., Gaasenbeek, M., Mesirov, J., Coller, H., Loh, M., Downing, J. and Caligiuri, M. (1999). Molecular classification of cancer: class discovery and class prediction by gene expression monitoring. *Science*, **286**, 531–536.

Goodfellow, I., Bengio, Y. and Courville, A (2016). *Deep Learning.* MIT Press.

Goodfellow, I., Pouget-Abadie, J., Mirza, M., Xu, B., Warde-Farley, D., Ozair, S., Courville, A. and Bengio, Y. (2014). Generative adversarial nets. In *Advances in Neural Information Processing Systems*, 2672–2680.

Grant, M. and Boyd, S. (2013). CVX: Matlab software for disciplined convex programming, version 2.0 beta. http://cvxr.com/cvx.

Grant, M. and Boyd, S. (2014). CVX: Matlab software for disciplined convex programming, version 2.1. http://cvxr.com/cvx.

Gravuer, K., Sullivan, J.J., Williams, P. and Duncan, R. (2008). Strong human association with plant invasion success for Trifolium introductions to New Zealand. *Proc. Natl. Acad. Sci.*, **105**(17), 6344–6349.

Greenshtein, E. and Ritov, Y. (2004). Persistence in high-dimensional predictor selection and the virtue of overparametrization. *Bernoulli*, **10**, 971–988.

Gu, Y. (2017). "Unconventional Regression for High-Dimensional Data". PhD Thesis, University of Minnesota.

Gu, Y., Fan, J., Kong, L., Ma, S. and Zou, H. (2018). ADMM for high-dimensional sparse penalized quantile regression. *Technometrics*, **60**, 319–331.

Gu, Y. and Zou, H. (2019). Sparse composite quantile regression in ultrahigh dimensions with tuning parameter calibration. Technical report, University of Minnesota.

Gunter, L. and Zhu, J. (2007). Efficient computation and model selection for the support vector regression. *Neural Computation*, **19**, 1633–1655.

Guo, J., Levina, E., Michailidis, G. and Zhu, J. (2010). Pairwise variable selection for high-dimensional model-based clustering. *Biometrics*, **66**, 793–804.

Guttman, L. (1954). Some necessary conditions for common-factor analysis. *Psychometrika*, **19**, 149–161.

Hájek, J. (1970). A characterization of limiting distributions of regular estimates. *Zeitschrift für Wahrscheinlichkeitstheorie und verwandte Gebiete*, **14**, 323–330.

Hájek, J. (1972). Local asymptotic minimax and admissibility in estimation. In: *Proceedings of the Sixth Berkeley Symposium on Mathematical Statistics and Probability*, **1**, 175–194.

Hall, P. and Miller, H. (2009). Using generalized correlation to effect variable selection in very high dimensional problems. *Jour. Comput. Graph. Statist.*, **18**, 533–550.

Hall, P., Park, B. and Samworth, R. (2008). Choice of neighbor order in nearest-neighbor classification. *Ann. Statist.*, **36**, 2135–2152.

Hall, P., Pittelkow, Y., and Ghosh, M. (2008). Theoretic measures of relative performance of classifiers for high dimensional data with small sample sizes. *Jour. Roy. Statist. Soc. B*, **70**, 158–173.

Hallin, M. and Liška, R. (2011). Dynamic factors in the presence of blocks. *Jour. Econ.*, **163**, 29–41.

Halmos, P.R. (2017). *Introduction to Hilbert Space and the Theory of Spectral Multiplicity.* Courier Dover Publications.

Han, F. and Liu, H. (2017). Statistical analysis of latent generalized correlation matrix estimation in transelliptical distribution. *Bernoulli*, **23**, 23–57.

Hancock, P., Burton, A. and Bruce, V. (1996). Face processing: human perception and principal components analysis. *Memory and Cognition*, **24**, 26–40.

Hand, D. J. (2006). Classifier technology and the illusion of progress. *Statist. Sci.*, **21**, 1–14.

Hannan, E. J., and Quinn, B. G. (1979). The determination of the order of an autoregression. *Jour. Roy. Statist. Soc. B*, **41**, 190–195.

Hao, N., Dong, B. and Fan, J. (2015). Sparsifying the Fisher linear discriminant by rotation. *Jour. Roy. Statist. Soc. B*, **77**, 827–851.

Hao, N. and Zhang, H.H. (2014). Interaction screening for ultrahigh dimensional data, *Jour. Ameri. Statist. Assoc.*, **109**, 1285–1301.

Härdle, W., Hall, P. and Ichimura, H. (1993). Optimal smoothing in single-index models. *Ann. Statist.*, **21**, 157–178.

Härdle, W., Liang, H. and Gao, J. T. (2000) *Partially Linear Models.* Springer Phisica-Verlag, Germany.

Hardt, M., Recht, B., and Singer, Y. (2015). Train faster, generalize better: Stability of stochastic gradient descent. *arXiv preprint arXiv:1509.01240.*

Hartigan, J. A. (1975). *Clustering Algorithms.* Wiley.

Hastie, T., Rosset, S., Tibshirani, R. and Zhu, J. (2004). The entire regularization path for the support vector machine. *Jour. Mach. Learn. Res.*, **5**, 1391–1415.

Hastie, T. and Stuetzle, W. (1989). Principal curves. *Jour. Ameri. Statist. Assoc.*, **84**, 502–516.

Hastie, T.J. and Tibshirani, R. (1990). *Generalized Additive Models.* Chapman and Hall, London.

Hastie, T. and Tibshirani, R. (1996). Discriminant adaptive nearest neighbor classification. *IEEE Pattern Recognition and Machine Intelligence*, **18**, 607–616.

Hastie, T., Tibshirani, R., Eisen, M., Brown, P., Ross, D., Scherf, U., Weinstein, J., Alizadeh, A., Staudt, L. and Botstein, D. (2000). 'Gene shaving' as a method for identifying distinct sets of genes with similar expression patterns. *Genome Biology*, **1**, 1–21.

Hastie, T.J., Tibshirani, R. and Friedman, J. (2009). *The Elements of Statistical Learning: Data Mining, Inference, and Prediction* (2nd ed). *Springer*, New York.

Hastie, T., Tibshirani, R., and Wainwright, M. (2015). *Statistical Learning with Sparsity. CRC Press*, New York.

He, K., Zhang, X., Ren, S. and Sun, J. (2016a). Deep residual learning for image recognition. In *Proceedings of the IEEE Conference on Computer Vision and Pattern Recognition*, 770–778.

He, K., Zhang, X., Ren, S. and Sun, J. (2016b). Identity mappings in deep residual networks. In *European Conference on Computer Vision*, 630–645. Springer.

He, X., Wang, L. and Hong, H.G. (2013). Quantile-adaptive model-free variable screening for high-dimensional heterogeneous data. *Ann. Statist.*, **41**, 342–369.

Heckman, N. (1986). Spline smoothing in a partly linear model. *J. Roy. Statist. Soc. Ser. B*, **48**, 244–248.

Hettmansperger, T. P. and McKean, J. W. (1998). *Robust Nonparametric Statistical Methods.* London: Arnold.

Hinton, G., Srivastava, N. and Swersky, K. (2012). Neural networks for machine learning. Lecture 6a: Overview of mini-batch gradient descent.

Hinton, G. E., Srivastava, N., Krizhevsky, A., Sutskever, I. and Salakhutdinov, R. R. (2012). Improving neural networks by preventing co-adaptation of feature detectors. *arXiv preprint arXiv:1207.0580.*

Hochreiter, S. and Schmidhuber, J. (1997). Long short-term memory. *Neural Computation*, **9**, 1735–1780.

Ho, T. (1998). The random subspace method for constructing decision forests. *IEEE Transactions on Pattern Analysis and Machine Intelligence*, **20**, 832–844.

Hochreiter, S. and Schmidhuber, J. (1997). Long short-term memory. *Neural Computation*, **9**, 1735–1780.

Hoeffding, W. (1963). Probability inequalities for sums of bounded random variables. *Jour. Ameri. Statist. Assoc.*, **58**, 13–30.

Hoerl A.E. (1962). Application of ridge analysis to regression problems. *Chemical Engineering Progress*, **58**, 54–59.

Hoerl, A.E. and Kennard, R.W. (1970). Ridge regression: Biased estimation for nonorthogonal problems. *Technometrics*, **42**, 55–67.

Hoffman, M.D., Blei, D.M., Wang, C., and Paisley, J. (2013). Stochastic variational inference. *Journal of Machine Learning Research*, **14**, 1303–1347.

Hofmann, T. (1999). Probabilistic latent semantic indexing. In *International ACM SIGIR conference*, 50-57.

Höfling, H. and Tibshirani, R. (2009). Estimation of sparse binary pairwise markov networks using pseudo-likelihoods. *Jour. Mach. Learn. Res.*, **10**, 883–906.

Holland, P. W. and Leinhardt, S. (1981). An exponential family of probability distributions for directed graphs. *Jour. Ameri. Statist. Assoc.*, **76**, 33–50.

Holland, P. W., Laskey, K. B., and Leinhardt, S.(1983). Stochastic block models: First steps. *Social networks*, **5**, 109-137.

Hotelling, H. (1933). Analysis of a complex of statistical variables into principal components. *Jour. of Educat. Psych.*, **24**, 417–441 and 498–520.

Hsu, D. and Kakade, S. M. (2013). Learning mixtures of spherical Gaussians: moment methods and spectral decompositions. In *Proceedings of the Fourth Conference on Innovations in Theoretical Computer Science.* ACM.

Huang, C.-C., Liu, K.Pope, R.M., et al. (2011). Activated TLR signaling in atherosclerosis among women with lower Framingham risk score: The multi-ethnic study of atherosclerosis. *PLoS One*, **6**, e21067.

Huang, G., Liu, Z., Van Der Maaten, L. and Weinberger, K. Q. (2017). Densely connected convolutional networks. In *Proceedings of the IEEE Conference on Computer Vision and Pattern Recognition*, 4700–4708.

Huang, J., Horowitz, J. and Ma, S. (2008a). Asymptotic properties of bridge estimators in sparse high-dimensional regression models. *Ann. Statist.*, **36**, 587–613.

Huang, J., Ma, S. and Zhang, C.-H. (2008b). Adaptive Lasso for sparse high-dimensional regression models. *Statistica Sinica*, **18**, 1603–1618.

Huang, J. and Zhang, C.-H. (2012). Estimation and selection via absolute penalized convex minimization and its multistage adaptive applications. *Jour. Mach. Learn. Res.*, **13**, 1809–1834.

Huber, P. J. (1964). Robust estimation of a location parameter. *Ann. Math. Statist.*, **35**, 73–101.

Huber, P. J. (1973). Robust regression: asymptotics, conjectures and Monte Carlo. *Ann. Statist.*, **1**, 799–821.

Hunter, D. and Lange, K. (2000a). Rejoinder to discussion of "Optimization transfer using surrogate objective functions." *Jour. Comput. Graph. Statist.* **9**, 52–59.

Hunter, D. and Lange, K. (2000b), Quantile regression via an MM algorithm, *Jour. Comput. Graph. Statist.*, **9**, 60–77.

Hunter, D. and Lange, K. (2004). A tutorial on MM algorithms. *The American Statistician*, 58, 30–37.

Hunter, D. and Li, R. (2005). Variable selection using MM algorithms. *Ann. Statist.*. **33**, 1617–1642

Iandola, F. N., Han, S., Moskewicz, M. W., Ashraf, K., Dally, W. J. and Keutzer, K. (2016). SqueezeNet: AlexNet-level accuracy with 50x fewer parameters and¡ 0.5 MB model size. *arXiv preprint arXiv:1602.07360.*

Ioffe, S. and Szegedy, C. (2015). Batch normalization: Accelerating deep network training by reducing internal covariate shift. *arXiv preprint arXiv:1502.03167.*

Ishwaran, H. and Rao, J. S. (2005). Spike and slab variable selection: Frequentist and Bayesian strategies. *Ann. Statist.*, **33**, 730–773.

Ising, E. (1925). Beitrag zur theorie des ferromagnetismus. *Z. Physik*, **31**, 53–258.

Jacobs, R. A., Jordan, M. I., Nowlan, S. J., and Hinton, G. E. (1991). Adaptive mixtures of local experts. *Neural Computation*, **3**, 79–87.

Jaeckel, L. A. (1972). Estimating regression coefficients by minimizing the dispersion of the residuals. *Ann. Math. Statist.*, **43**, 1449–1458.

Jain, P., Kakade, S. M., Kidambi, R., Netrapalli, P. and Sidford, A. (2017). Accelerating stochastic gradient descent. *arXiv preprint arXiv:1704.08227.*

James, G. and Radchenko, P. (2009). A generalized Dantzig selector with shrinkage tuning. *Biometrika*, **96**, 323–337.

James, G., Radchenko, P. and Lv, J. (2009). DASSO: Connections between the Dantzig selector and Lasso. *Jour. Roy. Statist. Soc. B*, **71**, 127–142.

James, W. and Stein, C. (1961). Estimation with quadratic Loss. In *Proc. Fourth Berkeley Symp. Math. Statist. Probab.*, **1**, 361–379. Univ. California Press, Berkeley.

Jason, L., Fithian, W. and Hastie, T. (2015). Effective degrees of freedom: a flawed metaphor. *Biometrika*, **102**, 479–485.

Javanmard, A. and Montanari, A. (2014). Confidence intervals and hypothesis testing for high-dimensional regression. *Jour. Mach. Learn. Res.*, **15**, 2869-2909.

Javanmard, A., Modelli, M., and Montanari, A. (2019). Analysis of a two-layer neural network via displacement convexity. *arXiv preprint arXiv:1901.01375.*

Javanmard, A. and Montanari, A. (2014a). Confidence intervals and hypothesis testing for high-dimensional regression. *Jour. Mach. Learn. Res.*, **15**, 2869–2909.

Javanmard, A. and Montanari, A. (2014b). Hypothesis testing in high-dimensional regression under the gaussian random design model: Asymptotic theory. *IEEE Trans. Inform. Theory*, **60**, 6522–6554.

Jeffers, J. (1967). Two case studies in the application of principal component. *Appl. Statist.*, **16**, 225–236.

Jia, J. and Yu, B. (2010). On model selection consistency of the Elastic Net when $p \gg n$. *Statist. Sinica*, **20**, 595–611.

Jiang, W. and Tanner, M. A. (1999). Hierarchical mixtures-of-experts for exponential family regression models: approximation and maximum likelihood estimation. *Ann. Statist.*, **27**, 987-1011.

Jin, J. (2015). Fast community detection by score. *Ann. Statist.* **43**, 57–89.

Jin, J. and Cai, T.T. (2007). Estimating the null and the proportion of non-null effects in large-scale multiple comparisons. *Jour. Ameri. Statist. Assoc.* **102**, 495–506.

Jin, J., Ke, Z.T. and Luo, S. (2017). Estimating network memberships by simplex vertex hunting. arXiv:1708.07852.

Jin, J. and Wang, W. (2016). Influential Features PCA for high dimensional clustering (with discussion). *Ann. Statist.*, **44**(6), 2323–2359.

Johnson, R. A. and D. W. Wichern (2007). *Applied Multivariate Statistical Analysis.* 6th Edition. Prentice Hall.

Johnstone, I.M. and Lu, A.Y. (2009). On consistency and sparsity for principal components analysis in high dimensions. *Jour. Ameri. Statist. Assoc.*, **104**, 682–693.

Johnstone, I. M. and Silverman, B. W. (2005). Empirical Bayes selection of wavelet thresholds. *Ann. Statist.*, **33**, 1700–1752.

Jolliffe, I. T. (1986). *Principal Component Analysis*, Springer Verlag, New York.

Jolliffe, I. T. (1986). Principal component analysis and factor analysis. In *Principal Component Analysis.* Springer, 115–128.

Jolliffe, I. T. (1995). Rotation of principal components: choice of normalization constraints. *Jour. Appl. Statist.*, **22**, 29–35.

Jolliffe, I. T. and Uddin, M. (2003). A modified principal component technique based on the lasso, *Jour. Comput. Graph. Statist.*, **12**, 531–547.

Jones, M., Marron, J. and Sheather, S. J. (1996). A brief survey of bandwidth selection for density estimation. *Jour. Ameri. Statist. Assoc.*, **91**, 401–407.

Jordan, M. I. and Jacobs, R. A. (1994). Hierarchical mixtures of experts and the EM algorithm. *Neural Computation*, **6**, 181-214.

Jordan, M. I., Lee, J.D., and Yang, Y. (2019). Communication-efficient distributed statistical inference. *Jour. Ameri. Statist. Assoc.*, **2019**, 668 - 681.

Journée, M., Nesterov, Y., Richtárik, P. and Sepulchre, R. (2010). Generalized power methods for sparse principal component analysis. *Jour. Mach. Learn. Res.*, **11**, 517–533.

Kai, B., Li, R. and Zou, H. (2010). Local CQR smoothing: An efficient and safe alternative to local polynomial regression. *Jour. Roy. Statist. Soc. B*, **72**, 49–69.

Kai, B., Li, R. and Zou, H. (2011). New efficient estimation and variable selection methods for semiparametric varying-coefficient partially linear models. *Ann. Statist.*, **39**, 305–332.

Kaufman, L. and Rousseeuw, P. (1990). *Finding Groups in Data: An Introduction to Cluster Analysis.* New York: Wiley.

Ke, Z. and Wang, M. (2019). A new SVD approach to optimal topic estimation. *arXiv preprint arXiv:1704.07016v2.*

Kendall, M.G. (1948). *Rank Correlation Methods.* Griffin, London.

Keshavan, R. H.; Montanari, A. and Oh, S. (2010). Matrix completion from a few entries. *IEEE Trans. Inform. Theory*, **56**, 2980–2998.

Khalili, A. and Chen, J. (2007). Variable selection in finite mixture of regression models. *Jour. Ameri. Statist. Assoc.*, **102**, 1025–1038.

Khalili, A., Chen, J. and Lin, S. (2011). Feature selection in finite mixture of sparse normal linear models in high dimensional feature space. *Biostatistics*, **12**, 156–172.

Kidambi, R., Netrapalli, P., Jain, P. and Kakade, S. (2018). On the insufficiency of existing momentum schemes for stochastic optimization. In *2018 Information Theory and Applications Workshop (ITA), IEEE*, 1–9.

Kim, Y., Choi, H., and Oh, H.S. (2008). Smoothly clipped absolute deviation on high dimensions. *Jour. Amer. Statist. Assoc.* **103**, 1665–1673.

Kingma, D. P. and Ba, J. (2014). Adam: A method for stochastic optimization. *arXiv preprint arXiv:1412.6980.*

Klaassen, C. A. (1987). Consistent estimation of the influence function of locally asymptotically linear estimators. *Ann. Statist.*, **15**, 1548–1562.

Kneip, A. and Sarda, P. (2011). Factor models and variable selection in high-dimensional regression analysis. *Ann. Statist.*, **39**, 2410–2447.

Knight, K. and Fu, W. (2000). Asymptotics for Lasso-type estimators. *Ann. Statist.*, **28**, 1356–1378.

Koenker, R. (2005). *Quantile Regression*, Econometric Society Monograph Series, Cambridge University Press.

Koenker, R., and Bassett, G. (1978). Regression quantiles. *Econometrica*, **46**, 33–50.

Koenker, R. and Geling, R. (2001). Reappraising medfly longevity: a quantile regression survival analysis. *Jour. Amer. Statist. Assoc.*, **96**, 458–468.

Koenker, R. and Hallock, K. (2001). Quantile regression. *Jour. Econ. Perspect.*, **15**, 143–156.

Koenker, R. and Ng, P. (2005). A Frisch-Newton algorithm for sparse quantile regression. *Acta Mathematicae Applicatae Sinica*, **21**, 225–236.

Koltchinskii, V. (2009). The dantzig selector and sparsity oracle inequalities. *Bernoulli*, **15**, 799–828.

Koltchinskii, V. and Lounici, K. (2017). Concentration inequalities and moment bounds for sample covariance operators. *Bernoulli*, **23**, 110–133.

Koltchinskii, V., Lounici, K. and Tsybakov, A. B. (2011). Nuclear-norm penalization and optimal rates for noisy low-rank matrix completion. *Ann. Statist.*, **39**, 2302–2329.

Krizhevsky, A., Sutskever, I. and Hinton, G. E. (2012). Imagenet classification with deep convolutional neural networks. *Advances in Neural Information Processing Systems*, 1097–1105.

Krzanowski, W. J. and Lai, Y. T. (1985). A criterion for determining the number of clusters in a data set. *Biometrics*, **44**, 23–34.

Kvam, V.M., Liu, P. and Si, Y. (2012). A comparison of statistical methods for detecting differentially expressed genes from RNA-seq data. *Ameri. Jour. Botany*, **99**, 248–256

Lam, C. and Fan, J. (2009). Sparsistency and rates of convergence in large covariance matrices estimation. *Ann. Statist.*, **37**, 4254–4278.

Lam, C. and Yao, Q. (2012). Factor modeling for high-dimensional time series: inference for the number of factors. *Ann. Statist.*, **40**, 694–726.

Lambert, D. (1992). Zero-inflated Poisson regression, with an application to defects in manufacturing. *Technometrics*, **34**, 1–13.

Lambert-Lacroix, S. and Zwald, L. (2011). Robust regression through the Huber's criterion and adaptive lasso penalty. *Electron. Jour. Stat.*, **5**, 1015–1053.

Langaas, M. and Lindqvist, B. (2005). Estimating the proportion of true null hypotheses, with application to DNA microarray data. *Jour. Roy. Statist. Soc. B*, **67**, 555–572.

Lange, K. (1995). A gradient algorithm locally equivalent to the EM algorithm. *Jour. Roy. Statist. Soc. B.*, **57**, 425–437.

Lange, K., Hunter, D.R. and Yang, I. (2000). Optimization transfer using surrogate objective functions (with discussion). *Jour. Comput. Graph. Statist.*, **9**, 1–20.

Lauritzen, S. L. (1996). *Graphical Models.* Clarendon Press, Oxford.

LeCam, L. (1960). Locally asymptotically normal families of distributions. *Univ. California Publ. Statist.*, **3**, 37–98.

Le Cam, L. (1969). *Théorie Asymptotique de la Décision Statistique.* Les Presses de l'Univérsite de Montreal.

Le Cam, L. (1972). Limits of experiments. In: *Proceedings of the Sixth Berkeley Symposium on Mathematical Statistics and Probability, Volume 1: Theory of Statistics.* The Regents of the University of California.

Le Cam, L. (1986). *Asymptotic Methods in Statistical Decision Theory.* Springer-Verlag.

LeCun, Y., Bengio, Y. and Hinton, G. (2015). Deep learning. *Nature*, **521**, 436–444.

LeCun, Y., Bottou, L., Bengio, Y. and Haffner, P. (1998). Gradient-based learning applied to document recognition. *Proceedings of the IEEE*, **86**, 2278–2324.

Lee, J. D., Liu, Q., Sun, Y. and Taylor, J. E. (2017). Communication-efficient sparse regression. *Jour. Mach. Learn. Res.*, **18**, 1–30.

Lee, J.D., Sun, D.L., Sun, Y., Taylor, J.E. (2016). Exact post-selection inference, with application to the lasso. *Ann. Statist.*, **44**, 907–927.

Lee, J. D., Sun, Y. and Taylor, J. E. (2015). On model selection consistency of regularized $M$-estimators. *Electron. Jour. Stat.*, **9**, 608–642.

Leeb, H. and Potscher, B. M. (2006). Can one estimate the conditional distribution of post-model-selection? *Ann. Statist.*, **34**, 2554–2591.

Leek, J.T. and Storey, J.D. (2008). A general framework for multiple testing dependence. *Proc. Natl. Acad. Sci.*, **105**, 18718–19723.

Lehmann, E. L. (1983), *Theory of Point Estimation.* Pacific Grove, CA: Wadsworth and Brooks/Cole.

Lehmann, E. L. and Romano, J. P. (2005). Generalizations of the familywise error rate. *Ann. Statist.*, **33**, 1138-1154.

Lehmann, E.L., Romano, J. P., and Shaffer, J. P. (2005). On optimality of stepdown and stepup multiple test procedures. *Ann. Statist.*, **33**, 1084–1108.

Lei, J. (2016). A goodness-of-fit test for stochastic block models. *Ann. Statist.*, **44**, 401–424.

Lei, J. and Rinaldo, A. (2015). Consistency of spectral clustering in stochastic block models. *Ann. Statist.*, **43**, 215–237.

Levina, E., Rothman, A.J. and Zhu, J. (2008). Sparse estimation of large covariance matrices via a nested lasso penalty. *Ann. Appl. Statist.*, **2**, 245–263.

Li, D., Ke, Y. and Zhang, W. (2015). Model selection and structure specification in ultra-high dimensional generalized semi-varying coefficient models. *Ann. Statist.*, **43** 2676–2706.

Li, D. and Li, R. (2016). Local composite quantile regression smoothing for Harris recurrent Markov processes. *Jour. Econ.*, **194**, 44–56.

Li, D. and Zou, H. (2016). SURE Information Criteria for Large Covariance Matrix Estimation and Their Asymptotic Properties. *IEEE Trans. Inform. Theory*, **62**, 2153–2169.

Li, G., Peng. H., Zhang J. and Zhu, L. (2012). Robust rank correlation based screening. *Ann. Statist.*, **40**, 1846–1877.

Li, H. and Gui, J. (2006). Gradient directed regularization for sparse Gaussian concentration graphs, with applications to inference of genetic networks. *Biostatistics*, **7**, 302–317.

Li, J., Zhong, W., Li, R. and Wu, R. (2014). A fast algorithm for detecting gene-gene interactions in genome-wide association studies. *Ann. Appl. Statist.*. **8**, 2292 – 2318.

Li, K.-C. (1987). Asymptotic optimality for $C_p$, $C_L$, cross-validation and generalized cross-validation: discrete index set. *Ann. Statist.*, **15**, 958–975.

Li, K.-C. (1991). Sliced inverse regression for dimension reduction (with discussion). *Jour. Amer. Statist. Assoc.*, **86**, 316–342.

Li, K.-C. (1992). On principal Hessian directions for data visualization and dimension reduction: another application of Stein's lemma. *Jour. Amer. Statist. Assoc.*, **87**, 1025–1039.

Li, Q. and Racine, J. (2007). *Nonparametric Econometrics: Theory and Practice.* Princeton University Press, Princeton.

Li, R., Liu, J. and Lou, L. (2017). Variable selection via partial correlation. *Statistica Sinica*, **27**, 983 -996.

Li, R., Zhong, W. and Zhu, L. (2012). Feature screening via distance correlation learning. *Jour. Ameri. Statist. Assoc.*, **107**, 1129–1139.

Li, Y., Swersky, K. and Zemel, R. (2015). Generative moment matching networks. *International Conference on Machine Learning*, 1718–1727.

Li, Y. and Zhu, J. (2008). L1-norm quantile regression. *Jour. Comput. Graph. Statist.*, **17**, 163–185.

Li, X., Lu, J., Wang, Z., Haupt, J. and Zhao, T. (2018). On tighter generalization bound for deep neural networks: CNNs, Resnets, and beyond. *arXiv preprint arXiv:1806.05159.*

Liang, Y., Balcan, M.-F. F., Kanchanapally, V. and Woodruff, D. (2014). Improved distributed principal component analysis. In *Advances in Neural Information Processing Systems*, 3113–3121

Liang, T. (2017). How well can generative adversarial networks (GAN) learn densities: A nonparametric view. *arXiv preprint arXiv:1712.08244.*

Lin, L., Sun, J. and Zhu, L. (2013). Nonparametric feature screening. *Comput. Statist. Data Anal.*, **67**, 162 – 174.

Lin, M., Chen, Q. and Yan, S. (2013). Network in network. *arXiv preprint arXiv:1312.4400.*

Lin, Y. (2002). Support vector machines and the Bayes rule in classification. *Comput. Statist. Data Anal.*, **6**, 259–275.

Lin, Y. (2004). A note on margin-based loss functions in classification. *Statist. Probab. Lett.*, 68, 73–82.

Lin, Y. and Jeon, Y. (2003). Discriminant analysis through a semiparametric model. *Biometrika*, **90**, 379–392.

Lin, Y. and Jeon, Y. (2006). Random forests and adaptive nearest neighbors. *Jour. Ameri. Statist. Assoc.*, **101**, 578–590.

Lin, Y. and Zhang, H. H. (2006). Component selection and smoothing in smoothing spline analysis of variance models. *Ann. Statist.*, **34**, 2272–2297.

Lindsay, B. G. (1988). Composite likelihood methods. In Statistical Inference from Stochastic Processes (Ithaca, NY, 1987). *Contemporary Mathematics* **80**, 221–239. Amer. Math. Soc., Providence, RI.

Liu, H., Han, F., Yuan, M., Lafferty, J., and Wasserman, L. (2012). High-dimensional semiparametric Gaussian copula graphical models. *Ann. Statist.*, **40**, 2293–2326.

Liu, H., Han, F., and Zhang, C. (2012). Transelliptical graphical models, in: *Advances in Neural Information Processing Systems*, **25**, 809–817.

Liu, H., and Wang, L. (2017). TIGER: a tuning-insensitive approach for optimally estimating Gaussian graphical models. *Electron. Jour. Stat.*, **11**, 241–294.

Liu, H., Wang, L., and Zhao, T. (2014). Sparse covariance matrix estimation with eigenvalue constraints. *Jour. Comput. Graph. Statist.*, **23**, 439–459.

Liu, H., Wang, X., Yao, T., Li, R. and Ye, Y. (2019). Sample average approximation with sparsity-inducing penalty for high-dimensional stochastic programming. *Mathematical Programming*, **78**, 69-108

Liu, H., Yao, T. and Li, R. (2016). Global solutions to folded concave penalized nonconvex learning. *Ann. Statist.*, **44**, 629 – 659.

Liu, H., Yao, T, Li, R. and Ye, Y. (2017). Folded concave penalized sparse linear regression: complexity, sparsity, statistical performance, and algorithm theory for local solutions. *Mathematical Programming Series A*, **166**, 207–240.

Liu, J., Li, R. and Wu, R. (2014). Feature selection for varying coefficient models with ultrahigh dimensional covariates. *Jour. Ameri. Statist. Assoc.*, **109**, 266–274.

Liu, W. and Shao, Q.-M. (2014). Phase transition and regularized bootstrap in large-scale $t$-tests with false discovery rate control. *Ann. Statist.*, **42**, 2003–2025.

Liu, Y., Shen, X., and Doss, H. (2005). Multicategory $\psi$-learning and support vector machine: Computational tools. *Jour. Comput. Graph. Statist.* , **14**, 219–36.

Lloyd, S. P. (1957). Least square quantization in PCM. Bell Telephone Laboratories Paper. Published in journal much later: Lloyd., S. P. (1982). "Least squares quantization in PCM". , *IEEE Trans. Inform. Theory*, **28**, 129–137.

Lockhart, R., Taylor, J., Tibshrani, R.J. and Tibshrani, R. (2014). A significance test for the lasso (with discussion). *Ann. Statist.*, **42**, 413–468.

Loh, P.-L. (2017). Statistical consistency and asymptotic normality for high-dimensional robust $M$-estimators. *Ann. Statist.*, **45**, 866–896.

Loh, P.-L. and Wainwright, M. (2014). Regularized M-estimators with non-convexity: Statistical and algorithmic theory for local optima. *Jour. Mach. Learn. Res.*, 1–56.

Loh, P.-L. and Wainwright, M. (2015). Regularized m-estimators with non-convexity: Statistical and algorithmic theory for local optima. *Jour. Mach. Learn. Res.*, **16**, 559–616.

Loh, P.-L. and Wainwright, M. (2017). Support recovery without incoherence: a case for nonconvex regularization. *Ann. Statist.*, **45**, 2455–2482.

Loh, W.-Y. (2009). Improving the precision of classification trees. *Ann. Appl. Statist.*, **3**, 1710–1737.

Lu, Z. (2009). Smooth optimization approach for sparse covariance selection. *SIAM Jour. Optim.*, **19**, 1807–1827.

Luce, R. D. (1959). *Individual choice behavior: A theoretical analysis.* John Wiley & Sons, Inc., New York.

Ludvigson, S.C. and Ng, S. (2009). Macro factors in bond risk premia. *Review of Financial Studies*, **22**, 5027–5067.

Luo, R., Wang, H., and Tsai, C. L. (2009). Contour projected dimension reduction. *Ann. Statist.*, **37**, 3743–3778.

Luo, X., Stefanski, L. A. and Boos, D. D. (2006). Tuning variable selection procedure by adding noise. *Technometrics*, **48**, 165–175.

Lv, J. and Fan, Y. (2009). A unified approach to model selection and sparse recovery using regularized least squares. *Ann. Statist.*, **37**, 3498–3528.

Ma, S., Li, R. and Tsai, C.-L. (2017). Variable screening via partial quantile correlation. *Jour. Ameri. Statist. Assoc.*, **112**, 650– 663.

Ma, Z. (2013). Sparse principal component analysis and iterative thresholding. *Ann. Statist.*, **41**, 772–801.

Maas, A. L., Hannun, A. Y. and Ng, A. Y. (2013). Rectifier nonlinearities improve neural network acoustic models. *Proc. ICML.*

Mai, Q., Yang, Y. and Zou, H. (2019). Multiclass sparse discriminant analysis. *Statistica Sinica*, **29**, 97–111.

Mai, Q. and Zou, H. (2013). The Kolmogorov filter for variable screening in high-dimensional binary classification. *Biometrika*, **100**, 229–234.

Mai, Q. and Zou, H. (2015). The fused Kolmogorov filter: A nonparametric model-free screening method. *Ann. Statist.*, **43**, 1471–1497.

Mai, Q. and Zou, H. (2015). Semiparametric sparse discriminant analysis in ultra-high dimensions. *Jour. Multivar. Anal.*, **135**, 175–188.

Mai, Q., Zou, H. and Yuan, M. (2012). A direct approach to sparse discriminant analysis in ultra-high dimensions. *Biometrika*, **99**, 29–42.

Mallot, S. G. and Zhang, Z. (1993). Matching pursuits with time-frequency dictionaries. *IEEE Transactions on Signal Processing*, **41**, 3397–3415.

Mallows, C.L. (1973). Some comments on $C_p$. *Technometrics*, **15**, 661–675.

Marron, S., Todd, M. and Ahn, J. (2007). Distance-weighted discrimination. *Jour. Ameri. Statist. Assoc.*, **102**, 1267–1271.

Mason, L., Baxter, J., Bartlett, P. and Frean, M. (1999). Boosting algorithms as gradient descent. *Proceedings of the 12th International Conference on Neural Information Processing Systems*, 512-518

Masse, K. (1997). Statistical models applied to the rating of sports teams. Technical report, Bluefield College.

McCallum, A. and Nigam, K. (1998). A comparison of event models for naive Bayes text classification. *AAAI/ICML-98 Workshop on Learning for Text Categorization*, 41–48.

McCracken, M.W. and Ng, S. (2015). FRED-MD: A monthly database for macroeconomic research. Working paper.

McCullagh, P. and Nelder, J. A. (1989). *Generalized Linear Models.* 2nd Edition. Chapman & Hall, London.

McLachlan, G. and Peel, D. (2000). *Finite Mixture Models.* John Wiley and Sons, New York.

Mei, S., Misiakiewicz, T. and Montanari, A. (2019). Mean-field theory of two-layers neural networks: dimension-free bounds and kernel limit. *arXiv preprint arXiv:1902.06015.*

Mei, S., Montanari, A. and Nguyen, P. (2018). A mean field view of the landscape of two-layer neural networks. *Proc. Natl. Acad. Sci.*, **115**, E7665–E7671.

Meier, L., van de Geer, S. , and Bühlmann, P. (2008). The group lasso for logistic regression. *Jour. Roy. Statist. Soc. B*, **70**, 53–71.

Meier, I., Geer, V. and Bühlmann, P. (2009). High-dimensional additive modeling. *Jour. Roy. Statist. Soc. B*, **71**, 1009–1030.

Meinshausen, N. and Bühlmann, P. (2010). Stability selection (with discussion). *Jour. Roy. Statist. Soc., B* **72**, 417-473.

Meinshausen, N. and Bühlmann, P. (2006). High dimensional graphs and variable selection with the Lasso. *Ann. Statist.*, **34**, 1436–1462.

Meinshausen, N. and Rice, J. (2006). Estimating the proportion of false null hypotheses among a large number of independently tested hypotheses. *Ann. of Statist.*, **34**, 373–393.

Meinshausen, N. and Yu, B. (2009). Lasso-type recovery of sparse representations for high-dimensional data. *Ann. Statist.*, **37**, 246–270.

Michie, D., Spiegelhalter, D. and Taylor, C. (1994). *Machine Learning, Neural and Statistical Classification*, first edn. Ellis Horwood.

Minsker, S. (2016). Sub-gaussian estimators of the mean of a random matrix with heavy-tailed entries. *arXiv preprint arXiv:1605.07129.*

Mitchell, T. J. and Beauchamp, J. J. (1988) Bayesian variable selection in linear regression. *Jour. Ameri. Statist. Assoc.*, **83**, 1023–1032.

Misra, J., Schmitt, W., Hwang, D., Hsiao, L., Gullans, S., and Stephanopoulos, G. (2002). Interative exploration of microarray gene expression patterns in a reduced dimensional space. *Genome Research*, **12**, 1112–1120.

Mitra, R. and Zhang, C.-H. (2014). Multivariate analysis of nonparametric estimates of large correlation matrices. *arXiv preprint arXiv:1403.6195.*

Mitra, R. and Zhang, C.-H. (2016). The benefit of group sparsity in group inference with de-biased scaled group lasso. *Electron. Jour. Stat.*, **10**, 1829–1873.

Molenaar, P. C. (1985). A dynamic factor model for the analysis of multivariate time series. *Psychometrika*, 50, 181-202.

Morgan, J. and Sonquist, J. (1963). Problems in the analysis of survey data, and a proposal. *Jour. Ameri. Statist. Assoc.*, 58, 415–434.

Muirhead, R. J. (1982). *Aspects of Multivariate Statistical Theory.* Wiley, New York.

Mullahy, J. (1986). Specification and testing of some modified count data models. *Jour. Econ.*, **33**, 341–365.

Nagalakshmi, U., Wang, Z., Waern, K., Shou, C., Raha, D., Gerstein, M. and Snyder, M. (2008). The transcriptional landscape of the yeast genome defined by RNA sequencing. *Science*, **320**, 1344–1349.

Narisetty, N.N., and He, X. (2014). Bayesian variable selection with shrinking and diffusing priors. *Ann. Statist.*, **42**, 789–817.

Negahban, S.N., Oh, S. and Shah, D. (2017). Rank centrality: ranking from pairwise comparisons. *Oper. Res.*, **65**, 266–287.

Negahban, S.N., Ravikumar, P., Wainwright, M. J. and Yu, B. (2012). A unified framework for high-dimensional analysis of $M$-estimators with decomposable regularizers. *Statist. Sci.*, **27**, 538–557.

Negahban, S.N. and Wainwright, M. J. (2011). Estimation of (near) low-rank matrices with noise and high-dimensional scaling. **39**, *Ann. Statist.*, 1069–1097.

Negahban, S. and Wainwright, M. J. (2012). Restricted strong convexity and weighted matrix completion: Optimal bounds with noise. *Jour. Mach. Learn. Res*, **13**, 1665–1697.

Nelder, J.A. (1966). Inverse polynomials, a useful group of multi-factor response functions. *Biometrics*, **22**, 128–141.

Nesterov, Y. (1983). A method of solving a convex programming problem with convergence rate $O(1/k^2)$. In *Soviet Mathematics Doklady*, **27**, 372–376.

Nesterov, Y. (2007). Gradient methods for minimizing composite objective function. CORE Discussion Papers 2007076, Université Catholique de Louvain, Center for Operations Research and Econometrics (CORE).

Neyshabur, B., Tomioka, R. and Srebro, N. (2015). Norm-based capacity control in neural networks. In *Conference on Learning Theory*, 1376–1401.

Ng, A., Jordan, M., and Weiss, Y. (2002). On spectral clustering: analysis and an algorithm. *Advances in Neural Information Processing Systems*, **14**, 849–856.

Ning, Y. and Liu, H. (2017). A general theory of hypothesis tests and confidence regions for sparse high dimensional models. *Ann. Statist.*, **45**, 158–195.

Ninomiya, Y. and Kawano, S. (2016). AIC for the Lasso in generalized linear models. *Electron. Jour. Stat.* , **10**, 2537– 2560.

Nowozin, S., Cseke, B. and Tomioka, R. (2016). f-gan: Training generative neural samplers using variational divergence minimization. *Advances in Neural Information Processing Systems*, 271–279.

Oberthuer, A., Berthold, F., Warnat, P., Hero, B., Kahlert, Y., Spitz, R., Ernestus, K., König, R., Haas, S., Eils, R., Schwab, M., Brors, B., Westermann, F., Fischer, M. (2006). Customized oligonucleotide microarray gene expression based classification of neuroblastoma patients outperforms current clinical risk stratification, *Journal of Clinical Oncology*, **24**, 5070–5078.

Ockenhouse, C. F., Hu, W. C., Kester, K. E., Cummings, J. F., Stewart, A., Heppner, D. G., Jedlicka, A. E., Scott, A. L., Wolfe, N. D., Vahey, M. and Burke, D. S. (2006). Common and divergent immune response signaling pathways discovered in peripheral blood mononuclear cell gene expression patterns in presymptomatic and clinically apparent malaria. *Infection and Immunity*, **74**, 5561–5573.

Olhede, S. C. and Wolfe, P. J. (2014). Network histograms and universality of block model approximation. *Proc. Natl. Acad. Sci.*, **111**, 14722–14727.

Onatski, A. (2010). Determining the number of factors from empirical distribution of eigenvalues. *Review of Economics and Statistics*. **92**, 1004–1016.

Onatski, A. (2012). Asymptotics of the principal components estimator of large factor models with weakly influential factors. *Jour. Econ.*, **168**, 244–258.

Osborne, M., Presnell, B., and Turlach, B. (2000a). A new approach to variable selection in least squares problems. *IMA Journal of Numerical Analysis*, **20**, 389–404.

Osborne, M.R., Presnell, B. and Turlach, B. (2000b). On the LASSO and its dual. *Jour. Comput. Graph. Statist.*, **9**, 319–337.

Owen, A.B. (2005). Variance of the number of false discoveries. *Jour. Roy. Statist. Soc. B*, **67**, 411–426.

Pan, R., Wang, H. and Li, R. (2016). Ultrahigh dimensional multiclass linear discriminant analysis by pairwise sure independence screening. *Jour. Ameri. Statist. Assoc.*, **111**, 169–179.

Pan, W. and Shen, X. (2007). Penalized model-based clustering with applications to variable selection. *Jour. Mach. Learn. Res*, **8**, 1145–1164.

Park, B., Mammen, E., Härdle, W. and Borzk, S. (2009). Time series modelling with semiparametric factor dynamics. *Jour. Ameri. Statist. Assoc.*, **104**, 284–298.

Park, M. Y. and Hastie, T. (2007). $L_1$-regularization path algorithm for generalized linear models. *Jour. Roy. Statist. Soc. B.*, **69**, 659–677.

Park, S.H. (1981). Collinearity and optimal restrictions on regression parameters for estimating responses. *Technometrics*, **23**, 289–295.

Park, T. and Casella. G. (2008). The Bayesian lasso. *Jour. Ameri. Statist. Assoc.*, **103**, 681–686.

Parikh, N. and Boyd, S. (2013). Proximal algorithms. *Foundations and Trends in Optimization*, **1**, 123–231.

Parzen, E. (1962). On estimation of a probability density function and mode. *Ann. Math. Statist.*, **33**, 1065–1076.

Paul, D. (2007). Asymptotics of sample eigen structure for a large dimensional spiked covariance model. *Statisti. Sinica*, **17**, 1617–1642.

Paul, D., Bair, E., Hastie, T., Tibshirani, R. (2008). "Pre-conditioning" for feature selection and regression in high-dimensional problems. *Ann. Statist.*, **36**, 1595–1618.

Paszke, A., Gross, S., Chintala, S., Chanan, G., Yang, E., DeVito, Z., Lin, Z., Desmaison, A., Antiga, L. and Lerer, A. (2017). Automatic differentiation in PyTorch. In *NIPS 2017 Autodiff Workshop: The Future of Gradient-based Machine Learning Software and Techniques*, Long Beach, CA, USA.

Pearson, K. (1901). On lines and planes of closest fit to systems of points in space. *Philosophical Magazine, Series 6*, **2**(11), 559–572.

Peng, B. and Wang, L. (2015). An iterative coordinate descent algorithm for high-dimensional nonconvex penalized quantile regression. *Jour. Comput. Graph. Statist.*, **79**, 871– 880.

Polyak, B. T. (1964). Some methods of speeding up the convergence of iteration methods. *USSR Computational Mathematics and Mathematical Physics*, **4**, 1–17.

Poultney, C., Chopra, S., LeCun, Y. et al. (2007). Efficient learning of sparse representations with an energy-based model. *Advances in Neural Information Processing Systems*, 1137–1144.

Pourahmadi, M. (2013). *High-Dimensional Covariance Estimation*, John-Wiley and Sons.

Price, A. L., Patterson, N. J., Plenge, R. M., Weinblatt, M. E., Shadick, N. A. and Reich, D. (2006). Principal components analysis corrects for stratification in genome-wide association studies. *Nature Genetics*, **38**, 904–909.

Qi, H. and Sun, D. (2006). A quadratically convergent Newton method for computing the nearest correlation matrix. *SIAM Jour. Matrix Anal. Appl.*, **28**, 360–385.

Qin, T. and Rohe, K. (2013). Regularized spectral clustering under the degree-corrected stochastic blockmodel. In *Advances in Neural Information Processing Systems*, 3120-3128.

Quinlan, J. (1993). *C4.5: Programs for Machine Learning.* Morgan Kaufmann, San Mateo, CA.

Raftery, A. E. (1995). Bayesian model selection in social research. *Sociological Methodology*, **25**, 111–164.

Raskutti, G., Wainwright, M. J. and Yu, B. (2011). Minimax rates of estimation for high–dimensional linear regression over $\ell_q$–balls. *IEEE Trans. Inform. Theory*, **57**, 6976–6994.

Ravikumar, P., Liu, H., Lafferty, J. and Wasserman, L. (2008). SpAM: Sparse Additive Models. *Advances in Neural Information Processing Systems*, **20**, 1201–1208

Ravikumar, P., Lafferty, J., Liu, H. and Wasserman, L. (2009). Sparse additive models, *Jour. Roy. Statist. Soc. B*, **71**, 1009-1030.

Ravikumar, P., Wainwright, M. J. and Lafferty, J. (2010). High-dimensional Ising model selection using $\ell_1$-regularized logistic regression. *Ann. Statist.*, **38**, 1287–1319.

Ravikumar, P., Wainwright, M. J., Raskutti, G. and Yu, B. (2011). High-dimensional covariance estimation by minimizing $\ell_1$-penalized log-determinant divergence. *Electron. J. Stat.*, **5**, 935–980.

Recht, B. (2011). A simpler approach to matrix completion. *Jour. Mach. Learn. Res.*, **12**, 3413-3430.

Reddi, S. J., Kale, S. and Kumar, S. (2019). On the convergence of Adam and beyond. *arXiv: 1904.09237.*

Ren, Z., Sun, T., Zhang, C.-H. and Zhou, H. H. (2015). Asymptotic normality and optimalities in estimation of large Gaussian graphical model. *Ann. Statist.*, **43**, 991-1026.

Rigollet, P., and Tsybakov, A. (2012). Estimation of covariance matrices under sparsity constraints. *arXiv preprint arXiv:1205.1210.*

Robbins, H. and Monro, S. (1951). A stochastic approximation method. *Ann. Math. Statist.*, **22**, 400–407.

Robinson, P.M. (1988). The stochastic difference between econometrics and statistics, *Econometrica*, **56**, 531–547.

Rohe, K., Chatterjee, S., and Yu, B. (2011). Spectral clustering and the high-dimensional stochastic block model. *Ann. Statist.*, **39**, 1878–1915.

Romano, Y., Sesia, M. and Candès, E. J. (2018). Deep knockoffs. *arXiv preprint arXiv:1811.06687.*

Rosset, S. and Zhu, J. (2007). Piecewise linear regularized solution paths. *Ann. Statist.*, **35**, 1012–1030.

Rosset, S., Zhu, J. and Hastie, T. (2004). Margin maximizing loss functions. *Advances in Neural Information Processing Systems.*

Rothman, A.J., Bickel, P.J., Levina, E. and Zhu, J. (2008). Sparse permutation invariant covariance estimation. *Electron. Jour. Stat.*, **2**, 494–515.

Rothman, A.J., Levina, E. and Zhu, J. (2009). Generalized thresholding of large covariance matrices. *Jour. Ameri. Statist. Assoc.*, **104**, 177–186.

Rotskoff, G. M. and Vanden-Eijnden, E. (2018). Neural networks as interacting particle systems: Asymptotic convexity of the loss landscape and universal scaling of the approximation error. *arXiv preprint arXiv:1805.00915.*

Rousseeuw, P.J. (1984). Least median of squares regression. *Jour. Ameri. Statist. Assoc.*, **24**, 676–694.

Rousseeuw, P.J. and Leroy, A. M. (1987). *Robust Regression and Outlier Detection.* Wiley.

Rousseeuw, P. J. and van Zomeren, B. C. (1990). Unmasking multivariate outliers and leverage points. *Jour. Ameri. Statist. Assoc.*, **85**, 633–639.

Rudelson, M. and Zhou, S. (2013). Reconstruction from anisotropic random measurements. *IEEE Trans. Inform. Theory*, **59**, 3434–3447.

Rumelhart, D. E., Hinton, G. E. and Williams, R. J. (1985). Learning internal representations by error propagation. Technical Report, California Univ San Diego La Jolla Inst for Cognitive Science.

Russakovsky, O., Deng, J., Su, H., Krause, J., Satheesh, S., Ma, S., Huang, Z., Karpathy, A., Khosla, A., Bernstein, M., Berg, A. C. and Fei-Fei, L. (2015). ImageNet large scale visual recognition challenge. *International Journal of Computer Vision (IJCV)*, **115**, 211–252.

Salimans, T., Goodfellow, I., Zaremba, W., Cheung, V., Radford, A. and Chen, X. (2016). Improved techniques for training GANs. *Advances in Neural Information Processing Systems*, 2234–2242.

Sarkar, S.K. (2002). Some results on false discovery rate in stepwise multiple testing procedures. *Ann. Statist.*, **30**, 239–257.

Sarkar, S.K. (2006). False discovery and false non discovery rates in single-step multiple testing procedures. *Ann. Statist.*, **34**, 394–415.

Sak, H., Senior, A. and Beaufays, F. (2014). Long short-term memory recurrent neural network architectures for large scale acoustic modeling. *In Fifteenth Annual Conference of the International Speech Communication Association.*

Schapire, R., Freund, Y., Bartlett, P. and Lee, W.S. (1998). Boosting the margin: a new explanation for the effectiveness of voting methods. *Ann. Statist.*, **26**, 1651–1686.

Scheinberg, K., Ma, S. and Goldfarb, D. (2010). Sparse inverse covariance selection via alternating linearization methods. In *Advances in Neural Information Processing Systems* 23, Lafferty, J., Williams, C.K.I., Shawe-Taylor, J., Zemel, R. and Culotta, A. eds. pp. 2101–2109.

Schick, A. (1986). On asymptotically efficient estimation in semiparametric models. *Ann. Statist.*, **14**, 1139–1151.

Schizas, I. D. and Aduroja, A. (2015). A distributed framework for dimensionality reduction and denoising. *IEEE Trans. Signal Process.*, **63**, 6379–6394.

Schmidt-Hieber, J. (2020). Nonparametric regression using deep neural networks with ReLU activation function. *arXiv preprint arXiv:1708.06633.*

Schölkopf, B., Smola, A. and Müller, K.R. (1999). Kernel principal component analysis, in B. Schölkopf, C. Burges and A. Smola (eds), *Advances in Kernel Methods: Support Vector Learning*, MIT Press, Cambridge, MA, USA, pp. 327–352.

Schwartzman, A. and Lin, X. (2011). The effect of correlation in false discovery rate estimation. *Biometrika*, **98**, 199–214.

Schwarz, G. (1978). Estimating the dimension of a model. *Ann. Statist.*, **6**, 461–464.

Sedghi, H., Janzamin, M. and Anandkumar, A. (2016). Provable tensor methods for learning mixtures of generalized linear models. *Artificial Intelligence and Statistics*, 1223–1231.

Shamir, O., Srebro, N. and Zhang, T. (2014). Communication-efficient distributed optimization using an approximate Newton-type method. *Proceedings of the 31st International Conference on Machine Learning*, 1000–1008.

Shao, J. (1997). An asymptotic theory for linear model selection. *Statistica Sinica*, **7**, 221–264.

Sheather, S. and Jones, M. (1991). A reliable data-based bandwidth selection method for kernel density estimation. *Jour. Roy. Statist. Soc. B*, **53**, 683–690.

Shen, D., Shen, H., Zhu, H. and Marron, J. S. (2016). The statistics and mathematics of high dimension low sample size asymptotics. *Statist. Sinica*, **26**, 1747–1770.

Shen, H. and Huang, J.Z. (2008). Sparse principal component analysis via regularized low rank matrix approximation. *Jour. Multivar. Anal.*, **99**, 1015–1034.

Shen, X., Pan, W. and Zhu, Y. (2012). Likelihood-based selection and sharp parameter estimation. *Jour. Ameri. Statist. Assoc.*, **107**, 223–232.

Shen, X., Tseng, G., Zhang, X. and Wong, W. (2003). On psi-learning. *Jour. Ameri. Statist. Assoc.*, **98**, 724–734.

Shendure, J. and Ji, H. (2008). Next-generation DNA sequencing. *Nature Biotechnology*, **26**, 1135–1145.

Shi, C., Song, R., Chen, Z. and Li, R. (2019). Linear hypothesis testing for high dimensional generalized linear models. *Ann. Statist.*, **47**, 2671– 2703.

Shi, C., Song, R., Lu, W. and Li, R. (2020). Statistical inference for high-dimensional models via recursive online-score estimation. *Jour. Ameri. Statist. Assoc.*. In press.

Shi, J. and Malik, J. (2000). Normalized cuts and image segmentation. *IEEE Transactions on Pattern Analysis and Machine Intelligence*, **22**, 888–905.

Shibata, R. (1981). An optimal selection of regression variables. *Biometrika*, **68**, 45–54.

Shibata, R. (1984). Approximate efficiency of a selection procedure for the number of regression variables. *Biometrika*, **71**, 43–49.

Sims, C.A. (1980). Macroeconomics and reality. *Econometrica*, **48**, 1–48.

Simonyan, K. and Zisserman, A. (2014). Very deep convolutional networks for large-scale image recognition. *arXiv preprint arXiv:1409.1556.*

Silver, D., Schrittwieser, J., Simonyan, K., Antonoglou, I., Huang, A., Guez, A., Hubert, T., Baker, L., Lai, M., Bolton, A., Chen, Y., Lillicrap, T., Hui, F., Sifre, L., van den Driessche, G., Graepel, T. and Hassabis, D. (2017). Mastering the game of Go without human knowledge. *Nature*, **550**, 354–359.

Silverman, B. (1998). *Density Estimation for Statistics and Data Analysis.* Chapman and Hall/CRC, London.

Singh, D., Febbo, P., Ross, K., Jackson, D., Manola, J., Ladd, C., Tamayo, P., Renshaw, A. A., D'Amico, A. V., Richie, J., Lander, E., Loda, M., Kantoff, P., Golub, T. and Sellers, W. (2002). Gene expression correlates of clinical prostate cancer behavior. *Cancer Cell*, **1**, 203–209.

Sirignano, J. and Spiliopoulos, K. (2018). Mean field analysis of neural networks. *arXiv preprint arXiv:1805.01053.*

Sjöstrand, K., Rostrup, E., Ryberg, C., Larsen, R., Studholme, C., Baezner, H., Ferro, J., Fazekas, F., Pantoni, L., Inzitari, D. and Waldemar, G. (2007). Sparse decomposition and modeling of anatomical shape variation. *IEEE Trans. Medical Imaging*, **26**, 1625–1635.

Song, R., Yi, F. and Zou, H. (2014). On varying-coefficient independence screening for high-dimensional varying-coefficient models, *Statist. Sinica*, **24**, 1735–1752

Speckman, P. (1988). Kernel smoothing in partial linear models. *Jour. Roy. Statist. Soc. B*, **50**, 413–436.

Speed, T.P. (2003). *Statistical Analysis of Gene Expression Microarray Data.* Chapman & Hall/CRC, New York.

Städler, N., Bühlmann, P., van de Geer, S. (2010). $\ell_1$-penalization for mixture regression models (with discussion). *Test*, **19**, 209–256.

Stein, C. (1956).Efficient nonparametric testing and estimation. In: *Proceedings of the Third Berkeley Symposium on Mathematical Statistics and Probability*, volume 1, pages 187–196. University of California Press.

Stein, C. (1981). Estimation of the mean of a multivariate normal distribution. *Ann*. Statist., **9**, 1135–1151.

Stock, J. H. and Watson, M. W. (2002a). Forecasting using principal components from a large number of predictors. *Jour. Ameri. Statist. Assoc.*, **97**, 1167–1179.

Stock, J. H. and Watson, M. (2002b). Macroeconomic forecasting using diffusion indexes. *Jour. Busi. Econ. Statist.*, **20**, 147–162.

Stock J.H., Watson, M.W. (2006). Forecasting with many predictors. In *Handbook of Economic Forecasting* (G. Elliott, C. Granger and A. Timmermann, eds.), Chapter 10, 515–554. Elsevier.

Stone, C.J. (1985). Additive regression and other nonparametric models. *Ann. Statist.*, **13**, 689–705.

Stone, C.J. (1994). The use of polynomial splines and their tensor products in multivariate function estimation (with discussion). *Ann. Statist.*, **22**, 118–184.

Stone, C.J., Hansen, M., Kooperberg, C. and Truong, Y.K. (1997). Polynomial splines and their tensor products in extended linear modeling (with discussion). *Ann. Statist.*, **25**, 1371–1470.

Stone, M. (1974). Cross-validatory choice and assessment of statistical predictions (with discussion). *Jour. Roy. Stat. Soc. B*, **36**, 111–147.

Storey, J.D. (2002) A direct approach to false discovery rates. *Jour. Roy. Stat. Soc. B*, **64**, 479–498.

Storey, J.D. (2003). The positive false discovery rate: A Bayesian interpretation and the q-value. *Ann. Statist.*, **23**, 2013–2035.

Storey J.D., Taylor J.E., and Siegmund, D. (2004). Strong control, conservative point estimation, and simultaneous conservative consistency of false discovery rates: A unified approach. *Jour. Roy. Stat. Soc. B*, **66**, 187-205.

Storey, J.D., and Tibshirani R. (2003). Statistical significance for genomewide studies. *Proc. Natl. Acad. Sci.*, **100**, 9440–9445.

Su, W. and Candés, E. J. (2016). Slope is adaptive to unknown sparsity and asymptotically minimax. *Ann. Statist.*, **44**, 1038–1068.

Sugar, C. and James, G. (2003). Finding the number of clusters in a data set: an information theoretic approach. *Jour. Ameri. Statist. Assoc.*, **98**, 750–763.

Sun, Q., Zhou, W.X., and Fan, J. (2020). Adaptive Huber regression. *Jour., Amer. Statist. Assoc.*, In press.

Sun, T. and Zhang, C.-H. (2010). Comments on: $\ell_1$-penalization for mixture regression models. *Test*, **19**, 270–275.

Sun, T. and Zhang, C.-H. (2012a). Comments on: Optimal rates of convergence for sparse covariance matrix estimation. *Statist. Sinica*, **22**, 1354–1358.

Sun, T. and Zhang, C.-H. (2012b). Scaled sparse linear regression. *Biometrika*, **99**, 879–898.

Sun, T. and Zhang, C.-H. (2013). Sparse matrix inversion with scaled Lasso. *Jour. Mach. Learn. Res.*, **14**, 3385–3418.

Sun, W. and Cai, T. (2009). Large-scale multiple testing under dependency. *Jour. Roy. Statist. Soc. B*, **71**, 393-424.

Sutskever, I., Martens, J., Dahl, G. and Hinton, G. (2013). On the importance of initialization and momentum in deep learning. *International Conference on Machine Learning*, 1139–1147.

Szegedy, C., Liu, W., Jia, Y., Sermanet, P., Reed, S., Anguelov, D., Erhan, D., Vanhoucke, V. and Rabinovich, A. (2015). Going deeper with convolutions. *Proceedings of the IEEE Conference on Computer Vision and Pattern Recognition*, 1–9.

Székely, G. J., Rizzo, M. L. and Bakirov, N. K. (2007). Measuring and testing dependence by correlation of distances. *Ann. Statist.*, **35**, 2769–2794.

Taylor, J. and Tibshirani, R.J. (2015). Statistical learning and selective inference. *Proc. Natl. Acad. Sci.*, **112**, 7629–7634.

Tibshirani, R. (1996). Regression shrinkage and selection via lasso. *Jour. Roy. Statist. Soc. B.*, **58**, 267–288.

Tibshirani, R. (1997). The lasso method for variable selection in the Cox model. *Statistics in Medicine*, **16**, 385-395.

Tibshirani, R., Hastie, T., Narasimhan, B. and Chu, G. (2002). Diagnosis of multiple cancer types by shrunken centroids of gene expression. *Proc. Natl. Acad. Sci.*, **99**, 6567–6572.

Tibshirani, R., Walther, G. and Hastie, T. (2001). Estimating the number of clusters in a data set via the gap statistic. *Jour. Roy. Statist. Soc. B*, **63**, 411–423.

Tibshirani, R.J. (2013). The lasso problem and uniqueness. *Electronic Journal of Statistics*, **7**, 1456-1490.

Tibshirani, R.J., Taylor, J., Lockhart, R. and Tibshirani, R. (2016). Exact post-selection inference for sequential regression procedures. *Jour. Ameri. Statist. Assoc.*, **111**, 600–620.

Tikhonov, A. N. (1943). On the stability of inverse problems. *Doklady Akademii Nauk SSSR*, **39**, 195–198.

Toh, K.C., Todd, M.J. and Tutuncu, R.H. (1999). SDPT3 — a Matlab software package for semidefinite programming, *Optimization Methods and Software*, **11**, 545–581.

Tropp, J. A. (2006). Just relax: convex programming methods for identifying sparse signals in noise. *IEEE Trans. on Information Theory*, **52**, 1030–1051.

Tropp, J.A. (2015). An introduction to matrix concentration inequalities. *Foundations and Trends in Machine Learning*, **8**, 1–230.

Tseng, P. (2001). Convergence of a block coordinate descent method for nondifferentiable minimization. *Jour. Optim. Theory Appl.*, **109**, 475–494.

Tseng, P. and Yun, S. (2009). A coordinate gradient descent method for nonsmooth separable minimization. *Math. Program.*, **117**, 387–423.

Turlach, B.A., Venables, W.N. and Wright, S.J. (2005). Simultaneous variable selection. *Technometrics*, **47**, 349–363.

van de Geer, S. (2007). The Deterministic Lasso. Technical Report 140, ETH Zurich, Switzerland.

van de Geer, S. (2008). High-dimensional generalized linear models and the Lasso. *Ann. Statist.*, **36**, 614–645.

van de Geer, S. and Bühlmann, P. (2009). On the conditions used to prove oracle results for the Lasso. *Electron. Jour. Stat.*, **3**, 1360–1392.

van de Geer, S., Bühlmann, P., Ritov, Y., and Dezeure, R. (2014). On asymptotically optimal confidence regions and tests for high-dimensional models. *Ann. Statist.*, **42**,1166–1202.

van de Geer, S. and Müller, P. (2012). Quasi-likelihood and/or robust estimation in high dimensions. *Statist. Sci.*, **27**, 469–480.

van de Geer, S., Bühlmann, P., Ritov, Y. A., and Dezeure, R. (2014). On asymptotically optimal confidence regions and tests for high-dimensional models. *Ann. Statist.*, **42**, 1166–1202.

Van der Vaart, A. W. (2000). *Asymptotic Statistics*, volume 3. Cambridge University Press.

van der Vaart, A.W. and Wellner, J.A. (1996). *Weak Convergence and Empirical Processes.* Springer, New York.

Vapnik, V. (1996). *The Nature of Statistical Learning*, Springer Verlag, New York.

Vershynin, R. (2010). Introduction to the non-asymptotic analysis of random matrices. *arXiv preprint arXiv:1011.3027.*

Vial, J.-P. (1982). Strong convexity of sets and functions. *Journal of Mathematical Economics*, **9**, 187–205.

Vincent, P., Larochelle, H., Bengio, Y. and Manzagol, P.-A. (2008). Extracting and composing robust features with denoising autoencoders. *Proceedings of the 25th International Conference on Machine Learning*, 1096–1103, ACM.

von Luxburg, U., Belkin, M., and Bousquet, O. (2008). Consistency of spectral clustering. *Ann. Statist.*, **36**, 555–586.

Vu, V. Q. and Lei, J. (2013). Minimax sparse principal subspace estimation in high dimensions. *Ann. Statist.*, **41**, 2905–2947.

Wager, S., Wang, S. and Liang, P. S. (2013). Dropout training as adaptive regularization. *Advances in Neural Information Processing Systems*, 351–359.

Wahba, G. (1990). *Spline Models for Observational Data*, Series in Applied Mathematics, Vol.59, SIAM, Philadelphia.

Wainwright, M.J. (2009). Sharp thresholds for noisy and high-dimensional recovery of sparsity using $\ell_1$-constrained quadratic programming (Lasso). *IEEE Transactions on Information Theory*, **55**, 2183–2202.

Wainwright, M. J. (2019). *High-dimensional Statistics: A Non-asymptotic Viewpoint.* Cambridge University Press, Cambridge, UK.

Wang, B. and Zou, H. (2016). Sparse distance weighted discriminant. *Jour. Comput. Graph. Statist.* , **25**, 826–838.

Wang, B. and Zou, H. (2018). Another look at distance-weighted discrimination. *Jour. Roy. Statist. Soc. B*, **80**, 177–198.

Wang, H. (2009). Forward regression for ultra-high dimensional variable screening, *Jour. Ameri. Statist. Assoc.*, **104**, 1512–1524.

Wang, H., Li, B. and Leng, C. (2009). Shrinkage tuning parameter selection with a diverging number of parameters. *Jour. Roy. Statist. Soc. B*, **71**, 671–683.

Wang, H., Li, G. and Jiang, G. (2007). Robust regression shrinkage and consistent variable selection through the LAD-Lasso. *Jour. Bus. and Econ. Statist.*, **25**, 347–355.

Wang, H., Li, R. and Tsai, C.-L. (2007). Tuning parameter selectors for the smoothly clipped absolute deviation method. *Biometrika*, **94**, 553–568.

Wang, J., Zhao, Q., Hastie, T. and Owen, A.B. (2017). Confounder adjustment in multiple hypothesis testing. *Ann. Statist.*, **45**, 1863–1894.

Wang, L. (2013). The $L_1$ penalized LAD estimator for high dimensional linear regression. *Jour. Multivar. Anal.*, **120**, 135–151.

Wang, L., Kim, Y. and Li, R. (2013). Calibrating nonconvex penalized regression in ultrahigh dimension. *Ann. Statist.*, **41**, 2505–2536.

Wang, L. and Li, R. (2009). Weighted Wilcoxon-type smoothly clipped absolute deviation method. *Biometrics*, **65**, 564–571.

Wang, L., Liu, J., Li, Y. and Li, R. (2017). Model-free conditional independence feature screening for ultrahigh dimensional data. *Sci. China Math.*, **60**, 551– 568.

Wang, L. and Shen, X. (2006). Multicategory support vector machines, feature selection and solution path. *Statist. Sinica*, **16**, 617–634.

Wang, L., Wu, Y. and Li, R. (2012). Quantile regression for analyzing heterogeneity in ultrahigh dimension. *Jour. Ameri. Statist. Assoc.*, **107**, 214–222.

Wang, L., Wu, Y. and Li, R. (2012). Quantile regression for analyzing heterogeneity in ultrahigh dimension. *Jour. Ameri. Statist. Assoc.*, **107**, 214 – 222.

Wang, L., Zhu, J., and Zou, H. (2008). Hybrid Huberized support vector machines for microarray classification and gene selection. *Bioinformatics*, **24**, 412–419.

Wang, S. and Zhu, J. (2008). Variable selection for model-based high-dimensional clustering and its application to microarray data. *Biometrics*, **64**(2), 440–448.

Wang, W. and Fan, J. (2017). Asymptotics of empirical eigen-structure for high dimensional spiked covariance. *Ann. Statist.*, **45**, 1342–1374.

Wang, Z., Liu, H. and Zhang, T. (2014). Optimal computational and statistical rates of convergence for sparse nonconvex learning problems. *Ann. Statist.*, **42**, 2164–2201.

Wasserman, L. and Roeder, K. (2009). High dimensional variable selection, *Ann. Statist.*, **37**, 2178 – 2201.

Wedderburn, R. W. M. (1974). Quasilikelihood functions, generalized linear models and the Gauss-Newton method. *Biometrika*, **61**, 439–447.

Wegkamp, M. and Zhao, Y. (2016). Adaptive estimation of the copula correlation matrix for semiparametric elliptical copulas. *Bernoulli*, **22**, 1184–1226.

Wei, Y and He, X. (2006). Conditional growth charts (with discussions). *Ann. Statist.*, **34**, 2069–2097.

Weisberg, S. (2005). *Applied Linear Regression*, 3rd Edition, John Wiley, Hoboken, New Jersey.

Whittaker, J. (2009). *Graphical Models in Applied Multivariate Statistics.* Wiley Publishing.

Wille, A., Zimmermann, P., Vranova, E., *et al.* (2004). Sparse graphical gaussian modeling of the isoprenoid gene network in arabidopsis thaliana. *Genome Biology*, **5** R92.

Witten, D. M. and Tibshirani, R. (2010). A framework for feature selection in clustering. *Jour. Ameri. Statist. Assoc.*, **105**, 713–726.

Witten, D. M. and Tibshirani, R. (2011). Penalized classification using Fisher's linear discriminant. *Jour. Roy. Statist. Soc. B*, **73**, 753–772.

Witten, D. M., Tibshirani, R., and Hastie, T. (2009). A penalized matrix decomposition, with applications to sparse principal components and canonical correlation analysis. *Biostatistics*, **10**(3), 515–534.

Wright, J., Ganesh, A., Rao, S., Peng, Y., and Ma, Y. (2009). Robust principal component analysis: Exact recovery of corrupted low-rank matrices via convex optimization. *Advances in Neural Information Processing Systems*, 2080–2088.

Wu, M., Zhang, L., Wang, Z., Christiani, D. and Lin, X. (2008). Sparse linear discriminant analysis for simultaneous testing for the significance of a gene set pathway and gene selection. *Bioinformatics*, **25**, 1145–1151.

Wu, W. B. and Pourahmadi, M. (2003). Nonparametric estimation of large covariance matrices of longitudinal data. *Biometrika*, **90**, 831–844.

Wu, Y. and Liu, Y. (2007). Robust truncated hinge loss support vector machines. *Jour. Ameri. Statist. Assoc.*, **102**, 974–983.

Wu, Y. and Liu, Y. (2009). Variable selection in quantile regression. *Statist. Sinica*, **19**, 801–817.

Wu, Y., Schuster, M., Chen, Z., Le, Q. V., Norouzi, M., Macherey, W., Krikun, M., Cao, Y., Gao, Q., Macherey, K. et al. (2016). Google's neural machine translation system: Bridging the gap between human and machine translation. *arXiv preprint arXiv:1609.08144*.

Xie, B., Pan, W. and Shen, X. (2008). Penalized model-based clustering with cluster-specific diagonal covariance matrices and grouped variables. *Electron. Jour. Stat.* , **2**,168–212.

Xie, H. and Huang, J. (2009). Scad-penalized regression in high-dimensional partially linear models. *Ann. Statist.*, **37**, 673–696.

Xu, C. and Chen, J. (2014). The sparse MLE for ultrahigh-dimensional feature screening, *Jour. Ameri. Statist. Assoc.*, **109**, 1257–1265.

Xu, P., Zhu, L. and Li, Y. (2014). Ultrahigh dimensional time course feature selection, *Biometrics*, **70**, 356–365.

Xue, L., Ma, S. and Zou, H. (2012). Positive definite $\ell_1$ penalized estimation of large covariance matrices. *Jour. Ameri. Statist. Assoc.*, **107**, 1480–1491.

Xue, L., and Zou, H. (2012). Regularized rank-based estimation of high-dimensional nonparanormal graphical models. *Ann. Statist.*, **40**, 2541-2571.

Xue, L. and Zou, H. (2014). Rank-based tapering estimation of bandable correlation matrices. *Statist. Sinica*, **24**, 83–100.

Xue, L. and Zou, H. (2014). Optimal estimation of sparse correlation matrices of semiparametric Gaussian copula. *Stat. Interface* , **7**, 201–209.

Xue, L., Zou, H. and Cai, T. (2012). Nonconcave penalized composite conditional likelihood estimation of sparse Ising models. *Ann. Statist.*, **40**, 1403–1429.

Yang, Y. and Zou, H. (2013). An efficient algorithm for computing the HHSVM and its generalizations. *Jour. Comput. Graph. Statist.* , **22**, 396–415.

Yang, Y. and Zou, H. (2015). A fast unified algorithm for solving group-lasso penalized learning problems. *Stat. Comput.*, **25**, 1129–1141

Ye, F. and Zhang, C.-H. (2010). Rate minimaxity of the lasso and Dantzig selector for the $\ell_q$ loss in $\ell_r$ balls. *Jour. Mach. Learn. Res.*, **11**, 3481–3502.

Yin, W., Osher, S., Goldfarb, D. and Darbon, J. (2008). Bregman iterative algorithms for L1-minimization with applications to compressed sensing. *SIAM Jour. Imaging Sci.*, **1**, 143–168.

Yu, G. and Bien, J. (2017). Learning local dependence in ordered data. *Jour. Mach. Learn. Res.*, **18**, 1–60.

Yu, Y., Wang, T., and Samworth, R. J. (2014). A useful variant of the Davis-Kahan theorem for statisticians. *Biometrika*, **102**, 315–323.

Yuan, M. (2010). High dimensional inverse covariance matrix estimation via linear programming. *Jour. Mach. Learn. Res.*, **11**, 2261–2286.

Yuan, M. and Lin, Y. (2006). Model selection and estimation in regression with grouped variables. *Jour. Roy. Statist. Soc. B*, **68**, 49–67.

Yuan, M. and Lin, Y. (2007). Model selection and estimation in the Gaussian graphical model. *Biometrika*, **94**, 19–35.

Yuan, M., and Lin, Y. (2007). On the non-negative garrote estimator. *Jour. Roy. Statist. Soc., B* **69**, 143–161.

Zhang, C.-H. (2005). Estimation of sums of random variables: examples and information bounds. *Ann. Statist.*, **33**, 2022–2041.

Zhang, C.-H. (2007). Information-theoretic Optimality of Variable Selection with Concave Penalty. Technical Report 2007-008, Department of Statistics, Rutgers University.

Zhang, C.-H. (2010). Nearly unbiased variable selection under minimax concave penalty. *Ann. Statist.*, **38**, 894–942.

Zhang, C.-H. (2011). Statistical inference for high-dimensional data. In: *Mathematisches Forschungsinstitut Oberwolfach: Very High Dimensional Semiparametric Models, Report No. 48/2011*, pages 28–31.

Zhang, C.-H. and Huang, J. (2008). The sparsity and bias of the Lasso selection in high-dimensional linear regression. *Ann. Statist.*, **36**, 1567–1594.

Zhang, C. -H., and Zhang, S. S. (2014). Confidence intervals for low dimensional parameters in high dimensional linear models. *Jour. Roy. Statist. Soc. B*, **76**, 217-242.

Zhang, C.-H. and Zhang, T. (2012). A general theory of concave regularization for high-dimensional sparse estimation problems. *Statist. Sci.*, **27**, 576–593.

Zhang, H. H. , Ahn, J., Lin, X., and Park, C. (2006). Gene selection using support vector machines with nonconvex penalty. *Bioinformatics*, **2**, 88–95.

Zhang, H. H. and Lu, W. (2007). Adaptive Lasso for Cox's proportional hazards model. *Biometrika*, **94**, 691–703.

Zhang, H.P. and Singer, B. (1999) *Recursive Partitioning in the Health Sciences*. Springer-Verlag: New York.

Zhang, N., Jiang, W., Lan, Y. (2019). Conditional variable screening via ordinary least squares projection. *arXiv preprint arXiv:1910.11291.*

Zhang, T. (2009). Some sharp performance bounds for least squares regression with $L_1$ regularization. *Ann. Statist.*, **37**, 2109–2144.

Zhang, T. (2010b). Analysis of multi-stage convex relaxation for sparse regularization. *Jour. Mach. Learn. Res.*, **11**, 1087–1107.

Zhang, T. (2011). Adaptive forward-backward greedy algorithm for learning sparse representations. *IEEE Trans. Inform. Theory*, **57**, 4689–4708.

Zhang, T. and Zou, H. (2014). Sparse precision matrix estimation via Lasso penalized D-Trace loss. *Biometrika*, **101**, 103–120.

Zhang, Y., Duchi, J., and Wainwright, M. (2015). Divide and conquer kernel ridge regression: A distributed algorithm with minimax optimal rates. *Jour. Mach. Learn. Res.*, **16**, 3299–3340.

Zhang, Y. and Li, R. (2011). Iterative conditional maximization algorithm for nonconcave penalized likelihood. in *Nonparametric Statistics and Mixture Models: A Festschrift in Honor of Thomas P. Hettmansperger* (Eds by D. R. Hunter, D. St. Richards and J. Rosenberger), 336-352. Singapore: World Scientific Publishing Co.

Zhao, A., Feng, Y., Wang, X. and Tong, X. (2016). Neyman-Pearson classification under high-dimensional settings. *Jour. Mach. Learn. Res.*, **17**, 1–39.

Zhao, D.S. and Li, Y. (2012). Principled sure independence screening for Cox models with ultra-high-dimensional covariates. *Jour. Multivar. Anal.*, **105**, 397–411.

Zhao, P. and Yu, B. (2006). On model selection consistency of Lasso. *Jour. Mach. Learn. Res.*, **7**, 2541–2567.

Zhao, Y., Levina, E., and Zhu, J. (2012). Consistency of community detection in networks under degree-corrected stochastic block models. *Ann. Statist.*, **40**, 2266–2292.

Zhong, W. and Zhu, L. P. (2015). An iterative approach to distance correlation based sure independence screening. *Jour. Stat. Comput. Simul.*, **85**, 2332 – 2345.

Zhou, T., Zhu, L., Xu, C. and Li, R. (2020). Model-free forward regression via cumulative divergence. *Jour. Ameri. Statist. Assoc.*, In press.

Zhu, L. P., Li, L., Li, R. and Zhu, L. X. (2011). Model-free feature screening for ultrahigh dimensional data. *Jour. Ameri. Statist. Assoc.*, **106**, 1464–1475.

Zhu, J., Rosset, S., Hastie, H. and Robert, T. (2003). 1-norm support vector machines. *Proceedings of the 16th International Conference on Neural Information Processing Systems*, 49–56.

Zhu, Z., Li, Y., and Song, Z. (2018). A convergence theory for deep learning via over-parameterization. *arXiv preprint arXiv:1811.03962.*

Zou, H. (2006). The adaptive Lasso and its oracle properties. *Jour. Amer. Statist. Assoc.*, **101**, 1418–1429.

Zou, H. (2007). An improved 1-norm svm for simultaneous classification and variable selection. *Proceedings of the Eleventh International Conference on Artificial Intelligence and Statistics*, 675–681.

Zou, H. (2008). A note on path-based variable selection in the penalized proportional hazards model. *Biometrika*, **95**, 241 - 247.

Zou, H. and Hastie, T. (2005). Regularization and variable selection via the elastic net. *Jour. Roy. Statist. Soc. B*, **67**, 301–320.

Zou, H., Hastie, T., and Tibshirani, R. (2006). Sparse principal component analysis. *Jour. Comput. Graph. Statist.*, *15*, 265–286.

Zou, H., Hastie, T., and Tibshirani, R. (2007). On the degrees of freedom of the Lasso, *Ann. Statist.*, **35**, 2173–2192.

Zou, H. and Li, R. (2008). One-step sparse estimates in nonconcave penalized likelihood models (with discussion). *Ann. Statist.*, **36**, 1509–1533.

Zou, H. and Yuan, M. (2008). Composite quantile regression and the oracle model selection theory. *Ann. Statist.*, **36**, 1108–1126.

Zou, H. and Zhang, H. (2009). On the adaptive elastic-net with a diverging number of parameters. *Ann. Statist.*, **37**, 1733–1751.

# Author Index

# Index

9781466510845